W9-ATK-940

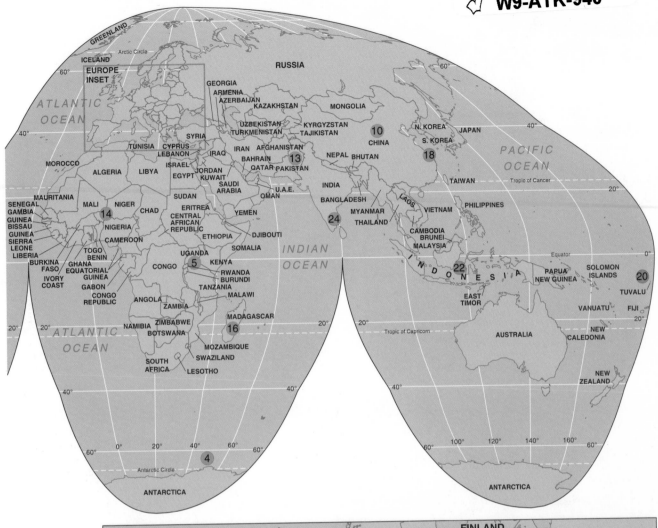

ENVIRONMENT

4/E

Peter H. Raven

Missouri Botanical Garden

Linda R. Berg

St. Petersburg College

JOHN WILEY & SONS, INC.

COVER IMAGE   ©*John Martin/Images.com*

ACQUISITIONS EDITOR   *Keri Witman*
DEVELOPMENTAL EDITOR   *Ellen Ford*
MARKETING MANAGER   *Clay Stone*
SENIOR PRODUCTION EDITOR   *Norine M. Pigliucci*
DESIGN DIRECTOR/COVER DESIGNER   *Madelyn Lesure*
ILLUSTRATION EDITOR   *Sandra Rigby*
PHOTO RESEARCHER   *Teri Stratford*
PHOTO EDITOR   *Hilary Newman*
PRODUCTION MANAGEMENT SERVICES   *Hermitage Publishing Services*
EDITORIAL ASSISTANT   *Maureen Powers*

This book was set in 10/12 Janson by Hermitage Publishing Services and printed and bound by Von Hoffmann Press. The cover was printed by Von Hoffmann Press.

♻ This book is printed on recycled paper containing 10% post-consumer waste.

To order books or for customer service please, call 1(800)-CALL-WILEY (225-5945).

ISBN 0-471-44452-9
WIE ISBN 0-471-45167-3

Printed in the United States of America

10 9 8 7 6 5 4 3 2

Preface

The challenge of creating and maintaining a sustainable environment is probably the single most pressing issue that will confront students throughout their lives. Today, environmental science is not only relevant to students' personal experiences but also vital to the future of the entire planet. As humans increasingly alter Earth's land, water, and atmosphere on local, regional, and global levels, the resulting environmental problems can seem insurmountable. Armed with the proper tools, however, students need not find these issues overwhelming. *Environment*, fourth edition, equips students with the most essential of these tools: a working knowledge of the ecological concepts that underlie environmental problems.

The overarching concept of **environmental sustainability** has never been more important to the field of environmental science than it is today. We have therefore made sustainability a central theme of *Environment* by integrating it throughout the text. We begin in Chapter 1 with a definition of environmental sustainability, a brief discussion of why experts in environmental science think human society is not operating sustainably, and an introduction to the concept of sustainable development. Sustainability is revisited throughout the text, particularly in the context of sustainable water use (Chapter 13), sustainable soil use (Chapter 14), sustainable manufacturing (Chapter 15), sustainable forest management (Chapter 17), and sustainable agriculture (Chapter 18). In Chapter 24 we conclude with an extended discussion of living sustainably.

From the opening pages, we acquaint students with current environmental issues—issues that have many dimensions and that defy easy solutions. We begin by examining the scientific, historical, ethical, governmental, and economic underpinnings of environmental science. This provides a conceptual foundation for students that they can then bring to bear on the rest of the material in the book. One of our principal goals is to convey to students an appreciation of the remarkable complexity and precise functioning of natural ecosystems. Thus, we next explore the basic ecological principles that govern the natural world and consider the many ways in which humans affect the environment. Later chapters examine in detail the effects of human activities, including overpopulation, energy production and consumption, depletion of natural resources, and pollution.

Although we avoid unwarranted optimism when presenting these problems—many are very serious indeed—we try to avoid the gloomy predictions of disaster so commonly presented by the media today. Instead, students are encouraged to take active, positive roles, using the practical and conceptual tools presented in this book, to meet the environmental challenges of today and tomorrow.

Environment, 4/e, integrates important information from many different fields, such as biology, geography, chemistry, geology, physics, economics, sociology, natural resources management, law, and politics. Because environmental science is an interdisciplinary field, this book is appropriate for use in environmental science courses offered by a variety of departments, including (but not limited to) biology, geology, geography, and agriculture.

All of the chapters have been painstakingly researched, and extraordinary efforts have been made to obtain the most recent data available. Both instructors and students will benefit from the book's **currency** because environmental issues and trends are continually changing.

This book is intended as an introductory text for undergraduate students, both science and nonscience majors. Although relevant to all students, *Environment*, 4/e, is particularly appropriate for those majoring in education, journalism, government and politics, and business, as well as the traditional sciences. We assume our students have very little prior knowledge of how ecosystems work, the dynamics of how matter and energy move through ecosystems, and how populations affect and are affected by ecosystems. These important ecological concepts and processes are presented in a straightforward, unambiguous manner.

CHARTING A COURSE FOR LEARNING

Learning environmental science is a challenging endeavor. A well-developed pedagogical plan that facilitates student mastery of the material has always been a hallmark of *Environment*.

Pedagogical features in the fourth editon include:

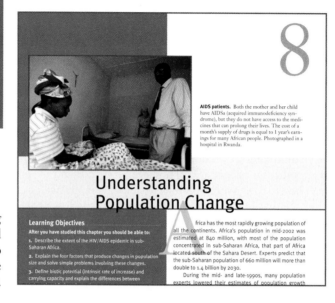

8

AIDS patients. Both the mother and her child have AIDSa (acquired immunodeficiency syndrome), but they do not have access to the medicines that can prolong their lives. The cost of a month's supply of drugs is equal to 1 year's earnings for many African people. Photographed in a hospital in Rwanda.

Understanding Population Change

Learning Objectives

After you have studied this chapter you should be able to:

1. Describe the extent of the HIV/AIDS epidemic in sub-Saharan Africa.

2. Explain the four factors that produce changes in population size and solve simple problems involving these changes.

3. Define biotic potential (intrinsic rate of increase) and carrying capacity and explain the differences between

Africa has the most rapidly growing population of all the continents. Africa's population in mid-2002 was estimated at 840 million, with most of the population concentrated in sub-Saharan Africa, that part of Africa located south of the Sahara Desert. Experts predict that the sub-Saharan population of 660 million will more than double to 1.4 billion by 2030.

During the mid- and late-1990s, many population experts lowered their estimates of population growth

Chapter Introductions illustrate certain concepts in the chapter with stories about some of today's most pressing environmental issues.

Learning Objectives at the beginning of the chapter indicate in behavioral terms what the student must be able to do to demonstrate mastery of the material in the chapter.

Case In Point features in the body of the chapter offer a wide variety of in-depth case studies that address important issues in the field of environmental science.

CASE-IN-POINT **The Arctic National Wildlife Refuge**

One of the oil industry's biggest obstacles to locating and extracting oil is public resistance based on the perceived threat to environmentally sensitive areas. Consider the proposed opening of the Arctic National Wildlife Refuge

Envirobriefs provide additional topical material about relevant environmental issues.

ENVIROBRIEF

Using Goats to Fight Fires

California has about 6,000 wildfires each year, and they are becoming increasingly expensive and dangerous to manage because so many people are building homes and living in fire-vulnerable chaparral. For one thing, the topography is so steep that firefighters often cannot use mechanized equipment but instead must be transported to fires by helicopters. Afraid that prescribed burns will get

MEETING THE CHALLENGE

Reclamation of Coal-Mined Land

The 1977 Surface Mining Control and Reclamation Act (SMCRA) has been extremely effective in protecting the environment. In the years since this law's passage, thousands of permits regulating the reclamation of active coal mines covering almost 2 million hectares (5 million acres), an area the size of New Jersey, have been issued. In addition, some of the most dangerous (from a health and safety viewpoint) abandoned coal mines have been reclaimed. The Office of Surfaces ing of the Department of the Interior gives annual a

Meeting the Challenge boxes profile environmental success stories.

YOU CAN MAKE A DIFFERENCE

Getting Around Town

Can you imagine getting around town without a car? How would you get to class, the grocery store, the laundromat? Hopping in your car for every errand seems like the natural thing to do. According to the American Automobile Association, American motorists drive an average of 10,100 miles annually and burn 507 gallons of gasoline in the process.

However, consider this: From the production of gasoline to the disposal of old automobiles, the car has a significant negative impact on the environment. Acid deposition and

You Can Make a Difference boxes suggest specific courses of action or lifestyle changes students can make to improve the environment.

3. Carpool to class, to work, to the grocery store, to social events. One car on the road is better than three or four.

4. Buy a good bicycle; it is less expensive than a car to buy and maintain, and it is great for local transportation. It is also good exercise.

5. Walk to class or work if you live within a mile or so. You will need to allow yourself a little extra time, but once you get into the habit, it is easy. Walking is good exercise, too.

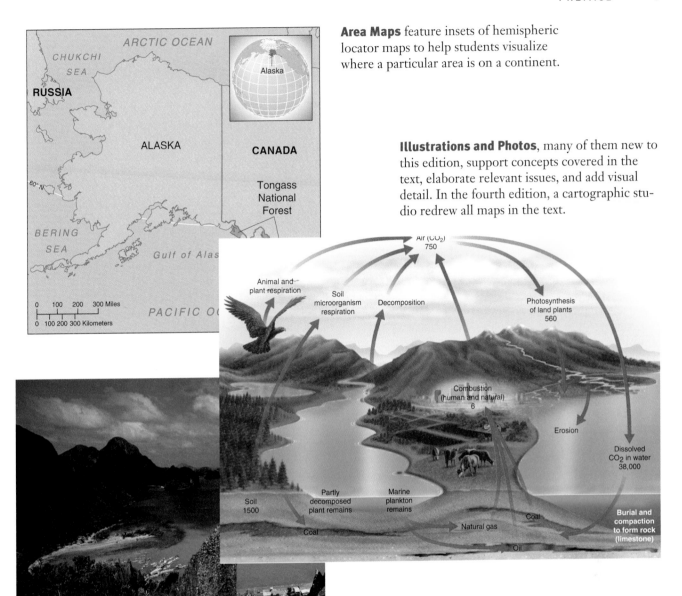

Area Maps feature insets of hemispheric locator maps to help students visualize where a particular area is on a continent.

Illustrations and Photos, many of them new to this edition, support concepts covered in the text, elaborate relevant issues, and add visual detail. In the fourth edition, a cartographic studio redrew all maps in the text.

Table 5.1	Ecosystem Services
Ecosystem	**Services Provided by Ecosystem**
Forests	Purify air and water; produce and maintain soil; absorb carbon dioxide; provide wildlife habitat; provide humans with wood and recreation
Freshwater systems (rivers and streams, lakes, and groundwater)	Moderate water flow; dilute and remove pollutants; provide wildlife habitat; provide humans with drinking and irrigation water, food, transportation corridors, electricity, and recreation
Grasslands	Purify air and water; produce and maintain soil; absorb carbon dioxide; provide wildlife habitat; provide humans with livestock and recreation
Coasts	Provide a buffer against storms; dilute and remove pollutants; provide wildlife habitat; provide humans with food, harbors and transportation routes, and recreation
Sustainable agricultural ecosystems*	Produce and maintain soil; absorb carbon dioxide; provide wildlife habitat for birds, insect pollinators, and soil organisms; provide humans with food and fiber crops

* Sustainable agriculture ecosystems are human-made and therefore inherently different from the other ecosystems. Sustainable agriculture is discussed in Chapter 18.

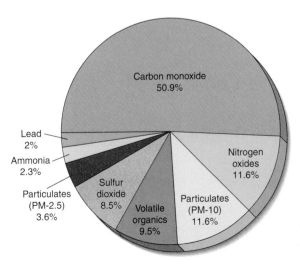

Tables and graphs, with data sources cited at the end of the text, summarize and organize important information.

Mini-Glossaries, located within many chapters, provide handy definitions and distinctions between closely related terms or potentially confusing new terms.

MINI-GLOSSARY

Köppen's Climate Zones

humid equatorial climate: Every month is warm; mean temperature is over 18°C (64°F).

dry climate: Evaporation exceeds precipitation in most months.

humid temperate climate: Distinct winter and summer seasons; winters are mild; mean temperature in coldest month is above 3°C (27°F).

Summaries with Selected Key Terms in outline form at the end of each chapter provide a review of the material presented. Boldfaced selected key terms are integrated within each summary, enabling students to study vocabulary words in the context of related concepts.

SUMMARY WITH SELECTED KEY TERMS

I. Science is a systematic process to investigate the natural world. Science seeks to reduce the apparent complexity of our world to general principles that can be used to solve problems or provide new insights.

A. The established processes that scientists use to solve problems are collectively called the **scientific method**. There are many variations of the scientific method, which basically

2. Any difference in the outcome between the control group and the experimental group must be the result of the variable.

C. Both inductive and deductive reasoning are used in the scientific method.

1. **Inductive reasoning** begins with specific examples and seeks to draw a conclusion or discover a unifying rule on

---

Thinking About the Environment questions, many new to this edition, encourage critical thinking and highlight important concepts and applications. All Learning Objectives presented at the beginning of the chapter are covered in the Thinking About the Environment questions.

THINKING ABOUT THE ENVIRONMENT

1. What is a biome? What two climate factors are most important in determining an area's characteristic biome?
2. How does vegetation change with increasing elevation and latitude?
3. What climate and soil factors produce each of the major terrestrial biomes?
4. Describe representative organisms of the forest biomes dis-

8. What environmental factors are most important in determining the kinds of organisms found in aquatic environments?
9. Distinguish between freshwater wetlands and estuaries, and between flowing-water and standing-water ecosystems.
10. What would happen to the organisms in a river with a fast

---

Take a Stand features appear at the end of every chapter and ask students to enter into a debate about an issue or controversy from the chapter. Students then visit our Web site to find links for researching the situation and tools for organizing their arguments. Take A Stand can be done individually or in small collaborative groups, and as an oral presentation, a writing assignment, or both.

TAKE A STAND

Visit our Web site at **http://www.wiley.com/college/raven** (select Chapter 8 from the Table of Contents) for links to more information about the controversies surrounding U.S. immigration policies. Consider the opposing views of supporters and opponents to current immigration policies, and debate the issues with your classmates. You will find tools to help you organize your research, analyze the data, think critically about

Stand activities can be done individually or as part of a team, as oral presentations, written exercises, or Web-based (e-mail) assignments.

Additional on-line materials relating to this chapter, including Student Quizzes, Activity Links, Useful Web Sites, Flash Cards, and more, can also be found on our Web site.

Suggested Reading provides current references for further learning.

2002 World Population Data Sheet, provided by the Population Reference Bureau, is folded into the text (inside the back cover) and is intended to be pulled out for classroom use. Chapter 8 provides a student assignment of questions about population based on the data sheet.

ORGANIZATION

Educators present the major topics of an introductory environmental science course in a variety of orders. We make no pretense that we have found the best way to organize the subject of environmental science. However, we have put our best efforts into writing the seven parts and their chapters so that they can be successfully presented in any number of sequences.

Part 1, Humans in the Environment, introduces environmental science and current environmental concerns. The three chapters in this section also develop the scientific process and examine how human endeavors such as economics, government policies, and environmental ethics affect the environment.

Part 2, The World We Live In, provides a detailed introduction to basic ecological principles. This part, which consists of four chapters, is organized around the ecosystem, which is the fundamental unit of ecology.

Part 3, A Crowded World, discusses the principles of population ecology and emphasizes the fact that human populations comply with the same principles of population ecology as other organisms. The two chapters in this part also examine urbanization and other sociological and cultural factors that affect human population growth.

Part 4, The Search for Energy, considers the environmental impact of the human quest for energy. This part, which consists of three chapters, discusses key issues associated with the use of fossil fuels, nuclear power, and renewable energy sources.

Part 5, Our Precious Resources, examines the overuse and abuse of our natural resources. The six chapters in this section explore resource issues and dilemmas involving water, soil, minerals, biological resources, land, and food production.

The five chapters in **Part 6, Environmental Concerns,** review the effects of local, regional, and global pollution. We discuss the problems associated with air pollution, acid deposition, global climate change, stratospheric ozone destruction, water and soil pollution, pesticide pollution, and solid and hazardous wastes.

Part 7, Tomorrow's World, concludes the book with a single chapter that presents the opinions of the authors on social responsibilities, identifying some of the most critical issues that must be grappled with today to assure a better tomorrow.

MAJOR CHANGES IN THE FOURTH EDITION

Issues in environmental science change rapidly, and the fourth edition of *Environment* is therefore a major revision. Overall, the authors meticulously updated and verified all facts and made a serious attempt to provide more

balance and eliminate bias. A complete list of all changes and updates to the fourth edition is too long to fit in the Preface, but one or two of the more important changes to each chapter follows:

Chapter 1, Our Changing Environment, provides a new Chapter Introduction on the Lewis Center of Oberlin College, a new Envirobrief on the invasion of jellyfish in the Gulf of Mexico, and a new Meeting the Challenge box on the 2002 World Summit. **Chapter 2, Using Science to Address Environmental Problems,** contains expanded coverage of toxicology, a new section on the precautionary principle, a new Table on the probability of risk of dying by selected causes, and a new Mini-Glossary of toxicology terms. **Chapter 3, Environmental History, Legislation, and Economics,** offers expanded coverage of both environmental history and environmental economics.

Chapter 4, Ecosystems and Energy, presents additional discussion on the concept of ecosystems, updates Vitousek's groundbreaking research on the human appropriation of land-based net primary productivity, and contains a new Envirobrief on unintended changes in food webs. **Chapter 5, Ecosystems and Living Organisms,** increases coverage of evolution, the mechanisms that underlie succession, and the concept of keystone species; it also has a new Table on ecosystem services. **Chapter 6, Ecosystems and the Physical Environment,** contains new sections on the sulfur cycle and on humans and the hydrologic cycle; in addition, much of the artwork in this chapter has been redrawn. **Chapter 7, Major Ecosystems of the World,** has a new Case-in-Point on the Chesapeake Bay, a new Envirobrief on using goats to fight fires, and an updated section on the Florida Everglades.

Chapter 8, Understanding Population Change, contains a new section on population boom-bust cycles and density dependence, as well as additional discussion of Malthus' ideas. **Chapter 9, Facing the Problems of Overpopulation,** has new discussions of ecological footprints and sustainable consumption, as well as a new Mini-Glossary of consumption terms.

Chapter 10, Fossil Fuels, contains a new Mini-Glossary of synfuel terms and a presentation of Bush's National Energy Policy. **Chapter 11, Nuclear Energy,** has updates on breeder reactors, fixing technical and safety problems in existing nuclear power plants, Chornobyl, and Yucca Mountain. **Chapter 12, Renewable Energy and Conservation,** contains a new Mini-Glossary on direct solar energy terms and a new Table of energy intensities in selected countries.

Chapter 13, Water: A Fragile Resource, updates the Case-in-Point on the declining condition of the Missouri River and contains a new Mini-Glossary of water terms. **Chapter 14, Soils and Their Preservation,** has new material on agroforestry, the 2001 soil study by the International Food Policy Research Institute, and sodbusting associated with the Grassland Reserve Program. **Chapter 15, Minerals: A Nonrenewable Resource,**

contains updates on diamonds and gold and on Copper Basin. **Chapter 16, Preserving Earth's Biological Diversity,** has a new section on Earth's 25 biodiversity hotspots, expanded coverage on the concepts that guide conservation biologists, and a new Envirobrief on when one rare species eats another. Coverage of forestry was significantly expanded in **Chapter 17, Land Resources and Conservation,** which also has a new Envirobrief on taking cattle "to the bank." **Chapter 18, Food Resources: A Challenge for Agriculture,** contains improved definitions of the principal types of agriculture in the body of the chapter as well as in a new Mini-Glossary; Chapter 18 also has a new section on food safety and updates on the use of low-dose antibiotics for livestock.

Chapter 19, Air Pollution, has a better explanation for how a temperature inversion occurs in the Los Angeles basin, a new Mini-Glossary of Air Pollutants, and significant updates on air pollution and human health, Mexico City, and clean cars. New updates on the IPCC Third Assessment Report, melting of glaciers and icecaps, rising sea levels, the effects of climate change on organisms and on human health, and the status of the Kyoto Protocol are found in **Chapter 20, Regional and Global Atmospheric Changes. Chapter 21, Water and Soil Pollution,** has new material on bacterial source testing, a 2002 study of organic pollutants in U.S. waterways, and groundwater pollution in Santa Monica, California. **Chapter 22, The Pesticide Dilemma,** has new paragraphs on resistance management for weeds and on the possible link between pesticides and Parkinson's disease, as well as an updated section on the global ban of persistent organic pollutants. New information on computer waste, the *Khian Sea* saga, and how PCBs bioaccumulate in food webs is found in **Chapter 23, Solid and Hazardous Wastes.**

Chapter 24, Tomorrow's World, now has pedagogical material, including new Learning Objectives, a new Summary with Selected Key Terms, new Thinking About the Environment questions, and new Suggested Readings; a new Table shows U.S. participation in selected international environmental treaties.

■ SUPPLEMENTS

The package accompanying *Environment*, 4/e includes several items developed specifically to augment students' understanding of environmental issues and concerns. Together, these ancillaries provide instructors and students with interesting and helpful teaching and learning tools and take full advantage of both electronic and print media.

For the Student and the Instructor

The newly expanded and revised *Environment*, 4/e, **Web site** coauthored by Chris Migliaccio, Miami-Dade Community College, Wolfson Campus, and Greg Ballinger, Miami-Dade Community College, Kendall Campus is located at **www.wiley.com/college/raven.** Some of the Student Resources at this site include Useful Web Sites; Take a Stand (see description above); Activity Links; Quizzes for student self-testing; and Flash Cards. Some of the Instructor Resources include an Image Bank of **all** of the line illustrations in the textbook in jpeg format; and the Instructor's Manual/Test Bank.

For the Student

Study Guide by John Aliff, Georgia Perimeter College. This innovative guide not only helps students learn the details of environmental science but also fosters critical thinking skills and scientific reasoning. Each chapter includes Learning Objectives, Vocabulary Review, Critical thinking Questions, Matching Questions, and Multiple Choice Questions. The Study Guide also includes suggested Web sites and further reading.

For the Instructor

Instructor's Manual/Test Bank by Wendy Ryan, Kutztown University. This resource contains lecture outlines, key terms, and test bank questions. The test bank consists of 40 to 50 challenging multiple-choice questions per chapter, as well as true/false, matching, short answer, and thought questions, with answers provided.

Computerized Test Bank. The easy-to-use test-generation program fully supports graphics, print tests, student answer sheets, and answer keys. The software's advanced features allow you to create an exam to your exact specifications.

Instructor's Resource™ CD-ROM contains all of the line illustrations from *Environment*, 4/e, in jpeg format so that the images can be used in a variety of ways: to include in PowerPoint™ presentations, to custom-make additional four-color overhead transparencies, to import to a Web site, and more.

Overhead Transparency Acetates. The full-color set includes 150 transparencies with figures from the text. Overhead transparencies have been reformatted with large-print labeling for easy viewing in any classroom.

John Wiley & Sons may provide complementary instructional aids and supplements or supplement packages to those adopters qualified under our adoption policy. Please contact your sales representative for more information. If as an adopter or potential user you receive supplements you do not need, please return them to your sales representative or send them to:

Attn: Returns Department
Heller Park Center
360 Mill Road
Edison, NJ 08817

ACKNOWLEDGMENTS

The development and production of *Environment*, 4/e, was a process involving interaction and cooperation among the author team and between the authors and many individuals in our home and professional environments. We are keenly aware of the valuable input and support from editors, colleagues, and students. We also owe our families a debt of gratitude for their understanding, support, and encouragement as we struggled through many revisions and deadlines.

The Editorial Environment

We are privileged to work on this edition with our new publisher, John Wiley and Sons, which is unsurpassed in the college textbook industry. Preparing this book has been an enormous undertaking, but working with the outstanding editorial and production staff at John Wiley and Sons has made it an enjoyable task. We thank our Publisher Kaye Pace and Editor Keri Witman for their support, enthusiasm, and ideas. Senior Developmental Editor Ellen Ford expertly guided us through the revision process, coordinated the final stages of development, and provided us with valuable suggestions before the project went into production. We thank Geraldine Osnato for overseeing and coordinating the development of the supplements, and Maureen Powers, for her editorial assistance.

We are grateful to Senior Marketing Manager Clay Stone for his superb marketing and sales efforts.

We thank Teri Stratford for her contribution as photo researcher, helping us find the wonderful photographs that enhance the text. We appreciate the artistic expertise of Maddy Lesure for a striking cover design, Production Editor Norine Pigliucci for her efforts in maintaining a smooth production process, and Sandra Rigby, who coordinated the art development and ensured that a consistent standard of quality was maintained throughout the illustration program. We greatly appreciate our Project Editor, Larry Meyer of Hermitage Publishing Services, for guiding us through the many deadlines of production.

Our colleagues and students have provided us with valuable input and have played an important role in shaping *Environment*, 4/e. We thank them and ask for additional comments and suggestions from instructors and students who use this text. You can reach us through our editors at John Wiley and Sons; they will see that we get your comments. Any errors can be corrected in subsequent printings of the book, and more general suggestions can be incorporated into future editions.

The Professional Environment

The success of *Environment*, 4/e, is due largely to the quality of the many professors and specialists who have read the manuscript during various stages of its preparation and provided us with valuable suggestions for improving it. We appreciate the efforts of Alan R. Berg, who was instrumental in researching and analyzing the data used in the fourth edition. Special thanks go to Mashood Ahmed Siddiqui, Assistant Director, Environmental Protection Agency, Pakistan, for his invaluable insights into environmental problems outside the United States and Canada. In addition, the reviewers of the first three editions made important contributions that are still part of this book. They are as follows:

Reviewers of the Fourth Edition

John Aliff, Georgia Perimeter College
Diana Anderson, Northern Arizona University
David Boose, Gonzaga University
Karen DeFries, Erie Community College
Michael L. Denniston, Georgia Perimeter College
Jean Dupon, Menlo College
Allen Lee Farrand, Bellevue Community College
David Hassenzahl, University of Nevada, Las Vegas
Solomon Isiorho, Indiana University/Purdue University at
    Fort Wayne
Thomas Justice, McLennan Community College
Penny Koines, University of Maryland
Allen Koop, Grand Valley State University
Alberto Mancinelli, Columbia University
Michael McCarthy, Eastern Arizona College
Christopher Migliaccio, Miami-Dade Community College
Mark Morgan, Rutgers University
James Morris, University of South Carolina
James Ramsay, University of Wisconsin Stevens Point
Sandy Rock, Bellevue Community College
Neil Sabine, Indiana University East
Mark Smith, Fullerton College
Ray Williams, Rio Hondo College

Reviewers of Earlier Editions

Sylvester Allred, Northern Arizona University
Lynne Bankert, Erie Community College
Richard Bates, Rancho Santiago Community College
George Bean, University of Maryland
Mark Belk, Brigham Young University
Bruce Bennett, Community College of Rhode Island–Knight
    Campus
Eliezer Bermúdez, Indiana State University
Patricia Beyer, Bloomsburg University of Pennsylvania
John Dryden Burton, McLennan Community College
Kelly Cain, University of Arizona
Ann Causey, Prescott College
Robert Chen, University of Massachusetts–Boston
Gary Clambey, North Dakota State University
George Cline, Jacksonville State University
Rebecca Cook, Lambuth University
Harold Cones, Christopher Newport College
Bruce Congdon, Seattle Pacific University
Armando de la Cruz, Mississippi State University
Laurie Eberhardt, Valparaiso University

Thomas Emmel, University of Florida
Donald Emmeluth, Fulton-Montgomery Community College
Margaret "Tim" Erskine, Lansing Community College
Kate Fish, EarthWays, St. Louis, Missouri
Steve Fleckenstein, Sullivan County Community College
Andy Friedland, Dartmouth College
Carl Friese, University of Dayton
Bob Galbraith, Crafton Hills College
Stan Guffey, University of Tennessee
John Harley, Eastern Kentucky University
Neil Harriman, University of Wisconsin–Oshkosh
Denny Harris, University of Kentucky
Leland Holland, Pasco-Hernando Community College
James Horwitz, Palm Beach Community College
Dan Ippolito, Anderson University
John Jahoda, Bridgewater State College
Jan Jenner, Talladega College
David Johnson, Michigan State University
Linda Johnston, University of Kansas
Patricia Johnson, Palm Beach Community College
Karyn Kakiba-Russell, Mount San Antonio College
Colleen Kelley, Northern Arizona University
Katrina Smith Korfmacher, Denison University
Karl F. Korfmacher, Denison University
Ericka Lawson, Baylor College of Medicine (formerly Columbia College)
Norm Leeling, Grand Valley State University
Joe Lefevre, State University of New York at Oswego
James Luken, Northern Kentucky University
Timothy Lyon, Ball State University
Jane M. Magill, Texas A&M University
Linda Matson McMurry, San Pedro Academy/Texas Department of Agriculture
George Middendorf, Howard University
Chris Migliaccio, Miami-Dade Community College
Gary Miller, University of North Carolina at Asheville
Michael Morgan, University of Wisconsin–Green Bay
James Morris, University of South Carolina
Andrew Neill, Joliet Junior College

Lisa Newton, Fairfield University
James T. Oris, Miami University, Ohio
Robert Paoletti, Kings College
Richard Pemble, Moorhead State University
Ervin Poduska, Kirkwood Community College
Lowell Pritchard, Emory University
James Ramsey, University of Wisconsin–Stevens Point
Elizabeth Reeder, Loyola College
Maralyn Renner, College of the Redwoods
C. Howard Richardson, Central Michigan University
W. Guy Rivers, Lynchburg College
Barbra Roller, Miami Dade Community College–South Campus
Richard Rosenberg, The Pennsylvania State University
Wendy Ryan, Kutztown University
Neil Sabine, Indiana University East
Frank Schiavo, San Jose State University
Jeffery Schneider, State University of New York at Oswego
Judith Meyer Schultz, University of Cincinnati
Lynda Swander, Johnson County Community College
Thomas Schweizer, Princeton Energy Resources International
Mark Smith, Victor Valley Community College
Nicholas Smith-Sebasto, University of Illinois
Joyce Solochek, Marquette University
Morris Sotonoff, Chicago State University
Steven Steel, Bowling Green State University
Allan Stevens, Snow College
Karen Swanson, William Paterson University
Eileen Van Tassel, Michigan State University
Jerry Towle, California State University–Fresno
Jack Turner, University of South Carolina at Spartanburg
Bruce Van Zee, Sierra College
Matthew Wagner, Texas A&M University
Phillip Watson, Ferris State University
Alicia Whatley, Troy State University
Jeffrey White, Indiana University
James Willard, Cleveland State University
Ray E. Williams, Rio Hondo College

About the Authors

Peter H. Raven, one of the world's leading botanists, has dedicated more than three decades to conservation and biodiversity as Director of the Missouri Botanical Garden in St. Louis, where he has cultivated a world-class institution of horticultural display, education, and research. Described by *Time* magazine as "Hero for the Planet," Dr. Raven champions research around the world to preserve endangered species and is a leading advocate for conservation and a sustainable environment.

Dr. Raven is Chair of the National Geographic Society's Committee for Research and Exploration, Chair of the Division of Earth and Life Studies of the National Research Council, and he is Chairman of the American Association for the Advancement of Science. He is the recipient of numerous prizes and awards, including the prestigious National Medal of Science in 2001, the highest award for scientific accomplishments in this country, Japan's International Prize for Biology, the Environmental Prize of the Institut de la Vie, the Volvo Environment Prize, the Tyler Prize for Environmental Achievement, and the Sasakawa Environment Prize. He also has held Guggenheim and MacArthur fellowships.

Dr. Raven received his Ph.D. from the University of California, Los Angeles, after completing his undergraduate work at the University of California, Berkeley.

Linda R. Berg is an award-winning teacher and textbook author. She received a B.S. in science education, M.S. in botany, and Ph.D. in plant physiology from the University of Maryland. Her research focused on the evolutionary implications of steroid biosynthetic pathways in various organisms. Her recent interests involve the Florida Everglades.

Dr. Berg taught at the University of Maryland–College Park, for 17 years, and is presently an Adjunct Professor at St. Petersburg College in Florida. She has taught introductory courses in environmental science, biology, and botany to thousands of students since 1972. At the University of Maryland, she received numerous teaching and service awards. Dr. Berg is also the recipient of many national and regional awards, including the National Science Teachers Association Award for Innovations in College Teaching, the Nation's Capital Area Disabled Student Services Award, and the Washington Academy of Sciences Award in University Science Teaching.

During her career as a professional science writer, Dr. Berg has authored or coauthored several leading college science textbooks. Her writing reflects her teaching style and love of science.

Brief Contents

Contents

PART TWO
THE WORLD WE LIVE IN

4 Ecosystems and Energy 63

5 Ecosystems and Living Organisms 81

6 Ecosystems and the Physical Environment 103

PART THREE
A CROWDED WORLD

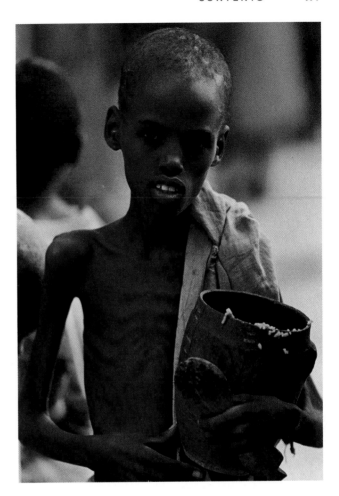

PART FOUR
THE SEARCH FOR ENERGY

10 Fossil Fuels 208

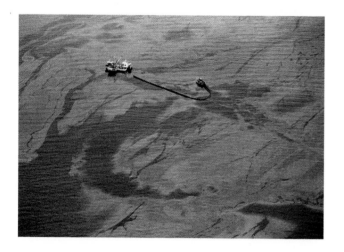

11 Nuclear Energy 234

PART FIVE

OUR PRECIOUS RESOURCES

15 Minerals: A Nonrenewable Resource   331

16 Preserving Earth's Biological Diversity   351

17 Land Resources and Conservation   379

18 **Food Resources: A Challenge for Agriculture 408**

PART SIX
ENVIRONMENTAL CONCERNS

19 **Air Pollution 436**

20 Regional and Global Atmospheric Changes 460

21 Water and Soil Pollution 487

PART SEVEN
TOMORROW'S WORLD

APPENDICES

Green architecture. The Adam Joseph Lewis Center for Environmental Studies at Oberlin College is one of the best examples of ecological design in the United States. It uses about 20% of the energy that a new building typically uses.

1

Our Changing Environment

Learning Objectives

After you have studied this chapter you should be able to:

1. Define environmental science and explain why environmental sustainability is an important concern of environmental science.

2. Summarize human population issues, including population size and level of consumption.

3. Describe the three factors that are most important in determining human impact on the environment and solve a problem using the *IPAT* equation.

4. Briefly describe some of the data that suggest that certain chemicals used by humans may also function as endocrine disrupters in animals, including humans.

5. Provide an overview of how human activities have affected the following: the Georges Bank fishery, tropical migrant birds, wolf populations in Yellowstone National Park, and invasive species such as comb jellies and zebra mussels.

6. Characterize human impacts on the global atmosphere, including stratospheric ozone depletion and climate warming.

7. Describe some of the consequences of tropical rainforest destruction.

8. Define environmental ethics and discuss distinguishing features of the Western and deep ecology worldviews.

The attractive building in the opening photo—the Adam Joseph Lewis Center for Environmental Studies at Oberlin College in Ohio—does not at first glance appear to be anything special, but it represents a vision implemented by dedicated students, faculty members, architects, and many others. Dedicated in 2000, the building is an award-winning example of **green architecture,** which encompasses environmental considerations such as energy conservation, improved indoor air quality, water conservation, and recycled or reused building materials. The Lewis Center is designed to work like a natural ecosystem such as a forest or pond—that is, it obtains its energy from sunlight, preserves biological diversity, and produces few wastes that cannot be recycled or reused.

An advanced-design, earth-coupled heat pump, which extracts heat from 24 geothermal wells in the ground, supplies some of the energy for heating and cooling the building. On the roof, an array of photovoltaic (PV) cells collects the sun's energy to meet most of the building's electricity needs. On sunny days, the PV cells produce more electricity than the building uses, so the excess is sold to the Ohio power grid. On cloudy winter days, the building buys the extra energy it needs from the grid.

Energy is saved by the triple-paned windows, which allow sunlight to enter the building but insulate against heat loss during winter and heat gain during summer. Classrooms in the building have motion sensors that detect when people are not in the room, triggering the energy-efficient lighting fixtures and ventilation systems to shut down.

The building's wastewater from sinks and toilets is cleansed and recycled by a so-called living machine. This organic system consists of a series of tanks that contain bacteria, aquatic organisms, and plants. These organisms interact to remove organic wastes, nutrients such as nitrogen and phosphorus, and disease-causing organisms from the water as it moves through the tanks. After it has been cleaned, the water is stored and reused in the building's toilets.

The acoustic panels in the auditorium are made from agricultural straw wastes, and the computer table in the resource center is constructed from a bowling alley lane from the old campus bowling alley. All new wood used in the building came from nearby forests that are certified to produce sustainably harvested wood. All the steel frames used in the building are composed of recycled steel. The carpets in the classrooms are installed in squares that are easy to remove. As the carpet wears out, it is not disposed of in a landfill. Instead, the squares are returned to the manufacturer, where they are recycled into "new" carpet.

The landscape surrounding the building is carefully designed to highlight a variety of natural and agricultural ecosystems. Because most of Oberlin College's students come from urban areas, these landscaped grounds supply an outdoor classroom that exposes students to the natural environment. Much of the area on which Oberlin College was built originally consisted of marshes and deciduous forests. A wetland that provides a haven for frogs, insects such as dragonflies, and songbirds has been constructed along one side of the building. Native forest trees are planted along another side of the building. Orchards of apple and pear trees, gardens, and a greenhouse are also part of the landscape. Near the orchard is a cistern that collects precipitation runoff from the roof for reuse.

The more efficient use of materials and energy in buildings is just one way that people have responded in recent years to pressing environmental concerns. In this chapter we examine several current environmental issues. Some of the problems are global in scope, such as destruction of Earth's ozone shield; others are more regional, such as the collapse of the Georges Bank fishery due to overfishing; and still others are local, such as the loss and reintroduction of wolves in Yellowstone National Park. Some problems are cases of upsetting the balance among native organisms by introducing foreign species; the invasion of the Great Lakes by zebra mussels is a problem of this sort. Other problems involve the destruction of entire ecosystems, such as tropical rain forests that are burned to provide land for growing crops or raising cattle.

We provide this overview of specific environmental problems to help you understand that any action, no matter how small, has environmental consequences. As you study this text, you will begin to recognize the trade-offs inherent in environmental decision making. *No perfect solutions exist for environmental problems.* Rather, you must weigh carefully all costs and benefits of each possible solution and select the option that is least damaging.

■ ENVIRONMENTAL SCIENCE

How humans can best live within Earth's environment is the theme of what is loosely called **environmental science**, the interdisciplinary study of humanity's relationship with other organisms and the nonliving physical environment. Environmental science is interdisciplinary because it uses and combines information from many disciplines, such as biology (particularly ecology), geography, chemistry, geology, physics, economics, sociology (particularly demography, the study of populations), cultural anthropology, natural resources management, agriculture, engineering, law, politics, and ethics.

Environmental science encompasses many complex and interconnected problems involving human population, Earth's natural resources, and environmental pollution. **Pollution** is any alteration of air, water, or soil that harms the health, survival, or activities of humans and other living organisms.

The Goals of Environmental Science

Environmental scientists try to establish general principles about how the natural world functions. They use these principles to develop viable solutions to environmental problems—solutions that are based as much as possible on scientific knowledge. Environmental problems are generally complex, however, and so our scientific understanding of them is often less complete than we would like it to be. Environmental scientists are often called on to reach a scientific consensus before the data are complete. As a result, they often make recommendations to elected officials based on probabilities rather than precise answers.

The science of **ecology**, a discipline of biology that studies the interrelationships between organisms and their environment, is a basic tool of environmental science, and so we present a detailed treatment of ecology (Chapters 4 through 7). Using what we have learned about ecology, we then directly address human population growth (Chapters 8 and 9) and three of the major results of that growth: the increasing need for energy (Chapters 10 to 12), depletion of resources (Chapters 13 to 18), and rising pollution (Chapters 19 to 23).

Many of the environmental problems we consider in this book are serious ones that must be addressed—if not by us, then unavoidably by our children. Environmental science is not, however, simply a "doom and gloom" listing of problems coupled with predictions of a bleak future. To the contrary, its focus, and our focus as individuals and as world citizens, is on identifying, understanding, and solving problems that we as a society have generated. A great deal is being done, and more must be done to address the problems of today's world.

Environmental Sustainability

One of the most important and most frequently used terms in environmental science is **environmental sustainability**. According to the Brundtland report, *Our Common Future*, published in 1987 by the World Commission on Environment and Development, environmental sustainability is the ability to meet humanity's current needs without compromising the ability of future generations to meet their needs. Sustainability implies that the environment can function indefinitely without going into a decline from the stresses imposed by human society on natural systems (such as fertile soil, water, and air) that maintain life (Figure 1.1). When the environment is used sustainably, humanity's present needs are met without endangering the welfare of future generations. Environmental sustainability applies at many levels, including individual, community, regional, national, and global levels.

Environmental sustainability is based in part on the following ideas:

1. We must consider the effects of our actions on the health and well-being of the natural environment, including all living things.
2. Earth's resources are not present in infinite supply. We must live within limits that let renewable resources such as fresh water regenerate for future needs.
3. We must understand *all* the costs to the environment and to society of products we consume.
4. We must each share in the responsibility for environmental sustainability.

Many experts in environmental science think that human society is not operating sustainably because of the following human behaviors:

1. We are using nonrenewable resources such as fossil fuels as if they were present in unlimited supplies.
2. We are using renewable resources such as fresh water and forests faster than they can be replenished naturally.
3. We are polluting the environment with toxins as if the capacity of the environment to absorb them is limitless.

Figure 1.1 Planet Earth. Earth's life support system and the sun supply us with six essential requirements: air, water, food, energy, shelter, and clothing.

4. Our numbers continue to grow despite Earth's finite ability to feed and sustain us and absorb our wastes.

If left unchecked, these activities may reach the point of environmental catastrophe, threatening the life-support systems of Earth to the extent that recovery is impossible.

At first glance, the issues may seem simple. Why don't we just stop the overconsumption, population growth, and pollution? The solutions are more complex than they may initially seem, in part because of various interacting ecological, societal, and economic factors. Our inadequate scientific understanding of how the environment works and how different human choices affect the environment is a major reason that problems of environmental sustainability are difficult to resolve. Because the effects of many interactions between the environment and humans are unknown or difficult to predict, we generally do not know if corrective actions should be taken before our scientific understanding is complete.

The key challenge, then, is to meet immediate human needs while protecting the environment for the long term. This book tries to present a balanced evaluation of the many causes and repercussions of environmental problems. We also provide the scientific foundation to evaluate these problems. However, environmental problems interact with numerous competing interests and cannot be solved by science alone (see *Meeting the Challenge: The 2002 World Summit*; also see Chapters 2 and 3).

MEETING THE CHALLENGE

The 2002 World Summit

In 1992 representatives from most of the world's countries met in Rio de Janeiro, Brazil, for a ground-breaking summit officially called the *U.N. Conference on Environment and Development*. Countries attending the conference examined environmental problems that cross international borders: pollution and deterioration of the planet's atmosphere and ocean, a decline in the number and kinds of organisms, and destruction of forests.

In addition, the 1992 participants adopted *Agenda 21*, a complex action plan of **sustainable development** for the 21st century in which future economic development, particularly in developing countries, will be reconciled with environmental protection (see figure). The goals of sustainable development are improved living conditions for all people while maintaining a healthy environment in which natural resources are not overused and excessive pollution is not generated. A serious application of the principles of environmental sustainability to development will require many changes in such fields as population policy, agriculture, industry, economics, and energy use. Agenda 21 recommended more than 2,500 actions to deal with our most urgent environmental, health, and social problems.

In 2002 the United Nations sponsored the *World Summit on Sustainable Development* in Johannesburg, South Africa, to assess the progress and failures in the 10 years since the 1992 summit. The 2002 summit considered issues such as population growth, poverty, the depletion of fresh water, food security, the use of unsustainable energy sources, and land degradation and habitat loss. Some of the world's problems have worsened in the past decade. The human population has grown by more than 795 million since 1992. The per-capita incomes in more than 80 countries are lower than they were

ten years ago. More than 4.7 million people die each year from water-related diseases, a fact that is tied to the unavailability of basic sanitation services for about 2.4 billion people. About one third of Earth's soil has been degraded so much that it cannot be used to grow food.

Thus, it appears that, with few exceptions, little progress has been made during the past decade in solving the world's most serious environmental problems through implementation of the conventions agreed on at Rio. The real world is operating much as it always has. Also, many governments have shifted their attention from serious environmental priorities to other challenges such as terrorism and worsening international tensions. Meanwhile, scientific warnings on important environmental problems such as global climate change have grown increasingly severe.

The 2002 summit reached conceptual agreements and a produced a timetable for meeting certain goals that should reduce poverty and environmental degradation. But these agreements can only work if the world's nations enforce them. The 2002 summit, like the 1992 summit, has little power to accomplish anything. Despite the lack of significant change at the international level since the 1992 Earth Summit, important environmental progress has been made at national, state, and local levels. Many countries, for example, have enacted more stringent air pollution laws, including the phasing out of leaded gasoline. The Worldwatch Institute says that more than 100 countries have created sustainable-development commissions. Corporations that promote environmentally responsible business practices have joined to form the World Business Council for Sustainable Development. Also, the World Bank has invested $8.5 billion in sustainable development projects around the world.

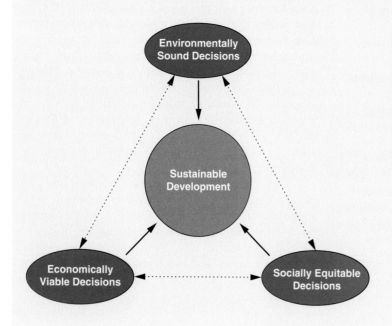

Sustainable development. Three factors—environmentally sound decisions, economically viable decisions, and socially equitable decisions—interact to promote sustainable development. Environmentally sound decisions do not harm the environment or deplete natural resources. Economically viable decisions consider *all* costs, including long-term environmental and societal costs. Socially equitable decisions reflect the needs of society and ensure that costs and benefits are shared equally by all groups.

Figure 1.2 Satellite view of North America at night. This image shows most major cities and metropolitan areas in the United States, Mexico, and Canada.

OUR IMPACT ON THE ENVIRONMENT

To gain some idea of the diversity of problems dealt with by today's environmental scientists, look at a newspaper. You will see many stories about the environment and our impact on it. In the pages that follow are a few environmental issues that have been important during the last several years. Each of these issues affects the quality of our lives, either directly or indirectly.

Increasing Human Numbers

Figure 1.2 is a portrait of about 421 million people. It is a satellite photograph of North America, including the United States, Mexico, and Canada, at night. The tiny specks of light represent cities, whereas the great metropolitan areas, such as New York along the northeastern seacoast, are ablaze with light.

The central problem of environmental science, the one that links all others together, is that there are many people in this picture—and soon there will be many more. According to the United Nations (U.N.), in 1950 only eight cities in the world had populations larger than 5 million, the largest being New York, with 12.3 million. By 2000 the largest city, Tokyo, Japan, had 26.4 million inhabitants, and the combined population of the world's 10 largest cities was 159.4 million (see Table 9.1). There are currently 16 megacities with populations greater than 10 million. Of the almost 400 cities worldwide that have a population of at least 1 million, 284 are in developing countries such as Brazil, India, and Indonesia. (See Chapter 8 for a discussion of highly developed and developing countries.)

In 1999 the human population as a whole passed a significant milestone: 6 billion individuals. Not only is this number incomprehensibly large, but our population has grown this large in a very brief span of time. In 1960 the human population was only 3 billion (Figure 1.3). By 1975 there were 4 billion people, and by 1987 there were 5 billion. The more than 6 billion people who currently inhabit our planet consume great quantities of food and water, use a great deal of energy and raw materials, and produce much waste.

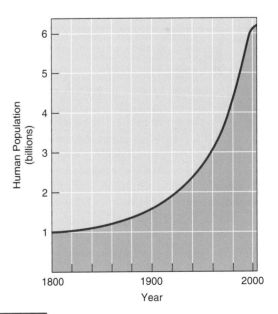

Figure 1.3 Human population numbers, 1800 to present. It took thousands of years for the human population to reach 1 billion (in 1800), 130 years to reach 2 billion (1930), 30 years to reach 3 billion (1960), 15 years to reach 4 billion (1975), 12 years to reach 5 billion (1987), and 12 years to reach 6 billion (1999).

Figure 1.4 Slum in Mumbai (Bombay), India. Many of the world's people live in extreme poverty. One trend associated with poverty is that poor people are increasingly moving from rural to urban areas. As a result, the number of poor people living in or around the fringes of cities is mushrooming.

Despite the vigorous involvement of most countries with family planning, population growth rates cannot be expected to change overnight. Several billion people will be added to the world in the 21st century, so even if we continue to be concerned about the overpopulation problem and even if our solutions are very effective, the coming decades may very well be clouded with tragedy.

The World Bank, which makes loans to developing countries for projects it thinks will encourage development, estimates that 2.8 billion—almost half of the world's people—are living in extreme **poverty**, unable to meet their basic needs for adequate food, clothing, shelter, education, or health (Figure 1.4). By one measure, poverty is defined as having a per-capita income of less than $1 per day, expressed in U.S. dollars adjusted for purchasing power; using this measure, 1.2 billion people currently live at this level of poverty. Poverty is associated with a low life expectancy, illiteracy, and inadequate access to health services, safe water, and balanced nutrition. According to the United Nations, 828 million people consume less than 80% of the recommended daily levels of food (calories).

The world population may stabilize by the end of the 21st century, given the family planning efforts that are currently under way. U.N. demographers have noticed a decrease in fertility rate worldwide to a current average of about three children per family, and the fertility rate is projected to continue to decline in coming decades. Population experts have made various projections for the world population at the end of the 21st century, from about 7.9 billion to 10.9 billion, based on how fast the fertility rate decreases.

No one knows whether Earth can support so many people indefinitely. Finding ways for it to do so represents one of the greatest challenges of our times. Among the tasks that we must accomplish is feeding a world population considerably larger than the present one without destroying the biological communities that support us. The quality of life available to our children and grandchildren will depend to a large extent on our ability to achieve this goal.

A factor as important as population *size* is a population's level of *consumption*. Inhabitants of the United States and other highly developed countries consume many more resources per person than do citizens of developing countries such as Nigeria, India, and Peru. Our high rate of resource consumption affects the environment at least as much as the explosion in population that is occurring in other parts of the world.

Population, Consumption, and Environmental Impact: A Simple Model   When you turn on the tap to brush your teeth in the morning, you probably do not think about where the water comes from or about the environmental consequences of removing it from a river or the ground. Likewise, most North Americans do not think about where the energy comes from when they flip on a light switch or start their car, van, or truck. Many of us do not realize that all of the materials that make up the products we use everyday come from Earth, nor do we grasp that these materials eventually are returned to Earth, much of it in sanitary landfills.

Such human impacts on the environment are difficult to assess. Since the 1970s, they have been estimated by using the three factors that are most important in determining environmental impact *(I):*

The number of people *(P).*

The affluence per person, which is a measure of the consumption or amount of resources used per person *(A).*

The environmental effects (resources needed and wastes produced) of the technologies used to obtain and consume the resources *(T).*

$$I = P \times A \times T$$

This model, first proposed by biologist Paul R. Ehrlich and physicist John P. Holdren, shows the mathematical relationship between environmental impacts and the forces that drive them. To determine the environmental impact of carbon dioxide (CO_2) emissions from motor vehicles, for example, multiply the population times the number of cars per person (affluence/consumption per person) times the average annual CO_2 emissions per year (technological impact). This model demonstrates that although increasing motor vehicle efficiency and developing cleaner technologies will reduce pollution and environmental degradation, a larger reduction will result if population and per-capita consumption are also controlled.

The *IPAT* equation, while useful, must be interpreted with care, in part because we often do not understand all of the environmental impacts of a particular technology. Motor vehicles, for example, are linked not only to global warming from CO_2 emissions but also to air pollution (tailpipe exhaust), water pollution (improperly disposed motor oil and antifreeze), stratospheric ozone depletion (from leakage of air conditioner coolants), and solid waste (disposal of automobiles in sanitary landfills). There are currently about 600 million motor vehicles on the planet, and the number is rising.

The three factors in the *IPAT* equation are always changing in relation to each other. For example, consumption of a particular resource may increase, but technological advances may decrease the environmental impact of the increased consumption. Consumer trends and choices may affect environmental impact. The average fuel economy of new cars and light trucks (sport utility vehicles, vans, and pickup trucks) in the United States declined from 22.1 miles per gallon in 1988 to 20.4 miles per gallon in 2001, in part because of the popularity of sport utility vehicles (SUVs). In addition to being about 29% less fuel efficient than cars, SUVs emit between 30% and 75% more emissions. Because of such trends and uncertainties, the *IPAT* equation is of limited usefulness when making long-term predictions.

The equation is valuable in that it helps to identify what we do not know or understand about consumption and its environmental impact. In 1997 the National Research Council organized by the National Academy of Sciences to advise the federal government on complex issues in science and technology, examined our knowledge about consumption in *Environmentally Significant Consumption: Research Directions*. It identified research areas that need to be addressed, such as: Which kinds of consumption have the greatest destructive impact on the environment? Which groups in society are responsible for the greatest environmental disruption? How can the activities of these environmentally disruptive groups be altered? It will take years to address such questions, but the answers should help decision makers in government and business formulate policies that will alter consumption patterns in an environmentally responsible way. Our ultimate goal should be to make consumption sustainable so that our current practices do not compromise the abil-

ity of future generations to use and enjoy the riches of our planet.

To summarize, as human numbers and consumption increase worldwide, so does humanity's impact on Earth, posing new challenges to us all. Success in achieving sustainability in population size and consumption will require the cooperation of all the world's peoples. (Human population issues are examined further in Chapters 8 and 9.)

Endocrine Disrupters

The activities of Earth's growing human population are known to have a profound, sometimes global, effect on the environment. Is it possible that some of the thousands of chemicals that humans produce and use could threaten the health of animals, including the human species? Unfortunately, the answer may be yes. Mounting evidence suggests that dozens of widely used industrial and agricultural chemicals are **endocrine disrupters** that mimic or interfere with the actions of the endocrine system (the body's hormones) in humans and wildlife. These chemicals include chlorine-containing industrial compounds known as PCBs and dioxins; the heavy metals lead and mercury; some pesticides such as DDT, kepone, dieldrin, chlordane, and endosulfan; and certain plastics and plastic additives such as phthalates.

In 2001 the Centers for Disease Control and Prevention (CDC) reported that they tested a diverse sample of the civilian U.S. population for the presence of 27 different environmental chemicals—heavy metals, nicotine from tobacco smoke, pesticides, and plastics, many of which are thought to be endocrine disrupters. Nearly every chemical was detected in every participant's body at higher levels than previously guessed. Because scientists had never actually measured 24 of the 28 chemicals in humans before, the study could not say if the levels of various environmental chemicals are increasing or decreasing. However, the study provides a good baseline of exposure of the U.S. population to these compounds against which future studies (the CDC plans to continue the tests yearly) can be compared. The study is also a first step toward determining if any of these chemicals is the hidden cause of modern illnesses or simply a benign by-product of modern society.

Hormones are chemical messengers produced by organisms in minute quantities to regulate their growth, reproduction, and other important biological functions. Some endocrine disrupters mimic the estrogens, a class of female sex hormones, and send false signals to the body that interfere with the normal functioning of the reproductive system. Because both males and females of many species, including humans, produce estrogen, endocrine disrupters that mimic estrogen can affect both sexes. Additional endocrine disrupters interfere with the endocrine system by mimicking hormones other than estrogen, such as androgens (male hormones such as testosterone) and thyroid hormones. Like hormones, endocrine disrupters are active at very low concentrations and, as a result, may cause significant health effects at relatively low doses.

Many endocrine disrupters appear to alter reproductive development in males and females of various animal species. Accumulating evidence indicates that fishes, frogs, birds, reptiles such as turtles and alligators, mammals such as polar bears and otters, and other animals exposed to these environmental pollutants exhibit reproductive disorders and are often left sterile.

A chemical spill in 1980 contaminated Lake Apopka, Florida's third largest lake, with DDT and other agricultural chemicals with known estrogenic properties. Male alligators living in Lake Apopka in the years following the spill had low levels of testosterone (an androgen) and elevated levels of estrogen. Their reproductive organs were often feminized or abnormally small. The mortality rate for eggs in this lake is still extremely high—only 40% hatch, and half of these die within 10 days (Figure 1.5 and Table 1.1).

Humans may be equally at risk from endocrine disrupters, as the number of reproductive disorders, infertility cases, and hormonally related cancers (such as testicular cancer and breast cancer) appears to be increasing. More than 60 studies since 1938 have reported that sperm counts in a total of nearly 15,000 men from many nations, including the United States, dropped by more

| Table 1.1 | The Population of Juvenile Alligators in Lake Apopka After a Chemical Spill in 1980 |

Year	Number of Alligators, Size 30 to 120 Centimeters, per Kilometer of Shoreline*
1980	21.9**
1981	21.8
1982	10.5
1983	3.2
1984	2.2
1985	1.8
1986	3.1
1987	2.1
1988	1.2
1989	0.4
1990	2.5
1991	3.0
1992	3.8
1993	5.2
1994	2.7

* Counts are adjusted for changes in water level.
** Chemical spill occurred just prior to the 1980 surveys.

than 50% between 1940 and 1990. Although only one sperm is needed to fertilize an egg, infertility increases as sperm counts decline. Scientists do not know if there is a link between this apparent decline and environmental factors, however.

Certain phthalates, ingredients of cosmetics, fragrances, nail polish, and common plastics used in a variety of food packaging, toys, and household products, have been implicated in birth defects and reproductive abnormalities. Animal studies in which rats are fed large doses of phthalates have shown that these compounds can damage fetal development, particularly of reproductive organs. In humans there may be a correlation between certain phthalates and the increase since the 1970s in the incidence of hypospadias, a birth defect in which the urethral opening (passageway for urine) is on the underside of the penis instead of at its tip. Certain phthalates have also been linked to an increased observance of premature breast development in young girls, typically between 6 and 24 months of age. Other phthalates are of negligible concern, according to a review by the National Toxicology Program's Center for the Evaluation of Risks to Human Reproduction.

Definite links between environmental endocrine disrupters and human health problems cannot be made at this time because of the limited number of human studies. Human exposure to endocrine-disrupting chemicals needs to be quantified so we know exactly how much of these chemicals affect various communities. Complicating such assessments is the fact that humans are also exposed to *natural*, hormone-mimicking substances in the plants we eat. Soy-based foods such as bean curd and soymilk, for example, contain natural estrogens.

Figure 1.5 Lake Apopka alligators. A young American alligator (*Alligator mississippiensis*) hatches from eggs taken by University of Florida researchers from Lake Apopka, Florida. Few Lake Apopka eggs actually hatch, and many of the young alligators that do hatch have abnormalities in their reproductive systems. This young alligator may not leave any offspring.

It is not known how the hundreds of different endocrine-disrupting chemicals interact. Sometimes two or more pollutants interact in such a way that their combined effects are more severe than the sum of their individual effects, a phenomenon known as **synergism**. The combined effects of two or more pollutants may also be less severe than the sum of their individual effects, a phenomenon known as **antagonism**. Our ignorance of possible synergisms and antagonisms among the many different endocrine-disrupting chemicals makes it very difficult to conclusively understand their individual and collective effects.

Congress amended the **Food Quality Protection Act** and the **Safe Drinking Water Act** in 1996 to require the U.S. Environmental Protection Agency to develop a plan and establish priorities to test thousands of chemicals for their potential to disrupt the endocrine system. In the first round of testing, chemicals are tested to see if they interact with any of five different endocrine receptors. (The body has specially shaped receptors on and in cells to which specific hormones attach. Once a hormone attaches to a receptor, it triggers other changes within the cell.) Chemicals testing positive—that is, binding to one or more types of receptors—are subjected to an extensive battery of tests to determine what specific damages, if any, they cause to reproduction and other biological functions. These tests, which may take decades to complete, should reveal the level of human and animal exposure to endocrine disrupters and the effects of this exposure.

The hypothesis that endocrine disrupters could affect the reproductive health of animals, including humans, was not widely recognized until the 1990s. *One lesson we can learn from this is that serious environmental problems sometimes take us by surprise. Another lesson is that the absence of scientific certainty about a chemical's effects does not guarantee its safety.* (The effects of toxic chemicals, including several endocrine disrupters, are discussed in Chapters 19, 21, 22, and 23.)

Closing the Georges Bank Fishery

In late 1994, the U.S. Commerce Department closed two large portions of Georges Bank, a 16,500-km^2 (6,600-mi^2) area off the coast of New England in the North Atlantic Ocean and once one of the world's richest fishing grounds. These closures were not a surprise to any observer of the fishing industry. Catches of cod, haddock, and yellowtail flounder had been steadily declining there for several decades (Figure 1.6). During that period, for example, the haddock population dropped by 75%, primarily due to overfishing. *Overfishing* means that fishes are harvested faster than they can replace themselves. As a result of the closure of the Georges Bank fishery, thousands of people lost their jobs.

Since the closures were declared, fish populations, particularly haddock and yellowtail flounder, have started

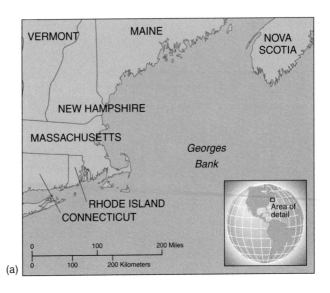

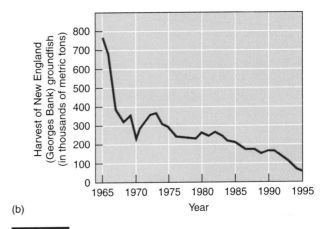

Figure 1.6 Georges Bank. (a) Location of Georges Bank off the coast of New England. (b) Annual harvests from northeastern U.S. coastal waters, including the Georges Bank. The harvest of groundfishes (such as cod, haddock, and flounder) from the Georges Bank fishing grounds declined steadily until two large portions were closed in December 1994. The closures are still in effect.

to increase. However, fish stocks are currently less than half their sustainable levels, according to the Marine Fish Conservation Network, a coalition of more than 150 national and regional environmental organizations, commercial and recreational fishing groups, and marine science groups. Fishery experts do not know how long it will take for the Georges Bank fishery to recover to the point that it can sustain large commercial harvests, but the evidence now indicates it is possible to restore most populations to higher levels of abundance than exist at present.

The closure of the Georges Bank fishery is not unique. The Alaskan salmon fishery declined during the 1960s and 1970s; Canada's Atlantic cod fishery along the Grand Banks closed in the 1990s; and in 1998 a moratorium on commercial salmon fishing was declared in the

North Atlantic. Fish shortages are occurring all over the globe as a result of overfishing. According to the National Research Council, 80% of 191 commercially important fish stocks in the United States are fully exploited or over-fished. The problem is particularly acute in the Pacific Northwest, Nova Scotia, and the Gulf of Mexico. World-wide, about 30% of fish species have been overfished.

Overfishing occurs because the world demand for fish has grown, and to meet the demand, there are more large fishing boats than ever before. In addition, more and more fishermen are using high-tech methods to find and harvest fishes. Advanced technologies include the use of sonar and Comsat satellite images to locate fishes. International fishing fleets roam the ocean, competing with one another, and as fish populations dwindle, the competition becomes more and more fierce. According to the United Nations, there are at least 100 ongoing disputes among nations over fishing rights; some of these disputes have led to violent clashes at sea.

In 1996 Congress reauthorized the **Magnuson–Stevens Fishery Conservation and Management Act**, which was originally enacted in 1976. The current law requires that the National Marine Fisheries Service and eight regional councils that regulate fishing under U.S. jurisdiction—that is, from the U.S. coast to 200 nautical mi (370 km) offshore—devise detailed strategies, such as strict quotas, to help declining marine fish populations recover.

The United States is not the only nation to recognize that declining fish catches are a serious issue. In 1995 the United Nations approved the **U.N. Fish Stocks Agreement**, the first international treaty to regulate marine fishing. The treaty went into effect in December 2001. At that time, 59 nations had signed the treaty, and 31 countries had ratified it. (The United States has ratified the treaty.) Because the problem continues to escalate, the United Nations has sponsored other overfishing pacts since 1995. A plan to reduce fishing capacity by capping the size of fishing fleets now and by reducing the number of fishing vessels, starting between 2003 and 2005, was drafted in 1999. Worldwide, there are thought to be about 4 million fishing vessels, with 40,000 of these larger than 100 tons and longer than 23.8 m (78 ft).

Governments will also have to reduce or remove subsidies that help support the fishing industry. (A **subsidy** is a form of government support given to a business or institution to promote the activity performed by that business or institution.) The U.N. Food and Agricultural Agency has determined that worldwide, the fishing industry annually spends at least $50 billion more than it earns because it is becoming more and more expensive to net fewer and fewer fish. Government subsidies, which encourage modernization and expansion of fishing fleets, help to make the fishing industry profitable.

One lesson that the closing of the Georges Bank fishery teaches us is that as technologies such as more effective fishing techniques advance, so does the impact we have on the environment. As a result, we must regulate our impacts, such as over-harvesting, more closely than ever. It is a theme you will encounter repeatedly as you proceed through the text. (The topic of declining fisheries is revisited in Chapter 18.)

Declining Bird Populations

Evidence is mounting that the populations of many birds that were once abundant have been dropping steadily for about two decades. Population declines have been noted across the North American continent, particularly in songbirds of forests, shrublands, and grasslands. Many songbirds are *tropical migrants* that spend the winter in Central and South America and the Caribbean and then migrate north to the United States and Canada to breed during the summer. Cerulean warblers, olive-sided fly-catchers, yellow-billed cuckoos, eastern wood pewees, and wood thrushes are examples of tropical migrants with declining numbers. The population of cerulean warblers, for example, declined by 51% from the 1960s to the 1990s, according to Breeding Bird Survey data.

Tropical migrants are faced with changing environments in both their winter and summer homes, and loss of habitat appears to be an important reason for their decline. These migratory birds are under stress from the burning of tropical rain forests in Central and South America as well as the fragmenting of forest habitat in North America to accommodate suburban development, agriculture, and logging. Forest **fragmentation** is the division of large, unbroken areas of forest into smaller areas by roads, fields, and cities,

Several studies have shown that fragmentation of northern forests increases the likelihood of reproductive failure in tropical migrants, many of which are insect eaters that nest only in forests. As a result of forest fragmentation, the nests are more likely to be located near a **forest edge**, the often-sharp boundary between the forest and residential neighborhoods, surrounding farm-lands, or clear-cut areas rather than deep in the forest. Nests that are within 100 m (328 ft) of a forest edge are more vulnerable to predation by small mammalian predators such as raccoons, opossums, snakes, and feral cats (domestic cats that have gone wild). Blue jays, American crows, and other egg-eating birds do most of their hunting along the forest edge. (Edges are discussed in the context of species richness in Chapter 5.)

Nest parasitism by brown-headed cowbirds is also more significant along the forest edge. Female cowbirds lay their eggs in the nests of other bird species and leave all parenting jobs to the hosts (Figure 1.7). In a 1989 to 1993 study that monitored 5,000 nests of tropical migrants, the majority of nests were parasitized in areas with less than 55% forest cover, and in many cases there were more cowbird eggs in the nests than eggs from the host species. Nest parasitism causes reproductive failure because the host parents provide food for the larger,

Figure 1.7 Nest parasitism. Note the cowbird (*Molothrus ater*) egg, which is speckled, in the nest of a wood thrush (*Hylocichla mustelina*). When the wood thrush nestlings hatch, they will have to compete for food with their larger, more aggressive foster nestling. More cowbird eggs are laid in songbird nests along the forest edge than deep within a forest.

more aggressive cowbird babies, while their own young starve.

Cowbirds avoid nesting or feeding in intact forests. Before Europeans settled in North America, cowbirds followed the migratory herds of bison across the Great Plains, foraging on seeds and insects stirred up by the bison. With the settlement of the continent by Europeans, many forests were cut to provide agricultural lands, and the cowbird expanded its range into fields and cattle pastures. The cowbirds began parasitizing new bird species that had not yet evolved ways to resist them. (Some birds, such as robins, gray catbirds, and Baltimore orioles, reject nests with cowbird eggs.)

An important conclusion we can make regarding declining bird populations is that human activities such as forest fragmentation often unintentionally benefit some species at the expense of others. (The effects of habitat fragmentation on organisms are discussed further in Chapter 16.)

Reintroducing Wolves to Yellowstone

Beginning in 1995 and 1996, one of the most controversial reintroductions of an endangered species took place in Yellowstone National Park and the central Idaho wilderness. Under a provision of the 1973 **Endangered Species Act (ESA)**, a small number of gray wolves (*Canis lupus*) were captured in Canada and released into Yellowstone National Park in Wyoming and wilderness area in Idaho by the U.S. Fish and Wildlife Service (FWS) (Figure 1.8). The two populations have thrived and increased to about 300 individuals. The ultimate goals of the reintroduction were to establish gray wolves as a functional part of the Yellowstone ecosystem and to remove the gray wolf from the endangered species list.

Wolves originally ranged from northern Mexico to Greenland, but they have been trapped, poisoned,

ENVIROBRIEF

Coffee That's for the Birds

Coffee drinkers may hear "bird friendly or unfriendly?" along with "regular or decaf?" at their favorite coffee spot. Many species of migratory songbirds, favorites among North American bird-lovers, are in decline, and it appears that Americans' coffee habits may play a role. In Central and South America, high-yield farms that cultivate coffee in full sunlight are rapidly replacing traditional coffee farms, known as shade plantations. Shade plantations are coffee plants grown in the shade of tropical rainforest trees. These trees support a vast diversity of songbird species that winter in the tropics—as many as 150 species in 5 hectares in one study—as well as large numbers of other vertebrates and insects. In contrast, sun plantations provide poor bird habitat. Because sun-grown varieties of coffee are treated with large inputs of chemical pesticides and fertilizers, they outperform the shade-grown varieties. About half of the region's shade plantations have been converted to sun plantations since the 1970s.

Although there is no direct proof that the reduction in the number of shade plantations has harmed songbird populations, the number of birds has dropped alarmingly during the same time period. Wood thrushes have declined by 40%, golden-winged warblers by 46%, and orchard orioles by 29%. Various conservation organizations and development agenicies such as the World Bank and USAID have initiated programs to certify coffee as "shade-grown," which would allow consumers the chance to support the preservation of tropical rain forest. Shade-grown coffee typically costs more than sun-grown coffee because it is hand picked, and only ripe berries are picked. In contrast both ripe and unripe berries of sun-grown coffee are picked when the coffee is harvested.

snared, and hunted to extinction in most places. To prevent wolves from attacking livestock, a federal program assisted ranchers in removing wolves from Yellowstone and the American Rocky Mountains in the 1930s. By 1960, the only wolves remaining in the lower 48 states were small populations in Minnesota. Under the protection offered by the ESA, however, small packs of wolves have spread from Canada into Montana and Washington and from Minnesota into Michigan and Wisconsin.

Many biologists and environmentalists support the wolf reintroduction because it has restored a more natural balance between predators and prey. The wolves' main prey are elk, deer, and an occasional bison. The packs of wolves that roam these wilderness areas have reduced Yellowstone's huge elk population, which, in the absence of natural predators, was at an all-time high. When elk numbers are not managed properly, they overgraze their habitat, and thousands starve during hard winters. Biologists are studying the dynamics of predator and prey as the wolf populations increase.

The pruning effect that wolves are having on elk will cause broader, long-term impacts. Yellowstone's elk population eventually could be reduced by as much as 20%,

Figure 1.8 Reintroduction of gray wolves into Yellowstone National Park. The reintroduction has caused many changes in plant and wildlife populations in Yellowstone.

which would relieve heavy grazing pressure on aspen, willow, and other plants and encourage a more lush and varied plant composition. Richer vegetation will support more herbivores such as beavers and snow hares, which in turn will support small predators such as foxes, badgers, and martens. The increase in beavers, which build dams across streams, will increase the amount of wetland habitat, on which numerous species depend.

The wolves are also affecting coyotes, which are potential prey for wolves. Wolf packs have decimated some coyote populations. A reduction in coyotes has allowed populations of the coyotes' prey, such as ground squirrels and chipmunks, to increase. Scavengers such as ravens, bald eagles, and grizzly bears have benefited from dining on scraps from wolf kills.

The reintroduction of wolves has had a positive impact on tourism in the Yellowstone area. There has been a surge in tourists, and surveys indicate that visitors want to see wolves more than any other animal in the park. Thousands of people have reported catching glimpses of wolves through binoculars or hearing their song.

The reintroduction of wolves did not occur without a fight. The FWS was given the green light only after holding more than 150 public hearings, conducting numerous scientific studies, and considering 160,000 comments and public testimonies. Two lobbying groups that represent farmers and ranchers filed suit challenging the program. Ranchers who live in the area were against the reintroduction because their livelihood depends on livestock being safe from predators.

Under a compromise developed to deal with this concern, ranchers are allowed to kill Yellowstone wolves that attack their cattle and sheep, and federal officers can remove any wolf that threatens humans or livestock. To interject flexibility into the ESA, which forbids killing an endangered species, the Yellowstone wolf was declared an "experimental nonessential" in the area. There are prob-

lems with this designation, however. How do ranchers determine if a wolf that kills livestock is from Yellowstone (and therefore can be killed) or is naturally occurring (and therefore cannot be killed because it is protected)? This concern led a U.S. District Court to rule in 1997 that the "experimental nonessential" designation violates the ESA because it threatens the natural wolf population. The judge ordered the removal of all wolves from the park, a ruling that was overturned by a U.S. appeals court in 2000.

The Defenders of Wildlife, an environmental organization that concentrates on protecting wild plant and animal species in their natural environments, has a wolf compensation fund that reimburses ranchers for full market value of livestock lost to wolves. However, the rancher must prove that a wolf made the kill, which means that the carcass must be examined within 24 hours, before it decomposes too much to evaluate the teeth marks.

The reintroduction of wolves to Yellowstone is a vivid lesson that environmental issues are often complicated by differing opinions about what should be done. Compromise is essential to solving environmental problems. (Chapter 16 expands on endangered species and the ESA.)

The Introduction of Invasive Species

Cargo-carrying ocean vessels carry water from their ports of origin in order to increase their ships' stability in the open ocean. When they reach their destinations, this water, known as *ballast water*, is discharged into local bays, rivers, or lakes. According to the U.S. Coast Guard, the United States annually receives more than 79 million tons of ballast water from foreign seas. Ballast water may contain clams, mussels, worms, small fishes, and crabs, in addition to millions of microscopic aquatic organisms. These organisms, if they become successfully established in their new environment, can threaten the area's aquatic environment and contribute to the possible extinction of native organisms. Foreign species whose introduction causes economic or environmental harm are called **invasive species.**

In 1982, an invasive species from the United States—a jellyfish-like organism called a comb jelly—arrived somewhere in the Black Sea. Carried there in the ballast water of a ship, the comb jelly quickly became established because it had an ample food supply (microscopic plankton) and no natural predators in its new environment. As its numbers increased, it consumed so much plankton that anchovies and other fishes were deprived of their food supply. The comb jelly's success has all but eliminated commercially important fishes in the Black Sea. Fish catches, once an important part of the local Russian economy, have declined dramatically since the late 1980s, as comb jellies became so dominant. Marine biologists say the establishment of comb jellies in the Black Sea is one of the most devastating biological invasions to occur in the past 50 years.

species. Each introduction seems a unique accident when it happens, but the overall problem is general and very serious because our highly mobile society and globilization of trade have facilitated the movement of foreign organisms.

The **Invasive Species Management Plan**, which went into effect in 2001, strengthens the **National Invasive Species Act of 1996** and coordinates the policies of dozens of U.S. federal agencies that deal with the problem of aquatic invasive species. One of the unintended consequences of international treaties such as the North American Free Trade Agreement is that they increase the risk of invasion (because they increase the number of international shipments). Meanwhile, since 1999 all ships that enter U.S. ports are required to empty their ballast tanks at sea and refill them with seawater; this move will hopefully help stem the tide of aquatic invasions.

One lesson we can learn from this is that all too often introduced species produce havoc in the environment and generate high economic costs. (Chapter 16 expands on the problems of introduced species.)

Figure 1.9 Zebra mussels. Zebra mussels (*Dreissena polymorpha*), shown here clogging a pipe, have caused billions of dollars in damage in addition to displacing native clams and mussels.

One of North America's greatest biological threats—and the continent's most economically damaging aquatic invader—is the zebra mussel, a native of the Caspian Sea that was probably introduced inadvertently through ballast water flushed into the Great Lakes by a foreign ship in 1985 or 1986. Since then, the tiny freshwater mussel, which clusters in extraordinary densities, has massed on hulls of boats, piers, buoys, and, most damaging of all, water intake systems (Figure 1.9). The zebra mussel's strong appetite for algae, phytoplankton, and zooplankton is also cutting into the food supply of native fishes, mussels, and clams, threatening their survival. By 1994, the zebra mussel had progressed from the Great Lakes into the Mississippi River. It is currently found in 19 states—as far south as New Orleans, as far north as Minnesota in the United States and Quebec in Canada, and as far east as the Hudson River in New York. The U.S. Coast Guard estimates that economic losses and control efforts cost the United States about $5 billion each year.

Unfortunately, invasions of exotic species are not unusual. The Great Lakes alone contain at least 139 introduced species. According to the National Research Council, more than 200 foreign species have been identified in the United States since 1980, and 25% of these have caused significant damage. Worldwide, most regions are estimated to contain 10% to 30% foreign

Invasion of the Jellyfish

Recently, dramatic increases in the populations of certain jellyfish have been noted in several parts of the world. The Bering Sea adjacent to Alaska and the Gulf of Mexico off Louisiana have been particularly hard hit. Scientists are concerned that jellyfish blooms (a "bloom" is the sudden appearance of a large number of individuals) may indicate imbalances in marine environments that may harm commercial fisheries. The Australian spotted jellyfish is an invasive species that was first noticed in the northern Gulf of Mexico in 2000 (see figure). It is not known how the spotted jellyfish got to the Gulf of Mexico from the tropical Pacific Ocean where it is native, nor is it known what conditions in the gulf are enabling it to thrive there. Scientists hypothesize that nutrient enrichment of the gulf's coastal waters may encourage an abundance of the phytoplankton on which the jellyfish feed. By eating much of the food that supports other aquatic organisms, a bloom of jellyfish could reduce local fish, crab, and shrimp populations.

Damage to the Atmosphere: Stratospheric Ozone Depletion

The swirling colors of Figure 1.10 are a computer-generated image of the South Pole, based on measurements taken from a satellite. This is a reconstruction of ozone levels in the **stratosphere**, the layer of the atmosphere that extends from 10 to 45 km (6.2 to 28 mi) above Earth's surface. As you can see, there is a large ozone "hole" over Antarctica, within which the ozone concentration has thinned. In September 2001, the ozone thinning covered an area larger than the size of North America. The quantity of ozone, measured by balloons released at the South Pole, dropped from 280 Dobson units, the normal level of ozone over the poles, to 99 Dobson units. (Dobson units are used to measure stratospheric ozone levels. They are named after British scientist Gordon Dobson, who studied ozone levels over Antarctica during the late 1950s.)

Ozone thinning was first hypothesized by two chemists at the University of California, Irvine on the basis of calculations they made in 1974 of the atmospheric effects of **chlorofluorocarbons** (**CFCs**). CFCs, a group of chemicals familiar as aerosol propellants in spray cans (now banned in most countries) and as the cooling agent Freon in refrigerators and air conditioners (now being phased out in most countries) are the main cause of ozone thinning.

British scientists reported the first direct observations of ozone thinning in 1985. The thinning is not a permanent feature but a seasonal phenomenon, evident only for a few months at the onset of the Antarctic spring in September. In 1990 the minimum ozone concentration was 50% lower than the minimum 10 years earlier. And by 1992 there was clear evidence that stratospheric ozone was also thinning over the Arctic.

Antarctic ozone thinning is related to a gradual thinning of stratospheric ozone worldwide. In the heavily populated midlatitudes of the planet, ozone levels have been decreasing for several decades. Why is this worrisome? Stratospheric ozone absorbs harmful ultraviolet radiation from sunlight. Without its protection, human skin cancers caused by ultraviolet radiation would become far more common. The increased ultraviolet radiation would also harm other species.

As a direct result of public awareness of the problem, strong national and international laws were eventually passed restricting the use of CFCs. Since 1987, more than 160 nations have agreed to a total ban on CFC production. For the United States and other highly developed countries, the phase-out date for CFCs and similar compounds was 1996, although existing stockpiles may be used after that date. Developing countries are on a different timetable and will phase out CFC production by 2006. Because CFCs are very stable, they can survive in the atmosphere 120 years or more, so the changes initiated by the release of CFCs may not be quickly reversible. However, evidence indicates that the area of ozone thinning over Antarctica will slowly become smaller each year. Full recovery of the stratospheric ozone layer is not expected until 2050.

An important lesson to learn from this story is that we often must deal with serious environmental problems on the basis of what we already know, even when there may be some gaps in the available scientific knowledge. Had we waited until all scientific evidence on ozone thinning had been gathered before we took any action, the problem would have been much worse than it is. (Stratospheric ozone depletion is examined further in Chapter 20.)

Global Climate Warming and Increasing Carbon Dioxide Levels

During the past two centuries, the level of CO_2 in the atmosphere has increased dramatically. Figure 1.11 shows the increase since 1958. The causes of the increase in atmospheric CO_2 are no mystery: The burning of fossil fuels such as coal, oil, and natural gas, and the clearing and burning of forests are major causes. Environmental scientists are increasingly concerned that the rising levels of CO_2 may change Earth's climate. Carbon dioxide levels rose from 315 parts per million (ppm) in 1958 to

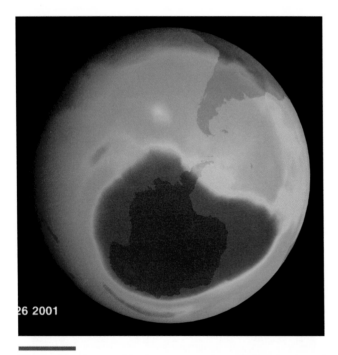

26 2001

Figure 1.10 Ozone thinning. A computer-generated image of part of the Southern Hemisphere, taken on September 26, 2001, reveals ozone thinning (purple area over Antarctica). The ozone hole is not stationary but moves about as result of air currents.

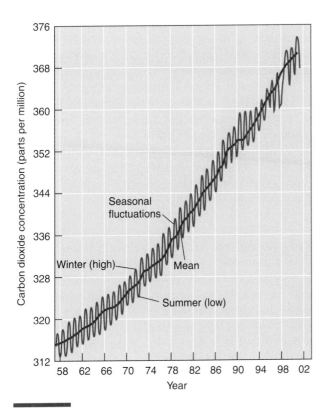

Figure 1.11 Carbon dioxide in the atmosphere, 1958 to 2001. Note the steady increase in the concentration of atmospheric CO_2 since 1958, when measurements began at the Mauna Loa Observatory. This location was selected because it is far from urban areas where factories, power plants, and motor vehicles emit CO_2. The seasonal fluctuation corresponds to winter (a high level of CO_2) and summer (a low level of CO_2) and is caused by greater photosynthesis in summer.

370.9 ppm in 2001 (latest data available). Carbon dioxide in the atmosphere allows solar radiation to pass through but does not allow heat to radiate into space. Instead, the heat is radiated back to Earth's surface. As CO_2 accumulates in the atmosphere, enough heat may be trapped to warm the planet gradually.

Observations confirm that our climate is changing rapidly, particularly in the past few decades. The 1980s, 1990s, and early 2000s saw some of the warmest years in weather records. Eight of the 10 warmest years have occurred since 1990 (the other two occurred in 1988 and 1987). Earth's temperature in 1998 was by far the warmest since 1860, and the second warmest temperature on record occurred in 2001.

Environmental scientists estimate that if trends are not changed, Earth's mean temperature could rise 1.4° to 5.8°C (2.5° to 10.4°F) by the end of the 21st century. This temperature increase would make the atmosphere warmer than tree ring and glacial ice analyses estimate it has been for the past 1,000 years. This warming could produce major shifts in patterns of rainfall and initiate

melting of the West Antarctic and Southern Greenland ice sheets, as did the last warm period 120,000 years ago. Such melting would cause the ocean level to rise, an alarming scenario, as it might put many major cities at least partly under water.

Definite conclusions about long-term trends are difficult to reach, but an increasing number of environmental scientists are concerned that human-induced global climate warming may well have started. The U.S. National Academy of Science has stated that global climate warming may be the most pressing international issue of the 21st century. In 2001 the U.N. Intergovernmental Panel on Climate Change (IPCC) released a report based on the consensus of hundreds of climate scientists worldwide. The IPCC declared that the human-produced increase in greenhouse gas concentrations are likely responsible for most of the observed warming in the last 50 years. Given the potential negative impacts of global warming, a do-nothing approach today is unwise and may be even foolhardy.

An international climate warming conference was held in Kyoto, Japan, in December 1997. Many countries came to the talks with conflicting positions, so it was impossible to reach a consensus about how to delay or prevent global warming. Nonetheless, an initial treaty known as the **Kyoto Protocol** was formulated. This stipulated that highly developed countries must cut their emissions of CO_2 and other gases that cause warming by an average of 5.2% by 2012; developing countries do not have to make comparable cuts. Unless the treaty is ratified and observed by many countries, CO_2 emissions are expected to continue to increase. (The United States signed the protocol in 1998 at a follow-up meeting in Buenos Aires, but the Bush administration withdrew the United States from that commitment.)

Energy experts at the U.S. Department of Energy say that much of the required U.S. emissions reductions could be accomplished by energy conservation measures and by an increase in energy efficiency. As much as 60 million metric tons of CO_2 emissions may be saved each year if cars are designed to get better gas mileage. Installing high-efficiency wind turbines to generate electricity in states such as Texas and California may save as much as 20 million metric tons of CO_2 emissions per year. (We say more about energy conservation and energy efficiency measures in Chapter 12 and global warming in Chapter 20.)

One important conclusion that we can make about climate warming is that the effects of human activities locally and regionally can have global repercussions, even changing the composition of the atmosphere. Another lesson is that deciding what to do to address global warming, the most devastating effects of which may not be felt for many decades, is a very challenging endeavor.

Destroying a Tropical Rain Forest

The fire seen in Figure 1.12 is in a tropical rain forest in the Amazon Basin of Brazil. The fire burned for weeks, turning thousands of acres of forest into charred stumps. This fire was set deliberately to clear the forest and produce grazing land, although it burned a much larger area than was originally intended. Many developing countries clear forests to make pasture or agricultural land. During the late 1990s rainforest loss in Brazil averaged about 2 million hectares (5 million acres) a year. In addition, forests are cleared in many parts of the world to meet the demand for timber or to provide firewood. (The poor in many developing countries have no alternative to wood for fuel.)

What are some of the environmental consequences of this particular forest being destroyed? Most obvious is the loss of hundreds of species of plants, animals, and microorganisms that lived in the forest. The worldwide decrease in biological diversity is one of the great concerns of scientists today.

Many tropical migrant songbirds are dependent on mature rain forests for part of their life cycle. When forests are destroyed, these birds are forced into more marginal land such as the scrubland that develops a few years after a rain forest has been cleared. Many of these birds are unable to build up the reserves needed for the long migration to North America. Studies indicate that even if these birds are able to make it to their northern homes, they are less likely to establish a nesting site and breed successfully.

Slash-and-burn agriculture, in which the forest is cut down, allowed to dry, and burned for cropland, also has adverse effects outside of the forest. The forest soil is no longer held in place by living trees and can be washed into surrounding streams, clogging them with sediment. Fish and other aquatic organisms may perish from the resulting sediment pollution, which clogs their gills and makes it difficult to find prey.

The burning of the rain forest may affect the quality and flow of drinking water for people living far from the forest. **Watersheds** are areas drained by river systems. When a watershed is covered by forest, the trees allow more water to be absorbed by the soil, which effectively filters out many contaminants. The forest regulates the flow of this water as it drains into rivers and their tributaries. With the clearing of the forest, most rainfall quickly washes into rivers without the benefits of filtration and flow regulation.

Rainforest clearing may contribute to global climate warming. Trees and other forest plants absorb CO_2 from the atmosphere and incorporate it into the carbon compounds found in wood and other plant tissues. Because forest trees can live for hundreds of years,

Figure 1.12 Fire in a tropical rain forest in Roraima, Brazil. This 1998 fire, originally set by ranchers and farmers to clear land, got out of control and burned almost 15% of Roraima.

they provide a long-term storage of CO_2. The combustion of the wood, however, releases this stored CO_2 into the atmosphere.

Clearing of the forest reduces the long-term potential for people living in the area to earn livelihoods from the forest. If harvested sustainably, tropical forests provide many nontimber resources, including latex for rubber, chicle for chewing gum, turpentine, cork, and various spices. In the future, a foreign company might purchase the right to **bio-prospect** in the forest—that is, to investigate the forest organisms as potential sources of chemical products, including drugs, flavorings, fragrances, and natural pesticides. Alternatively, the forest could be developed as a site for **ecotourism**, in which tourists pay to observe wildlife in natural settings. Sustainable timber practices and the promotion of nontimber products, including bio-prospecting and ecotourism, may provide people with more income than they would earn if the forest was cleared to grow crops or raise cattle.

One positive outcome of the Roraima fire and similar fires is that in 2000 Brazil implemented a fire prevention and control plan in the Amazon Basin that prohibits burning during the dry season. As a result, there has been a significant reduction in the number of accidental fires identified by satellite images throughout most of the heavily settled parts of the Amazon Basin.

The effects of forest loss are many and varied, from destruction of songbird habitat to exacerbated global warming. *One lesson we can apply to the forest clearing in Brazil is that environmental problems should not be studied as separate issues, because they are interconnected, often in surprisingly complex ways.* (We say more about forests and forest management in Chapter 17.)

■ THE ROLE OF ETHICS, VALUES, AND WORLDVIEWS IN ADDRESSING ENVIRONMENTAL PROBLEMS

Thus far in this chapter we have considered the scope of environmental science by examining several current issues. We now shift our attention to the views of different individuals and societies and how those views affect our ability to understand and solve environmental problems.

Ethics is the branch of philosophy that is derived through the logical application of human values. These **values** are the principles that an individual or society considers important or worthwhile. Values are not static entities but change as societal, cultural, political, and economic priorities change. Ethics helps us determine which forms of conduct are morally acceptable or unacceptable, right or wrong. Ethics plays a role in whatever types of human activities involve intelligent judgment and voluntary action. Whenever alternative, conflicting values occur, ethics helps us choose which value is better, or worthier, than other values.

Environmental ethics is a large and complex field of applied ethics that considers the moral basis of environmental responsibility and how far this responsibility extends. Environmental ethicists try to determine how we humans should relate to the natural environment. For example, what role should humans play in determining the fate of Earth's resources, including other species? How do we develop an environmental ethic that is acceptable in the short term for us as individuals but also in the long term for our species and the planet? These questions and others like them are difficult intellectual issues that involve political, economic, societal, and individual trade-offs.

Environmental ethics considers not only the rights of people living today, both individually and collectively, but also the rights of future generations. This aspect of environmental ethics is critical because the impacts of today's activities and technologies are changing the environment. In some cases these impacts may be felt for hundreds or even thousands of years. By defining our environmental ethics, we are in a better position to use science and technology for long-term environmental sustainability.

Worldviews

Each of us has a particular **worldview**—that is, a commonly shared perspective based on a collection of our basic values that helps us make sense of the world, understand our place and purpose in it, and determine right and wrong behaviors. These worldviews lead to behaviors and lifestyles that may or may not be compatible with environmental sustainability.

There are many worldviews; some worldviews share certain fundamental beliefs, whereas others are mutually exclusive. A worldview that is considered ethical in one society may be considered irresponsible or even sacrilegious in another. Here we present two extreme, competing **environmental worldviews,** what we'll call the Western worldview and the deep ecology worldview. These two worldviews, admittedly broad generalizations, are near opposite ends of a spectrum of worldviews relevant to global environmental problems, and each approaches environmental responsibility in a radically different way.

The traditional *Western worldview*, also known as the *expansionist worldview*, is human-centered and utilitarian. It mirrors the beliefs inherent in the 18th century **frontier attitude**, a desire to conquer and exploit nature as quickly as possible (see Chapter 3). The Western worldview includes human superiority and dominance over nature, the unrestricted use of natural resources, increased economic growth to manage an expanding industrial base, the inherent rights of individuals, and accumulation of wealth and unlimited consumption of goods and services to provide material comforts. According to the Western worldview, humans have a primary obligation to themselves and are therefore responsible for managing natural resources to benefit human society. Thus, any concerns about the environment are derived from human interests.

The *deep ecology worldview* is actually a diverse set of viewpoints that dates from the 1970s and is based on the work of **Arne Naess**, a Norwegian philosopher, and others, including ecologist **Bill Devall** and philosopher **George Sessions**. The principles of deep ecology as expressed by Arne Naess in *Ecology, Community, and Lifestyle*, include:

1. The flourishing of human and non-human life on Earth has intrinsic value. The value of non-human life forms is independent of the usefulness they may have for narrow human purposes.

2. Richness and diversity of life forms are values in themselves and contribute to the flourishing of human and non-human life on Earth.

3. Humans have no right to reduce this richness and diversity except to satisfy vital needs.

4. Present human interference with the non-human world is excessive, and the situation is rapidly worsening.

5. The flourishing of human life and cultures is compatible with a substantial decrease in the human population. The flourishing of non-human life requires such a decrease.

6. Significant change of life conditions for the better requires change in policies. These affect economic, technological, and ideological structures.

7. The ideological change is mainly that of appreciating life quality (dwelling in situations of intrinsic value) rather than adhering to a high standard of living. There will be profound awareness of the difference between big and great.

8. Those who subscribe to the foregoing points have an obligation directly or indirectly to participate in the attempt to implement the necessary changes.

Compared to the Western worldview, the deep ecology worldview represents a radical shift in how humans relate themselves to the environment. Deep ecology stresses harmony with nature, a spiritual respect for life, and the belief that humans and all other species have an equal worth. Ethically speaking, all forms of life have the right to exist, and humans are not different or separate from other organisms. Thus, humans are obligated both to themselves and to the environment. The deep ecology worldview advocates sharply curbing human population growth. It does not advocate returning to a society free of the technological advances we have today but instead proposes a significant rethinking of our use of current technologies and alternatives. It asks individuals and societies to share an inner spirituality connected to the natural world.

Ted Perry, a film scriptwriter, eloquently summarized the worldview that deep ecologists say must be embraced to help solve the many serious environmental problems of today. His quote is based on notes of a speech reportedly delivered in 1854 by Chief Sealth, chief of the Suquamish tribe in the Pacific Northwest, in response to President Franklin Pierce's offer to buy their land and provide them with a reservation:

How can you buy [the land,] ... or [we] sell the sky...? This idea is strange to us ... If we do not own the freshness of the air and the sparkle of the water, how can you buy them? ... This we know—the Earth does not belong to man, man belongs to the Earth. ... All things are connected like the blood which unites one family. Whatever befalls the Earth befalls the sons of the Earth. ... Man does not weave the web of life, he is merely a strand in it. Whatever he does to the web, he does to himself.

Most people today do not fully embrace either the Western worldview, which is anthropocentric, or the deep ecology worldview, which is biocentric. The planet's natural resources could not support its more than 6 billion humans if each consumed the high level of goods and services sanctioned by the Western worldview. On the other hand, if all humans were to adhere completely to the tenets of deep ecology, it would entail giving up some of the material comforts and benefits of modern technology. Further, the world as envisioned by the deep ecology worldview could only support a fraction of the existing human population (Figure 1.13). These worldviews, while not practical for widespread adoption, are useful to keep in mind as we examine various environmental issues in later chapters. In the meantime, you should think about your own worldview and discuss it

Figure 1.13 Embracing deep ecology. At one time or another, most of us yearn for a simpler life advocated by the tenets of deep ecology. However, there are far too many people and far too little land for us all to embrace this lifestyle. Photographed in Glencoe, Scotland.

with others. Listen carefully to their worldviews, which will probably be different from your own. Thinking leads to actions, and actions lead to consequences. What are the consequences of your particular worldview? *If the*

environment is to be sustainable for ourselves, other living organisms, and future generations, we need to develop and incorporate a longlasting, environmentally sensitive worldview into our culture.

SUMMARY WITH SELECTED KEY TERMS

I. Environmental science is the interdisciplinary study of humanity's relationship with other organisms and the nonliving physical environment.

A. Environmental science encompasses many complex and interconnected problems involving human numbers, Earth's natural resources, and environmental **pollution**.

B. Environmental science is interdisciplinary because it uses and combines information from many disciplines: biology, geography, chemistry, geology, physics, economics, sociology, cultural anthropology, natural resources management, agriculture, engineering, law, politics, and ethics.

C. Environmental sustainability is the ability to meet humanity's current needs without compromising the ability of future generations to meet their needs.

II. The world's human population surpassed 6 billion in 1999.

A. The increasing population is placing unsustainable stresses on the environment, as humans consume ever-increasing quantities of food and water, use more and more energy and raw materials, and produce enormous amounts of waste and pollution.

B. The World Bank estimates that 2.8 billion people are now living in extreme **poverty**, which means they are unable to meet their basic needs for adequate food, clothing, shelter, education, or health.

C. One model of environmental impact *(I)* has three factors: the number of people *(P)*; the affluence per person *(A)*, which is a measure of the consumption or amount of resources used per person; and the environmental effect of the technologies used to obtain and consume those resources *(T)*.

 1. $I = P \times A \times T$
 2. This model shows the mathematical relationship between environmental impacts and the forces that drive them.

III. Human activities include the release of materials that may harm the environment.

A. Many widely used chemicals are **endocrine disrupters** that may interfere with the actions of **hormones**, chemical messengers that regulate growth, reproduction, and other activities in the body.

 1. Studies of wildlife and laboratory animals suggest that certain abnormalities are linked to endocrine disrupters.
 2. The U.S. Environmental Protection Agency is testing thousands of chemicals for their potential to disrupt the endocrine system. These tests will determine the level of

human and animal exposure to endocrine disrupters and what effects this exposure causes.

B. Chlorofluorocarbons (**CFCs**) are chemicals that were once widely used as cooling agents in refrigerators and air conditioners and are still used somewhat today. When released into the atmosphere, CFCs drift into the **stratosphere**, the layer of the atmosphere from 10 to 45 km above Earth's surface, and cause ozone thinning.

 1. Although the production of CFCs has largely been phased out, existing CFCs can survive in the atmosphere 120 years or more.
 2. Full recovery of the stratospheric ozone layer is not expected until 2050.

C. Human production of carbon dioxide (CO_2) may alter global climate.

 1. Both the burning of fossil fuels (coal, oil, and natural gas) and the clearing of forests increase CO_2 levels.
 2. Global CO_2 levels rose from 315 ppm in 1958 to 370.9 ppm in 2001.
 3. If CO_2 trends are not changed, Earth's mean temperature could rise 1.4° to 5.8°C by the end of the 21st century.
 4. The **Kyoto Protocol** stipulates that highly developed countries must cut their emissions of CO_2 and other gases that cause warming. The United States has not committed to the Kyoto Protocol.

IV. Some other environmental challenges are caused by our attempts to manage organisms with which we share the planet.

A. The gray wolf, an endangered species, was reintroduced in Yellowstone National Park and the central Idaho wilderness where wolves had all been removed earlier in the 20th century.

 1. A provision of the 1973 **Endangered Species Act** allowed the reintroduction.
 2. Wolves have affected elk and other plant and animal populations in Yellowstone.

B. Humans are responsible for the introduction of foreign species that invade the environment, sometimes in dramatic fashion.

 1. Comb jellies from North America have all but eliminated commercial fishing in the Black Sea. Comb jellies eat microscopic plankton, thereby depriving fish of their food supply.
 2. Zebra mussels from the Caspian Sea have caused billions of dollars of damage to boats, piers, buoys, and water pipes in the Great Lakes. Zebra mussels have also threat-

ened native fishes, mussels, and clams by reducing their food supply.

V. Developing natural resources has adversely impacted the environment.

A. The Georges Bank, once one of the world's richest fishing grounds, has been overfished, and two areas have been closed to fishing since 1994. Fish stocks are currently less than half their sustainable levels.

 1. The Georges Bank is not unique, and fish shortages are occurring around the globe.

 2. In 1997 Congress reauthorized the **Magnuson–Stevens Fishery Conservation and Management Act**. It requires the National Marine Fisheries Service and eight regional councils to devise strict quotas and other strategies to help declining marine fish populations recover.

 3. The **U.N. Fish Stocks Agreement**, created in 1995, is the first international treaty to regulate marine fishing.

B. Tropical migrant songbird populations are declining.

 1. Tropical migrants spend the winter in Central and South America and the summer in North America.

 2. Their breeding habitats—the forested areas in North America—are increasingly **fragmented** by development. Because nests are more likely to be located near a **forest edge**, the nests are more vulnerable to **nest parasitism** by brown-headed cowbirds.

 3. Habitat fragmentation is also occurring in tropical migrants' winter homes.

C. All over the world, forests are being cleared to meet the demand for timber or firewood or for **slash-and-burn agriculture**.

 1. During the late 1990s, rainforest loss in Brazil averaged about 2 million hectares a year.

 2. Clearing of forest causes habitat loss, poor water quality (from erosion), watershed problems, global climate warming, and a reduced potential for local people to earn a sustainable livelihood from the forest.

VI. Ethics is the branch of philosophy that deals with human values.

A. Values are the principles that an individual or society considers important or worthwhile.

B. Environmental ethics is a field of applied ethics that considers the moral basis of environmental responsibility and how far this responsibility extends. Environmental ethicists try to determine how we should relate to nature.

C. Each of us has a particular **worldview**, or perspective based on a collection of our values that helps us understand our place in the world. There are many worldviews.

 1. The Western worldview, also known as the expansionist worldview, stresses human dominance over nature, the unrestricted use of natural resources, increased economic growth for an expanding human population, the inherent rights of individuals, and accumulation of wealth and unlimited consumption of goods and services to provide material comforts.

 2. The deep ecology worldview stresses spirituality with nature and advocates reducing human population growth and rethinking our use of current technologies. The deep ecology worldview is based partly on the work of the Norwegian **Arne Naess**.

THINKING ABOUT THE ENVIRONMENT

1. How does the Lewis Center at Oberlin College save energy? How does it reduce the amount of waste at local landfills? How does it conserve water? How does the building teach environmental ethics?

2. Explain the following Native American saying as it relates to the concept of environmental sustainability: We have not inherited the world from our ancestors; we have borrowed it from our children.

3. Do you think it is possible for the world to sustain its present population of more than 6 billion indefinitely? Why or why not?

4. Cite some of the evidence that environmental chemicals may be harming the endocrine systems of humans and wildlife.

5. Briefly discuss human involvement in one of the following: overfishing in the Georges Bank fishery, declining populations of tropical migrant birds, introduction of wolves into Yellowstone National Park, or invasion of zebra mussels in North America.

6. How do human effects on the global atmosphere, such as stratospheric ozone depletion and climate warming, make it more difficult for humans to reach the goal of environmental sustainability?

7. Describe some of the essential services that forests provide. How does destroying a forest affect these services?

8. State whether each of the following statements reflects the Western worldview, the deep ecology worldview, or both.

 a. Species exist to be used by humans.

 b. All organisms, humans included, are interconnected and interdependent.

 c. There is a unity between humans and nature.

 d. Humans are a superior species capable of dominating other organisms.

 e. Humans should protect the environment.

 f. Nature should be used, not preserved.

 g. Economic growth will help the Earth manage an expanding human population.

 h. Humans have the right to modify the environment to benefit society.

 i. All forms of life are intrinsically valuable and therefore have the right to exist.

 j. Humans are just one of the many kinds of organisms that inhabit the world.

*9. Use the *IPAT* equation to calculate the environmental impact in terms of CO_2 emissions per year at the beginning of the 21st century, when there were 6 billion people, an average of 0.1 motor vehicles per person, and 5.4 tons of CO_2 emitted by each car per year. Then make a similar calculation for the year 2050, based on these projections: a population of 10 billion people, 0.4 cars per person, and CO_2 emissions per vehicle similar to what we have today (that is, no technological improvements). How might we hold global CO_2 emissions from motor vehicles to 2000 levels in the year 2050?

*10. Use the data from Table 1.1 to graph the decline in the population of juvenile alligators in Lake Apopka after the pesticide spill in 1980. Refer to Appendix II for information and help in graphing.

* The solution appears in Appendix VII.

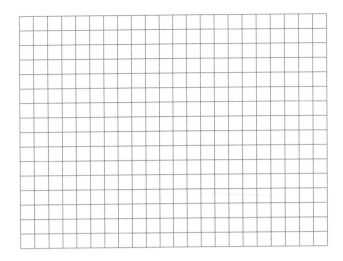

TAKE A STAND

Visit our Web site at **http://www.wiley.com/college/raven** (select Chapter 1 from the Table of Contents) for links to more information about the controversies surrounding the reintroduction of wolves into Yellowstone National Park. Consider the opposing views of those who seek to protect the wolves and of ranchers who wish to protect their livestock, and debate the issues with your classmates. You will find tools to help you organize your research, analyze the data, think critically about the issues, and construct a well-considered argument. Take a

Stand activities can be done individually or in a team, as oral presentations, written exercises, or Web-based (e-mail) assignments.

Additional on-line materials relating to this chapter, including a Student Testing Section with study aids and self-tests, Environmental News, Activity Links, Research Projects, and more, can also be found on our Web site.

SUGGESTED READING

Chadwick, D.H. "Return of the Gray Wolf." *National Geographic*, Vol. 193, No. 5 (May 1998). Controversy surrounds the reintroduction of the gray wolf to some of its former U.S. habitats.

Farrell, A., and M. Hart. "What Does Sustainability Mean?" *Environment*, Vol. 40, No. 9 (November 1998). It is possible to measure progress toward sustainability by using certain indicators.

Greene, K. "Bigger Populations Needed for Sustainable Harvests." *Science*, Vol. 296 (17 May 2002). The fishing industry, environmentalists, and scientists diverge on harvest rates needed to sustain a productive fishery on the Georges Bank.

Halweil, B. "Sperm Counts Are Dropping." *WorldWatch*, Vol. 12, No. 2 (March/April 1999). Many studies suggest that human sperm counts have been declining since the mid-20th century.

Halweil, B. "Why Your Daily Fix Can Fix More Than Your Head." *WorldWatch*, Vol. 15, No. 3 (May–June 2002). Shade-grown coffee helps preserve the rain forest, which is an important winter habitat for songbirds.

Harder, B. "Stemming the Tide." *Science News*, Vol. 161 (13 April 2002). Several techniques show promise in killing organisms in ballast water so they cannot gain a foothold in foreign waters.

Nepstad, D., et al. "Frontier Governance in Amazonia." *Science*, Vol. 295 (25 January 2002). This article highlights some of Brazil's promising trends in managing the tropical rain forest in its Amazon Basin.

Soroos, M.S. "Preserving the Atmosphere as a Global Commons." *Environment*, Vol. 40, No. 2 (March 1998). Examines the legal treaties and laws that address specific air pollution issues, including thinning of the ozone layer and global warming.

Temple, S.A. "Easing the Travails of Migratory Birds." *Environment*, Vol. 40, No. 1 (January–February 1998). This well-researched article examines population trends in tropical migrants and the role of humans in their problems.

Uhl, C., and A. Anderson. "Green Destiny: Universities Leading the Way to a Sustainable Future." *BioScience*, Vol. 51, No. 1 (January 2001). Universities should embrace a comprehensive ecological mission aimed at achieving sustainability in all aspects of univeristy operations.

U.S. Global Change Research Program. *Climate Change Impacts on the United States: The Potential Consequences of Climate Variability and Change.* Cambridge University Press: New York (2001). This massive report outlines the impacts of climate changes expected in the United States in the next 100 years.

Williams, T. "Lessons from Lake Apopka." *Audubon* (July–August 1999). An excellent review of Lake Apopka's environmental woes, which began when farmers started draining the lake in the 1940s.

Landscape elements and malaria risk in Mexico. Each village, surrounded by a 1-km buffer, is represented as a black ring; of these, the 10 villages predicted to be most at risk for malaria are shown as magenta circles. The remote-sensing model correctly identified 7 of the 10 villages (magenta circles with black dots) with the greatest abundance of mosquitoes and therefore the greatest risk of malaria transmission.

Using Science to Address Environmental Problems

Learning Objectives

After you have studied this chapter you should be able to

1. Outline the steps of the scientific method.

2. Distinguish between deductive and inductive reasoning.

3. Define risk assessment and explain how it helps determine adverse health effects.

4. Describe how a dose–response curve is used in determining the health effects of environmental pollutants.

5. Discuss the precautionary principle as it relates to the introduction of new technologies or products.

6. Explain how policy makers use cost–benefit analyses to help formulate and evalutate environmental legislation.

7. List and briefly describe the five stages of solving environmental problems.

8. Briefly describe the history of the Lake Washington pollution problem of the 1950s and how it was resolved.

9. Relate Garrett Hardin's description of the tragedy of the commons in medieval Europe to the global commons today.

Human health has improved significantly over the past several decades, but environmental factors remain a significant cause of human disease in many areas of the world. Epidemiologists, scientists who investigate the ecology of disease in a population, are establishing increasingly strong links between human health and human activities that alter the environment. The U.N. World Health Organization released a 1997 report that concluded that about 25% of disease and injury worldwide is related to environmental changes caused by humans. The environmental component of human health is sometimes direct and obvious, as when people drink unsanitary water and contract a waterborne disease agent that causes a condition such as diarrhea, which annually causes 4 million deaths worldwide, mostly in children.

The health effects of many human activities are complex and often indirect. The disruption of natural environments may give disease-causing agents an opportunity to break out of their isolation. Development activities such as cutting down forests, building dams, and agricultural expansion may bring more humans into contact with new or rare disease-causing agents. Alternatively, such projects may increase the spread of disease by increasing the population and distribution of disease-carrying organisms such as mosquitoes. Social factors may also contribute to disease epidemics. Human populations increasingly concentrate in large cities, permitting the rapid spread of infectious organisms among people. Global travel also has the potential to contribute to the rapid spread of disease as infected individuals move easily from one place to another.

Malaria, a disease transmitted to humans by mosquitoes, infects between 300 million and 500 million people worldwide and causes as many as 2.7 million deaths each year. About 60 species of *Anopheles* mosquitoes transmit the protozoan parasites that cause malaria. Each mosquito species has its own unique combination of environmental conditions (such as elevation, amount of precipitation, temperature, relative humidity, and availability of surface water) under which it thrives.

In some regions of the world, such as Peru and Bangladesh, the incidence of malaria is increasing, in part because of environmental changes. Areas of recently cleared forest tend to have small, temporary pools of water that provide ideal sites in which mosquitoes can breed. The incidence of malaria is demonstrably higher in parts of the Amazon where the forest has been cleared and the human population has expanded because of colonization. In the Amazon the construction of roads, which typically have drainage ditches on each side, has also benefited malaria-transmitting mosquitoes. It is anticipated that human-induced changes in world climate will affect the incidence of malaria by allowing expansion of the malaria-transmitting mosquito into areas that are not currently part of the mosquito's range. During the recent global warming trend, malaria has been noted at higher elevations in the tropics, which are warmer than they were previously.

One of the newest tools used by epidemiologists is remote-sensing data gathered by low-flying aircraft or satellites. The Center for Health Applications of Aerospace-Related Technologies (CHAART) is a branch of the National Aeronautics and Space Administration (NASA). In studying malaria in the coastal areas of southern Chiapas, Mexico, CHAART used satellite images of landscapes to predict which localities were most susceptible to malaria (see opening photo). They found that villages located near two landscape elements, transitional wetlands and unmanaged pastures, tended to have a greater incidence of malaria, and they successfully identified 7 of 10 villages with the highest mosquito abundance (and therefore greatest risk of malaria transmission). These villages were then targeted for interventions to reduce the population of mosquitoes.

In studying environmental problems that face the world today, from the increased incidence in diseases such as malaria to the pollution of freshwater lakes, it is important to remember that much can be done to improve our situation. *Environmental science is a problem-solving endeavor, with the role of identifying problems and suggesting and evaluating potential solutions.* Although the choice to implement a proposed solution is almost always a matter of public policy, environmental scientists play key roles in educating both government officials and the general public.

THE NATURE OF SCIENCE

The key to the successful solution of any environmental problem is rigorous scientific evaluation. It is important to understand clearly just what science is, as well as what it is not. Most people think of science as a body of knowledge—a collection of facts about the natural world. However, science is also a dynamic *process*, a systematic way to investigate the natural world. **Science** seeks to reduce the apparent complexity of our world to general principles, which can then be used to make predictions, solve problems, or provide new insights.

Scientists collect objective **data** (singular, *datum*), the information with which science works. Data are collected by observation and experimentation and then analyzed or interpreted. Scientific conclusions are inferred from the available data and are not based on faith, emotion, or intuition. A requirement of science is *repeatability*—that is, observations and experiments must produce consistent data when they are repeated.

Science is an ongoing enterprise, and scientific concepts must be reevaluated in light of newly discovered data. Thus, scientists can never claim to know the "final answer" about anything, because scientific understanding changes.

Several areas of human endeavor are not scientific. Ethical principles often have a religious foundation, and political principles reflect social systems. Some general principles, however, derive not from religion or politics but from the physical world around us. If you drop an apple, it will fall, whether or not you wish it to, and despite any laws you may pass forbidding it to do so. Science aims to discover and better understand the general principles that govern the operation of the natural world.

The Scientific Method

The established processes that scientists use to answer questions or solve problems are collectively called the **scientific method** (Figure 2.1). Although there are many variations of the scientific method, it basically involves five steps:

1. Recognize a question or unexplained occurrence in the natural world. After a problem is recognized, one determines what is already known about it by investigating the relevant scientific literature.

Figure 2.1 Scientific method. These steps provide the framework for scientific investigations. A simplified experiment is described on the right.

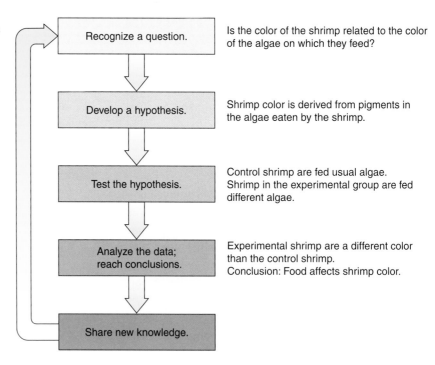

Recognize a question.

Is the color of the shrimp related to the color of the algae on which they feed?

Develop a hypothesis.

Shrimp color is derived from pigments in the algae eaten by the shrimp.

Test the hypothesis.

Control shrimp are fed usual algae. Shrimp in the experimental group are fed different algae.

Analyze the data; reach conclusions.

Experimental shrimp are a different color than the control shrimp.
Conclusion: Food affects shrimp color.

Share new knowledge.

2. Develop a **hypothesis**, or educated guess, to explain the problem. A good hypothesis makes a prediction that can be tested and possibly disproved. The same factual evidence can often be used to formulate several alternative hypotheses, each of which must be tested.

3. Design and perform an experiment to test the hypothesis. An experiment involves the collection of data by making careful observations and measurements (Figure 2.2). Much of the creativity in science is involved in designing experiments that help sort out the confusion caused by competing hypotheses. The scientific process never "proves" anything; instead, it disproves or falsifies alternative hypotheses until all that is left is the most plausible hypothesis.

4. Analyze and interpret the data to reach a conclusion. Does the evidence match the prediction stated in the hypothesis—that is, do the data support or refute the hypothesis? Does the hypothesis have to be modified or rejected on the basis of the observed data?

5. Share new knowledge with the scientific community. This is done by publishing articles in scientific journals or books and by presenting the information at scientific meetings. Sharing new knowledge with the scientific community permits other scientists to repeat the experiment or design new experiments that either verify or refute the work.

Although we have just described the scientific method as a linear sequence of events, science is rarely as straightforward or tidy as the scientific method implies. Good science involves *creativity*, not only in recognizing

questions and developing hypotheses but also in designing experiments. Because scientists try to expand our current knowledge, their work is in the realm of the unknown. Many creative ideas end up as dead ends, and there are often temporary setbacks or reversals of direction as scientific knowledge progresses. Scientific knowledge often expands haphazardly, with the "big picture" emerging slowly from confusing and sometimes contradictory details.

Scientific discoveries are often incorrectly portrayed in the media as "new facts" that have just come to light. At a later time, additional "new facts" that question the validity of the original study are reported. If one were to read the scientific papers on which such media reports are based, however, one would find that all the scientists involved made very tentative conclusions based on their data. Science progresses from uncertainty to less uncertainty, not from certainty to greater certainty. Thus, science is self-correcting over time, despite the fact that it never "proves" anything.

Inductive and Deductive Reasoning

Scientists use inductive and deductive reasoning. Discovering general principles by the careful examination of specific cases is called **inductive reasoning**. The scientist begins by organizing data into manageable categories and asking the question "What does this information have in common?" He or she then seeks a unifying explanation for the data. Inductive reasoning is the basis of modern experimental science.

As an example of inductive reasoning, consider the following:

Figure 2.2 Experimentation is an essential part of the scientific method. Here a field scientist makes observations that are critical to understanding rainforest plants. Photographed in Costa Rica.

Fact: Gold is a metal that is heavier than water.

Fact: Iron is a metal that is heavier than water.

Fact: Silver is a metal that is heavier than water.

Conclusion based on inductive reasoning: All metals are heavier than water.

Even if inductive reasoning makes use of correct data, the conclusion may be either true or false. As new data come to light, they may show that the generalization arrived at through inductive reasoning is false. Science has shown, for example, that the density of lithium, the lightest of all metals, is about half that of water. When one adds this information to the preceding list, a different conclusion must be formulated, in this case: *Most* metals are heavier than water. Inductive reasoning, then, produces new knowledge but is prone to error.

Science also makes use of **deductive reasoning**, which proceeds from generalities to specifics. Deductive

reasoning adds nothing new to knowledge, but it can make relationships among data more apparent. For example:

General rule: All birds have wings.

A specific example: Robins are birds.

Conclusion based on deductive reasoning: All robins have wings.

This is a valid argument. The conclusion that robins have wings follows inevitably from the information given. Scientists use deductive reasoning to determine the type of experiment or observations necessary to test a hypothesis.

The Importance of Prediction

A successful scientific hypothesis needs to be both valid and useful—it needs to tell you something you want to know. A hypothesis is most useful when it makes predictions, because the predictions provide a way to test the validity of the hypothesis. If your experiment refutes your prediction, then you must carefully recheck the entire experiment. If the prediction is still refuted, then you must reject the hypothesis. The more verifiable predictions a hypothesis makes, the more valid that hypothesis is. There is something very satisfying about a successful prediction, because the prediction being tested is generated by the hypothesis itself, and the result is not known ahead of time.

Experimental Controls

Most often, the processes we want to learn about are influenced by many factors. We call each factor that influences a process a **variable**. To evaluate alternative hypotheses about a given variable, it is necessary to hold all other variables constant so that we do not get misled or confused by them.

To test a hypothesis about a variable, we carry out two forms of the experiment in parallel. In the experimental group we alter the chosen variable in a known way. In the **control** group we do not alter that variable. We make sure that in all other respects the two groups are the same. We then ask, "What is the difference, if any, between the outcomes for the two groups?" Any difference that we see must be due to the influence of the variable that we changed, because all other variables remained the same. Much of the challenge of environmental science lies in designing control groups and in successfully isolating a single variable from all other variables.

Theories

A **theory** is an integrated explanation of numerous hypotheses, each of which has been supported by a large body of observations and experiments. A theory condenses and simplifies many data that previously appeared to be unrelated. A good theory grows as additional information becomes known. It predicts new data and suggests new relationships among a range of natural phenomena.

By demonstrating the relationships among classes of data, a theory simplifies and clarifies our understanding of the natural world. Theories are the solid ground of science, the explanations of which we are most sure. This definition contrasts sharply with the general public's use of the word *theory*, implying *lack* of knowledge, or a guess—as in "I have a theory about the assassination of John Kennedy." In this book, the word *theory* is always used in its scientific sense, to refer to a broadly conceived, logically coherent, and well-supported explanation.

Yet there is no absolute truth in science, only varying degrees of uncertainty. Science is continually evolving as new evidence comes to light, and therefore, its conclusions are always provisional or uncertain. It is therefore always possible that the results of a future experiment will contradict a prevailing theory and show it to be false.

Uncertainty, however, does not mean that scientific conclusions are invalid. For example, there is overwhelming evidence linking exposure to tobacco smoke and incidence of lung cancer. We cannot state with absolute certainty that every smoker will be diagnosed with lung cancer, but this uncertainty does not mean that there is no correlation between smoking and lung cancer. On the basis of the available evidence, we say that people who smoke have an *increased risk* of developing lung cancer.

Figure 2.3 Smoker. Despite known cancer risks—at least 30% of all estimated cancer deaths in the United States are caused by tobacco use—many people continue to smoke.

SCIENTIFIC DECISION MAKING AND UNCERTAINTY: AN ASSESSMENT OF RISKS

Each of us takes risks every day of our lives. A **risk** is the probability of harm (such as injury, disease, death, or environmental damage) occurring under certain circumstances. Risks exist for most human activities. Walking on stairs involves a small risk, but a risk nonetheless, because some people die from falls on stairs. Using household appliances is slightly risky, because some people die from electrocution when they operate appliances with faulty wiring or use appliances in an unsafe manner. Driving in an automobile or flying in a jet offers risks that are easier for most of us to recognize. Yet few of us hesitate to fly in a plane, and even fewer hesitate to drive in a car because of the associated risk.

Although we sometimes speak of percentages, probabilities of risk are always calculated as fractions. If a risk is certain to occur, its probability is 1; if it is certain *not* to occur, its probability is 0. Most probabilities of risk are some number between 0 and 1. For example, according to the American Cancer Society, in 2002 about 170,000 Americans who smoked died of cancer (Figure 2.3). This translates into a probability of risk of 0.00059, or about 6

Table 2.1	Probability of Risk of Dying by Selected Causes, 1998	
Cause of Death	*U.S. Deaths in 1998*	*Probability of Risk*
Cardiovascular disease	940,600	0.0035 or 3.5×10^{-3}
Cancer (all types)	541,500	0.0020 or 2.0×10^{-3}
Accidents (including motor vehicle)	97,800	0.00036 or 3.6×10^{-4}
Suicide	30,600	0.00011 or 1.1×10^{-4}
Homicide	18,300	0.00007 or 0.7×10^{-4}
Accidental falls	16,274	0.00006 or 0.6×10^{-4}
Accidental poisonings by drugs	9,838	0.000036 or 3.6×10^{-5}
Accidental drownings	3,964	0.000015 or 1.5×10^{-5}
Fire	3,255	0.000012 or 1.2×10^{-5}
Accidents by firearms	726	0.0000026 or 2.6×10^{-6}
Accidents (airplane)	692	0.0000025 or 2.5×10^{-6}

Table 2.2	The Four Steps of Risk Assessment for Adverse Health Effects
Step	*What It Answers*
1. Hazard identification	Does exposure to a substance cause an increased likelihood of an adverse health effect such as cancer or birth defects?
2. Dose–response assessment	What is the relationship between amount of exposure (the dose) and the seriousness of the adverse health effect?
3. Exposure assessment	How much, how often, and how long are humans exposed to the substance in question? For hazardous air pollutants, emissions are measured and analyzed to determine the relationship between emissions and concentrations in the environment. Where humans live relative to the emissions is also considered.
4. Risk characterization	What is the probability of an individual or population having an adverse health effect? Risk characterization combines and evaluates data from dose–response assessment and exposure assessment (steps 2 and 3).

people of every 10,000 Americans. (See Table 2.1 for probabilities of risk of dying in a given year by selected causes.)

Using statistical methods to quantify the risks involved in a particular action so that they can be compared and contrasted with other risks is known as **risk assessment**. The four steps involved in risk assessment for adverse health effects are summarized in Table 2.2. Once a risk assessment has been performed, its results are evaluated with relevant political, social, and economic considerations to determine whether a particular risk should be reduced or eliminated and, if so, what should be done. This evaluation, which includes the development and implementation of laws to regulate hazardous substances, is known as **risk management**.

Risk assessment helps us to estimate the probability that an event will occur and enables us to set priorities and manage risks in an appropriate way. As an example, consider a person who smokes a pack of cigarettes a day and drinks well water containing traces of the cancer-causing chemical trichloroethylene (in acceptable amounts as established by the Environmental Protection Agency, or EPA). Without knowledge of risk assessment, this person might buy bottled water in an attempt to reduce his or her chances of getting cancer. Based on risk assessment calculations, the annual risk from smoking is 0.00059, or 5.9×10^{-4}, whereas the annual risk from drinking water with EPA-accepted levels of trichloroethylene is 0.000000002, or 2.0×10^{-9}. This means that this person is almost *300,000 times* more likely to get cancer from smoking than to get it from ingesting such low levels of trichloroethylene. Knowing this, the person in our example would, we hope, be persuaded to stop smoking.

One of the most perplexing dilemmas of risk assessment is that people often ignore substantial risks but get extremely upset about minor risks. The average life expectancy of smokers is more than 8 years shorter than that of nonsmokers, and almost one third of all smokers die from diseases caused or exacerbated by their habit. Yet many people get much more upset over a one-in-a-million chance of getting cancer from pesticide residues

on food than they do over the relationship between smoking and cancer. Perhaps part of the reason for this attitude is that behaviors such as diet, smoking, and exercise are parts of our lives that we can control *if we choose to*. Risks over which most of us have no control, such as pesticide residues or nuclear wastes, tend to evoke more fearful responses.

DETERMINING THE HEALTH EFFECTS OF ENVIRONMENTAL POLLUTANTS

The human body is exposed to many kinds of chemicals in the environment. Both natural and synthetic chemicals are in the air we breathe, the water we drink, and the food we eat. *All* chemicals, even "safe" chemicals such as sodium chloride (table salt), are toxic if exposure is high enough. For example, a 1-year-old child will die from ingesting about 2 tablespoons of table salt; table salt is also harmful to people with heart or kidney disease.

The study of **toxicants**, chemicals with adverse effects on health, is known as **toxicology** (see "Mini-Glossary: Toxicology Terms"). It encompasses the effects of toxicants on living organisms and the mechanisms whereby they cause toxicity, as well as ways to prevent or minimize adverse effects, such as by developing appropriate handing or exposure guidelines.

The effects of toxicants following exposure can be immediate (acute toxicity) or prolonged (chronic toxicity). **Acute toxicity**, which ranges from dizziness and nausea to death, occurs immediately to within several days following a single exposure. In comparison, **chronic toxicity** generally produces damage to vital organs, such as the kidneys or liver, following a long-term, low-level exposure to chemicals. Toxicologists know far less about chronic toxicity than they do about acute toxicity, in part because the symptoms of chronic toxicity often mimic those of other chronic diseases.

We measure toxicity by the dose at which adverse effects are produced. A **dose** of a toxicant is the amount

Table 2.3	LD$_{50}$ Values for Selected Chemicals	
	Chemical	*LD$_{50}$ (mg/kg)**
	Aspirin	1,750.0
	Ethanol	1,000.0
	Morphine	500.0
	Caffeine	200.0
	Heroin	150.0
	Lead	20.0
	Cocaine	17.5
	Sodium cyanide	10.0
	Nicotine	2.0
	Strychnine	0.8

* Administered orally to rats.

that enters the body of an exposed organism. The **response** is the type and amount of damage caused by exposure to a particular dose. A dose may cause death (lethal dose) or cause harm but not cause death (sublethal dose). Lethal doses, which are usually expressed in milligrams of toxicant per kilogram of body weight, vary depending on the organism's age, sex, health, metabolism, and how the dose was administered (all at once or over a period of time). Lethal doses in humans are known for many toxicants because of records of homicides and accidental poisonings.

One way to determine acute toxicity is to administer various doses to populations of laboratory animals, measure the responses, and use these data to predict the chemical effects on humans. The dose that is lethal to 50% of a population of test animals is called the **lethal dose-50%**, or **LD$_{50}$**. It is usually reported in milligrams of chemical toxicant per kilogram of body weight. There is an inverse relationship between the LD$_{50}$ and the acute toxicity of a chemical: The smaller the LD$_{50}$, the more toxic the chemical, and, conversely, the greater the LD$_{50}$, the less toxic the chemical (Table 2.3). The LD$_{50}$ is determined for all new synthetic chemicals—thousands are produced each year—as a way of estimating their toxic potential. It is generally assumed that a chemical with a low LD$_{50}$ for several species of test animals is also very toxic in humans.

The **effective dose-50%**, or **ED$_{50}$**, is used for a wide range of biological responses, such as stunted development in the offspring of a pregant animal, reduced enzyme activity, or onset of hair loss. The ED$_{50}$ is the dose that causes 50% of a population to exhibit whatever response is under study.

A **dose–response curve** shows the effect of different doses on a population of test organisms (Figure 2.4). Scientists begin by testing the effects of high doses and then work their way down to a **threshold** level, the maximum dose that has no measurable effect (or, alternatively, the minimum dose that produces a measurable effect). It is assumed that doses lower than the threshold level will not have an effect on the organism and are therefore safe.

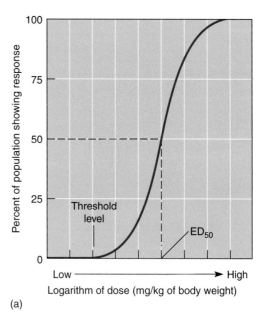

(a)

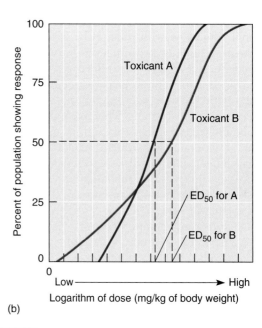

(b)

Figure 2.4 Dose–response curves. (a) This hypothetical dose–response curve demonstrates two assumptions of classical toxicology: first, that the biological response increases as the dose is increased; second, that there is a safe dose—that is, a level of the toxicant at which no response occurs. Harmful responses occur only above a certain threshold level. (b) Dose–response curves for two hypothetical toxicants, A and B. As you can see, dose–response curves have a variety of shapes. In this example, toxicant A has a lower effective dose-50% (ED$_{50}$) than toxicant B. However, at lower doses, toxicant B is more toxic than toxicant A.

A growing body of evidence, however, suggests that for certain toxicants there is no safe dose. A threshold does not exist for these chemicals, and even the smallest amount of the toxicant can cause a measurable response.

Toxicology Terms

toxicant: Chemical with adverse effects on health.

toxicology: Study of chemicals with adverse effects on health.

acute toxicity: Harmful effects that occur within a short period after exposure to a toxicant.

chronic toxicity: Harmful effects that occur after a long period of exposure to a toxicant.

dose: The amount of a toxicant that enters the body of an exposed individual.

response: The type and amount of damage caused by exposure to a particular dose of a toxicant.

LD_{50}: The dose of a particular toxicant that is lethal to 50% of a population of test animals.

ED_{50}: The dose of a particular substance that causes an observable response in 50% of a population.

dose–response curve: A graph that represents the relationship between the administered dose and the response produced.

threshold: The maximum dose that has no measurable effect.

Children and Chemical Exposure

Because they weigh substantially less than adults, children are more susceptible to chemicals. Consider a toxicant with an LD_{50} of 100 mg/kg. A potentially lethal dose for a child who weights 11.3 kg (25 lb) would be $100 \times 11.3 = 1130$ mg, which is equal to a scant $\frac{1}{4}$ teaspoon if the chemical is a liquid. In comparison, the potentially lethal dose for an adult who weighs 68 kg (150 lb) is 6,800 mg, or slightly less than 2 teaspoons. This exercise demonstrates that children must be protected from exposure to environmental chemicals because lethal doses are smaller for children than for adults.

Identifying Cancer-Causing Substances

Because of widespread concerns about cancer-inducing chemicals in the environment and because cancer is so feared, it traditionally was the only disease evaluated in chemical risk assessment. Environmental contaminants are known to be linked to several other serious diseases, such as birth defects, damage to the immune response, reproductive problems (recall the discussion of endocrine disrupters in Chapter 1), and damage to the nervous system or other body systems. Although cancer is not the only disease that is caused or aggravated by toxicants, we focus here on risk assessment as it relates to cancer. Noncancer hazards, such as diseases of the liver, kidneys, or nervous system, are assessed in ways that are conceptually similar to cancer risk assessment.

The most common method of determining whether a chemical causes cancer is to expose laboratory animals such as rats to extremely large doses of that chemical and see whether they develop cancer. This method is indirect and uncertain, however. For one thing, although humans and rats are both mammals, they are different organisms and may respond differently to exposure to the same chemical. (Even rats and mice, which are more similar to one another than are rats and humans, often respond differently to the same toxicant.)

Another problem is that the rats are exposed to massive doses of the suspected **carcinogen** (cancer-causing chemical, radiation, or virus) relative to their body size, whereas humans are usually exposed to much lower amounts. Researchers must use large doses in order to cause cancer in a small group of laboratory animals within a reasonable amount of time. Otherwise, such tests would take years, require thousands of test animals, and be prohibitively expensive to produce enough data to have statistically significant results.

It is assumed that one can extrapolate (work backward) from the huge doses of chemicals and the high rates of cancer they cause in rats to determine the rates of cancer that might be expected in humans exposed to lower amounts of the same chemicals. However, there is little evidence to indicate that extrapolating backward is scientifically sound. Even if you are reasonably sure that exposure to high doses of a chemical causes the same effects for the same reasons in both rats and humans, you cannot assume that these same mechanisms work at low doses in humans. The way the body metabolizes (breaks down) very small doses of a chemical is often not the same way it deals with very large doses; for example, the body may be able to break down carcinogens in small quantities by enzymatic action in the liver, but in the presence of an excessive amount of carcinogen, the enzymes are overwhelmed.

In short, extrapolating from one species to another and from high doses to low doses is filled with uncertainty and may overestimate or underestimate the danger. Despite these shortcomings, animal carcinogen studies do provide valuable information: A toxicant that does not cause cancer in laboratory animals at high doses is not likely to cause cancer in humans at lower levels found in the environment or in occupational settings.

Although scientists do not currently have a reliable way to determine if exposure to small amounts of a substance causes cancer in humans, the EPA is planning to change how toxic chemicals are evaluated and regulated. Methods are being developed that will give direct evidence of the risk involved in exposure to low doses of chemicals that cause cancer. Once implemented, these methods should be more accurate in assessing risk.

Epidemiological evidence, including studies of human groups accidentally exposed to high levels of suspected carcinogens, is also used to determine whether chemicals are carcinogenic. For example, in 1989 epidemiologists in Germany established a direct link between cancer and a group of chemicals called dioxins. They observed the incidence of cancer in workers exposed to high concentrations of dioxins during an accident at a chemical plant in 1953, and found unexpectedly high levels of cancers of both the digestive and respiratory tracts (see Chapter 23 for more recent studies involving dioxin and cancer).

Risk Assessment of Chemical Mixtures

Humans are frequently exposed to various combinations of chemical compounds. Such chemical mixtures are present in the air we breathe, the food we eat, and the water we drink. Cigarette smoke contains a mixture of chemicals, as does automobile exhaust. However, the vast majority of toxicology studies have been performed on single chemicals rather than chemical mixtures, and for good reason. Mixtures of chemicals can interact in a variety of ways, increasing the level of complexity in risk assessment, a field already complicated by many uncertainties. Moreover, toxicologists point out that there are simply too many chemical mixtures to evaluate all of them.

Chemical mixtures can interact by additivity, synergy, or antagonism. When a chemical mixture is **additive**, the effect is exactly what one would expect, given the individual effects of each component of the mixture. If a chemical with a toxicity level of 1 is mixed with a different chemical, also with a toxicity level of 1, the combined effect of exposure to the mixture is 2. Recall from Chapter 1 that a chemical mixture that is **synergistic** has a greater combined effect than would be expected; two chemicals, each with a toxicity level of 1, might have a combined toxicity of 3. An **antagonistic** interaction in a chemical mixture results in a smaller combined effect than would be expected; for example, the combined effect of two chemicals, each with toxicity levels of 1, might be 1.3.

If toxicological studies of chemical mixtures are lacking, how is risk assessment for chemical mixtures assigned? Toxicologists assign risk to mixtures by additivity—that is, by adding the known effects of each compound in the mixture. Such an approach sometimes overestimates or underestimates the actual risk involved, but it has been deemed the best approach currently available. The alternative—that of waiting for years or decades until numerous studies have been designed, funded, and completed—is unreasonable.

THE PRECAUTIONARY PRINCIPLE

You've probably heard the expression "An ounce of prevention is worth a pound of cure." This statement is the heart of a policy, known as the **precautionary principle**, that is advocated by many politicians and environmental activists: When a new technology or chemical product is suspected of threatening human health or the environment, precautionary measures should be taken even if there is scientific uncertainty about the scope of danger. The new technology or chemical should not be introduced until it can be demonstrated that the risks are small and that the benefits outweigh the risks.

The precautionary principle can also be applied to existing technologies when new evidence suggest they are more dangerous than they were originally thought to have been. For example, when observations and experiments suggested that chlorofluorocarbons (CFCs) harm the ozone layer in the stratosphere, the precautionary principle led to the production and use of these compounds being phased out. Additional studies supported this step (see Chapter 20).

To many people the precautionary principle is just common sense, given that science and risk assessment often cannot provide definitive answers to policy makers dealing with environmental and public health problems. The precautionary principle puts the burden of proof on the developers of the new technology or substance. They must prove it is safe beyond a reasonable doubt instead of society proving it is harmful after it has already been introduced. However, the precautionary principle does not require that developers provide absolute proof that their product is safe; such proof would be impossible to provide.

The precautionary principle has been incorporated into certain laws and decisions in many member countries of the European Union, and some laws in the United States have a precautionary tone. In October 2000 Christine Todd Whitman, then governor of New Jersey, said in a speech to the National Academy of Sciences,

Policy makers need to take a precautionary approach to environmental protection. … We must acknowledge that uncertainty is inherent in managing natural resources, recognize it is usually easier to prevent environmental damage than to repair it later, and shift the burden of proof away from those advocating protection toward those proposing an action that may be harmful.

The precautionary principle has generated much controversy. Some scientists fear that the precautionary principle challenges the role of science and endorses making decisions without the input of science. Some critics contend that its imprecise definition can reduce trade and limit technological innovations. For example, several European countries made precautionary decisions to ban beef from the United States and Canada because these countries use growth hormones to make the cattle grow faster (see Chapter 18). Europeans contend that the growth hormone might harm humans eating the beef, but the ban, which has been in effect for more than 10 years, is widely viewed as protecting their own beef industry. Another international controversy in which the precautionary principle has been involved is the introduction of genetically modified foods (see Chapter 18).

ECOLOGICAL RISK ASSESSMENT

Doing a risk assessment as it relates to human health is relatively easy compared to doing one for the environment. How does one assess cleanup options for a hazardous waste site, predict the effects of water pollution on the survival of endangered fish species, or determine if wildlife population declines are the result of natural trends or human actions? Yet the EPA and other federal and state environmental monitoring groups are increasingly trying to evaluate ecosystem health. While there is no formal method for **ecological risk assessments**, the EPA has established guidelines for estimating the probable effects of a wide range of human activities on ecosystems.

Such analyses are difficult because the effects may be felt on a wide scale, from individual animals or plants in a local area to ecological communities across a large region. Given the hazards and exposure levels of human-induced **environmental stressors** (human-induced changes that tax the environment), ecological effects can range from good to bad, or from acceptable to unacceptable. Because many ecological effects are incompletely understood or difficult to measure, using scientific knowledge in environmental decision making is filled with uncertainty. Despite these problems, there is a real need to quantify risk to the environment and to develop strategies to cope with the uncertainty.

The EPA is using ecological risk assessment to tackle complex environmental problems. For example, the EPA examined the cumulative effects of many natural and human-induced stressors on various plant and animal species in the Snake River ecosystem in Southern Idaho. The Snake River provides irrigation water for agriculture, and dams harness the water to generate electricity (Figure 2.5). These and other land use practices in the **watershed** (the area of land drained by the river) have resulted in a reduced river flow, elevated water temperature, and nutrient enrichment. Algae and aquatic weeds now grow in great profusion, and many fish and aquatic invertebrates are severely reduced in number. Ecological risk assessment is helping the EPA and other federal agencies, regional groups, state agencies, Native American tribes, local groups, and private individuals to set priorities to meet their common goal of managing and protecting the biological communities in the Snake River watershed.

Cost–Benefit Analysis of Risks

Before the benefits of scientific risk assessment were understood or widely appreciated, politicians and government agencies tended to respond to the environmental issues that received the most publicity. As data on actual risks became available, however, it was discovered that some highly publicized environmental problems are

Figure 2.5 Milner Dam of Snake River. Ecological risk assessment of the Snake River ecosystem will help to sustainably manage the river and its watershed, which are suffering from a variety of human-induced stressors, including dams.

astronomically expensive to correct and at the same time do not pose as much of a threat as many of the less-publicized problems. As a result, decision makers have increasingly adopted an approach known as cost–benefit analysis to address environmental problems, particularly those that involve human health and safety. In a **cost–benefit analysis**, the estimated cost is compared with potential benefits to determine how much expense society is willing to incur to derive the benefits.

Cost–benefit analysis is an important mechanism to help decision makers formulate and evaluate environmental legislation, but cost–benefit analysis is only as good as the data and assumptions on which it is based. Corporate estimates of the cost to control pollution are often many times higher than the actual cost turns out to be. During the debate over phasing out leaded gasoline in 1971, the oil industry predicted that the cost during the transition would be $7 billion per year, but the actual cost was less than $500 million per year.

Despite the wide range that often occurs between projected and actual costs, the cost portion of cost–benefit analysis is often easier to determine than are the health and environmental benefits. The cost of installing air pollution–control devices at factories is relatively easy to estimate, but how does one put a price tag on the benefits of a reduction in air pollution? What is the value of reducing respiratory problems in children and the elderly, two groups that are very susceptible to air pollution? How much is clean air worth?

Another problem with cost–benefit analysis is that the risk assessments on which such analyses are based are far from perfect. Scientists admit that even the best risk assessments are based on assumptions that, if changed, could substantially alter the estimated risk. Risk assessment by its very nature is an uncertain science.

To summarize, cost–benefit analyses and risk assessments are useful in evaluating and solving environmental problems, but decision makers must recognize the limitations of these methods when developing new government regulations.

A Balanced Perspective on Risks

Threats to our health, particularly from toxic chemicals in the environment, make big news. Many of these stories are more sensational than factual. If they were completely accurate, people would be dying left and right, whereas in fact, human health is generally better today than at any time in our history, and our life expectancy continues to increase rather than decline.

This does not mean that we should ignore chemicals that humans introduce into the environment. Nor does it mean we should discount the stories that are sometimes sensationalized by the news media. These stories serve an important role in getting the regulatory wheels of the government moving to protect us as much as possible from the dangers of our technological and industrialized world.

People should not expect no-risk foods, no-risk water, or no-risk anything else. Risk is inherent in all our actions and in everything in our environment. We do, however, have the right to be informed about the risks we face. We should not ignore small risks just because larger ones exist. However, it is extremely important that we have an adequate understanding of the nature and size of risks before deciding what actions are appropriate to avoid them.

ADDRESSING ENVIRONMENTAL PROBLEMS: AN OVERVIEW

Before we begin a detailed examination of the environmental problems that are discussed in remaining chapters of this text, it is useful to consider the many elements that contribute to solving environmental problems. How is information gathered, and at what point can conclusions be regarded as certain enough to warrant action? Who makes the decisions, and what are the trade-offs? Viewed simply, there are five stages in addressing an environmental problem (Figure 2.6).

1. **Scientific assessment.** The first stage of addressing any environmental problem is scientific assessment, the gathering of information. The problem is defined. Data are then collected, and experiments or simulations performed to construct a **model**, which is a formal statement that describes the behavior of a process. Such a model can be used to understand how the present situation developed from the past or to predict the future course of events. Models also help

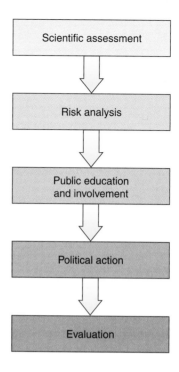

Figure 2.6 Addressing environmental problems. These five steps provide a framework for addressing environmental problems: scientific assessment, risk analysis, public education and involvement, political action, and evaluation.

us generate additional questions about environmental issues.

2. **Risk analysis.** Using the results of a scientific investigation, it is possible to analyze the potential effects of intervention—that is, what could be expected to happen if a particular course of action were followed, including any adverse effects the action might generate. In other words, risk-management goals are defined, and risks are examined. Then, remediation options are considered.

3. **Public education and involvement.** In a democracy, public awareness and endorsement is an essential part of addressing environmental problems. When choices can be made among alternative courses of action, the affected parties—that is, the public—must be informed. This involves explaining the problem, presenting all the available alternatives for action, and revealing the probable cost and results of each choice.

4. **Political action.** The affected parties, through their elected officials, select a course of action—that is, the risk-management strategy they will employ—and implement it. Ideally, such an action is based on the best available scientific evidence. During the political process, however, there are always differences of opinion about how this evidence should be inter-

preted when selecting a specific course of action. Some of these disagreements are based on economic or social considerations rather than on scientific evidence.

5. **Evaluation.** The results of any action taken should be carefully monitored, both to see if the environmental problem is being addressed and to improve the initial assessment and modeling of the problem. Thus, the success of the risk-management strategy is evaluated.

These five stages represent an ideal approach to systematically addressing environmental problems. In real life, addressing environmental problems is rarely so neat and tidy, particularly when the problem is exceedingly complex, of regional or global scale, or with higher costs and less obvious benefits for the money invested. Quite often, the public becomes aware of a problem and triggers discussion of remediation before the problem has been clearly identified. However, to demonstrate the five steps as they operate in an ideal situation, let us consider a relatively simple environmental problem that was recognized and addressed in the 1950s, pollution in Lake Washington. This problem, unlike many environmental issues we face today, was relatively easy to diagnose and solve.

CASE·IN·POINT Lake Washington

Lake Washington is a large (86 km² or 33 mi²), deep freshwater lake that forms the eastern boundary of the city of Seattle (Figure 2.7). During the first part of the 20th century, the Seattle metropolitan area expanded eastward toward the lake from the shores of Puget Sound, an inlet of the Pacific Ocean. As this expansion occurred, Lake Washington came under increasingly intense environmental pressures. Recreational use of the lake expanded greatly, and so did its use for waste disposal. Sewage arrangements in particular had a major impact on the lake.

Birth of an Environmental Problem Like many U.S. cities, Seattle is surrounded by suburbs, each with individual municipal governments. These suburbs expanded rapidly in the 1940s, generating an enormous waste disposal problem. Between 1941 and 1954, 10 suburban sewage treatment plants began operating at points around the lake, with a combined daily discharge of 75.7 million L (20 million gal) into Lake Washington. Each plant treated the raw sewage to break down the organic material within it and released the effluent (treated sewage) into the lake. By the mid-1950s, a great deal of treated sewage had been poured into the lake.

Gabriel Comita and George Anderson, doctoral students of W.T. Edmondson at the University of Washington in Seattle, were the first to note the effects of this discharge on the lake. Their studies of the lake's microscopic organisms in 1950 indicated that large masses of

Figure 2.7 Lake Washington. This large freshwater lake forms the eastern boundary of Seattle, Washington.

Oscillatoria, a filamentous cyanobacterium, were growing in the lake (Figure 2.8). The abundance of these long strings of photosynthetic bacterial cells in Lake Washington was unexpected. The growth of such large num-

Figure 2.8 Light micrograph of *Oscillatoria*. The filaments of this photosynthetic cyanobacterium are composed of chains of cells.

bers of cyanobacteria requires a plentiful supply of nutrients, and deepwater lakes such as Lake Washington do not usually have enough dissolved nutrients to support cyanobacterial growth. (*Low* levels of nutrients are desirable in freshwater lakes because they permit the controlled growth of photosynthetic organisms that are the base of the food web. When a body of water contains a *high* level of nutrients, the photosynthetic organisms are present in vast numbers, upsetting the natural balance in the lake.) The amount of filamentous cyanobacteria in Lake Washington's waters hinted that the lake was somehow changing, becoming richer in dissolved nutrients such as phosphorus.

In July 1955, a technical report by the Washington Pollution Control Commission sounded one of the first public alarms. Citing the work of Comita and Anderson, it concluded that the treated sewage effluent that was being released into the lake's waters was raising the lake's levels of dissolved nutrients to the point of serious pollu-

Figure 2.9 Nutrient enrichment. A eutrophic pond in Wisconsin is covered with the blue-green scum of filamentous cyanobacteria.

tion. Whereas primary treatment (see Chapter 21) followed by chlorination was ridding it of bacteria, it was not eliminating many chemicals, particularly phosphorus, a major component of detergents. In essence, the treated sewage was fertilizing the lake by enriching it with dissolved nutrients.

Eutrophication, the process of nutrient enrichment of freshwater lakes is well understood by ecologists (see Chapter 21). Eutrophication is undesirable because high nutrient levels contribute to the growth of filamentous cyanobacteria. These photosynthetic organisms need only three things in order to grow: light for photosynthesis, which they get from the sun; carbon atoms, which they get from carbon dioxide dissolved in water; and nutrients such as nitrogen and phosphorus, which were provided by the treated sewage. Without the nutrients, cyanobacteria cannot grow; supply excess nutrients, and soon mats of cyanobacteria form a green scum over the surface of the water and the water begins to stink from the odor of rotting organic matter (Figure 2.9).

Then the serious problem begins: The bacteria that decompose the masses of dead cyanobacteria multiply explosively, consuming vast quantities of oxygen in the process, until the lake's deeper waters become so depleted that they can no longer support other organisms that require oxygen to live. Fishes can no longer extract enough oxygen through their gills, and neither can the myriad of tiny invertebrates that populate freshwater lakes.

To Edmondson, the abundance of *Oscillatoria* in Lake Washington was a clear warning. On October 13, 1955, the University of Washington *Daily* ran a story, "Edmondson Announces Pollution May Ruin Lake," in which Edmondson explained the likely meaning of the large masses of *Oscillatoria*: The eutrophication of Lake Washington was demonstrably at an advanced stage, and unless it was reversed, it would soon destroy the water quality of the lake.

ENVIROBRIEF

No Quick Fix for the Salton Sea

The Salton Sea of southern California holds many dilemmas for scientists, conservationists, and policy makers. It was created by accident in 1905, when engineers were attempting to run irrigation canals from the Colorado River, and the entire river emptied into a shallow basin for 16 months. The 984-km² sea provides an unusual habitat—an inland marine environment. Because it was filled with salty irrigation runoff, its salinity levels are 25% higher than that of the ocean. Only certain hardy fish species, which were stocked beginning in the 1950s, can tolerate these stressful high-saline conditions. These fishes have experienced massive die-offs, probably caused by water pollution and high levels of nutrients from agricultural runoff.

As a desert oasis, the Salton Sea hosts millions of wintering birds and "stopover" migrants, including several threatened or endangered species. Prime location notwithstanding, the Salton Sea is considered by many to be a less than ideal bird habitat, or even hazardous to birds' health. Hundreds of thousands of birds have died there since 1992, apparently from such diseases as botulism, avian cholera, and Newcastle virus. Scientists suspect a yet unproven link between the environment's high salinity and the animals' susceptibility to disease.

Some scientists argue that it would be best to let the sea cycle naturally, even if that means it reaches salinities that eliminate fish species. Others insist on preserving the wildlife habitat at all costs by reducing the salinity, especially in the face of great wetland destruction elsewhere. To complicate matters further, California, which has long used more water from the Colorado River than allowed under an agreement with other states, is required by federal mandate to reduce its use significantly by the end of 2002. To get the water that it needs, California may have to use the agricultural runoff that is the main source of water for the Salton Sea. As a result, the Salton Sea would be begin to dry up and would be largely gone by 2030. Any water remaining would be too salty to support fish or other forms of life, so the huge bird populations that have come to rely on the Salton Sea would have no place to go.

Scientific Assessment   Scientific assessment of an environmental problem verifies that a problem exists and builds a sound set of observations on which to base a solution. Lake Washington's microscopic life had been the subject of a detailed study in 1933. Thus, when the telltale signs of pollution first appeared in 1950, Edmondson's students quickly detected changes from the previous study. Without the earlier study's careful analyses of the many microorganisms living in the lake, understanding the changes that were occurring would have been delayed or possibly missed entirely.

Edmondson examined and compared data from the earlier study of the lake and confirmed that there had indeed been a great increase in dissolved nutrients in the lake's water. Surmising that the added nutrients were the result of sewage treatment discharge into the lake by suburban communities, Edmondson formed the hypothesis that treated sewage was introducing so many nutrients into the lake that its waters were beginning to support the growth of cyanobacteria.

Edmondson's hypothesis made a clear prediction: The continued addition of phosphates and other nutrients to the lake would change its surface into a stinking mat of rotting cyanobacteria, unfit for swimming or drinking, and the beauty of the lake would be only a memory. Bolstering his prediction was the fact that lakes near other cities, such as Madison, Wisconsin, had deteriorated after receiving discharges of treated sewage.

Making a Model   Edmondson constructed a graphical model of the lake, which predicted that the decline could be reversed: If the pollution were stopped, the lake would clean itself at a predictable rate, reverting to its previous, unpolluted state within 5 years. (In freshwater lakes, iron reacts with phosphorus to form an insoluble complex that sinks to the bottom of the lake and is buried in the sediments. Thus, if additional phosphorus were not introduced into the lake from sewage effluent, the lake would slowly recover.) In April 1956, Edmondson outlined three steps that would be necessary in any serious attempt to save the lake:

1. Comprehensive regional planning by the many suburbs that ringed the lake;

2. Complete elimination of sewage discharge into the lake;

3. Research to identify the key nutrients that were causing the cyanobacteria to grow.

His proposal received widespread publicity in the Seattle area, and the stage was set to bring scientists and civic leaders together.

Risk Analysis   It is one thing to suggest that the addition of treated sewage to Lake Washington stop, and quite another to devise an acceptable remediation option.

Further treatment of sewage can remove some nutrients, but it may not be practical to remove all of them. The alternative is to dump the sewage somewhere else—but where? In this case, officials weighed their options and decided to discharge the treated sewage into Puget Sound. In their plan, a ring of sewers to be built around the lake would collect sewage treatment discharges, treat them further, and then transport them to be discharged at great depth into Puget Sound.

Why go to all the trouble and expense of treating the discharges further, if you are just going to dump them? And why bother discharging them deep under water? It is important that the solution to one problem not produce another. The plans to further treat the discharge and release it at great depth were formulated in an attempt to minimize the environmental impact of diverting Lake Washington's discharge into Puget Sound. It was assumed that sewage effluent would have less of an impact on the greater quantity of water in Puget Sound than on the much smaller amount of water in Lake Washington. Also, nutrient chemistry in marine water is different from that in fresh water. Puget Sound is naturally rich in nutrients, and phosphate does not control cyanobacterial growth there as it does in Lake Washington. The growth of photosynthetic bacteria and algae in Puget Sound is largely limited by tides, which mix the water and transport the tiny organisms into deeper water, where they cannot get enough light to grow rapidly.

Public Education and Involvement   Despite the technical bulletin published by the Washington Pollution Control Commission in 1955, local sanitation authorities were not convinced that urgent action was necessary. Public action required further education, and it was at this stage that scientists played a key role. Edmondson and other scientists wrote articles for the general public that contained concise explanations of what nutrient enrichment is and what problems it causes. The general public's awareness of the problem increased as local newspapers published these articles.

In December 1956, Edmondson wrote a letter in an effort to alert a committee established by the mayor of Seattle to examine regional problems affecting Seattle and its suburbs. Edmondson explained that even well-treated sewage would soon destroy the lake, and that Lake Washington was already showing signs of deterioration. He received an encouraging response and prepared for the committee a nine-page report of his scientific findings. After presenting his data showing that the mass of cyanobacteria varied in strict proportion to the amounts of nutrients being added to the lake, Edmondson posed a series of questions: "How has Lake Washington changed? What will happen if nothing is done to halt nutrient accumulation? Why not poison the cyanobacteria and then continue to discharge the effluent?" He then answered the questions and outlined two alternative courses of public action—do nothing, or stop

Environmental Literacy

Because responses to environmental problems depend on the public's awareness and understanding of the issues and the underlying scientific concepts involved, environmental education is critical to appropriate decision making. The emphasis on environmental education has grown dramatically over the years:

- Three international treaties supporting environmental education went into effect between 1975 and 1990.

- In 1990, 22 university presidents from 13 nations issued a declaration of their commitment to environmental education and research at their institutions. More than 150 university presidents from at least 38 countries have since followed suit.

- As of 1997, more than 30 states required some form of environmental education in primary and secondary schools.

The North American Association for Environmental Education has issued guidelines for educators to help them select materials such as textbooks and films that are based on sound scientific evidence and that present a balanced perspective on environmental problems. Fairness and accuracy are emphasized in these guidelines. However, a backlash against environmental education occurred during the late 1990s. Some conservative research groups criticized what they perceived as a biased presentation of environmental issues, particularly the promotion of environmental activism, in schools. At least one state (Arizona) overturned its environmental education law in response to these critics.

adding nutrients to the lake—and made a clear prediction about the consequences of each.

Political Action   Edmondson's report was widely circulated among local governments, but implementing its

proposals presented serious political problems because there was no public mechanism in place that would permit the many local suburbs to act together on regional matters such as sewage disposal. In late 1957 the state legislature passed a bill permitting a public referendum in the Seattle area on the formation of a regional government with six functions: water supply, sewage disposal, garbage disposal, transportation, parks, and planning. The referendum was defeated in March 1958, apparently because suburban voters felt that the plan was an attempt to tax them for the city's expenses. Understanding the urgency of Edmondson's proposals, an advisory committee immediately submitted to the voters a revised bill limited to sewage disposal. Over the summer there was widespread discussion of the lake's future, and when the votes were counted on September 9, 1958, the revised bill had passed by a wide margin.

At the time it was passed, the Lake Washington plan was the most ambitious and most expensive pollution control project in the United States. Every household in the area had to pay $2 a month in additional taxes for construction of a massive trunk sewer to ring the lake, collect all the effluent, treat it, and discharge it into Puget Sound.

Groundbreaking ceremonies for the new project were held in July 1961. Meanwhile, the lake had deteriorated further. Visibility declined from 4 m (12.3 ft) in 1950 to less than 1 m (3.1 ft) in 1962, the water being clouded with cyanobacteria. In 1963 the first of the waste treatment plants around the lake began to divert its effluent into the new trunk sewer. One by one, the others diverted theirs, until the last effluent was diverted in 1968. The lake's deterioration stopped by 1964, and then its condition began to improve (Figure 2.10).

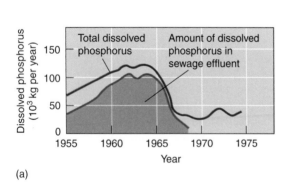

(a)

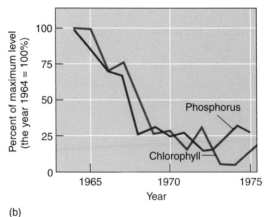

(b)

Figure 2.10 Nutrient additions to Lake Washington and cyanobacterial growth. (a) Dissolved phosphorus (dark maroon line) in Lake Washington from 1955 to 1975. Note how the level of dissolved phosphorus declined in the lake as the amount of phosphorus contributed by sewage effluent (shaded area) declined. (b) Cyanobacterial growth from 1965 to 1975, during Lake Washington's recovery, as measured indirectly by the amount of chlorophyll. Note that as the level of phosphorus dropped in the lake, the number of cyanobacteria (that is, the chlorophyll content) also declined.

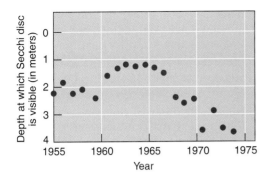

Figure 2.11 Recovery of water transparency in Lake Washington. Measurements were taken in Lake Washington in July or August from 1955 to 1975 using a Secchi disk, a round disk attached to a rope. The greater the water transparency, the deeper the Secchi disk can be lowered and still be visible. If the Secchi disk disappears at 1 m, it means the water is very cloudy from the growth of cyanobacteria. If the Secchi disk can be lowered over 3 m before disappearing from sight, few cyanobacteria are present.

Evaluation Edmondson's hypothesis about what was happening to the lake was correct. Water transparency returned to normal within a few years (Figure 2.11). *Oscillatoria* persisted until 1970, but eventually it disappeared. By 1975 the lake was back to normal.

Every environmental intervention is an experiment, and continued monitoring is necessary because environmental scientists work with imperfect tools. There is a great deal we do not know, and every added bit of information increases our ability to deal with future problems. The unanticipated always lurks just beneath the surface of any experiment carried out in nature. It was not anticipated, for example, that water transparency would continue to increase. By 1980, the lake was clearer than at any time in recent memory, with visibility exceeding 12 m (39.4 ft) at times. Today, the lake remains clear.

Before the recovery, the presence of filamentous cyanobacteria such as *Oscillatoria* had limited the population of a microscopic organism called *Daphnia* because cyanobacterial filaments clog *Daphnia*'s feeding apparatus. The disappearance of *Oscillatoria* and other filamentous cyanobacteria allowed the lake's *Daphnia* population to flourish and become dominant among the many kinds of invertebrates that live there. Because *Daphnia* are very efficient eaters of nonfilamentous algae, levels of these algae fell, too, so that the water became even clearer.

Another unanticipated change observed since 1988 is that the lake water has become increasingly alkaline. The cause of this chemical change in the lake's water is unknown at present, but it has been suggested that the development of land around the eastern side of the lake is involved. The kinds of chemicals draining into a lake from a natural drainage basin are quite different from those draining from storm drains, and it has been hypothesized that this change is responsible for the lake's altered chemistry. Additional scientific studies should help clarify the situation in Lake Washington. ∎

CASE·IN·POINT **The Tragedy of the Commons**

Garrett Hardin, a professor of human ecology at the University of California at Santa Barbara, writes about human environmental dilemmas. In 1968 he published a classic essay, "The Tragedy of the Commons," in the journal *Science* in which he contended that our inability to solve complex environmental problems is the result of a struggle between short-term individual welfare and long-term **environmental sustainability** and societal welfare.

Hardin used the commons to illustrate this struggle. In medieval Europe, the inhabitants of a village shared pastureland, called the commons, and each herder could bring animals onto the commons to graze. The more animals a herder brought onto the commons, the greater the advantage to that individual. When every herder in the village brought as many animals onto the commons as possible, however, the plants were killed from overgrazing, and the entire village suffered. Thus, the users inevitably destroyed the commons that they depended on.

Hardin said that one of the outcomes of the eventual destruction of the commons was private ownership of land, because when each individual owned a parcel of land, it was in that individual's best interest to protect the land from overgrazing. A second outcome Hardin considered was government ownership and management of such resources, because the government's authority could impose rules on users of the resource and thereby protect it.

Hardin's paper has stimulated a great deal of research since it was published in 1968. In general, scholars agree that degradation of the self-governing commons has sometimes occurred in the past and is sometimes occurring today. However, scholars now think that such destruction is not inevitable—that is, it is possible to sustainably manage common resources without privatization (individual ownership) or government management.

As one goes from local to regional to global commons, the challenges of sustainably managing resources increase in complexity. In today's world Hardin's parable has particular relevance at the global level. The commons are those parts of our environment that are available to everyone but for which no single individual has responsibility—shared resources such as the atmosphere, fresh water, forests, wildlife, and fisheries in the ocean. These modern-day commons, sometimes collectively called the **global commons**, are experiencing increasing environmental stress. Because no individual, jurisdiction, or country owns them, they are susceptible to overuse. Although their exploitation may benefit only a few, everyone on Earth must pay for the environmental cost of exploitation.

The world needs effective legal and economic policies to prevent the short-term degradation of our global commons and ensure the long-term well-being of our natural resources. There are no quick fixes, because solutions to global environmental problems are not as simple or short-term as are solutions to some local problems, such as Lake Washington. Most environmental ills are inextricably linked to other persistent problems, such as poverty, overpopulation, and social injustice—problems that are beyond the ability of a single nation to resolve. The large number of participants that must organize, agree on limits, and enforce rules complicates the development of global treaties to manage global commons. Cultural and economic differences among participants make finding solutions even more challenging.

Clearly, all people, businesses, and governments must foster a strong sense of **stewardship**, or shared responsibility for the sustainable care of our planet. Cooperation and commitment at the international level are essential if we are to alleviate poverty, stabilize the human population, and preserve our environment and its resources for future generations. ■

■ WORKING TOGETHER

The reversal of the pollution of Lake Washington is a particularly clear example of how environmental science can work to identify and address environmental problems. Many environmental problems facing us today are far more complex than Lake Washington, including those involving the global commons, and public attitudes are often different. Lake Washington's pollution problem was solved only because the many small towns involved in the problem cooperated in seeking a solution. Today, confrontation over an environmental problem frequently makes it difficult to reach an agreement. Even scientists disagree among themselves and call for additional research to help them arrive at a consensus. In such an atmosphere, politicians often compromise by adopting a "wait and see" approach.

Such delays are really a form of negative action because the consequences of many environmental problems are so serious that they must be acted on before a scientific consensus is reached. The need for additional scientific studies should not prevent us from taking action on such serious regional and global issues as stratospheric ozone depletion and global climate warming (see Chapter 20), pollution in the Great Lakes (see Chapter 21), and acid rain (see Chapter 20). We need to recognize the uncertainty inherent in environmental problems; consider a variety of possible approaches; weigh the cost, benefits, and probable outcomes of each; and set in motion a policy that is flexible enough to allow us to modify it as additional information becomes available. In the final analysis, then, environmental scientists identify a problem and often suggest a solution, but implementation depends on a political decision that is influenced by social and economic agendas as well as scientific evidence.

■ SUMMARY WITH SELECTED KEY TERMS

I. Science is a systematic process to investigate the natural world. Science seeks to reduce the apparent complexity of our world to general principles that can be used to solve problems or provide new insights.

A. The established processes that scientists use to solve problems are collectively called the **scientific method**. There are many variations of the scientific method, which basically involves five steps.

　1. A scientist recognizes and states the problem or unanswered question.

　2. The scientist develops a **hypothesis**, or educated guess, to explain the problem. Hypotheses are most useful when they make predictions that can be tested.

　3. An experiment is designed and performed to test the hypothesis.

　4. **Data**, the results obtained from the experiment, are analyzed and interpreted to reach a conclusion.

　5. The conclusion is shared with the scientific community.

B. A factor that influences a process is called a **variable**.

　1. A well-designed experiment has two parts, a **control** group and an experimental group, which differs from the control only with respect to the single variable being studied.

　2. Any difference in the outcome between the control group and the experimental group must be the result of the variable.

C. Both inductive and deductive reasoning are used in the scientific method.

　1. **Inductive reasoning** begins with specific examples and seeks to draw a conclusion or discover a unifying rule on the basis of those examples. Inductive reasoning provides new knowledge but is error-prone.

　2. **Deductive reasoning** operates from generalities to specifics and adds nothing new to knowledge, but it can make relationships among data more apparent.

D. A **theory** is an integrated explanation of numerous hypotheses, each of which is supported by a large body of observations and experiments.

II. An element of **risk**, the probability of harm, is inherent in everything we do.

A. Risk assessment, the estimation of risks for comparative purposes, helps us set priorities and manage risks. If a risk is certain to occur, its probability is 1; if it is certain *not* to occur, its probability is 0. Most probabilities of risk are fractions, that is, some number between 0 and 1.

B. Risk management is the determination of the need to reduce or eliminate a particular risk. Risk management is based on data from risk assessment as well as political, economic, and social considerations.

III. The study of **toxicants**, chemicals with adverse effects on health, is known as **toxicology**.

A. Acute toxicity occurs immediately to within several days following a single exposure. **Chronic toxicity** generally produces damage to vital organs, such as the kidneys or liver, following a long-term, low-level exposure to chemicals.

B. A **dose–response curve** shows the effect of different doses on a population of test organisms. Scientists test the effects of high doses and work their way down to a threshold level, a dose that has no measurable response. A threshold may not exist for certain toxicants.

C. The dose that is lethal to 50% of a population of test animals is called the **lethal dose-50%, or LD$_{50}$**.

D. Children must be protected from exposure to environmental chemicals because lethal doses are smaller for children than for adults.

E. The most common method of determining whether a chemical causes cancer is to expose laboratory animals such as rats to extremely large doses of that chemical and see whether they develop cancer.

1. This method is indirect and uncertain, in part because humans and rats may respond differently to exposure to the same chemical.

2. Another problem is that the rats are exposed to massive doses of the suspected **carcinogen** (cancer-causing chemical, radiation, or virus) relative to their body size, whereas humans are usually exposed to much lower amounts. Researchers must use large doses to cause cancer in a small group of laboratory animals within a reasonable amount of time.

IV. According to the **precautionary principle**, when a new technology or chemical product is suspected of threatening human health or the environment, precautionary measures should be taken even if there is scientific uncertainty about the scope of danger.

V. Ecological risk assessment is the relatively new process by which the ecological consequences of human activities are estimated.

A. Ecological risk assessment is helping the EPA and other groups to manage and protect the biological communities in

the Snake River **watershed**. Irrigation, hydroelectric power, and other land use practices have resulted in a reduced river flow, elevated water temperature, and nutrient enrichment. Algae and aquatic weeds now grow in great profusion, and many fish and aquatic invertebrates are severely reduced in number.

B. Cost–benefit analysis, in which estimated cost is compared with potential benefits, is increasingly used to determine how much expense society is willing to incur to derive an environmental benefit.

VI. Addressing environmental problems requires the application of approaches from a diversity of fields.

A. There are five stages in this process.

1. Scientific assessment involves identifying a potential environmental problem and collecting data to construct a model.

2. Risk analysis evaluates the potential effects of intervention.

3. Public education and involvement occurs when the results of scientific assessment and risk analysis are placed in the public arena.

4. Political action is the implementation of a particular risk-management strategy by elected or appointed officials.

5. Evaluation monitors the effects of the action that was taken.

B. Lake Washington exemplifies a successful approach to addressing a relatively simple environmental problem.

1. The pouring of treated sewage into Lake Washington had raised its level of nutrients to the point where the lake supported the growth of filamentous cyanobacteria.

2. Disposal of the sewage in another way solved the lake's pollution problem.

VII. In 1968 **Garrett Hardin** published a classic essay, "The Tragedy of the Commons," in the journal *Science*.

1. He contended that our inability to solve complex environmental problems is the result of a struggle between short-term individual welfare and long-term **environmental sustainability** and societal welfare.

2. In today's world Hardin's parable has particular relevance at the global level. The **global commons** are those parts of our environment available to everyone but for which no single individual has responsibility—shared resources such as the atmosphere, fresh water, forests, wildlife, and fisheries in the ocean.

▮ THINKING ABOUT THE ENVIRONMENT

1. Explain how human-induced changes in the environment may be related to an increased incidence in certain diseases.

2. Thomas Henry Huxley once wrote, "The great tragedy of science—the slaying of a beautiful hypothesis by an ugly fact." Explain what he meant, based on what you have learned about the nature of science in this chapter.

3. In the chapter, the term *model* is defined as a formal statement that describes a situation and that can be used to predict the future course of events. On the basis of this definition, is a model the same thing as a scientific hypothesis? Explain your answer.

4. People want scientists to give them precise, definitive answers to environmental problems. On the basis of what you have learned about the nature of science, explain why this is not possible.

5. When Sherlock Holmes amazed his friend Watson by determining the general habits of a stranger on the basis of isolated observations, what kind of reasoning was he using? Explain.

6. Select one of the two choices to complete the following sentence, and then explain your choice: The absence of scientific certainty about the health effects of an environmental pollutant *is/is not* synonymous with the absence of risk.

7. Explain how one would use the precautionary principle when a new product or chemical substance is developed.

8. What information does a cost–benefit analysis provide decision makers?

9. Place the following stages in addressing environmental problems in order and briefly explain each: evaluation, public education and involvement, risk analysis, scientific assessment, political action.

10. Although the Lake Washington case demonstrates the five components of addressing an environmental problem, the final outcome—dumping treated sewage into Puget Sound—is not an ideal, long-term solution. Explain why.

11. How is the collapse of the Georges Bank fishery, discussed in Chapter 1, an example of the tragedy of the commons? Name an additional example of a global commons other than those mentioned in the chapter.

*12. The annual death rate from sitting in a classroom with an asbestos ceiling is estimated at 0.05 people per million who are exposed to the risk. This risk is equivalent to 1 person per how many million people?

*13. The lethal dose of cyanide is generally 10.0 mg per kilogram of body weight. Calculate the lethal dose of cyanide, in grams, for a 100-lb woman. (*Hint:* Don't forget to convert the woman's weight to kilograms. Use Appendix III to help.)

*14. The LD_{50} of the insecticide parathion administered orally to rats is 20 mg/kg. How much parathion (in grams) would have to be fed to a rat that weighs 0.75 kg to give a 50% chance of killing it?

* Solutions to questions preceded by asterisks appear in Appendix VII.

TAKE A STAND

Visit our Web site at **http://www.wiley.com/college/raven** (select Chapter 2 from the Table of Contents) for links to more information about current environmental problems in both Lake Washington and Puget Sound. Find out about current political actions that are in progress to deal with these problems, and debate the issues with your classmates. You will find tools to help you organize your research, analyze the data, think critically about the issues, and construct a well-considered argument. Take a Stand activities can be done individually or as part of a team, as oral presentations, written exercises, or Web-based (e-mail) assignments.

Additional on-line materials relating to this chapter, including a Student Testing section with study aids and self-tests, Environmental News, Activity Links, Research Projects, and more, can also be found on our Web site.

SUGGESTED READING

Appell, D. "The New Uncertainty Principle." *Scientific American* (January 2001). This article highlights the increasing use of the precautionary principle.

Bower, B. "Objective Visions: Historians Track the Rise and Times of Scientific Objectivity." *Science News*, Vol. 154 (December 5, 1998). Scientific historians say that subjective impressions were considered a valid part of the scientific process until the 19th century.

Howard, V. "Synergistic Effects of Chemical Mixtures—Can We Rely on Traditional Toxicology?" *The Ecologist*, Vol. 27 (September–October 1997). Examines the complexities of evaluating the effects of chemical mixtures on human health.

Jensen, M.N. "Common Sense and Common-Pool Resources." *BioScience* (August 2000). This article reviews the historical impact and contemporary implications of Garrett Hardin's seminal paper.

Pickrell, J. "Aerial War Against Disease." *Science News* (April 6, 2002). Satellite data can be used to indicate where people are at greatest risk from certain diseases so that disease-control measures can be focused there.

Robbins, J. "Farms and Growth Threaten a Sea and Its Creatures." *New York Times* (April 2, 2002). Saving the Salton Sea is complicated by the fact that California needs the water that empties into the salty lake.

Tattersall, I. "Science Versus Religion? No Contest." *Natural History* (April 2002). This very short essay examines the nature of science.

Tyson, N.G. "Certain Uncertainties." *Natural History* (October 1998). In part 1 of a two-part essay, the author discusses how uncertainty is inherent in the scientific process.

Tyson, N.G. "Belly Up to the Error Bar." *Natural History* (November 1998). In part 2 of a two-part essay, the author considers how the scientific method minimizes human biases.

Clean water from the tap. The Safe Drinking Water Act, which ensures safe drinking water, is an unfunded mandate that requires local water systems to monitor their drinking water for contaminants.

Environmental History, Legislation, and Economics

Learning Objectives

After you have studied this chapter you should be able to:

1. Define *conservation* and distinguish between *conservation* and *preservation*.

2. Briefly outline the environmental history of the United States.

3. Describe the environmental contributions of the following people: John James Audubon, Henry David Thoreau, George Perkins Marsh, Theodore Roosevelt, Gifford Pinchot, John Muir, Franklin Roosevelt, Aldo Leopold, Wallace Stegner, Rachel Carson, and Paul Ehrlich.

4. Explain why the National Environmental Policy Act is the cornerstone of U.S. environmental law.

5. Relate how environmental impact statements provide such powerful protection of the environment.

6. Sketch a simple diagram that shows how economics is related to natural capital. Make sure you include sources and sinks.

7. Describe various approaches to pollution control, including command and control regulation and incentive-based regulation (that is, emissions charges and marketable waste-discharge permits).

8. Give two reasons why the national income accounts are incomplete estimates of national economic performance.

9. Distinguish among the following economic terms: *marginal cost of pollution, marginal cost of pollution abatement, optimum amount of pollution.*

10. Discuss some of the complexities of the "jobs versus the environment" issue in the Pacific Northwest.

11. Describe some of the environmental problems facing formerly communist governments in Central and Eastern Europe.

In 1995 Congress passed and the president signed into law a bill, the **Unfunded Mandate Review Act,** requiring the federal government to pay for expensive future programs it imposes on state and local governments. This law targets **unfunded mandates,** which are federal requirements imposed on states and local governments, and for which they must absorb all or most of the cost of implementation and enforcement. The unfunded mandate bill will probably have far-reaching impacts on future environmental legislation, as well as future legislation in education, transportation, health, and welfare.

Prior to passage of the unfunded mandate bill, Congress frequently passed directives to states and local governments without providing a way to pay for the cost of compliance. Congress must now determine the cost of proposed federal orders that it passes along to states and local governments, and the federal government must pay for any federal mandates that exceed $50 million. The law does not apply to *existing* environmental laws, such as those covering clean air or clean water, only future laws.

One of the reasons that state and local governments rebelled against unfunded mandates is that these federal regulations may be expensive and sometimes have questionable benefits. The Safe Drinking Water Act requires states to implement and enforce federal drinking water regulations. In one well-publicized example, cities in Ohio tested for 52 pesticides and other contaminants in drinking water, as required by the Safe Drinking Water

41

Act. One of these pesticides, the so-called pineapple pesticide (DBCP), was shown to be a probable cause of cancer in humans. Although it had originally been widely used on more than 40 crops prior to 1979, the Environmental Protection Agency restricted its use to only pineapples in 1979. In 1987, the pineapple pesticide was banned from all uses. Although traces of this pesticide had never been found in Ohio drinking water, Ohio cities were required to continue testing for it. The rules have since been modified so that the pineapple pesticide no longer has to be monitored in Ohio. (Although DBCP has been banned from use throughout the United States, it can still be detected in drinking water samples from at least 16 states, so the original requirement to monitor for it appears to be a reasonable one.)

Despite this example of regulatory inflexibility, most state and local governments are not against federal environmental mandates as much as the fact that they are *unfunded*. The issue, then, revolves around which level of government should assume the financial burden of oversight and implementation. The Safe Drinking Water Act sets national standards for public drinking water, thereby protecting human health, and nobody wants to drink pesticide-laced water. Some state and local governments say they cannot afford the Safe Drinking Water Act, despite its commendable goals, because it is only one of many unfunded mandates that they must pay for.

In this chapter we continue our examination of how environmental problems are addressed, a process that began in Chapter 2, which considered the role of science in environmental decision making. In Chapter 3 we first examine the environmental history of the United States. Then we examine the roles of government and economics in handling environmental issues.

CONSERVATION AND PRESERVATION OF RESOURCES

Resources are any part of the natural environment that are used to promote the welfare of people or other species. Examples of resources include air, water, soil, forests, minerals, and wildlife. **Conservation**, the sensible and careful management of natural resources, has been practiced in one form or another for thousands of years. Conservation should not be confused with preservation. Conservation allows the use of resources in a responsible manner—that is, without inflicting excessive environmental damage, so that resources are available not only for current needs but also for the needs of future generations. In contrast, **preservation** is concerned with setting aside undisturbed areas, maintaining them in a pristine state, and protecting them from human activities that might alter the "natural" state.

Humans have practiced conservation for thousands of years. Three thousand years ago, the Phoenicians terraced hilly farmland to prevent soil erosion. More than 2,000 years ago, the Greeks practiced crop rotation to maintain yields on farmlands, and the Romans practiced irrigation. These and other conservation techniques were gradually adopted and further refined by other Europeans. Conservation did not become a popular movement until the early 20th century, when expanding industrialization, coupled with enormous growth in human population, began to put increased pressure on the world's supply of natural resources.

ENVIRONMENTAL HISTORY OF THE UNITED STATES

From the establishment of the first permanent English colony at Jamestown, Virginia, in 1607, the first two centuries of U.S. history were a time of widespread environmental destruction. Land, timber, wildlife, rich soil, clean water, and other resources were cheap and seemingly inexhaustible. The European settlers did not dream that the bountiful natural resources of North America would one day become scarce. During the 1700s and early 1800s, most Americans had a **frontier attitude**, a desire to conquer and exploit nature as quickly as possible (see discussion of Western worldview in Chapter 1). Concerns about the depletion and degradation of resources occasionally surfaced, but efforts to conserve were seldom made because the vastness of the continent made it seem that there would always be enough resources.

Protecting Forests

The great forests of the Northeast were leveled within a few generations, and, shortly after the Civil War in the 1860s, loggers began deforesting the Midwest at an alarming rate. Within 40 years they deforested an area the size of Europe, stripping Minnesota, Michigan, and Wisconsin of virgin forest (Figure 3.1). By 1897 the sawmills of Michigan had processed 160 billion board feet of white pine, leaving less than 6 billion board feet standing in the whole state. There has been nothing like this unbridled environmental destruction since.

During the 19th century, many U.S. naturalists began to voice concerns about conserving natural resources. **John James Audubon** (1785–1851) painted lifelike portraits of birds and other animals in their natural surroundings. His paintings aroused widespread public interest in the wildlife of North America. **Henry David Thoreau** (1817–1862), a prominent U.S. writer, lived for 2 years on the shore of Walden Pond near Concord, Massachusetts. There he observed nature and contemplated how people could live in harmony with the natural world by economizing and simplifying their

Figure 3.1 Logging operations in 1884. This huge logjam occurred on the St. Croix River near Taylors Falls, Minnesota.

lives. **George Perkins Marsh** (1801–1882) was a farmer, linguist, and diplomat at various times during his life. Today he is most remembered for his book *Man and Nature*, which was published in 1864. *Man and Nature* provided one of the first discussions of humans as agents of global environmental change. Marsh was widely traveled, and *Man and Nature* was based in part on his observations of environmental damage in areas as geographically separate as the Middle East and his native Vermont.

In 1875 a group of public-minded citizens formed the American Forestry Association, with the intent of influencing public opinion against the wholesale destruction of America's forests. Sixteen years later, in 1891, the **General Revision Act** gave the president the authority to establish forest reserves on public (federally owned) land. Benjamin Harrison (1833–1901), Grover Cleveland (1837–1908), and **Theodore Roosevelt** (1858–1919) used this law to put 17.4 million hectares (43 million acres) of forest, primarily in the West, out of the reach of loggers.

In 1907 angry Northwest congressmen pushed through a bill rescinding the president's powers to establish forest reserves. Theodore Roosevelt, who made many important contributions to the conservation movement, responded by designating 21 new national forests that totaled 6.5 million hectares (16 million acres). He *then* signed the bill into law that would prevent him and future presidents from establishing future forest reserves.

Roosevelt appointed **Gifford Pinchot** (1865–1946) the first head of the U.S. Forest Service. Pinchot supported expanding the nation's forest reserves and managing forests scientifically, such as by harvesting trees only at the rate at which they can regrow. Today, national forests are managed for multiple uses, from biological habitats to recreation to timber harvest to cattle grazing (see Chapter 17).

Establishing and Protecting National Parks and Monuments

The world's first national park was established by Congress in 1872 after a party of Montana explorers reported on the natural beauty of the canyon and falls of the Yellowstone River; Yellowstone National Park now includes parts of Idaho, Montana, and Wyoming. In 1890 the Yosemite and Sequoia National Parks were established in California by the Yosemite National Park Bill, largely in response to the efforts of a single man, naturalist and writer **John Muir** (1838–1914) (Figure 3.2). Muir also founded the Sierra Club, a national conservation organization that is still active on a range of environmental issues.

In 1906 Congress passed the Antiquities Act, which authorized the president to set aside as national monuments sites, such as the Badlands in South Dakota, that had scientific, historic, or prehistoric importance. By 1916 there were 13 national parks and 20 national monuments, under the loose management of the U.S. Army. (Today there are 57 national parks and 74 national monuments under the management of the National Park Service; see Chapter 17.)

Some environmental battles involving the protection of national parks were lost. John Muir's Sierra Club fought such a battle with the city of San Francisco over its efforts to dam a river and form a reservoir in the Hetch Hetchy Valley (Figure 3.3), which lay within Yosemite National Park and was as beautiful as Yosemite Valley. In 1913 Congress voted to approve the dam.

But the controversy generated a strong sentiment that the nation's national parks should be better protected, and in 1916 Congress created the National Park Service to

Figure 3.2 President Theodore Roosevelt (left) and John Muir. Photo was taken on Glacier Point above Yosemite Valley, California.

(a)

(b)

Figure 3.3 Hetch Hetchy Valley in Yosemite. A view in Hetch Hetchy Valley (a) before and (b) after Congress approved a dam to supply water to San Francisco.

manage the national parks and monuments for the enjoyment of the public, "without impairment." It was this clause that gave a different outcome to another battle, fought in the 1950s between conservationists and dam builders over the construction of a dam within Dinosaur National Monument. No one could deny that to drown the canyon with 400 feet of water would "impair" it. This victory for conservation established the "use without impairment" clause as the firm backbone of legal protection afforded our national parks and monuments.

Conservation in the Mid-20th Century

During the Great Depression, the federal government financed many conservation projects to provide jobs for the unemployed. During his administration **Franklin Roosevelt** (1882–1945) established the Civilian Conservation Corps, which employed more than 175,000 men

to plant trees, make paths and roads in national parks and forests, build dams to control flooding, and perform other activities that protected natural resources.

During the droughts of the 1930s, windstorms carried away much of the topsoil in parts of the Great Plains, forcing many farmers to abandon their farms and search for work elsewhere (see Chapter 14). The so-called American Dust Bowl alerted the United States about the need for soil conservation and resulted in the formation of the Soil Conservation Service in 1935 by President Roosevelt.

Aldo Leopold (1886–1948) was a wildlife biologist and environmental visionary who was extremely influential in the conservation movement of the mid- to late-20th century (Figure 3.4). His textbook, *Game Management*, was published in 1933 and supported the passage of a 1937 act in which new taxes on sporting weapons and ammunition funded wildlife management and research. Leopold also wrote philosophically about humanity's relationship with nature and about the need to conserve wilderness areas in *A Sand County Almanac*, which was published in 1949. Leopold argued persuasively for a land ethic and the sacrifices that such an ethic requires. Leopold wrote:

> *All ethics so far evolved rest upon a single premise: that the individual is a member of a community of interdependent*

Figure 3.4 Aldo Leopold. Leopold's *A Sand County Almanac* is widely considered an environmental classic.

parts. His instincts prompt him to compete for his place in the community, but his ethics prompt him also to co-operate (perhaps in order that there may be a place to compete for).

The land ethic simply enlarges the boundaries of the community to include soils, waters, plants, and animals, or collectively, the land.

This sounds simple: Do we not already sing our love for and obligation to the land of the free and the home of the brave? Yes, but just what and whom do we love? Certainly not the soil, which we are sending helter-skelter downriver. Certainly not the waters, which we assume have no function except to turn turbines, float barges, and carry off sewage. Certainly not the plants, of which we exterminate whole communities without batting an eye. Certainly not the animals, of which we have already extirpated many of the largest and most beautiful species. A land ethic of course cannot prevent the alteration, management, and use of these "resources," but it does affirm their right to continued existence, and, at least in spots, their continued existence in a natural state.

Aldo Leopold had a profound influence on many American thinkers and writers, including **Wallace Stegner** (1909–1993), who penned his famous "Wilderness Essay" in 1962. Stegner's essay, written to a commission that was conducting a national inventory of wilderness lands, helped create support for the passage of the Wilderness Act of 1964. Stegner wrote:

Something will have gone out of us as a people if we ever let the remaining wilderness be destroyed; if we permit the last virgin forests to be turned into comic books and plastic cigarette cases; if we drive the few remaining members of the wild species into zoos or to extinction; if we pollute the last clean air and dirty the last clean streams and push our paved roads through the last of the silence, so that never again will Americans be free in their own country from the noise, the exhausts, the stinks of human and automotive waste…

We simply need that wild country available to us, even if we never do more than drive to its edge and look in. For it can be a means of reassuring ourselves of our sanity as creatures, a part of the geography of hope.

During the 1960s, public concern about pollution and resource quality began to increase, in large part due to marine biologist **Rachel Carson** (1907–1964). Carson wrote about interrelationships among living organisms, including humans, and the natural environment (Figure 3.5). Her most famous work, *Silent Spring*, was published in 1962. In it Carson wrote against the indiscriminate use of pesticides:

Pesticide sprays, dusts, and aerosols are now applied almost universally to farms, gardens, forests, and homes—nonselective chemicals that have the power to kill every insect, the "good" and the "bad," to still the song of birds and the leaping of fish in the streams, to coat the leaves with a deadly film, and to linger on in soil—all this though the intended

Figure 3.5 Rachel Carson. Carson's book, *Silent Spring*, heralded the beginning of the environmental movement.

target may be only a few weeds or insects. Can anyone believe it is possible to lay down such a barrage of poisons on the surface of the earth without making it unfit for all life? They should not be called "insecticides," but "biocides."

Silent Spring heightened public awareness and concern about the dangers of uncontrolled use of DDT and other pesticides, including poisoning birds and other wildlife and contaminating human food supplies. Ultimately, it led to restriction on the use of certain pesticides (see Chapter 22). Around this time, the media began to increase its coverage of environmental incidents, such as hundreds of deaths in New York City from air pollution (1963); closed beaches and fish kills in Lake Erie from water pollution (1965); and detergent foam in a creek in Pennsylvania (1966).

In 1968, when the population of Earth was "only" 3.5 billion people, ecologist **Paul Ehrlich** published *The Population Bomb*. In it he described the damage to Earth's life support system that was occurring to support such a huge population, including the depletion of essential resources such as fertile soil, groundwater, and other living organisms. Ehrlich's book raised the public's awareness of the dangers of overpopulation and triggered debates on how to deal effectively with population issues.

The Environmental Movement of the Late-20th Century

Until 1970 the voice of **environmentalists**, people concerned about the environment, was heard in the United States primarily through societies such as the Sierra Club and the National Wildlife Federation. There was no generally perceived environmental movement until the spring of 1970, when **Gaylord Nelson**, former senator of Wisconsin, urged Harvard graduate student **Denis Hayes** to organize the first nationally celebrated Earth Day. This event awakened U.S. environmental consciousness to population growth, overuse of resources, and pollution and degradation of the environment. On Earth Day 1970, an estimated 20 million people in the United States demonstrated their support of improvements in resource conservation and environmental quality by planting trees, cleaning roadsides and riverbanks, and marching in parades. (Figure 3.6 presents a timeline of selected environmental events that have occurred since Earth Day 1970.)

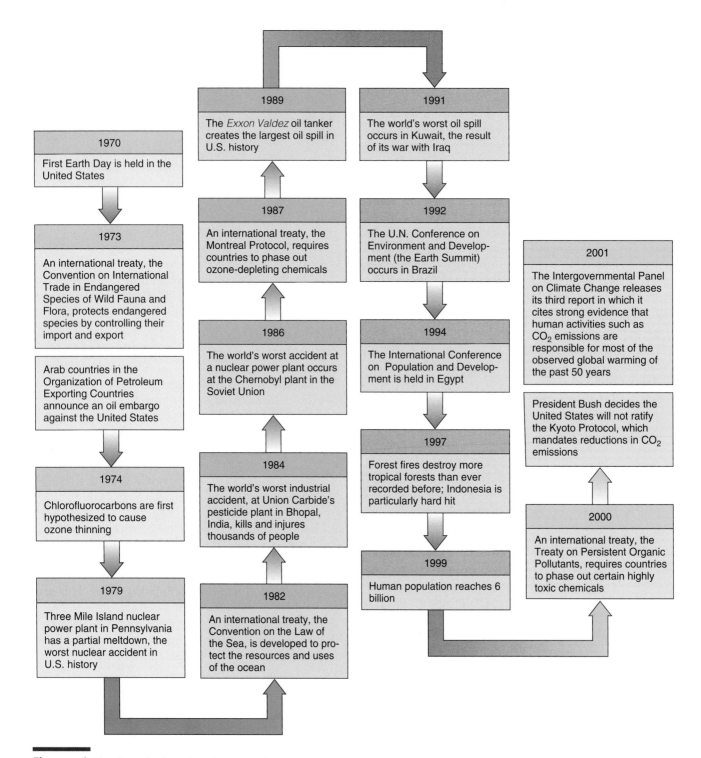

Figure 3.6 Timeline of selected environmental events, from 1970 to the present.

In the years that followed the first Earth Day, environmental awareness and the belief that individual actions could repair the damage humans were doing to Earth became a pervasive popular movement. Musicians such as Marvin Gaye, Joni Mitchell (a Canadian), and the group Alabama popularized environmental concerns. Many of the world's religions—such as Christianity, Judaism, Islam, Hinduism, Buddhism, Taoism, Shintoism, Confucianism, and Jainism—embraced environmental themes such as protecting endangered species and controlling global warming.

By Earth Day 1990, the movement had spread around the world, signaling the rapid growth in environmental consciousness (Figure 3.7). An estimated 200 million people in 141 nations demonstrated to increase public awareness of the importance of individual efforts ("Think globally, act locally"). The theme of Earth Day 2000, "Clean Energy Now!", reflected the dangers of global climate change and what individuals and communities could do: replace fossil fuel energy sources, which produce greenhouse gases, with solar electricity, wind

Figure 3.7 Earth Day. Schoolchildren join with environmentalists to celebrate Earth Day 1990 in Hong Kong.

power, and the like. However, by 2000 many environmental activists had begun to think that individual actions as espoused by Earth Day, while collectively important, are not as important as pressuring governments and large corporations to make environmentally friendly decisions.

U.S. ENVIRONMENTAL LEGISLATION

Galvanized by well-publicized ecological disasters such as the 1969 oil spill off the coast of Santa Barbara, California, and by overwhelming public support for the Earth Day movement, in 1970 the **Environmental Protection Agency (EPA)** was formed, and the **National Environmental Policy Act (NEPA)** was signed into law. A key provision of NEPA stated that the federal government must consider the environmental impact of a proposed federal action, such as financing highway or dam construction, when making decisions about that action. NEPA provides the basis for developing detailed **environmental impact statements (EISs)** to accompany every federal recommendation or proposal for legislation. These EISs are supposed to help federal officials make informed decisions. Each EIS must include the following:

1. The nature of the proposal and why it is needed.

2. The environmental impacts of the proposal, including short-term and long-term effects and any adverse environmental effects should the proposal be implemented.

3. Alternatives to the proposed action that will lessen the adverse effects. This part of the EIS generally concentrates on ways to mitigate the impact of the project.

NEPA also requires that public comments be solicited when preparing an EIS, which generally provide a broader perspective on the proposal and its likely effects.

NEPA established the **Council on Environmental Quality** to monitor the required EISs and report directly to the president. Because this council had no enforcement powers, the NEPA was originally thought to be innocuous, generally more a statement of good intentions than a regulatory policy. During the next few years, however, environmental activists took people, corporations, and the federal government to court to challenge their EISs or use them to block proposed development. The courts decreed that EISs had to be substantial documents that thoroughly analyzed the environmental consequences of anticipated projects on soil, water, and organisms and that EISs must be made available to the public. These rulings put very sharp teeth into the law—particularly the provision for public scrutiny, which placed intense pressure on federal agencies to respect EIS findings.

The NEPA revolutionized environmental protection in the United States. In addition to overseeing federal highway construction, flood and erosion control, military projects, and many other public works, federal agencies oversee nearly one third of the land in the United States. Federally owned holdings include extensive fossil fuel and mineral reserves as well as millions of hectares of public grazing land and public forests. Since 1970 very little has been done to any of them without some sort of environmental review. NEPA has also influenced environmental legislation in many states and other countries. At least 36 states have passed similar legislation requiring EISs for state-funded projects. Canada, Australia, France, New Zealand, and Sweden are some of the countries that now require EISs for government-sponsored projects.

Although almost everyone agrees that NEPA has been successful in helping federal agencies reduce adverse environmental impacts of their activities and projects, it has its critics. Environmentalists complain that EISs are sometimes incomplete or that reports are ignored when decisions are made. Other critics think the EISs delay important projects ("paralysis by analysis") because the EISs are too involved, take too long to prepare, and are often the targets of lawsuits.

Addressing New Environmental Problems with Government Policies

An extensive network of federal, state, and local environmental management agencies interacts with citizens, private consultants, environmental advocacy groups, and others in identifying and addressing new environmental problems. Additionally, many environmental problems are discovered as part of established pollution control programs that conduct routine monitoring programs or investigate complaints from citizens.

Once an environmental problem becomes widely recognized as needing to be addressed, the process of environmental regulation begins with a U.S. congressperson drafting legislation. When the legislation is passed and the president signs it, it usually goes to the EPA, which is given the job of translating the law's language into regulations that specify allowable levels of pollution. Before the regulation officially becomes law, several rounds of public comments allow affected parties to present their views; the EPA is required to respond to all of these comments. Then the Office of Management and Budget reviews the new regulations. Implementation and enforcement of the new law often fall to state governments, which must send the EPA detailed plans showing how they plan to achieve the goals of the new regulations. Currently, the EPA oversees thousands of pages of environmental regulations that affect individuals, corporations, local communities, and states. According to EPA estimates, during the late 1990s the cost of complying with these federal regulations was $210 billion per year, which amounts to about 2.6% of the U.S. gross domestic product.

Accomplishments of U.S. Environmental Legislation

During the period since Earth Day 1970, Congress has passed almost 40 major environmental laws that address a wide range of issues, such as endangered species, clean water, clean air, energy conservation, hazardous wastes, and pesticides (Table 3.1). These laws greatly increased federal regulation of pollution, creating a tough interlocking mesh of laws to improve environmental quality. Many environmental laws contain statutes that allow private citizens to take violators, whether they are private industries or government-owned facilities, to court for noncompliance. These **citizen suits** have contributed significantly in the enforcement of environmental legislation.

However, the laws are not perfect (recall the chapter introduction). Economists and industries have argued that many of the regulations make pollution abatement

Table 3.1	Some Important Federal Environmental Legislation

General
Freedom of Information Act of 1966
National Environmental Policy Act of 1969
National Environmental Education Act of 1990

Conservation of Energy
Energy Policy and Conservation Act of 1975
Northwest Power Act of 1980
National Appliance Energy Conservation Act of 1987
Energy Policy Act of 1992

Conservation of Wildlife
Fish and Wildlife Act of 1956
Anadromous Fish Conservation Act of 1965
Fur Seal Act of 1966
National Wildlife Refuge System Act of 1966
Species Conservation Act of 1966
Marine Mammal Protection Act of 1972
Marine Protection, Research, and Sanctuaries Act of 1972
Endangered Species Act of 1973
Federal Noxious Weed Act of 1974
Magnuson Fishery Conservation and Management Act of 1976
Whale Conservation and Protection Study Act of 1976
Fish and Wildlife Improvement Act of 1978
Fish and Wildlife Conservation Act of 1980
Fur Seal Act Amendments of 1983
Wild Bird Conservation Act of 1992
National Invasive Species Act of 1996

Conservation of Land
General Revision Act of 1891
Taylor Grazing Act of 1934
Soil Conservation Act of 1935
Multiple Use Sustained Yield Act of 1960 (re: national forests)
Wilderness Act of 1964
Land and Water Conservation Fund Act of 1965
Wild and Scenic Rivers Act of 1968
National Trails System Act of 1968
Coastal Zone Management Act of 1972
National Reserves Management Act of 1974
Forest and Rangeland Renewable Resources Act of 1974
Federal Land Policy and Management Act of 1976
National Forest Management Act of 1976
Soil and Water Resources Conservation Act of 1977
Surface Mining Control and Reclamation Act of 1977
Public Rangelands Improvement Act of 1978
Antarctic Conservation Act of 1978
Endangered American Wilderness Act of 1978

Alaska National Interest Lands Act of 1980
Coastal Barrier Resources Act of 1982
Emergency Wetlands Resources Act of 1986
North American Wetlands Conservation Act of 1989
California Desert Protection Act of 1994
Farm Security and Rural Investment Act of 2002
  (the latest version of the "farm bill," which has been
  amended and renamed every 5 years or so since the 1930s)

Air Quality and Noise Control
Noise Control Act of 1965
Clean Air Act of 1970
Quiet Communities Act of 1978
Asbestos Hazard and Emergency Response Act of 1986
Clean Air Act Amendments of 1990

Water Quality and Management
Refuse Act of 1899
Water Resources Research Act of 1964
Water Resources Planning Act of 1965
Clean Water Act of 1972
Ocean Dumping Act of 1972
Safe Drinking Water Act of 1974
National Ocean Pollution Planning Act of 1978
Water Resources Development Act of 1986
Great Lakes Toxic Substance Control Agreement of 1986
Water Quality Act of 1987 (amendment of Clean Water Act)
Ocean Dumping Ban Act of 1988

Control of Pesticides
Food, Drug, and Cosmetics Act of 1938
Federal Insecticide, Fungicide, and Rodenticide Act of 1947
Food Quality Protection Act of 1996

Management of Solid and Hazardous Wastes
Solid Waste Disposal Act of 1965
Resource Recovery Act of 1970
Hazardous Materials Transportation Act of 1975
Toxic Substances Control Act of 1976
Resource Conservation and Recovery Act of 1976
Low-Level Radioactive Policy Act of 1980
Comprehensive Environmental Response, Compensation, and Liability ("Superfund") Act of 1980
Nuclear Waste Policy Act of 1982
Superfund Amendments and Reauthorization Act of 1986
Marine Plastic Pollution Control Act of 1987
Oil Pollution Act of 1990
Pollution Prevention Act of 1990

unduly complex and expensive. Nor have the laws always worked as intended. The Clean Air Act of 1977 required coal-burning power plants to outfit their smokestacks with expensive "scrubbers" to remove sulfur dioxide from their emissions but made an exception for tall smokestacks (Figure 3.8). This loophole led directly to the proliferation of tall stacks that have since produced acid rain throughout the Northeast. The Clean Air Act Amendments of 1990, described in Chapter 19, go a long way toward closing this loophole.

Despite imperfections, environmental legislation has had overall positive effects. Since 1970:

Figure 3.8 Tall smokestacks. Such smokestacks, which emit sulfur dioxide from coal-burning power plants, were exempt from the requirement of pollution-control devices under the Clean Air Act of 1977.

■ Eleven national parks have been established, and the National Wilderness Preservation System now totals more than 42 million hectares (104 million acres).

■ Millions of hectares of farmland that are particularly vulnerable to erosion have been withdrawn from production, reducing soil erosion by more than 60%.

■ Many previously endangered species are better off than they were in 1970, and the American alligator, California gray whale, and bald eagle have recovered enough to be removed from the endangered species list. (However, dozens of other species, such as the manatee, ivory-billed woodpecker, and Kemp's sea turtle, have suffered further declines or extinction since 1970.)

Although we still have a long way to go, pollution control efforts have been particularly successful. Since 1970:

■ Lead levels in the air have dropped by 98% with the phaseout of leaded gasoline.

■ Hydrocarbon emissions from motor vehicles have declined from 10.3 million tons to 5.5 million tons.

■ Emissions of sulfur dioxide, carbon monoxide, and soot have been reduced by more than 30%.

■ Use of chlorofluorocarbons (CFCs) and other chemicals that deplete the ozone layer in the stratosphere has declined by more than 70% (but ozone thinning remains a problem).

■ Release of toxic chemicals into water and air from industrial sources has declined by 43%.

■ Fewer rivers and streams are in violation of water quality standards. (However, the number of fish-consumption advisories relating to specific toxins such as mercury or polychlorinated biphenyls [PCBs] has risen since the 1970s.)

■ The number of secondary sewage treatment facilities, in which bacteria break down organic wastes before the water is discharged into rivers and streams, has increased by 72% since the 1990s.

■ Certain toxic chemicals, such as dichlorodiphenyltrichloroethane (DDT), asbestos, and dioxins, have been banned from use in the United States.

In the 1960s and 1970s, pollution was often very obvious—witness the Cuyahoga River in Cleveland, Ohio, which burst into flames and burned for 8 days in 1969 from the oily pollutants floating on its surface. Legislators, the media, and the public typically viewed environmental problems such as a burning river as posing serious threats that required immediate attention without regard to the cost. Now that the most obvious pollution problems in the United States have been largely addressed, more and more people think we should consider the cost of cleaning up the environment together with the benefits of a cleaner environment. Thus, economics has become increasingly important in environmental legislation and policy making.

ENVIROBRIEF

Trading Turtle Safety

Recent trade rulings affecting sea turtles suggest that one nation's environmental interests may be irrelevant in the face of international free trade. To limit the destruction of endangered sea turtles, the United States requires its own shrimpers to employ turtle exclusion devices (TEDs) in their nets. A U.S. law enacted in 1989 extended this restriction, and by October 1996, trade courts had mandated that other nations must implement TED regulations if they were to continue trading with the United States. Four nations—Thailand, India, Malaysia, and Pakistan—joined to oppose the trade ban, and in April 1998, the World Trade Organization (WTO) supported their opposition. The WTO ruled that the United States must import shrimp from these nations, regardless of the danger to sea turtles. Apparently, free-trade issues win out over environmental protection. According to the WTO, environmental concerns cannot be battled through trade restrictions but must be resolved instead through multilateral agreements, which certainly may provide effective long-term solutions but are frustratingly slow to negotiate.

ECONOMICS AND THE ENVIRONMENT

Economics is the study of how people use their limited resources to try to satisfy their unlimited wants. Economists use specific analytical tools, including developing hypotheses, testing models (see Chapter 2), and analyzing observations and data, to try to understand the consequences of the ways in which people, businesses, and governments allocate their limited resources.

Seen through an economist's eyes, the world is one large marketplace, where resources are allocated to a variety of uses, and where goods—a car, a pair of shoes, a hog—and services—a haircut, a tour of a museum, an education—are consumed and paid for. In a free market, the price of a good is determined by its supply and by the demand for it. If something in great demand is in short supply, its price will be high. High prices encourage suppliers to produce more of a good or service, as long as the selling price is higher than the cost of producing the good or service. This interaction of consumer demand, producer's supply, prices, and costs underlies much of what happens in the U.S. economy, from the price of a hamburger to the cycles of economic expansion (increase in economic activity) and recession (slowdown in economic activity).

In a free-market system such as that of the United States, economists study the prices of goods and services and how those prices influence the amount of a given good or service that is produced and consumed. Like any scientists conducting experiments, economists try to predict the consequences of particular economic actions. When the actions involve economic development, their predictions may contribute to policy decisions that have significant environmental consequences.

Economies depend on the natural environment as sources for raw materials and sinks for waste products (Figure 3.9). A **source** is that part of the natural environment from which materials move, and a **sink** is that part of the natural environment that receives an input of materials. Both sources and sinks contribute to **natural capital**, which is Earth's resources and processes that sustain living organisms, including humans. Natural capital includes mineral resources, forests, soils, groundwater, clean air, wildlife, and fisheries. According to economists, natural capital is what the environment provides for our production and consumption. Resource degradation and pollution represent the overuse of natural capital: *Resource degradation* is the overuse of sources, and *pollution* is the overuse of sinks. Both resource degradation and pollution threaten our long-term economic future.

Natural Resources, the Environment, and the National Income Accounts

Much of our economic well-being flows from natural, rather than human-made, assets—our land, rivers, the ocean, natural resources such as oil and timber, and indeed the air that we breathe. Ideally, for the purposes of economic and environmental planning, the use and misuse of natural resources and the environment should be appropriately measured in the national income accounts.

National income accounts represent the total income of a nation for a given year. Two measures used in national income accounting are *gross domestic product* (*GDP*) and *net domestic product* (*NDP*). Both GDP and NDP provide estimates of national economic performance that are used to make important policy decisions.

Unfortunately, current national income accounting practices are misleading and incomplete because they do not incorporate environmental factors. There are at least two important conceptual problems with the way the national income accounts currently handle the economic use of natural resources and the environment. These problems involve natural resource depletion and the cost and benefits of pollution control.

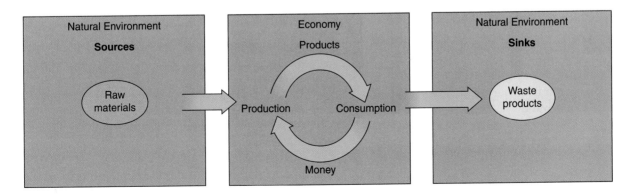

Figure 3.9 Economics and the environment. Goods and services (products) and money (to pay for the products) flow between businesses (production) and consumers (consumption). Economies depend on the natural environment to provide sources for raw materials and sinks for waste products.

Natural Resource Depletion　If a firm produces some product (output) but in the process wears out a portion of its plant and equipment, the firm's output is counted as part of GDP, but the depreciation of capital is subtracted in the calculation of NDP. Thus NDP is a measure of the net production of the economy, after a deduction for used-up capital. In contrast, when an oil company drains oil from an underground field, the value of the oil produced is counted as part of the nation's GDP, but no offsetting deduction to NDP is made to account for the fact that nonrenewable resources have been used up. In principle, the draining of the oil field should be considered a type of depreciation, and the net product of the oil company should be accordingly reduced. The same point applies to any other natural resource that is depleted in the process of production. To summarize, natural capital is a very large part of a country's economic wealth, and it should be treated the same as human-made capital.

The Cost and Benefits of Pollution Control　Imagine that a company has the following choices: It can produce $100 million worth of output and in the process pollute the local river by dumping its wastes. Alternatively, by using 10% of its workers to properly dispose of its wastes, it can avoid polluting but will only get $90 million of output. Under current national income accounting rules, if the firm chooses to pollute rather than not to pollute, its contribution to GDP will be larger ($100 million rather than $90 million), because the national income accounts attach no explicit value to a clean river. In an ideal accounting system, the economic cost of environmental degradation would be subtracted in the calculation of a firm's contribution to GDP, and activities that improve the environment—because they provide real economic benefits—would be added to GDP. To summarize, estimates of environmental damage should be subtracted from GDP.

Discussing the national income accounting implications of resource depletion and pollution may seem to trivialize these important problems. But in fact, because GDP and related statistics are used continually in policy analyses, abstract questions of measurement may often turn out to have significant real effects. Economic development experts have expressed concern that some poor countries, in attempting to raise measured GDP as quickly as possible, have done so in part by overexploiting their natural resources and impairing the environment. Conceivably, if "hidden" resource and environmental costs were explicitly incorporated into official measures of economic growth, these policies might be modified. Similarly, in industrialized countries political debates about the environment have sometimes emphasized the impact on conventionally measured GDP of proposed pollution control measures, rather than the impact on overall economic welfare. Better accounting for environmental quality might serve to refocus these debates to the more relevant question of whether, for any given environmental proposal, the benefits (economic and noneconomic) exceed the cost. An increasing number of economists, government planners, and scientists support replacing GDP and NDP with a more comprehensive measure of national income accounting that includes estimates of both natural resource depletion and the environmental costs of economic activities.

An Economist's View of Pollution

An important aspect of the operation of a free-market system is that the person consuming a product should be the one to pay for all the cost of producing that product. When consumption or production of a product has a harmful side effect that is borne by people not directly involved in the market exchange for that product, the side effect is called an **external cost** or **negative externality**. Because an external cost is usually not reflected in a product's market price—that is, is not borne by the buyer or seller—the market system generally does not operate in the most efficient way.

If an industry makes a product and, in so doing, also releases a pollutant into the environment, the product is bought at a price that reflects the cost of making it but *not* the cost of the damage to the environment by the pollutant. Because this damage is not included in the product's price and because the consumer may not be aware that the pollution exists or that it harms the environment, the cost of the pollution has no impact on the consumer's decision to buy the product. As a result, consumers of the product buy more of it than they would if its true cost, including the cost of pollution, were known or reflected in the selling price.

The failure to add the price of environmental damage to the cost of products generates a market force that increases pollution. From the perspective of economics, then, one of the root causes of the world's pollution problem is the failure to consider negative externalities in the pricing of goods. We now examine industrial pollution from an economist's viewpoint, as a policy-making failure. Keep in mind, however, that lessons about the economics of industrial pollution also apply to other environmental issues (such as resource degradation) where harm to the environment is a consequence of economic activity.

How Much Pollution Is Acceptable?　To come to grips with the problem of assigning a proper price to pollution, economists first attempt to answer the basic question "How much pollution should be allowed in our environment?" One can imagine two environmental extremes: a wilderness in which no pollution is produced and a "sewer" that is completely polluted from excess production of goods. In a wilderness, the environmental quality would be the highest possible, but many goods that are desirable, or even necessary, to humans would be scarce or nonexistent. In a sewer, millions and millions of goods could be produced, but environmental quality would be extremely low. In our world, a move toward a better environment almost always entails a cost in terms of goods.

How do we, as individuals, as a country, and as part of the larger international community, decide where we want to be between the two extremes of a wilderness and a sewer? Economists answer such questions by analyzing the marginal costs of environmental quality and of other goods (see "Mini-Glossary: The Economics of Pollution"). A **marginal cost** is the additional cost associated with one more unit of something. Two examples of marginal costs associated with pollution are the effects of pollution on human health and on organisms in the natural environment.

The trade-off between environmental quality versus more goods involves balancing two kinds of marginal costs: (1) the cost, in terms of environmental damage, of more pollution (the marginal cost of pollution) and (2) the cost, in terms of giving up goods, of eliminating pollution (the marginal cost of pollution abatement).

The Marginal Cost of Pollution   The **marginal cost of pollution** is the added cost for all present and future members of society of an additional unit of pollution. Determining the marginal cost of pollution is generally not an easy process. It involves assessing the risks

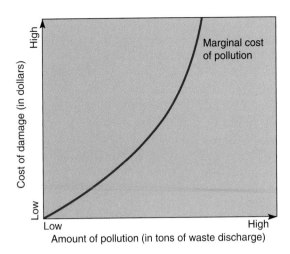

Figure 3.10 Marginal cost of pollution. This graph, represented by an upward sloping curve (from left to right), specifies the cost of damage associated with additional units of pollution. The marginal cost of pollution shows that as the level of pollution rises, the social cost (in terms of human health and a damaged environment) increases sharply.

associated with the pollution. (See Chapter 2 for a discussion of risk assessment.)

Bearing this in mind, let us consider a simple example involving the marginal cost of sulfur dioxide, a type of air pollution produced during the combustion of fuels containing sulfur. Sulfur dioxide is removed from the atmosphere as acid rain, which causes damage to the environment, particularly aquatic systems (see Chapter 20). Economists add up the harm done by each additional unit of sulfur dioxide pollution, such as another ton of sulfur dioxide added to the atmosphere. As the total amount of pollution increases, the harm done by each additional unit usually also increases, so that the curve showing the marginal cost of pollution slopes upward (Figure 3.10). At low pollution levels, the environment may be able to absorb the damage, so that the marginal cost of one added unit of pollution is near zero. As the quantity of pollution increases, the marginal cost rises, and at very high levels of pollution, the cost soars.

The Marginal Cost of Pollution Abatement   The **marginal cost of pollution abatement** is the added cost for all present and future members of society of reducing a given type of pollution by one unit. This cost tends to rise as the level of pollution falls (Figure 3.11). It is relatively inexpensive to reduce automobile exhaust emissions by half, but costly devices are required to reduce the remaining emissions by half again. For this reason, the curve showing the marginal cost of pollution abatement slopes downward. At high pollution levels, the marginal cost of eliminating one unit of pollution is low, but as more and more pollution is eliminated, the cost rises.

MINI-GLOSSARY

The Economics of Pollution

marginal cost of pollution: The cost, in environmental damage, of a unit of pollution that is emitted into the environment.

marginal cost of pollution abatement: The cost to dispose of a unit of pollution in a nonpolluting way.

optimum amount of pollution: The amount of pollution that is economically most desirable. It is determined by plotting two curves, the marginal cost of pollution and the marginal cost of pollution abatement. The point where the two curves meet is the optimum amount of pollution.

command and control regulation: Pollution control that works by setting pollution ceilings. Examples include the Clean Water and Clean Air Acts.

incentive-based regulation: Pollution control in which emission targets are established and industries are given incentives to reduce emissions. Examples include emission charges and marketable waste-discharge permits.

emission charge: Pollution control by charging the polluter for each unit of emissions—that is, by establishing a tax on pollution.

marketable waste-discharge permit: Pollution control by issuing permits allowing the holder to pollute a given amount. Holders are not allowed to produce more emissions than are sanctioned by their permits.

emission reduction credit (ERC): A waste-discharge permit that can be bought and sold by companies producing emissions. Companies have a financial incentive to reduce emissions because they can recover some or all of their cost of pollution abatement by the sale of the ERCs that they no longer need.

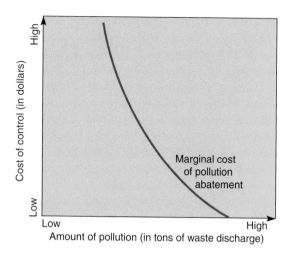

Figure 3.11 Marginal cost of pollution abatement. This graph, represented by a downward sloping curve (from left to right), specifies the cost of reducing pollution to a lower level so there is less damage. The marginal cost of pollution abatement shows that as more and more pollution is eliminated from the environment, the cost of removing each additional (marginal) unit of pollution increases.

How Economists Estimate the Optimum Amount of Pollution In Figure 3.12, the two marginal-cost curves from Figures 3.10 and 3.11 are plotted together on one graph, called a cost–benefit diagram. Economists use this diagram to identify the point at which the marginal cost of pollution equals the marginal cost of abatement—that

is, the point where the two curves intersect. As far as economics is concerned, this point represents an **optimum amount of pollution**.

If pollution exceeds the optimum amount of pollution (is to the right of the point of intersection), the harm done (measured on the marginal cost of pollution curve) exceeds the cost of reducing it (the marginal cost of pollution abatement curve), and it is economically efficient to reduce pollution (Figure 3.13a). On the other hand, if pollution is less than this amount (is to the left of the intersection point), then the marginal cost incurred to reduce it exceeds the marginal cost of the pollution (Figure 3.13b). In the latter situation, the gain in environmental quality is more than offset monetarily by the decrease in resources available for other uses. In such a situation, an economist would argue that the polluter is

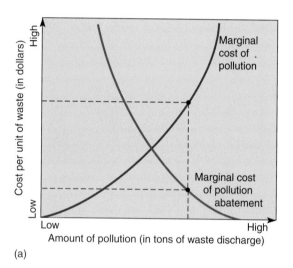

(a)

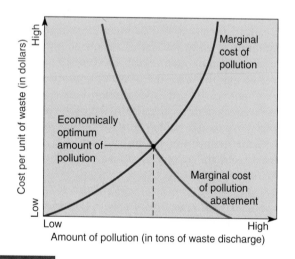

Figure 3.12 Cost–benefit diagram. Economists identify the optimum amount of pollution as the amount at which the marginal cost of pollution equals the marginal cost of pollution abatement (the point at which the two curves intersect). If more pollution than the optimum is allowed, the social cost will be unacceptably high. If less than the optimum amount of pollution is allowed, the pollution abatement cost will be unacceptably high.

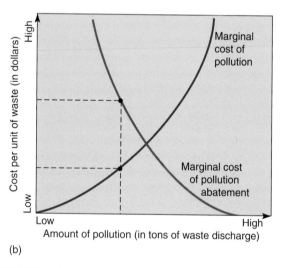

(b)

Figure 3.13 When pollution is less than optimal. (a) In this example, the harm caused by the pollution (measured on the "marginal cost of pollution" curve) exceeds the cost of reducing the pollution (measured on the "marginal cost of pollution abatement" curve). (b) In this example, the cost of reducing the pollution exceeds the harm caused by the pollution.

being charged too much and that the amount of permissible pollution should be increased.

Flaws in the Optimum Pollution Concept There are two major flaws in the economist's concept of optimum pollution. First, the true cost of environmental damage by pollution is difficult to determine. Second, the risks of unanticipated environmental catastrophe are not taken into account in assessing the potential environmental damage of pollution.

Economists usually measure the cost of pollution in terms of damage to property, damage to health in the form of medical expenses or time lost from work, and the monetary value of animals and plants killed. It is difficult, however, to place a value on quality of life or on damage to natural beauty. How much is a scenic river or the sound of a bird singing worth? And how does one assign a value to the extinction of a species? Also, when pollution covers a large area and involves millions of people—as acid rain does—assessing pollution cost is extremely complex.

In adding up pollution costs, economists do not take into account the possible disruption or destruction of the environment. As you will see in the next few chapters, the web of relationships within the environment is extremely intricate and may be more vulnerable to pollution damage than is initially obvious, sometimes with disastrous results. This is truly a case where the whole is much greater than the sum of its parts, and it is inappropriate for economists to simply add up the cost of lost elements in a polluted environment.

Economic Strategies for Pollution Control

Historically, most pollution control efforts in the United States have involved **command and control regulation**, which is the passage of laws that set limits on levels of pollution. Sometimes such laws state that a specific pollution control method must be used, such as catalytic converters in cars to decrease polluting emissions in exhaust. In other cases, a quantitative goal is set. The Clean Air Act Amendments of 1990 established a goal of a 60% reduction in nitrogen oxide emissions in passenger cars by the year 2003. Usually, all polluters must comply with the same rules and regulations regardless of their particular circumstances.

Economists have criticized command and control regulation for being more costly than necessary. They think command and control regulation sets environmental pollution levels much lower than the economically optimum level of pollution. Most economists, whether they have progressive or conservative views, prefer incentive-based regulation over command and control regulation. In **incentive-based regulation**, emission targets are established and industries are given incentives to reduce emissions. Incentive-based regulation is considered a market-oriented strategy because it seeks to use the economic forces of a free market to alleviate the pollution problem—that is, it depends on market incentives to reduce pollution and minimize the cost of control.

A very popular incentive-based regulation for controlling pollution, particularly in Europe, involves imposing an **emission charge** on polluters. In effect, this charge is a tax on pollution. Economists propose such "green taxes" to correct what they perceive as a distortion in the market caused by not including the environmental cost of driving an automobile or cutting down trees or polluting streams. Many European countries have restructured their taxes to take into account environmentally destructive products and activities. Beginning in 1999 Germany increased taxes on gasoline, heating oil, and natural gas while simultaneously lowering its income tax. One result was that carpooling increased by 25% in the first half of 2000. (However, there were concerns about whether Germany's energy-intensive industries could remain competitive with industries in countries without similar energy taxes.) Great Britain put into place a steadily increasing fuel tax increase from 1993 to 1999. One outcome was that the average fuel efficiency of heavy trucks increased by 13%, and fuel consumption dropped. The Netherlands has had a fuel tax in place since 1988 but its primary goal was to raise revenues, not encourage energy conservation. From 1996 to 1998 the Netherlands introduced a tax on natural gas, electricity, fuel oil, and heating oil that included incentives for energy efficiency; income taxes were decreased to offset the tax burden on individuals. Both electricity and fuel use has declined as a result. Finland implemented a carbon dioxide tax in 1990. In 1991 Sweden introduced taxes on carbon and sulfur while simultaneously reducing income taxes. These charges increase the cost of polluting or of overusing natural resources.

It is hoped that users react to the increase in cost from emission charges by decreasing pollution or decreasing consumption. In actual practice, this approach has not been notably successful, because people usually object to paying a tax on something they perceive as free. Even when it is suggested that the revenues from pollution taxes be used to finance rebates to the public in the form of income tax cuts, public response is usually lukewarm. Another problem with taxes on pollution is that they are almost always set too low to have much effect on the behavior of people or companies.

Many economists argue for a different incentive-based pollution control approach. The government sets a cap on pollution at an "acceptable" level and then issues a fixed number of **marketable waste-discharge permits**, each of which allows the holder to emit a specified amount of a given pollutant, such as sulfur dioxide. The marketable permits, called **emission reduction credits (ERCs)**, can be freely bought and sold by industries that produce that specific pollutant. Economists think this approach is an efficient way to move the market toward the point of optimum pollution. Companies that produce pollution in excess of what they are allowed must purchase additional ERCs, whereas those that pollute less than their allowances may sell their excess ERCs. Using marketable permits allows companies to determine what level of pollution and of investment in clean technologies and pollution abatement is economically appropriate for them.

The EPA used marketable permits to mandate lower levels of lead in gasoline. Refineries were assigned quotas of lead, which they could trade with each other. Since 1974 the EPA has also issued air pollution permits as marketable ERCs. Consider a company that wishes to expand its business and move into a new city. Like all communities in the United States, this city must comply with the air quality standards of the 1970 Clean Air Act. To produce goods in this city (and therefore, unavoidably, to produce pollution), the company must buy ERC polluting rights from established firms that have cut their own emissions.

The Clean Air Act Amendments of 1990 included a similar plan to cut acid rain–causing sulfur dioxide emissions by issuing tradable permits for coal-burning electricity utilities. These permits have been a very effective policy tool. Sulfur dioxide emissions are being reduced ahead of schedule at about 50% of the original projected cost.

We have examined the roles of government and economics in addressing environmental concerns. Now let us consider two very different environmental problems that are closely connected to both governmental and economic policies: management of old-growth forests of the Pacific Northwest and environmental destruction in Central and Eastern Europe.

Figure 3.14 Northern spotted owl (Strix occidentalis caurina). This rare species is found predominantly in old-growth forests in the Pacific Northwest, from northern California to southern British Columbia. Protecting the owl's habitat benefits many other species that also reside in the same environment.

CASE·IN·POINT Old-Growth Forests of the Pacific Northwest

Although legislative approaches have generally been effective in dealing with environmental problems, some issues are so contentious that they are difficult to resolve. One such controversy escalated during the late 1980s and 1990s in western Oregon, Washington, and northern California and continues today. At stake were thousands of jobs and the future of large tracts of old-growth (virgin) coniferous forest, along with the existence of organisms that depend on the forest.

One of these forest animals, the northern spotted owl, came to symbolize the confrontation because of its dependence on mature forests for habitat (Figure 3.14). These birds are specialists that usually need large areas of intact forest to supply them with food such as flying squirrels and red tree voles (a mouselike animal). Since 1990 the northern spotted owl has been listed as a threatened species under the Endangered Species Act.

Old-growth forests, by definition, have never been logged. They are very different from forests that were logged and have regrown. For one thing, many of the

redwoods and sequoias in old-growth forests are ancient, some as old as 2,000 years. Many trees have attained immense sizes. Some Douglas firs are 91 m (300 ft) tall and 4.9 m (16 ft) in diameter. Old-growth forests have trees at various ages, from seedlings and saplings to mature giants, whereas previously logged forests usually contain trees of the same age and size.

Because most of the forests in the United States have been logged at one time or another, the amount of old-growth forest represents a small fraction of the total forested area in the United States. Less than 10% of old-growth forests remain. This fraction is decreasing because old-growth forests continue to be logged. Most old-growth forests in the United States are found in the Pacific Northwest and Alaska.

To environmentalists, the old-growth forests are a national treasure to be protected and cherished. These stable forests provide the primary biological habitats for many species, including the northern spotted owl and 40 other endangered or threatened species. Provisions of the Endangered Species Act (ESA) require the government to protect the habitat of endangered species so that their numbers can increase. To enforce this law, in 1991 a court ordered the suspension of logging in about 1.2 million hectares (3 million acres) of federal forest where the owl lives.

The timber industry bitterly opposed the 1991 moratorium, stating that thousands of jobs would be lost if the northern spotted owl habitat were to be set aside. Many rural communities in the Pacific Northwest did not have diversified economies; timber was their main source of revenue. Thus, a major confrontation over the future of the old-growth forest ensued between the timber industry and environmentalists. Strong feelings were expressed on both sides—witness the bumper sticker reading "Save a logger, kill an owl."

Complexities of the Controversy The situation was more complex than simply jobs versus the owl, however. The timber industry was already declining in terms of its ability to support people in the Pacific Northwest. During the decade between 1977 and 1987, logging in Oregon's national forests increased by over 15%, whereas employment dropped by 15% (an estimated 12,000 jobs) during the same period. The main cause of this decline was automation of the timber industry. In addition, the timber industry in that region had not been operating **sustainably**—that is, the industry removed trees faster than the forest could regenerate. If the industry continued to log at its 1980s rates, most of the remaining old-growth forests would have disappeared within 20 years.

Timber is not as important to the economy of the Pacific Northwest as it used to be, a change that began long before the controversy over the northern spotted owl. Industry in the Pacific Northwest diversified, and by the late 1980s, the timber industry's share of the economy in Oregon and Washington was less than 4%. Although some had predicted economic disaster in 1991 when logging was blocked in the old-growth forests, Oregon's unemployment rate in 1994 was the lowest in 25 years. Oregon attracted several high-technology companies, which produced more jobs than had been lost from logging. (Of course, these new jobs were small consolation to unemployed loggers who were not trained in technology.)

The Political Solution to the Controversy In 1993, President Bill Clinton convened a much-publicized timber summit in Portland, Oregon, with members of all sides of the issue. The 1994 **Northwest Forest Plan** that arose from this summit represented a compromise between environmental and timber interests. Thanks to a healthy infusion of federal aid to the area, some timber workers were retrained for other kinds of careers. State programs, such as the Jobs for the Environment project funded by the state of Washington, also helped reduce unemployment. Hundreds of former loggers were employed by Jobs for the Environment and similar programs to restore watersheds and salmon habitats in the forests they used to harvest.

As a result of the Northwest Forest Plan, logging was resumed on federal forests in Washington, Oregon, and northern California. However, only about one fifth of the logging that occurred during the 1980s was permitted. The plan, along with previous congressional and administrative actions, reserved about 75% of federal timberlands to safeguard watersheds and provide protection for the northern spotted owl and several hundred other species, including salmon and other fishes.

As happens with many compromises, neither environmentalists nor timber-cutting interests were happy with the Northwest Forest Plan. Some environmentalists did not think the plan was scientifically sound and worried that it took too many risks with conservation values. Timber-cutting interests challenged the legality of the plan and tried to revoke or revise the laws that restricted logging. They asked Congress to pass legislation allowing a greater harvest of timber because of economic hardship.

Responding to powerful timber interests, a controversial bill that permitted **salvage logging** in national forests was signed into law in 1995. The law allowed loggers to cut dead trees and trees weakened by insects, disease, or fire, as well as "associated trees," healthy trees considered to be in danger of catching a disease. The law, which expired at the end of 1996, allowed loggers access to parts of the forest that had been declared off-limits by the Northwest Forest Plan as well as other national forests not normally open to logging. It also exempted timber companies from complying with provisions of the Clean Water Act and the ESA.

Forestry scientists and environmentalists opposed the salvage logging law because dead and diseased trees have important ecological roles in forests. Taxpayer advocates, such as the National Taxpayers Union, opposed the law because it cost taxpayers money. The U.S. Forest Service annually spends millions of dollars more to build and maintain roads through national forests than it earns from timber sales. Timber companies haul lumber on these roads. Taxpayer advocates consider road construction in national forests a **subsidy** to the timber industry because it did not pay for any of the road construction or repair costs.

In 1999 a federal judge ruled that the U.S. Forest Service and the Bureau of Land Management, the two government agencies overseeing logging of old-growth forests on federal land, had not adequately carried out the provisions of the 1994 Northwest Forest Plan. Specifically, the Northwest Forest Plan had required that the federal agencies complete extensive surveys of 77 local endangered and threatened species before granting the timber industry permission to log. The judge agreed with the environmental groups that brought the suit against the federal agencies that the agencies had failed to complete the surveys. The surveys were completed in 2001 and will be reviewed annually.

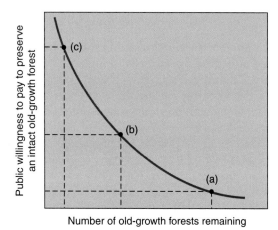

y-axis: Public willingness to pay to preserve an intact old-growth forest

x-axis: Number of old-growth forests remaining

Figure 3.15 Economic evaluation of old-growth forest. The public is willing to pay a great deal to preserve stands of intact old-growth forest when few of such forests remain.

An Economic Case for Preserving Old-Growth Forests
According to environmental economists, the old-growth forests that remain should not be logged because a cost–benefit analysis indicates the resulting loss to society is too great. Not only do taxpayers subsidize the timber industry but also, the financial benefits associated with preserving these forests, such as protection of species and increased recreation, outweigh the cost of preservation, such as loss of logging jobs and an increased price of wood.

Examine Figure 3.15, which shows the public's willingness to pay to preserve stands of intact old-growth forest. When many intact forests exist (see point *a*), the public value of a single forest is low, and the public is willing to allow the forest to be cut. As such forests become less abundant (see point *b*), the value of each old-growth forest increases. Point *c* represents the current situation in the United States. There are so few intact old-growth forests that the cost of losing even a single stand is unacceptably high. The public is therefore willing to pay a great deal to preserve any remaining old-growth forests.

CASE-IN-POINT Environmental Problems in Central and Eastern Europe

The fall of the Soviet Union and communist governments in Central and Eastern Europe during the late 1980s revealed a grim legacy of environmental destruction (Figure 3.16). Water was so poisoned from raw sewage and chemicals that it could not be used for industrial purposes, let alone for drinking. Unidentified chemicals leaked out of dump sites into the surrounding soil and water, while nearby, fruits and vegetables were grown in chemical-laced soil. Power plants emitted soot and sulfur dioxide into the air, producing a persistent chemical

haze. Buildings and statues eroded, and entire forests died because of air pollution and acid rain. Crop yields fell despite intensive use of chemical pesticides and fertilizers. One of the most polluted areas in the world was the "Black Triangle," which consists of the bordering regions of the former East Germany, northern Czech Republic, and southwest Poland.

How did this massive pollution affect human health? Many Central and Eastern Europeans suffered from asthma, emphysema, chronic bronchitis, and other respiratory diseases as a result of breathing the filthy, acrid air. By the time most Polish children were 10 years old, they suffered from chronic respiratory diseases or heart problems. The levels of cancer, miscarriages, and birth defects were also extremely high. Life expectancies are still lower than in other industrialized nations; in 2002 the average Eastern European lived to age 68, which was 10 years less than the average Western European.

The economic assumption behind communism was one of high production and economic self-sufficiency—regardless of environmental damages—and so pollution in communist-controlled Europe went largely unchecked. Meeting industrial production quotas always took precedence over environmental concerns, even though production was not carried out for profit. Because clean air, water, and soil had no economic value, it was acceptable to pollute. The governments supported heavy industry—power plants, chemicals, metallurgy, and large machinery—at the expense of the more environmentally benign service industries. As a result, Cen-

Figure 3.16 Pollution problems in former communist countries. This coal mine on the Mius River in Ukraine produces pollutants that seep through piles of waste into soil, surface water, and groundwater. Note the scummy layer of chemical wastes covering the river water. Thousands of these sites exist in former communist countries, the result of rapid expansion of industrialization without regard for the environment. It will take decades for the contamination from these sites to be cleaned up. Photographed in the late 1990s.

tral and Eastern Europe was overindustrialized, and because most of its plants were built during the post–World War II period, it lacked the pollution abatement equipment now required in factories in most industrialized countries.

In addition, communism did little to encourage resource conservation, which is a very effective way to curb pollution. Because energy subsidies and lack of competition allowed power plants to provide energy at prices far below its actual cost, neither industries nor individuals had a strong incentive to conserve energy.

Communism also took a toll on the environment because a government run by a single political party cannot be held accountable. Political opposition to any aspect of government operations, including environmental damage, was unthinkable, and people who wanted to scrutinize pollution information found that it was unavailable, having been classified as top secret. Only the collapse of communism allowed the citizens of Central and Eastern Europe to begin to assess the full extent of damage to their environments.

Clearly the current Central and Eastern European governments faced an intimidating task. While switching from communism to democracy with a free-market economy, they also faced the overwhelming responsibility of improving the environment, and any political system in transition will not have the environment as its first priority. Nonetheless, they tried to formulate environmental policies based on the experiences of the United States and Western Europe over the past several decades.

How long will the cleanup take? Experts predict that decades will be required to clean up the pollution legacy of communism. How much will it cost? The figures are staggering. It is estimated that improving the environment in what was formerly East Germany alone will cost up to $300 billion.

Meanwhile, the environment of countries in the former Soviet bloc is slowly improving. Some countries, such as Hungary, Poland, and the Czech Republic, have been relatively successful in moving toward a market economy. These countries are generating enough money to invest some of it in environmental cleanup. In the Czech Republic, 1.5% of the country's GDP was invested in environmental protection in 1999. The levels of air pollution (sulfur dioxide, nitrogen oxide, and particulates) are declining despite their increased economic growth. From 1990 to 2000 in the Czech Republic, sulfur dioxide emissions declined 86%, nitrogen oxide emissions from industries declined 69%, and particulate emissions declined 91%.

Other countries, such as Bulgaria, Romania, and Russia, have not had as smooth a socioeconomic transition to a free-market economy. Economic recovery has been slow in these countries, and severe budgetary problems have forced the environment to take a backseat to political and economic reform. ■

SUMMARY WITH SELECTED KEY TERMS

I. Conservation is the sensible and careful management of **resources** such as air, water, soil, forests, minerals, and wildlife. **Preservation** is concerned with setting aside undisturbed areas, maintaining them in a pristine state, and protecting them from human activities.

II. The first two centuries of U.S. history were a time of widespread environmental destruction. By the late 20th century, environmental awareness had become a pervasive popular movement.

A. During the 1700s and early 1800s, most Americans had a **frontier attitude**, a desire to conquer and exploit nature as quickly as possible.

B. During the 19th century, many U.S. naturalists began to be concerned about conserving natural resources.

   1. John James Audubon aroused widespread public interest in the wildlife of North America.

   2. Henry David Thoreau wrote about living in harmony with the natural world.

   3. George Perkins Marsh wrote about humans as agents of global environmental change.

C. The earliest conservation legislation revolved around protecting land—forests, parks, and monuments.

   1. The **General Revision Act** of 1891 gave the president the authority to establish forest reserves on public land.

     a. Benjamin Harrison, Grover Cleveland, and **Theodore Roosevelt** employed this law to put more than 17 million hectares aside as national forests.

     b. Theodore Roosevelt appointed **Gifford Pinchot** as the first head of the U.S. Forest Service. Pinchot supported expanding the nation's forest reserves and managing forests scientifically.

     c. Today, national forests have multiple uses, from biological habitats to recreation to timber harvest to cattle grazing.

   2. Yellowstone National Park, the world's first national park, was established in 1872.

     a. The Yosemite and Sequoia National Parks were established in California in 1890, largely in response to the efforts of naturalist **John Muir**, who also founded the Sierra Club.

     b. The Antiquities Act of 1906 authorized the president to set aside national monuments with scientific, historic, or prehistoric importance.

     c. The National Park Service was created in 1916 to manage U.S. national parks and monuments.

     d. Today there are 57 national parks and 74 national monuments.

D. Franklin Roosevelt established the Civilian Conservation Corps during the Great Depression. Workers planted trees, made paths and roads in national parks and forests, and built dams to control flooding.

E. The Dust Bowl during the 1930s called attention to the need for soil conservation. To address this issue, Franklin Roosevelt established the Soil Conservation Service.

F. Aldo Leopold was very influential in the conservation movement of the mid- to late 20th century. In *A Sand County Almanac*, published in 1949, he wrote about humanity's relationship with nature and about the need to conserve wilderness areas.

G. In 1962 **Wallace Stegner** wrote his famous "Wilderness Essay," which helped create support for the passage of the Wilderness Act of 1964.

H. During the 1960s, the public became concerned about pollution. **Rachel Carson** published *Silent Spring* in 1962, alerting the public about the dangers of uncontrolled pesticide use.

I. Paul Ehrlich published *The Population Bomb*, which raised the public's awareness of the dangers of overpopulation and triggered debates on how to deal effectively with population issues.

J. The first Earth Day occurred in 1970, when an estimated 20 million people in the United States demonstrated their support of improvements in resource conservation and environmental quality.

III. Since 1970 the federal government has addressed many environmental problems.

A. The **National Environmental Policy Act** (**NEPA**) was passed in 1970.

  1. By requiring **environmental impact statements** (**EISs**) that are open to public scrutiny, this law initiated serious environmental protection in the United States.

  2. NEPA established the **Council on Environmental Quality** to monitor the required EISs and report directly to the president.

B. Later laws, such as the Endangered Species Act, the Clean Air Act, and the Clean Water Act, have added to the environmental safety net, with some success. Many environmental laws contain statutes that allow **citizen suits**, in which private citizens can take violators, whether they are private industries or government-owned facilities, to court for noncompliance.

IV. Economics is the study of how people use their limited resources to try to satisfy their unlimited wants.

A. Economies depend on the natural environment as **sources** for raw materials and **sinks** for waste products.

  1. Both sources and sinks contribute to **natural capital**, which is Earth's resources and processes that sustain living organisms, including humans.

  2. Resource degradation and pollution represent the overuse of natural capital: Resource degradation is the overuse of sources, and pollution is the overuse of sinks.

B. National income accounts represent the total income of a nation for a given year.

  1. Many economists, government planners, and scientists support replacing gross domestic product (GDP) and net domestic product (NDP) with a more comprehensive measure of national income accounting that includes estimates of both natural resource depletion and the environmental cost of economic activities.

  2. Better accounting for environmental quality would help determine if, for a given environmental proposal, the benefits (economic and noneconomic) exceed the cost.

C. An **external cost**, or **negative externality**, is a harmful side effect of production or consumption of a product and is borne by individuals who are not directly involved in the market exchange for that product. Economists view pollution in a market economy as a failure in pricing.

D. From an economic point of view, the appropriate amount of pollution is a trade-off between harm to the environment and inhibition of development.

  1. The cost in environmental quality of a unit of pollution that is emitted into the environment is known as the **marginal cost of pollution**.

  2. The cost to dispose of a unit of pollution in a nonpolluting way is called the **marginal cost of pollution abatement**.

  3. Economists think the use of resources for pollution abatement should be increased only until the cost of abatement equals the cost of the damage done by the pollution. This results in the **optimum amount of pollution**—the amount that is economically most desirable.

  4. There are two main flaws in the economic approach to pollution: It is difficult to actually measure the monetary cost of pollution, and the risk of environmental disruption is rarely taken into account.

E. Government often controls pollution by nonincentive-based regulation such as **command and control regulation**, which imposes legal limits on amounts of permissible pollution.

F. Incentive-based regulations are market-oriented strategies that seek to use the economic forces of a free market to control pollution.

  1. An incentive-based **emission charge** controls pollution by charging the polluter for each unit of emissions—that is, by establishing a tax on pollution.

  2. Pollution can also be controlled by incentive-based **marketable waste-discharge permits**, which allow the holder to pollute a given amount. Holders of these **emission reduction credits** are not allowed to produce more emissions than are sanctioned by their permits, although they can buy and sell permits as needed.

V. The question of how much remaining old-growth forest should be logged in the Pacific Nowthwest remains a contentious issue.

A. Logging in the 1980s was not **sustainable**, and all remaining old-growth forests would have been lost had logging not been curtailed.

B. The 1994 **Northwest Forest Plan** represented a compromise between environmental and timber interests. Only about one fifth of the logging that occurred during the 1980s was permitted.

C. From 1995 to 1996, **salvage logging** was permitted in these forests, in response to pressure from powerful timber interests.

VI. Formerly communist countries in Central and Eastern Europe face an expensive cleanup from environmental damage inflicted during the communist era of high production and economic self-sufficiency.

THINKING ABOUT THE ENVIRONMENT

1. Discuss the pros and cons of the 1995 unfunded mandate law as it relates to future environmental legislation.

2. Was Theodore Roosevelt a conservationist or a preservationist? Explain your answer, making sure you clearly distinguish between conservation and preservation.

3. Briefly describe each of the following aspects of U.S. environmental history: protecting forests; establishing and protecting national parks and monuments; conservation in the mid-20th century; the environmental movement of the late 20th century.

4. Describe the environmental contributions of two of the following: John James Audubon, Henry David Thoreau, George Perkins Marsh, Theodore Roosevelt, Gifford Pinchot, John Muir, Franklin Roosevelt, Aldo Leopold, Wallace Stegner, Rachel Carson, and Paul Ehrlich.

5. The National Environmental Policy Act (NEPA) is sometimes called the "Magna Carta of environmental law." What is meant by such a comparison?

6. Why are environmental impact statements such a powerful tool of NEPA?

7. If you were a member of Congress, what legislation would you introduce to deal with each of the following problems?

   a. Poisons from a major sanitary landfill are polluting your state's groundwater.

   b. Acid rain from a coal-burning power plant in a nearby state is harming the trees in your state. Loggers and foresters are upset.

   c. There is a high incidence of cancer in the area of your state where heavy industry is concentrated.

8. How would an economist approach each of the problems listed in question 7? How would an environmentalist?

9. Draw a diagram showing how the following are related: natural environment, sources, sinks, economy, production, consumption, raw materials, and waste products.

10. Distinguish between command and control regulation and incentive-based regulation.

11. What are two problems with the national income accounts as they relate to the environment?

12. Draw a graph that shows the marginal cost of pollution, the marginal cost of pollution abatement, and the optimum amount of pollution.

13. Discuss the events that led to the Northwest Forest Plan of 1994.

14. Describe the extent of environmental destruction in formerly comunist countries.

*15. Interpret the graphs in Figure 3.13. In each case, is the amount of pollution indicated more or less than the economically optimum amount of pollution? Is the marginal cost of pollution higher or lower than if the pollution level was at its optimum? Is the cost of pollution abatement higher or lower than if the pollution level was at its optimum?

*16. The graph below shows two marginal cost of pollution abatement curves. In this hypothetical situation, technological innovations were developed between 1995 and 2000 that lowered the abatement cost. Which curve (a or b) corresponds to 1995, and which to 2000? Explain your answer.

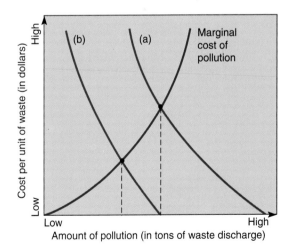

* Solutions to questions preceded by asterisks appear in Appendix VII.

TAKE A STAND

Visit our Web site at **http://www.wiley.com/college/raven** (select Chapter 3 from the Table of Contents) for links to more information about the controversies surrounding the old-growth forests in the Pacific Northwest. Consider the opposing views of loggers and other rural people and environmentalists and debate the issues with your classmates. You will find tools to help you organize your research, analyze the data, think critically about the issues, and construct a well-considered argu-

ment. Take a Stand activities can be done individually or as part of a team, as oral presentations, written exercises, or Web-based (e-mail) assignments.

Additional on-line materials relating to this chapter, including a Student Testing Section with study aids and self-tests, Environmental News, Activity Links, Research Projects, and more, can also be found on our Web site.

SUGGESTED READING

Abramovitz, J.N. "Putting a Value on Nature's 'Free' Services." *World Watch*, Vol. 11, No. 1 (January–February 1998). Most of the world's economy is based on the normal functioning of the environment's forests, grasslands, and rivers.

Austin, D. "The Green and the Gold: How a Firm's Clean Quotient Affects Its Value." *Resources*, Vol. 132 (summer 1998). Evidence suggests that companies can profit from reducing pollution that their activities produce.

Carson, R.L. *Silent Spring.* Boston: Houghton Mifflin, 1962. This classic book first alerted the public to the environmental dangers of pesticide use.

Darmstadter, J. "Greening the GDP: Is It Desirable? Is It Feasible?" *Resources*, Issue 139 (spring 2000). Summarizes *Nature's Numbers: Expanding the National Economic Accounts to Include the Environment*, which was recently published by the National Research Council's Committee on National Statistics.

Franklin, A. "Are Spotted Owls Doing Better?" *Environmental Review*, Vol. 8, No. 6 (June 2001). Alan Franklin, a conservation biologist, discusses the current status of northern spotted owls.

Fullerton, D., and R. Stavins. "How Economists See the Environment." *Nature*, Vol. 395 (October 1, 1998). This

commentary explains some of the misunderstandings between economists and ecologists about the environment.

Goodstein, E.S. *Economics and the Environment*, 3rd edition. New York: John Wiley & Sons, Inc., 2002. This excellent text considers all aspects of ecological economics, from environmental regulation to promoting clean technology.

Jackson, D.D. "A Sage for All Seasons." *Smithsonian* (September 1998). An intimate look at Aldo Leopold, whose book, *A Sand County Almanac*, is an environmental classic.

Kahn, J.R. *The Economic Approach to Environmental and Natural Resources*, 2nd edition. Fort Worth, TX: Dryden Press, 1998. Chapter 11 considers the controversy surrounding old-growth forests and discusses some of the economic studies relating to this highly polarized debate.

Leopold, A. *A Sand County Almanac.* New York: Ballantine Books, reprint edition, 1970. This environmental classic is a philosophical examination of man's interrelationship with nature.

Portney, P.R. "Counting the Cost: The Growing Role of Economics in Environmental Decisionmaking." *Environment*, Vol. 40, No. 2 (March 1998). Examines why economics is increasingly important in environmental policy debates and why this is a good thing.

Cordgrass in a Chesapeake Bay salt marsh.

Ecosystems and Energy

Learning Objectives

After you have studied this chapter you should be able to:

1. Define *ecology* and distinguish among the following ecological levels: population, community, ecosystem, landscape, and biosphere.

2. Define *energy,* and explain how it is related to work and to heat.

3. Use examples to contrast potential energy and kinetic energy.

4. State the first and second laws of thermodynamics, and discuss the implications of these laws as they relate to organisms.

5. Write summary reactions for photosynthesis and respiration, and contrast these two biological processes.

6. Describe the communities around hydrothermal vents and explain the source of energy that sustains them.

7. Summarize how energy flows through a food web, using the terms *producer, consumer,* and *decomposer.*

8. Explain some of the impacts humans have had on the Antarctic food web.

9. Draw and explain typical pyramids of numbers, biomass, and energy.

10. Distinguish between gross primary productivity and net primary productivity, and discuss human impact on the latter.

The environment often contains as astonishing assortment of organisms that interact with each other and are interdependent in a variety of ways. Consider for a moment a salt marsh in the Chesapeake Bay on the East Coast of the United States. This bay is one of the world's richest **estuaries,** semi-enclosed bodies of water found where fresh water from a river drains into the ocean. Estuaries are under the influence of tides and gradually change from unsalty fresh water to salty ocean water. In the Chesapeake Bay this change results in three distinct marsh communities: freshwater marshes at the head of the bay, brackish (moderately salty) marshes in the middle bay region, and salt marshes on the ocean side of the bay. Each community has its own characteristic organisms.

A Chesapeake Bay salt marsh consists of flooded meadows of cordgrass (*Spartina*; see photo). Few other plants are found because high salinity and twice-daily tidal inundations produce a challenging environment to which only a few plants have adapted. Nutrients such as nitrates and phosphates, much of them from treated sewage and agriculture, drain into the marsh from the land and promote rapid growth of both cordgrass and microscopic algae suspended in the water. These organisms are eaten directly by some animals, and when they die, their remains provide food for other salt marsh inhabitants.

A casual visitor to a salt marsh would observe two different types of animal life: insects and birds. Insects, particularly mosquitoes and horseflies, number in the millions. Birds nesting in the salt marsh include seaside sparrows, laughing gulls, and clapper rails. Study the salt marsh carefully and you will find it has numerous other species. Large numbers of invertebrates, such as shrimps, lobsters, crabs, barnacles, worms, clams, and snails, seek refuge in the water surrounding the cordgrass. Here they eat, hide from predators to avoid being eaten, and reproduce. Many invertebrates gather on the shore between the high-tide mark and the low-tide mark because food—detritus, algae, protozoa, and worms—is abundant there.

Almost no amphibians inhabit salt marshes (the salty water dries out their skin), but a few reptiles, such as the northern diamondback terrapin (a semi-aquatic turtle), have adapted. It spends its time basking in the sun or swimming in the water searching for food—snails, crabs, worms, insects, and fish. Although a variety of snakes abound in the dry areas adjacent to salt marshes, only the northern water snake, which preys on fish, is adapted to salty water.

The meadow vole is a small rodent that lives in the salt marsh. It constructs a nest of cordgrass on the ground above the high-tide zone. Meadow voles are excellent swimmers and scamper about the salt marsh day and night. Their diet consists mainly of insects and the leaves, stems, and roots of cordgrass.

The Chesapeake Bay marshes are an important nursery for numerous marine fishes—spotted sea trout, Atlantic croaker, striped bass, and bluefish, to name just a few. These fishes typically spawn (reproduce) in the open ocean, and the young then enter the estuary, where they eat smaller fish and invertebrates and grow into juveniles.

Add to all these visible plant and animal organisms the unseen microscopic world of the salt marsh, which includes countless numbers of protozoa, fungi, and bacteria, and you can begin to appreciate the complexity of a saltmarsh community.

WHAT IS ECOLOGY?

The concept of ecology was first developed in the 19th century by Ernst Haeckel, who also devised its name—*eco* from the Greek word for "house" and *logy* from the Greek word for "study." Thus, *ecology* literally means "the study of one's house." The environment—one's house—consists of two parts, the **biotic** (living) environment, which includes all organisms, and the **abiotic** (nonliving, or physical) environment, which includes such physical factors as living space, temperature, sunlight, soil, wind, and precipitation. **Ecology**, then, is the study of the interactions among organisms and between organisms and their abiotic environment.

The focus of ecology can be local or global, specific or generalized, depending on what questions the scientist is asking and trying to answer. One ecologist might determine the temperature or light requirements of a single species of oak, another might study all the organisms that live in a forest where the oak is found, and another might examine how nutrients flow between the forest and surrounding communities.

Ecology is the broadest field within the biological sciences, and it is linked to every other biological discipline. The universality of ecology also links subjects that are not traditionally part of biology. Geology and earth science are extremely important to ecology, especially when ecologists examine the physical environment of planet Earth. Chemistry and physics are also important. Because humans are biological organisms, all of our activities have a bearing on ecology. Even economics and politics have profound ecological implications, as was presented in Chapter 3.

How does the field of ecology fit into the organization of the biological world? As you may know, one of the characteristics of life is its high degree of organization (Figure 4.1). Atoms are organized into molecules, which are organized into cells. In multicellular organisms, cells are organized into tissues, tissues into organs such as a bone or stomach, organs into body systems such as the nervous system and digestive system, and body systems into individual organisms such as dogs, ferns, and so on.

Ecologists are most interested in the levels of biological organization that include or are above the level of the individual organism (see "Mini-Glossary: Ecology Terms"). Organisms occur in **populations**, which are members of the same species that live together in the same area at the same time. (A **species** is a group of similar organisms whose members freely interbreed with one another in the wild to produce fertile offspring; members of one species do not interbreed with other species of organisms.) A population ecologist might study a popula-

MINI-GLOSSARY

Ecology Terms

population: A group of organisms of the same species that live together in the same area at the same time.

community: All the populations of different species that are living together in the same area at the same time.

ecosystem: A community and its physical environment. Includes all the interactions among organisms and between organisms and their abiotic environment. These interactions form a complex network of energy flow and materials cycling.

landscape: A region that includes several ecosystems.

biosphere: The layer of Earth containing all living organisms. As an ecological system, the biosphere interacts with the land, the water, and the atmosphere.

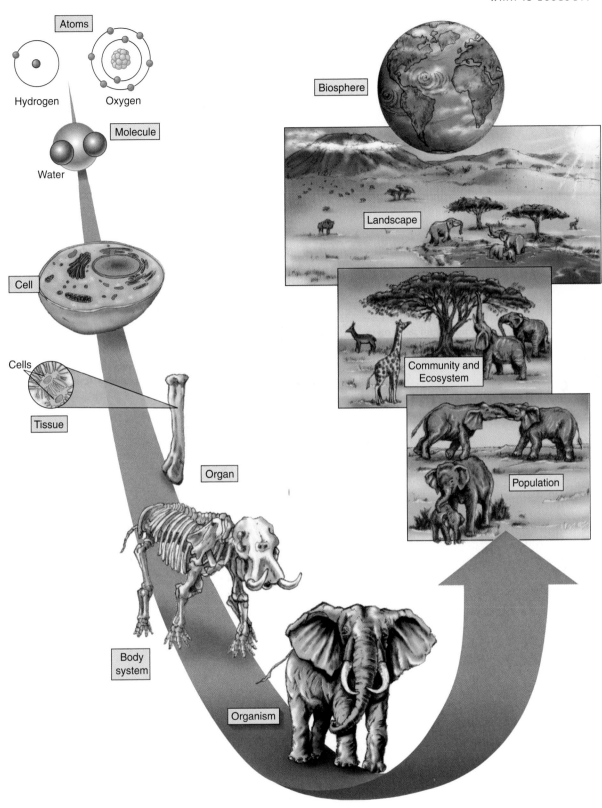

Figure 4.1 Levels of biological organization. Starting at the simplest level, atoms are organized into molecules, which are organized into cells. Cells are organized into tissues, tissues into organs, organs into body systems, and body systems into individual multicellular organisms. A group of individuals of the same species is a population. Populations of different species interact to form communities. A community and its abiotic environment are an ecosystem, whereas a region with several ecosystems is a landscape. The layer of Earth containing all living organisms comprises the biosphere. Ecologists study the highest levels of biological organization: individual organisms, populations, communities, ecosystems, landscapes, and the biosphere.

Figure 4.2 Coral reef community. Coral reef communities have the greatest number of species and are the most complex aquatic community. A coral reef community in the Indian Ocean off the coast of Maldives is shown.

tion of polar bears or a population of marsh grass. Populations, particularly human populations, are so important in environmental science that we devote two chapters, Chapters 8 and 9, to their study.

Populations are organized into communities. A **community** is a natural association that consists of all the populations of different species that live and interact together within an area at the same time. Different communities are characterized by the number and kinds of species that live there, along with their relationships with one another. A community ecologist might study how organisms interact with one another—including feeding relationships (who eats whom)—in an alpine meadow community or in a coral reef community (Figure 4.2).

Ecosystem is a more inclusive term than *community* because an **ecosystem** is a community together with its physical environment. Thus, an ecosystem includes not only all the biotic interactions of a community but also the interactions between organisms and their abiotic environment. In an ecosystem, all of the biological, physical, and chemical components of an area form an extremely complicated interacting network of energy flow and materials cycling. An ecosystem ecologist might examine how energy, nutrients, organic (carbon-containing) materials, and water affect the organisms living in a desert community or a coastal bay ecosystem.

The ultimate goal of ecosystem ecologists is to understand how ecosystems function. This is not a simple task, but it is important because ecosystem processes collectively regulate global cycles of water, carbon, nitrogen, and oxygen that are essential to the survival of humans and all other organisms (see Chapter 6). As humans increasingly alter ecosystems for their own uses,

the natural functioning of ecosystems is changed, and we need to know if these changes will affect the **sustainability** of our life-support system.

Landscape ecology is a relatively new subdiscipline in ecology that studies the connections among ecosystems found in a particular region. Consider a **landscape** consisting of a forest ecosystem located adjacent to a pond ecosystem. One connection between these two ecosystems might be great blue herons, which eat fish, frogs, insects, crustaceans, and snakes along the shallow water of the pond but often build nests and raise their young in the secluded treetops of the nearby forest. Landscapes, then, are based on larger land areas that include several ecosystems.

The layer of Earth that contains all living organisms is known as the **biosphere**. The organisms of the biosphere—Earth's communities, ecosystems, and landscapes—depend on one another and on the other realms of the Earth's physical environment: the atmosphere, hydrosphere, and lithosphere (Figure 4.3). The **atmosphere** is the gaseous envelope surrounding the Earth; the **hydrosphere** is the Earth's supply of water—liquid and frozen, fresh and salty; and the **lithosphere** is the soil and rock of Earth's crust. Ecologists who study the biosphere examine the complex global interrelationships among the Earth's atmosphere, land, water, and organisms.

The biosphere teems with life. Where do these organisms get the energy to live? And how do they harness this energy? We now examine the importance of energy to organisms, which can continue to survive only as long as the environment continuously supplies them with energy. We will revisit the importance of energy as it relates to human endeavors in many chapters throughout this text.

Figure 4.3 The northern tip of Palawan Island, Philippines, shows Earth's four realms. The atmosphere contains cumulus clouds, which indicate warm, moist air. The jagged spires of rock, formed from volcanic lava flows that have eroded over time, represent the lithosphere. The shallow water of the bay represents the hydrosphere. The biosphere includes the green vegetation and human settlement along the shore as well as the coral reefs visible as darker areas in the bay.

THE ENERGY OF LIFE

Energy is the capacity or ability to do work. In organisms, the biological work that requires energy includes processes such as growing, moving, reproducing, and maintaining and repairing damaged tissues. Energy exists in several different forms: chemical, radiant, heat, mechanical, nuclear, and electrical (see "Mini-Glossary: Forms of Energy"). Biologists generally express energy in units of work (**kilojoules, kJ**) or units of heat energy (**kilocalories, kcal**). One kilocalorie, which is the energy required to raise the temperature of 1 kg of water by 1°C, equals 4.184 kJ.

Energy can exist as stored energy—called **potential energy**—or as **kinetic energy**, the energy of motion (Figure 4.4). You can think of potential energy as an

MINI-GLOSSARY

Forms of Energy

chemical energy: Energy stored in the chemical bonds of molecules. For example, food contains chemical energy.

radiant, or solar, energy: Energy transported from the sun as electromagnetic waves.

heat energy: Thermal energy that flows from an object with a higher temperature (the heat source) to an object with a lower temperature (the heat sink).

mechanical energy: Energy in the movement of matter.

nuclear energy: Energy found within atomic nuclei (discussed in Chapter 11).

electrical energy: Energy that flows as charged particles.

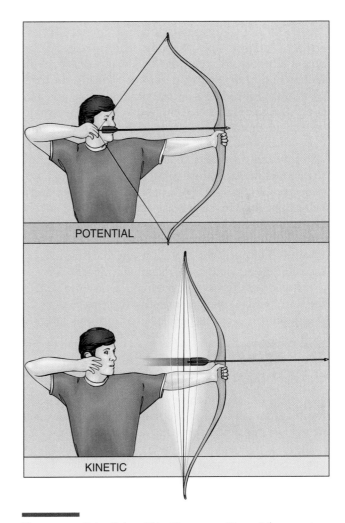

Figure 4.4 Potential and kinetic energy. Potential energy is stored in the drawn bow and is converted to kinetic energy as the arrow speeds toward its target.

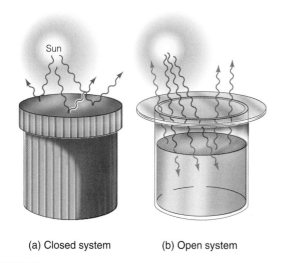

(a) Closed system   (b) Open system

Figure 4.5 Closed and open systems, with regard to energy.
(a) Energy is not exchanged between a closed system and its surroundings. (b) Energy is exchanged between an open system and its surroundings. Earth is an open system because it receives energy from the sun.

arrow on a drawn bow. When the string is released, this potential energy is converted to kinetic energy as the motion of the bow propels the arrow. Similarly, the cordgrass that a meadow vole eats has chemical potential energy, some of which is converted to kinetic energy and heat as the meadow vole swims in the salt marsh. Thus, energy can change from one form to another.

The study of energy and its transformations is called **thermodynamics**. When considering thermodynamics, scientists use the term *system* to refer to an object that is being studied. The rest of the universe other than the system being studied is known as the *surroundings*. A **closed system** is one that does not exchange energy with its surroundings, whereas an **open system** is one that can exchange energy with its surroundings (Figure 4.5).

Regardless of whether a system is open or closed, there are two laws about energy that apply to all things in the universe: the first and second laws of thermodynamics.

The First Law of Thermodynamics

According to the **first law of thermodynamics**, energy cannot be created or destroyed, although it can be transformed from one form to another. As far as we know, the energy present in the universe at its formation, approximately 15 billion to 20 billion years ago, equals the amount of energy present in the universe today. This is all the energy that can ever be present in the universe. Similarly, the energy of any system and its surroundings is constant. A system may absorb energy from its surroundings, or it may give up some energy into its surroundings, but the total energy content of that system and its surroundings is always the same.

As specified by the first law of thermodynamics, then, an organism cannot *create* the energy that it requires to live. Instead, it must capture energy from the environment to use for biological work, a process involving the transformation of energy from one form to another. In photosynthesis, plants absorb the radiant energy of the sun and convert it into the chemical energy contained in the bonds of carbohydrate (sugar) molecules. Similarly, some of that chemical energy may later be transformed by an animal that eats the plant into the mechanical energy of muscle contraction, enabling it to walk, run, slither, fly, or swim.

The Second Law of Thermodynamics

As each energy transformation occurs, some of the energy is changed to heat energy that is then released into the cooler surroundings. This energy can never again be used by any organism for biological work; it is "lost" from the biological point of view. However, it is not really gone from a thermodynamic point of view because it still exists in the surrounding physical environment. The use of food to enable us to walk or run does not destroy the chemical energy that was once present in the food molecules. After we have performed the task of walking or running, the energy still exists in the surroundings as heat energy.

The **second law of thermodynamics** can be stated most simply as follows: Whenever energy is converted from one form to another, some usable energy—that is, energy available to do work—is degraded into heat, a less-usable form that disperses into the environment. As a result, the amount of usable energy available to do work in the universe decreases over time.

It is important to understand that the second law of thermodynamics is consistent with the first law; that is, the total amount of energy in the universe is *not* decreasing with time. However, the total amount of energy in the universe that is available to do biological work is decreasing over time.

Less-usable energy is more diffuse, or disorganized. **Entropy** is a measure of this disorder or randomness; organized, usable energy has low entropy, whereas disorganized energy such as heat has high entropy. Entropy is continuously increasing in the universe in all natural processes. It may be that at some time, billions of years from now, all energy will exist as heat uniformly distributed throughout the universe. If that happens, the universe as a closed system will cease to operate because no work will be possible. Everything will be at the same temperature, so there will be no way to convert the thermal energy of the universe into usable mechanical energy.

Another way to explain the second law of thermodynamics, then, is that entropy, or disorder, in a system spontaneously tends to increase over time. (The word *spontaneously* in this context means that entropy occurs naturally rather than being caused by some external influence.)

As a result of the second law of thermodynamics, no process requiring an energy conversion is ever 100% efficient, because much of the energy is dispersed as heat, resulting in an increase in entropy. For example, an automobile engine, which converts the chemical energy of gasoline to mechanical energy, is between 20% and 30% efficient. That is, only 20% to 30% of the original energy stored in the chemical bonds of the gasoline molecules is actually transformed into mechanical energy, or work. In our cells, energy utilization for metabolism is about 50% efficient; the remaining energy is released to the surroundings as heat.

Organisms have a high degree of organization, and at first glance, they appear to refute the second law of thermodynamics. As organisms grow and develop, they maintain a high level of order and do not appear to become more disorganized. However, organisms are able to maintain their degree of order over time only with the constant input of energy. That is why plants must photosynthesize and animals must eat food.

Photosynthesis and Cellular Respiration

Photosynthesis is the biological process in which light energy from the sun is captured and transformed into the chemical energy of carbohydrate (sugar) molecules. Photosynthetic pigments such as **chlorophyll**, which is green and gives plants their green color, absorb radiant energy. This energy is used to manufacture a carbohydrate called glucose ($C_6H_{12}O_6$) from carbon dioxide (CO_2) and water (H_2O), with the liberation of oxygen (O_2):

$$6CO_2 + 12\,H_2O + \text{radiant energy} \rightarrow C_6H_{12}O_6 + 6H_2O + 6O_2$$

The chemical equation for photosynthesis is read as follows: 6 molecules of carbon dioxide plus 12 molecules of water plus light energy are used to produce 1 molecule of glucose plus 6 molecules of water plus 6 molecules of oxygen. (See Appendix I for a review of basic chemistry.)

Plants, some bacteria, and algae (photosynthetic aquatic organisms that range from single cells to seaweeds well over 50 m in length) perform photosynthesis, a process that is essential for almost all life. Photosynthesis provides these organisms with a ready supply of energy in carbohydrate molecules that they can use as the need arises. The energy can also be transferred from one organism to another—for instance, from plants to the organisms that eat plants (Figure 4.6). Oxygen, which is required by many organisms when they break down glucose or similar foods by cellular respiration, is a by-product of photosynthesis.

The chemical energy that plants store in carbohydrates and other molecules is released within cells of plants, animals, or other organisms through **cellular respiration**. In this process, molecules such as glucose are broken down in the presence of oxygen and water

Figure 4.6   Black-tailed prairie dog *(Cynomys ludovicianus).* The chemical energy produced by photosynthesis and stored in seeds and leaves is transferred to the black-tailed prairie dog as it eats.

into carbon dioxide and water, with the release of energy:

$$C_6H_{12}O_6 + 6O_2 + 6H_2O \rightarrow 6CO_2 + 12H_2O + \text{energy}$$

Cellular respiration makes the chemical energy stored in glucose and other food molecules available to the cell for biological work, such as moving around, courting, and growing new cells and tissues. All organisms, including green plants, respire to obtain energy. Some organisms, however, do not use oxygen for this process. Some types of bacteria that live in waterlogged soil, stagnant ponds, or animal intestines respire in the absence of oxygen.

CASE-IN-POINT   Life Without the Sun

The sun is the energy source for almost all of the biosphere. A notable exception was discovered in 1977, when an oceanographic expedition aboard the submersible research craft *Alvin* studied the Galapagos Rift, a deep cleft in the ocean floor off the coast of Ecuador. The expedition revealed, on the floor of the deep ocean, a series of **hydrothermal vents** where seawater apparently had penetrated and been heated by the hot rocks below. During its time within the Earth, the water had been

Figure 4.7 A hydrothermal vent ecosystem. Bacteria living in the tissues of these tube worms extract energy from hydrogen sulfide to manufacture organic compounds. Because these worms lack digestive systems, they depend on the organic compounds provided by the bacteria, along with materials filtered from the surrounding water. Also visible in the photograph are some filter-feeding clams (*yellow*) and a crab (*white*).

charged with inorganic compounds, including hydrogen sulfide (H_2S).

At the tremendous depth (greater than 2,500 m, or 8,202 ft) of the Galapagos Rift, there is no light for photosynthesis. But the hot springs support a rich and bizarre ecosystem that contrasts with the surrounding "desert" of the deep-ocean floor. Most of the species found in these oases of life were new to science. Giant, bloodred tube worms almost 3 m (10 ft) in length cluster in great numbers around the vents (Figure 4.7). Other animals around the hydrothermal vents include clams, crabs, barnacles, and mussels.

Scientists initially wondered what the ultimate source of energy for the species in this dark environment is. Most deep-sea ecosystems depend on the organic material that drifts down from surface waters; that is, they depend on energy derived from photosynthesis. But the Galapagos Rift ecosystem and other hydrothermal vent ecosystems are too densely clustered and too productive to be dependent on chance encounters with organic material from surface waters.

The base of the food web in these aquatic oases consists of certain bacteria that can survive and multiply in water so hot (exceeding 200°C, or 392°F) that it would not even remain in liquid form were it not under such extreme pressure. These bacteria function as producers, but they do not photosynthesize. Instead, they **chemosynthesize**, obtaining energy and making carbohydrate molecules from inorganic raw materials. Chemosynthetic bacteria possess enzymes (organic catalysts) that cause the inorganic molecule hydrogen sulfide to react with oxygen, producing water and sulfur or sulfate. Such chemical reactions provide the energy required to support these bacteria and other organisms in deep-ocean hydrothermal vents. Many of the Galapagos Rift animals consume the bacteria directly by filter feeding. Others, such as the giant tube worms, are supplied energy from chemosynthetic bacteria that live symbiotically inside their bodies. ◼

THE FLOW OF ENERGY THROUGH ECOSYSTEMS

We have seen that, with the exception of hydrothermal vent ecosystems, energy enters ecosystems as radiant energy (sunlight), some of which is trapped by plants during photosynthesis. The energy, now in chemical form, is stored in the bonds of organic molecules such as glucose. Animals obtain their energy by eating plants or by eating animals that ate plants. All organisms—plants, animals, and microorganisms—respire to obtain some of the energy in organic molecules. When these molecules are broken apart by cellular respiration, the energy becomes available to do work such as repairing tissues, producing body heat, or reproducing. As the work is accomplished, the energy escapes the organism and dissipates into the environment as heat (recall the second law of thermodynamics). Ultimately, this heat energy radiates into space. Thus, once energy has been used by an organism, it becomes unusable. The movement of energy just described—in a one-way direction through an ecosystem—is known as **energy flow**.

Producers, Consumers, and Decomposers

The organisms of an ecosystem can be divided into three categories on the basis of how they obtain nourishment: producers, consumers, and decomposers (Figure 4.8). Virtually all ecosystems contain representatives of all three groups, which interact extensively, both directly and indirectly, with one another.

Producers, also called **autotrophs** (Greek *auto*, "self," and *tropho*, "nourishment"), manufacture complex organic molecules from simple inorganic substances, generally carbon dioxide and water, usually using the energy of sunlight to do so. In other words, most produc-

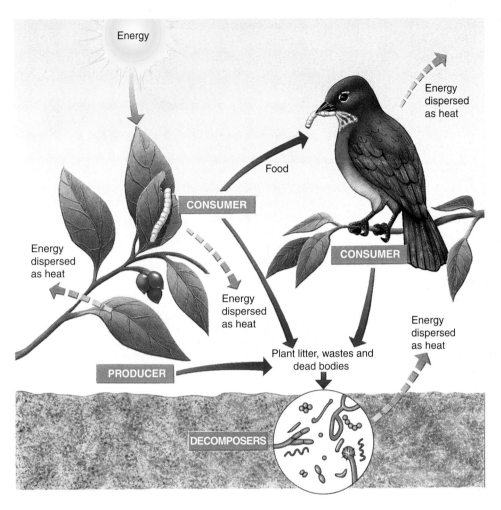

Energy

CONSUMER

Food

Energy
dispersed
as heat

CONSUMER

Energy
dispersed
as heat

Energy
dispersed
as heat

PRODUCER

Plant litter, wastes and
dead bodies

Energy
dispersed
as heat

DECOMPOSERS

Figure 4.8 Producers, consumers, and decomposers. During photosynthesis, producers use the energy from sunlight to make complex molecules from carbon dioxide and water. Consumers obtain energy when they eat producers or consumers that ate producers. Wastes and dead organic material supply decomposers with energy. During every energy transaction, some energy is lost to biological systems as it is dispersed as heat.

ers perform the process of photosynthesis. By incorporating the chemicals they manufacture into their own bodies, producers become potential food resources for other organisms. Whereas plants are the most significant producers on land, algae and certain types of bacteria are important producers in aquatic environments. In the saltmarsh ecosystem discussed in the chapter introduction, cordgrass, algae, and photosynthetic bacteria are the important producers.

Animals are **consumers**; that is, they use the bodies of other organisms as a source of food energy and bodybuilding materials. Consumers are also called **heterotrophs** (Greek *heter*, "different," and *tropho*, "nourishment"). Consumers that eat producers are called **primary consumers**, which usually means that they are exclusively **herbivores** (plant eaters). Rabbits and deer are examples of primary consumers, as is the marsh periwinkle, a type of snail that feeds on algae in the saltmarsh ecosystem.

Secondary consumers eat primary consumers, whereas **tertiary consumers** eat secondary consumers. Both secondary and tertiary consumers are flesh-eating **carnivores** that eat other animals. Lions and spiders are examples of carnivores, as are the northern diamondback terrapin and the northern water snake in the saltmarsh ecosystem. Other consumers, called **omnivores**, eat a

variety of organisms, both plant and animal. Bears, pigs, and humans are examples of omnivores; the meadow vole, which eats both insects and cordgrass in the saltmarsh ecosystem, is also an omnivore.

Some consumers, called **detritus feeders** or **detritivores**, consume **detritus**, which is organic matter that includes animal carcasses, leaf litter, and feces. Detritus feeders, such as snails, crabs, clams, and worms, are especially abundant in aquatic environments, where they burrow in the bottom muck and consume the organic matter that collects there. Marsh crabs are detritus feeders in the saltmarsh ecosystem. Earthworms are terrestrial (land-dwelling) detritus feeders, as are termites, beetles, snails, and millipedes. Detritus feeders work together with microbial decomposers to destroy dead organisms and waste products. An earthworm actually eats its way through the soil, digesting much of the organic matter contained there.

Decomposers, also called **saprotrophs** (Greek *sapro*, "rotten," and *tropho*, "nourishment"), are microbial heterotrophs that break down dead organic material and use the decomposition products to supply themselves with energy. They typically release simple inorganic molecules, such as carbon dioxide and mineral salts, that can then be reused by producers. Bacteria and fungi are important examples of decomposers. Sugar-metabolizing fungi first invade dead wood and consume the wood's simple carbohydrates, such as glucose and maltose. When these carbohydrates are exhausted, other fungi, often aided by termites with symbiotic bacteria in their guts, complete the digestion of the wood by breaking down cellulose, a complex carbohydrate that is the main component of wood.

Ecosystems such as the Chesapeake Bay salt marsh contain a balanced representation of all three ecological categories of organisms—producers, consumers, and decomposers—and all of these have indispensable roles in ecosystems. Producers provide both food and oxygen for the rest of the community. Consumers play an important role by maintaining a balance between producers and decomposers. Detritus feeders and decomposers are necessary for the long-term survival of any ecosystem because, without them, dead organisms and waste products would accumulate indefinitely. Without microbial decomposers, important elements such as potassium, nitrogen, and phosphorus would remain permanently in dead organisms and therefore be unavailable for use by new generations of organisms.

The Path of Energy Flow: Who Eats Whom in Ecosystems

In an ecosystem, energy flow occurs in **food chains**, in which energy from food passes from one organism to the next in a sequence (Figure 4.9). Each level, or "link," in a food chain is called a **trophic level** (recall that the Greek *tropho* means nourishment). The first trophic level is formed by producers (organisms that photosynthesize), the second level by primary consumers (herbivores), the third level by secondary consumers (carnivores), and so on. At every step in a food chain are decomposers, which respire organic molecules in the carcasses and body wastes of all members of the food chain.

Simple food chains rarely occur in nature, because few organisms eat just one kind of organism. More typically, the flow of energy and materials through an ecosystem takes place in accordance with a range of choices of food for each organism involved. In an ecosystem of average complexity, numerous alternative pathways are possible. A hawk eating a rabbit is a different energy pathway than a hawk eating a snake. Thus, a **food web**, which is a complex of interconnected food chains in an ecosystem, is a more realistic model of the flow of energy and materials through ecosystems (Figure 4.10).

The most important thing to remember about energy flow in ecosystems is that it is linear, or one-way. That is, energy can move along a food chain or food web from one organism to the next as long as it has not been used to do biological work. Once energy has been used by an organism, however, it is lost as heat and is unavailable for use by any other organism in the ecosystem.

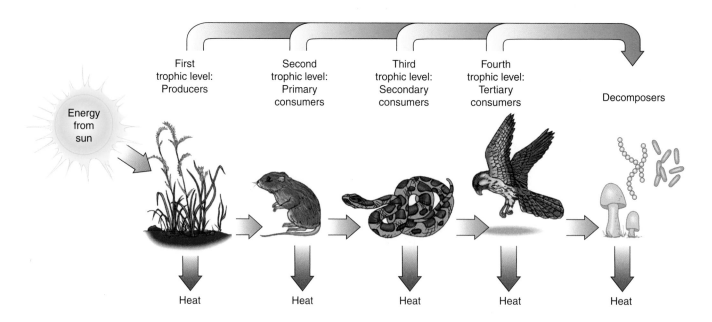

Figure 4.9 Energy flow through a food chain. Energy enters ecosystems from an external source (the sun), flows linearly—in a one-way direction—through ecosystems and exits as heat loss. Much of the energy acquired by a given level of the food chain is used for respiration at that level and escapes into the surrounding environment as heat. This energy, as stipulated by the second law of thermodynamics, is unavailable to the next level of the food chain.

①Pitch pine	⑤Eastern chipmunk	⑨Red-tailed hawk	⑬American robin	⑰Worms and ants	㉑Insect larvae
②White oak	⑥Eastern cottontail	⑩Eastern bluebird	⑭Woodpecker	⑱Moths	㉒Insects
③Barred owl	⑦Red fox	⑪Red-winged blackbird	⑮Red clover	⑲Deer mouse	㉓Fungi
④Gray squirrel	⑧White-tailed deer	⑫Blackberry	⑯Bacteria and fungi	⑳Spiders	

Figure 4.10 Food web at the edge of an eastern deciduous forest. This food web is greatly simplified compared to what actually happens in nature. Groups of species are lumped into single categories such as "spiders" and "fungi," many other species are not included, and numerous links in the web are not shown.

Unintended Changes in Food Webs

Sometimes the best human intentions can backfire. At the National Elk Refuge near Jackson, Wyoming, wildlife biologists made a management decision to feed the elk during the bitter cold winters, thereby preventing large population die-offs. As a result, the elk population has flourished in and around Jackson. But winter still causes natural die-offs, and the much larger elk population translates into many elk carcasses in the area. Ravens, black birds that are much larger than crows, are attracted to the elk carcasses, which are their main source of food during cold winter months. The number of ravens has increased dramatically because of their ample winter food supply.

In the spring, when the elk migrate northward to their summer ranges, the large raven population migrates with them. Ravens are omnivores and feed on anything they can find—berries, small lizards and snakes, insects, and songbird eggs. Because there are fewer elk carcasses in spring and summer months, the ravens are eating the eggs and babies of songbirds in much larger numbers than before. This nest predation has resulted in a heavy toll on songbird populations. Thus, feeding the elk at the National Elk Refuge has led to a decrease in songbirds outside the refuge.

CASE·IN·POINT ## How Humans Have Affected the Antarctic Food Web

Although the icy waters around Antarctica may seem to be an inhospitable environment, a complex food web is found there. The base of the food web is microscopic algae, which are present in vast numbers in the well-lit nutrient-rich water. These marine algae are eaten by a huge population of herbivores—tiny shrimplike animals called **krill** (Figure 4.11). Krill in turn support a variety of larger animals. One of the main consumers of krill is baleen whales, which filter krill out of the frigid water. Baleen whales include blue whales, humpback whales, and right whales. Krill are also consumed in great quantities by squid and fish. These in turn are eaten by other carnivores: toothed whales such as the sperm whale; elephant seals and leopard seals; king penguins and emperor penguins; and birds such as the albatross and petrel.

Humans have had an impact on the complex Antarctic food web, as they have on most other ecosystems. Before the advent of whaling, baleen whales consumed huge quantities of krill. During the past 150 years—until a 1986 global ban on hunting all large whales—whaling steadily reduced the number of large baleen whales in Antarctic waters. Some whale populations are so decimated that they are on the brink of extinction. As a result of fewer whales eating krill, more krill have been available for other krill-eating animals, whose populations have increased. Seals, penguins, and smaller baleen whales have replaced the large baleen whales as the main eaters of krill.

Now that commercial whaling is regulated, it is hoped that the number of large baleen whales will slowly increase, and that appears to be the case for many species. As of the late 1990s, the only whale population that did not appear to be growing in response to the moratorium on whaling was the southern blue whale. It is not known whether baleen whales will return to or be excluded from their former position of dominance in terms of krill consumption in the food web. Biologists will monitor changes in the Antarctic food web as the whale populations recover.

Recently, a human-related change has developed in the atmosphere over Antarctica that has the potential to cause far greater effects on the entire Antarctic food web—thinning of the ozone layer in the stratospheric region of the atmosphere. This ozone thinning allows more of the sun's ultraviolet radiation to penetrate to the Earth's surface. Ultraviolet radiation contains more energy than visible light. It is so energetic that it can break the chemical bonds of some biologically important molecules, such as deoxyribonucleic acid (DNA).

Scientists are concerned that ozone thinning over Antarctica may damage the algae that form the base of the food web in the southern ocean. A 1992 study confirmed that increased ultraviolet radiation is penetrating the surface waters around Antarctica and that algal productivity has declined by at least 6% to 12%, probably as a result of increased exposure to ultraviolet radiation. (The problem of stratospheric ozone depletion is discussed in detail in Chapter 20.)

Another human-induced change that may be responsible for declines in certain Antarctic populations is global warming. As the water has warmed in recent decades around Antarctica, less pack ice has formed during winter months. Large numbers of marine algae are

Figure 4.11 Antarctic krill (Euphausia superba). These tiny, shrimplike animals live in large swarms and eat photosynthetic algae in and around the pack ice. Whales, seals, penguins, and fish consume vast numbers of krill. Photographed in a tank on Palmer Station, Antarctica.

Fishing Down the Food Web

According to a 1998 study of changes in marine food webs over time, fish being caught today are significantly lower on the food chain than those caught 10 years ago. Dozens of fish species, mostly top predators such as cod, have been fished to commercial extinction, and today's commercial fishermen are catching smaller, bonier, less-valuable species. The drop in the average trophic level of commercial species indicates a loss of biological diversity and overall resource exhaustion caused by overfishing. Fishermen are literally fishing down the food web, now relying on species that eat plankton or invertebrates. If overfishing continues, even these species will no longer be commercially available. Many marine biologists, including the study's authors, stress that fishing down the food web is unsustainable. They argue that one way to reverse this movement down marine food webs—and to preserve diverse marine environments—is to create protected areas where no fishing is allowed.

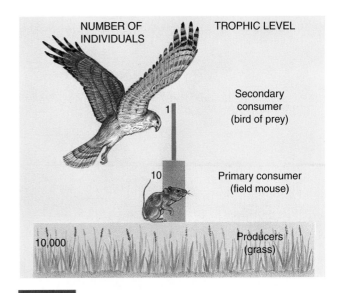

Figure 4.12 Pyramid of numbers. This pyramid is for a hypothetical area of temperate grassland. Based on the number of organisms found at each trophic level, a pyramid of numbers is not as useful as other ecological pyramids. It provides no information about biomass or energy relationships between one trophic level and the next. (Note that decomposers are not shown.)

found in and around the pack ice, providing a critical supply of food for the krill, which spawn in the area. Years with below-average pack ice cover means less algae, which mean less krill spawning. Scientists have demonstrated that low krill abundance coincides with unsuccessful breeding seasons in penguins and fur seals, all of which struggle to find food during warmer winters. Scientists are concerned that global warming may continue to decrease the amount of pack ice, which will reverberate through the food web. (Global warming, including the effect on Adelie penguins, is discussed in Chapter 20.)

To complicate matters further, some commercial fishermen have started to harvest krill, which is used to make fishmeal for aquaculture industries (discussed in Chapter 18). Because so many marine animals depend on krill for food, scientists worry that the human harvest of krill may endanger many animals higher on the food web. ∎

Ecological Pyramids

An important feature of energy flow is that as a result of the second law of thermodynamics, most of the energy going from one trophic level to the next in a food chain or food web dissipates into the environment. **Ecological pyramids** often graphically represent the relative energy values of each trophic level. There are three main types of pyramids—a pyramid of numbers, a pyramid of biomass, and a pyramid of energy.

A **pyramid of numbers** shows the number of organisms at each trophic level in a given ecosystem, with greater numbers illustrated by a larger area for that section of the pyramid (Figure 4.12). In most pyramids of numbers, the organisms at the base of the food chain are the most abundant, and each successive trophic level is occupied by fewer organisms. Thus, in African grasslands the number of herbivores, such as zebras and wildebeests,

is far greater than the number of carnivores, such as lions. Inverted pyramids of numbers, in which higher trophic levels have *more* organisms than lower trophic levels, are often observed among decomposers, parasites, tree-dwelling herbivorous insects, and similar organisms. One tree, for example, can provide food for thousands of leaf-eating insects. Pyramids of numbers are of limited usefulness because they do not indicate the biomass of the organisms at each level, and they do not indicate the amount of energy transferred from one level to another.

A **pyramid of biomass** illustrates the total biomass at each successive trophic level. **Biomass** is a quantitative estimate of the total mass, or amount, of living material; it indicates the amount of fixed energy at a particular time. Biomass units of measure vary: Biomass may be represented as total volume, as dry weight, or as live weight. Typically, pyramids of biomass illustrate a progressive reduction of biomass in succeeding trophic levels (Figure 4.13). On the assumption that there is, on the average, about a 90% reduction of biomass for each trophic level, 10,000 kg of grass should be able to support 1,000 kg of grasshoppers, which in turn support 100 kg of toads. (The 90% reduction in biomass is an approximation; actual field numbers for biomass reduction in nature vary widely.) By this logic, the biomass of toad eaters such as snakes could be, at the most, only about 10 kg. From this brief exercise, you can see that although carnivores may eat no vegetation, a great deal of vegetation is still required to support them.

A **pyramid of energy** illustrates the energy content, often expressed as kilocalories per square meter

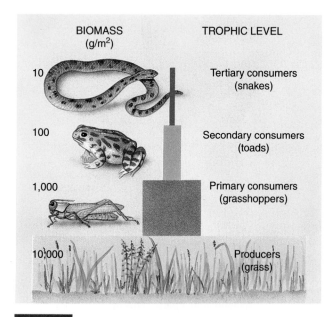

BIOMASS
(g/m²)

TROPHIC LEVEL

10 — Tertiary consumers
(snakes)

100 — Secondary consumers
(toads)

1,000 — Primary consumers
(grasshoppers)

10,000 — Producers
(grass)

Figure 4.13 Pyramid of biomass. This pyramid is for a hypothetical area of temperate grassland. Based on the biomass at each trophic level, pyramids of biomass generally have a pyramid shape with a large base and progressively smaller areas for each succeeding trophic level. (Note that decomposers are not shown.)

per year, of the biomass of each trophic level (Figure 4.14). These pyramids, which always have large energy bases and get progressively smaller through succeeding trophic levels, show that most energy dissipates into the environment when going from one trophic level to the

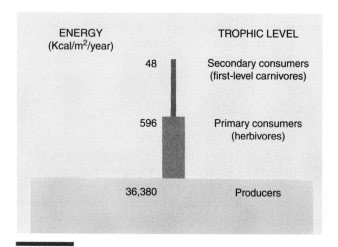

ENERGY
(Kcal/m²/year)

TROPHIC LEVEL

48 — Secondary consumers
(first-level carnivores)

596 — Primary consumers
(herbivores)

36,380 — Producers

Figure 4.14 Pyramid of energy. This pyramid represents energy flow, which is the functional basis of ecosystem structure. It indicates how much energy is present at each trophic level and how much is transferred to the next trophic level. Note the substantial loss of usable energy from one trophic level to the next. (Note that decomposers are not shown. Also, the 36,380 kcal/m²/year for the producers is gross primary productivity, or GPP, which will be discussed shortly.)

next. Less energy reaches each successive trophic level from the level beneath it because some of the energy at the lower level is used by those organisms to perform work, and some of it is lost. (Remember, no biological process is ever 100% efficient.) Energy pyramids explain why there are so few trophic levels: *Food webs are short because of the dramatic reduction in energy content that occurs at each trophic level.* (The eating habits of humans as they relate to food chains and trophic levels are discussed in Chapter 18 in "You Can Make a Difference: Vegetarian Diets.")

Productivity of Producers

The **gross primary productivity (GPP)** of an ecosystem is the *rate* at which energy is captured during photosynthesis.[1] Thus, GPP is the total amount of photosynthetic energy captured in a given period of time. Of course, plants must respire to provide energy for their own life processes, and cellular respiration by plants acts as a drain on photosynthesis. Energy that remains in plant tissues after cellular respiration has occurred is called **net primary productivity (NPP)**. That is, NPP is the amount of biomass found in excess of that broken down by a plant's cellular respiration for normal daily activities for survival. NPP represents the *rate* at which this organic matter is actually incorporated into plant tissues for growth.

$$
\begin{array}{ccc}
\text{Net primary} & \text{Gross primary} & \\
\text{productivity} & \text{productivity} & \text{Plant} \\
\text{(plant growth} = \text{(total} - \text{respiration} \\
\text{per unit area} & \text{photosynthesis} & \text{(per unit area} \\
\text{per unit time)} & \text{per unit area} & \text{per unit time)} \\
& \text{per unit time)} &
\end{array}
$$

Only the energy represented by NPP is available for consumers, and of this energy only a portion is actually used by them. Both GPP and NPP can be expressed as energy per unit area per unit time (kilocalories of energy fixed by photosynthesis per square meter per year) or in terms of dry weight (grams of carbon incorporated into tissue per square meter per year).

Ecosystems differ strikingly in their productivities (Figure 4.15 and Table 4.1). On land, tropical rain forests have the highest productivity, probably because of their abundant rainfall, warm temperatures, and intense sunlight. As you might expect, tundra with its harsh, cold winters and deserts with their lack of precipitation are the least productive terrestrial ecosystems.

---

[1] Gross and net primary productivities are referred to as *primary* because plants occupy the first trophic level in food webs.

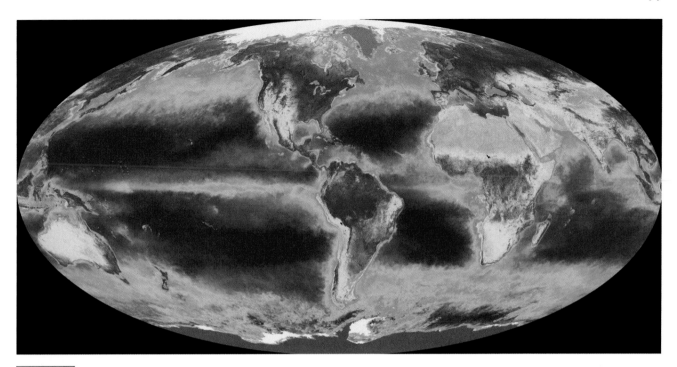

Figure 4.15 A measure of Earth's primary productivity. The data are from a satellite launched as part of NASA's Mission to Planet Earth in 1997. The satellite measured the amount of plant life on land as well as the concentration of phytoplankton (algae) in the ocean. On land, the most productive areas, such as tropical rain forests, are dark green, whereas the least productive ecosystems (deserts) are yellow and orange. In the ocean and other aquatic ecosystems, the most productive regions are red, followed by orange, yellow, green, and blue (the least productive). Data are not available for the gray area.

Wetlands, swamps and marshes that connect terrestrial and aquatic environments, are extremely productive. The most productive aquatic ecosystems are algal beds, coral reefs, and estuaries. The lack of available nutrient minerals in some regions of the open ocean makes them extremely unproductive, equivalent to aquatic deserts.

| Table 4.1 | Net Primary Productivities (NPP) for Selected Ecosystems |

Ecosystem	Average NPP (g dry matter/m²/year)
Swamp and marsh	3,000
Tropical rain forest	2,200
Temperate evergreen forest	1,300
Temperate deciduous forest	1,200
Savanna	900
Boreal (northern) forest	800
Woodland and shrubland	700
Agricultural land	650
Temperate grassland	600
Lake and stream	400
Arctic and alpine tundra	140
Desert and semidesert scrub	90
Extreme desert (rock, sand, ice)	3

ENVIROBRIEF

More Is Sometimes Less

Conventional ecological theory held that the more productive the habitat, the greater the biodiversity it supported. Now ecologists are seeing a recurring pattern worldwide: Habitats become more productive and diverse as resources increase, but at some point, diversity actually *declines* with increasing productivity. For example, the resource-poor depths of the abyssal plain of the Atlantic Ocean have higher species richness than the productive shallow waters near the coasts; intermediate depths exhibit the greatest diversity. Scientists have little solid data to help explain the pattern, which holds true with rodents in Israel, birds in South America, and large mammals in Africa. Mathematical ecosystem models, however, suggest that a patchy distribution of resources reduces competition and allows the coexistence of a greater variety of organisms. (Recall from Chapter 2 that a **model** is a formal statement that describes the behavior of a process, thereby helping us to understand how the present situation developed from the past or to predict the future course of events.) The bad news for global biodiversity is that we are constantly enriching our environment with inputs from fossil fuels, fertilizer, and human and livestock sewage. This continual enrichment may make the Earth's ecosystems increasingly productive, a shift that could cost the world a substantial loss of biodiversity. (Factors that affect species richness, the number of species within an ecosystem, are discussed in Chapter 5.)

(Earth's major aquatic and terrestrial ecosystems are discussed in Chapter 7.)

Human Impact on Net Primary Productivity Humans consume far more of the Earth's resources than any of the other millions of animal species. Peter Vitousek and colleagues at Stanford University calculated in 1986 how much of the global NPP is appropriated for the human economy and therefore not transferred to other organisms. When both direct and indirect human impacts are accounted for, humans are conservatively estimated to use 32% of the annual NPP of land-based ecosystems and 40% of the annual land-based NPP using the "most reasonable" definition of human appropriation. Essentially, humans' use of global productivity is competing with other species' needs for energy. Our use of so much of the world's productivity may contribute to the loss of many species, some potentially useful to humans,

through extinction. Clearly, at these levels of consumption of the Earth's resources, human population growth becomes a serious threat to the planet's ability to support its occupants.

Vitousek's influential work was revisited in 2001 by Stuart Rojstaczer and colleagues at Duke University. Rojstaczer used contemporary data sets, many of which are satellite-based and are more accurate measures than the data Vitousek had available for his ground-breaking research. Rojstaczer's mean value for his conservative estimate of annual land-based NPP appropriation by humans was 32%, like Vitousek's, although Rojstaczer arrived at that number using different calculations. It must be emphasized that both Vitousek's and Rojstaczer's numbers are estimates, not actual values. The take-home message is simple, however: If we want our planet to operate sustainably, we must share terrestrial photosynthesis products—that is, NPP—with other organisms.

SUMMARY WITH SELECTED KEY TERMS

I. The study of the relationships between organisms (the **biotic** environment) and the nonliving (**abiotic**) environment is called **ecology**. Ecologists study individual organisms, populations, communities/ecosystems, landscapes, and the biosphere.

A. A **population** is all the members of the same **species** that live together in a particular area at a particular time.

B. A **community** is all the populations of different species living in the same area at the same time.

C. An **ecosystem** is a community and its abiotic environment. In an ecosystem, all of the biological, physical, and chemical components of an area form a complicated interacting network of energy flow and materials cycling.

D. **Landscape ecology** considers the connections among ecosystems found in a particular region. Landscapes are larger land areas than individual ecosystems.

E. The **biosphere** is the layer of Earth that contains all living organisms. The organisms of the biosphere depend on one another and on the other realms of the Earth's physical environment: the **atmosphere** (gaseous envelope), **lithosphere** (soil and rock), and **hydrosphere** (water).

II. Energy is the capacity to do work. Energy can be transformed from one form to another.

A. Potential energy is stored energy; **kinetic energy** is energy of motion.

B. Energy can be conveniently measured as **heat energy**; the unit of heat energy is the **kilocalorie (kcal)**.

III. The study of energy and its transformations is known as **thermodynamics**.

A. The **first law of thermodynamics** states that energy can be converted from one form to another, but it can neither be created nor destroyed. The first law explains why organisms

cannot produce energy but must continuously capture it from the surroundings.

B. The **second law of thermodynamics** states that disorder (**entropy**) continually increases in the universe as usable energy is converted to heat, a lower-quality, less-usable form.

1. The second law explains why no process requiring energy is ever 100% efficient.

2. In every energy transaction, some energy is dissipated as heat, which contributes to entropy.

IV. All life depends on a continuous input of energy.

A. Plants, algae, and some bacteria capture radiant energy during **photosynthesis** and incorporate some of it into carbohydrate molecules. The chemical formula that summarizes photosynthesis is: $6CO_2 + 12H_2O + $ radiant energy $\rightarrow C_6H_{12}O_6 + 6O_2 + 6H_2O$.

B. All organisms obtain the energy in carbohydrate and other molecules by **cellular respiration**, in which molecules such as glucose are broken down with the release of energy. The chemical formula that summarizes cellular respiration is: $C_6H_{12}O_6 + 6O_2 + 6H_2O \rightarrow 6CO_2 + 12H_2O + $ energy.

C. Although the sun is the energy source for almost all ecosystems, **hydrothermal vents** are an exception. The base of the food web in hydrothermal vents consists of certain bacteria that **chemosynthesize**, obtaining energy and making carbohydrate molecules from inorganic raw materials.

V. Energy flow through an ecosystem is linear, from the sun to producer to consumer to decomposer.

A. Much of this energy is converted to less-usable heat as the energy moves from one organism to another, as stipulated in the second law of thermodynamics.

B. Organisms in ecosystems assume the roles of producer, consumer, and decomposer.

   1. Producers, or **autotrophs**, are the photosynthetic organisms that are potential food resources for other organisms. Producers include plants, algae, and some bacteria.

   2. Consumers, which feed on other organisms, are almost exclusively animals. **Primary consumers**, or **herbivores**, feed on plants; **secondary consumers**, or **carnivores**, feed on primary consumers; **detritus feeders**, or **detritivores**, feed on **detritus**, dead organic material.

   3. Microbial **decomposers**, or **saprotrophs**, feed on the components of dead organisms and organic wastes, degrading them into simple inorganic materials that can then be used by producers to manufacture more organic material. Both consumers and decomposers are **heterotrophs**.

C. Trophic levels may be expressed in **food chains** or, more realistically, **food webs**, which show the many pathways that energy can take among the producers (the first trophic level), primary consumers (the second trophic level), secondary consumers (the third trophic level), and so on within an ecosystem.

D. Ecological pyramids express the progressive reduction in numbers of organisms (**pyramid of numbers**), biomass (**pyramid of biomass**), and energy (**pyramid of energy**) found in successively higher trophic levels.

E. Gross primary productivity (**GPP**) of an ecosystem is the rate at which organic matter is produced by photosynthesis. **Net primary productivity** (**NPP**) expresses the rate at which some of this matter is incorporated into plant bodies.

   1. NPP is less than GPP because of the losses resulting from cellular respiration by plants.

   2. Scientists have estimated how much of the global NPP is appropriated for the human economy and therefore not transferred to other organisms. When both direct and indirect human impacts are accounted for, humans are conservatively estimated to use 32% of the annual NPP of land-based ecosystems.

THINKING ABOUT THE ENVIRONMENT

  1. Draw a food web containing organisms found in a Chesapeake Bay salt marsh.

  2. What is the difference between a community and an ecosystem? Between an ecosystem and a landscape?

  3. How are the following forms of energy significant to organisms in ecosystems: (a) radiant energy, (b) mechanical energy, (c) chemical energy, (d) heat?

  4. Use the water stored behind a dam to explain the concepts of potential and kinetic energy.

  5. When coal is burned in a power plant, only 3% of the energy in the coal is converted into light in a regular lightbulb. What happens to the other 97% of the energy? Explain your answer using the laws of thermodynamics.

  6. Distinguish between photosynthesis and cellular respiration. Which organisms perform each process?

  7. Why do deep-sea organisms cluster around hydrothermal vents? What is their energy source?

  8. Why is the concept of a food web generally preferred over that of a food chain?

  9. Could a balanced ecosystem be constructed that contained only producers and consumers? Only consumers and decomposers? Only producers and decomposers? Explain the reasons for your answers.

10. How have humans affected the Antarctic food web?

11. Suggest a food chain that might have an inverted pyramid of numbers—that is, greater numbers of organisms at higher trophic levels than at lower trophic levels.

12. Is it possible to have an inverted pyramid of energy? Why or why not?

13. Relate the pyramid of energy to the second law of thermodynamics.

*14. The NPP for a particular river ecosystem is measured at 8,833 kcal/m^2/year. Respiration by the aquatic producers is estimated as 11,977 kcal/m^2/year. Calculate the GPP for this ecosystem.

*15. Graph the information given in Table 4.1 on page 77. Would it be more appropriate to construct a line graph or a bar graph? Why?

* Solutions to questions preceded by asterisks appear in Appendix VII.

TAKE A STAND

Visit our Web site at **http://www.wiley.com/college/raven/** (select Chapter 4 from the Table of Contents) for links to more information about the environmental issues surrounding Antarctica. Consider the opposing views of Japanese whalers and environmentalists who oppose whaling, and debate the issues with your classmates. You will find tools to help you organize your research, analyze the data, think critically about the issues, and construct a well-considered argument. Take a

Stand activities can be done individually or as part of a team, as oral presentations, as written exercises, or as Web-based (e-mail) assignments.

Additional on-line materials relating to this chapter, including Student Quizzes, Activity Links, Useful Web Sites, Flash Cards, and more, can also be found on our Web site.

SUGGESTED READING

Dybas, C.L. "Undertakers of the Deep." *Natural History* (November 1999). The body of a dead whale supports a diverse community of organisms on the ocean floor for many years.

Field, C. "Human Appropriation of Natural Systems." *Environmental Review*, Vol. 9, No. 3 (March 2002). This prominent scientist who specializes in global ecology talks about human use of photosynthesis products (i.e., NPP) for food, fibers, and grazing by domesticated animals.

Hinrichs, R.A. *Energy: Its Use and the Environment*, 3rd ed. Philadelphia: Harcourt College Publishers, 2002. The focus of this introductory text is the physical principles behind energy use and its effects on the environment.

Johnson, K. "The Vital Link." *Living Planet* (summer 2001). The health of many marine species in the Southern Ocean around Antarctica is based largely on the population size of small, shrimplike krill.

Lutz, R.A., and R.M. Haymon. "Rebirth of a Deep-Sea Vent." *National Geographic*, Vol. 186, No. 5 (November 1994). Undersea lava flows recently buried some hydrothermal vent ecosystems off the west coast of Mexico. The author follows their quick rebirth as nearby vent species migrated into the area.

Ross, J.F. "Hardly a Mouse or a Molecule Moves Here without Being Noticed." *Smithsonian* (July 1996). The Smithsonian Environmental Research Center studies how the Chesapeake Bay functions as an ecological unit.

Schoen, D. "Primary Productivity: The Link to Global Health." *BioScience*, Vol. 47, No. 8 (September 1997). Examines an important question that scientists are currently addressing: How terrestrial primary productivity will respond to climate warming.

Tyson, P. "Neptune's Furnace." *Natural History* (June 1999). Describes an expedition to raise black smoker chimneys from the ocean floor so their geology and biology can be studied further.

Invasion of water hyacinth in Lake Victoria. The water hyacinth (*Eichhornia crassipes*) has choked much of Lake Victoria's shoreline with thick vegetation. Many small ports are closed, and boats cannot move.

Ecosystems and Living Organisms

Learning Objectives

After you have studied this chapter you should be able to:

1. Explain the four premises of evolution by natural selection as proposed by Charles Darwin.

2. Describe ecological succession and distinguish between primary and secondary succession.

3. Discuss an example of a keystone species.

4. Explain symbiosis and distinguish among mutualism, commensalism, and parasitism.

5. Define *predation* and describe the effects of natural selection on predator–prey relationships.

6. Define *competition* and distinguish between *intraspecific* and *interspecific competition*.

7. Describe the factors that contribute to an organism's ecological niche and distinguish between *fundamental niche* and *realized niche*.

8. Give several examples of limiting factors and discuss how they might affect an organism's ecological niche.

9. Relate the concepts of *competitive exclusion* and *resource partitioning*.

10. Summarize the main determinants of species richness in a community and describe factors associated with high species richness.

11. Give several examples of ecosystem services.

ntil relatively recently, the world's second largest freshwater lake, Lake Victoria in East Africa, was home to about 400 species of small, colorful fishes known as cichlids (pronounced *sik'lids*). (Recall from Chapter 4 that a **species** is a group of similar organisms whose members freely interbreed with one another in the wild and do not interbreed with other species of organisms.) The cichlid species in Lake Victoria had remarkably different eating habits. Some grazed on algae; some consumed dead organic material at the bottom of the lake; and others ate insects, shrimp, or other cichlid species. These fishes, which thrived throughout the lake ecosystem, provided much-needed protein to the diets of 30 million humans living in the lake's vicinity.

Today, the aquatic community is very different in Lake Victoria. More than half of the cichlids and other native fish species are now extinct. As a result of the disappearance of most of the algae-eating cichlids, the algal population has increased explosively. When these algae die, their decomposition uses up the dissolved oxygen in the water. The bottom zone of the lake, once filled with cichlids, is empty because it contains too little dissolved oxygen. Any fishes venturing into the oxygen-free zone suffocate. Local fishermen, who once caught and ate hundreds of different types of fishes, now catch only a few types.

The most important reason for the destruction of Lake Victoria's delicate ecological balance was the deliberate introduction of the Nile perch, a large and voracious predator, into the lake in 1960. Proponents of the introduction thought that the successful establishment of the Nile perch would stimulate the local economy and help the fishermen. For about 20 years, as its population slowly increased, the Nile perch did not appear to have an appreciable effect on the lake. But in 1980, fishermen noticed that they were harvesting increasing quantities of Nile perch and decreasing amounts of native fishes. By 1985, most of the annual catch was Nile perch, which was experiencing a population explosion fueled by an abundant food supply—the cichlids.

Pollution is also a factor in the cichlids' disappearance. Introduction of the Nile perch has changed the surrounding forest as fishermen cut down the trees to create firewood to dry the very large Nile perch they caught. As nearby forests were cut, soil erosion made the water of Lake Victoria turbid (cloudy). Agricultural practices in the area contributed fertilizer as well as sediment pollution from soil erosion, and both of these pollutants increased the turbidity of the water. Water transparency in deep lake water decreased from an average of 6.8 m (22.3 ft) in the 1920s to 2.2 m (7.2 ft) in the 1990s. Water transparency in shallow areas along the shoreline decreased from 3 m (9.8 ft) to 1.5 m (4.9 ft) during the 1990s.

The Lake Victoria story is far from finished. In 1990 the water hyacinth, a South American plant, invaded Lake Victoria. This floating plant has spread along the shoreline so rapidly that fishing is impossible in many sections of the lake. Small ports and harbors have had to close (see opening figure). In ways not yet understood, the invasion of the water hyacinth appears to have increased the incidence of snail- and mosquito-borne diseases, threatening the health of millions of people living in the region. The World Bank has established the Lake Victoria Environmental Management Program to control the spread of the water hyacinth, but progress has been slow.

Regardless of whether an ecosystem is as large as Lake Victoria or as small as a roadside drainage ditch, its organisms are continually interacting with one another and adapting to changes in the environment. This chapter is concerned with making sense of community structure and diversity by finding common patterns and processes in a wide variety of communities.

◼ EVOLUTION: HOW POPULATIONS CHANGE OVER TIME

Where did the original 400 species of cichlids in Lake Victoria come from? They, like all of Earth's remarkable number of species living today, descended from earlier species by a process known as evolution. **Evolution** is a genetic change in a population of organisms that occurs over time. The concept of evolution dates back to the time of Aristotle (384–322 B.C.), but Charles Darwin (1809–1882), a 19th-century naturalist, proposed the mechanism of evolution that is still accepted by the scientific community today. As you will see, the environment plays a crucial role in Darwin's theory of evolution.

It occurred to Darwin that over time, inherited traits favorable to survival in a given environment would tend to be preserved, whereas unfavorable ones would be eliminated. The result would be **adaptation**, evolutionary modification that improves the chances of survival and reproductive success of the population in a given environment. Eventually the accumulation of modifications might result in a new species.

Darwin proposed the theory of evolution by natural selection in his monumental book, *The Origin of Species by Means of Natural Selection*, which was published in 1859 (Figure 5.1). Since that time, scientists have accumulated an enormous body of observations and experiments that support Darwin's theory. Although biologists still do not agree completely on some aspects by which evolutionary changes occur, the concept that evolution by natural selection has taken place is now well documented.

Natural Selection

Darwin's mechanism of evolution is by **natural selection**, in which better-adapted individuals—those with a combination of genetic traits better suited to environmental conditions—are more likely to survive and reproduce, increasing their proportion in the population. Over time, the population changes as favorable traits increase in frequency in successive generations, and as unfavorable traits decrease or disappear. Natural selection consists of four observations about the natural world: overproduction, variation, limits on population growth, and differential reproductive success.

1. **Overproduction.** Each species produces more offspring than will survive to maturity. Natural populations have the reproductive potential to increase their numbers continuously over time. For example, if each breeding pair of elephants produces 6 offspring during a 90-year life span, in 750 years a single pair of elephants will have given rise to a population of 19 million! Yet elephants have not overrun the planet.

2. **Variation.** The individuals in a population exhibit variation. Each individual has a unique combination of traits, such as size, color, and ability to tolerate harsh environments. Some traits improve the chances of an individual's survival and reproductive success, whereas others do not. It is important to remember that the variation necessary for evolution by natural selection must be inherited so that it can be passed to offspring.

3. **Limits on population growth, or a struggle for existence.** There is only so much food, water, light,

Warbler Finch
(Insects)

Small Tree Finch
(Insects, fruits, soft seeds)

Medium Tree Finch
(Insects, soft seeds)

Large Cactus Finch
(Insects, seeds)

Vegetarian Finch
(Soft seeds, fruits, leaves)

Large Tree Finch
(Insects)

Woodpecker Finch
(Insects)

Cactus Finch
(Moderate seeds)

Mangrove Finch
(Mostly insects)

Small Ground Finch
(Soft seeds)

Sharp Beaked Ground Finch
(Insects, seeds, nectar)

Medium Ground Finch
(Moderate seeds)

Cocos Island Finch
(Insects, seeds, nectar)

Large Ground Finch
(Hard seeds)

Figure 5.1 Darwin's finches.
Charles Darwin was a ship's naturalist on a 5-year voyage around the world. During an extended stay in the Galápagos Islands off the coast of Ecuador, he studied the plants and animals of each island, including the 14 species of finches. Each finch species was specialized for a particular lifestyle different from those of the others and different from finches on the South American mainland. Darwin came to realize that the 14 species of Galapagos finches descended from a single common ancestor— one or a small population of finches that originally colonized the Galápagos from the South American mainland.

growing space, and so on available to a population, and organisms compete with one another for the limited resources available to them. Because there are more individuals than the environment can support, not all of the offspring will survive to reproductive age. Other limits on population growth include predators and diseases.

4. **Differential reproductive success.** Those individuals that possess the most favorable combination of characteristics (those that make individuals better adapted to their environment) are more likely to survive, reproduce, and pass their traits to the next generation. Because offspring tend to resemble their parents, the next generation obtains the parents' heritable traits. Sexual reproduction is the key to natural selection: The best-adapted individuals are those that reproduce most successfully, whereas less-fit individuals die prematurely or produce fewer or inferior offspring. Over time, enough changes may accumulate in geographically separated populations (often with slightly different environments) to produce new species.

A vast body of evidence supports evolution, including observations from the fossil record, comparative anatomy, biogeography (the study of the geographical locations of organisms), and molecular biology. In addition, evolutionary hypotheses are tested experimentally. On the basis of these kinds of evidence, most biologists accept the principles of evolution by natural selection but have recently scrutinized some of its details. For example, what is the role of chance in evolution? How rapidly do new species evolve? These questions have arisen in part from a reevaluation of the fossil record and in part from discoveries in molecular aspects of inheritance. Such critical analyses are an integral part of the scientific process because they stimulate additional observation and experimentation, along with reexamination of previous evidence. As discussed in Chapter 2, science is an ongoing process, and information obtained in the future may require modifications to certain parts of the theory of evolution by natural selection.

The Kingdoms of Life

Biologists try to make sense of the remarkable diversity of life that has evolved on Earth by arranging organisms into logical groups. For hundreds of years, biologists regarded organisms as falling into two broad categories—plants and animals. With the development of microscopes, however, it became increasingly obvious that many organisms did not fit very well into either the plant kingdom or the animal kingdom. For example, bacteria have a *prokaryotic* cell structure: They lack a nuclear envelope and other internal cell membranes. This feature, which separates bacteria from all other organisms, is far more fundamental than the differences between plants and animals, which have similar cell structures. Hence, it became clear that bacteria were neither plants nor animals. Furthermore, certain microorganisms such as *Euglena*, which is both motile (has a hairlike flagellum for movement) and photosynthetic, seem to possess characteristics of both plants and animals.

These and other considerations eventually led to a six-kingdom system of classification used by many biologists today: Archaebacteria, Eubacteria, Protista, Fungi, Plantae, and Animalia (Figure 5.2). Bacteria, with a prokaryotic cell structure, are now divided into two very distinct groups on the basis of significant biochemical differences. The archaebacteria frequently live in oxygen-deficient environments and are often adapted to harsh conditions; these include hot springs, salt ponds, and hydrothermal vents (see Chapter 4). The thousands of remaining kinds of bacteria are collectively known as the eubacteria.

The remaining four kingdoms are composed of organisms with a *eukaryotic* cell structure. Eukaryotic cells have a high degree of internal organization, containing nuclei, chloroplasts (in photosynthetic cells), and

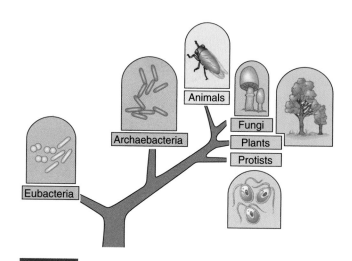

Figure 5.2 The six-kingdom system of classification.

mitochondria. Eukaryotes that are unicellular or relatively simple multicellular organisms, such as algae, protozoa, slime molds, and water molds, are classified as members of the kingdom Protista.

In addition to the kingdom Protista, there are three specialized groups of multicellular organisms—fungi, plants, and animals—that evolved independently from different groups of Protista. The kingdoms Fungi, Plantae, and Animalia differ from one another in—among other features—their types of nutrition. Fungi secrete digestive enzymes into their food and then absorb the predigested nutrients. Plants use radiant energy to manufacture food molecules by photosynthesis. Animals ingest their food, then digest it inside their bodies.

Although the six-kingdom system is a definite improvement over the two-kingdom system, it is not perfect. Most of its problems concern the kingdom Protista, which includes some organisms that may be more closely related to members of other kingdoms than to certain other protists. For example, green algae are protists that are clearly similar to plants but do not appear to be closely related to other protists, such as slime molds and red algae.

BIOLOGICAL COMMUNITIES

As you may recall from Chapter 4, the term *community* has a far broader sense in ecology than in everyday speech. For the biologist, a **community** is an association of different populations of organisms that live and interact together in the same place at the same time.

Communities are extraordinarily complex because the organisms that comprise a community are interdependent in a variety of ways. Species compete with one another for food, water, living space, and other resources. (Used in this context, a **resource** is anything from the

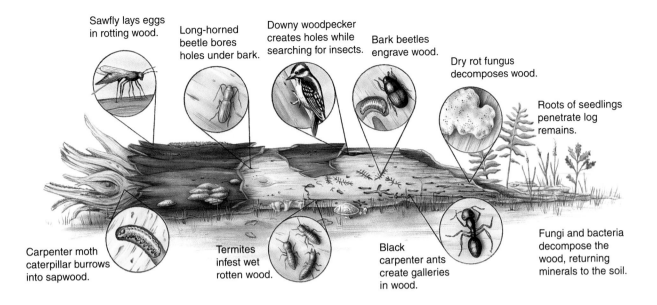

Sawfly lays eggs in rotting wood.

Long-horned beetle bores holes under bark.

Downy woodpecker creates holes while searching for insects.

Bark beetles engrave wood.

Dry rot fungus decomposes wood.

Roots of seedlings penetrate log remains.

Carpenter moth caterpillar burrows into sapwood.

Termites infest wet rotten wood.

Black carpenter ants create galleries in wood.

Fungi and bacteria decompose the wood, returning minerals to the soil.

Figure 5.3 A rotting log community. Detritivores and decomposers consume a rotting log, which teems with a variety of fungi, animals, and plants.

environment that meets a particular species' needs.) Some organisms kill and eat other organisms. Some species form intimate associations with one another, while other species seem only distantly connected. As discussed in Chapter 4, each organism plays one of three main roles in community life: producer, consumer, or decomposer. The unraveling of the many positive and negative, direct and indirect interactions of organisms living together as a community is one of the goals of community ecologists.

Communities vary greatly in size, lack precise boundaries, and are rarely completely isolated. They interact with and influence one another in countless ways, even when the interaction is not readily apparent. Furthermore, communities are nested within one another like Chinese boxes—that is, there are communities within communities. A forest is a community, but so is a rotting log in that forest. Insects, plants, and fungi invade a fallen tree as it undergoes a series of decay steps (Figure 5.3). First, wood-boring insects and termites forge paths through the bark and wood. Later, other insects, plant roots, and fungi follow and enlarge these openings. Mosses and lichens that establish on the log's surface trap rainwater and extract nutrient minerals, and fungi and bacteria speed decay, thus providing nutrients for other inhabitants. As decay progresses, small mammals burrow into the wood and eat the fungi, insects, and plants.

Organisms exist in an abiotic (nonliving) environment that is as essential to their lives as are their interactions with one another. Minerals, air, water, and sunlight are just as much a part of a honeybee's environment as the flowers that it pollinates and from which it takes nectar. A biological community and its abiotic environment together comprise an **ecosystem**. (The abiotic environment of ecosystems is considered in Chapter 6.)

From the Little Acorn … Comes Lyme Disease?

In a study illustrating how indirect interactions between species determine population dynamics, an intricate and fascinating relationship emerged among white-footed mice, acorn production, gypsy moth population growth, and the potential occurrence of Lyme disease. Large-scale experiments conducted in oak forests of the northeastern United States linked bumper acorn crops to booming mouse populations. Because the mice eat gypsy moth pupae, these high-acorn conditions also led to low populations of gypsy moths. This ecological outcome helps the oaks because gypsy moths cause serious defoliation, but the prospects for human health may not be as bright. Abundant acorns apparently also attract tick-bearing deer; the ticks' offspring feed on the mice. The mice often carry the Lyme disease–causing bacterium, which may be transmitted to the maturing ticks and in turn to humans bitten by infected ticks. The study's findings suggest the possibility of predicting which years pose the greatest potential threat of Lyme disease to humans—based on when oaks are most productive. Awareness of these community relationships might also affect forest management: What's good for the trees—a decline in gypsy moths—may not be so good for human health down the road.

Succession: How Communities Change over Time

A community of organisms does not spring into existence full-blown but develops gradually, through a sequence of species. The process of community development over time, which involves species in one stage being replaced by different species, is called **succession**. An area is initially colonized by certain organisms that are replaced over time by others, which themselves may be replaced much later by still others (see "Mini-Glossary: Succession").

The actual mechanisms that underlie sucession are not clear. In some cases, it may be that an earlier species modifies the environment in some way, thereby making it more suitable for a later species to colonize. It is also possible that earlier species are able to exist because there is little competition from other species. Later, as more invasive species arrive, the original species are outcompeted and displaced.

Ecologists initially thought that succession inevitably led to a stable and persistent community, such as a forest, known as a *climax community*. But more recently, the traditional view has fallen out of favor. The apparent stability of a "climax" forest is probably the result of how long trees live relative to the human life span. It is now recognized that mature "climax" communities are not in a state of permanent equilibrium but rather in a state of continual disturbance. Over time, a mature community changes in species composition and in the relative abundance of each species, despite the fact that it retains an overall uniform appearance.

Succession is usually described in terms of the changes in the species composition of the plants growing in an area, although each successional stage also may have its own characteristic kinds of animals and other organisms. The time involved in ecological succession is on the scale of tens, hundreds, or thousands of years, not the millions of years involved in the evolutionary time scale.

Primary Succession   **Primary succession** is the change in species composition over time in an environment that has not previously been inhabited by organisms. No soil exists when primary succession begins. Bare rock surfaces, such as recently formed volcanic lava and rock scraped clean by glaciers, are examples of sites where primary succession may take place (Figure 5.4).

Although the details vary from one site to another, on bare rock lichens are often the most important element in the **pioneer community**, which is the initial community that develops during primary succession. Lichens secrete acids that help to break the rock apart, beginning the process of soil formation. Over time, mosses and drought-resistant ferns may replace the lichen community, followed in turn by tough grasses and herbs. Once enough soil accumulates, grasses and herbs may be replaced by low shrubs, which in turn would be replaced by forest trees in several distinct stages. Primary succession on bare rock from a pioneer community to a forest community often occurs in this sequence: lichens → mosses → grasses → shrubs → trees.

Primary Succession on Krakatoa   Primary succession can take hundreds or thousands of years to occur. The Indonesian island of Krakatoa has provided scientists with a perfect long-term study of primary succession in a tropical rain forest. In 1883 a volcanic eruption destroyed all life on the island. Biologists have surveyed the ecosystem in the more than 100 years since the devastation to document the return of life forms. As of the 1990s, biologists had found that the progress of primary succession was extremely slow, in part because of Krakatoa's isolation. Both Java and Sumatra are nearby, but many species are limited in their ability to disperse over water. A portion of Krakatoa's forest might have only one tenth the number of tree species found in undisturbed tropical rain forests on nearby islands. The lack of plant diversity has in turn limited the number of colonizing animal species. In a forested area of Krakatoa where zoologists would expect more than 100 butterfly species, there are only 2 species.

Primary Succession on Sand Dunes   Some lake and ocean shores have extensive sand dunes that are deposited by wind and water. At first these dunes are blown about by the wind. The sand dune environment is severe, with high temperatures during the day and low temperatures at night. The sand may also be deficient in certain nutrient minerals needed by plants. As a result, few plants can tolerate the environmental conditions of a sand dune.

Henry Cowles developed the concept of succession in the 1880s when he studied succession on sand dunes around the shores of Lake Michigan, which had been gradually shrinking since the last ice age. The shrinking lake exposed new sand dunes that displayed a series of stages in the colonization of the land.

Grasses are common pioneer plants on sand dunes around the Great Lakes. As the grasses extend over the

---

MINI-GLOSSARY

Succession

succession: A process of community development that involves a changing sequence of species.

pioneer community: The first organisms to colonize (or recolonize) an area.

primary succession: Ecological succession in an environment that has not previously been inhabited; no soil is present initially.

secondary succession: Ecological succession in a previously inhabited environment that was exposed to some type of disturbance; soil is already present.

(a)

(b)

(c)

Figure 5.4 Primary succession on glacial moraine. During the past 200 years, glaciers have retreated in Glacier Bay, Alaska. Although these photos were not taken in the same area, they show some of the stages of primary succession on glacial moraine (rocks, gravel, and sand deposited by a glacier). (a) After the glacier's retreat, the barren landscape is initially colonized by lichens, then mosses and small shrubs. (b) At a later time, dwarf trees and shrubs colonize the area. (c) Still later, spruces dominate the community.

surface of a dune, their roots hold it in place, helping to stabilize it. At this point, mat-forming shrubs may invade, further stabilizing the dune. Later, shrubs may be replaced by poplars (cottonwoods), which years later are replaced by pines and finally by oaks. Because the soil fertility remains low, other forest trees rarely replace oaks. A summary of how primary succession on sand dunes around the Great Lakes might proceed is in this sequence: grasses → shrubs → poplars (cottonwoods) → pine trees → oak trees.

Secondary Succession **Secondary succession** is the change in species composition that takes place after some disturbance destroys the existing vegetation; soil is already present. An open area caused by a forest fire and abandoned farmland are common examples of sites where secondary succession occurs.

During the summer of 1988, wildfires burned approximately one third of Yellowstone National Park. This natural disaster provided a valuable chance for biologists to study secondary succession in areas that had been protected forests. After the conflagration, gray ash covered the forest floor, and most of the trees, although standing, were charred and dead. Secondary succession in Yellowstone has occurred rapidly since 1988. Less than one year later, in the spring of 1989, trout lily and other herbs sprouted and covered much of the ground. By 1998, a young forest of knee-high to shoulder-high lodgepole pines dominated the area. Douglas fir seedlings also began appearing in 1998. Biologists continue to monitor the changes in Yellowstone as secondary succession takes place.

Secondary succession on abandoned farmland has been studied extensively. Although it takes more than 100 years for secondary succession to occur at a single site, it is possible for a single researcher to study old field succession in its entirety by observing different sites in the same area that have been undergoing succession for different amounts of time. The biologist may examine county tax records to determine when each field was abandoned.

Abandoned farmland in North Carolina is colonized by a predictable succession of communities (Figure 5.5). The first year after cultivation ceases, the field is dominated by crabgrass. During the second year, horseweed, a larger plant that outgrows crabgrass, is the dominant species. Horseweed does not dominate more than 1 year, however, because decaying horseweed roots inhibit the growth of young horseweed seedlings. In addition, horseweed does not compete well with other plants that become established in the third year. During the third year after the last cultivation, other weeds—broomsedge, ragweed, and aster—become established. Typically, broomsedge outcompetes aster, because broomsedge is drought-tolerant, whereas aster is not.

In years 5 to 15, the dominant plants in an abandoned field are pines such as shortleaf pine and loblolly

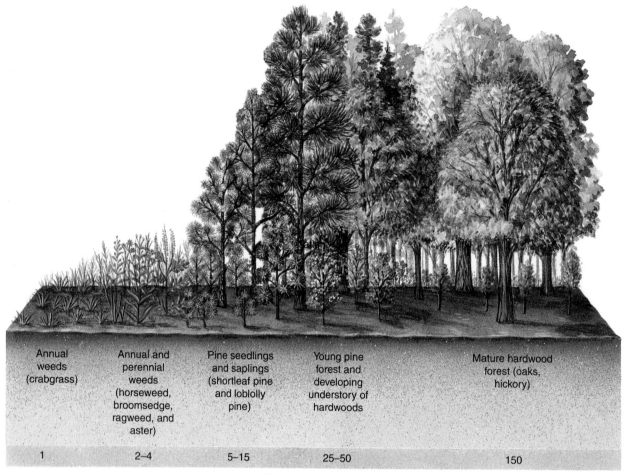

Annual weeds (crabgrass)	Annual and perennial weeds (horseweed, broomsedge, ragweed, and aster)	Pine seedlings and saplings (shortleaf pine and loblolly pine)	Young pine forest and developing understory of hardwoods	Mature hardwood forest (oaks, hickory)
1	2–4	5–15	25–50	150

Years After Cultivation

Figure 5.5 Secondary succession on an abandoned field in North Carolina.

pine. Through the buildup on the soil of litter, such as pine needles and branches, pines produce conditions that cause the earlier dominant plants to decline in importance. Over time, pines give up their dominance to hardwoods such as oaks. The replacement of pines by oaks depends primarily on the environmental changes produced by the pines. The pine litter causes soil changes, such as an increase in water-holding capacity, that are necessary for young oak seedlings to become established. In addition, hardwood seedlings are more tolerant of shade than are young pine seedlings. Secondary succession on abandoned farmland in the southeastern United States proceeds in this sequence: crabgrass → horseweed → broomsedge and other weeds → pine trees → hardwood trees.

Animal Life During Old Field Succession As secondary succession in North Carolina proceeds, a progression of animal life follows the changes in vegetation.

Although a few animals—such as the short-tailed shrew—are found in all stages of abandoned farmland succession, most animals appear with certain stages and disappear with others. During the crabgrass and weed stages of secondary succession, the open fields support grasshoppers, meadow voles, cottontail rabbits, and birds such as grasshopper sparrows and meadowlarks. As young pine seedlings become established, animals of open fields give way to animals common in mixed herbaceous and shrubby habitats. Now white-tailed deer, white-footed mice, ruffed grouse, robins, and song sparrows are common, whereas grasshoppers, meadow mice, grasshopper sparrows, and meadowlarks disappear. As the pine seedlings grow into trees, animals of the forest replace those that are common in mixed herbaceous and shrubby habitats. Cottontail rabbits give way to red squirrels, and ruffed grouse, robins, and song sparrows are replaced by warblers and veeries. Thus, each stage of succession supports its own characteristic animal life.

KEYSTONE SPECIES

Certain species are more crucial to the maintenance of their ecosystem than others. Such species, called **keystone species**, are vital in determining the nature and structure of the entire ecosystem—that is, its species composition and its ecosystem functioning. Other species of a community depend on or are greatly affected by the keystone species. Keystone species are usually not the most abundant species in the ecosystem. Although present in relatively small numbers, keystone species exert a profound influence on the entire ecosystem because they often affect the available amount of food, water, or some other resource.

Identifying and protecting keystone species are crucial goals of conservation biologists because if a keystone species disappears from an ecosystem, many other organisms in that ecosystem may become more common, become rare, or even disappear. One example of a keystone species is a top predator such as the gray wolf. Where wolves were hunted to extinction, the populations of deer and other herbivores increased explosively (recall the reintroduction of wolves to Yellowstone, discussed in Chapter 1). As these herbivores overgrazed the vegetation, many plant species that could not tolerate such grazing pressure disappeared. Many smaller animals such as insects also were lost from the ecosystem because the plants that they depended on for food were now less abundant. Thus, the disappearance of the wolf resulted in an ecosystem with considerably less biological diversity.

Because different species of fig trees collectively produce a continuous crop of fruits, they appear to be keystone species in tropical rain forests. Fruit-eating monkeys, birds, bats, and other vertebrates of the forest do not normally consume large quantities of figs in their diets. During the time of year when other fruits are less plentiful, however, fig trees become very important in sustaining fruit-eating vertebrates. Should the fig trees disappear, most of the fruit-eating vertebrates would also be eliminated. Thus, protecting fig trees in such tropical rainforest ecosystems is an important conservation goal because it increases the likelihood that monkeys, birds, bats, and other vertebrates will survive.

Some scientists think the concept of keystone species, although of potential usefulness in conservation biology, should be abandoned because it is too problematic. For one thing, most of the information we have about keystone species is anecdotal. Few long-term scientific studies have been performed to identify keystone species and to determine the nature and magnitude of their effects on the ecosystems they inhabit.

INTERACTIONS AMONG ORGANISMS

No organism exists independently of other organisms. The producers, consumers, and decomposers of an ecosystem interact with one another in a variety of complex ways, and each forms associations with other organisms. Three main types of interactions occur among species in an ecosystem: symbiosis, predation, and competition.

SYMBIOSIS

Symbiosis is any intimate relationship or association between members of two or more species. Usually, one species lives in or on another species. The partners of a symbiotic relationship, called **symbionts**, may benefit from, be unaffected by, or be harmed by the relationship.

Coevolution

Symbiosis is the result of **coevolution**, the interdependent evolution of two interacting species. Flowering plants and their animal pollinators have a symbiotic relationship that is an excellent example of coevolution. Because plants are rooted in the ground, they lack the mobility that animals have when mating. Many flowering plants rely on animals to help them mate. Bees, beetles, hummingbirds, bats, and other animals transport the male reproductive units, called pollen, from one plant to another, in effect giving plants mobility. How has this come about?

During the millions of years over which these associations developed, flowering plants evolved several ways to attract animal pollinators. One of the rewards for the pollinator is food—nectar (a sugary solution) and pollen. Plants often produce food that is precisely correct for one type of pollinator. For example, the nectar of flowers that are pollinated by bees usually contains 30% to 35% sugar, the concentration that bees need to make honey. Bees will not visit flowers with lower sugar concentrations in their nectar. Bees also use pollen to make beebread, a nutritious mixture of nectar and pollen that is eaten by their larvae.

Plants also possess a variety of ways to get the pollinator's attention, most of which involve colors and scents. Showy petals visually attract the pollinator much as a neon sign or golden arches attract a hungry person to a restaurant. Scents are also an effective way to attract pollinators. Insects have a well-developed sense of smell, and many insect-pollinated flowers have strong scents, which may also be very pleasant perfumes to humans. A few specialized kinds of flowers have unpleasant odors. The carrion plant produces flowers that smell like rotting flesh. Carrion flies move from one flower to another, looking for a place to deposit their eggs, and in the process pollen is transferred from one flower to another.

During the time plants were acquiring specialized features to attract pollinators, the animal pollinators coevolved specialized body parts and behaviors that enabled them to both aid pollination and obtain nectar and pollen as a reward. Coevolution is responsible for the hairy bodies of bumblebees, which catch and hold the sticky pollen for transport from one flower to another. Coevolution is also responsible for the long, curved beaks of certain honeycreepers, Hawaiian birds that insert their beaks into tubular flowers to obtain nectar (Figure 5.6).

Animal behavior has also coevolved. The flowers of certain orchids resemble female wasps in coloring and shape. The resemblance between one orchid species and female wasps is so strong that male wasps mount the flowers and attempt to copulate with them. During this misdirected activity, a pollen sac usually attaches to the back of the wasp. When the frustrated wasp departs and attempts to copulate with another orchid flower, pollen grains are transferred to the flower.

The thousands, or even millions, of symbiotic associations that result from coevolution fall into three categories: mutualism, commensalism, and parasitism (see "Mini-Glossary: Symbiosis").

Figure 5.6 Coevolution. The gracefully curved bill of the 'I'iwi, one of the Hawaiian honeycreepers, enables it to sip nectar from flowers of the lobelia. The 'I'iwi bill fits perfectly into the long, tubular lobelia flowers.

Mutualism: Sharing Benefits

Mutualism is a symbiotic relationship in which both partners benefit. The association between nitrogen-fixing bacteria of the genus *Rhizobium* and legumes, which are plants such as peas, beans, and clover, is an example of mutualism. Nitrogen-fixing bacteria live in nodules in the roots of legumes and supply the plants with all of the nitrogen they need. The legumes supply sugar to their bacterial symbionts.

Another example of mutualism is the association between reef-building coral animals and microscopic algae. These symbiotic algae, which are called **zooxanthellae** (pronounced *zoh-zan-thel´ee*, live inside cells of

---

MINI-GLOSSARY

Symbiosis

symbiosis: Any intimate relationship between two or more species. Includes mutualism, commensalism, and parasitism.

mutualism: A symbiotic relationship in which both partners benefit.

commensalism: A symbiotic relationship in which one partner benefits and the other partner is unaffected.

parasitism: A symbiotic relationship in which one partner—the parasite—obtains nutrients at the expense of the other—the host.

the coral, where they photosynthesize and provide the animal with carbon and nitrogen compounds as well as oxygen. Zooxanthellae have a stimulatory effect on the growth of corals, causing calcium carbonate skeletons to form around their bodies much faster when the algae are present. The corals, in turn, supply their zooxanthellae with waste products such as ammonia, which the algae use to make nitrogen compounds for both partners.

Mycorrhizae (*my-kor-rye´zee*) are mutualistic associations between fungi and the roots of about 80% of all plants. The fungus, which grows around and into the root as well as into the surrounding soil, absorbs essential minerals, especially phosphorus, from the soil and provides them to the plant. In return, the plant provides the fungus with food produced by photosynthesis. Plants grow more vigorously in mycorrhizal relationships (Figure 5.7), and they are better able to tolerate **environmental stressors** such as drought and high soil

Figure 5.8 Commensalism. Epiphytes are small plants that grow attached to the bark of a tree. Photo was taken in St. Johns, Virgin Islands.

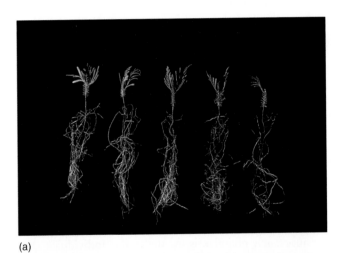

(a)

(b)

Figure 5.7 Mutualism. Western red cedar (*Thuja plicata*) seedlings respond to mycorrhizal fungi. (a) These seedlings were grown in low phosphorus in the absence of the fungus. (b) Comparable seedlings were grown in low phosphorus with the fungus.

temperatures. Indeed, some plants cannot maintain themselves under natural conditions if the fungi with which they normally form mycorrhizae are not present.

Commensalism: Taking Without Harming

Commensalism is a type of symbiosis in which one organism benefits and the other one is neither harmed nor helped. One example of commensalism is the relationship between two kinds of insects, silverfish and army ants. Certain kinds of silverfish move along in permanent association with the marching columns of army ants and share the plentiful food caught in their raids. The army ants derive no apparent benefit or harm from the silverfish.

Another example of commensalism is the relationship between a tropical tree and its **epiphytes**, smaller plants, such as mosses, orchids, and ferns that live attached to the bark of the tree's branches (Figure 5.8). The epiphyte anchors itself to the tree but does not obtain nutrients or water directly from the tree. Its location on the tree enables it to obtain adequate light, water (as rainfall dripping down the branches), and required nutrient minerals (washed out of the tree's leaves by rainfall). Thus, the epiphyte benefits from the association, whereas the tree is apparently unaffected.

Parasitism: Taking at Another's Expense

Parasitism is a symbiotic relationship in which one member, the *parasite*, benefits and the other, the *host*, is adversely affected. The parasite obtains nourishment from its host, but although a parasite may weaken its host, it rarely kills it. (A parasite would have a difficult life if it kept killing off its hosts!) Some parasites, such as ticks, live outside the host's body; other parasites, such as tapeworms, live within the host. Parasitism is a very successful lifestyle; more than 100 parasites live in or on the human species alone!

Since the 1980s, wild and domestic honeybees in the United States have been dying off. Although **habitat fragmentation** and pesticide use have contributed to the problem, tracheal mites (Figure 5.9) and larger varroa mites are a major reason for the honeybee decline. The number of commercial colonies has fallen by about 50% during the past several decades. In the United States honeybees annually pollinate more than 90 food, fiber, and seed crops valued at $14.6 billion and produce about $124 million of honey, so their decline is a major threat to U.S. agriculture. Beginning in 1999, beekeepers in several States, in cooperation with the U.S. Department of Agriculture, have tested Russian honeybees for pollination and honey production. Preliminary results indicate that the Russian bees are very resistant to mites and can tolerate cold winters better than American bees.

When a parasite causes disease and sometimes the death of a host, it is called a **pathogen**. Crown gall disease, which is caused by a bacterium, occurs in many different kinds of plants and results in millions of dollars' worth of damage each year to ornamental and agricultural plants. Crown gall bacteria, which also live on detritus (organic debris) in the soil, enter plants through small wounds such as those caused by insects. They cause galls (tumorlike growths), often at a plant's crown—that is, between the stem and the roots, at or near the soil surface. Although plants seldom die from crown gall disease, they are weakened, grow more slowly, and often succumb to other pathogens.

Many parasites do not cause disease. Humans can become infected with the beef tapeworm by eating undercooked beef that is infested with immature tape-

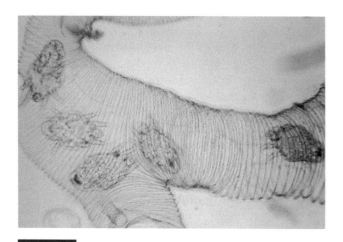

Figure 5.9 Parasitism. Tiny tracheal mites live in the breathing tubes of honeybees and suck their blood, weakening and eventually killing them. Since the mid-1990s, these mites have devastated many of North America's wild and domestic honeybee populations.

worms. Once the tapeworm is inside the human digestive system, it attaches itself to the wall of the small intestine and grows rapidly by absorbing nutrients that pass through the small intestine. A single beef tapeworm that lives in a human digestive tract does not usually cause any noticeable symptoms, although some weight loss may be associated with a multiple infestation.

PREDATION

Predation is the consumption of one species, the *prey*, by another, the *predator*. Predation includes both animals eating other animals and animals eating plants. Predation has resulted in an "arms race," with the coevolution of predator strategies—more efficient ways to catch prey—and prey strategies—better ways to escape the predator. A predator that is more efficient at catching prey exerts a strong selective force on its prey, and over time the prey species may evolve some sort of countermeasure that reduces the probability of being captured. The countermeasure acquired by the prey in turn may act as a strong selective force on the predator.

We now consider several adaptations that are related to predator–prey interactions. These include predator strategies (pursuit and ambush) and prey strategies (plant defenses and animal defenses). Keep in mind as you read these descriptions that such strategies are not "chosen" by the respective predators or prey. New traits arise randomly in a population as a result of changes in genetic material, or deoxyribonucleic acid (DNA). Some new traits may be beneficial, whereas others may be harmful or have no effect at all. As a result of natural selection, beneficial strategies, or traits, persist in a population because such characteristics make the individuals that possess them well suited to thrive and reproduce. In contrast, characteristics that make the individuals that possess them poorly suited to their environment tend to disappear in a population.

Pursuit and Ambush

The brown pelican sights its prey—a fish—while in flight. Less than 2 seconds after diving into the water at a speed as great as 72 km (45 mi) per hour, it has its catch (Figure 5.10). Killer whales, which hunt in packs, often herd salmon or tuna into a cove so that they are easier to catch. Any trait that increases hunting efficiency, such as the speed of a brown pelican or the intelligence of killer whales, favors predators that pursue their prey. Because these carnivores must be able to process information quickly during the pursuit of prey, their brains are generally larger, relative to body size, than those of the prey they pursue.

Ambush is another effective way to catch prey. The goldenrod spider is the same color as the white or yellow flowers in which it hides. This camouflage prevents

Figure 5.10 Predation. A brown pelican (*Pelecanus occidentalis*) alights in the water, having just caught a fish. One effective predator strategy is speed of pursuit.

unwary insects that visit the flower for nectar from noticing the spider until it is too late. Predators that are able to *attract* prey are particularly effective at ambushing. For example, a diverse group of deep-sea fishes called anglerfish possess rodlike luminescent lures close to their mouths to attract prey.

Plant Defenses Against Herbivores

Plants cannot escape predators by fleeing, but they possess adaptations that protect them from being eaten. The presence of spines, thorns, tough leathery leaves, or even thick wax on leaves discourages foraging herbivores from grazing. Other plants produce an array of protective chemicals that are unpalatable or even toxic to herbivores. The active ingredients in such plants as marijuana, opium poppy, tobacco, and peyote cactus may discourage the foraging of herbivores. The nicotine found in tobacco is so effective at killing insects, for example, that it is a common ingredient in many commercial insecticides.

Milkweeds are an excellent example of the evolutionary arms race between plants and herbivores. Milkweeds produce alkaloids and cardiac glycosides, chemicals that are poisonous to all animals except for a small group of insects. During the course of evolution, these insects acquired the ability to either tolerate or metabolize the milkweed toxins. As a result, they can eat milkweeds without being poisoned. These insects avoid competition from other herbivorous insects because few other insects are able to tolerate milkweed toxins. Predators also learn to avoid these insects, which accumulate the toxins in their tissues and are usually brightly colored to announce that fact.

Defensive Adaptations of Animals

Many animals, such as woodchucks, flee from predators by running to their underground burrows. Others have mechanical defenses, such as the barbed quills of a porcupine and the shell of a pond turtle. Some animals live in groups—a herd of antelope, colony of honeybees, school of anchovies, or flock of pigeons. Because the group has so many eyes, ears, and noses watching, listening, and smelling for predators, this social behavior decreases the likelihood of a predator catching one of them unaware.

Chemical defenses are also common among animal prey. The South American poison arrow frog has poison glands in its skin. Its bright **warning coloration** prompts avoidance by experienced predators (Figure 5.11). Snakes or other animals that have tried to eat a poisonous frog do not repeat their mistake! Other examples of warning coloration occur in the striped skunk, which sprays acrid chemicals from its anal glands, and the bombardier beetle, which sprays harsh chemicals at potential predators.

Some animals hide from predators by blending into their surroundings. Such camouflage is often enhanced by the animal's behavior. There are many examples of camouflage. Certain caterpillars resemble twigs so closely that you would never guess they are animals until they move. Pipefish are almost perfectly camouflaged in green eelgrass. The pygmy seahorse resembles gorgonian coral so closely that the little seahorse was not discovered until

Figure 5.11 Warning coloration. The poison arrow frog (*Dendrobates tinctorius*) advertises its poisonous nature with its conspicuous warning coloration to avoid would-be predators. Photo was taken in Guyana, a small South American country.

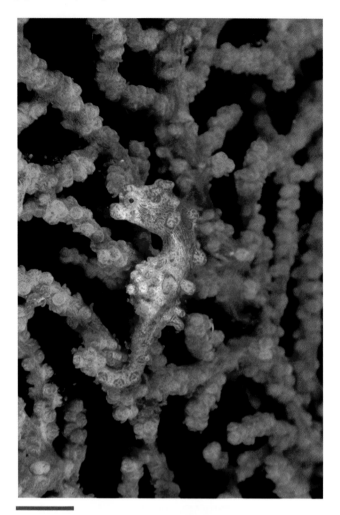

Figure 5.12 Camouflage. Pygmy seahorses (*Hippocampus bargibanti*), about the size of small fingernails, are virtually invisible in and around gorgonian coral (*Muricella* sp.). Photographed off the coast of Sulawesi, Indonesia.

1970 when a zoologist found it in the gorgonian coral he had placed in an aquarium (Figure 5.12). Such camouflage has been preserved and accentuated by evolution.

COMPETITION

Competition occurs when two or more individuals attempt to use an essential common resource such as food, water, shelter, living space, or sunlight. Because resources are often in limited supply in the environment, their use by one individual decreases the amount available to others. If a tree in a dense forest grows taller than surrounding trees, it is able to absorb more of the incoming sunlight. Less sunlight is therefore available for nearby trees that are shaded by the taller tree.

Competition can occur among individuals within a population (**intraspecific competition**) or between different species (**interspecific competition**). Intraspecific competition is discussed in Chapter 8.

Ecologists traditionally assumed that competition is the most important determinant of both the number of species found in a community as well as the size of each population. Today ecologists recognize that competition is only one of many interacting biotic and abiotic factors that affect community structure. Furthermore, competition is not always a straightforward, direct interaction. A variety of flowering plants live in a young pine forest and presumably compete with conifers for such resources as soil moisture and soil nutrient minerals. Their relationship, however, is more complex than simple competition. The flowers produce nectar that is consumed by some insect species that also prey on needle-eating insects, thereby reducing the number of insects feeding on pines. It is therefore difficult to assess the overall effect of flowering plants on pines. If the flowering plants were removed from the community, would the pines grow faster because they were no longer competing for necessary resources? Or would pine growth be inhibited by the increased presence of needle-eating insects (caused by fewer omnivorous insects)?

Short-term experiments in which one competing plant species is removed from a forest community often have demonstrated an improved growth for the remaining species. However, very few studies have tested the

ENVIROBRIEF

Credible Science in Biosphere 2

Biosphere 2 is artificial self-contained ecosystem that began operating in 1991 in the Arizona desert. It is a 1.27-hectare (3-acre) greenhouse that contains a miniature world, including a tropical rain forest, savanna, desert, saltwater lagoon, mangrove estuary, and agricultural regions. (Biosphere 1 is planet Earth.) After failing to support eight nonscientist inhabitants in a 2-year test of survival and self-sufficiency in a sealed environment, the Biosphere 2 facility's credibility suffered from a barrage of criticism from the scientific community. Many critics challenged the original idea of Biosphere 2—that of a holistic, self-contained prototype of a Martian space colony—as being too theatrical and lacking in application of the scientific process. Although many scientists recognized the tremendous potential of Biosphere 2 as a unique site for conducting environmental studies with potentially global applications, they were frustrated that such experiments were not allowed.

In 1994 Biosphere 2 started a new direction when owner Edward Bass, a Texas billionaire, announced the appointment of a new scientific director and the start of a nonprofit joint venture with Columbia University in New York. The new arrangement opened the door for undergraduate students from Columbia and other institutions to study at Biosphere 2 and for outside scientists to propose and carry out ecosystem experiments within the facility. Scientists must contend with the structure's limitations, such as difficulties in conducting control or duplicate experiments, but most scientists are excited by the challenge. Ongoing projects examine the responses of different organisms, from ocean-dwelling corals to cottonwood trees, to increased levels of carbon dioxide. This information may help us understand what might be in store for planet Earth as carbon dioxide levels continue to rise in the 21st century (see Chapter 20).

long-term effects on forest species of the removal of one competing species. These long-term effects may be subtle, indirect, and difficult to assess. They may lower or negate the negative effects of competition for resources.

THE ECOLOGICAL NICHE

We have seen that a diverse assortment of organisms inhabits each community and that these organisms obtain nourishment in a variety of ways. We have also examined some of the ways that species interact to form interdependent relationships within the community. Now we examine the way of life of a given species in its ecosystem. An ecological description of a species typically includes (1) whether it is a producer, consumer, or decomposer; (2) the kinds of symbiotic associations it forms; (3) whether it is a predator and/or prey; and (4) what species it competes with. Other details are needed, however, to provide a complete picture.

Every organism is thought to have its own role within the structure and function of an ecosystem; we call this role its **ecological niche**. An ecological niche, which is difficult to define precisely, takes into account *all* aspects of the organism's existence—that is, all physical, chemical, and biological factors that the organism needs to survive, remain healthy, and reproduce. Among other things, the niche includes the local environment in which an organism lives—that is, its **habitat**. An organism's niche also encompasses what it eats, what eats it, what organisms it competes with, and how it interacts with and is influenced by the abiotic components of its environment, such as light, temperature, and moisture. The niche represents the totality of an organism's adaptations, its use of resources, and the lifestyle to which it is suited. Thus, a complete description of an organism's ecological niche involves numerous dimensions.

The ecological niche of an organism may be much broader potentially than it is in actuality. Put differently, an organism is potentially capable of using much more of its environment's resources or of living in a wider assortment of habitats than it actually does. The potential, idealized ecological niche of an organism is its **fundamental niche**, but various factors such as competition with other species usually exclude it from part of its fundamental niche. Thus, the lifestyle that an organism actually pursues and the resources that it actually uses make up its **realized niche**.

An example may help clarify the distinction between fundamental and realized niches. The green anole, a lizard native to Florida and other southeastern states, perches on trees, shrubs, walls, or fences during the day and waits for insect and spider prey (Figure 5.13a). In past years these little lizards were widespread in Florida. Several years ago, however, a related species, the brown anole, was introduced from Cuba into southern Florida and quickly became common (Figure 5.13b). Suddenly the green anoles became rare, appar-

(a)

(b)

Figure 5.13 Effect of competition on an organism's realized niche. (a) The green anole (*Anolis carolinensis*) is native to Florida. (b) The brown anole (*Anolis sagrei*) was introduced in Florida. (c) The fundamental niches of the two lizards initially overlapped. Species 1 is the green anole, and species 2 is the brown anole. (d) The brown anole was able to outcompete the green anole, restricting its niche.

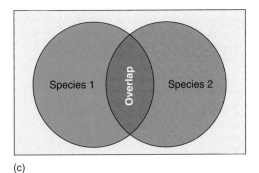

(c)

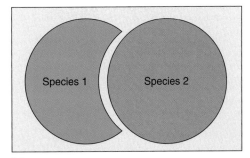

(d)

ently driven out of their habitat by competition from the slightly larger brown lizards. Careful investigation disclosed, however, that green anoles were still around. They were now confined largely to the vegetation in wetlands and to the foliated crowns of trees, where they were less easily seen.

The habitat portion of the green anole's fundamental niche includes the trunks and crowns of trees, exterior house walls, and many other locations. Where they became established, brown anoles were able to drive green anoles out from all but wetlands and tree crowns, so that the green anole's realized niche became much smaller as a result of competition (Figure 5.13c, d). Because natural communities consist of numerous species, many of which compete to some extent, the complex interactions among species produce the realized niche of each.

Limiting Factors

What factors actually determine the niche of a species? An ecological niche is basically determined by the total of a species' structural, physiological, and behavioral adaptations. Such adaptations determine the tolerance an organism has for environmental extremes. If any feature of its environment lies outside the bounds of its tolerance, then the organism cannot live there. Just as you would not expect to find a cactus living in a pond, you would not expect water lilies in a desert.

The environmental factors that help determine an organism's ecological niche can be extremely difficult to identify. For this reason, the concept of ecological niche is largely abstract, although some of its dimensions can be experimentally determined. Any environmental resource that, because it is scarce or unfavorable, restricts the ecological niche of an organism is called a **limiting factor**.

Most limiting factors that have been investigated are simple variables such as the mineral content of soil, extremes of temperature, and amount of precipitation. Such investigations have disclosed that any factor that

exceeds an organism's tolerance, or that is present in quantities smaller than the minimum required, limits the occurrence of that organism in an ecosystem (Figure 5.14). By their interaction, such factors help to define an organism's ecological niche.

Limiting factors often affect only one part of an organism's life cycle. Although adult blue crabs can live in brackish (slightly salty) water, they cannot become permanently established there because their larvae (immature forms) require water with a higher concentration of dissolved salt. Similarly, the ring-necked pheasant, a popular game bird, has been introduced widely in North America but does not survive in the southern United States. The adult birds do well, but the eggs cannot develop properly in the warm southern temperatures.

Competitive Exclusion and Resource Partitioning

When two species are very similar, as are the green and brown anoles, their fundamental niches may overlap. However, many ecologists think that no two species can indefinitely occupy the same niche in the same community because **competitive exclusion** eventually occurs. In competitive exclusion, one species is excluded from a portion of a niche by another as a result of competition between species—that is, interspecific competition. Although it is possible for species to compete for some necessary resource without being total competitors, two species with absolutely identical ecological niches cannot coexist. Coexistence *can* occur, however, if the overlap in the two species' niches is reduced. In the lizard example, direct competition between the two species was reduced as the brown anole excluded the green anole from most of its former habitat until the only places that remained open to it were wetland vegetation and tree crowns.

The initial evidence that competition between species determines an organism's realized niche came from a series of experiments conducted by the Russian biologist G.F. Gause in 1934. In one study Gause grew populations of two species of *Paramecium* (a type of pro-

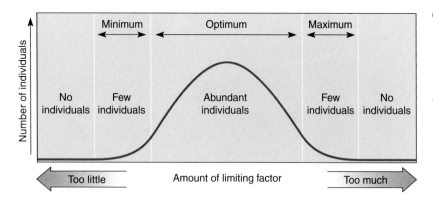

Figure 5.14 Limiting factor. Any environmental resource that is present in excess of an organism's tolerance or in quantities smaller than the minimum required limits the occurrence of that organism.

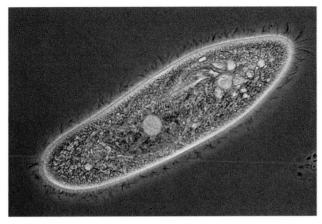

(a)

Figure 5.15 Interspecific competition. (a) *Paramecium* is a complex unicellular protist. Gause studied competition between two species of *Paramecium*. The top graph (b) shows how a population of *P. aurelia* grows in separate culture (that is, in a single-species environment), whereas (c) shows a separate culture of *P. caudatum*. The bottom graph (d) shows how these two species grow in a single culture, in competition with each other. *P. aurelia* outcompetes *P. caudatum* and drives it to extinction.

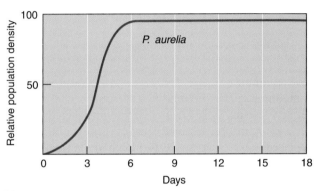

(b)

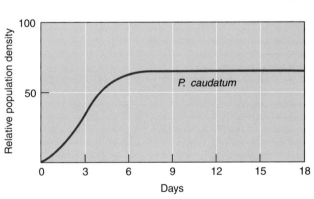

(c)

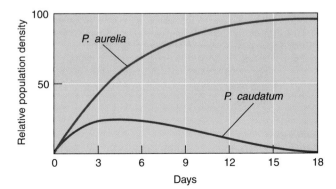

(d)

tist), *P. aurelia* and the larger *P. caudatum* (Figure 5.15). When grown in separate test tubes, the population of each species quickly increased to a high level, which was maintained for some time thereafter. When grown together, however, only *P. aurelia* thrived. *P. caudatum* dwindled and eventually died out. Under different environmental conditions, *P. caudatum* prevailed over *P. aurelia*. Gause interpreted this to mean that although one set of conditions favored one species, a different set favored the other. Nonetheless, because both species were so similar, given time, one or the other would eventually triumph at the other's expense.

Competition, then, has adverse effects on all species that use a limited resource and may result in competitive exclusion of one or more species. It therefore follows that natural selection should favor individuals of each species that avoid or at least reduce competition by **resource partitioning**, in which coexisting species' niches differ from each other in one or more ways. Evidence of resource partitioning in animals is well documented and includes studies in tropical forests of Central and South America that demonstrate little overlap in the diets of fruit-eating birds, primates, and bats that coexist in the same habitat. Although fruits are the primary food for several hundred bird, primate, and bat species, the wide variety of fruits available has allowed fruit eaters to specialize, thereby reducing competition.

Resource partitioning also may include timing of feeding, location of feeding, nest sites, and other aspects of an organism's ecological niche. Robert MacArthur's study of five North American warbler species is a classic example of resource partitioning (Figure 5.16). Although it initially appeared that their niches were nearly identical, MacArthur determined that individuals of each species spend most of their feeding time in different portions of the spruces and other conifer trees they frequent. They also move in different directions through the canopy, consume different combinations of insects, and nest at slightly different times.

We have seen that an organism's ecological niche takes into account all aspects of that organism's existence. Now we examine the number of niches available in different communities, which in turn affects the number of species those communities contain.

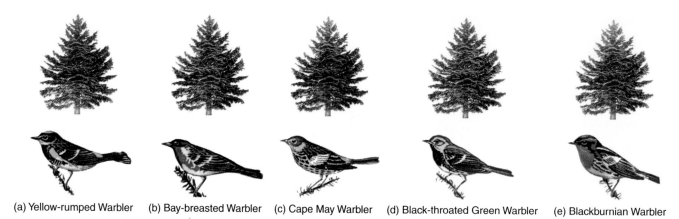

(a) Yellow-rumped Warbler   (b) Bay-breasted Warbler   (c) Cape May Warbler   (d) Black-throated Green Warbler   (e) Blackburnian Warbler

Figure 5.16 Resource partitioning. Each warbler species spends at least half its foraging time in the designated area (brown section) of a spruce tree, thereby reducing competition for foraging sites.

SPECIES RICHNESS

Species richness, the number of species present in a community, varies greatly from one community to another. Tropical rain forests and coral reefs are examples of communities with extremely high species richness. In contrast, geographically isolated islands and mountaintops exhibit low species richness.

What determines the number of species in a community? There seems to be no single answer, but several factors appear to be significant, such as the abundance of potential ecological niches, closeness to the margins of adjacent communities, geographical isolation, dominance of one species over others, habitat stress, and geological history.

Species richness is related to the *abundance of potential ecological niches*. An already complex community offers a greater variety of potential ecological niches than does a simple community (Figure 5.17). It may become even more complex if organisms potentially capable of filling those niches evolve or migrate into the community. Thus, it appears that species richness is self-perpetuating.

Species richness is usually greater *at the margins of adjacent communities* than in their centers. This is because an **ecotone**—a transitional zone where two or more communities meet—contains all or most of the ecological niches of the adjacent communities as well as some niches that are unique to the ecotone. The change in species composition produced at ecotones is known as the **edge effect**.

Species richness is inversely related to the *geographical isolation* of a community. Isolated island communities tend to be much less diverse than are communities in similar environments found on continents. This is due partly to the difficulty encountered by many species in reaching and successfully colonizing the island (recall the

discussion of primary succession on Krakatoa). Also, sometimes species become locally extinct as a result of random events. In isolated environments such as islands or mountaintops, extinct species cannot be readily replaced. Isolated areas are also likely to be small and to possess fewer potential ecological niches.

Species richness is reduced when any *one species enjoys a decided position of dominance within a community* so that it is able to appropriate a disproportionate share of available resources, thus crowding out, or outcompet-

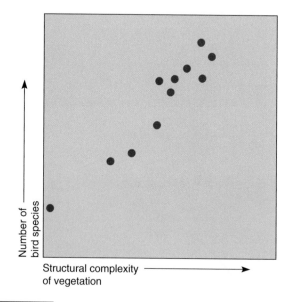

Figure 5.17 Effect of community complexity on species richness. A community in which the vegetation is structurally complex—a forest, for example—provides birds with more kinds of food and hiding places than does a structurally simpler community, such as grassland.

ing, other species. Ecologist James H. Brown of the University of New Mexico has studied species competition and diversity in long-term experiments conducted since 1977 in the Chihuahuan desert of southeastern Arizona. In one experiment, the removal of three dominant species, all kangaroo rats, from several plots resulted in an increased diversity of other rodent species. (The scientists enclosed their study areas with fencing and then cut holes in the fencing that would allow smaller rodents to come and go but exclude kangaroo rats.) This increase was attributed both to less competition for food and also to an altered habitat, because the abundance of grass species increased dramatically after the removal of the kangaroo rats.

Generally, species richness is inversely related to the *environmental stress of a habitat*. Only those species capable of tolerating extreme environmental conditions can live in an environmentally stressed community. Thus, the species richness of a highly polluted stream is low compared to that of a nearby pristine stream. Similarly, the species richness of high-latitude (further from the equator) communities exposed to harsh climates is less than that of lower-latitude (closer to the equator) communities with milder climates. Although the equatorial countries of Colombia, Ecuador, and Peru occupy only 2% of the Earth's land, they contain an astonishing 45,000 native plant species. The continental United States and Canada, with a significantly larger land area, possess a total of 19,000 native plant species. Ecuador alone contains more than 1,300 native bird species, twice as many as the United States and Canada combined.

Species richness is greatly affected by *geological history*. Tropical rain forests are thought to be very old, stable communities that have undergone few climate changes in the entire history of the Earth. During this time, myriad species evolved in tropical rain forests, having experienced few or no abrupt climate changes that might have led to their extinction. In contrast, glaciers have repeatedly altered temperate and arctic regions during Earth's history. An area recently vacated by glaciers will have a low species richness because few species will as yet have had a chance to enter it and become established.

Species Richness, Ecosystem Services, and Community Stability

Ecologists and conservationists have long debated whether the extinction of species threatens the normal functioning and stability of ecosystems. This question is of great practical concern because ecosystems supply human societies with many environmental benefits, such as clean air to breathe, clean water to drink, and fertile soil in which to grow crops (Table 5.1). Conservationists maintain that ecosystems with greater species richness are better able to supply such **ecosystem services** than are ecosystems with lower species richness.

Traditionally, most ecologists assumed that **community stability**—the ability of a community to withstand environmental disturbances—is a consequence of community complexity. That is, a community with considerable species richness is thought to function better and therefore be more stable than a community with less

Table 5.1 Ecosystem Services	
Ecosystem	*Services Provided by Ecosystem*
Forests	Purify air and water; produce and maintain soil; absorb carbon dioxide; provide wildlife habitat; provide humans with wood and recreation
Freshwater systems (rivers and streams, lakes, and groundwater)	Moderate water flow; dilute and remove pollutants; provide wildlife habitat; provide humans with drinking and irrigation water, food, transportation corridors, electricity, and recreation
Grasslands	Purify air and water; produce and maintain soil; absorb carbon dioxide; provide wildlife habitat; provide humans with livestock and recreation
Coasts	Provide a buffer against storms; dilute and remove pollutants; provide wildlife habitat; provide humans with food, harbors and transportation routes, and recreation
Sustainable agricultural ecosystems*	Produce and maintain soil; absorb carbon dioxide; provide wildlife habitat for birds, insect pollinators, and soil organisms; provide humans with food and fiber crops

* Sustainable agriculture ecosystems are human-made and therefore inherently different from the other ecosystems. Sustainable agriculture is discussed in Chapter 18.

species richness. According to this view, the greater the species richness, the less critically important any single species should be. With many possible interactions within the community, it appears unlikely that any single disturbance could affect enough components of the system to make a significant difference in its functioning.

Evidence for this hypothesis can be found in the fact that destructive outbreaks of pests are more common in cultivated fields, which are low-diversity communities, than in natural communities with greater species richness. As another example, the almost complete loss of the American chestnut tree to the chestnut blight fungus had little ecological impact on the moderately diverse Appalachian woodlands of which it used to be a part.

Ongoing studies by David Tilman of the University of Minnesota and John Downing of the University of Iowa have strengthened the link between species richness and community stability. In their initial study, reported in the journal *Nature* in 1994, they established and monitored 207 plots of Minnesota grasslands for seven years. During the study period, Minnesota's worst drought in 50 years occurred (1987–1988). The biologists found that those plots with the greatest number of plant species lost less ground cover and recovered faster than did species-poor plots. Further research published in 1996 in the journal *Ecology* supported these conclusions and showed a similar effect of species richness on community stability during nondrought years. Similar work by almost three dozen ecologists at eight grassland sites in Europe, published in 1999 in the journal *Science*, also support the link between species richness and ecosystem functioning.

Although species richness appears to increase community stability, populations of individual species within a diverse community may vary significantly from year to year. It may seem paradoxical that instability within populations of individual species relates to the stability of the entire community. When one considers all the interactions among the organisms in a community, however, it is obvious that some species benefit at the expense of others. If one species declines during a disturbance such as a drought, other species that compete with it may flourish during that period. Thus, it appears that ecosystems with greater species richness are more likely to contain species that are resistant to any given disturbance.

SUMMARY WITH SELECTED KEY TERMS

I. Evolution can be defined as a genetic change in a population of organisms that occurs over time.

A. Charles Darwin proposed the theory of evolution by **natural selection**, which is based on four premises:

  1. Each species produces more offspring than will survive to maturity.

  2. The individuals in a population exhibit inheritable variation in their traits.

  3. Organisms compete with one another for the resources needed to survive.

  4. Those individuals with the most favorable combination of traits are most likely to survive and reproduce, passing their genetic characters on to the next generation.

B. A vast body of evidence supports evolution, including observations from the fossil record, comparative anatomy, biogeography, and molecular biology. In addition, evolutionary hypotheses are tested experimentally.

C. Biologists try to make sense of the diversity of life that has evolved on Earth by classifying organisms into groups. Currently, many biologists use a six-kingdom system of classification.

II. A biological **community** consists of a group of organisms that interact and live together in the same place at the same time. A living community and its abiotic environment constitute an **ecosystem**.

A. Succession is the orderly replacement of one community by another.

B. Primary succession begins in an environment that has not previously been inhabited. No soil exists when primary succession begins. Two examples are primary succession on recently formed volcanic lava and on sand dunes along lake shores.

C. Secondary succession begins in an area where there was a preexisting community and a well-formed soil. Two examples are open areas caused by a forest fire and abandoned farmland.

III. Within an ecosystem, **keystone species** help determine the species composition and functioning of the entire ecosystem. Although present in relatively small numbers, keystone species often affect the available amount of food, water, or some other **resource** for much of the community.

IV. Symbiosis is any intimate association between two or more species. Symbiosis is the result of **coevolution**.

A. In **mutualism**, both partners benefit. Three examples of mutualism are nitrogen-fixing bacteria and legumes, zooxanthellae and corals, and mycorrhizae (fungi and roots of plants).

B. In **commensalism**, one organism benefits and the other is unaffected. Two examples of commensalism are silverfish and army ants, and epiphytes and larger plants.

C. In **parasitism**, one organism (the parasite) benefits while the other (the host) is harmed. One example of parasitism is honeybees and mites. Some parasites are **pathogens** that cause disease. One example of a pathogen is the bacterium that causes crown gall disease in many plants.

V. Predation is the consumption of one species (the prey) by another (the predator).

A. During coevolution between predator and prey, the predator evolves more efficient ways to catch prey, and the prey evolves better ways to escape the predator.

B. Two effective predator strategies are pursuit and ambush.

C. Adaptations that protect plants from being eaten include spines; thorns; tough, leathery leaves; and protective chemicals.

D. Strategies that help animals avoid being killed and eaten include flight, association in groups, and camouflage, as well as a variety of mechanical and chemical defenses. Animals that possess chemical defenses often exhibit **warning coloration**.

VI. Competition occurs when two or more individuals attempt to use an essential common resource such as food, water, shelter, living space, or sunlight.

A. Competition can occur among individuals within a population (**intraspecific competition**) or between species (**interspecific competition**).

B. Interactions among species competing for the same resources may be subtle, indirect, and difficult to assess.

VII. An organism's **ecological niche** includes its distinctive lifestyle and its role in the community. The ecological niche takes into account all aspects of the organism's existence—that is, all of the physical, chemical, and biological factors that the organism needs to survive, remain healthy, and reproduce.

A. Organisms are potentially able to exploit more resources and play a broader role in the life of their community than they actually do.

1. The potential ecological niche of an organism is its **fundamental niche**, whereas the niche an organism actually occupies is its **realized niche**.

2. Interspecific competition is an important biological determinant of a species' realized niche.

B. Many ecologists think that no two species can occupy the same niche in the same community for an indefinite period of time.

1. Limiting factors (environmental resources such as the mineral content of soil, temperature extremes, and amount of precipitation) tend to restrict an organism's ecological niche.

2. In **competitive exclusion**, one species excludes another as a result of competition for limited resources. Gause's experiment with two species of *Paramecium* is a classic study of competitive exclusion.

3. Some species reduce competition by **resource partitioning**, in which they evolve differences in resource use. MacArthur's study of North American warblers is a classic example of resource partitioning.

VIII. Community complexity is expressed in terms of **species richness**, the number of species within a community.

A. Species richness is often great when there are many potential ecological niches, when the area studied is at the margins of adjacent communities, when the community is not isolated or severely stressed, when one species does not dominate others, and when communities have a long history.

B. Ecosystems with greater species richness may be better able to supply **ecosystem services**, environmental benefits such as clean air, clean water, and fertile soil.

THINKING ABOUT THE ENVIRONMENT

1. During mating season, male giraffes slam their necks together in fighting bouts to determine which male is stronger and can therefore mate with females. Explain how long necks may have evolved under this scenario, using Darwin's theory of evolution by natural selection.

2. Describe an example of secondary succession. Begin your description with the specific disturbance that preceded it.

3. Some biologists think that protecting keystone species would help preserve biological diversity in an ecosystem. Explain.

4. What type of symbiotic relationship—mutualism, commensalism, or predation—do you think exists between the pygmy seahorse and the gorgonian coral pictured in Figure 5.12? Explain your answer.

5. Describe how evolution has affected predator–prey relationships.

6. What are the two main kinds of competition that ecologists recognize?

7. Why is a realized niche usually narrower, or more restricted, than a fundamental niche?

8. What portion of the human's fundamental niche are we occupying today? Do you think our realized niche has changed over the past 200 years? Why or why not?

9. Explain how the concept of fundamental and realized niche is applicable to the rodent experiment in the Chihuahuan desert of Arizona.

10. What is the most likely limiting factor for plants and animals in deserts? Explain your answer.

11. Describe two determinants of species richness and give an example of each.

12. Describe some ecosystem services provided by a forest.

*13. Examine and interpret the figure shown below, to answer the following questions.

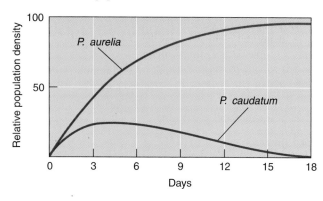

a. When (approximately what day) does the population of *P. aurelia* exceed that of *P. caudatum*?

b. When does the population of *P. caudatum* peak?

c. When does the population of *P. aurelia* peak?

* The solution to this question appears in Appendix VII.

TAKE A STAND

Visit our Web site at **http://www.wiley.com/college/raven** (select Chapter 5 from the Table of Contents) for links to more information about the Biosphere 2 project. You will find tools to help you organize your research, analyze the data, think critically about the issues, and construct a well-considered argument. Take a Stand activities can be done individually or as part of a team, as oral presentations, written exercises, or Web-based (e-mail) assignments.

Additional on-line materials relating to this chapter, including Student Quizzes, Activity Links, Useful Web Sites, Flash Cards, and more, can also be found on our Web site.

SUGGESTED READING

Baskin, Y. "Yellowstone Fires: A Decade Later." *BioScience*, Vol. 49, No. 2 (February 1999). Yellowstone National Park is recovering rapidly from the wildfires that burned one third of it in 1988.

Bond, W. "Keystone Species—Hunting the Snark?" *Science*, Vol. 292 (April 6, 2001). Examines the concept of keystone species and cites the long-term study of kangaroo rats by Brown as the type of research needed to identify keystone species.

Douglas, K. "Hot Spots." *New Scientist* (April 4, 1998). This article examines the complex reasons why there are so many species in the tropics.

Foster, D. "Small Matters." *Smithsonian* (May 2002). Explores the fascinating symbiotic relationship between leafcutter ants and the fungi that the ants grow in subterranean gardens.

Holloway, M. "The Glass House in the Desert." *Scientific American* (Januaray 2002). Biosphere 2, located north of Tucson, Arizona, attracts scientists, students, and tourists.

Knapp, A.K., et al. "The Keystone Role of Bison in North American Tallgrass Prairie." *BioScience*, Vol. 49, No. 1 (January 1999). The presence of grazing bison increases biological diversity in North American grasslands.

Lovett, R.A. "Mount St. Helens, Revisited." *Science*, Vol. 288 (June 2, 2000). Scientists continue to study the recovery of Mount St. Helens from its massive eruption in 1980, changing scientific thinking about succession in the process.

Moffett, W. "Ants and Plants: Friends and Foes." *National Geographic*, Vol. 195, No. 5 (May 1999). Examines some of the fascinating interactions among plants, ants, and other organisms.

Souder, W. "Study of Plants Makes a Case for Biodiversity." *Washington Post* (April 16, 2001). Experimental plots that have a greater richness of plant species outproduce plots in which species richness is low. The author evaluates the significance of this experiment.

Planet Earth, as viewed from space.
According to the Gaia hypothesis, Earth is capable of self-maintenance.

6

Ecosystems and the Physical Environment

Learning Objectives

After you have studied this chapter you should be able to:

1. Diagram the carbon, nitrogen, phosphorus, sulfur, and hydrologic cycles.

2. Describe how humans have influenced the carbon, nitrogen, phosphorus, sulfur, and hydrologic cycles.

3. Summarize the effects of solar energy on Earth's temperature, including the influence of albedos of various surfaces.

4. Discuss the roles of solar energy and the Coriolis effect in the production of global air and water flow patterns.

5. Define *El Niño–Southern Oscillation* (*ENSO*) and *La Niña* and describe some of their effects.

6. Distinguish between weather and climate and give three causes of regional precipitation differences.

7. Contrast tornadoes and tropical cyclones.

8. Define *plate tectonics* and explain its relationship to earthquakes and volcanic eruptions.

Almost completely isolated from everything in the universe but sunlight, Earth has often been compared to a vast spaceship with a life-support system consisting of the organisms that inhabit it. These living things modify the composition of gases in the atmosphere, transfer energy, and recycle waste products with great efficiency.

Broadly speaking, Earth is capable of self-maintenance of its complex biological-physical systems. Organisms interact with the abiotic environment to produce and maintain the chemical composition of the atmosphere, the global temperature, the ocean's salinity, and other characteristics. Thus, Earth's environment and organisms are intimately linked to one another and work together as a *homeostatic mechanism*. (Biological systems have homeostatic mechanisms such as feedback loops to help maintain a steady state or constant environment.)

The series of hypotheses that Earth's organisms adjust the environment to keep it habitable for life has been called, collectively, the **Gaia hypothesis.** (*Gaia* is derived from the Greek *Gaea*, who is Earth personified as a goddess.) James Lovelock, a British chemist, and Lynn Margulis, a U.S. biologist, first proposed the Gaia hypothesis in the early 1970s. Since then, the original hypothesis—that planet Earth behaves like a living organism—has been modified to emphasize that substantial interactions between organisms and the physical environment can lead to global self-regulation. The hypothesis requires no foresight or planning on the part of living organisms. The field of study based on the Gaia hypothesis is called **geophysiology.**

As an example of the Gaia hypothesis, consider Earth's temperature. It is generally accepted that the temperature has remained relatively constant at a level suitable for life over the 3.5 to 4 billion years that life has existed. Yet there is evidence that the sun has been heating up during that time. Why has Earth's temperature not increased? Gaia proponents hypothesize that Earth has remained the same temperature because the level of atmosphere-warming carbon dioxide (CO_2) has dropped during that time. This happened because the living Earth compensated for increased sunlight by "fixing" CO_2 into calcium carbonate shells of countless billions of marine phytoplankton (microscopic algae). As the phytoplankton died, their shells sank to the ocean floor, thus removing CO_2 from the system. This planetary temperature mechanism is an example of a **negative feedback loop** between the abiotic environment and organisms, which mutually interact to regulate and stabilize Earth's temperature. Negative feedback loops work to restore the normal values of a variable, thereby stabilizing a system. Many feedback loops operate between organisms and the physical environment.

Many scientists are reluctant to accept all aspects of the original Gaia hypothesis. Evolutionary biologists do not see any selective advantage for organisms whose activities contribute to global stability. Virtually everyone agrees that the environment modifies organisms and organisms modify the environment; these parts of the Gaia hypothesis are easily testable, and much evidence supports them. Moreover, the dependence of organisms on the physical environment is indisputable. However, the Gaia hypothesis has probably been most important because it has stimulated research that has helped us begin to understand Earth as a self-regulating system.

◼ THE CYCLING OF MATERIALS WITHIN ECOSYSTEMS

In Chapter 4 we learned that energy flows in one direction through an ecosystem. In contrast, matter, the material of which organisms are composed, moves in numerous cycles from one part of an ecosystem to another—that is, from one organism to another and from living organisms to the abiotic environment and back again (Figure 6.1). We call these cycles of matter **biogeochemical cycles** because they involve biological, geological, and chemical interactions.

Five different biogeochemical cycles of matter—carbon, nitrogen, phosphorus, sulfur, and water—are representative of all biogeochemical cycles. These five cycles are particularly important to organisms, as these materials are used to make the chemical compounds of cells. Carbon, nitrogen, and sulfur are elements that form gaseous compounds, whereas water is a compound that readily evaporates; thus, these four cycles have components that can move over large distances of the atmosphere with relative ease. Phosphorus, however, is an element that does not form gaseous compounds, and as a result, only local cycling occurs easily.

The Carbon Cycle

Proteins, carbohydrates, and other molecules essential to life contain carbon, so organisms must have carbon available to them. Carbon makes up approximately 0.037% of the atmosphere as a gas, CO_2. It is present in the ocean in several forms: dissolved carbon dioxide—that is, carbonate (CO_3^{2-}) and bicarbonate (HCO_3^-); other forms of dissolved inorganic carbon; and dissolved organic carbon from decay processes. Carbon is also present in rocks such as limestone. The global movement of carbon between the abiotic environment, including the atmosphere and ocean, and organisms is known as the **carbon cycle** (Figure 6.2).

During photosynthesis, plants, algae, and certain bacteria remove carbon dioxide from the air and *fix*, or incorporate, it into complex chemical compounds such as sugar (see Chapter 4). Plants use sugar to make other compounds. Thus, photosynthesis incorporates carbon from the abiotic environment into the biological compounds of producers. Those compounds are usually used as fuel for cellular respiration by the producer that made them, by a consumer that eats the producer, or by a decomposer that breaks down the remains of the producer or consumer. Thus, carbon dioxide is returned to the atmosphere by the process of cellular respiration. A similar carbon cycle occurs in aquatic ecosystems between aquatic organisms and carbon dioxide dissolved in the water.

Sometimes the carbon in biological molecules is not recycled back to the abiotic environment for some time. A large amount of carbon is stored in the wood of trees, where it may stay for several hundred years or even longer. In addition, millions of years ago, vast coal beds formed from the bodies of ancient trees that did not decay fully before they were buried. Similarly, the organic compounds of unicellular marine organisms probably gave rise to the underground deposits of oil and natural gas that accumulated in the geological past. Coal, oil, and natural gas, called **fossil fuels** because they formed from the remains of ancient organisms, are vast deposits of carbon compounds—the end products of photosynthesis that occurred millions of years ago (see Chapter 10).

The carbon in coal, oil, natural gas, and wood can be returned to the atmosphere by the process of burning, or **combustion.** In combustion, organic molecules are rap-

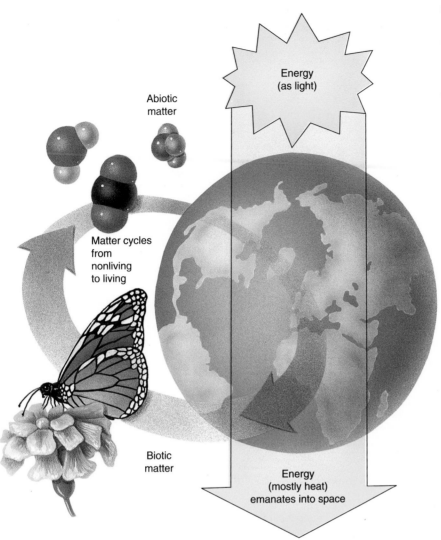

Figure 6.1 Energy and matter.
Although energy flows one way through ecosystems, matter continually cycles from the abiotic to the biotic components of ecosystems and back again.

idly oxidized—that is, combined with oxygen—and thus converted into carbon dioxide and water, with an accompanying release of heat and light.

An even greater amount of carbon that leaves the carbon cycle for millions of years is incorporated into the shells of marine organisms. When these organisms die, their shells sink to the ocean floor and are covered by sediments, forming seabed deposits several kilometers thick. The deposits are eventually cemented together to form a sedimentary rock called limestone. The crust is dynamically active, and over millions of years, sedimentary rock on the bottom of the sea floor may lift to form land surfaces. The summit of Mount Everest, for example, is composed of sedimentary rock. When the process of geological uplift exposes limestone, it slowly erodes away by chemical and physical weathering processes. This returns the carbon to the water and atmosphere, where it is available to participate in the carbon cycle once again.

The Carbon Cycle and Global Warming Human activities are increasingly disturbing the balance of the carbon

cycle. Before 1850, the advent of the Industrial Revolution, the global carbon cycle was in a steady state. Enormous amounts of carbon moved to and from the atmosphere, ocean, and terrestrial ecosystems, but these movements within the global carbon cycle just about cancelled out one another.

Since 1850, our industrial society has required a lot of energy, and humans have obtained this energy by burning increasing amounts of fossil fuels—coal, oil, and natural gas. This trend, along with a greater combustion of wood as a fuel and the burning of large sections of tropical forests, has released CO_2 into the atmosphere at a rate greater than the natural carbon cycle can handle.

Less than half of the CO_2 emitted by the combustion of fossil fuels and other human activities remains in the atmosphere. The rest is taken up, at least temporarily, by the ocean and by vegetation on land, both of which act as carbon "sinks." One of the great unanswered mysteries of climate science is quantifying exactly where the "missing" CO_2 that leaves the atmosphere goes. Scientists need to understand and quantify this aspect of the global

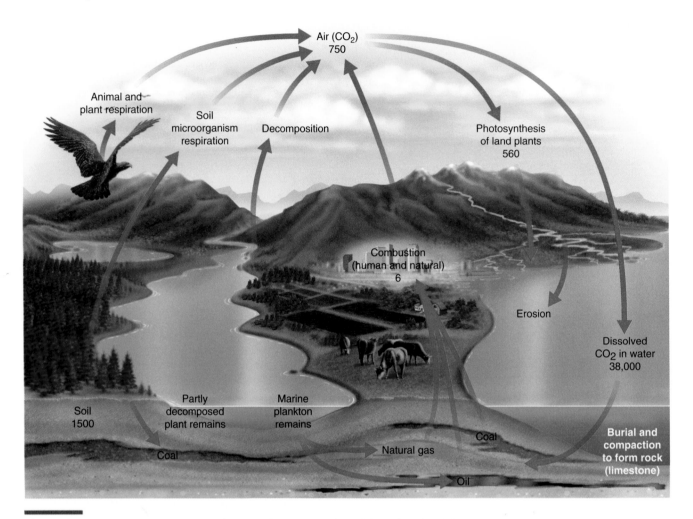

Figure 6.2 Simplified model of carbon cycle. Sedimentary rocks and fossil fuels hold almost all of Earth's estimated 10^{23} g of carbon. The values shown for some of the active pools in the global carbon budget are expressed as 10^{15} g of carbon. For example, the soil contains an estimated $1,500 \times 10^{15}$ g of carbon.

carbon cycle so they can make more accurate predictions of future trends in atmospheric CO_2 levels.

The level of atmospheric CO_2 increased particularly dramatically during the last half of the 20th century (see Figure 1.11), and this rise of CO_2 in the atmosphere may cause human-induced changes in climate called global warming. Global warming could result in a rise in sea level, changes in precipitation patterns, the death of forests, the extinction of organisms, and problems for agriculture. It could force the displacement of thousands or even millions of people, particularly from coastal areas. A more thorough discussion of the potential impacts of global warming is found in Chapter 20.

The Nitrogen Cycle

Nitrogen is crucial for all organisms because it is an essential part of biological molecules such as proteins and nucleic acids (for example, deoxyribonucleic acid [DNA]). At first glance it would appear that a shortage of

nitrogen for organisms is impossible. The atmosphere is 78% nitrogen gas (N_2), a two-atom molecule. But atmospheric nitrogen is so stable that it does not readily combine with other elements. Therefore, atmospheric nitrogen must first be broken apart before the nitrogen atoms can combine with other elements to form proteins and nucleic acids.

There are five steps in the **nitrogen cycle**, in which nitrogen cycles between the abiotic environment and organisms: nitrogen fixation, nitrification, assimilation, ammonification, and denitrification (Figure 6.3 and "Mini-Glossary: The Nitrogen Cycle"). Bacteria are exclusively involved in all of these steps except assimilation.

The first step in the nitrogen cycle, biological **nitrogen fixation**, is the conversion of gaseous nitrogen (N_2) to ammonia (NH_3). The process gets its name from the fact that nitrogen is *fixed* into a form that organisms can use. Combustion, volcanic action, lightning discharges, and industrial processes, all of which supply enough energy to break up atmospheric nitrogen, also fix considerable nitro-

MINI-GLOSSARY

The Nitrogen Cycle

nitrogen fixation: The conversion of atmospheric nitrogen to ammonia, performed by nitrogen-fixing bacteria, including cyanobacteria.

nitrification: The conversion of ammonia or ammonium to nitrate, performed by nitrifying bacteria.

assimilation: The conversion of inorganic nitrogen (nitrate, ammonia, or ammonium) to the organic molecules of organisms.

ammonification: The conversion of organic nitrogen (biological molecules containing nitrogen) to ammonia and ammonium ions, performed by ammonifying bacteria.

denitrification: The conversion of nitrate to nitrogen gas, performed by denitrifying bacteria.

gen. Nitrogen-fixing bacteria, including cyanobacteria, carry out nitrogen fixation in soil and aquatic environments. Nitrogen-fixing bacteria employ an enzyme called **nitrogenase** to split atmospheric nitrogen and combine the resulting nitrogen atoms with hydrogen. Because nitrogenase functions only in the absence of oxygen, the bacteria that use nitrogenase must insulate the enzyme from oxygen by some means. Some nitrogen-fixing bacteria live beneath layers of oxygen-excluding slime on the roots of certain plants. Other important nitrogen-fixing bacteria, *Rhizobium*, live inside special swellings, or nodules, on the roots of legumes such as beans or peas and some woody plants. The relationship between *Rhizobium* and its host plants is mutualistic: The bacteria receive carbohydrates from the plant, and the plant receives nitrogen in a form that it can use. In aquatic environments most of the nitrogen fixation is done by cyanobacteria. Filamentous cyanobacteria have special oxygen-excluding cells that function as the sites of nitrogen fixation.

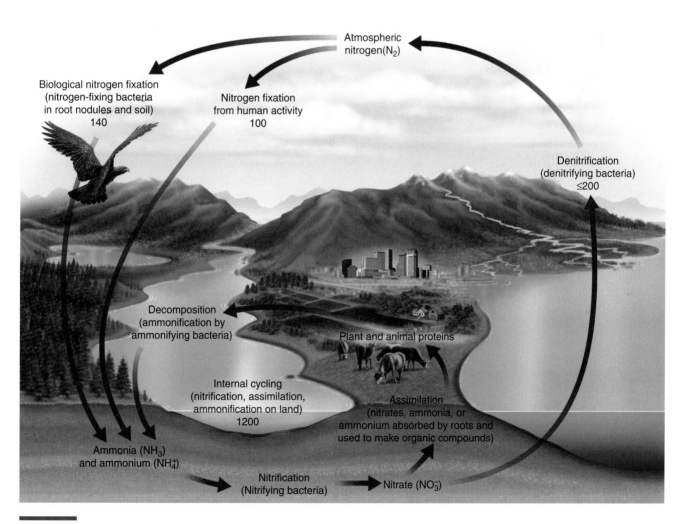

Figure 6.3 Simplified model of nitrogen cycle. The atmosphere holds the largest pool of nitrogen, 3.9×10^{21} g. The values shown for some of the active pools in the global nitrogen budget are expressed as 10^{12} g of nitrogen per year. For example, each year humans fix an estimated 100×10^{12} g of nitrogen.

The conversion of ammonia (NH_3) or ammonium (NH_4^+, formed when water reacts with ammonia) to nitrate (NO_3^-) is called **nitrification**. Soil bacteria perform nitrification, which is a two-step process. First soil bacteria convert ammonia or ammonium to nitrite (NO_2^-). Then another soil bacterium oxidizes nitrite to nitrate. The process of nitrification furnishes these bacteria, collectively called nitrifying bacteria, with energy.

Plant roots absorb nitrate (NO_3^-), ammonia (NH_3), or ammonium (NH_4^+), and **assimilate** the nitrogen of these molecules into plant proteins and nucleic acids. When animals consume plant tissues, they also assimilate nitrogen by taking in plant nitrogen compounds (amino acids) and converting them to animal compounds (proteins).

The conversion of biological nitrogen compounds into ammonia (NH_3) and ammonium ions (NH_4^+) is known as **ammonification**. Ammonification begins when organisms produce nitrogen-containing waste products such as urea (in urine) and uric acid (in the wastes of birds). These substances, plus the nitrogen compounds that occur in dead organisms, are decomposed, releasing the nitrogen into the abiotic environment as ammonia. The bacteria that perform this process both in the soil and in aquatic environments are called ammonifying bacteria. The ammonia produced by ammonification enters the nitrogen cycle and is once again available for the processes of nitrification and assimilation.

The reduction of nitrate (NO_3^-) to gaseous nitrogen (N_2) is called **denitrification**. Denitrifying bacteria reverse the action of nitrogen-fixing and nitrifying bacteria by returning nitrogen to the atmosphere as nitrogen gas. Denitrifying bacteria prefer to live and grow where there is little or no free oxygen. For example, they are found deep in the soil near the water table, an environment that is nearly oxygen-free.

Human-Induced Changes to the Global Nitrogen Cycle

Human activities have disturbed the balance of the global nitrogen cycle. During the 20th century, humans more than doubled the amount of fixed nitrogen entering the global nitrogen cycle. (Fixed nitrogen refers to nitrogen that has been chemically combined with hydrogen, oxygen, or carbon.) The excess nitrogen is seriously altering many terrestrial and aquatic ecosystems.

Large quantities of nitrogen fertilizer, both ammonia and nitrate, are produced from nitrogen gas for agriculture. According to the U.N. Food and Agriculture Organization, 141 million tons of nitrogen fertilizer were used worldwide in 2000. The increasing use of fertilizer has resulted in higher crop yields (see Chapter 18).

In 1997, Peter Vitousek and seven other ecologists published an extensive review of scientific research on the negative environmental impacts of human-produced nitrogen. One serious problem with nitrogen is that it is extremely mobile and is easily transferred from the land to rivers to the ocean. Thus, the overuse of commercial fertilizer on the land can cause water quality problems that may help explain long-term declines in many coastal fisheries. The amount of nitrate or ammonium in most aquatic ecosystems is usually in limited supply and therefore limits the growth of algae. Rain washes fertilizer into rivers and lakes, where it stimulates the growth of algae, some of which are toxic. As these algae die, their decomposition by bacteria robs the water of dissolved oxygen, which in turn causes other aquatic organisms, including many fishes, to die of suffocation.

Nitrates from fertilizer can also leach (dissolve and wash down) through the soil and contaminate **groundwater** (fresh water stored in underground caverns and porous layers of rock). Many people drink groundwater, and groundwater contaminated by nitrates is dangerous to drink, particularly for infants and small children. The effects of nitrate contamination of groundwater are discussed in Chapter 21.

Another human activity that affects the nitrogen cycle is the combustion of fossil fuels. When fossil fuels are burned, the nitrogen locked in organic compounds in the fuel is chemically altered and transferred to the atmosphere. In addition, the high temperatures of combustion convert some atmospheric nitrogen (N_2) to **nitrogen oxides,** trace gases in the atmosphere that are produced by chemical interactions between nitrogen and oxygen (see Chapter 19). Automobile exhaust is one of the main sources of nitrogen oxides.

Nitrogen oxides exacerbate several serious environmental problems. Nitrogen oxides are a necessary ingredient in the production of **photochemical smog**, a mixture of several air pollutants that can injure plant tissues, irritate eyes, and cause respiratory problems in humans. Nitrogen oxides also react with water in the atmosphere to form nitric acid (HNO_3) and nitrous acid (HNO_2). When these and other acids leave the atmosphere as **acid deposition**, they cause the pH of surface waters (lakes and streams) and soils to decrease. Acid deposition has been linked to declining animal populations in aquatic ecosystems. On land, acid deposition alters soil chemistry so that certain essential nutrient minerals such as calcium and potassium wash out of the soil and are therefore unavailable for plants. Nitrous oxide (N_2O), one of the nitrogen oxides, retains heat in the atmosphere (like CO_2) and so promotes global warming. Nitrous oxide also contributes to the depletion of ozone in the stratosphere. (See Chapter 19 for a discussion of photochemical smog; Appendix I for a discussion of acids and pH; and Chapter 20 for a discussion of acid deposition, global warming, and stratospheric ozone depletion.)

The Phosphorus Cycle

In the **phosphorus cycle**, phosphorus, which does not form compounds in the gaseous phase and therefore does

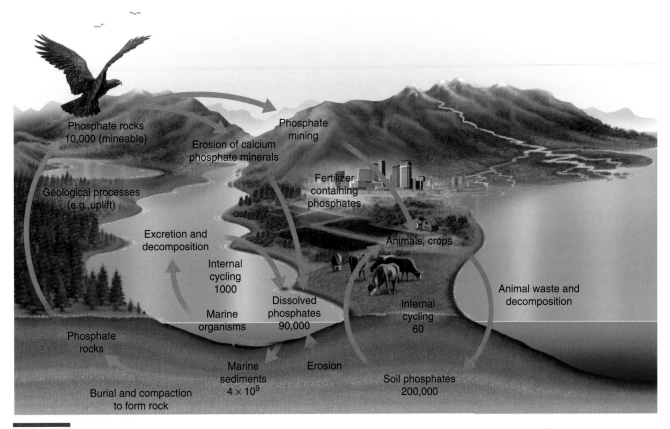

Figure 6.4 Simplified model of phosphorus cycle. The values shown in the figure for the global phosphorus budget are expressed as 10^{12} g of phosphorus per year. For example, each year an estimated 60×10^{12} g of phosphorus cycles from the soil to terrestrial organisms and back again.

not enter the atmosphere, cycles from the land to sediments in the ocean and back to the land (Figure 6.4). As water runs over rocks containing phosphorus, it gradually wears away the surface and carries off inorganic phosphate (PO_4^{3-}) molecules.

The erosion of phosphorus-containing rocks releases phosphorus into the soil, where it is taken up by plant roots in the form of inorganic phosphates. Once in cells, phosphates are used in a variety of biological molecules, including nucleic acids and ATP (adenosine triphosphate, an organic compound important in energy transfer reactions in cells). Animals obtain most of their required phosphate from the food they eat, although in some localities drinking water may contain a substantial amount of inorganic phosphate. Phosphorus released by decomposers becomes part of the soil's pool of inorganic phosphate that can be reused by plants. Thus, like carbon and nitrogen, phosphorus moves through the food web as one organism consumes another.

Phosphorus cycles through aquatic communities in much the same way it does through terrestrial communi-

ties. Dissolved phosphorus enters aquatic communities through absorption and assimilation by algae and plants, which are then consumed by plankton and larger organisms. A variety of fishes and mollusks eat these in turn. Ultimately, decomposers that break down wastes and dead organisms release inorganic phosphorus into the water, where it is available to be used by aquatic producers again.

Phosphate can be lost from biological cycles. Some phosphate is carried from the land by streams and rivers to the ocean, where it can be deposited on the sea floor and remain for millions of years. The geological process of uplift may someday expose these sea floor sediments as new land surfaces, from which phosphate will once again be eroded.

A very small portion of the phosphate in the aquatic food web finds its way back to the land. A few fishes and aquatic invertebrates are eaten by sea birds, which may defecate on land where they roost. Guano, the manure of sea birds, contains large amounts of phosphate and nitrate. Once on land, the roots of plants may absorb these minerals. The phosphate contained in guano may enter terrestrial food webs in this way, although the amounts involved are quite small.

Humans and the Phosphorus Cycle   Humans affect the natural cycling of phosphorus by accelerating its long-term loss from the land. Corn grown in Iowa, which contains phosphate absorbed from the soil, may be used to fatten cattle in an Illinois feedlot. Part of the phosphate absorbed by the roots of the corn plants thus ends up in the feedlot wastes, which may eventually wash into the Mississippi River. Beef from the Illinois cattle may be consumed by people living far away—in New York City, for instance. Hence, more of the phosphate ends up in human wastes and is flushed down toilets into the New York City sewer system. Sewage treatment rarely removes phosphates, and so they cause water quality problems in rivers and lakes (see Chapters 2 and 21). To compensate for the steady loss of phosphate from their land, farmers must add phosphate fertilizer to their fields. That fertilizer is produced in Florida, Tennessee, and several other states from the large deposits of phosphate rock that are mined there.

In natural ecosystems, very little phosphorus is lost from the cycle, but few ecosystems today are in a natural state. Phosphorus loss from the soil is accelerated by land-denuding practices, such as the clearcutting of forests, and by erosion of agricultural and residential lands. For practical purposes, phosphorus that washes from the land into the ocean is permanently lost from the terrestrial phosphorus cycle (and from further human use), for it remains in the ocean for millions of years.

The Sulfur Cycle

Our understanding of how the global **sulfur cycle** works is filled with uncertainty. Most of the planet's sulfur is found underground in sedimentary rocks and minerals, which over time erode to release sulfur-containing compounds into the ocean (Figure 6.5). Sulfur gases enter the atmosphere from natural sources in both the ocean and land. Seaspray delivers sulfates (SO_4^{2-}) into the air, as do forest fires and dust storms (desert soils are rich in calcium sulfate, $CaSO_4$). Volcanoes release both hydrogen sulfide (H_2S), a poisonous gas with an unforgettable smell of rotten eggs, and sulfur oxides (SO_x). Sulfur oxides include sulfur dioxide (SO_2), a choking, acrid gas, and sulfur trioxide (SO_3).

Sulfur gases comprise a minor part of the atmosphere and are not long-lived because atmospheric sulfur compounds are reactive. Hydrogen sulfide reacts with oxygen to form sulfur oxides, and sulfur oxides (SO_x) react with water to form sulfuric acid (H_2SO_4). Although the total amount of sulfur compounds present in the atmosphere at any given time is relatively small, the total annual movement of sulfur to and from the atmosphere is substantial.

A tiny fraction of global sulfur is present in living organisms, where it is an essential component of proteins. Plant roots absorb sulfate (SO_4^{2-}) and assimilate it by incorporating the sulfur into plant proteins. When animals consume plant tissues, they also assimilate sulfur by taking in plant proteins and converting them to animal proteins. In the ocean, certain marine algae release large amounts of dimethyl sulfide, or DMS (CH_3SCH_3) into the atmosphere, where it helps condense water into droplets in clouds. In the atmosphere dimethyl sulfide is converted to sulfate, and most of the sulfur in DMS is deposited into the ocean as sulfate.

As in the nitrogen cycle, bacteria drive the sulfur cycle. In freshwater wetlands, tidal flats, and flooded soils, which are oxygen-deficient, certain bacteria convert sulfates to hydrogen sulfide gas, which is released into the atmosphere, or to metallic sulfides, which are deposited as rock. In the absence of oxygen, other bacteria perform an ancient type of photosynthesis that uses hydrogen sulfide instead of water. Where oxygen is present, different bacteria oxidize sulfur compounds to sulfates.

Humans and the Sulfur Cycle   Coal and, to a lesser extent, oil contain sulfur. When these fuels are burned in power plants, factories, and motor vehicles, sulfur dioxide (SO_2), a major cause of acid deposition, is released into the atmosphere. Sulfur dioxide is also released during the smelting of sulfur-containing ores of such metals as copper, lead, and zinc. Although sulfur oxides are also produced by natural sources, human inputs of sulfur oxides are substantial. Ice core samples taken in Greenland record a large increase in sulfur deposition from the atmosphere to the land since the beginning of the Industrial Revolution. Pollution abatement, such as scrubbing smokestack gases to remove sulfur oxides, has reduced the amount of sulfur emissions in highly developed countries in recent years, but the global level continues to increase.

Human activities also affect the sulfur cycle when sulfur is mined and used to produce sulfuric acid, which has many industrial uses, including the manufacture of fertilizers, processing of metals, refining of petroleum, and the manufacture of other chemicals. Increasingly, humans are making use of bacterial reactions with sulfur compounds to remove metals from low-grade ores.

The Hydrologic Cycle

Life on Earth would be impossible without water, which makes up a substantial part of the mass of most organisms. All life forms, from bacteria to plants and animals, use water as a medium for chemical reactions as well as for the transport of materials within and among cells.

Water continuously circulates from the ocean to the atmosphere to the land and back to the ocean. It provides a renewable supply of purified water for terrestrial organisms. This complex cycle, known as the **hydro-**

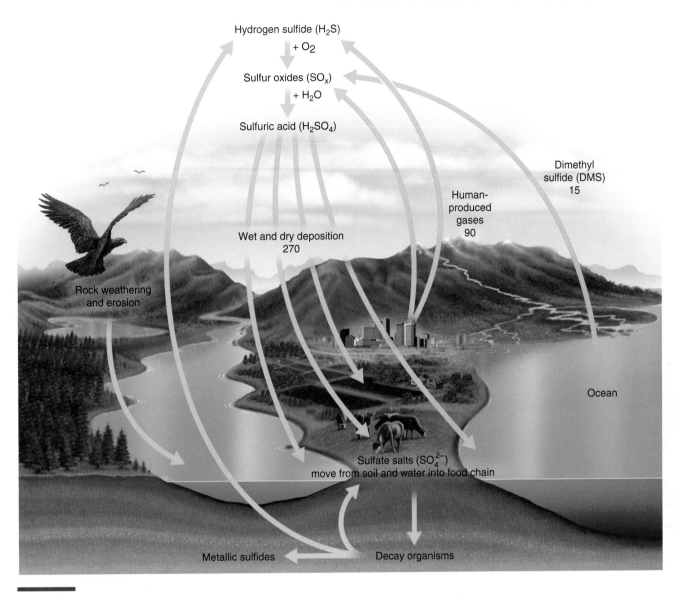

Figure 6.5 Simplified model of sulfur cycle. The largest pool of sulfur on the planet is sedimentary rocks, which contain an estimated $7,440 \times 10^{18}$ g of sulfur. The second largest pool is the ocean, which contains $1,280 \times 10^{18}$ g of sulfur. The values shown in the figure for the global sulfur budget are expressed in units of 10^{12} g of sulfur per year. For example, the ocean emits an estimated 15×10^{12} g of sulfur per year as the gas dimethyl sulfide.

logic cycle, results in a balance between water in the ocean, on the land, and in the atmosphere (Figure 6.6). Water moves from the atmosphere to the land and ocean in the form of precipitation—rain, snow, sleet, or hail. When water evaporates from the ocean surface and from soil, streams, rivers, and lakes on land, it forms clouds in the atmosphere. In addition, **transpiration**, the loss of water vapor from land plants, adds water to the atmosphere. Roughly 97% of the water absorbed from the soil by a plant is transported to the leaves, where it is lost by transpiration.

Water may evaporate from land and reenter the atmosphere directly. Alternatively, it may flow in rivers and streams to coastal **estuaries** where fresh water meets the ocean. The movement of water from land to rivers, lakes, wetlands and, ultimately, the ocean is called **runoff**, and the area of land being drained by runoff is called a **watershed**. Water also percolates, or seeps, downward through the soil and rock to become groundwater. Groundwater may reside in the ground for hundreds to many thousands of years, but eventually it supplies water to the soil, to vegetation, to streams and rivers, and to the ocean.

Regardless of its physical form—solid, liquid, or vapor—or location, every molecule of water eventually moves through the hydrologic cycle. Tremendous quan-

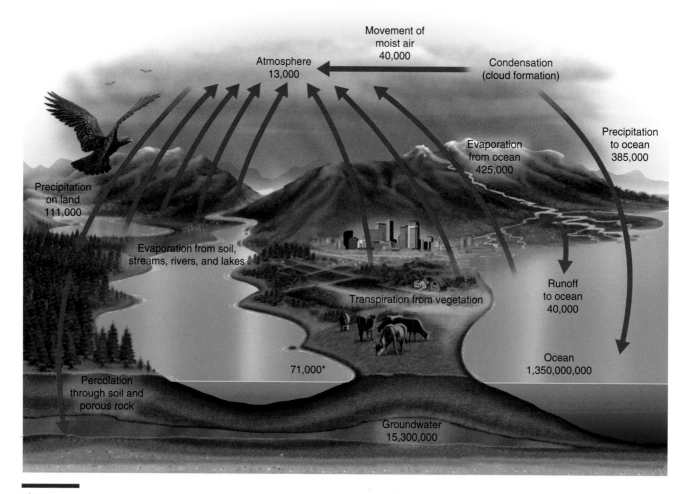

Figure 6.6 Simplified model of hydrologic cycle. Estimated values for pools in the global water budget are expressed as cubic kilometers, and the values for movements (associated with arrows) are in cubic kilometers per year. The starred value (71,000 km³ per year) includes both transpiration from plants and evaporation from soil, streams, rivers, and lakes.

tities of water are cycled annually between Earth and its atmosphere. The volume of water entering the atmosphere each year is estimated at about 389,500 cubic km (95,000 cubic mi). Approximately three fourths of this water reenters the ocean directly as precipitation over water; the remainder falls on land.

Humans and the Hydrologic Cycle A major concern of human society has always been obtaining and guaranteeing an adequate water supply. Humans alter the hydrologic cycle by building engineering structures such as levees and dams to obtain water for domestic and industrial use, to irrigate crops, and to produce hydroelectricity. Reducing flood damage and dealing with droughts continues to occupy our attention. (We discuss water quantity issues in Chapter 13 and water pollution concerns in Chapter 21.)

A 2001 paper published in the journal *Science* by researchers at Scripps Institution of Oceanography sug-

gests a further human-induced effect on the hydrologic cycle: It appears that air pollution may be weakening the global hydrologic cycle. The concern is **aerosols**, tiny particles of air pollution consisting mostly of sulfates, nitrates, carbon, mineral dusts, and fly ash. These aerosols are produced largely from fossil fuel combustion and the burning of forests. Once in the atmosphere, aerosols enhance the scattering and absorption of sunlight in the atmosphere and cause brighter clouds to form. Both the clouds and the light-scattering effect in the atmosphere cause a warming of the atmosphere and a threefold reduction in the amount of solar radiation reaching Earth's surface, including the ocean. Also, clouds formed in aerosols are less likely to release their precipitation. As a result, scientists think aerosols may affect the availability and quality of water in some regions during the 21st century.

We have seen how living things depend on the abiotic environment to supply energy (see Chapter 4) and essential materials (in biogeochemical cycles). We now consider five aspects of the physical environment that also affect organisms: solar radiation, the atmosphere, the ocean, weather and climate, and internal planetary processes.

SOLAR RADIATION

The sun makes life on Earth possible. It warms the planet, including the atmosphere, to habitable temperatures. Without the sun's energy, the temperature would approach absolute zero (–273°C), and all water would be frozen, even in the ocean. The sun powers the hydrologic cycle, carbon cycle, and other biogeochemical cycles and is the primary determinant of climate. The sun's energy is captured by photosynthetic organisms, which use it to make the food molecules required by almost all forms of life. Most of our fuels—wood, oil, coal, and natural gas—represent solar energy captured by photosynthetic organisms. Without the sun, almost all life would cease.

The sun's energy is the product of a massive nuclear fusion reaction (see Chapter 11) and is emitted into space in the form of electromagnetic radiation—especially visible light and infrared and ultraviolet radiation, which are not visible to the human eye. Approximately one billionth of the total energy released by the sun strikes our

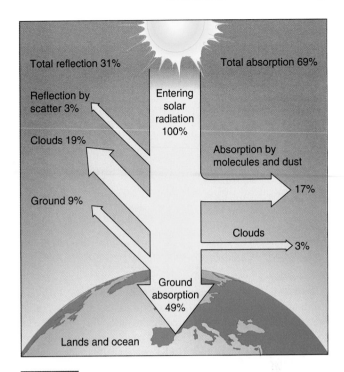

Figure 6.7 Fate of solar radiation that reaches Earth. Most of the energy produced by the sun never reaches Earth. The solar energy that does reach Earth warms the planet's surface, drives the hydrologic cycle and other biogeochemical cycles, produces our climate, and powers almost all life through the process of photosynthesis.

atmosphere, and of this tiny trickle of energy, a minute part operates the biosphere.

About 31% of the solar radiation that falls on Earth is immediately reflected away by clouds and, to a lesser extent, by surfaces, especially snow, ice, and the ocean (Figure 6.7). **Albedo** is the proportional reflectance of Earth's surface. Glaciers and ice sheets have high albedos and reflect 80% to 90% of the sunlight hitting their surfaces. At the other extreme, asphalt pavement and buildings have low albedos and reflect 10% to 15%, whereas the ocean and forests reflect only about 5%.

As shown in Figure 6.7, the remaining 71% of the solar radiation that falls on Earth is absorbed and runs the hydrologic cycle, drives winds and ocean currents, powers photosynthesis, and warms the planet. Ultimately, however, all of this energy is lost by the continual radiation of long-wave infrared (heat) energy into space.

Temperature Changes with Latitude

The most significant local variation in Earth's temperature is produced because the sun's energy does not reach all places uniformly. A combination of Earth's roughly spherical shape and the tilt of its axis produces a great deal of variation in the exposure of the surface to the energy delivered by sunlight.

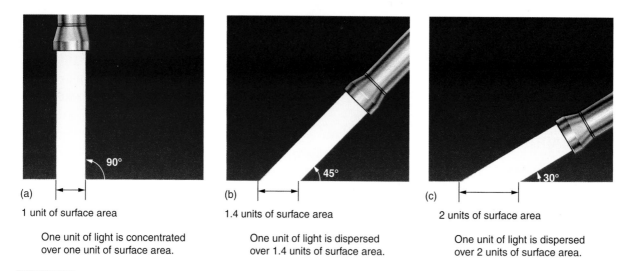

(a) 1 unit of surface area

One unit of light is concentrated over one unit of surface area.

(b) 1.4 units of surface area

One unit of light is dispersed over 1.4 units of surface area.

(c) 2 units of surface area

One unit of light is dispersed over 2 units of surface area.

Figure 6.8 Solar intensity and latitude. The angle at which the sun's rays strike Earth varies from one geographical location to another due to Earth's spherical shape and its inclination on its axis. (a) Sunlight (represented by the flashlight) that shines vertically near the equator is concentrated on Earth's surface. (b, c) As one moves toward the poles, the light hits the surface more and more obliquely, spreading the same amount of radiation over larger and larger areas.

The principal effect of the tilt is on the angles at which the sun's rays strike different areas of the planet at any one time (Figure 6.8). On the average, the sun's rays hit vertically near the equator, making the energy more concentrated and producing higher temperatures. Near the poles the sun's rays hit more obliquely, and, as a result, their energy is spread over a larger surface area. Also, rays of light entering the atmosphere obliquely near the poles must pass through a deeper envelope of air than those entering near the equator. This causes more of the sun's energy to be scattered and reflected back to space, which in turn further lowers temperatures near the poles. Thus, because the solar energy that reaches polar regions is less concentrated, temperatures are lower.

Temperature Changes with Season

Seasons are primarily determined by Earth's inclination on its axis. Earth's inclination on its axis is 23.5 degrees from a line drawn perpendicular to the orbital plane. During half of the year (March 21 to September 22) the Northern Hemisphere tilts *toward* the sun, and during the other half (September 22 to March 21) it tilts *away* from the sun (Figure 6.9). The orientation of the Southern Hemisphere is just the opposite at these times. Summer in the Northern Hemisphere corresponds to winter in the Southern Hemisphere.

THE ATMOSPHERE

The atmosphere is an invisible layer of gases that envelops Earth. Oxygen (21%) and nitrogen (78%) are the predominant gases in the atmosphere, accounting for about 99% of dry air. Other gases, including argon, carbon dioxide, neon, and helium, make up the remaining 1%. In addition, water vapor and trace amounts of various air pollutants, such as methane, ozone, dust particles, microorganisms, and chlorofluorocarbons (CFCs), are present in the air. The atmosphere becomes less dense as it extends outward into space; as a result of gravity, most of its mass is found near Earth's surface.

The atmosphere performs several ecologically important functions. It protects Earth's surface from most of the sun's ultraviolet radiation and x-rays and from lethal amounts of cosmic rays from space. Without this shielding by the atmosphere, life as we know it would cease to exist. While the atmosphere protects Earth from high-energy radiation, it allows visible light and some infrared radiation to penetrate, and they warm the surface and the lower atmosphere. This interaction between the atmosphere and solar energy is responsible for weather and climate.

Organisms depend on the atmosphere, but they also maintain and, in certain instances, modify its composition. Atmospheric oxygen is thought to have increased to its present level as a result of millions of years of photosynthesis. A balance between oxygen-producing photosynthesis and oxygen-using respiration maintains the current level of oxygen.

Layers of the Atmosphere

The atmosphere is composed of a series of five concentric layers—the troposphere, stratosphere, mesosphere, thermosphere, and exosphere (Figure 6.10). These layers

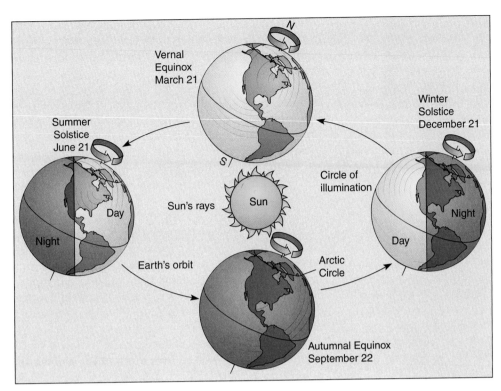

Figure 6.9 Progression of seasons. Earth's inclination on its axis remains the same as it travels around the sun. Thus, the sun's rays hit the Northern Hemisphere obliquely during its winter months and more directly during its summer. In the Southern Hemisphere, the sun's rays are oblique during its winter, which corresponds to the Northern Hemisphere's summer. At the equator, the sun's rays are approximately vertical on March 21 and September 22.

vary in altitude and temperature with latitude and season. The **troposphere**, the layer of atmosphere closest to Earth's surface, extends up to a height of approximately 10 km (6.2 mi). The temperature of the troposphere decreases with increasing altitude by about –6°C for every kilometer. Weather, including turbulent wind, storms, and most clouds, occurs in the troposphere.

In the next layer of atmosphere, the **stratosphere**, there is a steady wind but no turbulence. There is little water, and the temperature is more or less uniform (–45°C to –75°C) in the lower stratosphere; commercial jets fly here. The stratosphere extends from 10 to 45 km (6.2 to 28 mi) above Earth's surface and contains a layer of ozone that is critical to life because it absorbs much of the sun's damaging ultraviolet radiation (see Chapter 20). The absorption of ultraviolet radiation by the ozone layer heats the air, and so temperature increases with increasing altitude in the stratosphere.

The **mesosphere**, the layer of atmosphere directly above the stratosphere, extends from 45 to 80 km (28 to 50 mi) above Earth's surface. Temperatures drop steadily in the mesosphere to the lowest in the atmosphere—as low as –138°C.

The **thermosphere**, which extends from 80 to 500 km (50 to 310 mi), is characterized by steadily rising temperatures. Gases in the extremely thin air of the thermosphere absorb x-rays and short-wave ultraviolet radiation. This absorption drives the few molecules present in the thermosphere to great speeds, raising their temperature in the process to 1,000°C or more. The aurora, a colorful display of lights in dark polar skies, is produced when

charged particles from the sun hit oxygen or nitrogen molecules in the thermosphere. The thermosphere is important in long-distance communication because it reflects outgoing radio waves back toward Earth without the aid of satellites.

The outermost layer of the atmosphere, the **exosphere**, begins about 500 km (310 mi) above Earth's surface. The exosphere continues to thin until it converges with interplanetary space.

Atmospheric Circulation

In large measure, differences in temperature caused by variations in the amount of solar energy reaching different locations on Earth drive the circulation of the atmosphere. The very warm surface near the equator heats the air that is in contact with it, causing this air to expand and rise (Figure 6.11). As the warm air rises, it cools, and then it sinks again. Much of it recirculates almost immediately to the same areas it has left, but the remainder of the heated air splits and flows in two directions, toward the poles. The air chills enough to sink to the surface at about 30 degrees north and south latitudes. This descending air splits and flows over the surface in two directions. Similar upward movements of warm air and its subsequent flow toward the poles occur at higher latitudes, farther from the equator. At the poles the cold polar air sinks and flows toward the lower latitudes, generally beneath the sheets of warm air that simultaneously flow toward the poles. The constant motion of air transfers heat from the equator toward the

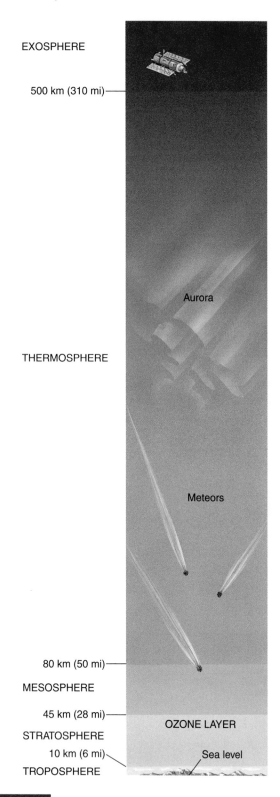

EXOSPHERE

500 km (310 mi)

THERMOSPHERE

Aurora

Meteors

80 km (50 mi)

MESOSPHERE

45 km (28 mi)

OZONE LAYER

STRATOSPHERE

10 km (6 mi)

Sea level

TROPOSPHERE

Figure 6.10 Layers of atmosphere. The troposphere is the layer closest to Earth's surface. The layer above the troposphere is the stratosphere, followed by the mesosphere and thermosphere. The outermost layer of the atmosphere, the exosphere, has no distinct boundary separating it from interplanetary space.

poles, and as the air returns, it cools the land over which it passes. This continuous turnover does not equalize temperatures over the surface of Earth, but it does moderate them.

Surface Winds In addition to global circulation patterns, the atmosphere exhibits complex horizontal movements that are commonly referred to as **winds**. The nature of wind, with its turbulent gusts, eddies, and lulls, is difficult to understand or predict. It results in part from differences in atmospheric pressure and from the rotation of Earth.

The gases that constitute the atmosphere have weight and exert a pressure that is, at sea level, about 1013 millibars (14.7 lb per in.^{2}). Air pressure is variable, however, changing with altitude, temperature, and humidity. Winds tend to blow from areas of high atmospheric pressure to areas of low pressure, and the greater the difference between the high- and low-pressure areas, the stronger the wind.

Earth's rotation also influences the direction of wind. Earth rotates from west to east, which causes moving air to be deflected from its path and swerve to the right of the direction in which it is traveling in the Northern

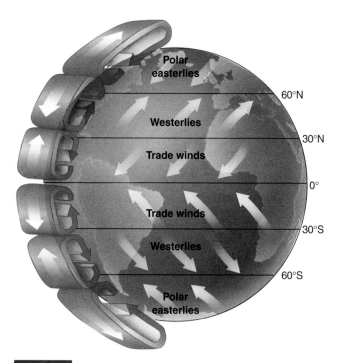

Polar easterlies

60°N

Westerlies

30°N

Trade winds

0°

Trade winds

30°S

Westerlies

60°S

Polar easterlies

Figure 6.11 Atmospheric circulation and heat exchange. Atmospheric circulation transports heat from the equator to the poles (*left side of figure*). The greatest solar energy input occurs at the equator, heating air most strongly in that area. The air rises and travels to the poles but is cooled in the process so that much of it descends again at around 30 degrees latitude in both hemispheres. At higher latitudes, the patterns of movement are more complex.

Hemisphere and to the left of the direction in which it is traveling in the Southern Hemisphere. This tendency is known as the **Coriolis effect**.

The Coriolis effect can be visualized by imagining that you and a friend are sitting about 10 ft apart on a merry-go-round that is turning clockwise (Figure 6.12). Suppose you throw a ball directly to your friend. By the time the ball reaches the place where your friend was, he or she is no longer in that spot. The ball will have

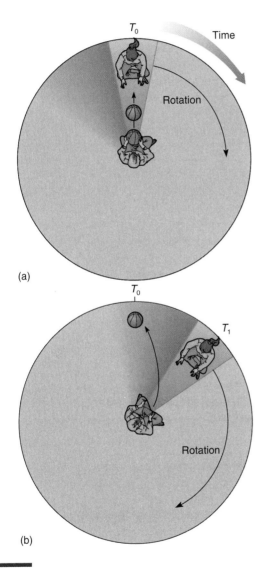

(a)

(b)

Figure 6.12 Coriolis effect. A merry-go-round (seen from above) can be used to demonstrate the Coriolis effect. The center of the merry-go-round corresponds to the South Pole, and the outer edge to the equator. (a) If, while sitting on the merry-go-round, you throw a ball to a friend (also on the merry-go-round) at time zero T_0, when the merry-go-round is rotating clockwise, the ball at time T_1 (b) appears to curve to the left instead of going straight. Because of Earth's rotation, winds curve to the left in the Southern Hemisphere and to the right in the Northern Hemisphere.

swerved far to the left of your friend. This is how the Coriolis effect works in the Southern Hemisphere.

To visualize how the Coriolis effect works in the Northern Hemisphere, imagine you and your friend are sitting on the same merry-go-round, only this time it is moving counterclockwise. Now when you throw the ball, it will swerve far to the right of your friend.

The atmosphere has three **prevailing winds**—major surface winds that blow more or less continually (see Figure 6.11). Prevailing winds that blow from the northeast near the North Pole or from the southeast near the South Pole are called **polar easterlies**. Winds that blow in the midlatitudes from the southwest in the Northern Hemisphere or from the northwest in the Southern Hemisphere are called **westerlies**. Tropical winds that blow from the northeast in the Northern Hemisphere or from the southeast in the Southern Hemisphere are called **trade winds**.

THE GLOBAL OCEAN

The global ocean is a huge body of salt water that surrounds the continents and covers almost three fourths of Earth's surface. It is a single, continuous body of water, but geographers divide it into four sections that are separated by the continents: the Pacific, Atlantic, Indian, and Arctic Oceans. The Pacific Ocean is the largest by far: It covers one third of Earth's surface and contains more than half of Earth's water.

Patterns of Circulation in the Ocean

The persistent prevailing winds blowing over the ocean produce mass movements of surface ocean water

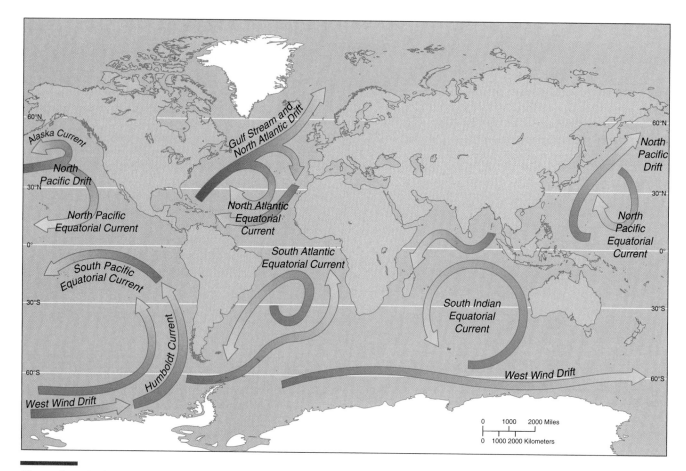

Figure 6.13 Surface ocean currents. The basic pattern of ocean currents is caused largely by the action of winds. The main ocean current flow—clockwise in the Northern Hemisphere and counterclockwise in the Southern Hemisphere—results partly from the Coriolis effect.

known as **currents** (Figure 6.13). The prevailing winds generate *circular* ocean currents called **gyres**. In the North Atlantic, the tropical trade winds tend to blow toward the west, whereas the westerlies in the midlatitudes blow toward the east. This helps establish a clockwise gyre in the North Atlantic. Thus, surface ocean currents and winds tend to move in the same direction, although there are many variations on this general rule.

The Coriolis effect influences the paths traveled by surface ocean currents. Earth's rotation from west to east causes surface ocean currents to swerve to the right in the Northern Hemisphere, helping establish a circular, clockwise pattern of water currents. In the Southern Hemisphere, ocean currents swerve to the left, thereby moving in a circular, counterclockwise pattern.

The position of land masses also affects ocean circulation. As you can see in Figure 6.14, the ocean is not distributed uniformly over the globe: There is clearly more water in the Southern Hemisphere than in the Northern

Hemisphere. Therefore, the circumpolar (around the pole) flow of water in the Southern Hemisphere is almost unimpeded by land masses.

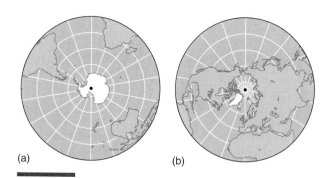

(a) (b)

Figure 6.14 Ocean and land masses in the Northern and Southern Hemispheres. (a) The Southern Hemisphere as viewed from the South Pole. (b) The Northern Hemisphere as viewed from the North Pole. Ocean currents are freer to flow in a circumpolar manner in the Southern Hemisphere.

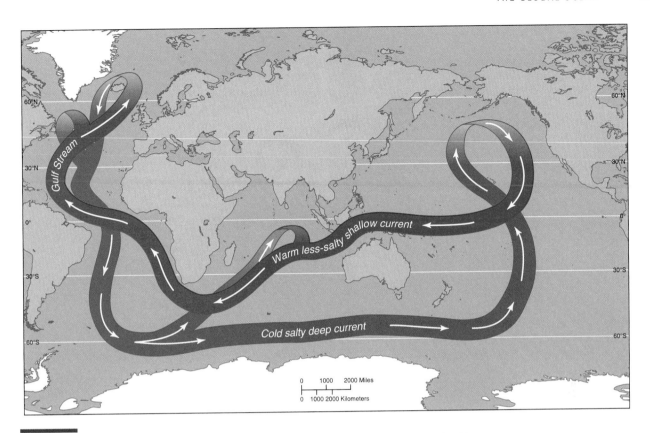

Figure 6.15 Ocean conveyor belt. This loop consists of both warm, shallow water and cold, deep water. The ocean conveyor belt is responsible for warming Europe and may also affect global climate.

Vertical Mixing of Ocean Water

The varying **density** (mass per unit volume) of seawater affects deep ocean currents. Water that is colder or saltier is denser than warmer, less salty water. (The density of water increases with decreasing temperature down to 4°C.) Thus, colder, salty ocean water sinks and flows under warmer, less salty water, generating currents far below the surface. Deep ocean currents often travel in different directions and at different speeds than do surface currents, in part because the Coriolis effect is more pronounced at greater depths. Figure 6.15 shows the present circulation of shallow and deep currents, known informally as the *ocean conveyor belt.*

The ocean conveyor belt affects regional and possibly global climate. As the Gulf Stream and North Atlantic Drift (see Figure 6.13) push into the North Atlantic, it delivers an immense amount of heat from the tropics to Europe. As this shallow current transfers its heat to the atmosphere, the water becomes denser and sinks. The deep current flowing southward in the North Atlantic is, on average, 8°C cooler than the shallow current flowing northward.

Scientific evidence from seafloor sediments and Greenland ice indicates that the ocean conveyor belt is not unchanging but can shift from one equilibrium state to another in a relatively short period of time (a few years to a few decades). The present ocean conveyor belt reorganized between 11,000 and 12,000 years ago. During this period of about 1,000 years, heat transfer to the North Atlantic stopped, and both North America and Europe experienced conditions of intense cold. Global temperatures also dropped during this time. The exact causes and effects of such large shifts in climate are not currently known, but scientists are concerned that human activities may unintentionally affect the link between the ocean conveyor belt and global climate.

Ocean Interactions with the Atmosphere

The ocean and the atmosphere are strongly linked, with wind from the atmosphere affecting the ocean currents and heat from the ocean affecting atmospheric circulation. One of the best examples of the interaction between ocean and atmosphere is the **El Niño–Southern Oscillation (ENSO)** event. ENSO is a periodic warming of surface waters of the tropical East Pacific that alters both ocean and atmospheric circulation patterns and results in unusual weather in areas far from the tropical Pacific (Figure 6.16). Normally, westward-blowing trade winds

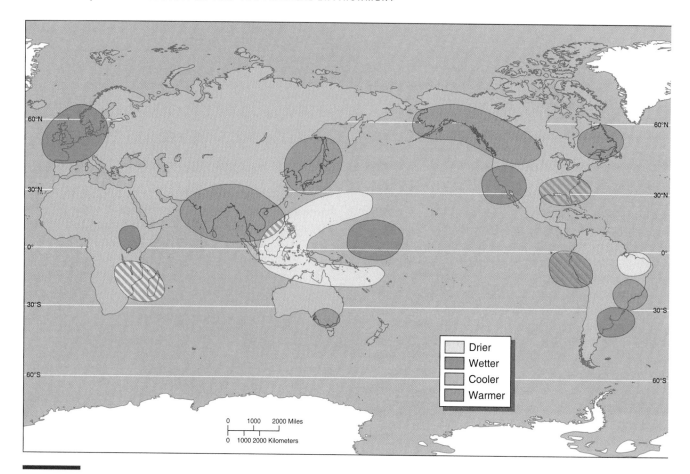

Figure 6.16 Climate patterns associated with ENSO. El Niño–Southern Oscillation (ENSO) events can drastically alter the climate in many areas remote from the Pacific Ocean. As a result of ENSO, some areas are drier, some wetter, some cooler, and some warmer than usual. Typically, northern areas of the contiguous United States are warmer during the winter, whereas southern areas are cooler and wetter.

restrict the warmest waters to the western Pacific near Australia. Every 3 to 7 years, however, the trade winds weaken, and the warm mass of water expands eastward to South America, increasing surface temperatures in the East Pacific. Ocean currents, which normally flow westward in this area, slow down, stop altogether, or even reverse and go eastward. The phenomenon is called *El Niño* (Spanish, "the boy child") and refers to the Christ child because the warming usually reaches the fishing grounds off Peru just before Christmas. Most ENSOs last from 1 to 2 years.

ENSO has a devastating effect on the fisheries off South America. Normally, the colder, nutrient-rich deep water is about 40 m (130 ft) below the surface and **upwells** (comes to the surface) along the coast, partly in response to strong trade winds (Figure 6.17). During an ENSO event, the colder, nutrient-rich deep water is about 152 m (500 ft) below the surface in the Eastern Pacific, and the warmer surface temperatures and weak trade winds prevent upwelling. The lack of nutrients in the water results in a severe decrease in the populations

of anchovies and many other marine fishes. During the 1982–83 El Niño, one of the worst ever recorded, the anchovy population decreased by 99%. However, other species such as shrimp and scallops thrive during an ENSO event.

ENSO alters global air currents, directing unusual weather to areas far from the tropical Pacific. By one estimate, the 1997–98 ENSO, the strongest on record, caused more than 20,000 deaths and $33 billion in property damages worldwide. It resulted in heavy snows in parts of the western United States and ice storms in eastern Canada. This ENSO was also responsible for the torrential rains that flooded Peru, Ecuador, California, Arizona, and Western Europe. In addition, this ENSO was linked to droughts in Texas, Australia, and Indonesia. Indonesia was particularly hurt by an ENSO-related environmental crisis for which humans were mostly responsible: The 1997–98 drought in Indonesia was the worst in 50 years. It exacerbated fires, many of which were deliberately set by large, multinational companies to clear land for agriculture (such as rice and oil palm

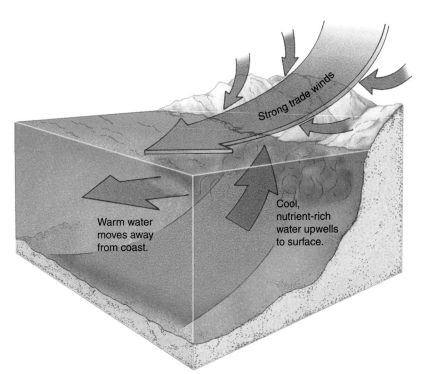

Figure 6.17 Upwelling. Coastal upwelling, where deeper waters come to the surface, occurs in the Pacific Ocean along the South American coast. Upwelling provides nutrients for large numbers of microscopic algae, which in turn support a complex food web. Coastal upwelling weakens considerably during years with El Niño–Southern Oscillation (ENSO) events, temporarily reducing fish populations.

plantations). The fires burned out of control and destroyed an area as large as the state of New Jersey.

La Niña El Niño is not the only periodic ocean temperature event to affect the tropical Pacific Ocean. The condition known as **La Niña** (Spanish, "the little girl") occurs when the surface water temperature in the eastern Pacific Ocean becomes unusually cool, and westbound trade winds become unusually strong. La Niña often occurs after an El Niño event and is considered to be part of the natural oscillation. During the spring of 1998, the surface water of the Eastern Pacific cooled by 6.7°C (12°F) in just 20 days. Like ENSO, La Niña affects weather patterns around the world, but its effects are more difficult to predict. In the contiguous United States, La Niña typically causes wetter than usual winters in the Pacific Northwest, warmer weather in the Southeast, and drought conditions in the Southwest. Atlantic hurricanes are stronger and more numerous during a La Niña event.

WEATHER AND CLIMATE

Weather refers to the conditions in the atmosphere at a given place and time; it includes temperature, atmospheric pressure, precipitation, cloudiness, humidity, and wind. Weather changes from one hour to the next and from one day to the next.

Climate comprises the average weather conditions that occur in a place over a period of years. The two most important factors that help determine an area's climate are temperature—both average temperature and temperature extremes—and precipitation—both average precip-

itation and seasonal distribution. Other climate factors include wind, humidity, fog, and cloud cover. Lightning is an important aspect of climate in some areas because it starts fires (see Chapter 7 introduction).

Day-to-day variations, day-to-night variations, and seasonal variations in climate factors are important dimensions of climate that affect organisms. Latitude, elevation, topography, vegetation, distance from the ocean, and location on a continent or other land mass influence temperature, precipitation, and other aspects of climate. Unlike weather, which changes rapidly, climate changes slowly, over hundreds or thousands of years.

Earth has many different climates, and because each is relatively constant for many years, organisms have adapted to them. The many different kinds of organisms on Earth are here in part because of the large number of different climates—from cold, snow-covered polar climates to tropical climates where it is hot and rains almost every day. A German botanist and climatologist, Wladimir Köppen, developed the most widely used system for classifying climates in the early part of the 20th century (see "Mini-Glossary: Köppen's Climate Zones"). He based his scheme on the observation that various types of vegetation are associated with different climates, particularly temperature and precipitation (see Chapter 7). Figure 6.18 shows a world climate map modified from Köppen. Note that there are six climate zones—humid equatorial, dry, humid temperate, humid cold, cold polar, and highland climate—and that each is subdivided into climate types. For example, the three types of humid temperate climates are no dry season, dry winter, and dry summer.

MINI-GLOSSARY

Köppen's Climate Zones

humid equatorial climate: Every month is warm; mean temperature is over 18°C (64°F).

dry climate: Evaporation exceeds precipitation in most months.

humid temperate climate: Distinct winter and summer seasons; winters are mild; mean temperature in coldest month is above 3°C (27°F).

humid cold climate: Distinct winter and summer seasons; winters are cold; mean temperature in coldest month is below 3°C (27°F).

cold polar climate: Distinct winter and summer seasons; winters are long and extremely cold; mean temperature in the warmest month is below 10°C (50°F).

highland climate: Climate changes as elevation increases and from one side of the mountain to the other; characteristic of high plateaus and mountains.

Precipitation

Precipitation refers to any form of water, such as rain, snow, sleet, and hail, that falls from the atmosphere. Precipitation varies from one location to another and has a profound effect on the distribution and kinds of organisms present. One of the driest places on Earth is in the Atacama Desert in Chile, where the average annual rainfall is 0.05 cm (0.02 in.). In contrast, Mount Waialeale in Hawaii, Earth's wettest spot, receives an average annual precipitation of 1,200 cm (472 in.).

Differences in precipitation depend on several factors. The heavy rainfall of some areas of the tropics results mainly from the equatorial uplift of moisture-laden air. High surface-water temperatures cause the evaporation of vast quantities of water from tropical parts of the ocean, and prevailing winds blow the resulting moist air over land masses. Heating of the air by a land surface that has been warmed by the sun causes moist air to rise. As it rises, the air cools, and its moisture-holding ability decreases (cool air holds less water vapor than warm air). When the air reaches its saturation point—that is, when it cannot hold any additional water vapor—

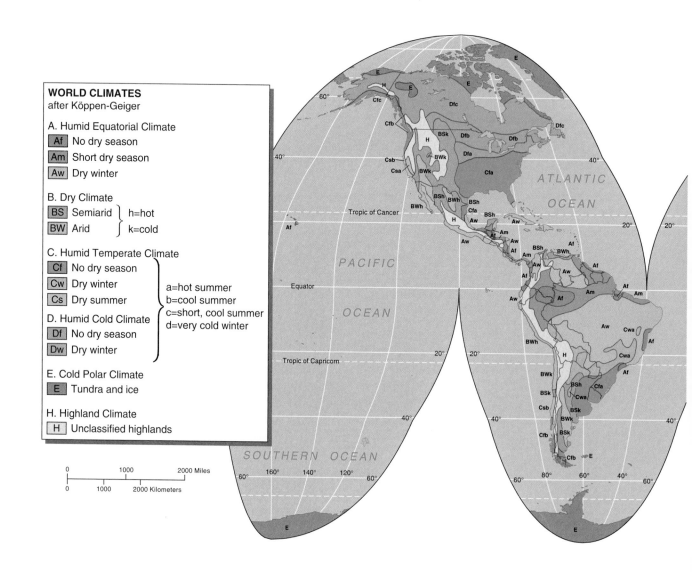

WORLD CLIMATES
after Köppen-Geiger

A. Humid Equatorial Climate
- Af No dry season
- Am Short dry season
- Aw Dry winter

B. Dry Climate
- BS Semiarid   } h=hot
- BW Arid   } k=cold

C. Humid Temperate Climate
- Cf No dry season
- Cw Dry winter
- Cs Dry summer

a=hot summer
b=cool summer
c=short, cool summer
d=very cold winter

D. Humid Cold Climate
- Df No dry season
- Dw Dry winter

E. Cold Polar Climate
- E Tundra and ice

H. Highland Climate
- H Unclassified highlands

clouds form and water is released as precipitation. The air eventually returns to Earth on both sides of the equator near the Tropics of Cancer and Capricorn (latitudes 23.5 degrees north and 23.5 degrees south). By then most of its moisture has precipitated, and the dry air returns to the equator. This dry air makes little biological difference over the ocean, but its lack of moisture over land produces some of the great tropical deserts, such as the Sahara Desert.

Air is also dried by long journeys over land masses. Near the windward (the side from which the wind blows) coasts of continents, rainfall may be heavy. However, in the temperate zones—the areas between the tropics and the polar zones—continental interiors are usually dry, because they are far from the ocean that replenishes water in the air passing over it.

Rain Shadows   Moisture is removed from humid air by mountains, which force air to rise. The air cools as it gains altitude, clouds form, and precipitation occurs—primarily on the windward slopes of the mountains. As the air mass moves down on the other side of the mountain, it is warmed, thereby lessening the chance of precip-

itation of any remaining moisture. This situation exists on the west coast of North America, where precipitation falls on the western slopes of mountains that are close to the coast. The dry land on the side of the mountains away from the prevailing wind—in this case, east of the mountain range—is called a **rain shadow** (Figure 6.19).

Tornadoes

A **tornado**, or twister, is a powerful, rotating funnel of air associated with severe thunderstorms. Tornadoes form when a mass of cool, dry air collides with warm, humid air, producing a strong updraft of spinning air on the underside of a cloud. The spinning funnel is called a tornado when it descends from the cloud and touches the ground. Wind velocity in a strong tornado may reach 480 km per hour (300 mi per hour). Tornadoes range from 1 m to 3.2 km (2 mi) in width. They last from several seconds to as long as 7 hours and travel along the ground from several meters to more than 320 km (200 mi).

On a local level, tornadoes have more concentrated energy than any other kind of storm. They can destroy buildings, bridges, and freight trains and even blow the

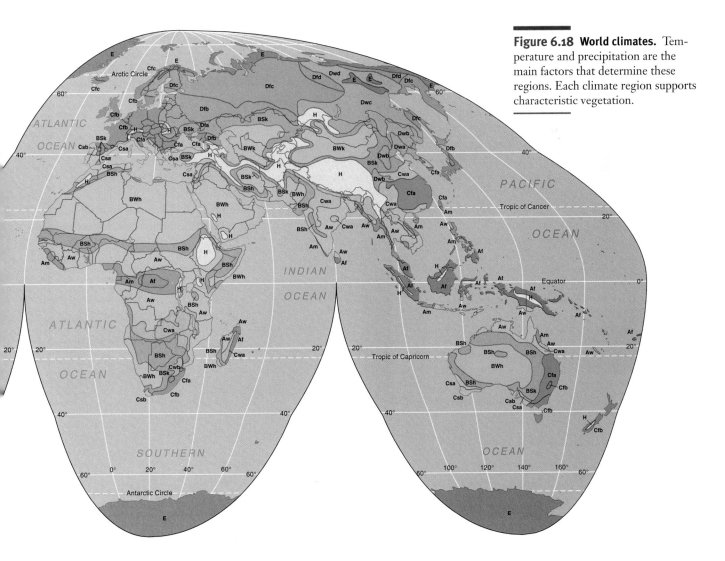

Figure 6.18 World climates. Temperature and precipitation are the main factors that determine these regions. Each climate region supports characteristic vegetation.

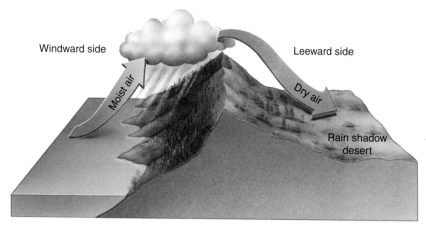

Windward side

Leeward side

Moist air

Dry air

Rain shadow desert

Figure 6.19 Rain shadow. A rain shadow refers to arid or semiarid land that occurs on the far side (leeward side) of a mountain. Prevailing winds blow warm, moist air from the windward side. Air cools as it rises, releasing precipitation so that dry air descends on the leeward side. Such a rain shadow exists east of the Cascade Range in Washington State.

water out of a river or small lake, leaving it empty. Tornadoes also kill people; more than 10,000 people in the United States died in tornadoes during the 20th century. Although tornadoes occur in other countries, the United States, which is known as the severe-storm capital of the world, has more tornadoes—typically about 1,000 per year—than anywhere else. They are most common in the spring months throughout the Great Plains and Midwestern states (especially Texas, Oklahoma, and Kansas), as well as states along the Gulf of Mexico coast (especially Florida).

Tropical Cyclones

Tropical cyclones are giant, rotating tropical storms with winds of at least 119 km per hour (74 mi per hour); the most powerful have wind velocities greater than 250 km per hour (155 mi per hour). They form as strong winds pick up moisture over warm surface waters of the tropical ocean and start to spin as a result of Earth's rotation. The spinning causes an upward spiral of massive clouds as air is pulled upward. Known as *hurricanes* in the Atlantic, *typhoons* in the Pacific, and *cyclones* in the Indian

Ocean, tropical cyclones are most common during summer and autumn months when ocean temperatures are warmest. With their spiral of clouds measuring about 800 km (500 mi) in diameter, tropical cyclones are easy to recognize in satellite photographs (Figure 6.20).

Tropical cyclones are destructive when they hit land, not so much from strong winds as from resultant *storm surges*, waves that rise as much as 7.5 m (25 ft) above the ocean surface. Storm surges can cause property damage and loss of life. Some hurricanes produce torrential rains. Hurricane Mitch, which hit the Atlantic coast of Central America in October 1998, caused more than 10,000 deaths during the flooding and landslides that followed. Mitch, which caused extensive damage in Honduras and Nicaragua, was the deadliest hurricane to occur in the Western Hemisphere in at least 200 years.

Some years produce more hurricanes than others. The 1995 hurricane season in the Atlantic Ocean was one of the busiest in the 20th century, with 19 named tropical storms, 11 of which became hurricanes. (Tropical storms have wind speeds between 63 to 117 km per hour [39 and 73 mi per hour], whereas hurricanes have wind speeds of 74 mi per hour [119 km per hour] or greater.) Some of

Figure 6.20 Hurricane Andrew. This satellite image shows Hurricane Andrew as it approached Florida in August 1992. Andrew devastated South Florida, causing 40 deaths and $30 billion in damage. It was the most expensive hurricane in U.S. history.

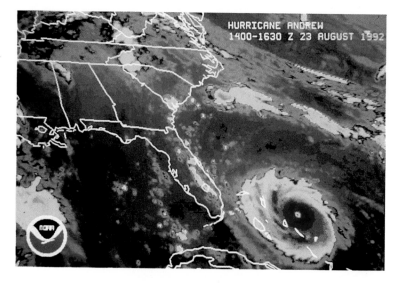

HURRICANE ANDREW
1400-1630 Z 23 AUGUST 1992

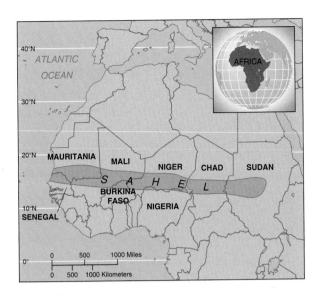

Figure 6.21 The Sahel region in Africa.

the factors that influence hurricane formation in the North Atlantic include precipitation in western Africa and water temperatures in the eastern Pacific Ocean. A wetter than usual rainy season in the western Sahel region of Africa (Figure 6.21) appears to translate into

more hurricanes, as does a dissipation of ENSO, which results in cooler water temperatures in the Pacific.

INTERNAL PLANETARY PROCESSES

Earth's crust is composed of seven large plates, plus a few smaller ones, that float on the mantle (the layer of hot, soft rock lying beneath the crust and above the core) (Figure 6.22). The land masses are situated on some of these plates. As the plates move horizontally across Earth's surface, the continents change their relative positions. The movement of these crustal plates is called **plate tectonics**.

Any area where two plates meet, called a **plate boundary**, is a site of intense geological activity (Figure 6.23). Earthquakes and volcanoes are common in such a region. Both the San Francisco area, noted for its earthquakes, and the volcano Mount Saint Helens in Washington State are situated where two plates meet. Where land masses are on the boundary between two plates, mountains may form. The Himalayas formed when the plate carrying India rammed into the plate carrying Asia. When two plates grind together, one of them sometimes descends under the other, in a process known as **subduction**. When two plates move apart, a ridge of molten

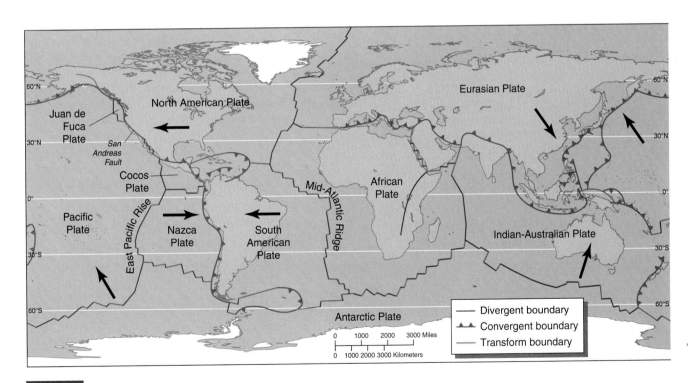

Figure 6.22 **Plates and plate boundary locations.** There are seven major independent plates that move horizontally across Earth's surface: African, Eurasian, Indian-Australian, Antarctic, Pacific, North American, and South American. Arrows show the directions of plate movements. The three types of plate boundaries are explained in Figure 6.23.

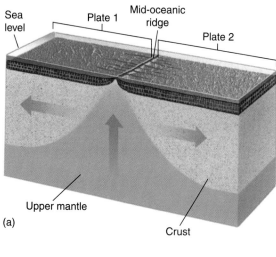

(a)

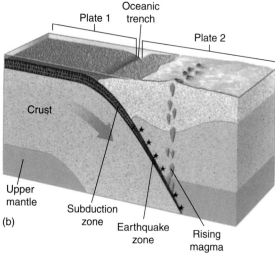

(b)

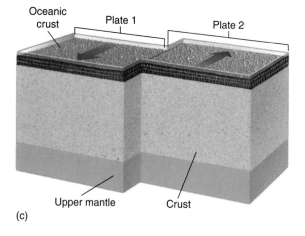

(c)

Figure 6.23 Plate boundaries. All three types of plate boundaries occur both in the ocean and on land. (a) Two plates move apart at a divergent plate boundary. (b) When two plates collide at a convergent plate boundary in the seafloor, subduction may occur. Convergent collision can also form a mountain range (not shown). (c) At a transform plate boundary, plates move horizontally in opposite but parallel directions. On land, such a boundary is often evident as a long, thin valley due to erosion along the fault.

rock from the mantle wells up between them; the ridge continually expands as the plates move farther apart. The Atlantic Ocean is growing as a result of the buildup of lava along the mid-Atlantic ridge, where two plates are separating.

Volcanoes

The movement of crustal plates on the hot, soft rock of the outer mantle is responsible for most volcanic activity. In places, the rock reaches the melting point, forming pockets of molten rock called **magma**. When one plate slides under or away from an adjacent plate, magma may rise to the surface, often forming volcanoes. Magma that reaches the surface is called **lava**.

Volcanoes occur at three locations: subduction zones, spreading centers, and above hot spots. Subduction zones around the Pacific Basin have given rise to hundreds of volcanoes around Asia and the Americas known as the "ring of fire." Plates that are spreading apart also form volcanoes. Iceland is a volcanic island that formed along the mid-Atlantic ridge. The Hawaiian Islands also have a volcanic origin, but they did not form at plate boundaries. This chain of volcanic islands formed as the Pacific plate moved over a **hot spot**, a rising plume of magma that flowed from an opening in the crust.

The largest volcanic eruption in the 20th century occurred in 1991 when Mount Pinatubo in the Republic of the Philippines exploded (see Figure 20.4). Despite the evacuation of more than 200,000 people, 338 deaths occurred, mostly from the collapse of buildings under the thick layer of wet ash that blanketed the area. The volcanic cloud produced when Mount Pinatubo erupted extended upward some 30 km (48 mi). We are used to hearing about human activities affecting climate, but many significant natural phenomena, including volcanoes, also affect global climate. The magma and ash that were ejected into the atmosphere when Mount Pinatubo erupted blocked much of the sun's warmth and caused a slight cooling of global temperatures for a year or so (see Chapter 20).

Earthquakes

Forces inside Earth sometimes push and stretch rocks in the crust. The rocks absorb this energy for a time, but eventually, as the energy accumulates, the stress is too great and the rocks suddenly shift or break. The energy—released as **seismic waves**, vibrations that spread through the rocks rapidly in all directions—causes one of the most powerful events in nature, an earthquake. Most earthquakes occur along **faults**, fractures in the crust along which rock moves forward and backward, up and down, or from side to side. Fault zones are often found at plate boundaries.

The site where an earthquake begins, often far below the surface, is called the **focus**. Directly above the focus, at Earth's surface, is the earthquake's **epicenter**. When seismic waves reach the surface, they cause the ground to shake. Buildings and bridges collapse, and roads break. One of the instruments used to measure seismic waves is a seismograph, which helps seismologists (scientists who study earthquakes) determine where an earthquake started, how strong it was, and how long it lasted. One way the magnitude of energy released by an earthquake can be measured is by the *Richter scale*, invented by Charles Richter, a California seismologist, in 1935. Each unit on the Richter scale represents about 30 times more released energy than the unit immediately below it. As an example, a magnitude 8 earthquake is 30 times more powerful than a magnitude 7 earthquake and 900 times more powerful than a magnitude 6 earthquake. The Richter scale makes it easy to compare earthquakes, but it tends to underestimate the energy of very large quakes.

Although the public is used to hearing about the Richter scale, seismologists typically do not use it. There are several ways to measure the magnitude of an earthquake, just as there are several ways to measure the size of a person (for example, height, weight, and amount of body fat). Most seismologists use a more precise scale called the *moment magnitude scale* to measure earthquakes, especially those larger than magnitude 6.5 on the Richter scale. The moment magnitude scale calculates the total energy released by a quake.

More than 1 million earthquakes are recorded by seismologists each year. Some of these are major, but most are too small to be felt, equivalent to readings of about 2 on the Richter scale. A magnitude 5 earthquake usually causes property damage. Every 5 years or more, a great earthquake occurs with a reading of 8 or higher. Such quakes usually cause massive property destruction and kill large numbers of people. Few people die directly because of the seismic waves; most die because of building collapses or fires started by ruptured gas lines.

Landslides and tsunamis are some of the side effects of earthquakes. A landslide is an avalanche of rock, soil, and other debris that slides swiftly down a mountainside. A 1970 earthquake in Peru resulted in a landslide that buried the town of Yungay and killed 17,000 people. A

tsunami, a giant sea wave caused by an underwater earthquake or volcanic eruption, can sweep through the water at more than 750 km (450 mi) per hour. Although a tsunami may be only about 1 m high in deep ocean water, it can build to a wall of water 30.5 m (100 ft)—as high as a 10-story building—when it comes ashore, often far from where the original earthquake triggered it. Tsunamis have caused thousands of deaths, particularly along the Pacific coast. Although the Pacific Tsunami Warning System monitors submarine earthquakes and warns people of approaching tsunamis, deaths still occur because there is so little time to respond. A 1998 tsunami in Papua New Guinea killed at least 2,200 people. The first wave reached the northern shore of this country about 5 to 10 minutes after the underwater earthquake occurred. Debris at the tops of palm trees showed that waves were more than 14 m (46 ft) high.

One of the most geologically active places in North America is California's San Andreas Fault, which runs parallel to the California coast from the Mexican border to northern California, a length of more than 1,100 km (700 mi). The Pacific plate (the west side of the San Andreas Fault) is sliding northward relative to the North American plate (the east side) at a rate of about 3.5 cm (1.4 in.) per year. In 1906, much of San Francisco, which is located near the San Andreas Fault, was destroyed by a magnitude 8.3 earthquake and the fire, caused by ruptured gas lines, that followed it. In 1989, a magnitude 6.9 earthquake along the San Andreas Fault in Loma Prieta, about 90 km south of San Francisco, killed 67 people and caused $6 billion in damage to the San Francisco Bay area.

Not all earthquakes occur at plate boundaries. Some occur on smaller faults that crisscross the large plates—the major earthquakes that damaged Northridge, California, in 1994 and Kobe, Japan, in 1995, for example. Such earthquakes pose a seismic hazard that is difficult to evaluate because major quakes occur along a given small fault only every 1,000 to 5,000 years, in contrast to a larger fault line, which may have a major quake every century or so.

SUMMARY WITH SELECTED KEY TERMS

I. In contrast to energy, which moves in one direction through ecosystems, matter is cyclical. All materials vital to life are continually recycled from the abiotic environment to organisms and back to the environment.

II. Biogeochemical cycles are cycles of matter such as carbon, nitrogen, phosphorus, sulfur, and water.

A. Carbon dioxide (CO_2) is the important gas of the **carbon cycle**.

1. Carbon enters plants, algae, and cyanobacteria as CO_2, which is incorporated into organic molecules by photosynthesis.
2. Cellular respiration by plants, by animals that eat plants, and by decomposers returns CO_2 to the atmosphere, making it available for producers again.
3. **Combustion** and weathering also return CO_2 to the atmosphere.
4. The level of atmospheric CO_2 has increased dramatically

during the last half of the 20th century and beginning of the 21st century. This increase may cause human-induced changes in climate called global warming.

B. There are five steps in the **nitrogen cycle**. Bacteria drive the nitrogen cycle.

  1. Nitrogen fixation is the conversion of nitrogen gas to ammonia.

  2. Nitrification is the conversion of ammonia or ammonium to nitrate, one of the main forms of nitrogen used by plants.

  3. Assimilation is the biological conversion of nitrates, ammonia, or ammonium into proteins and other nitrogen-containing compounds by plants; the conversion of plant proteins into animal proteins is also part of assimilation.

  4. Ammonification is the conversion of organic nitrogen to ammonia and ammonium ions.

  5. Denitrification converts nitrate to nitrogen gas.

  6. Humans have more than doubled the amount of fixed nitrogen entering the global nitrogen cycle. The excess nitrogen is contributing to water quality problems, air pollution, acid deposition, global warming, and depletion of stratospheric ozone.

C. The **phosphorus cycle** has no biologically important gaseous compounds.

  1. Phosphorus erodes from rock as inorganic phosphates and is absorbed from the soil by the roots of plants.

  2. Plants incorporate phosphorus into organic compounds. Animals obtain the phosphorus they need from their diets. Decomposers release inorganic phosphate into the environment.

  3. Phosphorus can be lost from terrestrial cycles for millions of years when it washes into the ocean and is deposited on the sea floor.

D. In the **sulfur cycle**, most sulfur occurs as rocks or as sulfur dissolved in the ocean.

  1. Sulfur-containing gases, which include hydrogen sulfide, sulfur oxides, and dimethyl sulfide (DMS) comprise a minor part of the atmosphere and are not long-lived.

  2. A tiny fraction of sulfur is present in the proteins of living organisms.

  3. Bacteria drive the sulfur cycle.

  4. Human activities have greatly increased sulfur emissions, which contribute to air pollution and acid deposition.

E. The **hydrologic cycle**, which continuously renews the supply of water that is essential to life, involves an exchange of water among the land, the atmosphere, and organisms.

  1. Water enters the atmosphere by evaporation and **transpiration** and leaves the atmosphere as precipitation.

  2. On land, water filters through the ground or runs off to lakes, rivers, and the ocean. **Groundwater** is stored in underground caverns and porous layers of rock. The movement of surface water from land to rivers, lakes, wetlands, and, ultimately, the ocean is called **runoff**.

  3. Aerosols, tiny particles of air pollution produced from fossil fuel combustion and the burning of forests, enhance the scattering and absorption of sunlight in the atmosphere and cause brighter clouds to form. Scientists think aerosols may weaken the hydrologic cycle.

III. Sunlight is the primary (almost sole) source of energy available to the biosphere.

A. Of the solar energy that reaches Earth, 31% is immediately reflected away and the remaining 69% is absorbed.

B. Albedo is the proportional reflectance of Earth's surface. Glaciers and ice sheets have high albedos, and the ocean and forests have low albedos.

C. Ultimately, all absorbed solar energy is radiated into space as infrared (heat) radiation.

D. A combination of Earth's roughly spherical shape and the tilt of its axis concentrates solar energy at the equator and dilutes solar energy at the poles.

  1. The tropics are therefore hotter and less variable in climate than are temperate and polar areas.

  2. Seasons are primarily determined by the inclination of Earth's axis.

IV. The atmosphere protects Earth's surface from most of the sun's ultraviolet radiation and x-rays as well as from lethal amounts of cosmic rays from space.

A. Visible light and some infrared radiation penetrate the atmosphere to warm the surface and lower part of the atmosphere.

B. Atmospheric heat transfer from the equator to the poles produces a movement of warm air toward the poles and a movement of cool air toward the equator, thus moderating the climate.

C. In addition to these global circulation patterns, the atmosphere exhibits complex horizontal movements called **winds** that result in part from differences in atmospheric pressure and from the rotation of Earth (the **Coriolis effect**).

V. The global ocean is a single, continuous body of water that surrounds the continents and covers almost three fourths of Earth's surface.

A. Surface ocean **currents** result largely from prevailing winds. Other factors that contribute to ocean currents include the Coriolis effect, the position of land masses, and the varying **density** of water.

B. Deep ocean currents often travel in different directions and at different speeds than do surface currents, in part because the Coriolis effect is more pronounced at greater depths.

  1. The present circulation of shallow and deep currents, known informally as the ocean conveyor belt, affects regional and possibly global climate.

  2. Evidence indicates that the ocean conveyor belt can switch from one equilibrium state to another.

C. The **El Niño–Southern Oscillation** (**ENSO**) event is a periodic warming of surface waters of the tropical East Pacific that alters both ocean and atmospheric circulation patterns and results in unusual weather in areas far from the tropical Pacific. During a **La Niña** event, surface water in the eastern Pacific becomes unusually cool.

VI. An area's **climate** comprises the average **weather** conditions that occur there over a period of years.

A. Temperature (both average temperature and temperature extremes) and precipitation (both average precipitation and seasonal distribution) are the two most important factors that help determine an area's climate.

B. Precipitation is greatest where warm air passes over the ocean, absorbing moisture, and is then cooled, such as when mountains force humid air upward. Deserts develop in the **rain shadows** of mountain ranges or in continental interiors.

C. A **tornado** is a powerful, rotating funnel of air associated with severe thunderstorms.

D. A **tropical cyclone** is a giant, rotating tropical storm with high winds. Tropical cyclones are called hurricanes in the Atlantic, typhoons in the Pacific, and cyclones in the Indian Ocean.

VII. Earth's crust consists of seven large plates, plus a few smaller ones, that float on the mantle.

A. As the plates move horizontally, the continents change their relative positions. The movement of the crustal plates is called **plate tectonics**.

B. **Plate boundaries** are sites of intense geological activity, such as mountain building, volcanoes, and earthquakes.

THINKING ABOUT THE ENVIRONMENT

1. Why might industrial polluters suppose that the Gaia hypothesis gives them permission to pollute the air, water, and soil indefinitely?

2. What is a biogeochemical cycle? Why is the cycling of matter essential to the continuance of life?

3. Describe how organisms participate in each of these biogeochemical cycles: carbon, nitrogen, phosphorus, and sulfur.

4. Diagram the hydrologic cycle.

5. What human activities alter the carbon cycle? The nitrogen cycle?

6. How does the sun affect temperature at different latitudes? Why?

7. What basic forces determine the circulation of the atmosphere? Describe the general directions of atmospheric circulation.

8. How do ocean currents affect climate on land?

9. Describe the El Niño–Southern Oscillation (ENSO) and some of its global effects.

10. Distinguish between weather and climate. What are the two most important climate factors?

11. What are some of the environmental factors that produce areas of precipitation extremes, such as rain forests and deserts?

12. Distinguish between tornadoes and tropical cyclones.

13. What are crustal plates and plate boundaries? Where are earthquakes and volcanoes commonly located, and why?

14. Evaluate the area where you live with respect to natural dangers. Is there a threat of possible earthquakes, volcanic eruptions, hurricanes, tornadoes, or tsunamis?

*15. Convert the temperature range in the lower stratosphere ($-45°C$ to $-75°C$) to degrees Fahrenheit.

* The solution to this question appears in Appendix VII.

TAKE A STAND

Visit our Web site at **http://www.wiley.com/college/raven** (select Chapter 6 from the Table of Contents) for links to more information about human-induced changes in the global nitrogen cycle. Consider the opposing views of farmers who use nitrogen fertilizers and ecologists who study the effects of excess nitrate on water quality and coastal fisheries, and debate the issues with your classmates. You will find tools to help you organize your research, analyze the data, think critically about the issues, and construct a well-considered argument. Take a Stand activities can be done individually or as a team, as oral presentations, written exercises, or Web-based (e-mail) assignments.

Additional on-line materials relating to this chapter, including Student Quizzes, Activity Links, Useful Web Sites, Flash Cards, and more, can also be found on our Web site.

SUGGESTED READING

Ackerman, J. "New Eyes on the Oceans." *National Geographic*, Vol. 198, No. 4 (October 2000). Scientists are using advanced technologies to help them understand marine systems.

Beardsley, T. "Dissecting a Hurricane." *Scientific American*, Vol. 282, No. 3 (March 2000). Highlights the work of scientists who fly into hurricanes to study them.

Bennet, E., and S.R. Carpenter. "P Soup." *World Watch*, Vol. 15, No. 2 (March–April 2002). Humans are having an increasing impact on the global phosphorus cycle.

Crutzen P.J., and V. Ramanathan. "The Ascent of Atmospheric Sciences." *Science*, Vol. 290 (October 13, 2000). Human influence on atmospheric processes is now occurring at the global level.

Gurnis, M. "Sculpting the Earth from Inside Out." *Scientific American*, Vol. 284, Vol. 3 (March 2001). How the mantle shapes Earth's surface.

Koenig, R. "Researchers Target Deadly Tsunamis." *Science*, Vol. 293 (August 17, 2001). Scientists are using computer models, improved maps of the ocean floor, and new sensory equipment to understand the causes of tsunamis.

Nierenberg, D. "Toxic Fertility." *World Watch*, Vol. 14, No. 2 (March–April 2001). Humans are dominating the global nitrogen cycle.

Perkins, S. "Pinning Down the Sun–Climate Connection." *Science News*, Vol. 159 (January 20, 2001). Scientists are beginning to understand some of the complexities involved in the influence of the sun on Earth's climate.

Perkins, S. "Tornado Alley, USA." *Science News*, Vol. 161 (May 11, 2002). Researchers have developed a map that shows where tornadoes are the greatest concern.

Suplee, C. "El Niño, La Niña: Nature's Vicious Cycle." *National Geographic*, Vol. 195, No. 3 (March 1999). Examines some of the effects of variations in global atmosphere and ocean circulation patterns caused by El Niño and La Niña.

Vesilind, P.J. "Once and Future Fury: California's Volcanic North." *National Geographic*, Vol. 200, No. 4 (October 2001). Three tectonic plates collide in northern California, triggering volcanoes such as Mount Shasta.

Wofsy, S.C. "Where Has All the Carbon Gone?" *Science*, Vol. 292 (June 22, 2001). Understanding carbon sinks for excess CO_2 is essential to coming to terms with the effect of increased CO_2 on global climate change.

Wolfe, D.W. "Out of Thin Air." *Natural History* (September 2001). Biological nitrogen fixation is essential to life, including human life, on Earth.

August 2000 wildfire in Montana.

Major Ecosystems of the World

Learning Objectives

After you have studied this chapter you should be able to:

1. Define *biome* and discuss how biomes are related to climate.

2. Explain the similarities and the changes in vegetation observed with increasing elevation and increasing latitude.

3. Briefly describe the nine major terrestrial biomes, giving attention to the climate, soil, and characteristic plants and animals of each.

4. Relate at least one human effect on each of the biomes discussed.

5. Summarize the important environmental factors that affect aquatic ecosystems.

6. Briefly describe the various freshwater, estuarine, and marine ecosystems, giving attention to the environmental characteristics and representative organisms of each.

7. Relate at least one human effect on each of the aquatic ecosystems discussed.

8. Outline the environmental history of the Florida Everglades.

Wildfires are fires started by lightning. They are an important environmental force in many geographical areas. Those areas most prone to wildfires have wet seasons followed by dry seasons. Vegetation that grows and accumulates during the wet season dries out enough during the dry season to burn easily. When lightning hits the ground, it ignites the dry organic material, and wind spreads the fire through the area. At the peak of a wildfire season in the American West, where extensive lands are prone to wildfires, hundreds of new wildfires can break out each day. In 2000, which was a particularly bad year for wildfires, more than 92,000 fires burned about 3 million hectares (7.4 million acres) in the United States.

Fires have several effects on the environment. First, combustion frees the nutrient minerals that were locked in dry organic matter. The ashes remaining after a fire are rich in potassium, phosphorus, calcium, and other nutrient minerals essential for plant growth. Thus, vegetation flourishes after a fire. Second, fire removes plant cover and exposes the soil, which stimulates the germination of seeds that require bare soil and encourages the growth of shade-intolerant plants. Third, fire can cause increased soil erosion because it removes plant cover, leaving the soil more vulnerable to wind and water.

Grasses adapted to wildfire have underground stems and buds, which are unaffected by a fire sweeping over them. After the aerial parts have been killed by fire, the underground parts send up new sprouts. Fire-adapted trees such as bur oak and ponderosa pine have a thick bark that is resistant to fire. In contrast, fire-sensitive

131

trees such as many hardwoods have a thin bark. Certain pines such as jack and lodgepole pine depend on fire for successful reproduction, because the heat of the fire opens the cones so that seeds can be released.

Fires were a part of the natural environment long before humans appeared, and many terrestrial ecosystems have adapted to fire. African savannas, California chaparrals, North American grasslands, and pine forests of the southern United States are some fire-adapted ecosystems. Fire helps maintain grasses as the dominant vegetation in grasslands by removing fire-sensitive hardwood trees.

The influence of fire on plants became even more pronounced once humans appeared. Because humans deliberately and accidentally set fires, fires became more frequent. Humans set fires for many reasons: to provide the grasses and shrubs that many game animals require; to clear the land for agriculture and human development; and in times of war, to reduce enemy cover.

Humans also try to prevent fires, and sometimes this effort can have disastrous consequences. When fire is excluded from a fire-adapted ecosystem, organic litter accumulates. As a result, when a fire does occur, it burns hotter and is much more destructive and therefore ecologically unhelpful. During the past several years, the fires in Arizona, Colorado, New Mexico, and Utah were caused in part by decades of suppressing fires in the region. Prevention of fire also converts grassland to woody vegetation and facilitates the invasion of fire-sensitive trees into fire-adapted forests.

Another complexity introduced into the human-fire equation is the so-called *wildland–urban interface fire*, or *intermix fire*, which occurs where suburban homes are built in close proximity to areas naturally prone to wildfires. Developers typically clear very small lots around the homes they're building, and they often build wood-frame homes. Fire management experts say that short-term changes, such as replacing wooden roofs and clearing the brush around houses would help control intermix fires. Long-term changes to reduce the destruction of property by wildfires include modifications of land use planning and building codes to restrict building in areas that are fire-prone.

Humans sometimes conduct *prescribed burns*, a tool of ecological management in which the organic litter is deliberately burned under controlled conditions before it accumulates to dangerous levels. Prescribed burns are also used to suppress fire-sensitive trees, thereby maintaining the natural fire-adapted ecosystem. However, prescribed burns do not always accomplish their goal of reducing fire risk. In 2000, a huge New Mexico fire that started out as a prescribed burn got out of control and burned 18,600 hectares (46,000 acres). Fire management experts agree that prescribed burns are helpful in reducing the damage in certain areas but will not prevent fires. There are always going to be wildfires.

EARTH'S MAJOR BIOMES

In Chapter 6 we learned that Earth has many different **climates,** which are based primarily on temperature and precipitation differences (see Figure 6.18). Characteristic organisms have adapted to each climate. A **biome** is a large, relatively distinct terrestrial region characterized by similar climate, soil, plants, and animals, regardless of where it occurs in the world (Figure 7.1). Because it is so large in area, a biome encompasses many interacting ecosystems. In terrestrial ecology, a biome is considered the next level of ecological organization above those of community, ecosystem, and landscape.

Near the poles, temperature is generally the overriding climate factor, whereas in temperate and tropical regions, precipitation becomes more significant than temperature (Figure 7.2). Light is relatively plentiful in biomes, except in certain environments such as the rainforest floor. Other abiotic factors to which certain biomes are sensitive include temperature extremes as well as rapid temperature changes, fires, floods, droughts, and strong winds. Elevation also affects biomes: Changes in vegetation with increasing elevation resemble the changes in vegetation observed in going from warmer to colder climates.

Vertical Zonation: The Distribution of Vegetation on Mountains

Hiking up a mountain is similar to traveling toward the North Pole with respect to the major ecosystems encountered (Figure 7.3). This elevation-latitude similarity occurs because as one climbs a mountain, the temperature drops just as it does when one travels north. The types of organisms living on the mountain change as the temperature changes.

The base of a mountain in Colorado, for example, might be covered by deciduous trees, which shed their leaves every autumn. At higher elevations, where the climate is colder and more severe, one might find a coniferous forest called subalpine forest, which resembles the northern boreal forest. Higher still, where the climate is very cold, a kind of tundra occurs, with vegetation composed of grasses, sedges, and small tufted plants; it is called alpine tundra to distinguish it from arctic tundra. At the very top of the mountain, a permanent ice or snow cap might be found, similar to the nearly lifeless polar land areas.

There are important environmental differences between high elevations and high latitudes, however, that affect the types of organisms found in each place. Alpine tundra typically lacks permafrost and receives more pre-

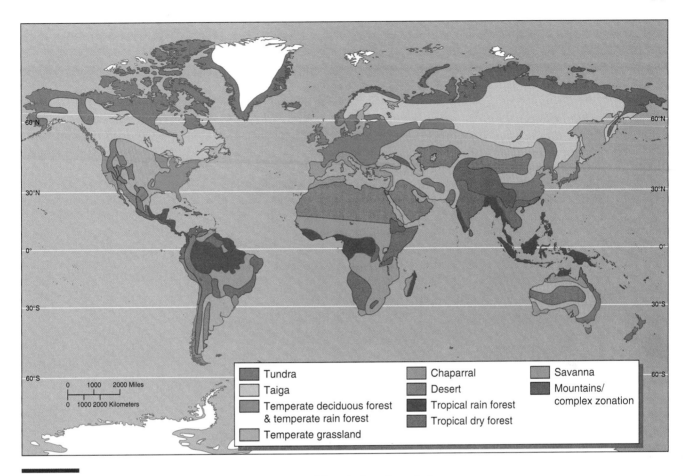

Figure 7.1 Distribution of the world's terrestrial biomes. Although sharp boundaries are shown in this highly simplified map, biomes actually grade together at their boundaries. Biomes generally correspond to the climate zones discussed in Chapter 6 (see Figure 6.18).

cipitation than arctic tundra. Also, high elevations of temperate mountains do not have the great extremes in day length that are associated with the changing seasons in biomes at high latitudes. Furthermore, the intensity of solar radiation is greater at high elevations than at high latitudes. At high elevations the sun's rays pass through less atmosphere, which results in a greater exposure to ultraviolet (UV) radiation—that is, less UV is filtered out by the atmosphere—than occurs at high latitudes.

We now consider nine major biomes and how humans are affecting them: tundra, taiga, temperate rain forest, temperate deciduous forest, temperate grassland, chaparral, desert, savanna, and tropical rain forest.

Tundra: Cold Boggy Plains of the Far North

Tundra (also called **arctic tundra**) occurs in the extreme northern latitudes wherever the snow melts seasonally (Figure 7.4). The Southern Hemisphere has no equivalent of the arctic tundra because it has no land in the corresponding latitudes. A similar ecosystem located in the

higher elevations of mountains, above the tree line, is called **alpine tundra** to distinguish it from arctic tundra.

Arctic tundra has long, harsh winters and very short summers. Although the growing season, with its warmer temperatures, is short (from 50 to 160 days depending on location), the days are long. Above the Arctic Circle, the sun does not set at all for many days in midsummer, although the amount of light at midnight is one tenth that at noon. There is little precipitation (10 to 25 cm, or 4 to 10 in., per year) over much of the tundra, with most of it falling during the summer months.

Tundra soils tend to be geologically young, because most of them were formed when glaciers retreated after the last Ice Age. (Glacier ice, which occupied about 29% of the Earth's land during the last Ice Age, began retreating about 17,000 years ago. Today, glacier ice occupies about 10% of the land.) These soils are usually nutrient-poor and have little organic litter such as dead leaves and stems, animal droppings, and remains of organisms. Although the soil melts at the surface during the summer, tundra has a layer of **permafrost**, permanently frozen ground that varies in depth and thickness. Permafrost is

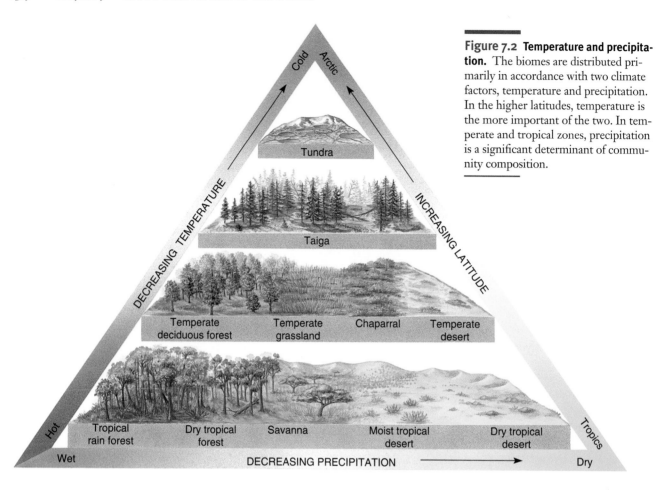

Figure 7.2 Temperature and precipitation. The biomes are distributed primarily in accordance with two climate factors, temperature and precipitation. In the higher latitudes, temperature is the more important of the two. In temperate and tropical zones, precipitation is a significant determinant of community composition.

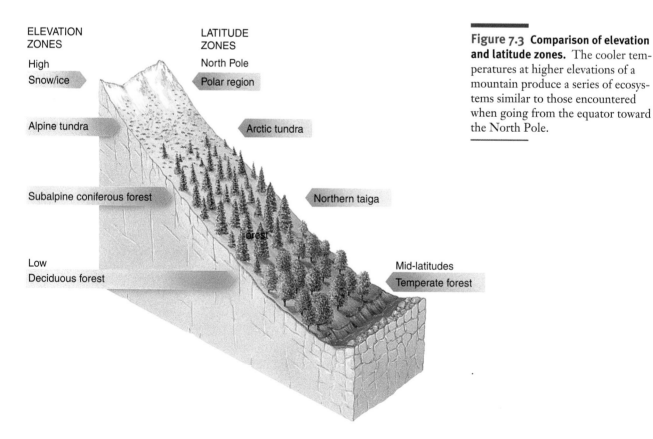

Figure 7.3 Comparison of elevation and latitude zones. The cooler temperatures at higher elevations of a mountain produce a series of ecosystems similar to those encountered when going from the equator toward the North Pole.

Figure 7.4 Arctic tundra. A caribou buck stands in the Alaskan tundra. Because of its short growing season and permafrost, only small, hardy plants grow in the northernmost biome that encircles the Arctic Ocean.

most extensive in northern Canada and Siberia. Because permafrost interferes with drainage, the thawed upper zone of soil is usually waterlogged during the summer. Permafrost limits the depth to which roots can penetrate, thereby preventing the establishment of most woody species. The limited precipitation in combination with low temperatures, flat topography (surface features), and the permafrost layer produces a landscape of broad, shallow lakes and ponds, sluggish streams, and bogs.

Low **species richness** and low **primary productivity** characterize tundra. Few plant species occur, but individual species often exist in great numbers. Tundra is dominated by mosses; lichens, such as reindeer moss; grasses; and sedges, which are grasslike plants. No readily recognizable trees or shrubs grow except in sheltered locations, although dwarf willows, dwarf birches, and other dwarf trees are common. As a rule, tundra plants seldom grow taller than 30 cm (12 in.).

The year-round animal life of the tundra includes lemmings, voles, weasels, arctic foxes, snowshoe hares, ptarmigan, snowy owls, and musk oxen. In the summer, caribou migrate north to the tundra to graze on sedges, grasses, and dwarf willow. Dozens of bird species also migrate north in summer to nest and feed on abundant insects. Mosquitoes, blackflies, and deerflies survive the winter as eggs or pupae and occur in great numbers during summer weeks. There are no reptiles or amphibians except for wood frogs, which are sometimes found in tundra ponds.

Tundra regenerates very slowly after it has been disturbed. Even casual use by hikers can cause damage. Long-lasting injury, likely to persist for hundreds of years, has been done to large portions of the arctic tundra as a result of oil and natural gas exploration and military

use (see Chapter 10, "Case in Point: The Arctic National Wildlife Refuge").

Taiga: Evergreen Forests of the North

Just south of the tundra is the **taiga**, or **boreal forest**, which stretches across North America and Eurasia, covering approximately 11% of the Earth's land (Figure 7.5). A biome comparable to the taiga is not found in the

Figure 7.5 Taiga, or boreal forest. These coniferous forests occur in cold regions of the Northern Hemisphere adjacent to the tundra. Photographed in Yukon, Canada.

Southern Hemisphere. Winters are extremely cold and severe, although not as harsh as in the tundra. The growing season of the taiga is somewhat longer than that of the tundra. Taiga receives little precipitation, perhaps 50 cm (20 in.) per year, and its soil is typically acidic, mineral-poor, and characterized by a deep layer of partly decomposed pine and spruce needles at the surface. Permafrost is patchy and, where found, is often deep under the soil. Taiga has numerous ponds and lakes in water-filled depressions that were dug in the ground by grinding ice sheets during the last Ice Age.

Black and white spruces, balsam fir, eastern larch, and other conifers (cone-bearing evergreens) dominate the taiga, although deciduous trees (trees that shed their leaves in autumn), such as aspen and birch, may form striking stands. Conifers have many drought-resistant adaptations, such as needlelike leaves with a minimal surface area for water loss. Such an adaptation enables conifers to withstand the "drought" of the northern winter months (roots cannot absorb water when the ground is frozen). Being evergreen, conifers can resume photosynthesis as soon as warmer temperatures return.

The animal life of the taiga consists of some larger species such as caribou, which migrate from the tundra to the taiga for winter; wolves; bears; and moose. However, most mammals are medium-sized to small, including rodents, rabbits, and fur-bearing predators such as lynx, sable, and mink. Most species of birds are abundant in the summer but migrate to warmer climates for winter. Insects are abundant, but there are few amphibians and reptiles except in the southern taiga.

Most of the taiga is not well suited to agriculture because of its short growing season and mineral-poor soil. However, the taiga yields lumber, pulpwood for paper products, animal furs, and other forest products. Currently, taiga is the world's primary source of industrial wood and wood fiber, and extensive logging of certain boreal forests has occurred. (See Chapter 17 for a discussion of deforestation of this biome.)

Temperate Rain Forest: Lush Temperate Forests

A coniferous **temperate rain forest** occurs on the northwest coast of North America. Similar vegetation exists in southeastern Australia and in southern South America. Annual precipitation in this biome is high, from 200 to 380 cm (80 to 152 in.), and this is augmented by condensation of water from dense coastal fogs. The proximity of temperate rain forest to the coastline moderates the temperature so that the seasonal fluctuation is narrow; winters are mild and summers are cool. Temperate rain forest has relatively nutrient-poor soil, although its organic content may be high. Cool temperatures slow the activity of bacterial and fungal decomposers. Thus, needles and large fallen branches and trunks accumulate on the ground as litter that takes many years to decay and release nutrient minerals to the soil.

The dominant vegetation in the North American temperate rain forest is large evergreen trees such as western hemlock, Douglas fir, western red cedar, Sitka spruce, and western arborvitae (Figure 7.6). Temperate rain forests are rich in epiphytic vegetation, which are smaller plants that grow on the trunks and branches of large trees. Epiphytes in this biome are mainly mosses, club mosses, lichens, and ferns, all of which also carpet the ground. Deciduous shrubs such as vine maple grow wherever a break in the overlying canopy occurs. Squirrels, wood rats, mule deer, elk, numerous bird species, and several species of amphibians and reptiles are common temperate rainforest animals.

Temperate rain forest is a rich wood producer, supplying us with lumber and pulpwood. It is also one of the world's most complex ecosystems in terms of species richness. Care must be taken to avoid overharvesting the original old-growth (never logged) forest, however, because such an ecosystem takes hundreds of years to develop. When the logging industry harvests old-growth forest, it typically replants the area with a monoculture of trees (that is, a single species) that it can harvest in 40- to 100-year cycles. Thus, the old-growth forest ecosystem, once harvested, never has a chance to redevelop. A small fraction of the original old-growth temperate rain forest in Washington, Oregon, and northern California remains untouched. These stable forest ecosystems provide biological habitats for many species, including about 40 endangered and threatened species. The issues surrounding old-growth forests of the Pacific Northwest were explored in Chapter 3.

Figure 7.6 Temperate rain forest. This temperate biome is characterized by large amounts of precipitation. Photographed in Olympic National Park in Washington state.

Temperate Deciduous Forest: Broad-Leaved Trees That Shed Their Leaves

Seasonality (hot summers and cold winters) is characteristic of the **temperate deciduous forest**, which occurs in temperate areas where precipitation ranges from about 75 to 126 cm (30 to 50 in.) annually. Typically, the soil of a temperate deciduous forest consists of a topsoil rich in organic material and a deep, clay-rich lower layer. As organic materials decay, mineral ions are released. Ions that are not absorbed by the roots of living trees **leach** (filter) into the clay, where they may be retained.

The temperate deciduous forests of the northeastern and mideastern United States are dominated by broad-leaved hardwood trees, such as oak, hickory, and beech, that lose their foliage annually (Figure 7.7). In the southern areas of the temperate deciduous forest, the number of broad-leaved evergreen trees, such as magnolia, increases. The trees of the temperate deciduous forest form a dense canopy that overlies saplings and shrubs.

Temperate deciduous forests originally contained a variety of large mammals, such as puma, wolves, and bison, which are now absent, plus deer, bears, and many small mammals and birds. In Europe and North America,

Figure 7.7 Temperate deciduous forest. The broad-leaf trees that dominate this biome are deciduous and will shed their leaves before winter. Photographed during autumn in Pennsylvania.

logging and land clearing for farms, tree plantations, and cities have removed much of the original temperate deciduous forest. Where it has been allowed to regenerate, temperate deciduous forest is often in a seminatural state that is highly modified by humans for recreation, livestock foraging, timber harvest, and other uses. Many forest organisms have successfully become reestablished in these returning forests.

Worldwide, deciduous forests were among the first biomes to be converted to agricultural use. In Europe and Asia, many soils that originally supported deciduous forests have been cultivated by traditional agricultural methods for thousands of years without a substantial loss in fertility. During the 20th century, however, intensive agricultural practices were widely adopted; these, along with overgrazing and deforestation, have contributed to the degradation of some agricultural lands. Most damage to farmland has been done since the end of World War II. We say more about land degradation in Chapters 14, 17, and 18.

Grasslands: Temperate Seas of Grass

Summers are hot, winters are cold, and rainfall is often uncertain in **temperate grasslands**. Annual precipitation averages 25 to 75 cm (10 to 30 in.). In grasslands with less precipitation, nutrient minerals tend to accumulate in a well-defined layer just below the topsoil. These nutrient minerals tend to wash out of the soil in areas with more precipitation. Grassland soil has considerable organic material because the aboveground portions of many grasses die off each winter and contribute to the organic content of the soil, while the roots and rhizomes (underground stems) survive underground. The roots and rhizomes eventually die and also contribute to the soil's organic material. Many grasses are sod formers—that is, their roots and rhizomes form a thick, continuous underground mat.

Moist temperate grasslands, also known as tallgrass prairies, occur in the United States in parts of Illinois, Iowa, Minnesota, Nebraska, Kansas, and other Midwestern states (Figure 7.8). Although few trees grow except near rivers and streams, grasses grow in great profusion in the deep, rich soil. Periodic wildfires help maintain grasses as the dominant vegetation in grasslands. Tallgrass prairies were formerly dominated by several species of grasses that, under favorable conditions, grew as tall as a person on horseback. The land was covered with herds of grazing animals, such as bison and pronghorn elk. The principal predators were wolves, although in sparser, drier areas coyotes took their place. Smaller animals included prairie dogs and their predators (foxes, black-footed ferrets, and various birds of prey), grouse, reptiles such as snakes and lizards, and great numbers of insects.

Shortgrass prairies are temperate grasslands that receive less precipitation than the moist temperate grasslands just described but more precipitation than deserts.

Figure 7.8 Moist temperate grassland. This tallgrass prairie contains a profusion of grasses and other herbaceous flowering plants. Photographed in Iowa on one of the virgin tracts of tallgrass prairie owned by the Nature Conservancy.

In the United States, shortgrass prairies occur in the eastern half of Montana, the western half of South Dakota, and parts of other Midwestern states. Grasses that grow knee high or lower dominate shortgrass prairies. The plants grow in less abundance than in the moister grasslands, and occasionally some bare soil is exposed. Native grasses of shortgrass prairies are drought-resistant.

The North American grassland, particularly the tallgrass prairie, was so well suited to agriculture that little of it now remains. More than 90% has vanished under the plow, and the remaining prairie is so fragmented that almost nowhere can we see even an approximation of what European settlers saw when they migrated into the Midwest. Today, the tallgrass prairie is considered North America's rarest biome. It is not surprising that the North American Midwest, the Ukraine, and other moist temperate grasslands became the breadbaskets of the world, because they provide ideal growing conditions for crops such as corn and wheat, which are also grasses.

Chaparral: Thickets of Evergreen Shrubs and Small Trees

Some hilly temperate environments have mild winters with abundant rainfall combined with very dry summers. Such **mediterranean climates**, as they are called, occur not only in the area around the Mediterranean Sea but also in the North American Southwest, southwestern and southern Australia, central Chile, and southwestern South Africa. In mountain slopes of southern California, this mediterranean-type community is known as **chaparral**. Chaparral soil is thin and often not very fertile. Frequent fires occur naturally in this environment, particularly in late summer and autumn.

Chaparral vegetation looks strikingly similar in different areas of the world, even though the individual species are different. Chaparral is usually dominated by a dense growth of evergreen shrubs but may contain short, drought-resistant pine or scrub oak trees that grow 1 to 3 m (3.3 to 9.8 ft) tall (Figure 7.9). During the rainy winter season, the environment may be lush and green, but the plants lie dormant during the hot, dry summer. Trees and shrubs often have hard, small, leathery leaves that resist water loss. Many plants are also specifically fire-adapted and grow best in the months following a fire. Such growth is possible because fire releases nutrient minerals that were present in aboveground parts of the plants that burned. The underground parts of some plants and the seeds of many others are not killed by the fire, and with the new availability of essential nutrient minerals, the plants sprout vigorously during winter rains. Mule deer, wood rats, chipmunks, lizards, and many species of birds are common animals of the chaparral.

The fires that occur at irregular intervals in California chaparral are quite costly to humans when they consume expensive homes built on the hilly chaparral landscape. Unfortunately, efforts to prevent the naturally occurring fires sometimes backfire. Denser, thicker vegetation tends to accumulate over several years; then, when a fire does occur, it is much more severe. Removing the chaparral vegetation, whose roots hold the soil in place, can also cause problems—witness the mudslides that sometimes occur during winter rains in these areas.

Figure 7.9 Chaparral. Chaparral vegetation consists mainly of drought-resistant evergreen shrubs and small trees. Hot, dry summers and mild, rainy winters characterize the chaparral. Photographed in the Santa Monica Mountains, California.

Using Goats to Fight Fires

California has about 6,000 wildfires each year, and they are becoming increasingly expensive and dangerous to manage because so many people are building homes and living in fire-vulnerable chaparral. For one thing, the topography is so steep that firefighters often cannot use mechanized equipment but instead must be transported to fires by helicopters. Afraid that prescribed burns will get out of control, local governments increasingly are trying an effective, low-tech method to reduce the fuel load: During the 6-month fire season, goats are clearing hills around Oakland and Berkeley, Monterey, and Malibu.

A herd of 350 goats can denude an entire acre of heavy brush in about a day, but their use entails a lot of advance organization and support. Before goats can clear hazardous dry fuels from surrounding hillsides, botanists must walk the terrain to put fences around any small trees or other plants that are rare or endangered; fencing keeps the goats from eating those plants. A portable home site for goatherds is then installed, as are electric fencing and water troughs for the goats. The goatherds typically use dogs to help herd the goats.

Goats have turned out to be an excellent tool for fire management because they preferentially browse woody shrubs and thick undergrowth—the exact fuel that causes disastrous fires. Fires that have occurred in areas after goats have browsed there have been much easier to contain. Clearly, local governments are happy, not annoyed, to "get one's goat."

Deserts: Arid Life Zones

Deserts are dry areas found in both temperate (*cold deserts*) and subtropical or tropical regions (*warm deserts*). The low water vapor content of the desert atmosphere results in daily temperature extremes of heat and cold, so that a major change in temperature occurs in a single 24-hour period. Deserts vary greatly, depending on the amount of precipitation they receive, which is generally less than 25 cm (10 in.) per year. A few deserts are so dry that virtually no plant life occurs in them, as for example, the African Namib Desert and the Atacama Desert of northern Chile and Peru. As a result of sparse vegetation, desert soil is low in organic material but is often high in mineral content, particularly the salts sodium chloride ($NaCl$), calcium carbonate ($CaCO_3$), and calcium sulfate ($CaSO_4$). In some regions, such as areas of Utah and Nevada, the concentration of certain soil minerals reaches toxic levels for many plants.

Plant cover is sparse in deserts, so much of the soil is exposed. Both perennials (plants that live for more than 2 years) and annuals (plants that complete their life cycles in one growing season) occur in deserts. However, annuals are common only after rainfall. Plants in North American deserts include cacti, yuccas, Joshua trees, and sagebrushes (Figure 7.10). Desert plants tend to have reduced leaves or no leaves, an adaptation that conserves water. In cacti such as the giant saguaro, photosynthesis is carried out by the stem, which also expands accordion-style to store water; the leaves are modified into spines, which discourage herbivores. Other desert plants shed their leaves for most of the year, growing only during the brief moist season. Many desert plants are provided with spines, thorns, or toxins to resist the heavy grazing pressure often experienced in this food- and water-deficient environment.

Desert animals tend to be small. During the heat of the day, they remain under cover or return to shelter periodically, whereas at night they come out to forage or hunt. In addition to desert-adapted insects, there are a few desert-adapted amphibians (frogs and toads) and many specialized desert reptiles, such as the desert tortoise, desert iguana, Gila monster, and Mohave rattlesnake. Desert mammals include rodents such as gerbils and jerboas in African and Asian deserts and kangaroo rats in North American deserts. There are also mule deer and jackrabbits in these deserts, oryxes in African deserts, and kangaroos in Australian deserts. Carnivores such as the African fennec fox and some birds of prey, especially

Figure 7.10 Desert. Inhabitants of deserts are strikingly adapted to the demands of their environment. The moister deserts of North America contain large columnar cacti such as the giant saguaro (*Carnegiea gigantea*), which grows 15 to 18 m (50 to 60 ft) tall. The smaller cacti, called chollas (*Opuntia* sp.), have a dense covering of barbed spines.

owls, live on the rodents and jackrabbits. During the driest months of the year, many desert insects, amphibians, reptiles, and mammals tunnel underground, where they remain inactive.

Humans have altered North American deserts in several ways. Off-road vehicles damage desert vegetation, which sometimes takes years to recover. Tank crews that trained for World War II in California's Mojave Desert left tracks in 1940 that are still visible today. People who drive across the desert in four-wheel-drive vehicles also inflict environmental damage. When the top layer of desert soil is disturbed, erosion occurs more readily, and less vegetation grows to support native animals. Another problem in deserts is that certain cacti and desert tortoises are rare as a result of poaching. Houses, factories, and farms built in desert areas require vast quantities of water, which must be imported from distant areas. Increased groundwater consumption by many desert cities has caused groundwater levels to drop. Aquifer depletion in U.S. deserts is particularly critical in southern Arizona and southwestern New Mexico. See Chapter 13 for a discussion of water problems in this region.

Savanna: Tropical Grasslands

The **savanna** biome is a tropical grassland with widely scattered clumps of low trees (Figure 7.11). Savanna is found in areas of low rainfall or seasonal rainfall with prolonged dry periods. The temperatures in tropical savannas vary little throughout the year, and seasons are regulated by precipitation, not by temperature as they are in temperate grasslands. Annual precipitation is 85 to 150 cm (34 to 60 in.). Savanna soil is somewhat low in essential nutrient minerals, in part because it is strongly leached—that is, nutrient minerals have filtered out of the topsoil. Savanna soil is often rich in aluminum, which resists leaching, and in places the aluminum reaches levels that are toxic to many plants. Although the African savanna is best known, savanna also occurs in South America, western India, and northern Australia.

Savanna is characterized by wide expanses of grasses interrupted by occasional trees such as acacia, which bristle with thorns that provide protection against herbivores. Both trees and grasses have fire-adapted features, such as extensive underground root systems, that enable them to survive seasonal droughts as well as periodic fires.

Spectacular herds of hoofed mammals—such as wildebeest, antelope, giraffe, zebra, and elephants—occur in the African savanna. Large predators, such as lions and hyenas, kill and scavenge the herds. In areas of seasonally varying rainfall, the herds and their predators may migrate annually.

Savannas are rapidly being converted to rangeland for cattle and other domesticated animals, which are replacing the big herds of wild animals. The problem is particularly acute in Africa, which has the most rapidly growing human population of any continent. In some places, severe overgrazing and use of existing trees for firewood have contributed to the conversion of marginal savanna to desert (see Chapter 17).

Tropical Rain Forests: Lush Equatorial Forests

Tropical rain forests occur where temperatures are warm throughout the year and precipitation occurs almost daily. The annual precipitation of a tropical rain forest is typically from 200 to 450 cm (80 to 180 in.).

Figure 7.11 Savanna. Tropical grasslands such as this one, with widely scattered Acacia trees, support large herds of grazing animals and their predators. These are swiftly vanishing under pressure from pastoral and agricultural land use. Photographed in Tanzania.

Much of this precipitation comes from locally recycled water that enters the atmosphere by transpiration (loss of water vapor from plants) of the forest's own trees.

Tropical rain forest commonly occurs in areas with ancient, highly weathered, mineral-poor soil. Little organic matter accumulates in such soils: Because temperatures are high year round, bacteria, fungi, and detritus-feeding ants and termites decompose organic litter quite rapidly. Roots and mycorrhizae quickly absorb nutrient minerals from the decomposing material. Thus, the nutrient minerals of tropical rain forests are tied up in the vegetation rather than the soil. Tropical rain forests are found in Central and South America, Africa, and southeast Asia.

Tropical rain forest is very productive—that is, the plants capture a lot of energy by photosynthesis. Despite the scarcity of nutrient minerals in the soil, abundant solar energy and precipitation stimulate high productivity.

Of all the biomes, the tropical rainforest is unexcelled in species richness and variety. No single species dominates this biome. A person can often travel for hundreds of meters without encountering two members of the same species of tree. Local factors, such as varying soil fertility and topography, affect the composition of rainforest species. Valleys have more plant diversity than hills, for example.

The trees of tropical rain forests are typically evergreen flowering plants (Figure 7.12a). Their roots are often shallow and concentrated near the surface in a mat only a few centimeters (an inch or so) thick. The root mat catches and absorbs almost all nutrient minerals released from leaves and litter by decay processes. Swollen bases or braces called buttresses hold the trees upright and aid in the extensive distribution of the shallow roots.

A fully developed tropical rain forest has at least three distinct stories, or layers, of vegetation. The topmost story consists of the crowns of occasional, very tall trees, some 50 m (164 ft) or more in height, which are exposed to direct sunlight. The middle story, which reaches a height of 30 to 40 m (100 to 130 ft), forms a continuous canopy of leaves that lets in very little sunlight to support the sparse understory. Only 2% to 3% of

(a)

(b)

Figure 7.12 Tropical rain forest. (a) A broad view of tropical rain forest vegetation along a riverbank in Southeast Asia. Except at riverbanks, tropical rain forests have a closed canopy that admits little light to the rainforest floor. (b) Thick, epiphyte-covered lianas grow into the canopy using a tree trunk for support. Photographed in Costa Rica.

the light bathing the forest canopy reaches the forest understory. Smaller plants that are specialized for life in the shade, as well as the seedlings of taller trees, comprise the understory. The vegetation of tropical rain forests is not dense at ground level except near stream banks or where a fallen tree has opened the canopy.

Tropical rainforest trees support extensive epiphytic communities of plants such as ferns, mosses, orchids, and bromeliads. Epiphytes grow in crotches of branches, on bark, or even on the leaves of their hosts, but they use their host trees primarily for physical support, not for nourishment.

Because little light penetrates to the understory, many of the plants living there are adapted to climb already established host trees rather than to invest their meager photosynthetic resources in building the cellulose tissues of their own trunks. Lianas (woody tropical vines), some as thick as a human thigh, twist up through the branches of the huge rainforest trees (Figure 7.12b). Once in the canopy, lianas grow from the upper branches of one forest tree to another, connecting the tops of the trees together and providing a walkway for many of the canopy's residents. They and herbaceous vines provide nectar and fruit for many tree-dwelling animals.

Not counting bacteria and other soil-dwelling organisms, about 90% of tropical rainforest organisms live in the upper canopy. Rainforest animals include the most abundant and varied insects, reptiles, and amphibians on Earth. The birds, often brilliantly colored, are also varied, with some specialized to consume fruit (parrots, for example) and others to consume nectar (hummingbirds and sunbirds, for example). Most rainforest mammals, such as sloths and monkeys, live only in the trees and never climb down to the ground, although some large, ground-dwelling mammals, including elephants, are also found in rain forests.

Unless strong conservation measures are initiated soon, human population growth and industrial expansion in tropical countries may spell the end of tropical rain forests by the middle of the 21st century. Biologists know that many rainforest organisms will become extinct before they have even been identified and scientifically described. The ecological impacts of tropical rainforest destruction are discussed extensively throughout this text (see Chapter 17, for example).

AQUATIC ECOSYSTEMS

Not surprisingly, aquatic life zones are different in almost all respects from terrestrial life zones. Recall that in biomes, temperature and precipitation are the major determinants of plant and animal inhabitants, and light is relatively plentiful, except in certain habitats such as the rainforest floor. Significant environmental factors in aquatic ecosystems are very different. Generally speaking, temperature is less important in watery environments because the water itself tends to moderate temperature. Water is obviously not an important limiting factor in aquatic ecosystems.

The most fundamental division in aquatic ecology is probably between freshwater and saltwater environments. **Salinity**, which is the concentration of dissolved salts such as sodium chloride (NaCl) in a body of water, affects the kinds of organisms present in aquatic ecosystems, as does the amount of dissolved oxygen. Water greatly interferes with the penetration of light, so floating aquatic organisms that photosynthesize must remain near the water's surface, and vegetation attached to the bottom can grow only in shallow water. In addition, low levels of essential nutrient minerals limit the number and distribution of organisms in certain aquatic environments. Other abiotic determinants of species composition in aquatic ecosystems include temperature, pH, and presence or absence of waves and currents.

Aquatic ecosystems contain three main ecological categories of organisms: free-floating plankton, strongly swimming nekton, and bottom-dwelling benthos. **Plankton** are usually small or microscopic organisms that are relatively feeble swimmers. For the most part, they are carried about at the mercy of currents and waves. They are unable to swim far horizontally, but some species are capable of large daily vertical migrations and are found at different depths of water at different times of the day or at different seasons. Plankton are generally subdivided into two major categories, phytoplankton and zooplankton. **Phytoplankton**, which are free-floating photosynthetic cyanobacteria and algae, are producers that form the base of most aquatic food webs. **Zooplankton** are nonphotosynthetic organisms that include protozoa (animal-like protists), tiny shrimplike crustaceans, and the larval (immature) stages of many animals. In aquatic food webs, zooplankton feed on algae and are in turn consumed by newly hatched fish and other small aquatic organisms. **Nekton** are larger, more strongly swimming organisms such as fishes, turtles, and whales. **Benthos** are bottom-dwelling organisms that fix themselves to one spot (sponges, oysters, and barnacles), burrow into the sand (worms, clams, and echinoderms), or simply walk about on the bottom (crawfish, aquatic insect larvae, and brittle stars).

FRESHWATER ECOSYSTEMS

Freshwater ecosystems include rivers and streams (flowing-water ecosystems), lakes and ponds (standing-water ecosystems), and marshes and swamps (freshwater wetlands). Specific abiotic conditions and characteristic organisms distinguish each freshwater ecosystem. Although freshwater ecosystems occupy a relatively small portion (about 2%) of Earth's surface, they have an important role in the hydrologic cycle: They assist in recycling precipitation that flows as surface runoff to the ocean (see hydrologic cycle in Chapter 6). Large bodies of fresh water also help moderate daily and seasonal temperature fluctuations on nearby land. Freshwater habitats also provide homes for large numbers of species.

Rivers and Streams: Flowing-Water Ecosystems

Many different conditions exist along the length of a river or stream (Figure 7.13). The nature of a **flowing-water ecosystem** changes greatly between its source (where it begins) and its mouth (where it empties into another body of water). Certain parts of the stream may be shaded by surrounding forest, whereas other parts may be exposed to direct sunlight. Headwater streams (the small streams that are the sources of a river) are usually shallow, cold, swiftly flowing, and therefore highly oxygenated. In contrast, rivers downstream from the headwaters are wider and deeper, cloudy (they contain suspended particulates), not as cold, slower flowing, and therefore less oxygenated. Along parts of a river or stream, groundwater wells up through sediments on the bottom; this local input of water moderates the water temperature so that summer temperatures are cooler and winter temperatures are warmer than in adjacent parts of the flowing-water ecosystem.

The kinds of organisms found in flowing-water ecosystems vary greatly from one stream to another, depending primarily on the strength of the current. In streams with fast currents, the inhabitants may have adaptations such as suckers to attach themselves to rocks so that they are not swept away. The larvae of blackflies attach themselves with a suction disk located on the end of their abdomen. Some stream inhabitants, such as immature water-penny beetles, may have flattened bodies to enable them to slip under or between rocks. The water-penny beetle larva (immature form) gets its common name from its flattened, nearly circular shape. Alternatively, inhabitants such as fish may be streamlined and muscular enough to swim in the current. Organisms in large, slow-moving streams and rivers do not need such adaptations, although they are typically streamlined like most aquatic organisms to lessen resistance when moving through water. Where the current is slow, organisms of the headwaters are replaced by organisms characteristic of ponds and lakes.

Unlike other freshwater ecosystems, streams and rivers depend on the land for much of their energy. In headwater streams, up to 99% of the energy input comes from detritus, such as leaves carried from the land into streams and rivers by wind or surface runoff. Downstream, rivers contain more producers and therefore have a slightly lower dependence on detritus as a source of energy than do the headwaters.

Human activities have several adverse impacts on rivers and streams, including water pollution and the effects of dams, which are built to contain or divert the water of rivers or streams. Pollution alters the physical environment and changes the biotic component down-

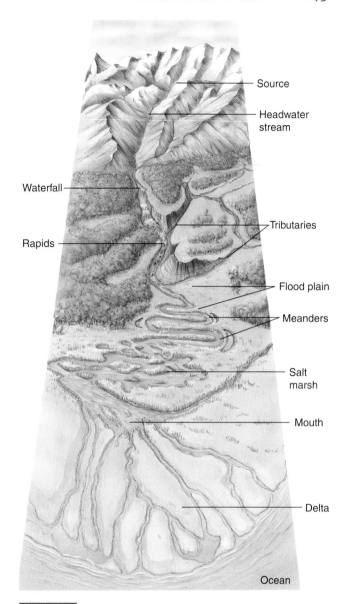

Figure 7.13 Features of a typical river. The river begins at a source, often high in mountains and fed by melting snows or glaciers. Headwater streams flow downstream rapidly, often over rocks (as rapids) or bluffs (as waterfalls). Along the way, smaller streams called tributaries feed into the river, adding to its flow. As the river's course levels out, the river flows more slowly and winds from side to side, forming bends called meanders. The flood plain is the relatively flat area on either side of the river that is subject to flooding. Some flood plains are quite large: The Mississippi River's flood plain is up to 130 km (80 mi) wide. Near the ocean, the river may form a salt marsh where fresh water from the river and salt water from the ocean mix. The delta is a fertile, low-lying plain at the river's mouth that forms from sediments deposited by the slow-moving river as it empties into the ocean.

stream from the pollution source (see Chapter 21). A dam causes water to back up, flooding large areas of land and forming a reservoir, which destroys terrestrial habitat (see Chapters 12 and 13). Below the dam, the once-powerful river is often reduced to a relative trickle, altering aquatic ecosystems.

Lakes and Ponds: Standing-Water Ecosystems

Standing-water ecosystems are characterized by zonation. A large lake has three zones: the littoral, limnetic, and profundal zones (Figure 7.14). The **littoral zone** is a shallow-water area along the shore of a lake or pond. Emergent vegetation, such as cattails and bur reeds, plus several deeper-dwelling aquatic plants and algae, live in the littoral zone. The littoral zone is the most productive section of the lake—that is, photosynthesis is greatest here—in part because it receives nutrient inputs from surrounding land that stimulate the growth of plants and algae. Animals of the littoral zone include frogs and their tadpoles; turtles; worms; crayfish and other crustaceans; insect larvae; and many fishes, such as perch, carp, and bass. Surface dwellers such as water striders and whirligig beetles are found in the quieter areas.

The **limnetic zone** is the open water beyond the littoral zone—that is, away from the shore; it extends down as far as sunlight penetrates to permit photosynthesis. The main organisms of the limnetic zone are microscopic phytoplankton and zooplankton. Larger fishes also spend most of their time in the limnetic zone, although they may visit the littoral zone to feed and reproduce. Owing to its depth, less vegetation grows here than in the littoral zone.

The deepest zone, the **profundal zone**, is beneath the limnetic zone of a large lake; smaller lakes and ponds typically lack a profundal zone. Because light does not penetrate effectively to this depth, plants and algae do not live here. Food drifts into the profundal zone from the littoral and limnetic zones. Bacteria decompose dead organisms that reach the profundal zone, using up oxygen and liberating the nutrient minerals contained in the organic material. These nutrient minerals are not effectively recycled, because there are no producers to absorb them and incorporate them into the food web. As a result, the profundal zone tends to be both mineral-rich and anaerobic (without oxygen), with few organisms other than anaerobic bacteria occupying it.

Thermal Stratification and Turnover in Temperate Lakes

The marked layering of large temperate lakes caused by how far light penetrates is accentuated by **thermal stratification**, in which the temperature changes sharply with depth. Thermal stratification occurs because the summer sunlight penetrates and warms surface waters, making them less dense. (The density of water is greatest at 4°C. Both above and below this temperature, water is less dense.) In the summer, cool and therefore denser water remains at the lake bottom, separated from the warm and therefore less-dense water above by an abrupt temperature transition called the **thermocline** (Figure 7.15a). Seasonal distribution of temperature and oxygen (more oxygen dissolves in water at cooler temperatures) affects the distribution of fish in the lake.

In temperate lakes, falling temperatures in autumn cause a mixing of the layers of lake water known as the **fall turnover** (Figure 7.15b). (Because there is little seasonal temperature variation in the tropics, such turnovers are not common there.) As the surface water cools, its density increases and eventually it displaces the less-dense, warmer, mineral-rich water beneath. The warmer water then rises to the surface where it, in turn, cools and sinks. This process of cooling and sinking continues until the lake reaches a uniform temperature throughout.

When winter comes, the surface water cools below 4°C, its temperature of greatest density, and if it is cold enough, ice forms. Ice, which forms at 0°C, is less dense than cold water; thus, ice forms on the surface, and the

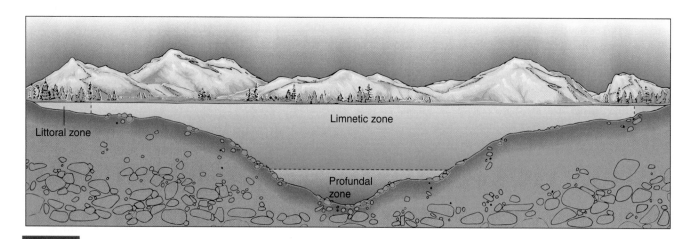

Figure 7.14 Zonation in a large lake. A lake is a standing-water ecosystem surrounded by land. The littoral zone is the shallow-water area around the lake's edge. The limnetic zone is the open, sunlit water away from the shore. The profundal zone, under the limnetic zone, is below where light penetrates.

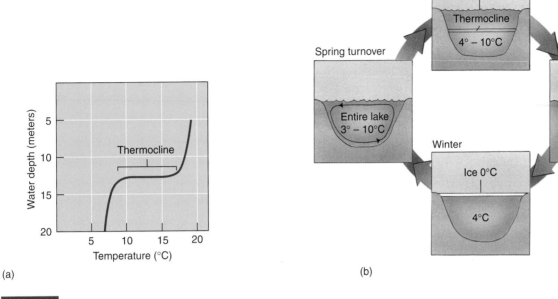

(a)

(b)

Figure 7.15 Thermal stratification in a temperate lake. (a) Temperature differences by depth during the summer. There is an abrupt temperature transition, called the thermocline, between the upper warm layer and the bottom cold layer. (b) During fall and spring turnovers, a mixing of upper and lower layers of water brings oxygen to the oxygen-depleted depths of the lake and nutrient minerals to the mineral-deficient surface water.

water on the lake bottom is warmer than the ice on the surface.

In the spring, a **spring turnover** occurs as ice melts and the surface water reaches 4°C. Surface water again sinks to the bottom, and bottom water returns to the surface, resulting in a mixing of the layers. As summer arrives, thermal stratification occurs once again.

The mixing of deeper, nutrient-rich water with surface, nutrient-poor water during the fall and spring turnovers brings essential nutrient minerals to the surface and oxygenated water to the bottom. The sudden presence of large amounts of essential nutrient minerals in surface waters encourages the development of large algal and cyanobacterial populations, which form temporary **blooms** (population explosions) in the fall and spring. (Harmful algal blooms such as red tides are discussed in Chapter 21.)

Marshes and Swamps: Freshwater Wetlands

Freshwater wetlands are usually covered by shallow water for at least part of the year and have characteristic soils and water-tolerant vegetation. Wetland soils are waterlogged for variable periods and are therefore anaerobic. Most wetland soils are rich in accumulated organic materials, in part because anaerobic conditions discourage decomposition. Freshwater wetlands include marshes, in which grasslike plants dominate, and swamps, in which woody trees or shrubs dominate (Fig-

ure 7.16). Freshwater wetlands also include hardwood bottomland forests (lowlands along streams and rivers that are periodically flooded), prairie potholes (small, shallow ponds that formed when glacial ice melted at the

Figure 7.16 Freshwater swamp. Freshwater swamps are inland areas permanently saturated or covered by water and dominated by trees, such as bald cypress (*Taxodium distichum*, shown). A floating carpet of tiny aquatic plants covers the water surface. Photographed in northeast Texas.

end of the last Ice Age), and peat moss bogs (peat-accumulating wetlands where sphagnum moss dominates).

Wetland plants, which are highly productive, provide enough food to support a wide variety of organisms. Wetlands are valued as a wildlife habitat for migratory waterfowl and many other bird species, beaver, otters, muskrats, and game fishes. Wetlands help to control flooding by acting as holding areas for excess water when rivers flood their banks. The floodwater stored in wetlands then drains slowly back into the rivers, providing a steady flow of water throughout the year. Wetlands also serve as groundwater recharging areas. One of their most important roles is to help cleanse and purify water by trapping and holding pollutants in the flooded soil. Such important environmental functions as these are known as **ecosystem services**.

At one time wetlands were considered wastelands, areas that needed to be filled in or drained so that farms, housing developments, and industrial plants could be built on them. Wetlands are also breeding places for mosquitoes and therefore were viewed as a menace to public health. Today, however, the crucial ecosystem services that wetlands provide are widely recognized, and wetlands have some legal protection, although they are still threatened by agriculture, pollution, engineering (dams), and urban and suburban development.

ENVIROBRIEF

The Value of Nature

Services provided by nature, termed ecosystem services, include all ecosystem processes that sustain human life, such as air and water purification; waste decomposition; flood mitigation; crop pollination by insects, bats, and other species; pest control by insect-eating birds and other species; soil renewal; and conservation of biodiversity. Ecosystem services often lead to the production of ecosystem goods—food, timber, medicines, fuels, and so on. Though humans have essentially depended on ecosystem services for millennia, they have only recently begun to disrupt them on a global scale. By taking ecosystem services for granted, the modern world threatens to lose them. In a dramatic and unconventional effort to reverse this devaluation, a team of researchers used economic market principles to price 17 ecosystem services for 16 biomes. In simple terms, their goal was to determine how much the world would have to pay for the goods and services that nature provides for free. Their answer? An estimated $33 trillion annually, about twice as much as the global gross national product. Many ecologists consider even this a conservative estimate, because certain biomes and ecosystem services were omitted. Wetlands and coastal and oceanic areas, followed by forests, proved most valuable.

Some ecologists and conservationists are offended by the dollar estimates, arguing that the Earth should be valued for moral and not financial reasons. The study's supporters point out that people already assign value to nature all the time in choices they make. Some economists criticize the study's methods and suggest estimates are inflated. The study's authors recognize its flaws but insist that however imperfect the estimates, assigning value raises awareness of Earth's worth to humans.

ESTUARIES: WHERE FRESH WATER AND SALT WATER MEET

Where the ocean meets the land, there may be one of several kinds of ecosystems: a rocky shore, a sandy beach, an intertidal mud flat, or a tidal estuary. An **estuary** is a coastal body of water, partly surrounded by land, with access to the open ocean and a large supply of fresh water from a river. Water levels in an estuary rise and fall with the tides, whereas salinity fluctuates with tidal cycles, the time of year, and precipitation. Salinity also changes gradually within the estuary, from unsalty fresh water at the river entrance to salty ocean water at the mouth of the estuary.

Because estuaries undergo significant daily, seasonal, and annual variations in such physical factors as temperature, salinity, and depth of light penetration, estuarine organisms must be able to tolerate these changing conditions.

Estuaries are among the most fertile ecosystems in the world, often having a much greater productivity than either the adjacent ocean or the fresh water upriver. This high productivity is brought about by four factors. One, nutrients are transported from the land into rivers and creeks that flow into the estuary. Two, tidal action promotes a rapid circulation of nutrients and helps remove waste products. Three, a high level of light penetrates the shallow water. Four, the presence of many plants provides an extensive photosynthetic carpet and also mechanically traps detritus, forming the base of a detritus food web. Many commercially important fishes and shellfish spend their larval stages in estuaries among the protective tangle of decaying plants.

Temperate estuaries usually contain **salt marshes**, shallow wetlands dominated by salt-tolerant grasses (see the Chapter 4 introductory photograph). Salt marshes have often appeared to be worthless, empty stretches of land to uninformed people. As a result, they have been used as dumps and become severely polluted or have been filled with dredged bottom material to form artificial land for residential and industrial development. A large part of the estuarine environment has been lost in this way, along with many of its ecosystem services, such as biological habitats, sediment and pollution trapping, groundwater supply, and storm buffering (salt marshes absorb much of the energy of a storm surge and thereby prevent flood damage elsewhere).

Mangrove forests, the tropical equivalent of salt marshes, cover perhaps 70% of tropical coastlines (Figure 7.17). Like salt marshes, mangrove forests provide valuable ecosystem services. Their interlacing roots are breeding grounds and nurseries for several commercially important fishes and shellfish, such as blue crabs, shrimp, mullet, and spotted sea trout. Mangrove branches are nesting sites for many species of birds, such as pelicans, herons, egrets, and roseate spoonbills. Mangrove roots

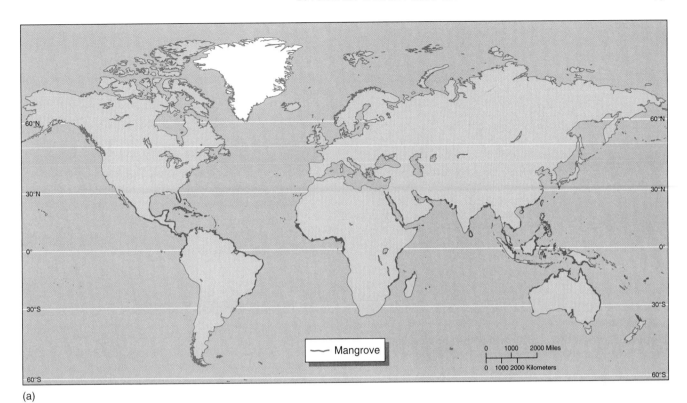

(a)

(b)

Figure 7.17 Mangroves. (a) Distribution of mangrove forests worldwide. (b) Red mangroves (*Rhizophora mangle*) have stilt-like roots that support the tree. These roots grow into deeper water as well as into mudflats that are often exposed by low tide. Many animals live in the complex root systems of mangrove forests. Photographed at low tide along the coast of Florida, near Miami.

stabilize the submerged soil, thereby preventing coastal erosion and providing a barrier against the ocean during storms. Scientists at the Mangrove Ecosystem Research Center in Vietnam have evidence that mangroves are more effective than concrete sea walls in dissipating wave energy and controlling floodwater from tropical storms. Mangroves are under assault from coastal development, unsustainable logging, and aquaculture (see Figure 18.16). Some countries, such as the Philippines, Bangladesh, and Guinea-Bissau, have lost 70% or more of their mangrove forests.

CASE·IN·POINT The Chesapeake Bay

The Chesapeake Bay is the largest estuary in the United States (6,475 km², or 2,500 mi²). It is also the most productive, annually yielding tons of seafood such as oysters, blue crabs, shad, striped bass, and white perch. For example, although the Chesapeake Bay's annual harvest of blue crabs has declined in recent years, the bay remains the largest source of blue crabs in the world. Nine rivers and about 150 headwater streams from six different states empty their waters into the Chesapeake Bay (Figure 7.18). This fresh

Figure 7.18 Watershed and airshed of the Chesapeake Bay. Water from six states drains into the bay, and air pollution from an even larger area is deposited in the bay's watershed.

water, which contains nutrient minerals and silt, mixes with the salty water inbound from the Atlantic Ocean.

The Chesapeake Bay area is home to more than 17 million people. The bay has suffered in recent years from deteriorating water quality due to pollution associated with the motor vehicles, farms, industries, and homes of all these people. Consider nitrogen pollution. Polluted water that empties into the bay contains nitrogen from agricultural runoff, sewage treatment plants, electric power companies, and motor vehicles. Most of this pollution remains in the bay, and only a small amount washes into the ocean.

The long process of restoring and protecting the bay and its resources is underway. The **Chesapeake Bay Program** is a partnership of federal and state agencies with goals such as sound land use, restoration of water quality and habitats, and management of fisheries. Its overall goal is to have the bay restored by 2010. Various state agencies have reduced the pollution load, both nutrients and sediments, going into the bay and restricted coastal development. For example, the level of phosphorus in bay waters has declined since the 1980s, in spite of the fact that the human population in the region increased. In 1999 an agreement to preserve 30,750 hectares (76,000 acres) of Chesapeake Bay watershed (that is, forests and wetlands) was announced. New oyster reef habitats are being successfully created and have been naturally colonized by oysters.

Many problems remain, however. Huge factory farms grow poultry in the watershed, and the runoff from these farms, which is rich in nitrogen and phosphorus, encourages excessive algal growth that disrupts the natural balance between producers and consumers. Excessive algal growth results in a much lower level of dissolved oxygen in the water, which harms fishes and other animals. Excessive nutrients can also result in outbreaks of toxic algae such as *Pfiesteria*, which received international attention in 1997 when it was detected in three rivers that empty into the Chesapeake Bay. The *Pfiesteria* outbreak caused sores on the bodies of fish and health problems for those who harvested the fish. Scientists studying the problem recommended that decreasing the amount of nutrients entering the water would lessen the risk of toxic outbreaks. In 1998, the state of Maryland passed a law requiring farmers to adopt nutrient management plans to control the pollution coming from their lands.

Despite the many challenges, the Chesapeake Bay Program is committed to the preservation and restoration of the bay and its watershed. ∎

MARINE ECOSYSTEMS

Although lakes and the ocean are comparable in many ways, there are also many differences. The depths of even the deepest lakes do not approach those of the ocean abysses, which have extremely deep areas that extend more than 6 km (3.6 mi) below the sunlit surface. Tides and currents exert a profound influence on the ocean. Gravitational pulls of both the sun and the moon usually produce two high tides and two low tides each day along the ocean's coastlines, but the height of those tides varies with the season, local topography, and phases of the moon (a full moon causes the highest tides).

The immense marine environment is subdivided into several zones: the intertidal zone, benthic (ocean floor) environment, and pelagic (ocean water) environment (Figure 7.19 and "Mini-glossary: The Marine Environ-

MINI-GLOSSARY

The Marine Environment

intertidal zone: The area of shoreline between low and high tides.

pelagic environment: The open ocean environment; divided into neritic and oceanic provinces.

neritic province: Open ocean that overlies the ocean floor from the shoreline to a depth of 200 m.

oceanic province: Open ocean that overlies the ocean floor at depths greater than 200 m.

euphotic zone: The surface layer of the ocean—that is, the upper part of the pelagic environment—where enough light penetrates to support photosynthesis; extends from the surface to a maximum depth of 150 m.

benthic environment: The ocean bottom or floor.

abyssal benthic zone: That part of the ocean floor that extends from a depth of 4,000 to 6,000 m.

hadal benthic zone: That part of the ocean floor that extends from 6,000 m to the bottom.

usually do not have any notable adaptations to survive drying out or exposure.

A rocky shore provides a fine anchorage for seaweeds and marine animals but is exposed to wave action when immersed during high tides and to drying and temperature changes when exposed to air during low tides (Figure 7.20). A typical rocky-shore inhabitant has some way of sealing in moisture, perhaps by closing its shell, if it has one, plus a powerful means of anchoring itself to the rocks. Mussels have tough, threadlike anchors secreted by a gland in the foot, and barnacles have special glands that secrete a tightly bonding glue that hardens under water. Rocky-shore intertidal algae, such as rockweed (*Fucus*), usually have thick, gummy coats, which dry out slowly when exposed to air, and flexible bodies not easily broken by wave action. Some rocky-shore organisms hide in burrows or under rocks or crevices at low tide, and some small crabs run about the splash line, following it up and down the beach.

The Benthic Environment: Seagrass Beds, Kelp Forests, and Coral Reefs

The **benthic environment** is the ocean floor. Most of the ocean floor consists of sediments (mostly sand and mud) in which many marine animals, such as worms and clams, burrow. Bacteria are common in marine sediments, but until recently it was assumed that bacteria did not extend very far into the sediments. In the 1990s bacteria were reported in ocean sediments more than 500 m (1625 ft) below the ocean floor at several different sites in the Pacific Ocean.

The **abyssal benthic zone** is that part of the benthic environment that extends from a depth of 4,000 to 6,000 m (2.5 to 3.7 mi), whereas the **hadal benthic zone** extends from 6,000 m to the bottom. ("Case in Point: Life Without the Sun" in Chapter 4 provided a description of the unusual organisms in the abyssal zone.) Here we describe benthic communities in shallow ocean waters that are particularly productive—seagrass beds, kelp forests, and coral reefs.

Sea grasses are flowering plants that have adapted to complete submersion in salty ocean water. They only occur in shallow water, to depths of 10 m (33 ft) where they receive enough light to photosynthesize efficiently. Extensive beds of sea grasses occur in quiet temperate, subtropical, and tropical waters; no sea grasses occur in polar waters. Eelgrass is the most widely distributed sea grass along the coasts of North America. The most common sea grasses in the Caribbean Sea are manatee grass and turtle grass (Figure 7.21). Sea grasses have a high primary productivity and are therefore ecologically important in shallow marine areas. Their roots and rhizomes help stabilize the sediments, reducing surface erosion. Sea grasses provide food and habitat for many marine organisms. In temperate waters, ducks and geese eat sea grasses, whereas in tropical waters, manatees, green tur-

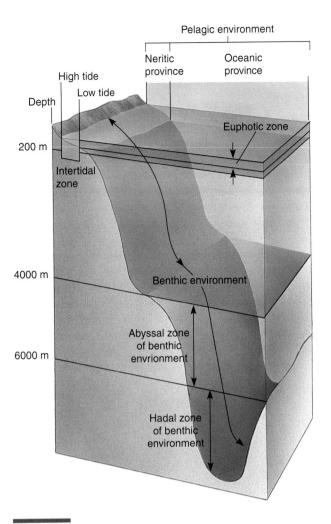

Figure 7.19 Zonation in the ocean. The ocean has three main life zones: the intertidal zone, benthic environment, and pelagic environment. The pelagic environment consists of the neritic and oceanic provinces. (The slopes of the ocean floor are not as steep as shown; they are exaggerated to save space.)

ment"). The pelagic environment is in turn divided into two provinces based on the distance from shore, the neritic province and the oceanic province.

The Intertidal Zone: Transition Between Land and Ocean

The area of shoreline between low and high tides is called the **intertidal zone**. Although the high levels of light and nutrients, together with an abundance of oxygen, make the intertidal zone a biologically productive habitat, it is also a very stressful one. If an intertidal beach is sandy, inhabitants must contend with a constantly shifting environment that threatens to engulf them and gives them scant protection against wave action. Consequently, most sand-dwelling organisms, such as mole crabs, are continuous and active burrowers. Because they are able to follow the tides up and down the beach, they

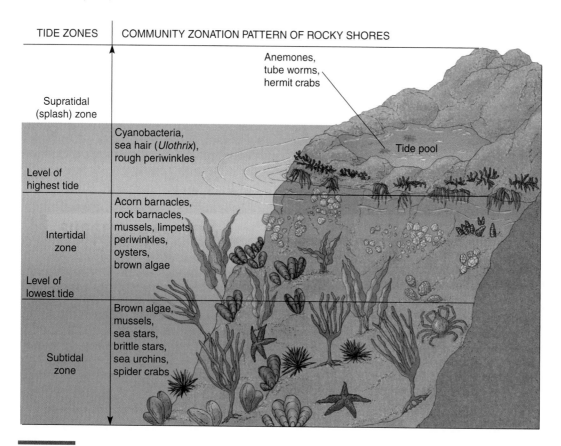

Figure 7.20 Zonation along a rocky shore. Three zones are shown: the supratidal (splash) zone, the intertidal zone, and the subtidal zone (part of the benthic environment). Representative organisms are given for each of these zones. Although exposed to wave action when submerged and to drying and seasonal heat and freezing when exposed to the air, the intertidal zone is crowded with many species. The subtidal zone is rarely exposed to the atmosphere except during spring tides. A greater diversity of organisms lives here than at higher levels on the rocky shore.

Figure 7.21 Sea grass bed. Turtle grasses (*Thalassia* sp.) have a large number of epiphytes attached to their leaves. Note the white sea anemone in the middle of the bed of turtle grass. These underwater meadows are ecologically important for shelter and food for many organisms. Photographed off the coast of Florida.

tles, parrot fish, sturgeon fish, and sea urchins eat them. These herbivores consume only about 5% of the sea grasses. The remaining 95% eventually enter the detritus food web and are decomposed by bacteria when the sea grasses die. The bacteria are in turn consumed by a variety of animals such as mud shrimp, lugworms, and mullet (a type of fish).

Kelps, which may reach lengths of 60 m (200 ft), are the largest brown algae (Figure 7.22). Kelps are common in cooler temperate marine waters of both Northern and Southern Hemispheres. They are especially abundant in relatively shallow waters (depths of about 25 m, or 82 ft) along rocky coastlines. Kelps are photosynthetic and are therefore the primary food producers for the kelp "forest" ecosystem. Kelp forests also provide habitats for many marine animals. Tube worms, sponges, sea cucumbers, clams, crabs, fishes, and sea otters find refuge in the algal fronds. Some animals eat the fronds, but kelps are mainly consumed in the detritus food web. Bacteria that decompose the kelp remains provide food for sponges,

Figure 7.22 Kelp forest. These underwater forests are ecologically important because they support many kinds of aquatic organisms. Photographed off the coast of California.

tunicates, worms, clams, and snails. The diversity of life supported by kelp beds almost rivals that found in coral reefs.

Coral reefs, which are built from accumulated layers of calcium carbonate ($CaCO_3$), are found in warm (usually greater than 21°C), shallow seawater (Figure 7.23). The living portions of coral reefs must grow in shallow waters where light penetrates. Many coral reefs are composed principally of red coralline algae that require light for photosynthesis. Other coral reefs consist of colonies of millions of tiny coral animals, which also require light for the large number of symbiotic algae, known as **zooxanthellae**, that live and photosynthesize in their tissues. Although species of coral exist without zooxanthellae,

only those with zooxanthellae build reefs. In addition to obtaining food from the zooxanthellae that live inside them, coral animals capture food at night with stinging tentacles that paralyze zooplankton and small animals that drift nearby. The waters in which coral reefs are found are often poor in nutrients. Yet other factors are favorable for a high productivity, including the presence of zooxanthellae, a favorable temperature, and year-round sunlight.

Coral reefs grow slowly, as coral organisms build on the calcium carbonate remains of countless organisms before them. On the basis of their structure and underlying geological features, there are three kinds of coral reefs: fringing reefs, atolls, and barrier reefs (Figure

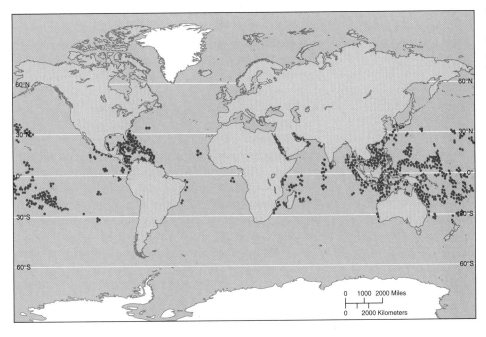

Figure 7.23 Distribution of coral reefs. There are more than 6,000 coral reefs worldwide.

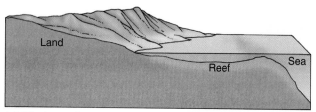

(a) Fringing reef

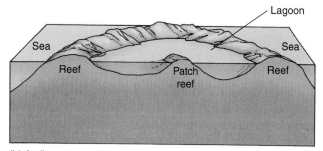

(b) Atoll

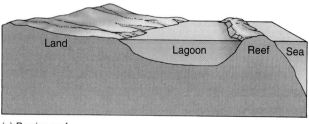

(c) Barrier reef

Figure 7.24 Types of coral reefs. (a) Fringing reef. (b) Atoll. (c) Barrier reef.

7.24). The most common type of coral reef is the fringing reef. A **fringing reef** is directly attached to the shore of a volcanic island or continent and has no lagoon associated with it. An **atoll** is a circular coral reef that surrounds a central lagoon of quiet water. An atoll forms on top of the cone of a submerged volcano island. More than 300 atolls are found in the Pacific and Indian Oceans, whereas the Atlantic Ocean has few atolls because the water there is generally too cool for coral animals. A lagoon of open water separates a **barrier reef** from the nearby land. The world's largest barrier reef is the Great Barrier Reef, which is nearly 2,000 km (more than 1,200 mi) in length and up to 100 km (62 mi) across. It extends along the northeastern coast of Australia. The second largest barrier reef is found in the Caribbean Sea off the coast of Belize.

Coral reef ecosystems are the most diverse of all marine environments (see Figure 4.2). They contain hundreds of species of fishes and invertebrates, such as giant clams, snails, sea urchins, sea stars, sponges, flat-

worms, brittle stars, sea fans, shrimp, and spiny lobsters. The Great Barrier Reef occupies only 0.1% of the ocean's surface, but 8% of the world's fish species live there. Many fishes are brightly colored to advertise the fact that they are poisonous. The conspicuous lion fish, for example, defends itself from potential predators with numerous spiny projections that contain a powerful poison. The complex multitude of relationships and interactions that occur at coral reefs is comparable only to tropical rain forests among terrestrial ecosystems. As in the rain forest, competition is intense, particularly for light and space to grow.

Many unusual relationships occur at coral reefs. Certain tiny fishes swim over and even inside the mouths of larger fishes to remove potentially harmful parasites. Fishes sometimes line up at these cleaning stations, waiting their turn to be serviced. As another example, *Podillopora* corals have a remarkable defense mechanism that protects them from coral browsers such as the crown-of-thorns sea star. Tiny crabs live among the branches of the coral. When a sea star crawls onto the coral, the crabs swarm over it, nipping off its tube feet and eventually killing it if it does not retreat. (Since the late 1960s, population explosions of the crown-of-thorns sea stars have overrun many coral reefs, causing extensive devastation of the coral. In most cases, the number of sea stars has eventually declined to normal levels. It is not known what factors cause these outbreaks.)

Coral reefs are ecologically important because they both provide a habitat for a wide variety of marine organisms and protect coastlines from shoreline erosion. They also provide humans with seafood, pharmaceuticals, and recreational/tourism dollars. Although coral formations are important ecosystems, they are being degraded and destroyed. Of 109 countries with large reef formations, 90 are damaging them. According to the first global assessment of reefs jointly published in 1998 by the U.N. Environment Program and several leading conservation organizations, 27% of the world's coral reefs are at high risk. Coral reefs off southeastern Asia, which contain the most species of all coral reefs, are the most threatened of any region.

Human Impacts on Coral Reefs Throughout the tropical western Atlantic, 30% to 50% of coral species are either rare or endangered. In some areas, silt washing downstream from clearcut inland forests has smothered reefs under a layer of sediment. High salinity resulting from the diversion of fresh water to supply the growing human population is thought by some scientists to be killing Florida reefs. Overfishing, pollution from sewage discharge and agricultural runoff, oil spills, boat groundings, fishing with dynamite or cyanide, hurricane damage, disease, coral bleaching, land reclamation, tourism, and the mining of corals for building material are also taking a heavy toll.

Marine biologists are especially concerned about the more than 1 million scuba divers and snorkelers that visit coral reefs each year. Most divers and snorkelers are unaware of how vulnerable coral reefs and their associated organisms are to human interactions. Many reef dwellers are killed or injured simply by touching or squeezing them. When a diver or snorkeler accidentally kicks or grabs the reef, pieces break off. Divers also stir up the bottom sediments, which suffocate the coral animals.

Since the late 1980s, corals in the tropical Atlantic and Pacific have suffered extensive bleaching, a phenomenon in which stressed corals expel their zooxanthellae, the symbiotic algae that supply nutrients to the corals and give the reef its color. Scientists are slowly gaining ground on an understanding of what causes bleaching. They suspect several environmental stressors, the most likely being warmer seawater temperatures (water only about 1 degree above average can lead to bleaching). Many scientists attribute recent oceanic warming, with record sea temperatures in 1998 and the largest die-off of corals in history (more than 70% losses in some areas), to El Niño effects, global warming, or a combination of the two. Other potential stressors are pollution or disease.

Bleaching, however, may not be the fatal blow once believed. Scientists are learning that the coral–zooxanthellae relationship is highly complex and flexible:

- Natural variation in zooxanthellae density, coupled with seasonal environmental fluctuations, may explain some bleaching episodes.
- Corals may lose 75% of their zooxanthellae seasonally without harming the reef.
- Corals may hold a "secret reserve" of zooxanthellae, not immediately apparent, that allows them to recover when bleached.
- Corals can take in any of several different zooxanthella species, perhaps allowing them to be rescued by one zooxanthella species when abandoned by another.

In short, corals may be able to bounce back from even vast bleaching episodes and thus may adapt to climate shifts. Prospects for corals remain uncertain, however, as the incidence of coral diseases is on the rise, and estimates of the percentage of reefs beyond recovery continue to climb. In 2002 biologists listed the top 10 coral reefs that are most vulnerable to human activities. They are: Philippines; Gulf of Guinea, off the west coast of Africa; Sunda Islands, Indonesia; southern Mascarene Islands (Indian Ocean); eastern coastal waters of South Africa; northern Indian Ocean; waters of southern Japan, Taiwan, and southern China; Cape Verde Islands off the west coast of Africa; western Caribbean; and the Red Sea and Gulf of Aden.

The Neritic Province: Shallow Waters Close to Shore

The **neritic province** is open ocean that overlies the ocean floor from the shoreline to a depth of 200 m (650 ft). Organisms that live in the neritic province are all floaters or swimmers. The upper reaches of the pelagic environment comprise the **euphotic zone**, which extends from the surface to a maximum depth of 150 m (488 ft) in the clearest open ocean water. Sufficient light penetrates the euphotic zone to support photosynthesis. Large numbers of phytoplankton, particularly diatoms in cooler waters and dinoflagellates in warmer waters, produce food by photosynthesis and are thus the base of food webs. Zooplankton (including tiny crustaceans, jellyfish, comb jellies, protists such as foraminiferans, and larvae of barnacles, sea urchins, worms, and crabs) feed on phytoplankton. Zooplankton are consumed by plankton-eating nekton such as herring, sardines, squid, baleen whales, and manta rays. These in turn become prey for carnivorous nekton such as sharks, tuna, porpoises, and toothed whales. Nekton are mostly confined to the shallower neritic waters (less than 60 m, or 195 ft, deep), near their food.

The Oceanic Province: Most of the Ocean

The average depth of the ocean is 4,000 m (more than 2 mi). The **oceanic province** is the part of the pelagic environment that overlies the ocean floor at depths greater than 200 m (650 ft). It is the largest marine environment, comprising about 75% of the ocean's water. Most of the oceanic province is loosely described as the "deep sea." Cold temperatures, high hydrostatic pressure, and an absence of sunlight characterize all but the surface waters of the oceanic province. These environmental conditions are uniform throughout the year.

Most organisms of the deep waters of the oceanic province depend on **marine snow**, organic debris that drifts down into their habitat from the upper, lighted regions of the oceanic province. Organisms of this little-known realm are filter feeders, scavengers, or predators. Many are invertebrates, some of which attain great sizes. The giant squid measures up to 18 m (59 ft) in length, including its tentacles. Fishes of the deep waters of the oceanic province are strikingly adapted to darkness and scarcity of food. An organism that encounters food infrequently needs to eat as much as possible when food is available. The rare gulper eel has huge jaws that enable it to swallow large prey (Figure 7.25). Adapted to drifting or slow swimming, animals of the oceanic province are often characterized by reduced bone and muscle mass. Many of these animals have light-producing organs that enable them to locate one another for mating or food capture. The dragonfish has a pocket of red light shining from beneath each eye. Because other

Figure 7.25 Oceanic province.
The gulper eel (*Saccopharynx laven-bergi*) is highly specialized for its deep-sea habitat. It uses its "trap-door" jaws to swallow prey as large as itself. The tail, of which only a small portion is shown, makes up most of the length of a gulper eel's body. Gulper eels grow to 1.8 m (6 ft). Shown is a live specimen, photographed in an aquarium aboard ship after being captured at a depth of 1,500 m (4,921 ft) off southern California.

species living in the ocean's depths are unable to see red light, the dragonfish is able to detect organisms in its surroundings without being seen. (Most animals in the ocean depths produce a blue-green light.)

The Impact of Human Activities on the Ocean

Because the ocean is so vast, it is hard to visualize that human activities could affect it, much less harm it. Such is the case, however. Development of resorts, cities, industries, and agriculture along coasts alters or destroys many coastal ecosystems, including mangrove forests, salt marshes, seagrass beds, and coral reefs. Coastal and marine ecosystems receive pollution from land, from rivers emptying into the ocean, and from atmospheric contaminants that enter the ocean via precipitation. Disease-causing viruses and bacteria from human sewage contaminate shellfish and other seafood and pose an increasing threat to public health. Millions of tons of trash, including plastic, fishing nets, and packaging materials find their way into coastal and marine ecosystems; some of this trash entangles and kills marine organisms. Less visible contaminants of the ocean include fertilizers, pesticides, heavy metals, and synthetic chemicals from agriculture and industry.

Offshore mining and oil drilling pollute the neritic province with oil and other contaminants. Millions of ships dump oily ballast and other wastes overboard in the neritic and oceanic provinces. Fishing is highly mechanized, and new technologies can remove every single fish in a targeted area of the ocean (see Figure 18.14). Scallop dredges and shrimp trawls are dragged across the benthic

environment, destroying entire communities with a single swipe.

These problems have been studied for many years, and conservation groups and government agencies have made numerous recommendations to protect and manage the ocean's resources. The following are a few ideas that have been proposed:

1. All countries should develop strict policies to protect their coastal and marine ecosystems.
2. Governments should use existing scientific knowledge to develop immediate plans to manage fish stocks; in addition, governments should support scientific research to expand our knowledge in this and other critical fields of marine science.
3. Environmental education should be part of the curriculum in every country and should include a strong marine component.
4. Protected coastal and marine areas should be established and carefully monitored around the world.
5. Reduction of marine pollution from land-based sources (sewage discharges, agricultural pollutants, and industrial effluents) should be a top priority of every government.
6. Pollution from ships and offshore installations should be prohibited, and this prohibition should be effectively enforced.

Most countries recognize the importance of the ocean to life on this planet, but few have the resources or programs to protect and manage the ocean effectively. The actions just listed will require global, regional, and local cooperation; billions of dollars; and years of effort by millions of people if they are to be realized.

INTERACTION OF LIFE ZONES

Although we have discussed terrestrial and aquatic life zones as discrete entities, none of them exists in isolation. When parts of the Amazon rain forest flood annually, fishes leave the stream beds and swim all over the forest floor, where they play a role in dispersing the seeds of many species of plants. And in the Antarctic, where waters are much more productive than land areas, the many seabirds and seals form a link between marine and terrestrial environments. Although the ocean supports these animals, their waste products, cast-off feathers, and the like, when deposited on land, support whatever lichens, insects, and microorganisms live there.

Some inhabitants of terrestrial and aquatic life zones cover great distances—in the case of migratory fishes and birds, even global distances. Many young albacore tuna migrate from the California coast across the Pacific Ocean to Japan. Some species of flycatchers spend their summers in Canada and the United States and their winters in Central and South America. Like the flycatchers, many other migratory birds commonly spend critical parts of their life cycles in entirely different countries, which can make their conservation difficult (see Chapter 1). It does little good, for instance, to protect a songbird in one country if the inhabitants of the next put it in the cooking pot as soon as it arrives. Such large-scale interactions among various ecosystems make ecological concepts difficult for many people and nations to grasp and apply.

CASE·IN·POINT **The Everglades**

Throughout this chapter, we have provided glimpses of how human activities are altering and in some cases destroying biomes and aquatic ecosystems. The Everglades is an excellent example of a severely altered ecosystem that, depending on human policies and intervention, may be partially restored during the first half of the 21st century.

The Everglades in the southernmost part of Florida is a vast expanse of predominantly sawgrass wetlands dotted with small islands of trees. At one time 80 km (50 mi) wide, 160 km (100 mi) long, and 15 cm to 0.9 m (6 in. to 3 ft) deep, the "river of grass" drifted south in a slow-moving sheet of water from Lake Okeechobee to Florida Bay (Figure 7.26). The Everglades is a haven for wildlife, including alligators, snakes, panthers, otters, raccoons,

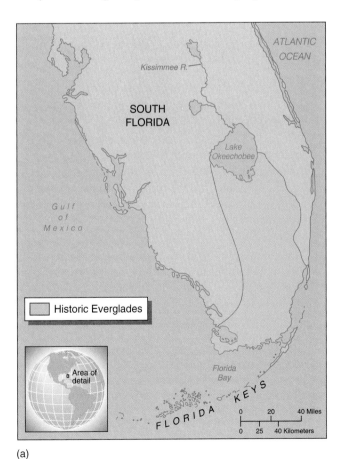

(a)

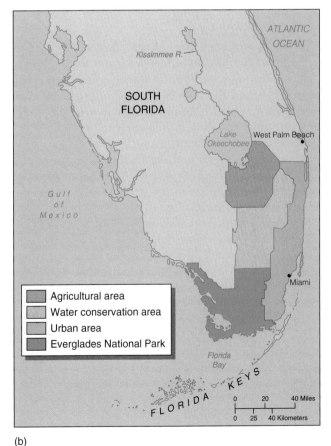

(b)

Figure 7.26 Florida Everglades. (a) Extent of the original Everglades. (b) The Everglades today.

and thousands of wading birds and birds of prey—great blue herons, snowy egrets, great white herons, roseate spoonbills, and osprey, to name just a few. The region's natural wonders were popularized in an environmental classic, *The Everglades: River of Grass*, written by Marjory Stoneman Douglas in 1947. In it Douglas voiced concern about what would happen to the Everglades if development and commercial interests were allowed to drain and develop it. The southernmost part of the Everglades is now protected as a national park that has also been given three international designations: an International Biosphere Reserve, a World Heritage Site, and a Wetland of International Importance.

South of the Everglades is Florida Bay, a shallow estuary dotted with many tiny islands, or keys. Florida Bay and the Florida Keys are greatly affected by the water leaving the Everglades. Both Florida Bay and the Everglades are important nurseries for commercially important fishes, shrimp, lobsters, and stone crabs. With all the wildlife and recreational opportunities in the Everglades and Florida Bay, it is not surprising that the local economy relies heavily on tourism and commercial fishing.

The Everglades today is about half its original size of 1.6 million hectares (4 million acres), and it has many serious environmental problems. Most water bird populations are down by 90% in recent decades, and the area is now home to 50 endangered or threatened species.

How the Everglades Was Almost Destroyed Let us examine a brief history of the Everglades as an illustration of how misguided human activities can cause more harm than good. Basically, two problems override all others in the Everglades today—it receives too little water, and the water it receives is polluted with nutrient minerals from agricultural runoff.

During heavy rains, Lake Okeechobee historically flooded its banks, creating wetlands that provided biological habitat and helped to recharge the Everglades. However, when a hurricane hit the lake in 1928 and more than 1,800 people died, the Army Corps of Engineers built the 12.2-m (40-ft) high Hoover Dike along the eastern 240 km (150 mi) of the lake. The Hoover Dike, which was completed in 1932, stopped the flooding, but it also prevented the water in Lake Okeechobee from recharging the Everglades. Four canals built by the Everglades Drainage District effectively drained 214,000 hectares (530,000 acres) of land south of Lake Okeechobee, which was converted to farmland. The fertilizers and pesticides used in this area eventually make their way to the Everglades, where they alter native plant communities. Phosphorus in fertilizer is particularly harmful because it encourages the growth of non-native cattails that overrun the native sawgrasses.

After several tropical storms caused flooding damage in South Florida in 1947, Congress authorized the Army Corps of Engineers to construct an extensive system of canals, levees, and pump stations to prevent flooding, provide drainage, and supply water to South Florida. These structures, which divert excess water to the Atlantic Ocean rather than the Everglades, stopped the periodic floods in South Florida. However, the drier lands produced by the levees, canals, and pumps encouraged accelerated urban growth, particularly along the east coast, and the expansion of agriculture, primarily sugar growing, into the southernmost parts of the Everglades.

Thus, more than 70 years of engineering projects have reduced the quantity of water flowing into the Everglades, and the water that does enter is polluted from agricultural runoff. Urbanization has also contributed to the Everglades' problems by causing **habitat fragmentation**. Like the Everglades, Florida Bay is showing signs of environmental decline. Because it receives so little water from the Everglades, Florida Bay's water has become saltier. Blooms of cyanobacteria sometimes cover as much as 40% of the bay. Its sea grasses, fishes, and sponge populations have all declined, and tourism and commercial fishing have suffered.

Restoration of the Everglades Florida's unique river of grass will never return completely to its original natural condition because there are too many sugar plantations and too many cities in the region. The Everglades can be partially restored, however, and in 1996 the state of Florida and the U.S. government began planning a massive restoration project to undo decades of human interference. The plan has three parts:

1. Farmers will be forced to clean up their runoff so that the amount of phosphorus entering the Everglades will be reduced.

2. At least 16,180 hectares (40,000 acres) of agricultural land located at the southern end of the Everglades Agricultural Area below Lake Okeechobee will be bought. This land will be converted to marshes that will filter and further clean the agricultural runoff of the remaining 267,000 hectares (660,000 acres) of farmland before it reaches the Everglades.

3. The U.S. Army Corps of Engineers and the South Florida Water Management District will undertake a massive project to reengineer the area's entire system of canals, levees, and pumps so that a more natural flow of water will be restored to the Everglades.

Restoration plans have not come without bitter debates that have pitted the state's tourism, fishing, and environmental interests against the region's urban developers and sugar farmers, who are prosperous and politically powerful. One reason the sugar industry is so prosperous is that the federal government provides it with price supports, or subsidies, that prevent U.S. consumers from purchasing less expensive imported sugar. As a result, U.S. consumers pay sugar prices that are at least two times higher than average world sugar prices.

In 1997 the U.S. government and the state of Florida agreed to purchase more than 20,250 hectares (more than 50,000 acres) of sugarcane fields close to Everglades National Park. The **Comprehensive Everglades Restoration Plan**, unveiled in 1998, will eventually supply clean, fresh water to the Everglades and to the fast-growing population of South Florida. (The population in South Florida is currently 6 million and is projected to be 15 million by the year 2050.) The plan involves drilling 333 wells into the Florida Aquifer, a porous layer of limestone about 305 m (1,000 ft) underground. The wells, which would be located near Lake Okeechobee, would pump excess water produced during the rainy season underground, where it could be stored until it is needed during dry spells. Roads through the area must be raised so the water can flow under them.

In 2001 the plan was criticized by an independent National Academy of Sciences panel of prominent scientists, who examined the significant uncertainties of using aquifer storage, which is largely untested. In 2002 the panel issued a second report with more criticisms of the plan; this report said the plan will probably not help transform the cloudy waters of Florida Bay into the fisherman's paradise it once was (as recently as the 1970s). Obviously, a great deal of scientific research must be done before the restoration, including the aquifer storage and recovery plan, can be completed. Many questions remain about how best to restore a more natural water flow, repel invasions of foreign species, and reestablish the native biological diversity.

The Army Corps of Engineers and the South Florida Water Management District, a state agency, have been charged with overseeing the project, which will take more than 20 years and cost $8 billion. As the Everglades is restored to a scaled-down version of what it originally was, it is hoped that water quality in Florida Bay and the Florida Keys will also improve. ◼

SUMMARY WITH SELECTED KEY TERMS

I. A **biome** is a large, relatively distinct terrestrial region with characteristic climate, soil, plants, and animals, regardless of where it occurs. A biome encompasses many interacting ecosystems. Temperature and precipitation are important abiotic factors that influence biome distribution.

A. Tundra, the northernmost biome, is characterized by a permanently frozen layer of subsoil called the **permafrost**, by low-growing vegetation adapted to extreme cold, and by a very short growing season. Low **species richness** and low **primary productivity** characterize tundra.

B. The **taiga**, or **boreal forest**, lies south of the tundra and is dominated by coniferous trees that are adapted to the cold winters, a short growing season, and acidic, mineral-poor soil.

C. Temperate rain forest, such as occurs on the northwest coast of North America, receives very high precipitation and is dominated by large conifers.

D. Temperate deciduous forest occurs where precipitation is relatively high and soils are rich in organic matter. Broad-leaf trees that lose their leaves seasonally dominate this biome.

E. Temperate grassland typically possesses deep, mineral-rich soil and has moderate but uncertain precipitation. Temperate grassland is well suited to growing grain crops.

F. Thickets of small-leaf evergreen shrubs and trees and a **Mediterranean climate** of wet, mild winters and dry summers characterize the **chaparral**.

G. Desert, found in both temperate (cold desert) and subtropical or tropical regions (warm desert) with low levels of precipitation, contains organisms with specialized water-conserving adaptations.

H. Tropical grassland, called **savanna**, has widely scattered trees interspersed with grassy areas. It occurs in tropical areas with low or seasonal rainfall with prolonged dry periods.

I. A tropical environment with mineral-poor soil and very high rainfall that is evenly distributed through the year characterizes **tropical rain forest**. Tropical rain forest has high species richness and high productivity.

II. In aquatic ecosystems, important environmental factors include **salinity**, amount of dissolved oxygen, and availability of light for photosynthesis.

A. Aquatic life is ecologically divided into **plankton** (free-floating), **nekton** (strongly swimming), and **benthos** (bottom-dwelling).

1. The microscopic **phytoplankton** are photosynthetic and are the base of the food web in most aquatic communities.

2. Zooplankton are nonphotosynthetic organisms that include protozoa, tiny crustaceans, and the immature stages of many animals.

B. Freshwater ecosystems include flowing-water (rivers and smaller streams), standing-water (lakes and ponds), and freshwater wetlands.

1. In **flowing-water ecosystems**, the water flows in a current, which is swifter in headwaters than downstream. Flowing-water ecosystems have few phytoplankton and depend on detritus from the land for much of their energy.

2. Large **standing-water ecosystems** (freshwater lakes) are divided into zones on the basis of water depth.

a. The marginal **littoral zone** contains both emergent vegetation and algae and is very productive.

b. The **limnetic zone** is open water away from the shore that extends down as far as sunlight penetrates to permit photosynthesis. Organisms in the limnetic zone include phytoplankton, zooplankton, and larger fishes.

c. The deep, dark **profundal zone** holds little life other than bacterial decomposers.

    d. Large temperate lakes exhibit **thermal stratification**, in which the temperature changes sharply with depth.

  3. Freshwater wetlands, lands that are transitional between freshwater and terrestrial ecosystems, are usually covered by shallow water and have characteristic soils and vegetation. They are highly productive areas that perform many valuable **ecosystem services**.

C. An **estuary** is a coastal body of water, partly surrounded by land, with access to the ocean and a large supply of fresh water from rivers.

  1. Water levels in an estuary rise and fall with the tides.

  2. Salinity fluctuates with tidal cycles, the time of year, and precipitation. Within the estuary, salinity also changes gradually, from unsalty fresh water at the river entrance to salty ocean water at the estuary's mouth.

  3. Estuaries are very productive, in part because they receive a high input of nutrients from the adjacent land. Temperate estuaries usually contain **salt marshes**, whereas tropical estuaries are lined with **mangrove forests**.

  4. The largest, most productive estuary in the United States is the Chesapeake Bay. The **Chesapeake Bay Program** is overseeing the restoration and protection of the bay.

D. Four important marine environments are the intertidal zone, benthic environment, neritic province, and oceanic province.

  1. The **intertidal zone** is the shoreline area between low and high tides. It is a very productive area. Organisms of the intertidal zone possess adaptations to resist wave action and the extremes of being covered by water (high tide) and exposed to air (low tide).

  2. The **benthic environment** is the ocean floor.

    a. Sea grasses are flowering plants that have adapted to complete submersion in salty ocean water. They have a high primary productivity and are ecologically important in shallow marine areas.

    b. Kelps, the largest brown algae, are common in cooler temperate marine waters, particularly in relatively shallow waters along rocky coastlines. Kelp forests provide food and habitats for many marine animals.

    c. Coral reefs are important benthic communities in shallow tropical ocean waters. Coral reefs, which have high species richness and high productivity, are classified as **fringing reefs**, **barrier reefs**, and **atolls**.

  3. The **neritic province** is open ocean that overlies the ocean floor from the shoreline to a depth of 200 m. Organisms that live in the neritic province are all floaters or swimmers. Phytoplankton in the **euphotic zone** are the base of the food web.

  4. The **oceanic province** is that part of the open ocean that overlies the ocean floor at depths greater than 200 m. The uniform environment beneath the surface waters of the oceanic province is dark, cold, and under high pressure. Animal inhabitants are predators or scavengers that subsist on detritus that drifts in from surface waters.

III. The Everglades in the southernmost part of Florida is a vast expanse of sawgrass wetlands dotted with small islands of trees. Called the "river of grass," it is a slow-moving sheet of water that drifts from Lake Okeechobee to Florida Bay.

A. The Everglades today is about half its original size and has many serious environmental problems. Most water bird populations are down by 90% in recent decades.

B. Engineering projects have reduced the quantity of water flowing into the Everglades, and the water that does enter is polluted from agricultural runoff. Urbanization has caused **habitat fragmentation**.

C. Although the Everglades will never return completely to its original natural condition, it can be partially restored. The **Comprehensive Everglades Restoration Plan** is a massive restoration project that will eventually supply fresh water to the Everglades and to the fast-growing population of South Florida.

THINKING ABOUT THE ENVIRONMENT

1. What is a biome? What two climate factors are most important in determining an area's characteristic biome?

2. How does vegetation change with increasing elevation and latitude?

3. What climate and soil factors produce each of the major terrestrial biomes?

4. Describe representative organisms of the forest biomes discussed in the text: taiga, temperate deciduous forest, temperate rain forest, and tropical rain forest.

5. In which biome do you live? If your biome does not match the description given in this book, how do you explain the discrepancy?

6. Which biomes are best suited for agriculture? Explain why each of the biomes you did not specify is less suitable for agriculture.

7. What human activities are harmful to deserts? To grasslands? To forests?

8. What environmental factors are most important in determining the kinds of organisms found in aquatic environments?

9. Distinguish between freshwater wetlands and estuaries, and between flowing-water and standing-water ecosystems.

10. What would happen to the organisms in a river with a fast current if a dam were built? Explain your answer.

11. List and briefly describe the four main marine environments.

12. Which aquatic ecosystem is often compared to a tropical rain forest? Why?

13. Explain how coral reefs are vulnerable to human activities.

14. What are some of the challenges associated with restoration of the Florida Everglades?

*15. According to the U.N. Environment Program, marine pollution sources include 44% runoff and discharges from land, 33% airborne emissions from land, 12% shipping and accidental spills, 10% ocean dumping, and 1% off-shore mining and oil and gas drilling. Use a protractor to draw a pie chart that reflects the percentages of each pollution source.

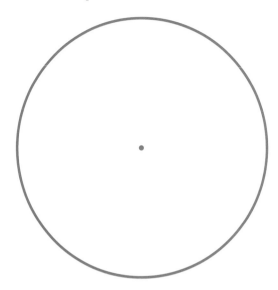

*16. The following three figures show the average precipitation (blue bars) and the average temperature (black line) each month at a particular location in North America. Study the figures, which are called climographs. Determine which figure most likely corresponds to a temperate desert, which to a temperate deciduous forest, and which to a temperate grassland. Explain your answers.

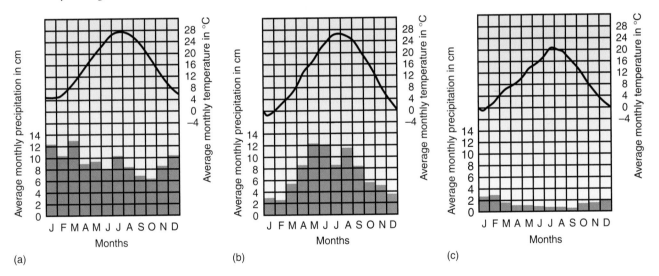

(a)   (b)   (c)

* Solutions to questions preceded by asterisks appear in Appendix VII.

TAKE A STAND

Visit our Web site at **http://www.wiley.com/college/raven** (select Chapter 7 from the Table of Contents) for links to more information about the restoration of the Florida Everglades. Consider the opposing views of sugar farmers and conservationists, and debate the issue with your classmates. You will find tools to help you organize your research, analyze the data, think critically about the issues, and construct a well-considered argument. Take a Stand activities can be done individually or as part of a team, as oral presentations, written exercises, or Web-based (e-mail) assignments.

Additional on-line materials relating to this chapter, including Student Quizzes, Activity Links, Useful Web sites, Flash Cards, and more, can also be found on our Web site.

SUGGESTED READING

Douglas, M.S. *The Everglades: River of Grass.* Sarasota, FL: Pineapple Press, 1947 (revised ed., 1988). The classic book that informed the public of the Everglades' unique natural and cultural history.

Earle, S. "Does It Matter What We Do to the World's Oceans?" *Environmental Review*, Vol. 6 (November 1999). This is the complete text of Sylvia Earle's marvelous speech to the Ecological Society of America. Her love for the ocean and all living things in it is contagious.

Kiester, E. "Getting Their Goats." *Smithsonian* (October 2001). California communities are using goats to lessen the dangers from wildfires in the chaparral.

Mastny, L. "A Worldwatch Addendum." *World Watch*, Vol. 14, No. 3 (May–June 2001). Examines the status of coral reefs around the world.

Montaigne, F. "Boreal: The Great Northern Forest." *National Geographic*, Vol. 201, No. 6 (June 2002). The northern forest of Canada, Scandinavia, and Russia is under threat from logging, mining, road building, and hydroelectric development.

Pain, S. "Lair of the Dragon." *New Scientist*, Vol. 161 (March 27, 1999). Many fish in the dark depths of the ocean have light-producing organs to help them hunt and find prospective mates.

Perkins, S. "A Nation Aflame." *Science News*, Vol. 159 (February 24, 2001). An excellent review of the impact of wildfires in the United States.

Poston, L. "Rags to Riches." *Living Planet* (Fall 2001). Beautiful photographs highlight this essay on the Florida Everglades.

Samuels, S.H. "In Iowa, Restitching the Torn Fragment of the Prairie." *New York Times* (September 21, 1999). The Neal Smith National Wildlife Refuge near Prairie City, Iowa, is a tallgrass prairie restored from former farmland.

Valiela, I., J.L. Bowen, and J.K. York. "Mangrove Forests: One of the World's Threatened Major Tropical Environments." *BioScience*, Vol. 51, No. 10 (October 2001). In the past 20 years, more than 35% of the world's mangrove forests have been lost.

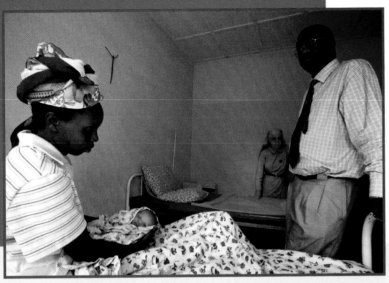

AIDS patients. Both the mother and her child have AIDS (acquired immune deficiency syndrome), but they do not have access to the medicines that can prolong their lives. The cost of a month's supply of drugs is equal to 1 year's earnings for many African people. Photographed in a hospital in Rwanda.

Understanding Population Change

Learning Objectives

After you have studied this chapter you should be able to:

1. Describe the extent of the HIV/AIDS epidemic in sub-Saharan Africa.

2. Explain the four factors that produce changes in population size and solve simple problems involving these changes.

3. Define biotic potential (intrinsic rate of increase) and carrying capacity and explain the differences between J-shaped and S-shaped growth curves.

4. Distinguish between density-dependent and density-independent factors that affect population size and give examples of each.

5. Describe some of the density-dependent factors that may affect boom-or-bust population cycles.

6. Describe Type I, Type II, and Type III survivorship curves.

7. Summarize the history of human population growth.

8. Identify Thomas Malthus, relate his ideas on human population growth, and explain why he may or may not be wrong.

9. Explain why it is impossible to answer precisely the question of how many people the Earth can support—that is, Earth's carrying capacity for humans.

10. Explain how highly developed and developing countries differ in population characteristics such as infant mortality rate, total fertility rate, replacement-level fertility, and age structure.

11. Briefly relate the history and controversies of U.S. immigration.

Africa has the most rapidly growing population of all the continents. Africa's population in mid-2002 was estimated at 840 million, with most of the population concentrated in sub-Saharan Africa, that part of Africa located south of the Sahara Desert. Experts predict that the sub-Saharan population of 660 million will more than double to 1.4 billion by 2030.

During the mid- and late-1990s, many population experts lowered their estimates of population growth rates in sub-Saharan Africa. These lower projections were not based on declining birth rates. Tragically, the estimates were lowered because acquired immune deficiency syndrome (AIDS), caused by the human immunodeficiency virus (HIV), is ravaging an increasing number of countries in the sub-Saharan region. The U.N. Program on AIDS estimates that more than 28.5 million of the world's 40 million people who are now infected with HIV/AIDS live in sub-Saharan Africa. Ten percent of the adult sub-Saharan population, ages 15 to 49, are infected. Botswana, Zimbabwe, and Swaziland in southern Africa have the highest prevalence of HIV/AIDS in the world: More than 33% of the adult population in each of these countries is infected. Even more tragic is the fact that at the end of 2001, about 11 million African children living in the region had lost one or both parents to AIDS. By 2010, that number is expected to double. Many of these AIDS orphans do not have adequate education, health care, or nutrition.

161

In developing countries, including those in sub-Saharan Africa, almost all people infected with HIV/AIDS die, usually within 10 years of infection, because they cannot afford the high cost of potent antiviral drugs. Countries in sub-Saharan Africa are experiencing high death rates from HIV/AIDS, which is now the leading cause of death in Africa. In 2001, 2.2 million people died of AIDS in sub-Saharan Africa. (Three million people died worldwide of AIDS that year.)

The high mortality from HIV/AIDS has caused life expectancies to decline in many African countries. The average life expectancy in Botswana, for example, declined from a high of about 60 years in the late 1980s to 39 years in 2002. Africa's AIDS crisis has many repercussions beyond a reduced life expectancy, however. The HIV/AIDS epidemic threatens the economic stability and social support networks of the region and has overwhelmed health care systems. Labor shortages are common, and foreign investments have slowed because many investors are wary of the high rate of infection among workers.

The HIV/AIDS epidemic is but one factor affecting the human population. In this chapter, we focus on the dynamics of population change characteristic of all organisms and then describe the current state of the human population. Chapter 9 examines the consequences of continued human population growth and explores ways to limit the expanding world population.

PRINCIPLES OF POPULATION ECOLOGY

Because the size of the human population is central to so many environmental problems and their solutions, it is important that we understand how populations increase or decrease. We can understand much about the biological principles that affect the growth of human populations by first studying populations of other species.

Individuals of a given species are part of a larger organization known as a population. Populations exhibit characteristics distinctive from those of the individuals of which they are composed. Some of the features discussed in this chapter that are characteristic of populations but not of individuals are population density, birth and death rates, growth rates, and age structure.

Population ecology deals with the number of individuals of a particular species that are found in an area and how and why those numbers increase or decrease over time. Population ecologists try to determine the population processes that are common to all populations. They study how a population responds to its environment, such as how individuals in a population compete for food or other resources, and how predation, disease,

and other environmental pressures affect the population. Population growth, whether of bacteria or maples or giraffes, cannot increase indefinitely because of such environmental pressures.

Additional aspects of populations that are important to environmental science are their reproductive success or failure (that is, extinction) and how populations affect the normal functioning of communities and ecosystems. Scientists in applied disciplines, such as forestry, agronomy (crop science), and wildlife management, must possess an understanding of population ecology in order to manage populations of economic importance, such as, forests, field crops, game animals, and fishes. An understanding of the population dynamics of endangered species plays a key role in efforts to prevent their slide to extinction. An understanding of the population dynamics of pest species plays a key role in efforts to prevent their increase to levels that cause significant economic or health impacts.

Population Density

By itself, the size of a population tells us relatively little. Population size is only meaningful when the boundaries of the population are defined. Consider the difference between 1,000 mice in 100 hectares (250 acres) and 1,000 mice in 1 hectare (2.5 acres). Often a population is too large to study in its entirety. Such a population is examined by sampling a part of it and then expressing the population in terms of density. Examples include the number of dandelions per square meter of lawn, the number of water fleas per liter of pond water, and the number of cabbage aphids per square centimeter of cabbage leaf. **Population density**, then, is the number of individuals of a species per unit of area or volume at a given time.

Different environments vary in the population density of any species that they can support. This density may also vary in a single habitat (local environment) from season to season or year to year. As an example, consider red grouse populations in northwest Scotland at two locations only 2.5 km (1.6 mi) apart (Figure 8.1). At one location the population density remained stable during a 3-year period, but at the other site it almost doubled in the first 2 years and then declined to its initial density in the third year. The reason was likely a difference in habitat. The area where the population density increased had been experimentally burned, and young heather shoots produced during the 2 years following the burn provided nutritious food for the red grouse. Population density, then, may be determined in large part by external factors in the environment.

How Do Populations Change in Size?

Populations of organisms, whether they are sunflowers, eagles, or humans, change over time. On a global scale, this change is due to two factors: the rate at which indi-

Figure 8.1 Red grouse. Red grouse are game birds whose populations are managed for hunting. Photographed in Deeside, Scotland.

viduals produce offspring (the birth rate) and the rate at which organisms die (the death rate) (Figure 8.2a). In humans, the **birth rate** *(b)* is usually expressed as the number of births per 1,000 people per year, and the **death rate** *(d)* as the number of deaths per 1,000 people per year (see "Mini-Glossary: Basic Population Terms").

The rate of change (increase or decrease), or **growth rate** *(r)*, of a population is the birth rate *(b)* minus the

---

death rate *(d)*. Growth rate is also referred to as **natural increase** in human populations.

$$r = b - d$$

As an example, consider a hypothetical human population of 10,000 in which there are 200 births per year (that is, by convention, 20 births per 1,000 people) and 100 deaths per year (that is, 10 deaths per 1,000 people).

$$r = \underbrace{20/1000}_{b} - \underbrace{10/1000}_{d}$$

$$r = 0.02 - 0.01 = 0.01, \text{ or } 1\% \text{ per year}$$

If organisms in the population are born faster than they die, *r* is a positive value, and population size increases. If organisms in the population die faster than they are born, *r* is a negative value, and population size decreases. If *r* is equal to zero, births and deaths match, and population size is stationary despite continued reproduction and death.

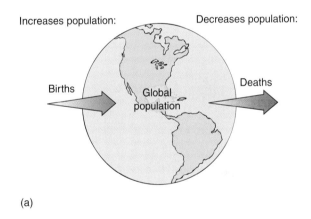

(a)

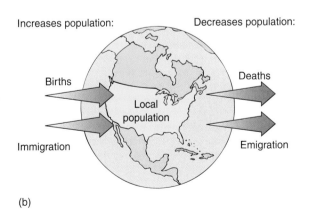

(b)

Figure 8.2 Factors that interact to change population size. (a) On a global scale, the change in a population is due to the number of births and deaths. (b) In local populations, such as the population of the United States, the number of births, deaths, immigrants, and emigrants affect population size.

MINI-GLOSSARY

Basic Population Terms

birth rate: The number of births per 1,000 individuals.

death rate: The number of deaths per 1,000 individuals.

growth rate: The natural increase of a population, expressed as percent per year.

dispersal: The movement of individuals among populations.

immigration: The dispersal of individuals into a population from another area or country.

emigration: The dispersal of individuals from a population, bound for another area or country in which to live.

In addition to birth and death rates, **dispersal,** which is movement from one region or country to another, must be considered when changes in populations on a local scale are examined. There are two types of dispersal: **immigration** *(i),* by which individuals enter a population and thus increase its size, and **emigration** *(e),* by which individuals leave a population and thus decrease its size. The growth rate of a local population must take into account birth rate *(b),* death rate *(d),* immigration *(i),* and emigration *(e)* (Figure 8.2b). The growth rate equals (the birth rate minus the death rate) plus (immigration minus emigration):

$$r = (b - d) + (i - e)$$

For example, the growth rate of a population of 10,000 that has 100 births (by convention, 10 per 1,000), 50 deaths (5 per 1,000), 10 immigrants (1 per 1,000), and 100 emigrants (10 per 1,000) in a given year would be calculated as follows:

$$r = \underbrace{(10/1000}_{b} - \underbrace{5/1000)}_{d} + \underbrace{(1/1000}_{i} - \underbrace{10/1000)}_{e}$$

$$r = (0.010 - 0.005) + (0.001 - 0.010)$$
$$r = 0.005 - 0.009 = -0.004, \text{ or } -0.4\% \text{ per year}$$

Maximum Population Growth

The maximum rate at which a population could increase under ideal conditions is known as its **biotic potential** or **intrinsic rate of increase.** Different species have different biotic potentials. A particular species' biotic potential is influenced by several factors. These include the age at which reproduction begins, the fraction of the life span during which an individual of that species is capable of reproducing, the number of reproductive periods per lifetime, and the number of offspring produced during each period of reproduction. These factors, called *life history characteristics,* determine whether a particular species has a large or a small biotic potential.

Generally, larger organisms, such as blue whales and elephants, have the smallest biotic potentials, whereas microorganisms have the greatest biotic potentials. Under ideal conditions (that is, an environment with unlimited resources), certain bacteria can reproduce by dividing in half every 30 minutes. At this rate of growth, a single bacterium would increase to a population of more than 1 million in just 10 hours (Figure 8.3a), and the population from a single individual would exceed 1 billion in 15 hours! If we plot the population number versus time, the graph has a J shape that is characteristic of **exponential population growth,** the accelerating population growth that occurs when optimal conditions allow a constant reproductive rate (Figure 8.3b). When a population grows exponentially, the larger that population

Time (hours)	Number of bacteria
0	1
0.5	2
1.0	4
1.5	8
2.0	16
2.5	32
3.0	64
3.5	128
4.0	256
4.5	512
5.0	1,024
5.5	2,048
6.0	4,096
6.5	8,192
7.0	16,384
7.5	32,768
8.0	65,536
8.5	131,072
9.0	262,144
9.5	524,288
10.0	1,048,576

(a)

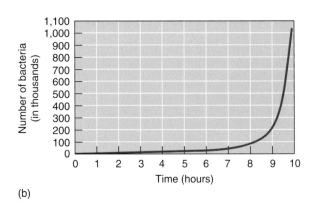

(b)

Figure 8.3 Exponential population growth. (a) When bacteria divide at a constant amount during each time period, their numbers increase exponentially. This set of figures assumes a zero death rate, but even if a certain percentage of each generation of bacteria died, exponential population growth would still occur; it would just take longer to reach the very high numbers. (b) When these data are graphed, the curve of exponential population growth has a characteristic J shape.

gets, the faster it grows. It doubles, then doubles again, then again, but each time the doubling occurs in a *shorter* time period.

Regardless of which species one is considering, whenever the population is growing at its biotic potential, population size plotted versus time gives a curve of the same shape. The only variable is time. It may take longer for a lowland gorilla population than for a bacterial population to reach a certain size (because gorillas do not reproduce as rapidly as bacteria), but both populations will always increase exponentially as long as their growth rates remain constant.

Environmental Resistance and Carrying Capacity

Certain populations may exhibit exponential population growth for a short period of time. Exponential population growth has been experimentally demonstrated in bacterial and protist cultures and in certain insects. However, organisms cannot reproduce indefinitely at their biotic potentials, because the environment sets limits, which are collectively called **environmental resistance**. Environmental resistance includes such unfavorable environmental conditions as the limited availability of food, water, shelter, and other essential resources (resulting in increased competition) as well as limits imposed by disease and predation. Using the earlier example, bacteria would never be able to reproduce unchecked for an indefinite period of time because they would run out of food and living space, and poisonous body wastes would accumulate in their vicinity. With crowding, bacteria would also become more susceptible to parasites (high population densities facilitate the spread of infectious organisms such as viruses among individuals) and predators (high population densities increase the likelihood of a predator catching an individual). As the environment deteriorated, their birth rate (*b*) would decline and their death rate (*d*) would increase. The environmental conditions might worsen to a point where *d* would exceed *b*, and the population would decrease. The number of individuals in a population, then, is controlled by the ability of the environment to support it. As the number of individuals in a population increases, so does environmental resistance, which acts to limit population growth. Environmental resistance is an excellent example of a **negative feedback loop** (see Chapter 6).

Over longer periods of time, the rate of population growth may decrease to nearly zero. This leveling out occurs at or near the limit of the environment's ability to support a population. The **carrying capacity (K)** represents the largest population that can be maintained for an indefinite period by a particular environment, assuming there are no changes in that environment. In nature, the carrying capacity is dynamic and changes in response to environmental changes. An extended drought, for example, could decrease the amount of vegetation growing in an area, and this change, in turn, would lower the carrying capacity for deer and other herbivores in that environment.

When a population influenced by environmental resistance is graphed over longer periods of time (Figure 8.4), the curve has a characteristic S shape. The curve shows the population's initial exponential increase (note the curve's J shape at the start, when environmental resistance is low), followed by a leveling out as the carrying capacity of the environment is approached. Although the S curve is an oversimplification of how most populations change over time, it does appear to fit some populations that have been studied in the laboratory, as well as a few that have been studied in nature. For example, G.F. Gause, a Russian ecologist who conducted experiments during the 1930s, grew a population of a single species, *Paramecium caudatum*, in a test tube (see Figure 5.15). He supplied a limited amount of food (bacteria) daily and replenished the media occasionally to eliminate the buildup of metabolic wastes. Under these conditions, the population of *P. caudatum* increased exponentially at first, but then their growth rate declined to zero, and the population size leveled off.

A population rarely stabilizes at *K* (carrying capacity), as shown in Figure 8.4, but may temporarily rise higher than *K*. It will then drop back to, or below, the carrying capacity. Sometimes a population that overshoots *K* will experience a *population crash*, an abrupt decline from high to low population density. Such an abrupt change is commonly observed in bacterial cultures, zooplankton, and other populations whose resources have been exhausted.

The carrying capacity for reindeer, which live in cold northern habitats, is determined largely by the availability of winter forage (Figure 8.5). In 1910 a small herd of 26 reindeer was introduced on one of the Pribilof Islands of Alaska. The herd's population increased exponentially for about 25 years until there were approximately 2,000 reindeer, many more than the island could support, particularly in winter. The reindeer overgrazed the vegetation until the plant life was almost wiped out. Then, in slightly over a decade, as reindeer died from starvation, the number of reindeer plunged to 8, which was about one third the size of the original introduced population

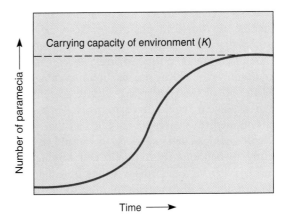

Figure 8.4 Population growth as the carrying capacity is approached. In many laboratory studies, including Gause's investigation with *Paramecium caudatum*, exponential population growth slows as the carrying capacity of the environment is approached. This produces a curve with a characteristic S shape.

Figure 8.5 Reindeer. The population of reindeer on one of the Pribilof Islands in the Bering Sea, off the coast of Alaska, experienced rapid growth followed by a sharp decline when the excess of reindeer damaged the environment.

and less than 1% of the population at its peak. Recovery of arctic and subarctic vegetation after overgrazing by reindeer can take 15 to 20 years, during which time the carrying capacity for reindeer is greatly reduced.

REPRODUCTIVE STRATEGIES

Each species has a lifestyle that is uniquely adapted to its reproductive patterns. Many years pass before a young magnolia tree flowers and produces seeds, whereas wheat grows from seed, flowers, and dies in a single season. A mating pair of black-browed albatrosses produces a single chick every year, whereas a mating pair of gray-headed albatrosses produces a single chick biennially (every other year). Biologists try to understand the adaptive consequences of these various life history strategies.

Imagine an organism possessing the "perfect" life history strategy that ensures continual reproduction at the maximum biotic potential. In other words, this hypothetical organism produces the maximum number of offspring, and all of these offspring survive to reproduce. Such an organism would have to reach reproductive maturity immediately after it was born so that it could begin reproducing at a very early age. It would reproduce frequently throughout its long life and produce large numbers of offspring each time. Further, it would have to provide care for all of its young in order to ensure their total survival.

In nature, such an organism does not exist, because if an organism puts all its energy into reproduction, it could not expend any energy toward ensuring its own survival. Animals must use energy to hunt for food, and plants need energy to grow taller than surrounding plants to obtain adequate sunlight. Nature, then, requires organisms to make trade-offs in the expenditure of energy.

Individuals, if they are to be successful, must do what is required to survive as individuals and as populations (by reproducing). If they allocate all their energy for reproduction, none is available for the survival of the individual, and the individual dies. If they allocate all their energy for the individual, none is available for reproduction, and there are no further generations.

Thus, each species has its own life history strategy—its own reproductive characteristics, body size, habitat requirements, migration patterns, and behaviors—that represent a series of trade-offs reflecting this energy compromise. Although many different life history strategies exist, some ecologists recognize two extremes with respect to reproductive characteristics, *r*-selected species and *K*-selected species. Keep in mind as you read the following descriptions of *r* selection and *K* selection that these concepts, while useful, oversimplify most life histories. Many species possess a combination of *r*-selected and *K*-selected traits, as well as traits that cannot be classified as either *r*-selected or *K*-selected.

Populations described by the concept of *r selection* have traits that contribute to a high population growth rate. Recall that *r* designates the growth rate. Because such organisms have a high *r*, they are known as *r strategists* or *r-selected species*. Small body size, early maturity, short life span, large broods, and little or no parental care are typical of many *r* strategists, which are usually opportunists found in variable, temporary, or unpredictable environments where the probability of long-term survival is low. Some of the best examples of *r* strategists are insects such as mosquitoes and common weeds such as the dandelion (Figure 8.6a).

In populations described by the concept of *K selection*, traits maximize the chance of surviving in an environment where the number of individuals (N) is near the carrying capacity (K) of the environment. These organisms, called *K strategists or K-selected species*, do not produce large numbers of offspring. They characteristically have long life spans with slow development, late reproduction, large body size, and a low reproductive rate. Redwood trees are classified as *K* strategists. Animals that are *K* strategists typically invest in parental care of their young. *K* strategists are found in relatively constant or stable environments, where they have a high competitive ability.

Tawny owls are *K* strategists that pair-bond for life, with both members of a pair living and hunting in separate, well-defined territories (Figure 8.6b). Their reproduction is regulated in accordance with the resources, especially the food supply, present in their territories. In an average year, 30% of the birds do not breed at all. If food supplies are more limited than initial conditions had indicated, many of those that breed fail to incubate their eggs. Rarely do the owls lay the maximum number of eggs that they are physiologically capable of laying, and breeding is often delayed until late in the season, when the rodent populations on which they depend have become large.

(a)

(b)

Figure 8.6 **Reproductive strategies.** (a) Dandelions (*Taraxacum officinale*) are *r* strategists, annuals that mature early and produce many small seeds. A dandelion population fluctuates from year to year but rarely approaches the carrying capacity of its environment. (b) Tawny owls (*Strix aluco*) are *K* strategists and maintain a fairly constant population size at or near the carrying capacity. They mature slowly, delay reproduction, and have a relatively large body size.

Thus, the behavior of tawny owls ensures better reproductive success of the individual and leads to a stable population at or near the carrying capacity of the environment. Starvation, an indication that the tawny owl population has exceeded the carrying capacity, rarely occurs.

Survivorship

Ecologists construct *life tables* for plants and animals that show the likelihood of survival for individuals at different times during their lives. Insurance companies also use life tables to determine how much policies should cost; life

tables show the relationship between a client's age and the likelihood that the client will survive to pay enough insurance premiums to cover the cost of the policy.

Survivorship is the probability that a given individual in a population will survive to a particular age. Figure 8.7 is a graph of the three main survivorship curves that ecologists recognize. In type I survivorship, as exemplified by humans and elephants, the young (that is, prereproductive) and those at reproductive age have a high probability of living. The probability of survival decreases more rapidly with increasing age; that is, deaths are concentrated later in life.

In type III survivorship, the probability of death is greatest early in life, and those individuals that avoid early death subsequently have a high probability of survival. In animals, type III survivorship is characteristic of many fish species and oysters. Young oysters have three free-swimming larval stages before adulthood, when they settle down and secrete a shell. These larvae are extremely vulnerable to predation, and few survive to adulthood.

In type II survivorship, which is intermediate between types I and III, the probability of survival does not change with age. The probability of death is likely across all age groups, resulting in a linear decline in survivorship. This constancy probably results from essentially random events that cause death with little age bias. Although this relationship between age and survivorship is rare, some lizards have a type II survivorship.

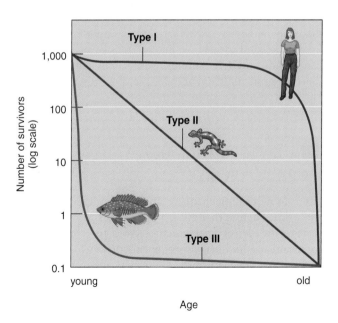

Figure 8.7 **Survivorship.** These generalized survivorship curves represent the ideal survivorships of species in which death is greatest in old age (type I), spread evenly across all age groups (type II), and greatest among the young (type III). The survivorship of most organisms can be compared to these curves.

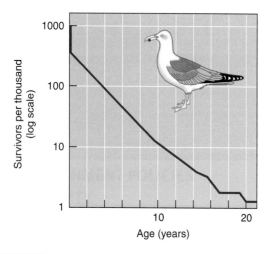

Figure 8.8 Survivorship for a herring gull population. This survivorship curve reveals type III survivorship as chicks and type II survivorship as adults.

The three survivorship curves are generalizations, and few populations exactly fit one of the three. Some species have one type of survivorship curve early in life and another type as adults. Herring gulls have a type III survivorship curve early in life and a type II curve as adults (Figure 8.8). Note that with most, death occurs almost immediately after hatching, despite the protection and care given to the chicks by the parent bird. Herring gull chicks die from predation or attack by other herring gulls, inclement weather, infectious disease, or starvation following the death of the parent. Once the chicks become independent, their survivorship increases dramatically, and death occurs at about the same rate throughout their remaining lives. As a result, few or no herring gulls die from the degenerative diseases of "old age" that cause death in most humans.

FACTORS THAT AFFECT POPULATION SIZE

Certain natural mechanisms appear to influence population size. Factors that affect population size fall into two categories, density-dependent factors (the "regulatory" mechanisms) and density-independent factors (the "nonregulatory" mechanisms). These two sets of factors vary in importance from one species to another, and in most cases probably interact simultaneously in complex ways to determine the size of a population.

Density-Dependent Factors

Certain environmental factors have a greater influence on a population when its density is greater. If a change in population density alters how an environmental factor affects that population, then the environmental factor is said to be a **density-dependent factor.** As population density increases, density-dependent factors tend to slow population growth by causing an increase in death rate and/or a decrease in birth rate. The effect of these density-dependent factors on population growth increases as the population density increases—that is, density-dependent factors affect a larger proportion, not just a larger number, of the population. Density-dependent factors can also affect population growth when population density declines, by decreasing death rate and/or increasing birth rate. Density-dependent factors, then, tend to regulate a population at a relatively constant size that is near the carrying capacity of the environment. (Keep in mind, however, that the carrying capacity of the environment frequently changes.)

Predation, disease, and competition are examples of density-dependent factors. As the density of a population increases, predators are more likely to find an individual of a given prey species. When population density is high, the members of a population encounter one another more frequently, and the chance of their transmitting infectious disease organisms increases. As population density increases, so does competition for resources such as living space, food, cover, water, minerals, and sunlight. The opposite effects occur when the density of a population decreases. Predators are less likely to encounter individual prey, parasites are less likely to be transmitted from one host to another, and competition among members of the population for resources such as living space and food declines.

Most studies of density dependence have been conducted in laboratory settings where all density-dependent (and density-independent) factors except one are controlled experimentally. But populations in natural settings are exposed to a complex set of variables that continually change. As a result, in natural communities it can be difficult to evaluate the relative effects of different density-dependent and density-independent factors.

Ecologists from the University of California, Davis noted that few spiders occur on tropical islands inhabited by lizards, whereas more spiders and more species of spiders are found on lizard-free islands. Deciding to study these observations experimentally, the researchers selected 12 very small Caribbean islands, 4 with lizards and 8 without; all contained web-building spiders. They introduced a small population of lizards onto four of the lizard-free islands. After 7 years, spider population densities were higher in the lizard-free islands than in islands with lizards. Moreover, the islands without lizards had more species of spiders. Therefore, we might conclude that lizards control spider populations. But even in this relatively simple experiment, the results may also be explained by a combination of two density-dependent factors, predation (lizards eat spiders) and competition (lizards compete with spiders for insect prey—that is, both spiders and lizards eat insects). In this experiment, the effects of the two density-dependent factors in determining spider population size cannot be evaluated separately. Additional experiments are needed to determine the relative roles of competition and predation.

CASE·IN·POINT Predator–Prey Dynamics on Isle Royale

Wildlife biologists have studied the populations of moose and wolves on Isle Royale since the 1950s. During the early 1900s, a small herd of moose wandered across the ice of frozen Lake Superior to an island, Isle Royale. In the ensuing years, until 1949, moose became successfully established, although the population experienced oscillations. The story became more complex beginning in 1949, when a few Canadian wolves wandered across the frozen lake and discovered abundant moose prey on the island. The wolves also remained and became established.

Moose have been the wolf's primary winter prey on Isle Royale. Wolves hunt in packs by encircling a moose and trying to get it to run so it can be attacked from behind. (The moose is the wolf's largest, most dangerous prey. A standing moose is more dangerous than one that is running because when standing, it can kick and slash its attackers with its hooves.)

Wildlife biologists, such as Rolf Peterson of Michigan Technological University, have studied the effects of both density-dependent and density-independent factors on the Isle Royale populations of moose and wolves. They found that the two populations fluctuated over the years (Figure 8.9). Generally, as the population of wolves decreased, the population of moose increased. The reverse was also true—as the population of wolves increased, the population of moose decreased. Although the overall effect of wolves was to reduce the population of moose, the wolves did not eliminate moose from the island. Studies indicate that wolves primarily feed on the very old and very young in the moose population. Healthy moose in their peak reproductive years are not eaten.

A new episode in the Isle Royale story began during the late 1980s and early 1990s. The wolf population plunged, from 50 animals in 1980 to a low of 12 animals

in 1989, possibly the result of a deadly disease, canine parvovirus. Analysis of their blood revealed the presence of antibodies to canine parvovirus, confirming that the wolves had been exposed to this disease.

As expected, the moose population increased as the wolf population declined; in 1995 there were more than 2,400 moose on Isle Royale, many more than the island's vegetation could support. The moose overgrazed the island, particularly mountain ash and aspen, their preferred food. Lack of food in combination with a particularly bad winter (1995–96) caused hundreds of moose to die; only 500 moose were counted in the 1997 survey. A year later, in 1998, the wolf population declined to 14 individuals. Lack of vulnerable prey may have caused the wolf decline, but scientists must again check for another disease outbreak. ■

Density Dependence and Boom-or-Bust Population Cycles Lemmings are small rodents that live in the colder parts of the Northern Hemisphere. Lemming populations undergo dramatic increases in their population, followed by crashes over fairly regular time intervals—every 3 to 4 years (Figure 8.10). This cyclic oscillation is often described as a *boom-or-bust* cycle. Other species such as snowshoe hares and red grouse also undergo cyclic population fluctuations.

The actual causes of such population oscillations are poorly understood, but many hypotheses involve density-dependent factors. For example, the population density of lemming predators such as weasels, arctic foxes, and jaegers (arctic birds that eat lemmings) may increase in response to the increasing density of prey. As more predators consume the abundant prey, the prey population declines. Another possibility is that a huge prey population overwhelms the food supply. Recent studies on lemming population cycles suggest that lemming populations crash because they eat all the vegetation in the area, not because predators eat them.

Parasites may also cause such cyclic fluctuations. Red grouse populations (see Figure 8.1) that live in the wild or in managed wildlife preserves exhibit population oscil-

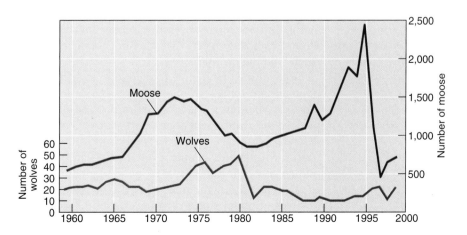

Figure 8.9 Wolf and moose populations on Isle Royale. Data from aerial surveys indicate that in 1999 there were 25 wolves and 750 moose.

(a)

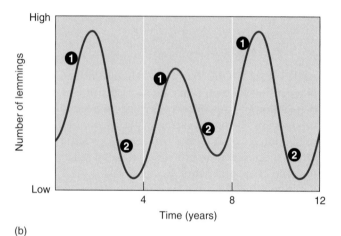

(b)

─────────

Figure 8.10 Lemming population oscillations. (a) The brown lemming lives in the arctic tundra. (b) This hypothetical diagram shows the cyclic nature of lemming population oscillations, which are not well understood. At the parts of the curve labeled 1, the population is increasing, and density-dependent factors are increasingly severe; as a result, the population peaks and begins to decline. At the parts of the curve labeled 2, the population is declining, and density-dependent factors are increasingly relaxed; as a result, the population bottoms out and begins to increase.

─────────

lations. Studies have shown that reproduction in red grouse is related to the presence of parasitic roundworms living in the birds' intestines. When adult birds are infected with worms, they do not breed successfully, and the population crashes. Biologists were able to eliminate population oscillations in several red grouse populations by catching the birds and treating them with a medicine that causes the worms to be ejected from their bodies.

Density-Independent Factors

Any environmental factor that affects the size of a population but is not influenced by changes in population density is called a **density-independent factor**. Such factors are typically abiotic. Random weather events that reduce population size serve as density-independent factors. A killing frost, severe blizzard, hurricane, or fire may cause extreme and irregular reductions in a population regardless of its size and thus might be considered largely density independent.

Consider a density-independent factor that influences mosquito populations in arctic environments. These insects produce several generations per summer and achieve high population densities by the end of the season. A shortage of food does not seem to be a limiting factor for mosquitoes, nor is there any shortage of ponds in which to breed. Instead, winter puts a stop to the skyrocketing mosquito population. Not a single adult mosquito survives winter, and the entire population must grow afresh the next summer from the few eggs and hibernating larvae that survive. Thus, the timing and severity of winter weather is a density-independent factor that affects arctic mosquito populations.

Density-dependent and density-independent factors are often interrelated. Social animals, for example, are often able to resist dangerous weather conditions by means of their collective behavior, as in the case of sheep huddling together in a snowstorm. In this case it appears that the greater the population density, the better their ability to resist the environmental stress of a density-independent event (that is, the snowstorm).

THE HUMAN POPULATION

Now that we have examined some of the basic concepts of population ecology, we can apply those concepts to the human population. Examine Figure 8.11, which shows the world increase in population since 1800. Now look back at Figure 8.3b and compare the two curves. The characteristic J curve of exponential population growth shown in Figure 8.11 reflects the decreasing amount of time it has taken to add each additional billion people to our numbers. It took thousands of years for the human population to reach 1 billion, a milestone that took place around 1800. It took 130 years to reach 2 billion (in 1930), 30 years to reach 3 billion (in 1960), 15 years to reach 4 billion (in 1975), 12 years to reach 5 billion (in 1987), and 12 years to reach 6 billion (in 1999). The United Nations projects that the human population will reach 7 billion by 2013.

One of the first people to recognize that the human population cannot continue to increase indefinitely was Thomas Malthus, a British economist (1766–1834). He pointed out that human population growth is not always desirable—a view contrary to the beliefs of his day and to those of many people even today—and that the human population is capable of increasing faster than its food supply. He maintained that the inevitable consequences of population growth are famine, disease, and war. Since Malthus's time, the human population has

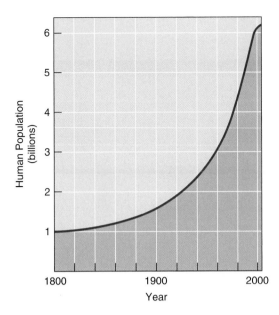

Figure 8.11 Human population numbers, 1800 to present.
The human population has been increasing exponentially. Population experts predict that the population will level out during the 21st century, possibly forming the S curve observed in other species.

grown from about 1 billion to more than 6 billion. It appears that Malthus was proved wrong. Our population has been able to grow so dramatically because scientific advances have enabled us to improve the productivity of the land so that in most places food production has kept pace with population growth. The reason Malthus's ideas may ultimately be proved correct, however, is that we do not know if our increase in food production is *sustainable*. Have we achieved this increase in food production at the environmental cost of reducing the ability of the land to meet the needs of future populations?

Current Population Numbers

Our world population was 6.2 billion in 2002 and increased by approximately 78 million from 2001 to 2002. This increase is not due to an increase in the birth rate (*b*). In fact, the world birth rate has actually *declined* slightly during the past 200 years. The increase in population is due instead to a dramatic decrease in the death rate (*d*), which has occurred primarily because greater food production, better medical care, and improvements in water quality and sanitation practices have increased the life expectancies for a great majority of the global population. From about 1920 to 2000, the death rate in Mexico fell from approximately 40 to 5 per 1,000 individuals, whereas the birth rate dropped from approximately 40 to 26 per 1,000 individuals (Figure 8.12). Because Mexico's birth rate currently remains much greater than

its death rate, it has a high annual growth rate (*r*), 2.1% in 2002:

$$r = b - d$$
$$r = 26/1{,}000 - 5/1{,}000$$
$$r = 0.026 - 0.005 = 0.021, \text{ or } \textbf{2.1\% per year}$$

Projecting Future Population Numbers

The human population has reached a turning point. Although our numbers continue to increase, the world growth rate (*r*) has declined slightly over the past several years, from a peak of 2.2% per year in the mid-1960s to 1.3% per year in 2002. Population experts at the United Nations and the World Bank have projected that the growth rate will continue to decrease slowly until zero population growth is attained. Thus, exponential growth of the human population will end, and the J curve may be replaced by the S curve. It is projected that **zero population growth**—when the birth rate equals the death rate—will occur toward the end of the 21st century (see "Mini-Glossary: Population Terms Commonly Applied to Humans")

The United Nations periodically publishes population projections for the 21st century. The latest (2000) U.N. figures available forecast that the human population will be between 7.9 billion (their "low" projection) and 10.9 billion (their "high" projection) in the year 2050, with 9.3 billion thought to be "most likely" (Figure 8.13). The estimates vary depending on fertility changes, particularly in less developed countries because that is where almost all of the growth will take place.

Such population projections are "what if" exercises: Given certain assumptions about future tendencies in the

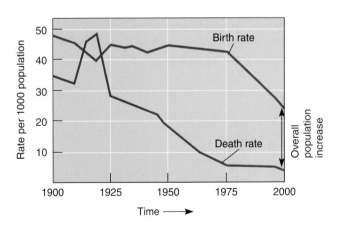

Figure 8.12 Birth and death rates in Mexico, 1900 to 2000.
Both birth and death rates generally declined in Mexico during the 20th century. Because the death rate declined much more than the birth rate, Mexico experienced a high growth rate. (The high death rate prior to 1925 was caused by the Mexican Revolution.)

MINI-GLOSSARY

Population Terms Commonly Applied to Humans

zero population growth: The condition when a population is no longer increasing (or decreasing) because the birth rate equals the death rate.

infant mortality rate: The number of infant deaths per 1,000 live births.

doubling time: The number of years it will take a population to double in size, given its current growth rate.

replacement-level fertility: The number of children a couple must have to "replace" themselves.

total fertility rate: The average number of children born to each woman during her lifetime.

birth rate, death rate, and migration, an area's population can be calculated for a given number of years into the future. Population projections indicate the changes that may occur, but they must be interpreted with care because they vary depending on what assumptions have been made. In projecting that the world population will be 7.9 billion (their low projection) in the year 2050, U.N. population experts assume that the average number of children born to each woman in all countries will have declined to 1.5 in the 21st century. In 2002 the average number of children born to each woman on Earth was 2.8. If the decline to 1.5 does not occur, our population could be significantly higher. If the average number of children born to each woman declines only to 2.5 instead of 1.5, the 2050 population will be 10.9 billion (the U.N. high projection). Small differences in fertility, then, produce large differences in population forecasts.

The main unknown factor in any population growth scenario is Earth's *carrying capacity*. According to Dr. Joel Cohen, who has a joint professorship in populations at Rockefeller University and Columbia University, most published estimates of how many people Earth can support range from 4 billion to 16 billion. These estimates vary widely depending on what assumptions are made about standard of living, resource consumption, technological innovations, and waste generation. If we want all people to have a high level of material well-being equivalent to the lifestyles common in highly developed countries, then the Earth will clearly be able to support far fewer humans than if everyone lives just above the subsistence level. The Earth's carrying capacity for humans, then, is not decided simply by natural environmental constraints. Human choices and values have to be factored into the assessment.

It is also not clear what will happen to the human population when the carrying capacity is approached. Optimists suggest that the human population will stabilize because of a decrease in the birth rate. Some experts take a more pessimistic view and predict that the widespread degradation of our environment caused by our

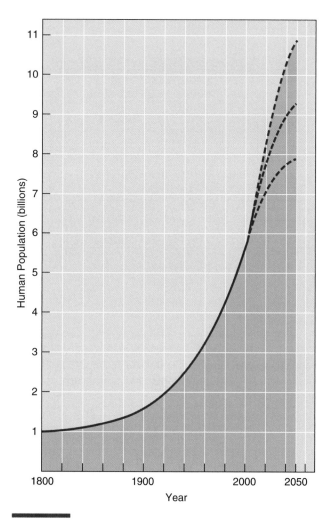

Figure 8.13 Population projections to 2050. In 2000 the United Nations made three projections, each based on different fertility rates.

ever-expanding numbers will make the Earth uninhabitable for humans as well as for other species. These experts contend that a massive wave of human suffering and death will occur. This view does not mean that we will go extinct as a species but that there will be severe hardships for a very large number of people. Some experts think that the human population has already exceeded the carrying capacity of the environment, a potentially dangerous situation that threatens our long-term survival as a species.

DEMOGRAPHICS OF COUNTRIES

Whereas world population figures illustrate overall trends, they do not describe other important aspects of the human population story, such as population differences from country to country (Table 8.1). **Demographics,** the applied branch of sociology that deals with population statistics, provides interesting information on

Table 8.1	The World's Ten Most Populous Countries	
Country	**2002 Population (in millions)**	**Population density (per mi²)**
China	1280.7	347
India	1049.5	827
United States	287.4	77
Indonesia	217.0	295
Brazil	173.8	53
Pakistan	143.5	467
Russia	143.5	22
Bangladesh	133.6	2403
Nigeria	129.9	364
Japan	127.4	873

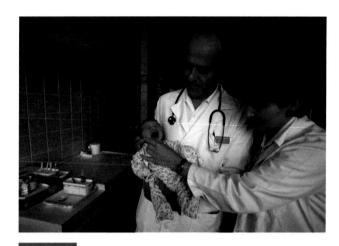

Figure 8.14 A doctor examines a newborn infant. Medical advances have contributed to a decline in infant deaths, particularly in highly developed countries.

the populations of various countries. As you probably know, not all countries have the same rates of population increase. Countries can be classified into two groups—highly developed and developing—depending on growth rates, degree of industrialization, and relative prosperity (Table 8.2).

Highly developed countries (also called **developed countries**), such as the United States, Canada, France, Germany, Sweden, Australia, and Japan, have low rates of population growth, and are highly industrialized relative to the rest of the world. Highly developed countries have the lowest birth rates in the world. Indeed, some countries such as Germany have birth rates just below those needed to sustain their populations and are thus declining slightly in numbers. Highly developed countries also have very low **infant mortality rates** (the number of infant deaths under age 1 per 1,000 live births) (Figure 8.14). The infant mortality rate of the United States was 6.6 in 2002, compared with a world rate of 54. Highly developed countries also have longer life expectancies (77 years in the United States versus 67 years worldwide) and high average per-capita GNI PPPs ($34,100 in the United States versus $7,140 worldwide). Per-capita GNI PPP is gross

national income (GNI) in purchasing power party (PPP) divided by midyear population. It indicates the amount of goods and services an average citizen of that particular region or country could buy in the United States.

Developing countries fall into two subcategories, moderately developed and less developed. Mexico, Turkey, Thailand, and most South American nations are examples of **moderately developed countries**. Their birth rates and infant mortality rates are higher than those of highly developed countries, but they are declining. Moderately developed countries have a medium level of industrialization, and their average per-capita GNI PPPs are lower than those of highly developed countries. **Less developed countries** (**LDCs**) include Bangladesh, Niger, Ethiopia, Laos, and Cambodia. These countries have the highest birth rates, the highest infant mortality rates, the shortest life expectancies, and the lowest average per-capita GNI PPPs in the world.

Table 8.2	Comparison of 2002 Population Data in Developed and Developing Countries		
	Developed	**Developing**	
	(Highly Developed) United States	**(Moderately Developed) Brazil**	**(Less Developed) Ethiopia**
Fertility rate	2.1	2.2	5.9
Projected population change, 2002–2050	+44%	+42%	+155%
Infant mortality rate	6.6 per 1,000	33 per 1,000	97 per 1,000
Life expectancy at birth	77 years	69 years	52 years
Per-capita GNI PPP (2000; U.S. $)*	$34,100	$7,300	$660
Women using modern contraception	72%	70%	6%

* GNI PPP = gross national income in purchasing power parity.

One way to express the population growth of a country is to determine its **doubling time**, the amount of time it would take for its population to double in size, assuming that its current growth rate does not change. A simplified formula for doubling time (t_d) is $t_d = 70/r$. The actual formula involves calculus and is beyond the scope of this text.

A country's doubling time can usually identify it as a highly, moderately, or less developed country: The shorter the doubling time, the less developed the country. At rates of growth in 2002, the doubling time is 30 years for Laos, 28 years for Ethiopia, 47 years for Turkey, 87 years for Thailand, and 175 years for France.

It is also instructive to examine **replacement-level fertility**—that is, the number of children a couple must produce in order to "replace" themselves. Replacement-level fertility is usually given as 2.1 children. The number is greater than 2.0 because some infants and children die before they reach reproductive age. Worldwide, the **total fertility rate**—the average number of children born to each woman—is currently 2.8, which is well above the replacement level.

Demographic Stages

Demographers recognize four demographic stages, based on their observations of Europe as it became industrialized and urbanized (Figure 8.15). These stages converted Europe from relatively high birth and death rates to relatively low birth and death rates. Because all highly developed and moderately developed countries with more advanced economies have gone through this **demographic transition**, demographers generally assume that the same progression will occur in less developed countries as they become industrialized.

In the first stage, called the **preindustrial stage**, birth and death rates are high, and population grows at a modest rate. Although women have many children, the infant mortality rate is high. Intermittent famines, plagues, and wars also increase the death rate, so the population grows slowly or temporarily declines. If we use a European country such as Finland to demonstrate the four demographic stages, we can say that Finland was in the first demographic stage from the time of its first human settlements until the late 1700s.

As a result of improved health care and more reliable food and water supplies that accompany the beginning of an industrial society, the second demographic stage, called the **transitional stage**, is characterized by a lowered death rate. Because the birth rate is still high, the population grows rapidly. Finland in the mid-1800s was in the second demographic stage.

The third demographic stage, the **industrial stage**, is characterized by a decline in birth rate and takes place at some point during the industrialization process. The decline in birth rate slows population growth despite a relatively low death rate. For Finland, this occurred in the early 1900s.

Low birth and death rates characterize the fourth demographic stage, sometimes called the **postindustrial stage**. In countries that are heavily industrialized, people are better educated and more affluent; they tend to desire smaller families and take steps to limit family size. The population grows very slowly or not at all in the fourth demographic stage. This is the situation in such highly developed countries and groups of countries as the United States, Canada, Australia, Japan, and Europe, including Finland.

Why has the population stabilized in more than 30 highly developed countries that are in the fourth demographic stage? The reasons are complex. The decline in birth rate has been associated with an improvement in living standards. However, it is difficult to say whether improved socioeconomic conditions have resulted in a

Figure 8.15 Demographic transition. The demographic transition consists of four demographic stages through which a population progresses as its society becomes industrialized. Note that the death rate declines first, followed by a decline in the birth rate.

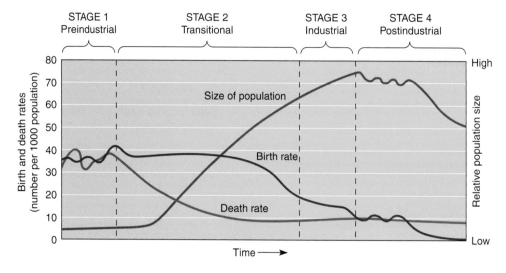

Table 8.3	Fertility Changes in Selected Developing Countries	
	Total Fertility Rate*	
Country	**1960–65**	**2002**
Bangladesh	6.7	3.3
Brazil	6.2	2.2
China	5.9	1.8
Egypt	7.1	3.5
Guatemala	6.9	4.6
India	5.8	3.2
Kenya	8.1	4.4
Mexico	6.8	2.9
Nepal	5.9	4.1
Nigeria	6.9	5.8
Thailand	6.4	1.8

* Total fertility rate = average number of children born to each woman during her lifetime.

decrease in birth rate, or a decrease in birth rate has resulted in improved socioeconomic conditions. Perhaps both are true. Another reason for the decline in birth rate in highly developed countries is the increased availability of family planning services. Other socioeconomic factors that influence birth rate are increased education, particularly of women, and urbanization of society. These factors are considered in greater detail in Chapter 9.

Once a country reaches the fourth demographic stage, is it correct to assume it will continue to have a low birth rate indefinitely? The answer is that we do not know. Low birth rates may be a permanent response to the socioeconomic factors that are a part of an industrialized, urbanized society. On the other hand, low birth rates may be a response to socioeconomic factors such as the changing roles of women in highly developed

countries. It could be that unforeseen changes in the socioeconomic status of women and men in the future may again change birth rates. No one knows for sure.

The population in many developing countries is beginning to approach stabilization. The fertility rate must decline in order for a population to stabilize; see Table 8.3 and note the general decline in total fertility rates in selected developing countries from the 1960s to 2002. The total fertility rate in developing countries has decreased from an average of 6.1 children per woman in 1970 to 3.1 in 2002.

Although the fertility rates in these countries have declined, it should be remembered that they still exceed replacement-level fertility. Consequently, the populations in these countries are still increasing. Also, even when fertility rates equal replacement-level fertility, population growth will still continue for some time. To understand why this is so, we now examine the age structure of various countries.

Age Structure of Countries

In order to predict the future growth of a population, it is important to know its **age structure**, the number and proportion of people at each age in a population. The number of males and number of females at each age, from birth to death, can be represented in an **age structure diagram** (Figure 8.16).

Predicting Population Changes Using Age Structure Diagrams The overall shape of an age structure diagram indicates whether the population is increasing, stable, or shrinking. The age structure diagram of a country with a very high growth rate, based on a high fertility rate—for example, Nigeria or Bolivia—is shaped like a pyramid (Figure 8.17a). Because the largest percentage of the population is in the pre-repro-

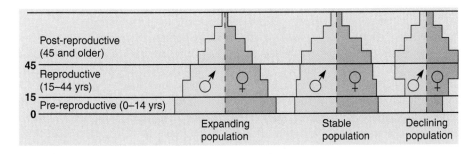

Figure 8.16 Generalized age structure diagrams. These diagrams show an expanding population, a stable population, and a population that is decreasing in size. Each diagram is divided vertically in half, the left side representing the males in a population and the right side the females. The bottom third of each diagram represents pre-reproductive humans (between 0 and 14 years of age); the middle third, reproductive humans (15 to 44 years); and the top third, post-reproductive humans (45 years and older). The widths of these segments are proportional to the population sizes—a broader width implies a larger population.

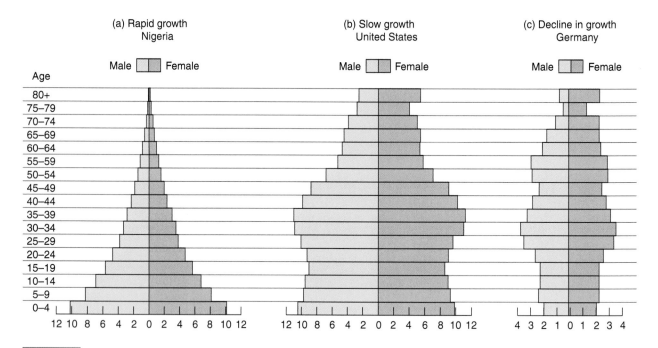

Figure 8.17 Age structure diagrams. Shown are countries with (a) fast (Nigeria), (b) slow (United States), and (c) declining (Germany) population growth. These age structure diagrams indicate that less developed countries such as Nigeria have a much higher percentage of young people than highly developed countries. As a result, less developed countries are projected to have greater population growth than highly developed countries.

ductive age group (that is, 0 to 14 years of age), the probability of future population growth is very great. A positive **population growth momentum** exists because when all these children mature, they will become the parents of the next generation, and this group of parents will be larger than the previous group. Thus, even if the fertility rate of such a country has declined to replacement level (that is, couples are having smaller families than their parents did), *the population will continue to grow for some time.* Population growth momentum, which can be either positive or negative but is usually discussed in a positive context, explains how the future growth of a population is affected by its present age distribution.

In contrast, the more tapered bases of the age structure diagrams of countries with slowly growing, stable, or declining populations indicate that a smaller proportion of the population will become the parents of the next generation (Figure 8.17b and c). The age structure diagram of a stable population, one that is neither growing nor shrinking, demonstrates that the numbers of people at pre-reproductive and reproductive ages are approximately the same. Also, a larger percentage of the population is older—that is, post-reproductive—than in a rapidly increasing population. Many countries in Europe have stable populations.

In a population that is shrinking in size, the prereproductive age group is *smaller* than either the reproductive or post-reproductive group. Russia, Bulgaria, and Germany are examples of countries with slowly shrinking populations.

Worldwide, 30% of the human population is under age 15 (Figure 8.18). When these people enter their reproductive years, they have the potential to cause a large increase in the growth rate. Even if the birth rate does not increase, the growth rate will increase simply because there are more people reproducing.

Most of the world population increase since 1950 has taken place in developing countries as a result of the younger age structure and the higher-than-replacement-level fertility rates of their populations. In 1950, 66.8% of the world's population was in developing countries in Africa, Asia (minus Japan), and Latin America. Between 1950 and 2000, the world's population more than doubled in size, but most of that growth occurred in developing countries. As a reflection of this, in 2002 the number of people in developing countries had increased to 80.7% of the world population. Most of the population increase that will occur during the 21st century will also take place in developing countries, largely the result of their younger age structures. These countries are least able to support such growth.

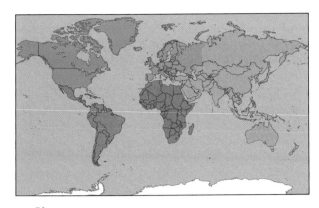

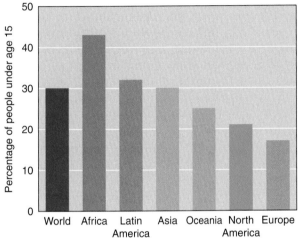

Figure 8.18 Percentages of the population under age 15 for various regions in 2002. The higher this percentage, the greater the potential for population growth when people in this group reach their reproductive years.

DEMOGRAPHICS OF THE UNITED STATES

The United States has the largest population of all highly developed countries, and compared to the global population, U.S citizens have a high level of material well-being. However, a persuasive argument could be made that the United States is one of the most overpopulated countries in the world, because of overconsumption by its individuals (see Chapter 9). Moreover, the population of the United States continues to grow significantly, in part because people from developing countries immigrate to the United States to try for a better life. Population experts and environmental scientists try to predict what the increasing population size and high consumption level of the United States will mean to future generations, both in the United States and worldwide. Should the United States have a formal population policy that stipulates desirable family size and economic prosperity?

Should we—or can we—control immigration? Is it ethical to curtail individual rights of the present generation in order to assure a better future for generations that will follow us?

Currently, the United States does not have a formal population policy, and the closest it has come to such a policy occurred under President Nixon's leadership, when a commission was established to examine U.S. population growth. The commission spent 3 years gathering data and listening to testimonies from experts and the public before concluding in 1972 that the United States would gain no substantial benefits from continued population growth. The commission therefore recommended that the United States should try to stabilize its population. Despite this recommendation, the U.S. population has continued to increase.

The United States has one of the highest rates of population increase of all the highly developed countries. The U.S. population increased by 19.7 million from 1997 to 2002. This translates to a 0.6% annual increase in 2002, which is greater than most highly developed countries—for example, Western Europe's annual increase in 2002 was 0.1% and Japan's was 0.2%. These figures take only birth and death rates into account; immigration is not considered.

Immigration has a greater effect on population size in the United States than in many other nations. Since 1992, the United States has accepted about 1 million legal immigrants annually. The number of unauthorized migrants who gain access to the United States and are not deported is not known with any certainty, but this

---

ENVIROBRIEF

A Look at U.S. Population

Here are some facts about the changing and complex population within the United States, based on 2000 data (latest available).

- The three states with the greatest number of residents are California (33.9 million), Texas (20.9 million), and New York (19 million).

- The two states with the highest population densities are New Jersey (1,134 people per square mile of land area) and Rhode Island (1,003 people per square mile). The two states with the lowest population densities are Alaska (1 person per square mile) and Wyoming (5 people per square mile).

- Between 1990 and 2000, Nevada had the highest growth rate (66.3% increase), with Arizona second (40% increase). The District of Columbia had the largest population decline (–5.7%).

- In 2000, Utah had the highest percentage of its population below the age of 18 (32%) whereas Florida had the highest percentage of its population over age 65 (18%).

- In 2025, the 3 states with the greatest number of residents are projected to be California (49.3 million), Texas (27.2 million), and Florida (20.7 million).

number is estimated by the International Organization for Migration to be about 300,000 per year. The total population of unauthorized migrants living in the United States in 2001 was estimated to be 7.8 million. Assuming that recent immigration levels continue, the U.S. Census Bureau estimates that the U.S. population will reach 404 million by 2050.

CASE-IN-POINT U.S. Immigration

Before we examine the history and scope of immigration in the United States, it is useful to understand that human migration across international borders is a worldwide phenomenon that has escalated dramatically during the past few decades. According to the U.N. Population Fund, more than 150 million people currently live outside their country of birth (Figure 8.19). There are many reasons for the increase in international migration. People migrate in search of jobs or an improved standard of living (the most important reason); to escape war or persecution for their race, religion, nationality, or political opinions; or to join other family members who have already migrated.

The increase in the human population during the 20th century is an underlying factor that has contributed to the current surge in migration, which population experts predict will continue to increase during the next 30 years. Deteriorating environmental conditions brought on by population growth may contribute to international migration. Population growth is also responsible for the addition of millions of people to the workforce each year, and many countries' economies do not create enough new jobs to accommodate these people. At the same time, a few other countries, such as Saudi Arabia, do not have a large enough workforce to accommodate available jobs.

History of Immigration in the United States   Prior to 1875, there were no immigration laws in the United States, and thus no such thing as unauthorized migrants. In 1875, however Congress passed a law denying convicts and prostitutes entrance to the United States. In 1882 the Chinese Exclusion Act was passed, and in 1891 the Bureau of Immigration was established. Thus by the late 1800s a policy of selective exclusion was officially established in the United States and began to shape the population of this country.

During the early 20th century, Congress set numerical restrictions on immigration, including quotas allowing only a certain number of people from each foreign country to immigrate. This severely restricted entrance to the United States by people from Asia, southern and Eastern Europe, and Mexico. With these stronger laws, the influx of unauthorized migrants began to increase. U.S. immigration policy relaxed during World War II, when labor shortages made it possible for workers from Mexico, Barbados, Jamaica, and British Honduras to gain temporary residence in the United States.

In 1952 the **Immigration and Nationality Act** was passed, and although it has been revised since then (it is now called the **Immigration Reform and Control Act**, or **IRCA**), it is still the basic immigration law in effect. Amendments passed in 1965 abolished national quotas and gave three groups of people priority when immigrating to the United States: those with family members already living in the United States, those who can fill vacant jobs (employment-based preference), and those who are refugees seeking asylum.

The Current Wave of Immigrants to the United States   In the past, waves of immigrants entering the United States were racially and ethnically homogeneous. At the beginning of the 20th century, the vast majority of immigrants came from southern and eastern parts of Europe;

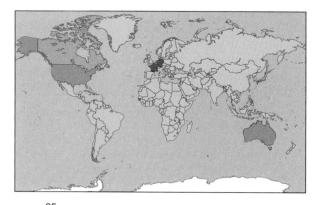

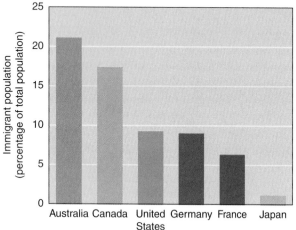

Figure 8.19 Immigrant (foreign-born) populations in selected countries in 2000. Although the percentage of immigrants to the United States is not the highest in the world, more foreign-born immigrants come to the United States than to any other country. Currently, 25 million nonnative immigrants live in the United States, according to the International Organization for Migration.

ENVIROBRIEF

Diverting the Flood of Refugees

The world's refugees face a complicated future not easily defined by demographic statistics. According to the U.N. High Commissioner for Refugees, the number of refugees internationally is actually declining, at 19.8 million in 2002, down from the historic high of 27.4 million in 1995. Though this trend appears positive, it is tempered by a dramatic increase in internal displacement—the more than 21 million people who are forced to move and subjected to persecution or violence but do not cross a border into another country. For example, 500,000 to 750,000 Afghanis are internally displaced.

Repatriation movements (that is, people returning to their country of birth at the end of a war or some other national crisis) in nations such as Cambodia, Mozambique, Rwanda, and Central and Eastern Europe also explain the decline in refugees since the early 1990s.

Many people attempting to flee their homelands are not welcomed with open arms. The Western world's reluctance to embrace refugees stems from several sources:

- Economic problems at home.

- The high costs of processing asylum applications.

- Skepticism over whether refugees are actually fleeing political persecution and are not simply "economic migrants."

Failure to address the high demand for asylum leads to a lower quality of life for prospective refugees:

- Refugee camps are growing more violent.

- Human trafficking (charging asylum-seekers money to illegally transport them across borders) is on the rise, with its accompanying financial exploitation, human rights abuses, and violence.

- Less developed nations take their cue from highly developed ones, refusing to accept refugees when more prosperous nations will not.

most were young and poor, often barely able to afford travel fares. In contrast, typical immigrants to the United States today are a more diverse group. Four of every five legal immigrants come from either Asia or Latin America. The top five countries from which legal U.S. immigrants migrate are Mexico, the Philippines, Vietnam, the Dominican Republic, and China. About 70% of these new arrivals are poor and have few skills but the remainder are college graduates and may practice their professions, such as medicine, engineering, and computer science, in the United States. More than 4% of legal immigrants are senior citizens, many of whom live with their adult children.

Immigration as a Controversial Environmental Issue
U.S. environmental groups have not reached a consensus on immigration. Some environmentalists think the population of the United States is already too large. According to this viewpoint, the presence of legal immigrants and unauthorized migrants is undesirable because it contributes greatly to pollution and resource depletion as these people adopt the affluent, high-consumption lifestyle of "typical" U.S. citizens. These groups argue that the United States must reduce both birth rates and immigration rates to bring about sustainability.

Other environmentalists point out that the United States is not the only country to take in immigrants. They argue that it is morally unacceptable to deny people a chance for a better life in the United States. The United States can absorb the environmental consequences of immigrants, who often have simpler lifestyles than most U.S. citizens (at least for the first generation), more easily than their home countries can. Because census figures indicate that immigrants tend to have smaller families than if they had stayed in their countries of origin, immigration helps to reduce the rate of global population growth. Further, immigrants provide U.S. environmental groups with a much-needed global consciousness, a "we're in this together" viewpoint that is easy to overlook when simply considering U.S. environmental issues.

SUMMARY WITH SELECTED KEY TERMS

I. Population ecology is the branch of biology that deals with the number of individuals of a particular species that are found in an area and how and why those numbers change over time.

A. Populations have certain properties that individual organisms lack, such as population density, birth and death rates, growth rates, and age structure.

B. Population density is the number of individuals of a species per unit of area or volume at a given time.

C. On a global scale the **growth rate (r)**, the rate of change in a population's size, is due to two factors, the **birth rate (b)** and the **death rate (d)**.

1. On a global scale (when **dispersal** is not a factor), $r = b - d$.

2. Populations increase in size as long as the number of births is greater than the number of deaths.

3. Growth rate is referred to as **natural increase** in human populations.

D. In addition to birth and death rates, dispersal must be considered when examining changes in local populations.

1. **Emigration (e)**, the number of individuals leaving an area, and **immigration (i)**, the number of individuals entering an area, affect a local population's size and growth rate.

2. For a local population (where dispersal is a factor), $r = (b - d) + (i - e)$.

II. Biotic potential, or intrinsic rate of increase, is the maximum rate at which a species or population could increase under ideal conditions.

A. Although populations with a constant reproductive rate exhibit **exponential population growth** for limited periods of time (the J curve), eventually the growth rate decreases to around zero or becomes negative.

B. Population size is modified by limits set by the environment, which are collectively called **environmental resistance**. Environmental resistance includes the limited availability of food, water, shelter, and other resources, as well as limits imposed by disease and predation.

C. The **carrying capacity (K)** of the environment is the largest population that can be maintained for an indefinite time by a particular environment.

D. The S curve shows an initial lag phase (when the population is small), followed by an exponential phase, followed by a leveling phase as the carrying capacity of the environment is reached. The S curve is an oversimplification of how most populations change over time.

III. Each species has its own life history strategy.

A. An *r* strategy emphasizes a high growth rate. These organisms often have small body sizes, high reproductive rates, and short life spans, and they typically inhabit variable environments.

B. A *K* strategy results in maintenance of a population near the carrying capacity of the environment. These organisms often have large body sizes, low reproductive rates, and long life spans, and they typically inhabit stable environments.

C. Many species possess a combination of *r*-selected and *K*-selected traits, as well as traits that are characteristic of neither *r* selection nor *K* selection.

D. There are three general types of **survivorship** curves.
  1. In type I survivorship, death is greatest in old age.
  2. In type III survivorship, death is greatest among the young.
  3. In type II survivorship, death is spread evenly across all age groups.

IV. Environmental factors limit population growth.

A. **Density-dependent factors** regulate population growth by affecting a larger proportion of the population as population density rises. Predation, disease, and competition are examples.

B. **Density-independent factors** limit population growth but are not influenced by changes in population density. Hurricanes and fires are examples.

C. Cyclic population oscillations are poorly understood but may involve density-dependent factors. Examples of animals with such boom-or-bust cycles include lemmings, snowshoe hares, and red grouse.

V. The principles of population ecology apply to humans.

A. The world population was 6.2 billion in 2002 and increased by 78 million from 2001 to 2002.
  1. Although our numbers continue to increase, the growth rate *(r)* has declined slightly over the past several years, from a peak in the mid-1960s of 2.2% per year to 1.3% per year in 2002.
  2. **Demographers,** scientists who study population statistics, project that the world population will become stationary *(r = 0)*—that is, **zero population growth**—by the end of the 21st century.

B. **Highly developed countries** have the lowest birth rates, lowest **infant mortality rates,** longest life expectancies, and highest per-capita GNI PPPs. **Developing countries** have the highest birth rates, highest infant mortality rates, shortest life expectancies, and lowest per-capita GNI PPPs.

C. The **age structure** of a population greatly influences population dynamics.
  1. It is possible for a country to have **replacement-level fertility** and still experience population growth if the largest percentage of the population is in the pre-reproductive years.
  2. Such an age structure causes a positive **population growth momentum**.

D. The United States has one of the highest rates of population increase (0.6% in 2002) of all the highly developed countries.
  1. Immigration has a greater effect on population size in the United States than in many other countries.
  2. The **Immigration Reform and Control Act (IRCA),** the basic immigration law in effect in the United States, gives three groups of people priority when migrating to the United States: those with family members living in the United States, those who can fill vacant jobs, and those who are refugees seeking asylum.

THINKING ABOUT THE ENVIRONMENT

some of the factors that have contributed to the
se in the incidence of AIDS in sub-Saharan

ch of the following on population
igration, and emigration.

3. Explain how biotic potential and/or carrying capacity produce the J-shaped and S-shaped population growth curves.

4. Draw a graph to represent the *long-term* growth of a population of bacteria cultured in a test tube containing a nutrient medium that is *not* replenished.

5. Give three examples each of density-dependent and density-independent factors that affect population growth.

6. Explain how the spread of human diseases such as the Black Death, tuberculosis, and AIDS is related to high population densities found in urban environments. Are such diseases density-dependent or density-independent?

7. What is a boom-or-bust population cycle? Describe density-dependent factors that may influence such cyclic population oscillations.

8. Draw the three main survivorship curves and relate them to *r* selection and *K* selection in animals.

9. Describe human population growth for the past 200 years.

10. Who was Thomas Malthus, and what were his views on human population growth?

11. When determining Earth's carrying capacity for humans, why is it not enough to just consider human numbers? What else must be considered?

12. What is infant mortality rate? How does it affect life expectancy?

13. Distinguish between the total fertility rate and replacement-level fertility.

14. If all the women in the world suddenly started bearing children at replacement-level fertility rates, would the population stop increasing immediately? Why or why not?

15. Use the Population Reference Bureau's *World Population Data Sheet*, a poster located inside the back cover of this text, to obtain what percentage of the population is younger than 15 and what percentage is older than 65 for two countries, Bolivia and Austria. On the basis of this information, which of these countries will have the highest population growth momentum over the next two decades? Why?

16. Should the rapid increase in world population be of concern to the average citizen in the United States? Why or why not?

17. Should the United States increase or decrease the number of legal immigrants? Present arguments in favor of both sides.

*18. If a population of 100,000 has 2,700 births per year and 900 deaths per year, what is its growth rate in percent per year?

*19. Use the Population Reference Bureau's *World Population Data Sheet* inside the back cover of this text to obtain the population and growth rate (natural increase, annual percentage) of India. Assuming that the growth rate remains constant, what will India's population be in 2005? In what year will India's population be double that of its 2002 population?

* Solutions to questions preceded by asterisks appear in Appendix VII.

WORLD POPULATION DATA SHEET ASSIGNMENT

Use the Population Reference Bureau's *World Population Data Sheet* from inside the back cover to answer the following questions.

1. What is the estimated world population in *billions*?

2. What is the current birth rate for the world?

3. Which continent has the highest birth rate, and what is it?

4. Which country or countries has/have the highest birth rate, and what is it?

5. Which continent has the lowest birth rate, and what is it?

6. Which country or countries has/have the lowest birth rate, and what is it?

7. What is the current death rate for the world?

8. Which country or countries has/have the highest death rate, and what is it?

9. How is natural increase (annual, %) calculated?

10. Which country or countries does not/do not have a positive natural increase (i.e., are either stable or decreasing in population)?

11. What is an infant mortality rate?

12. Which country or countries has/have the world's highest

13. Which country or countries has/have the world's lowest infant mortality rate, and what is it?

14. What is a total fertility rate?

15. Which country or countries has/have the world's highest total fertility rate, and what is it?

16. Which country or countries has/have the world's lowest total fertility rate, and what is it?

17. Which country or countries has/have the highest percentage of the population under age 15, and what is it?

18. Which country or countries has/have the highest percentage of the population over age 65, and what is it?

19. Which country or countries has/have the longest life expectancy, and what is it?

20. Which country or countries has/have the shortest life expectancy, and what is it?

21. Which country or countries has/have the highest percentage of the population living in urban areas, and what is the percentage?

22. Which country or countries has/have the highest per-capita GNI PPP, and what is it?

23. Which country or countries has/have the lowest per-capita

TAKE A STAND

Visit our Web site at **http://www.wiley.com/college/raven** (select Chapter 8 from the Table of Contents) for links to more information about the controversies surrounding U.S. immigration policies. Consider the opposing views of supporters and opponents to current immigration policies, and debate the issues with your classmates. You will find tools to help you organize your research, analyze the data, think critically about the issues, and construct a well-considered argument. Take a

Stand activities can be done individually or as part of a team, as oral presentations, written exercises, or Web-based (e-mail) assignments.

Additional on-line materials relating to this chapter, including Student Quizzes, Activity Links, Useful Web Sites, Flash Cards, and more, can also be found on our Web site.

SUGGESTED READING

Bongaarts, J. "Demographic Consequences of Declining Fertility." *Science*, Vol. 282 (October 16, 1998). A nice overview of trends in fertility rates in different regions of the world.

Brown, L., and B. Halweil. "Breaking Out or Breaking Down." *WorldWatch*, Vol. 12 (September–October 1999). The upward trend in life expectancy has reversed in certain of the world's poorer countries.

Doyle, R. "Assembling the Future." *Scientific American*, Vol. 286, No. 2 (February 2002). International migrants are now the major contributor to demographic change in many countries, including the United States.

Ezzell, C. "Care for a Dying Continent." *Scientific American*, Vol. 282, No. 5 (May 2000). Documents the extent of the AIDS crisis in Zimbabwe.

Ferguson, A.R.B. "Perceiving the Population Bomb." *World Watch*, Vol. 14, No. 4 (July–August 2001). The hard facts about human fertility and population growth are considered.

Knodel, J. "Deconstructing Population Momentum." *Population Today*, Vol. 27 (March 1999). An explanation of how population momentum affects the dynamics of population growth.

Martin, P., and E. Midgley. "Immigration to the United States." *Population Bulletin*, Vol. 54 (June 1999). Examines current U.S. immigration policies and patterns.

Perkins, S. "And Counting…" *Science News*, Vol. 161 (February 23, 2002). The 2000 U.S. census is the most accurate one to date.

Tuljapurkar, S. "Taking the Measure of Uncertainty." *Nature*, Vol. 387 (June 19, 1997). A discussion of the strengths and shortcomings of the methods used by demographers to forecast populations.

Wattenberg, B.J. "'Overpopulation' Turns Out to Be Overhyped." *Wall Street Journal* (March 4, 2002). This author presents a view about human population that is quite different from that of most population experts.

Wilson, E.O. "The Bottleneck." *Scientific American*, Vol. 286, No. 2 (February 2002). This excerpt from Wilson's recently published book, *The Future of Life*, is thought-provoking and informative.

2002 World Population Data Sheet. Washington, D.C.: Population Reference Bureau, 2002. A chart published annually that provides current population data for all countries: birth rates, death rates, infant mortality rates, total fertility rates, and life expectancies, as well as other pertinent information. See the poster located inside the back cover of this text.

Family planning. Women at the Dakalaia Health Clinic in Egypt learn about family planning and birth control.

Facing the Problems of Overpopulation

Learning Objectives

After you have studied this chapter you should be able to:

1. Relate human population size to each of the following: hunger; natural resources and the environment; and economics.

2. Distinguish between people overpopulation and consumption overpopulation.

3. Describe the concept of ecological footprints.

4. Outline some of the complexities associated with the concept of sustainable consumption.

5. Define urbanization and describe trends in the distribution of people in rural and urban areas.

6. Explain how cities are analyzed from an ecosystem perspective.

7. Describe some of the problems associated with the rapid growth rates of large urban areas.

8. Relate total fertility rates to each of the following: cultural values; social and economic status of women; and the availability of family planning services.

9. Compare the ways that the governments of China, India, Mexico, and Nigeria have tried to slow human population growth.

10. Outline some of the steps that both governments and individuals can take to achieve global population stabilization.

ike many developing countries, Egypt's population of 71.2 million has more than tripled since 1950, and because the majority of the land in Egypt is uninhabitable desert, most Egyptians live in crowded strips along the Nile River. Yet despite its population growth, Egypt's **total fertility rate (TFR)**, which is the average number of children born to each woman, declined to 3.5 in 2002, down from 7.0 during the 1960s.

Egypt's successful national family planning program, established in 1965, became particularly effective in the late 1980s when its administrators decided to communicate through religion and the media. (At that time, Egypt's TFR was 5.3.) Teams of religious leaders and health care workers started holding neighborhood meetings to educate people about birth control. About 54% of married Egyptian women currently use modern methods of contraception.

Traditionally, Islamic leaders have opposed any form of birth control as an impediment against God's wish to create life. However, the Prophet Mohammed's words have recently been reinterpreted by Islamic scholars in Egypt and several other Islamic nations as sanctioning birth control because it contributes to the social welfare of Muslims. Moreover, the Grand Mufti, the chief cleric and highest interpreter of Islamic law in Egypt, approves of most forms of birth control.

Television entered the campaign, with a popular soap opera called *And the Nile Flows On* that promotes the desirability of marrying at a later age and the benefits of small families. The hero in *And the Nile Flows On* is a handsome, well-educated Muslim cleric who debunks the idea that birth control is sinful. The villain is the local mullah, a religious leader whose wife is a midwife and who therefore profits from every birth in the village. In addition to the soap opera, contests, such as *Wedding of the Month*, ask prospective brides questions about birth control. The winner's wedding is broadcast on national television and is attended by television celebrities. Similar programs on the radio award cash prizes to those who correctly answer the most questions about Egypt's population problem.

St. Lucia, a small island nation in the Caribbean, and Kenya are examples of other developing countries that have produced radio and television soap operas that educate people about family planning, contraceptive use, and sexually transmitted diseases such as acquired immune deficiency syndrome (AIDS). Partly in response to such programs, St. Lucia's TFR has declined from 3.8 in 1990 to 2.0 in 2002, whereas Kenya's TFR has declined from 6.7 in 1990 to 4.4 in 2002.

Many other countries are experiencing fertility declines, but we are far from achieving population stabilization. In this chapter we examine world hunger, underdevelopment, poverty, resource consumption, environmental problems, and urban problems, all of which are aggravated by rapid population growth. We then discuss factors that influence the TFR, including cultural traditions, the social and economic status of women, and the availability of family planning services. If population growth is slowed and resource consumption per person is decreased, the world will be in a better position to tackle many of its serious environmental problems.

THE HUMAN POPULATION EXPLOSION

Most people would agree that all people in all countries should have access to the basic requirements of life: a balanced diet, clean water, decent shelter, and adequate clothing. As the 21st century proceeds, however, it will become increasingly difficult to meet these basic needs, especially in countries that have not achieved population stabilization. About 81% of the world's population lives in less developed countries. If their rate of growth in 2002 were to continue, 126 of these countries would double their populations by the year 2050.

Demographers at the Population Institute consider rapid population growth in the poorest countries of the world to be the most urgent global population problem. It is likely that the social, political, and economic problems resulting from continued population growth in these countries will affect other countries that have already achieved stabilized populations and high standards of living (see Chapter 8 for discussions of immigrants and refugees). For these reasons, population growth is of concern to the entire world community, regardless of where it is occurring.

As our numbers increase during the 21st century, environmental degradation, hunger, persistent poverty, economic stagnation, urban deterioration, and health issues will continue to challenge us (see Chapter 1 for a discussion of poverty). Already the need for food for the increasing numbers of people living in environmentally fragile arid lands, such as parts of sub-Saharan Africa, has led to overuse of the land for grazing and crop production. (*Sub-Saharan Africa* refers to all African countries located south of the Sahara Desert.) When land overuse occurs in combination with an extended period of drought, these formerly productive lands may decline in agricultural productivity—that is, their **carrying capacity** may decrease. Although it is possible to reclaim such arid lands, efforts to do so are made difficult by the large numbers of people and their animal herds trying to live off the land (see Chapter 17 for a discussion of rangeland degradation and desertification).

You may recall from Chapter 8 that when the carrying capacity of the environment is reached, the population may stabilize or crash because of a decrease in the birth rate, an increase in the death rate, or a combination of both. No one knows whether the Earth can sustainably support 10 billion humans or even the 6.2 billion humans that we currently have. It is impossible to quantify the carrying capacity for humans in any meaningful way, in part because our impact on natural resources and the environment involves more than the number of humans. To estimate carrying capacity for humans, we must make certain assumptions about our *quality* of life. Do we assume that everyone in the world should have the same standard of living as average U.S. citizens currently do? If so, then the Earth would be able to support a fraction of the humans it could support if everyone in the world had only the barest minimum of food, clothing, and shelter needed to survive. Also, we do not know what future types of technology might occur that would completely alter Earth's sustainable population size. It may be that we have already reached or overshot our carrying capacity and that the numerous environmental problems we are experiencing will cause the world population increase to come to a halt or even decline precipitously.

On a national level, developing countries have the largest rates of population increase and often have the fewest resources to support their growing numbers. If a country is to support its human population, it must have either the agricultural land to raise enough food for those people or enough of other natural resources, such as minerals or oil, to provide buying power to purchase food.

Population and World Hunger

Many of the world's people—more than 800 million—do not get enough food to thrive, and in certain areas of the world people, especially children, still starve to death (Chapter 18). According to the U.N. Food and Agricultural Organization, 86 countries are considered low-income and food-deficient. South Asia and sub-Saharan Africa are the two regions of the world with the greatest food insecurity (Figure 9.1).

The cause of world hunger, however, is anything but clear.[1] Agriculture currently produces enough food for everyone to have enough if it were distributed evenly. The more than 800 million people who are undernourished are unable to grow or afford food. Experts agree that population, world hunger, poverty, and environmental problems are interrelated, but they do not agree about the most effective way to stop world hunger.

Those who assert that population growth is the root cause of the world's food problem point out that those countries with some of the highest total fertility rates are also the ones with the greatest food shortages. Some of the more extreme members of this group argue that it is imperative to reduce population growth, even through drastic measures such as the establishment of world population quotas. Under such a system, a country that exceeded its assigned population size would not be eligible for relief from the international community during times of food shortages.

Some people think the way to tackle world food problems is not by controlling population growth but by promoting the economic development of countries that are unable to produce adequate food for their people. They presume that development would provide the appropriate technology for the people living in those countries to increase their food production. Also, once a country becomes more developed, its total fertility rate should decline, helping to lessen the population problem (recall the Chapter 8 discussion of fertility rates in highly developed countries).

A third group of people maintains that neither controlling population growth nor enhancing economic development alone will solve world food problems. They argue that the inequitable distribution of resources is the primary cause of world hunger. According to this view, there are enough resources, land, and technologies to produce food for all humans, but people on the lower end of the socioeconomic scale in many countries do not have access to the resources they need to support themselves. In other words, the principal cause of hunger is that people are unable to afford food. (Poverty and hunger are revisited in Chapter 18.)

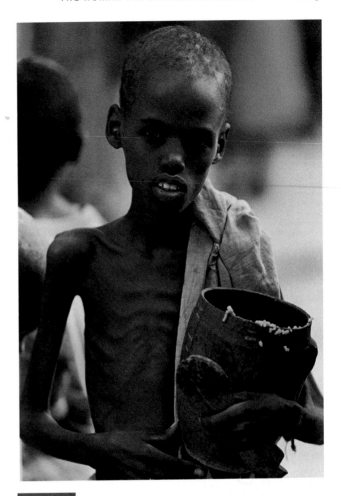

Figure 9.1 Hunger. A young Somali boy receives rice and beans from a Red Cross food distribution center. Overpopulation is not the only reason that there is not enough food for all the world's peoples. In Somalia, other underlying causes include drought and continuing civil strife.

These differing viewpoints indicate that the relationship between world hunger and population may be affected by economic development as well as by poverty and the uneven distribution of resources. Regardless of whether population growth is the main cause of world hunger, however, it is clear that world food problems are exacerbated by population pressures.

Consider sub-Saharan Africa, which is plagued with food insecurity and periodic famine. If this region were to continue its current 2002 rate of growth, it would double its population every 28 years. The U.N. Food and Agricultural Organization estimates that Africa will have to increase its food production by 300% to feed its population in 2050. Alternatively, many African countries will have to import additional food, which they currently cannot afford to pay for out of their own export earnings. Thus, a real danger exists that sub-Saharan Africa's food supply will be overwhelmed by its population growth. Slowing population growth in sub-Saharan countries and other developing countries in Latin America and Asia

---

[1] Famines are a notable exception. A famine can usually be attributed to bad weather, insect outbreaks, war (which causes a breakdown in social and political institutions), or some other disaster (see Chapter 18).

will provide time to increase food production, expand economic development (providing the purchasing power to buy food), and conserve natural resources (which will not have to be exploited as much to pay for food imports).

Economic Effects of Continued Population Growth

The relationship between economic development and population growth is difficult to evaluate. It goes without saying that population growth affects economic development *and* economic development affects population growth. The degree to which each affects the other is unclear, however. Some economists have argued that population growth stimulates economic development and technological innovation. Other economists hold that a rapidly expanding population hampers developmental efforts. Most major technological advances are now occurring in countries where population growth is low to moderate, an observation that seems to support the latter point of view.

In 1986 the National Research Council[2] examined whether large increases in population are a deterrent to economic development. Their panel of experts took into account the complex interactions among global problems such as underdevelopment, hunger, poverty, environmental problems, and population growth. While concluding that population stabilization alone would not eliminate other world problems, the panel determined that for most of the developing world, economic development would profit from slower population growth. Thus, population stabilization would not guarantee higher living standards but would probably promote economic development, which in turn would raise the standard of living. In 1992 the U.S. National Academy of Sciences took a stronger stance and issued a statement with the Royal Society of London that if population growth continues at currently predicted levels, much of the world will experience irreversible environmental degradation and continued poverty.

Debt in Developing Countries   If a country's standard of living is to be raised, its economic growth must be greater than its population growth. If a population doubles every 40 years, then its economic goods and services must more than double during that time. Until recently, many developing nations were able to realize economic growth despite increases in population, largely because of financial assistance, usually in the form of loans from banks and governments of highly developed nations, or

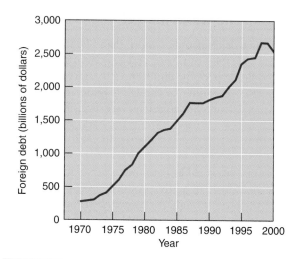

Figure 9.2   External debt in developing countries, 1970 to 2000. (Based on dollar value in the year 2000.)

from multilateral institutions such as the World Bank and the International Monetary Fund (IMF).

However, it has become increasingly difficult for many developing countries to continue raising their standards of living, because the tremendous debts they have accumulated while funding past development preclude future loans. For example, sub-Saharan African countries are overwhelmed by a massive foreign debt, estimated by the World Bank as 70% of the entire region's output in goods and services. This debt burden is severely impeding economic development.

As of 2000, developing nations owe more than $2.5 trillion to highly developed nations, foreign banks, and multilateral institutions (Figure 9.2). Because many of their loans were in default during the 1980s and 1990s, much of this debt has been restructured—that is, reduced and renegotiated so that the debtor nations will be able to repay it. Some creditors have even forgiven debts. According to the U.S. General Accounting Office, the United States canceled $2.3 billion in loans to developing countries between 1990 and 1997. However, the very poorest nations still have significant unrestructured debts.

POPULATION, RESOURCES, AND THE ENVIRONMENT

The relationships among population growth, use of natural resources, and environmental degradation are complex. We address the details of resource management and environmental problems in this and later chapters, but for now, we can make two useful generalizations: (1) The resources that are essential to an individual's survival are small, but a rapidly increasing number of people (as we see in developing countries) tends to overwhelm and

---

[2] The National Research Council is a private, nonprofit society of distinguished scholars. It was organized by the National Academy of Sciences to advise the federal government on complex issues in science and technology.

deplete a country's soils, forests, and other natural resources (Figure 9.3a). (2) In highly developed nations, individual resource demands are large, far above requirements for survival. In order to satisfy their desires rather than their basic needs, many people in more affluent nations exhaust resources and degrade the global environment through extravagant consumption and "throwaway" lifestyles (Figure 9.3b).

Types of Resources

When examining the effects of population on the environment, it is important to distinguish between the two types of natural resources: nonrenewable and renewable. **Nonrenewable resources**, which include minerals (such as aluminum, tin, and copper) and fossil fuels (coal, oil, and natural gas), are present in limited supplies and are

depleted by use. Natural processes do not replenish nonrenewable resources within a reasonable period of time on the human time scale. Fossil fuels, for example, take millions of years to form.

In addition to a nation's population, several other factors affect how nonrenewable resources are used—including how efficiently the resource is extracted and processed and how much of it is required or consumed by different groups of people. People in the United States and other highly developed nations tend to consume most of the world's nonrenewable resources. Nonetheless, the inescapable fact is that Earth has a finite supply of nonrenewable resources that sooner or later will be exhausted. In time, technological advances may enable us to find or develop substitutes for nonrenewable resources. Therefore, slowing the rate of population growth will help us buy time to develop such alternatives.

(a)

(b)

Figure 9.3 Consumption of natural resources. (a) Shown is a typical Indian family, from Ahraura Village, India, with their possessions. The rapidly increasing number of people in developing countries overwhelms their natural resources, even though individual resource requirements may be low. (b) Shown is a typical U.S. family, from Pearland, Texas, with all their possessions. People in highly developed countries consume a disproportionate share of natural resources.

Some examples of **renewable resources** are trees, fishes, fertile agricultural soil, and fresh water. Nature replaces these resources fairly rapidly (on a scale of days to decades), and they can be used forever as long as they are not overexploited in the short term. In developing countries, forests, fisheries, and agricultural land are particularly important renewable resources because they provide food. Indeed, many people in developing countries are subsistence farmers, able to harvest just enough food so that they and their families can survive.

Rapid population growth can cause renewable resources to be overexploited. For example, large numbers of poor people must grow crops on land that is inappropriate for farming—such as mountain slopes or tropical rain forests. Although this practice may provide a short-term solution to the need for food, it does not work in the long run, because when these lands are cleared for farming, their agricultural productivity declines rapidly and severe environmental deterioration occurs. Renewable resources, then, are *potentially* renewable. They must be used in a sustainable way—that is, in a manner that gives them time to replace or replenish themselves.

The effects of population growth on natural resources are particularly critical in developing countries. The economic growth of developing countries is often tied to the exploitation of their natural resources, often for export to highly developed countries. Developing countries are faced with the difficult choice of exploiting natural resources to provide for their expanding populations in the short term (that is, to pay for food or to cover debts) or conserving those resources for future generations. It is instructive to note that the economic growth and development of the United States and of other highly developed nations came about through the exploitation—and in some cases the destruction—of their resources. Continued growth and development in highly developed countries now relies significantly on the importation of these resources from less-developed countries.

Poverty is tied to the effects of population pressures on natural resources and the environment. Poor people in developing countries find themselves trapped in a vicious circle of poverty. They use environmental resources unwisely for short-term gain (that is, to survive), but this exploitation degrades the resources and diminishes long-term prospects of economic development.

Population Size and Resource Consumption

Whereas it is true that resource issues are clearly related to population size (more people use more resources), an equally if not more important factor is a population's *resource consumption*. **Consumption** is the human use of materials and energy (see "Mini-Glossary: Consumption Terms"). Consumption is both an economic and a social act. Consumption provides the consumer with a sense of

---

> ### MINI-GLOSSARY
>
> **Consumption Terms**
>
> **consumption:** The human use of materials and energy.
>
> **people overpopulation:** A situation in which there are too many people in a given area, resulting in pollution, environmental degradation, and resource depletion, even though each individual consumes few resources.
>
> **consumption overpopulation:** A situation in which each individual in a population consumes too large a share of resources, resulting in pollution, environmental degradation, and resource depletion.
>
> **ecological footprint:** The average amount of land and ocean needed to supply an individual with food, energy, water, housing, transportation, and waste disposal.
>
> **sustainable consumption:** The use of goods and services that satisfy basic human needs and improve the quality of life but that also minimize the use of nonrenewable and renewable resources so they are available for future generations.

identity as well as status among peers. The media, including the advertising industry, promote consumption as a way to achieve happiness. We are encouraged to spend, to consume.

People in highly developed countries are extravagant and wasteful consumers; their use of resources is greatly out of proportion to their numbers. A single child born in a highly developed country such as the United States causes a greater impact on the environment and on resource depletion than do a dozen or more children born in a developing country. Many natural resources are needed to provide the automobiles, air conditioners, disposable diapers, cell phones, videocassette recorders, computers, clothes, newspapers, athletic shoes, furniture, boats, and other "comforts" of life in highly developed nations. Yet such consumer goods represent a small fraction of the total materials and energy needed to produce and distribute these goods. According to the Worldwatch Institute, a private research institution in Washington, D.C., Americans collectively consume almost 10 billion tons of materials every year. Thus, the disproportionately large consumption of resources by the United States and other highly developed countries affects natural resources and the environment as much as or more than the population explosion in the developing world.

People Overpopulation and Consumption Overpopulation

A country is *overpopulated* if the level of demand on its resource base results in damage to the environment. If we compare human impact on the environment in developing and highly developed countries, we see that a country can be overpopulated in two ways. **People overpopula-**

tion occurs when the environment is worsening because there are too many people, even if those people consume few resources per person. People overpopulation is the current problem in many developing nations.

In contrast, **consumption overpopulation** occurs when each individual in a population consumes too large a share of resources. The effect of consumption overpopulation on the environment is the same as that of people overpopulation—pollution and degradation of the environment. Many affluent, highly developed nations, including the United States, suffer from consumption overpopulation: *Highly developed nations represent only 20% of the world's population, yet they consume significantly more than half of its resources.* According to the Worldwatch Institute, highly developed nations account for the lion's share of total resources consumed:

- 86% of aluminum used
- 76% of timber harvested
- 68% of energy produced
- 61% of meat eaten
- 42% of the fresh water consumed

These nations also generate 75% of the world's pollution and waste.

Mathis Wackernagel and colleagues at the Universidad Anáhuac de Xalapa in Mexico compared the ecological impact of individuals living in 52 large nations. In this study, each person has an **ecological footprint**, an average amount of productive land and ocean needed to supply that person food, energy, water, housing, transportation, and waste disposal. In developing nations, such as India and Nigeria, the ecological footprint is about 1 hectare (2.5 acres). In the United States, the ecological footprint is about 9.6 hectares (23.7 acres). If all 6.2 billion people in the world had the same lifestyle and level of consumption as an average American, and assuming no changes in technology, we would need 4 additional planets the size of Earth (Figure 9.4).

New Consumers in Developing Countries   As developing nations increase their economic growth and improve their standard of living, more and more people in those countries are able to purchase consumer goods. By the early 2000s, more new cars were sold annually in Asia than in North America and Western Europe combined. These new consumers may not consume at the high level of the average consumer in a highly developed nation, but their consumption has increasingly adverse effects on the environment. For example, air pollution caused by automotive traffic in urban centers in developing countries is bad and getting worse every year. Millions of dollars are lost because of health problems caused by air pollution in these cities. One of society's challenges is to provide new consumers (as well as ourselves) with less polluting, less consuming forms of transportation.

Sustainable Consumption

Consumption overpopulation reflects the growing idea that much of the world's environmental and resource

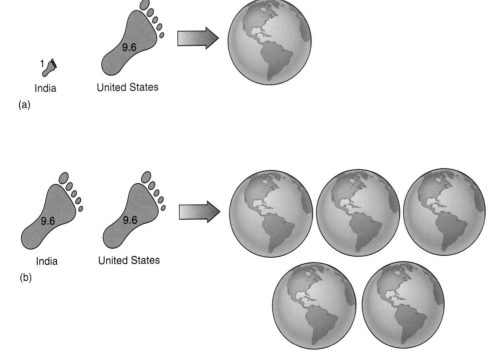

Figure 9.4 Ecological footprints. (a) In developing nations such as India, about 1 hectare is needed to meet the resource requirements of an average person, whereas the ecological footprint of each individual in a highly developed country such as the United States is almost 10 hectares. (b) If everyone in the world had the same level of consumption as the average American, it would take the resources and land area of five planet Earths.

problems stem from the lifestyles of people living in highly developed nations. We interpret *lifestyle* broadly, to include goods and services bought for food, clothing, housing, travel, recreation, and entertainment. In evaluating consumption overpopulation, all aspects of the production, use, and disposal of these goods and services must be taken into account, including the environmental costs of resource extraction, production, transport, trade, and waste management. Such an analysis provides a sense of what it means to consume sustainably versus unsustainably.

Sustainable consumption is the use of goods and services that satisfy basic human needs and improve the quality of life but that also minimize the use of resources so they are available for future use. Like sustainable development (see Chapter 1), sustainable consumption is a complex concept that forces us to address whether our present actions undermine the long-term ability of the environment to meet the needs of future generations. Factors that affect sustainable consumption include population, economic activities, technology choices, social values, and government policies.

At the global level, sustainable consumption requires the eradication of poverty, which in turn requires that poor people in developing countries increase their consumption of certain essential resources. For their increased consumption to be sustainable, however, the consumption patterns of people in highly developed countries must change.

Widespread adoption of sustainable consumption will not be easy. It will require major changes in the consumption patterns and lifestyles of most people in highly developed countries. Some examples of promoting sustainable consumption include switching from motor vehicles to public transport and bicycles and developing durable, repairable, recyclable products.

Sustainable consumption is not a popular idea with policy makers, politicians, economists, or even with consumers such as you and me. It contains an inherent threat to "business as usual," but it also offers new and exciting opportunities. Many scientists and population experts increasingly advocate that we adopt sustainable consumption before it is forced on us by an environmentally degraded, resource-depleted world.

POPULATION AND URBANIZATION

The geographical distribution of people in rural areas, towns, and cities significantly influences the social, environmental, and economic aspects of population growth Throughout recent history, people have increasingly migrated to cities. When Europeans first settled in North America, the majority of the population consisted of farmers in rural areas. Today, approximately 25% of the people in the United States are involved in agriculture, and 75% of the U.S. population lives in cities. The process in which people increasingly move from rural areas to densely populated cities is known as **urbanization**. The process of urbanization also involves the transformation of rural areas into urban areas.

How many people does it take to make an urban area or city? The answer varies from country to country; it can be anything from 100 homes clustered in one place to a population of 50,000 residents. A population of 250 qualifies as a city in Denmark, whereas in Greece a city has a population of 10,000 or more. According to the U.S. Bureau of the Census, a location with 2500 or more people qualifies as an urban area.

One important distinction between rural and urban areas is not how many people live there but how people make a living. Most people residing in a rural area have occupations that involve harvesting natural resources—such as fishing, logging, and farming. In urban areas, most people have jobs that are not directly connected with natural resources.

Cities have grown at the expense of rural populations for several reasons. With advances in agriculture, including the increased mechanization of farms, fewer farmers support an increased number of people. Consequently, there are fewer employment opportunities for people in rural settings. Cities have traditionally provided more jobs because cities are the sites of industry, economic development, and educational and cultural opportunities.

Characteristics of the Urban Population

Every city is unique in terms of its size, climate, culture, and economic development (Figure 9.5). Although there is no such thing as a typical city, certain traits are common to city populations in general. One of the basic characteristics of city populations is that they are far more heterogeneous with respect to race, ethnicity, religion, and socioeconomic status than populations in rural areas. People living in urban areas tend to be younger than in the surrounding countryside. The young age structure of cities is due not to a higher birth rate but instead to the influx of many young adults from rural areas.

Urban and rural areas often have different proportions of males and females. Cities in developing nations tend to have more males. In cities in Africa, for example, males migrate to the city in search of employment, while females tend to remain in the country and tend the farm and their children. Cities in highly developed countries, however, often have a higher ratio of females to males. Women in rural areas often have little chance of employment after they graduate from high school, and so they move to urban areas.

The City as an Ecosystem

Many urban sociologists use an ecosystem approach to better understand how cities function and how they

Figure 9.5 Vancouver, Canada. This city is unique in its beautiful location on the Pacific Ocean. It is Canada's most important port city on the Pacific coast. Noted for its lovely parks and gardens, Vancouver has mild weather with a wet winter season. The Greater Vancouver region has a population of more than 2 million, including a large immigrant population, particularly from China, India, the Philippines, Taiwan, and the Republic of Korea.

change over time. Recall from Chapter 4 that to ecologists an *ecosystem* is an interacting system that encompasses a biological community and its nonliving, physical environment. Urban ecologists study urban trends and patterns in the context of four variables: population, organization, environment, and technology; they use the acronym POET to refer to these variables.

Population refers to the number of people; the factors that change this number (births, deaths, immigration, and emigration); and the composition of the city by age, sex, and ethnicity. *Organization* refers to the social structure of the city, including its economic policies, method of government, and social hierarchy. *Environment* refers both to the natural environment, such as if the city is situated by a river or in a desert, and to the city's physical infrastructure, including its roads, bridges, and buildings. Environment also includes changes to the natural environment that are caused by humans—air and water pollution, for example. *Technology* refers to human inventions that directly affect the urban environment. Examples of technology include aqueducts, which carry water long distances to cities in arid environments, and air conditioning, which allows people to live in comfort in hot, humid cities. The four variables (POET) do not function independently of one another. Just as in natural ecosystems, they are interrelated, often in complex ways.

Environmental Problems Associated with Urban Areas
The concept of the city as an ecosystem would be incomplete without considering the effects the city has on its natural environment. Growing urban areas affect land-use patterns and destroy or fragment wildlife habitat by suburban sprawl that encroaches into former forest, wetlands, desert, or agricultural land in rural areas. For example, portions of Chicago, Boston, and New Orleans are former wetlands. Most cities have blocks and blocks of *brownfields*, areas of abandoned, vacant factories, warehouses, and residential sites that may be contaminated from past uses. Meanwhile, the suburbs continue to expand outward, swallowing natural areas and farmland. (See Chapter 17 for additional discussion of urbanization and its accompanying suburban sprawl.)

Most workers in U.S. cities have to commute dozens of miles through traffic-congested streets, from suburbs where they live to downtown areas where they work. Because development is so spread out in the suburbs, automobiles are a necessity to accomplish everyday chores. This heavy dependence on motor vehicles as our primary means of transportation increases air pollution and causes other environmental problems.

Cities affect water flow by covering the rainfall-absorbing soil with buildings and paved roads. Storm sewage systems are built to handle the runoff from rainfall, which is polluted with organic wastes (garbage, animal droppings, and such), motor oil, lawn fertilizers, and heavy metals. In most cities across the United States, urban runoff is cleaned up in sewage treatment plants before being discharged into nearby waterways. In many cities, however, high levels of precipitation can overwhelm the sewage treatment plant and result in the release of untreated urban runoff. When this occurs, the polluted runoff contaminates water far beyond the boundaries of the city.

The high density of automobiles, factories, and commercial enterprises in urban areas causes a buildup of airborne emissions, including particulate matter (dust), sulfur oxides, carbon oxides, nitrogen oxides, and volatile organic compounds. Urban areas in developing nations have the worst air pollution in the world. In Mexico City, the air is so polluted that schoolchildren are not permitted to play outside during much of the school year. Although progress has been made in reducing air pollution in highly developed nations, the atmosphere in many of their cities often contains higher levels of pollutants than are acceptable based on health standards.

Cities are also warmer than the surrounding countryside. The heat released by human activities is retained by the paved streets and buildings and slowly released into the atmosphere, creating an **urban heat island**. The atmosphere over cities is also cloudier and produces more precipitation than the surrounding countryside. (We consider these various environmental issues in greater detail in Chapters 10, 13, 17, 19, and 21.)

Environmental Benefits of Urbanization Although it may appear from our previous discussion that the concentration of people into cities has an overall harmful effect on the environment, urbanization also has the potential to provide tangible environmental benefits that in many cases outweigh the negative aspects. A well-planned city actually benefits the environment by reducing pollution and preserving rural areas. A solution to urban growth is **compact development**, in which cities are designed so that tall, multiple-unit residential buildings are close to shopping and jobs, all of which are connected by public transportation. Dependence on motor vehicles and their associated pollution is reduced as people walk, cycle, or take public transit such as buses or light rails to work and shop. With compact development, fewer parking lots and highways are needed, so there is more room for parks, open space, housing, and businesses. Thus, compact development makes a city more livable, and more people *want* to live there.

Portland, Oregon, provides a good example of compact development. Portland has limited urban sprawl by developing effective land use policies that dictate where and how growth will occur. The city looks inward to brownfields rather than outward to the suburbs for new development sites. Since 1975, Portland's population has grown by about 50%, yet the urbanized area has increased by only 2%. (In contrast, the population of the Chicago metropolitan area grew by 4% from 1970 to 1990, yet its urbanized area increased by 46%.) Portland's metropolitan transportation system incorporates a 33-mile light-rail line, bike lanes, 102 bus routes, and walkways in addition to the automobile (Figure 9.6). Employers are encouraged to provide bus passes to their employees instead of paying for parking. As an indication of its success in reducing the importance of the automobile, a 1999 study showed that 39% of all adults in the Portland metropolitan region use public transportation at least twice a month, and an average of 252,700 people use public transportation daily. Another recent study found that during the 1990s, a decade of rapid population growth, the number of vehicle miles traveled by automobiles did not increase. In 2000, *Money Magazine* said that Portland is the most livable city in the United States.

CASE·IN·POINT **Curitiba, Brazil**

Livable cities are not restricted to highly developed countries. Curitiba, a city of more than 2.5 million peo-

Figure 9.6 Portland's public transportation system. The light-rail line (*right*) and buses (*left*) eliminate 164,000 car trips each day, easing traffic congestion and reducing air pollution. The public transportation system provides service for the three counties in the Portland metropolitan area. It accommodates people with disabilities and provides free parking at Park and Ride lots along most light-rail routes.

ple in Brazil, provides a good example of compact development in a moderately developed country. Curitiba's city officials and planners have had notable successes in public transportation, traffic management, land use planning, waste reduction and recycling, and community livability. The city developed an inexpensive, efficient mass transit system that uses clean, modern buses that run in high-speed bus lanes. High-density development was largely restricted to areas along the bus lines, encouraging population growth where public transportation was already available. About 72% of the commuters use mass transportation. Since 1974 Curitiba's population has more than doubled, yet traffic has declined by 30%. Because Curitiba does not rely on automobiles as much as do comparable cities, it has less traffic congestion and significantly cleaner air, both of which are major goals of compact development. Instead of streets crowded with vehicular traffic, the center of Curitiba is a *calcadao*, or "big sidewalk," that consists of 49 downtown blocks of pedestrian walkways that are connected to bus stations, parks, and bicycle paths.

Over several decades, Curitiba purchased and converted flood-prone properties along rivers in the city to a series of interconnected parks that are crisscrossed by bicycle paths. This move reduced flood damage and increased the per-capita amount of "green space" from 0.5 m² in 1950 to 50 m² today, a significant accomplishment considering Curitiba's rapid population growth during the same period.

Another example of Curitiba's creativity is its labor-intensive Garbage Purchase program, in which poor people exchange filled garbage bags for bus tokens, surplus food (eggs, butter, rice, and beans), or school notebooks. This program encourages garbage pickup from the unplanned shantytowns (where garbage trucks are unable to drive) that surround the city. Curitiba supplies more services to these unplanned settlements than most cities do. It tries to provide water, sewer, and bus service to them, the bus service allows the settlers to seek employment in the city.

These changes did not happen overnight. Like Curitiba, however, most cities can be carefully reshaped over a period of several decades to make better use of space and to reduce dependence on motor vehicles. City planners and local and regional governments are increasingly adopting measures that will provide the benefits of compact development in the future. ■

Urbanization Trends

Urbanization is a worldwide phenomenon. According to the Population Reference Bureau, 47% of the world population currently lives in urban areas, and sometime before 2010, that number will increase to more than 50%. (As mentioned earlier in the chapter, the definition of *urban* varies from country to country. Generally, however, towns with populations of 2,000 or greater are considered urban.) The percentage of people living in cities compared with rural settings currently is greater in highly developed countries than in developing countries. In 2002, urban inhabitants comprised 75% of the total population of highly developed countries but only 40% of the total population of developing countries.

Although proportionately more people still live in rural settings in developing countries, urbanization there has been increasing rapidly. Currently, most urban growth in the world is occurring in developing countries whereas highly developed countries are experiencing little urban growth. Illustrating the greater urban growth of developing nations is the fact that most of the world's largest cities are in developing countries. In 1950, only 3 of the 10 largest cities were in developing countries. Shanghai, Buenos Aires, and Calcutta. In 2000, 7 of the 10 largest cities were in developing countries: Mexico City, São Paulo, Mumbai (Bombay), Calcutta, Shanghai, Dhaka, and Delhi (Table 9.1). According to the United Nations, there are almost 400 cities worldwide with a population of at least 1 million inhabitants, and 284 of these cities are in developing countries.

Mexico City illustrates the rapidity of urban growth. According to the U.N. Population Division, in 1950 Mexico City had a population of 2.9 million and was the world's 19th largest city. By 2000, its population had increased to 18.1 million, and it was the world's second largest city, surpassed only by Tokyo. Mexico City continues to increase in population by approximately 1,000 new immigrants from overpopulated, economically stagnated rural areas each day.

Urbanization is increasing in highly developed nations, too, but at a much slower rate. Consider the United States as representative of highly developed nations. Here, most of the migration to cities occurred during the past 150 years, when an increased need for industrial labor coincided with a decreased need for agricultural labor. The growth of U.S. cities over such a long period of time was typically slow enough to allow important city services such as water, sewage, education, and adequate housing to keep pace with the influx of people from rural areas.

In contrast, the recent faster pace of urban growth in developing nations has outstripped the limited capacity of many cities to provide basic services. It has also overwhelmed their economic growth (although cities still offer more job possibilities than rural areas). Consequently, cities in developing nations are generally faced with more serious challenges than are cities in highly developed countries. These challenges include substandard housing (slums and squatter settlements; see Figure 9.7); poverty; exceptionally high unemployment; heavy pollution; and inadequate or nonexistent water, sewage, and waste disposal. Rapid urban growth also strains school, medical, and transportation systems.

Cities in highly developed and developing countries share some problems, such as homelessness. Every country, even a highly developed country such as the United States, has city-dwelling people who lack shelter. Although estimates vary widely, most urban scholars estimate that the United States has a total of between 300,000 and 500,000 homeless people on any given night. Urban problems such as homelessness are usually more pronounced in the cities of developing nations, however. In Calcutta, India, perhaps 250,000 homeless people sleep in the streets each night (Figure 9.8).

As an indication of the growing seriousness of urban problems, the U.N. Conference on Human Settlements, which was held in Istanbul, Turkey, in 1996, considered urban issues, such as poverty, crime, and the potential of epidemics in densely populated cities. A major theme of

---

Table 9.1   The World's Ten Largest Cities

1950	*2000*	*2005 projections*
New York, USA, 12.3*	Tokyo, Japan, 26.4	Tokyo, Japan, 26.8
London, UK, 8.7	Mexico City, Mexico, 18.1	São Paulo, Brazil, 19.6
Tokyo, Japan, 6.9	São Paulo, Brazil, 18.0	Mexico City, Mexico, 18.9
Paris, France, 5.4	New York, USA, 16.7	Mumbai (Bombay), India, 18.3
Moscow, USSR, 5.4	Mumbai (Bombay), India, 16.1	New York, USA 17.1
Shanghai, China, 5.3	Los Angeles, USA, 13.2	Dhaka, Bangladesh, 15.9
Essen, Germany, 5.3	Calcutta, India, 13.1	Delhi, India, 15.3
Buenos Aires, Argentina, 5.0	Shanghai, China, 12.9	Calcutta, India, 14.3
Chicago, USA, 4.9	Dhaka, Bangladesh 12.5	Los Angeles, USA, 13.8
Calcutta, India, 4.4	Delhi, India, 12.4	Jakarta, Indonesia, 13.2

* Population in millions.

Figure 9.7 Squatter settlement. The substandard, poor-quality housing seen in this section of Port au Prince, Haiti, is common in cities of developing nations. These houses are built out of whatever material their owners can salvage and patch together. Squatters illegally occupy the land that they build on and therefore cannot require the city to supply them with services such as clean water, sewage treatment, paved roads, or police and fire protection. According to the United Nations, squatter settlements house about one third of the entire urban population in most developing countries.

the conference was that cities could be made livable. To demonstrate that theme, the conference gave 12 awards for urban development projects that have achieved the greatest improvements in human settlements.

Is urbanization related to the rate at which the population grows? Urbanization appears to be a factor in *decreasing* total fertility rates, perhaps because family planning services, including access to contraceptives, are more readily available in urban settings. We now examine various factors that influence the total fertility rate.

■ REDUCING THE TOTAL FERTILITY RATE

Dispersal (moving from one place to another) used to be a solution for overpopulation, but not today. As a

species, we have expanded our range throughout Earth, and few habitable areas remain that have the resources to adequately support a major increase in human population. Nor is increasing the death rate an acceptable means of regulating population size. Clearly, the way to control our expanding population is by reducing the number of births. Cultural traditions, women's social and economic status, and family planning all influence the total fertility rate (TFR).

Culture and Fertility

The values and norms of a society—what is considered right and important and what is expected of a person—constitute a part of that society's culture. Gender is an important part of culture. Different societies have different gender expectations—that is, varying roles that men

Figure 9.8 Homelessness. Entire families of desperately poor "pavement people" eat, sleep, and raise their families on a street in Calcutta, India. Homelessness, a serious problem in the cities of highly developed countries, is an even greater problem in cities of developing nations.

and women are expected to fill. In parts of Latin America men are expected to do the agricultural work, whereas in sub-Saharan Africa women do most of the agricultural work. With respect to fertility and culture, a couple is expected to have the number of children determined by the traditions of their society.

High TFRs are traditional in many cultures. The motivations for having many babies vary from culture to culture, but overall *a major reason for high TFRs is that infant and child mortality rates are high.* In order for a society to endure, it must continue to produce enough children who survive to reproductive age. Thus, if infant and child mortality rates are high, TFRs must also be high to compensate. Although world infant and child mortality rates have been decreasing, it will take longer for culturally imbedded fertility levels to decline. Parents must have enough confidence that the children they already have will survive before they stop having additional babies. Another reason for the lag in fertility decline is cultural: Changing anything that has been traditional, including large family size, usually takes a long time.

Higher TFRs in some developing countries are also due to the important economic and societal roles of children. In some societies, children usually work in family enterprises such as farming or commerce, contributing to the family's livelihood (Figure 9.9). The International Labor Organization estimates that worldwide, 100 million to 200 million children under the age of 15 work; more than 95% of these children live in developing countries. When these children become adults, they provide support for their aging parents. In contrast, children in highly developed countries have less value as a source of labor, because they attend school and because human labor is required less in an industrialized society. Further, highly developed countries provide many social services for the elderly, so the burden of their care does not fall entirely on their offspring.

Many cultures place a higher value on male children than on female children. In these societies, a woman who bears many sons achieves a high status; thus, there is a social pressure to have male children that keeps the TFR high.

Religious values are another aspect of culture that affects TFRs. Several studies done in the United States point to differences in TFRs among Catholics, Protestants, and Jews. In general, Catholic women have a higher TFR than either Protestant or Jewish women, and women who do not follow any religion have the lowest TFR of all. However, it is difficult to conclude that the observed differences in TFRs are the result of religious differences alone. Other variables, such as ethnicity (certain religions are associated with particular ethnic groups) and residence (certain religions are associated with urban or with rural living), complicate any generalizations that might be made.

Figure 9.9 Children at work. These children are gathering firewood near the Ethiopian village of Meshal. In developing countries, total fertility rates are high partly because children contribute to the family by working. Young children gather firewood, carry water, and work in the fields with their parents. Older children often have wage-producing jobs that add to the family income.

The Social and Economic Status of Women

Gender inequality exists in most societies. Women do not have the same rights, opportunities, or privileges as men. Sons are more highly valued than daughters are, so girls are often kept at home to work rather than being sent to school. In most developing countries, a higher percentage of women are illiterate than men (Table 9.2). Note in the table, however, that definite progress has been made in recent years in increasing literacy in both women and men and in narrowing the gender gap. Younger women and men are less likely to be illiterate than older women and men within a given country. Fewer women than men attend secondary school (high school). In some African countries only 2% to 5% of girls are enrolled in secondary school. Worldwide, some 90 million girls are not given the opportunity to receive a primary (elementary school) education. Laws, customs, and lack of education often limit women to low-skilled,

Table 9.2 Percent Illiteracy of Men and Women in Selected Developing Countries, 2000

Country	Ages 15 to 24		Ages 25+	
	Women	Men	Women	Men
Bangladesh	62	41	87*	63*
Brazil	10	15	25	22
China	9	3	42	17
Egypt	46	29	78	50
Ethiopia	52	48	89*	74*
India	38	22	81*	50*
Kenya	14	8	54	26
Mexico	5	4	20	13
Philippines	3	4	9	7

* 1990 data are the latest available.

low-paying jobs. In such societies, marriage is usually the only way for a woman to achieve social influence and economic security.

Evidence is accumulating that *the single most important factor affecting high TFRs may be the low status of women in many societies.* A significant way to tackle population growth, then, is to improve the social and economic status of women. We say more about this later, but for now we examine how marriage age and educational opportunities, especially for women, affect fertility.

Marriage Age and Fertility The TFR is affected by the average age at which women marry, which in turn is determined by the laws and customs of the society in which they live. Women who marry are more apt to bear children than women who do not marry, and the earlier a woman marries, the more children she is likely to have.

The percentage of women who marry and the average age at marriage vary widely among different societies, but there is always a correlation between marriage age and TFR. Consider Sri Lanka and Bangladesh, two developing countries in South Central Asia. In Sri Lanka the average age at marriage is 25, and the average number of children born per woman is 2.0, which means it will take 58 years for the population to double (at its current rate of natural increase). In contrast, in Bangladesh the average age at marriage is 17, the average number of children born per woman is 3.3, and the doubling time is 32 years.

Educational Opportunities and Fertility In nearly all societies women with more education tend to marry later and have fewer children. Figure 9.10 shows the total fertility rates of women in the United States with different education levels. Providing women with educational opportunities delays their first childbirth, thereby reducing the number of childbearing years and increasing the amount of time between generations. Education also

opens the door to greater career opportunities and may change women's lifetime aspirations. In the United States, it is not uncommon for a woman in her thirties or forties to give birth to her first child, after establishing a career.

Studies in dozens of countries show a strong correlation between the average amount of education women receive and TFR. For example, women in Botswana with a secondary education have an average of 3.1 children each, women with a primary education have 5.1 children each, and women with no formal education have 5.9 children each.

Education increases the probability that women will know how to control their fertility, and it provides them with knowledge to improve the health of their families, which results in a decrease in infant and child mortality. A study in Kenya showed that 10.9% of children born to women with no education died by age 5, as compared with 7.2% of children born to women with a primary education and 6.4% of children born to women with a secondary education. Education also increases women's options, opening doors to other careers and ways of achieving status besides having babies.

Education may have an indirect effect on TFR as well. Children who are educated have a greater chance of improving their living standards, partly because they have more employment opportunities. Parents who recognize this may be more willing to invest in the education of a few children than in the birth of many children whom they cannot afford to educate. The ability of bet-

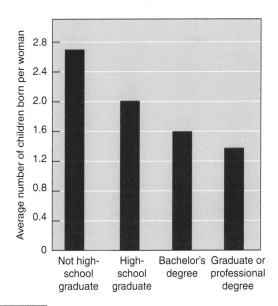

Figure 9.10 Education and fertility. The number of children a woman has varies with the amount of education she has received. Shown are the U.S. total fertility rates for 35- to 44-year-old women with differing levels of education in 1995 (latest data available).

ENVIROBRIEF

Successful Shots

Preventing and controlling common childhood diseases reduces the mortality of infants and children, which in turn is associated with lowered birth rates. Immunizing children is therefore an important part of efforts to stabilize the human population. Since the 1980s, international immunization programs have enjoyed tremendous success:

- The U.N. Expanded Program for Immunizations administers vaccines against the six major childhood diseases: tuberculosis, diphtheria, pertussis, tetanus, polio, and measles.

- By 1995, more than 80% of infants from developing nations were immunized, up from 25% in the 1980s and the less than 5% in the mid-1970s.

- Since the mid-1990s, the numbers have not changed appreciably, although there have been improvements in specific countries and with selected diseases.

- Researchers strive for further immunization success at lower cost by developing vaccines that require fewer doses and maintain their potency at tropical temperatures.

ter-educated people to earn more money may be one of the reasons why smaller family size is associated with increased family income.

Family Planning Services

Socioeconomic factors may encourage people to want smaller families, but *reduction in fertility will not become a reality without the availability of health and family planning services.* The governments of most countries recognize the importance of educating people about basic maternal and child health care. Developing countries that have had success in significantly lowering TFRs credit many of these results to effective family planning programs. Prenatal care and proper birth spacing make

women healthier. In turn, healthier women give birth to healthier babies, leading to fewer infant deaths. The percentage of women using family planning services to limit family size has increased from less than 10% in the 1960s to more than 50% during the 1990s. During that time, however, the actual number of women who are not using family planning has increased due to population growth.

Family planning services provide information on reproductive physiology and contraceptives, as well as the actual contraceptive devices, to those who wish to control the number of children they produce or to space their children's births. Family planning programs are most effective when they are designed with sensitivity to local social and cultural beliefs. Family planning services do not try to force people to limit their family sizes but rather attempt to convince people that small families (and the contraceptives that promote small families) are acceptable and desirable. The major birth control methods in use today are shown in Table 9.3.

Contraceptive use is strongly linked to lower total fertility rates (Figure 9.11). Research has shown that 90% of the decrease in fertility in 31 developing countries was a direct result of the increased knowledge and availability of contraceptives. In highly developed countries, where total fertility rates are at replacement levels or lower, the percentage of married women of reproductive age who use contraceptives is often greater than 70%. Fertility declines have been noted in developing countries where contraceptives are readily available. During the 1970s, 1980s, and 1990s, use of contraceptives in East Asia and many areas of Latin America increased significantly, and these regions experienced a corresponding decline in birth rate. In areas where contraceptive use remained low, such as parts of Africa, there was little or no decline in birth rate.

Family planning centers provide information and services primarily to women. As a result, in the male-dominated societies of many developing countries, such

Figure 9.11 Contraceptive use and total fertility rate, 2002. Note that greater contraceptive use among married women of reproductive age correlates with a lower fertility rate.

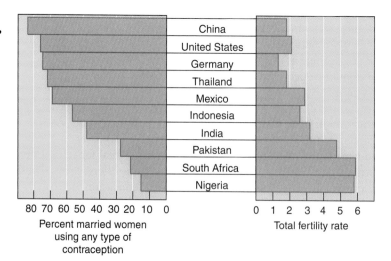

Percent married women using any type of contraception

Total fertility rate

Table 9.3	Birth Control Methods			
Method	**Failure Rate***	**Mode of Action**	**Advantages**	**Disadvantages**
Oral contraceptives	0.3; 5	Inhibit ovulation; may also affect endometrium and cervical mucus and prevent implantation	Highly effective; regulate menstrual cycle	Minor discomfort in some women; because of potential for thromboembolism or hypertension, should not be used by women over age 35 who smoke
Injectable contraceptives	About 1	Inhibit ovulation	Effective; long-lasting	Irregular menstral bleeding; fertility may not return for 6 to 12 months after use discontinued
Progesterone implantation	About 1	Inhibits ovulation	Effective; long-lasting	Irregular menstrual bleeding in some women; requires small incision in arm
Intrauterine device (IUD)	1; 1	Not known; probably stimulates inflammatory response and prevents implantation	Provides continuous protection; highly effective for several years	Cramps; increased menstrual flow; spontaneous expulsion; increased risk of pelvic inflammatory disease and infertility; not recommended for women who have not had a child
Spermicides (foams, jellies, creams)	3; 20	Chemically kill sperm	No known side effects; can be used with a condom or diaphragm to improve efficacy	Messy to use; must be applied before intercourse
Contraceptive diaphragm (with jelly)[†]	3; 14	Diaphragm mechanically blocks entrance to cervix; jelly is spermicidal	No side effects	Must be prescribed (and fitted) by physician; must be inserted prior to intercourse and left in place for several hours afterward
Condom, male	2.6; 14	Mechanically prevents sperm from entering vagina	No side effects; some protection against STDs,** including human immuno-deficiency virus	Interruption of foreplay to put it on; slightly decreased sensation for male; could break
Rhythm[‡]	13; 19	Abstinence during fertile period	No known side effects	Not very reliable
Withdrawal (coitus interruptus)	9; 22	Male withdraws penis from vagina prior to ejaculation	No side effects	Not reliable; sperm in the fluid secreted before ejaculation may be sufficient for conception
Sterilization Tubal ligation	0.04	Prevents ovum from leaving uterine tube	Most reliable method	Requires surgery; considered permanent
Vasectomy	0.15	Prevents sperm from leaving scrotum	Most reliable method	Requires surgery, considered permanent
Chance (no contraception)	About 90			

*The lower figure is the failure rate of the method; the higher figure is the rate of method failure plus failure of the user to use the method correctly. Based on number of failures per 100 women who use the method per year in the United States.
** STDs are sexually transmitted diseases.
† The failure rate is lower when the diaphragm is used with spermicides.
‡ There are several variations of the rhythm method. For those who use the calendar method alone, the failure rate is about 35. However, if the body temperature is taken daily and careful records are kept (temperature rises after ovulation), the failure rate can be reduced. When women use some method to help determine the time of ovulation, and have intercourse *only* more than 48 hours *after* ovulation, the failure rate can be reduced to about 7.

services may not be as effective as they could otherwise be. Polls of women in developing countries reveal that many who say they do not want additional children still do not practice any form of birth control. When asked why they do not use birth control, these women frequently respond that their husbands or in-laws want additional children.

GOVERNMENT POLICIES AND FERTILITY

The involvement of governments in childbearing and childrearing is well established. Laws determine the minimum age at which people may marry and the amount of education that is compulsory. Governments may allot portions of their budgets to family planning services, education, health care, old-age security, or incentives for smaller or larger family size. The tax structure, including additional charges or allowances based on family size, also influences fertility.

In recent years, the governments of at least 78 developing countries—41 in Africa, 19 in Asia, and 18 in Latin America and the Caribbean—have recognized the need to limit population growth and have formulated policies such as economic rewards and penalties designed to achieve this goal. Most countries sponsor family planning projects, many of which are integrated with health care, education, economic development, and efforts to improve women's status. Many of these projects are supported by the U.N. Fund for Population Activities.

Population control measures have been instituted in many developing countries (recall the chapter introduction). Here we examine those measures in China, India (the two most populous nations), Nigeria, and Mexico. We then examine population concerns in Europe as an interesting comparison.

CASE·IN·POINT China's Controversial Family Planning Policy

China, with a mid-2002 estimated population of 1.28 billion people, has the largest population in the world, and its population exceeds the combined populations of all highly developed countries. Recognizing that its rate of population growth had to decrease or the quality of life for everyone in China would be compromised, the Chinese government in 1971 began to pursue birth control seriously. It urged couples to marry later, increase spacing between children, and limit the number of children to two.

In 1979 China instigated an aggressive plan to push China into the third demographic stage. (Recall from Chapter 8 that the third demographic stage is charac-

terized by a decline in the birth rate along with a relatively low death rate.) Announcements were made of incentives to promote later marriages and one-child families (Figure 9.12). Local jurisdictions were assigned the task of reaching this goal. A couple who signed a pledge to limit themselves to a single child might be eligible for such incentives as medical care and schooling for that child, cash bonuses, preferential housing, and retirement funds. Penalties were also instituted, including fines and the surrender of all of these privileges if a second child was born.

China's aggressive plan brought about the most rapid and drastic reduction in fertility in the world, from 5.8 births per woman in 1970 to 2.1 births per woman in 1981. However, it has been controversial and unpopular because it compromised individual freedom of choice. In some instances, social pressures from the community induced women who were pregnant with a second child to get an abortion.

Moreover, based on the disproportionate number of male versus female babies reported born in recent years, it is suspected that many expectant parents determine the sex of their fetus and abort it if it is a female. In addition, hundreds of thousands of newborn baby girls were killed or abandoned by their parents who, when required to conform to the one-baby policy, wanted a boy. In China, sons carry on the family name and traditionally provide old-age security for their parents. Thus, sons are valued more highly than daughters are. As a result of the increased ratio of male to female births, demographers project that by the middle of the 21st century, marriageable males will outnumber marriageable females by 1 million.

In 1984 the one-child family policy was relaxed in rural China, where 70% of all Chinese live. China's recent population control program has relied on education, publicity campaigns, and fewer penalties to achieve its goals. China trains population specialists at institu-

Figure 9.12 China's one-child family policy. A billboard campaign in China promotes the one-child family. Note that the child is a male rather than a female.

tions such as the Nanjing College for Family Planning Administrators. In addition, thousands of secondary school teachers have been taught how to integrate population education into the curriculum. China's total fertility rate in 2002 was 1.8, and the Chinese government continues its policy of strict population control. ■

CASE-IN-POINT India's Severe Population Pressures

India is the world's second most populous nation, with an estimated mid-2002 population of 1.05 billion (Figure 9.13). In the 1950s it became the first country to establish government-sponsored family planning. Unlike China, India did not experience immediate results from its efforts to control population growth, in part because of the diverse cultures, religions, and customs in different regions of the country. Indians speak 14 main languages and more than 700 other languages and dialects, which makes communicating a broad program of family planning education difficult. Like China, the Indian culture is biased toward male children. Husbands often desert wives who do not produce sons. Custom dictates that when a husband dies, the sons rather than the daughters support their mother. These and other factors promote a "large family" psychology in Indian society.

In 1976 the Indian government became more aggressive. It introduced incentives to control population growth and controversial programs of compulsory sterilization in several states. If a man had three or more living children, he was compelled to obtain a vasectomy. Compulsory sterilization was a failure; it had little effect on the birth rate and was exceedingly unpopular. It may have

been partly responsible for the fact that Indira Gandhi, who was prime minister beginning in 1966, was voted out of office in 1977. Even after compulsory sterilization was abandoned, population experts continued to criticize India's family planning policy because it still put too much emphasis on sterilizations, which are almost exclusively used by mature couples who have already had several children and are seeking a long-term birth control procedure.

In recent years, India's family planning policies have taken a different approach—to educate young couples, who need to delay childbirth and space their young if the TFR is going to reach its target of 2.1. India has also attempted to integrate economic development and family planning projects. For example, adult literacy and population education programs have been combined. Multimedia advertisements and education have been used to promote voluntary birth control, and contraceptives have been made more available. India has also emphasized improving health services to lower infant and child mortality rates and reduce the rate of female infanticide. These changes have had an effect: India's total fertility rate has declined from 5.3 in 1980 to 3.2 in 2002.

Despite these relatively modest gains, India's total fertility rate remains above replacement level, and the nation is facing severe problems. Population pressure has caused the deterioration of India's environment in the past few decades, and 40% of Indians live below the official poverty level. Population experts predict that by 2050, India's population will exceed that of China. These numbers can only exacerbate India's poverty, environmental degradation, and economic underdevelopment. ■

Figure 9.13 Mass wedding. This gathering of 104 couples and their families, which took place in Jaipur, India, on February 18, 1996, symbolizes the population pressures occurring in India.

CASE-IN-POINT **Mexico's Young Age Structure**

Mexico, with a mid-2002 estimated population of 102 million, is the second most populous nation in Latin America. (Brazil, with a population of 174 million, is the most populous.) Mexico has a tremendous potential for growth because 33% of its population is less than 15 years of age. Even with a low birth rate, population growth momentum will cause the population to continue to increase in the future because of the large number of young women having babies.

Traditionally, the Mexican government supported rapid population growth, but in the late 1960s the government became alarmed at how rapidly the population was increasing. In 1974 the Mexican government instigated several measures to reduce population growth, such as educational reform, family planning, and health care. The time was right for such changes, as many Mexican women were already ignoring the decrees of the progrowth government and the Roman Catholic Church and purchasing contraceptives on the black market. Mexico has had great success in reducing its fertility level, from 6.7 births per woman in 1970 to 2.9 births per woman in 2002.

Mexico's goal includes not only population stabilization but also balanced regional development. Its urban population is 74% of its total population, and most of these people live in Mexico City. Although Mexico is largely urbanized compared with other developing countries, its urban-based industrial economy has not been able to absorb the large number of people in the work force. (According to the Mexico office of the International Labor Organization, about 1.3 million new workers join Mexico's labor force each year.) Because unemployment is so high, many Mexicans have migrated, both legally and illegally, to the United States (see Chapter 8).

Mexico's recent efforts at population control include multimedia campaigns. As in Egypt, popular television and radio soap operas carry family planning messages such as "Small families live better." Booklets on family planning are distributed, and population education is being integrated into the public school curriculum. Social workers receive training in family planning as part of their education.

CASE-IN-POINT **Much to Be Done in Nigeria**

Nigeria is part of the sub-Saharan region that currently has the world's most rapid population growth. Nigeria has the largest population of any African country: In mid-2000 its population was estimated at 130 million people. Nigeria's TFR today (5.8) is basically what it was in the early 1980s—6.0 births per woman. Nigeria has a huge reproductive potential because 44% of the population is less than 15 years of age. Today only about 9% of married women use a modern method of contraception. Furthermore, less than half of all Nigerian women in their reproductive years (ages 15 to 49) have any knowledge of contraception. The average life expectancy in Nigeria is 52 years, which is low in part because of high infant and child mortality rates.

The Nigerian government has recognized that its economic goals are more likely to be attained if its rate of population growth decreases. In 1986 Nigeria developed a national population policy that integrates population and development projects. The plan involves improving health care, including training nurses and other health care professionals. Population education is being used to encourage later marriages (currently, half of all women are married by age 17) and birth spacing.

CASE-IN-POINT **Population Concerns in Europe**

Population has stabilized in Europe, where most countries have lower-than-replacement-level total fertility rates (TFRs), and several countries have experienced a slight decline in population. As a result of fewer births, the proportion of elderly people in the European population is increasing. Population control has thus become a controversial issue in Europe, with two opposing viewpoints emerging.

Those who favor population growth are called **pronatalists**. Pronatalists think that the vitality of their region is at risk because of declining birth rates. They are concerned that the decrease in population might result in a loss of economic growth. They presume that their countries' positions in the world will weaken and that their cultural identity will be diluted by immigrants from non-European countries. (It is well established that declining birth rates and increased immigration to Europe have resulted in increased diversity. The current wave of immigrants to Europe is primarily from North Africa and the Middle East.) Pronatalists are also concerned about the possibility that pension and old age security systems will be overwhelmed by the large numbers of elderly in Europe unless a larger work force is available to contribute to those systems. The most outspoken pronatalists assert that women should marry young and have many children for the good of society. Pronatalists favor government policies that provide incentives for larger families (such as paid maternity and paternity leaves, easily available child care, and monetary "baby bonuses") and penalties for smaller families.

The opponents of pronatalists do not view European society as declining. Instead, they contend that European influence is increasing, because they judge power in economic terms rather than in terms of population. Generally, they are not as opposed to immigration, and they maintain that Europe should not be making a concerted effort to increase its TFRs when overpopulation is such a serious problem in much of the world. Further, they point out that technological innovations have eliminated many jobs in Europe, and the consequential unemployment would only be made worse by an increase in the labor force caused by a rise in birth rate. Opponents of pronatalists also question whether the elderly are a burden to society. They argue that the elderly are not the only ones the work force supports. Their view is that the cost of providing for the young, especially for many years of education, cancels out any perceived benefit from an increased birth rate.

THE GLOBAL SUMMIT ON POPULATION AND DEVELOPMENT

The third and most recent U.N. International Conference on Population and Development was held in Cairo in 1994 to address population issues and the role of government in controlling population growth. The two previous U.N. population conferences, in 1974 and 1984, focused on setting national and global population targets and meeting those targets through family planning.

Departing from this tradition, the 1994 conference drafted a 20-year World Program of Action that provides specific measures to be addressed at the *local* level. It focuses on individuals' reproductive rights, empowerment of women, and reproductive health. *Reproductive rights* refer to the rights of individuals to make informed decisions about their fertility and reproductive health. *Empowerment of women* is concerned with improving women's status through education and economic opportunities; it is thought that such empowerment will give women more choices and therefore greater control over their reproductive rights. *Reproductive health* includes access to affordable contraception, modern obstetrical practices, and improved maternal and child heath care, as well as control of sexually transmitted diseases. Most of the more than 500,000 women who die each year as a result of pregnancy and childbirth do not have access to medical care during pregnancy and childbirth. Such health care is either not available or unaffordable for the women.

The conference also recognized a wide range of issues that had been largely ignored during previous conferences: the responsibilities of men in family planning and raising children (Figure 9.14), violence against women, female genital mutilation, unsafe abortions, and prenatal sex selection.

Despite the bitter debates over contentious issues such as abortion, the 1994 plan was approved by 180 nations. Although the plan was not a treaty and was therefore not legally binding, it was expected to set the agenda for national and international programs to slow population growth. Local laws, cultural values, and religious beliefs were to be respected in adopting the provisions of the plan.

Figure 9.14 Men and family planning. The importance of male involvement in family planning and child rearing has been overlooked in the past, when family planning agencies focused almost exclusively on women. Shown is a father assuming family responsibilities in Tijuana, Mexico.

The World Program of Action was further supported by the Fourth World Conference on Women's Rights, which met in 1995 in Beijing, China. The conference emphasized that expanding women's access to health care, education, and employment is important not only in its own right but also because this access will slow population growth.

The World Program of Action included an initial annual budget of $17 billion beginning in the year 2000. This amount sounds very high, but it is less than what the world spends per week on military armaments. Two thirds of this cost was to be paid by developing countries, and one third was to be donated by highly developed nations. As of 2002 the allocations by individual governments have been but a fraction of what is needed. Highly developed countries—particularly the United States, which is the largest international donor—have not paid their share.

Reproduction and Women—Different Countries, Different Rights

One of the many issues raised at the U.N. population summit in Cairo in 1994 was the reproductive rights of women, and particularly the vastly different government policies that pressure or force women to adhere to specific reproductive plans:

- Urban women in China (total fertility rate [TFR] = 1.8) are limited to one child through allegedly enforced contraception (intrauterine device [IUD] insertion), sterilization, or abortion.

- In Japan (TFR = 1.3), oral contraceptives (birth control pills) are illegal, in part because it is feared that their widespread use will further lower the TFR. The only legal birth control options available to Japanese couples are condoms and abortion, although many Japanese women take high-dose birth control pills for "menstrual disorders." The Japanese government may soon lift its ban on oral contraceptives.

- In certain European countries (Germany [TFR = 1.3], Hungary [TFR = 1.3], Luxembourg [TFR = 1.8], and Portugal [TFR = 1.5]) and in Japan (TFR = 1.3), women are strongly encouraged, with cash payments or promises of benefits such as payment of obstetric expenses and child care to bear several children.

- Women in Bangladesh (TFR = 3.3) and India (TFR = 3.2) are encouraged to limit reproduction to no more than two children, and many undergo substandard tubal ligations.

- Saudi Arabian (TFR = 5.7) and Libyan (TFR = 3.7) couples do not have legal access to contraceptives.

- Abortions are illegal in most Latin American nations (average TFR = 2.7), and many maternal deaths are caused by botched illegal abortions.

- Russian women (TFR = 1.3) are forced to rely primarily on abortions as birth control because quality contraceptives are often unavailable.

According to the U.N. Population Fund, underfunding of the World Program of Action could result in

- 122 million to 220 million unintended pregnancies, which result in 57 million to 104 million births and 43 million to 88 million additional abortions.

- 65,000 to 117,000 maternal deaths.

- 5.2 million to 9.3 million infant and child deaths.

ACHIEVING POPULATION STABILIZATION

In this chapter we have considered how global problems—including hunger, poverty, economic underdevel-

opment, and environmental issues—are exacerbated by an increase in population and by overconsumption. Population stabilization and reduction of overconsumption are critically important if we are to effectively tackle these serious problems. But what kinds of policies will help achieve population stabilization?

Developing countries should increase the amount of money that they allot to public health (to reduce infant and child mortality rates) and family planning services, particularly the dissemination of information on affordable, safe, and effective methods of birth control. Governments should take steps to increase the average level of education, especially of women, and women must be given more employment opportunities. National population policies will not work without cultural awareness, respect for individual religious beliefs, and local community involvement and acceptance; community leaders can provide feedback on the effectiveness of programs.

Highly developed countries can help developing nations achieve a decrease in population growth by providing financial support for the U.N. Population Fund, which supports international family planning efforts, and by supporting the research and development of new birth control methods. Providing funding for development projects that increase the incomes of poor people in developing countries would help alleviate the hunger and poverty that are associated with population growth.

Most important, highly developed nations need to face their own population problems, particularly the environmental costs of consumption overpopulation by affluent people. Policies should be formulated that support reducing the use of resources, increasing the reuse and recycling of materials, and expunging the throwaway mentality. These policies will also show developing nations that highly developed countries are serious about the issues of overpopulation at home as well as abroad.

On a personal level, individuals in highly developed countries should examine their own consumptive practices and take steps to reduce them. Individual efforts to reduce material consumption—sometimes called *voluntary simplicity*—are effective not only for their collective effects but also because they often influence additional people to reduce unnecessary consumption. Before purchasing a product, individuals should ask if it is *really* needed. A material lifestyle—two or three cars in the garage, steaks on the grill, a television in every room, and similar luxury items—may provide short-term satisfaction. Such a lifestyle is not as fulfilling in the long term, however, as learning about the world, interacting with others in meaningful ways, and contributing to the betterment of family and community. In the end, a rich life is measured not by what was owned but by what was done for others.

SUMMARY WITH SELECTED KEY TERMS

I. Many human problems, such as hunger, resource depletion, environmental problems, underdevelopment, poverty, and urban problems, are exacerbated by the rapid increase in population.

A. The countries with the greatest food shortages also have some of the highest **total fertility rates (TFRs)**.

B. Economic development, poverty, and the inequitable distribution of resources influence the relationship between world hunger and population growth.

C. Underdevelopment and poverty are associated with high TFRs. Most economists think that slowing population growth promotes economic development.

II. In developing countries, individual resource demands are small, but rapidly increasing populations deplete natural resources. In highly developed countries, individual resource demands are large and deplete natural resources.

A. **Nonrenewable resources** are present in a limited supply and cannot be replenished in a reasonable period. Slowing the rate of population growth will give more time to find substitutes for nonrenewable resources as they are depleted.

B. **Renewable resources** are replaced by natural processes and can be used forever, provided they are used in a sustainable way. Overpopulation causes renewable resources to be overexploited. When this happens, renewable resources become nonrenewable.

C. **Consumption** is the human use of materials and energy.

   1. **People overpopulation** occurs when the state of the environment is worsening because there are too many people, even if those people consume few resources per person. Developing countries have people overpopulation.

   2. **Consumption overpopulation** occurs when each individual consumes too large a share of resources, thereby degrading the environment. Highly developed countries have consumption overpopulation.

D. A person's **ecological footprint** is the average amount of productive land and ocean needed to supply that person with food, energy, water, housing, transportation, and waste disposal.

   1. The ecological footprint of a person in a developing nation is about 1 hectare.

   2. The ecological footprint of a person in a highly developed country is about 9.6 hectares.

E. **Sustainable consumption** is the use of goods and services that satisfy basic human needs and improve the quality of life but that minimize the use of resources so they are available for future generations.

III. As a nation develops economically, the proportion of the population living in cities increases.

A. The process by which people increasingly move from rural areas to densely populated cities is known as **urbanization**.

B. In developing nations, most people live in rural settings, but their rates of urbanization are rapidly increasing. This makes it difficult to provide city dwellers with basic services such as housing, water, sewage, and transportation systems.

C. A solution to urban growth is **compact development**, in which cities are designed so that multiple-unit residential buildings are close to shopping and jobs, all of which are connected to public transportation.

   1. Portland, Oregon, is a good example of compact development in a highly developed country.

   2. Curitiba, Brazil, is a good example of compact development in a moderately developed country.

D. Urbanization appears to be a factor in decreasing population growth.

IV. The relationships between TFR and cultural, social, governmental, and economic factors are intricate.

A. A combination of four factors is thought to be primarily responsible for high TFRs: high infant and child mortality rates, the important economic and societal roles of children in some cultures, the low status of women in many societies, and a lack of health and family planning services.

B. The single most important factor affecting high TFRs is the low status of women in many societies.

V. The governments of many developing countries are trying to limit population growth.

A. In 1979 China began a coercive, one-child family policy to reduce TFR. The program, which has been successful but unpopular, was relaxed in rural China in 1984. Education and publicity campaigns are also used today.

B. India's government has sponsored family planning since the 1950s. In 1976 India introduced compulsory sterilization in several states, but this coercive policy failed. Education and publicity campaigns are the focus of its family planning efforts today.

C. Mexico's government has sponsored family planning since 1974. Mexico's TFR continues to drop, largely in response to education, improved health care, economic development, and publicity campaigns.

D. Nigeria put a national population policy into place in 1986 that includes education, improved health care, and economic development. These efforts have been largely ineffective, and Nigeria's TFR remains almost as high as it was in the early 1980s.

VI. The 1994 U.N. International Conference on Population and Development drafted a 20-year World Program of Action that focuses on reproductive rights, empowerment of women, and reproductive health. The program is currently underfunded.

THINKING ABOUT THE ENVIRONMENT

1. How is human population growth related to world hunger? To natural resource depletion and environmental degradation? To economic development?

2. Explain how a single child born in the United States can have a greater effect on the environment and natural resources than a dozen or more children born in a developing country.

3. How does a rapidly expanding population affect nonrenewable resources? Renewable resources?

4. What is your ecological footprint?

5. What is sustainable consumption? Are you a sustainable consumer? Why or why not?

6. Keep a list of the natural resources you use in a single day. How would this list compare to a similar list made by a poor person in a developing country such as India?

7. Although the United States ranks third in population number, some experts contend that it is the most overpopulated country in the world. Explain their rationale.

8. Given what you have learned about consumption overpopulation, do you think it is reasonable that in order for developing countries to have a larger share of the world's natural resources, highly developed countries must adjust to a smaller share themselves? Explain your answer.

9. What is urbanization? Which countries are the most urbanized? The least urbanized? What is the urbanization trend today in largely rural nations?

10. What are some of the problems brought on by rapid urban growth in developing countries?

11. What is the relationship between fertility rate and marriage age? Between fertility and educational opportunities for women?

12. Briefly describe and compare the successes and failures of China, India, Mexico, and Nigeria in slowing human population growth.

13. What is family planning? Is family planning effective in reducing fertility rates?

14. Explain the rationale behind this statement: It is better for highly developed countries to spend millions of dollars on family planning in developing countries now than to have to spend billions of dollars on relief efforts later.

15. Discuss this statement: The current human population crisis causes or exacerbates all environmental problems.

16. Discuss some of the ethical issues associated with overpopulation. Is it ethical to have more than two children? Is it ethical to consume so much in the way of material possessions? Is it ethical to try to influence a couple's decision about family size?

*17. The United States produces 30 times more of the pollutant carbon dioxide per capita than India, but India's population is much larger than that of the United States. Refer to the Population Reference Bureau's *World Population Data Sheet*, a poster located inside the back cover of this text, to obtain the populations of the United States and India. Use these data to determine which country adds more carbon dioxide to the atmosphere. Calculate how many times more.

* The solution to this question appears in Appendix VII.

TAKE A STAND

Visit our Web site at **http://www.wiley.com/college/raven** (select Chapter 9 from the Table of Contents) for links to more information about the controversy of debt in developing countries. Consider the opposing views of proponents and opponents of the Highly Indebted Poor Countries (HIPC) initiative, started by the World Bank and the International Monetary Fund (IMF) to reduce the debt burden of developing countries, and debate the issues with your classmates. You will find tools to help you organize your research, analyze the data, think critically about the issues, and construct a well-considered argument. Take a Stand activities can be done individually or as part of a team, as oral presentations, written exercises, or Web-based (e-mail) assignments.

Additional on-line materials relating to this chapter, including Student Quizzes, Activity Links, Useful Web Sites, Flash Cards, and more, can be found on our Web site.

SUGGESTED READING

Chen, D.D.T. "The Science of Smart Growth." *Scientific American*, Vol. 283, No. 6 (December 2000). Examines why cities have suburban sprawl and efforts to combat it.

Cincotta, R.P., and B.B. Crane. "The Mexico City Policy and U.S. Family Planning Assistance." *Science*, Vol. 294 (October 19, 2001). President Bush has imposed restrictions on U.S. funding for family planning in developing countries.

Doyle, R. "Sprawling into the Third Millenium." *Scientific American*, Vol. 284, No. 3 (March 2001). A short essay on suburban sprawl.

Frimm, N.B., et al. "Integrated Approach to Long-Term Studies of Urban Ecological Systems." *BioScience*, Vol. 50, No. 7 (July 2000). Examines the long-term ecological studies of Phoenix and Baltimore that were discussed in the text.

Hwang, A. "Exportable Righteousness, Expendable Women." *WorldWatch*, Vol. 15, No. 1 (January–February 2002). International family planning programs are vulnerable to political pro-life politics. Regardless of your political views, the article raises some interesting ethical dilemmas.

Jordan, M. "Among Poor Villagers, Female Infanticide Still Flourishes in India." *The Wall Street Journal* (May 9, 2000). The female-to-male ratio in India declined dramatically during the 20th century, possibly because of female infanticide.

Knickerbocker, B. "Taking Stock of Our Stuff." *The Christian Science Monitor* (February 11, 1999). An examination of the huge quantity of "stuff" (about 200 pounds per person per day) we consume in our everyday lives. This article also considers how individuals and corporations are beginning to address the issue of overconsumption.

Koerner, B.I. "Cities That Work." *U.S. News and World Report* (June 8, 1998). Highlights some of the promising features of six cities, including Vancouver, Canada; Minneapolis, United States; Chattanooga, United States; Curitiba, Brazil; Tilburg, the Netherlands; and Melbourne, Australia.

Potts, M. "The Unmet Need for Family Planning." *Scientific American* (January 2000). Many people in developing countries still do not have adequate access to contraceptives.

Sachs, J.D., A.D. Mellinger, and J.L. Gallup. "The Geography of Poverty and Wealth." *Scientific American*, Vol. 284, No. 3 (March 2001). The poorest nations have tropical climates and lack of access to sea trade.

Walfish, D. "National Count Reveals Major Societal Changes." *Science*, Vol. 292 (June 8, 2001). China's most recent census indicates its population is changing: It is older, better educated, and more transient.

Yunus, M. "The Grameen Bank." *Scientific American* (November 1999). Discusses an effective way to eliminate poverty—by offering small loans to women in developing countries to help them get started in business.

2002 World Population Data Sheet. Washington, D.C.: Population Reference Bureau, 2002. A chart published annually that provides current population data for all countries: birth rates, death rates, infant mortality rates, total fertility rates, and life expectancies, as well as other pertinent information. See the poster located inside the back cover of this text.

Long lines at a filling station, 1973. Gas rationing and long lines at filling stations were the result of the oil embargo imposed by OPEC in 1973. Photographed in California.

Fossil Fuels

Learning Objectives

After you have studied this chapter you should be able to

1. Briefly explain U.S. dependence on foreign oil.

2. Compare per-capita energy consumption in highly developed and developing countries.

3. Describe the processes that formed coal, oil, and natural gas.

4. Discuss the advantages and disadvantages, including environmental problems, of using coal.

5. Discuss the advantages and disadvantages, including environmental problems, of using oil and natural gas.

6. Summarize the continuing controversy surrounding the Arctic National Wildlife Refuge.

7. Distinguish among the five kinds of synfuels (tar sands, oil shales, gas hydrates, liquid coal, and coal gas) and briefly consider the environmental implications of using synfuels.

8. Relate three reasons that the United States needs a comprehensive national energy strategy. Briefly describe the National Energy Policy of the George W. Bush administration.

Dependence on energy resources from other countries has been dramatically demonstrated several times in the past few decades. In 1973, Israel and surrounding Arab countries went to war. Because the United States and many other industrialized countries supported Israel, the Organization of Petroleum Exporting Countries (OPEC), which consists largely of Arab countries, restricted oil shipments to the United States, precipitating an energy crisis. Total oil consumption was cut by only 5% in response to the restrictions, but the United States was thrown into a panic when it realized its fuel supplies were neither completely reliable nor endlessly cheap.

The embargo resulted in an economic recession that was triggered in part by escalating prices for gasoline and home heating oil. Long lines at filling stations were commonplace in many states, and in some states motorists were restricted to buying gasoline every other day. Car sales dropped, and people who did buy cars generally purchased the more fuel-efficient foreign makes. Because cars get better mileage at moderate speeds, federal law reduced the freeway speed limit to 55 miles per hour as an energy conservation measure.

The OPEC oil embargo of 1973 was not the only oil crisis Americans faced during the 1970s. In 1979 crude oil (petroleum) prices skyrocketed from $13 to $34 a barrel[1]

---

[1] Unless noted otherwise, all energy facts cited in this chapter were obtained from the Energy Information Administration (EIA), the statistical agency of the U.S. Department of Energy (DOE). Because EIA and DOE were not created until 1975, pre-1975 data cannot be verified and are not provided in this chapter.

because of an oil shortage touched off by the Iranian revolution. Lines returned at gasoline stations, and higher prices contributed to another economic recession.

Beginning in 1973, the Nixon, Ford, and Carter administrations passed a series of laws to help the United States cope with the problems of oil dependence. For example, President Ford signed the Energy Policy and Conservation Act into law in December 1975. One provision of this law mandated the establishment of the **Strategic Petroleum Reserve** to protect against the economic damage caused by interruptions in petroleum supplies. This emergency supply of up to 1 billion barrels of oil is stored in underground salt caverns along the coast of the Gulf of Mexico. The Strategic Petroleum Reserve is an important part of U.S. energy security policy.

Supported by Presidents Nixon, Ford, and Carter, the United States embarked on a massive campaign to reduce U.S. dependence on oil and to conserve energy. Americans turned down their thermostats during winter months and up during the summer months and began carpooling to work. Owners installed additional insulation in homes and older buildings Technological advances paved the way for more efficient automobiles, lighting fixtures, appliances, and buildings.

By the 1980s, the oil scares of the seventies were largely forgotten as oil prices declined. Americans purchased more cars, both domestic and foreign, than ever before, and gasoline was so cheap and abundant that its consumption increased. Between 1985 and 1989, daily oil imports increased from 3.2 million barrels (134 million gallons) to 5.8 million barrels (244 million gallons). At the same time, domestic oil production in the United States declined.

During the 1990s the United States continued to depend on foreign energy supplies and was even willing to go to war to ensure that supply remained dependable—witness the 1990 war in the Persian Gulf, which has been at least partially attributed to our dependence on foreign oil. During the conflict, President George H.W. Bush released oil reserves from the Strategic Petroleum Reserve to minimize cost increases due to a temporary disruption in supply. Despite this effort, the increase in energy costs precipitated by the crisis contributed to an economic recession, the only one in the 1990s.

The world's oil market has changed since the 1970s. For one thing, there are more non-OPEC suppliers, including countries in Latin America and around the North Sea. Because energy is currently cheap and abundant, per-capita energy consumption in the United States is as high today as during its 1973 peak. On average, U.S. citizens are driving larger, more powerful motor vehicles (for example, sport utility vehicles, or SUVs) and living in larger homes with more energy-using features such as high ceilings, central air conditioning, and Jacuzzis. Congress repealed the national speed limits in 1995, and some states have set limits as high as 75 miles per hour.

But the United States relies more than ever on foreign oil. Currently, 55.5% of the oil used in the United States is imported, up from 35% in 1973, and the Persian Gulf region provides the United States with 12.9% of its oil, almost three times the 1973 level. Thus, the United States remains as vulnerable as ever to an energy crisis. Moreover, the world's oil supplies will not remain abundant indefinitely. When supplies eventually tighten, knowledgeable experts predict the United States will desperately scramble to return to energy conservation measures.

In this chapter we examine the various fossil fuels—nonrenewable fuels such as petroleum, natural gas, and coal—that supply us with most of our energy. We also discuss the environmental problems associated with their production and use. The chapter concludes with a discussion of the U.S. National Energy Policy.

■ ENERGY CONSUMPTION IN HIGHLY DEVELOPED AND DEVELOPING COUNTRIES

Human society depends on energy. We use it to warm our homes in winter and cool them in summer; to grow, store, and cook our food; to light our homes; to extract and process natural resources for manufacturing items we use daily; and to power various forms of transportation. Many of the conveniences of modern living depend on a ready supply of energy.

A conspicuous difference in per-capita energy consumption exists between highly developed and developing nations (Figure 10.1). As you might expect, highly developed nations consume much more energy per person than developing nations. Although only 20% of the world's population lived in highly developed countries in 2000,[2] these people used 60% of the commercial energy consumed worldwide. That means that each person in highly developed countries uses approximately eight times as much energy as each person in developing countries. (Keep in mind, however, that world commercial energy use does not take into account the use of charcoal and firewood, which meet the energy needs of a substantial portion of the population in many developing countries.)

A comparison of energy requirements for food production clearly illustrates the energy consumption differences between developing and highly developed

---

[2] Energy data take several years to compile, and preliminary data are revised several times during this period. As a result, as we go to press, the latest energy data available are from 2000.

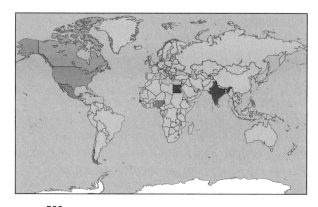

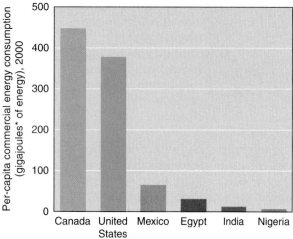

*1 gigajoule = 1 billion joules

Figure 10.1 Annual per-capita commercial energy consumption in selected countries, 2000. Note that energy consumption per person in highly developed nations is much greater than it is in developing countries.

countries. Farmers in developing nations rely on their own physical energy or the energy of animals to plow and tend fields. In contrast, agriculture in highly developed countries involves many energy-consuming machines, such as tractors, automatic loaders, and combines. Additional energy is required to produce the fertilizers and pesticides widely used in industrialized agriculture. The larger energy input is one of the reasons that the agricultural productivity of highly developed countries is greater than that of developing countries.

According to the EIA, world energy consumption has increased every year since 1982, with most of the increase occurring in developing countries. From 1999 to 2000, for example, energy consumption increased worldwide by about 1.9%. One of the goals of developing countries is to improve their standard of living. One way to achieve this is through economic development, a process usually accompanied by a rise in per-capita energy consumption. Furthermore, the world's energy requirements will continue to increase during the 21st

century, as the human population continues to increase. Most of this population growth will be in developing countries (see Chapter 8).

In contrast, the population in highly developed nations is more stable, and many energy experts think that those nations' per-capita energy consumption may be at or near saturation. Also, it is possible that additional energy demands may be more than compensated for by increased **energy efficiency**, the use of less energy to accomplish the same task, of such items as appliances, automobiles, and home insulation.

Figure 10.2 shows how energy is used in the United States. Industry, which encompasses the production of chemicals, minerals, food, and additional energy resources, accounts for about 42% of the energy we consume. Another 33% of our consumed energy is used to make buildings comfortable with heating, air conditioning, lighting, and hot water. The remaining 25% of energy we consume provides transportation, primarily for motor vehicles.

FOSSIL FUELS

Energy is obtained from a variety of sources, including fossil fuels, nuclear reactors (see Chapter 11), and solar and other alternative energy sources (see Chapter 12). Today, most of the energy required in North America is supplied by fossil fuels: coal, oil, and natural gas. A **fossil fuel** is composed of the partially decayed remnants of organisms. Most of the fossil fuels that we use today were formed millions of years ago.

Fossil fuels are **nonrenewable resources**—that is, the Earth's crust has a finite, or limited, supply of them, and that supply is depleted by use. Although coal and other fossil fuels are still forming by natural processes today, they are forming too slowly (on a scale of millions of years) to replace the fossil fuel reserves we are using. Because fossil fuel formation does not keep pace with use, when our present supply of fossil fuels has been used up, we will have to make a transition to other, more sustainable forms of energy (see Chapter 12).

How Fossil Fuels Were Formed

Three hundred million years ago, the climate of much of Earth was mild and warm, and plants grew year round. Vast swamps were filled with plant species that have long since become extinct. Many of these plants—horsetails, ferns, and club mosses—were large trees (Figure 10.3).

Plants in most environments decay rapidly after death, owing to the activities of decomposers such as bacteria and fungi. As the ancient swamp plants died, either from old age or from storm damage, they fell into the swamp, where they were covered by water. Their watery grave prevented the plants from decomposing much;

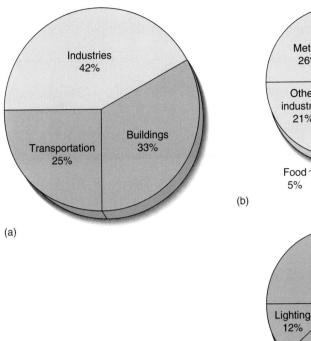

(a)

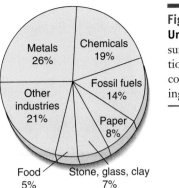

(b)

Figure 10.2 Energy consumption in the United States. (a) Overall, industries consume 42%, buildings 33%, and transportation 25%. (b, c, d) Breakdown of energy consumption by (b) industries, (c) buildings, and (d) transportation.

(c)

(d)

wood-rotting fungi cannot act on plant material where oxygen is absent, and anaerobic bacteria, which thrive in oxygen-deficient environments, do not decompose wood very rapidly. Over time, more and more dead plants piled up. As a result of periodic changes in sea level, layers of sediment (mineral fragments deposited by gravity) accumulated, forming layers that covered the plant material. Aeons passed, and the heat and pressure that accompanied burial converted the nondecomposed plant material into a carbon-rich rock called coal, and the layers of sediment into sedimentary rock. Much later, geological upheavals raised these layers so that they were nearer the Earth's surface.

Oil was formed when large numbers of microscopic aquatic organisms died and settled in the sediments. As these organisms accumulated, their decomposition depleted the small amount of oxygen that was present in the sediments. The resultant oxygen-deficient environ-

ment prevented further decomposition. Over time, the dead remains were covered and buried deeper in the sediments. Although we do not know the basic chemical reactions that produce oil, the heat and pressure caused by burial presumably aided in the conversion of these remains to the mixture of **hydrocarbons** (molecules containing carbon and hydrogen) known as oil.

Natural gas, composed primarily of the simplest hydrocarbon, **methane**, was formed in essentially the same way as oil, only at higher temperatures (typically greater than 100°C). Over millions of years, as the remains of organisms were converted to oil or natural gas, the sediments covering them were transformed to sedimentary rock.

Deposits of oil and natural gas are often found together. Because oil and natural gas are less dense than rock, they tend to move upward through porous rock layers and accumulate in pools beneath nonporous or imper-

Figure 10.3 Reconstruction of a Carboniferous swamp. The plants of the Carboniferous period, 360 million to 286 million years ago, included giant ferns, horsetails, and club mosses that formed our present-day coal deposits.

meable rock layers. Almost all oil and natural gas deposits are found in rocks that are less than 400 million years old.

COAL

Although coal had been used as a fuel for centuries, it was not until the 18th century that it began to replace wood as the dominant fuel in the Western world, and since then it has had a significant impact on human history. It was coal that powered the steam engine and supplied the energy needed for the Industrial Revolution, which began in the mid-18th century. Today coal is used primarily by utility companies to produce electricity and, to a lesser extent, by heavy industries. For example, coal generally supplies the energy needed to melt iron during its conversion to steel.

Coal occurs in different grades, largely as a result of the varying amounts of heat and pressure to which it was exposed during formation. Coal that was exposed to high heat and pressure during its formation is drier, is more compact (and therefore harder), and has a higher heating value (that is, a higher energy content). Lignite, subbituminous, bituminous coal, and anthracite are the four most common grades of coal (Table 10.1).

Lignite is a soft coal, brown or brown-black in color with a soft, woody texture. It is moist and produces little heat compared with other types of coal. Lignite is often used to fuel electric power plants. Sizable deposits of it are found in the western states, and the largest producer of lignite in the United States is North Dakota.

Subbituminous coal is a grade of coal intermediate between lignite and bituminous. Like lignite, subbituminous coal has a relatively low heat value and sulfur con-

Table 10.1	A Comparison of Different Kinds of Coal					
Type of Coal	Color	Water Content (%)	Relative Sulfur Content	Carbon Content (%)	Average Heat Value (BTU/pound)	2000 Cost at Mine for 2,000 lbs of Coal
Lignite	Dark brown	45	Medium	30	6,000	$11.41
Subbituminous coal	Dull black	20–30	Low	40	9,000	$7.12
Bituminous coal	Black	5–15	High	50–70	13,000	$24.15
Anthracite	Black	4	Low	90	14,000	$40.90

tent. Because its sulfur content is low, many coal-fired electric power plants in the United States burn subbituminous coal. It is found primarily in Alaska and a few western states, such as Montana and Wyoming.

Bituminous coal, the most common type, is also called **soft coal** even though it is harder than lignite and subbituminous. It is dull to bright black with dull bands. Much bituminous coal contains sulfur, a chemical element that causes severe environmental problems (discussed shortly) when the coal is burned in the absence of pollution-control equipment. Bituminous coal is nevertheless used extensively by electric power plants because it produces a lot of heat. In the United States, bituminous coal deposits are found in the Appalachian region, near the Great Lakes, in the Mississippi valley, and in central Texas.

The highest grade of coal, **anthracite** or **hard coal**, was exposed to extremely high temperatures during its formation. It is a dark, brilliant black and burns most cleanly—that is, it produces the fewest pollutants per unit of heat released—of all the types of coal because it is not contaminated by large amounts of sulfur. Anthracite also has the highest heat-producing capacity of any grade of coal. Anthracite in the United States has been largely depleted, but most of the remaining deposits are located east of the Mississippi River, particularly in Pennsylvania.

Coal is usually found in seams, underground layers that vary from 2.5 cm (1 in.) to more than 30 m (100 ft) in thickness. Because they are easily located, geologists think that most, if not all, major coal deposits have probably been identified. Scientists working with coal are therefore concerned less about finding new deposits than about the safety and environmental problems associated with coal.

Coal Reserves

Coal, the most abundant fossil fuel in the world, is found primarily in the Northern Hemisphere (Figure 10.4). The largest coal deposits are in the United States, Russia, China, Australia, India, Germany, and South Africa. The United States has 25% of the world's coal supply in its massive deposits. According to the World Resources Institute, known world coal reserves could last for more than 200 years at the present rate of consumption. Additional coal resources that are currently too expensive to develop have the potential to provide enough coal to last for 1,000 or more years (at current consumption rates). For example, some coal deposits are buried more than 5,000 feet inside the Earth's crust. Drilling a shaft that deep would cost considerably more than the current price of coal would justify.

Coal Mining

The two basic types of coal mines are surface and subsurface (underground) mines. The type of mine chosen

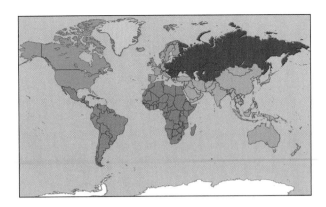

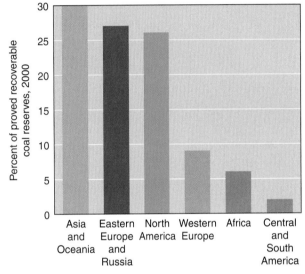

Figure 10.4 Distribution of coal deposits. Data are presented as percentages of the 2000 estimated recoverable reserves—that is, of coal known to exist that can be recovered under present economic conditions with existing technologies. The majority of the world's coal deposits are located in the Northern Hemisphere.

depends on surface contours and on the location of the coal bed relative to the surface. If the coal bed is within 30 m (100 ft) or so of the surface, **surface mining** is usually done (Figure 10.5). In one type of surface mining, known as strip mining, a trench is dug to extract the coal, which is scraped out of the ground and loaded into railroad cars or trucks. Then a new trench is dug parallel to the old one, and the overburden from the new trench is put into the old trench, creating a hill of loose rock known as a **spoil bank**. Digging the trenches involves using bulldozers, giant power shovels, and wheel excavators to remove the ground covering the coal seam. Approximately 60% of the coal mined in the United States is obtained by surface mining.

When the coal is deeper in the ground or runs deep into the Earth from an outcrop on a hillside, it is mined underground. **Subsurface mining** accounts for approximately 40% of the coal mined in the United States.

Figure 10.5 Surface coal mine near Douglas, Wyoming. The overlying vegetation, soil, and rock is stripped away, and then the coal is extracted out of the ground.

Surface mining has several advantages over subsurface mining: It is usually less expensive, safer for miners, and generally allows a more complete removal of coal from the ground. However, surface mining disrupts the land much more extensively than subsurface mining and has the potential to cause several serious environmental problems.

Safety Problems Associated with Coal

Although we usually focus on the environmental problems caused by mining and burning coal, there are also significant human safety and health risks in the mining process itself. Underground mining is a hazardous occupation. According to the DOE, during the 20th century more than 90,000 American coal miners died in mining accidents, although the number of deaths per year declined significantly in the latter part of the century. Miners also have an increased risk of cancer and **black**

lung disease, a condition in which the lungs are coated with inhaled coal dust so that the exchange of oxygen between the lungs and the blood is severely restricted. It is estimated that these diseases are responsible for the deaths of at least 2,000 miners in the United States each year.

Environmental Impacts Associated with the Mining Process

Coal mining, especially surface mining, can have substantial effects on the environment. Prior to passage of the 1977 **Surface Mining Control and Reclamation Act (SMCRA)**, abandoned surface coal mines were usually left as large open pits or trenches. Cliffs of excavated rock called *highwalls*, some more than 30 m (100 ft) high, were left exposed. Acid and toxic mineral drainage from such mines, along with the removal of topsoil, which was buried or washed away by erosion, prevented most plants from naturally recolonizing the land. Streams were polluted with sediment and **acid mine drainage**, produced when rainwater seeps through iron sulfide minerals exposed in mine wastes and carries sulfuric acid to nearby lakes and streams. Dangerous landslides occurred on hills that were unstable due to the lack of vegetation.

Surface-mined land can be restored to prevent such degradation and to make the land productive for other purposes, although restoration is expensive and technically challenging. Prior to passage of the SMCRA, few restorations of surface-mined land took place, and many of these "restorations" were half-hearted. The SMCRA requires coal companies to restore areas that have been actively surface mined, beginning in 1977. The SMCRA protects the environment by requiring permits and inspections of active coal mining operations and reclamation sites (see "Meeting the Challenge: Reclamation of Coal-Mined Land"). Reclamation begins during the mining process, not after surface mining ends.

The SMCRA also prohibits coal mining in sensitive areas such as national parks, wildlife refuges, wild and scenic rivers, and sites listed on the National Register of Historic Places. In addition, the SMCRA stipulates that surface-mined land abandoned prior to 1977 (more than 0.4 million hectares, or 1 million acres) gradually be restored, using money from a tax that coal companies pay on currently mined coal. According to the U.S. Office of Surface Mining, between 1977 and 1997 more than $1.5 billion was spent reclaiming the most dangerous abandoned mine lands; approximately two thirds of this money was spent in four states—Pennsylvania, Kentucky, West Virginia, and Wyoming. However, so many abandoned coal mines exist that it is doubtful that they can all be restored.

Mountaintop Removal One of the most land-destructive types of surface mining is mountaintop removal

MEETING THE CHALLENGE

Reclamation of Coal-Mined Land

The 1977 Surface Mining Control and Reclamation Act (SMCRA) has been extremely effective in protecting the environment. In the years since this law's passage, thousands of permits regulating the reclamation of active coal mines covering almost 2 million hectares (5 million acres), an area the size of New Jersey, have been issued. In addition, some of the most dangerous (from a health and safety viewpoint) abandoned coal mines have been reclaimed. The Office of Surface Mining of the Department of the Interior gives annual awards to those coal companies that have developed outstanding reclamation sites, such as the one shown in the figure.

Sometimes this reclaimed land is used for biological habitat, such as rangeland for wild grazing animals in western states or wetlands in midwestern and eastern states. The Eastern Kentucky Regional Airport is located on a reclaimed mining site, as is a tree farm in Garrett County, Maryland. Other post-mining land uses include camping and recreational areas such as golf courses, farmlands, sanitary landfills, cemeteries, and housing developments.

The SMCRA stipulates that coal-mined land must be reclaimed as soon as possible. Old highwalls are *backfilled* (covered with excavated rock from current mining operations) and *recontoured*—that is, graded to resemble the original premining terrain. Then topsoil is spread, and the area is prepared for planting of seeds or small plants.

Soil erosion is a problem until the area has vegetation established, particularly on steeper slopes, so sediment and erosion-control structures are constructed. Many sites contain *sedimentation ponds*, which confine sediment-laden runoff and allow the sediment to settle out before the water leaves the area. Water from nearby streams is monitored during mining and reclamation to ensure that it remains free of heavy metals and acid. In some places, wells are dug to monitor groundwater quality as well.

Identifying acid and toxic materials such as selenium, excavating, and safely disposing of them prevents soil and water contamination by these materials. In arid regions such as Wyoming, the toxic *spoils* (excavated rock) may be placed at the bottom of a nearby mine pit, which is covered with additional rock and topsoil, then planted. In moister regions, acid-producing rock is kept "high and dry," because water on the mine floor would cause acid mine drainage.

Because excavated rock often occupies more space than unexcavated rock, excess spoil usually remains after the mine

Reclaimed coal-mined land. Prior to reclamation, this 45-acre site in West Virginia had huge piles of coal refuse, dangerous highwalls, and 11 abandoned underground mine openings. All health and safety hazards have been removed, and the land is similar to what it was before coal mining occurred. Photographed near Harding, West Virginia.

is regraded as much as possible to its original contours. This material is often placed in a nearby valley, which is eventually covered with topsoil and replanted.

The permanent establishment of plant cover on land affected by coal mining is required by the SMCRA. The plants selected for revegetation are usually native to the area and thus adapted to the climate. Generally, seeds are planted, but greater success is often obtained when seedlings are hand-planted (a more expensive technique). Coal mining companies must maintain the vegetation for a minimum of 5 years in the East and 10 years in the arid West, after which time the vegetation is considered successfully reestablished. During this period, the mining companies continue to monitor water quality to ensure that minimum standards are met.

(Figure 10.6). A $100 million machine called a **dragline**, a huge shovel with a 20-story-high arm, takes enormous chunks out of a mountain, eventually removing the entire mountaintop in order to reach the coal located below. According to Environmental Media Services, between 15% and 25% of the mountaintops in southern West Virginia have already been leveled by mountaintop removal. The valleys and streams between the mountains are gone as well, filled with debris from the mountaintops. At the current rate of mountaintop removal, environmentalists say that half the peaks of southern West Virginia will be gone by 2020. Mountaintop removal is also occurring in parts of Kentucky, Pennsylvania, Tennessee, and Virginia. The SMRCA specifically exempts mountaintop removal. In 1977, when the SMCRA was passed, existing technology for

Figure 10.6 Mining for coal by mountaintop removal. This photo was taken in Raleigh County, West Virginia.

mountaintop removal could take out the top seam of coal along a mountain ridge. Today, mountaintop removal can take out as many as 16 seams of coal, from the ridge to the base of the mountain. A 1999 court decision to limit mountaintop removal was upheld in 2002 by a federal judge who said disposing of mountaintop waste rock in valleys violates the Clean Water Act because it buries streams.

Environmental Impacts Associated with Burning Coal

Burning any fossil fuel releases carbon dioxide, CO_2, into the atmosphere. This carbon that we are releasing into the atmosphere in a few short centuries had been stored in the Earth's crust for millions of years. You may recall from the discussion of the carbon cycle in Chapter 6 that a natural equilibrium exists between the CO_2 in the atmosphere and the CO_2 dissolved in the ocean. Currently we are releasing so much CO_2 into the atmosphere through our consumption of fossil fuels that the Earth's CO_2 equilibrium has been disrupted. Because the concentration of CO_2 in the atmosphere is increasing and CO_2 prevents heat from escaping from the planet, global temperature may be affected. An increase of a few degrees in global temperature caused by higher levels of CO_2 and other greenhouse gases may not seem very serious at first glance, but a closer look reveals that such an increase is potentially harmful. A global rise in temperature may cause polar ice to melt, raising sea levels and flooding coastal areas. This would increase coastal erosion and put many coastal buildings (and the people that live in them) at higher risk from violent storms. Other serious environmental consequences of global climate warming are considered in

Chapter 20. The CO_2 problem is made more severe by the burning of coal than by the burning of other fossil fuels, because coal burning releases more CO_2 per unit of heat energy produced.

Coal burning generally contributes more air pollutants (other than CO_2) than does burning either oil or natural gas. A lot of coal contains mercury that is released into the atmosphere during combustion. This mercury moves readily from the atmosphere to water and land. As mercury accumulates in the environment, it harms humans as well as wildlife. (See Chapter 21 for a discussion of the human health hazards of elevated levels of mercury.) In the United States, coal-burning electric power plants currently produce one third of all airborne mercury emissions.

Much bituminous coal contains sulfur and nitrogen that, when burned, are released into the atmosphere as sulfur oxides (SO_2 and SO_3) and nitrogen oxides (NO, NO_2, and N_2O). Both sulfur oxides and the nitrogen oxides NO and NO_2 form acids when they react with water. These reactions result in a type of air pollution known as **acid deposition**, in which acid falls from the atmosphere to the surface as precipitation (**acid precipitation**) or as dry acid particles. The combustion of coal is partly responsible for acid deposition, which is particularly prevalent downwind from coal-burning electric power plants. Normal rain is slightly acidic (pH 5.6), but in some areas acid precipitation has been measured at a pH of 2.1, which is equivalent to that of lemon juice. (See Appendix I for a review of pH.) Acidification of lakes and streams has resulted in the decline of aquatic animal populations and has also been linked to some of the forest decline that has been documented worldwide (Figure 10.7). Acid precipitation and forest decline are discussed in greater detail in Chapter 20.

Figure 10.7 Dead trees enveloped in acid fog on Mt. Mitchell, North Carolina. Forest decline was first documented in Germany and Eastern Europe. More recently, it has been observed in the eastern United States, particularly at higher elevations. Acid deposition is an important factor contributing to forest decline.

Although it is relatively easy to identify and measure pollutants in the atmosphere, it is difficult to trace their exact origins. They are transported and dispersed by air currents and are often altered as they react chemically with other pollutants in the air. Even so, it is clear that some nations suffer the damage of acid deposition caused by pollutants produced in other countries, and as a result acid deposition has become an international issue.

Using Technology to Make Coal a Cleaner Fuel

It is possible to reduce sulfur emissions associated with the combustion of coal by installing desulfurization systems, or **scrubbers**, to clean the power plants' exhaust. As the polluted air passes through a scrubber, chemicals in the scrubber react with the pollution and cause it to precipitate (settle) out. Modern scrubbers remove 98% of the sulfur and 99% of the particulate matter in smokestacks. Desulfurization systems are expensive; they cost about $50 to $80 per installed kilowatt or about 10% to 15% of the construction costs of a coal-fired electric power plant.

In lime scrubbers, a chemical spray of water and lime neutralizes acidic gases such as sulfur dioxide, which remain behind as a calcium sulfate sludge that itself becomes a disposal problem (see Figure 19.10d). A large power plant may produce enough sludge annually to cover 2.6 km^2 (1 mi^2) of land 0.3 m (1 ft) deep. Although many power plants currently dispose of the sludge in landfills, some have found markets for the material. In **resource recovery**, the sludge is treated as a marketable product rather than as a polluted emission. Some utilities have begun selling calcium sulfate from scrubber sludge to wallboard manufacturers. (Wallboard is traditionally manufactured from gypsum, a mineral composed of calcium sulfate.) Other companies are using **fly ash**, the ash from the chimney flues, to make a lightweight concrete that could substitute for wood in the building industry. Some farmers have started applying one type of sludge (calcium sulfate) as a soil conditioner. Plants grow better because calcium sulfate increases the water-holding capacity of the soil—that is, the calcium sulfate acts like a sponge; calcium sulfate also neutralizes acids in some soils.

The **Clean Air Act Amendments of 1990** required the nation's 111 dirtiest coal-burning power plants to cut sulfur dioxide emissions by 1995 (or by 1997 if they committed to buying scrubbers). Compliance resulted in a total annual decrease of 3.8 million metric tons nationwide by 1995. This represented a significant portion of the total amount of sulfur dioxide emitted in the United States each year (15.7 million metric tons in 1993, before reductions mandated by the Clean Air Act Amendments went into effect). In the second phase of the Clean Air Act Amendments, more than 200 additional power plants made SO$_2$ cuts by the year 2000, resulting in a total annual decrease of 10 million metric tons nationwide. A nationwide cap on SO$_2$ emissions from coal-burning power plants was imposed after 2000. Utilities also cut nitrogen oxide emissions by 2.6 million tons per year, beginning in 1995, out of 7.2 million tons per year total.

New methods for burning coal (called **clean coal technologies**) are being developed that will not contaminate the atmosphere with sulfur oxides and will significantly reduce nitrogen oxide production. Clean coal technologies include fluidized-bed combustion and coal gasification and liquefaction (considered shortly, in the discussion of synfuels). It is important to recognize, however, that these technologies have little impact on reducing CO$_2$ emissions.

Fluidized-bed combustion mixes crushed coal with particles of limestone in a strong air current during combustion (Figure 10.8). Because fluidized-bed combustion takes place at a lower temperature than regular coal burning, fewer nitrogen oxides are produced. (Higher temperatures cause atmospheric nitrogen and oxygen to combine, forming nitrogen oxides.) Also, because the sulfur in coal reacts with the calcium in limestone to form calcium sulfate, which then precipitates out, sulfur is

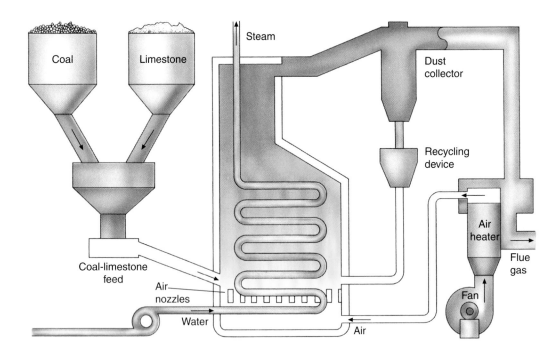

Figure 10.8 Fluidized-bed combustion of coal. Crushed coal and limestone are suspended in air. As the coal burns, limestone neutralizes most of the sulfur dioxide in the coal. The heat generated during combustion is used to convert water to steam, which can be used to power various industrial processes.

removed from the coal *during* the burning process, so scrubbers are not needed to remove it *after* combustion.

In the United States several large power plants are testing fluidized-bed combustion, and a few small power plants that use this technology are already in commercial operation. The Clean Air Act Amendments of 1990 provide incentives for utility companies to convert to clean coal technologies, such as fluidized-bed combustion. The cost of installing fluidized-bed combustion compares favorably with the cost of installing desulfurization systems.

Fluidized-bed combustion is also more efficient than traditional coal burning—that is, it produces more heat from a given amount of coal—and therefore reduces CO_2 emissions per unit of electricity produced. If improvements of this technology were developed and adopted widely by coal-burning power plants, fluidized-bed combustion could significantly reduce the amount of CO_2 released into the atmosphere by as early as 2020. Pressurized fluidized-bed combustion is currently being developed as a way to reduce not only CO_2 but also nitrogen and sulfur oxides. By operating fluidized-bed combustion under very high pressures, complete combustion of coal can take place at unusually low temperatures. Sulfur emissions are removed as calcium sulfate, and very few nitrogen oxides form because of the low temperatures. Pressurized fluidized-bed combustion is more expensive than regular fluidized-bed combustion, however, because it requires a costly pressurized vessel.

OIL AND NATURAL GAS

Although coal was the most important energy source in the United States during the early 1900s, oil and natural gas became increasingly important, beginning in the 1940s. This change occurred largely because oil and natural gas are easier to transport and because they burn cleaner than coal. In 2000, oil and natural gas supplied approximately 63% of the energy used in the United States. In comparison, other U.S. energy sources included coal (23%), nuclear power (8.2%), and hydropower (3.1%). Globally in 2000, oil and natural gas provided 61.5% of the world's energy. In comparison, other major energy sources used worldwide included coal (23.7%), hydroelectric power (7%), and nuclear power (65%) (Figure 10.9).

Petroleum, or **crude oil**, is a liquid composed of hundreds of hydrocarbon compounds. During petroleum refining, the compounds are separated into different products—such as gases, gasoline, heating oil, diesel oil, and asphalt—based on their different boiling points (Figure 10.10). Oil can be used to produce **petrochemicals**, compounds that are used in the production of such diverse products as fertilizers, plastics, paints, pesticides, medicines, and synthetic fibers.

In contrast to petroleum, natural gas contains only a few different hydrocarbons: methane and smaller amounts of ethane, propane, and butane. Propane and

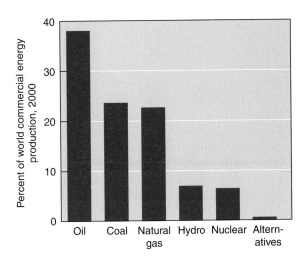

Figure 10.9 World commercial energy sources, 2000. Note the overwhelming importance of oil, coal, and natural gas as commercial energy sources. Alternatives include geothermal, solar, wind, and wood and waste electric power.

butane are separated from the natural gas, stored in pressurized tanks as a liquid called **liquefied petroleum gas**, and used primarily as fuel for heating and cooking in rural areas. Methane is used to heat residential and commercial buildings, to generate electricity in power plants, and for a variety of purposes in the organic chemistry industry. Methane is usually distributed by being pumped through pressurized pipelines or shipped as a solid in refrigerated tankers. (At very low temperatures, natural gas becomes a solid.)

Use of natural gas is increasing in three main areas—generation of electricity, transportation, and commercial cooling. One example of an increasingly popular approach that uses natural gas is **cogeneration**, in which natural gas is used to produce both electricity and steam; the heat of the exhaust gases provides the energy to make steam for water and space heating (see Chapter 12). Cogeneration systems that use natural gas are able to provide electricity cleanly and efficiently.

Natural gas as a fuel for trucks, buses, and automobiles offers significant environmental advantages over gasoline or diesel: Natural gas vehicles emit 80% to 93% fewer hydrocarbons, 90% less carbon monoxide, 90% fewer toxic emissions, and almost no soot. According to the EIA, as of 1999 more than 96,000 vehicles that use compressed natural gas for fuel are in use in the United States. Most of these are fleet vehicles. The city of Los Angeles is the largest operator of natural gas–powered transit buses in North America.

Natural gas can efficiently fuel residential and commercial air cooling systems. One example is the use of natural gas in a desiccant-based (air-drying) cooling system, which is ideal for supermarkets, where humidity control is as important as temperature control. Restau-

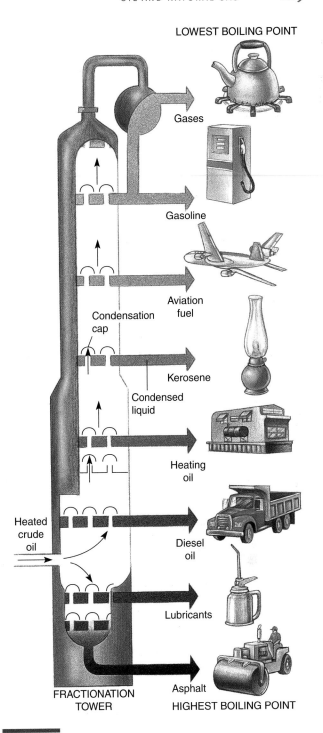

Figure 10.10 Petroleum refining. Crude oil is separated into a variety of products based on their different boiling points. After being heated, they are separated in a fractionation tower, which may be 30.5 m (100 ft) tall. The lower the boiling point, the higher the compounds rise in the tower.

rants are also important users of natural gas–powered desiccant-based cooling systems.

The main disadvantage of natural gas is that deposits are often located far from where the energy could be used. Because it is a gas and is therefore less dense than a liquid, natural gas costs four times more to transport through pipelines than crude oil. Currently, many oil well locations simply burn off the natural gas they emit or, concerned with CO_2 emissions, reinject it into the ground. Efforts are under way to improve the chemical conversion of natural gas into a liquid fuel such as methanol that would be easy to transport and store. (Methanol fuel is discussed further in Chapter 12.)

Geological Exploration for Oil and Natural Gas

Exploration is continually under way in search of new oil and natural gas deposits, which are usually found together under one or more layers of rock. Oil and natural gas deposits are usually discovered indirectly by the detection of **structural traps**, geological structures that tend to trap any oil or natural gas that is present (Figure 10.11). (Recall that oil and natural gas tend to migrate upward until they reach an impermeable rock layer.)

The upward folding of sedimentary rock **strata** (layers) may occur because of plate tectonic movements. Sometimes the strata that arch upward include both porous and impermeable rock. If impermeable layers overlie porous layers, it is possible that any oil or natural gas present from a source rock such as shale will work its way up through the porous rock to accumulate under the impermeable layer.

Many important oil and natural gas deposits (for example, oil deposits known to exist in the Gulf of Mexico) have been found in association with **salt domes**, underground columns of salt. Salt domes develop when extensive salt deposits form at the Earth's surface because of the evaporation of water. All surface water contains dissolved salts. The salts dissolved in ocean water are so concentrated that they can be tasted, but even fresh water contains some dissolved material. If a body of water lacks a passage to the ocean, as an inland lake often does, the salt concentration in the water gradually increases. (The

Great Salt Lake in North America is an example of a salty inland body of water that formed in this way. Although three rivers empty into the Great Salt Lake, water escapes from the lake only by evaporation, accounting for its high salinity—four times higher than ocean water).

If such a lake were to dry up, a massive salt deposit would remain. Layers of sediment may eventually cover such deposits and convert to sedimentary rock after millions of years. Because salt is less dense than rock, the rock layers settle, and the salt deposit tends to rise in a column—a salt dome. The ascending salt dome, together with the rock layers that buckle over it, provides a trap for oil or natural gas.

Geologists use a variety of techniques to identify structural traps that might contain oil or natural gas. One method is to drill test holes in the surface and obtain rock samples. Another method is to produce an explosion at the surface and measure the echoes of sound waves that bounce off rock layers under the surface. These data can be interpreted to determine whether or not structural traps are present. It should be emphasized, however, that many structural traps do not contain oil or natural gas.

Three-dimensional seismology is a new technology that maps oil fields three-dimensionally, enabling geologists to have a higher rate of success when drilling. Another new technology that improves oil recovery is horizontal drilling. Traditional oil wells are vertical and cannot veer off to follow the contours of underground formations that contain oil. Wells dug with horizontal drilling can follow contours, and they generally yield three to five times as much oil as vertical wells.

Even with the new technologies, searching for oil and natural gas is very expensive. It costs millions of dollars just to do the basic geological analyses to find structural traps. And once oil or natural gas has been located, drilling and operating the wells cost additional millions.

Reserves of Oil and Natural Gas

Although oil and natural gas deposits exist on every continent, their distribution is uneven, and a disproportionate share of total oil deposits are clustered relatively close to each other. Enormous oil fields containing more than half of the world's total estimated reserves are situated in the Persian Gulf region, which includes Iran, Iraq, Kuwait, Oman, Qatar, Saudi Arabia, Syria, United Arab Emirates, and Yemen (Figure 10.12). In addition, major oil fields are known to exist in Venezuela, Mexico, Russia, Kazakhstan, Libya, and the United States (in Alaska and the Gulf of Mexico).

Almost half of the world's proved recoverable reserves of natural gas are located in two countries, Russia and Iran (Figure 10.13). Because the United States has more deposits of natural gas than Western Europe, use of natural gas is more common in North America. Canada and the United States also extract **coal bed methane**, a form of natural gas associated with coal deposits.

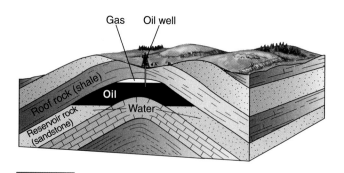

Figure 10.11 Structural trap. The most important of several kinds of structural traps is shown. These traps form when sedimentary rock strata buckle, or fold upward. Oil and natural gas seep through porous reservoir rock such as sandstone and collect under nonporous layers such as a roof of shale. Natural gas accumulates on top of oil, which in turn floats on groundwater.

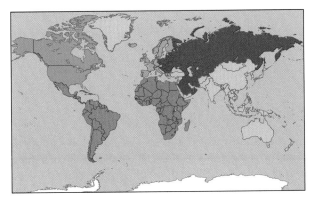

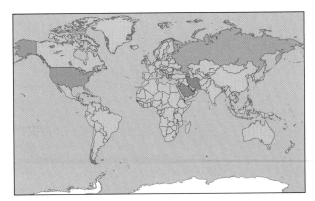

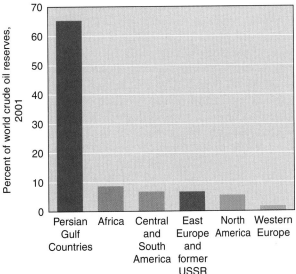

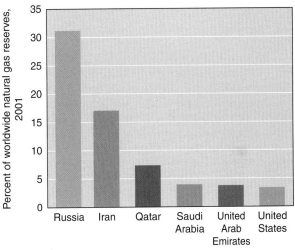

Figure 10.12 Distribution of oil deposits. The Persian Gulf region contains huge oil deposits in a relatively small area, whereas other regions have very few. Data are presented as regional percentages of the January 1, 2001, world estimate of crude oil reserves.

Figure 10.13 Six countries with the greatest natural gas deposits. Data are presented as percentages of the January 1, 2001, world estimate of natural gas reserves. Note that Russia and Iran together possess 48% of the world's natural gas deposits. (Qatar and United Arab Emirates, which are not shown on the map, are tiny countries east of Saudi Arabia.)

It is unlikely that major new oil fields will be discovered in the continental United States, which has been explored for oil more extensively than any other country. Since the 1980s, the success rate of searches for new oil fields has declined, as has the amount of exploration.

There is reason to believe that large oil deposits exist under the **continental shelves**, the relatively flat underwater areas that surround continents, and in deepwater areas adjacent to the continental shelves. Despite problems, such as storms at sea and the potential for major oil spills, many countries engage in offshore drilling for this oil. New technologies, such as platforms the size of football fields, enable oil companies to drill several thousand feet for oil, making seafloor oil fields that were once considered inaccessible open for tapping. As many as 18 billion barrels (756 billion gallons) of oil and natural gas may exist in the deep water of the Gulf of Mexico, just off the continental shelf from Texas to Alabama. Continental shelves off the coasts of western Africa and Brazil are also

promising. The oil industry is currently developing remote-controlled robots that can install and maintain underwater equipment and pipelines.

Environmentalists generally oppose opening the outer continental shelves for oil and natural gas exploration because of the threat a major oil spill would pose to marine and coastal environments. Coastal industries, including fishing and tourism, also oppose oil and natural gas exploration in these areas.

How Long Will Oil and Natural Gas Supplies Last?
Because there are so many unknowns, it is difficult to say with certainty how long it will be before the world runs out of oil and natural gas. We do not know how many additional oil and natural gas reserves will be discovered, nor do we know if or when technological breakthroughs will allow us to extract more fuel from each deposit. The answer to how long these fuels will last also depends on whether world consumption of oil and natural gas increases, remains the same, or decreases. Economic fac-

Offshore Rig Debate

When aging offshore oil platforms have reached the end of their usefulness, what should be done with them? Should they be completely dismantled, a complex task that costs millions of dollars per platform, or should underwater parts be left behind to serve as artificial reefs? For years, oil companies have left the bases of obsolete oil platforms anchored in the floor of the Gulf of Mexico. (They cut the legs off at a depth safe for ships.) Now many of the rigs off the coast of Southern California are nearing retirement age, and environmentally conscious Californians are debating their future. Oil companies want to leave the underwater legs behind because they save substantial money in removal costs. Moreover, the vertical underwater reefs provide habitat for a large number of marine organisms.

Researchers at the University of California at Santa Barbara studied a single underwater rig in the Santa Barbara Channel and catalogued more than 25,000 fishes in the vicinity of the reef, as well as thousands of mussels, crabs, and starfish. Many environmentalists and fishermen however, think the rigs should be completely removed because their presence could provide oil companies with a justification for not cleaning up oil seepage in the area as extensively as they should. Environmentalists are concerned that oil companies will say they cannot clean the area because the cleanup could jeopardize the artificial reefs.

tors influence oil and natural gas availability and consumption. As reserves are exhausted, prices increase, which can drive down consumption and stimulate greater energy efficiency, the search for additional deposits, and the use of alternative energy sources.

Despite adequate oil supplies for the near future, at some point world oil production will peak and then enter a decline that will last for a period of several decades. Most experts think we will begin to have serious problems with oil supplies sometime during the 21st century. Many geologists estimate that oil production will peak between 2010 and 2020. They point out that about 80% of current production comes from oil fields that were discovered before 1973, and that most of these fields have already started to decline in production. These analysts say the world must move quickly to develop alternative energy sources because the global demand for energy will only continue to increase.

Industrial analysts and economists are more optimistic. They suggest that improving technology will allow us to extract more oil out of old oil fields, obtain new oil from fields that were formerly unreachable (such as beneath deep ocean waters), and produce oil from natural gas, coal, and synfuels (discussed later in the chapter). Even with these technological advances, however, most industry analysts predict that oil production will peak sometime between 2050 and 2100.

Natural gas is more plentiful than oil. Experts estimate that readily recoverable reserves of natural gas, if

converted into a liquid fuel, would be equivalent to between 500 billion and 770 billion barrels of crude oil, enough to keep production rising for at least 10 years after conventional supplies of petroleum have begun to decline.

Global Oil Demand and Supply

One difficult aspect of the oil market is that the world's major oil producers are not its major oil consumers. According to the EIA, in 2000 North America and Western Europe consumed 50.6% of the world's total petroleum, yet these same countries produced only 28.9% of the world's crude oil. In contrast, the Persian Gulf region consumed 5.9% of the world's petroleum and produced 26.7% of the world's crude oil.

The United States currently imports more than one half of its oil, which produced a 2001 balance of payments deficit of $94 billion. This oil bill is expected to increase. The U.S. Department of Energy and other knowledgeable experts project that the United States will be importing almost 100% of its oil by 2015.

Because the Persian Gulf region has much higher proven reserves than other countries, the imbalance between oil consumers and oil producers will probably worsen in the years to come. At current rates of production, North America's oil reserves will run out decades before those of the Persian Gulf nations, which have

Energy for China

Faced with explosive economic growth, China has seen its energy demands soar, forcing it to become a net importer of oil, beginning in 1993. Current trends indicate that oil imports will make up 40% of China's total consumption by 2010. Because it is a net importer of oil, China has tried to bring its domestic cost of gasoline more in line with oil costs on the world market. As a result, gasoline prices increased steadily during the early 2000s. Attempting to limit its dependence on oil imports, the Chinese government wants to develop domestic oil sources and to substitute other fuels for oil. Thus far, however, China's oil supplies have proved less than promising, and coal is the only major alternative fuel under production.

With its fossil fuel consumption—and the accompanying greenhouse gas emissions—so steadily on the rise, China dominates international concerns over global climate warming. In the early 2000s, China's per-capita motor vehicle ownership was low, and buses and other forms of mass transit were the most widely used types of transportation. Because of projected increases in motor vehicle ownership in coming decades, however, China's 2025 projected carbon dioxide emissions are 3.2 billion tons per year, compared with current global carbon dioxide emissions of 6.15 billion tons per year. However, China can justify its increased energy consumption and emissions as products of fair economic development. For example, China's projected ownership of motor vehicles in 2020 is only 52 vehicles per thousand people, which is about one-fifteenth what the U.S. level was in 1999.

65% of the known world oil reserves and may be able to produce oil at current rates for perhaps a century. This dependence of the United States and other countries on Middle Eastern oil has potential international security implications as well as economic impacts.

Environmental Problems Associated with Oil and Natural Gas

Two sets of environmental problems are associated with the use of oil and natural gas: the problems that result from burning the fuels (combustion) and the problems involved in obtaining them in order to burn them (production and transport). We have already mentioned the CO_2 emissions that are a direct result of the combustion of fossil fuels. As with coal, the burning of oil and natural gas produces CO_2. Every gallon of gasoline you burn in your automobile releases an estimated 9 kg (20 lb) of CO_2 into the atmosphere. As CO_2 accumulates in the atmosphere, it insulates the planet, preventing planetary heat from radiating back into space. The global climate may be warming more rapidly now than it did during any of the warming periods following the ice ages. The environmental impact of rapid global climate warming could be catastrophic (see Chapter 20).

Another negative environmental impact of burning oil is acid deposition. Although oil does not produce appreciable amounts of sulfur oxides, it does produce nitrogen oxides, mainly through gasoline combustion in automobiles, which contributes approximately half the nitrogen oxides released into the atmosphere. (Coal combustion is responsible for the other half.) Nitrogen oxides contribute to acid deposition and the formation of photochemical smog (see Chapters 19 and 20).

The burning of natural gas, on the other hand, does not pollute the atmosphere as much as the burning of oil. Natural gas is a relatively clean, efficient source of energy that contains almost no sulfur, a contributor to acid deposition. In addition, natural gas produces far less CO_2, fewer hydrocarbons, and almost no particulate matter, as compared to oil and coal.

One of the concerns in oil and natural gas production is the environmental damage that may occur during their transport, which is often over long distances by pipelines or by ocean tankers. A serious spill along the route creates an environmental crisis, particularly in aquatic ecosystems, where the oil slick can travel.

The 1989 Alaskan Oil Spill   On March 24, 1989, the supertanker *Exxon Valdez* hit Bligh Reef and spilled 260,000 barrels (10.9 million gallons) of crude oil into Prince William Sound along the coast of Alaska, creating the largest oil spill in U.S. history. As it spread, the black, tarry gunk eventually covered thousands of square kilometers of water (Figure 10.14) and contaminated hundreds of kilometers of shoreline. According to the U.S. Fish and Wildlife Service and the Alaska Department of Environ-

mental Conservation, more than 30,000 birds (sea ducks, loons, cormorants, bald eagles, and other species) and between 3,500 and 5,500 sea otters died as a result of the spill. The area's killer whale and harbor seal populations declined, and salmon migration was disrupted. Throughout the area, there was no fishing season that year.

Within hours of the spill, scientists began to arrive on the scene to advise both Exxon Corporation and the government on the best way to try to contain and clean up the spill. But it took much longer for any real action to be taken. Eventually, nearly 12,000 workers took part in the cleanup; their activities included mechanized steam cleaning and rinsing, which killed additional shoreline organisms such as barnacles, clams, mussels, eelgrass, and rockweed.

In late 1989, Exxon declared the cleanup "complete." But it left behind contaminated shorelines, particularly rocky coasts, marshes, and mudflats; continued damage to some species of birds (such as the common loon and harlequin duck), fishes (such as murrelet and rockfish), and mammals (such as the harbor seal); and a reduced commercial salmon catch, among other problems. Multiple studies are still being conducted to determine the status of the cleanup. In 1991 Exxon agreed to pay Alaska a settlement of $1 billion, and in 1994, a jury awarded 34,000 fishermen and other Alaskans an additional $5 billion in punitive damages; Exxon is appealing the punitive damages. Additional lawsuits have been filed and will take years to settle. The final cost to Exxon is expected to exceed $10 billion.

One positive outcome of the disaster was passage of the **Oil Pollution Act** of 1990. This legislation establishes liability for damages to natural resources resulting from a catastrophic oil spill, including a trust fund that pays to clean up spills when the responsible party is unable to; a tax on oil provides money for the trust fund. The Oil Pollution Act also requires double hulls on all oil tankers that enter U.S. waters by 2015. (Had the *Exxon Valdez* possessed a double hull, the disaster probably would not have occurred because only the outer hull might have broken.) Because tankers are extremely expensive to build, however, most oil tankers will have single hulls until the 2015 deadline.

The 1991 Persian Gulf Oil Spill   The world's most massive oil spill occurred in 1991 during the Persian Gulf War, when about 6 million barrels (250 million gallons) of crude oil—more than 20 times the amount of the *Exxon Valdez* spill—were deliberately dumped into the Persian Gulf. Many oil wells were also set on fire, and lakes of oil spilled into the desert around the burning oil wells. Cleanup efforts along the coastline and in the desert were initially hampered by the war. In 2001 Kuwait began a massive remediation project to clean up its oil-contaminated desert. However, progress is slow, and environmentalists fear that it may take a century or more for the area to completely recover.

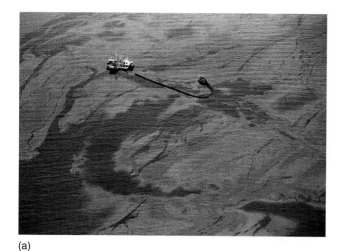

(a)

(b)

(c)

Figure 10.14 Alaskan oil spill, 1989. (a) An aerial view of the massive oil slick in Prince William Sound. (b) Workers try to clean the rocky shoreline of Eleanor Island, Alaska, several months after the spill. (c) The extent of the spill (*black arrows*). Water currents caused it to spread rapidly for hundreds of kilometers. Countless animals, such as sea otters and ocean birds, died.

CASE·IN·POINT **The Arctic National Wildlife Refuge**

One of the oil industry's biggest obstacles to locating and extracting oil is public resistance based on the perceived threat to environmentally sensitive areas. Consider the proposed opening of the Arctic National Wildlife Refuge to oil exploration. This controversy has been a major environment-versus-economy conflict off and on since 1980. On one side are those who seek to protect rare and fragile natural environments; on the other side are those whose higher priority is the development of some of the last major U.S. oil supplies.

The refuge, which has been referred to as "America's Serengeti," is home to an extremely diverse community of organisms, including polar bears, arctic foxes, peregrine falcons, musk oxen, Dall sheep, wolverines, and snow geese. It also contains the calving area for a large migrating herd of caribou. The Porcupine caribou herd, named after the Porcupine River in Canada where the herd winters, contains more than 150,000 head. Dominant plants in this coastal plain of tundra include mosses, lichens, sedges, grasses, dwarf shrubs, and small herbs. Under a thin upper layer of soil is the permafrost layer, which contains permanently frozen water.

Although it is biologically rich, the tundra is an extremely fragile ecosystem, in part because of its harsh climate. The organisms living here have adapted to their environment, but they live "on the edge." Any additional stress has the potential to harm or even kill them. Thus, arctic organisms are particularly vulnerable to human activities.

History of the Arctic National Wildlife Refuge In 1960 Congress declared a section of northeastern Alaska protected because of its distinctive wildlife. In 1980 Con-

Figure 10.15 Arctic National Wildlife Refuge. Located in the northeastern part of Alaska, the refuge is situated close to the Trans-Alaska Pipeline, which begins at Prudhoe Bay and extends south to Valdez. The National Petroleum Reserve–Alaska is also shown.

gress expanded this wilderness area to form the Arctic National Wildlife Refuge—7.3 million hectares (18 million acres) of untouched northern forests, tundra, wetlands, and glaciers (Figure 10.15). The Department of the Interior was given permission to conduct a study of the potential for oil discoveries in the area, but exploration and development could proceed only with congressional approval. The U.S. Geological Survey estimates there are 7.7 billion barrels of oil within the refuge. However, the amount of oil that it is economically feasible to obtain depends on the price of oil. At $24 per barrel, 2 billion barrels could be recovered.

Pressure to open the refuge to oil development subsided for about 5 years following the Alaskan oil spill, when public sentiments were strongly against oil companies. However, in the mid-1990s, pro-development interests became more vocal, partly because in 1994, for the first time in its history, the United States imported *more* than half the oil it used. Although a 1995 study by the Department of the Interior concluded that oil drilling in the refuge would harm the area's ecosystem, both the Senate and the House of Representatives passed measures to allow it. (President Clinton vetoed the bill in the fall of 1995.)

At the beginning of the 21st century, the issue appeared to have reached a stalemate. In 2001, President George W. Bush announced that he supports opening the refuge to oil drilling as part of his comprehensive energy policy (discussed later in the chapter). However,

after a contentious debate in Congress, the Senate voted in 2002 against opening the refuge for oil development. Pro-development interests also want to open a 4.6-million-acre section of Alaska *west* of the refuge. This piece of federal land was designated the National Petroleum Reserve–Alaska in 1923. Until recently, it was not considered economically feasible to extract the 9.3 billion barrels estimated to be beneath this land, but technological improvements have made it more attractive. In 1998 the U.S. Department of the Interior opened part of the National Petroleum Reserve–Alaska to oil and gas leasing but excluded other areas that are critical to wildlife or to native people living there.

Support for Oil Exploration in the Refuge Supporters cite economic considerations as the main reason for drilling for oil in the refuge. They point out that the United States is spending a large proportion of its energy budget to purchase foreign oil. Development of domestic oil would help to improve the balance of trade and make us less dependent on foreign countries for our oil.

The oil companies are eager to develop this particular site because it is near Prudhoe Bay, where large oil deposits are already being tapped. Prudhoe Bay has a sprawling industrial complex to support oil production, including roads, pipelines, gravel pads, and storage tanks. The Prudhoe Bay oil deposits peaked in production in 1985 and have declined in productivity since then. As a result, the oil industry is looking for sites that can make use of the infrastructure already in place.

Supporters of oil exploration in the refuge argue that it will have little lasting impact on the environment or on organisms. They say that there has been little environmental disruption or contamination of Prudhoe Bay by the oil industry; further they point out that the number of caribou in that area has actually increased.

Opposition to Oil Exploration in the Refuge Conservationists think that oil exploration poses permanent threats to the delicate balance of nature in the Alaskan wilderness, in exchange for a very temporary oil supply. They reason that the money that would be spent drilling for oil would be better used for research into alternative, renewable energy sources and energy conservation—a more permanent solution to the energy problem. They further argue that "drain America first" policies will only increase our future dependence on foreign oil supplies.

Opponents of oil exploration also refute supporters' claims that Prudhoe Bay has been developed with little environmental damage. Studies, such as one conducted by the U.S. Fish and Wildlife Service, document considerable habitat damage and declining numbers of wolves and bears in the Prudhoe Bay area. (Top predators are usually more susceptible to environmental disruption than are organisms occupying lower positions in a food web.) Biologists directly attribute the increase in the caribou population to the decline in the number of wolves and bears. The oil industry and conservationists *do* agree on one point: It is not financially practical to restore developed areas in the Arctic to their natural states. Thus, development in the Arctic causes permanent changes in the natural environment. ■

SYNFUELS AND OTHER POTENTIAL FOSSIL FUEL RESOURCES

Synthetic fuels, or **synfuels**, are used in place of oil or natural gas (see "Mini-Glossary: Synfuels"). They are derived from coal and other naturally occurring sources and may be liquid or gaseous. Synfuels include tar sands, oil shales, gas hydrates, liquefied coal, and coal gas. Although synfuels are more expensive to produce than fossil fuels, they may become more important as fossil fuel reserves decline.

Tar sands, also called **oil sands**, are underground sand deposits permeated with a thick, asphalt-like oil known as *bitumen*. The bitumen in tar sands deep in the ground cannot be pumped out unless it is heated underground with steam to make it more fluid. If tar sands are close to the Earth's surface, however, they can be surface-mined. Once bitumen is obtained from tar sands, it must be refined like crude oil. World tar sand reserves are estimated to contain half again as much fuel as world oil reserves. Major tar sands are found in Venezuela and in Alberta, Canada, where an estimated 300 billion barrels of oil occur in tar sands; this estimate is greater than the estimated oil reserves in Saudi Arabia. An energy company in Canada is currently mining, extracting, and refining tar sands to produce almost 30 million barrels of oil a year.

"Oily rocks" were discovered by western American pioneers whose rock hearths caught fire and burned. **Oil shales** are rocks containing a mixture of hydrocarbons known collectively as *kerogen*. In order to yield their oil, these sedimentary rocks, called oil shales, must be crushed and heated, and the kerogen must be refined after it is mined. Because the mining and refinement of oil shales require the expenditure of a great deal of energy, it is not yet cost efficient to process oil shales. Large oil shale deposits are located in Australia, Estonia, Brazil, Sweden, the United States, and China. Wyoming, Utah, and Colorado have the largest deposits in the United States. Like tar sands, oil shale reserves may contain half again as much fuel as world oil reserves.

Gas hydrates, also called **methane hydrates**, are reserves of ice-encrusted natural gas located deep underground in porous rock. Massive deposits have been identified in the arctic tundra, deep under the permafrost, and in the deep ocean sediments of the continental slope and ocean floor. Until recently, the U.S. oil industry was not particularly interested in extracting natural gas from gas hydrates because of the expense involved. Several U.S. oil companies are currently developing methods to extract gas hydrates and hope the extraction will be economically feasible by 2015 or so. Countries with lots of gas hydrates (for example, Russia) or with few conventional fossil fuel deposits (for example, India and Japan) have established national gas-hydrates programs. In a pilot program in Siberia, Russian scientists successfully removed natural gas from hydrate deposits by pumping methanol (an alcohol) into the hydrate region. Natural gas that is associated with ice readily dissolves in methanol and can then be pumped out of the ground.

Coal has been used to produce a nonalcohol liquid fuel similar to oil. The liquid fuel, which can be cleaned before burning, is less polluting than solid coal. This process, called **coal liquefaction**, was first developed before World War II, but its expense prevented it from replacing gasoline production. Although research in the 1980s resulted in a series of technological improvements that lowered the cost of coal liquefaction, it is still not cost-competitive.

Another synfuel is a gaseous product of coal. Coal gas has been produced since the 19th century. As a matter of fact, it was the major fuel used for lighting and heating in American homes until oil and natural gas replaced it during the 20th century. Production of the combustible gas methane from coal is called **coal gasification** (Figure 10.16). A promising coal gasification technique was developed at Stanford University, and several demonstration power plants that convert coal into gas have been constructed in the United States. One advantage of coal gas over solid coal is that it burns almost as cleanly as natural gas. Because sulfur is removed during coal gasification, scrubbers are not needed when coal gas is burned. Like other synfuels, coal gas is currently more expensive to produce than fossil fuels.

MINI-GLOSSARY

Synfuels

synthetic fuels: Liquid or gaseous fuels synthesized from coal and other naturally occurring sources; also called *synfuels*.

tar sands: Underground sand deposits permeated with a thick, asphalt-like oil known as bitumen; also called *oil sands*.

oil shales: Rocks containing a mixture of hydrocarbons known as kerogen.

gas hydrates: Ice-encrusted natural gas located underground in porous rock; also called *methane hydrates*.

coal liquefaction: Process by which solid coal is converted to a synthetic liquid fuel.

coal gasification: Process by which solid coal is converted to a synthetic gaseous fuel.

Environmental Impacts of Synfuels

Although synfuels are promising energy sources, they have many of the same undesirable effects as fossil fuels.

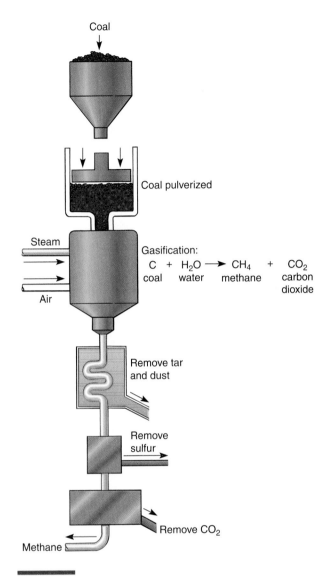

Coal

Coal pulverized

Steam

Gasification:

$$C + H_2O \longrightarrow CH_4 + CO_2$$
coal   water   methane   carbon dioxide

Air

Remove tar and dust

Remove sulfur

Remove CO_2

Methane

Figure 10.16 Coal gasification. Shown is one method of coal gasification, in which the combustible gas methane is generated from coal.

Their combustion releases enormous quantities of CO_2 and other pollutants into the atmosphere, thereby contributing to global warming and air pollution. Some synfuels, such as coal gas, require large amounts of water during production and would therefore be of limited usefulness in arid areas, where water shortages are already commonplace. Also, enormously large areas of land would have to be surface mined in order to recover the fuel in tar sands and oil shales.

AN ENERGY STRATEGY FOR THE UNITED STATES

The United States has a comprehensive energy policy for several reasons: (1) The supply of fossil fuels is lim-ited; (2) the production, transport, and use of fossil fuels pollute the environment; (3) our heavy dependence on foreign oil makes us economically vulnerable. Because of the complex nature of energy issues, any policy that is adopted by our political leaders must have many approaches. Although there is no way to completely eliminate our vulnerability to disruptions in foreign oil supplies and to oil price increases, we can lessen the effects of such events through a comprehensive energy policy. Such a strategy must provide us with a secure supply of energy, encourage us to use less energy, and protect the environment. The following elements could be included in a comprehensive national energy policy.

Objective 1: Increase Energy Efficiency and Conservation

Over the past several decades, energy efficiency in the United States has improved significantly, but despite gains, the United States uses more energy than any other country. There is room for great improvement on all fronts, from individuals conserving heating oil by weatherproofing their homes, to groups of commuters conserving gasoline by carpooling, to corporations developing more energy-efficient products. The automobile industry could be required to increase the average new-car gasoline mileage, for example.

Reinstating the 55-mile-per-hour speed limit in areas where the limit is now as high as 75 mph would conserve gasoline. Fuel consumption increases by approximately 50% if a car is driven at 75 mph rather than 55 mph. Fuel consumption increases by 30% at 65 mph rather than 55 mph. In addition, federal financial support for transportation could be shifted from highway construction to public transportation. We say more about energy conservation in Chapter 12. (Also see "You Can Make a Difference: Getting Around Town.")

One way to encourage energy conservation is to eliminate government subsidies that keep energy prices artificially low. *When prices reflect the true costs of energy, including the environmental costs incurred by its production, transport, and use, energy will be used more efficiently.* Gasoline prices in the United States do not reflect the true cost of gasoline and are unrealistically low. During 2002, Western Europeans paid about two and one half times more for gasoline than U.S. citizens did (Table 10.2). It has been demonstrated many times that the price of gasoline affects the level of gasoline consumption: Lower prices encourage greater consumption. Over the next few years, a more realistic price for gasoline could be introduced to encourage people to buy fuel-efficient automobiles, carpool, and use public transportation (Table 10.3).

YOU CAN MAKE A DIFFERENCE

Getting Around Town

Can you imagine getting around town without a car? How would you get to class, the grocery store, the laundromat? Hopping in your car for every errand seems like the natural thing to do. According to the American Automobile Association, American motorists drive an average of 10,100 miles annually and burn 507 gallons of gasoline in the process.

However, consider this: From the production of gasoline to the disposal of old automobiles, the car has a significant negative impact on the environment. Acid deposition and global warming are just two of the problems caused by the combustion of gasoline. Vehicle exhaust, photochemical smog, and chronic low-level exposure to toxins are all health threats to car owners and to those who live in areas with a high density of cars (see Chapter 19). Dumping of engine oil, fumes from the burning of tires and batteries, and automobile junkyards threaten both our health and our environment.

There is something you can do about it. Granted, you may not be able to give up your car entirely, but you can cut down on its use wherever possible. For example, try the following:

1. Ride the bus or the train whenever possible. Think about how jammed the road would be if all 50 people on a bus were driving individual cars!

2. Consider whether you need to drive to accomplish a task. Sometimes a phone call can be substituted for a trip in the automobile.

3. Carpool to class, to work, to the grocery store, to social events. One car on the road is better than three or four.

4. Buy a good bicycle; it is less expensive than a car to buy and maintain, and it is great for local transportation. It is also good exercise.

5. Walk to class or work if you live within a mile or so. You will need to allow yourself a little extra time, but once you get into the habit, it is easy. Walking is good exercise, too.

6. Modify your driving habits to save gasoline. Minimize braking, and do not let your engine idle for more than 1 minute. Keep your car well tuned, replace air and oil filters often, and keep your tires inflated at the recommended pressure. Remove any unnecessary weight from your car. All of these measures help boost gasoline mileage.

7. When you purchase a motor vehicle, use the EPA's official mileage ratings (http://www.fueleconomy.gov) as one of the criteria to help you select a model. For example, using the mid-2002 price ($1.39) for a gallon of regular gasoline, you will spend $306 to drive 15,000 miles in the 2002 Honda Insight, which gets 68 miles per gallon for highway driving. Driving that same distance in a 2002 GMC Yukon Denali, which gets 15 miles per gallon for highway driving, will cost you $1,390.

Table 10.2	Comparison of Gasoline Prices in Selected Countries (Including Taxes)

Country	Premium Gasoline Price (Dollars per Gallon, July 29, 2002)
United States	$1.59
Belgium	$3.66
France	$3.79
Germany	$3.89
Italy	$3.90
Netherlands	$4.26
United Kingdom	$4.35

Table 10.3	Comparison of Energy Input for Different Kinds of Transportation

Method of Transportation	Energy Input (in BTUs)* per Person, per Mile
Automobile (driver only)	6,530
Rail	3,534
Carpool	2,230
Vanpool	1,094
Bus	939

* BTU stands for British thermal unit, an energy unit equivalent to 252 calories or 1,054 joules.

CASE·IN·POINT Energy Subsidies and the Real Price of Fuel

The price you pay for gasoline at the fuel pump is not its actual price. Its real cost is subject to a national energy pricing policy, which attempts to stabilize energy prices and reduce our dependence on sources of foreign energy supplies. Most governments practice some type of energy pricing policy, often in the form of energy subsidies or energy taxes.

A **subsidy** is government support to a business or other economic institution that the government thinks is beneficial to the economy. Subsidies include monetary payments, public financing, tax benefits, and tax exemptions. Energy is heavily subsidized and therefore appears cheaper than it really is. An energy subsidy can be calculated as the difference between the world market price and the domestic market price for a particular fuel. If gasoline trades for $1.25 per gallon on the world market but costs only $1.00 within a particular country, then that country's government is subsidizing energy at a cost of $.25 per gallon.

Energy subsidies are often aimed at reducing the price of fuel for consumers, with the intent of stimulating economic growth. Low costs for consumers encourage a high use of energy, which in turn accelerates the depletion of a nonrenewable energy resource. When a government removes its energy subsidies, consumers have to pay more for fuel, which encourages them to use energy more efficiently so they can save money. The resulting decrease in energy use reduces a country's dependence on costly fuel imports and lessens the harmful environmental impacts of fuel production and consumption.

Whereas subsidies reduce the cost of fuel for consumers, energy taxes *increase* the cost. Energy taxes serve to raise a government's revenues. In addition, they encourage consumers to conserve energy, thereby helping to reduce a country's dependence on foreign energy supplies. Energy taxes reflect some of the hidden costs of energy consumption, such as pollution control and cleanup.

According to Norman Myers in *Perverse Subsidies: Tax Dollars Undercutting Our Economies and Environments Alike*, U.S. energy subsidies are roughly estimated at $32 billion. (This is a rough estimate because it is difficult for analysts to agree on exactly what constitutes a subsidy.) Fossil fuels received approximately 60% of this amount, nuclear energy approximately 30%, and renewable and nonpolluting sources of energy, plus energy conservation and improved energy efficiency, less than 10%. Within fossil fuels, natural gas received less than either coal or oil, despite the fact that natural gas is a cleaner fuel.

Objective 2: Secure Future Fossil Fuel Energy Supplies

A comprehensive national energy strategy could include the environmentally sound and responsible development of domestically produced fossil fuels, especially natural gas. There are two types of opposition to this element of a national energy strategy; one is economic and the other is environmental. Some think it is better to deplete foreign oil reserves while prices are reasonable and save domestic supplies for the future. Most economists argue against this view, however, because of the U.S. trade deficit. We do not currently finance our oil imports by exporting goods and services of equal value. Many environmentalists oppose the development and increased use of domestic fossil fuels, largely because of the environmental problems already discussed.

Everyone, environmentalists included, recognizes the need for a dependable energy supply. Securing a future supply of fossil fuels is a *temporary* strategy, however, because fossil fuels are nonrenewable resources that will eventually be depleted, regardless of how efficient our use or how much we conserve. Having a secure energy supply for the short term will, however, allow us time to develop alternative energy sources for the long term.

Objective 3: Develop Alternative Energy Sources

Research and development must be expanded for all possible alternatives to fossil fuels, especially renewable energy sources such as solar and wind energy (see Chapter 12). Our long-term energy policy goal could be to shift to energy sources that are less harmful to the environment.

Who should pay for the research costs of improving energy conservation and developing alternative forms of energy? The answer is that we all should share in these costs because we will all share in the benefits. The proceeds of a gasoline tax have been suggested as a means of financing programs to achieve a sustainable energy future. Some policy makers have suggested a tax of as much as 50 cents per gallon.

Objective 4: Accomplish the First Three Objectives Without Further Damaging the Environment

The environmental costs of using a particular energy source must be weighed against its benefits when it is considered as a practical component of an energy policy. If domestic supplies of fossil fuels are developed with as much attention to the environment as possible, they will help reduce our dependence on foreign oil. One suggestion has been to add a 5-cent tax on each barrel of domestically produced oil to establish a reclamation fund for undoing some of the environmental damage caused by mining and production of oil and natural gas.

How Politics Influences the National Energy Policy

The nation's energy policy reflects the varied political views of the president, Congress, and the American public. It is also subject to major changes with every change in the U.S. presidential office. The current National Energy Policy, developed by President George W. Bush, is summarized in Table 10.4. It has five components: (1) modernize conservation; (2) modernize our energy infrastructure; (3) increase energy supplies; (4) accelerate the protection and improvement of the environment; and (5) increase our nation's energy security.

Table 10.4   **Highlights of President Bush's National Energy Policy**

Modernize Conservation

1. Increase funding for renewable energy and energy efficiency research and development programs.
2. Create an income tax credit for people who purchase hybrid and fuel cell vehicles.
3. Extend the Energy Star efficiency program to include schools, retail buildings, health care facilities, and homes, and extend the Energy Star labeling program to additional products and appliances.
4. Fund the Intelligent Transportation Systems program, the Fuel Cell Powered Transit Bus program, and the Clean Buses program.
5. Review the Corporate Average Fuel Economy (CAFE) standards.

Modernize Our Energy Infrastructure

1. Expedite permits for energy-related projects on a national basis.
2. Grant authority for rights-of-way for electricity transmission lines, to create a national transmission grid.
3. Enact electricity legislation that encourages new electricity generation.
4. Improve the reliability of the interstate transmission system.

Increase Energy Supplies

1. Open the Arctic National Wildlife Refuge (ANWR) to exploration and production. Examine the potential of oil and natural gas development on other federal lands.
2. Earmark $1.2 billion from the leasing of ANWR to fund research into renewable energy resources.
3. Fund research on clean coal technology and on electricity from biomass cofired with coal.
4. Provide for the safe expansion of nuclear energy by establishing a national repository for nuclear waste and by streamlining the licensing of nuclear power plants.

Accelerate the Protection and Improvement of the Environment

1. Enact legislation to establish a flexible, market-based program to reduce and cap emissions of sulfur dioxide, nitrogen oxides, and mercury from electric power generators.
2. Earmark royalties from oil and gas exploration in ANWR to fund land conservation efforts.

Increase Our Nation's Energy Security

1. Prepare for potential energy-related emergencies.
2. Expand cross-border energy investments, oil and gas pipelines, and electricity grid connections with Canada and Mexico.
3. Expedite permits for a gas pipeline route from Alaska to the lower 48 states.

SUMMARY WITH SELECTED KEY TERMS

I. Energy consumption increased worldwide by about 1.9% from 1999 to 2000.

A. Most of the increase occurred in developing countries, which use more energy as they improve their standard of living.

B. Highly developed nations consume much more energy per person than developing countries.

II. Oil, natural gas, and coal are **fossil fuels** that formed several hundred million years ago from plant and animal remains.

A. Fossil fuels are **nonrenewable resources**. The Earth has a finite supply of fossil fuels; this supply is depleted by use.

B. No fossil fuel is a perfect energy source. Their combustion produces several pollutants—in particular, large amounts of CO_2, a greenhouse gas that prevents heat from escaping from the Earth.

III. Coal was formed when partially decomposed plant material was exposed to heat and pressure for aeons.

A. There are several grades of coal.

   1. **Lignite** is soft coal that is low in sulfur and produces less heat than other grades.

   2. **Subbituminous coal** is a grade of coal intermediate between lignite and bituminous coal. It has a low heat value and sulfur content.

   3. **Bituminous coal**, also called **soft coal**, is usually high in sulfur and produces a lot of heat.

   4. **Anthracite**, also called **hard coal**, is low in sulfur and produces more heat than the other types of coal.

B. Coal is present in greater quantities than oil or natural gas (known world reserves could last for more than 200 years at the present rate of consumption) but its use has a greater potential to harm the environment.

1. **Subsurface mining**, or underground coal mining, can be dangerous and unhealthful.

2. **Surface mining** disturbs large areas of land, which is expensive to restore. **Strip mining** is a type of surface mining.

3. In the United States surface mining accounts for 60%, and subsurface mining accounts for 40% of the coal mined.

C. Many environmental problems are associated with coal mining and combustion.

1. Surface mining destroys existing vegetation and topsoil. The **Surface Mining Control and Reclamation Act (SMCRA)** requires that surface-mined lands be restored to make the land productive for other uses.

2. Coal produces more CO_2 emissions per unit of heat than do other fossil fuels. CO_2 accumulates in the atmosphere, where it prevents planetary heat from radiating into space. There is currently no way to reduce or eliminate CO_2 emissions when coal is burned.

3. Burning soft coals that contain sulfur contributes to **acid deposition**. There are several ways to control sulfur emissions, including **fluidized-bed combustion** (a cleaner coal-burning process) and the installation of **scrubbers** (desulfurization systems) in smokestacks. The sludge produced by scrubbers must be disposed of or sold **(resource recovery)**.

IV. Oil and natural gas formed when large numbers of microscopic aquatic organisms died and settled in oxygen-deficient sediments.

A. Oil and natural gas deposits occur in association with **structural traps** such as upward foldings of rock strata and **salt domes** (underground columns of salt).

B. Although world oil and natural gas reserves are large, they will probably be depleted during the 21st century.

C. More than half the world's total estimated reserves of oil are located in the Persian Gulf region. Russia and Iran have almost half of the world's proved recoverable reserves of natural gas.

D. Many environmental problems are associated with obtaining and burning oil and, to a lesser extent, natural gas.

1. An accidental oil spill during transport, often over long distances by pipelines or ocean tankers, creates an environmental crisis. The 1989 Alaskan oil spill was the largest accidental spill in U.S. history. The 1991 Persian Gulf oil spill was the largest in the world.

2. Oil exploration and extraction are a threat to environmentally sensitive areas. Whether or not to drill in the Alaska National Wildlife Refuge has been an ongoing controversy since 1980.

3. CO_2 emissions that are released when oil and natural gas are burned may contribute to global climate warming.

4. Production of nitrogen oxides when oil is burned contributes to acid deposition.

V. **Synthetic fuels**, or **synfuels**, are naturally occurring liquid or gaseous fuels that substitute for oil or natural gas.

A. Synfuels include **tar sands**, **oil shales**, **gas hydrates**, **liquid coal**, and **coal gas**.

B. Synfuels are currently more expensive to produce than oil or natural gas. However, they may be used more in the future, when oil and natural gas reserves decline.

C. Synfuels have many of the environmental drawbacks associated with traditional fossil fuels.

VI. The goal of a national energy strategy should be to ensure adequate energy supplies without harming the environment.

A. Such a policy can be accomplished by increasing energy efficiency, securing future energy supplies, improving energy technology, and avoiding harm to the environment.

B. President George W. Bush's National Energy Policy includes the following objectives: modernize conservation; modernize our energy infrastructure; increase energy supplies; accelerate the protection and improvement of the environment; and increase our nation's energy security.

THINKING ABOUT THE ENVIRONMENT

1. How does U.S. dependence on foreign oil affect our energy security?

2. How does per-capita energy consumption compare in highly developed countries and developing countries?

3. What are fossil fuels, and how are coal, oil, and natural gas formed?

4. What are the advantages and disadvantages of using coal as an energy resource?

5. It has been suggested that the Industrial Revolution was concentrated in the Northern Hemisphere largely because coal is located there. What is the relationship between coal and the Industrial Revolution?

6. Several politicians have commented that the United States is "the OPEC of coal." What does this mean?

7. What are the advantages and disadvantages of using oil and natural gas as energy resources?

8. On the basis of what you have learned about coal, oil, and natural gas, which fossil fuel do you think the United States should exploit in the short term (during the next 20 years)? Explain your rationale.

9. In your estimation, which fossil fuel has the greatest potential for the 21st century? Why?

10. Which of the negative environmental impacts associated with fossil fuels is most serious? Why?

11. Which major consumer of oil is most vulnerable to disruption in the event of another energy crisis: electric power generation, motor vehicles, heating and air conditioning, or industry? Why?

12. What is the controversy surrounding the Arctic National Wildlife Refuge? Do you think oil drilling should be permitted there? Why or why not?

13. Why are synfuels promising but not a panacea for our energy needs?

14. What are the main points in President George W. Bush's National Energy Policy?

*15. Draw a graph showing daily per-capita gasoline consumption (in gallons) from 1950 to 2002. Use the information provided in Table 10.A. You will need to convert barrels of gasoline to gallons (1 barrel = 42 gallons) and then determine the per-capita gasoline consumption. (Divide gasoline consumption in gallons per day by the population; round to the nearest tenth of a gallon.)

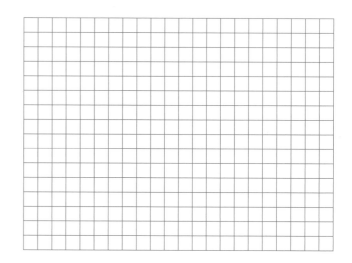

Table 10.A Per-Capita Gasoline Consumption in the United States, 1950 to 2002

Year	Gasoline (Million Barrels/Day)	Gasoline (Million Gallons/Day)	Population (Millions)	Gasoline Consumption (Gallons/Day/Person)
1950	2.72	114.24	151.3	0.8
1955	3.66		165.1	
1960	4.13		179.3	
1965	4.59		193.5	
1970	5.78		203.2	
1975	6.67		215.5	
1980	6.58		226.5	
1985	6.83		238.7	
1990	7.23		248.7	
1995	7.79		263.2	
2000	8.36		275.6	
2002	9.27		287.4	

* The solution to this question appears in Appendix VII.

TAKE A STAND

Visit our Web site at **http://www.wiley.com/college/raven** (select Chapter 10 from the Table of Contents) for links to more information about the controversies surrounding the Arctic National Wildlife Refuge. Consider the opposing views of those who seek to protect rare and fragile natural environments and those who wish to develop the last major oil supplies in the United States and debate the issues with your classmates. You will find tools to help you organize your research, analyze the data, think critically about the issues, and construct a well-considered argument. Take a Stand activities can be done individually or as part of a team, as oral presentations, written exercises, or Web-based (e-mail) assignments.

Additional on-line materials relating to this chapter, including Student Quizzes, Activity Links, Useful Web Sites, Flash Cards, and more, can also be found on our Web site.

SUGGESTED READING

Chiles, J.R. "Getting in Deep." *Smithsonian* (November 2001). The oil industry is probing reserves located more than a mile beneath the surface of the Gulf of Mexico.

Deffeyes, K.S. *Hubbert's Peak: The Impending World Oil Shortage.* Princeton, NJ: Princeton University Press, 2001. The author, a geophysicist, calculates that world oil production will peak sometime around 2004 to 2008.

Fouda, S.A. "Liquid Fuels from Natural Gas." *Scientific American* (March 1998). Recent technological advances have made it possible to convert natural gas to a liquid fuel.

George, R.L. "Mining for Oil." *Scientific American* (March 1998). The oil in Canadian tar sands can now be economically extracted, making it a viable substitute for dwindling oil reserves.

Gibbs, W.W. "The Arctic Oil and Wildlife Refuge." *Scientific American*, Vol. 284, No. 5 (May 2001). The North Slope of Alaska is the center of a bitter debate over oil versus wildlife.

Jehl, D. "Energy Disruptions Brighten Future of Coal, a Fossil of a Fuel." *New York Times* (June 16, 2001). Bush's focus on energy security over environmental protection has allowed utilities to reconsider coal as the energy source of choice.

Lipton, M. "The Fight for the Soul of Coal Country." *New York Times* (May 17, 2002). Mountaintop removal has ignited a controversy over mining jobs versus the beautiful environment of West Virginia.

Mitchell, J.G. "In the Wake of the Spill: Ten Years After *Exxon Valdez*." *National Geographic* (March 1999). The environmental damage caused by the 1989 Alaskan oil spill is improving, but certain negative effects are still lingering.

Mitchell, J.G. "Oil Field or Sanctuary?" *National Geographic*, Vol. 200, No. 2 (August 2001). Examines the issue over whether the Arctic National Wildlife Refuge should be opened to drilling.

Suess, E., G. Bohrmann, J. Greinert, and E. Lausch. "Flammable Ice." *Scientific American* (November 1999). In addition to being a potential source of future energy supplies, methane hydrates in marine sediments may contribute to the stability of continental slopes. There is also concern that warming temperatures in the ocean could trigger the release of methane that would further exacerbate global warming.

11

Nuclear power plant at Salem, New Jersey. There are three nuclear reactors here. Salem I opened in 1977, Salem II in 1981, and Hope Creek in 1986. Nuclear power meets almost 50% of New Jersey's electricity needs.

Nuclear Energy

Learning Objectives

After studying this chapter you should be able to

1. Distinguish between *nuclear energy* and *chemical energy* and define *radioactive elements* and *radioactive decay*.

2. Summarize how mined uranium becomes fuel for nuclear power plants.

3. Diagram a pressurized water reactor and show its three water circuits.

4. Contrast conventional nuclear fission, breeder nuclear fission, and fusion.

5. Describe the two major nuclear power plant accidents discussed in the text (Three Mile Island and Chornobyl).

6. Distinguish between low-level and high-level radioactive wastes.

7. Relate the pros and cons of permanent storage of high-level radioactive wastes at Yucca Mountain.

8. Explain what happens to nuclear power plants after they are closed.

W hen nuclear energy was discovered during the 20th century, human history was forever changed. Lauded as a potential source of energy for humanity, nuclear energy was also denounced as a weapon that could destroy human civilization. This dichotomous view continues to permeate issues involving nuclear energy today. The main question that we wish to address throughout this chapter is what role, if any, nuclear energy may have in providing us with electrical energy in the 21st century. To consider this question effectively, we begin our discussion with a brief history of the scientific development of nuclear energy.

Nuclear reactions have the potential to release a vast amount of energy, primarily as heat. In this process, a small amount of the mass of an atom is transformed into a large amount of energy. Albert Einstein first hypothesized in 1905 that mass and energy are related in his now-famous equation, $E = mc^2$, in which energy (E) is equal to mass (m) times the speed of light (c) squared. Einstein's research was based on the work of earlier scientists. In 1896 French physicist Henri Becquerel discovered that uranium-containing minerals spontaneously and continually give off invisible rays of energy—that is, radiation. Beginning in 1898, British physicist Ernest Rutherford conducted a series of experiments that determined that radiation consists of high-energy particles. In 1911, his work and the work of others led Rutherford to the conclusion that each atom contains a nucleus in which most of its mass is located.

Scientists began to study what happens when the nuclei of atoms are bombarded with high-energy particles. In 1919 Rutherford bombarded nitrogen nuclei with alpha particles (positively charged radiation), changing the nitrogen into oxygen in the process! Attempts to bombard heavier atoms with alpha particles failed, however, because the larger nuclei repelled the positively charged particles. In 1938 two German scientists, Otto Hahn and Fritz Strassmann, succeeded in bombarding uranium with neutrons, which are atomic particles that lack a charge. During this reaction, some of the uranium nuclei split into two smaller nuclei of barium and krypton. Additional neutrons were also emitted, along with energy.

Scientists quickly realized the implications of splitting atomic nuclei. If the emitted neutrons could in turn be made to bombard other uranium nuclei, a chain reaction would be started. Using Albert Einstein's equation, scientists determined that 0.45 kg (1 lb) of uranium could release as much energy as 7,300 metric tons (8,000 tons) of TNT, enough to make a very powerful bomb. The development of nuclear weapons was the first application of nuclear energy attempted by several nations because, beginning in 1939, Europe was at war. Einstein fled to the United States to avoid persecution by the Nazis during World War II, and he informed President Roosevelt that the Germans were working on a nuclear bomb. Roosevelt established the top-secret Manhattan Project to build such a bomb. The scientists began to experiment with chain reactions and to produce enough enriched uranium and plutonium for a bomb. In July 1945, the world's first atomic bomb was exploded in the desert at Alamogordo, New Mexico. Shortly thereafter, in August 1945, the United States dropped two atomic bombs on Japan.

Following World War II, the United States, Soviet Union, United Kingdom, France, and Canada began to design nuclear reactors that would harvest nuclear energy for peaceful purposes (see figure). The world's first large-scale nuclear power plant opened at Calder Hall in England in 1956, whereas the first commercial nuclear power plant in the United States opened at Shippingport, Pennsylvania, in 1957. Canada's first large-scale plant opened in 1962 at Rolphton, Ontario.

INTRODUCTION TO NUCLEAR PROCESSES

As a way to obtain energy, nuclear processes are fundamentally different from the combustion that produces energy from fossil fuels. Combustion is a chemical reaction. In ordinary chemical reactions, atoms of one element do not change into atoms of another element, nor does any of their mass (matter) change into energy. The energy released in combustion and other chemical reactions comes from changes in the chemical bonds that hold the atoms together. Chemical bonds are associations between electrons, so ordinary chemical reactions involve the rearrangement of electrons (see Appendix I).

In contrast, nuclear energy involves changes within the *nuclei* of atoms; small amounts of matter from the nucleus are converted into very large amounts of energy. There are two different reactions that release nuclear energy: fission and fusion. In **fission**, the process that nuclear power plants use, larger atoms of certain elements are split into two smaller atoms of different elements. In **fusion**, the process that powers the sun and other stars, two smaller atoms are combined to make one larger atom of a different element. In each case, the mass of the end product(s) is less than the mass of the starting material(s) because a small quantity of the starting material is converted to energy.

Nuclear reactions produce 100,000 times more energy per atom than do chemical reactions such as combustion. In nuclear bombs this energy is released all at once, producing a tremendous surge of heat and power that destroys everything in its vicinity. When nuclear energy is used to generate electricity, the nuclear reaction is *controlled* to produce smaller amounts of energy in the form of heat, which can then be converted to electricity.

Atoms and Radioactivity

All atoms are composed of positively charged protons, negatively charged electrons, and electrically neutral neutrons (Figure 11.1). Protons and neutrons, which have approximately the same mass, are clustered in the center of the atom, making up its nucleus. Electrons, which possess little mass in comparison with protons and neutrons, orbit around the nucleus in distinct regions. Electrically neutral atoms possess identical numbers of positively charged protons and negatively charged electrons.

The **atomic mass** of an element is equal to the sum of protons and neutrons in the nucleus. Each element contains a characteristic number of protons per atom, called its **atomic number**. In contrast, the number of neutrons in each atom of a given element may vary, resulting in atoms of one element with different atomic masses. Forms of a single element that differ in atomic mass are known as **isotopes**. For example, normal hydrogen, the lightest element, contains one proton and no neutrons in the nucleus of each atom. The other two isotopes of hydrogen are **deuterium**, which contains one proton and one neutron per nucleus, and **tritium**, which contains one proton and two neutrons per nucleus. Many isotopes are stable, but some are unstable; the unstable ones, called **radioisotopes**, are said to be **radioactive** because they spontaneously emit **radiation**, a form of energy consisting of particles (see "Mini-Glossary:

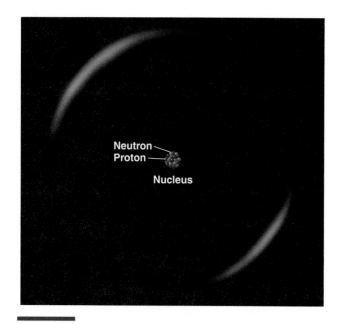

Figure 11.1 Atomic structure. Atoms contain a nucleus made of positively charged particles (protons) and particles with no charge (neutrons). Circling the nucleus is a "cloud" of small negatively charged particles called electrons.

Nuclear Energy Terms"). The only radioisotope of hydrogen is tritium.

As a radioactive element emits radiation, its nucleus changes into the nucleus of a different element, one that is more stable. This process is known as **radioactive decay**. For example, the radioactive nucleus of one isotope of uranium, U-235, decays over time into lead (Pb-207). (Uranium-235 is an isotope of uranium with an

MINI-GLOSSARY

Nuclear Energy Terms

nuclear energy: The energy released by nuclear fission or fusion.

fission: The splitting of an atomic nucleus into two smaller fragments, accompanied by the release of a large amount of energy.

fusion: The joining of two lightweight atomic nuclei into a single heavier nucleus, accompanied by the release of a large amount of energy.

isotope: One of two or more forms of an element that have the same atomic number but different atomic masses. Some isotopes are radioactive.

radioactive decay: Radioactivity; the emission of energetic particles or rays from unstable atomic nuclei. Naturally occurring radiation includes positively charged alpha particles; negatively charged beta particles; and high-energy, electromagnetic gamma rays.

radioisotope: A radioactive isotope of an element; an unstable isotope that releases radiation.

radioactive half-life: The amount of time it takes for half of the atoms in a radioactive substance to decay.

Table 11.1	Some Common Radioactive Isotopes Associated with the Fission of Uranium
Radioisotope	**Half-Life (years)**
Iodine-131	0.02 (8.1 days)
Xenon-133	0.04 (15.3 days)
Cerium-144	0.80
Ruthenium-106	1.00
Krypton-85	10.40
Tritium	12.30
Strontium-90	28.00
Cesium-137	30.00
Radium-226	1,600.00
Plutonium-240	6,600.00
Plutonium-239	24,400.00
Neptunium-237	2,130,000.00

atomic mass of 235.) Each radioisotope has its own characteristic rate of decay. The period of time required for one half of the total amount of a radioactive substance to change into a different material is known as its **radioactive half-life**. There is enormous variation in the half-lives of different radioisotopes (Table 11.1). The half-life of iodine (I-132) is only 2.4 hours, the half-life of tritium is 12.3 years, and the half-life of an isotope of uranium (U-234) is 250,000 years.

CONVENTIONAL NUCLEAR FISSION

Uranium ore, the mineral fuel used in conventional nuclear power plants, is a nonrenewable resource present in limited amounts in sedimentary rock in the Earth's crust. The processes involved in producing the fuel used in nuclear reactors and in disposing of radioactive wastes (also called nuclear wastes) are collectively called the **nuclear fuel cycle** (Figure 11.2).

Substantial deposits of uranium are found in Australia (20.4% of known world reserves), Kazakhstan (18.2%), the United States (10.6%), Canada (9.9%), and South Africa (8.9%).[1] In the United States, uranium is found in Wyoming, Texas, Colorado, New Mexico, and Utah. Uranium ore contains three isotopes, U-238 (99.28%), U-235 (0.71%), and U-234 (less than 0.01%). Because U-235, the isotope that is used in conventional fission reactions, is such a minor part (0.71%) of uranium ore, uranium ore must be refined after mining to increase the concentration of U-235 to about 3%. This refining process, which can be very energy-intensive, is known as **enrichment**.

After enrichment, the uranium fuel used in a **nuclear reactor** is processed into small pellets of uranium dioxide

---

[1] Unless noted otherwise, all energy facts cited in this chapter were obtained from the Energy Information Administration (EIA), the statistical agency of the U.S. Department of Energy (DOE).

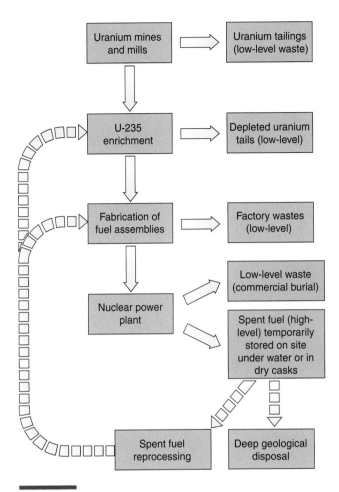

Figure 11.2 Nuclear fuel cycle, from mining to disposal. The *left* side of the figure shows how mined uranium becomes fuel for nuclear power plants. The *right* side shows radioactive wastes that must be handled and disposed of during the cycle. Broken lines indicate steps that are not currently occurring. In 1976 the United States stopped reprocessing (reusing spent fuel) for economic and political reasons. During the late-1990s, Europe's nuclear reprocessing business was essentially closed for the same reasons. Deep geological disposal of spent fuel is currently under study in several countries, including the United States.

(a)

(b)

Figure 11.3 Uranium fuel. (a) Uranium dioxide pellets, held in a gloved hand, contain about 3% uranium-235, which is the fission fuel in a nuclear reactor. Each pellet contains the energy equivalent of one ton of coal. (b) The uranium pellets are loaded into long fuel rods, which are grouped into square fuel assemblies (shown).

(Figure 11.3a), each of which contains the energy equivalent of a ton of coal. The pellets are placed in **fuel rods,** closed pipes, often as long as 3.7 m (12 ft). The fuel rods are then grouped into square **fuel assemblies,** generally of 200 rods each (Figure 11.3b). A typical nuclear reactor contains about 250 fuel assemblies.

In nuclear fission, U-235 is bombarded with neutrons (Figure 11.4). When the nucleus of an atom of U-235 is struck by and absorbs a neutron, it becomes unstable and splits into two smaller atoms, each approximately half the size of the original uranium atom. In the fission process, two or three neutrons are also ejected from the uranium atom. They collide with other U-235 atoms, generating a chain reaction as those atoms are split and more neutrons are released to collide with additional U-235 atoms.

The fission of U-235 releases an enormous amount of heat, which is used in a nuclear power plant to transform water into steam. The steam, in turn, is used to generate electricity. Production of electricity is possible because the fission reaction is controlled. (Recall that nuclear bombs make use of uncontrolled fission reac-

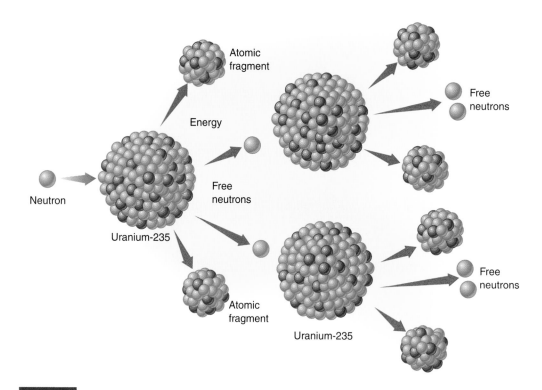

Figure 11.4 Nuclear fission. Starting at the left side of the figure, neutron bombardment of a uranium-235 (U-235) nucleus causes it to split into two smaller atomic fragments and several free neutrons. The free neutrons bombard nearby U-235 nuclei, causing them to split and release still more free neutrons in a process called a chain reaction. Many different pairs of smaller atomic fragments are produced during the fission of U-235.

tions. Even if the control mechanism in a nuclear power plant failed, a bomb-like nuclear explosion cannot take place in a nuclear power reactor because nuclear fuel only has 3% U-235, whereas bomb-grade material is 20% U-235.) Fission reactions in the reactor of a nuclear power plant can be started or stopped, increased or decreased, thus allowing the desired amount of heat energy to be produced.

How Electricity Is Produced from Nuclear Energy

A typical nuclear power plant has four main parts: the reactor core, the steam generator, the turbine, and the condenser (Figure 11.5). Fission occurs in the **reactor core**, and the heat produced by nuclear fission is used to produce steam from liquid water in the **steam generator**. The **turbine** uses the steam to generate electricity, and the **condenser** cools the steam, converting it back to a liquid.

The reactor core contains the fuel assemblies. Above each fuel assembly is a **control rod** made of a special metal alloy that is capable of absorbing neutrons. The plant operator signals the control rod to move either up out of or down into the fuel assembly. If the control rod is out of the fuel assembly, free neutrons collide with the

fuel rods, and fission of uranium takes place. If the control rod is completely lowered into the fuel assembly, it absorbs the free neutrons, and fission of uranium no longer occurs. By exactly controlling the placement of the control rods, the plant operator can produce the exact amount of fission required.

A typical nuclear power plant has three water circuits. The **primary water circuit** (orange circuit in Figure 11.5) heats water, using the energy produced by the fission reaction. This circuit is a closed system that circulates water under high pressure through the reactor core, where it is heated to about 293°C (560°F). Because it is under such high pressure, this superheated water cannot expand to become steam and so remains in a liquid state.

From the reactor core, the very hot water circulates to the steam generator, where it boils water held in a **secondary water circuit** (blue circuit in Figure 11.5), converting the water to steam. The pressurized steam goes to and turns the turbine, which in turn spins a generator to produce electricity. After it has turned the turbine, the steam in the secondary water circuit goes to a condenser, where it is converted to a liquid again. The cooling is necessary to obtain the pressure differential that helps turn the turbine blades.

A **tertiary water circuit** (green circuit in Figure 11.5) provides cool water to the condenser, which cools

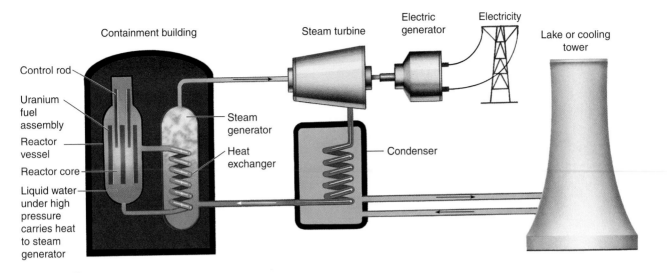

Figure 11.5 Pressurized water reactor. Fission of uranium-235 that occurs in the reactor vessel produces heat, which is used to produce steam in the steam generator. The steam drives a turbine to generate electricity. The steam then leaves the turbine and is pumped through a condenser before returning to the steam generator. Pumping hot water from the condenser to a lake or massive cooling tower controls excess heat. After it is cooled, the water is pumped back to the condenser. Approximately two thirds of all nuclear power plants in the United States are of this type.

the spent steam in the secondary water circuit. As the water in the tertiary water circuit is heated, it moves from the condenser to a lake or **cooling tower**, where it is cooled before circulating back to the condenser.

Safety Features of Nuclear Power Plants

The reactor core where fission occurs is surrounded by a huge steel potlike structure called a **reactor vessel**—a safety feature designed to prevent the accidental release of radiation into the environment. The reactor vessel and steam generator are placed in a **containment building**, an additional line of defense against accidental radiation leaks. Containment buildings have steel-reinforced concrete walls 0.9 to 1.5 m (3 to 5 ft) thick and are built to withstand severe earthquakes, the high winds of hurricanes and tornadoes, and even planes crashing into them.

BREEDER NUCLEAR FISSION

Uranium ore is mostly U-238, which is not fissionable and is therefore a waste product of conventional nuclear fission. In **breeder nuclear fission**, however, U-238 is converted to plutonium, Pu-239, a human-made isotope that is fissionable. Breeder reactors can use U-235, Pu-239, or thorium (Th-232) as fuel. Some of the neutrons that are emitted in breeder nuclear fission are used to produce additional plutonium from U-238 (Figure 11.6). A breeder reactor thus makes more fissionable fuel than it uses.

Because breeder fission can use U-238, it has the potential to generate much larger quantities of energy

from uranium ore than traditional nuclear fission can. If the U-238 stored in radioactive waste sites across the United States could be **reprocessed** and used in breeder reactors, it would supply the entire country with electricity for at least 100 years! When one adds the uranium reserves in the ground to these radioactive waste stockpiles, breeder fission has the potential to supply the entire country with electrical energy for several centuries.

Although breeder fission sounds promising, many safety problems have yet to be resolved. For reasons too complex to consider here, breeder fission reactors use liquid sodium rather than water as a coolant. Sodium is a highly reactive metal that reacts explosively with water and burns spontaneously in air at the high temperatures maintained in the breeder reactor. Thus, a leak in a breeder reactor's plumbing system is very dangerous. Should a leak be large enough to cause the loss of much of the liquid sodium coolant, the temperature within the reactor might get high enough to cause an uncontrolled nuclear fission reaction—that is, a small nuclear bomb-like explosion. The force of this explosion would almost certainly rip open the containment building, releasing radioactive materials into the atmosphere.

Public and governmental distrust of breeder reactors is even greater than their misgivings about conventional fission, in part because plutonium is used not only in breeder nuclear fission but also in nuclear weapons (discussed later in this chapter). The plutonium used for India's first nuclear test bomb in 1974 came from its breeder-reactor development program.

The United States performed the first breeder reactor experiments but abandoned its breeder reactor devel-

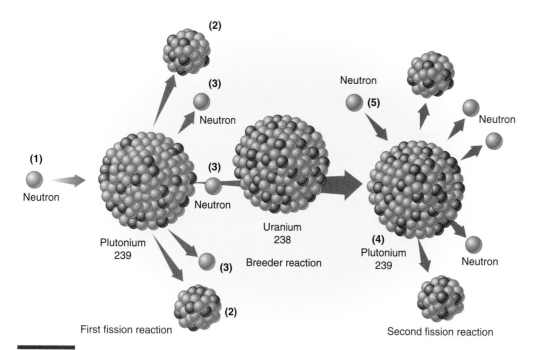

Figure 11.6 Breeder nuclear fission. A neutron (1) from a previous fission reaction bombards plutonium-239 (Pu-239), causing it to split into smaller radioactive fission fragments (2). Additional neutrons (3) are ejected in the process, some of which may collide with uranium-238 to form Pu-239. (4). Neutron bombardment (5) of this Pu-239 molecule causes it to split, and the process continues.

opment program in 1977, during the administration of President Carter. As a result of public opposition and economic factors (uranium turned out to be more plentiful and therefore cheaper than originally thought), other countries have slowly followed suit. Germany built a breeder reactor, completed in 1991, and then abandoned it. England's breeder reactor, which was built in 1974, was closed in 1994. France's commercial-scale breeder reactor, Superphénix, was built in 1985 but was plagued with safety problems and operated for only a few months between 1986 and 1996. In 1997 the French government announced that based on economic reasons, it was closing the Superphénix permanently. Similarly, a series of technical problems and accidents forced the closure of Japan's Monju breeder reactor in 1995. As of 2002, only four breeder reactors remain active around the world, and two of those are scheduled for closure in the next few years.

IS NUCLEAR ENERGY A CLEANER ALTERNATIVE THAN COAL?

One of the reasons proponents of nuclear energy argue for the widespread adoption of nuclear energy is that it has less of an environmental impact than fossil fuels, particularly coal (Table 11.2). They point out that the combustion of coal to generate electricity is responsible for

more than one third of the air pollution in the United States. As discussed in Chapter 10, coal is an extremely dirty fuel, especially because we have used up most of our reserves of cleaner-burning coal. Today most coal-burning power plants burn soft coal that produces sulfur-containing emissions that interact with moisture in the atmosphere to form acid precipitation. In addition, the combustion of coal releases carbon dioxide, a greenhouse gas that traps solar heat in our atmosphere and may be causing global warming.

In comparison, nuclear energy emits very few pollutants into the atmosphere. It does, however, generate radioactive waste, such as **spent fuel**, the fuel elements that were irradiated in a nuclear reactor. Nuclear power plants also produce other radioactive wastes, such as radioactive coolant fluids and gases in the reactor. As we will discuss shortly, spent fuel and other radioactive wastes are highly radioactive and therefore very dangerous. The extreme health and environmental hazards caused by this waste require that special measures be taken in its storage and disposal.

Opponents of nuclear energy contend that the fact that coal is a dirty fuel is not so much an argument in favor of nuclear energy as it is an argument in favor of a cleaner alternative to coal. They point out that pollution control devices could significantly lessen the air pollution produced by coal-burning power plants. As provisions of the Clean Air Act Amendments of 1990 have gone into

| Table 11.2 | Comparison of Environmental Impacts of 1,000-MWe Coal and Nuclear Power Plants* | |

Impact	*Coal*	*Nuclear (Conventional Fission)*
Land use	17,000 acres	1,900 acres
Daily fuel requirement	9,000 tons/day	3 kg/day
Availability of fuel, based on present economics	A few hundred years	100 years, maybe longer (much longer with breeder fission)
Air pollution	Moderate to severe, depending on pollution controls	Low
Climate change risk (carbon dioxide emissions)	Severe	No risk
Radioactive emissions, routine	1 curie	28,000 curies
Water pollution	Often severe at mines	Potentially severe at nuclear waste disposal sites
Risk from catastrophic accidents	Short-term local risk	Long-term risk over large areas
Link to nuclear weapons	No	Yes
Annual occupational deaths	0.5 to 5	0.1 to 1

* Impacts include extraction, processing, transportation, and conversion. Assumes coal is strip-mined. (A 1,000-MWe utility, at a 60% load factor, produces enough electricity for a city of 1 million people.)

effect, more and more coal-fired power plants have installed pollution control equipment or adopted more efficient designs such as fluidized-bed combustion (see Chapter 10).

The replacement of coal-burning power plants with nuclear power plants does not significantly lessen the threat of global warming, because only 15% of greenhouse gases come from power plants in the first place. Most greenhouse gases are produced by automobile emissions and industrial processes, both of which are unaffected by nuclear power. Also, the uranium-mining through uranium-enrichment steps in the nuclear fuel cycle require the combustion of fossil fuels, meaning that nuclear energy indirectly contributes a small amount to the greenhouse effect.

■ IS ELECTRICITY PRODUCED BY NUCLEAR ENERGY CHEAP?

In 1999, 104 nuclear power plants supplied about 23% of U.S. electricity. Beginning in the late-1950s, the price of electricity produced by nuclear energy was more expensive than conventionally produced electricity. However, U.S. government subsidies, such as funding for research and development and for public management of the production and disposal of nuclear materials, made up the difference. Many energy analysts predicted in the 1960s and 1970s that the cost of nuclear power would decrease over time and would be cheaper than fossil fuels by the year 2000. Generally speaking, this did not happen, and most analysts today think natural gas will be the cheapest fuel in coming decades.

Proponents of nuclear energy usually point to France, which obtains 77% of its electricity from nuclear energy, because electricity generated by French nuclear

power plants is 27% less expensive than that generated from coal. The lower cost of nuclear power can be explained in part because France uses the same design for all its nuclear power plants. Nuclear energy proponents believe there is no reason why the United States could not achieve the same economic efficiency. Opponents of nuclear power dismiss the example of France because the French government, like the governments of other countries that use nuclear power, heavily subsidizes the nuclear industry. Among other things, the French government funds research and development, radioactive waste disposal, and insurance coverage against accidents in its nuclear power plants.

The true costs of nuclear energy, like those of coal and other fossil fuels, are not always obvious in utility bills, whether in France or in the United States. You may recall from Chapter 10 that Norman Myers estimates in *Perverse Subsidies: Tax Dollars Undercutting Our Economies and Environments Alike* that U.S. tax dollars provide about $9.6 billion in nuclear energy subsidies annually. In an increasingly competitive power market, the generation of electricity using nuclear energy is very expensive when all costs, including those that are subsidized by the government, are taken into account.

The Cost of Building a Nuclear Power Plant

In the United States, no nuclear power plants have been ordered since 1976. One of the reasons electrical utilities will not commit to building new nuclear power plants is their high cost, which translates into higher prices paid by utility customers for their electricity. Traditionally designed nuclear power plants are large and take years to plan and build. In comparison, electric plants powered by fossil fuels, especially natural gas, can be much smaller and therefore easier and less expensive to build. Also, the regulatory process by which permits are obtained to build a nuclear power plant is cumbersome and adds to the expense.

Although the initial cost estimates for building a nuclear power plant are high, the actual costs are usually much higher than the forecasts. Cost overruns, which are borne by the utility and its customers, occur partly because the slow permitting process makes construction fall far behind schedule. Consider the Seabrook nuclear reactor in Seabrook, New Hampshire. Seabrook obtained its operating license from the Nuclear Regulatory Commission (NRC) in 1990, after numerous delays had put its construction 11 years behind schedule. The plant cost more than $6 billion, 12 times the original estimate.

Fixing Technical and Safety Problems in Existing Plants

Before the late 1990s, the price of electricity was averaged so that the higher cost of nuclear power–produced electricity was partly paid for by conventionally produced electricity. During the late 1990s, some state governments deregulated the U.S. electricity market, a move that opened electricity to competition by giving businesses and residential customers choices about where to buy their electricity. In addition, as a result of deregulation, power companies began to purchase power from the least expensive source and charge consumers based on that price. (Electricity deregulation is discussed further in Chapter 12.)

In the deregulated market, the low operating costs of some existing nuclear power plants continued to make them attractive sources of electric power. However, other nuclear power plants, particularly older ones, had much higher fuel and labor costs, and price competition led to the closing of certain nuclear power plants that produced expensive electricity. In 1998 Northeast Utilities shut down a nuclear power plant in Waterford, Connecticut, and Commonwealth Edison shut down two plants in Zion, Illinois. These three plants had been off-line—that is, closed temporarily—for several years prior to 1998 while a series of technical and safety problems were addressed. (As nuclear power plants age, such problems become more complex and expensive.) In all three cases, the utility companies decided that in the early climate of deregulation, it was cheaper to close the plants permanently than to fix the problems.

In the early 2000s, amid widely publicized electricity shortages in several states, the market price of electricity soared, and nuclear power became more attractive economically. Power companies began to think about increasing the amount of electricity they generate by nuclear power. Several power companies asked the NRC if they could increase electricity production at nuclear power plants currently in operation. Also, the NRC began to study the feasibility of reopening previously closed nuclear power plants. Currently, fixing many technical and safety problems in existing plants is economically possible.

CAN NUCLEAR ENERGY DECREASE OUR RELIANCE ON FOREIGN OIL?

International crises, such as the oil embargo of the early 1970s and the Persian Gulf crisis in the early 1990s, occasionally threaten the supply of oil to the United States. Some supporters of nuclear energy point out that our dependence on foreign oil would be lessened if all oil-burning power plants were converted to nuclear plants. This claim is not as convincing as it seems, however, because oil is responsible for generating only about 3% of the electricity in the United States; we rely on oil primarily for automobiles and for heating buildings. Thus, the replacement of electricity generated by oil with electricity generated by nuclear energy would do little in the short term to lessen our dependence on foreign oil, because we would still need oil for heating buildings and driving automotive vehicles.

However, technological advances could change nuclear power's potential contribution in the future. If electric heat pumps and electric motor vehicles become more common in the future, nuclear power plants have the potential to heat buildings and power automobiles and thus could decrease our reliance on foreign oil.

SAFETY IN NUCLEAR POWER PLANTS

According to the International Atomic Energy Agency (IAEA), 438 conventional nuclear power plants are in operation worldwide, and accidents have occurred in some of them. Although conventional nuclear power plants cannot explode like atomic bombs, accidents can happen in which dangerous levels of radiation might be released into the environment and result in human casualties. At high temperatures the metal encasing the uranium fuel melts, releasing radiation; this is called a **meltdown**. Also, the water that is used in a nuclear reactor to transfer heat can boil away during an accident, contaminating the atmosphere with radioactivity.

The probability that a major accident will occur is considered low by the nuclear industry, but public perception of the risk is high because it is an involuntary risk, unlike smoking or driving in an automobile (see Chapter 2 discussion of risk). Also, the consequences of such an accident are drastic and life threatening, both immediately and long after the accident has occurred. We now consider two major accidents, one in the United States (Three Mile Island) and the other at Chornobyl in Ukraine.

CASE·IN·POINT Three Mile Island

The most serious nuclear reactor accident in the United States occurred in 1979 at the Three Mile Island power plant in Pennsylvania, the result of human error after the cooling system failed. A 50% meltdown of the reactor core took place. Had there been a complete meltdown of the fuel assembly, dangerous radioactivity would have been emitted into the surrounding countryside. Fortunately, the containment building kept almost all the radioactivity released by the core material from escaping. Although a small amount of radiation entered the environment, there were no substantial environmental damages and no immediate human casualties. A study conducted within a 10-mile radius around the plant 10 years after the accident concluded that cancer rates were in the normal range and that there was no association between cancer rates and radiation emissions from the accident. Numerous other studies have failed to link abnormal health problems (other than increased stress) to the accident.

The seriousness of the situation at Three Mile Island elevated public apprehension about nuclear power. It took 12 years and $1 billion for Three Mile Island to be repaired and reopened. Although the reactor that was involved in the accident at Three Mile Island was destroyed, a second reactor that was undamaged during the accident is currently in operation. In the aftermath of the accident, public wariness prompted construction delays and cancellations of several new nuclear power plants across the United States.

On the positive side, the accident at Three Mile Island reduced the complacency that had been commonplace in the nuclear industry. New safety regulations were put in place, including more frequent safety inspections, new risk assessments, and improved emergency and evacuation plans for nuclear power plants and surrounding communities. The Institute of Nuclear Power Operations was created to promote safety and improve the training of nuclear operators. ■

CASE·IN·POINT Chornobyl

The worst accident ever to occur at a nuclear power plant took place at the Chornobyl plant, located in the former Soviet Union in what is now Ukraine, on April 26, 1986. One or possibly two explosions ripped a nuclear reactor apart and expelled large quantities of radioactive material into the atmosphere (Figure 11.7). The effects of this accident were not confined to the area immediately surrounding the power plant: significant amounts of radioisotopes quickly spread across large portions of Europe. The Chornobyl accident affected and will continue to affect many nations.

The first task faced after the accident was to contain the fire that had broken out after the explosion and pre-

(a)

(b)

Figure 11.7 Chornobyl. (a) The arrow indicates the site of the explosion. The upper part of the reactor was completely destroyed. (b) Since the accident, the damaged area has been completely entombed in a concrete-and-metal sarcophagus.

vent it from spreading to other reactors at the power plant. Local firefighters, many of whom later died from exposure to the high levels of radiation, battled courageously to contain the fire. In addition, 116,000 people who lived within a 30-km (18.5-mi) radius around the plant had to be quickly evacuated and resettled. Ultimately, more than 170,000 people had to permanently abandon their homes.

Once the danger from the explosion and fire had passed, the radioactivity at the power station had to be cleaned up and contained so that it would not spread. Dressed in protective clothing, workers were transported to the site in radiation-proof vehicles, and initially the radioactivity was so high that they could stay in the area for only a few minutes at a time. There are few photographs of the cleanup because camera film was quickly ruined by the radiation. After the initial cleanup, the damaged reactor building was encased in 300,000 tons of concrete. Then the surrounding countryside had to be decontaminated: Highly radioactive soil was removed, and buildings and roads were scrubbed down to remove the radioactive dust that had settled on them.

Although cleanup in the immediate vicinity of Chornobyl is finished, the people in Ukraine face many long-term problems. Much of the farmland and forests are so contaminated that they cannot be used for more than a century. Loss of agricultural production is one of the largest costs for the local economy. Inhabitants in many areas of Ukraine still cannot drink the water or consume locally produced milk, meat, fish, fruits, or vegetables.

Mothers do not nurse their babies because their milk is contaminated by radioactivity. The Ukrainian Scientific Center for Radiation Medicine has continually monitored the health of approximately 80,000 Chornobyl patients.

In the investigation that ensued after the accident, it became apparent that there had been two fundamental causes. First, the design of the nuclear reactor was flawed—the reactor was not housed in a containment building and was extremely unstable at low power. This type of reactor, called an RBMK reactor, is not used commercially in either North America or Western Europe because nuclear engineers consider it too unsafe. Russia, Ukraine, and several adjacent countries still have 15 RBMK reactors in operation, although safety improvements have been implemented since the Chornobyl explosion. The International Chornobyl Conference, which convened in Vienna in 1996, concluded that additional safety improvements are needed, however.

Second, human error contributed greatly to the disaster. Many of the Chornobyl plant operators lacked scientific or technical understanding of the plant they were operating, and they made several major mistakes in response to the initial problem. As a result of the disaster at Chornobyl, the Soviet government developed a retraining program for operators at all the nuclear power plants in the country. In addition, safety features were added to existing reactors.

One of the disquieting consequences of Chornobyl was the lack of predictability of the course taken by spreading radiation (Figure 11.8). Chornobyl's radiation

Figure 11.8 Radioactive fallout from Chornobyl. Parts of Ukraine, Belarus, Sweden, Norway, France, Italy, and Switzerland received heavy radioactive fallout from the accident.

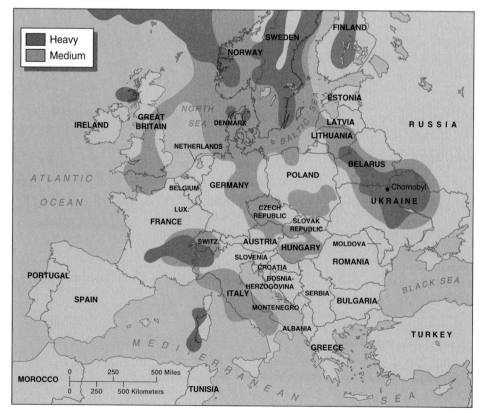

cloud unevenly dumped radioactive fallout over some areas of Europe and Asia, leaving other areas relatively untouched and making it difficult to plan emergency responses for a possible future nuclear accident.

As the health effects of the accident at Chornobyl are monitored over the years, the death toll is rising. In 1999, the Ukrainian health minister said the death toll to date was nearly 170,000. Western scientists regard such assessments with caution, in part because an increase in intensive screening for diseases that occurred after the accident has also uncovered previously undiagnosed diseases that were unrelated to the accident.

Nearly 400,000 adults and more than 1 million children currently receive government aid for health problems related to Chornobyl. In addition, many people in nearby countries received dangerous doses of radiation. An increase in the frequency of birth defects and mental retardation in newborns has been documented in Belarus, Ukraine, and certain other parts of Europe, along with an increased incidence of leukemia in infants and of thyroid cancer and immune abnormalities in children exposed to the Chornobyl fallout (Figure 11.9). Breast cancer, stomach cancer, and other organ cancers are not expected to appear in large numbers until 20 years or more after the accident. Psychological injuries to those living under the cloud of Chornobyl are also being assessed because many of these individuals are suffering from debilitating mental stress, anxiety, and depression.

Since 1992, the Ukrainian government has spent more than $4 billion for Chornobyl-related projects such as compensation for the victims. Ukraine has shut down all of the reactors at Chornobyl, and it is developing alternative energy sources, including two nuclear reactors that are currently under construction. In Ukraine and nearby countries that were formerly part of the Soviet Union, about 57 reactors do not fully comply with Western safety standards, although many have incorporated significant safety features in recent years.

Recent inspections of Chornobyl have revealed numerous potential safety hazards. The most serious threat is posed by cracks in the concrete sarcophagus encasing the radioactive fuel in the destroyed reactor. Because radioactive dust leaking from the cracks could contaminate both air and groundwater, Ukraine and international donor countries are currently spending $750 million on repairs to stabilize the sarcophagus. They also plan to erect an enormous structure to completely cover the sarcophagus.

The world has learned much from this nuclear disaster. Most countries now take nuclear power more seriously, hoping to prevent accidents. Safety features that are commonplace in North American and Western European reactors are being incorporated into new nuclear power plants around the world. In addition, nuclear engineers learned a great deal from the cleanup and entombment of Chornobyl; this knowledge will be useful when

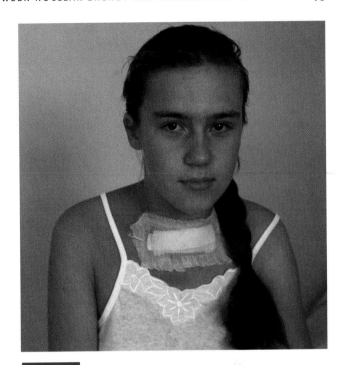

Figure 11.9 Health consequence of Chornobyl. Shown is a 14-year-old who is recovering from thyroid cancer. A significant increase in thyroid cancer in children and adolescents occurred within a few years after the accident. Children are more sensitive to thyroid cancer than adults because the thyroid gland is very active during childhood. Fortunately, if the thyroid gland is removed, thyroid cancer has a very high survival rate. Patients must take thyroid-hormone replacement therapy for the rest of their lives.

old nuclear power plants are dismantled. Doctors learned more about effective treatment of people who have been exposed to massive doses of radiation. In the years to come, health researchers will learn more about the relationship between cancer and radiation as they follow the health of the thousands who were exposed to radiation from Chornobyl.

THE LINK BETWEEN NUCLEAR ENERGY AND NUCLEAR WEAPONS

Fission is involved in both the production of electricity by nuclear energy and the destructive power of nuclear weapons. Uranium-235 and plutonium-239 are the two fuels commonly used in atomic fission weapons. As you know, plutonium is produced in breeder reactors. It is also possible to reprocess spent fuel from conventional fission reactors to make plutonium for breeder reactors or nuclear weapons.

Many countries are using or contemplating using nuclear energy to generate electricity. The possession of nuclear power plants gives these countries access to the fuel needed for nuclear weapons (by reprocessing spent fuel). Responsible world leaders are concerned about the consequences of terrorist groups and states of concern building nuclear weapons. These concerns have caused many people to shun nuclear energy, particularly breeder fission, and to seek alternatives that are not so intimately connected with nuclear weapons.

In the mid-1990s there were about 270 metric tons (298 tons) of weapons-grade plutonium worldwide, much of it surplus from scrapped nuclear weapons. An additional 930 metric tons (1026 tons) of plutonium wastes existed from commercial reactors; this amount is expected to more than double by 2010. Storing plutonium is a security nightmare because it only takes several kilograms to make a nuclear bomb as powerful as the ones that destroyed the Japanese cities of Nagasaki and Hiroshima in World War II.

Of special concern to international security is Russia's political instability and economic turmoil, which increase the chance that terrorist groups could steal plutonium and enriched uranium and use them to make nuclear weapons. During the late 1990s, guards and other nuclear personnel often did not receive paychecks for months, making them vulnerable to bribes. According to the EIA, the former Soviet Union has enough highly enriched uranium and plutonium stored at about 40 different sites to make 40,000 nuclear bombs! The U.S. Department of Energy has entered into a program with Russia's Atomic Energy Ministry to help Russia maintain nuclear security. The program includes training for Russian guards and the installation of security equipment to monitor bomb materials.

The United States plans to get rid of more than 50 tons of surplus plutonium from the dismantling of its nuclear warheads. After reviewing dozens of possible solutions, the Clinton administration decided to pursue a $2 billion program that includes two options. One, plutonium will be combined with highly radioactive waste and then vitrified into glass logs (discussed shortly). Two, some plutonium will be converted to a mixed oxide, or MOX, and then "burned" as a fuel in commercial nuclear power reactors. The MOX fuel will not be burned until about 2007 because a special facility must first be built to process the plutonium into MOX. Both options would make the plutonium so radioactive that it could not be safely handled to construct a bomb. The National Academy of Sciences and many nuclear scientists and arms-control experts have endorsed the two-option Clinton plan, although all agree that there is no perfect solution to such a complex problem. The George W. Bush administration has announced it supports option two as the only approach for plutonium disposition; it plans to suspend all work on vitrification.

Additionally, the United States wants to ensure that Russia dismantles its nuclear weapons and destroys the surplus plutonium. To meet that goal, the United States has agreed "in principle" with Russia to buy about 50 metric tons of its weapons-grade uranium, convert it to MOX, and burn it in commercial nuclear power reactors in the United States.

RADIOACTIVE WASTES

Radioactive wastes are classified as either low-level or high-level. **Low-level radioactive wastes** are radioactive solids, liquids, or gases that give off small amounts of ionizing radiation. Produced by nuclear power plants, university research labs, nuclear medicine departments in hospitals, and industries, low-level radioactive wastes include glassware, tools, paper, clothing, and other items that have been contaminated by radioactivity. The **Low-Level Radioactive Policy Act**, passed in 1980, specified that all states are responsible for the waste they generate, and it encouraged states to develop facilities to handle low-level radioactive wastes by 1996. There are three sites currently accepting waste for the entire country—in Washington state, South Carolina, and Utah. New technologies to compact low-level radioactive waste have dramatically reduced the volume being disposed of in these nuclear dumps.

High-level radioactive wastes are radioactive solids, liquids, or gases that initially give off large amounts of ionizing radiation. Examples of high-level radioactive wastes produced during nuclear fission include the reactor metals (fuel rods and assemblies), coolant fluids, and air or other gases found in the reactor. High-level radioactive wastes are also generated during the reprocessing of spent fuel. Produced by nuclear power plants and nuclear weapons facilities, high-level radioactive wastes are among the most dangerous hazardous wastes created by humans.

Fuel rods, which absorb neutrons, thereby forming radioisotopes, can be used for only about 3 years, after which they become highly radioactive spent fuel. In 1999 the United States had more than 37,000 metric tons of spent fuel stored temporarily at more than 100 nuclear power plants around the country (Table 11.3). As the radioisotopes in spent fuel decay, they produce considerable heat, are extremely toxic to organisms, and remain radioactive for thousands of years. Their dangerous level of radioactivity requires that they be handled in special ways. Secure storage of these materials must be guaranteed for thousands of years, until they can decay sufficiently to be safe.

Clearly, the safe disposal of radioactive wastes is one of the main difficulties that must be overcome if nuclear energy is to realize its potential in the 21st century. Many people question whether we can safely guarantee the storage of wastes that must be isolated from organisms for millennia. High-level radioactive wastes must be stored in an isolated area where there is minimal possibility they can contaminate the environment. The storage site must also have geological stability and little or no

| Table 11.3 | Nuclear Power Plants and Waste in Selected Countries | |

Country	Number of Reactors, 2002	Spent Fuel Inventories, 1998*
Argentina	2	2,000
Canada	14	24,500
France	59	14,800
Japan	54	6,100
South Africa	2	400
South Korea	16	3,800
Sweden	11	3,800
United Kingdom	33	7,800
United States	104	37,667

* Metric tons of radioactive waste.

water flowing nearby, which might transport the waste away from its original site.

What are the best sites for the long-term storage of high-level radioactive wastes? Many scientists recommend storing the wastes in stable rock formations deep in the ground. Another suggestion for the long-term storage of radioactive wastes is aboveground mausoleums, which would be built in remote locations. If we built mausoleums, however, we would not be able to simply store the wastes and forget about them. Mausoleums would have to have adequate security to guarantee their safety. Other long-term possibilities that have been considered include storage in Antarctic ice sheets and burial in the seabed (beneath the ocean floor). Because ocean disposal has the potential to harm the marine environment, international agreements currently prohibit ocean disposal of both high-level and low-level radioactive wastes.

Most experts today support underground geological disposal in rock formations. The selection of these sites is complicated by people's reluctance to have radioactive wastes stored near their homes.

Meanwhile, radioactive wastes continue to accumulate. In the United States, there are 72 "temporary" sites at which domestic radioactive wastes have been stored for decades. Most commercially operated nuclear power plants store their spent fuel in huge indoor pools of water on-site. None of these plants was designed for long-term storage of spent fuel. Because they have nowhere to send their spent fuel when on-site storage areas are full, these plants must either expand their on-site storage—an expensive proposition because of legal battles that usually occur before expansion is granted—or shut down. As of 1999, nine commercially operated plants had expanded on-site storage by building aboveground, air-cooled concrete and steel casks (Figure 11.10).

CASE-IN-POINT Yucca Mountain

In 1982 the passage of the **Nuclear Waste Policy Act** put the burden of developing permanent sites for civilian and military radioactive wastes on the federal govern-

ENVIROBRIEF

Human Nature and Nuclear Energy

The NIMBY response lessens the promise of nuclear energy. *NIMBY* stands for "not in my backyard." As soon as people hear that a nuclear power plant or a radioactive waste disposal site may be situated nearby, the NIMBY response rears its head. Part of the reason NIMBYism is so prevalent in nuclear energy issues is that, despite the assurances given by experts that a site will be safe, no one can guarantee *complete* safety, with no possibility of an accident. (Recall from the discussion of risk assessment in Chapter 2 that *nothing* is risk-free.)

A sister response to NIMBY is the NIMTOO response which stands for "not in my term of office." Politicians who wish to get reelected are sensitive to their constituents' concerns and are not likely to support the construction of a nuclear power plant or radioactive waste disposal site in their districts.

Given human nature, the NIMBY and NIMTOO responses are not surprising. But emotional reactions, however reasonable they might be, do little to constructively solve complex problems. Consider the disposal of radioactive waste. There is universal agreement that we need to safely isolate radioactive waste until it can decay enough to cause little danger. But NIMBY and NIMTOO, with their associated demonstrations, lawsuits, and administrative hearings, prevent us from effectively dealing with radioactive waste disposal. Every potential disposal site is near someone's home, in some politician's state.

Most people agree that our generation has the responsibility to dispose of radioactive waste generated by the nuclear power plants we have already built. Only we want to put it in someone else's state, in someone else's backyard. Arguing against any disposal scheme that is proposed will simply result in letting the waste remain where it is now ... at existing nuclear power plants. Although this may be the only *politically* acceptable solution, it is unacceptable from an environmental viewpoint.

Figure 11.10 Storage casks for spent fuel at the Prairie Island nuclear power plant in Minnesota. Each cask temporarily holds 17.6 tons of spent fuel assemblies.

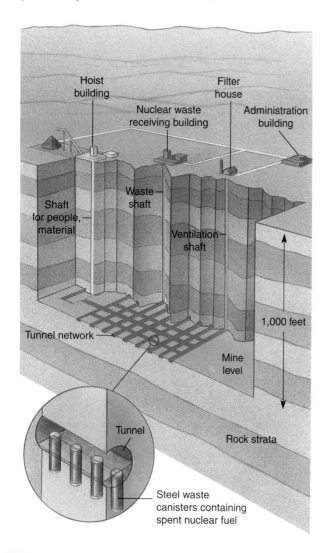

Hoist
building

Filter
house

Nuclear waste
receiving building

Administration
building

Shaft
for people,
material

Waste
shaft

Ventilation
shaft

1,000 feet

Tunnel network

Mine
level

Tunnel

Rock strata

Steel waste
canisters containing
spent nuclear fuel

Figure 11.11 Yucca Mountain. This radioactive waste facility will be a huge complex of interconnected tunnels located in dense volcanic rock 305 m (1,000 ft) beneath the mountain. Canisters containing radioactive waste will be stored in the tunnels.

ment and required the first site to be operational by 1998. Since then, the deadline for completion of an operational high-level radioactive waste repository has been postponed from 1998 to 2010 at the earliest. In a 1987 amendment to this bill, Congress identified Yucca Mountain in Nevada as the only candidate for a permanent underground storage site for 77,000 tons of high-level radioactive wastes from commercially operated power plants (Figure 11.11). (Beginning in 1999, low-level radioactive wastes from the manufacture of nuclear weapons are being stored permanently in deep underground salt beds near Carlsbad, New Mexico.)

Since 1983 the U.S. Department of Energy has spent billions of dollars conducting feasibility studies on Yucca Mountain's geology. Results indicate that the site is safe, at least from volcanic eruptions and earthquakes. How-

ever, study of the suitability of Yucca Mountain has been mired in scientific, management, cost, public opposition, and scheduling controversies, and Nevadans have opposed the selection of their state for a radioactive waste site. In 2002 Congress finally approved the choice of Yucca Mountain as the U.S. nuclear-waste repository, despite the state of Nevada's opposition.

The Yucca Mountain site, some 145 km (90 mi) northwest of Las Vegas, is controversial in part because it is near a volcano (its last eruption may have been as recent as 20,000 years ago) and active earthquake fault lines. The possibility of a volcanic eruption at Yucca Mountain is currently considered very remote (1 chance in 10,000 during the next 10,000 years). Concerns that earthquakes might possibly disturb the site and raise the water table, which could result in radioactive contamination of air and groundwater, were examined when a magnitude 5.6 earthquake occurred in 1992 about 20 km (about 12 mi) from Yucca Mountain. Scientists were already monitoring water table elevation, and they measured a 1-meter change caused by the earthquake. Because the water table is some 800 m (2,625 ft) beneath the mountain crest, however, water elevation changes due to earthquake activity are not considered a serious problem.

Transporting high-level wastes from nuclear reactors and weapons complex sites by truck, rail, or air is a major concern of opponents of the Yucca Mountain site (Figure 11.12). The typical shipment would travel an average of 2,300 miles and 43 states would have these dangerous materials passing through on their way to Yucca Mountain. Eight states—Illinois, Indiana, Iowa, Kansas, Missouri, Nebraska, Utah, and Wyoming—would contain major transportation corridors to Yucca Mountain. How-

Figure 11.12 Transport of nuclear waste. Shipments to Yucca Mountain, Nevada, will travel an average of 2,300 miles.

ever, proponents of Yucca Mountain point out that leaving nuclear waste at multiple sites spread across the United States poses a greater risk of theft and, possibly, human health problems.

High-Level Radioactive Liquid Waste

High-level liquid wastes are dangerously unstable and very hard to monitor. For that reason, they must be converted to solid form before they can be stored at Yucca Mountain. The U.S. government is planning to store high-level liquid wastes in enormous glass or ceramic logs 10 ft high and 2 ft in diameter that are themselves contained in stainless steel canisters. The glass or ceramic contains boron, which absorbs neutrons and so will prevent the logs from exploding. Solidifying liquid waste into solid glass or ceramic logs, known as **vitrification**, is an established practice in Germany, where the logs are stored underground in a salt mine, and France, where they are stored in a surface facility. The United States began producing glass logs in South Carolina (the Savannah River site) and New York (the West Valley site) in 1996, although it does not yet have a permanent disposal site for these hazardous logs.

Radioactive Wastes with Relatively Short Half-Lives

Some radioactive wastes are produced directly from the fission reaction. Uranium-235, the reactor fuel, may split in several different ways, forming smaller atoms, many of which are radioactive. Most of these, including krypton-85 (half-life, 10.4 years), strontium-90 (half-life, 28 years), and cesium-137 (half-life, 30 years), have relatively short-term radioactivity. In 300 to 600 years they will have decayed to the point where they are safe.

The safe storage of fission products with relatively short half-lives is of concern because fission produces larger amounts of these materials than of the materials with extremely long half-lives. Also, health concerns exist because many of the shorter-lived fission products mimic essential nutrients and tend to concentrate in the body, where they continue to decay, with harmful effects (see discussion of bioaccumulation in Chapter 22). One of the common fission products, strontium-90, is chemically similar to calcium. If strontium-90 were to be accidentally released into the environment from radioactive waste that had not been stored properly, it could be incorporated into human and animal bones and teeth in place of calcium. In like manner, cesium-137 replaces potassium in the body and accumulates in muscle tissue, and iodine-131 concentrates in the thyroid gland.

Decommissioning Nuclear Power Plants

Nuclear power plants are licensed to operate for a maximum of 40 years, although some nuclear power plants

ENVIROBRIEF

A Nuclear Waste Nightmare

Over the past three decades, Soviet (and now Russian) practices for radioactive waste disposal have often violated international standards:

- Billions of gallons of liquid radioactive wastes have been pumped directly underground, without being stored in protective containers. Russian officials claim that layers of clay and shale at the sites, which are all near major rivers, prevent leakage, but they also admit that some wastes have already leaked more than expected.

- Highly radioactive wastes were dumped into the ocean in amounts double the combined total of dumped wastes from 12 other nuclear nations. The wastes disposed in the ocean include the nuclear reactors from 17 submarines and an icebreaker. Six of these reactors contained live fuel and are considered the most dangerous of the wastes.

- Both underground injection and underwater dumping of radioactive wastes continue because Russia lacks alternatives for nuclear waste processing. Potential health and environmental hazards associated with these wastes are unknown because so little data exist for these types of long-term storage.

with good operating records have asked the NRC to extend their operating licenses past the 40-year cutoff. Several of these reactors have been granted 20-year extensions.

As nuclear power plants age, certain critical sections, such as the reactor vessel, become brittle or corroded. At the end of their operational usefulness, however, nuclear power plants cannot simply be abandoned or demolished, because many parts have become contaminated with radioactivity.

Three options exist when a nuclear power plant is closed: storage, entombment, and decommissioning. If an old plant is put into **storage,** it is simply guarded by the utility company for 50 to 100 years, during which time some of the radioactive materials decay. This decrease in radioactivity makes it safer to dismantle the plant later, although accidental leaks during the storage period are still a concern.

Entombment, permanently encasing the entire power plant in concrete, is not considered a viable option by most experts because the tomb would have to remain intact for at least 1,000 years. It is likely that accidental leaks would occur during that time. Also, we cannot guarantee that future generations would inspect and maintain the "tomb."

The third option for the retirement of a nuclear power plant is to **decommission,** or dismantle, the plant immediately after it closes. The workers who dismantle the plant must wear protective clothing and masks. Some portions of the plant are too "hot" (radioactive) to be safely dismantled by workers, although advances in robotics may make it feasible to tear down these sections. As the plant is torn down, small sections of it are transported to a permanent storage site.

Several small nuclear power plants have been decommissioned. Shippingport, the nation's first commercial nuclear power plant, was dismantled in 1989 and transported by barge more than 12,872 km (8,000 mi) from its working site in Pennsylvania to Hanford Nuclear Reservation in Washington state (see "Case in Point: Hanford Nuclear Reservation" in Chapter 23). The decommissioning of a large nuclear power plant will not be possible, however, until there are permanent storage sites for all the radioactive pieces.

Decommissioning nuclear power plants is the responsible thing to do once a plant is no longer operable. There are risks, however, including dangers to workers during the decommissioning process and accidental discharges of radiation into the environment during either dismantling or transport of radioactive debris to a permanent site.

According to the IAEA, worldwide, 95 nuclear power plants were permanently retired as of 2002, and many nuclear power plants are nearing retirement age; in 2002 approximately 132 operational plants were 25 years old or older. During the 21st century, we may find that we are paying more in our utility bills to close old plants than we are to have new plants constructed. For example, decommissioning the Yankee Rowe, a small nuclear reactor in Massachusetts that was shut down in 1991, will cost some $370 million; the plant's construction costs in 1960 were $186 million (in today's dollars). As the utility industry is deregulated, difficult questions will arise about who pays for the retirement of these plants.

■ FUSION: NUCLEAR ENERGY FOR THE FUTURE?

The atomic reaction that powers the stars, including our sun, is fusion. In fusion, two lighter atomic nuclei are brought together under conditions of high heat and pressure in such a way that they combine, producing a larger nucleus. The energy produced by fusion is considerable; it makes the energy produced by the burning of fossil fuels seem trifling by comparison: 30 mL (1 oz) of fusion fuel has the energy equivalent of 266,000 L (70,000 gal) of gasoline.

Isotopes of hydrogen are the fuel for fusion. In one type of fusion reaction, the nuclei of deuterium and tritium combine to form helium (Figure 11.13), releasing huge amounts of heat energy in the process; in a fusion power plant, the heat would be converted to electricity. Deuterium, also called heavy hydrogen, is present in water and is relatively easy to separate from normal hydrogen. Tritium, which is radioactive, is not found in nature; this human-made hydrogen isotope can be

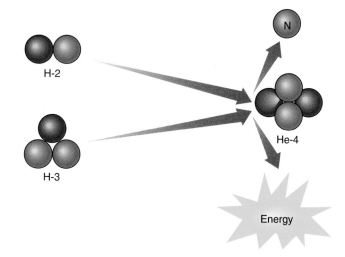

Figure 11.13　Fusion. In one possible fusion reaction, a deuterium nucleus fuses with a tritium nucleus, producing a helium-4 nucleus, a neutron, and considerable energy. Neutrons formed during the reaction would bombard lithium, splitting it into helium and tritium. Thus, additional tritium fuel would be generated.

formed during the fusion reaction by bombarding another element, lithium, with neutrons. The isotope of lithium that is required, Li-6, is found in seawater and in certain types of surface rocks.

Supporters of nuclear energy view fusion as the best possible form of energy, not only because its fuel, hydrogen, is available in virtually limitless supply, but also because fusion will produce no high-level radioactive waste. (Fusion will produce low-level waste, however.) Unfortunately, many technological difficulties have been encountered in efforts to stage a controlled fusion reaction. It takes phenomenally high temperatures (millions of degrees) to make atoms fuse. To date, the best fusion experiments generate about one third of the energy that must be supplied to heat the fuel.

Another challenge is confining the fuel. At extremely high temperatures, a gas separates into negative electrons and positive nuclei. This superheated, ionized gas, known as **plasma**, has a tendency to expand. Confinement of the plasma is necessary so that the nuclei are close enough to one another to fuse, but a regular container doesn't work because as soon as the nuclei hit the container walls, they lose so much energy that they cannot fuse. Several technological approaches to confine plasma are under investigation. One approach that has been studied for decades involves using magnetic confinement to circulate the plasma in a doughnut-shaped chamber called a *tokamak*. Research-sized tokamak reactors are relatively small and fit inside a large room.

Fusion research is currently being conducted in several countries, particularly in Europe, Japan, Russia, Canada, and the United States. Fusion research is very expensive. Since the 1960s, the United States has invested more than $10 billion in its fusion energy program. While progress has been made, there is no guarantee that fusion will ever be successful at producing commercial electric power.

THE FUTURE OF NUCLEAR POWER

In an effort to promote nuclear energy, nuclear and utility executives in the early 1990s developed a plan addressing the safety and economic issues associated with nuclear power. They envisioned building a series of "new generation" nuclear reactors designed to be 10 times safer than current reactors. They also gave serious consideration to the financial risks involved in building a nuclear power plant. According to their plan, costs could be held in line by standardizing nuclear power plants rather than custom-building each one. When France standardized its plant designs, the costs were lowered considerably. The plan also calls for building schedules to be improved and the regulatory process to be streamlined.

Several new designs have already received certification by the NRC. The new fission reactors are considered by nuclear experts to be much safer than the nuclear reactors currently operating in the United States. Some of the new reactors would not have a meltdown even in a worst-case scenario. Because the fuel cannot melt, radiation cannot be released into the environment during the plant's operation. Some of the designs use helium rather than water to turn the turbines and cool the system. Helium, a gas that is much less corrosive than steam, is less likely to cause leaks by corroding pipes. However, the new generation of nuclear power plants, although smaller, simpler in design, less expensive to build, and safer to operate, will still produce high-level radioactive wastes and have a potential link to nuclear weapons.

Although some countries in Asia are currently building nuclear power plants, the United States and most other countries are not. Most energy analysts do not think the United States will build a new generation of nuclear power plants before 2020, if ever. Many questions involving economics, safety, management of radioactive waste, and nuclear safeguards will have to be addressed in a satisfactory way before the public will support a new expansion of nuclear power. Although nuclear power has the potential to help us mitigate global climate change, the future of nuclear power in the 21st century is not yet known.

SUMMARY WITH SELECTED KEY TERMS

I. Nuclear energy harnesses reactions in the nucleus of an atom.

A. In **fission**, atomic nuclei are split apart, whereas in **fusion**, atomic nuclei are combined. A small amount of mass is converted to a large amount of energy, as first hypothesized by Albert Einstein in 1905.

B. Both fission and fusion result in a significant release of energy, in comparison to the chemical combustion of fossil fuels.

II. The processes involved in producing the fuel for use in nuclear reactors and in disposing of radioactive wastes are collectively called the **nuclear fuel cycle**.

A. Australia, Africa, and North America contain substantial deposits of uranium.

B. **Enrichment** is that part of the nuclear fuel cycle in which uranium ore is refined to increase the concentration of uranium-235 (U-235), the isotope of uranium that is used as fuel.

III. Fission is the process used in today's nuclear power plants to generate electricity.

A. In a nuclear power reaction, U-235 is bombarded with neutrons, which split the nucleus into two smaller atoms plus additional neutrons. These neutrons, in turn, collide with additional U-235 atoms. The fission of U-235 releases heat that converts water to steam, which is used to generate electricity.

B. A typical nuclear power plant contains a **reactor core**, where fission occurs; a **steam generator**; a **turbine**; and a **condenser**.

1. The reactor core contains about 250 **fuel assemblies**, each of which consists of 200 **fuel rods**. Fuel rods contain pellets of uranium dioxide.

2. Above each fuel assembly is a **control rod** that can move into or out of the fuel assembly, thereby producing the amount of fission required.

C. Safety features include a steel **reactor vessel** and a **containment building** built of steel-reinforced concrete.

IV. In **breeder nuclear fission**, both U-235 and plutonium-239 (Pu-239) are split to release energy.

A. The large quantity of U-238 in uranium fuel is converted into Pu-239; thus, breeder reactors make more fuel than they use.

B. Many safety problems have to be resolved if breeder fission is to be used.

V. Supporters say nuclear energy is better than alternatives because it is less polluting and more economical, and its fuel, uranium, is plentiful. Opponents of nuclear energy refute these arguments.

VI. Problems associated with nuclear power include questions about safety in nuclear power plants, particularly **meltdowns**, accidents where high temperatures cause the metal encasing the uranium fuel to melt.

A. Radioactive fallout from the 1986 accident at the Chornobyl power plant resulted in widespread environmental pollution as well as serious local contamination.

B. The Chornobyl accident stimulated a world reassessment of nuclear energy.

VII. The increase in global supplies of weapons-grade plutonium and plutonium wastes from commercial reactors threatens international security because it increases the chance that certain nations and terrorist groups could use them to make nuclear weapons.

VIII. In all countries that use nuclear power, permanent waste disposal sites are urgently needed to house spent fuel, which is highly radioactive, as well as radioactive parts of dismantled power plants.

A. **Low-level radioactive wastes** are radioactive solids, liquids, or gases that give off small amounts of ionizing radiation.

B. **High-level radioactive wastes** are radioactive solids, liquids, or gases that initially give off large amounts of ionizing radiation.

C. The United States has selected Yucca Mountain in Nevada for a permanent storage site for high-level radioactive wastes from commercially operated nuclear power plants.

D. Nuclear power plants that have reached the end of their operational usefulness are contaminated with radioactivity and must be put into **storage**, **entombed**, or **decommissioned**. A nuclear power plant that is decommissioned is dismantled, with sections transported to a permanent storage site.

IX. Commercial fusion as a source of energy is many years from becoming a reality, but fusion has the theoretical potential to produce unlimited energy with very little radioactive waste.

X. New fission reactor designs are safer, smaller, simpler in design, and less expensive to build than the reactors currently operating in the United States. However, like current reactors, the new generation of nuclear reactors would produce high-level radioactive waste.

THINKING ABOUT THE ENVIRONMENT

1. Explain the difference between chemical energy and nuclear energy.

2. What is the nuclear fuel cycle?

3. How does a pressurized water reactor produce electricity? What are its three water circuits?

4. Distinguish between conventional nuclear fission, breeder nuclear fission, and fusion.

5. Compare the environmental effects of coal combustion and conventional nuclear fission for the generation of electricity.

6. Breeder reactors produce more fuel than they consume. Does this mean that if we use breeder reactors, we will have a perpetual supply of plutonium for breeder fission? Why or why not?

7. Describe the nuclear power plant accidents at Three Mile Island in Pennsylvania and Chornobyl in Ukraine.

8. Can catastrophic accidents at nuclear power plants be prevented in the future? Why or why not?

9. What is low-level radioactive waste, and how is it disposed of in the nuclear fuel cycle? What is high-level radioactive waste, and how is it currently stored?

10. How does the disposal of radioactive wastes pose technical problems? Political problems?

11. What are some of the advantages of storing high-level radioactive waste at Yucca Mountain? What are some of the disadvantages?

12. Why is decommissioning nuclear power plants such a major task?

13. Are you in favor of the United States developing additional nuclear power plants to provide us with electricity in the 21st century? Why or why not?

*14. Uranium-235 (U-235) has an atomic mass of 235 and an atomic number of 92. Calculate the number of protons and neutrons in an atom of U-235.

*15. Assume that you start with 1 kg of uranium-234 (U-234), which has a half-life of 250,000 years.

    a. How many grams of U-234 will remain after 250,000 years?

    b. How many years will it take for 750 g of U-234 to decay?

    c. How many grams of U-234 will remain after 1 million years?

* Solutions to questions preceded by asterisks appear in Appendix VII.

TAKE A STAND

Visit our Web site at **http://www.wiley.com/college/raven** (select Chapter 11 from the Table of Contents) for links to more information about the controversy involving the transport of high-level radioactive waste to Yucca Mountain. Consider the opposing viewpoints of those who think it is safer to use Yucca Mountain for the permanent storage of wastes from around the country and those who think the transport of these wastes makes a single site at Yucca Mountain too dangerous. You will find tools to help you organize your research, analyze the data, think critically about the issues, and construct a well-considered argument. Take a Stand activities can be done individually or as part of a team, as oral presentations, written exercises, or Web-based (e-mail) assignments.

Additional on-line materials relating to this chapter, including Student Quizzes, Activity Links, Useful Web Sites, Flash Cards, and more, can also be found on our Web site.

SUGGESTED READING

Christensen, J. "New Questions Plague Nuclear Waste Storage Plan." *New York Times* (August 10, 1999). This article is an overview of the continuing controversy about the storage of high-level radioactive waste at Yucca Mountain, Nevada. Includes an interesting timeline ("The Quest to Bury Nuclear Waste") that extends from 1954, when the Atomic Energy Act was passed, to the year 622,010, when nearby residents may be exposed to peak radiation levels of 85 millirems per year.

Flavin, C., and N. Lenssen. "Nuclear Power Nears Its Peak." *WorldWatch* (July–August 1999). The authors explain why nuclear energy is rapidly becoming a thing of the past. This article should be read with the suggested reading by Yost, which takes the opposite view.

Lake, J.A., R.G. Bennet, and J.F. Kotek. "Next-Generation Nuclear Power." *Scientific American*, Vol. 286, No. 1 (January 2002). The authors develop a case for expanding our use of nuclear power and describe some of the features that will make newer nuclear power plants safer and more economical.

Long, M.E. "Half Life: The Lethal Legacy of America's nuclear Waste." *National Geographic*, Vol. 202, No. 1 (July 2002). An excellent, well-researched article about the nuclear waste problem in the United States.

Makhijani, A. "Plutonium End Game: Stop Reprocessing, Start Immobilizing." *Science for Democratic Action*, Vol. 9, No. 2 (February 2001). This journal, published by the Institute for Energy and Environmental Research, is admittedly antinuclear, but the author, who has a Ph.D. in nuclear physics, build a well-documented case for the dangers involved in reprocessing spent nuclear fuel.

Stone, R. "Living in the Shadow of Chornobyl." *Science*, Vol. 292 (April 20, 2001). An examination of the health effects that have occurred in the 15 years since the accident at the Chornobyl nuclear power plant.

Stone, R. "Nuclear Trafficking: A Real and Dangerous Threat." *Science*, Vol. 292 (June 1, 2001). A chilling account of the illegal trafficking of weapons-grade nuclear materials from the former Soviet Union.

Von Hippel, F.N. "Plutonium and Reprocessing of Spent Nuclear Fuel." *Science*, Vol. 293 (September 28, 2001). An excellent review of the history and tradeoffs involved in commercial reprocessing, which continues in several European countries.

Von Hippel, F., and S. Jones. "The Slow Death of the Fast Breeder." *The Bulletin of the Atomic Scientists* (September–October 1997). The impacts of closing most breeder nuclear reactors around the world include safely disposing of huge inventories of plutonium.

Yost, M. "Three Mile What?" *The Wall Street Journal* (September 13, 1999). The authors explain why nuclear energy is making a comeback. This article should be read with the suggested reading by Flavin and Lenssen, which takes the opposite view.

Wind farm. The 6,000 wind turbines tower over hillsides in Altamont, California. These turbines generate 1 billion to 1.2 billion kWh of electricity per year.

Renewable Energy and Conservation

Learning Objectives

After you have read this chapter you should be able to

1. Distinguish between active and passive solar energy and describe how each is used.

2. Contrast the advantages and disadvantages of solar thermal electric generation and photovoltaic solar cells in converting solar energy into electricity.

3. Define biomass, explain why it is an example of indirect solar energy, and outline its advantages and disadvantages as a source of energy.

4. Describe the locations that can make optimum use of wind energy and of hydropower. Compare the potential of wind energy and hydropower.

5. Describe two renewable energy sources that are not direct or indirect results of solar energy.

6. Distinguish between energy conservation and energy efficiency and give examples of each.

7. Define *cogeneration* and give an example of a large-scale cogeneration system.

n recent years **wind farms,** arrays of nonpolluting wind turbines, have sprung up in many open landscapes, the rotors of each turbine turning to generate electricity when the wind blows. Germany, the United States, Spain, Denmark, and India are just a few of the countries that have built large clusters of windmills to generate electricity.

Wind is not the only alternative energy technology emerging. In the Philippines, wood plantations are being managed for present and future wood-fired power plants. Brazil has significantly reduced its reliance on imported oil by converting sugarcane crops into alcohol fuels. India, Mexico, and Kenya are manufacturing photovoltaic solar cells. Japan, Germany, and Italy are installing solar rooftops on thousands of buildings. Hungary and Mexico use increasingly large amounts of geothermal energy. In the United States, at least six power plants generate electricity from the combustion of old tires. As these examples show, most countries are trying new approaches in the endless quest for energy.

Alternatives to fossil fuels and nuclear power are receiving a great deal of attention these days, and for good reason. Alternative energy sources are not only renewable, but they cause fewer environmental problems than fossil fuels or nuclear power. We have seen that there are several concerns about using fossil fuels for energy. Reserves of oil, gas, and even coal are limited and will eventually be depleted, and burning these fossil fuels for energy has negative environmental consequences such as global warming, air pollution, acid rain, and oil spills. Some people have suggested that nuclear energy can be used when we run out of fossil fuels, but as we saw in Chapter 11, several serious problems are associated with the use of nuclear fission. In addition, uranium, the fuel for nuclear fission, is a nonrenewable resource.

Given the rapidly expanding world population, future energy needs will probably demand the exploitation of most energy sources. The recognized need for a long-term solution has prompted world interest in **renewable energy sources,** those sources that are replenished by natural processes so that they can be used indefinitely. Among the most attractive renewable energy sources is **solar energy.** *Direct solar energy* can be used to heat water, heat buildings, and generate electricity. Wind, biomass (such as wood, agricultural wastes, and fast-growing plants), and hydropower (the energy of flowing water) are all examples of *indirect solar energy.* Currently, the nonsolar renewable energy source that is most widely used is geothermal energy—the heat of the Earth.

Despite the promise of renewable forms of energy, we humans still rely primarily on economically inexpensive but environmentally costly fossil fuels to power our society. Renewables currently make up a small fraction of the world's energy, largely because the economics of renewables do not yet make them competitive with traditional forms of energy. Nonetheless, substantial technological improvements to renewables have occurred since the 1990s. These improvements have lowered the cost of many renewables to within striking distance of fossil fuels and nuclear energy. In many places wind energy generates electricity at prices that are competitive with coal and nuclear power. Worldwide during the 1990s and early 2000s, renewables such as wind power, photovoltaics, and geothermal energy grew faster than coal, oil, natural gas, or nuclear power.

Most energy analysts predict the transition from fossil fuels and nuclear power to renewable energy sources will take place during the 21st century. Large energy companies, such as Shell, Arco, and British Petroleum (BP), recognize the vast potential market for renewables and

are investing in them. BP, for example, has a solar division with more than 700 employees.

During the transition to renewables, we must use our energy resources with conservation in mind. We must continue to find ways to enhance energy efficiency and decrease energy waste. In many ways, energy conservation is our single most important long-term energy solution because it helps meet the increasing energy needs of the expanding human population. This chapter examines the various renewable energy sources and considers the importance of energy conservation and energy efficiency.

DIRECT SOLAR ENERGY

The sun produces a tremendous amount of energy, most of which dissipates into outer space. Only a very small portion is radiated to the Earth. Solar energy is different from fossil and nuclear fuels because it is perpetually available; we will run out of solar energy only when the sun's nuclear fire burns out. Solar energy is dispersed over the Earth's entire surface rather than concentrated in highly localized areas, as are coal, oil, and uranium deposits. In order to make solar energy useful, we must collect it.

Recall from Chapter 6 that solar radiation varies in intensity depending on the latitude, season of the year, time of day, and degree of cloud cover. Areas at lower latitudes—that is, closer to the equator—receive more solar radiation annually than do latitudes closer to the North and South Poles (Table 12.1). More solar radiation is received during summer than during winter, because the sun is directly overhead in the summer and lower on the horizon in winter. Solar radiation is more intense when the sun is high in the sky (noon) than when it is low in the sky (dawn or dusk). Clouds both scatter incident light and absorb some of the sun's energy, thereby reducing its intensity. Because of its lack of cloud cover and lower latitude, the southwestern United States receives the greatest amount of solar radiation annually, whereas the Northeast receives the least (Figure 12.1).

Although the technology exists to use solar energy directly, it has not been adopted widely, largely because

Table 12.1	Variation in Solar Radiation for Selected U.S. Cities*				
City	*Latitude*	*December*	*March*	*June*	*September*
Miami, FL	26°	1,292	1,829	1,992	1,647
Los Angeles, CA	34°	912	1,641	2,259	1,892
Dodge City, KS	38°	874	1,566	2,400	1,842
Washington, D.C.	38°	632	1,255	2,081	1,446
East Lansing, MI	42°	380	1,086	1,914	1,303
Seattle, WA	47°	218	917	1,724	1,129

* Solar radiation on a horizontal surface in $Btu/ft^2/day$.

Figure 12.1 Solar energy distribution over the United States.
This map shows the average daily total of solar energy (on an annual basis) that would be received on a solar collector that tilts to compensate for latitude. The units are in megajoules per square meter. The Southwest is the best area in the United States for year-round solar energy collection.

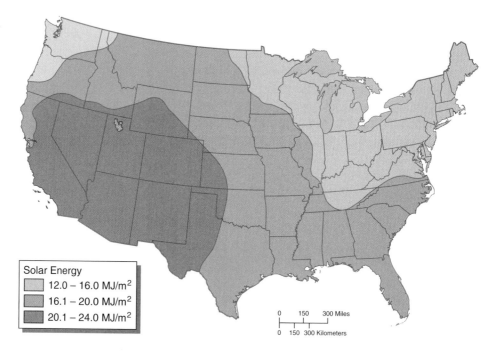

Solar Energy
- 12.0 – 16.0 MJ/m²
- 16.1 – 20.0 MJ/m²
- 20.1 – 24.0 MJ/m²

0   150   300 Miles
0   150   300 Kilometers

the initial costs associated with converting to solar power are high. However, the long-term energy savings of solar power may offset the high start-up costs. Trapping the sun's energy using current technology is also inefficient, meaning that relatively little of the sun's energy that hits the solar panels (collecting devices) is actually used. With new technological developments, the efficiency of solar energy collection is increasing, making it a more cost-effective alternative source of energy. The use of solar energy is projected to increase in the future, in both highly developed and developing countries (see Mini-Glossary: Direct Solar Energy Terms).

For example, an estimated 0.5 million people in the rural areas of Africa, Central America, India, and China are currently using solar cookers. Recent designs for solar cookers transmit solar light into the cooker, and the glass cover does not transmit out the infrared wavelengths (heat) that would normally escape. Pots containing the food to be cooked are placed inside the box on a black metal plate. The solar cooker can reach a temperature of 177°C (350°F) and can be used to boil, bake, simmer, and sauté foods. In average sunlight, a person can cook a full meal in 2 to 4 hours.

MINI-GLOSSARY

Direct Solar Energy Terms

direct solar energy: Technologies that make use of direct radiant energy from the sun.

passive solar heating: A system that captures solar energy directly to heat buildings without the need for pumps or fans to distribute the collected heat.

active solar heating: A system that captures solar energy directly, primarily to heat water; a liquid heated by the solar collector is pumped to a heat exchanger, where the water is heated.

solar thermal electric generation: A system that captures solar energy directly to generate electricity; the sun's energy is concentrated by mirrors or lenses onto a fluid-filled pipe, and the heated fluid is used to generate electricity.

photovoltaic (PV) solar cells: A system that captures solar energy directly to generate electricity; wafers or thin-film devices generate electricity when solar energy is absorbed.

solar-generated hydrogen: A system that uses PV-generated or wind-generated electricity to produce hydrogen fuel from water.

Heating Buildings and Water

You have probably noticed that in winter or summer, the air inside a car that is sitting in the sun with its windows rolled up becomes much hotter than the surrounding air. Similarly, the air inside a greenhouse remains warmer than the outside air during cold months. (Greenhouses usually require additional heating in cold climates, but far less than might be expected.) This kind of warming occurs partly because the material—such as glass—that envelops the air inside the enclosure is transparent to visible light but impenetrable to heat. Thus, visible light from the sun penetrates the glass and warms the surfaces of objects inside, which in turn give off **infrared radiation**—invisible waves of heat energy. Because infrared radiation cannot penetrate glass, heat does not escape, and the area surrounded by glass grows continuously warmer.

In **passive solar heating**, solar energy is used to heat buildings without the need for pumps or fans to distribute the collected heat. Certain design features can be put to use in a passive solar heating system to warm buildings in winter and help them remain cool in summer (Figure

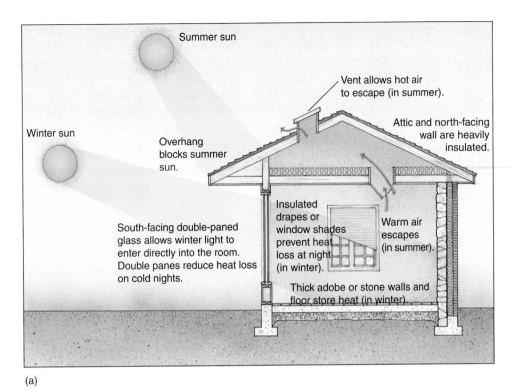

(a)

(b)

Figure 12.2 Passive solar heating designs. (a) Several passive designs are incorporated into this home. (b) A solar sunspace can be added to existing homes.

12.2). In the Northern Hemisphere large south-facing windows receive more total sunlight during the day than windows facing other directions. The sunlight entering through the windows provides heat that is then stored in floors and walls made of concrete or stone, or in containers of water. This stored heat can be transmitted throughout the building naturally by convection, which is the circulation that occurs because warm air rises and cool air sinks. Buildings with passive solar heating systems must be well insulated so that accumulated heat does not escape. Depending on the building's design and location, passive heating can save as much as 50% of heating costs. Currently, about 7% of new homes built in the United States have passive solar features.

In **active solar heating**, a series of collection devices mounted on a roof or in a field is used to gather solar energy. The most common collection device is a flat solar panel or plate of black metal (which absorbs the sun's energy) enclosed in an insulated box (Figure 12.3). Active solar heating is used primarily for heating water, either for household use or for swimming pools. The heat absorbed by the solar collector is transferred to a liquid inside the panel, which is then pumped to the heat exchanger, where the heat is transferred to water that will be stored in the hot water tank. Because approximately 8% of the energy consumed in the United States goes toward heating water, active solar heating has the potential to supply a significant amount of the nation's energy demand.

The use of active solar energy for space heating, currently not as common as for heating water, may become more important when diminishing supplies of fossil fuels force their prices higher. Active solar heating of buildings, now costlier than more conventional forms of space heating, would then become more competitive. Solar air conditioning, although technically feasible, is expensive and is not widely available. Some commercial buildings have installed solar air conditioning, however.

Solar Thermal Electric Generation

Electricity can be produced by several different systems that collect incident sunlight and concentrate it, using mirrors or lenses, to heat a working fluid to high temper-atures. Known as **solar thermal electric generation**, these systems can only use sunlight arriving directly from the sun. In one such system, trough-shaped mirrors, guided by computers, track the sun for optimum efficiency, center sunlight on oil-filled pipes, and heat the oil to 390°C (735°F) (Figure 12.4). The hot oil is circulated to a water storage system and used to change water into superheated steam, which turns a turbine to generate electricity. Alternatively, the heat can be used in industrial processes or for desalinization—that is, removal of salt from water—and water purification.

Solar thermal systems often have a backup—usually natural gas—that is available to generate electricity at night and during cloudy days when solar power is not operating. The backup increases the system's value to the electricity grid system. (Electricity produced in power stations across the United States is fed into a *grid*, a network of cables that carry electricity where it is needed.) The world's largest solar thermal system of this type is currently operating in the Mojave Desert in southern California.

The solar power tower is a solar thermal system with a tall tower surrounded by hundreds of mirrors. The computer-controlled mirrors move to follow the sun, focusing solar radiation on a central receiver at the top of the tower. There a circulating liquid—molten salt—is heated by the concentrated sunlight, and the heat is used to produce steam for generating electricity. Because molten salt retains heat, some of the heat may be stored to be used for electricity generation during the night, when solar energy is unavailable. Solar power towers have been tested in the United States, several European countries, and Japan; the United States is no longer working on this technology.

Figure 12.3 Active solar water heating. Solar collectors are mounted on the roof of a building. Each solar panel is a box with a black metal base and glass covering. Sunlight enters the glass and warms the pipes and the liquid that is flowing through them. The liquid that is heated in solar collectors is used to heat water, which is further heated to required temperatures by a backup heater that uses electricity or natural gas. Solar domestic water heating can provide a family's hot water needs year-round.

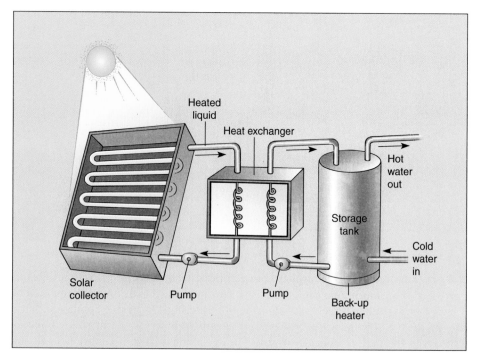

(a)

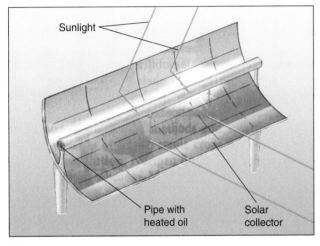

(b)

Figure 12.4 Solar thermal electric generation. (a) A solar thermal plant in California uses troughs to focus sunlight on a fluid-filled tube, as shown in (b). The heated oil is pumped to a water tank where it generates steam used to produce electricity. For simplicity, arrows show sunlight converging on several points; sunlight actually converges on the pipe throughout its length.

Solar thermal energy systems are inherently more efficient than other solar technologies because they concentrate the sun's energy. With improved engineering, manufacturing, and construction methods, solar thermal energy may become cost-competitive with fossil fuels. For example, electricity currently costs $.10 to $.15 per kilowatt-hour (kWh) at the Solar Electric Generating System in southern California. (Table 12.2 compares the generating costs of electricity using different energy sources, including solar thermal.) In addition, the environmental benefits of solar thermal plants are significant because they do not produce air pollution or contribute to acid rain or global warming.

Photovoltaic Solar Cells

It is possible to convert sunlight directly into electricity by using **photovoltaic (PV) solar cells**. Photovoltaic cells are wafers or thin films of solid state materials, such as silicon or gallium arsenide, that are treated with certain metals in such a way that they generate electricity—that is, a flow of electrons—when solar energy is absorbed (Figure 12.5). Photovoltaic solar cells are arranged on large panels that are set up to absorb sunlight, even on cloudy or rainy days.

PVs generate electricity with no pollution and minimal maintenance. They can be used on any scale, from small, portable modules to large, multimegawatt power plants. Our current PV solar cell technology, although used to power satellites, uncrewed airplanes, highway signals, watches, and calculators, has a few limitations that prevent the cells' widespread use to generate electricity. Photovoltaic solar cells are only about 10% to 15% efficient at converting solar energy to electricity (although efficiencies are steadily improving), and the number of solar panels needed for large-scale use requires a great deal of land. At current efficiencies several thousand acres of solar panels would be required to absorb enough solar energy to produce the electricity generated by a single large conventional power plant.

One of the main benefits of PV devices for utility companies is that they can be purchased in small modular units that become operational in a short amount of time. A utility company can purchase PV elements to increase its generating capacity in small increments, rather than committing a billion dollars or more and a decade or more of construction for a massive conventional power plant. Used in this supplementary way, the PV units can provide the additional energy needed, for example, to power irrigation pumps on hot, sunny days.

In remote areas that are not served by electrical power plants, such as rural areas of developing countries, it is more economical to use PV solar cells for electricity than to extend power lines. Photovoltaics are the energy choice to pump water, refrigerate vaccines, grind grain, charge batteries, and supply rural homes with lighting. According to the Institute for Sustainable Power, more than 1 million households in the developing countries of Asia, Latin America, and Africa have installed PV solar cells on the roofs of their homes. A PV panel the size of

Table 12.2	2002 Generating Costs of Electric Power Plants

Energy Source	Generating Costs (cents per kilowatt hour)
Hydropower	4–10
Biomass	6–8
Geothermal	3–8
Wind	4–5
Solar thermal	10–15
Photovoltaics	15–25
Natural gas	3–4
Coal	4–5
Nuclear power	10–15

(a)

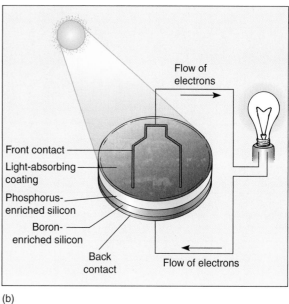

(b)

Figure 12.5 Photovoltaic cells. (a) Arrays of photovoltaic (PV) cells collect solar energy used to generate electricity at the Intercultural Center of Georgetown University. About 10% of the electricity needed by the school is supplied by its PV system. (b) Photovoltaic cells contain silicon and other materials. Sunlight excites electrons, causing them to be ejected from silicon atoms. Useful electricity is generated when the ejected electrons flow out of the PV cell through a wire.

two pizza boxes can supply a rural household with enough electricity for five lights, a radio, and a television.

The cost of manufacturing PV modules has steadily declined over the past 25 years, from an average factory price of almost $90 per watt in 1975 to about $3.50 per watt in 2002, according to *PV News*. Also, the cost of producing electricity from PVs has steadily declined from 1970 to the present. Despite this progress, in 2002 the cost was still about $.15 to $.25 per kilowatt-hour.

Future technological progress may make PVs economically competitive with electricity produced by conventional energy sources. The production of "thin-film" solar cells, which are much cheaper to manufacture, has decreased costs. Thin-films can be produced as flexible sheets that can be incorporated into building materials, such as roofing shingles, tiles, and window glass (Figure 12.6). More than 120,000 Japanese homes had PV solar-energy roofing installed in the past few years. The Mil-

Figure 12.6 Solar shingles. These thin-film solar cells look much like conventional roofing materials.

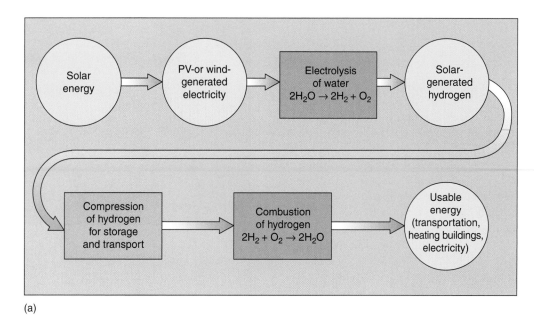

(a)

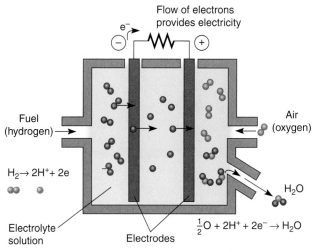

(b)

Figure 12.7 Solar-generated hydrogen. (a) Solar electricity can be used to split water in a process called electrolysis. This produces hydrogen gas, which represents a chemical form of solar energy. Following gas compression, pipelines can transport hydrogen to users. When burned in the presence of oxygen, hydrogen produces usable energy and water. PV = photovoltaic. (b) Cross-section of a hydrogen fuel cell.

lion Solar Roofs Initiative, sponsored by the U.S. government, plans to have solar-energy roofing on one million buildings by 2010.

Solar-Generated Hydrogen

Electricity generated by PVs or wind energy can be used to split water into the gases oxygen and hydrogen. (As discussed shortly, wind is an indirect form of solar energy.)

Hydrogen can also be produced using conventional energy sources such as fossil fuels and nuclear power, but then, of course, the environmental and security problems of these forms of energy are not avoided. For that reason, we limit this discussion to hydrogen fuel production using solar electricity (Figure 12.7a).

Hydrogen is a clean fuel; it produces water and heat when it is burned and produces no sulfur oxides, carbon monoxide, hydrocarbon particulates, or CO_2 emissions. It does produce some nitrogen oxides. (Some of the heat produced when hydrogen is burned chemically combines nitrogen and oxygen, both of which are present in the atmosphere.) However, the nitrogen oxides are produced in amounts that are fairly easy to control. Hydrogen has the potential to provide energy for transportation in the form of hydrogen-powered electric automobiles as well as for heating buildings and producing electricity.

It may seem wasteful to use electricity generated from solar energy to make hydrogen, which can then be used to generate more electricity. However, electricity that is generated by existing photovoltaic cells cannot be stored long-term; it must be used immediately. Hydrogen offers a convenient way to store solar energy as chemical energy. It can be transported by pipeline, possibly less expensively than electricity can be transported by wire.

Production of hydrogen from PV electricity currently has an efficiency of 8%, which means that only 8% of the solar energy absorbed by the PV cells is actually converted into the chemical energy of hydrogen fuel. Scientists are working to improve the efficiency, and in 1998 scientists at the National Renewable Energy Laboratory announced a breakthrough that allowed them to convert about 12.5% of the solar energy to hydrogen fuel. Continued advances should decrease costs and make solar-generated hydrogen fuel commercially viable. A pilot plant that makes solar-generated hydrogen fuel is being tested in Saudi Arabia under the joint sponsorship of Germany and Saudi Arabia.

A **fuel cell** is an electrochemical cell similar to a battery (Figure 12.7b). Fuel cells differ from batteries in that fuel cells produce power as long as they are supplied with fuel, whereas batteries store a fixed amount of energy. Also, fuel cell reactants (hydrogen and oxygen) are supplied from external reservoirs, whereas battery reactants are contained within the battery. When hydrogen and oxygen react in a fuel cell, water forms and energy is produced as an electrical current. The major car makers are currently developing automobiles powered by hydrogen fuel cells (see "Meeting the Challenge: Clean Cars, Clean Fuels" in Chapter 19). Iceland plans to build the world's first fleet of fuel cell buses, obtaining its hydrogen fuel by using existing geothermal and hydroelectric resources.

INDIRECT SOLAR ENERGY

There are renewable energy sources that indirectly use the sun's energy. Combustion of **biomass**—wood and other organic matter—is an example of indirect solar energy, because green plants, which use solar energy for photosynthesis, produce the energy contained in biomass. Windmills can harness **wind energy**—surface air currents that are caused by the solar warming of air. The damming of rivers and streams to generate electricity is a type of **hydropower**—the energy of flowing water—which exists because of the hydrologic cycle that is driven by solar energy (see Chapter 6).

Biomass Energy

Biomass, one of the oldest fuels known to humans, consists of such materials as wood, fast-growing plant and algal crops, crop wastes, sawdust and wood chips, and animal wastes (Figure 12.8). Biomass contains chemical energy, the source of which can be traced to radiant energy from the sun, which was used by photosynthetic organisms to form the organic molecules of biomass. Biomass is a renewable form of energy as long as it is managed properly.

(a)

(b)

Figure 12.8 Biomass. (a) Firewood is the major energy source for most of the developing world. (b) Animal dung is an important energy source in Ethiopia and certain other countries.

Biomass fuel, which can be a solid, liquid, or gas, is burned to release its energy. Solid biomass such as wood is burned directly to obtain energy. Biomass—particularly firewood, charcoal (wood that has been turned into coal by partial burning), animal dung, and peat (partly decayed plant matter found in bogs and swamps)—supplies a substantial portion of the world's energy. At least half of the human population relies on biomass as their main source of energy. In developing countries, wood is the primary fuel for cooking. In the United States, biomass accounts for about 3% of total U.S. energy production.[1] Biomass in the form of low-cost residues from sawmills, paper mills, and agricultural industries is burned in power plants to generate about 7.6 gigawatts (GWe) of electricity.

---

[1] Unless noted otherwise, all energy facts cited in this chapter were obtained from the Energy Information Administration (EIA), the statistical agency of the U.S. Department of Energy (DOE).

It is possible to convert biomass, particularly animal wastes, into **biogas**. Biogas, which is usually composed of a mixture of gases (mostly methane), can be stored and transported easily like natural gas. It is a clean fuel, the combustion of which produces fewer pollutants than either coal or biomass. In India and China, several million family-sized **biogas digesters** use microbial decomposition of household and agricultural wastes to produce biogas that is used for cooking and lighting (Figure 12.9a). When biogas conversion is complete, the solid remains are removed from the digester and used as fertilizer. Although the technology for biogas digesters is relatively simple, the conditions inside the digester, such as the moisture level and pH, must be carefully monitored if the bacteria are to produce biogas at an optimum level.

Biogas has the potential to be used in fuel cells to generate electricity. A pilot program at Greater Boston's main sewage treatment plant began producing electricity from biogas (methane) in 1997. The methane is produced from sewage sludge in large biogas digesters and then burned in a methane fuel cell to produce enough electricity for 150 homes. Like hydrogen fuel cells discussed earlier in the chapter, methane fuel cells produce relatively few pollutants.

Biomass can also be converted to liquid fuels, especially **methanol** (methyl alcohol) and **ethanol** (ethyl alcohol), which can then be used in internal combustion engines (Figure 12.9b). Mixing gasoline with 10% ethanol produces a cleaner-burning mixture known as gasohol.

Although some U.S. energy companies convert sugar cane, corn, or wood crops to alcohol, others are interested in the commercial conversion of agricultural and municipal wastes into ethanol. Costs are high, however, and few companies have successfully invested in waste-to-ethanol. A company (BC International) in Louisiana plans to convert *bargasse*, the fibrous residues left after the juices have been squeezed out of sugar cane, into 20 million gallons of ethanol per year, beginning in 2004. Their conversion process makes use of genetically altered bacteria and enzymes. Several companies are currently building plants that convert biomass (one using cornstalks, one using rice straw, and one digesting sewage sludge) to ethanol. Currently, the profitability of ethanol is only possible because of heavy government subsidies, which effectively reduce ethanol's cost. Companies planning to pursue waste-to-ethanol processes acknowledge that they need government subsidies initially but think they eventually will be able to compete with gasoline without government help.

A major disadvantage of alcohol fuels, whether they are produced from biomass, natural gas, or coal, is that as much as 30% to 40% of the energy in the starting material is lost in the conversion to alcohol. For biomass, this energy loss does not take into account the additional energy required to grow, harvest, and process the biomass. Thus, the overall efficiency for converting biomass to alcohol is even lower.

(a)

(b)

Figure 12.9 Liquid and gaseous fuels from biomass. (a) A concrete biogas digester (circular structure in ground) in Chu Zhang Village, near Beijing. The 135 digesters in this village of 600 people offer the benefit of improved sanitation as well as fuel for cooking and lighting. Bacteria convert crop residues and human and animal wastes to methane gas. Note the tubing that leads directly from a digester to an individual home. A 6- to 8-m^3 digester can usually satisfy the energy requirements of a five-person family, including cooking three meals and providing 2 hours of lighting with a biogas lamp. (b) A methanol-powered bus on a street in Denver, Colorado.

Advantages of Biomass Use Biomass is attractive as a source of energy because it reduces dependence on fossil fuels and because it can often make use of wastes, thereby reducing our waste disposal problem (see Chapter 23). For example, the Mesquite Lake Resource Recovery Project in southern California burns cow manure in special furnaces to generate electricity for thousands of homes. The manure is too salty and contaminated with weed seeds to be used as fertilizer, so its use as an energy source helps solve the problem of its disposal.

Biomass is usually burned to produce energy, so the pollution problems caused by fossil fuel combustion, particularly carbon dioxide emissions, are not completely absent in biomass combustion. However, the low levels of sulfur and ash produced by biomass combustion compare favorably with the levels produced when bituminous coal is burned. It is possible to offset the CO_2 that is released into the atmosphere from biomass combustion by increasing tree planting. As trees photosynthesize, they absorb atmospheric CO_2 and lock it up in organic molecules that make up the body of the tree, thereby providing a carbon "sink." Thus, if biomass is regenerated to replace the biomass used, there is no net contribution of CO_2 to the atmosphere and to global warming.

Disadvantages of Biomass Use Some problems are associated with use of biomass, especially from plants. For one thing, biomass production requires land and water. The use of agricultural land for energy crops competes with the growing of food crops, so shifting the balance toward energy production might decrease food production, contributing to higher food prices. For this reason, some scientists are interested in the commercial development of certain desert shrubs that produce oils that could be used for fuel. The shrubs do not require prime agricultural land, although care would have to be taken to ensure that the desert soils were not degraded or eroded by overuse.

As mentioned earlier, at least half of the world's population relies on biomass as its main source of energy. Unfortunately, in many areas people burn wood faster than they replant trees. Intensive use of wood for energy has resulted in severe damage to the environment, including soil erosion (see Chapter 14), deforestation and desertification (see Chapter 17), air pollution (see Chapters 19 and 20), and degradation of water supplies (see Chapter 21).

Crop residues, another category of biomass that includes cornstalks, wheat stalks, and wood wastes at paper mills and sawmills, are increasingly being used for energy. At first glance, it may seem that crop residues, which normally remain in the soil after harvest, would be a good source of energy if they were collected and burned. After all, they are just waste materials that will eventually decompose. As it turns out, however, crop residues left in and on the ground prevent erosion by helping to hold the soil in place. Also, their decomposition serves to enrich the soil by making the minerals that were originally locked up in the plant residues available for new plant growth. If all crop residues were removed from the ground, the soil would eventually be depleted of minerals, and its future productivity would decline. For-

est residues, which remain in the soil after trees are harvested, fill similar ecological roles.

Sweden was embroiled in a biomass controversy in the late 1990s as it phased out all of its nuclear power plants. The Swedish government had examined alternative energy options and decided to expand the use of biomass. (Sweden's long, dark winters prevent it from using solar power directly, and a law passed in 1987 prevents further development of hydropower.) Some biologists opposed using larger areas of land to grow biomass crops because of the threat to biological diversity. The scientists were concerned that the large-scale removal of natural vegetation to plant biomass crops could lead to habitat loss for many native bird species, including several endangered species. In response to this opposition campaign, Sweden cut its projections for future biomass use in half. Sweden's energy controversy is a useful reminder that the use of all forms of energy, even renewables, has environmental impacts.

Wind Energy

During the 1990s and early 2000s, wind energy became the world's fastest growing source of energy. Wind, which results from the warming of the atmosphere by the sun, is an indirect form of solar energy in which the radiant energy of the sun is transformed into mechanical energy—the movement of air molecules. Wind is sporadic over much of the Earth's surface, varying in direction and magnitude; and, like direct solar energy, wind power is a highly dispersed form of energy. Harnessing wind energy to generate electricity has great potential, however, and wind is becoming increasingly important in supplying our energy needs (see the chapter opener and Figure 12.10).

As turbines have become larger and more efficient, costs for wind power have declined rapidly—from \$.40 per kilowatt-hour in 1980 to \$.04 to \$.05 per kilowatt-hour in 2002. Wind power is currently the most cost-competitive of all forms of renewable energy. New technological advances, such as turbines that use variable speed operation, suggest that wind energy could become an important global source of electricity during the first few decades of the 21st century. Denmark, one of the world leaders in wind power, now generates 18% of its electricity using wind energy. Other leading wind energy countries include Germany, which produces more than one third of the world's total wind power; the United States; Spain; and India.

Harnessing wind energy is most profitable in rural areas that receive fairly continual winds, such as islands, coastal areas, mountain passes, and grasslands. The world's largest concentration of wind turbines is currently located in the Tehachapi Pass at the southern end of the Sierra Nevada mountain range in California. In the continental United States, some of the best locations for large-scale electricity generation from wind energy

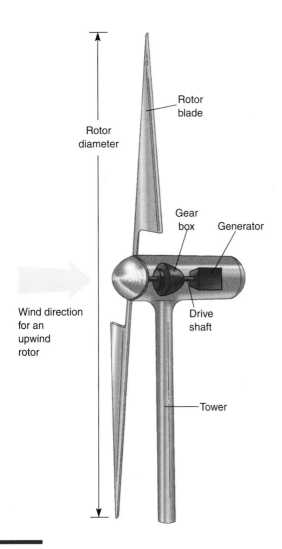

Figure 12.10 Wind energy. This basic wind turbine design has a horizontal axis (*horizontal* refers to the orientation of the drive shaft). Airflow causes the turbine's blades to turn 15 to 60 revolutions per minute (rpm). As the blades turn, gears within the turbine spin the drive shaft 1,200 to 1,800 rpm. This spinning powers the generator, which sends electricity through underground cables to a nearby utility. Many wind turbines are more than 10 stories tall and have 80-foot-long blades. Wind turbine technology is advancing rapidly, and many changes in design are anticipated. (The tower is not drawn to scale and is much taller than depicted.)

are on the Great Plains. The 10 states with the greatest wind energy potential, according to the American Wind Energy Association, are North Dakota, Texas, Kansas, South Dakota, Montana, Nebraska, Wyoming, Oklahoma, Minnesota, and Iowa. Current U.S. wind power projects are under way in these and many other states.

The use of wind power does not cause major environmental problems, although one initial concern was reported bird kills. The California Energy Commission estimated that several hundred birds, many of them raptors (birds of prey), turned up dead in the vicinity of the 7,000 turbines at Altamont Pass in California during a 2-year study; most had collided with the turbines. Studies later determined that Altamont Pass is a major bird migration pathway. Technical "fixes," such as painted blades and anti-perching devices to discourage raptors from roosting on the towers, were implemented at Altamont Pass, and future wind farm sites will be chosen away from bird routes.

Because it produces no waste, wind is a clean source of energy. It produces no emissions of sulfur dioxide, carbon dioxide, or nitrogen oxides. Every kilowatt-hour of electricity generated by wind power rather than fossil fuels prevents 1 to 2 lb of the greenhouse gas CO_2 from entering the atmosphere.

A concern with increasing our use of wind power is aesthetics: Wind machines change the character of the landscape. Fortunately, most of the locations that are appropriate for large-scale wind power are not densely populated, and lease payments from wind power producers will probably help offset landowners' aesthetic concerns. Combining wind farms with cattle grazing, as is done in Altamont, California, or with raising corn or other crops, as is done in Iowa, is a very productive and profitable use of land. Several countries, including Ireland and the United States, plan to build offshore wind farms, in part because ocean winds are strong. Also, offshore wind farms are not visible to most people.

Hydropower

The sun's energy drives the hydrologic cycle, which encompasses precipitation, evaporation from land and water, transpiration from plants, and drainage and runoff. As water flows from higher elevations back to sea level, we can harness its energy. Unlike the sun's energy, which is highly dispersed, hydropower is a more concentrated energy. The potential energy of water held back by a dam is converted to kinetic energy as the water falls down a penstock, where it turns turbines to generate electricity (Figure 12.11).

Currently, hydropower generates approximately 19% of the world's electricity, making it the form of solar energy in greatest use. The 10 countries with the greatest hydroelectric production are, in decreasing order, Canada, the United States, Brazil, China, Russia, Norway, Japan, India, France, and Sweden. In the United States, approximately 2,100 hydropower plants produce 9% of its electricity. Highly developed countries have already built dams at most of their potential sites, but this is not the case in many developing nations. There—particularly in undeveloped, unexploited parts of Africa and South America—hydropower represents a great potential source of electricity.

Impacts of Dams One of the problems associated with hydropower is that building a dam changes the natural flow of a river. A dam causes water to back up, flooding large areas of land and forming a reservoir, which

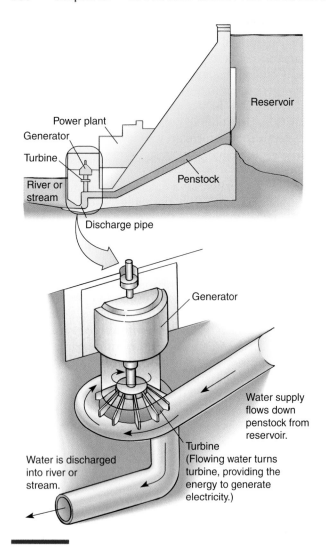

Figure 12.11 Hydroelectric power. A controlled flow of water released down the penstock turns a turbine, which generates electricity.

destroys plant and animal habitats. Native fishes are particularly susceptible to dams because the original river ecosystem is so altered. The migration of spawning fish is also disrupted (see "Case in Point: The Columbia River" in Chapter 13). Below the dam, the once-powerful river is reduced to a relative trickle. The natural beauty of the countryside is affected, and certain forms of wilderness recreation are made impossible or less enjoyable, although water sports in the reservoir are made possible by dams.

At least 200 large dams around the world have been associated with reservoir-induced seismicity, earthquakes that occur during and after the filling of a large reservoir behind a dam. The larger the reservoir and the faster it is filled, the greater the intensity of seismic activity. An area does not have to be seismically active to have earthquakes induced by reservoirs.

In arid regions, the creation of a reservoir results in greater evaporation of water, because the reservoir has a larger surface area in contact with the air than the stream or river did. As a result, serious water loss and increased salinity of the remaining water may occur (see Chapter 13).

One of the most serious consequences of large dams is that they involuntarily displace people in the areas that are flooded by reservoirs. Villages, churches and temples, burial grounds, farmlands, and archaeological sites are often inundated. The Three Gorges Project, the largest dam ever built, is currently under construction on the Yangtze River in China. This dam, which is scheduled for completion in 2009 or 2010, will create a reservoir that is 663 km (412 mi) long. Filling of the reservoir will displace almost 2 million people, the largest number for any dam project. More than 1,200 temples and other cultural relic sites are also endangered.

If a dam breaks, people and property downstream may be endangered. In addition, waterborne diseases such as schistosomiasis may spread throughout the local population. **Schistosomiasis**, a tropical disease caused by a parasitic worm, can damage the liver, urinary tract, nervous system, and lungs. It is estimated that as much as half the population of Egypt suffers from this disease, largely as a result of the Aswan Dam, built on the Nile River in 1902 to control flooding but used since 1960 to provide electrical power. (The large reservoir behind the dam provides habitat for the worm, which spends part of its life cycle in the water. The worms infect humans during bathing, swimming, or walking barefoot along water banks, or by drinking infected water.)

The environmental and social impacts of a dam may not be acceptable to the people living in a particular area. Laws have been passed to prevent or restrict the building of dams in certain locations. In the United States, the **Wild and Scenic Rivers Act** prevents the hydroelectric development of certain rivers, although the number of rivers protected by this law is less than 1% of the nation's total river systems. Norway and Sweden have similar laws.

Dams cost a great deal to build but are relatively inexpensive to operate. A dam has a limited life span, usually 50 to 200 years, because over time the reservoir fills in with silt until it cannot hold enough water to generate electricity. This trapped silt, which is rich in nutrients, is prevented from enriching agricultural lands downstream. The gradual depletion of agricultural productivity downstream from the Aswan Dam in Egypt is well documented. Egypt now relies on heavy applications of chemical fertilizers to maintain fertility of the Nile River valley and its delta.

Ocean Waves

Ocean waves are produced by winds, which are caused by the sun, so wave energy is considered an indirect form of solar energy. Like other types of flowing water, wave power has the potential to turn a turbine, thereby generating electricity. Norway, Great Britain, Japan, and several other countries are investigating the production of electricity from ocean waves. Only one commercial wave power station is currently operational, on the west coast of Islay, a Scottish island.

The wave power station uses a simple technology (Figure 12.12). Essentially, the concrete, hollow power-plant box is sunk into a gully off the coast to catch waves. As each new wave enters the chamber (about every 10 seconds), the rising water in the chamber pushes air into a vent that contains a turbine, causing the turbine to spin. In turn, the spinning turbine drives a generator. When the wave recedes, it draws the air back into the chamber, and the moving air continues to drive the turbine.

Although a great deal of power can be obtained from the ocean, wave power plants need to be able to harness energy at a rate that is efficient and productive. More testing and improvements in design are needed, and engineers cautiously warn the public about past failures of wave power stations. For example, the world's first commercial wave power station (Osprey), located off the coast of northern Scotland, sank and was destroyed in 1995 during a storm, less than 1 month after it opened.

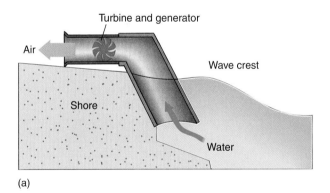

(a)

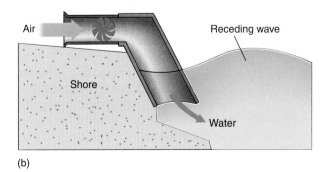

(b)

Figure 12.12 Wave energy. (a) As a wave crests, the rising water in the hollow chamber pushes air that spins a turbine, generating electricity. (b) When the wave recedes, it draws air back into the chamber, which continues to drive the turbine.

After the Osprey failure, scientists and engineers revised their designs and found backers for commercial projects. As a result, by the early 2000s, 15 or more commercial wave power stations were planned or under construction. Proponents of wave power predict that electricity from these stations will be cheaper than that produced by new coal-fired or nuclear power plants.

Ocean Thermal Energy Conversion

In the future it may be possible to generate power using **ocean temperature gradients**, the differences in temperature at various ocean depths. There may be as much as a 24°C difference between warm surface water and very cold, deeper ocean water. Ocean temperature gradients, which are greatest in the tropics, are the result of solar energy warming the surface of the ocean.

The generation of electricity from ocean temperature gradients is known as **ocean thermal energy conversion** (**OTEC**). Warm surface water is pumped into the power plant, where it heats a liquid, such as ammonia, to the boiling point. (Liquid ammonia has a very low boiling point, –33°C.) The ammonia steam drives a turbine to generate electricity as ammonia is cooled by the very cold water brought up from the ocean depths. Alternatively, warm seawater can be made to boil in a vacuum chamber, forming steam to drive the turbine as the vapor is cooled and condensed back into a liquid by cold seawater.

Hawaii has been a prime location for pioneering work in OTEC. In addition to generating a small amount of electricity, OTEC prototypes supported several spin-off projects. After the cold water from the ocean depths was used in the power plant, it was still colder than surface ocean water, so it was used to air-condition buildings. Then it was piped to nearby aquaculture facilities to provide clean, nutrient-laden seawater for growing algae, fishes, and crustaceans. The first commercial OTEC plant is under construction at the Natural Energy Laboratory of Hawaii Authority on the island of Hawaii.

Although OTEC is technologically possible, the potential impact of bringing massive quantities of cold water to the surface in a tropical area needs to be considered carefully before it is adopted on a large scale. Such water properties as dissolved gases, turbidity (cloudiness), nutrient levels, and salinity gradients (differences in salt concentrations) would be altered along with the temperature, and these changes would probably have a profound effect on marine organisms.

Because the temperature differences in OTEC (24°C) are not as great as the temperature differences generated in a conventional power plant (500°C), OTEC must pump a lot of water to generate power. So much energy is required to pump this water that OTEC is not very efficient (only 3% to 4%), and costs are high.

■ OTHER RENEWABLE ENERGY SOURCES

Tidal energy and geothermal energy are renewable energy sources that are not direct or indirect results of solar energy. **Tidal energy**, which is caused by changes in water level between high and low tides, has been exploited to generate electricity on a very limited scale. **Geothermal energy** is the naturally occurring heat within the Earth. This heat can be used for space heating and to generate electricity.

Tidal Energy

Tides, the alternate rising and falling of the surface waters of the ocean and seas that generally occur twice each day, are the result of the gravitational pull of the moon and the sun. Normally, the difference in water level between high and low tides is about 0.5 m (1 or 2 ft). However, certain coastal regions with narrow bays have extremely large differences in water level between high and low tides. The Bay of Fundy in Nova Scotia has the largest tides in the world, with up to 16 m (53 ft) difference between high and low tides.

By building a dam across a bay, it is possible to harness the energy of large tides to generate electricity. In one type of system, the dam's floodgates are opened as high tide raises the water on the bay side. Then the floodgates are closed. As the tide falls, water flowing back out to the ocean over the dam's spillway is used to turn a turbine and generate electricity.

Currently, power plants that make use of tidal power are in operation in France, Russia, China, and Canada. However, tidal energy cannot become a significant resource worldwide, because few geographical locations have large enough differences in water level between high and low tides to make power generation feasible. The most promising locations for tidal power in North America include the Bay of Fundy in Nova Scotia, Passamaquoddy Bay in Maine, Puget Sound in Washington, and Cook Inlet in Alaska.

Other problems associated with tidal energy include the high cost of building a tidal power station and potential environmental problems associated with tidal energy in estuaries, coastal areas where river currents meet ocean tides. Because the mixing of fresh and salt waters creates a nutrient-rich environment, estuaries are among the most productive aquatic environments in the world. Fishes and countless invertebrates migrate there to spawn. Building a dam across the mouth of an estuary would prevent these animals from reaching their breeding habitats. Estuaries are also popular sites for recreation, which would be severely curtailed by a tidal dam.

Geothermal Energy

Geothermal energy, the natural heat within the Earth, arises from the ancient heat within the Earth's core, from friction where continental plates slide over one another, and from the decay of radioactive elements. The amount of geothermal energy is enormous. Scientists estimate that just 1% of the heat contained in the uppermost 10 km of the Earth's crust is equivalent to 500 times the energy contained in all of Earth's oil and natural gas resources.

Geothermal energy is typically associated with volcanism. Large underground reservoirs of heat exist in areas of geologically recent volcanism. As groundwater in these areas travels downward and is heated, it becomes buoyant and rises until it is trapped by an impermeable layer in the Earth's crust, forming a **hydrothermal reservoir**. Hydrothermal reservoirs contain hot water and possibly also steam, depending on the temperature and pressure of the fluid. Some of the hot water or steam may escape to the surface, creating hot springs or geysers.

Hydrothermal reservoirs are tapped by drilling wells similar to those used for extracting oil and natural gas. The hot fluid is brought to the surface and can be used to supply heat directly or to generate electricity. Hot springs have been used for thousands of years for bathing, cooking, and heating buildings. Hydrothermal reservoirs can be used to generate electricity (Figure 12.13). The fluid is brought up a well to the surface, and the resulting steam is expanded through a turbine to spin a generator, creating electricity. The electricity generated by these power stations is inexpensive and reliable, and geothermal energy requires a very small area of land (as compared to coal, nuclear, or other renewable energy sources).

The United States is the world's largest producer of geothermal electricity. Electric power is currently produced at 17 different geothermal fields in California Nevada, Utah, and Hawaii. The world's largest geothermal power plant is The Geysers, a geothermal field in northern California that provides electricity for 1.7 million homes. Other important producers of geothermal energy include the Philippines, Italy, Japan, Mexico, Indonesia, and Iceland.

Iceland, a country that possesses minimal energy resources other than geothermal energy and hydropower, is a good example of the benefits of geothermal energy. Situated on the mid-Atlantic ridge, a boundary between two continental plates, Iceland is an island of intense volcanic activity and consequently has considerable geothermal resources. Iceland uses geothermal energy to generate electricity and to heat two thirds of its homes. In addition, most of the fruits and vegetables required by the people of Iceland are grown in geothermally heated greenhouses.

Some argument exists as to whether geothermal energy is renewable. As a source of heat for geothermal energy, the planet is inexhaustible on a human time scale. However, the water used to transfer the heat to the surface is not inexhaustible. Some geothermal applications

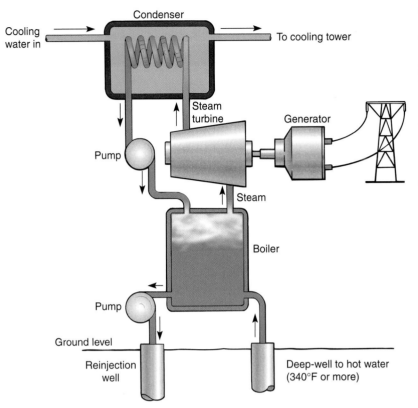

Figure 12.13 Geothermal energy. Shown is one design for a geothermal power plant. Steam separated from hot water that was pumped from underground is used to turn a turbine and generate electricity. After its use, the steam is condensed and pumped back into the ground. By reinjecting spent water into the ground, geothermal energy remains a renewable energy source because the cooler, reinjected water can be reheated and used again.

recirculate all the water back into the underground reservoir, ensuring many decades of heat extraction from a given reservoir. Other geothermal applications consume a portion of the water in the process, leading to the eventual depletion of the water in the underground reservoir. In these cases, better fluid management practices are helping to ensure longevity for these reservoirs. A good example is The Geysers geothermal field, where after nearly 40 years of production, experts estimate that about half the water has been depleted but that about 95% of the heat remains in the rock. Improved fluid management will probably help the field to remain productive for several decades more.

Geothermal energy is considered environmentally benign compared to conventional fossil fuel–based energy technologies. It emits only a fraction of the air pollutants and entails minimal land use. The most common environmental hazard associated with geothermal energy is the emission of hydrogen sulfide (H_2S) gas, which comes from the large amounts of dissolved minerals and salts found in the steam or hot water. Some geothermal reservoirs contain H_2S in quantities that require mitigation to meet air quality standards. Air pollution control methods are highly effective but do increase the energy cost. A lesser concern associated with geothermal energy is that the surrounding land may subside, or sink, as the water from hot springs and their connecting underground reservoirs is removed. Although experience has shown that this is not a problem at most geothermal fields, it has occurred at a few.

Geothermal Energy from Hot, Dry Rock   Conventional use of geothermal energy relies on hydrothermal reservoirs—that is, on groundwater—to bring the heat to the surface. These geothermal resources, however, are limited geographically and represent only a small fraction of total geothermal energy. Scientists in Australia, Europe, Japan, and the United States are studying how to extract some of the vast amount of geothermal energy stored in hot, dry rock. Scientists at the Los Alamos National Laboratory in New Mexico pioneered the concept of geothermal energy from hot, dry rock and succeeded in demonstrating its feasibility. They drilled a well into the hot, dry rock, used hydraulic pressure to fracture the rock, and then circulated water into the fractured area to make an artificial underground reservoir. When the pressurized water returned to the surface by way of a second well, it turned to steam, which drove an electricity-generating turbine. Current technology is available to create such systems, but it is very expensive. If technology can be improved to make hot, dry geothermal energy economically attractive, it could greatly expand the extent and use of geothermal resources.

Heating and Cooling Buildings with Geothermal Energy Increasingly, geothermal energy is employed to heat and cool commercial and residential buildings. **Geothermal heat pumps (GHPs)** take advantage of the difference in temperature between the Earth's surface and subsurface at depths from 1 m to about 100 m. Because the ground continually absorbs and retains much of the sun's energy, underground temperatures fluctuate only slightly and are much cooler in summer and warmer in winter than air temperatures. Geothermal heat pumps (GHPs) have an underground arrangement of pipes containing circulating fluids to extract natural heat in winter, when the Earth acts as a heat source, and transfer excess heat underground in summer, when the Earth acts as a heat sink. The hundreds of feet of pipe form a ground loop that feeds into a heat pump, which directs the flow of heated or cooled air. Geothermal heating systems can be modified to also provide supplemental hot water.

Though they have been available for many years, GHPs have not been widely used, as their installation can be expensive and complicated. However, with the growth of green architecture efforts (see Chapter 1) and rising fuel costs, commercial and residential use of the systems is on the rise. The system's benefits include low operating costs, which may be half those of conventional systems, and high efficiency. The Environmental Protection Agency (EPA) estimates GHPs to be the most efficient heating system available, two to three times more efficient than other heating methods, and producing the lowest carbon dioxide emissions. Currently, more than 250,000 U.S. homes have GHPs.

ENERGY SOLUTIONS: CONSERVATION AND EFFICIENCY

Human requirements for energy will continue to increase, if only because the human population is growing. In addition, energy consumption continues to increase as developing countries raise their standard of living. We must therefore place a high priority not only on developing alternative energy sources but also on energy conservation and energy efficiency. **Energy conservation** is moderating or eliminating wasteful or unnecessary energy-consuming activities, whereas **energy efficiency** is using technology to accomplish a particular task with less energy. As an example of the difference between energy conservation and energy efficiency, consider gasoline consumption by automobiles. Energy conservation measures to reduce gasoline consumption would include carpooling and lowering driving speeds, whereas energy efficiency measures would include designing and manufacturing more fuel-efficient automobiles. Both conservation and efficiency accomplish the same goal, that of saving energy.

Many energy experts consider energy conservation and energy efficiency to be the most promising energy "sources" available to humans, because they not only save energy for future use but also buy us time to explore new energy alternatives. Energy conservation and efficiency can cost less than development of new sources or supplies of energy and improve the productivity of the economy. The adoption of energy-efficient technologies generates new business opportunities, including the research, development, manufacture, and marketing of those technologies.

In addition to economic benefits and energy resource savings, there are important environmental benefits from greater energy efficiency and conservation. For example, using more energy-efficient appliances could cut our CO_2 emissions by millions of tons each year, thereby slowing global climate warming. Energy conservation and energy efficiency also reduce air pollution, acid precipitation, and other environmental damage related to energy production and consumption.

Energy Consumption Trends and Economics

A country's or region's total energy consumption divided by its gross domestic product gives one measure of its **energy intensity** (Table 12.3). Lower energy intensity implies that the economy is more energy efficient. According to the U.S. Department of Energy (DOE), if U.S. energy intensity had remained at its 1970 level, the United States would be spending an additional $150 billion to $200 billion on energy each year. Despite gains in efficiency, the energy intensities of the United States and Canada are still considerably higher than the energy intensities of Japan and Europe. Although the U.S. economy has become more energy efficient, total energy consumption has still increased in recent years, partly the result of a decline in energy prices and partly due to an increase in the population.

Energy Trends in Developing Countries   Per-capita consumption of energy in developing nations is substantially less than it is in industrialized countries (see Figure 10.1), although the greatest increase in energy consumption today is occurring in the developing nations. As these countries boost their economic development, their energy demands increase. This is partly because the "new" industrial and agricultural processes being adopted in developing countries are often older, less expensive technologies that are also less energy efficient. Also, the burgeoning populations in developing countries contribute to rising energy demands.

Table 12.3	1997 Energy Intensities for Selected Countries

Country	Energy Intensity*
Japan	1.07
Germany	1.49
France	1.60
United States	2.77
Canada	4.02

* Total energy consumption in metric tons of oil equivalent per gross domestic product in U.S. dollars $\times 10^{-4}$.

Developing countries are faced with the need for economic development and the need to control environmental degradation. At first glance, these two goals appear to be mutually exclusive. However, both goals can be realized by the adoption of new technologies now being developed in industrialized nations to achieve greater energy efficiency. For example, it would cost Brazil $44 billion to build power plants to meet its projected electricity needs for the near future; this cost could be avoided by investing $10 billion in more efficient refrigerators, lighting, and electric motors. The energy efficiency approach in Brazil would not only cause fewer environmental problems but also foster and expand the growth of manufacturing industries devoted to energy-efficient products.

Energy-Efficient Technologies

The development of more efficient appliances, automobiles, buildings, and industrial processes has helped reduce energy consumption in highly developed countries. Compact fluorescent lightbulbs, introduced in the late 1980s, produce light of comparable quality but require 25% of the energy used by regular incandescent bulbs and last nine times longer. The energy-efficient bulbs are expensive (they cost about $15 each), but they more than pay for themselves in energy savings. (The payback period for a light that is on 12 hours a day is 1 year.) Standard long-tube fluorescent bulbs have also become more efficient. New condensing furnaces require approximately 30% less fuel than conventional gas furnaces. "Superinsulated" homes use 70% to 90% less heat than do homes insulated by standard methods (Figure 12.14).

The **National Appliance Energy Conservation Act** (**NAECA**) sets national appliance efficiency standards for refrigerators, freezers, washing machines, clothes dryers, dishwashers, room air conditioners, and ranges/ovens (including microwaves). Refrigerators built in 2001, for example, consume 75% less energy than comparable models built in the mid-1970s. This translates to an average saving to the consumer of $135 per year. When 150 million refrigerators are operating at the 2001 standard, the annual electrical energy saved will be equivalent to the output of about 32 nuclear power plants!

The NAECA requires appliance manufacturers to provide Energy Guide labels on all new appliances. These yellow labels provide estimates of the annual operating costs and efficiency levels. Consumers who use this information to buy energy-efficient appliances save hundreds of dollars on utility bills.

Automobile efficiency has improved dramatically since the mid-1970s as a result of the use of lighter materials and designs that reduce air drag. The U.S. average fuel efficiency of new passenger cars doubled between the mid-1970s and the mid-1980s, although it has declined

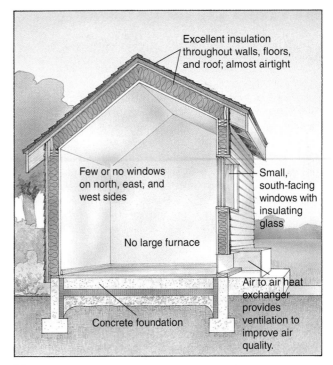

Excellent insulation throughout walls, floors, and roof; almost airtight

Few or no windows on north, east, and west sides

Small, south-facing windows with insulating glass

No large furnace

Air to air heat exchanger provides ventilation to improve air quality.

Concrete foundation

(a)

(b)

Figure 12.14 Superinsulated buildings. (a) Some of the characteristics of a superinsulated home, which is so well insulated and airtight that it does not require a furnace in winter. Heat from the inhabitants, lightbulbs, stove, and other appliances provides almost all the necessary heat. (b) A superinsulated office building in Toronto, Canada, has south-facing windows with insulating glass. The building is so well insulated that it uses no furnace.

A Starring Role in Energy Efficiency

The use of energy-efficient appliances instead of conventional models can reduce air pollution, often at lower energy costs. But how does a consumer know which models to buy? The Energy Star labeling program, a partnership effort among the EPA, DOE, product manufacturers, retailers, and local utilities, provides detailed energy efficiency information to appliance buyers. Since 1992, the program has assigned the Energy Star label to 29 different types of home and office products, such as dishwashers, refrigerators, clothes washers, air conditioners, and computers and monitors. Recommended models, which are widely available at conventional retailers, earn the rating by demonstrating a level of energy efficiency significantly greater than minimum government standards. Many retailers and manufacturers identify their Energy Star products, and local utilities often offer rebates on their purchase.

Because of the program's success, the EPA is considering expanding its labeling efforts to other commercial and residential products such as ice machines, vending machines, and traffic lights. A complete listing of Energy Star–qualified products, as well as tips on determining energy efficiency of specific appliances, is available at the Energy Star Web site, **http://www.energystar.gov**.

since then. A trend that has reduced energy efficiency is the popularity of minivans, sport utility vehicles, and light trucks, all of which have lower average gas mileages than sedan-type automobiles. Despite the reduction in energy efficiency caused by recent consumer preferences, significant gains could easily be implemented. Using current technology, automobiles with fuel efficiencies of 60 to 65 mpg could be routinely manufactured within the next decade or so, assuming there is consumer demand for these vehicles.

During the past few decades, many industries have improved their energy efficiencies. New aircraft are much more fuel-efficient than older models, and technological improvements in the papermaking industry make it possible to use less energy to manufacture paper today than was used just a few years ago. The energy savings from such improvements in efficiency translate into greater profits for the companies employing them.

Cogeneration   One energy technology with a bright future is **cogeneration**, which is the production of two useful forms of energy from the same fuel. Cogeneration is also known as **combined heat and power** (**CHP**). Typically, CHP involves the generation of electricity, and then the steam produced during this process is used rather than wasted. In CHP, the overall conversion efficiency (that is, the ratio of useful energy produced to fuel energy used) is very high because some of what would usually be waste heat is used.

Cogeneration can be done very cost effectively on a small scale. Modular CHP systems enable hospitals, hotels, restaurants, factories, and other businesses to harness steam that would otherwise be wasted. In a typical CHP system, electricity is produced in a traditional manner—that is, some type of fuel provides heat to form steam from water. Normally, the steam used to turn the electricity-generating turbine would be cooled before being pumped back to the boiler to be reheated. In cogeneration, after the steam is used to turn the turbine, it supplies energy to heat buildings, cook food, or operate machinery before it is cooled and pumped back to the boiler as water (Figure 12.15).

Cogeneration can also be accomplished on a large scale. One of the largest CHP systems in the United States is a "combined cycle" natural gas plant completed in 1995 in Oswego, New York, that produces electricity for the local utility. It uses natural gas turbines to generate electricity. The exhaust gases, which are 1,000°C (1,832°F) to 1,200°C (2,192°F), are then used to produce high-pressure steam that is used by a nearby industry. The overall efficiency of the Oswego cogeneration system is 54%, as compared to 33% efficiency in a typical fossil fuel power plant.

Figure 12.15 Cogeneration. In this example of a cogeneration system, fuel combustion occurs in a boiler (1). The heat produced is used to make steam, which turns a turbine (2) that generates electricity in a generator (3). The electricity that is produced (4) is used in-house or sold to a local utility. Cogeneration involves using the waste heat (leftover steam) from electricity generation to do useful work, such as cooking, space heating, or operating machinery (5). Any residual steam that is not used is piped into a condenser (6) and recycled back to the boiler.

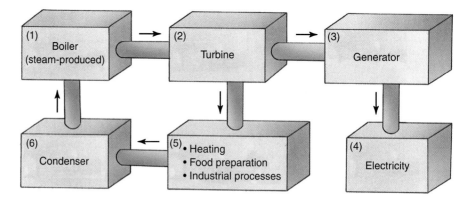

Energy Savings in Commercial Buildings   Energy costs often account for 30% of a company's operating budget. Unlike cars, which are traded in every few years, buildings are usually used for 50 or 100 years, so a company housed in an older building normally does not have the benefits of new energy-saving technologies. It makes good economic sense for these businesses to invest in energy improvements, which often pay for themselves in a few years (Table 12.4). More than 20,000 schools, hospitals, commercial office buildings, retail outlets, multifamily residential buildings, and other facilities across the United States have had energy-efficient upgrades that have provided energy savings of $.20 to $2.00 per square foot.

To get businesses to install new energy-efficient technologies, energy-services companies, which specialize in designing and installing energy-efficient technologies, offer their assistance in such a way that the business makes little or no financial outlay. Here is how it works. An energy-services company makes a detailed assessment of how a business can improve its energy efficiency. In developing its proposal, the energy-services company guarantees a certain amount of energy savings. It also provides the funding to accomplish the improvements, which may be as simple as fine-tuning existing heating, ventilation, and air conditioning systems or as major as replacing all existing windows and lights. The reduction in utility costs is used to pay the energy-services company, but once the bill is paid, the business benefits from all additional energy savings. (See "You Can Make a Difference: Saving Energy at Home" for energy-saving suggestions you can use in your own home.)

Electric Power Companies and Energy Efficiency

Changes in the regulations governing electric utilities have enabled utilities to make more money by generating *less* electricity. Such programs provide incentives to save

ENVIROBRIEF

Netting the Benefits of Home Energy Production

Environmentally aware homeowners have often paid a large economic price when they install renewable energy sources such as solar panels or wind-driven generators, known for their "clean" energy production. Such systems can be expensive to purchase and install, and they only provide electricity intermittently, when weather conditions permit. During unfavorable conditions, residents must purchase electricity from their local utility.

In a growing trend, some utility companies are permitting homeowners who produce their own energy to "net meter" their electricity. Any excess energy they generate is supplied to the utility's power grid, allowing their electrical meters to run backward. Net metering offsets energy costs over a billing period by essentially providing individuals the full retail price for the electricity they generate. By itself, net metering does not turn residential energy generation into an inexpensive prospect, but it certainly makes it more affordable, thus encouraging the additional development of renewable energy for homes. Allowed to "bank" their energy to withdraw it later, consumers save money while contributing to a cleaner environment.

More than half of the individual states allow regulated net metering, and a Congressional bill introduced in March 2001, the Home Energy Generation Act, would, if passed, amend the Federal Power Act to make net metering available nationally.

energy and thereby reduce power plant emissions that contribute to environmental problems.

Traditionally, to meet future power needs, electric utilities planned to build new power plants or purchase additional power from alternative sources. Now they often avoid these massive expenses by **demand-side management**, in which they help electricity consumers save energy. Some utilities support energy conservation and efficiency by offering cash awards to consumers who install energy-efficient technologies. When voters in California decided to close the nuclear power plant Ran-

Table 12.4   **Energy-Efficiency Upgrades in Selected Commercial Buildings**

Project	Energy Payback Time*	Unexpected Benefits Attributed to Project**
Energy-efficient lighting (post office in Nevada)	6 years	6% increase in mail sorting productivity
Energy-efficient (metal-halide) lighting (aircraft assembly plant in Washington)	2 years	Up to 20% better quality control
Energy-efficient lighting (drafting area of utility company in Pennsylvania)	About 4 years	25% lower absenteeism; 12% increase in drawing productivity
Energy-efficient lighting and air conditioning (office building in Wisconsin)	0 years (paid for by utility rebates); energy savings estimated at 40%	16% increase in worker productivity
Energy-saving daylighting, passive solar heating, heat recovery system (bank in Amsterdam)	3 months	15% lower absenteeism

* How long it takes for energy savings to cover the cost of the project.
** Lighting quality as well as lighting efficiency is improved, resulting in greater worker comfort.

YOU CAN MAKE A DIFFERENCE

Saving Energy at Home

The average household spends $1,500 each year on utility bills. This cost could be reduced considerably by investments in energy-efficient technologies.

When buying a new home, a smart consumer should demand energy efficiency. Although a more energy-efficient house might cost more, depending on the technologies employed, the improvements usually pay for themselves in two or three years. Any time spent in the home after the payback period means substantial energy savings. Energy efficiency has become an essential element of design codes nationwide and will almost certainly be an important part of future home designs.

Some energy-saving improvements, such as thicker wall insulation, are easier to install while the home is being built. Other improvements can be made in older homes to enhance energy efficiency and, as a result, reduce the cost of heating the homes. Examples include installing thicker attic insulation, installing storm windows and doors, caulking cracks around windows and doors, replacing inefficient furnaces and refrigerators, and adding heat pumps.

Many of the same improvements also provide energy savings when a home is air-conditioned. Additional cooling efficiency is achieved by insulating the air conditioner ducts, especially in the attic; buying an energy-efficient air conditioner; and shading the south and west sides of a house with deciduous trees. Window shades and awnings on south- and west-facing windows can also help reduce the heat a building gains from its environment. Ceiling fans can supplement air conditioners by making a room feel comfortable at a higher thermostat setting. Make sure, however, that your ceiling fan is set to draw warm air *toward* the ceiling in the summer, and reverse this setting in the winter.

Other energy savings in the home include replacing incandescent bulbs with energy-efficient compact fluorescent lightbulbs; installing a programmable thermostat, which can cut heating and air-conditioning costs up to 33%; lowering the temperature setting on water heaters to 140°F (with a dishwasher) or 120°F (without); and installing low-flow shower heads and faucet aerators to reduce the amount of hot water used (see figure).

How does a homeowner learn which improvements will result in the most substantial energy savings? In addition to reading the many articles on energy efficiency that appear in newspapers and magazines, a good way to learn about your home is to have a comprehensive energy audit done. Most local utility companies can send an energy expert to your home to perform an audit for little or no charge. The audit will determine the total energy consumed and where thermal losses are occurring (through the ceiling, floors, walls, or windows). On the basis of this assessment, the energy expert will then make recommendations about how you can reduce your heating and cooling bills.

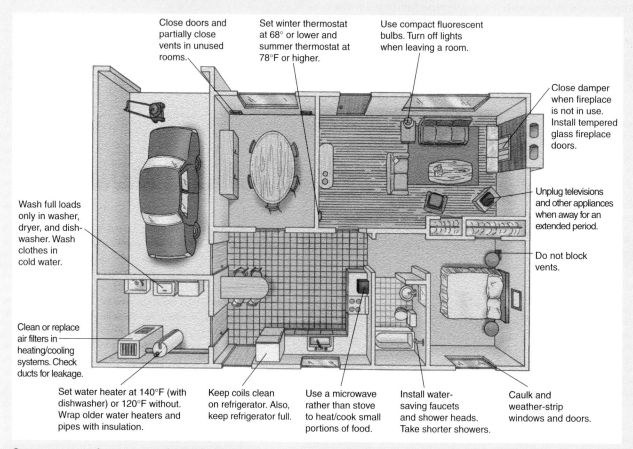

Some energy-saving measures for the home.

cho Seco, the Sacramento Municipal Utility District paid customers to buy more efficient refrigerators and plant trees to shade their houses, thereby lowering air conditioning costs. These efforts helped the utility company to reduce the demand for electricity.

Some utilities also give customers energy-efficient compact fluorescent lightbulbs, air conditioners, or other appliances. They then charge slightly higher rates or a small leasing fee, but the greater efficiency results in savings for both the utility company and the consumer. The utility company makes more money from selling less electricity because it does not have to invest in additional power generation to meet increased demand. The consumer saves because the efficient lightbulbs or appliances use less energy, which more than offsets the higher rates.

Electricity Deregulation   The U.S. electricity industry, which consists of about 5,000 power plants and 254,000 km (158,000 mi) of transmission lines, traditionally had hundreds of monopolies regulated by state public utility commissions. Now, however, the electricity industry is changing to one in which companies compete with one another for customers across state lines. Competitive utilities are thought to be more likely than regulated monopolies to adopt technological advances that increase energy efficiency and thereby lower costs. Deregulation depends on competitive market forces for investment, operation, and reliability management of power plants and transmission lines.

California was the first state to become deregulated, in 1996. Since then, 23 states and the District of Columbia have adopted electricity deregulation in some form or another. During the winter of 2001, California had a power crisis in which there were rolling blackouts and a hefty increase in the price of electricity. Deregulation was initially blamed for the crisis in California, but the actual reasons were complex and related to an imbalance in the supply of and demand for electricity. During the next 10 years, U.S. demand for electricity is expected to increase by about 25%. To prevent regional shortages such as occurred in California, the U.S. electric power system will have to increase electricity supply and reliability. According to the Federal Energy Regulatory Commission, 190,000 megawatts of generating capacity are currently under construction.

Energy Conservation at the Individual Level

Simple measures such as lowering your thermostat during the winter and raising it during the summer, turning off lights when you leave a room, and driving more slowly result in small energy (and cost!) savings. You also contribute to energy conservation by making use of carpools or public transportation. The cumulative effect of many people taking similar measures is substantial.

Energy conservation and efficiency are sound ideas for all of us. Using more energy, even from renewables, impacts the environment. Energy saved today will be available for our children and grandchildren. The energy we save now will help to slow down climate warming and environmental degradation so that our consumption will not become an overwhelming burden on future generations. The energy we save now will give us additional time to develop and improve alternative energy sources.

SUMMARY WITH SELECTED KEY TERMS

I. Renewable energy sources are increasingly being examined with our future energy needs in mind.

A. Renewable energy sources generally cause less environmental impact than fossil fuels or nuclear energy.

B. Many forms of renewable energy are dispersed and therefore require costly energy collection equipment.

II. Solar energy, the radiant energy derived from the sun, is used directly to heat water, heat buildings, and generate electricity. Solar energy is also used indirectly as, for example, wind energy used to generate electricity.

III. Direct solar energy is projected to become increasingly important in the future.

A. Direct solar energy can be used either actively or passively to heat buildings and water.

　1. Buildings designed for **passive solar heating** distribute the collected heat without the need for pumps or fans. Currently, about 7% of new homes built in the United States have passive solar features.

　2. In **active solar heating**, a series of collection devices gather solar energy for heating water and, to a lesser extent, space heating.

B. Solar thermal electric generation produces electricity by concentrating sunlight onto a fluid. The heated fluid changes water to steam, which turns a turbine to generate electricity.

C. Photovoltaic (PV) solar cells convert solar energy directly into electricity.

　1. Photovoltaic cells are wafers or thin films of solid state materials that are treated with certain metals such that they absorb solar energy and generate electricity.

　2. Photovoltaics generate electricity with no pollution and minimal maintenance.

　3. Photovoltaics are only about 10% to 15% efficient at converting solar energy to electricity.

D. Solar-generated electricity can be used to split water into hydrogen and oxygen.

1. Solar-generated hydrogen is a clean fuel that can provide energy for transportation as well as for generating electricity.

2. A **fuel cell** is an electrochemical cell in which the reactants (in this case, hydrogen and oxygen) are supplied from external reservoirs. Motor vehicles powered by hydrogen fuel cells are being developed.

IV. Forms of indirect solar energy include biomass, wind, hydropower, ocean waves, and ocean thermal energy conversion.

A. Biomass consists of wood, agricultural wastes, fast-growing plants, and other organic materials that are the product of photosynthesis.

1. Biomass can be burned directly to produce heat or electricity or converted to gas (**biogas**) or liquid fuels (**methanol** and **ethanol**).

2. Biomass is already being used for energy on a large scale, particularly in developing nations. India and China have several million **biogas digesters** that produce biogas from household and agricultural wastes.

B. Wind energy results from the warming of the atmosphere by the sun.

1. Harvesting wind energy to generate electricity has great potential because it is currently the most cost-competitive of all forms of solar energy.

2. Harnessing wind energy is most profitable in areas that receive fairly continual winds, such as islands, coastal areas, mountain passes, and grasslands.

C. Hydropower uses the energy of flowing water to generate electricity. Hydropower is an indirect form of solar energy, because the sun drives the hydrologic cycle.

1. Currently, hydropower produces about 19% of the world's electricity.

2. Both environmental and social problems are associated with hydropower. These include destruction of wildlife habitat, reservoir-induced seismicity, greater evaporation of water, displacement of people, and destruction of villages and farmland.

D. Ocean waves are an indirect form of solar energy, in which wave power turns a turbine, thereby generating electricity.

E. Ocean thermal energy conversion, which makes use of the temperature gradient of the ocean, may be used to generate electricity in the future.

V. Renewable energy sources that are not the direct or indirect result of solar energy include tidal energy and geothermal energy.

A. Tidal energy, caused by the change in water level between high and low tides, is being used to generate electricity on a very limited scale.

B. Geothermal energy can be obtained from **hydrothermal reservoirs** of heated water near the Earth's surface.

1. The established technology for extracting geothermal energy from heated areas of Earth's crust involves drilling wells and bringing the steam or hot water to the surface.

2. Geothermal energy from hot, dry rocks, if developed, could make geothermal energy more widely available.

3. **Geothermal heat pumps** take advantage of the difference in temperature between the Earth's surface and subsurface. Geothermal heat pumps use the ground as a heat source to provide warmth in winter and as a heat sink to provide coolness in the summer.

VI. Promising alternative energy "sources" are energy conservation and improved energy efficiency.

A. Energy efficiency is using technology to accomplish a particular task with less energy.

1. Increased energy efficiency results in energy savings that help conserve Earth's conventional fuel supplies.

2. One measure of **energy intensity** is a country's total energy divided by its gross domestic product. Lower energy intensity implies greater energy efficiency for a particular standard of life.

3. **Cogeneration**, the production of two forms of energy (usually, electricity and useful heat) from the same fuel, increases energy efficiency. Cogeneration is also known as **combined heat and power (CHP)**.

B. Energy conservation involves moderating or eliminating wasteful or unnecessary energy-consuming activities.

▮ THINKING ABOUT THE ENVIRONMENT

1. Distinguish between *active* and *passive solar energy.*

2. Explain the following statement: Unlike fossil fuels, solar energy is not resource-limited but is technology-limited.

3. Distinguish between *solar thermal* and *photovoltaic (PV) solar cells* to produce electricity.

4. Biomass is considered an example of indirect solar energy because it is the result of photosynthesis. Given that plants are the organisms that photosynthesize, why are animal wastes also considered biomass?

5. One advantage of the various forms of renewable energy, such as solar thermal and wind energy, is that they cause no net increase in atmospheric carbon dioxide. Is this true for biomass? Why or why not?

6. Cite the advantages and disadvantages of using wind to produce electricity and of using hydropower to produce electricity.

7. Some energy experts refer to the Great Plains states as "the Saudi Arabia of wind power." Explain what the reference means.

8. Why is it easier to obtain energy from a small river with a steep grade than from the vast ocean currents?

9. Japan wishes to make use of solar power, but it does not

have extensive tracts of land for building large solar power plants. Which solar technology do you think is best suited to Japan's needs? Why?

10. Give an example of how one or more of the alternative energy sources discussed in this chapter could have a negative effect on each of the following aspects of ecosystems:

   a. Soil preservation

   b. Natural water flow

   c. Production of foods used by wild plant and animal populations

   d. Preservation of the diversity of organisms found in an area

11. What are the pros and cons of using geothermal energy to produce electricity? Of using tidal power to produce electricity?

12. Explain how energy conservation and efficiency are major "sources" of energy.

13. What is cogeneration, and what are its advantages?

14. Evaluate which forms of energy other than fossil fuels and nuclear power have the greatest potential where you live.

15. List energy conservation measures that you could adopt for each of the following aspects of your life: washing laundry, lighting, bathing, cooking, buying a car, driving a car.

*16. Examine the data in Table 12.A on global wind power generating capacity from 1990 to 2001. Calculate the percent increase over the preceding year and place your answers in Table 12.A.

Table 12.A	Global Wind Power Generating Capacity, 1990 to 2001	
Year	**Capacity (megawatts)**	**Percent Increase over Preceding Year**
1990	1,930	11.6
1991	2,170	
1992	2,510	
1993	2,990	
1994	3,490	
1995	4,780	
1996	6,070	
1997	7,640	
1998	10,150	
1999	13,930	
2000	18,100	
2001 (preliminary)	24,800	

* The solution to this question appears in Appendix VII.

TAKE A STAND

Visit our Web site at **http://www.wiley.com/college/raven** (select Chapter 12 from the Table of Contents) for links to more information about the controversy surrounding subsidies for alternative energy development. Consider the views of proponents and opponents, and debate the issue with your classmates. You will find tools to help you organize your research, analyze the data, think critically about the issues, and construct a well-considered argument. Take a Stand activities can be done individually or as part of a team, as oral presentations, written exercises, or Web-based (e-mail) assignments.

Additional on-line materials relating to this chapter, including Student Quizzes, Activity Links, Useful Web sites, Flash Cards, and more, can also be found on our Web site.

SUGGESTED READING

Crossen, C. "How Much Power Do You Use?" *Wall Street Journal* (August 16, 2001). Examines energy use in the United States.

Dunn, S. "The Hydrogen Experiment." *World Watch*, Vol. 13, No. 6 (November–December 2000). Iceland is committed to becoming the first in the world to use hydrogen as an energy carrier.

Ezzell, C. "The Himba and the Dam." *Scientific American*, Vol. 284, No. 6 (June 2001). How an African tribe's way of life will be destroyed by a proposed dam.

Flavin, C., and M, O'Meara. "Solar Power Markets Boom." *World Watch*, Vol. 11, No. 5 (September–October 1998). Some of the exciting developments in the world of photovoltaics.

Gibb, W.W. "The Power of Gravity." *Scientific American*, Vol. 287, No. 1 (July 2002). How electricity is generated by the Hoover Dam.

Hart, D. "Fuelling the Future." *New Scientist: Inside Science* (June 16, 2001). The fuel cell may help us consume the vast amount of energy we require without harming the planet.

Löfstedt, R. "Sweden's Biomass Controversy: A Case Study of Communicating Policy Issues." *Environment*, Vol. 40, No. 4 (May 1998). An in-depth analysis of the controversy surrounding Sweden's decision to expand its use of biomass.

Pearce, F. "Catching the Tide." *New Scientist* (June 20, 1998). Covers the history and potential future of tidal power.

Turner, J.A. "A Realizable Renewable Energy Future." *Science*, Vol. 285 (July 30, 1999). The author builds a strong case for subsidizing renewable energy technologies as a means of controlling carbon dioxide emissions, which are linked to global climate warming.

Weiss, P. "Oceans of Electricity." *Science News*, Vol. 159 (April 14, 2001). New technologies are being tried to convert wave energy into electricity.

Williams, W. "Blowing Out to Sea." *Scientific American*, Vol. 286, No. 3 (March 2002). Offshore wind farms may solve some of the problems associated with wind power.

Landsat image of the San Francisco Bay area.

Water: A Fragile Resource

Learning Objectives

After you have studied this chapter you should be able to

1. Draw a simple diagram of a water molecule, indicating the regions of partial positive and partial negative charges and how hydrogen bonds form between adjacent water molecules.

2. Describe *surface water* and *groundwater*, using the following terms in your descriptions: *wetland, runoff, drainage basin, unconfined* and *confined aquifer*, and *water table*.

3. Explain how humans exacerbate property damage caused by floods using the upper Mississippi River basin as an example.

4. Relate some of the problems caused by overdrawing surface water, aquifer depletion, and salinization of irrigated soil.

5. Relate the background behind each of the following U.S. water problems: Mono Lake, the Colorado River basin, and the Ogallala Aquifer.

6. Briefly describe each of the following international water problems: drinking water problems, population growth and water problems, the Rhine River basin, the Aral Sea, and potentially volatile international situations over water rights.

7. Contrast the benefits and drawbacks of dams and reservoirs, using the Columbia River to provide specific examples.

8. Give examples of how water can be conserved by agriculture, industry, and individual homes and buildings.

The San Francisco Bay and its delta, a 4,144-km² (1,600-mi²) estuary that drains 45% of California's land area, is both environmentally and economically important. Home to more than 750 plant and animal species, it is the largest, most productive U.S. estuary on the Pacific coast (see figure). It is also highly developed. The cities of San Francisco (the peninsula on the upper left, 1), Oakland (upper right, opposite San Francisco, 2), and San Jose (bottom right, 3) are evident in this Landsat image. The Sacramento and San Joaquin Rivers that flow into the bay and its delta provide drinking water for two thirds of the state population—more than 20 million people. The rivers also supply irrigation water for more than 7 million acres of farmland that produce 45% of the nation's vegetables and fruits. In addition, many industries in the region require a plentiful supply of water.

The bay is smaller than it was 100 years ago, the result of tidal marshes being filled and diked to provide land for farms, homes, and industries. Today there is less than 130 km² of marshland remaining, as compared to 2,200 km² in 1850.

Years of water diversion for municipal and agricultural uses, along with water pollution and the introduction of exotic species, have harmed the delta. Water diversions mean that less fresh water flows through the delta toward the Pacific Ocean. As a result, more salt water intrudes from the ocean. Some municipalities have experienced salty tap water—a violation of the Clean Water Act—and populations of certain fish species have plummeted. When several species, notably the delta smelt and chinook salmon, were declared endangered, the diversion of water became a violation of the Endangered Species Act.

Various federal and state agencies and competing interest groups such as water users and environmentalists have argued for years over the extent of the bay's environmental problems and who is to blame. Developing a viable solution was impossible in such a confrontational atmosphere. In 1994, compromise and cooperation among the various groups during a year of intense negotiations initiated by the Clinton administration resulted in a historic agreement, called the CALFED Bay-Delta Program, to protect the San Francisco Bay and its delta.

Since 1994, the CALFED Bay-Delta Program has successfully specified the environmental problems and water management difficulties in San Francisco Bay and its delta: ecosystem health and restoration, water quality, water supply reliability, and levee system integrity. (Levees are embankments built alongside rivers to prevent flooding of adjacent lands during periods of high precipitation. Concern exists that the levee system might fail during a flood.) The program is developing a comprehensive plan to address these problem areas. The project is expected to take three decades and cost more than $10 billion to complete.

In this chapter we examine the complexities of water use and management. We begin with a discussion of the importance of water resources.

THE IMPORTANCE OF WATER

The view of planet Earth from outer space reveals that it is different from other planets in the Solar System. Earth is a predominantly blue planet because of the water that covers three fourths of its surface. Water has a tremendous effect on our planet: It helps shape the continents, it moderates our climate, and it allows organisms to survive.

Life on planet Earth would be impossible without water. All life forms, from simple unicellular bacteria to complex multicellular plants and animals, contain water. Humans are composed of approximately 70% water by body weight. We depend on water for our survival as well as for our convenience: We drink it, cook with it, wash with it, travel on it, and use an enormous amount of it for agriculture, manufacturing, mining, energy production, and waste disposal.

Although the Earth has plenty of water, about 97% of it is salty. The fresh water that exists is distributed unevenly, resulting in serious regional water supply problems. In regions where fresh water is in short supply, such as deserts, obtaining it is critically important. Because the use of water for one purpose decreases the amount available for other purposes, serious conflicts often arise over how water should be used. Even regions with readily available fresh water have problems, however, and maintaining the quality and quantity of water is important.

Worldwide, freshwater use is increasing. This is in part because the human population is expanding and in part because, on the average, each person is using more water. According to the U.N. Economic and Social Council, during the 20th century water use grew at more than twice the rate of population growth. A 1996 report published in the journal *Science* estimated that humans use 54% of all fresh water flowing in rivers and streams. An increasing number of countries are experiencing water shortages as population growth and human activities place increasing demands on a limited water supply. A 1998 report from the Johns Hopkins University School of Public Health predicted that by 2025 more than one third of the human population will live in areas where there is not enough fresh water for drinking and irrigation.

To meet the growing need for water, we try to augment our supply by building dams to create **reservoirs** (artificial lakes in which water is stored for later use) and by diverting river water. In many areas, the quantity of water is not as critical as its quality, and steps must be taken to ensure a supply of clean water. All of these efforts to obtain and maintain a steady supply of clean water involve considerable expense.

This chapter examines some of the ecological processes and human activities that affect the availability of water. Water quality, including water pollution, is such a significant issue that it is covered separately, in Chapter 21.

Properties of Water

Water is composed of molecules of H_2O, each consisting of two atoms of hydrogen and one atom of oxygen. Water exists in any of three forms: solid (ice), liquid, and vapor (water vapor or steam). Water molecules are polar—that is, one end of the molecule has a positive electrical charge, and the other end has a negative charge (Figure 13.1). The negative (oxygen) end of one water molecule is attracted to the positive (hydrogen) end of another water molecule, forming a **hydrogen bond** between the two molecules. Hydrogen bonds are the basis for many of water's physical properties, including its high melting/freezing point (0°C, 32°F) and high boiling point (100°C, 212°F). Because most of the Earth has a temperature between 0°C and 100°C, most water exists in the liquid form organisms need.

Water absorbs a great deal of solar heat without its temperature rising substantially. It is this high heat capacity that allows the ocean to have a moderating influence on climate, particularly along coastal areas. Another consequence of water's high heat capacity is that the

(a) Polar nature of water molecule

(b) Hydrogen bonding of water molecules
due to their polarity

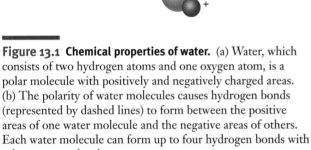

Figure 13.1 Chemical properties of water. (a) Water, which consists of two hydrogen atoms and one oxygen atom, is a polar molecule with positively and negatively charged areas. (b) The polarity of water molecules causes hydrogen bonds (represented by dashed lines) to form between the positive areas of one water molecule and the negative areas of others. Each water molecule can form up to four hydrogen bonds with other water molecules.

Water 965 grams

Salts 35 grams

All others
0.20 gram

Potassium (K$^+$)
0.38 gram

Calcium (Ca^{2+})
0.42 gram

Magnesium (Mg^{2+})
1.3 grams

Sulfate (SO$_4^{2-}$)
2.7 grams

Chloride (Cl$^-$)
19.3 grams

Sodium (Na$^+$)
10.7 grams

Figure 13.2 Chemical composition of 1 kg (2.2 lb) of seawater. Seawater contains a variety of dissolved salts, which are present as ions.

ocean does not experience the wide temperature fluctuations that are common on land.

Water must absorb a lot of heat before it **vaporizes,** or changes from a liquid to a vapor. (2,250 joules [540 calories] of energy are required to convert 1 g of water from a liquid to a vapor.) When it does evaporate, it carries the heat, called its heat of vaporization, with it into the atmosphere. Thus, evaporating water has a cooling effect. That is why your body is cooled when perspiration evaporates from your skin.

Water can also **sublimate,** or change from a solid to a vapor without going through the liquid phase. The preservation of various products, such as food, blood plasma, vaccines, and antibiotics, by freeze-drying takes advantage of the fact that water can sublimate.

Water is sometimes called the "universal solvent," and although this is an exaggeration, many materials do dissolve in water. In nature, water is never completely pure, because it contains dissolved gases from the atmosphere and dissolved mineral salts from the land. Seawater contains a variety of dissolved salts, including sodium chloride, magnesium chloride, magnesium sulfate, cal-

cium sulfate, and potassium chloride (Figure 13.2). Water's dissolving ability has a major drawback: Many of the substances that dissolve in water also cause water pollution.

Water partially obeys the general physical rule that heat expands and cold contracts. As water cools, it contracts and becomes denser until it reaches 4°C (39°F), the temperature at which it is the densest. When the temperature of water falls *below* 4°C, it becomes less dense. Thus, ice (at 0°C) floats on the denser, slightly warmer, liquid water. Because of this, water freezes from the top down rather than from the bottom up, and aquatic organisms can survive beneath a frozen surface.

The Hydrologic Cycle and Our Supply of Fresh Water

Water continuously circulates through the abiotic environment, from the ocean to the atmosphere to the land and back to the ocean, in a complex cycle known as the **hydrologic cycle** (see Chapter 6). The result is a balance among water in the ocean, on the land, and in the atmosphere. The hydrologic cycle continually renews the supply of fresh water on land, which is essential to terrestrial organisms.

Approximately 97% of the Earth's water is in the ocean and contains a high amount of dissolved salts (Figure 13.3). Seawater is too salty for human consumption and for most other uses. For example, if you watered your garden with seawater, your plants would die. Most fresh water is unavailable for easy human consumption because it is frozen as polar or glacial ice or is in the atmosphere or soil. Lakes, creeks, streams, rivers, and groundwater account for only a small portion—about 0.03%—of the Earth's fresh water.

Surface water is fresh water found on Earth's surface in streams and rivers, lakes, ponds, reservoirs, and **wetlands**, areas of land that are covered with water for at least part of the year. Surface waters are replenished by the **runoff** of precipitation from the land and are therefore considered a renewable, although finite, resource. A **drainage basin** or **watershed** is the area of land that is drained by a single river or stream. Watersheds range in size from less than 1 km² for a small stream to a huge area of the continent for major river systems such as the

Table 13.1	The World's Ten Largest Watersheds		
Watershed	**Region**	**Area of Watershed (thousand km²)**	**Population Density per km², Late 1990s**
Amazon	South America	6,144	4.3
Congo	Africa	3,807	14.5
Nile	Africa	3,255	42.7
Mississippi	North America	3,202	21.5
Ob	Asia	3,028	191.9
Parana	South America	2,583	23.5
Yenisey	Asia	2,499	2.3
Lena	Asia	2,307	1.3
Niger	Africa	2,262	31.2
Yangtze	Asia	1,722	223.7

Mississippi River. Table 13.1 lists the world's 10 largest watersheds.

Earth contains underground formations that collect and store water in the ground. This water originates as rain or melting snow that seeps into the soil and finds its way down through cracks and spaces in sand, gravel, or rock until it is stopped by an impenetrable layer; there it accumulates as **groundwater**. Groundwater flows through permeable sediments or rocks slowly—typically covering distances of several millimeters to a few meters per day. Eventually it is discharged into rivers, wetlands, springs, or the ocean. Thus, surface water and groundwater are interrelated parts of the hydrologic cycle.

The underground caverns and porous layers of sand, gravel, or rock in which groundwater is stored are called **aquifers** (Figure 13.4). Aquifers can be either unconfined

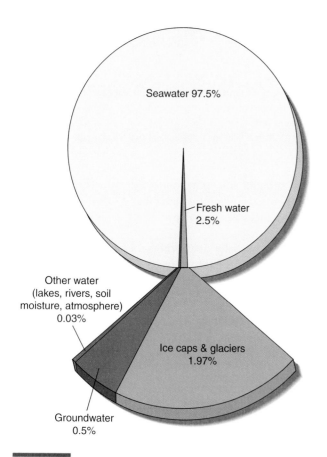

Figure 13.3 Distribution of water. Although three fourths of the Earth's surface is covered with water, substantially less than 1% is available for humans. Most water is salty, frozen, or inaccessible in the soil and atmosphere.

Labels in figure:
Seawater 97.5%
Fresh water 2.5%
Other water (lakes, rivers, soil moisture, atmosphere) 0.03%
Ice caps & glaciers 1.97%
Groundwater 0.5%

MINI-GLOSSARY

Water Terms

surface water: Fresh water found in streams, rivers, lakes, ponds, reservoirs, and wetlands.

runoff: The movement of fresh water from precipitation and snowmelt to rivers, lakes, wetlands, and, ultimately, the ocean.

drainage basin: The area of land drained by a single river or river system; also called *watershed*.

groundwater: The supply of fresh water under Earth's surface that is stored in aquifers.

aquifers: Underground caverns and porous layers of sand, gravel, or rock in which groundwater is stored.

unconfined aquifer: An aquifer whose upper surface coincides with the water table.

water table: The upper surface of the saturated zone of groundwater.

confined aquifer: An aquifer bounded by impermeable rock layers; also called *artesian aquifer*.

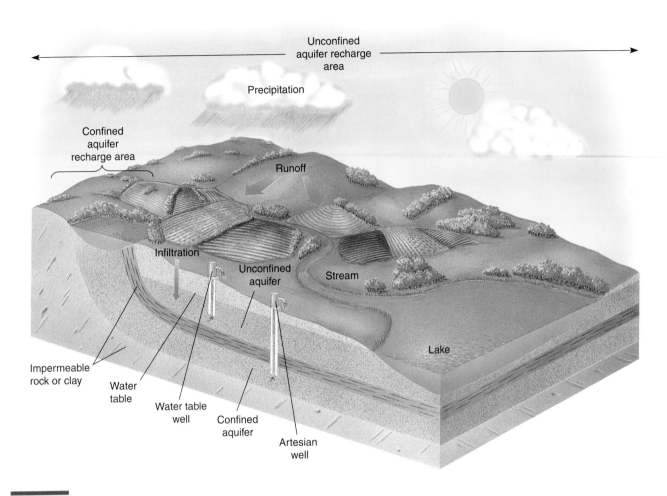

Figure 13.4 Groundwater. Excess surface water seeps downward through soil and porous rock layers until it reaches impermeable rock or clay. Groundwater that is recharged by surface water directly above is known as an unconfined aquifer. In a confined aquifer, groundwater is stored between two impermeable layers and is often under pressure. Artesian wells, which produce water from confined aquifers, often do not require pumping because of this pressure.

or confined. In **unconfined aquifers**, the layers of rock above are porous and allow surface water directly above them to seep downward, replacing the aquifer contents. The upper limit of an unconfined aquifer, below which the ground is saturated with water, is the **water table**. The water table varies in depth depending on amount of precipitation occurring in an area. In deserts, the water table may be far below the surface. In contrast, lakes, streams, and wetlands occur where the water table intersects with the surface. When a well goes dry, the water table has dropped below the depth of the well.

A **confined aquifer**, or **artesian aquifer**, is a groundwater storage area between impermeable layers of rock. The water in a confined aquifer is trapped and often under pressure. Its recharge area (the land from which water percolates to replace groundwater) may be hundreds of kilometers away.

Most groundwater is considered a nonrenewable resource because it has taken hundreds or even thou-

sands of years to accumulate, and usually only a small portion of it is replaced each year by percolation of precipitation. The recharge of confined aquifers is particularly slow.

HOW WE USE WATER

Water consumption varies among countries. In some countries, acute water shortages limit water use per person to several gallons per day. In highly developed countries that possess abundant water resources, per-capita water use may be as high as several *hundred* gallons per day. This amount encompasses agricultural and industrial uses as well as direct individual consumption.

The greatest user of water worldwide is agriculture, for irrigation. Irrigation accounts for 68.3% of the world's total water consumption, industry for 23.1%, and domestic and municipal use for only 8.6% (Table 13.2).

Table 13.2	Water Usage (in Cubic Kilometers), Mid-1990s*		
Region	**Irrigation**	**Industry**	**Domestic/ Municipal**
Africa	127.7	7.3	10.2
Asia	1,388.8	147.0	98.0
Australia-Oceania	5.7	0.3	10.7
Europe	141.1	250.4	63.7
North and Central America	298.1	255.5	54.8
South America	62.7	24.4	19.1
World total	2,024.1	684.9	256.5
World total, as percent	68.3	23.1	8.6

* Data were compiled from individual countries and represent estimates done at varying times. Most countries have data produced sometime during the 1990s.

Irrigation

Arid lands, or deserts, are fragile ecosystems in which plant growth is limited by lack of precipitation. **Semiarid** lands receive more precipitation than deserts but are subject to frequent and prolonged droughts. (Figure 6.18 shows where arid and semiarid lands occur worldwide.) Farmers can increase the agricultural productivity of arid and semiarid lands with irrigation. Most crops can be grown in the desert if enough water is supplied to the soil.

Irrigation of arid and semiarid lands has become increasingly important worldwide in efforts to produce enough food for burgeoning populations (Figure 13.5). Since 1955 the amount of irrigated land has more than tripled, to 274 million hectares (677 million acres) worldwide in 1999, the most recent year for which global data are available. Asia has more agricultural land under irrigation than do other continents, with China, India, and

Figure 13.5 Agricultural use of water. Center-pivot irrigation produces massive green circles. Each circular shape is the result of a long irrigation pipe that extends along the radius from the circle's center to its edge and slowly rotates, spraying the crop. Photographed in Nebraska.

Saving Water by Xeriscaping

The landscaped yard shown in the photograph does not have the usual grass-covered lawn and water-hungry flowers, trees, and shrubs. Instead, this property in Berkeley, California, has been **xeriscaped** (from the Greek *xero*, meaning dry)—that is, landscaped with rocks and plants that require little water. People in western states and in south Florida are increasingly xeriscaping their properties to conserve water. Xeriscaping has been shown to decrease household water consumption by up to 60%, reducing water bills and the energy-intensive maintenance that is required to mow, edge, and weed expansive lawns. Many homeowners do not wish to eliminate lawns entirely but instead reduce the size of grassy areas and xeriscape the remaining land. The stone mulch spread around the base of the drought-resistant plants keeps the soil moist, reduces weed growth, and protects the soil from erosion when precipitation does occur. Dozens of attractive flower, tree, and shrub species are adapted to thrive with little water and so are perfect candidates for xeriscaping.

Pakistan accounting for most of it. It is projected that water use for irrigation will continue to increase in the 21st century, particularly in Asia, but at a slower rate than in the last half of the 20th century.

WATER RESOURCE PROBLEMS

Water resource problems fall into three categories: too much, too little, and poor quality/contamination. (Chapter 21 addresses the third category.) Floods and droughts are part of natural climate variations and cannot be prevented. Human activities sometimes exacerbate their seriousness, however. Humans often court disaster when they make environmentally unsound decisions, such as building in an area that is prone to flooding.

Too Much Water

Many ancient civilizations—ancient Egypt, for example—developed near rivers that periodically spilled over, inundating the surrounding land with water. When the water receded, a thin layer of sediment that was rich in organic matter remained and enriched the soil. These civilizations flourished partly because of their agricultural productivity, which in turn was the result of floods replenishing the soil's nutrients.

Modern floods can cause widespread destruction of property and sometimes loss of life. In the United States, floods are responsible for approximately 9 out of 10 disaster declarations by the president. Today's floods are more disastrous in terms of property loss than those of the past because humans often remove water-absorbing plant cover from the soil and construct buildings on **flood plains**, areas bordering a river that are subject to flooding. These activities increase the likelihood of both floods and flood damage.

Forests, particularly on hillsides and mountains, provide nearby lowlands with some protection from floods by trapping and absorbing precipitation. When woodlands are cut down, particularly if they are clear-cut, the area cannot hold water nearly as well. Heavy rainfall then results in rapid runoff from the exposed, barren hillsides. This not only causes soil erosion but also puts lowland areas at extreme risk of flooding.

When a natural area—that is, an area undisturbed by humans—is inundated with heavy precipitation, the plant-protected soil absorbs much of the excess water. What the soil cannot absorb runs off into the river, which may then spill over its banks onto the flood plain. However, because rivers meander, the flow is slowed, and the swollen waters rarely cause significant damage to the surrounding area. (See Figure 7.13 for a diagram of a typical river, including its flood plain.)

When an area is developed for human use, much of the water-absorbing plant cover is removed. Buildings and paved roads do not absorb water, so runoff, usually in the form of storm sewer runoff, is significantly greater (Figure 13.6). People who build homes or businesses on the flood plain of a river will most likely experience flooding at some point.

It is easier and more economical to ban or restrict development in a flood plain than to build on it and then try to prevent flooding by building such structures as retaining walls and levees. Increasingly, local governments, both in the United States and in the rest of the world, zone flood plains to curtail development. Consider the January 1997 floods in California, which killed 8 people, destroyed more than 16,000 homes, and caused $1.6 billion in damage. During the floods, many expensive levees failed (Figure 13.7a). Rather than rebuild levees adjacent to the rivers to try to prevent floods (Figure 13.7b), river experts recommended that California try a different approach in which rivers are allowed to occupy part of the flood plain during a flood (Figure 13.7c). Smaller levees are built some distance from the river's edge. The new approach would be less expensive, result in less damage during floods, and provide some of the natural benefits of floods, such as improved habitat for

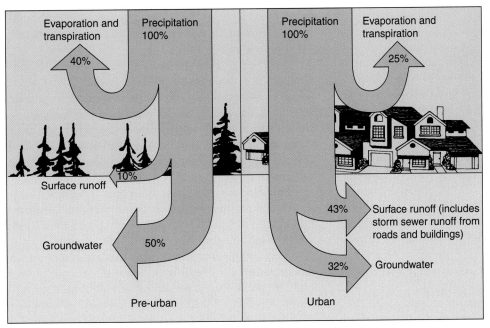

(a)   (b)

Figure 13.6 How development changes the natural flow of water. Shown is the fate of precipitation in Ontario, Canada (a) before and (b) after urbanization. After Ontario was developed, surface runoff increased substantially, from 10% to 43%.

Figure 13.7 Flood management. (a) A levee broke and water surged onto the flood plain (*right*) from the Cosumnes River (*to left of levee*) near Wilton, California, in 1997. (b) To control flooding, many rivers are channelized (straightened and deepened), with high levees adjacent to the river. This flood control method is expensive and may or may not be successful in preventing floods. (c) River managers today recommend letting rivers meander naturally through much of their flood plains, with smaller levees set back some distance from the rivers. The flood plains absorb much of the river's water, forming a buffer between the river and developed areas. When a flooding river spills over its banks, it creates a wetland that is an important wildlife habitat.

(a)

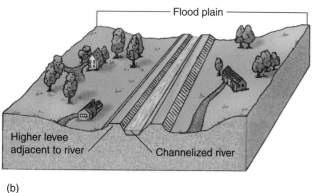

(b)

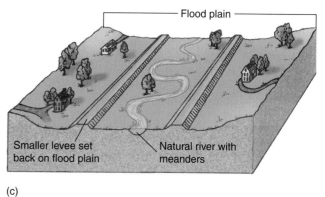

(c)

waterfowl and other wildlife and replenishment of the soil in the flood plain.

CASE·IN·POINT The Floods of 1993

The Mississippi River, which drains 31 states and 2 Canadian provinces, is one of the world's largest rivers. During the summer of 1993, the Mississippi and its tributaries flooded, spreading over 9.3 million hectares (23 million acres) of flood plains and engulfing farms and towns in nine Midwestern states (Figure 13.8). Fifty people were killed, and damage to property was estimated in excess of $12 billion. More than 70,000 people lost their homes, and 8.7 million acres of farmland were damaged. Large quantities of pesticides and other agricultural chemicals that washed off the fields were carried to the Gulf of Mexico. Floodwaters carried the zebra mussel, an exotic species that has caused havoc wherever it has spread (see Chapter 1), to new habitats.

The flood, considered by many experts to be the worst in U.S. history, was caused by above-average precipitation during the first 6 months of 1993, followed by

prolonged rain in the summer. However, the damage was exacerbated by three practices: draining wetlands; building on flood plains; and constructing levees to hold back floodwaters.

For the past hundred years or so, people in the Midwest drained wetlands to produce farmland or land on which to build homes. The ability of wetlands to moderate floods was simply unrecognized. It was probably no coincidence that Missouri, Illinois, and Iowa, the three states damaged the most by the 1993 floods, had each drained and developed more than 85% of their original wetlands.

Hundreds of levees have been built along the Mississippi and its tributaries to hold floodwaters back from the flood plain. Although levees may save lives and property where they are built, they cause floodwaters upstream to surge, damaging farms and towns that are less protected downstream.

The 1993 floods rekindled an old debate: Should the government rebuild damaged levees, or should it help relocate people away from the flood plain and restore the land to a more natural state so that it can

Figure 13.8 Floods of 1993. Nine states experienced flooding (green area) during the summer of 1993.

better face the next flood? Valmeyer and Hartsburg, two towns affected by the flooding, represent the two different strategies. The 900 residents of Valmeyer, Illinois, voted to use federal funds to relocate their town on a nearby hill. The town's old site on the flood plain became a park and wetland. In contrast, the 131 residents of Hartsburg, Missouri, decided to stay in their original location on the flood plain. The levees that were built to protect the town and that failed during the 1993 floods were rebuilt, and homes were repaired. Of course, Hartsburg remains vulnerable to flooding should its levees fail again.

The Flood Plain Management Task Force, assembled by the White House to reassess the national flood policy in the aftermath of the 1993 floods, released its report in 1994. According to the report, people need to make wiser use of flood plains and rely less on levees to control flooding. (Two thirds of the levees along the Mississippi River and its tributaries were damaged or destroyed by the floods of 1993.) Some flood plains need to be restored to their natural condition, which means that the towns built on them need to be moved to higher ground.

The authors of the report concluded that the presence of more wetlands would not have prevented flooding of this magnitude because the wetlands would have become saturated with water. The presence of more wetlands in the affected area would have indirectly reduced property damage from the floods of 1993, however, by keeping people and their property off the flood plain.

This observation is significant because the nation's flood plains continue to be developed in many places.

Meanwhile, flooding continues along the upper Mississippi River and its tributaries as a result of seasonal snowmelt and rainfall. In 2001, for example, many communities recorded the second highest water levels ever in floods in parts of southern Wisconsin, northern Iowa, and southern Minnesota. ◾

Too Little Water

Approximately 40% of the world's population lives in arid or semiarid lands, primarily in Asia and Africa. These people spend substantial amounts of time and effort obtaining water. Each day they may have to walk many miles to a stream or river and carry back the heavy water.

Population growth in arid and semiarid regions intensifies the problem of water shortage. More people need more food, so additional water resources must be diverted for irrigation. Also, the immediate need for food prompts people to remove natural plant cover in order to grow crops on marginal lands, which are subject to frequent drought and subsequent crop losses. Their livestock overgraze the small amount of plant cover in natural pastures. As a result of the lack of plants, the soil cannot absorb the water as well when the rains do come, and runoff is greater. Because the soil is not replenished by the precipitation that does fall, crop productivity is poor and the people are forced to cultivate food crops on additional marginal land.

Overdrawing Surface Waters   Removing too much fresh water from a river or lake can have disastrous consequences in local ecosystems. Humans can remove perhaps 30% of a river's flow without greatly affecting the natural ecosystem. In some places, however, considerably more than that amount is withdrawn for human use. In the arid American Southwest, it is not unusual for 70% or more of surface water to be removed.

When surface water is overdrawn, wetlands dry up. Natural wetlands play many roles, such as serving as a breeding ground for many species of birds (see Chapter 17). Estuaries, where rivers empty into seawater, become saltier when surface waters are overdrawn (recall the chapter introduction), and this change in salinity reduces the productivity that is associated with estuaries.

Aquifer Depletion   **Aquifer depletion**, the removal by humans of more groundwater than can be recharged by precipitation or melting snow, lowers the water table. Prolonged aquifer depletion drains an aquifer dry, effectively eliminating it as a water resource. In addition, aquifer depletion from porous rock causes **subsidence**, or sinking, of the land on top. Some areas of the San Joaquin Valley in California have sunk almost 10 m (33 ft) in the past 50 years because of aquifer depletion.

The limestone bedrock of Florida erodes as water moves through it, sometimes causing a **sinkhole**, a large surface cavity or depression where an underground cave roof has collapsed. Sinkholes occur more frequently when droughts or excessive pumping of water cause a lowering of the water table.

Saltwater intrusion, the movement of seawater into a freshwater aquifer, can occur along coastal areas when groundwater is depleted faster than it can be replenished (Figure 13.9). Well water in such areas can eventually become too salty for human consumption or other uses. Once it occurs, saltwater intrusion is difficult to reverse.

Salinization of Irrigated Soil Although irrigation improves the agricultural productivity of arid and semi-arid lands, it sometimes causes salt to accumulate in the soil, a phenomenon called **salinization** (Figure 13.10). In a natural scenario, as a result of precipitation runoff, rivers, carry salt away. Irrigation water, however, nor-

Figure 13.10 Salinization of irrigated soil. Irrigation water contains dissolved mineral salts. As the water evaporates from the soil's surface, the salts are left behind and gradually accumulate as a visible white powder, making the soil unfit for agriculture. Photographed in Montana.

mally soaks into the soil and does not run off the land into rivers, so when it evaporates, the salt remains behind and accumulates in the soil. Salty soil results in a decline in productivity and, in extreme cases, renders the soil completely unfit for crop production. Chapter 21 discusses the problem of soil salinization in greater detail.

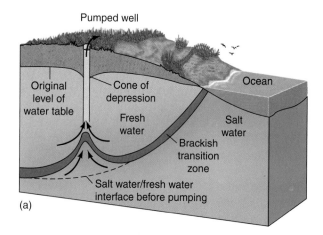

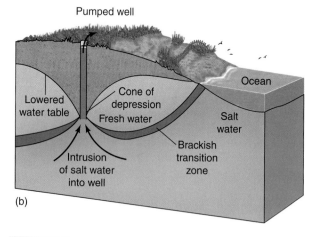

Figure 13.9 Saltwater intrusion. Normally, fresh groundwater overlies salty groundwater. The well supplies fresh water as long as pumping is not excessive. However, the removal of large amounts of fresh groundwater (a) has caused the brackish transition zone to migrate so that the well now draws up salty groundwater (b) that is unfit to drink.

WATER PROBLEMS IN THE UNITED STATES

Compared with many countries, the United States has a plentiful supply of fresh water. According to the U.S. Geological Survey (USGS), approximately 15.9 trillion L (4.2 trillion gal) of water fall as precipitation per day in the continental United States. Approximately two thirds of this water returns to the atmosphere through evaporation and transpiration, leaving roughly 5.3 trillion L (1.4 trillion gal) per day in the soil, surface waters, and groundwater (Figure 13.11). Most of this remaining water, approximately 4.9 trillion L (1.3 trillion gal) per day, makes its way into the ocean without ever being used. The relatively small amount of water that we borrow for our own purposes eventually returns to rivers and streams—for example, as treated or untreated sewage or as industrial wastes.

Despite the overall abundance of fresh water in the United States, many areas have severe water shortages because of geographical and seasonal variations (Figure 13.12). Arid and semiarid areas of the western United States normally receive little rainfall. Annual fluctuations

U.S. Water Use Trends

During the 20th century, per-capita water use in the United States increased, even as the population grew, at least until 1980. This trend of increased per-capita demand on water resources changed, beginning in 1980. According to the U.S. Geological Survey (USGS), water use in the United States actually declined by nearly 10% during the period from 1980 to 1995, whereas the population grew by 16%. In 1995, a nation with 37 million more people than in 1980 consumed 38 billion fewer gallons of water per day, which means that per-capita water consumption was about 20% less. Increased public awareness of water as a limited resource has accelerated more efficient water use. Newer appliances use less water, and low-flush toilets and water-efficient fixtures are now required in new construction.

Industrial water use, at 29 billion gal a day, is the lowest it has been since record keeping began in 1950. Agriculture consistently places heavy demands on water resources, but it, too, is gaining in efficiency. New irrigation techniques deliver water closer to the ground than in previous delivery systems, so less is lost to wind and evaporation.

These USGS data are only gathered every 5 or so years and therefore may reflect short-term trends. Still, the evidence is in that at least in the short term, the increase in water use in the United States has come to a halt.

in precipitation also occur. Droughts, higher-than-average precipitation rates, and other natural conditions cause problems in water availability throughout the country. Human activities sometimes exacerbate the difficulties.

Surface Water

The use of U.S. surface water for agriculture, industry, and personal consumption since the 1960s has caused many water supply and quality problems. Some U.S. regions that have grown in population during this period—for example, California, Nevada, Arizona, and Florida—have had correspondingly greater burdens on their water supplies. If water consumption in these and other areas continues to increase, the availability of surface waters could become a serious regional problem, even in places that have never before experienced water shortages.

Nowhere in the country are water problems as severe as they are in the West and Southwest. Much of this large region is arid or semiarid and receives less than 50 cm (about 20 in.) of precipitation annually. The West and Southwest consume an average of 44% of their renewable water, as compared to an average consumption of 4% of renewable water elsewhere in the United States (Figure 13.13). Historically, water in the West was used primarily for irrigation. However, with the rapid expansion of population in that region during the past 25 years, municipal, commercial, and industrial uses now compete heavily with irrigation for available water.

Until recently, the development of new sources of water met expanding water needs in the West and Southwest. Water was diverted from distant sources and transported via **aqueducts** (large conduits) to areas that needed it. As long ago as 1913, Los Angeles started bringing in water from the Owens Valley, an area of California 400 km (250 mi) north, along the east side of the Sierra Nevada. Dams were built and water-holding basins created to ensure a year-round supply. These solutions are no longer viable because the closest, most practical water sources have already been used and because the public, which has come to expect inexpensive water, opposes paying for costly solutions such as dams and aqueducts.

Mono Lake   Removing too much surface water can have serious environmental repercussions. Mono Lake, a salty lake in eastern California, is a striking example of the effects of removing too much surface water. Rivers and streams that are largely formed from snowmelt in the Sierra Nevada range replenish Mono Lake. Evaporation provides the only natural outflow from the lake. Over time, Mono Lake is slowly becoming saltier as rivers deposit dissolved salts (recall that fresh water contains some salt) and as water, but not salt, is removed by evaporation.

Beginning in 1941, much of the surface water that would naturally feed Mono Lake was diverted to Los Angeles, 442 km (275 mi) away. Over time, Mono Lake's water level subsided about 14 m (46 ft), and its salinity

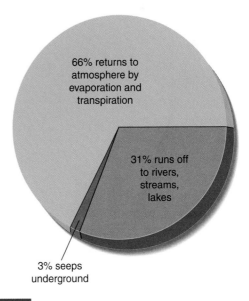

Figure 13.11 U.S. water budget, excluding Alaska and Hawaii. Approximately 66% of the precipitation that falls on the continental United States evaporates back into the atmosphere almost immediately. Most of the rest (about 31%) flows in rivers and streams to the ocean or to Canada or Mexico. Only about 3% of the precipitation seeps into underground aquifers.

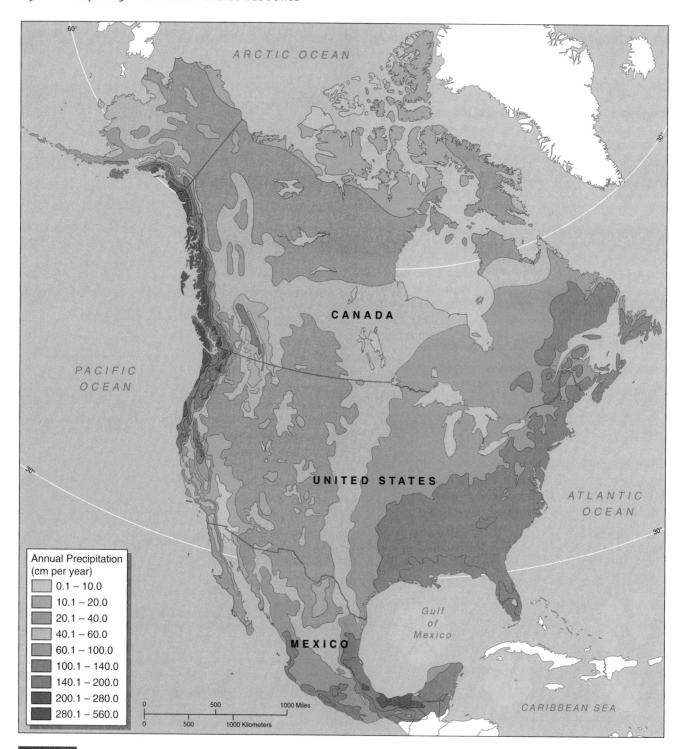

Figure 13.12 Average annual precipitation in North America. Note the arid and semiarid regions of the western United States.

increased dramatically (Figure 13.14a and b). These changes adversely affected brine shrimp and alkali fly populations, as well as more than 80 species of ducks, geese, and other water birds that feed on brine shrimp and alkali flies. In addition, winds whipped up dust storms from the exposed lakebed, posing a health hazard and a violation of federal air pollution standards.

A court order halted water diversions from Mono Lake in 1989, and California began a review of the Mono Lake situation. In 1994 the state of California worked out an agreement on Mono Lake water rights between the Los Angeles water authority and environmental groups such as the National Audubon Society. Mono Lake will be allowed to return to about 72% of its original volume,

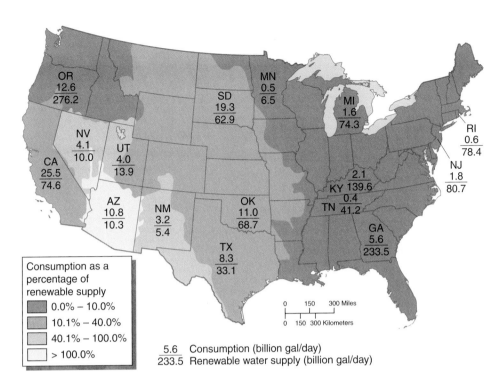

Figure 13.13 U.S. water consumption as a percentage of the renewable supply. The pale and light blue areas have severe water problems. Note that consumption actually exceeds renewable supply in the Southwest (the Colorado River basin).

Consumption as a percentage of renewable supply

- 0.0% – 10.0%
- 10.1% – 40.0%
- 40.1% – 100.0%
- > 100.0%

$\frac{5.6}{233.5}$ Consumption (billion gal/day)
Renewable water supply (billion gal/day)

(a)

(b)

(c)

Figure 13.14 Mono Lake, California. These three shots were taken from approximately the same location. As Los Angeles diverted water from Mono Lake, the lake level subsided, exposing towers of calcium carbonate known as tufa. (a) Two tufa were exposed by the lowered lake in 1962. (b) The same tufa were on dry land when this shot was photographed in 1994. (c) In 1997 the water level had risen somewhat (*see background behind the left tufa*). Now that water is no longer being diverted, the lake should refill to about 72% of its original volume by about 2015.

which should be reached by around 2015 (Figure 13.14c). As a result of the agreement, the National Audubon Society expects that hundreds of thousands of migratory and nesting birds will return to the lake's shores to nest.

The city of Los Angeles is using state funds to develop water conservation and reclaimed water projects to replace water supplies from Mono Lake. **Reclaimed water** is treated wastewater that is reused in some way, such as for irrigation, manufacturing processes that require water for cooling, wetland restoration, or groundwater recharge. By 2015 California's reclaimed water projects will produce enough water to make up for the water lost from Mono Lake.

The Colorado River Basin   One of the most serious water supply problems in the United States is in the Colorado River basin. The river's headwaters are formed from snowmelt in Colorado, Utah, and Wyoming, and major tributaries—often collectively called the upper Colorado—extend throughout these states. The lower Colorado River runs through part of Arizona and then along the border between Arizona and both Nevada and California. It then crosses into Mexico and empties into the Gulf of California.

The Colorado River provides water for 20 million people, including the cities of Denver, Las Vegas, Salt Lake City, Albuquerque, Phoenix, Los Angeles, and San Diego. It also provides irrigation water for 3.5 million acres of fruit, vegetable, and field crops worth $1.5 billion per year. The Colorado River has 49 dams, 11 of which produce electricity by hydropower. More than 30 Native American tribes live along the Colorado River and claim rights to some of its water. The river provides $1.25 billion per year in revenues from almost 30 million people who use it for recreation.

An international agreement with Mexico, along with federal and state laws, severely restricts the use of the Colorado's waters. The most important of all the state treaties is the 1922 Colorado River Compact. It stipulates an annual allotment of 7.5 million acre-feet of water each to the upper Colorado (Colorado, Utah, and Wyoming) and the lower Colorado (California, Nevada, Arizona, and New Mexico). (Each acre-foot equals 320,000 gal, enough for eight people for 1 year.)

Traditionally, the upper Colorado region appropriated little of the water to which it was entitled, because it had few people and little development. This made more water available to the faster-developing lower Colorado region, but it also gave that area a false impression of the size of its water supply. Recent population growth in the upper Colorado region is now threatening the lower Colorado region's water supply. Further, people in the states through which the lower Colorado flows take so much water that the remainder is insufficient to meet Mexico's needs as set forth by international treaty (Figure 13.15). To compound the problem, as more and more water is used, the lower Colorado becomes increasingly salty as it

Figure 13.15   Colorado River bed in San Luis Rio Colorado, Mexico. As a result of diversion for irrigation and other uses in the United States, the Colorado River often dries up before reaching the Gulf of California in Mexico.

flows toward Mexico; in places, the Colorado River is saltier than the ocean. To help meet the low salinity requirements and water amounts stipulated by water treaties between the United States and Mexico, the United States built a huge desalting plant in Arizona (discussed later in the chapter).

Groundwater

Roughly half the population of the United States uses groundwater for drinking. Many large cities, including Tucson, Miami, San Antonio, and Memphis, have municipal well fields and depend entirely or almost entirely on groundwater for their drinking water. In addition, many rural homes have private wells for their water supply. Groundwater is also used for industry and agriculture. Approximately 40% of the water used for irrigation in the United States comes from groundwater. Owing to increased groundwater consumption since the 1950s, groundwater levels have dropped in many areas of heavy use across the United States. Aquifer depletion is particularly critical in three regions: southern Arizona, California, and the High Plains, a band of states extending from Montana and North Dakota south to Texas (Figure 13.16). Groundwater has been overdrawn for irrigation in these arid and semiarid areas.

In certain coastal areas of Louisiana and Texas, the removal of too much groundwater has resulted in the intrusion of salt water from the Gulf of Mexico. Saltwater intrusion from the Pacific Ocean has occurred along

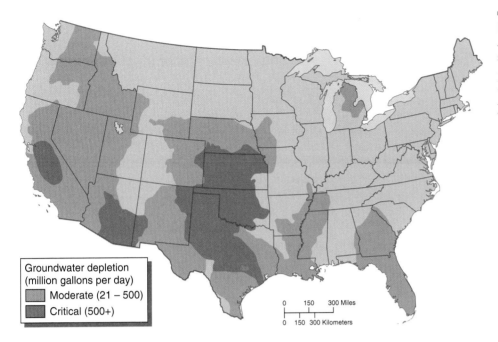

Figure 13.16 Aquifer depletion.
Aquifer depletion is a widespread problem in the United States, particularly in the High Plains, California, and southern Arizona.

Groundwater depletion
(million gallons per day)
Moderate (21 – 500)
Critical (500+)

parts of the California coast, along coastal areas of Puget Sound in Washington state, and in certain areas of Hawaii. Florida and many coastal regions in the Northeast and mid-Atlantic states also have saltwater intrusion.

The quality of an area's groundwater varies from fresh to salty, depending on the soil and rock characteristics of the area as well as the age of the water and its rate of recharge. Generally, groundwater is saltier in areas that have high evaporation rates and where salty water infiltrates, or penetrates, through rocks such as carbonate and sandstone, as it does in parts of North Dakota and Minnesota. Certain areas in the United States have poor groundwater quality because of naturally localized high concentrations of fluoride or arsenic. Some areas of Nevada and Utah, for example, have unusually high levels of toxic minerals in their groundwater as a result of natural conditions, rather than pollution caused by humans. Chapter 21 discusses groundwater contamination caused by pollution.

The Ogallala Aquifer The High Plains cover 6% of U.S. land but produce more than 15% of its wheat, corn, sorghum, and cotton and almost 40% of its livestock. To achieve this productivity, it requires approximately 30% of the irrigation water used in the United States. Farmers on the High Plains rely on water from the **Ogallala Aquifer**, the largest groundwater deposit in the world (Figure 13.17).

In some areas farmers are drawing water from the Ogallala Aquifer as much as 40 times faster than nature replaces it. The depletion of the Ogallala has lowered the water table by more than 30 m (100 ft) in some places. Where this has occurred, higher pumping costs have made it too expensive to irrigate. The amount of irrigated Texas farmland had declined by 11% in recent

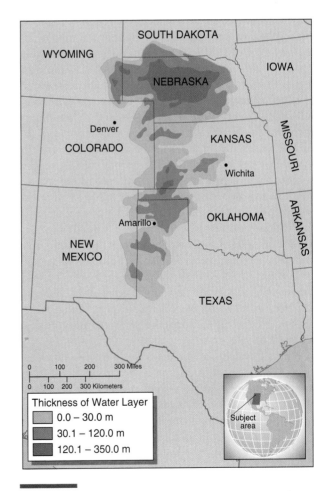

Thickness of Water Layer
0.0 – 30.0 m
30.1 – 120.0 m
120.1 – 350.0 m

Figure 13.17 Ogallala Aquifer. This massive deposit of groundwater lies under eight Midwestern states, with extensive portions in Texas, Kansas, and Nebraska. Water in the Ogallala Aquifer takes hundreds or even thousands of years to be renewed after it is withdrawn to grow crops and raise cattle.

years. When farmers revert to dry-land farming in these semiarid regions, they risk economic and ecological ruin during droughts (see discussion of the Dust Bowl in Chapter 14). Ogallala water is unevenly distributed underground. Areas where the Ogallala is shallowest have experienced recent population declines as farms fail during dry spells. Those areas where the Ogallala is deepest, however, may not experience water shortages during the 21st century. Most hydrologists (scientists who deal with water supplies) predict that groundwater will eventually drop in *all* areas of the Ogallala to a level that is uneconomical to pump. Their goal is to postpone that day through water conservation, including the use of water-saving irrigation systems.

GLOBAL WATER PROBLEMS

Data on global water availability and use indicate that overall, the amount of fresh water on the planet is adequate to meet human needs, even taking population growth into account. These data do not, however, consider the distribution of water resources in relation to human populations. Citizens of Bahrain, a tiny island nation in the Persian Gulf, for example, have *no* freshwater supply and must rely completely on desalinization (removing salt) of salty ocean water for their fresh water.

Large variations in per-capita water use exist from country to country and from continent to continent, depending on the size of the human population and the available water supply. South America and Asia are the two continents with the greatest total water supply. Together they receive more than one half of the world's renewable fresh water (by precipitation). Although South America has more available water per person than Asia does, it does not have the potential to support as many people as its water supply would suggest. That is because most of the precipitation received by South America falls in the Amazon River basin, which has poor soil and is therefore unsuitable for large-scale agriculture. In contrast, because most of the precipitation in Asia falls on land suitable for agriculture, the water supply can support more people.

Humans need an adequate supply of water year round. Global water supply is complicated by the fact that **stable runoff**, the portion of runoff from precipitation that is available *throughout* the year, can be low despite the fact that total runoff is quite high. India has a wet season—June to September—during which 90% of its annual precipitation occurs. Most of the water that falls during India's wet season quickly drains away into rivers and is unavailable during the rest of the year. Thus, India's stable runoff is low.

Variation in annual water supply is an important factor in certain areas of the world. The African Sahel region, just south of the Sahara Desert (see Figure 6.21), has wet years and dry years, and the lack of water during the dry years limits human endeavors during the wet years.

Drinking-Water Problems

Many developing countries have insufficient water to meet the most basic drinking and household needs of their people. Many people have to travel great distances to secure the water they need. This practice consumes large amounts of time, particularly for women and children, and tends to perpetuate poverty.

The World Health Organization (WHO) estimates that 1.4 billion people lack access to safe drinking water, and about 2.9 billion are without access to a satisfactory means of domestic wastewater and fecal waste disposal. These people risk disease because sewage or industrial wastes contaminate the water they consume (see Chapter 21). WHO also estimates that 80% of human illness results from insufficient water supplies and poor water quality caused by lack of sanitation. Although many developing countries have installed or are installing public water systems, population increases tend to overwhelm efforts to improve the water supply.

The United States and other highly developed countries are involved in efforts to improve water quality and supply in countries with critical water problems. The U.S. Agency for International Development (AID) manages projects in areas vulnerable to prolonged drought, such as the Sahel region in Africa. AID has assisted in well digging and other measures that alleviate the effects of drought in the Sahel. In addition to contributions by individual governments, both the United Nations and the World Bank[1] sponsor water management projects in developing countries.

Population Growth and Water Problems

As the world's population continues to increase, global water problems will become more serious. Even in Asia, which has the world's largest available water resources, population growth is outstripping water supplies. In India, where 20% of the world's population has access to 4% of the world's fresh water, approximately 8,000 villages have no local water. The water supply to some Indian cities—Madras, for example—has been so severely depleted that water is rationed from a public tap. Extracting groundwater faster than its rate of recharge has caused water tables to fall over 4 m (13 ft) in parts of Punjab and Haryana.

Water supplies are also precarious in much of China, owing to population pressures. One third of the wells in Beijing have gone dry, and the water table continues to drop. Compounding the problem, northern China, with

---

[1] The World Bank makes loans to developing countries for projects it thinks will lessen poverty and encourage development.

a population more than twice the population of the United States, suffered a decade-long drought during the 1990s. Much of the water in the Yellow River is diverted for irrigation, leaving downstream areas with little or no water. During the last 15 years of the 20th century, the Yellow River ran dry hundreds of kilometers inland before it reached the Yellow Sea.

Pakistan also faced an extended drought during the 1990s and early 2000s. The water shortages resulted in clashes between certain provinces over water use, particularly of the Indus River, the lifeline in Pakistan. The inability to produce enough food in agricultural regions has resulted in an increase in poverty.

Mexico is facing the most serious water shortages of any country in the Western Hemisphere. The main aquifer supplying Mexico City is dropping by as much as 3.5 m (11.2 ft) per year.

Shortages in global water supplies may also affect humans by limiting the amount of food they can grow. Recall that the main use of fresh water is for irrigation. As the growing human population depletes freshwater supplies, less water will be available for crops. Local or even widespread famines from water shortages are a very real possibility.

Sharing Water Resources Among Countries

Global water supply is complicated by the fact that surface water is often an international resource. Three fourths of the world's 200 or so major watersheds are shared between at least two nations. Management of rivers that cross international boundaries requires international cooperation.

The Rhine River Basin   The river basin for the Rhine River in Europe is in five countries—Switzerland, Germany, France, Luxembourg, and the Netherlands (Figure 13.18). The river basin is highly developed and densely populated; about 50 million people live in the basin. Traditionally, Switzerland, Germany, and France used water from the Rhine for industrial purposes and then discharged polluted water back into the river. The Dutch then had to clean up the water so they could drink it. Today, these countries recognize that international cooperation is essential if the supply and quality of the Rhine River are to be conserved and protected.

In 1950 the five countries in the Rhine River basin formed the International Commission for Protection of the Rhine (ICPR) to deal with water issues relating to the Rhine River. Owing to initial lack of political support, little was accomplished for several decades. River quality began to improve in the mid-1970s, largely in response to international reports on the river's poor condition. In 1986 a severe chemical spill in Switzerland dumped 30 tons of dyes, herbicides, fungicides, insecticides, and mercury into the river, killing half a million fish and ruining water supplies from Switzerland to the Netherlands.

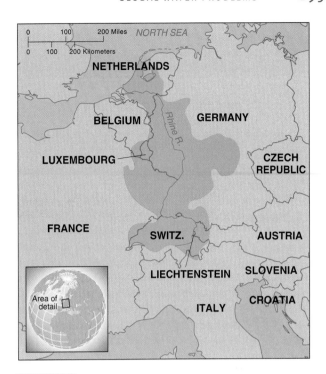

Figure 13.18 Rhine River basin. The Rhine River drains five European countries—Switzerland, Germany, France, Luxembourg, and the Netherlands. (The green area represents the drainage basin.) Water management of such a river requires international cooperation.

The spill galvanized the ICPR, which initiated a 15-year Rhine Action Plan. The main sources of pollution have since been eliminated, and the water in the Rhine River today is almost as pure as drinking water. Long-absent fishes have returned to the river, including Atlantic salmon, which returned in 1990 after a 30-year absence. The ICPR is currently working on bank restoration, flood control, and cleaning up remaining pollutants.

The Aral Sea   The Aral Sea in Kazakstan, part of the former Soviet Union, is suffering from the same problem as Mono Lake in California. Like Mono Lake, the Aral Sea has no outflow other than evaporation. In the 1950s, communist authorities began diverting water from the Amu Darya and the Syr Darya, the two rivers that feed into the Aral Sea, to irrigate desert areas surrounding the lake. By the early 1980s, irrigation had diverted more than 95% of the Aral Sea's inflow.

Since 1960, the Aral Sea, which was once the world's fourth largest freshwater lake, has declined in area by more than 50%. Its total volume is down by 80%. Much of its biological diversity has disappeared—all 24 fish species originally found there are gone. The satellite photos in Figure 13.19 demonstrate the shrinking of the Aral Sea.

About 35 million people live in the Aral Sea's watershed. Millions of them have developed health problems

Figure 13.19 Aral Sea. The satellite images show the Aral Sea in (a) 1976 and (b) 1997.

(a)

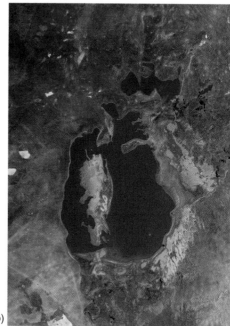

(b)

ranging from tuberculosis to severe anemia, and their death rate from respiratory illnesses is among the world's highest. Kidney disease and various cancers are also on the rise. International health experts have begun to assess which of these medical problems are due to the Aral Sea's environmental problems. Toxic salt storms caused by winds that whip the salt on the receding shoreline into the air may be responsible for many of these chronic conditions. (The constituents of the toxic salt have yet to be fully analyzed.) Since the 1950s, such storms have increased 60-fold. The salt is carried by the wind hundreds of kilometers from the Aral Sea, and where it is deposited it reduces the productivity of the land.

Immediately following the breakup of the Soviet Union in 1991, plans to save the Aral Sea faltered as responsibility for its rescue shifted from Moscow to the five central Asian republics that share the Aral basin: Uzbekistan, Kazakstan, Kyrgyzstan, Turkmenistan, and Tajikistan. In 1994 the five republics established a joint fund to prevent the complete disappearance of the Aral Sea. In addition, the World Bank and the U.N. Environment Program approved a grant to the five republics to help address the environmental problems of the area. Some money will be used to line the irrigation systems so less water will be wasted (and therefore more will flow into the Aral Sea).

There is virtually no chance of the Aral Sea returning to its former size. Water experts say that it would take 50 years to refill the lake, assuming that all irrigation diversions were stopped. Because the fishing industry collapsed, agriculture is the mainstay of the economy, so irrigation must continue or the countries' economies will be ruined.

To make a grim situation even worse, in 1988 the Soviets buried hundreds of tons of anthrax bacterial endospores on an island in the Aral Sea. Anthrax is an infectious disease deadly in humans and other warm-blooded animals. The bacterial endospores were part of highly secret biological warfare studies conducted by the former Soviet Union.

Although the anthrax endospores were treated with bleach before being buried, soil tests indicate that some of the endospores are still alive. Because the island on which the endospores were dumped has been enlarging as the lakeshore has been contracting, the island will join the mainland sometime soon. Uzbekistan and Kazakstan have asked the United States to help decontaminate this site, a process that experts say will be prohibitively expensive.

Potentially Volatile International Water Situations
Because about 150 major river watersheds and many underground aquifers cross national boundaries, the 21st century may well see countries facing one another in armed conflict over water rights. Tensions are high along the Mekong River basin, shared by Laos, Thailand, and Vietnam. Likewise, the Indus River basin is shared between Pakistan and India, two countries that have never been friendly, while India and Bangladesh quarrel over the Ganges River. Slovakia and Hungary both depend on the Danube River, whereas tensions are rising among Turkey, Syria, and Iraq for water from the Tigris–Euphrates River basin. According to U.N. Secretary General Kofi Annan, developing cooperative international agreements on shared water resources is an urgent global issue.

An example of a troublesome international spot is the Jordan River, which supplies water to Israel, Jordan, the West Bank, and Gaza Strip. These Middle Eastern countries are experiencing large population increases that could make their water situation even more critical. Water use is also increasing because of additional agricultural and economic activity in the region. According to a collaborative study among the U.S. National Academy of Sciences, the Israel Academy of Sciences and Humanities, the Palestine Academy for Sciences and Technology, and the Royal Scientific Society of Jordan, the future outlook for water supplies in the region is one of significant water stress. Participants urged their respective governments to cooperate with one another in developing and expanding water conservation measures such as reusing wastewater and installing more efficient irrigation.

Northeastern Africa also has a serious water-use situation: the Nile River. Ten nations share the Nile River basin, including Egypt, Ethiopia, and Sudan. The nations of Ethiopia and Sudan, in particular, are expanding their use of the Nile River's flow to meet the demands of their rapidly growing populations. Because almost all of Egypt's water supply comes from the Nile, the actions of Ethiopia and Sudan could imperil Egypt's freshwater supply at a time when its population is also increasing. The United Nations engineered an international water-use agreement among the Nile River countries to help diffuse this potentially dangerous water situation, and the ten nations in the region recently formed the Nile Basin Association to review past agreements and develop future ones.

WATER MANAGEMENT

People have always considered water different from other resources. Resources such as coal or gold may be owned privately and sold as free-market goods, but we view water as public property.

Historically, both in much of the United States and in many other countries, water rights were bound with land ownership. As more and more users compete for the same water, however, state or provincial governments are increasingly making allocation decisions. Some countries have separated land and water ownership so that water rights are sold separately.

Because rivers usually flow through more than one governmental jurisdiction, jurisdictions or states must develop agreements with each other about the management of a river or other shared water resource. Such interstate cooperation permits comprehensive rather than piecemeal management. In addition, these arrangements allow the water to be divided fairly between the jurisdictions, which then apportion their respective shares to individual users according to an established set of priorities.

Groundwater management is more complicated, in part because the extent of local groundwater supplies is not known. Some states manage groundwater where demand exceeds supply. Groundwater management includes issuing permits to drill wells, limiting the number of wells in a given area, and restricting the amount of water that may be pumped from each well.

The price of water varies, depending on how it is used. Historically, domestic use is most expensive and agricultural use is least expensive. In any case, the consumer rarely pays directly for the *entire* cost of water, which includes its transportation, storage, and treatment. State and federal governments heavily subsidize water costs, so we pay for some of the cost of water indirectly, through taxes. Increasingly, state and local governments are considering adjustments to the price of water as a mechanism to help ensure an adequate supply of water. Raising the price of water to users so that it reflects the actual cost generally promotes a more efficient use of water.

Providing a Sustainable Water Supply

The main goal of water management is to provide a sustainable supply of high-quality water. **Sustainable water use** means that humans can use water resources wisely without harming the essential functioning of the hydrologic cycle or the ecosystems on which humans depend, so that water is available for future generations.

Water supplies can be obtained by building dams, diverting water, or removing salt from seawater or salty groundwater. Additional water supplies can also be obtained by conservation, which includes reusing water, recycling water, and improving water use efficiency.

Dams and Reservoirs Dams ensure a year-round supply of water in areas that have seasonal precipitation or snowmelt. Dams confine water in reservoirs, from which the flow is regulated (Figure 13.20). Dams have other benefits, including the generation of electricity (recall the discussion of dams and hydroelectric power in Chapter 12). They provide flood control for areas downstream, because a reservoir can hold a large amount of excess water during periods of heavy precipitation and then release it gradually. Some of the reservoirs formed by dams also have recreational benefits: People swim, boat, and fish in them. Many people, however, feel that the drawbacks of dams, including the environmental costs, far outweigh any benefits they provide.

In recent years scientists have come to understand many of the ways that dams alter river ecosystems, both upstream and downstream. Heavy deposition of sediment occurs in the hydrologic environment upstream of the dam, and the water that passes over the dam does not have its normal sediment load. As a result, the river floor downstream of the dam is scoured, producing a deep-cut channel that is a poor habitat for aquatic organisms.

The Glen Canyon Dam, built in 1963, has profoundly affected the Colorado River in the Grand Canyon National Park. Prior to the dam's construction, powerful spring floods deposited beaches and sandbars that provided nesting sites for birds and shallow waters

Figure 13.20 Grand Coulee Dam on the Columbia River. Shown are the dam and part of its reservoir, the Franklin D. Roosevelt Lake. Dams help to regulate water supply, storing water that is produced in times when precipitation is plentiful to be used during dry periods. The many beneficial uses of dams include electricity generation and flood control, but they also destroy the natural river habitat and are expensive to build.

for breeding fishes. The regulated flow of water since the Glen Canyon Dam was constructed changed the ecosystem, to the detriment of some of the Grand Canyon's wildlife. The Bureau of Reclamation tried to rectify some of the changes that have occurred to the river by releasing an additional 117 billion gallons of water during a 1-week period in the spring of 1996. The experimental flood, although small in comparison to some of the natural floods of the past, rebuilt some 50 beaches and sandbars that had disappeared since 1963 and enlarged most of the existing ones. It also killed some of the exotic vegetation that had taken over the area and partly restored fish spawning habitats. Although it will take scientists years to analyze the full effects of the experimental flood, preliminary evidence indicates that rivers controlled by dams would benefit from periodic floods.

Sometimes dams are torn down when their presence adversely affects fish populations. In 1998 the Federal Energy Regulatory Commission ordered the destruction of the 160-year-old Edwards Dam on the Kennebec

River in Maine because it prevented 10 migratory fish species such as striped bass, Atlantic salmon, sturgeon, shad, and herring from going upstream (above the dam) to spawn. In 1999 Oregon announced its intention to remove two 90-year-old dams in the Sandy River basin east of Portland to restore the natural habitat of chinook salmon and steelhead trout, both of which are listed as threatened species. The removal of the dams, which will occur in 2006 at the earliest, will result in 161 km (100 mi) of unobstructed river, from the Pacific Ocean to Mount Hood.

CASE·IN·POINT The Columbia River

The impact of dams on natural fish communities is well illustrated by the Columbia River, the fourth largest river in North America. Its watershed, which covers an area the size of France, includes seven states and two Canadian provinces. Such a large, complex river system has multiple uses. There are more than 100 dams within the Columbia River system, 19 of which are major generators of inexpensive hydroelectric power (Figure 13.21). The Columbia River system supplies municipal and industrial water to several major urban areas, including Boise, Portland, Seattle, and Spokane. More than 1.2 million hectares (3 million acres) of agricultural land are irrigated with the Columbia's waters. Commercial ships navigate 805 km (500 mi) of the river. Recreational uses of the river include boating, windsurfing, and swimming. The Columbia River system also offers sport and commercial fishing for salmon, steelhead trout, and other fishes.

As is often the case in natural resource management, a particular use of the Columbia River system may have a negative impact on other uses. The dam impoundments along the Columbia River that generate electricity and control floods have adversely affected fish populations, particularly salmon. Salmon are migratory fish that spawn in the upper reaches of freshwater rivers and streams. The young offspring, called smolts, migrate to the ocean, where they spend most of their adult lives. Salmon complete their life cycle by returning to their place of birth to reproduce and die.

The salmon population in the Columbia River system is only a fraction of what it was before the watershed was developed. While several factors have contributed to their decline, the many dams that impede salmon migrations are widely considered the most significant. In addition, salmon have been overfished in the Pacific Ocean. Logging around salmon spawning streams contributes sediment pollution that degrades the salmon's habitat. Also, because streams in logged areas are no longer shaded, the water temperature becomes too hot for the developing salmon eggs.

Several projects to rebuild salmon populations were implemented during the 1980s and early 1990s, but none have been particularly effective. Many of the dams had

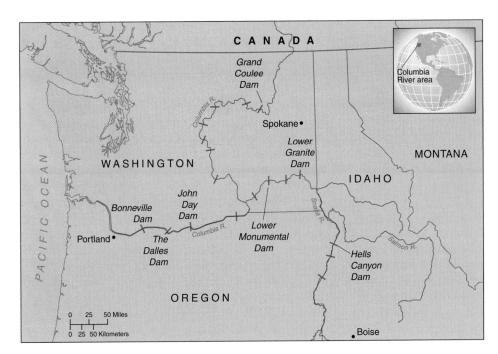

Figure 13.21 Major dams along the Columbia River and its tributaries.

already installed fish ladders, structures that allow some of the adult salmon to bypass the dams and continue their upstream migration (Figure 13.22).

The *downstream* migration of the young smolts is quite perilous: Of every five smolts that start the trip, only about one makes it to the Pacific Ocean. Prior to extensive dam building, smolts made the downstream trip to the Pacific in a few days, carried by the heavy spring flush of water from rains and melting snow. Now the same trip may take several months, with many of the fish becoming stranded in the tranquil waters of reservoirs behind the dams. Many of the smolts are consumed by large predators in the reservoirs or destroyed by the turbine blades as they fall over the dams.

To increase the number of salmon, several hatcheries have been built upstream of the dams, in tributaries of the Columbia. Young fish produced at these hatcheries are released to imprint the "smell" of the streams, enabling them to return there as adults to reproduce. Unfortunately, this effort has not reestablished natural spawning, in part because hatchery fish appear to be genetically incompatible with wild populations.

To protect some of the remaining natural salmon habitats, several streams in the Columbia River system have been designated off limits for dam development. Special underwater screens and passages are being installed at dams to steer the smolts away from the turbine blades. Trucks and barges transport some of the young fish around dams, while others swim safely over the dam because the electrical generators are periodically turned off to allow passage.

One of the more interesting approaches is the establishment of a "water budget" for the young fish. Extra water, simulating spring snowmelt, is released from the

dams to help wash the smolts downstream. Conservationists prefer the water-budget approach to trucking and barging, which have been done for almost 20 years without much success. According to scientific studies, less than 0.5% of barged salmon return to their spawning grounds. However, other interest groups are opposed to increasing water flow for salmon. Farmers do not want the water "wasted" during spring months when water is plentiful; they want it saved for irrigation during summer

Figure 13.22 Fish ladder. This ladder is located at the Bonneville Dam along the Oregon side of the Columbia River. Fish ladders help migratory fishes to bypass dams in their migration upstream.

months. The hydropower industry is also opposed because they want to save the water to generate electricity during the winter months, their time of peak demand.

In 1999 the National Marine Fisheries Service (NMFS) extended the protection of the Endangered Species Act to all species of salmon and steelhead trout found in Northwest rivers from the Canadian border to northern California, and eastward to Montana. This action is the largest implementation of the Endangered Species Act ever, and it affects not only public lands but also private lands, from rural areas to major cities such as Portland, Oregon.

Initial public opinion on the action to save the salmon, a symbol of the Northwest's natural heritage, was remarkably supportive. Resistance may increase, however, as the full implications begin to be felt during the first decade or so of the 21st century. Farmers may have to fence off streams so that trees and other vegetation will grow there and provide shade that cools the water for the fish. Farmers may also have to irrigate their fields less to prevent farm chemicals from polluting rivers. To prevent soil erosion from polluting streams and rivers, timber companies will probably not be allowed to log near streams or on steep slopes (the soil smothers salmon eggs). The construction industry will probably have to undergo thorough environmental reviews before construction projects begin, and development near stream banks will probably be prohibited. These actions may increase housing costs in the area. City residents will probably have restrictions on washing their cars and watering their lawns (to save water) and using pesticides and fertilizers (to avoid polluting the water that eventually drains into streams). The cost of domestic water will probably increase to cover the expense of building new sewage treatment plants to improve water quality in the Columbia and Willamette Rivers.

The most controversial proposal that the NMFS is considering is to tear down four dams on the lower Snake River, a tributary of the Columbia River. Scientists say this action would restore the natural flow and be the single most effective way to help restore salmon populations. Wild salmon from the Snake River have declined by almost 90% since the dams were installed during the 1960s and 1970s. Many biologists, environmentalists, and Native American tribes strongly endorse the proposal, whereas hydroelectric companies, area farmers, and many other area residents bitterly oppose it. ■

CASE-IN-POINT The Missouri River

The Missouri River flows from Montana to St. Louis, Missouri, where it joins the Mississippi River and flows on to the Gulf of Mexico (see Figure 13.8). The Missouri River, which is 3,968 km (2,466 mi) long, is the longest river in the United States; it drains about one sixth of the country. It contains North America's three largest reservoirs (Lake Sakakawea in North Dakota, Lake Oahe in South Dakota, and Fort Peck Reservoir in Montana) and the world's largest compacted-earth dam (Fort Peck Dam in Montana). In fact, the Missouri River has six dams built by the Army Corps of Engineers that provide both benefits and problems for people living along the river.

Since 1987 the Corps has opened up the northern dams to protect downstream navigation, including the shipping of 2 million tons of cargo each year. In addition, people who live downstream count on the river water for irrigation, electrical power, and individual water consumption. But the area along the northern Missouri River depends on the river for its multimillion-dollar fishing and tourism industry. Farmers want the river reined in with additional dikes and levees to protect their crops on the flood plains from damage during floods, whereas environmentalists want the river restored to its natural state as much as possible. Native Americans with claims to water rights along portions of the river want to use the water in a variety of ways, from generating hydroelectric power to irrigating cropland.

North Dakota, South Dakota, and Montana sued the Army Corps of Engineers for discharging water in an "arbitrary and capricious way." They won the decision, which was to have kept the Corps from releasing water from one of the large reservoirs in North Dakota, but the U.S. Court of Appeals reversed the ruling, and water flowed through the dams again.

The battle over water rights has become more heated and entangled as those who live upstream and those who live downstream fight to protect their interests. Although each side says it is willing to share, neither side seems willing to give up much. South Dakota wants the Corps to shorten the navigation season during years of drought so that upstream states will have enough water. But residents downstream counter that their needs should be favored over those of the fishing and tourism industry up north.

Complicating the upstream versus downstream issue is a growing confrontation between environmentalists and farmers. Environmentalists want to change the overall management plan for the Missouri River because 51 of its 67 native fish populations are in a dramatic decline. According to a 2002 report issued by the environmental group American Rivers, the Missouri River is the nation's most threatened river. This designation mirrors a 2002 report from the National Academy of Sciences that the river will continue to deteriorate unless its natural flow is significantly restored and the river is reconnected to its flood plain. Farmers and the barge industry bitterly oppose restoring the largely channelized river.

The Missouri River Basin Association, a coalition of eight river-basin states and two dozen Native American tribes, has the unenviable job of working with the Corps to meet the demands of the various competing interest groups as they decide the river's future. The association recognizes that the river does not belong to anyone, and they have to somehow prioritize uses of the river in a way

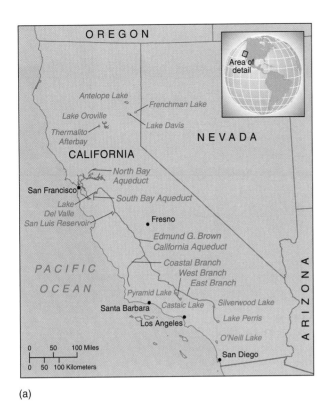

(a)

(b)

Figure 13.23 Water diversion in Southern California. Largely desert, Southern California relies on water diversion for the water needs of its millions of inhabitants. (a) The California Water Project includes 1,042 km (648 mi) of aqueducts to transfer large quantities of water to Southern California. This map also shows some of the main reservoirs of the California Water Project. (b) An aqueduct near Fresno, California.

that will at least partially satisfy the environmental groups, farmers, hydroelectric producers, Native Americans, and fishing and tourism interests.

Water Diversion Projects   One way to increase the natural supply of water to a particular area is to divert water from areas where it is in plentiful supply by pumping water through a system of aqueducts. Much of Southern California receives its water supply via aqueducts from northern California (Figure 13.23). Water from the Colorado River is also diverted into Southern California by aqueducts.

Large-scale water diversion projects are controversial and expensive. The Central Arizona Project, which pumps water 540 km (336 mi) from the Colorado River to Phoenix and Tucson, was completed at a cost of almost $4 billion. As we saw earlier, a river or other body of water is damaged when a major portion of its water is diverted. Pollutants, which would have been diluted in the normal river flow, reach higher concentrations when much of the flow has been removed. Fishes and other organisms may decline in number and diversity. Although no one denies that people must have water, opponents of water diversion projects contend that serious water conservation efforts would eliminate the need for additional large-scale water diversion.

Desalinization   Seawater and salty groundwater can be made fit to drink through the removal of salt, called **desalinization** (or **desalination**), by several methods. In **distillation**, salt water is heated until the water evaporates, leaving behind a crust of salt. The water vapor is

then condensed to produce fresh water. Another method, called **reverse osmosis**, involves forcing salt water through a membrane that is permeable to water but not to salt. About 97% of the salt can be removed from water by this process.

Desalinization is expensive because it requires a large input of energy, although recent advances in reverse osmosis technology have increased its efficiency so that it requires much less energy than distillation. Other expenses involved in desalinization projects include the cost of transporting the desalinized water from the site of production to where it will be used. During the early 2000s, the cost of desalinization, excluding transport costs, varied from $0.25 to $3.00 per cubic meter depending on what technology was used and how salty the water was to begin with. Removing salt from seawater costs three to five times more than removing salt from brackish water.

The disposal of salt produced by desalinization is also a concern. Simply dumping it back into the ocean, particularly near coastal areas that are highly productive, could cause a localized increase in salinity that would harm marine organisms.

Desalinization is a huge industry in North Africa and the Middle East, particularly in Saudi Arabia, because it is cost-competitive with alternative methods of obtaining fresh water in those arid regions. The United States is also a major producer of desalted water—for example, at the Yuma Desalting Plant in Arizona, which is the world's largest reverse osmosis desalting plant. The Yuma plant treats salty drainage water from nearby agricultural lands. After the water is desalted, it is added to the Colorado

River so that Mexico receives enough water (and water of a low enough salinity) from the United States.

WATER CONSERVATION

We have seen how population and economic growth have placed an increased demand on our water supply. Today there is more competition than ever among water users with different priorities, and water conservation measures are necessary to guarantee sufficient water supplies. Most water users use more water than they really need, whether it is for agricultural, industrial, or direct personal consumption. With incentives, these users will lower their rates of water consumption. Many studies have shown that higher prices for water provide the motivation to conserve water.

Reducing Agricultural Water Waste

Irrigation generally makes inefficient use of water. Traditional irrigation methods, which have been practiced for more than 5,000 years, involve flooding the land or diverting water to fields through open channels. Water flow must be increased in order to guarantee that the far end of the field or higher elevations of the field receive water. Less than 50% of the water applied to the soil by flood irrigation is absorbed by plants; the rest usually evaporates into the atmosphere.

One of the most important innovations in agricultural water conservation is **microirrigation**, also called

Figure 13.24 Microirrigation. Cutaway view of soil shows a small tube at the root line. Tiny holes in the tube deliver a precise amount of water directly to plant roots, eliminating much of the waste associated with traditional methods of irrigation. Photographed in Fresno, California.

drip or **trickle irrigation**, in which pipes with tiny holes bored in them convey water directly to individual plants (Figure 13.24). This reduces the water needed to irrigate crops by a substantial amount, usually 40% to 60%. Microirrigation also reduces the amount of salt left in the soil by irrigation water.

Another important water-saving measure in irrigation is the use of lasers to level fields, which allows water to be distributed more evenly. As a laser beam sweeps across a field, a field grader receives the beam and scrapes the soil, leveling it. Because farmers must use extra water to ensure that plants on higher elevations of a field receive enough, laser leveling of a field reduces the water required for irrigation.

The use of sound water management principles in agriculture reduces water consumption. Traditionally, farmers have been allotted specific amounts of water at specific times, with a "use it or lose it" philosophy. This approach encourages waste. Instead, a field's water needs can be carefully monitored, by measuring rainfall and determining soil moisture, to determine when irrigation should be scheduled and how much water should be applied. These water management strategies effectively reduce overall water consumption.

Although advances in irrigation technology are improving the efficiency of water use, many challenges remain. For one thing, sophisticated irrigation techniques are prohibitively expensive. Few farmers in highly developed countries, let alone subsistence farmers in developing nations, can afford to install them.

ENVIROBRIEF

Where Water Conservation Is Academic

Extensive efforts to increase water efficiency at the University of California at Santa Barbara (UCSB) have produced dramatic results: a nearly 50% reduction in campus water use between 1987 and 1994, even as the student population grew, and a total savings of $3.7 million attributed to efficiency improvements. Campus officials initiated the conservation program when the university began approaching its legal water allocation, as specified by the local water district. Audits of water use were assessed to determine appropriate conservation measures, such as leak repairs, improved irrigation systems using reclaimed water, new plumbing devices, and upgraded water-use appliances (dishwashers, washing machines).

The program's campus-wide acceptance and overall success are partially attributed to the extensive involvement of staff and students in planning and carrying out conservation efforts. Cost savings have been passed along to students and others in the campus community, and the university has transformed its community image from that of water hog to water conserver. Water conservation efforts similar to those in UCSB's program could be implemented at other higher education facilities, military bases, apartment complexes, and vacation resorts.

Reducing Water Waste in Industry

Electric power generators and many industries require water in order to function (recall from Chapters 10 to 12 that power plants heat water to form steam, which turns the turbines). In the United States, five major industries—chemical products, paper and pulp, petroleum and coal, primary metals, and food processing—consume almost 90% of industrial water. Water use by these industries does not include water used for cooling purposes.

Stricter pollution control laws provide some incentive for industries to conserve water. Industries usually reduce their water use, and therefore their water treatment costs, by recycling water. The National Steel Corporation plant in Granite City, Illinois, for example, recycles approximately two thirds of the 62 million gallons of water it uses daily. The used water is cleaned up before being discharged into a lake that spills into the Mississippi River.

It is likely that water scarcity, in addition to more stringent pollution control requirements, will encourage further industrial recycling. The potential for industries to conserve water by recycling is enormous.

Reducing Municipal Water Waste

Like industries, regions and cities can reduce their water consumption by recycling or reusing water before it is discharged. Individual homes and other buildings can be modified to collect and store "gray water"—water that has already been used in sinks, showers, washing machines, and dishwashers (Figure 13.25). The gray water can then be recycled to flush toilets, wash the car, or sprinkle the lawn.

In contrast to water recycling, *wastewater reuse* occurs when water is collected and treated before being redistributed for use. Israel probably has the world's most highly developed system of treating and reusing municipal wastewater. Israel does this out of necessity because all of its possible freshwater sources have already been tapped. The reclaimed water is used for irrigation, which allows higher-quality fresh water to be channeled to cities. Used water contains pollutants, but most of these are nutrients from treated sewage and are therefore beneficial to crops.

Automated systems to purify and reuse wastewater have been developed and are cost-competitive with fresh water. In Tokyo, for example, the wastewater in the Mitsubishi office building is purified and reused.

In addition to recycling and reuse, cities can decrease water consumption through other conservation measures. These include consumer education (both children and adults need to be taught the importance of using our limited supply of fresh water wisely), the use of water-saving household fixtures, and the development of economic incentives to save water (see "You Can Make a Difference: Conserving Water at Home"). These measures have been used successfully to pull cities through dry spells; they are effective because individuals are willing to conserve for the common good during water crisis periods.

Figure 13.25 Recycling water. Individual homes and buildings can be modified to collect and store "gray water," water that has already been used in sinks, showers, washing machines, and dishwashers. This "gray water" can be used when clean water is not required—for example, in flushing toilets, washing the car, and sprinkling the lawn.

Increasingly, however, cities are examining ways to encourage individual water conservation methods all the time. The installation of water meters in residences in Boulder, Colorado, reduced water consumption by one third. Before the installation, homeowners were charged a flat fee, regardless of their water use. For many apartment dwellers, water use is included in the rent; charging each apartment for its water use provides the incentive to use water more efficiently. In addition to installing water meters, a city might encourage water conservation by offering a rebate to any homeowner who installs a conserving device such as a water-saving toilet. Cities can help reduce municipal water consumption by passing building codes that specify installation of such water conservation fixtures as low-flush toilets and water-saving faucets and showerheads. The water supply systems (pipes and water mains) in many urban areas are old and leaky; repairing them would improve the efficiency of water use.

Cities also promote water conservation by increasing the price of water to reflect its true cost. As water prices rise, people learn very quickly to conserve water. For example, charging more for water during dry periods encourages individuals to conserve water. Although the average cost of water to consumers rose during the 1990s and early 2000s, many U.S. cities still did not charge consumers what the water actually cost them.

YOU CAN MAKE A DIFFERENCE

Conserving Water at Home

The average U.S. citizen uses 295 L (78 gal) of water per day at home (see figure). Many appliances, such as dishwashers, garbage disposals, and washing machines, need water. The growth of suburbs, with their expansive landscaping that requires watering, is also responsible for increased water use.

As a water user, you have a responsibility to use water carefully and wisely. The cumulative effect of many people practicing personal water conservation measures has a significant impact on overall water consumption. You can practice these yourself. The bathroom is a good place to start because most of the water used in an average home is for showers, baths, and flushing toilets.

1. Install water-saving showerheads and faucets to cut down significantly on water flow. Low-flow showerheads, for example, reduce water flow from 5 to 9 gal per minute to 2.5 gal per minute. Replacing one old showerhead brings a home $30 to $50 each year in water and energy savings. You can also save water by replacing washers on leaky faucets.

2. Install a low-flush toilet or use a water displacement device in the tank of a conventional toilet. Low-flush toilets require only 2 gal or less per flush, compared with 5 to 9 gal for conventional toilets. To save water with a conventional toilet, fill an empty plastic laundry bottle with water and place it in the tank to displace some of the water. Do not put the bottle where it will interfere with the flushing mechanism; also, do not add bricks to the tank, because they dissolve over time and can cause costly plumbing repairs.

3. An important way to conserve water at home is to fix leaky toilets. A toilet with a silent leak can waste 30 to 50 gal of water each day.

4. If you are in the market for a washing machine, high-efficiency washing machines require less water than traditional models. Also, wash full loads of clothes, and use the shortest cycle.

5. Modify your personal habits to conserve water. Avoid leaving the faucet running. For example, allowing the faucet to run while shaving consumes an average of 20 gal of water; you will use only 1 gal if you simply fill the basin with water or run the water only to rinse your razor. You may save as much as 10 gal of water a day by wetting your toothbrush and then turning off the tap while you brush your teeth, as opposed to running the water during the entire process. Also, most of us take longer showers than we need. Time yourself the next time you take a shower, if it is 10 minutes or longer, you can work on reducing your shower time.

6. Surprisingly, you will save water by using a dishwasher, which typically consumes about 12 gal per run, instead of washing dishes by hand with the tap running—but only if you run the dishwasher with a full load of dishes. That 12 gal of water is used regardless of whether the dishwasher is full or half empty.

Remember that wasting water costs you money. Conserving water at home reduces your water bill *and* heating bill: If you are using less hot water, you are also using less energy to heat that water.

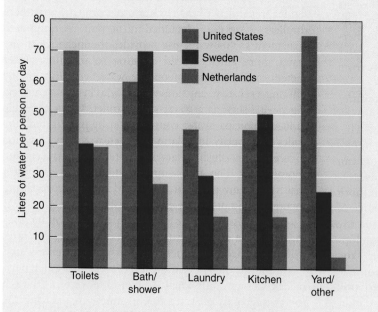

Residential water use in three highly developed countries. Data, in liters of water per person per day, are for the United States, Sweden, and the Netherlands.

SUMMARY WITH SELECTED KEY TERMS

I. Water has many special properties, including a high heat capacity and a high dissolving ability. Many of the properties of water are the result of its polarity, which causes **hydrogen bonding** between water molecules.

II. Although a large portion of the Earth is covered by water, only 2.5% of it is available as fresh water.

A. Surface water is fresh water found in streams, rivers, ponds, lakes, and **wetlands**, areas of land covered with water for at least part of the year.

  1. Surface waters are replenished by **runoff** of precipitation from the land.

  2. The area of land drained by a single river or stream is its **drainage basin** or **watershed**.

B. Groundwater occurs in **aquifers**, underground caverns and porous layers of rock.

  1. Unconfined aquifers have porous layers of rock above them. The upper limit of an unconfined aquifer is the **water table**.

  2. Confined aquifers have impermeable layers of rock above them. The water in a confined aquifer is trapped and is often under pressure.

III. Flooding occurs as a result of too much fresh water entering a particular area.

A. Flood damage is exacerbated by the deforestation of hillsides and mountains and by the development of **flood plains**, areas bordering a river that are subject to flooding.

B. The 1993 flood that occurred in nine Midwestern states is considered by many experts to be the worst flood in U.S. history.

IV. Fresh water is not evenly distributed, and some areas are barely able to support human life because of the shortage of water.

A. Aquifer depletion is the removal by humans of more groundwater than can be recharged by precipitation or melting snow.

  1. Aquifer depletion from porous rocks can cause **subsidence**, or sinking land. In some areas, **sinkholes** occur when droughts or excessive pumping of water cause a lowering of the water table.

  2. Saltwater intrusion is the movement of seawater into a freshwater aquifer.

B. Irrigation sometimes causes **salinization**, or accumulation of salt in the soil. Salinization results in a decline in productivity of agricultural soil.

V. Many areas in the United States have severe water shortages.

A. Population growth in such states as California, Nevada, Arizona, and Florida has placed great demands on water supplies.

B. Mono Lake in eastern California had surface water diverted to Los Angeles, lowering its water level and increasing its salinity.

  1. In 1994 California decided that Mono Lake would be returned almost to its original level.

  2. Los Angeles will develop **reclaimed water**, treated wastewater that is reused in some way, to replace water supplies from Mono Lake.

C. The Colorado River basin is seriously overdiverted for human consumption.

  1. The lower Colorado supplies water to Tucson and Phoenix in Arizona and to San Diego and Los Angeles in California.

  2. Recent population growth in the upper Colorado region (Colorado, Utah, and Wyoming) threatens the lower Colorado's water supply.

  3. The amount of water available from the Colorado River for Mexico, as stipulated by an international treaty, is insufficient.

D. The **Ogallala Aquifer** on the High Plains is the largest groundwater deposit in the world.

  1. The depletion of the Ogallala Aquifer for irrigation has lowered the water table in some places by more than 30 m.

  2. Water experts predict that groundwater in the Ogallala Aquifer will eventually drop to a level that is uneconomical to pump.

VI. The distribution of water resources in relation to human populations exacerbates global water problems.

A. People in many developing countries lack access to safe drinking water and wastewater disposal.

B. Population growth is outstripping water supplies in countries such as India, China, and Mexico.

C. The Aral Sea has been overdiverted for irrigation of farmland.

  1. Once the world's fourth largest freshwater lake, the Aral Sea has declined in area by more than 50%.

  2. Airborne salt from the dry lake bed may be adversely affecting the health of millions of people living near the Aral Sea.

D. International tensions over water rights could result in armed conflicts. Of particular concern are the Mekong River, Indus River, Ganges River, Tigris–Euphrates River, Jordan River, and Nile River.

VII. The long-term goal of water management is to provide a sustainable supply of high-quality water. **Sustainable water use** means that humans can use water resources into the future without harming the functioning of the hydrologic cycle or ecosystems.

A. Dams ensure a year-round supply of water in areas that have seasonal precipitation or snowmelt.

  1. Many people think the drawbacks of dams outweigh any benefits they provide.

  2. The Columbia River has more than 100 dams. The river is used for shipping, hydroelectric power, municipal and industrial water, recreational sports, and commercial fishing. Dam impoundments along the Columbia River have adversely affected salmon populations.

3. The Missouri River has been embroiled in a heated battle over water rights, particularly use of the water stored in upstream reservoirs.

B. Water diversion is sometimes used to increase the supply of water to a particular area.

1. Water is transported via conduits called **aqueducts**.

2. The Central Arizona Project pumps water from the Colorado River to Phoenix and Tucson. Water diversion concentrates pollutants in the remaining water and harms fishes and other organisms.

C. **Desalinization** is the removal of salt from seawater or salty groundwater.

1. One method of desalinization is **distillation**, heating salt water until water evaporates, leaving behind salt; the water vapor is then condensed.

2. A more energy-efficient method of desalinization is **reverse osmosis**, forcing salt water through a membrane that is permeable to water but not to salt.

D. Water conservation measures can make the present supply adequate, as opposed to trying to increase the available amount of water.

1. Agriculture (that is, irrigation) is the single largest user of water. Although irrigation increases agricultural productivity, it can contribute to water pollution, soil salinization, and depletion of water supplies. New agricultural techniques, including **microirrigation** and field leveling, can significantly cut agricultural water consumption.

2. After agriculture, the two largest users of water are industry and domestic/municipal water use. Water conservation, including recycling and reuse, can reduce both industrial and municipal water consumption.

THINKING ABOUT THE ENVIRONMENT

1. Why is the water accord discussed in the chapter introduction considered a model for water resource planning?

2. Diagram hydrogen bonding between water molecules and explain how it affects the properties of water.

3. Discuss the dissolving ability of water as it relates to ocean salinity, and to water pollution.

4. Distinguish between *surface water, runoff, drainage basin, groundwater, aquifer,* and *water table.*

5. Are our water supply problems largely the result of too many people? Give reasons why you support or refute this idea.

6. How has development along the upper Mississippi River exacerbated property damage during periods of flooding?

7. What are some of the problems associated with overdrawing surface water? With aquifer depletion?

8. Explain the problem of overdrawing surface water, using Mono Lake as an example.

9. Briefly describe the water scarcity problem in the Colorado River basin.

10. Explain the problem of aquifer depletion, using the Ogallala Aquifer as an example.

11. Briefly describe the complexity of international water use, using the Rhine River or the Aral Sea as an example.

12. List the major pros and cons of dams and reservoirs, using the Columbia River as an example.

13. Imagine you are a water manager for a Southwestern metropolitan district with a severe water shortage. What strategies would you use to develop a sustainable water supply?

14. How is water used in agriculture? Discuss two ways agricultural water can be used more sustainably.

15. Which industries consume the most water? Discuss one way in which water can be used more sustainably by industry.

16. Outline a brief water conservation plan for your own personal daily use. How do you think you could use water more sustainably?

17. Explain how water resource problems might contribute to economic or political instability.

18. Should housing be allowed on the flood plain of a river? Should taxpayers provide federal disaster assistance for those who choose to live on flood plains? Explain your answers.

*19. Repairing leaky faucets and toilets can save water at home. Calculate how many gallons of water are wasted each year from a leak of one drop per second (13,140 drops equals 1 gal).

* The solution to this question appears in Appendix VII.

TAKE A STAND

Visit our Web site at **http://www.wiley.com/college/raven** (select Chapter 13 from the Table of Contents) for links to more information about the controversy surrounding dam removal on the Snake River to help salmon species recover. Consider the views of proponents and opponents, and debate the issue with your classmates. You will find tools to help you organize your research, analyze the data, think critically about the issues, and construct a well-considered argument. Take a

Stand activities can be done individually or as part of a team, as oral presentations, written exercises, or Web-based (e-mail) assignments.

Additional on-line materials relating to this chapter, including Student Quizzes, Activity Links, Useful Web sites, Flash Cards, and more, can also be found on our Web site.

SUGGESTED READING

Alley, W.M., et al. "Flow and Storage in Groundwater Systems." *Science*, Vol. 296 (June 14, 2002). Understanding the dynamic nature of groundwater is essential to the sustainable development of groundwater resources.

Cohn, J.P. "Resurrecting the Dammed: A Look at Colorado River Restoration." *BioScience*, Vol. 51, No. 12 (December 2001). The Colorado River Indian Tribe is using part of its water allotment to restore aquatic habitat in the Colorado River.

Fischeti, M. "Drowning New Orleans." *Scientific American*, Vol. 285, No. 4 (October 2001). A major hurricane could flood New Orleans, killing thousands of people. The author examines how a proposed engineering project could save the city.

French, T. "Where the Sea Used to Be." *St. Petersburg Times* (December 2, 2001). A well-researched article about the shrinking Aral Sea.

Gleick, P.H. "Making Every Drop Count." *Scientific American*, Vol. 284, No. 2 (February 2001). This water expert examines the question "Will we have enough clean water to satisfy all the world's needs?"

Harden, B. "Dams, and Politics, Channel Flow of the Mighty Missouri." *New York Times* (May 5, 2002). The many special interests involved in the management of the Missouri River have created a political nightmare.

Kemper, S. "A Flap Over Water." *Smithsonian* (September 2001). A drought in the Klamath basin along the California–Oregon border has caused a dispute about water rights.

Montaigne, F. "A River Dammed." *National Geographic*, Vol. 1999, No. 4 (April 2001). An excellent review of the complexity of water management of the mighty Columbia River.

Perkins, S. "Crisis On Tap?" *Science News*, Vol. 162 (July 20, 2002). Pollution and population growth are harming the planet's water resources.

Postel, S. "Growing More Food with Less Water." *Scientific American*, Vol. 284, No. 2 (February 2001). This water expert argues persuasively that we must expand irrigation to feed our growing population, but at the same time we must reduce water waste associated with irrigation.

Yardley, J. "For Texas Now, Water and Not Oil Is Liquid Gold." *New York Times* (April 16, 2001). Should private companies sell groundwater? This article examines the political debate about water rights to aquifers in Texas.

Section of a shelter-forest near Dunhuang, China. Millions of trees defend China against raging dust storms from the Gobi Desert. Shown are pines and poplars developed by Chinese scientists to grow rapidly in poor soil. Note the agricultural land protected by the Great Green Wall.

Soils and Their Preservation

Learning Objectives

After you have studied this chapter you should be able to

1. Identify the factors involved in soil formation.

2. List the four components of soil and give the ecological significance of each.

3. Briefly describe soil texture and soil acidity.

4. Explain the impacts of soil erosion and mineral depletion on plant growth and on other resources such as water.

5. Describe the American Dust Bowl and explain how a combination of natural and human-induced factors caused this disaster.

6. Define *sustainable soil use* and summarize how conservation tillage, crop rotation, contour plowing, strip cropping, terracing, and shelterbelts help to minimize erosion and mineral depletion of the soil.

7. Discuss the basic process of soil reclamation.

8. Briefly describe the provisions of the Farm Bill regarding the Conservation Reserve Program and the Grasslands Reserve Program.

During its long history, China has had to contend with both invaders and devastating dust storms sweeping across its northern border. The Great Wall of China, a centuries-old earth and stone structure some 2,400 km (1,500 mi) long, was built across northern China to defend it against Mongolian raids. Although China's Great Wall was only partially successful in keeping out invaders, a more recently constructed "wall" has been very effective in controlling the dust storms from the expanding Gobi Desert. The new wall, often called the Great Green Wall, consists of some 300 million trees that were planted beginning in the 1950s. This wall of shelter-forest extends for some 4,800 km (3,000 mi) and is 800 km (500 mi) wide in places. The Great Green Wall and other shelter-forests in China's deserts now exceed 100,000 km² in area.

An average dust storm in northern China transports as much as 100 million tons of soil particles for hundreds or even thousands of kilometers. The dust damages crops, grounds air traffic, and causes many other inconveniences. The Great Green Wall, however, has reduced both the frequency and severity of dust storms because the trees slow the wind's velocity, and the moist forest floor discourages additional soil from being picked up and carried by the wind. During the 1950s Beijing, which is 480 km (300 mi) downwind of the Gobi Desert, experienced 10 to 20 major dust storms per year. By the 1970s, the annual number had declined to fewer than five, and during the 1990s, even a single dust storm per year was unusual.

Although climate was responsible for forming China's deserts, human activities probably caused them to enlarge. Removal of forests, overcollection of firewood, and overgrazing by sheep and goats on semiarid steppes (shortgrass prairies) loosened the fertile soil and exacerbated the dust storms coming out of the Gobi Desert before the Great Green Wall was planted. At least 8% of China's land area is desert, and **desertification,** the progressive degradation of grassland and other productive lands into unproductive desert, is an ongoing problem in China. (See Chapter 17 for a more detailed discussion of land degradation and desertification.) Planting trees to slow or stop the encroachment of desert, gradually improve soil quality, and reduce the severity of dust storms caused by wind erosion is a long-term Chinese goal. In 1998, for example, students and other volunteers planted a green wall of almost 2,000 trees in the En Ge Bei Desert in Inner Mongolia.

In this chapter you will learn about soil as a valuable natural resource on which humans depend for food. Many human activities cause or accentuate soil problems, such as erosion and nutrient mineral depletion. The goals of soil conservation are to minimize soil erosion and maintain soil fertility so that this resource can be used in a sustainable fashion. Conserving our soil resources is critical to human survival because more than 99% of our food comes from the land.

Table 14.1	Essential Elements For Plant Growth

Element	Source
Carbon	Air (as CO_2)
Hydrogen	Water
Oxygen	Water, air (as O_2)
Nitrogen	Soil
Phosphorus	Soil
Calcium	Soil
Magnesium	Soil
Sulfur	Soil
Potassium	Soil
Chlorine	Soil
Iron	Soil
Manganese	Soil
Copper	Soil
Zinc	Soil
Molybdenum	Soil
Boron	Soil

WHAT IS SOIL?

Soil is the relatively thin surface layer of the Earth's crust consisting of mineral and organic matter that has been modified by the natural actions of agents such as weather, wind, water, and organisms. It is easy to take soil for granted. We walk on and over it throughout our lives but rarely stop to think about how important it is to our survival.

Vast numbers and kinds of organisms, mainly microorganisms, inhabit soil and depend on it for shelter, food, and water. Plants anchor themselves in soil, and from it they receive essential nutrient minerals and water. Most of the 16 different elements essential for plant growth are obtained directly from the soil (Table 14.1). Terrestrial plants could not survive without soil, and because we depend on plants for our food, humans could not exist without soil, either.

How Soils Are Formed

Soil is formed from rock, called parent material, that is slowly broken down, or fragmented, into smaller and smaller particles by biological, chemical, and physical **weathering processes** in nature. It takes a very long time, sometimes thousands of years, for rock to disintegrate into finer and finer mineral particles. To form 2.5 cm (1 in.) of topsoil may require between 200 and 1,000 years. Time is also required for organic material to accumulate in the soil. Soil formation is a continuous process that involves interactions between the Earth's solid crust and the biosphere. The weathering of parent material beneath already-formed soil continues to add new soil. The thickness of soil varies from a thin film on very young lands, near the North and South Poles and on the slopes near the tops of mountains, to more than 3 m (10 ft) on very old lands, such as certain forests.

Organisms and climate both play essential roles in weathering, sometimes working together. When plant roots and other organisms that live in the soil respire, they produce carbon dioxide, CO_2, which diffuses into the soil and reacts with soil water to form carbonic acid, H_2CO_3. Soil organisms such as lichens also produce other kinds of acids. These acids etch tiny cracks in the rock; water then seeps into these cracks. If the parent material is located in a temperate climate, the alternate freezing and thawing of the water during the winter causes the cracks to enlarge, breaking off small pieces of rocks. Small plants can then become established and send their roots into the larger cracks, fracturing the rock further.

Topography, a region's surface features, such as the presence or absence of mountains and valleys, is also involved in soil formation. Steep slopes often have very little or no soil on them because soil and rock are continually transported down the slopes by gravity; runoff from precipitation tends to amplify erosion on steep slopes. Moderate slopes and valleys, on the other hand, may encourage the formation of deep soils.

SOIL COMPOSITION

Soil is composed of four distinct parts: mineral particles, which make up about 45% of a "typical" soil; organic matter (about 5%); water (about 25%); and air (about 25%). Soil occurs in layers, each of which has a certain composition and special properties. The plants, animals, fungi, and microorganisms that inhabit soil interact with it, and nutrient minerals are continually cycled from the soil to organisms, which use them in their biological processes. When the organisms die, bacteria and other soil organisms decompose the dead material, returning the nutrient minerals to the soil.

Components of Soil

The mineral portion, which comes from weathered rock, constitutes most of what we call soil. It provides anchorage and essential nutrient minerals for plants, as well as pore space for water and air. Because different rocks are composed of different minerals, soils vary in mineral composition and chemical properties. Rocks rich in aluminum form acidic soils, whereas rocks that contain silicates of magnesium and iron form soils that may be deficient in calcium, nitrogen, and phosphorus. Also, soils formed from the same kind of parent material may not develop in the same way because other factors such as weather, topography, and organisms differ.

The age of a soil affects its mineral composition. In general, older soils are more weathered and lower in certain essential nutrient minerals. Large portions of Australia, South America, and India have old, infertile soils. In contrast, in geologically recent time, glaciers passed across much of the Northern Hemisphere, pulverizing bedrock and forming fertile soils.[1] Essential nutrient minerals are readily available in these geologically young soils and in young soils formed in areas of volcanic activity.

Litter (dead leaves and branches on the soil's surface), animal dung, and dead remains of plants, animals, and microorganisms in various stages of decomposition constitute the organic portion of soils. Microorganisms, particularly bacteria and fungi, gradually decompose this material. During decomposition, essential nutrient mineral ions are released into the soil, where they may be bound by soil particles or absorbed by plant roots. Organic matter increases the soil's water-holding capacity by acting much like a sponge. For these reasons gardeners often add organic matter to soils, especially sandy soils, which are naturally low in organic matter.

The black or dark brown organic material that remains after much decomposition has occurred is called

Figure 14.1 Humus. Humus is partially decomposed organic material, primarily from plant and animal remains. Soil that is rich in humus has a loose, somewhat spongy structure with several properties, such as increased water-holding capacity, that are beneficial for plants and other organisms living in it.

humus (Figure 14.1 and "Mini-Glossary: Soil Terms"). Humus, which is not a single chemical compound but a mix of many organic compounds, binds nutrient mineral ions and holds water. On average, humus persists in agricultural soil for about 20 years. Certain components of humus, however, may persist in the soil for hundreds of years. Although humus is somewhat resistant to further decay, a succession of microorganisms gradually reduces it to carbon dioxide, water, and nutrient minerals. Detritus-feeding animals such as earthworms, termites, and ants also help to break down humus.

Soil has numerous pore spaces around and among the soil particles. The pore spaces occupy roughly 50% of a soil's volume and are filled with varying proportions of water (called **soil water**) and air (called **soil air**) (Figure 14.2); both are necessary to produce a moist but

MINI-GLOSSARY

Soil Terms

humus: Partly decomposed organic material in the soil; brown or black in color.

leaching: Movement of water and dissolved material downward through the soil.

illuviation: Deposition of a material into a lower soil layer from a higher layer as a result of leaching.

sand: Large (0.05- to 2-mm diameter) inorganic particles in the soil; coarse enough to feel gritty.

silt: Medium-sized (0.002- to 0.05-mm diameter) inorganic particles in the soil; too small to feel gritty.

clay: Small (< 0.002-mm diameter) inorganic particles in the soil; too small to settle out of suspension.

---

[1] The Pleistocene epoch, which began approximately 2 million years ago, was marked by four periods of glaciation. At their greatest extent these ice sheets covered nearly 4 million square miles of North America, extending south as far as the Ohio and Missouri Rivers.

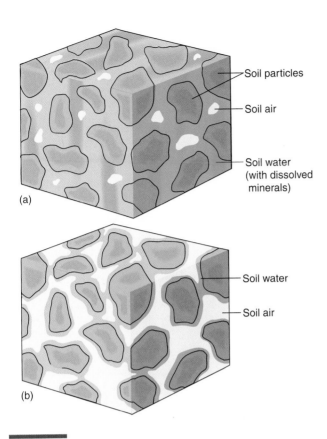

Soil particles

Soil air

Soil water
(with dissolved
minerals)

(a)

Soil water

Soil air

(b)

Figure 14.2 **Pore space.** (a) In a wet soil, most of the pore space is filled with water. (b) In a dry soil, a thin film of water is tightly bound to soil particles, and most of the pore space is occupied by soil air.

aerated soil that sustains plants and other soil-dwelling organisms. Generally speaking, water is held in the smaller pores (less than 0.05 mm in diameter), while air is found in the larger pores. After a prolonged rain, almost all of the pore spaces may be filled with water, but water drains rapidly from the larger pore spaces, drawing air from the atmosphere into those spaces.

Soil water originates as precipitation, which drains downward, or as groundwater, which rises upward from the water table (see Chapter 13). Soil water contains low concentrations of dissolved nutrient mineral salts that enter the roots of plants as they absorb the water. Water not bound to soil particles or absorbed by roots percolates (moves down) through the soil, carrying dissolved nutrient minerals with it. The removal of dissolved materials from the soil by water percolating downward is called **leaching**. The deposition of leached material in the lower layers of soil is known as **illuviation**. Iron and aluminum compounds, humus, and clay are some illuvial materials that can gather in the subsurface portion of the soil. Some substances completely leach out of the soil because they are so soluble that they migrate right down to the groundwater. It is also possible for water to move *upward* in the soil, transporting dissolved materials with it, such as when the water table rises.

Soil air contains the same gases as atmospheric air, although they are usually present in different proportions. Generally, as a result of cellular respiration by soil organisms, there is more carbon dioxide and less oxygen in soil air than in atmospheric air. (Recall from Chapter 4 that cellular respiration uses oxygen and produces carbon dioxide.) Among the important gases in soil air are oxygen, required by soil organisms for cellular respiration; nitrogen, used by nitrogen-fixing bacteria (see Chapter 6); and carbon dioxide, involved in soil weathering. As mentioned earlier in the chapter, carbon dioxide dissolves in water to form carbonic acid, a weak acid that accelerates the weathering process during soil formation.

Soil Horizons

A deep vertical slice, or section, through many soils reveals that they are organized into distinctive horizontal layers called **soil horizons**. A **soil profile** is a vertical section from surface to parent material, showing the soil horizons (Figure 14.3).

The uppermost layer of soil, the **O-horizon**, is rich in organic material. Plant litter accumulates in the O-horizon and gradually decays. In desert soils the O-horizon is often completely absent, but in certain organically rich soils it may be the dominant layer.

Just beneath the O-horizon is the topsoil, or **A-horizon**, which is dark and rich in accumulated organic matter and humus. The A-horizon has a granular texture and is somewhat nutrient-poor owing to the gradual loss of many nutrient minerals to deeper layers by leaching. In some soils, a heavily leached **E-horizon** develops between the A- and B-horizons.

ENVIROBRIEF

The Future of Everglades Soil

How are Everglades soils different from the vast majority of the world's soils? Most Everglades soils only contain 15% or less of mineral particles, but they have large amounts of organic matter. In fact, Everglades soils are formed primarily of organic material, heaps of partly decomposed saw grass that has piled up for centuries over a hard layer of limestone bedrock. The organic material never decomposed completely because it was buried under water. When Everglades soil is exposed to oxygen, however, the organic material quickly decomposes to carbon dioxide and water.

In the early 20th century, most areas of the Everglades had a soil layer as thick as 3.7 m (12 ft). In the Everglades Agricultural Area, where sugar and other crops have been grown for decades, the soil level has subsided dramatically. Scientists estimated that by the year 2000, 45% of the land in this area had less than 0.3 m (1 ft) of soil, which is too thin to grow sugar and most other crops economically. The loss of Everglades soil is unavoidable. As the ground is drained and worked to raise crops, the soil is exposed to air, and decomposition is accelerated. (Other aspects of the Everglades are discussed in Chapter 7.)

(a)

(b)

Figure 14.3 Soil profile. (a) A "typical" soil profile. Each horizon has its own chemical and physical properties. (b) This particular soil, located on a farm in Virginia, has no O-horizon because it is used for agriculture; the surface litter that would normally comprise the O-horizon was plowed into the A-horizon. The shovel gives an idea of the relative depths of each horizon.

The **B-horizon**, the light-colored subsoil beneath the A-horizon, is often a zone of illuviation in which nutrient minerals that leached out of the topsoil and litter accumulate. It is typically rich in iron and aluminum compounds and clay.

Beneath the B-horizon is the **C-horizon**, which contains weathered pieces of rock and borders the unweathered solid parent material. The C-horizon is below the extent of most roots and is often saturated with groundwater.

Soil Organisms

Although soil organisms are usually hidden underground, their numbers are huge. Millions of microorganisms, including bacteria, fungi, algae, microscopic worms, and protozoa, may inhabit just one teaspoon of fertile agricultural soil. Many other organisms also colo-

nize the soil ecosystem, including plant roots, insects such as termites and ants, earthworms, moles, snakes, and groundhogs (Figure 14.4). Most numerous in soil are bacteria, which number in the hundreds of millions per gram of soil. Scientists have identified about 170,000 species of soil organisms, but thousands remain to be identified. Moreover, scientists know very little about the roles of soil organisms, in part because it is hard to study their activities under natural conditions.

Soil organisms provide several essential **ecosystem services** (see Chapter 7), such as maintaining soil fertility by decaying and cycling organic material, preventing soil erosion, breaking down toxic materials, cleansing water, and affecting the composition of the atmosphere. Many of these services are discussed in greater detail in this chapter.

Worms are important organisms living in soil. Earthworms, probably one of the most familiar soil inhabitants, ingest soil and obtain energy and raw materi-

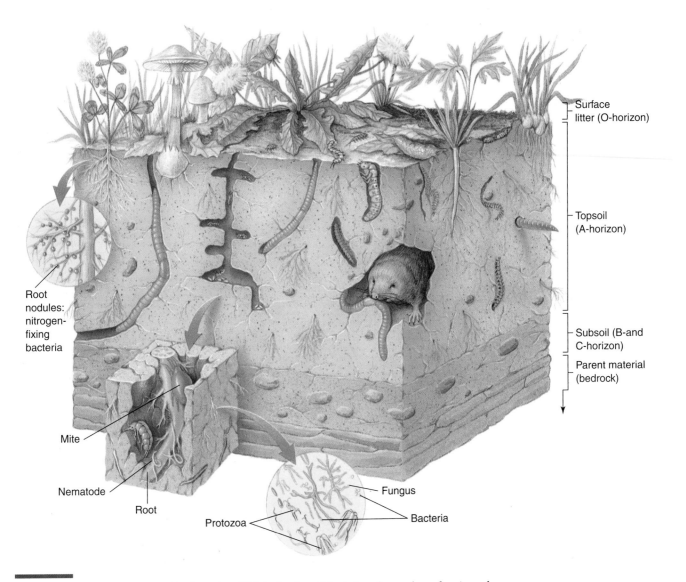

Root nodules: nitrogen-fixing bacteria

Mite

Nematode

Root

Protozoa

Fungus

Bacteria

Surface litter (O-horizon)

Topsoil (A-horizon)

Subsoil (B-and C-horizon)

Parent material (bedrock)

Figure 14.4 Soil organisms. The diversity of life in fertile soil includes plants, algae, fungi, earthworms, flatworms, roundworms, insects, spiders and mites, bacteria, and burrowing animals such as moles and groundhogs. (Soil horizons are not drawn to scale.)

als by digesting some of the compounds that make up humus. **Castings**, bits of soil that have passed through the gut of an earthworm, are deposited on the soil surface. In this way, nutrient minerals from deeper layers in the soil are brought to upper layers. Earthworm tunnels serve to aerate the soil, and the worms' waste products and corpses add organic material to the soil.

Ants live in the soil in enormous numbers, constructing tunnels and chambers that help to aerate it. Members of soil-dwelling ant colonies forage on the surface for bits of food, which they carry back to their nests. Not all of this food is eaten, however, and its eventual decomposition helps increase the organic matter in the soil. Many ants are also indispensable in plant reproduction because they bury seeds in the soil.

Plants are greatly affected by the properties of soil, although most plants can tolerate a wide range of soil types. In turn, the types of plants that grow in soil affect it. As a result of the complex interactions among plants, climate, and soil, it is hard to specify cause and effect in their relationships. Are the plants growing in a certain locality because of the soil that is found there, or is the soil's type determined by the plants?

One very important symbiotic relationship in the soil occurs between fungi and the roots of vascular plants. These associations, called **mycorrhizae**, enable plants to absorb adequate amounts of essential nutrient minerals from the soil. The threadlike body of the fungal partner, called a **mycelium**, extends into the soil well beyond the roots. Nutrient minerals absorbed from

Soil Fungi and Plant Diversity

Scientists have yet to determine the specific factors that influence plant diversity, but two teams of researchers caution that the importance of fungi in shaping plant communities should not be ignored. Researchers in Switzerland and Ontario together reported that plant diversity in a given location is strongly influenced by the variety of fungi present there. The fungi that were studied form associations with plant roots (mycorrhizae) and provide nutrient minerals for the majority of plant species. In controlled experiments simulating both European and U.S. landscapes, replicates of the same plant community were grown in the presence of from 1 to 14 fungal species. The scientists observed that plant communities that were grown in the presence of 8 or more fungal species maintained the greatest diversity. When plants were grown with fewer kinds of fungi, the structure and composition of plant communities fluctuated considerably, depending on the number and composition of fungal species present. Researchers think that fungal diversity thus plays a significant role in determining plant diversity, as well as ecosystem variability and productivity.

the soil by the fungus are transferred to the plant, while food produced by photosynthesis in the plant is delivered to the fungus. Mycorrhizal fungi have been demonstrated to enhance the growth of plants (see Figure 5.7). When mycorrhizal fungi are absent from the soil following a natural or human-induced disturbance, the reestablishment of certain tree species is retarded.

Nutrient Cycling

In a balanced ecosystem, the relationships among soil and the organisms that live in and on it ensure soil fertility. Essential nutrient minerals such as nitrogen and phosphorus are cycled from the soil to organisms and back again to the soil (Figure 14.5 and Chapter 6). Decomposition is part of **nutrient cycling**, which is the pathways of various nutrient minerals or elements from the environment through organisms and back to the environment. Bacteria and fungi decompose plant and animal detritus and wastes, transforming large organic molecules to small inorganic molecules, including carbon dioxide, water, and nutrient minerals; the nutrient minerals are then released into the soil to be used again.

Abiotic (nonliving) processes are also involved in nutrient cycling. Although leaching causes some nutrient minerals to be lost from the soil ecosystem to groundwater, the weathering of the parent material replaces much or all of them. In addition, dusts carried in the atmosphere for hundreds or thousands of kilometers help replace nutrient minerals in certain soils. Hawaiian rainforest soils, for example, receive dust inputs from central Asia, a distance that is more than 6,000 km (3,730 mi) away.

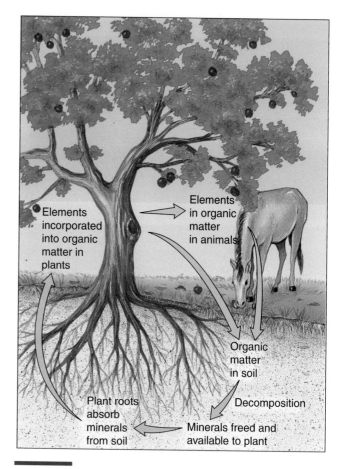

Figure 14.5 Nutrient cycling. In a balanced ecosystem, nutrient minerals cycle from the soil to organisms and then back again to the soil.

PHYSICAL AND CHEMICAL PROPERTIES OF SOIL

Texture and acidity are two parameters that help to characterize soils.

Soil Texture

The texture, or structural characteristic, of a soil is determined by the percentages (by weight) of different-sized inorganic mineral particles of sand, silt, and clay that it contains. The size assignments for sand, silt, and clay give soil scientists a way to classify soil texture. Particles larger than 2 mm in diameter, called gravel or stones, are not considered soil particles because they do not have any direct value to plants. The largest soil particles (0.05 to 2 mm in diameter) are called **sand**, medium-sized particles (0.002 to 0.05 mm in diameter) are called **silt**, and small particles (less than 0.002 mm in diameter) are called **clay** (Figure 14.6). Sand particles are large enough to be seen easily with the eye; silt par-

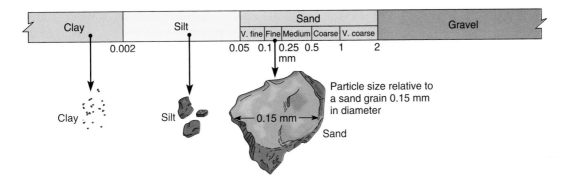

Figure 14.6 Relative sizes of soil particles.

ticles are about the size of flour particles and are barely visible with the eye; and clay particles can only be seen under an electron microscope. A soil's texture affects many of that soil's properties, which, in turn, influence plant growth. Clay is particularly important in determining many soil characteristics because clay particles have the greatest surface area of all soil particles. If the surface areas of about 450 g (1 lb) of clay particles were laid out side by side, they would occupy 1 hectare (2.5 acres).

Soil minerals are often present in charged forms called **ions**. Mineral ions may be positively charged (K^+, for example) or negatively charged (NO_3^-, for example).

Each clay particle has predominantly negative electrical charges on its outer surface that attract and reversibly bind positively charged mineral ions (Figure 14.7). Many of these mineral ions, such as potassium (K^+) and magnesium (Mg^{2+}), are essential for plant growth and are "held" in the soil for plant use by their interactions with clay particles. In contrast, negatively charged mineral ions are usually not held as tightly in the soil and are often washed out of the root zone.

Soil always contains a mixture of different-sized particles, but the proportions vary from soil to soil. A **loam**, which is an ideal agricultural soil, has an optimum combination of different soil particle sizes: It contains approxi-

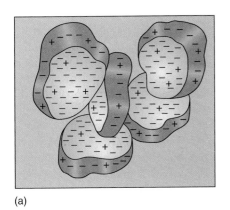

(a)

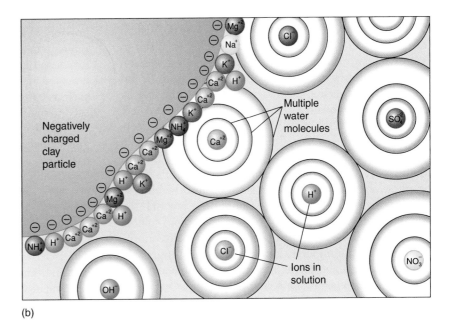

(b)

Figure 14.7 Availability of nutrient minerals. (a) The outer surfaces of clay particles are predominantly negatively charged. (b) In this close-up of part of a clay particle and the thin film of water around it, note the large number of positively charged ions attracted to the surface of the clay particle. (Circles represent the numerous water molecules surrounding the individual dissolved mineral ions.)

Table 14.2 Soil Properties Affected by Soil Texture

Soil Property	Soil Texture Type		
	Sandy Soil	Loam	Clay Soil
Aeration	Excellent	Good	Poor
Drainage	Excellent	Good	Poor
Nutrient mineral-holding capacity	Low	Medium	High
Water-holding capacity	Low	Medium	High
Workability (tillage)	Easy	Moderate	Difficult

mately 40% each of sand and silt, and about 20% of clay. Generally, the larger particles provide structural support, aeration, and permeability to the soil, whereas the smaller particles bind together into aggregates, or clumps, and hold nutrient minerals and water (Table 14.2). Soils with larger proportions of sand are not as desirable for most plants because they do not hold mineral ions or water as well. Plants grown in such soils are more susceptible to mineral deficiencies and drought. Soils with larger proportions of clay are also not as desirable for most plants because they provide poor drainage and often do not contain enough oxygen. Clay soils used for agriculture tend to get compacted, which reduces the number of pore spaces that can be filled by water and air.

Soil Acidity

Soil acidity is measured using the pH scale, which extends from 0 (extremely acidic) through 7 (neutral) to 14 (extremely alkaline). (See Appendix I for a discussion of pH.) The pH of most soils ranges from 4 to 8, but some soils are outside this range. The soil of the Pygmy Forest in Mendocino County, California, is extremely acidic, with a pH of 2.8 to 3.9. At the other extreme, certain soils in Death Valley, California, have a pH of 10.5.

Plants are affected by soil pH partly because the solubility of certain nutrient minerals varies with differences in pH. Soluble mineral elements can be absorbed by the plant, whereas insoluble forms cannot. At a low pH, the aluminum and manganese in soil water are more soluble and are sometimes absorbed by the roots in toxic concentrations. Certain mineral salts essential for plant growth, such as calcium phosphate, become less soluble and thus less available to plants at a higher pH.

Soil pH greatly affects the leaching of nutrient minerals. An acidic soil has a reduced ability to bind positively charged ions to it. As a result, certain nutrient mineral ions that are essential for plant growth, such as potassium (K^+), are leached more readily from acidic soil. The optimum soil pH for most plant growth is 6.0 to 7.0, because most nutrient minerals needed by plants are available in that pH range.

Soil pH affects plants and, in turn, is influenced by plants and other soil organisms. Litter composed of the needles of conifers contains acids that leach into the soil, lowering its pH. The decomposition of humus and the cellular respiration by soil organisms also decrease the pH of soil. (The effects of acid rain on soil are discussed in Chapter 20.)

MAJOR SOIL GROUPS

Variations in climate, local vegetation, parent material, underlying geology, topography, and soil age throughout the world result in thousands of soil types. These soils differ in color, depth, mineral content, acidity, pore space, and other properties. In the United States alone, as many as 17,000 different soil types have been identified. Here we focus on five soil types that are very common: spodosols, alfisols, mollisols, aridisols, and oxisols (Figure 14.8).

Regions with colder climates, ample precipitation, and good drainage typically have soils called **spodosols**, with very distinct layers. A spodosol usually forms under a coniferous forest and has an O-horizon of acidic litter composed primarily of needles; an ash-gray, acidic, leached E-horizon; and a dark brown, illuvial B-horizon. Spodosols do not make good farmland because they are too acidic and are nutrient-poor because of leaching.

Temperate deciduous forests grow on **alfisols**, soils with a brown to gray-brown A-horizon. Precipitation is great enough to wash much of the clay and soluble nutrient minerals out of the A- and E-horizons and into the B-horizon. When the deciduous forest is intact, soil fertility is maintained by a continual supply of plant litter such as leaves and twigs. When the soil is cleared for farmland, however, fertilizers (which contain nutrient minerals such as nitrogen, potassium, and phosphorus) must be used to maintain fertility.

Mollisols, found primarily in temperate, semiarid grasslands, are very fertile soils. They possess a thick, dark brown to black A-horizon that is rich in humus. Some soluble nutrient minerals remain in the upper layers, because precipitation is not great enough to leach them into lower layers. Most of the world's grain crops are grown on mollisols.

Aridisols are found in arid regions of all continents. The lack of precipitation in these deserts precludes much leaching, and the lack of lush vegetation precludes the accumulation of much organic matter. As a result, aridisols do not usually have distinct layers of leaching and illuviation. Some aridisols provide rangeland for grazing animals, and crops can be grown on aridisols if water is supplied by irrigation.

Oxisols, which are low in nutrient minerals, exist in tropical and subtropical areas with ample precipitation. Little organic material accumulates on the forest floor (O-horizon) because leaves and twigs are rapidly decom-

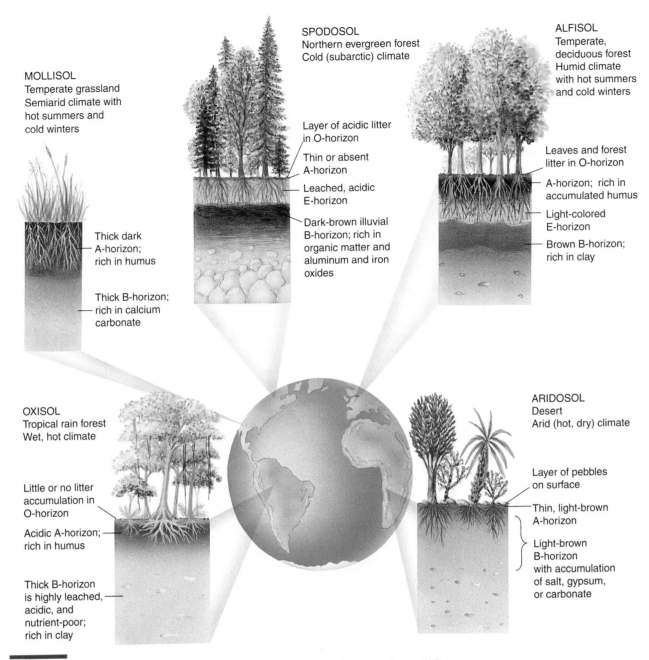

Figure 14.8 Major soil types. Five types are depicted, along with the vegetation and climate that are characteristic of each. The various climates and vegetation help determine the different soil types.

posed. The A-horizon is enriched with humus derived from the rapidly decaying plant parts. The B-horizon, which is quite thick, is highly leached, acidic, and nutrient-poor. Oddly enough, tropical rain forests, with their lush vegetation, grow on oxisols. Most of the nutrient minerals in tropical rain forests are locked up in the vegetation rather than in the soil. As soon as plant and animal remains touch the forest floor, they promptly begin to decay, and plant roots quickly reabsorb the nutrient minerals. Even wood, which may take years to be decomposed in temperate soils, is decomposed in a matter of

months in tropical rain forests, largely by subterranean termites.

SOIL PROBLEMS

Human activities often cause or exacerbate soil problems, including erosion and mineral depletion of the soil, both of which occur worldwide. Such activities do not promote **sustainable soil use**, in which humans use soil resources wisely, without a reduction in the amount or fertility of soil, so that it is productive for future generations. Soil that is used in a sustainable way is able to renew itself by natural processes year after year. (Soil pollution is discussed in Chapter 21.)

Soil Erosion

Water, wind, ice, and other agents promote **soil erosion**, the wearing away or removal of soil from the land. Water and wind are particularly effective in moving soil from one place to another. Rainfall loosens soil particles, which can then be transported away by moving water (Figure 14.9). Wind loosens soil and blows it away, particularly if the soil is barren and dry. Soil erosion is a natural process that can be greatly accelerated by human activities.

Erosion reduces the amount of soil in an area and therefore limits the growth of plants. Erosion also causes a loss of soil fertility, because essential nutrient minerals and organic matter that are part of the soil are also removed. As a result of these losses, the productivity of eroded agricultural soil drops, and more fertilizer must be used to replace the nutrient minerals lost to erosion.

Soil erosion is a national and international problem that rarely makes the headlines. To get a feeling for how serious the problem is, consider that the U.S. Department of Agriculture estimates that about 4.6 billion tons of soil are eroded from U.S. croplands and rangelands each year. Approximately one fifth of U.S. cropland is considered vulnerable to soil erosion damage.

Humans often accelerate soil erosion with poor soil management practices. Here we consider soil erosion caused by poor agricultural practices, but it is important to realize that agriculture is not the only culprit. Removal of natural plant communities, such as during the construction of roads and buildings, and unsound logging practices, such as clearcutting large forested areas, also accelerate soil erosion.

Soil erosion has an impact on other natural resources as well. Sediment that gets into streams, rivers, and lakes affects water quality and fish habitats. If the sediment contains pesticide and fertilizer residues, they further pollute the water. Also, when forests are removed within the watershed of a hydroelectric power facility, accelerated soil erosion can cause the reservoir behind the dam to fill in with sediment much faster than usual. This process results in a reduction of electricity production at that facility.

Sufficient plant cover limits the amount of soil erosion. Leaves and stems cushion the impact of rainfall, and roots help to hold the soil in place. Although soil erosion is a natural process, abundant plant cover makes it negligible in many natural ecosystems.

Figure 14.9 Soil erosion caused by water. Branching gullies, the most serious form of erosion, will continue to advance unless checked by some type of erosion control. Photographed in Colorado

CASE·IN·POINT **The American Dust Bowl**

Semiarid lands, such as the Great Plains of North America, have low annual precipitation and are subject to periodic droughts. Prairie grasses, the plants that grow best in semiarid lands, are adapted to survive droughts. Although the aboveground portions of the plant may die, the root systems can survive several years of drought. When the rains return, the root systems send up new leaves. Soil erosion is minimal because the dormant but living root systems hold the soil in place and resist the assault by wind and water.

The soils of semiarid lands are often of high quality, due largely to the accumulation over many centuries of a thick, rich humus. These lands are excellent for grazing and for growing crops on a small scale. Problems arise, however, when large areas of land are cleared for crops or when animals overgraze the land (see Chapter 17). The removal of the natural plant cover opens the way for climatic conditions to "attack" the soil, and it gradually deteriorates from the onslaught of hot summer sun, occasional violent rainstorms, and wind. If a prolonged drought occurs under such conditions, disaster can strike.

The effects of wind on soil erosion were vividly experienced over a wide region of the central United States during the 1930s (Figure 14.10). Throughout the late 19th and early 20th centuries, much of the native grasses had been removed to plant wheat. Then, between 1930 and 1937, the semiarid lands stretching from Oklahoma and Texas into Canada received 65% less annual precipitation than was normal. The rugged prairie grasses that had been replaced by crops could have survived these conditions, but not the wheat. The prolonged drought caused crop failures, which left fields barren and particularly vulnerable to wind erosion.

Winds from the west swept across the barren, exposed soil, causing dust storms of incredible magnitude (Figure 14.11). Topsoil from Colorado, Texas, Oklahoma, and other prairie states was blown eastward for hundreds of kilometers. Women hanging out clean laundry in Georgia went outside later to find it dust-covered. Bakers in New York City and Washington, D.C. had to keep freshly baked bread away from open windows so it would not get dirty. The dust even discolored the Atlantic Ocean several hundred kilometers off the coast. On April 14, 1937, known as Black Sunday, the most severe dust storm in U.S. history darkened the sky and blotted out the sun.

The Dust Bowl occurred during the Great Depression, and ranchers and farmers quickly went bankrupt. Many abandoned their dust-choked land and dead livestock and migrated west to the promise of California. The plight of these dispossessed farmers is movingly portrayed in the novel *The Grapes of Wrath*, written in 1939 by John Steinbeck, for which he won the Pulitzer Prize in 1940.

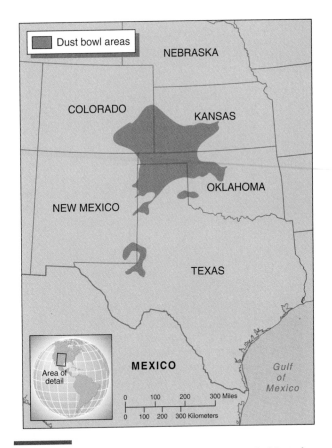

Figure 14.10 Location of the American Dust Bowl. More than 30 million hectares (74 million acres) of land in the Great Plains were damaged during the Dust Bowl years. Shaded parts of Colorado, Kansas, Oklahoma, Texas, and New Mexico suffered the most extensive damage.

When the rains finally arrived in the Great Plains, many areas were too devastated by erosion to support agriculture. In the years following the Dust Bowl, the U.S. Soil Conservation Service planted fence rows of shrubs and trees (to slow the winds) and seeded many badly eroded areas with native prairie grasses. Agriculture was able to return largely because irrigation protected against crop loss caused by droughts. We saw in Chapter 13, however, that the Ogallala Aquifer, which provides irrigation water for this region, is being drained faster than its rate of replacement. When farmers are forced to abandon irrigation and revert to dryland farming, only very careful management practices will prevent Dust Bowl conditions from returning during the next drought.

Although the United States no longer has a dust bowl, the Great Plains are still subject to droughts and soil erosion. For example, a severe 4-year drought in the early 2000s ruined crops in parts of Montana, Wyoming, Colorado, Kansas, Texas, and New Mexico. Dust storms (brownouts) reoccurred on the High Plains, reminding people of the "Dirty Thirties." ■

(a)

(b)

Figure 14.11 The Dust Bowl years. (a) A family running for shelter during a 1936 dust storm in Cimarron County, Oklahoma. (b) An abandoned Oklahoma farm in 1937. Total devastation was often the aftermath of dust storms.

Nutrient Mineral Depletion

In a natural ecosystem, essential nutrient minerals cycle from the soil to organisms, particularly plants, which absorb them through their roots. When those organisms die and microorganisms decompose them, the essential nutrient minerals are released into the soil, where they become available for use by organisms again. An agricultural system disrupts this pattern of nutrient cycling when the crops are harvested. Much of the plant material, containing nutrient minerals, is removed from the cycle, so it fails to decay and release its nutrient minerals back to the soil. Thus, over time, soil that is farmed inevitably loses its fertility (Figure 14.12).

Mineral Depletion in Tropical Rainforest Soils In tropical rain forests, the climate, the typical soil type, and the removal by humans of the natural forest community result in a particularly severe type of mineral depletion. Soils found in tropical rain forests are somewhat nutrient-poor because the nutrient minerals are stored primarily in the vegetation. Any nutrient minerals that are released as dead organisms decay in the soil are promptly reabsorbed by plant roots and their mutualistic fungi. If this did not occur, the heavy rainfall would quickly leach the nutrient minerals away. Nutrient reabsorption by vegetation is so effective that tropical rainforest soils can support luxuriant plant growth despite the relative infertility of the soil, as long as the forest remains intact.

When the forest is cleared, whether to sell the wood or to make way for crops or rangeland, its efficient nutrient cycling is disrupted. Removal of the vegetation that so effectively stores the forest's nutrient minerals allows them to leach out of the system. Crops can be grown on these soils for only a few years before the small mineral reserves in the soil are depleted. When cultivation is abandoned, the forest eventually returns to its original state, provided that there is a nearby forest to serve as a source of seeds. The regrowth of forest is a very slow process, however. (Chapter 17 discusses aspects of deforestation other than soil degradation.)

Soil Problems in the United States

Every 5 years the Natural Resources Conservation Service (NRCS), which was formerly called the Soil Conservation Service, measures the rate of soil erosion at thousands of sites across the United States. It also uses satellite data and models to estimate annual soil erosion. These measurements and estimates indicate that erosion is a serious threat to cultivated soils in many regions throughout the United States, particularly in parts of southern Iowa, northern Missouri, western and southern Texas, and eastern Tennessee. The good news is that soil erosion on all U.S. croplands declined by about 38% between 1982 and 1997, as calculated in December 2000, although it is still at significant levels.

Water erosion is particularly severe in the midwestern grain belt along the Mississippi and Missouri Rivers, as well as in the central valley of California and in the hilly Palouse River region of the Pacific Northwest. The NRCS estimates that about 25% of agricultural lands in the United States are losing topsoil faster than it can be regenerated by natural soil-forming processes. This loss

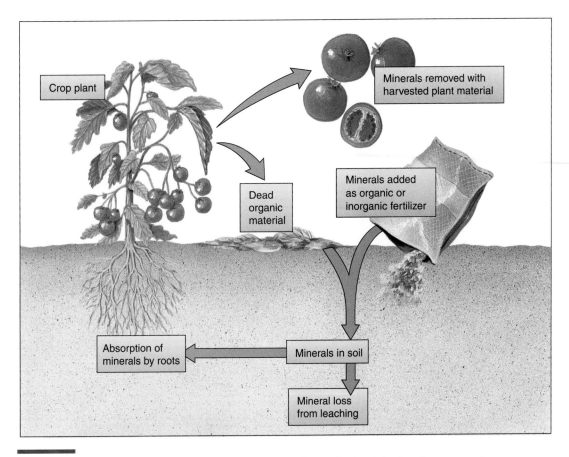

Figure 14.12 Mineral depletion in agricultural soil. As plant and animal detritus decomposes in natural ecosystems, nutrient minerals are cycled back to the soil for reuse. In agriculture, much of the plant material is harvested. Because the nutrient minerals in the harvested portions are unavailable to the soil, the nutrient cycle is broken, and fertilizer must be added periodically to the soil.

is often so gradual that even farmers fail to notice it. A severe rainstorm may wash away 1 mm (0.04 in.) of soil, which seems insignificant until the cumulative effects of many storms are taken into account. Twenty years of soil erosion amounts to the loss of about 2.5 cm (1 in.) of soil, an amount that could take hundreds of years to replace by natural soil-forming processes.

World Soil Problems

Soil erosion and mineral depletion are significant problems worldwide. Although estimates vary widely depending on what assumptions are made, soil erosion results in an annual loss of as much as 75 billion metric tons (83 billion tons) of topsoil around the world. Soil erosion is greatest in certain parts of Asia, Africa, and Central and South America (Figure 14.13). In India and China, soil experts estimate that erosion causes an annual loss of as much as 6.6 billion metric tons and 5.5 billion metric tons of soil, respectively. These two countries have 13% of the world's total land area from which they must feed 2.33 billion people—more than 37% of the world's human population.

It is estimated that more than 1 billion people depend on agricultural soils that are not productive enough to adequately support them. A combination of factors has caused this situation, including unsound farming methods, extensive soil erosion, and desertification. Along with these factors, the needs of a rapidly expanding population exacerbate soil problems worldwide.

The first global assessment of soil conditions, released in 1992, summarized a 3-year study of global soil degradation sponsored by the U.N. Environment Program. It reported that 1.96 billion hectares (4.84 billion acres) of soil—an area equal to 17% of the Earth's total vegetated surface—have been degraded since World War II. Eleven percent of the Earth's vegetated surface—an area the size of China and India combined—has been degraded so badly that it will be very costly, or in some cases impossible, to reclaim it. Soil degradation is primarily attributed to poor agricultural practices, overgrazing, and deforestation.

In 2001 the International Food Policy Research Institute issued the results of a study of world agricultural lands using satellite maps. The institute found widespread damage to soil quality. Only 16% of the world's

Figure 14.13 Erosion in the Baringo District of Kenya. Excessive withdrawal of water in this semi-arid region caused most of the plants on the hillside to die. Without plants to hold the soil in place, it eroded away.

farmland does not have soil fertility problems such as poor drainage, aluminum toxicity, acidity, salinity, low soil nutrients, or depletion of organic matter. North America, with 29% of its soil free of problems, has the largest share of good soil. In contrast, parts of Asia have as little as 6% of farmland without soil quality problems.

The African Sahel Asia and Africa have the largest land areas with extensive soil damage, and in both places the problem is compounded by rapid population growth. Consider the Sahel, a broad band of semiarid land that stretches across Africa just south of the Sahara Desert and includes all or parts of many countries (see Figure 6.21). The Sahel normally experiences periodic droughts, but for the past 30 years there has been a sustained rainfall deficit—that is, substantially less precipitation than normal. During droughts, the soil cannot support as many crops or grazing animals. Despite the drought, the Sahelians must use their land to grow crops and animals

for food or they will starve. The soil is so overexploited under these circumstances that it is able to support fewer and fewer people, and the day is approaching when the Sahel could become unproductive desert (Figure 14.14). As noted in the chapter introduction, the degradation of once-fertile rangeland or forest into nonproductive desert caused partly by soil erosion, forest removal, over-cultivation, and overgrazing is known as desertification. To reclaim the land would require restricting its use for many years so it could recover. If these measures were taken, however, the Sahelians would have no means of obtaining food.

Fortunately, organizations such as the International Centre for Research in Agroforestry, established in Nairobi, Kenya, are developing techniques to lessen environmental degradation of the Sahel and other tropical areas. One of their research aims is to use

Figure 14.14 Onset of human-induced desertification. Cattle in Burkina Faso have eaten all the ground cover; the trees that remain will probably be stripped of branches to feed the hungry cattle. Overexploitation of the Sahel, a semiarid region south of the Sahara Desert, is increasing the amount of unproductive desert area.

agroforestry, agricultural and forestry techniques that encompass the planting of trees and crops together, to improve soil fertility in degraded soils. For example, nitrogen-fixing acacia trees might be intercropped with traditional crops such as millet and sorghum. The acacia trees fix nitrogen, thereby improving soil fertility. When the leaves fall off the trees, they gradually decompose, returning mineral nutrients to the soil. The leaf layer also improves the soil's ability to hold moisture (less moisture evaporates from leaf-covered soil). The result of agroforestry techniques is higher crop yields. (Agroforestry is also discussed in the Chapter 20 introduction.)

SOIL CONSERVATION AND REGENERATION

Although soil degradation may be caused or accelerated by agriculture, good soil conservation practices promote sustainable soil use. Conservation tillage, crop rotation, contour plowing, strip cropping, terracing, and shelterbelts all help to minimize erosion and mineral depletion of the soil. Land that has been badly damaged by soil erosion and mineral depletion can be successfully restored, but it is a costly, time-consuming process.

Conservation Tillage

Conventional methods of tillage, or working the land, include spring plowing, in which the soil is cut and turned in preparation for planting seeds; and harrowing, in which the plowed soil is leveled, seeds are covered, and weeds are removed. Conventional tillage prepares the land for crops, but in removing all plant cover, it greatly increases the likelihood of soil erosion (Figure 14.15a). Fields that are conventionally tilled also contain less organic material and generally hold less water than does undisturbed soil.

Since the early 1980s, an increasing number of farmers have adopted a new approach called **conservation tillage**, in which residues from previous crops are left in the soil, partially covering it and helping to hold it in place. Several types of conservation tillage have been developed to fit different areas of the country and different crops. One of these, called **no-tillage**, leaves the soil undisturbed over the winter. During planting special machines cut a narrow furrow in the soil for seeds (Figure 14.15b). Conservation tillage is one of the fastest growing trends in U.S. agriculture. During the 2000 growing season, 36.3% of U.S. farmland was planted using conservation tillage, and the remaining 63.7% was planted using conventional tillage.

In addition to reducing soil erosion, conservation tillage increases the organic material in the soil, which in turn improves the soil's water-holding capacity. Decom-

(a)

(b)

Figure 14.15 Conventional and conservation tillage. (a) Conventional tillage leaves almost no plant residues on the soil surface. If heavy spring rains come before the young plants become established, significant soil loss can occur. (b) Decaying residues from the previous year's crop (rye) surround young soybean plants in a field in Iowa. Conservation tillage reduces soil erosion as much as 70% because plant residues from the previous season's crops are left in the soil.

posing organic matter releases nutrient minerals more gradually than when conventional tillage methods are employed. Farmers who adopt no-tillage save on fuel costs, machinery wear and tear, and labor time when they do not plow their land. However, use of conservation tillage requires new equipment, new techniques, and greater use of herbicides to control weeds. Research to develop alternative methods of weed control for use with conservation tillage is underway. (Chapter 18 discusses sustainable agriculture, which includes conservation tillage and the other soil conservation practices presented in this chapter.)

Crop Rotation

Farmers who practice effective soil conservation measures often use a combination of conservation tillage and **crop rotation**, the planting of a series of different crops in the same field over a period of years. When the same crop is grown continuously, pests for that crop tend to accumulate to destructive levels, so crop rotation lessens damage by insects and disease. Also, many scientific studies have shown that continuously growing the same crop over a period of years depletes the soil of certain essential nutrient minerals faster and makes the soil more prone to erosion. Crop rotation is therefore effective in maintaining soil fertility and in reducing soil erosion.

A typical crop rotation would be corn → soybeans → oats → alfalfa. Soybeans and alfalfa, both members of the legume family, increase soil fertility through their association with bacteria that fix atmospheric nitrogen into the soil. Thus, soybeans and alfalfa help produce higher yields of the grain crops that they alternate with in crop rotation.

Contour Plowing, Strip Cropping, and Terracing

Hilly terrain must be cultivated with care because it is more prone to soil erosion than flatland. Contour plowing, strip cropping, and terracing help control erosion of farmland with variable topography.

In **contour plowing**, fields are plowed and planted in curves that conform to the natural contours of the land, rather than in straight rows. Furrows run around, rather than up and down, hills. **Strip cropping**, a special type of contour plowing, produces alternating strips of different crops along natural contours (Figure 14.16a). For example, alternating a row crop such as corn with a closely sown crop such as wheat reduces soil erosion. Even more effective control of soil erosion is achieved when strip cropping is done in conjunction with conservation tillage.

Farming is undesirable on steep slopes, but if it must be done, **terracing** produces level areas and thereby reduces soil erosion (Figure 14.16b). Nutrient minerals and soil are retained on the horizontal platforms instead of being washed away. Soils are preserved in a somewhat

(a)

(b)

Figure 14.16 Strip cropping and terracing. (a) Strip cropping is evident in this well-managed farm. Quite often crop rotations in such strips include a legume, which reduces the need for nitrogen fertilizers. Photographed in Wisconsin. (b) Terracing hilly or mountainous areas, such as in the Luzon rice fields, Philippines, reduces the amount of soil erosion. However, some slopes are so steep they are totally unsuitable for agriculture. Such areas should be left covered by natural vegetation to prevent extensive erosion.

similar manner in low-lying areas that are diked to make rice paddies. The water forms a shallow pool, retaining sediments and nutrient minerals.

Preserving Soil Fertility

The two main types of fertilizer are organic and commercial inorganic. *Organic fertilizers* include such natural materials as animal manure, crop residues, bone meal, and compost. Organic fertilizers are chemically complex, and their exact compositions vary. The nutrient minerals in organic fertilizers become available to plants only as the organic material decomposes. For that reason, organic fertilizers are slow-acting and long-lasting. (For two very different discussions of compost, see "You Can Make a Difference: Practicing Environmental Principles" and "Meeting the Challenge: Municipal Solid Waste Composting."

YOU CAN MAKE A DIFFERENCE

Practicing Environmental Principles

You can maintain and improve the soil in your own lawn and garden by using compost and mulch. Gardeners often dispose of grass clippings, leaves, and other plant refuse by either bagging it for garbage collection or burning it. But these materials do not have to be treated like wastes; they can be a valuable resource for making **compost**, a natural soil and humus mixture that improves not only soil fertility but also soil structure. Grass clippings, leaves, weeds, sawdust, coffee grounds, ashes from the fireplace or grill, shredded newspapers, potato peels, and eggshells are just some of the materials that can be **composted**, or transformed by microbial action to compost.

To make a compost heap, spread a 6- to 12-inch layer of grass clippings, leaves, or other plant material in a shady area, sprinkle it with an organic garden fertilizer or a thin layer of farm animal manure, and cover it with several inches of soil. Add layers as you collect more organic debris. Water the mixture thoroughly, and turn it over with a pitchfork each month to aerate it (see Figure a). Although it is possible to make compost by just heaping it on the open ground in layers, it is more efficient to construct an enclosure, which allows temperatures to build from the heat generated by microbial action. (Organic material that is decomposing efficiently in a compost heap has a really hot core.) An enclosed compost heap is also less likely to attract animals.

When the compost is uniformly dark in color, is crumbly, and has a pleasant, "woodsy" odor, it is ready to use. The time it takes for decomposition will vary from 1 to 6 months depending on the climate, the materials you are using, and how often you turn it and water it.

Whereas compost is mixed into soil to improve the soil's fertility, **mulch** is placed on the surface of soil, around the bases of plants (see Figure b). Mulch helps control weeds and increases the amount of water in the upper levels of the soil by reducing evaporation. It lowers the soil temperature in the summer and extends the growing season slightly by providing protection against cold in the fall. Mulch also decreases erosion by lessening the amount of precipitation runoff.

Although mulches can consist of inorganic materials such as plastic sheets or gravel, natural mulches of compost, grass clippings, straw, chopped corncobs, or shredded bark have the added benefit of increasing the organic content of the soil. Grass clippings are a very effective mulch when placed around the bases of garden plants because they mat together, making it difficult for weeds to become established. You must replace grass mulches often, however, because they decay rapidly. Some gardeners prefer mulches of more expensive materials such as shredded bark, because they take longer to decompose and are more attractive.

(a)

(b)

Compost and mulch. (a) A compost heap. Composts form faster when the heap is located in a shady spot, kept damp, and aerated (turned over) frequently. (b) Mulches discourage the growth of weeds and help keep the soil damp. Organic mulches such as this shredded bark have the added benefit of gradually decaying, thereby increasing soil fertility.

MEETING THE CHALLENGE

Municipal Solid Waste Composting

The sanitary landfill, a modern replacement for the city dump, has been the recipient of most of the nation's solid waste for the past several decades, but it is not a long-term solution to waste disposal. For one thing, many landfills are filling up, and it is unlikely that enough replacement sites can be found (see Chapter 23). NIMBYism—the **not in my** backyard syndrome—frequently takes hold when government officials attempt to find new places for sanitary landfills.

Much of the bulky waste in sanitary landfills—paper, yard refuse, food wastes, and such—is organic and given the opportunity, could decompose into compost. However, in sanitary landfills, little of this material breaks down. Rapid and complete decomposition requires the presence of oxygen, and in a sanitary landfill, garbage is buried under a layer of soil so that little oxygen is available.

Of the several options for decreasing the quantity of trash and garbage in sanitary landfills (discussed in Chapter 23), one—**municipal solid waste composting**—received serious attention in the United States only since the 1990s. Municipal solid waste composting is the large-scale composting of the entire organic portion of a community's garbage. Because approximately two thirds of all household garbage is organic (paper, yard wastes, food wastes, and wood), municipal solid waste composting substantially reduces demand for sanitary landfills.

Numerous city and county governments are currently composting leaves and yard wastes in an effort to reduce the amount of solid waste sent to landfills. Although this endeavor alone is undeniably beneficial, municipal solid waste composting encompasses much more than yard wastes. It also involves composting food wastes, paper, and anything else in the solid waste stream that is organic.

Initial composting occurs quickly—in 3 to 4 days—because conditions such as moisture and the carbon–nitrogen ratio are continually monitored and adjusted (by adding water or fertilizer, for example) for maximum decomposition (see Figure 23.8). The decay process is carried out by billions of bacteria and fungi, which convert the organic matter into carbon dioxide, water, and humus. So many decomposers eat, reproduce, and die in the compost heap that the drum heats up, killing off potentially dangerous organisms such as disease-causing bacteria. When the material emerges, it is placed outside for several months to cure, during which time additional decomposition occurs. Finally it is sold as compost.

The potential market for compost is largely unexploited. Professional nurseries, landscapers, greenhouses, and golf courses use compost. The largest potential market, however, is the U.S. farmer. Tons of compost could be used to reclaim the 167 million hectares (413 million acres) of badly eroded farmland in the United States; 59 metric tons (65 tons) of compost would be needed to apply 1 inch of compost to a single acre. Compost could also improve the fertility of badly eroded rangeland, forestland, and strip mines. There appears to be no shortage of markets for compost. Thus, it appears that, should certain technical problems be resolved, composting on a large scale could become economically feasible.

Technical problems include concerns over the presence of pesticide residues and heavy metals in the compost. Pesticides sprayed on urban and suburban landscapes would naturally find their way into compost material on leaves, grass clippings, and other yard wastes. However, several scientific studies indicate that most pesticides are either decomposed by bacteria and fungi during composting or broken down by the high temperatures in the compost heap.

More troubling is the concern over heavy metals, such as lead and cadmium. Heavy metals can enter compost from sewage sludge, which may contain industrial wastewater, or consumer products such as batteries. (Sewage sludge is often added to compost because it is a rich source of nitrogen for the decomposing microorganisms.) Two ways to reduce heavy metal contamination in municipal compost are sorting out heavy metal sources before everything is dumped into the composting drum and requiring industries to pretreat their industrial wastewater before it gets to the sewage treatment facility.

Commercial inorganic fertilizers are manufactured from chemical compounds, and thus their exact compositions are known. Because they are soluble, they are immediately available to plants. However, commercial inorganic fertilizers are available in the soil for only a short period of time because they quickly leach away.

It is environmentally sound to avoid or limit the use of manufactured fertilizers, for several reasons. First, because of their high solubility, commercial inorganic fertilizers are very mobile and often leach into groundwater or surface runoff, polluting the water (see Chapter 21). Second, manufactured fertilizers do not improve the water-holding capacity of the soil as organic fertilizers do. Another advantage of organic fertilizers is that in ways that are not yet completely understood, they change the types of organisms that live in the soil, sometimes suppressing the microorganisms that cause certain plant diseases. Commercial inorganic fertilizers are also a source of nitrogen-containing gases (nitrous and nitric oxides) that are air pollutants (see Chapters 19 and 20). Finally, the production of commercial inorganic fertilizers requires a great deal of energy, which is largely obtained from our declining reserves of fossil fuels.

It is also economically advantageous to limit the use of commercial inorganic fertilizer. A 1998 study in Mexico that was reported in the journal *Science* demonstrated that adding less fertilizer later in the crop cycle than farmers normally did maintained expected crop yields. It also resulted in significant cost savings for farmers (an increase of 12% to 17% in after-tax profits).

Soil Reclamation

It is possible to reclaim land that is badly damaged from erosion. The United States has largely reversed the effects of the 1930s Dust Bowl, and China has reclaimed badly eroded land in Inner Mongolia (northern China). Soil reclamation involves two steps: (1) stabilizing the land to prevent further erosion and (2) restoring the soil to its former fertility. In order to stabilize the land, the bare ground is seeded with plants; they eventually grow to cover the soil, holding it in place. After the Dust Bowl, land in Oklahoma and Texas was seeded with drought-resistant native grasses. One of the best ways to reduce the effects of wind on soil erosion is by planting **shelterbelts**, rows of trees that lessen the impact of wind (Figure 14.17; also recall the Great Green Wall of China discussed in the chapter introduction).

The plants that have been established to stabilize the land start to improve the quality of the soil almost immediately, as dead portions are converted to humus. The humus holds nutrient minerals in place and releases them a little at a time; it also improves the water-holding capacity of the soil.

Restoration of soil fertility to its original level is a slow process, however. During the soil's recovery, use of the land must be restricted: It cannot be farmed or grazed. Disaster is likely if the land is put back to use before the soil has completely recovered. But restriction of land use for a period of several to many years is sometimes very difficult to accomplish. How can a government tell landowners that they may not use their own land? How can land use be restricted when people's livelihoods and maybe even their lives depend on it?

Figure 14.17 Shelterbelts surrounding kiwi orchards. Trees protect the delicate fruits from the wind and reduce the ability of the wind to pick up soil from farmland. Photographed in North Island, New Zealand.

SOIL CONSERVATION POLICIES IN THE UNITED STATES

During the late 1920s and early 1930s, Hugh H. Bennett, a soil scientist in the U.S. Department of Agriculture, spoke out about the dangers of soil erosion. A report he published in 1928, entitled "Soil Erosion: A Natural Menace," was largely ignored until the disastrous effects of the Dust Bowl years focused attention on the fact that soil is a valuable natural resource. The **Soil Conservation Act of 1935** authorized the formation of the Soil Conservation Service (now called the Natural Resources Conservation Service); its mission is to work with U.S. citizens to conserve natural resources on private lands. To that end, the NRCS assesses soil damage and develops policies to improve and sustain soil resources.

Historically, farmers have been more likely to practice soil conservation during hard financial times and periods of agricultural surpluses, both of which translate into lower prices for agricultural products. When prices are high, with a good market for agricultural products, farmers have more incentive to put every parcel of land into production, including marginal, highly erodible lands. During times when the farm economy has been strong, federal soil conservation programs have actually contributed to production on marginal lands by relying on voluntary rather than mandatory compliance to encourage soil conservation practices. The federal government has traditionally used incentives rather than penalties for noncompliance.

The **Food Security Act (Farm Bill) of 1985** contained provisions for two main soil conservation programs, a conservation compliance program and the Conservation Reserve Program. The conservation compliance program requires farmers with highly erodible land to develop and adopt a 5-year conservation plan for their farms that includes erosion-control measures. If they do not comply, they lose federal agricultural subsidies such as price supports.

The **Conservation Reserve Program (CRP)** is a voluntary program that pays farmers an average of $50 per acre per year to stop producing crops on highly erodible farmland. It requires planting native grasses or trees on such land and then "retiring" it from further use for 10 years. This land may not be grazed, nor may the grass be harvested for hay during that period.

The CRP has benefited the environment. Annual loss of soil on CRP lands that have been planted with grasses or trees has been reduced from an average of 7.7 metric tons of soil per hectare (8.5 tons per acre) to 0.6 metric ton per hectare (0.7 ton per acre). Because the vegetation is not disturbed once it is established, it pro-

vides biological habitat. Small and large mammals, birds of prey, and ground-nesting birds such as ducks have increased in number and kind on CRP lands. The reduction in soil erosion has also improved water quality and enhanced fish populations in surrounding rivers and streams. About 30 million acres are now enrolled in the program.

According to environmental groups, "sod-busting" may be an indirect and undesirable outcome of the CRP. Here's how it happens: A farmer enrolls environmentally sensitive cropland in the CRP, thereby retiring the land, and then plows native prairie to produce *new* cropland. Conservationists are concerned because little of the native prairie remains (see Chapter 7). They lobbied to have the CRP amended to prevent sod-busting. In response, a new conservation program, the **Grasslands Reserve Program,** was included in the 2002 Farm Bill. This program will pay farmers to protect up to 2 million acres of virgin and improved pastureland for a period of at least 10 years.

SUMMARY WITH SELECTED KEY TERMS

I. The formation of **soil,** the thin surface layer of the Earth's crust, involves interactions among parent material, climate, organisms, time, and **topography** (a region's surface features).

A. Biological, chemical, and physical **weathering processes** slowly break parent material into smaller and smaller particles.

B. Soil is composed of inorganic nutrient minerals, organic materials, soil air, and soil water.

    1. Humus is the dark brown or black organic material that remains after much decomposition has occurred.

    2. Soil water that is not absorbed by roots moves downward, **leaching** dissolved nutrient minerals from the soil.

C. Soil is often organized into layers called **soil horizons.**

D. Soil organisms such as plants, algae, fungi, worms, insects, spiders, and bacteria are important not only in forming soil, but also in cycling nutrient minerals.

    1. Earthworms bring nutrient minerals from deeper soil layers to upper layers when they deposit **castings** on the soil surface.

    2. Mycorrhizae are symbiotic relationships between soil-dwelling fungi and the roots of vascular plants.

    3. In a balanced ecosystem, **nutrient cycling** occurs in which the nutrient minerals removed from the soil by plants are returned when plants or the animals that eat plants die and are decomposed by soil microorganisms.

E. The texture of a soil depends on the relative amounts of **sand, silt,** and **clay** in it. The soil properties of texture and acidity affect a soil's water-holding capacity and nutrient availability, which in turn determine how well plants grow.

II. Sustainable soil use is the wise use by humans of soil resources without a reduction in the amount or fertility of soil, so that it is productive for future generations.

A. Soil erosion is the removal of soil from the land by the actions of water, wind, ice, or other agents.

    1. Soil erosion is a natural process that is often accelerated by human activities such as farming on arid land and clearing forests.

    2. The Dust Bowl that occurred in the western United States during the 1930s is an example of accelerated wind erosion caused by human exploitation of marginal land for agriculture.

B. Mineral depletion occurs in all soils that are farmed. It is a particularly serious problem when tropical rain forests are removed, because the nutrient minerals in the soil are quickly leached out.

C. Desertification is the degradation of once-fertile rangeland or forest into nonproductive desert; it is caused partly by soil erosion, overgrazing, overcultivation, and forest removal.

D. In 1992, the first global assessment of soil conditions reported that 17% of the Earth's total vegetated surface area had been degraded since World War II. A 2001 study of world agricultural lands using satellite maps found that only 16% of the world's farmland does not have soil fertility problems.

III. Soil conservation practices can minimize soil erosion and mineral depletion in healthy soils and help restore damaged soils.

A. In **conservation tillage** residues from previous crops are left in the soil, partially covering it and helping to hold it in place. This practice increases the organic matter and improves the soil's water-holding capacity but requires greater use of herbicides to control weeds.

B. **Crop rotation** is the planting of a series of different crops in the same field over a period of years. Crop rotation prevents depletion of certain nutrient minerals that occurs when the same crop is grown year after year.

C. In **contour plowing**, fields are plowed and planted in curves that conform to the natural contours of the land. **Strip cropping** is a special type of contour plowing with alternate strips of different crops planted along natural contours.

D. **Terracing** can be used to help control erosion in mountainous terrain that is being cultivated.

E. The use of organic fertilizers (animal manure, crop residues, bone meal, and compost) is preferred over commercial inorganic fertilizers because organic fertilizers are slow acting and long lasting and may suppress microorganisms that cause certain plant diseases.

F. Soil that has been badly damaged by erosion can be reclaimed.

   1. Providing plant cover and **shelterbelts**, rows of trees that lessen the impact of wind, help stabilize the land.

   2. Restricting land use until the soil has time to recover helps restore soil fertility.

IV. The Natural Resources Conservation Service works with U.S. citizens to conserve natural resources on private lands.

A. The **Conservation Reserve Program**, which is voluntary, pays farmers to stop producing crops on highly erodible farmland.

B. The **Grasslands Reserve Program**, a new program funded in the 2002 Farm Bill, pays farmers to protect virgin and improved pastureland for a period of at least 10 years.

THINKING ABOUT THE ENVIRONMENT

1. Explain the roles of weathering, organisms, climate, and topography in soil formation.

2. What are the four components of soil, and how is each important?

3. What is soil texture? How do the presence of various-sized particles (sand, silt, and clay) affect soil characteristics?

4. Give an example of how plants affect soil pH. Give an example of how soil pH affects plants.

5. Charles Darwin once wrote that the land is plowed by earthworms. Explain.

6. Describe three ways in which nutrient minerals can be lost from the soil.

7. It could be said that unlike other communities, tropical rain forests live *on* the soil rather than in it. What does this statement imply about tropical soils?

8. The American Dust Bowl is sometimes portrayed as a "natural" disaster brought on by drought and high winds. Present a case for the point of view that this disaster was not caused by nature as much as by humans.

9. Which soil horizons are most prone to erosion? What is the significance of your answer?

10. Where does eroded soil go after it is transported by water, wind, or ice?

11. What is sustainable soil use?

12. Distinguish among conservation tillage, crop rotation, contour plowing, strip cropping, terracing, and shelterbelts as methods of sustainable soil use.

13. Conservation tillage has many benefits, including reduction of soil erosion. However, certain pests that cause plant disease can reside in the plant residues left on the ground with conservation tillage. Knowing that disease-causing organisms are often quite specific for the plants they attack, recommend a way to control such disease organisms. Base your answer on the soil conservation methods discussed in this chapter.

14. Some Everglades farmers are growing rice, which allows them to keep the soil flooded. Explain why keeping Everglades soil flooded helps to preserve it.

15. How can degraded soils be reclaimed?

16. What is the Conservation Reserve Program? The Grasslands Reserve Program?

17. How is human overpopulation related to world soil problems?

18. President F.D. Roosevelt once sent a letter to the state governors in which he said, "A nation that destroys its soils, destroys itself." Explain.

*19. In the chapter you learned that a severe rainstorm may wash away 1 mm of soil and that such losses add up to 2.5 cm over a 20-year period. If 1 mm of topsoil coming off a land area of 1 hectare weighs 13.0 metric tons, how much topsoil (in metric tons per hectare) is lost in 20 years? Convert your answer to English units (tons per acre) using the conversions in Appendix IV.

* The solution to this question appears in Appendix VII.

TAKE A STAND

Visit our Web site at **http://www.wiley.com/college/raven** (select Chapter 14 from the Table of Contents) for links to more information about the environmental implications of the 2002 Farm Bill. Debate with your classmates whether the new Grasslands Reserve Program will be an effective deterrent against sod-busting. You will find tools to help you organize your research, analyze the data, think critically about the issues,

and construct a well-considered argument. Take a Stand activities can be done individually or as part of a team, as oral presentations, written exercises, or Web-based (e-mail) assignments.

Additional on-line materials relating to this chapter, including Student Quizzes, Activity Links, Useful Web Sites, Flash Cards, and more, can also be found on our Web site.

SUGGESTED READING

Egan, T. "Dry High Plains are Blowing Away, Again." *New York Times* (May 3, 2002). A severe drought over much of the High Plains has created Dust Bowl conditions.

Pimentel, D., et al. "Environmental and Economic Costs of Soil Erosion and Conservation Benefits." *Science*, Vol. 267 (February 24, 1995). A professor and several graduate students at Cornell University researched and wrote this sobering review of the global environmental threat of soil erosion.

Richter, D.D., and D. Markewitz. "How Deep Is Soil?" *Bio-Science*, Vol. 45, No. 9 (October 1995). Soil—that part of Earth's crust that has been modified by weathering—is much deeper than was previously hypothesized.

Rimmer, D. "Ultimate Interface." *New Scientist*, Vol. 160, No. 2160 (November 14, 1998). This excellent introduction to basic soil science also includes a discussion of soil's effects on climate warming. The article is found in the "Inside Science, No. 115" part of the journal.

Sanchez, P.A. "Soil Fertility and Hunger in Africa." *Science*, Vol. 295 (March 15, 2002). Highlights research done by the International Center for Research in Agroforestry on improving soil fertility.

Wall, D.H., and J.C. Moore. "Interactions Underground: Soil Biodiversity, Mutualism, and Ecosystem Processes." *Bio-Science*, Vol. 49, No. 2 (February 1999). The authors consider the contributions of species richness and mutualistic relationships to ecosystem functioning in soils.

Webster, D. "Alashan: China's Unknown Gobi." *National Geographic*, Vol. 201, No. 1 (January 2002). A fascinating account of Chinese life in the Gobi Desert, including soil issues such as dust storms and soil fertility.

Wilken, E. "Assault of the Earth." *World Watch*, Vol. 8, No. 2 (March/April 1995). Examines the serious problems of soil degradation.

15

Acid mine drainage from the Leadville mining district in the central Colorado Rockies. For more than 100 years, this area has been mined for silver, gold, lead, and zinc, which occur as sulfide deposits. Shown is the characteristic orange-red acid runoff that contains sulfuric acid contamined with lead, arsenic, cadmium, silver, and zinc. The runoff here drains into the Arkansas River from snowmelt and precipitation runoff.

Minerals: A Nonrenewable Resource

Learning Objectives

After you have studied this chapter you should be able to

1. Explain the difference between high-grade ores and low-grade ores, and between metallic and nonmetallic minerals.

2. Describe several natural processes by which minerals are concentrated in the Earth's crust.

3. Briefly describe how mineral deposits are discovered, extracted, and processed.

4. Relate the environmental impacts of mining and refining minerals and explain how mining lands can be restored.

5. Contrast the consumption of minerals by developing countries and by industrialized nations such as the United States and Canada.

6. Summarize the conservation of minerals by reuse, recycling, and changing our mineral requirements.

7. Explain how sustainable manufacturing and dematerialization can contribute to mineral conservation.

8. Sketch a diagram showing the interrelationships within the industrial ecosystem in Kalundborg, Denmark.

The **General Mining Law** of 1872 was established to encourage settlement in the sparsely populated western states. It allows companies or individuals, regardless of whether they are U.S. citizens or foreign, to stake mining claims on federal land. They can then purchase the land for $2.50 to $5 an acre, extract the valuable hardrock minerals such as gold, silver, copper, lead, or zinc, and keep all the profits. In contrast, 12.5% of profits on lumber, coal, oil, and natural gas obtained from federal land is paid to the government.

In 1995 the General Mining Law permitted a company (ASARCO) to obtain federal land in Arizona that contains copper and silver reserves worth an estimated $2.9 billion; the company paid $1,745 for this land. The Congressional Budget Office has determined that the federal treasury loses an estimated $150 million in annual revenue from the law's inequitable provisions. Although Congress put a hold on such sales, several hundred pending applications worth almost $16 billion in minerals were filed before the 1996 deadline.

The General Mining Law contains no provisions for environmental protection such as the replacement of topsoil and vegetation, or the reestablishment of biological habitat. As a result, hardrock mining has left a legacy of ravaged land, poisoned water, and lifeless ecosystems throughout the West that will cost billions to repair. Acid draining from tailings, which are loose rocks produced during the mining process, has made many streams and rivers totally lifeless (see Appendix I for a discussion of acids and pH).

More than 50 of the estimated 100,000 to 500,000 abandoned mines in the United States have been designated Superfund sites. The federal government—that is, U.S. taxpayers—will finance the cleanup of these sites (see Chapter 23 for a more thorough discussion of Superfund sites). For example, after a mining company extracted $105 million of gold from a mine in Summitville, Colorado, it declared bankruptcy in 1993, leaving behind an environmental disaster. Now a Superfund site, the Summitville mine will cost the federal government more than $140 million to clean up. Cleanup of all Superfund mining sites will cost an estimated $12.5 billion to $17.5 billion.

In 1872, the same year that President Ulysses S. Grant signed the General Mining Law, Yellowstone was designated the first U.S. national park. During the 1990s a proposed 200-acre mine site located in Montana less than 5 km from the Yellowstone border and within the watershed that drains into Yellowstone threatened its pristine condition. A Canadian company (Noranda, Inc.) had the rights to this land, which is surrounded by federal lands designated wilderness, and they planned to establish the Crown Butte gold, silver, and copper mine there. Opponents of the mine, including the superintendent of Yellowstone, worried that some of the pollution from the mine would contaminate Clark's Fork of the Yellowstone River and Yellowstone Park. Because Yellowstone is considered a national treasure, President Clinton stopped the mine by presidential order in 1996 and negotiated with the Canadian company. In 1997 Noranda agreed sell the property to the U.S. government for about $65 million.

In 2000 Congress enacted new mining regulations to protect taxpayers and the environment from the worst abuses of the General Mining Law, but when George W. Bush became president, he moved to weaken the new rules in response to pressure from the mining industry. The issue always comes back to whether the jobs that mining provides are worth the environmental damage caused by mining.

In this chapter we consider the distribution and abundance of minerals as well as the environmental damage caused by obtaining and processing them. We also examine our options for the future, when mineral reserves become depleted.

USES OF MINERALS

Minerals, elements or compounds of elements that occur naturally in the Earth's crust, are such an integral part of our daily lives that we often take them for granted (Figure 15.1, Table 15.1). Steel, an essential building material, is a blend of iron and other metals. Beverage cans, aircraft, automobiles, and buildings all contain aluminum. Copper, which readily conducts electricity, is used for electrical and communications wiring. The concrete used in buildings and roads is made from sand and gravel, as well as cement, which contains crushed limestone. Sulfur, a component of sul-

(a)

(b)

(c)

Figure 15.1 Examples of minerals. (a) Concrete highways are made of sand, gravel, and crushed limestone. (b) Table salt is a nonmetallic mineral. (c) Copper, a metallic mineral, is often shaped into wire for electrical equipment or sheets for roofing, gutters, and downspouts.

Table 15.1   Some Important Minerals and Their Uses

Mineral	Type	Some Uses
Aluminum (Al)	Metal element	Structural materials (airplanes, automobiles), packaging (beverage cans, toothpaste tubes), fireworks
Borax ($Na_2B_4O_7$)	Nonmetal	Diverse manufacturing uses—glass, enamel, artificial gems, soaps, antiseptics
Chromium (Cr)	Metal element	Chrome plate, pigments, steel alloys (tools, jet engines, bearings)
Cobalt (Co)	Metal element	Pigments, alloys (jet engines, tool bits), medicine, varnishes
Copper (Cu)	Metal element	Alloy ingredient in gold jewelry, silverware, brass, and bronze; electrical wiring, pipes, cooking utensils
Gold (Au)	Metal element	Jewelry, money, dentistry, alloys
Gravel	Nonmetal	Concrete (buildings, roads)
Gypsum ($CaSO_4 \cdot 2H_2O$)	Nonmetal	Plaster of Paris, soil treatments
Iron (Fe)	Metal element	Basic ingredient of steel (buildings, machinery)
Lead (Pb)	Metal element	Lead pipes, solder, battery electrodes, pigments
Magnesium (Mg)	Metal element	Alloys (aircraft), firecrackers, bombs, flashbulbs
Manganese (Mn)	Metal element	Steel, alloys (steamship propellers, gears), batteries, chemicals
Mercury (Hg)	Liquid metal element	Thermometers, barometers, dental inlays, electric switches, streetlights, medicine
Molybdenum (Mo)	Metal element	High-temperature applications, lamp filaments, boiler plates, rifle barrels
Nickel (Ni)	Metal element	Money, alloys, metal plating
Phosphorus (P)	Nonmetal element	Medicine, fertilizers, detergents
Platinum (Pt)	Metal element	Jewelry, delicate instruments, electrical equipment, cancer chemotherapy, industrial catalyst
Potassium (K)*	Metal element	Salts used in fertilizers, soaps, glass, photography, medicine, explosives, matches, gunpowder
Common salt (NaCl)	Nonmetal	Food additive, raw material for synthetics
Sand (largely SiO_2)	Nonmetal	Glass, concrete (buildings, roads)
Silicon (Si)	Metal element	Electronics, solar batteries, ceramics, silicones
Silver (Ag)	Metal element	Jewelry, silverware, photography, alloys
Sulfur (S)	Nonmetal element	Insecticides, rubber tires, paint, matches, papermaking, photography, rayon, medicine, explosives
Tin (Sn)	Metal element	Cans and containers, alloys, solder, utensils
Titanium (Ti)	Metal element	Paints; manufacture of aircraft, satellites, and chemical equipment
Tungsten (W)	Metal element	High-temperature applications, light bulb filaments, dentistry
Zinc (Zn)	Metal element	Brass, metal coatings, electrodes in batteries, medicine (zinc salts)

* Potassium, which is very reactive chemically, is never found free in nature; it is always combined with other elements.

furic acid, is an indispensable industrial mineral with many applications in the chemical industry. It is used to make plastics and fertilizers and to refine oil. Other important minerals include platinum, mercury, manganese, and titanium.

Human need and desire for minerals have influenced the course of history. Phoenicians and Romans explored Britain in a search for tin. One of the first metals to be used by humans, tin came into its own during the Bronze Age (3500 to 1000 BC), when tin and copper were combined to produce a tougher and more durable alloy known as bronze. The desire for gold and silver was directly responsible for the Spanish conquest of the New World. A gold rush in 1849 led to the settlement of California. More recently, the lure of gold in Amazonian and Indonesian rain forests has contributed to their destruction.

Earth's minerals are elements or (usually) compounds of elements and have precise chemical compositions. **Sulfides** are mineral compounds in which certain elements are combined chemically with sulfur, and **oxides** are mineral compounds in which elements are combined chemically with oxygen.

Rocks are naturally formed aggregates, or mixtures, of minerals and have varied chemical compositions. An **ore** is rock that contains a large enough concentration of a particular mineral that the mineral can be profitably mined and extracted. **High-grade ores** contain relatively large amounts of particular minerals, whereas **low-grade ores** contain lesser amounts.

Minerals can be metallic or nonmetallic. **Metals** are minerals such as iron, aluminum, and copper, which are malleable, lustrous, and good conductors of heat and electricity. **Nonmetallic minerals**, such as sand, stone,

Mineral Terms

minerals: Elements and compounds that occur naturally in the Earth's crust. Minerals are either metals or nonmetals.

metals: Minerals that are malleable, lustrous, and good conductors of heat and electricity. Examples are gold, copper, and iron.

nonmetals: Minerals that are nonmalleable, nonlustrous, and poor conductors of heat and electricity. Examples are sand, salt, and phosphates.

rock: A mixture of minerals that has varied chemical concentrations.

ore: Rock that contains a large enough concentration of a particular mineral for the mineral to be profitably mined and extracted. High-grade ores contain relatively large amounts of the desired mineral, and low-grade ores, relatively small amounts.

salt, and phosphates, lack these characteristics (see "Mini-Glossary: Mineral Terms").

MINERAL DISTRIBUTION AND FORMATION

Certain minerals, such as aluminum and iron, are relatively abundant in the Earth's crust. Others, including copper, chromium, and molybdenum, are relatively scarce. Abundance does not necessarily mean that the mineral is easily accessible or profitable to extract, however. It is possible, for instance, that you have gold and other expensive minerals in your own backyard. However, unless the concentrations are large enough to make them profitable to mine, they will remain there.

Like other natural resources, mineral deposits in the Earth's crust are distributed unevenly. Some countries have extremely rich mineral deposits, whereas others have few or none. Although iron is widely distributed in the Earth's crust, Africa has less than the other continents. Many copper deposits are concentrated in North and South America, particularly in Chile, whereas most of Asia has a relatively small amount of copper. The distribution of nickel is surprising in that a substantial portion of the world's known supply is found in the tiny island nation of Cuba. Much of the world's tin is in China and Indonesia, and most of the chromium reserves are in South Africa. We discuss the international implications of the unequal distribution of important minerals later in the chapter.

Formation of Mineral Deposits

Concentrations of minerals within the Earth's crust are apparently caused by several natural processes, including magmatic concentration, hydrothermal processes, sedimentation, and evaporation.

As magma (molten rock) cools and solidifies deep in the Earth's crust, it often separates into layers, with the heavier iron- and magnesium-containing rock[1] settling on the bottom and the lighter silicates (rocks containing silicon) rising to the top. Varying concentrations of minerals are often found in the different rock layers. This layering, which is thought to be responsible for some deposits of iron, copper, nickel, chromium, and other metals, is called **magmatic concentration**.

Hydrothermal processes involve groundwater that has been heated in the Earth. This water seeps through cracks and fissures and dissolves certain minerals in the rocks. The minerals are then carried along in the hot water solution. The dissolving ability of the water is greater if chlorine or fluorine is present, because these elements react with many metals (such as copper) to form salts (copper chloride, for example) that are soluble in water. When the hot solution encounters sulfur, a common element in the Earth's crust, a chemical reaction between the metal salts and the sulfur produces metal sulfides. Because metal sulfides are not soluble in water, they form deposits by settling out of the solution. Hydrothermal processes are responsible for deposits of minerals such as gold, silver, copper, lead, and zinc.

The chemical and physical weathering processes that break rock into finer and finer particles are important not only in soil formation (as we saw in Chapter 14) but in the production of mineral deposits. Weathered particles can be transported by water and deposited as sediment on riverbanks, deltas, and the sea floor in a process called **sedimentation**. During their transport, certain minerals in the weathered particles dissolve in the water. They later settle out of solution. When the warm water of a river meets the cold water of the ocean, settling occurs because less material dissolves in cold water than in warm water. Important deposits of iron, manganese, phosphorus, sulfur, copper, and other minerals have been formed by sedimentation.

Significant amounts of dissolved material can accumulate in inland lakes and in seas that have no outlet or only a small outlet to the ocean. If these bodies of water dry up by **evaporation**, a large amount of salt is left behind. Over time, it may be covered with sediment and incorporated into rock layers. Significant deposits of common table salt, borax, potassium salts, and gypsum have been formed by evaporation.

HOW MINERALS ARE FOUND, EXTRACTED, AND PROCESSED

The process of making mineral deposits available for human consumption occurs in several steps. First, a par-

---

[1] Pure magnesium is a relatively lightweight element. Rock usually contains magnesium in the form of magnesium oxide, which is heavier.

ticular mineral deposit is located. Second, the mineral is extracted from the ground by mining. Third, the mineral is processed, or refined, by concentrating it and removing impurities. During the fourth and final step, the purified mineral is used to make a product.

Discovering Mineral Deposits

Geologists employ a variety of instruments and measurements to help locate valuable mineral deposits. Aerial or satellite photography sometimes discloses geological formations that are associated with certain types of mineral deposits. Aircraft and satellite instruments that measure the Earth's magnetic field and gravity can reveal certain types of deposits. Geological knowledge of the Earth's crust and how minerals are formed is used to estimate locations of possible mineral deposits. Once these sites are identified, geologists drill or tunnel for mineral samples and analyze their composition. Seismographs, which are used to detect earthquakes, also provide valuable clues about mineral deposits.

Deposits on the ocean floor cannot be estimated until detailed three-dimensional maps of the sea floor are produced, usually with the aid of depth-measuring devices. Sophisticated computer analysis is necessary to evaluate the complex data recorded by such devices.

Extracting Minerals

The depth of a particular deposit determines whether surface or subsurface mining will be used. In **surface mining**, minerals are extracted near the surface, whereas in **subsurface mining**, minerals that are too deep to be removed by surface mining are extracted. Surface mining is more common because it is less expensive than subsurface mining. However, because even surface min-

ENVIROBRIEF

Diamonds Under the Tundra

Diamonds had never been found in large quantities in the Western Hemisphere until 1989, when Canadian geologists pinpointed the site of a whole cluster of "pipes," the veins that carry diamonds to the surface from great depths. The deposits, likely worth billions of dollars, are in Canada's Northwest Territories, in a wild sub-Arctic region called the Barren Lands. Now the area bustles with the activity of the more than 250 companies that had staked out 53 million acres for exploration by the mid-1990s.

North America's first diamond mine opened in the Northwest Territories in 1998, and a second mine is under construction. The region became a prime target for diamond discovery when seismologists determined that the rocks were the world's oldest; diamonds are found only in ancient deposits. Geologists' new understanding of the characteristics of indicator minerals associated with diamonds led them to the pipes' specific locations. At least a few of the pipes are extremely rich in diamonds, producing 3 carats per metric ton of mined material, and estimates of the potential deposits run as high as 1000 pipes. Environmentalists are understandably concerned about the possible impact mining will have on the isolated area, which supports a caribou herd of 325,000. But jobs in the region are scarce and potential profits are huge, so drilling for diamonds is under way.

eral deposits occur in rock layers beneath the Earth's surface, the overlying layers of soil and rock, called **overburden**, must first be removed, along with the vegetation growing in the soil. Then giant power shovels scoop the minerals out.

There are two kinds of surface mining, open-pit and strip mining. Iron, copper, stone, and gravel are usually extracted by **open-pit surface mining**, in which a giant hole is dug (Figure 15.2). Large holes that are formed by

Figure 15.2 Open-pit surface mining.
Shown is an open-pit copper mine near Tucson, Arizona.

open-pit surface mining are called quarries. In **strip mining**, a trench is dug to extract the minerals. Then a new trench is dug parallel to the old one; the overburden from the new trench is put into the old trench, creating a hill of loose rock known as a **spoil bank**.

Subsurface mining, which is done underground and is more complex, may be done with a shaft mine or a slope mine. A **shaft mine** is a direct vertical shaft to the vein of ore. The ore is broken up underground and then hoisted through the shaft to the surface in buckets. A **slope mine** has a slanting passage that makes it possible to haul the broken ore out of the mine in cars rather than hoisting it up in buckets. Sump pumps keep the subsurface mine dry, and a second shaft is usually installed for ventilation.

Subsurface mining disturbs the land less than surface mining, but it is more expensive and more hazardous for miners. There is always a risk of death or injury from explosions or collapsing walls, and prolonged breathing of dust in subsurface mines can result in lung disease.

Processing Minerals

Processing minerals often involves **smelting**, which is melting the ore at high temperatures to help separate impurities from molten metal. Purified copper, tin, lead, iron, manganese, cobalt, or nickel smelting is done in a blast furnace. Figure 15.3 shows a blast furnace used to smelt iron. Iron ore, limestone rock, and coke (modified

coal used as an industrial fuel) are added at the top of the furnace, while heated air or oxygen is added at the bottom. The iron ore reacts with coke to form molten iron and carbon dioxide. The limestone reacts with impurities in the ore to form a molten mixture called slag. Both molten iron and slag collect at the bottom, but slag floats on molten iron because it is less dense than iron. Note the vent near the top of the iron smelter for exhaust gases. If air pollution control devices are not installed, many dangerous gases are emitted during smelting.

ENVIRONMENTAL IMPLICATIONS

There is no question that the extraction, processing, and disposal of minerals harm the environment. Mining disturbs and damages the land, and processing and disposal of minerals pollute the air, soil, and water. As noted in the discussion of coal in Chapter 10, pollution can be controlled and damaged lands can be fully or partially restored, but these remedies cost money. Historically, the environmental cost of extracting, processing, and disposal of minerals has not been incorporated into the actual price of mineral products to consumers (see Chapter 3).

Most highly developed countries have regulatory mechanisms in place to minimize environmental damage from mineral consumption, and many developing nations are in the process of putting them in place. Such mechanisms include policies to prevent or reduce pollution, restore mining sites, and exclude certain recreational and wilderness sites from mineral development.

Mining and the Environment

Mining, particularly surface mining, disturbs huge areas of land. In the United States, current and abandoned metal and coal mines occupy an estimated 9 million hectares (22 million acres). Because any vegetation that had grown there is destroyed by mining, this land is particularly prone to erosion, with wind erosion causing air pollution and water erosion polluting nearby waterways and damaging aquatic habitats.

Open-pit mining of gold and other minerals uses huge quantities of water. As miners dig deeper into the ground to obtain the ore, they eventually hit the water table and have to pump out the water to keep the pit dry. In northern Nevada, scientists from the U.S. Geological Survey surveyed several wells and measured a drop in the water table of as much as 305 m (1,000 ft). This drop, which took place during the 1990s, was linked to gold mining in the region. (Nevada has numerous gold mines that provide the United States with half of its gold.) At the same time, the Western Shoshone tribe living in the area noticed that springs that are spiritually important to their culture have begun to dry up. The region's farmers and ranchers are also concerned that

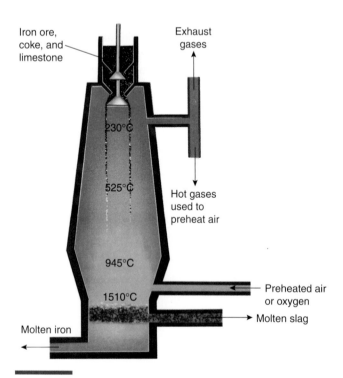

Figure 15.3 Blast furnace. Such towerlike furnaces are used to separate metal from impurities in the ore. The energy for smelting comes from a blast of heated air.

gold mining is depleting the groundwater they need for irrigation. Environmentalists and others would like the mines to reinject the water into the ground after they have pumped it out.

Mining also affects water quality. According to the Worldwatch Institute, mining has contributed to the contamination of at least 19,000 km (11,800 mi) of streams and rivers in the United States. Rocks rich in minerals often contain high concentrations of heavy metals such as arsenic and lead. When rainwater seeps through sulfide minerals exposed in mine wastes, sulfuric acid is produced that in turn dissolves other toxic substances, such as lead, arsenic, and cadmium, in the spoil banks of coal and metal ore mines. These acids and highly toxic substances, known as **acid mine drainage**, are washed into soil and water including groundwater by precipitation runoff. When such acids and toxic compounds make their way into nearby lakes and streams, they adversely affect the numbers and kinds of aquatic life. Rapid drainage during thunderstorms or a spring snowmelt produces sudden "toxic pulses" of poisonous water that are particularly harmful to waterfowl, fish, and other wildlife in the watershed.

Cost–Benefit Analysis of Mine Development Environmental economists suggest that before a decision is made to develop a mine, a cost–benefit analysis should be performed. It should consider the benefits of the mine in dollar terms versus the benefits, in dollar terms, of preserving the land intact for wildlife habitat, watershed protection, and recreation. This analysis should take into account that, over time, the benefits of the mine will decline as the mineral ore is exhausted, whereas the benefits of the natural environment will likely increase, in part because natural areas are becoming rarer as the number of developed areas increases. When a cost–bene-

Table 15.2	Ore and Waste Production for Selected Minerals	
Mineral*	**Amount of Mined Ore (Million Tons)**	**Percentage of Ore That Becomes Waste During Refining†**
Iron	25,503	60
Copper	11,026	99
Gold‡	7,235	99.99
Zinc	1,267	99.95
Lead	1,077	97.5
Aluminum	856	70

* 1995 data, except as noted.
† Data do not include the overburden of rock and soil that originally covered the ore deposits.
‡ 1997 data.

fit analysis of this type is performed, it often shows that present and future benefits of preserving the land are greater than present and future benefits of developing a mine.

Environmental Impacts of Refining Minerals

On average, approximately 80% or more of mined ore consists of impurities that become wastes after processing (Table 15.2). These wastes, called **tailings**, are usually left in giant piles on the ground or in ponds near the processing plants (Figure 15.4). The tailings contain toxic materials such as cyanide, mercury, and sulfuric acid. Left exposed in this way, these toxic substances can contaminate the air, soil, and water. Heavy metals in mine tailings at the Bunker Hill Superfund site in northern Idaho for example, have leached (washed) into the south fork of the Coeur d'Alene River, killing fishes and waterfowl.

Figure 15.4 Copper ore tailings from a strip mining operation in southern Illinois. Toxic materials from mine tailings that are left in mountainous heaps can pollute the air, soil, and water.

Smelting plants have the potential to emit large amounts of air pollutants during mineral processing. One impurity in many mineral ores is sulfur. Unless expensive pollution control devices have been added to smelters, the sulfur escapes into the atmosphere, where it forms sulfuric acid. The environmental implications of the resulting acid precipitation are discussed in Chapter 20. The pollution control devices are the same as those used when sulfur-containing coal is burned—scrubbers and electrostatic precipitators (see Figure 19.10).

Other contaminants found in many ores include the heavy metals lead, cadmium, arsenic, and zinc. These elements have the potential to pollute the atmosphere during the smelting process. Cadmium, for example, is found in zinc ores, and emissions from zinc smelters are a major source of environmental cadmium contamination. In humans, cadmium is linked to high blood pressure; diseases of the liver, kidneys, and heart; and certain types of cancer. (The health effects of lead, another common emission from smelters, are discussed in Chapter 21.) In addition to airborne pollutants, smelters emit hazardous liquid and solid wastes that can cause soil and water pollution. Pollution control devices can prevent such hazardous emissions, however.

One of the most significant environmental impacts in mineral production is the great amount of energy required to mine and refine minerals, particularly if they are being refined from low-grade ore. Gold is currently being extracted from low-grade ores in Nevada. For every 0.9 metric ton (2,000 pounds) of rock that is dug and crushed, as little as 0.7 g (0.025 oz) of gold is refined. Huge amounts of energy are required to dig and crush the countless tons of rock. Most of this energy is obtained by burning fossil fuels, which depletes energy reserves and produces carbon dioxide and other air pollutants (see Chapters 10 and 20).

ENVIROBRIEF

Not-So-Precious Gold

Annual global gold production in 1999 was 2,540,000 metric tons, up from 1,105,000 metric tons in 1980. Demand for gold is booming throughout the world, and the environment is suffering a heavy blow from the increased mining. The waste from mining and processing ore is enormous: Six tons of waste are produced to yield enough gold to make two wedding rings. A technology called cyanide heap leaching allows for profitable mining when minuscule amounts of gold are present, producing up to 3 million pounds of waste for every pound of gold produced. The highly toxic cyanide threatens waterfowl and fishes, as well as underground drinking water supplies. Small-scale miners use other extraction methods with destructive side effects: soil erosion, production of silt that clogs streams and threatens aquatic organisms, and contamination from mercury used to extract the gold. The threats of gold mining do not end when the gold is carried away: If not disposed of properly, mining wastes cause such long-term problems as acid mine drainage and heavy-metal contamination.

CASE·IN·POINT ## Copper Basin, Tennessee

Copper Basin, Tennessee, in the southeast corner of Tennessee, near its borders with Georgia and North Carolina, is a historical example of environmental degradation caused by smelting. Until relatively recently, the Copper Basin area progressed from lush forests to a panorama of red, barren hills baking in the sun (Figure 15.5). Few plant or animal species could be found—just 125 km² (50 mi²) of hills with deep ruts gouged into them. The ruined land had a stark, otherworldly appearance. How did this situation develop?

During the middle of the 19th century, copper ore was discovered near Ducktown in southeastern Tennessee. Copper mining companies extracted the ore from the ground and dug vast pits to serve as open-air smelters. They cut down the surrounding trees and burned them in the smelters to produce the high temperatures needed for the separation of copper metal from other contaminants in the ore. The ore contained great quantities of sulfur, which reacted with oxygen in the air to form sulfur dioxide. As sulfur dioxide from the open-air smelters billowed into the atmosphere, it reacted with water, forming sulfuric acid that fell as acid precipitation.

As a result of deforestation and acid precipitation, ecological ruin of the area occurred in a few short years. Any plants attempting a comeback after removal of the forests were quickly killed by the acid precipitation, which acidified the soil. Because plants no longer covered the soil and held it in place, soil erosion cut massive gullies in the gently rolling hills. Of course, the forest animals disappeared with the plants, which had provided their shelter and food. The damage did not stop here: Soil eroding from the Copper Basin, along with acid precipitation, ended up in the Ocoee River, killing its entire aquatic community.

Beginning in the 1920s and 1930s, several government agencies, including the Tennessee Valley Authority and the U.S. Soil Conservation Service, tried to replant a portion of the area. They planted millions of loblolly pine and black locust trees as well as shorter ground-cover grasses and legume plants that tolerate acid conditions, but most of the plants died. The success of such efforts was marginal until the 1970s, when land reclamation specialists began using new techniques such as application of seed and time-released fertilizer by helicopter. These plants had a greater survival rate, and as they became established, their roots held the soil in place. Leaves dropping to the ground contributed organic material to the soil. The plants provided shade and food for animals such as birds and field mice, which slowly began to return.

Today more than two thirds of the Copper Basin has one sort of vegetation or another. Reclamation continues under a 2001 agreement among the state of Tennessee, the U.S. Environmental Protection Agency, and OXY USA, Inc., with the goal to have the entire area under

Figure 15.5 Environmental devastation near Ducktown, Tennessee. Air pollution from a copper smelter in Tennessee killed the vegetation, and then water erosion carved gullies into the hillsides. This unrestored section of Copper Basin was photographed in the early 1980s. (The telephone pole at right provides a sense of scale.)

plant cover early in the 21st century. Of course, the return of the complex forest ecosystem that originally covered the land before the 1850s will take at least a century or two.

Plant scientists and land reclamation specialists have learned a lot from the Copper Basin, and they will put this knowledge to use in future reclamation projects around the world. ◼

Restoration of Mining Lands

When a mine is no longer profitable to operate, the land can be reclaimed, or restored to a seminatural condition,

as has been done with approximately two thirds of the Copper Basin in Tennessee. The goals of reclamation include preventing further degradation and erosion of the land, eliminating or neutralizing local sources of toxic pollutants, and making the land productive for purposes other than mining (Figure 15.6; also see "Meeting the Challenge: Reclamation of Coal-Mined Land," in Chapter 10). Restoration can also make such areas visually attractive.

A great deal of research is available on techniques of restoring lands that have been degraded by mining, called **derelict lands**. Restoration involves filling in and grading the land to its natural contours, then planting

Figure 15.6 Restoration of mining lands. Part of a phosphate mine near Fort Meade, Florida, has been reclaimed and is currently used as a pasture (*background*). The unrestored area that remains is in the foreground. Restoration of mining lands makes them usable once again or, at the very least, stabilizes them so that further degradation does not occur.

vegetation to hold the soil in place. The establishment of plant cover is not as simple as throwing a few seeds on the ground. Often the topsoil is completely gone or contains toxic levels of metals, so special types of plants that can tolerate such a challenging environment must be used. According to experts, the main limitation on the restoration of derelict lands is not lack of knowledge but lack of funding.

Reclamation of areas that were surface mined for coal is required by the **Surface Mining Control and Reclamation Act** of 1977. No federal law is in place to require restoration of derelict lands produced by mines other than coal mines, however. Recall from the chapter introduction that the General Mining Law makes no provision for reclamation.

Creative Approaches to Cleaning Up Mining Areas

Although wetlands are widely known to provide beneficial wildlife habitats, few people realize the potential of wetlands to help clean up former mining lands. Wetlands tend to trap sediments and pollutants that enter them from upstream areas, so that the quality of water resources located downstream from wetlands is improved. Although a single wetland provides these benefits, a series of wetlands constructed in the affected drainage basin is much more effective.

Consider the area around Butte, Montana, where copper was mined for 100 years. This area comprises the largest Superfund site in the United States. Its soil and water are contaminated with copper, zinc, nickel, cadmium, and arsenic. Many cleanup technologies are being developed and tested in Butte, including the design and construction of artificial wetlands. As contaminated water seeps into the wetland, bacteria consume the sulfur in the acid mine drainage, making the water less acidic. As the water becomes more basic, zinc and copper precipitate (settle out of solution) and enter the sediments. Constructed wetlands typically take 50 to 100 years to neutralize the acid enough for aquatic life to return to rivers and streams downstream from acid mine drainage. This time estimate is based on observations of more than 800 wetland systems that have been constructed at coal mining sites in Appalachia, the region in the eastern United States that encompasses the central and southern Appalachian Mountains.

Creating and maintaining wetlands is not cheap, although it is cost effective when compared to increasing the water's basicity by using lime. Recently, scientists at Butte have tried an inexpensive approach that involves cow manure, a "resource" that is readily available in Montana. Dumping cow manure onto mining lands causes the pH of the mine drainage water to increase as bacteria consume the manure. Toxic materials precipitate out of the basic water, just like in the artificial wetlands, improving the water quality.

Phytoremediation, the use of specific plants to absorb and accumulate toxic materials such as nickel from the soil, is also being tried to remove heavy metals from former mining lands. Although most plants do not tolerate soils rich in nickel, some plants, such as twist flower (*Streptanthus polygaloides*), thrive on it. This species is a hyperaccumulator, a plant that absorbs high quantities of a metal and stores it in its cells. The plants can be grown on nickel-contaminated land, harvested, and hauled to a hazardous waste site for disposal. Alternatively, the plants can be burned, and nickel obtained from the ashes. Thus, phytoremediation has great potential, not only to decontaminate mining and other hazardous waste sites but also to extract valuable metals from soil in an environmentally benign way. (See Chapter 23 for further discussion of phytoremediation.)

MINERAL RESOURCES: AN INTERNATIONAL PERSPECTIVE

The economies of industrialized countries require the extraction and processing of large amounts of minerals to make products. Most of these highly developed countries rely on the mineral reserves in developing countries, having long since exhausted their own supplies. As developing countries become more industrialized, their own mineral requirements increase correspondingly, adding further pressure to a nonrenewable resource. In fact, more minerals have been consumed since World War II than were consumed in the previous 5,000 years from the beginning of the Bronze Age to the middle of the 20th century.

We have seen that mining in the United States has caused many serious environmental problems. The problems in developing countries that rely on mining for a significant part of their economies are as great or greater than the ones faced by highly developed countries. The governments in developing nations lack the financial resources and political will to deal with acid mine drainage and other serious environmental problems caused by hardrock mining. To complicate the issue, foreign companies often have significant mining interests in developing countries. Spain, Great Britain, France, Germany, Japan, Russia, and the United States have been involved at various times during the past two centuries in mining (some would say exploiting) ores containing tin, zinc, copper, lead, and so on in Bolivia. The mining district of Bolivia currently faces a catastrophic environmental nightmare from decades of mining abuse. Yet the Bolivian government does not address the issue because mining is the predominant industry in Bolivia. (Recall from the chapter introduction how reluctant the federal government has been to deal with mining issues, and hardrock mining is a relatively minor part of the U.S. economy.)

U.S. and World Use

At one time, most of the highly developed nations had rich resource bases, including abundant mineral deposits that enabled them to industrialize. In the process of industrialization, they largely depleted their domestic reserves of minerals so that they must increasingly turn to developing countries. This is particularly true for Europe, Japan, and, to a lesser extent, the United States.

As with the consumption of other natural resources, there is a large difference in consumption of minerals between highly developed and developing countries. The United States and Canada, which have about 5.1% of the world's population, consume about 25% of many of the world's metals (Figure 15.7). It is too simplistic, however, to divide the world into two groups, the mineral consumers (highly developed countries) and the mineral producers (developing countries). For one thing, four of the world's top five mineral producers are highly developed countries: the United States, Canada, Australia, and the Russian Federation. South Africa, a moderately developed, middle-income country, is the other mineral producer in the top five. Furthermore, many developing countries lack any significant mineral deposits.

Because industrialization increases the demand for minerals, developing countries that at one time met their mineral needs with domestic supplies become increasingly reliant on foreign supplies as development occurs. South Korea is one such nation. During the 1950s it exported iron, copper, and other minerals. South Korea experienced dramatic economic growth from the 1960s to the present and, as a result, must now import iron and copper to meet its needs.

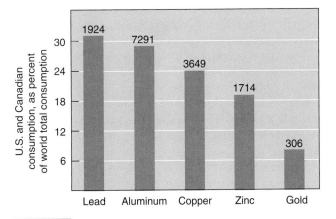

Figure 15.7 North American consumption of selected metals. The heavily industrialized United States and Canada, which have about 5.1% of the world's population, consume a disproportionate share of many of the world's metals. Actual quantities consumed by the United States and Canada are given at the top of each bar, in thousand metric tons. Data are from 2000, the latest available.

Distribution Versus Consumption

Chromium, a metallic element, provides a useful example of global versus national distribution and consumption. Chromium is used to make vivid red, orange, yellow, and green pigments for paints; chrome plate; and, combined with other metals, certain types of hard steel. There is no substitute for chromium in many of its important applications, including jet engine parts. Therefore, industrialized nations that lack significant chromium deposits, such as the United States, must import essentially all of their chromium. South Africa is one of only a few countries with significant deposits of chromium. Zimbabwe and Turkey also export chromium. Although world reserves of chromium are adequate for the immediate future, the United States and several other industrialized countries are utterly dependent on a few countries for their chromium supplies.

Many industrialized nations have stockpiled strategically important minerals to reduce their dependence on potentially unstable suppliers. The United States and others have stockpiles of such metals as titanium, tin, manganese, chromium, platinum, and cobalt, mainly because these metals are critically important to industry and defense. These stockpiles are large enough to provide strategic metals for a period of approximately three years.

Will We Run Out of Important Minerals?

In order to address this question, we must first examine how large the global supply of various minerals is. **Mineral reserves** are mineral deposits that have been identified and are currently profitable to extract. In contrast, **mineral resources**, deposits of low-grade ores, are potential sources of minerals that are currently unprofitable to extract but may be profitable to extract in the future. Mineral resources also include estimates of as-yet-unidentified deposits of minerals that may be confirmed in the future. The combination of a mineral's reserves and resources is called its **total resources** or its **world reserve base**.

Estimates of mineral reserves and resources fluctuate with economic, technological, and political changes. If the price of a mineral on the world market falls, certain borderline mineral reserves may slip into the mineral resource category; increasing prices may restore them to the mineral reserve category. When new technological methods decrease the cost of extracting ores, deposits that have been ranked in the mineral resource category are reclassified as mineral reserves. If the political situation in a country becomes so unstable that mineral reserves cannot be mined, they are reclassified as mineral resources; another change in the political situation at a later time may cause the minerals to be placed in the mineral reserve list again.

Table 15.3 Life Expectancies of Identified Economic World Reserves of Selected Minerals

Mineral	1999* Reserves (Thousand Metric Tons)	Life Expectancy (Years)†
Aluminum (bauxite ore)	25,000,000	81
Copper	340,000	22
Iron ore	74,000,000,000	65
Lead	64,000	17
Nickel	46,000	30
Tin	8,000	28
Zinc	190,000	20

* Latest data available.
† Based on a 2% growth rate in primary production.

Table 15.3 shows the life expectancies for selected minerals. These numbers are calculated by dividing the estimate of mineral reserves by the annual production. Life expectancies are often meaningless because it is extremely difficult to forecast future mineral supplies. In the 1970s, projections of escalating demand and impending shortages of many important minerals were commonplace. There are three reasons that none of these shortages actually materialized. One, new discoveries of major deposits occurred in recent decades—iron and aluminum deposits in Brazil and Australia, for example. Two, plastics, synthetic polymers, ceramics, and other materials replaced metals in many products. Three, a global economic slump resulted in a lower consumption of minerals. Today on the world market there is even a glut of some minerals, which has caused their value to spiral downward. However, there is always the possibility that changes in the world economic situation will contribute to future mineral shortages.

Economic factors aside, the prediction of future mineral needs is also difficult because it is impossible to know when or if there will be new discoveries of mineral reserves or replacements for minerals (such as the development of plastics). It is also impossible to know when or if new technological developments will make it economically feasible to extract minerals from low-grade ores.

With these reservations in mind, most experts currently think mineral supplies, both metallic and non-metallic, will be adequate during the 21st century. During that period, however, several important minerals—silver, copper, mercury, tungsten, and tin, for example—may become increasingly scarce.

Another reasonable projection is that the prices of even relatively plentiful minerals, such as iron and aluminum, will increase during your lifetime. The depletion of large, rich, and easily accessible deposits of these metals will require mining and refining of low-grade ores, which will be more expensive.

INCREASING OUR MINERAL SUPPLIES

As a resource becomes scarce, efforts intensify to discover new supplies, to conserve existing supplies of that resource, and to develop new substitutes for it. Although many reserves have been discovered and exploited, others that are as yet unknown may be found. In addition, the development of advanced mining technologies may make it possible to exploit known resources that are too expensive to develop using existing techniques.

Locating and Mining New Deposits

Many known mineral reserves have not yet been exploited. Although Indonesia is known to have many rich mineral deposits, its thick forests and mosquitoes that carry the malaria parasite have made accessibility to these deposits very difficult. Both northern and southern polar regions have had little mineral development. This is due in part to a lack of technology for mining in frigid environments. Normal offshore drilling rigs cannot be used in Antarctic waters, because the shifting ice formed during its harsh winter would tear the rigs apart. As new technologies become available, increasing pressure will be exerted to mine in northern Canada, Siberia, and Antarctica.

Plans are afoot to exploit some of the rich mineral deposits in Siberia, although new technologies will have to be developed to make this feasible. Some of the ore deposits in Siberia have unusual combinations of minerals (for example, potassium combined with aluminum) that cannot be separated using existing technology.

Is there a possibility that currently unknown mineral deposits will be discovered at some future time? The U.S. Geological Survey thinks that undiscovered mineral deposits may exist, particularly in developing countries where detailed geological surveys have not been performed. It is likely that a detailed survey of the western portion of South America, along the Andes Mountains, will reveal significant mineral deposits. Geologists also presume that minerals will be found in the Amazon Basin, although in many ways the rain forest and thick overlying alluvial layers of the river basin make these deposits as inaccessible as those in Antarctica. Examination of certain areas deep in the rain forest to assess the likelihood of deposits being present is hampered by logistical problems. Also, as in other regions, mining in the Amazon Basin would pose a grave environmental threat.

Geologists consider it likely that deep deposits buried 10 km or more in the Earth's crust will someday be discovered and exploited. The special technology required to mine deep deposits is not yet available.

Minerals in Antarctica

To date, no substantial mineral deposits have been found in Antarctica, although smaller amounts of valuable minerals have been discovered. Geologists think it likely that major deposits of valuable metals and oil are present and that they will be discovered in the future. Nobody owns Antarctica, and many nations have been involved in negotiations on the future of this continent and its possible mineral wealth.

The **Antarctic Treaty**, an international agreement that has been in effect since 1961, limits activity in Antarctica to peaceful uses such as scientific studies. Twenty-six nations are voting members to the Antarctic Treaty. During the 1980s, nearly a decade of delicate negotiations resulted in a pact that would have permitted exploitation of Antarctica's minerals. The pact, called the Convention on the Regulation of Antarctic Mineral Resource Activities, also required unanimous agreement in order to be ratified. In 1989, several countries refused to support the pact because of concerns that *any* mineral exploitation would damage Antarctica's environment. As a result of these concerns, an international agreement, known as the **Environmental Protection Protocol to the Antarctic Treaty**, or the **Madrid Protocol**, was established. The protocol, which went into effect in January 1990, includes a moratorium on mineral exploration and development for a minimum of 50 years. It designates Antarctica and its marine ecosystem as a "natural reserve dedicated to peace and society."

Why be concerned about Antarctica's environment? For one thing, polar regions are extremely vulnerable to human activities. Even scientific investigations and tourists, with their trash, pollution, and noise, have negatively affected the wildlife, such as emperor penguins, leopard seals, and blue whales, along Antarctica's coastline. No one doubts that large-scale mining operations would wreak havoc on such a fragile environment.

Maintaining Antarctica in a pristine state is also important because this continent plays a pivotal role in regulating many aspects of the global environment, such as global changes in sea level. By studying the natural environment of Antarctica, scientists gain valuable insights into such important environmental issues as global warming and stratospheric ozone depletion (see Chapter 20).

Minerals from the Ocean

The mineral reserves of the ocean may also provide us with future supplies. Minerals could be extracted from seawater. Alternatively, the sea floor may be mined where minerals have accumulated in the loose ocean sediments or near underwater volcanoes. Mining the seabed has environmental implications for marine ecosystems where minerals are extracted, as well as for land, where the minerals must be processed.

Seawater, which covers approximately three fourths of our planet, contains many different dissolved minerals. The total amount of minerals available in seawater is staggeringly high, but their concentrations are very low. Currently, sodium chloride (common table salt), bromine, and magnesium can be profitably extracted from seawater. It may be possible in the future to profitably extract other minerals from seawater and concentrate them, but current mineral prices and technology make this impossible now.

Large deposits of minerals lie on the ocean floor. **Manganese nodules**—small rocks the size of potatoes that contain manganese and other minerals, such as copper, cobalt, and nickel—are widespread on the ocean floor, particularly in the Pacific (Figure 15.8). According to the Marine Policy Center at the Woods Hole Oceanographic Institute, the estimates of these reserves are quite large. The Pacific Ocean may contain as much as 1.4 million metric tons (1.5 million tons) of these minerals. However, dredging manganese nodules from the ocean

Figure 15.8 Manganese nodules on the ocean floor. These potato-sized nodules have enticed miners, but it is not currently commercial feasibility to obtain them. Photographed in the Pacific Ocean.

floor would adversely affect sea life, and the current market value for these minerals would not cover the expense of obtaining them using existing technology. Further, it is not clear which country has the legal right to these minerals, which are in international waters. Despite these concerns, many experts think that deep-sea mining will be feasible in a few decades, and several industrial nations such as the United States have staked out claims in a region of the Pacific known for its large number of nodules. To date, however, none have been mined.

Such potential exploitations of the ocean floor are controversial. Many people think it is inevitable that minerals will be mined from the floor of the deep sea, but others think the seabed should be declared off limits because of the potential ecological havoc that mining could cause on the diverse life forms inhabiting the ocean floor. Sea urchins, sea cucumbers, sea stars, acorn worms, sea squirts, sea lilies, and lamp shells are but a few of the animals known to inhabit the seabed environment. In addition to ecological problems, mining the loose sediments of the deep sea also poses legal questions regarding which country owns the minerals.

These problems have been grappled with since the 1960s, when industrialized countries first expressed an interest in removing manganese nodules from the ocean floor. In 1982 their interest triggered the formation of an international treaty called the **U.N. Convention on the Law of the Sea** (UNCLOS), which became effective in 1994 after more than 60 countries ratified it. As of 2002, 138 countries had ratified this treaty. The United States and many other industrialized countries objected to some of the provisions relating to deep seabed mining beyond national jurisdictions and refused to sign or ratify the convention. As a result, a 1994 agreement that focused on mining issues was developed; this agreement revised the objectionable provisions. The revised agreement became effective in 1996, again without U.S. ratification. As we go to press, the U.S. Congress has not ratified either the convention or the agreement. Both are pending in the Senate Committee on Foreign Relations.

It seems likely that ocean mining will become technologically feasible and profitable sometime during the 21st century. In 1997 an Australian company became the first to claim mineral deposits in territorial waters. By 2005, the company plans to be mining high-grade ores originally produced by magmatic concentration at underwater volcanic sites in the South Pacific Ocean off the coast of Papua New Guinea. (Papua New Guinea has granted a permit to the Australian mining company and will receive royalties from it.) The provisions of UNCLOS are not binding for territorial waters, only for international waters, so there is no legal reason the seabed mining cannot be done. Ecologists are concerned about the environmental impact of this type of seabed mining because the volcanic hot springs associated with the mineral deposits teem with wildlife. The seabed miners acknowledge that some organisms will die during the mining process, but say they will be careful to avoid widespread destruction of the area.

Advanced Mining Technologies

We have already mentioned that special technologies will be needed to mine minerals in inaccessible areas such as polar regions and deposits deep in the ground. Making use of large, low-grade mineral deposits throughout the world will also require the development of special techniques. As minerals grow scarcer, economic and political pressure to exploit low-grade ores will increase. Obtaining high-grade metals from low-grade ores is an expensive proposition, in part because a great deal of energy must be expended to obtain enough ore. Future technology may make such exploitation more energy-efficient, thereby reducing costs.

Even if advanced technology makes obtaining minerals from low-grade ores feasible, other factors may limit exploitation of this potential source. In arid regions, the vast amounts of water required during the extraction and processing of minerals may be the limiting factor. Also, the environmental costs may be too high, because obtaining minerals from low-grade ores causes greater land disruption and produces far more pollution than does the development of high-grade ores.

Biomining In some cases microorganisms can be used to extract minerals from low-grade ores. Microorganisms have proved very efficient for copper mining, allowing the U.S. copper industry to become internationally competitive. When mixed with sulfuric acid, a bacterium called *Thiobacillus ferrooxidans* promotes a chemical reaction that leaches copper into an acidic solution, releasing larger quantities of the metal more efficiently than traditional methods.

Other important applications of biomining are also emerging. Treating low-grade gold ores with bacteria such as *Thiobacillus* allows a 90% recovery of gold, compared to 75% recovery for the more expensive and energy-intensive conventional methods. Phosphates, used primarily for fertilizers and additives in some manufactured goods, have traditionally been extracted by inefficient burning at high temperatures or by wasteful acid treatment processes. New biological processes can extract phosphates at room temperature.

■ EXPANDING OUR SUPPLIES BY SUBSTITUTION AND CONSERVATION

Because much of our civilization's technology depends on minerals, and because certain minerals may be unavailable or quite limited in the future, our society should extend existing mineral supplies as far as possible through substitution and conservation.

Finding Mineral Substitutes

The substitution of more abundant materials for scarce minerals is an important goal of manufacturing. The search for substitutes is driven in part by economics; one effective way to cut production costs is to substitute an inexpensive or abundant material for an expensive or scarce one. In recent years, plastics, ceramic composites, and high-strength glass fibers have been substituted for scarcer materials in many industries.

Earlier in the 20th century, tin was a critical metal for can-making and packaging industries; since then, other materials have been substituted for tin, including plastic, glass, and aluminum. The amounts of lead and steel used in telecommunications cables have decreased dramatically during the past 35 years, while the amount of plastics has had a corresponding increase. In addition, glass fibers have replaced copper wiring in telephone cables.

Although substitution can extend our mineral supplies, it is not a cure-all for dwindling resources. Certain minerals have no known substitutes. Platinum catalyzes many chemical reactions that are important in industry. So far, no other substance has been found that possesses the catalyzing abilities of platinum.

Figure 15.9 Recycling of scrap metal. The metal in these old, discarded motor vehicles will be fashioned into new products.

Mineral Conservation

Our mineral supplies can be extended by conservation, which includes both reuse and recycling. The **reuse** of items such as beverage bottles, which can be collected, washed, and refilled, is one way to extend mineral resources. In **recycling**, used items such as beverage cans and scrap iron are collected, remelted, and reprocessed into new products. In addition to the introduction of specific conservation techniques such as reuse and recycling, public awareness and attitudes about resource conservation can be modified to encourage low waste.

Reuse   When the same product is used over and over again, as when beverage containers are collected, washed, and refilled, both mineral consumption and pollution are reduced. The benefits of reuse are greater than those of recycling (see Chapter 23). To recycle a glass bottle requires crushing it, melting the glass, and forming a new bottle. Reuse of a glass bottle simply requires washing it, which obviously expends less energy than recycling. Reuse is a national policy in Denmark, where non-reusable beverage containers are prohibited.

Several countries and states have adopted beverage container deposit laws, which require consumers to pay a deposit, usually a nickel or dime, for each beverage bottle or can they purchase. The deposit is refunded when the container is returned to the retailer or to special redemption centers. Unredeemed deposits are generally used to provide revenue for environmental programs such as hazardous waste cleanups. In addition to encouraging reuse and recycling, thereby reducing mineral resource consumption, beverage container deposit laws save tax money by reducing litter and solid waste. Countries that have adopted beverage container deposit laws include the Netherlands, Germany, Norway, Sweden, and Switzerland. Parts of Canada and the United States[2] also have deposit laws.

Recycling   A large percentage of the products made from minerals—such as cans, bottles, chemical products, electronic devices, and batteries—is typically discarded after use. The minerals in some of these products—batteries and electronic devices, for instance—are difficult to recycle. Minerals in other products, such as paints containing lead, zinc, or chromium, are lost through normal use.

However, we have the technology to recycle many other mineral products. Recycling of certain minerals is already a common practice throughout the industrialized world. Significant amounts of gold, lead, nickel, steel, copper, silver, zinc, and aluminum are recycled (Figure 15.9).

Recycling has several advantages in addition to extending mineral resources. It saves unspoiled land from the disruption of mining, reduces the amount of solid waste that must be disposed (see Chapter 23), and reduces energy consumption and pollution. Recycling an aluminum beverage can saves the energy equivalent of about

---

[2] Eleven states (CA, CT, DE, HI, IA, MA, ME, MI, NY, OR, and VT) have beverage container deposit systems.

180 mL (6 oz) of gasoline. Recycling aluminum also reduces the emission of aluminum fluoride, a toxic air pollutant produced during the processing of aluminum ore.

About 55% of the aluminum cans in the United States are currently recycled. The aluminum industry, local governments, and private groups have established thousands of recycling centers across the country. It takes approximately 6 weeks for a can that has been returned to be melted, re-formed, filled, and put back on a supermarket shelf. Clearly, even more recycling is possible. It may be that today's sanitary landfills will become tomorrow's mines, as valuable minerals and other materials are extracted from them.

Changing Our Mineral Requirements We can reduce our mineral consumption by becoming a low-waste society. U.S. citizens have developed a "throwaway" mentality in which damaged or unneeded articles are discarded (Figure 15.10). This attitude has been encouraged by industries looking for short-term economic profits, even though the long-term economic and environmental costs of such an attitude are high. Products that are durable and repairable enable us to consume fewer resources. Laws such as those requiring a deposit on beverage containers also reduce consumption by encouraging reuse and recycling.

Figure 15.10 Throwaway mentality of industrial society. Many of these discarded materials can be recycled, and some could easily have been repaired and reused.

The throwaway mentality has also been evident in manufacturing industries. Traditionally, industries consumed raw materials and produced not only goods but also a large amount of waste that was simply discarded (Figure 15.11a). Increasingly, however, manufacturers are finding that the waste products from one manufacturing process can be used as raw materials in another indus-

Figure 15.11 Mineral flow in an industrial society. (a) The traditional flow of minerals is a one-way direction from Earth to solid waste production. Massive amounts of solid waste are produced at all steps, from mining the mineral to discarding the used-up product. (b) The flow of minerals in a low-waste society is more complex, with sustainable manufacturing, consumer reuse, and consumer recycling practiced at intermediate steps.

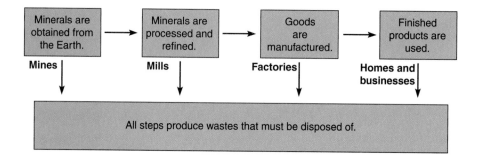

(a)

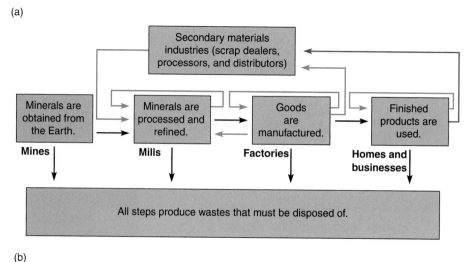

(b)

Key: ——— Sustainable manufacturing ——— Consumer reuse ——— Consumer recycling

(a)   (b)

Figure 15.12 Dematerialization. (a) The first general-purpose electronic calculator/computer was built at the University of Pennsylvania in 1946. It weighed 27 metric tons (30 tons), occupied 167 m² of floor space, and required six full-time technicians to operate. (b) A modern computer that has more applications than the historic computer is a fraction of its size.

try. By selling these "wastes," industries gain additional profits and lessen the amounts of materials that must be disposed.

The chemical and petrochemical industries are among the first businesses to have pioneered the minimization of wastes by converting such wastes into useful products. For example, some chemical companies buy a variety of used aluminum wastes from other companies and convert the aluminum in the wastes to aluminum sulfate, a chemical used to treat municipal water supplies (see Chapter 21). Such minimization of waste by industry is known as **sustainable manufacturing** (see Figure 15.11b and "Meeting the Challenge: Industrial Ecosystems"). Sustainable manufacturing requires that companies provide information about their waste products to other industries so that any potential waste recovery can be implemented. However, many companies are reluctant to reveal the kinds of wastes they produce because their competitors may be able to deduce valuable trade secrets from the nature of their wastes. This difficulty will have to be overcome if sustainable manufacturing is to be fully implemented.

Dematerialization As products evolve, they tend to become lighter in weight and often smaller. Washing machines manufactured in the 1960s were much heavier than comparable machines manufactured today. The same is true of other household appliances, automobiles, and electronic items (Figure 15.12). This decrease in the weight of products over time is called **dematerialization**.

Although dematerialization gives the appearance of reducing consumption of minerals and other materials, it can sometimes have the opposite effect. Products that are smaller and lighter may also be of lower quality. Because repairing broken, lightweight items is difficult and may cost more than the original products, retailers and manufacturers encourage consumers to replace rather than repair the items. Thus, although the weight of materials being used to make each item has decreased, the number of such items being used in a given period of time may actually have increased.

MEETING THE CHALLENGE

Industrial Ecosystems

Traditional industries operate in a one-way, linear fashion: natural resources from the environment → products → wastes dumped back into environment. However, natural resources such as minerals and fossil fuels are present in finite amounts, and the environment has a limited capacity to absorb waste. The field of **industrial ecology** has emerged to address these issues. An extension of the concept of sustainable manufacturing, industrial ecology seeks to use resources efficiently and regards "wastes" as potential products. Industrial ecology tries to create **industrial ecosystems** that compare in many ways to natural ecosystems.

Consider a pioneering industrial ecosystem in the Danish town of Kalundborg that consists of an electric power plant, an oil refinery, a pharmaceutical plant, a wallboard factory, a sulfuric acid producer, a cement manufacturer, fish farming, horticulture (greenhouses), and area homes and farms. At first glance, these entities appear to have little in common. However, they are linked to one another in complex ways that resemble a food web in a natural ecosystem (see figure). In this industrial ecosystem, the wastes produced by one company are sold to another company as raw materials for their processes, in a manner analogous to nutrient cycling in nature (see Chapter 14). Just as the interactions among producers, consumers, and decomposers in a natural ecosystem are complex, so too the interactions among the various components of Kalundborg's industrial ecosystem are intricate.

The coal-fired electric power plant originally cooled its waste steam and released it into the local fjord. The steam is now supplied to the oil refinery and the pharmaceutical plant, and additional surplus heat produced by the power plant warms greenhouses, the fish farm, and area homes. The need for 3,500 oil-burning home heating systems has been eliminated as a result.

Surplus natural gas from the oil refinery is sold to the power plant and the wallboard factory. The power plant now saves 27,200 metric tons (30,000 tons) of coal each year by burning the less expensive natural gas. Before selling the natural gas, the oil refinery removes excess sulfur from it, which is required by air pollution control laws. This sulfur is sold to a company that uses it to manufacture sulfuric acid.

To meet environmental regulations, the power plant installed pollution control equipment to remove sulfur from its coal smoke. This sulfur, in the form of calcium sulfate, is sold to the wallboard plant—some 72,530 metric tons (80,000 tons) annually—and used as a substitute for gypsum, which is calcium sulfate that occurs naturally in the Earth's crust and is mined. The fly ash produced by the power plant goes to the cement manufacturer to be used for road building.

Local farmers use the sludge from the fish farm as a fertilizer for their fields. The fermentation vats at the pharmaceutical plant also generate a high-nutrient sludge used by local farmers. Most pharmaceutical companies discard this sludge because it contains living microorganisms, but the Kalundborg plant heats the sludge to kill the microorganisms, thus converting a waste material into a commodity.

All of these interactions did not spring into existence at the same time; each represents a separately negotiated deal. It took a decade to develop the entire industrial ecosystem. Although these examples of industrial cooperation were initiated for economic reasons, each has distinct environmental benefits, from energy conservation to a reduction of pollution.

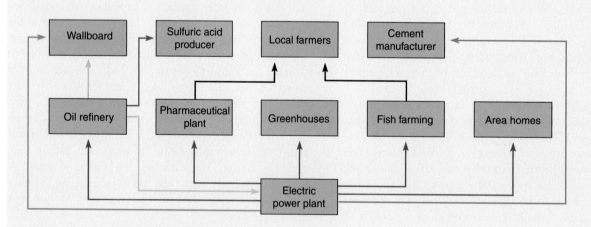

THE KALUNDBORG INDUSTRIAL ECOSYSTEM

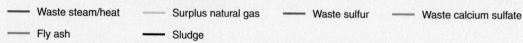

— Waste steam/heat   — Surplus natural gas   — Waste sulfur   — Waste calcium sulfate

— Fly ash   — Sludge

Industrial ecosystem in Kalundborg, Denmark. During the production of energy, food, and other products, resource recovery is maximized and waste production is minimized.

SUMMARY WITH SELECTED KEY TERMS

I. Minerals, which are essential to our industrial society, are naturally occurring elements and compounds. **Rocks** are mixtures of minerals; **ores** are rocks that contain a large enough concentration of a particular mineral to profitably mine and extract it.

A. High-grade ores contain relatively large amounts of particular minerals, whereas **low-grade ores** contain lesser amounts.

B. Minerals may be **metallic** such as iron, aluminum, and copper, or **nonmetallic** such as phosphates, salt, and sand.

C. There are several natural processes by which mineral deposits are formed.

1. **Magmatic concentration** is the formation of mineral deposits as liquid magma solidifies.

2. **Hydrothermal processes** form mineral deposits as hot water dissolves minerals from rock, and the solution seeps through cracks and fissures until the minerals settle out of solution.

3. In **sedimentation,** certain minerals dissolve in water and later settle out of solution.

4. In **evaporation,** dissolved materials in lakes with no outlet to the ocean can form mineral deposits when the water evaporates and the materials settle out of solution.

II. Earth's crust has an uneven distribution of minerals, which are often found in concentrated, highly localized deposits.

A. Mineral reserves are deposits that have been identified and are profitable to extract, whereas **mineral resources** are deposits of low-grade ores that may or may not be profitable to extract in the future.

B. The combination of a mineral's reserves and resources is called its **total resources** or **world reserve base**.

C. It is extremely difficult to estimate a mineral's total resources accurately because mineral consumption changes over time, and economic factors influence all aspects of mineral consumption.

III. Steps in converting a mineral deposit into a usable product include locating the deposit, mining, and processing or refining.

A. If the deposit is near the surface, it is extracted by **open-pit surface mining** or **strip mining**.

B. Subsurface mines, either **shaft mines** or **slope mines**, are used to obtain minerals located deep in the Earth's crust.

C. Processing minerals often involves **smelting**, melting the ore at high temperatures to help separate impurities from molten metal.

IV. Any form of mineral extraction and processing has negative effects on the environment.

A. Surface mining disturbs the land more than subsurface mining, but subsurface mining is more expensive and dangerous. Acids and other toxic compounds in the spoil banks of mines can cause **acid mine drainage**, which can harm surface water, soil, and groundwater.

B. Mineral processing can cause air, soil, and water pollution.

1. **Tailings**, the impurities that make up about 80% of mined ore, are often left in giant piles near processing plants.

2. Toxic materials such as mercury, cyanide, and sulfuric acid can leach into the environment from tailings.

3. Unless pollution control devices are used, smelting plants may emit large amounts of air pollutants during mineral processing.

C. Derelict lands, which are extensively damaged due to mining, can be restored to prevent further degradation and to make the land productive for other purposes.

1. Land reclamation is extremely expensive.

2. No federal law is in place to require restoration of derelict lands produced by mines other than coal mines.

V. Highly developed nations consume a disproportionate share of the world's minerals, but as developing countries industrialize, their need for minerals increases.

A. The richest concentrations of minerals in highly industrialized countries have largely been exploited. As a result, these nations have increasingly turned to developing countries for the minerals they require.

B. Sometimes highly developed nations must rely on potentially unstable developing nations for strategically important minerals.

VI. Mineral supplies can be extended if new deposits are identified. Also, advanced mining technology may make it possible to profitably extract minerals from inaccessible regions or from low-grade ores.

VII. Substitution and conservation extend mineral supplies.

A. Manufacturing industries continually try to substitute more common, less expensive minerals for those that are scarce and expensive.

B. Mineral conservation includes **reuse**, in which a product is collected and used over again, and **recycling**, in which discarded products are collected and reprocessed into new products. In addition to conserving minerals, reuse and recycling cause less pollution and save energy when compared to the extraction and processing of virgin ores.

C. If our mineral supplies are to last and if our standard of living is to remain high, consumers must decrease consumption. To accomplish this, manufacturers must make high-quality, durable products that can be repaired.

1. **Sustainable manufacturing** is the minimization of waste during industrial processes, including reuse and recycling.

2. **Industrial ecology** is an extension of the concept of sustainable manufacturing in which resources are used efficiently and "wastes" are regarded as potential products.

3. **Dematerialization** is the decrease in size and weight of a product as a result of technological improvements over time. Dematerialization only reduces consumption if products are durable and easily and inexpensively repaired.

THINKING ABOUT THE ENVIRONMENT

1. Distinguish between high-grade and low-grade ores.

2. Distinguish among the following ways in which mineral deposits may form: magmatic concentration, hydrothermal processes, sedimentation, and evaporation.

3. Explain why it is difficult to obtain an accurate appraisal of our total mineral resources.

4. Explain how mineral deposits are discovered.

5. Distinguish between *surface* and *subsurface mining*, between *open-pit* and *strip mines*, between *shaft* and *slope mines*.

6. What is smelting? What are tailings?

7. Discuss several harmful environmental effects of mining and processing minerals.

8. Explain why it is more environmentally damaging to obtain minerals from low-grade ores than to extract them from high-grade ores.

9. Historically, the cost of environmental damage arising from mining and processing minerals has not been included in the price of consumer products. Do you think it should be? Why or why not?

10. How does mineral consumption differ between industrialized and developing countries?

11. Distinguish between *reuse* and *recycling*.

12. Outline the benefits of beverage container deposit laws.

13. What is sustainable manufacturing, and whom does it benefit?

14. Some people in industry argue that the planned obsolescence of products, which means they must be replaced often, creates jobs. Others think that the production of smaller quantities of durable, repairable products would generate jobs and stimulate the economy. Explain each viewpoint.

15. How does the industrial ecosystem at Kalundborg resemble a natural ecosystem? What are some of the environmental benefits of Kalundborg's industrial ecosystem?

*16. According to Figure 15.7, in 2000, North America consumed 7,291,000 metric tons of aluminum, which represents 29% of global consumption. Calculate global consumption of aluminum in metric tons for 2000.

* The solution to this question appears in Appendix VII.

TAKE A STAND

Visit our Web site at **http://www.wiley.com/college/raven** (select Chapter 15 from the Table of Contents) for links to more information about the controversy surrounding mining the ocean floor. Consider the views of proponents and opponents, and debate the issue with your classmates. You will find tools to help you organize your research, analyze the data, think critically about the issues, and construct a well-considered argument. Take a Stand activities can be done individually or as part of a team, as oral presentations, written exercises, or Web-based (e-mail) assignments.

Additional on-line materials relating to this chapter, including Student Quizzes, Activity Links, Useful Web Sites, Flash Cards, and more, can also be found on our Web site.

SUGGESTED READING

Broad, W.J. "Undersea Treasure and Its Off Guardians." *New York Times* (December 30, 1997). Volcanic areas of the sea floor contain rich deposits of minerals that may be mined in the near future.

Brown, S., and P. Dickie. "Golden Rules." *Living Planet*, Issue 3 (Spring 2001). Examines the environmental price of gold.

Cockburn, A. "Diamonds." *National Geographic*, Vol. 201, No. 3 (March 2002). Describes the world of diamonds, including the link between diamond mines and war.

Hodges, C.A. "Mineral Resources, Environmental Issues, and Land Use." *Science*, Vol. 268 (June 2, 1995). Provides an overview of current and projected mineral use, and of environmental issues associated with mineral extraction.

Jacobson, M. "Guyana's Gold Standard." *Natural History* (September 1998). Discusses the economic importance and environmental toll of one of South America's largest open-pit gold mines.

Kaiser, J. "Turning Engineers into Resource Accountants." *Science*, Vol. 285 (July 30, 1999). The field of industrial ecology makes good business sense because it reduces wastes of materials and energy.

Kneese, Allen V. "Industrial Ecology and 'Getting the Prices Right.'" *Resources*, Vol. 130 (Winter 1998). How reforms in the market could accelerate the goals of industrial ecology.

Lambrecht, B. "Pollution Solution: Move the Town." *St. Louis Post Dispatch* (June 3, 2002). Lead contamination from an old mine has left Picher, Oklahoma, with a host of environmental problems.

McClure, R., and A. Schneider. "The General Mining Act of 1872 Has Left a Legacy of Riches and Ruin." *Seattle Post* (June 11, 2001). An in-depth history of the General Mining Act.

Robbins, J. "Butte Breaks New Ground to Mop Up a World-Class Mess." *New York Times* (July 21, 1998). Scientists are studying a variety of ways to clean up mine pollution in Butte, Montana, site of a former hardrock-mining complex.

Watkins, T.H. "Hard Rock Legacy." *National Geographic*, Vol. 199, No. 3 (March 2000). The mining industry is trying to clean up its act.

Nesting pair of bald eagles.

Preserving Earth's Biological Diversity

Learning Objectives

After you have studied this chapter you should be able to

1. Define *biological diversity* and distinguish among genetic diversity, species richness, and ecosystem diversity.

2. Discuss five important ecosystem services provided by biological diversity.

3. Contrast threatened, endangered, and extinct species, and list four characteristics common to many endangered species.

4. Define *biodiversity hotspots* and explain where most of the world's biodiversity hotspots are located.

5. Describe four human causes of species endangerment and extinction and tell which cause is the most important.

6. Define *biotic pollution* and explain how invasive species endanger native species.

7. Define *conservation biology* and compare in situ conservation and ex situ conservation.

8. Distinguish between *conservation biology* and *wildlife management*.

9. Discuss steps that can be taken to slow down or stop the decline in biological diversity.

he American bald eagle—the symbol of the United States and an emblem of strength—was a common sight throughout colonial North America. More recently, however, the bald eagle fell on hard times. Its numbers dropped precipitously to only 417 nesting pairs in the lower 48 states in 1963, and it was in danger of extinction.

Several factors contributed to its decline. As European settlers pushed across North America, they cleared many thousands of square kilometers of forest near lakes and rivers, destroying the bald eagle's habitat. Eagles were hunted for sport and because it was thought they had a significant impact on commercially important fishes. In fact, bounties were offered for dead bald eagles as recently as 1952. In addition, eagles' numbers dwindled because they could not reproduce at high enough levels to ensure their population growth or their survival. This reproductive failure was the direct result of ingesting food contaminated with the pesticide DDT (dichlorodiphenyltrichloroethane). DDT caused the eagles' eggs to be so thin-shelled that they cracked open before the embryos could mature and hatch (see Chapter 22). Mercury, lead, and selenium were other environmental pollutants that harmed bald eagles.

351

Banning the use of DDT in the United States in 1972 started the recovery efforts for the bald eagle, which was listed as an endangered species following enactment of the Endangered Species Act (ESA) in 1973. Conservation efforts involving the U.S. Fish and Wildlife Service, other federal agencies, state and local governments, Native American tribes, conservation organizations, universities, corporations, and individuals have helped the bald eagle make a remarkable comeback. In the mid-1970s, the first eagles to be bred in captivity were released in nature. In addition to raising birds in captive-breeding programs, biologists also removed eagle eggs from their nests in nature, raised the baby eagles in wildlife refuges, and returned them to nature. Removal of eggs helped increase the number of eagles, because nesting eagles commonly lay additional eggs to replace those that were removed.

As a result of continuing efforts, the number of nesting pairs in the continental United States increased to more than 6,000 pairs in 2002. (Nesting pairs are counted because bald eagles mate for life.) In 1994 the bald eagle was removed from the endangered list and transferred to the less critical threatened list. In 1999, the U.S. Fish and Wildlife Service (FWS) proposed that it be removed from the threatened list, an action that should be approved sometime soon.

Today the bald eagle symbolizes more than a country, for the bald eagle demonstrates that our biological heritage can be preserved if enough people care to do something about it. In this chapter we examine the importance of all forms of life and consider extinction, which has become an increasing threat to so many organisms. Finally, we explore what can be done to preserve our biological resources and save at least some of the many endangered species from disappearing forever.

HOW MANY SPECIES ARE THERE?

A **species** is a group of more or less distinct organisms that are capable of interbreeding with one another in the wild but do not interbreed with other organisms. We do not know exactly how many species of organisms exist. In fact, biologists in recent decades have begun to realize how very little we know about Earth's diverse organisms. Scientists estimate there may be as few as 5 million to 10 million or as many as 100 million species. About 1.8 million species of organisms have been scientifically named and described to date, including about 270,000 plant species, 45,000 vertebrate animal species, and some 950,000 insect species. About 10,000 new species are identified each year.

The variation among organisms is referred to as **biological diversity** or **biodiversity**, but the concept includes much more than simply the number of species,

Figure 16.1 Genetic diversity in corn. The variation in corn kernels and ears is evidence of the genetic diversity in the species *Zea mays*.

called **species richness**. Biological diversity occurs at all levels of biological organization, from populations to ecosystems (Chapter 4). It takes into account **genetic diversity**, the genetic variety *within* all populations of that species (Figure 16.1). Biological diversity also includes **ecosystem diversity**, the variety of interactions among organisms in natural communities. For example, a forest community with its trees, shrubs, vines, herbs, insects, worms, vertebrate animals, fungi, bacteria, and other microorganisms has greater ecosystem diversity than does a cornfield. Ecosystem diversity also encompasses the variety of ecosystems found on Earth: the forests, prairies, deserts, coral reefs, lakes, coastal estuaries, and other ecosystems of our planet.

WHY WE NEED ORGANISMS

Humans depend on the contributions of thousands of species for their survival. In primitive societies, these contributions are direct: plants, animals, and other organisms are the sources of food, clothing, and shelter. In industrialized societies most people do not hunt for their morning breakfasts or cut down trees for their shelter and firewood; nevertheless, we still depend on organisms.

Although all societies make use of many different kinds of plants, animals, fungi, and microorganisms, most species have never been evaluated for their potential usefulness. There are approximately 270,000 known plant species, but perhaps 250,000 of them have never been evaluated with respect to their industrial, medicinal, or agricultural potential. The same is true of most of the millions of microorganisms, fungi, and animals. Most people do not think of insects as an important biological resource, but insects are instrumental in several impor-

Pollinators in Decline

Two thirds of flowering plants depend on insects for successful pollination—with nearly a third of human food crops pollinated by bees—but scientists in recent years have documented sharp declines in pollinator populations worldwide. In North America, domesticated honeybees have declined by about 50% during the past several decades. The declines in bees and other pollinators are attributed to several likely threats—disease, habitat alteration through such activities as agriculture and grazing, introduction of competing pollinator species, and insect mortality suffered from the use of pesticides.

Threats to pollinator species—and therefore to the plants they pollinate—are made more serious by the complex nature of the plant–pollinator relationship. This mutualistic relationship can be highly specific, meaning that the extinction of one pollinating insect species may lead to the extinction of its dependent plant species. Pollination requirements are not known for most wild plant species, or even for many crop plants. Scientists and resource managers increasingly stress the importance of managing and protecting wild pollinators. The future success of many wild plants and food crops depends on our learning more about the biology of important pollinators and guarding these species from additional environmental threats.

tant ecological and agricultural processes, including pollination of crops, weed control, and insect pest control. In addition, many insects produce unique chemicals that may have important applications for human society. Bacteria and fungi provide us with foods, antibiotics and other medicines, and important biological processes such as nitrogen fixation (see Chapter 6). Biological diversity represents a rich, untapped resource for future uses and benefits, and many as-yet-unknown species may someday provide us with products. A reduction in biological diversity decreases this treasure prematurely and permanently.

Ecosystem Services and Species Richness

You may recall from Chapters 4, 5, 6, and 7 that the living world functions much like a complex machine. Each ecosystem is composed of many separate parts, the functions of which are organized and integrated to maintain the ecosystem's overall performance. The activities of all organisms are interrelated; we are bound together and dependent on one another and on the physical environment, often in subtle ways (Figure 16.2). When one species declines, other species that are linked to it may decline or increase in number.

Plants, animals, fungi, and microorganisms are instrumental in many environmental processes without which humans could not exist. Forests are not just a potential source of lumber; they provide watersheds from which we obtain fresh water, they reduce the number and severity of local floods, and they prevent soil erosion.

Many species of flowering plants depend on insects to transfer pollen for reproduction. Animals, fungi, and microorganisms help to keep the populations of various species in check so that the numbers of one species do not increase enough to damage the stability of the entire ecosystem. Soil dwellers, from earthworms to bacteria, develop and maintain soil fertility for plants. Bacteria and fungi perform the crucial task of decomposition, which allows nutrients to cycle in the ecosystem. All of these are **ecosystem services**, important environmental functions that organisms within ecosystems provide. Ecosystem services maintain the living world, including human societies, and we are completely dependent on these basic support services.

You might think that the loss of some species from an ecosystem would not endanger the rest of the organisms, but this is far from true. Imagine trying to assemble an automobile if some of the parts were missing. You might be able to piece it all together so that it resembled a car,

Figure 16.2 Role of alligators in the environment. The American alligator (*Alligator mississippiensis*) plays an integral, but often subtle, role in its natural ecosystem. It helps maintain populations of smaller fishes by eating the gar, a fish that preys on them. Alligators dig underwater holes that other aquatic organisms use during periods of drought when the water level is low. The nest mounds they build are enlarged each year and eventually form small islands that are colonized by trees and other plants. In turn, the trees on these islands support heron and egret populations. The alligator habitat is maintained in part by underwater "gator trails," which help to clear out aquatic vegetation that might eventually form a marsh. Photographed in Merrit Island, Florida.

but it probably would not run as well. Similarly, the removal of organisms from a community makes an ecosystem run less smoothly. If enough species are removed, the entire ecosystem will change. Species richness within an ecosystem provides the ecosystem with resilience, the ability to recover from environmental changes or disasters (see the discussion of species richness and community stability in Chapter 5).

Genetic Reserves

The maintenance of a broad genetic base is critical for each species' long-term health and survival. Consider economically important crop plants. During the 20th century, plant scientists developed genetically uniform, high-yielding varieties of important food crops such as wheat. It quickly became apparent, however, that genetic uniformity resulted in increased susceptibility to pests and disease.

By crossing the "super strains" with more genetically diverse relatives, disease and pest resistance can be reintroduced into such plants. A corn blight fungus that ruined the corn crop in the United States in 1970 was brought under control by crossing the cultivated, highly uniform U.S. corn varieties with genetically diverse ancestral varieties from Mexico. When some of the genes from Mexican corn were incorporated into the U.S. varieties, the latter became resistant to the corn blight fungus. (The global decline in domesticated plant and animal varieties is discussed in Chapter 18.)

Scientific Importance of Genetic Diversity

Genetic engineering, the incorporation of genes from one organism into an entirely different species (see Chapter 18), makes it possible to use the genetic resources of organisms on a much wider scale than had earlier been possible. The gene for human insulin, for example, has been engineered into bacteria. These bacteria subsequently become tiny chemical factories, manufacturing at a relatively low cost the insulin required in large amounts by diabetics. Genetic engineering, which has been available only since the mid-1970s, has already begun to provide us with new vaccines, more productive farm animals, and agricultural products with longer shelf life or other desirable characteristics.

Although we have the skills to transfer genes from one organism to another, we do not have the ability to *make* genes that encode for specific traits. Genetic engineering depends on a broad base of genetic diversity from which it can obtain genes. It has taken hundreds of millions of years for **evolution** to produce the genetic diversity found in organisms living on our planet today (see Chapter 5). This diversity may hold solutions not only to problems we have today but to problems we have not even begun to imagine. It would be very unwise to allow such an important part of our heritage to disappear.

Medicinal, Agricultural, and Industrial Importance of Organisms

The genetic resources of organisms are vitally important to the pharmaceutical industry, which incorporates into its medicines many hundreds of chemicals derived from organisms. From extracts of cherry and horehound for cough medicines to certain ingredients of periwinkle and mayapple for cancer therapy, derivatives of plants play important roles in the treatment of illness and disease (Figure 16.3). Many of the natural products taken directly from marine organisms, such as tunicates, red algae, mollusks, corals, and sponges, are promising anticancer or antiviral drugs. The AIDS (acquired immune deficiency syndrome) drug AZT (azidothymidine), for example, is a synthetic derivative of a compound from a sponge. The 20 best-selling prescription drugs in the United States are either natural products, natural products that have been slightly modified chemically, or manufactured drugs that were originally obtained from organisms.

The agricultural importance of plants and animals is indisputable, because we must eat to survive. However, the number of different kinds of foods we eat is limited when compared with the total number of edible species. There are probably many species that are nutritionally superior to our common foods. Quinoa, a plant long cultivated as food in the Andes Mountains in South America, looks and tastes somewhat like rice but has a much higher concentration of protein and is more nutritionally balanced. Winged beans are a tropical legume from

Figure 16.3 Medicinal value of the rosy periwinkle. The rosy periwinkle (*Catharanthus roseus*) produces chemicals that are effective against certain cancers. Drugs from the rosy periwinkle have increased the chance of surviving childhood leukemia from about 5% to more than 95%.

Figure 16.4 Agricultural value of the winged bean. Many edible plants have not been used to any great extent, including the winged bean (*Psophocarpus tetragonolobus*), a tropical legume that is nutritionally superior to many other foods.

Southeast Asia and Papua New Guinea (Figure 16.4). Because the seeds of the winged bean contain large quantities of protein and oil, they may be the tropical equivalent of soybeans. Almost all parts of the plant are edible, from the young, green fruits to the starchy storage roots.

Modern industrial technology depends on a broad range of genetic material from organisms, particularly plants, that are used in many products. Plants supply us with oils and lubricants, perfumes and fragrances, dyes, paper, lumber, waxes, rubber and other elastic latexes, resins, poisons, cork, and fibers. Animals provide wool, silk, fur, leather, lubricants, waxes, and transportation, and they are important in medical research. The armadillo is used for research in Hansen's disease (leprosy) because it is one of only two species known to be susceptible to that disease (the other species is humans).

Insects secrete a large assortment of chemicals that represent a wealth of potential products. Certain beetles produce steroids with birth-control potential, and fire-flies produce a compound that may be useful in treating viral infections. Centipedes secrete a fungicide over the eggs of their young that could help control the fungi that attack crops. Because biologists estimate that perhaps 90% of all insects have not yet been identified, insects represent a very important potential biological resource.

Aesthetic, Ethical, and Spiritual Value of Organisms

Organisms not only contribute to human survival and physical comfort, they also provide recreation, inspiration, and spiritual solace. Our natural world is a thing of beauty largely because of the diversity of living forms found in it. Artists have attempted to capture this beauty in drawings, paintings, sculpture, and photography, and it has inspired poets, writers, architects, and musicians to create works reflecting and celebrating the natural world.

The strongest ethical consideration involving the value of organisms is how humans perceive themselves in relation to other species. Traditionally, many human cultures have viewed themselves as superior beings, subduing and exploiting other forms of life for their benefit. An alternative view is that organisms have intrinsic value in and of themselves and that as stewards of the life forms on Earth, we humans should watch over and protect their existence (see Chapter 1 for a discussion of environmental ethics).

ENDANGERED AND EXTINCT SPECIES

Extinction, the death of a species, occurs when the last individual member of a species dies. Extinction is an irreversible loss: Once a species is extinct it can never reappear. Biological extinction is the eventual fate of all species, much as death is the eventual fate of all individuals. Biologists estimate that for every 2,000 species that have ever lived, 1,999 of them are extinct today.

During the span of time in which organisms have occupied Earth, there has been a continuous, low-level extinction of species, known as **background extinction**. At certain periods in the Earth's history, maybe five or six times, there has been a second kind of extinction, **mass extinction**, in which numerous species disappeared during a relatively short period of geological time. The course of a mass extinction episode may have taken millions of years, but that is a short time compared with the age of the Earth, estimated at 4.6 billion years.

The causes of background extinction and past mass extinctions are not well understood, but it appears that biological and environmental factors were involved. A major climate change could have triggered the mass extinction of species. Marine organisms are particularly vulnerable to temperature changes; if the Earth's temperature changed by just a few degrees, it is likely that many marine species would have become extinct. It is also possible that mass extinctions of the past were triggered by catastrophes, such as the collision of the Earth and a large asteroid or comet. The impact could have forced massive quantities of dust into the atmosphere, blocking the sun's rays and cooling the planet.

Extinctions Today

Although extinction is a natural biological process, it can be greatly accelerated by human activities. The burgeon-

ing human population has forced us to spread into almost all areas of the Earth. Whenever humans invade an area, the habitats of many organisms are disrupted or destroyed, which can contribute to their extinction. The dusky seaside sparrow, a small bird that was found only in the marshes of St. Johns River in Florida, became extinct in 1987, largely due to human destruction of its habitat.

Currently, the Earth's biological diversity is disappearing at an unprecedented rate (Figure 16.5). Conservation biologists estimate that species are presently

Figure 16.5 Representative endangered or extinct species. Declining biological diversity is a serious problem. Officials at the U.S. Fish and Wildlife Service estimate that more than 500 U.S. species have gone extinct during the past 200 years. Of these, roughly 250 have gone extinct since 1980.

becoming extinct at a rate at 100 to 1,000 times the natural rate of background extinctions. The 1995 U.N. Global Biodiversity Assessment, based on the work of about 1,500 scientists from around the world, estimated that more than 31,000 plant and animal species are currently threatened with extinction. More recent surveys indicate that these estimates are probably too conservative. The first World Conservation Union Red List of Threatened Plants, issued in 1997 and based on 20 years of data collection and analysis around the world, lists about 34,000 species of plants currently threatened with extinction.

Endangered and Threatened Species

The legal definition of an **endangered species**, as stipulated in the ESA, is a species in imminent danger of extinction throughout all or a significant portion of its range. (The area in which a particular species is found is its **range**.) A species is endangered when its numbers are so severely reduced that it is in danger of becoming extinct without human intervention.

A species is defined as **threatened** when extinction is less imminent but the population of a particular species is quite low. The legal definition of a threatened species is one that is likely to become endangered in the foreseeable future, throughout all or a significant portion of its range.

Endangered and threatened species represent a decline in biological diversity, because as their numbers decrease, their genetic variability is severely diminished. Long-term survival and evolution depend on genetic diversity, so its loss adds to the risk of extinction for endangered and threatened species, as compared to species that have greater genetic variability.

Characteristics of Endangered Species Many endangered species share certain characteristics that seem to have made them more vulnerable to extinction. Some of these characteristics are having an extremely small (localized) range; requiring a large territory; living on islands; having low reproductive success, often the result of a small population size or low reproductive rates; needing specialized breeding areas; and having specialized feeding habits.

Many endangered species have a very limited natural range, which makes them particularly prone to extinction if their habitat is altered. The Tiburon mariposa lily is found nowhere in nature except on a single hilltop near San Francisco. Development of that area would almost certainly cause the extinction of this species.

Species that require extremely large territories in order to survive may be threatened with extinction when all or part of their territory is modified by human activity. The California condor, a scavenger bird that lives off of carrion and requires a large, undisturbed territory—hundreds of square kilometers—in order to find adequate food, is slowly recovering from being on the brink of extinction. In 1983 the California condor population reached a low of 22 birds, and during the 5-year period from 1987 to 1992, it was no longer found in nature (Figure 16.6). A program to reintroduce zoo-bred California condors into the Los Padres National Forest, located about 100 miles northwest of Los Angeles, began in 1992. The release of condors in Big Sur, California, began in 1997. By 2001, 54 condors soared over California and adjacent areas in Arizona. The California condor story is not yet an unqualified success, however. The wild condor population is not self-sustaining, in part because many of the released birds are too young to reproduce. During the spring 2002 nesting season, three chicks were raised in the wild. There are high hopes that more chicks will be hatched in 2003, when the oldest members of the Big Sur flock begin nesting.

Figure 16.6 California condor. California condors (*Gymnogyps californianus*), which are scavengers, require large territories in order to obtain enough food. They can cover hundreds of miles a day by soaring effortlessly on thermal air currents. These birds, which have up to a 3-meter (10-foot) wingspan, are critically endangered largely because development has reduced the size of their wilderness habitat. Photographed in the San Diego Wild Animal Park.

Many island species that are **endemic** to certain islands (that is, they are not found anywhere else in the world) are endangered. These organisms often have small populations that cannot be replaced by immigration should their numbers be destroyed. Because they evolved in isolation from competitors, predators, and disease organisms, they have few defenses when such organisms are introduced, usually by humans, to their habitat. It is not surprising that of the 171 bird species that have become extinct in the past few centuries, 155 of them lived on islands.

In ecological terms, *island* refers not only to any land mass surrounded by water but also to any isolated habitat that is surrounded by an expanse of unsuitable territory. Accordingly, a small patch of forest surrounded by agricultural and suburban lands is considered an island (Figure 16.7). **Habitat fragmentation**, the breakup of large areas of habitat into small, isolated patches (that is, islands), is a major threat to the long-term survival of many species. (National parks are discussed as islands in Chapter 17.)

In order for a species to survive, its members must be present within their range in large enough numbers for males and females to mate. The minimum population density and size that ensure reproductive success vary from one type of organism to another. However, for all organisms, if the population density and size fall below a critical minimum level, the population declines, becoming susceptible to extinction.

Endangered species often share other characteristics. Some have low reproductive rates. The female blue whale produces a single calf every other year, whereas no more than 6% of swamp pinks, an endangered species of small flowering plant, produce flowers in a given year. Some endangered species breed only in very specialized areas: The green sea turtle lays its eggs on just a few beaches.

Highly specialized feeding habits can also endanger a species. In nature, the giant panda eats only bamboo. Periodically all of the bamboo plants in a given area flower and die together; when this occurs, panda populations face starvation. Like many other endangered species, giant pandas are also endangered because their habitat has been fragmented into small islands, and there are few intact habitats where they can survive. China's 1,100 wild giant pandas live in 24 isolated habitats that occupy a small fraction of their historic range.

Where Is Declining Biological Diversity the Greatest Problem?

Although declining biological diversity is a concern throughout the United States, it is most serious in the states of Florida, California, and Hawaii, according to a 1995 study by Defenders of Wildlife entitled *Endangered Ecosystems: A Status Report on America's Vanishing Habitat and Wildlife*. Hawaii has lost hundreds of species and has more species listed as endangered than any other state. At least two thirds of Hawaii's native forests are gone.

As serious as declining biological diversity is in the United States, it is even more serious abroad, particularly in tropical rain forests.

Tropical Rain Forests   Tropical rain forests are found in South and Central America, central Africa, and Southeast Asia. Although only 7% of the Earth's surface is covered by tropical rain forests, as many as 50% of the Earth's species inhabit them.

Ecosystem loss and degradation are occurring in many places around the world, but tropical rain forests are being destroyed faster than almost all other ecosystems. Using remote sensing surveys, scientists have determined that approximately 1% of tropical rain

Figure 16.7 Habitat fragmentation. Roads and agricultural lands effectively isolate the scattered remnants of forest.

forests are being cleared or severely degraded each year. The forests are making way for human settlements, banana plantations, oil and mineral explorations, and other human activities.

Tropical rain forests are home to thousands or even millions of the world's species. Many species in tropical rain forests are endemic; the clearing of tropical rain forests therefore contributes to their extinction. These species are important in their own right, but the mass extinction that is currently taking place in tropical rain forests has indirect ramifications as far away as North America. Birds that migrate from North America to Central America and the Caribbean have been declining in numbers. Not all migratory birds are declining at the same rate, however. Those birds that winter in tropical rain forests are declining at a much greater rate than birds that winter in tropical grasslands or tropical dry woodlands. Thus, tropical deforestation is also affecting organisms of the temperate region.

Tropical rain forests provide important ecosystem services that help to maintain their ecosystem. The forest itself generates much of the rainfall in tropical rain forests. If half of the existing rain forest in the Amazon region of South America were to be destroyed, precipitation in the remaining forest would decrease. As the land became drier, organisms adapted to moister conditions would be replaced by organisms able to tolerate the drier conditions. Many of the original species, being endemic and unable to tolerate the drier conditions, would become extinct.

Perhaps the most unsettling outcome of tropical deforestation is its disruptive effect on the evolutionary process. In the Earth's past, mass extinctions were followed during the next several million years by the formation of many new species to replace those that died out. For example, after the dinosaurs became extinct, ancestral mammals evolved into the variety of running, swimming, flying, and burrowing mammals that exist today. The evolution of a large number of related species from an ancestral organism is called **adaptive radiation**. In the past, tropical rain forests have supplied the base of ancestral organisms from which adaptive radiations could occur. By destroying tropical rain forests, we may be reducing or eliminating nature's ability to replace its species through adaptive radiation.

A few countries, primarily developing nations, hold most of the biological diversity that is so ecologically and economically important to the entire world. The situation is complicated by the fact that these countries are least able to afford the protective measures needed to maintain biological diversity. International cooperation will clearly be needed to preserve our biological heritage. (Tropical rain forests are also discussed in Chapters 7 and 17.)

Earth's Biodiversity Hotspots In the 1980s ecologist Norman Myers of Oxford University coined the term **biodiversity hotspots** to describe relatively small areas of land that contain an exceptional number of endemic species and are at high risk from human activities. In 2000, using plants as their criteria, Myers and ecologists at Conservation International identified 25 hotspots around the world (Figure 16.8). As many as 44% of all

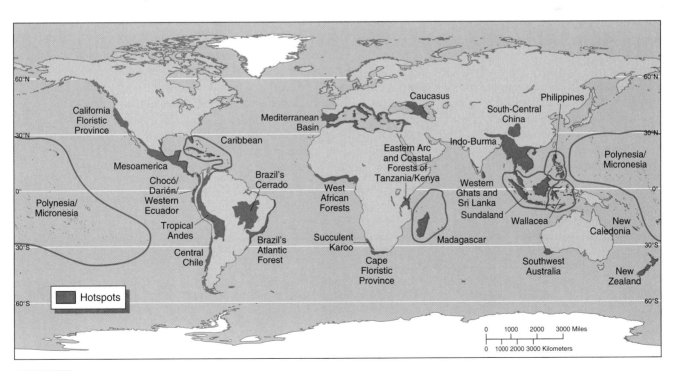

Figure 16.8 Earth's 25 biodiversity hotspots. These hotspots are rich in endemic species and at great risk from human activities.

species of vascular plants live within the hotspots. Interestingly, these 25 hotspots for plants also contain 29% of the world's endemic bird species, 27% of endemic mammal species, 38% of endemic reptile species, and 53% of endemic amphibian species. Many humans—nearly 20% of the world's population—also live in the hotspots. Fifteen of the 25 hotspots are tropical, and 9 are mostly or solely islands.

Myers and his colleagues recommend that conservation planners focus on preserving land in these hotspots to reduce the mass extinction of species that is currently under way. Not all biologists agree, however. Some critics think that concentrating most of our efforts on the 25 biodiversity hotspots causes us to neglect species living in other habitats, such as deserts, grasslands, tundra, and temperate forests, all of which are also at risk.

HUMAN CAUSES OF SPECIES ENDANGERMENT

Scientists generally agree that the single greatest threat to biological diversity is loss of habitat. The spread of invasive species, pollution, and overexploitation are also important. Table 16.1 summarizes an extensive study of the human threats to biological diversity in the United States. These data show the overall importance of habitat loss and degradation. Some interesting variations among the groups of organisms are also apparent. Pollution is a very serious threat to fishes, reptiles, and invertebrates but a relatively minor threat to plants.

Habitat Destruction, Fragmentation, and Degradation

Most species facing extinction today are endangered because of the destruction, fragmentation, or degradation of habitats by human activities (Figure 16.9). We demolish or alter habitats when we build roads, parking lots, bridges, and buildings; clear forests to grow crops or graze domestic animals; and log forests for timber. We

drain marshes to build on aquatic habitats, thus converting them to terrestrial ones, and we flood terrestrial habitats when we build dams with their reservoirs. Exploration and mining of minerals, including fossils fuels, disrupt the land and destroy habitats. Habitats are altered by outdoor recreation, including off-road vehicles, hiking off-trail, golfing, skiing, and camping. Because most organisms are utterly dependent on a particular type of environment, habitat destruction reduces their biological range and ability to survive.

As the human population has grown, the need to provide increased amounts of food for this population has resulted in a huge conversion of natural lands into croplands and permanent pastures. According to the U.N. Food and Agricultural Organization, total agricultural lands currently occupy 38% of the Earth's land area, excluding Antarctica. Agriculture has also had a major impact on aquatic ecosystems because of the diversion of water for irrigation. (Habitat loss is discussed throughout the text, such as in Chapters 10, 13, 15, and 17.)

Africa provides a vivid example of the conflict between humans and other species such as elephants over land use. African elephants are nomads that require a lot of natural landscape in which to forage for the hundreds of kilograms of food that each consumes daily. In southern Africa people are increasingly pushing into the elephants' territory to grow crops and graze farm animals. The elephants often trample and devour crops, ruining a year's growth of crops in a single night; they have even killed people. Farmers in the area are not permitted by law to shoot at or kill elephants because they are a protected species. (Before elephants were listed as a protected species, their numbers had declined precipitously because of over-hunting by ivory hunters.) A 1999 study in the journal *Conservation Biology* examined 25 African regions that had both wild areas and human settlements. The authors found that when the density of people reaches a certain level, the elephants migrate out of the area. The problem is that the wild areas to which elephants can move are steadily shrinking. One of the great challenges is finding a way to allow people and elephants to coexist in an increasingly crowded world.

Table 16.1 Percentages of Imperiled U.S. Species That Are Threatened by Various Human Activities

Activity	All Species (1,880)*	Plants (1,055)	Mammals (85)	Birds (98)	Reptiles (38)	Fishes (213)	Invertebrates (331)
Habitat loss/degradation	85†	81	89	90	97	94	87
Exotic species	49	57	27	69	37	53	27
Pollution	24	7	19	22	53	66	45
Overexploitation	17	10	45	33	66	13	23

* Numbers in parentheses are the total number of species evaluated. The "All Species" category represents about 75% of imperiled species in the United States.
† Because many of the species are affected by more than one human activity, the percentages in each column do not add up to 100.

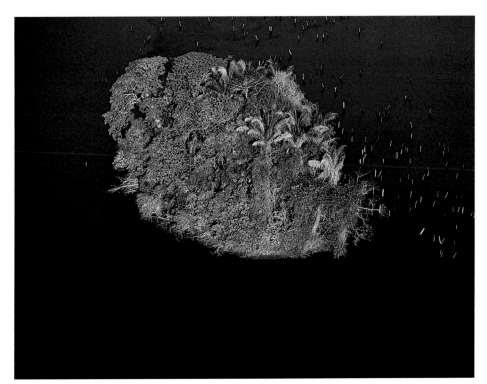

Figure 16.9 Habitat destruction. This tiny island is located in the Panama Canal. It was once a hilltop in a forest that was flooded when the Panama Canal was constructed.

Invasive Species

Biotic pollution, the introduction of a foreign species into an ecosystem in which it did not evolve, often upsets the balance among the organisms living in that area. The foreign species may compete with native species for food or habitat or may prey on them. Generally, an introduced competitor or predator has a greater negative effect on local organisms than do native competitors or predators. Foreign species whose introduction causes economic or environmental harm are called **invasive species** (see Chapter 1).

Although invasive species may be introduced into new areas by natural means, humans are usually responsible for such introductions, either knowingly or unknowingly. The blue water hyacinth was deliberately brought from South America to the United States because it has lovely flowers. Today it has become a nuisance in Florida waterways, clogging them so that boats cannot easily move and crowding out native species.

Islands are particularly susceptible to the introduction of invasive species. Less than 50 years ago, the brown tree snake was accidentally introduced in Guam, an island in the West Pacific (Figure 16.10). Thought to have arrived from the Solomon Islands on a U.S. Navy ship shortly after the end of World War II, the brown tree snake thrived and is now estimated to number about 3 million. It started consuming rainforest birds in large numbers, and as a result, 9 of the 12 native species of forest birds are extinct in nature. The Guam rail, a flightless bird endemic to Guam, numbered about 80,000 in 1968 but was extinct in nature by 1986; today

Figure 16.10 Biotic pollution. The brown tree snake (*Boiga irregularis*), which reaches 1.5 to 2.4 meters (5 to 8 ft) in length as an adult, is an efficient predator. Since its accidental introduction on Guam, where this photograph was taken, this nocturnal snake with bulging eyes has driven nine rainforest birds to extinction. Hawaii has mounted a campaign to prevent the snake's accidental introduction to its islands. At least seven brown tree snakes that hitched a ride in the wheel wells of planes have been discovered there and killed.

it exists only in small captive populations safely isolated from the snakes. The snakes have decimated Guam's small reptiles and small mammals and caused other problems as well. Although the venom does not harm adult humans, it is mildly poisonous to infants and children. The snakes sometimes crawl into people's beds and bite them while they are sleeping. Snakes have also caused many power outages by climbing utility poles and shorting electrical lines.

In the mid-1980s an aggressive aquarium-bred strain of alga known as *Caulerpa* was accidentally released into the Mediterranean Sea when a seaside aquarium cleaned out its tanks. The alga has spread like a dense carpet over more than 13,000 hectares (32,120 acres) of the Mediterranean seabed, crowding out biologically diverse seafloor communities of native sea grasses, sponges, corals, sea fans, anemones, sea stars, and lobsters. *Caulerpa* is also toxic to many Mediterranean species. Other regions far from the Mediterranean are concerned that *Caulerpa* could cause havoc to their marine ecosystems. It is now well established in southeast Australia. The United States banned the importation of *Caulerpa* under the **Federal Noxious Weed Act**. However, it was discovered in 2000 off the coast of California. An eradication effort is under way, but it is too soon to tell if it will be successful.

Pollution

Human-produced acid rain, ozone depletion, and climate warming degrade even wilderness habitats that are "totally" natural and undisturbed. Acid rain is thought to have contributed to the decline of large stands of forest trees and the biological death of many freshwater lakes. Because ozone in the upper atmosphere shields the ground from a large proportion of the sun's harmful ultraviolet (UV) radiation, ozone depletion in the upper atmosphere represents a very real threat to all terrestrial life. Climate warming, which is caused in part by an increase in atmospheric carbon dioxide released when fossil fuels are burned, is another threat. Such habitat modifications particularly reduce the biological diversity of species with extremely narrow and rigid environmental requirements.

Organisms are also affected by other types of pollutants, such as industrial and agricultural chemicals, organic pollutants from sewage, acid wastes seeping from mines, and thermal pollution from the heated wastewater of industrial plants. California condors are very susceptible to lead poisoning, which they get when they consume bullets inside carrion. (See Chapters 10, 15, and 19 through 23 for additional discussion of the adverse effects of pollutants on ecosystems.)

Overexploitation

Sometimes species become endangered or extinct as a result of deliberate efforts to eradicate or control their numbers. Many of these species prey on game animals or livestock. Ranchers, hunters, and government agents have reduced populations of large predators such as the wolf and grizzly bear. Predators of game animals and livestock are not the only animals vulnerable to human control efforts. Some animals are killed because their lifestyles cause problems for humans. The Carolina parakeet, a beautiful green, red, and yellow bird endemic to the southern United States, was extinct by 1920, exterminated by farmers because it ate fruit and grain crops.

Prairie dogs and pocket gophers were poisoned and trapped so extensively by ranchers and farmers that between 1900 and 1960, they disappeared from most of their original geographical range. As a result of sharply decreased numbers of prairie dogs, the black-footed ferret, the natural predator of these animals, became endangered. By the winter of 1985–86, only 10 ferrets were known to exist, 4 in Meeteetse, Wyoming, and 6 in captivity. A successful captive-breeding program enabled biologists to release black-footed ferrets in sections of South Dakota, Wyoming, and Montana, beginning in 1991. Black-footed ferrets have successfully reproduced in the wild since being reintroduced, but a recent outbreak of disease (plague) in black-tailed prairie dogs spread to the ferrets in Wyoming, killing the colony there. Recovery efforts for these charismatic animals continue in Colorado, Montana, and New Mexico, although they remain vulnerable to human development of the prairie habitat.

Unregulated hunting, also called overhunting, was a factor contributing to the extinction of certain species in the past but is now strictly controlled in most countries. The passenger pigeon was one of the most common birds in North America in the early 1800s, but a century of overhunting resulted in its extinction in the early 1900s. Unregulated hunting was one of several factors in the near extinction of the American bison. Bison were decimated by the U.S. Army, which killed bison to disrupt the food supply of the Plains Indians, and also by commercial hunters, who killed bison for their hides and tongues (considered a choice food) as well as meat for work crews of railroad companies.

Illegal commercial hunting, also known as poaching, endangers many larger animals, such as the tiger, cheetah, and snow leopard, whose beautiful furs are quite valuable (Figure 16.11). Rhinoceroses are slaughtered primarily for their horns, which are used for ceremonial dagger handles in the Middle East and for purported medicinal purposes in Asian medicine. Bears are killed for their gallbladders, which are used in Asian medicine to treat ailments from indigestion to heart ailments. Endangered American turtles are captured and exported illegally to China, where they are killed for food; wild turtles are thought to confer longevity and wisdom when they are eaten. Caimans (reptiles similar to crocodiles) are killed for their skins, which are made into shoes and handbags. Although these animals are legally protected,

Figure 16.11 Illegal trade in snow leopard pelts. Snow leopards are poached for their beautiful pelts as well as their bones, which are ground up and used in traditional Asian medicine. Photographed in Kashgar, Xinjiang, China.

ENVIROBRIEF

Solving Crimes Involving Organisms

The U.S. Fish and Wildlife Service Forensics Laboratory in Ashland, Oregon, is the only facility in the world that employs sophisticated technology to determine whether endangered species are being illegally killed or otherwise exploited. Investigators who specialize in such varied fields as forensics, criminal investigation, serology, morphology, and genetics investigate more than 900 cases, both in the United States and around the world, each year. They solve crimes such as poaching of elk, "head-hunting" of walruses for their tusk ivory, the sale of bear gallbladders to Asian medical markets, and the sale in upscale department stores of leather belts or purses made of skins of rare animals. The laboratory focuses on identifying protected animal species, determining causes of death, and linking crimes to their perpetrators. Investigative techniques employed are those available to most modern crime labs: autopsy, ballistics, conventional tissue typing fiber analysis, and DNA fingerprinting.

the demand for their products on the black market has caused them to be hunted illegally. For example, the American black bear's gallbladder can fetch $800 to $3,000 on the black market. In West Africa, poaching has contributed to the decline in lowland gorilla and chimpanzee populations. The meat (called *bushmeat*) of these rare primates and other protected species such as anteaters, elephants, and mandrill baboons is sold to satisfy the sophisticated palates of consumers in urban restaurants.

Commercial harvest is the collection of a live organism from nature. Commercially harvested organisms end up in zoos, aquaria, biomedical research laboratories, circuses, and pet stores. Several million birds are commercially harvested each year for the pet trade, but unfortunately many of them die in transit, and many more die from improper treatment after they are in their owners' homes.

At least 40 parrot species are now threatened or endangered, in part because of unregulated commercial trade. Although it is illegal to capture endangered animals from nature, there is a thriving black market, mainly because collectors in the United States, Europe, and Japan are willing to pay extremely large amounts to obtain rare tropical birds (Figure 16.12). Imperial Amazon macaws, for example, fetch up to $20,000 each. The United States passed the **Wild Bird Conservation Act** of 1992 that imposed a moratorium on importing rare bird species. Poaching data collected before and after 1992

Figure 16.12 Illegal animal trade. These hyacinth macaws (*Anodorhynchus hyacinthus*) were seized in French Guiana in South America as part of the illegal animal trade there.

indicate a drop in poaching rates after the law went into effect. Europe and Japan have not yet passed such a law.

Animals are not the only organisms threatened by excessive commercial harvest. Many unique and rare plants have been collected from nature to the point that they are endangered. These include carnivorous plants, wildflower bulbs, certain cacti, and orchids. On the other hand, carefully monitored and regulated commercial use of animal and plant resources creates an economic incentive to ensure that these resources do not disappear.

CASE-IN-POINT Disappearing Frogs

In all the examples of species in trouble that are mentioned in this chapter, frogs and other amphibians deserve special notice for several reasons. One is that amphibians are very sensitive indicators of environmental problems, and scientists increasingly perceive amphibians to be bellwether species. **Bellwether species**, also known as **sentinel species**, are organisms that provide an early warning of environmental damage that has the potential to affect other species. Amphibians also merit attention because recent precipitous declines in amphibian populations are not just a local problem. These declines are occurring around the world, although there are some regions that are not affected at this time. A recent discovery in some areas of high percentages of frogs and other amphibians with several kinds of deformities adds another layer of complexity to the amphibian crisis.

Amphibians, represented by about 5,000 species of frogs, toads, and salamanders, are survivors. These tough little organisms, which typically spend part of their life in the water and part on land, have existed as a group for more than 350 million years. Despite this evolutionary resilience, amphibians are remarkably sensitive environmental indicators of conditions in both aquatic and terrestrial ecosystems. Amphibians lay gelatinous and unprotected eggs in ponds and other pools of standing water. The water is also where tadpoles undergo metamorphosis, maturing into adult frogs that are able to live on land. As adults, frogs breathe primarily through their extremely permeable skin. This moist, absorptive skin also makes them very susceptible to environmental contaminants.

Amphibian Decline Since the 1970s, many of the world's frog populations have dwindled or disappeared. North America, Central and South America, and Australia have all experienced dramatic population declines. At least 14 species of Australian rainforest frogs have become endangered or extinct since 1980. In the United States, as many as 38% of its 242 native amphibian species are declining in numbers. Worldwide, 32 amphibian species have gone extinct in the last few decades, and about 200 species are in decline.

In assessing the possible causes of these declines, researchers have observed that the declines are not lim-ited to areas with obvious habitat destruction, such as drainage of wetlands where frogs live, or degradation from pollutants. Some remote, pristine locations also show dramatic declines in amphibians: Populations of all seven native species of frogs and toads in Yosemite National Park have declined. Biologists are not certain what is causing these mysterious declines, and it appears that no single factor is responsible. Potential factors for which there is strong evidence include pollutants, increased UV radiation, infectious diseases, and global climate warming.

Agricultural chemicals have been implicated in amphibian declines in California's Sierra Nevada Mountains. Frog populations on the eastern slopes are relatively healthy, but about eight species are declining on the western slopes, where prevailing winds carry residues of 15 different pesticides from the Central Valley, a huge agricultural region. Other areas in which agricultural chemicals may be contributing to amphibian decline include the eastern shore of Maryland and Ontario, Canada.

Another possible culprit may be increased UV radiation caused by ozone thinning. Amphibians possess an enzyme that allows them to repair DNA damage caused by natural UV radiation. Species suffering declines appear to be limited in their ability to repair such cellular damage. Researchers at Oregon State University exposed the eggs of three frog species to natural radiation. Egg survival was high for the Pacific tree frog, which had the greatest enzyme activity and is not in decline. Egg survival was much less (only 45% to 65%) for the Western toad and Cascades frog, both of which are declining. Egg survival for these species increased dramatically, however, when eggs were shielded from UV radiation.

Infectious diseases may explain some of the declines. In Australia, data suggest that a fungus called a chytrid is responsible for massive die-offs of more than 12 species, 4 of which are probably now extinct. Laboratory studies in the United States have shown that the chytrid can kill healthy frogs. This fungus may also be responsible for some of the frog declines observed in Central America and the United States.

Golden toads and about 20 other frog species that have disappeared in mountain areas of Costa Rica may have succumbed to climate warming. In recent years increasing global temperatures have reduced moisture levels in the cloud forests of Costa Rica's central highlands. The organisms in these tropical mountain forests depend on the regular formation of clouds and mist, particularly during the dry winter season. As the equatorial Pacific Ocean surface has warmed in recent years, a decline in the frequency of dry-season mist has been observed and correlated with declines in the abundance of many animal species, including the frogs that disappeared.

Amphibian Deformities In 1995 schoolchildren in Minnesota made a chilling discovery while they were on a field trip to a local pond. The children found that almost half of the leopard frogs they caught were deformed. (Less than 1% of frogs exhibit deformities in healthy frog populations.) Deformities include frogs with extra legs, extra toes, eyes located on the shoulder or back, deformed jaws, bent spines, missing legs, missing toes, and missing eyes (Figure 16.13). Deformed frogs usually die early, before they can reproduce. Predators can easily catch frogs with extra or missing legs. Since the children's discovery made worldwide headlines, 42 states have reported abnormally large numbers of deformities in 36 amphibian species. Canada has also reported frog deformities.

Many possible causes have been investigated and shown to produce amphibian deformities during development. These include exposure to chemicals such as pesticides, to increased UV light due to thinning of the ozone layer, and to parasites.

Several pesticides are known to affect normal development in frog embryos. At concentrations normally found in the environment, atrazine, the most commonly used herbicide in the United States, disrupts the sexual development of male frogs, turning them into hermaphrodites with both male and female characteristics. Although their appearance is unchanged, the male frogs are unable to reproduce successfully. Lab experiments using traces of the pesticide S-methoprene and its breakdown products have caused frog deformities that are similar to abnormalities found in ponds in Vermont that are close to where S-methoprene is used to control mosquitoes and fleas.

Laboratory tests have also demonstrated that infecting tadpoles with a parasitic flatworm, called a trematode, causes the adults that develop from the tadpoles to exhibit limb deformities like the ones observed in field sites in Santa Clara County, California.

No single factor explains the deformities found in all locations. Like amphibian declines, all frog deformities are not caused by the same environmental stressor. Furthermore, multiple stressors, such as habitat loss, disease, and air and water pollution, may interact synergistically with one another to cause deformities. An amphibian that is stressed by pesticide residues, high UV, or drought may be more susceptible to a parasite.

Amphibians are a special example of the kinds of injuries humans are inflicting on biological diversity. How do we respond to the problem? What strategies should we develop to cope with declining biological diversity? The field of conservation biology addresses these concerns. ■

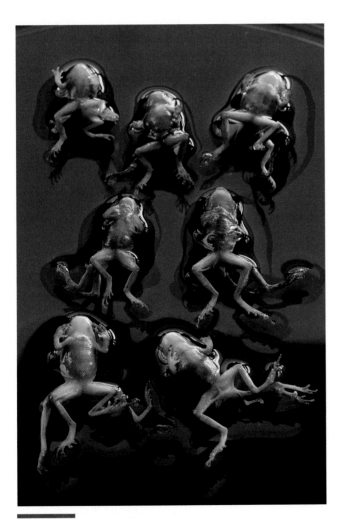

Figure 16.13 Frog deformities. Pollution, ultraviolet radiation, and parasites have been implicated in the recent widespread increase in developmental abnormalities in amphibians.

CONSERVATION BIOLOGY

Conservation biology is the scientific study of how humans impact organisms and of the development of ways to protect biological diversity. Conservation biology is a broad discipline that ranges from safeguarding populations of endangered species to preserving entire ecosystems and landscapes.

Several concepts guide conservation biologists when they work to safeguard species and the ecosystems in which they live. For example, a single large area of habitat, which has the potential to support large populations, is more effective at safeguarding species than several habitat fragments, each with the potential to support small populations. Also, a large area of habitat typically has the potential to support greater species richness than several habitat fragments. Another concept is that it is better if areas of habitat for a given species are located close together rather than far apart. If an area of habitat is isolated from other areas, individuals of a species may not be able to effectively disperse between them. Because the presence of humans adversely affects many species, another concept is that areas of habitat that lack roads or are inaccessible to humans are better than human-accessible areas. Virtually all conservation biologists think it is more effective and, ultimately, more economical to preserve intact ecosystems in which many species live than to work on preserving individual species one at a time. Conservation biologists typically assign a higher priority to preserving areas that are more biologically diverse than other areas (recall the earlier discussion of biodiversity hotspots).

Conservation biology includes two problem-solving techniques that are used to save organisms from extinction: in situ and ex situ conservation. **In situ conservation**, which includes the establishment of parks and reserves, concentrates on preserving biological diversity *in nature*. With increasing demands on land, however, in situ conservation cannot guarantee the preservation of all types of biological diversity. Sometimes only ex situ conservation can save a species. **Ex situ conservation** involves conserving biological diversity in *human-controlled settings*. The breeding of captive species in zoos and the seed storage of genetically diverse plant crops are examples of ex situ conservation.

Protecting Habitats

Conservation biologists maintain that protecting animal and plant habitats—that is, conserving and managing the ecosystem as a whole—is the single best way to protect biological diversity. Many nations are beginning to appreciate the need to protect their biological heritage and have set aside areas for biological habitats. Ecuador, Venezuela, Denmark, and the Dominican Republic are the four countries that have established protected areas totaling more than 30% of their land. Austria, Germany, New Zealand, Slovakia, Bhutan, and Belize have more than 20% of their land areas protected.

There are currently more than 3,000 national parks, sanctuaries, refuges, forests, and other protected areas throughout the world. These encompass some 1 billion hectares, an area almost as large as Canada. Some of these areas have been set aside to protect specific endangered species. The world's first such refuge was established in 1903 at Pelican Island, Florida, to protect the brown pelican. Today the U.S. National Wildlife Refuge System has land set aside in more than 500 refuges. Although the bulk of the protected land is in Alaska, refuges exist in all 50 states.

Many protected areas have multiple uses that sometimes conflict with the goal of preserving species. National forests may be open for logging, grazing, and mineral extraction. The mineral rights to many refuges are privately owned, and some refuges have had military exercises conducted on them.

Protected areas are not always effective in preserving biological diversity, particularly in developing countries where biological diversity is greatest, because there is little money or expertise to manage them. According to the Worldwatch Institute, the guards in Brazil's national parks are each responsible for an average of 6,000 km², an area larger than the state of Delaware. Such "paper parks," where governments are unwilling or unable to enforce conservation laws, are even more vulnerable to logging, farming, mining, and poaching than are protected areas in the United States and other highly developed countries.

Another shortcoming of the world's protected areas is that many are in lightly populated mountain areas, tundra, and the driest deserts, places that often have spectacular scenery but relatively few kinds of species. In contrast, ecosystems in which biological diversity is greatest often receive little protection. A high concentration of endemic species occurs in the Philippines and in Madagascar, yet these island nations have each set aside less than 2% of their land. Protected areas are urgently needed in tropical rain forests, the tropical grasslands and savannas of Brazil and Australia, and dry forests that are widely scattered around the world. Desert organisms are underprotected in northern Africa and Argentina, and many islands and temperate river basins need protection.

Many scientists and conservation organizations think that a minimum of 10% of Earth's total land area should be placed in national parks or other nature reserves and strictly protected. However, as the human population continues to increase and as global development accelerates the fragmentation of natural habitats, it is unlikely that 10% of Earth's land will be set aside.

Restoring Damaged or Destroyed Habitats

Although preserving habitats is an important part of conservation biology, the realities of our world, including the fact that the land-hungry human population continues to increase, dictate a variety of other conservation measures. Scientists can reclaim disturbed lands and convert them into areas with high biological diversity. **Restoration ecology**, in which the principles of ecology are used to help return a degraded environment as close as possible to its former state, is an important part of in situ conservation.

The most famous example of restoration ecology has been carried out since 1934 by the University of Wisconsin–Madison Arboretum (Figure 16.14). During that time, several distinct natural communities of Wisconsin were carefully developed on damaged agricultural land. These communities include a tallgrass prairie, a dry prairie, and several types of pine and maple forests.

Restoration of disturbed lands not only creates biological habitats but also has additional benefits such as the regeneration of soil that has been damaged by agriculture or mining (see Chapters 10, 14, and 15). The disadvantages of restoration include the expense and the amount of time it requires to restore an area. Even so, restoration is an important aspect of conservation biology, as it is thought that restoration will deter much extinction.

(a)

Figure 16.14 Prairie restoration. The University of Wisconsin–Madison Arboretum pioneered restoration ecology. (a) The restoration of the prairie was at an early stage in November 1935. (b) The prairie as it looks today. This picture was taken at approximately the same location as the 1935 photograph.

(b)

Zoos, Aquaria, Botanical Gardens, and Seed Banks

Zoos, aquaria, and botanical gardens often play a critical role in saving species that are on the brink of extinction. Eggs may be collected from nature, or the remaining few wild animals may be captured and bred in zoos, aquaria, and other research environments.

Special techniques, such as artificial insemination and embryo transfer, are used to increase the number of wild animal offspring. In **artificial insemination**, sperm collected from a suitable male of a rare species is used to artificially impregnate a female, perhaps located in another zoo in a different city or even in another country. In **embryo transfer**, a female of a rare species is treated with fertility drugs, which cause her to produce multiple eggs. Some of these eggs are collected, fertilized with sperm, and surgically implanted into a female of a related but less rare species, who later gives birth to offspring of the rare species (Figure 16.15). In late 1999, the first transfer of a frozen embryo from an endangered African wildcat to a common house cat was announced. Another technique involves hormone patches, which are being developed to stimulate reproduction in endangered birds; the patch is attached under the female bird's wing.

Scientists have learned how to successfully breed the endangered whooping crane in captivity should the species become extinct in the wild. The captive population now totals 131 cranes. Three wild whooping crane populations currently exist—in Texas, Florida (nonmigratory), and Wisconsin–Florida (migratory). In 2002 the total number of wild cranes was 284. Scientists are trying to establish additional flocks in the wild by soft-releasing cranes (to establish a nonmigratory flock) or flying them behind an ultra-light (to establish a migratory flock). These efforts have not yet led to additional self-sustaining flocks.

Attempting to save a species on the brink of extinction is very expensive; therefore, only a small proportion of endangered species can be saved by these methods. Zoos, aquaria, and botanical gardens do not have the space to mount efforts to save *all* endangered species. This means that conservation biologists must prioritize which species to attempt to save. Traditionally, the public has supported efforts to save large, charismatic species such as pandas, bald eagles, and black-footed ferrets. Smaller, less "attractive" endangered species, many of which are important providers of ecosystem services, have largely been ignored. Clearly, it is more cost-effective to maintain existing natural habitats so that species will never become endangered in the first place.

Figure 16.15 Bongo calf and its surrogate mother. This young bongo (*Tragelaphus eurycerus*) calf was transferred as an embryo to the uterus of a female eland (*Taurotragus oryx*), where it completed development. The bongo, a shy, elusive species inhabiting dense bamboo forests and deep jungle in equatorial Africa, is an endangered species, primarily because of habitat fragmentation and overhunting. The larger and more common eland, a different but related species, inhabits plains and open brush from Kenya to South Africa and Angola.

Reintroducing Endangered Species to Nature The ultimate goal of captive-breeding programs is to produce offspring in captivity and then release them into nature so that wild populations can be restored. However, only 1 of every 10 reintroductions using animals raised in captivity is successful. What guarantees that a reintroduced population will survive?

Whether such reintroductions actually succeed has been scientifically studied only in recent years. The Hawaiian goose called the nene (pronounced *nay-nay*), was down to 30 individuals in the 1960s. It was reintroduced to the islands of Hawaii (the Big Island) and Maui beginning in the 1970s, but although hundreds of birds have been released, a self-sustaining nene population has not developed on either island. Apparently, some of the same factors that originally caused the nene's extinction in nature, including habitat destruction and nonnative predators such as the Indian mongoose, are responsible for the failure of the reintroduced birds. (On Kauai a group of captive nenes escaped during a hurricane and appear to be successfully reestablishing themselves. Kauai is less developed than Maui or the Big Island and does not have the Indian mongoose.)

Now before attempting a reintroduction, conservation biologists make a feasibility study. This includes determining (1) what factors originally caused the species to become extinct in nature, (2) whether these factors still exist, and (3) whether any suitable habitat still remains.

If the animal to be reintroduced is a social animal, a small herd is usually released together. This is accomplished by first placing the herd in a large, semi-wild enclosure that is somewhat protected from predators but requires the herd to obtain its own food. When the herd's behavior begins to resemble the behavior of wild herds, it is released.

Sometimes it is impossible to teach critical survival skills to animals raised in captivity. The effort to reintroduce captive-raised thick-billed parrots (Figure 16.16) to the Chiricahua Mountains of Arizona was canceled in 1993 because all of the 88 birds released between 1986 and 1993 died or disappeared. Wild thick-billed parrots are loud, sociable birds whose flocking instinct contributes to their survival because individual birds act as sentinels and loudly announce the presence of danger, such as hawks or other birds of prey, to the group. Those parrots raised in captivity lacked such social behavior, and despite efforts to teach them to stay together, they separated from the flock after they were released. Future releases will have to wait on the availability of wild-caught, thick-billed parrots from Mexico.

Once animals are released, they must continue to be monitored. If any animals die, their cause of death is

Figure 16.16 Captive-raised thick-billed parrots. These thick-billed parrots (*Rhynchopsitta pachyrhyncha*) were unsuccessfully reintroduced to southeastern Arizona, where they have not been observed since 1922. Their extinction was due to habitat loss (logging), overhunting, and shifts in climate, with the resulting changes in food availability.

determined in order to search for ways to prevent unnecessary deaths in future reintroductions.

Seed Banks   More than 100 seed collections called **seed banks** exist around the world and collectively hold more than 3 million samples at very low temperatures (Figure 16.17). They offer the advantage of storing a large amount of plant genetic material in a very small space. Seeds stored in seed banks are safe from habitat destruction, climate warming, and general neglect. There have even been some instances of seeds from seed banks being used to reintroduce to nature a plant species that had become extinct.

Some disadvantages to seed banks exist, however. First, many types of plants, such as avocados and coconuts, cannot be stored as seeds. The seeds of these plants do not tolerate being dried out, which is a necessary step before the seeds can be sealed in moisture-proof containers for storage at $-18°C$ ($-0.4°F$). Second, seeds do not remain alive indefinitely and must be germinated periodically so that new seeds can be collected. Growing, harvesting, and returning seeds to storage is the most expensive aspect of storing plant material in seed banks. (Cryopreservation in liquid nitrogen at $-160°C$, or $-256°F$, is a new method being developed for certain kinds of seeds. Seeds stored at this temperature are able to survive for longer periods of time than seeds stored at warmer temperatures.) Because accidents such as fires or power failures can result in the permanent loss of the genetic diversity represented by the seeds, biologists typically subdivide seed samples and store them in several different seed banks.

Perhaps the most important disadvantage of seed banks is that plants stored in this manner remain stagnant in an evolutionary sense. They do not evolve in response to changes in their natural environments. As a result, they may be less fit for survival when they are reintroduced into nature.

Despite their shortcomings, seed banks are increasingly viewed as an important method of safeguarding seeds for future generations. Other international efforts to preserve plant genetic diversity are also being implemented. For example, some farmers may be paid to set aside some of their land for cultivating local varieties of crops, thereby preserving genetic diversity in agriculturally important plants.

Conservation Organizations

Conservation organizations are an essential part of the effort to maintain biological diversity, and we provide basic information about many such organizations in Appendix V, How to Make a Difference. These groups help to educate policy makers and the public about the importance of biological diversity. In certain instances they serve as catalysts by galvanizing public support for important biodiversity preservation efforts. They also provide financial support for conservation projects, from basic research to the purchase of land that is a critical habitat for a particular organism or group of organisms.

The World Conservation Union (IUCN)[1] assists countries with hundreds of conservation biology projects. It and other conservation organizations are currently assessing how effective established wildlife refuges are in maintaining biological diversity (Figure 16.18). In addition, IUCN and the World Wildlife Fund[2] have identified major conservation priorities by determining which biomes and ecosystems are not represented by protected areas. IUCN maintains a data bank on the status of the world's species; its material is published in *The IUCN Red Data Books* about organisms and habitats.

Figure 16.17 Seeds from a seed bank. Shown are small vials and packets of seeds from the seed bank in Svalbard, Norway.

CONSERVATION POLICIES AND LAWS

In 1973 the **Endangered Species Act (ESA)** was passed in the United States, authorizing the FWS to protect endangered and threatened species in the United States

---

[1] Formerly called the International Union for Conservation of Nature and Natural Resources, the World Conservation Union still goes by the acronym IUCN.

[2] Outside of the United States, Canada, and Australia, the World Wildlife Fund is known as the World Wide Fund for Nature.

Figure 16.18 Minimum critical size of ecosystems project.
When a protected area is set aside, it is important to know
what the minimum size of that area must be so that it will not
be affected by encroaching species from surrounding areas.
Shown are 1-hectare and 10-hectare plots of a long-term study
that is being conducted on the effects of habitat fragmentation
on Amazonian rain forest in Brazil by the World Wildlife
Fund and Brazil's National Institute for Amazon Research.
Plots with an area of 100 hectares are also under study, along
with identically sized sections of intact forest, which are used
as controls. Preliminary data indicate that the smaller forest
fragments are unable to maintain their ecological integrity. For
example, large trees often die or are damaged from exposure to
wind and weather.

and abroad. Many other countries now have similar legislation. The ESA conducts a detailed study of a species to
determine if it should be listed as endangered or threatened. Since the passage of the ESA, 1,260 species in the
United States have been listed as endangered or threatened (Table 16.2). The ESA provides legal protection to
listed species so that their danger of extinction is
reduced. This act makes it illegal to sell or buy any product made from an endangered or threatened species.

The ESA also requires the FWS to select critical
habitats and design a detailed recovery plan for each
species listed. The recovery plan includes an estimate of
the current population size, an analysis of what factors
contributed to its endangerment, and a list of activities
that will be used to help the population recover. Currently, there are 976 recovery plans. Some recovery plans
cover more than one species, whereas a few species have
two or more recovery plans for different parts of their
ranges. The recent reintroduction of wolves to Yellowstone National Park, discussed in Chapter 1, is part of
that species' recovery plan.

The ESA was updated in 1982, 1985, and 1988. It is
considered one of the strongest pieces of U.S. environmental legislation, in part because species are designated
as endangered or threatened entirely on biological
grounds. Currently, economic considerations cannot
influence the designation of endangered or threatened
species. Biologists generally agree that as a result of passage of the ESA in 1973, fewer species became extinct
than would have had the law never been passed.

The ESA has also been one of the most controversial
pieces of environmental legislation. The ESA does not
provide compensation for private property owners who
suffer financial losses because they cannot develop their
land if a threatened or endangered species lives there.
The ESA has also interfered with some federally funded
development projects.

The ESA was scheduled for congressional reauthorization in 1992 but has been entangled since then in
political wrangling between conservation advocates and
those who support private property rights. Conservation
advocates think the ESA does not do enough to save
endangered species, whereas those who own land on
which rare species live think the law goes too far and
infringes on property rights. Another contentious issue is
over the financial cost of the law. The FWS estimates
that federal and state governments spent $348 million in
1995 (latest data available), a figure that critics say is too
much, given the little bit of environmental gain that the
ESA accomplishes.

Some critics—notably business interests and private
property owners—view the ESA as an impediment to
economic progress. To protect the habitat of the northern spotted owl, the timber industry was blocked from
logging old-growth forests in certain parts of the Pacific
Northwest (see Chapter 3).

Those who defend the ESA point out that of 34,000
past cases of endangered species versus development,

Table 16.2   **U.S. Organisms Listed as Endangered or
Threatened, 2002**

Type of Organism	Number of Endangered Species	Number of Threatened Species
Mammals	65	9
Birds	78	14
Reptiles	14	22
Amphibians	12	9
Fishes	71	44
Snails	21	11
Clams	62	8
Crustaceans	18	3
Insects	35	9
Spiders	12	0
Flowering plants	568	144
Conifers	2	1
Ferns and other plants	24	2

only 21 cases could not be resolved by some sort of compromise. When the black-footed ferret was reintroduced on the Wyoming prairie, it was classified as an "experimental, nonessential species" so that its reintroduction would not block ranching and mining in the area. Thus, the ferret release program obtained the support of local landowners, support that was deemed crucial to its survival in nature.

This type of compromise is crucial to the success of saving endangered species, because according to the U.S General Accounting Office, more than 90% of endangered species live on at least some privately owned lands. Critics of the ESA think the law should be changed so that private landowners are given economic incentives to help save endangered species living on their lands. For example, tax cuts for property owners who are good land stewards could make the presence of endangered species on their properties an asset instead of a liability.

Defenders of the ESA agree that it is not perfect. Few endangered species have recovered enough to be delisted—that is, removed from protection of the ESA. (Table 16.3 shows U.S. species that have recovered enough to be delisted.) However, the FWS says that hundreds of listed species are stable or improving; they expect as many as several dozen additional species to be delisted in the next decade or so.

The ESA is geared more to saving a few popular or unique endangered species rather than the much larger number of less glamorous species that perform valuable ecosystem services. More than half the 1995 FWS funding for the ESA was used to help just 10 species, despite the fact that 1,260 species are currently listed. Yet it is the less glamorous organisms such as plants, fungi, and insects that play central roles in ecosystems and contribute most to their functioning.

Conservationists would like to see the ESA strengthened in such a way as to manage whole ecosystems and maintain complete biological diversity rather than attempt to save endangered species as isolated entities. This approach offers collective protection to many declining species rather than to single species.

Table 16.3   U.S. Species* That Are No Longer Classified as Endangered or Threatened

Organism	Year of Delisting
Robbins' cinquefoil (a plant)	2002
Aleutian Canada goose	2001
American peregrine falcon	1999
Pacific gray whale	1994
Arctic peregrine falcon	1994
Rydberg milk vetch (a plant)	1989
American alligator	1987
Brown pelican	1985

* Delisted species that have recovered.

ENVIROBRIEF

When One Rare Species Eats Another

What should be done when a rare species preys on an endangered species? At least two examples of this situation have occurred in recent years. Rice Island in the Columbia River was home to about 10% of the world's Caspian terns, which are declining in number worldwide. They nested on the island and ate fish, including highly endangered young salmon smolts, which accounted for 74% of their diet. Biologists were able to successfully lure the terns to another island where the smolts are not so concentrated by setting up bird models and broadcasting birdcalls. A study of the terns' diet on their new island home revealed they are still eating salmon—but salmon smolts account for only about 6% of their diet.

In New Mexico, rare mountain lions prey on highly endangered desert bighorn sheep. Because 75% of sheep deaths analyzed during a recent study appear to have been caused by mountain lions, state wildlife managers decided to adopt new hunting regulations that allow hunters to kill more lions. Opponents of the new regulations argue that most mountain lions eat mule deer, which are quite common, and only a few lions kill bighorn sheep. The opponents are concerned that mountain lions will be overhunted. They say that it would be better to kill only lions that have developed a taste for sheep, although this recommendation is expensive to carry out. In this case, more research is needed to understand the complex relationships between mountain lions and their prey as well as to determine an accurate population count of the elusive cats and what number can be sustainably killed.

Habitat Conservation Plans

The 1982 amendment of the ESA provided a way to resolve conflicts between protection of endangered species and development interests on private property: habitat conservation plans (HCPs). HCPs vary greatly, from small projects to regional conservation and development plans.

Habitat conservation plans allow a landowner to "take" (injure, kill, or modify the habitat of) a rare species if the "taking" does not threaten the survival or recovery of the threatened or endangered species on that property. If a landowner sets aside land as habitat for the rare species, he or she then has the right to develop a different part of the property without threat of legal action by the FWS. Conservationists are wary of HCPs because HCPs do not provide any promise of recovery of rare species. Conservationists are concerned that HCPs may actually contribute to a species' extinction.

The U.S. Biological Resources Discipline

The Biological Resources Discipline (BRD) was created in 1993 by combining science programs from seven different bureaus of the Department of the Interior.[3] The

---

[3] The BRD was originally named the National Biological Service.

mission of the BRD, now part of the U.S. Geological Service, is to provide information and technologies to manage and conserve biological resources on federal lands, which represent 22% of all U.S. lands. To accomplish this mission, basic science is needed to describe the current status of the nation's organisms and to monitor any changes. In 1998, the BRD, working with other agencies within the U.S. Department of the Interior and the U.S. Geological Survey, published a report entitled *Status and Trends of the Nation's Biological Resources*. This report, based on the contributions of almost 200 scientists, is the first comprehensive assessment of plants, animals, and ecosystems in the United States. It shows how U.S. biological resources are changing, including the establishment of more than 6,500 invasive species, and details the conservation problems that need to be addressed.

International Conservation Policies and Laws

The **World Conservation Strategy**, a plan designed to conserve biological diversity worldwide, was formulated in 1980 by the IUCN, the World Wildlife Fund, and the U.N. Environment Program. In addition to conserving biological diversity, the World Conservation Strategy seeks to preserve the vital ecosystem processes on which all life depends for survival and to develop sustainable uses of organisms and the ecosystems that they comprise. The most recent version of the World Conservation Strategy, published in 1991 and entitled *Caring for the Earth*, is presented in detail in Chapter 24.

The biological diversity treaty produced by the 1992 Earth Summit has 186 participating nations and is now considered binding. Under the conditions of the treaty, each signatory nation must inventory its own biodiversity and develop a **national conservation strategy**, a detailed plan for managing and preserving the biological diversity of that specific country.

The exploitation of endangered species can be somewhat controlled through legislation. At the international level, 160 countries participate in the Convention on International Trade in Endangered Species of Wild Flora and Fauna (CITES), which went into effect in 1975. Originally drawn up to protect endangered animals and plants considered valuable in the highly lucrative international wildlife trade, CITES bans hunting, capturing, and selling of endangered or threatened species and regulates trade of organisms listed as potentially threatened. Unfortunately, enforcement of this treaty varies from country to country, and even where enforcement exists, the penalties are not very severe. As a result, illegal trade in rare, commercially valuable species continues.

The goals of CITES often stir up controversy over such issues as who actually owns the world's wildlife and whether global conservation concerns take precedence over competing local interests. These conflicts often highlight socioeconomic differences between wealthy consumers of CITES products and poor people who trade the endangered organisms. The case of the African elephant bears out these controversies. Listed as an endangered species since 1989 to halt the slaughter of elephants driven by the ivory trade, the species seems to have recovered in southern Africa (Namibia, Botswana, and Zimbabwe). When elephant populations grow too large for their habitat, they root out and knock over so many smaller trees that the forest habitat can support fewer other species. Organizations such as the Humane Society in the United States are developing a birth control vaccine to reduce the number of elephant births. However, the African people living near the elephants want to be able to cull the herd periodically, so they can sell elephant meat, hides, and ivory for profit. In 1997, CITES transferred elephant populations in Namibia, Botswana, and Zimbabwe to a less restrictive, potentially threatened listing to allow trade of stockpiled ivory to Japan. Some conservationists oppose the resumption of any ivory trade, although others think that conservation goals are best attained by cooperating with local hunters and traders rather than by treating these groups as outlaws.

WILDLIFE MANAGEMENT

Wildlife management is an applied field of conservation biology that focuses on the continued productivity of plants and animals. Wildlife management includes the regulation of hunting and fishing and the management of food, water, and habitat. Wildlife management programs often have different priorities than conservation biology. Traditional wildlife management generally focuses on maintaining the population of a specific species, whereas conservation biology generally focuses on managing a community to ensure biological diversity in general. Wildlife managers use their knowledge of conservation biology to protect species that are endangered or of economic importance. They regulate an area by population control and habitat manipulation.

The natural predators of many game animals have largely been eliminated in the United States. As a result of the disappearance of predators such as wolves, the populations of animals such as squirrels, ducks, and deer sometimes exceed the carrying capacity of their environment (see Chapter 8). When this occurs, the habitat deteriorates, and many animals starve to death.

Sport hunting can effectively control overpopulation of game animals, provided restrictions are observed to prevent overhunting. Laws in the United States determine the time of year and length of hunting seasons for various species, as well as the number, sex, and size of each species that may be harvested.

Wildlife managers also affect a particular species by manipulating the plant cover, food, and water supplies of its habitat. Because different animals predominate in different stages of ecological succession (see Chapter 5), controlling the stage of ecological succession of an area's vegetation encourages the presence of certain animals and discourages others. Quail and ring-necked pheasant are found in grassy, open areas that are characteristic of early-succession stages. Moose, deer, and elk predominate in partially open forest, such as an abandoned field or meadow adjacent to a forest; the field provides food, and the forest provides protective cover. Other animals, such as grizzly bear and bighorn sheep, require undisturbed vegetation. Wildlife managers control the stage of succession with techniques such as planting certain types of vegetation, burning the undergrowth with controlled fires, and building artificial ponds.

Management of Migratory Animals

International agreements are established to protect migratory animals. Ducks, geese, and shorebirds spend their summers in Canada and their winters in the United States and Central America. During the course of their annual migrations, which usually follow established routes called **flyways**, they must have areas in which to rest and feed. Wetlands, the habitat of these animals, also must be protected in both their winter and summer homes.

CASE-IN-POINT Arctic Snow Geese

The Arctic snow goose has become a major challenge for wildlife managers because its population expanded rapidly during the last two decades of the 20th century. This goose breeds in large colonies along coastal salt marshes of the Arctic during the short Arctic summer. The population migrates south during autumn and traditionally wintered in salt marshes along the Texas and Louisiana coasts. The snow goose has successfully expanded its winter range into Arkansas, Mississippi, Oklahoma, New Mexico, and northern Mexico, largely because the geese are able to obtain seeds and other food from agricultural lands. Because snow geese have been so successful in expanding their winter range, more of the adults survive to return to the Arctic. Their adaptability to human-induced changes in the environment has enabled them to avoid density-dependent factors (i.e., lack of food during winter months) that would normally keep the population in check. The huge population of snow geese has damaged much of the Arctic's fragile coastal ecosystem as the geese forage there for food (the geese eat a variety of plants and insects).

Wildlife managers want to avoid a massive die-off of geese in the Arctic, a catastrophe that is unavoidable if the goose population is not brought under control. To reduce population numbers, U.S. and Canadian wildlife managers have increased the "taking" of snow geese by sport hunters. Animal rights groups such as the Humane Society of the United States, however, oppose hunting and argue that human-induced changes in the natural ecosystem are the cause of the goose population explosion. They think that farmers in the winter range of the geese should modify their agricultural methods to reduce the amount of food available for the geese. They acknowledge that this approach would be difficult to implement because of the impact it would have on so many farmers. Wildlife managers are looking at other options if increased sport hunting is not sufficient to reduce the number of snow geese. For example, the geese could be commercially harvested for human consumption. Wildlife managers could also modify certain practices at wildlife refuges to make the refuges less inviting to snow geese. ◼

Management of Aquatic Organisms

Fishes with commercial or sport value must be managed to ensure that they are not overexploited to the point of extinction. Freshwater fishes such as trout and salmon are managed in several ways. Fishing laws regulate the time of year, size of fish, and maximum allowable catch. Natural habitats are maintained to maximize population size. Ponds, lakes, and streams may be restocked with young hatchlings from hatcheries.

Traditionally, the ocean's resources have been considered common property, available to the first people to exploit them. As a result, many marine fishes have been severely reduced in numbers by commercial fishing. (Chapters 1 and 18 discuss this dwindling resource.)

During the 19th and 20th centuries, many whale species were harvested to the point of **commercial extinction**, meaning that so few remained that it was unprofitable to hunt them. Although commercially extinct species still have living representatives, their numbers are so reduced that they are endangered. In 1946 the International Whaling Commission set an annual limit on killed whales for each whale species in an attempt to secure sustainable whale populations. Unfortunately, these limits were set too high, resulting in further population declines during the next 20 years. Conservationists began to call for a global ban on commercial whaling; such a moratorium went into effect in 1986.

Scientists have since monitored whale populations and concluded that the ban is working overall. The populations of most whales, such as humpbacks and bowheads, appear to be growing. One species, the gray whale, has recovered sufficiently to be removed from the endangered species list and to be reclassified as only threatened (Figure 16.19). The North Atlantic right whale and southern blue whales, however, are still poised on the brink of extinction. In 1994 the International Whaling Commission established the Southern Ocean Whale Sanctuary in Antarctic waters, where many of the world's great whales feed and reproduce. This vast sanctuary, which bars commercial hunting, would remain should the current ban on whaling ever be lifted.

Figure 16.19 Gray whale. The gray whale (*Eschrichtius gibbosus*), found in the Pacific Ocean, has recovered sufficiently to be delisted from the endangered/threatened species lists. Photographed off Baja California in Mexico.

Despite international pressure, Japan and Norway do not honor either the global ban on commercial whaling or the Southern Ocean Whale Sanctuary. Japan has justified its continuing whale harvests by saying that the whales are harvested for "scientific purposes," although the whale meat from these harvests has been sold in Japanese markets and restaurants.

In an ironic twist, Japanese scientists detected dangerous levels of heavy metals and other pollutants in whale tissue and, in 1999, suggested that whale meat was unsafe to consume. The pollutants, which are present at low concentrations in ocean water, **bioaccumulate**, or increase in concentration, in the whale's bodies, at up to 70,000 times the level in surrounding seawater (see Chapter 22). Many Japanese supermarkets have now removed whale meat from their shelves.

WHAT CAN WE DO ABOUT DECLINING BIOLOGICAL DIVERSITY?

Although our children and grandchildren are faced with inheriting a biologically impoverished world, we should view this problem as a challenge. People who are dedicated to preserving our biological heritage can reverse the trend toward extinction. It is important to realize that you do not have to be a biologist to make a contribution; some of the most important contributions come from outside the biological arena. Following is a partial list of actions that can be taken to help maintain the biological diversity that is our heritage.

Increase Public Awareness

The consciousness of both the public and legislators must be increased so that they understand the impor-

tance of biological diversity. A political commitment to protect organisms is necessary because no immediate or short-term economic benefit is obtained from conserving species. This commitment must take place at all political levels, from local to international. Law making will not ensure the protection of organisms without strong public support. Thus, increasing public awareness of the benefits of biological diversity is critical.

Providing publicity on species conservation issues costs money. Private funds raised by organizations such as the Sierra Club, the Nature Conservancy, and the World Wildlife Fund support such endeavors, but clearly more money is needed. As an individual, you can help preserve biological diversity by joining and actively supporting conservation organizations.

Support Research in Conservation Biology

Before an endangered species can be saved, its numbers, range, ecology, biological nature, and vulnerability to changes in its environment must be determined. Basic research provides this information. We cannot preserve a given species effectively until we know how large a protected habitat must be established and what characteristics are essential in its design.

There are acute shortages of trained specialists in tropical forestry, conservation genetics, taxonomy, resource management, and similar disciplines. Many young people who are interested in these careers have selected others because of the dearth of funding for such research. The funding covers training and salaries of skilled personnel, research equipment and supplies, and miscellaneous expenses such as transportation costs.

As an individual, you can inform local and national politicians of your desire to have conservation research funded with tax dollars. When more funds are available, colleges, universities, and other research institutions will be able to justify adding faculty and research positions. As a result, more young people with interests in conservation will be able to obtain the necessary education and training for careers in this important field.

Support the Establishment of an International System of Parks

A worldwide system of protected parks and reserves that includes every major ecosystem must be established. The protected land would provide humans with other benefits in addition to the preservation of biological diversity. It would safeguard the watersheds that supply us with water, and it would serve as a renewable source of important biological products in areas with multiple uses. It would also provide people with unspoiled lands for aesthetic and recreational enjoyment. In addition to the establishment of new parks and reserves, particularly in developing nations, parks and reserves in highly developed nations must be expanded. As an individual, you can help establish parks by writing to national lawmakers.

Control Pollution

The establishment of parks and refuges will not be enough to prevent biological impoverishment if we continue to pollute Earth, because it is impossible to protect parks and refuges from threats such as acid rain, ozone depletion, and climate warming. Strong steps must be taken to curb the toxins we dump into the air, soil, and water—not only for human health and well-being but also for the well-being of the organisms that are so important to ecosystem stability. Specific recommendations on how you as an individual can help reduce pollution are discussed in Chapters 19 through 23.

Provide Economic Incentives to Landowners and Other Local People

Economic incentives encourage the preservation of biological diversity. Economic incentives are particularly critical because developing nations in the tropics, the repositories of most of the Earth's genetic diversity, do not have much money to spend on conservation. Their governments are consumed with human problems such as overpopulation, disease, and crushing foreign debts.

One way to help such countries appreciate the importance of the biological resources they possess is to allow them to charge fees for the use of genetic material. Much of the money earned could be used to help alleviate human problems. And some of the money generated by genetic resources could be used to provide protection for organisms, thus preserving biological diversity for continued, sustained exploitation.

The Suriname Biodiversity Prospecting Initiative, announced in 1993 by Conservation International, was the first such agreement of its kind. This enterprise pays money to local people of Suriname, a small country in South America, for new drugs delivered from the plants that they identify in their rain forests (Figure 16.20). With support from the U.S. government, universities, and various nonprofit organizations, the initiative represents the first partnership between indigenous people and a private pharmaceutical company (Bristol-Myers Squibb). Other agreements have followed. The National Cancer Institute has guaranteed the government of Malaysia part of the profits of any drugs obtained from organisms in Malaysian rain forests. Similar bioprospecting agreements exist between Merck and Costa Rica; Novo and Nigeria; and Glaxo and Ghana. It is essential that most of the money provided by these ventures find its way to the local level so landowners and other people appreciate the value of their biological resources.

Promoting ecotourism is another way that people can benefit financially by protecting their biological resources. Ecotourism is a type of tourism in which peo-

Figure 16.20 Suriname Biodiversity Prospecting Initiative. An ethnobotanist consults with a Tirio Indian in Suriname about the uses of a rainforest plant. Ethnobotany, the study of traditional uses of plants by indigenous people, helps pharmaceutical companies identify medicinal plants. Using this knowledge provides a shortcut in deciding which plants to test. Studies show that plants identified by shamans and traditional plant users are more likely to have medicinal value than randomly collected plants.

ple pay to visit natural environments and view native species. When done correctly, ecotourism conserves natural areas and improves the well-being of local people. Protection of coral reefs such as those off the coasts of Thailand and the Caribbean island of Bonaire has enabled ecotourism businesses that cater to scuba divers and snorkelers to thrive. Tours of Costa Rica's rain forests provide that country with its main earnings of foreign exchange. Kenya, Costa Rica, and Australia are currently considered the world leaders in ecotourism.

A third way of providing economic incentives to developing nations is for highly developed countries to forgive or reduce debts owed by such nations. In exchange, the developing countries would agree to protect their biological diversity. Such forgiveness of debts provides a tangible reward for preserving a nation's species. The United States arranged a debt-for-nature swap with Madagascar. It agreed to purchase $1 million of Madagascar's national debt in exchange for government support of local conservation efforts, including the protection of endangered species. Madagascar is known for its unique plants and animals. Twenty-three of Madagascar's 32 lemur species are currently in danger of extinction.

Another way to provide economic incentives to landowners would be for governments to cost-share habitat improvements for wildlife and to reduce the property taxes on such wildlife habitat. For example, the state of Texas is 97% privately owned. Essentially, all wildlife management is done through the cooperation of ranchers and other landholders. Laws in Texas allow land

that is appraised for agricultural production to become wildlife habitat without changing the tax valuation (agricultural lands have lower taxes than lands used for other purposes). Other incentives include payments to landowners who protect endangered species by restoring native plant communities.

Once again, you can help in the formulation of such policies. Let your lawmakers know where you stand. Join and support conservation groups. Campaign to preserve our biological heritage for future generations.

SUMMARY WITH SELECTED KEY TERMS

I. Biological diversity, the number and variety of organisms, encompasses **genetic diversity** (the variety within a species), **species richness** (the number of species), and **ecosystem diversity** (variety within and among ecosystems).

II. Organisms are an important natural resource.

A. Organisms perform essential **ecosystem services**. Bacteria and fungi, for example, perform the important task of decomposition.

B. Genetic reserves are used for domesticated plant and animal breeding, both through traditional breeding methods and through genetic engineering.

C. Organisms are sources of medicinal, agricultural, and industrial products.

D. Organisms give us recreation, inspiration, and aesthetic enjoyment.

III. When the last individual member of a species dies, it is said to be extinct. **Extinction** represents a permanent loss in biological diversity; once an organism is extinct, it can never exist again.

IV. An organism whose numbers are severely reduced so that it is in danger of extinction throughout all or a significant part of its **range** is said to be **endangered**. When extinction is less imminent but its numbers are quite low, a species is **threatened**.

A. Endangered and threatened species often have limited natural ranges and low population densities.

B. Endangered and threatened species may also have low reproductive rates or very specialized eating or reproducing requirements.

C. Many island species are endangered. **Habitat fragmentation**, the breakup of large areas of habitat into small, isolated patches (that is, islands), is a major threat to the long-term survival of many species.

D. **Biodiversity hotspots** are relatively small areas of land that contain an exceptional number of **endemic** species and are at high risk from human activities.

   1. Norman Myers and Conservation International have identified 25 biodiversity hotspots around the world.

   2. Fifteen of these hotspots are tropical, and 9 are mostly or entirely islands.

V. Human activities contribute to a reduction in biological diversity.

A. Habitat destruction is the most significant cause of declining biological diversity because it reduces a species' biological range and ability to survive.

B. **Biotic pollution** is the introduction of a foreign species into an area where it is not native.

   1. **Invasive species** are foreign species whose introduction causes economic or environmental harm.

   2. Islands are particularly susceptible to biotic pollution.

C. Pollution, such as acid rain, ozone-depleting compounds, and climate-warming atmospheric pollutants, is thought to contribute to the decline of many species.

D. Species that are over-hunted or harvested by people are sometimes at risk of endangerment or extinction.

   1. Efforts to eradicate a pest or predator have caused the endangerment or extinction of some species.

   2. Illegal commercial hunting (poaching) has a great impact on species endangerment and extinction, particularly for certain larger animals.

   3. **Commercial harvest** is the removal of live organisms from nature. Illegal commercial trade of rare animals and plants supplies a thriving black market, but regulated commercial harvest may sustain species from an economic standpoint.

VI. **Conservation biology** is the scientific study and protection of biological diversity.

A. Efforts to preserve biological diversity in nature are known as **in situ conservation**.

   1. Such efforts include establishing parks, wildlife sanctuaries and refuges, and other protected areas.

   2. **Restoration ecology**, in which the principles of ecology are used to help return a degraded environment as close as possible to its former state, is also part of in situ conservation.

B. **Ex situ conservation**, which includes captive breeding and storing genetic material, occurs in human-controlled settings. Zoos, aquaria, botanical gardens, and **seed banks** are examples of ex situ conservation.

C. Conservation organizations help to educate policy makers and the public about the importance of biological diversity. The World Conservation Union (IUCN) is assessing the effectiveness of established wildlife refuges in maintaining biological diversity.

VII. The **Endangered Species Act** (**ESA**) authorizes the U.S. Fish and Wildlife Service to protect from extinction endangered and threatened species, both in the United States and abroad. Many other countries have similar legislation.

A. The ESA does not include any economic considerations, such as providing compensation for private property owners who suffer financial losses as a result of obeying the law.

B. Habitat conservation plans are intended to help resolve ESA conflicts between conservation and development interests on private lands.

VIII. The IUCN, World Wildlife Fund, and U.N. Environment Program developed the **World Conservation Strategy**, which is designed to conserve biological diversity worldwide.

IX. Wildlife management is an applied field of conservation biology that focuses on the continued productivity of plants and animals.

A. Wildlife management includes the regulation of hunting and fishing and the management of food, water, and other habitat components.

B. Wildlife management programs often have different priorities than conservation biology. Traditional wildlife management tends to focus on maintaining the population of a specific species, whereas conservation biology focuses on managing a community to ensure biological diversity in general.

X. There are several ways to reverse the trend of declining biological diversity.

A. Both the public and lawmakers must become more aware of the importance of our biological heritage.

B. Funding must be found for additional research in both basic and applied fields relating to conservation biology.

C. A worldwide system of protected parks and reserves should be established, hopefully encompassing a minimum of 10% of the land area.

D. Pollution, which is damaging to both humans and other organisms, must be brought under control.

E. Economic incentives must be developed and expanded to help private landowners around the world manage their properties. Developing nations that are the repositories of much of the world's biological diversity must realize the value of their living resources through economic benefits.

THINKING ABOUT THE ENVIRONMENT

1. What is biological diversity?
2. Is biological diversity a renewable or nonrenewable resource? Why could it be seen both ways?
3. What are ecosystem services? Give at least five important ecosystem services provided by living organisms.
4. If we preserve species solely on the basis of their potential economic value—as a source of a novel drug, for example—does this mean that they lose their "value" once we have been able to capitalize on a newly discovered chemical? Why or why not?
5. Distinguish between *extinct*, *endangered*, and *threatened species*.
6. Give four characteristics that are common to many endangered species.
7. What are biodiversity hotspots? Where are most biodiversity hotspots located?
8. What are the four main causes of species endangerment and extinction? Give an example of a specific organism that has been harmed by each cause. Which cause do biologists consider most important?
9. What is biotic pollution?
10. Why are frogs and other amphibians considered bellwether species?
11. What is conservation biology?
12. Distinguish between *in situ* and *ex situ conservation* and give an example of each.
13. What is wildlife management? How are the goals of wildlife management different from those of conservation biology?
14. Why is the Arctic snow goose such a challenge for U.S. wildlife managers? For Canadian wildlife managers?
15. Does being pro-nature mean that you are also antidevelopment?
16. If you had the assets and authority to take any measure to protect and preserve biological diversity, but could take only one, what would it be?
17. According to this chapter, one of the ways in which you can help the world preserve its biological diversity is by talking about the problem, even to your friends. How will talking about biological impoverishment contribute to its reversal?
18. The most recent version of the World Conservation Strategy includes stabilizing the human population. How would stabilizing the human population affect biological diversity?
*19. Annual global trade in animals is conservatively estimated at $10 billion, and one third of that is illegal because it involves rare or endangered species. Calculate how much illegal trade, in millions of dollars, is done each day.

* The solution to this question appears in Appendix VII.

TAKE A STAND

Visit our Web site at **http://www.wiley.com/college/raven** (select Chapter 16 from the Table of Contents) for links to more information about the complexities of the elephant versus people issue in Africa. Consider the opposing views of conservationists and farmers, and debate the issue with your classmates. You will find tools to help you organize your research, analyze the data, think critically about the issues, and construct a well-considered argument. Take a Stand activities can be done individually or as part of a team, as oral presentations, written exercises, or Web-based (e-mail) assignments.

Additional on-line materials relating to this chapter, including Student Quizzes, Activity Links, Useful Web Sites, Flash Cards, and more, can also be found on our Web site.

SUGGESTED READING

Brooke, J. "Whalers and Their Foes Enlist Scientists." *New York Times* (June 4, 2002). Examines the controversial scientific whaling program sponsored by Japan.

Cohn, J. "Saving the California Condor." *BioScience*, Vol. 49, No. 11 (November 1999). Provides a fascinating account of what has been learned during the captive breeding and rearing of the highly endangered California condor.

Dobson, A., and A. Lyles. "Black-footed Ferret Recovery." *Science*, Vol. 288 (May 12, 2000). An overview of the recovery of the charismatic black-footed ferret, once though to be extinct in the United States.

Eliot, J.L. "Bald Eagles Come Back from the Brink." *National Geographic*, Vol. 202, No. 1 (January 2002). Bald eagle populations are increasing in the United States.

Gibbons, J.W., et al. "The Global Decline of Reptiles, Déjà vu Amphibians." *BioScience*, Vol. 50, No. 8 (August 2000). Reptile species are declining worldwide.

Gibbs, W.W. "On the Termination of Species." *Scientific American*, Vol. 285, No. 5 (November 2001). Ecologists' warnings of the current mass extinction are falling on deaf ears.

Hearn, J. "Unfair Game." *Scientific American*, Vol. 284, No. 6 (June 2001). The bushmeat trade is causing many African wildlife populations to decline.

Knight, J. "If They Could Talk to the Animals…" *Nature*, Vol. 414 (November 15, 2001). Animal conservation projects sometimes fail because the biologists do not know enough about the species' behavior.

Marinelli, J. "At Risk." *Natural History* (May 1999). Hundreds of U.S. plant species are listed as endangered or extinct.

Mestel, R. "Drugs from the Sea." *Discover* (March 1999). Discusses the potential of obtaining future drugs from ocean organisms.

Milius, S. "Are They Really Extinct?" *Science News*, Vol. 161 (March 16, 2002). It's very difficult to determine that a species is really extinct, as remnant populations may be surviving somewhere.

Morell, V. "The Fragile World of Frogs. *National Geographic*, Vol. 199, No. 5 (May 2001). This beautifully illustrated article highlights frog diversity and some of the threats that are affecting frog populations worldwide.

Pittman, C. "Fledgling Flier" *St. Petersburg Times* (May 25, 2002). The fascinating account of a baby whooping crane born in the wild in Florida—the first in many decades.

Simpson, R.D. "The Price of Biodiversity." *Issues in Science and Technology* (spring 1999). Examines some of the complex economic realities of bioprospecting and ecotourism.

Stone, R. "Hawaii: Hanging by a Thread." *Discover* (February 2000). Many of Hawaii's native plant and animal species are under siege from alien species and habitat loss.

Withgott, J. "California Tries to Rub Out the Monster of the Lagoon." *Science*, Vol. 295 (March 22, 2002). California is attempting to eradicate the invasive alga *Caulerpa* from its waters.

Mesquite Flat Dunes in Death Valley National Park.

Land Resources and Conservation

Learning Objectives

After you have studied this chapter you should be able to

1. Relate at least five ecosystem services provided by natural areas.

2. Summarize current land ownership in the United States.

3. Describe the following federal lands, stating which government agency administers each and current issues of concern: wilderness areas, national parks, national wildlife refuges, national forests, public rangelands, and national marine sanctuaries.

4. Define *deforestation* and relate the main causes of tropical deforestation.

5. Define *desertification* and explain its relationship to overgrazing.

6. Describe the current threats to freshwater and coastal wetlands, and explain why the definition of wetlands is controversial.

7. Discuss trends in U.S. agricultural lands, such as encroachment of suburban sprawl.

8. Contrast the views of the wise-use movement and the environmental movement regarding the use of federal lands.

he **California Desert Protection Act,** signed into law in 1994, created two of the United States' newest national parks—Death Valley and Joshua Tree—as well as the Mojave National Preserve in Southern California. Once considered desert wasteland, the parks and preserve protect at least 2,000 plant and animal species, varied archeological sites, 90 mountain ranges, and sand dunes as high as 213 m (700 ft).

Death Valley, which was named in 1849 by a group of gold prospectors who lost a companion there, has a rich mining history. Death Valley National Park was originally designated a National Monument by President Hoover in 1933. Mountain ranges surround the valley, which is 86 m (282 ft) below sea level at its lowest point, the lowest point in the entire Western Hemisphere. During summer, the valley floor is the hottest and one of the driest places in North America. Part of the valley contains beautiful sand dunes (see figure). The infrequent rainfall, which amounts to a few cloudbursts during the winter months, provides enough moisture for spectacular blooms of thousands of acres of wildflowers. Despite its name, Death Valley National Park abounds with hundreds of plant and animal species.

Joshua Tree National Park was originally designated a National Monument by President Franklin Roosevelt in

1936. It is named for the Joshua tree, an unusual desert plant that has a stout trunk, distinctive branches, and stiff, sharply pointed leaves. The national park contains two very different deserts, which are located at different elevations. Creosote bush, ocotillo, and cholla cactus dominate the Colorado Desert, located at the eastern side of the park, below 914 m (3,000 ft). The Mojave Desert, with its Joshua trees and unusual rock formations, is found at the western part of the park, where higher elevations make it cooler and moister.

The Mojave National Preserve contains more than 648,000 hectares (1.6 million acres) of some of the world's most variable desert environments. Geologically, it includes mountains, broad valleys, sand dunes, and exposed lava beds with volcanic cinder cones and craters. Botanically, the preserve contains several kinds of cactus, creosote bush, Joshua trees, mesquite, juniper, and pinyon pine woodlands. Many wildflowers bloom briefly following infrequent winter rains. Animals, which are mostly small and nocturnal, include the endangered desert tortoise.

In a decidedly modern twist on the role of national protected areas, the parks and the preserve do not function as simple nature sanctuaries but instead incorporate private interests into management of the wilderness. Existing mining and cattle ranching will continue to operate, recreational vehicles have been given some road access, and hunting is permitted in the preserve, all as a result of Congressional amendments to support the private property rights of the region's residents.

Although the "marriage" of private users and the National Park Service management will very likely experience tensions, it represents a solution to years of debate over environmental issues and the needs of human desert dwellers. Residents' concerns have been considered, yet most of the area will be protected. Habitat for the threatened desert tortoise will be preserved, and motorcycle races will no longer destroy dunes. Future development, mining, and road construction will be prohibited so that generations of people can appreciate the beauty of the desert wilderness.

IMPORTANCE OF NATURAL AREAS

Most of Earth's land area has a low density of humans. These sparsely populated areas, known as **nonurban** or **rural lands**, include forests, grasslands, deserts, and wetlands. Most people living in rural areas have jobs directly connected with natural resources—such as farming or logging. The many **ecosystem services** that are performed by rural lands enable the majority of humans to live in concentrated urban environments. Maintaining parcels of undisturbed land adjacent to agricultural and urban areas provides vital ecosystem services such as wildlife habitat, flood and erosion control, and groundwater recharge. Undisturbed land also breaks down pollutants and recycles wastes. Natural environments provide homes for organisms. One of the best ways to maintain biological diversity and to protect endangered and threatened species is by preserving or restoring the natural areas to which these organisms are adapted.

Undisturbed rural lands are ecosystems used by scientists as a benchmark, or point of reference, to determine the impact of human activity. Geologists, zoologists, botanists, ecologists, and soil scientists are some of the scientists who use undisturbed rural lands for scientific inquiry. These areas provide perfect settings for educational experiences not only in science but also in history, because they can be used to demonstrate the way the land was when humans originally settled here.

Unspoiled natural areas are important for their recreational value, providing places for hiking, swimming, boating, rafting, sport hunting, and fishing. Wild areas are important to the human spirit. Forest-covered mountains, rolling prairies, barren deserts, and other undeveloped areas not only are aesthetically pleasing but also help us to recover from the stresses of urban and suburban living. We can escape the tensions of the civilized world by retreating, even temporarily, to the solitude of natural areas.

CURRENT LAND USE IN THE UNITED STATES

About 55% of the land in the United States is privately owned by citizens, corporations, and nonprofit organizations, and about 3% by Native American tribes. The rest is owned by the federal government (about 35% of U.S. land) and by state and local governments (about 7% of U.S. land). Government-owned land encompasses all types of ecosystems, from tundra to desert, and includes land that contains important resources such as minerals and fossil fuels, land that possesses historical or cultural significance, and land that provides critical biological habitat. Most federally owned land is in Alaska and 11 western states (Figure 17.1). It is managed primarily by four agencies, three in the U.S. Department of the Interior—the Bureau of Land Management (BLM), the Fish and Wildlife Service (FWS), and the National Park Service (NPS)—and one in the Department of Agriculture—the U.S. Forest Service (USFS) (Table 17.1).

WILDERNESS

Wilderness encompasses regions where the land and its community of organisms have not been greatly disturbed by human activities and where humans visit but do not

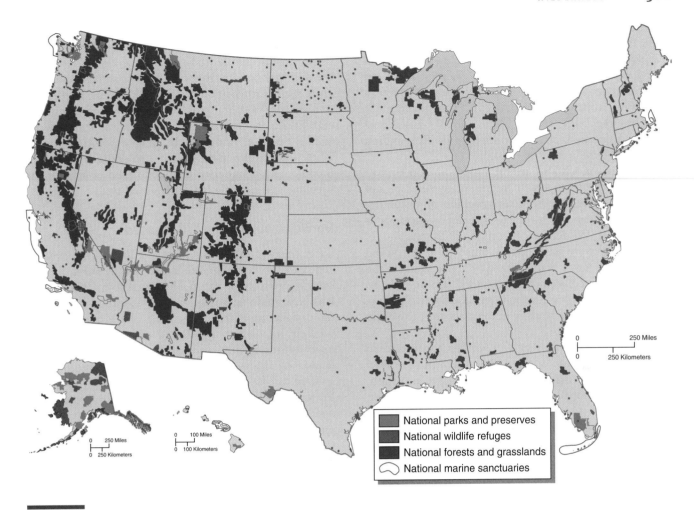

Figure 17.1 Selected federal lands. Shown are national parks and preserves, national wildlife refuges, national forests and grasslands, and national marine sanctuaries in the United States. Note the preponderance of federal lands in Western states and Alaska. Other federal lands, such as military installations and research facilities, are not included.

permanently inhabit. The 88th Congress recognized that increased human population and expansion into undeveloped areas might result in a future in which no lands occurred in their natural condition. Accordingly, the **Wilderness Act** of 1964 authorized the U.S. government to set aside federally owned land that retains its primeval character and lacks permanent improvements or human habitation, as part of the National Wilderness Preservation System. These federal lands range in size from tiny islands to portions of national parks (42% of wilderness areas are in national parks), national forests (33% of wilderness areas), and national wildlife refuges (22% of wilderness areas). The Big Gum Swamp wilderness in Florida is only 5,528 hectares (13,660 acres),

Table 17.1   Administration of Federal Lands

Agency	Land Held	Area in Millions of Hectares (Acres)
Bureau of Land Management (Dept. of Interior)	National resource lands	109 (270)
U.S. Forest Service (Dept. of Agriculture)	National forests	77 (191)
U.S. Fish and Wildlife Service (Dept. of Interior)	National wildlife refuges	37 (92)
National Park Service (Dept. of Interior)	National Park System	34 (84)
Other—includes Department of Defense, Corps of Engineers (Dept. of the Army), and Bureau of Reclamation (Dept. of Interior)	Remaining federal lands	29 (72)

whereas the Selway–Bitterroot Wilderness in Idaho is more than 530,000 hectares (1.3 million acres). Areas designated as wilderness are given the highest protection of any federal lands. These areas are to remain natural and unchanged so that they will be unimpaired for future generations to enjoy (Figure 17.2). Four government agencies—the NPS, USFS, FWS, and BLM—oversee the 630 wilderness areas that encompass 40.3 million hectares (102 million acres) of land.

Although mountains are the most common land to be safeguarded by this system, representative examples of other ecosystems have been set aside, including tundra, desert, and wetlands. More than one half of the lands in the National Wilderness Preservation System lie in Alaska, and western states contain much of the remaining lands. Because few sites untouched by humans exist in the eastern states, requirements were modified in 1975 so that the wilderness designation could be applied to certain federally owned lands where forests are recovering from logging.

Millions of people visit U.S. wilderness areas each year, and some areas are overwhelmed by this traffic: Eroded trails, soil and water pollution, litter and trash, and human congestion predominate over quiet, unspoiled land. Government agencies now restrict the number of people allowed into each wilderness area at one time so that the wilderness is not seriously affected by human use. Some of the most popular wilderness areas may require more intensive future management, such as the development of trails, outhouses, cabins, and campsites. These amenities are not encountered in true wilderness, posing a dilemma between wilderness preservation and human use and enjoyment of wild lands.

Limiting the number of human guests in a wilderness area does not control all factors that threaten wilderness. Exotic species that invade wilderness have the potential to upset the balance among native species. The white pine blister rust, an exotic fungus that can kill white pine trees, has invaded the wilderness in the northern Rocky Mountains. Wilderness managers are concerned that declining white pine populations could harm the population of grizzly bears in the region. (Pine seeds are a major part of the grizzlies' diet.) The Wilderness Act specifies both the preservation of natural conditions and the avoidance of intentional ecological management. In this example, however, the only way to preserve as much as possible of the original wilderness may be to intentionally manipulate the white pine population by breeding and planting fungus-resistant trees.

Do We Have Enough Wilderness?

Large tracts of wilderness, most of it in Alaska, have been added to the National Wilderness Preservation System since the passage of the Wilderness Act in 1964. People who view wilderness as a nonrenewable resource support the designation of wilderness areas. They think it is particularly important to preserve additional land in the lower 48 states, where currently less than 2% of the total land area is specified as wilderness. Increasing the amount of federal land in the National Wilderness Preservation System is usually opposed by groups who operate businesses on public lands (such as timber, mining, ranching, and energy companies) and by their political representatives.

The National Wild and Scenic Rivers System

The **Wild and Scenic Rivers Act** was passed in 1968 to protect rivers with outstanding beauty, recreational value, unique geologic features, important fish and wildlife, or historical value for the benefit of present and future generations. The NPS maintains the National Rivers Inventory and oversees management plans for wild and scenic rivers. About 170 river segments totaling about 18,180 km (11,300 mi) are currently protected by this act, with others being considered for inclusion. More than half the wild and scenic rivers are located in Alaska and the Pacific Northwest. Wild and scenic rivers represent less than 1% of the nation's total river systems.

Rivers that have been designated wild and scenic are not wilderness areas, but they have little or no development along their banks, and most have no dams. Camping, swimming, boating, sport hunting, and fishing are permitted, but development of the shoreline is prohibited. Mining claims are permitted, however.

NATIONAL PARKS

In 1872 Congress established the world's first national park, Yellowstone National Park, in federal lands in the territories of Montana and Wyoming. The purpose of the park was to protect this land of great scenic beauty and biological diversity in an unimpaired condition for

Figure 17.2 Wilderness areas. These hikers are enjoying the Selway–Bitterroot Wilderness in Idaho.

(a)

(b)

Figure 17.3 National parks. (a) Yosemite National Park in California. This winter view shows the Merced River flowing past the rock formation known as El Capitan. (b) Summer crowd at Grand Canyon National Park in Arizona. The popularity of certain national parks threatens to overwhelm them.

present and future generations. The National Park System was originally composed of such large, scenic areas in the West as Yellowstone, Grand Canyon, and Yosemite Valley (Figure 17.3a). Today, however, the National Park System has more cultural and historical sites—battlefields and historically important buildings

and towns—than places of scenic wilderness. Additions to the National Park System are made through acts of Congress, although the president has the authority to establish national monuments on federally owned lands.

The NPS was created in 1916 as a new federal bureau in the Department of Interior and given the responsibility to administer the national parks and monuments. The NPS currently administers 385 different sites, 57 of which are national parks. The total acreage administered by the NPS is 34.1 million hectares (84.3 million acres).

Many of these sites have been purchased with money provided by the **Land and Water Conservation Fund Act** of 1965. Urban parks, such as the Golden Gate National Recreation Area near San Francisco, have also been established. The most recent additions to the National Park System are, in 2001, the Virgin Islands Coral Reef National Monument, Governor's Island National Monument in New York, and Minidoka Internment National Monument in Idaho.

Because the NPS believes that knowledge and understanding increase enjoyment, one of the primary roles of the NPS is to teach people about the natural environment, management of natural resources, and history of a site by providing nature walks and guided tours of its parks. Exhibits along roads and trails, evening campfire programs, museum displays, and lectures are other educational tools.

The popularity and success of U.S. national parks (Table 17.2) have led many other nations to follow the example of the United States in establishing national parks. Today about 1,200 national parks exist in more than 100 countries. As in the United States, these parks usually have multiple roles, from providing biological habitat to facilitating human recreation.

Table 17.2   **The Ten Most Popular National Parks**

National Park	Number of Recreation Visitors in 2001 (millions)
Great Smoky Mountains (North Carolina and Tennessee)	9.2
Grand Canyon (Arizona)	4.1
Olympic (Washington)	3.4
Yosemite (California)	3.4
Rocky Mountain (Colorado)	3.1
Cuyahoga Valley (Ohio)	3.1
Yellowstone (Wyoming, Montana, and Idaho)	2.8
Grand Teton (Wyoming)	2.5
Acadia (Maine)	2.5
Zion (Utah)	2.2
Total visitors to the National Park System	279.9

CASE-IN-POINT **A National Park in Africa**

Deforestation is occurring at an unprecedented rate throughout the tropics. In most of West Africa the forests have disappeared. Cameroon, a West African country, is fortunate in that its government is committed to forest preservation. Korup National Park, established in 1986, has been extensively surveyed and has the richest biological diversity in Africa. It contains more than 400 tree species, the most of any comparable area in Africa; 50 mammal species, including almost 25% of the primate species found in Africa; and more than 320 bird species.

Korup National Park is a special project for wildlife conservation group World Wildlife Fund (WWF), in cooperation with the government of Cameroon and other international organizations. The WWF helps to protect and maintain the park, as well as integrate the park in regional development plans. To alleviate poverty, the WWF offers people in nearby villages the opportunity to receive training in local technical colleges. Educational training also discourages the killing of animals in the park, which is the park's most serious problem. When villagers have alternative sources of income, they are less likely to poach animals. Recently, an antipoaching partnership was formed between the local people and the park authorities. Most villagers now inform the park manager when they find poachers within the park.

Long-term plans for Korup National Park include many aspects of conservation and sustainable utilization of forest resources. The six villages now located in the park will be relocated on fertile land outside park boundaries. Local people are being educated about environmental issues and the importance of conservation. Some of these people are employed as park staff members, including game guards. A scientific research program has been established to help develop sustainable management practices for the park. Tourism, which depends on preservation of the area's unique biological diversity, helps provide income for the local economy. Korup National Park is a model of land conservation that other developing countries, in cooperation with conservation organizations and foreign aid, can emulate. ◼

Threats to U.S Parks

Some national parks are overcrowded (Figure 17.3b). All of the problems plaguing urban areas are found in popular national parks during peak seasonal use, including crime, vandalism, litter, traffic jams, and pollution of the soil, water, and air. In addition, thousands of resource violations, from cutting live trees and collecting plants, minerals, and fossils to defacing historical structures with graffiti, are investigated in national parks each year. Park managers have had to reduce visitor access to environmentally fragile park areas that have been degraded from overuse.

Many people think that more funding is needed to maintain and repair existing parks. Facilities at some of the largest, most popular parks, such as Yosemite, the Grand Canyon, and Yellowstone, were last upgraded some 30 years ago. Entrance fees account for about $132 million of the $2.2 billion a year the Park Service spends. Although steps have been taken to make parks more self-sufficient, they still depend on general tax revenues to pay for their operations.

Some national parks have imbalances in wildlife populations that involve declining populations of many species of mammals, including bears, white-tailed jackrabbits, and red foxes. Grizzly bears in national parks in the western United States are threatened. Grizzly bears require large areas of wilderness, and the human presence in national parks may influence their status. More important, the parks may be too small to support grizzlies. Fortunately, so far grizzly bears have survived in sustainable numbers in Alaska and Canada.

Other mammal populations, notably elk, have been allowed to proliferate. Elk in Yellowstone National Park's northern range have increased from a population of 3,100 in 1968 to between 10,000 and 20,000 in the late 1990s. Ecologists have documented that the elk have overgrazed the entire ecosystem, reducing the abundance of native vegetation, such as willow and aspen, and seriously eroding stream banks. The reintroduction of wolves to Yellowstone should reduce the elk population over time (see Chapter 1).

National parks are also affected by human activities beyond their borders. Pollution does not respect park boundaries (recall the Chapter 15 discussion of a proposed gold mine near Yellowstone National Park). Also, parks are increasingly becoming islands of natural habitat surrounded by human development. Development on the borders of national parks limits the areas in which wild animals may range, forcing them into isolated populations. Ecologists have found that when environmental stressors occur, several small "island" populations are more likely to become threatened than a single large population occupying a sizable range (see Chapter 16).

Natural Regulation

A controversial park management policy known as natural regulation was introduced in Yellowstone National Park and many other U.S. national parks in 1968. **Natural regulation** involves letting nature take its course most of the time, with corrective actions undertaken as needed to adjust for changes caused by pervasive human activities. Under natural regulation, the population of the elk herd in Yellowstone is allowed to fluctuate naturally because of varying weather conditions and predator populations such as wolves. Park managers do not try to maintain the herd at a consistent level by culling or artificially propagating elk. Because fires are an integral part of the Yellowstone ecosystem, wildfires in the park are not suppressed unless they threaten people or buildings. Park managers intervene to control the invasion of exotic species, however. They are removing lake trout from Yellowstone Lake because this species, presumably introduced by fishermen, threatens the natural population of cutthroat trout.

Some critics of natural regulation think that park managers should intervene more to control the elk population. They point out that the current population of elk is harming the rest of the ecosystem, and they do not think the growing wolf population and harsh winters are enough to reduce elk numbers to an acceptable level. Thus, most of the controversy over natural regulation involves what kinds and how much of human intervention are necessary to maintain the park in pristine condition.

WILDLIFE REFUGES

The National Wildlife Refuge System, which was established in 1903 by President Theodore Roosevelt, is the most extensive network of lands and waters committed to wildlife habitat in the world. The National Wildlife Refuge System contains more than 535 refuges, with at least 1 in each of the 50 states, and encompasses 38.4 million hectares (95 million acres) of land. The various refuges represent all major ecosystems found in the United States, from tundra to temperate rain forest to desert, and are home to some of North America's most endangered species, such as the whooping crane. The mission of the National Wildlife Refuge System, which is administered by the FWS, is to preserve lands and waters for the conservation of fishes, wildlife, and plants of the United States. Wildlife-dependent activities, such as hunting, fishing, wildlife observation, photography, and environmental education, are permitted on parts of some wildlife refuges as long as they are compatible with scientific principles of fish and wildlife management.

FORESTS

Forests, important ecosystems that provide many goods and services that support human society, occupy less than one third of the Earth's total land area. Timber harvested from forests is used for fuel, construction materials, and paper products. Forests also supply nuts, mushrooms, fruits, and medicines. Forests provide employment for millions of people worldwide. They offer recreation and spiritual sustenance to an increasingly crowded world.

Forests provide a variety of beneficial ecosystem services. Forests influence local and regional climate conditions. If you walk into a forest on a hot summer day, you will notice that the air is cooler and moister than it is outside the forest. This is the result of a biological cooling process called **transpiration**, in which water from the

soil is absorbed by roots, transported through plants, and then evaporated from their leaves and stems. Transpiration also provides moisture for clouds, eventually resulting in precipitation (Figure 17.4). Thus, forests help maintain local and regional precipitation.

Forests play an essential role in regulating global biogeochemical cycles, such as those for carbon and nitrogen. Photosynthesis by trees removes large quantities of heat-trapping carbon dioxide from the atmosphere and fixes it into carbon compounds. Forests thus act as carbon "sinks" that help mitigate global warming (see Chapter 20). At the same time, oxygen that almost all organisms require for cellular respiration is released into the atmosphere.

Tree roots hold vast tracts of soil in place, reducing erosion and mudslides. Forests protect watersheds because they absorb, hold, and slowly release water; this

Up to 75% water recycled by transpiration and evaporation

25% or more water seeps into ground or runs off to rivers, streams, and lakes

Figure 17.4 Role of forests in the hydrologic cycle. Forests return most of the water that falls as precipitation to the atmosphere by transpiration. In contrast, when an area is deforested, almost all precipitation is lost as runoff.

moderation of water flow provides a more regulated flow of water downstream, even during dry periods, and helps to control floods and droughts. Forest soils remove impurities from water, improving its quality. In addition, forests provide a variety of essential habitats for many organisms, such as mammals, reptiles, amphibians, fish, insects, lichens and fungi, mosses, ferns, conifers, and numerous kinds of flowering plants. (The importance of tropical forests, particularly tropical rain forests, as the repositories of most of the world's biological diversity was discussed in Chapter 16).

Forest Management

When forests are managed for timber production, their species composition and other characteristics are altered. Specific varieties of commercially important trees are planted, and those trees that are not as commercially desirable are thinned out or removed. Traditional forest management often results in low-diversity forests. In the southeastern United States, many tree plantations of young pine that are grown for timber and paper production are all the same age and are planted in rows a fixed distance apart (Figure 17.5). These "forests" are essentially **monocultures**—areas covered by one crop, like a field of corn. Herbicides are sprayed to kill shrubs and herbaceous plants between the rows. One of the disadvantages of monocultures is that they are more prone to damage by insect pests and disease-causing microorganisms. Consequently, pests and diseases must be controlled in managed forests, usually by applying insecticides and fungicides. Because managed forests contain few kinds of food, they cannot support the variety of organisms typically found in natural forests. Tree plantations have the potential to benefit remaining natural forests, however, provided that remaining forests are conserved and protected and that the plantations themselves do not replace natural forests.

In recognition of the many ecosystem services performed by forests, a new method of forest management, known as **ecologically sustainable forest management** or, simply, **sustainable forestry,** is evolving. This broader approach seeks not only to conserve forests for the long-term commercial harvest of timber and nontimber forest products but also to sustain biological diversity by providing an improved habitat for a variety of species; to prevent soil erosion and improve soil conditions; and to preserve watersheds that produce clean water. Effective sustainable forest management involves cooperation among environmentalists, loggers, farmers, indigenous people, and local, state, and federal governments. To achieve these goals, forestry operations are designed that maintain a mix of forest trees, by age and species.

When logging occurs using sustainable forestry principles, unlogged areas are set aside as sanctuaries for organisms, along with **wildlife corridors**, which are protected zones that connect isolated unlogged areas. The

Figure 17.5 Tree plantation. This intensively managed pine plantation is a monoculture, with trees of uniform size and age. Such plantations supplement harvesting of trees in wild forests to provide the United States with the timber it requires. According to a Forest Service report, the United States annually consumes nearly 20% more wood than it produces. Photographed somewhere in the southern U.S. pulpwood region, from Alabama to Georgia.

purposes of wildlife corridors are to provide escape routes should they be needed and to allow animals to migrate so they can interbreed. (Small, isolated populations that are inbred may have a higher risk of extinction.) Wildlife corridors are also thought to allow large animals such as the Florida panther to maintain large territories. Some scientists question the effectiveness of wildlife corridors, although recent research on wildlife corridors in fragmented landscapes suggests that wildlife corridors help certain wildlife populations to persist. Additional scientific research is needed to resolve the question for all endangered species.

The actual methods of ecologically sustainable forest management that distinguish it from traditional forest management are gradually being developed. These vary from one forest ecosystem to another, in response to different ecological, cultural, and economic conditions. For example, in Mexico many co-management projects

Ecologically Certified Wood

Homebuilders and homeowners are increasingly interested in "green" wood for flooring and other building materials. Such wood is ecologically certified by a legitimate third party, such as the Mexico-based Forest Stewardship Council (FSC), to have come from a forest managed with environmentally sound and socially responsible practices. Although these areas remain a small percentage of total forests, by early 1999 the FSC had certified as well managed more than 31 million acres in 27 countries. Certification is based on sustainability of timber resources, socioeconomic benefits provided to local people, and forest ecosystem health, which includes such considerations as preservation of wildlife habitat and watershed stability.

Green timber has proved profitable for its national distributors, who have seen sales increases of 20% to 30%. Often, the consumer pays no additional premium, or only slightly more, for ecologically certified wood, which has become so popular that demand threatens to exceed supply.

Green forestry has its detractors. Traditional forestry organizations are skeptical about the reliability of FSC investigations and the economic viability of this type of forestry. Trade experts caution that any governmental efforts to specify the purchase of certified timber could violate world free trade agreements. Still, green timber seems to be gaining on all fronts, pleasing business owners and consumers alike and offering greater promise of conservation in managed forests.

involving communities that are economically dependent on forests are developing. Because trees have such long life spans, scientists and forest managers of the future will judge the results of today's efforts.

Harvesting Trees According to the U.N. Food and Agricultural Organization, 3.38 billion cubic meters of roundwood (logs for fuelwood, timber, paper, and other products) were harvested in 1999 (the latest available data). The five countries with the greatest tree harvests are the United States, Canada, Russia, China, and Brazil; these countries currently produce more than half the world's roundwood. About 55% of harvested wood is burned directly as fuelwood or used to make charcoal. (Partially burning wood in a large kiln from which air is excluded converts the wood into charcoal.) Most fuelwood and charcoal are used in developing countries (discussed shortly). Highly developed countries consume more than three fourths of the remaining 45% of harvested wood for paper and wood products.

Loggers harvest trees in several ways—by selective cutting, shelterwood cutting, seed tree cutting, and clearcutting (Figure 17.6). **Selective cutting**, in which mature trees are cut individually or in small clusters while the rest of the forest remains intact, allows the forest to regenerate naturally. The trees left by selective cutting produce seeds that germinate to fill the void. Selective cutting has fewer negative effects on the forest environ-

ment than other methods of tree harvest, but it is not as profitable in the short term because timber is not removed in great enough quantities.

The removal of all mature trees in an area over a period of time is known as **shelterwood cutting**. In the first year of harvest, undesirable tree species and dead or diseased trees are removed. The forest is then left alone for perhaps a decade, during which the remaining trees continue to grow, and new seedlings become established. During the second harvest, many mature trees are removed but some of the largest trees are left to shelter the young trees. The forest is then allowed to regenerate on its own for perhaps another decade. A third harvest removes the remaining mature trees, but by this time a healthy stand of younger trees is replacing the mature ones. Little soil erosion occurs with this method of tree removal, even though more trees are removed than in selective cutting.

In **seed tree cutting**, almost all trees are harvested from an area; a scattering of desirable trees is left behind to provide seeds for the regeneration of the forest.

Clearcutting is the removal of all trees from an area (Figure 17.7). After the trees have been removed by clearcutting, the area is either allowed to reseed and regenerate itself naturally or is planted with one or more specific varieties of trees. Timber companies prefer clearcutting because it is the most cost-effective way to harvest trees. Clearcutting in small patches can actually benefit some wildlife species, such as deer and certain songbirds. These species thrive in the regrowth of trees and shrubs that follows removal of the overhead canopy. However, clearcutting over wide areas is ecologically unsound. It destroys biological habitats and increases soil erosion, particularly on sloping land. For example, in 1996 hundreds of mudslides from steep hillsides that had been clear-cut occurred in Oregon following heavy rains; properties and roads were damaged, and several people were killed. Sometimes the land is so degraded from clearcutting that reforestation does not take place; whereas lower elevations are usually regenerated successfully, higher elevations are often difficult. Obviously, the recreational benefits of forests are lost when clearcutting occurs.

Deforestation

The most serious problem facing the world's forests is **deforestation**, which is the temporary or permanent clearance of large expanses of forest for agriculture or other uses. The World Commission on Forests, formed following the Earth Summit in 1992, released its first report in 1999 after 3 years of research and public hearings around the world. It concluded that Earth's forests are shrinking *each year* by 15 million hectares (37 million acres). A study by the U.N. Food and Agricultural Organization (FAO), released in 2001, concluded that forest loss is lower than the estimate given by the World Com-

(a) SELECTIVE CUTTING

(b) SHELTERWOOD CUTTING

(c) SEED TREE CUTTING

(d) CLEAR-CUTTING

Figure 17.6 Systems of tree harvesting. (a) In selective cutting, the older, mature trees are selectively harvested from time to time, and the forest regenerates itself naturally. (b) In shelterwood cutting, less desirable and dead trees are harvested. As younger trees mature, they produce seedlings, which continue to grow as the now-mature trees are harvested. (c) Seed tree cutting involves the removal of all but a few trees, which are allowed to remain, providing seeds for natural regeneration. (d) In clearcutting, all trees are removed from a particular site. Clearcut areas may be reseeded or allowed to regenerate naturally.

mission on Forests. The FAO says that forests are shrinking by about 9 million hectares (22.2 million acres) each year. They say their estimate is lower because it takes into account the recent planting of extensive tree plantations in India and China.

Causes of the forest destruction include fires caused by drought and land clearing practices, expansion of agriculture, construction of roads in forests, tree harvests, and insects and diseases. When forests are converted to other land uses, they no longer make valuable contributions to the environment or to the people who depend on them. Forest destruction, particularly in the tropics, threatens indigenous people whose cultural and physical survival depends on the forests.

Deforestation results in decreased soil fertility through rapid leaching of the essential mineral nutrients found in most forest soils. Uncontrolled soil erosion, particularly on steep deforested slopes, can affect the production of hydroelectric power as silt builds up behind

Figure 17.7 Clearcutting. This aerial view of a privately-owned forest in Washington state shows a large patch of clear-cut forest. Clearcutting is the most common but most controversial type of logging. The lines are roads built to haul away the logs.

dams. Increased sedimentation of waterways caused by soil erosion can also harm downstream fisheries. In drier areas, deforestation contributes to the formation of deserts (discussed shortly). When a forest is removed, the total amount of surface water that flows into rivers and streams actually increases. However, because this water flow is no longer regulated by the forest, the affected region experiences alternating periods of flood and drought.

Deforestation contributes to the extinction of many species. Many tropical species, in particular, have very limited ranges within a forest, so they are especially vulnerable to habitat modification and destruction. Migratory species, including birds and butterflies, also suffer from deforestation.

Deforestation is thought to induce regional and global climate changes. Trees release substantial amounts of moisture into the air; about 97% of the water that roots absorb from the soil is evaporated directly into the atmosphere. This moisture falls back to the earth in the hydrologic cycle (see Chapter 6). When a large forest is removed, rainfall may decline and droughts may become more common in that region. Studies suggest that the local climate has become drier in parts of Brazil where tracts of the rain forest have been burned. Where deforestation has occurred in Central America, the nearby cloud forests have lost much of their moisture-providing clouds. Temperatures may also rise slightly in a deforested area because there is less evaporative cooling from the trees.

Deforestation may contribute to an increase in global temperature by causing a release of carbon originally stored in the trees into the atmosphere as carbon dioxide, which enables the air to retain heat. The carbon in forests is released immediately if the trees are burned, or more slowly when unburned parts decay. If trees are harvested and logs are removed, roughly one half of the forest carbon remains as dead materials (branches, twigs, roots, and leaves) that decompose, releasing carbon dioxide. When an old-growth forest is harvested, researchers estimate that it takes about 200 years for the replacement forest to accumulate the amount of carbon that was stored in the original forest.

Tropical Forests and Deforestation

There are two types of tropical forests: tropical rain forests and tropical dry forests. In places where the climate is warm and very moist throughout the year—with about 200 or more cm (at least 79 in.) of precipitation annually—**tropical rain forests** prevail. Tropical rain forests are found in Central and South America, Africa, and Southeast Asia, but almost half of them are in just three countries: Brazil, Democratic Republic of the Congo, and Indonesia (Figure 17.8).

In other tropical areas where annual precipitation is less but is still enough to support trees, including regions subjected to a wet season and a prolonged dry season, **tropical dry forests** occur. During the dry season, tropical trees shed their leaves and remain dormant, much as temperate trees do during the winter. India, Kenya, Zimbabwe, Egypt, and Brazil are a few of the countries that have tropical dry forests.

Most of the remaining undisturbed tropical forests, which lie in the Amazon and Congo river basins of South America and Africa, are being cleared and burned at a rate that is unprecedented in human history. Tropical forests are also being destroyed at an extremely rapid rate in southern Asia, Indonesia, Central America, and the Philippines.

Exact figures on rates of tropical forest destruction are unavailable. The FAO released its most recent assessment of tropical deforestation in *State of the World's Forests: 1999*. A total of 117 tropical countries were evaluated in this study. The FAO estimated an average annual loss in forests of 0.7% per year from 1990 to 1995, and

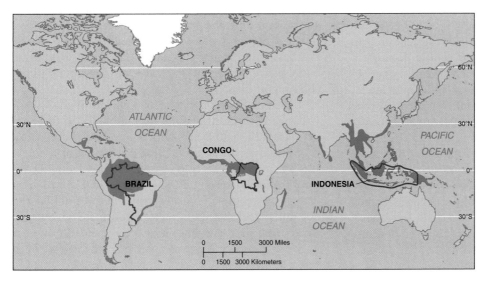

Figure 17.8 Distribution of tropical rain forests. Rain forests (*green areas*) are located in Central and South America, Africa, and Southeast Asia. Much of the remaining forested area is highly fragmented.

some areas, such as continental Southeast Asia, experienced a loss of forests estimated at 1.6% per year. If this rate of deforestation, which represents an annual loss of 12.6 million hectares (31.1 million acres), continues, tropical forests will be all but gone by the first half of the 22nd century.

In an encouraging move, in 1999 Brazil agreed to set aside up to 10% of the Amazonian rain forest for conservation. The World Bank and the World Wildlife Fund will provide financial and technical assistance. Deciding which lands to protect, how to manage them, and how to ensure their conservation are some of the issues that must be decided in this historic agreement.

Why are Tropical Forests Disappearing?

Several studies show a strong statistical correlation between population growth and deforestation. More people need more food, and so the forests are cleared for agricultural expansion. Tropical deforestation is a complex problem, however, that cannot be attributed simply to population pressures. The main causes of deforestation vary from place to place, and a variety of economic, social, and governmental factors interact to cause deforestation. Government policies sometimes provide incentives that favor the removal of forests. The Brazilian government opened the Amazonian frontier for settlement, beginning in the late 1950s, by constructing the Belem–Brasilia Highway, which cut through the Amazon Basin (Figure 17.9). Sometimes economic conditions encourage deforestation. The farmer who converts more forest to pasture can maintain a larger herd of cattle, which is a good hedge against inflation.

Keeping in mind that tropical deforestation is a complex problem, three agents—subsistence agriculture, commercial logging, and cattle ranching—are thought to be the most immediate causes of deforestation. Other reasons for the destruction of tropical forests include mining, particularly when ore smelters burn charcoal produced from rainforest trees, and the development of hydroelectric power, which inundates large areas of forest.

Subsistence Agriculture Subsistence agriculture, in which a family produces just enough food to feed itself, accounts for perhaps 60% of tropical deforestation. In many developing countries where tropical rain forests occur, the majority of people do not own the land they live and work on. In Brazil 5% of the farmers own 70% of the land. Most subsistence farmers have no place to go except into the forest, which they clear to grow food. Land reform in Brazil, Madagascar, Mexico, the Philippines, Thailand, and many other countries would make the land owned by a few available to everyone, thereby easing the pressure on tropical forests by subsistence farmers. This scenario is unlikely, however, because wealthy landowners have more economic and political clout than impoverished landless peasants.

Figure 17.9 Satellite photograph of human settlements along a portion of road in Brazil's tropical rain forest. Numerous smaller roads extend perpendicularly from the main roads. As the farmers settle along the roads, they clear out more and more forest (*dark green*) for their croplands and pastures (*tans and pinks*).

Subsistence farmers often follow loggers' access roads until they find a suitable spot. They first cut down the trees and allow them to dry, then they burn the area and plant crops immediately after burning; this is known as **slash-and-burn agriculture** (discussed further in Chapter 18). The yield from the first crop is often quite high because the nutrients that were in the burned trees are now available in the soil. However, soil productivity declines at a rapid rate, and subsequent crops are poor. In a very short time, the people farming the land must move to a new part of the forest and repeat the process. Cattle ranchers often claim the abandoned land for grazing, because land that is not rich enough to support crops can still support livestock.

Slash-and-burn agriculture done on a small scale, with plenty of forest to shift around in so that there are periods of 20 to 100 years between cycles, is sustainable. The forest regrows rapidly after a few years of farming. But when *millions of people* try to obtain a living in this way, the land is not allowed to lie uncultivated long enough to recover. Globally, between 200 million and 600 million subsistence farmers are estimated to obtain a living from slash-and-burn agriculture, and the number is growing rapidly. Moreover, there is only half as much forest available today as there was 50 years ago.

Commercial Logging   About 20% of tropical deforestation is the result of commercial logging. Vast tracts of tropical rain forests, particularly in Southeast Asia, are being harvested for export abroad. Most tropical countries allow commercial logging to proceed at a much faster rate than is sustainable. Unmanaged tropical deforestation does not contribute to economic development; rather, it depletes a valuable natural resource faster than it can be regenerated for sustainable use.

Cattle Ranching and Agriculture for Export   Approximately 12% of tropical deforestation is carried out to provide open rangeland for cattle. Cattle ranching, which employs relatively few local people (ranching does not require much labor), is particularly important in Latin America. Some of the beef raised on these ranches, which are often owned by foreign companies, is exported to highly developed countries, although much is consumed locally. After the forests are cleared, cattle can graze on the land for perhaps 20 years, after which time the soil fertility is depleted. When this occurs, shrubby plants, known as **scrub savanna**, take over the range.

A considerable portion of forestland cleared for plantation-style agriculture, often owned by foreign corporations, produces crops such as citrus fruits and bananas for export. Plantation-style agriculture is sustainable on forest soil as long as fertilizers and other treatments are applied.

Why Are Tropical Dry Forests Disappearing?

Tropical dry forests are also being destroyed at an alarming rate, primarily for fuel (Figure 17.10). Wood—perhaps half of the wood consumed worldwide—is used as heating and cooking fuel by much of the developing world. The unsustainable use of wood has led to a fuelwood crisis in many developing countries. The 2 billion or so people that the FAO says cannot get enough fuelwood to meet basic needs such as boiling water and cooking are at risk of waterborne infectious diseases (see Table 21.1 for a list of some human diseases transmitted by polluted water). Often the wood cut for fuel is converted to charcoal, which is then used to power steel, brick, and cement factories. Charcoal production is extremely wasteful: 3.6 metric tons (4 tons) of wood produce enough charcoal to fuel an average-sized iron smelter for only 5 minutes.

Boreal Forests and Deforestation

Although tropical forests have been depleted extensively, they are not the only forests being destroyed. Extensive deforestation due to the logging of certain boreal forests began in the late 1980s and early 1990s. **Boreal forests** occur in Alaska, Canada, Scandinavia, and northern Russia. Coniferous evergreen trees such as spruce, fir, cedar, and hemlock dominate these northern forests. Boreal forests comprise the world's largest biome, covering about 11% of the Earth's land area.

Boreal forests, which are harvested primarily by clearcut logging, are currently the primary source of the world's industrial wood and wood fiber. The annual loss of boreal forests has been estimated to encompass an area twice as large as the Amazonian rain forests of Brazil.

About 1 million hectares (2.5 million acres) of Canadian forests are logged annually, and most of Canada's forests are under logging tenures. (*Tenures* are agreements between provinces and companies that give companies the right to cut timber.) Canada is the world's biggest timber exporter. On the basis of current harvest quotas, however, logging in Canada appears to be unsustainable. According to the World Resources Institute, the Canadian government promotes but does not always implement sustainable forest policies.

Figure 17.10 Deforestation for fuel.
Indian women gather firewood in the Ranthambore National Park buffer zone. Note how the branches have been trimmed off the trees in the background. About 340,000 hectares (840,000 acres) of India's forests disappear per year, and almost 90% of wood removed from forests is burned as fuel. Fortunately, as many as 10 million rural Indians now use biogas (discussed in Chapter 12) instead of wood, and the number is projected to grow.

Extensive tracts of Siberian forests in Russia are also harvested, although exact estimates are unavailable. Alaska's boreal forests are at risk because the U.S. government may increase logging on public lands in the future.

Forests in the United States

In recent years, temperate forests in the Rocky Mountains, Great Lakes region, and New England and other eastern states have generally been holding steady or even expanding. In Vermont, for example, the amount of land covered by forests has increased from 35% in 1850 to 80% today. Expanding forests are the result of secondary succession on abandoned farms (see Chapter 5), commercial planting (tree plantations) on both private and public lands, and government protection. Although the returning forests generally do not have the biological diversity of virgin stands, many forest organisms have successfully become reestablished in regenerated forests. The good news about these returning forests must be tempered with the fact that additional forestland will probably have to be harvested during the 21st century to meet the increased demand for lumber, paper, and other wood products caused by an increase in the U.S. population.

Slightly more than one half of U.S. forests are privately owned, and three fourths of these private lands are in the northeastern and midwestern parts of the country (Figure 17.11). Many of these private forest owners are under economic pressure, such as high property taxes, to subdivide the land and develop tracts of it for housing or shopping malls. Projected conversion of forests to agricultural, urban, and suburban lands over the next 40 years will probably have the greatest impact in the South, where more than 85% of forest is privately owned and logging is largely unregulated.

The Forest Legacy Program is a provision of the 1990 Farm Bill that helps private landowners protect environmentally important forestlands from development. Here's how the program, which is managed by the U.S. Department of Agriculture, works. A willing landowner sells some or all ownership rights, such as the right to develop the land, to the U.S. government, which then holds a **conservation easement**, a legal agreement that protects the forest property from development for a specified number of years. The landowner usually continues to live and/or work on the property and can sell those rights not already sold to the government to other private individuals. All future property owners must abide by the provisions of the conservation easement, however.

U.S. National Forests According to the USFS, the United States has 155 national forests encompassing 77 million hectares (191 million acres) of land, mostly in Alaska and western states. The USFS manages most national forests, with the remainder overseen by the BLM. National forests were established for multiple uses—to provide for U.S. citizens maximum benefits of natural resources such as fish, wildlife, and timber. Multiple uses include timber harvest; livestock forage; water resources and watershed protection; mining; hunting, fishing, and other forms of outdoor recreation; and habitat for fishes and wildlife. Recreation, which increased dramatically in national forests during the 1990s, ranges from camping at designated campsites to backpacking in the wilderness. Visitors also swim, boat, picnic, and observe nature in national forests. With so many different possible uses of national forests, conflicts inevitably arise, particularly between timber interests and those who wish to preserve the trees for other purposes.

Road building is a particularly contentious issue in national forests, in part because the USFS builds taxpayer-funded roads that allow private logging companies access to the forest so they can remove timber. This money is not reimbursed by the price the USFS charges for timber concessions. The BLM also subsidizes logging operations on lands it manages. About 697,000 km (433,000 mi) of logging roads have been built throughout U.S. national forests. Road building in national forests can be environmentally destructive if improper construction accelerates soil erosion and mudslides (particularly on steep terrain) and causes water pollution in streams. Biologists are also concerned that so many roads fragment wildlife habitat and provide entries for disease organisms and exotic species.

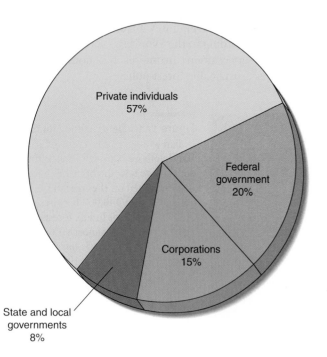

Figure 17.11 Forest ownership in the United States. Most forests are privately owned.

Another contentious issue in national forests is clearcutting. We now examine the clearcutting debate and its most recent developments in the Tongass National Forest, which is located along Alaska's southeastern coast.

CASE-IN-POINT Tongass National Forest

Despite its northern location, the Tongass National Forest is one of the world's few temperate rain forests (Figure 17.12; also see Chapter 7 for a description of temperate rain forests). It is one of the wettest places in the United States. This moisture supports old-growth forest of giant Sitka spruce, yellow cedar, and western hemlock, some of which are 700 years old. Stream banks are lined with bent willows, ferns, mosses, and other vegetation. This forest, the largest in the National Forest System, provides habitat for a wealth of wildlife, such as grizzly bears and bald eagles.

The Tongass is also a prime logging area because a single large Sitka spruce may yield as much as 10,000 board feet of high-quality timber. The logging industry forms the basis of much of the local economy. Regeneration of mature forest after it has been clearcut is very slow, on the order of several centuries. On Vancouver Island to the south of the Tongass, the trees in temperate rain forest that was clearcut in 1911 have regrown to only 20% of their original size.

As in most national forests, it is expensive to log in the Tongass. To cover the high costs of operating, timber interests such as pulp mills have always relied on obtaining the timber from the federal government at below-market prices. This right was granted in 1954 by a 50-year contract that expired in the 1990s. In 1990 some members of Congress tried to pass the Tongass Timber Reform Act to force timber interests to pay market prices, but the legislation was bitterly opposed by other members of Congress. The compromise agreement, reached in 1997, provided timber to the mills at market prices. As a result of this legislation, clearcut logging continued in the Tongass, but at lower rates than in the past.

In 1999, the 1997 Tongass Land Management Plan was modified after several dozen appeals were filed against the plan. The 1999 decision, also known as the Modified 1997 Forest Plan, protects an additional 100,000 acres of old growth forest from development. This brings the total protected area in the Tongass to 234,000 acres. (The total area of the Tongass National Forest is 17 million acres.) The modified plan increases timber harvest rotations from 100 years to 200 years in specially designated wildlife areas. This change reduces the impact of forest fragmentation and protects the Sitka black-tailed deer population, which is used for food by native tribes. The 1999 decision specifies that road density will be reduced to protect the habitat of wolves, bears, and other wildlife. It also reduces the maximum quantity of timber harvested on a sustainable basis from

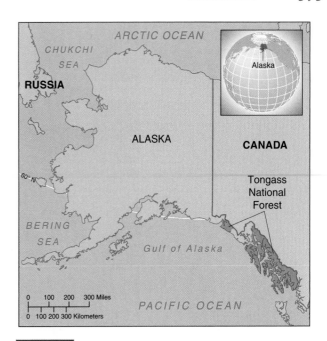

Figure 17.12 Alaska's Tongass National Forest. This temperate rain forest (*light green area*) is in southeastern Alaska along the Pacific Ocean.

267 million board feet (the 1997 value) to 187 million board feet. This quantity is considered sufficient to meet the requirements of timber operators in the region.

The USFS officially adopted the Roadless Area Conservation rule in 2000. A series of lawsuits challenging the rule ensued, initiated mainly by logging, mining, and gas and oil interests. In May 2001, a U.S. district judge blocked implementation of the roadless rule, and plans to log several areas of Tongass National Forest that lack roads proceeded. When President Bush's administration did not intervene to support the federal rule, several environmental groups filed suit on behalf of the federal government. In 2002 the Bush administration filed a brief defending the Roadless Area Conservation rule, in response to additional lawsuits filed by industries wanting to extract resources from other federal lands that are roadless. The take-home message from all of this legal wrangling is that the USFS, like many other government agencies, is a political organization that takes its lead from current presidential policies. Changes in administrations often leave the USFS and other government agencies floundering as they strive to implement rulings from a previous administration that are no longer supported by the current administration. ■

RANGELANDS

Rangelands are grasslands, in both temperate and tropical climates, that serve as important areas of food production for humans by providing fodder for livestock such as

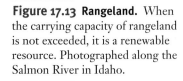

Figure 17.13 Rangeland. When the carrying capacity of rangeland is not exceeded, it is a renewable resource. Photographed along the Salmon River in Idaho.

cattle, sheep, and goats (Figure 17.13). Rangelands may also be mined for minerals and energy resources, used for recreation, and preserved for biological habitat and for soil and water resources. The predominant vegetation of rangelands includes grasses, forbs (small herbaceous plants other than grasses), and shrubs.

Rangeland Degradation and Desertification

Grasses, the predominant vegetation of rangelands, have a **fibrous root system**, in which many roots form a diffuse network in the soil to anchor the plant. Plants with fibrous roots hold the soil in place quite well, thereby reducing soil erosion. Grazing animals eat the leafy shoots of the grass, and the fibrous roots continue to develop, allowing the plants to recover and regrow to their original size.

Carefully managed grazing is beneficial for grasslands. Because the vegetation is adapted to grazing, the removal of mature vegetation by grazing animals stimulates rapid regrowth. At the same time, the hooves of grazing animals disturb the soil surface enough to allow rainfall to more effectively reach the root systems of grazing plants. Several studies around the world have reported that moderate levels of grazing encourage greater plant diversity.

The **carrying capacity** of a rangeland is the maximum number of animals the rangeland plants can sustain over an indefinite period of time without deterioration of the rangeland. When the carrying capacity of a rangeland is exceeded, grasses and other plants are **overgrazed**; that is, grazing animals continue to consume the plants in a particular area so that they cannot recover, and they may die. When plants die, the ground is left barren, and the exposed soil is susceptible to erosion. Sometimes

other plants that can tolerate the depleted soil invade an overgrazed area. In parts of Texas Hill Country that were overgrazed, the lush grasses were replaced by junipers, which are not good forage food.

Most of the world's rangelands lie in semiarid areas that have natural extended periods of drought. Under normal conditions, native grasses in these drylands can survive a severe drought: The aerial portion of the plant dies back, but the extensive root system remains alive and holds the soil in place. When the rains return, the roots develop new shoots.

When overgrazing occurs in combination with an extended period of drought, however, once-fertile rangeland can be converted to desert. The reduced grass cover caused by overgrazing allows winds to erode the soil. Even when the rains return, degradation is so extensive that the land may not recover. Land **degradation** is the natural or human-induced process that decreases the future ability of the land to support crops or livestock. Water erosion removes the little bit of remaining topsoil and the sand that is left behind forms dunes. This progressive degradation, which induces unproductive desert-like conditions on formerly productive rangeland (or tropical dry forest), is called **desertification** (Figure 17.14). It reduces the agricultural productivity of economically valuable land, forces many organisms out, and threatens endangered species.

There is still much that scientists do not understand about desertification and the processes related to it. We do not know if the declines in productivity caused by land degradation and desertification are permanent. We also do not know the extent to which desertification results from natural fluctuations in climate versus population pressures and human activities. Currently, many ecologists who study desertification do not think that human

Figure 17.14 Overgrazing. Overgrazing can contribute to the degradation of rangeland into nonproductive desert, as occurred in this arid region in western New Mexico and eastern Arizona.

activities are the main cause of desert encroachment. Several years of satellite data support the hypothesis that natural climate variation is the primary factor responsible for the shifting boundary between the southern edge of the Sahara Desert and the African Sahel (see Figure 6.21). Ecologists acknowledge that human activities and overpopulation contribute to degradation in the Sahel but say that fluctuations in climate are causing the desert to advance southward into the Sahel (during drier years) and then retreat (during wetter years). Ecologists therefore prefer to use the term *degradation* rather than desertification when referring to the effects of human activities in arid grasslands, whereas international policy makers prefer the term *desertification.*

About 30% of the human population lives in the Sahel or other rangelands that border deserts. One of the consequences of desertification/degradation is a decline in agricultural productivity, and studies have shown that when an area cannot feed its people, they emigrate. For example, there are more Senegalese in certain regions of France than in their native villages in Africa. According to the U.N. Environment Program, about 135 million people worldwide are in danger of displacement as a result of desertification. The World Bank, which estimates that desertification costs $42.3 billion per year in economic losses, is beginning to emphasize programs with sustainable agricultural systems for dryland areas.

Rangelands in the United States

Rangelands make up approximately 30% of the total land area in the United States and occur mostly in the western states. Of this, approximately one third is publicly owned and two thirds is privately owned. Excluding Alaska, there are at least 89 million hectares (220 million acres)

of public rangelands. The BLM, which is guided by the **Taylor Grazing Act** of 1934, the **Federal Land Policy and Management Act** of 1976, and the **Public Rangelands Improvement Act** of 1978, manages most of the approximately 69 million hectares (170 million acres) of public rangelands. The USFS manages an additional 20 million hectares (50 million acres).

Overall, the condition of public rangelands in the United States has slowly improved since the low point of the Dust Bowl in the 1930s (see Chapter 14). Much of this improvement can be attributed to fewer livestock being permitted to graze the rangelands after the passage of the Taylor Grazing Act in 1934. Better livestock management practices, such as controlling the distribution of animals on the range through fencing or herding, as well as scientific monitoring, have contributed to rangeland recovery. But restoration has been slow and costly, and more is needed. Rangeland management includes seeding in places where plant cover is sparse or absent, conducting controlled burns to suppress shrubby plants, constructing fences to allow rotational grazing, controlling exotic weeds, and protecting habitats of endangered species. Most livestock operators today use public rangelands in a way that results in their overall improvement.

Issues Involving U.S. Rangelands

The federal government distributes permits to private livestock operators that allow them to use public rangelands for grazing in exchange for a fee. (According to the BLM, in 2002 the monthly grazing fee was $1.43 per cow. The comparable cost on private land is more than $11.) The permits are held for many years and are not open to free-market bidding by the general public—that is, only ranchers who live in the local area are allowed to obtain grazing permits.

Some environmental groups are concerned about the ecological damage caused by overgrazing of public rangelands and want to reduce the number of livestock animals allowed to graze. They want public rangelands to be managed primarily for other uses, such as biological habitat, recreation, and scenic value, rather than for livestock grazing. To accomplish this goal, they would like to be able to purchase grazing permits and then set aside the land. (See "Meeting the Challenge: The Trout Creek Mountain Working Group" for an example of an unusual coalition among ranchers, environmentalists, and government officials.)

Conservative economists have joined environmentalists in criticizing the management of federal rangelands. According to the BLM, in 2000, taxpayers contributed at least $53 million more than the grazing fees collected on public rangelands. This money is used to maintain the rangelands, including installing water tanks and fences, and to fix the damage done by overgrazing. The National Taxpayers Union and other freemarket groups want grazing fees to cover all costs of

MEETING THE CHALLENGE

The Trout Creek Mountain Working Group

Confrontations among Western ranchers trying to preserve their way of life, environmentalists trying to protect the environment, and officials trying to administer state and federal laws are common news items. In southeastern Oregon, however, an unlikely partnership of 25 to 35 ranchers, environmentalists, and government officials makes up the Trout Creek Mountain Working Group. The coalition was organized in 1988 to improve the habitat of a threatened species, the Lahontan cutthroat trout, in an area of federally owned rangeland in southeastern Oregon. The Trout Creek Mountain Working Group's goal is to develop long-term strategies that will provide for both the ecological health of the land and the economic well-being of people living in the area. An important concept of the working group is that ranchers and other people are a part of the natural ecosystem.

The group adopted a broad approach, known as **ecosystem management**, to execute their task. Ecosystem management is a long-term, flexible collaboration among scientists, private landowners, various levels of government, and business leaders to sustain or restore natural systems and their functions. It is based on the growing understanding that organisms, including the Lahontan cutthroat trout, cattle, and humans, are interrelated and interdependent with one another and with their abiotic environment (soil, water, and air). Ecosystem management takes into account that ecosystems do not recognize political boundaries and may span federal, state, county, and private lands. It also recognizes that

the economic well-being of human communities is inextricably linked to the sustainable use of natural resources and environmental processes. The ecosystem management process takes a broad view because it is both geographically comprehensive and interdisciplinary (it takes social, economic, scientific, and political factors into account).

After a species has been declared threatened or endangered, the Endangered Species Act requires that federal officials design a plan to aid the species' recovery. For the Lahontan cutthroat trout, the focus was ecosystem management of **riparian areas**, the thin patches of vegetation along streams that interface between terrestrial and aquatic habitats. This requirement meant closing the streams to the anglers and limiting cattle from foraging and drinking along stream banks. Cattle trample and eat the smaller vegetation along a bank and may increase the rate of soil erosion if not managed properly. Excess cattle feces pollute the water where the fish live. Cattle also like to eat aspen and willow shoots that grow along the edges of stream banks, and when these trees are grazed, the stream is shaded less, which in turn increases the water temperature. Because trout are adapted to cold temperatures, warmer temperatures decrease the trout's reproductive success.

Although members of the Trout Creek Mountain Working Group were initially suspicious of one another, they continued to work together and gradually developed common goals and devised compatible solutions. The ranchers agreed

---

maintaining herds on publicly owned rangelands. These groups estimate that the federal government could save an estimated $250 million over 5 years if this were done.

Wild Horses and Burros Another rangeland issue involves the estimated 44,000 wild horses and burros that roam on public rangelands in 10 western states. These animals are not native to North America but were introduced from Europe by Spaniards. Because they have come to symbolize America's pioneer history, the wild horses and burros are protected. At the same time, they are very destructive to the natural ecosystem and must be managed so that they do not contribute to its deterioration. The goal is to have no more than 26,000 horses and burros roaming U.S. rangelands. The annual rate of increase of these animals is high (in good years, herds increase in size by as much as 40%) because they have no natural predators.

To prevent horse and burro populations from exceeding the carrying capacity of their rangeland home, the BLM started the Adopt-a-Horse program, in which between 5,000 and 9,000 wild horses and burros are annually removed from the range and given away.

However, the Adopt-a-Horse program is expensive to operate, and there are more wild horses and burros available for adoption than there are people willing to adopt them. More than 6,000 unadopted animals currently live in government-maintained corrals. In 1997 the media discovered that thousands of the horses that had been "adopted" were sold to slaughterhouses, prompting a public outcry from animal lovers. Animal scientists have developed a vaccine that provides 1 year of contraception for female horses, deer, and burros, and they are currently trying to develop a 3-year vaccine. Vaccine trials on western horses have been conducted for several years, and it may be that a vaccine will be used widely once it is approved by the U.S. Food and Drug Administration.

WETLANDS

Wetlands are lands that are transitional between aquatic and terrestrial ecosystems (see Figure 7.16). As discussed in Chapter 7, wetlands are usually covered by shallow water for at least part of the year and have characteristic

to withdraw about 3,000 cattle from environmentally sensitive parts of the stream for 3 years, from 1989 to 1992, a concession that was hard for them because it reduced their profits. As part of the compromise, environmentalists and government officials agreed to open parts of the stream to some cattle, beginning in 1992.

Today, the condition of the riparian areas, which are shared by both cattle and trout, has improved. Willow, aspen, alder, wild rose, and other plants are becoming reestablished along the stream banks (see figure). These plants are reducing sediments and modifying the water temperature of the stream, thereby providing the habitat necessary for trout. As a result of the improved ecosystem health of riparian areas, the population of Lahontan cutthroat trout is increasing.

The work of the Trout Creek Mountain Working Group is far from over, and additional difficult decisions will have to be made. The group continues to monitor the stream ecosystem and make adjustments as needed, taking into account both environmental and economic issues. For example, the group recently considered the development of a prescribed burn program for mountain pastures; the restoration of angling (catch and release) for Lahontan cutthroat trout (based on their recent population increase); and the closure of a 3-mile segment of road near Willow Creek so that the stream could reclaim its natural flood plain and freely meander.

Recovered riparian area. This stream, managed by the Trout Creek Mountain Working Group, is ideal trout habitat.

soils and water-tolerant vegetation. People used to think that the only benefit of wetlands was to provide habitat for migratory waterfowl and other wildlife. In recent years, however, the many ecosystem services that wetlands perform have come to be appreciated, and estimates of their economic value have increased accordingly. Wetlands help recharge groundwater and also reduce damage from flooding because they hold excess water when rivers flood their banks. Wetlands improve water quality by trapping and holding nitrates and phosphates from fertilizers, and they help to cleanse water that contains sewage, pesticides, and other pollutants. Wetlands provide habitat for many species listed as endangered or threatened; these include up to one half of all fish species, one third of all bird species, and one sixth of all mammal species that are listed on the U.S. endangered or threatened species lists. Freshwater wetlands produce many commercially important products, including wild rice, blackberries, cranberries, blueberries, and peat moss. They are also sites for fishing, hunting, boating, bird watching, photography, and nature study.

Wetlands are increasingly threatened by drainage for agriculture or mosquito control; dredging for navigation; channelization and construction of dams, dykes, or seawalls for flood control; filling in for solid waste disposal, road building, and residential and industrial development; conversion for aquaculture (discussed in Chapter 18); mining for gravel, phosphate, and fossil fuels; and logging. In the United States, wetlands have been steadily shrinking by an estimated 23,675 hectares (58,500 acres) per year since 1985. This represents a slower rate of loss than in previous years. Between 1975 and 1985, the annual loss of wetlands was about 120,000 hectares (300,000 acres). In the contiguous 48 states, more than one half of the more than 89.4 million hectares (221 million acres) of wetlands that originally existed during colonial times have been lost; only 42 million hectares (104 million acres) remain. Most of the loss since the 1950s has been the result of farmers' converting wetlands to cropland. The greatest percentage declines in wetlands have occurred in the agricultural states of Ohio, Indiana, Iowa, and California. The greatest total acreage declines in wetlands have occurred in Florida, Louisiana, and Texas.

The loss of wetlands is legislatively controlled by a section of the 1972 Clean Water Act. This legislation,

Taking Cattle to the Bank

More than a century ago, western rangelands in New Mexico consisted of occasional trees widely interspersed with open areas of grasses. Overgrazing and fire suppression have changed much of the western landscape from the native grasses that grew so prolifically to forests thick with trees and woody shrubs. The woody plants predominate now because the grasses were so overgrazed by cattle that they would not support the periodic fires that used to burn through the area, killing off the woody vegetation. As a result of the loss of grassy areas, the cattle that graze today are crowded onto unsustainably small areas of rangeland, further damaging the grasses and stream banks.

An innovative partnership is working to restore and preserve native grassland species on federal rangelands and to help ranchers at the same time. The Valle Grande Grass Bank in New Mexico is a partnership between the Conservation Fund, the Northern New Mexico Stockman's Association, the U.S. Forest Service, and National Forest grazing permittees in northern New Mexico. It allows ranchers in the area to graze their cattle on a stretch of land known as a "grass bank." Here's how it works: First, the Conservation Fund bought a 240-acre ranch and acquired a grazing permit for 36,000 acres of good-quality federal grazing land, which the organization named the Valle Grande Grass Bank. The Conservation Fund then made an offer to local ranchers that would benefit both the ranchers and the rangelands they lease from the federal government. During the time their cattle are grazing at the grass bank, the ranchers work with the Forest Service to improve their home pastures. The ranchers remove trees and woody shrubs, let native grasses grow enough to support a fire, conduct prescribed burns, and install fences around vulnerable riparian areas to allow stream banks to recover from the onslaught of cattle hooves. Ranchers typically participate in the grass bank for several years so that the new vegetation on their section of federal rangeland has a chance to become well established.

which has been up for renewal since 1997, does a reasonably good job of protecting coastal wetlands but a poor job of protecting inland wetlands, which is where most wetlands are. The **Emergency Wetlands Resources Act** of 1986 authorized the FWS to inventory and map U.S. wetlands. These maps have many uses, such as planning for drinking water supply protection, siting of development projects, flood plain planning, and development of endangered species recovery plans. Currently, the National Wetlands Inventory of the FWS has mapped 90% of the lower 48 states and 34% of Alaska.

Since the administration of President George H.W. Bush, the United States has attempted to prevent any new net loss of wetlands by conserving existing wetlands and restoring those that have been lost. Development of wetlands is allowed only if a corresponding amount of previously converted wetlands is restored. However, not all wetland restorations have been successful, and there is no routine tracking of compliance. Furthermore, there are many things we do not understand about wetland dynamics (Figure 17.15). The policy of no net loss of

wetlands has been only partially successful, and wetlands loss has continued, although at a slower rate.

The wetlands policy is complicated by two factors: (1) confusion and dissent about the definition of wetlands, which was not spelled out in the original Clean Water Act, and (2) the question of who owns wetlands. In 1989 a team of government scientists developed a comprehensive, scientifically correct definition of wetlands. (The definition is very technical and is beyond the scope of this

Figure 17.15 A reconstructed wetland. The Sweetwater Marsh (upper left) was constructed along San Diego Bay in 1984 by the California Department of Transportation, which was legally required to do so when it destroyed a similar marsh during road construction. One of the main purposes of the reconstructed marsh was to provide habitat for the light-footed clapper rail, an endangered species. The clapper rail never became established in the reconstructed marsh. Ecologists determined that the marsh grass in Sweetwater Marsh was too short for the bird to use for nesting. That species of marsh grass grows significantly taller in natural marshes, where the sediments retain enough nitrogen to fertilize it. The reconstructed marsh's sediments, which were dredged from San Diego Bay's shipping channel and obtained from an old urban dump, are too sandy to retain nitrogen.

text.) It provoked an outcry from farmers and real estate developers, who perceived it as a threat to their property values. Largely in response to their criticisms, politicians attempted to narrow the definition of wetlands several times during the 1990s, removing marginal wetlands that were not as wet as swamps or marshes. This narrower definition, which allowed for economic reality, compromised decades of wetlands research and excluded approximately one half of existing wetlands in the United States from protection. In 1992 Congress asked the National Research Council to help settle the issue because it had become so controversial. The council published its wetlands study in 1995, urging Congress to put the wetlands debate back on a more scientific footing. The council recommended that because shallow wetlands or wetlands that are only intermittently wet perform the same ecosystem services as swamps and marshes, they should be regulated by the same principles. Ironically, the council study generated an even greater debate in Congress. To further complicate the issue, a series of federal court cases involving wetlands protection resulted in conflicting decisions during the late-1990s. It appears that Congress will have to decide what wetlands it wishes to protect when it eventually renews the Clean Water Act.

The federal government owns less than 25% of wetlands in the lower 48 states; the remaining 75% is privately owned. This means private citizens control whether wetlands are protected and preserved or developed and destroyed. Because of the traditional rights of private land ownership in the United States, landowners resent the federal government's telling them what they may or may not do with their properties. Property-rights advocates in Congress side with landowners in thinking the government has overprotected wetlands. It is therefore important that private landowners become informed of the environmental importance of wetlands and the critical need to maintain wetlands where feasible. Although some private owners recognize the value of wetlands and voluntarily protect them, others are constrained by economic realities that dictate what they must do with their land. The federal government is examining proposals such as tax incentives and the outright purchase of wetlands to encourage their conservation.

Congress authorized the establishment of the Wetlands Reserve Program (WRP) under the **Food Security Act** of 1985 and its amendments, the 1990, 1996, and 2002 Farm Bills. The WRP is a voluntary program that seeks to restore and protect 433,000 hectares (1.07 million acres) of privately owned freshwater wetlands that have previously been drained, such as those drained for conversion to cropland. Participants are offered financial incentives to restore the wetlands, and they can establish conservation easements to protect the wetlands. If a property owner establishes a permanent easement, the WRP pays all restoration costs and an amount equal to the agricultural value of the land. As with other conservation easements, the landowner continues to control access to the land. The Natural Resources Conservation Service administers the WRP, which is funded annually by Congress and is therefore subject to budget cuts.

Coastlines

Coastal wetlands, also called saltwater wetlands, provide food and protective habitats for many aquatic animals (see Figure 7.17). They could be considered the ocean's nurseries because so many different marine fishes and shellfish spend the first parts of their lives there. In addition to being highly productive areas, coastal wetlands protect coastlines from erosion and reduce damage from hurricanes.

Historically, tidal marshes and other coastal wetlands have been regarded as wasteland, good only for breeding large populations of mosquitoes. Coastal wetlands throughout the world have been drained, filled in, or dredged out to turn them into "productive" endeavors such as industrial parks, housing developments, and marinas. Agriculture also contributes to the demise of coastal wetlands, which are drained for their rich soil. Other human endeavors that cause the destruction of coastal wetlands include fish farming (see Chapter 18) and timber harvesting.

In the United States, people have belatedly recognized the importance of coastal wetlands and have passed some legislation to slow their destruction. For example, intact coastal wetlands help moderate the effects of tides more inexpensively than engineering structures such as retaining walls. Sandy beaches along intact coastal wetlands are alternatively washed away and replenished by wave action. When retaining seawalls are built parallel to the beach, the beach rapidly erodes away, including beach on properties adjacent to the wall (Figure 17.16). Most legislation protects beaches at the shoreline from seawalls by *rolling easements*, which prioritize public access to the shore over property owners' rights to build walls. Few laws protect shoreline along bays and sounds, however, and walls are rapidly replacing these tidal shorelines in many areas of the United States. Because sea level may rise during the 21st century because of global warming (see Chapter 20), these walls will become even more commonplace unless states enact legislation against their construction.

Coastal Demographics   Many coastal areas are overdeveloped, highly polluted, and overfished. Although more than 50 countries have coastal management strategies, their focus is narrow and usually deals only with the economic development of a thin strip of land that directly borders the ocean. Coastal management plans generally do not integrate the management of both land and offshore waters, nor do they take into account the main reason for coastal degradation—human numbers. Perhaps as many as 3.8 billion people—about two thirds of the world's population—live within 150 km (93 mi) of a

Figure 17.16 Seawalls and beach erosion. To protect their property from rising water levels and storm surges, many property owners along coasts build seawalls, which erode beaches between the wall and the water. Such walls accelerate beach erosion on adjacent properties where no seawall exists. Seawalls also prevent public access to beaches.

coastline. Demographers project that three fourths of all humans—perhaps as many as 6.4 billion—will live in that area by 2025. Many of the world's largest cities are situated in coastal areas, and these cities are currently growing more rapidly than noncoastal cities.

The United States is not immune to environmental destruction along its coasts. A 2002 report from the Pew Oceans Commission estimated that more than 27 million additional people will settle along U.S. coasts over the next 15 years. Already, 14 of the 20 largest U.S. cities and 19 of the country's 20 most densely populated counties lie along coasts. These population statistics do not take into account the additional effects of seasonal visitors to coastal resorts. The report says that 14% of U.S. coasts were developed in 1997, and the percentage is expected to rise to 25% by 2025.

If the world's natural coastal areas are not to become urban sprawl or continuous strips of tourist resorts during the 21st century, coastal management strategies must be developed that take into account projections of human population growth and distribution. Such comprehensive management plans will be difficult to formulate and execute because they must regulate coastal development and prevent resource degradation, both on land and in offshore waters. The key to successful planning is local community involvement. If the public understands the importance of natural coastal areas, people may become committed to the sustainable development of coastlines.

National Marine Sanctuaries The United States has 12 national marine sanctuaries to minimize human impacts and protect unique natural resources and historical sites along the Atlantic, Pacific, and Gulf of Mexico coasts. Home to many diverse organisms, these sanctuaries include kelp forests off the coast of California, coral reefs in the Florida Keys, fishing grounds along the continental shelf, and deep submarine canyons, as well as shipwrecks and other sites of historical value.

The National Marine Sanctuary Program, which is part of the National Oceanic and Atmospheric Administration, administers the sanctuaries. Like many federal lands, they are managed for multiple purposes, including conservation, recreation, education, mining of some resources, scientific research, and ship salvaging. Commercial fishing is permitted in most of them, although several "no-take" zones exist where all fishing and collecting of biological resources is banned. Several recent studies have shown that these no-take zones for fish promote population increases of individual species as well as species richness (the no-take reserves typically contain 20% to 30% more species than nearby areas that are fished). The reserves also have a spillover effect—that is, they boost populations of fish surrounding their borders. Fishing groups and some politicians, however, are opposed to no-take reserves because they threaten public access to the seas.

AGRICULTURAL LANDS

The United States has more than 121 million hectares (300 million acres) of **prime farmland**, land that has the soil type, growing conditions, and available water to produce food, forage, fiber, and oilseed crops. Certain areas of the country have large amounts of prime farmland. For example, 90% of the Corn Belt, a corn-growing region in the Midwest encompassing parts of six states, is considered prime farmland. Not all prime farmland is used to grow crops; approximately one third contains roads, pastures, rangelands, forests, feedlots, and farm buildings.

Traditionally, farming was a family business. However, larger agribusiness conglomerates that can operate more efficiently are rapidly replacing the family farm. The U.S. Department of Agriculture reported in the 2000 Census of Agriculture that there were 2.2 million farms in the United States, as compared to 6.4 million in 1920. In the same time period, the average farm size increased from 60 hectares (148 acres) to 176 hectares (434 acres).

There is considerable concern that much of our prime agricultural land is falling victim to urbanization and suburban sprawl by being converted to parking lots, housing developments, and shopping malls. Although most agricultural land is not in danger of urban intrusion, much prime farmland is. Unquestionably, both natural ecosystems and agricultural lands adjacent to urban areas are being developed (Figure 17.17). In certain areas of the United States and Canada, loss of rural land is a significant problem (Table 17.3). According to the American Farmland Trust, more than 162,000

Figure 17.17 Suburban spread onto agricultural land. Homes and businesses occupy land that was once cornfields in York County, Pennsylvania. Loss of farmland to urban and suburban development is a problem in certain areas of the United States.

hectares (400,000 acres) of prime U.S. farmland are lost each year.

The 1996 Farm Bill included funding for the establishment of a national Farmland Protection Program. (Twenty-five states and several local jurisdictions also have farmland protection programs.) This voluntary program helps farmers keep their land in agriculture. The farmers sell conservation easements that prevent them from converting their land to nonagricultural uses. The easements are in effect from a minimum of 30 years to forever. As with other conservation easements, the farmers retain full rights to use their property for agricultural purposes. Funding of the Farmland Protection Program is subject to annual appropriations from Congress.

Table 17.3   The Top 12 U.S. Farm Areas Threatened by Population Growth and Urban/Suburban Spread

Farm Areas (in order of priority)

California's Central Valley
South Florida
California's coastal region
Mid-Atlantic Chesapeake (Maryland to New Jersey)
North Carolina Piedmont
Puget Sound Basin (Washington)
Chicago–Milwaukee–Madison metro (Illinois/Wisconsin)
Willamette Valley (western Oregon)
Twin Cities metro (Minnesota)
Western Michigan (Lake shore)
Shenandoah and Cumberland Valleys (Virginia to
    Pennsylvania)
Hudson River and Champlain Valleys (New York to Vermont)

SUBURBAN SPRAWL AND URBANIZATION

In Chapter 9 we discussed **urbanization**, the concentration of humans in cities, relative to human population growth. Urbanization and its accompanying suburban sprawl obviously affect land use (Figure 17.18). Prior to World War II, jobs and homes were concentrated in cities, but during the 1940s and 1950s, jobs and homes began to move from urban centers to the suburbs. New housing and industrial and office parks were built on rural land surrounding the city, along with a suburban infrastructure that included new roads, schools, and the like. Further development extended the edges of the suburbs, cutting deeper and deeper into the surrounding rural land and causing environmental problems such as loss of wetlands, loss of biological habitat, air pollution, and water pollution. Meanwhile, people who remained in the city and older suburbs found themselves the victims of declining property values and increasing isolation from suburban jobs. This pattern of land use, which continues to the present, has increased the economic disparity between older neighborhoods and newer suburbs.

There is an urgent need for regional planning involving government and business leaders, environmentalists, inner-city advocates, suburbanites, and farmers to determine where new development should take place and where it should not. Also, most metropolitan areas need to make more efficient use of land that has already been developed, including central urban areas.

The Greater Atlanta area is an excellent example of the rapid spread of urban sprawl. During the 1990s tract houses, strip malls, business parks, and access roads

Figure 17.18 **Expansion of the urban fringe.** This development, near Las Vegas, Nevada, was built in the mid-1990s.

replaced an average of 500 acres of surrounding farms and forests *each week*. The Greater Atlanta area now extends more than 175 km (110 mi) across, almost twice the distance it was in 1990. Atlanta now has 1,750 km² (700 mi²) of sprawl.

U.S. voters have grown increasingly concerned about the unrestricted growth. In the 1998 elections, 240 state and municipal areas had initiatives to prevent sprawl and preserve undeveloped land; 170 of these passed. At least 11 states now have comprehensive, statewide growth-management laws. Maryland, for example, has a "smart growth" plan that protects open space in highly developed areas while promoting growth in areas that could be helped by growth.

LAND USE

Many environmental concerns converge in the matter of land use. Pollution, population issues, preservation of our biological resources, mineral and energy requirements, and production of food are all tied to land use. Overriding all of these concerns are economic factors, such as the way that privately owned land is taxed. Sometimes forest or agricultural land located near urban and suburban areas is taxed as potential urban land. Because of the higher taxes on this land, its owners fall under greater pressure to sell it, which ultimately hastens its development. However, if such land is taxed as forest or farmland, the lower taxes are an incentive for owners to hold onto the land and maintain it in its undeveloped condition. Thus, land use is largely controlled by economic factors.

Public Planning of Land Use

Examine the use of land where you live. You may be surrounded by high rises and factories or by tree-lined streets interspersed with open parkland. Regardless of your surroundings, it is likely that they got that way by accident. Most areas have a land-use plan that includes zoning, but rarely do land-use plans take into account all aspects of land as a resource both before and after development. The philosophy of most land-use plans is that development is good because it increases the tax base, even though the revenue from these taxes is usually consumed providing services to the developed area.

Land-use decisions are complex because they have multiple effects. If a tract of land is to be developed for housing, then roads, sewage lines, and schools must be built nearby to accommodate the influx of people. This usually results in the opening of restaurants and shopping areas, which take up more land.

Public planning of land use must take into account all repercussions of the proposed land use, not just its immediate effects. It is helpful to begin with an inventory of the land, including its soil type, topography, types of organisms, endangered or threatened species, and historical or archaeological sites.

At this stage, the public planning commission attempts to understand the value of the land *as it currently exists*, as well as its potential value after any proposed change. In addition to providing people with open space for recreation and mental health, undeveloped land provides ecosystem services that must be recognized. All of these benefits should be compared with the possible economic benefits of development. In the long term, the best use of land may not be the use that provides immediate economic gain.

If the land will ultimately be developed, the development plan should be comprehensive. It should indicate which areas will remain open space, which will remain agricultural, and which will be zoned for high-, medium-, and low-density housing.

Management of Federal Lands

How do we best manage the legacy of federal lands? Should federal lands be managed under multiple uses, or should they be preserved so they benefit U.S. citizens for generations to come? These questions have divided many Americans into two groups: those who wish to exploit resources on federal lands now (a coalition of several hundred grassroots organizations known collectively as the wise-use movement) and those who wish to preserve the resources on federally owned lands (a coalition of several hundred grassroots organizations known collectively as the environmental movement). Keep in mind as you read the following paragraphs that these descriptions of wise-use and environmental movements are mainstream; certain groups associated with one or the other of these movements may not support all of the goals that are listed.

People who support the **wise-use movement** think that the government has too many regulations protecting the environment and that property owners should have more flexibility to use natural resources. They believe that a primary purpose of federal lands is to enhance economic growth. Some of their goals include the following:

1. Put all national forests under timber management, including old-growth forests.

2. Permit mining and commercial development of wilderness areas, wildlife refuges, and national parks, where appropriate.

3. Allow unrestricted development of wetlands.

4. Change the Endangered Species Act so that economic factors are considered along with scientific ones (recall from Chapter 16 that the definition of threatened and endangered species is currently based on scientific information only).

5. Sell parts of resource-rich federal lands to private interests, such as mining, oil, coal, ranching, and timber groups, to be managed for sustainable resource extraction.

Many of the organizations that embrace the wise-use movement have environmentally friendly names. The National Wetlands Coalition, for example, consists primarily of real estate developers and energy companies who wish to drain and develop wetlands. Similarly logging companies support the American Forest Resource Alliance. (Appendix V lists many organizations that are environmentally friendly.)

In contrast to the wise-use movement, the **environmental movement** views federal lands as a legacy of U.S. citizens. They think that

1. The primary purpose of public lands is to protect biological diversity and ecosystem integrity.

ENVIROBRIEF

Preserving the DOE's Natural Labs

The U.S. Department of Energy's weapons labs are surrounded by vast restricted areas designated as National Environmental Research Parks. For five decades these protected habitats, totaling 2 million acres, have served as natural ecology labs, allowing scientists unparalleled access to uninhabited areas for long-term studies of preserved landscapes and sheltered wildlife. Now, faced with budget cuts, the Department of Energy (DOE) has begun selling portions of these buffer zones, considering them inessential or even inappropriate to their peacetime efforts. The labs represent several ecosystems that are rare elsewhere, such as tallgrass prairie in northern Illinois, shrub-steppe in Washington state, and southern pine forests dominated by mature longleaf pine and wiregrass in South Carolina. Plant and animal species are abundant; Tennessee's Oak Ridge Research Park, for example, supports more than 1,100 species of vascular plants, including 26 rare species, and 315 wildlife species, including 20 rare or endangered animals.

The DOE has been disposing of research park acreage for several years, by selling it to local governments for development opportunities, or by transferring land to other federal agencies. A 1997 inspector general report recommended divesting about one fourth of the parklands. Ecologists and conservation groups oppose the divestment, pleading the importance of these unique areas to conducting applied research and to protecting plant and animal habitats. One proposed solution is for Congress to designate the areas as national monuments, preserving the parks for their scientific and educational resources.

2. Those who extract resources from public lands should pay U.S. citizens compensation equal to the fair market value of the resource and not be subsidized by taxpayers.

3. Those who use public lands should be held accountable for any environmental damage they cause.

CONSERVATION OF OUR LAND RESOURCES

Globally, humans use an estimated 38% of the world's total land area for agriculture—that is, for raising crops and livestock (Figure 17.19). Another 33% of the land surface consists of urban areas and of rock, ice, tundra, and desert—areas considered unsuitable for long-term human use. This leaves 29% of the land surface as natural ecosystems, such as forests, that could potentially be developed for human purposes. We have seen, however, that these natural ecosystems provide many valuable ecosystem services that are important to human survival.

Our ancestors considered natural areas as an unlimited resource to exploit. They appreciated prairies as valuable agricultural lands and forests as immediate sources of lumber and eventual farmlands. This outlook

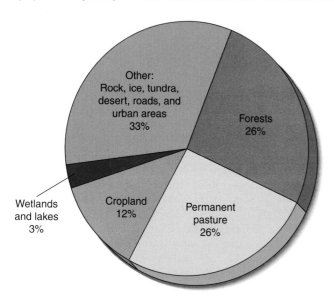

Figure 17.19 World land use. Note that 38% of the world's total land area is used for agriculture (cropland and pastureland).

Table 17.4	The Top 22 Most Endangered Ecosystems in the United States

Ecosystems (in order of priority)

South Florida landscape
Southern Appalachian spruce-fir forests
Longleaf pine forests and savannas
Eastern grasslands, savannas, and barrens
Northwestern grasslands and savannas
California native grasslands
Coastal communities in lower 48 states and Hawaii
Southwestern riparian communities
Southern California coastal sage scrub
Hawaiian dry forest
Large streams and rivers in lower 48 states and Hawaii
Cave and karst systems
Tallgrass prairie
California riparian communities and wetlands
Florida scrub
Shrublands and grasslands of the Intermountain West
Ancient eastern deciduous forest
Ancient forests of the Pacific Northwest
Ancient red and white pine forests of the Great Lake states
Ancient ponderosa pine forests
Midwestern wetlands
Southern forested wetlands

was practical as long as there was more land than people needed. But as the population increased and the amount of available land decreased, it was necessary to consider land as a limited resource. Increasingly, the emphasis has shifted from exploitation to preservation of the remaining natural areas.

Although all types of ecosystems must be conserved, several are in particular need of protection. The National Biological Service (now the Biological Resources Discipline) and the Defenders of Wildlife (see Appendix V) commissioned studies that developed a numerical ranking of the most endangered ecosystems in the United States. They used four criteria:

1. The area lost or degraded since Europeans colonized North America;

2. The number of present examples of a particular ecosystem, or the total area;

3. An estimate of the likelihood that a given ecosystem will lose a significant area or be degraded during the next 10 years;

4. The number of threatened and endangered species living in that ecosystem.

Table 17.4 lists the 22 most endangered U.S. ecosystems based on these criteria. As these ecosystems are lost and degraded, the organisms that compose them decline in number and in genetic diversity. Conservation strategies that set aside ecosystems are the best way to preserve an area's biodiversity.

As you have seen in this chapter, government agencies, private conservation groups, and private citizens have begun to set aside natural areas for permanent preservation. Such activities ensure that our children and grandchildren will inherit a world with wild places and other natural ecosystems.

SUMMARY WITH SELECTED KEY TERMS

I. Nonurban, or **rural lands** include wilderness, forests, grasslands, and wetlands. Natural areas provide many **ecosystem services**, including watershed management, soil erosion protection, climate regulation, and wildlife habitat.

II. Private citizens, corporations, and nonprofit organizations own about 55% of land in the United States. The federal government owns about 35%, state and local governments about 7%, and Native American tribes about 3%.

III. Areas that have not been greatly disturbed by human activities are called **wilderness**.

A. Public lands designated as part of the National Wilderness Preservation System are protected from development.

B. The National Park Service, Forest Service, Fish and Wildlife Service, and Bureau of Land Management administer the 630 wilderness areas.

C. Problems with federally designated wilderness areas include overuse of some areas and the introduction of exotic species.

IV. The National Park System administers 385 different sites, including 57 national parks.

A. National parks have multiple roles, including recreation and ecosystem preservation.

B. The National Park Service administers the national parks.

C. Problems at national parks include overuse, high operating costs, imbalances in wildlife populations, and development of lands adjoining park boundaries, which create "island" populations.

D. Natural regulation, letting nature take its course most of the time, has been the management policy of Yellowstone National Park and many other national parks since 1968.

V. The National Wildlife Refuge System contains habitat for the conservation of fishes, wildlife, and plants of the United States.

A. There are more than 535 national wildlife refuges.

B. The Fish and Wildlife Service administers the national wildlife refuges.

VI. Forests provide many ecosystem services, as well as commercially important timber.

A. Forests that are replanted for commercial harvest have little species richness. Tree plantations are essentially **monocultures**.

B. Ecologically sustainable forest management, or **sustainable forestry**, is a nontraditional method of forest management that seeks not only to conserve forests for timber harvest but also to sustain biological diversity, prevent soil erosion, and preserve watersheds.

   1. When logging occurs, unlogged areas are set aside as sanctuaries for organisms.

   2. Wildlife corridors are protected zones that connect isolated unlogged areas.

C. Selective cutting, shelterwood cutting, seed tree cutting, and clearcutting are tree harvesting methods.

   1. In **selective cutting**, mature trees are cut individually or in small clusters, while the rest of the forest remains intact.

   2. In **shelterwood cutting**, mature trees are removed from a forest in several harvests over a period of time. This method allows mature trees to shelter young trees.

   3. In **seed tree cutting**, almost all trees are harvested from an area; the few remaining trees provide seeds for the regeneration of the forest.

   4. In **clearcutting**, all trees are removed from an area.

D. The greatest problem facing world forests today is **deforestation**, the temporary or permanent removal of forests for agriculture or other uses.

   1. Deforestation causes decreased soil fertility and increased soil erosion; impairs watershed functioning; contributes to the extinction of species; and may induce regional and global climate changes.

   2. Tropical forests (**tropical rain forests** and **tropical dry forests**) are destroyed to provide temporary agricultural

land; obtain timber, particularly for highly developed nations; provide open rangeland for cattle; and supply firewood and charcoal for fuel.

   3. Extensive deforestation of **boreal forests** began in the late 1980s and early 1990s. Boreal forests are currently the world's primary source of industrial wood and wood fiber.

E. Slightly more than one half (57%) of U.S. forests are privately owned; the federal government owns 20%.

   1. Privately owned forests can be protected through the Forest Legacy Program, in which the landowner grants a **conservation easement** to the U.S. government.

   2. The United States has 155 national forests.

   3. The Forest Service manages most national forests, with the remainder overseen by the Bureau of Land Management.

   4. National forests have multiple uses, including timber harvest, livestock forage, water resources and watershed protection, mining, recreation, and habitat for fishes and wildlife.

   5. Issues in national forests include confrontations over multiple uses, building of logging roads with general tax revenues, and clearcutting, particularly of old-growth forest.

VII. Rangelands are grasslands in which grasses, forbs, and shrubs predominate.

A. Cattle, sheep, goats, and other domestic mammals often graze rangelands. Provided the number of grazing animals is balanced with the rangeland's **carrying capacity**, the rangeland remains a renewable resource.

B. Overgrazing can result in barren, exposed soil that is susceptible to erosion.

   1. Such land **degradation** decreases the future ability of the land to support crops or livestock.

   2. Progressive land degradation contributes to **desertification**, the development of unproductive desertlike conditions on formerly productive land.

C. About two thirds of U.S. rangelands are privately owned; the federal government owns much of the remainder.

   1. The Bureau of Land Management manages most of these lands, with the remainder overseen by the Forest Service.

   2. The federal government issues permits that allow private livestock operators to use public rangelands for grazing. Environmental groups want public rangelands to be managed for other uses, such as wildlife habitat, recreation, and scenic value.

VIII. Wetlands are transitional areas between aquatic and terrestrial ecosystems and may be either freshwater or coastal (saltwater).

A. Wetlands provide habitat for many organisms, purify natural bodies of water, and recharge groundwater.

B. Despite their many ecosystem services, wetlands are often drained or dredged for other purposes. In the United States, wetlands are primarily converted to agricultural land.

C. About 75% of wetlands are privately owned.

D. The U.S. policy of preventing any net loss of wetlands is complicated by the definition of wetlands and by property-rights advocates.

E. The Wetlands Reserve Program provides incentives for private landowners to restore and protect wetlands by establishing conservation easements.

IX. Coastlines include tidal marshes and other coastal wetlands.

A. Ecosystem services provided by coastal wetlands include food and habitat for many wildlife species and protection from coastal erosion and storm damage.

B. Failing to recognize the importance of coastal wetlands has led to their destruction and development, including draining and/or dredging.

C. Many coastal areas are over-developed, highly polluted, and overfished. About two thirds of the world's population lives within 150 km of a coastline.

D. The United States has 12 national marine sanctuaries to protect unique natural resources and historical sites along the Atlantic, Pacific, and Gulf of Mexico coasts.

    1. The National Marine Sanctuary Program, under the National Oceanic and Atmospheric Administration, administers the sanctuaries.

    2. The sanctuaries are managed for multiple purposes.

X. Agricultural lands are former forests or grasslands that have been plowed for cultivation.

A. The United States has more than 121 million hectares of **prime farmland**, land with the proper soil type, growing conditions, and available water to produce food, forage, fiber, and oilseed crops.

B. Increasingly, farming involves large tracts of land by agribusiness conglomerates, which may operate more cost-effectively than family farmers on small pieces of land do.

C. In certain areas, agricultural lands are threatened by expanding urban and suburban areas.

XI. The enlightened view of the interaction between humans and land is that humans are stewards of the land, managing it to achieve sustainable use.

A. **Urbanization** (the concentration of humans in cities) and its accompanying suburban sprawl affect land use.

B. It is possible to become involved in local land-use planning by becoming familiar with land-use policies and laws.

C. U.S. citizens are divided over whether federal lands should be exploited for multiple uses now or conserved for future generations.

    1. The **wise-use movement** thinks that the government has too many regulations that protect the environment and that a primary purpose of federal lands is to enhance economic growth.

    2. The **environmental movement** thinks that federal lands are a legacy for U.S. citizens that should be preserved to protect biological diversity and ecosystem integrity.

THINKING ABOUT THE ENVIRONMENT

1. Give at least five ecosystem services provided by nonurban lands. Why is it difficult to assign economic values to many of these benefits?

2. What percentage of land in the United States is privately owned? What percentage is public land owned by the federal government?

3. What are the six main types of federally owned land in the United States? What uses are permitted on each type of land? What are current issues of concern for each type?

4. Do you think additional federal lands should be added to the wilderness system? Why or why not?

5. Suppose a valley contains a small city surrounded by agricultural land. The valley is encircled by mountain wilderness. Explain why the preservation of the mountain ecosystem would help support both urban and agricultural land in the valley.

6. How would a park manager of a national park that adheres to natural regulation probably respond to a lightning-induced forest fire? To the invasion of a noxious exotic weed species?

7. What is deforestation and why is it a serious global environmental problem?

8. Describe at least four causes of tropical deforestation.

9. What are the environmental effects of clearcutting on steep mountain slopes? On tropical rainforest land?

10. Some analysts think that an effective way to combat the influx of unwanted immigrants to such countries as the United States is to curb deforestation. Explain the connection.

11. Distinguish between rangeland degradation and desertification. Why do ecologists prefer the term *degradation* when referring to the effects of range overuse by humans?

12. How do the activities of the Trout Creek Mountain Working Group qualify as ecosystem management?

13. What is a wetland? Why is the scientific definition of wetlands controversial?

14. List at least five causes of wetland destruction.

15. Describe current wetland protection policies in the United States and give their strengths and weaknesses.

16. What is urban sprawl?

17. Explain how certain tax and zoning laws can increase the conversion of prime farmland to urban and suburban development.

18. How do the wise-use and environmental movements differ regarding their views on the use of public lands?

19. Should private landowners have control over what they wish to do to their land? How would you as a landowner handle land use decisions that may affect the public? Present arguments for both sides of this issue.

20. Explain how economic growth and sustainable use of natural resources are compatible goals.

*21. Not counting agriculture, about 0.8% of Earth's total surface area of 14.7 billion hectares is settled—that is, developed for homes, highways, industrial parks, shopping malls, and the like. Calculate the total land area, in hectares, that is settled. Convert your answer to acres, using Appendix III to help you.

* The solution to this question appears in Appendix VII.

TAKE A STAND

Visit our Web site at **http://www.wiley.com/college/raven** (select Chapter 17 from the Table of Contents) for links to more information about the management of federal lands. Consider the opposing views of the wise-use and environment movements, and debate the issue with your classmates. You will find tools to help you organize your research, analyze the data, think critically about the issues, and construct a well-considered argument. Take a Stand activities can be done individually or as part of a team, as oral presentations, written exercises, or Web-based (e-mail) assignments.

Additional on-line materials relating to this chapter, including Student Quizzes, Activity Links, Useful Web Sites, Flash Cards, and more, can also be found on our Web site.

SUGGESTED READING

Carlton, J. "Canada, Timber Firms Agree on Rainforest Pact." *Wall Street Journal* (April 4, 2001). Some timber companies are curbing logging in the temperate rain forests of British Columbia.

Conniff, R. "Swamps of Jersey: The Meadowlands." *National Geographic*, Vol. 199, No. 2 (February 2001). New Jersey's wetlands are recovering from decades of abuse.

Dean, C. *Against the Tide: The Battle for America's Beaches.* New York: Columbia University Press, 1999. Describes the hard choices that must be made to protect the U.S. shoreline, given scientific uncertainties and political and economic considerations.

Godwin, P. "Without Borders: Uniting Africa's Wildlife Reserves." *National Geographic*, Vol. 200, No. 3 (September 2001). The countries in southern Africa are working to protect their shared environment.

Jensen, M.N. "Can Cows and Conservation Mix?" *BioScience*, Vol. 51, No. 2 (February 2001). Ranchers and environmentalists are beginning to agree about how to manage public rangelands.

Jordan, M. "Brazilian Mahogany: Too Much in Demand." *Wall Street Journal* (November 14, 2001). Illegal mahogany trade threatens the Brazilian rain forest.

Jordan, M. "From the Amazon to Your Armrest." *Wall Street Journal* (May 1, 2001). Some companies are trying to give people living in rain forests some economic alternatives to slash-and-burn agriculture.

Kaiser, J. "Bringing Science to the National Parks." *Science*, Vol. 288 (April 7, 2000). The Natural Resource Challenge is a program that strives to provide the scientific information needed to manage national parks.

Malusa, J. "Forest on the Rocks." *Natural History* (March 1999). Clearcutting threatens a remarkable habitat in Alaska's Tongass National Forest.

Mitchell, J.G. "The American Dream." *National Geographic*, Vol. 200, No. 1 (July 2001). Urban sprawl comes with high taxes, pollution, and traffic problems.

Mitchell, J.G. "The Big Open." *National Geographic*, Vol. 200, No. 2 (August 2001). Highlights public lands managed by the Bureau of Land Management.

Raloff, J. "Underwater Refuge." *Science News*, Vol. 159 (April 28, 2001). There are benefits to expanding coastal no-fishing zones in the national marine sanctuaries.

Walnut Acres Farm in Penn's Creek, Pennsylvania.

Food Resources: A Challenge for Agriculture

Learning Objectives

After you have studied this chapter you should be able to

1. Identify the main components of human nutritional requirements and differentiate among undernourished, malnourished, and overnourished conditions.

2. Define *world grain carryover stocks* and explain how they are a measure of world food security.

3. Describe the beneficial and harmful effects of domestication on crop plants and livestock.

4. Contrast industrialized agriculture with subsistence agriculture and describe three kinds of subsistence agriculture.

5. Relate the benefits and problems associated with the green revolution.

6. Describe current food safety issues.

7. Describe the environmental impacts of industrialized agriculture.

8. Define *sustainable agriculture* and contrast sustainable agriculture with industrialized agriculture.

9. Identify the potential benefits and problems with genetic engineering.

10. Contrast fishing and aquaculture and relate the environmental challenges of each activity.

n 1946 farmers Paul and Betty Keene began producing organic foods at Walnut Acres in central Pennsylvania. The Keenes kept their farming methods natural at a time when other farmers were adopting highly mechanized operations that depended on commercial inorganic fertilizers and pesticides for crops and on hormones and antibiotics for livestock. (Commercial inorganic fertilizers are manufactured mixtures of plant nutrient minerals that increase the productivity of the soil, whereas pesticides control insects, weeds, and disease-causing organisms on crops. Hormones regulate certain livestock animals' bodily functions, such as increasing weight or milk output, and antibiotics control livestock infections in crowded conditions.) At Walnut Acres, the soil is fertilized with animal manure and plant wastes, and beneficial insects are used to control pests. Crop rotations and careful crop selections also help control insect pests and maintain soil fertility. Every fourth or fifth year they plant a cover crop, such as alfalfa or clover, and let the soil rest; the cover crop is tilled into the soil before the next growing season.

When the Keenes first started farming at Walnut Acres, they had 108 acres. Today Walnut Acres, one of the oldest, most respected organic farms in the United States, consists of more than 1,000 acres. The farm includes a cannery, bakery, mill, warehouse, and retail store. Its mail-order organic food company is the largest in the country.

In 2001, organic farming, once on the agricultural fringe, exceeded more than $9 billion annually in the United States and represented a small but growing portion of the U.S. food supply. According to the Organic Farming Research Foundation and the Organic Trade Association, about 5,000 U.S. farmers practice certified organic methods on 607,000 hectares (1.5 million acres). Anyone who sells "organic" food is required by law to be certified.

What are organic foods? According to guidelines established by the **Organic Food Production Act** in 1990, *organic foods* are crops that are grown in soil that has been free of commercial inorganic fertilizers and pesticides for at least 3 years. (Organically grown food may contain traces of pesticides introduced by such sources as irrigation water or rain.) If the land that the crops are grown on has been inspected, private or state agencies can label it *certified organic*, which specifies the highest standards. Cattle and chickens that are labeled *free range* or *naturally raised* are raised in open pastures or fields rather than in cramped pens. These animals are not treated with antibiotics or hormones. As specified in the Organic Food Production Act, the U.S. Department of Agriculture (USDA) developed standards for certification of organically grown foods. These national standards, which went into effect in 2002, replaced standards that varied from state to state. The final standards prohibit the use of genetically engineered crops and livestock, sewage sludge (a type of fertilizer), and ionizing radiation (to kill microorganisms on processed foods).

Because organic farming is more labor intensive and produces food on smaller quantities of scale than conventional agriculture, organic food is 5% to 30% more expensive. Although the appeal of organically grown food is undeniable, it must be remembered that the chemicals used in conventional agriculture allowed U.S. agriculture to become the most efficient and productive in the world. However, many environmental problems are associated with conventional agriculture, and there are concerns about whether conventional agriculture is sustainable.

The production of food in an environmentally sustainable way is one of the principal challenges facing humanity today. In this chapter we examine the magnitude and nature of the world's food problems. Most farmers in developing countries produce barely enough food to feed themselves and their families, with little left over as a reserve. Highly developed countries, for their part, have energy-intensive agricultural methods that produce high yields of food but cause serious environmental problems such as soil erosion and pollution. Overshadowing and exacerbating all food problems is the increase in the human population. The production of adequate food for the world's people will be an impossible goal until population growth is brought under control.

HUMAN NUTRITIONAL REQUIREMENTS

The foods humans eat are composed of several major types of biological molecules that are necessary to maintain health: carbohydrates, proteins, and lipids. **Carbohydrates**, such as sugars and starches, are important primarily because they are metabolized readily by the body in **cellular respiration**. In this process, the energy of these biological molecules is transferred to a molecule called adenosine triphosphate (ATP). The body uses the energy in ATP to contract muscles, produce heat, repair damaged tissues, grow, fight off infections, and reproduce. In other words, carbohydrates supply the body with the energy required to maintain life.

Proteins are large, complex molecules composed of repeating subunits called **amino acids**. Proteins perform several critical roles in the body. When plant and animal proteins in food are digested, or broken down into amino acids, the body then absorbs these amino acids, which may be reassembled in different orders to form human proteins. A substantial part of the human body, from hair and nails to muscles, is made up of protein. Some proteins serve as *enzymes*, catalysts that regulate the thousands of different chemical reactions that take place in living cells. In addition, proteins may be metabolized in cellular respiration to release energy.

There are approximately 20 different amino acids required for human nutrition. The human body manufactures 10 to 11 of these for itself, using starting materials such as carbohydrates. However, human cells lack the ability to synthesize the other amino acids, called **essential amino acids**. (The essential amino acids are isoleucine, leucine, lysine, methionine, phenylalanine, threonine, tryptophan, valine, histidine, and, in children, arginine.) These must be obtained from food.

Lipids are a diverse group of biological molecules that includes fats and oils. Like carbohydrates and proteins, lipids are metabolized by cellular respiration to provide the body with a high level of energy. Pound for pound, lipids deliver more energy when metabolized than either carbohydrates or proteins. Lipids have several other important roles in the body. Some lipids are hormones, and others are essential components of cell membranes.

In addition to carbohydrates, proteins, and lipids, we require minerals, vitamins, and water in our diets. **Minerals** are inorganic elements, such as iron, iodine, and calcium, that are essential for the normal functioning of the human body. Minerals are ingested in the form of salts dissolved in food and water. **Vitamins** are complex biological molecules that are required in very small quantities by living cells. Vitamins help to regulate metabolism and the normal functioning of the human body. Whereas plants synthesize most vitamins, humans and other animals must obtain vitamins from food.

WORLD FOOD PROBLEMS

In 1998 the U.N. Food and Agricultural Organization (FAO) estimated that 828 million people lacked access to the food needed to be healthy and to lead productive lives. Most of these people live in rural areas of the poorest developing countries. Currently, 86 countries are considered *low-income, food-deficient countries*—that is, countries that cannot produce enough food or afford to import enough food to feed the entire population. The two regions of the world with the greatest food insecurity are South Asia, with an estimated 270 million hungry people, and sub-Saharan Africa, with an estimated 175 million. (Sub-Saharan Africa refers to all African countries located south of the Sahara Desert.) Today sub-Saharan Africa produces less food per person than it did in 1950.

The average adult human must consume enough food to get approximately 2,600 kilocalories (kcal) per day. (The average man requires 3,000 kcal per day, whereas the average woman requires 2,200 kcal per day.) If a person consumes less than this over an extended period of time, his or her health and stamina decline, even to the point of death. People who receive fewer calories than needed are **undernourished**. Worldwide, an estimated 182 million children under the age of 5 suffer from undernutrition and are seriously underweight for their age, according to the World Health Organization (WHO). This figure represents one third of all children under 5 in developing countries.

The total number of calories consumed is not the only measure of good nutrition, however. People can receive enough calories in their diets but still be **malnourished** because they are not receiving enough of specific, essential nutrients such as proteins, vitamin A, iodine, or iron. For example, a person whose primary food is rice can obtain enough calories, but a diet of rice lacks sufficient amounts of proteins, lipids, minerals, and vitamins to maintain normal body functions. Adults suffering from malnutrition are more susceptible to disease and have less strength to function productively than those who are well fed. In addition to poor physical development and increased disease susceptibility, children who are malnourished do not grow or develop normally. Because malnutrition affects cognitive development, malnourished children do not do as well in school as children who are well fed. Currently, WHO estimates that more than 3 billion people worldwide—the greatest number in history—are malnourished. In addition, more than half the deaths in children less than 5 years old in developing countries are associated with malnutrition.

The two most common diseases of malnutrition are marasmus and kwashiorkor. **Marasmus** (from the Greek word *marasmos*, meaning "a wasting away") is progressive emaciation caused by a diet low in both total calories and protein. Marasmus is most common among children in their first year of life—particularly children of very poor families in developing nations. Symptoms include a pronounced slowing of growth and extreme atrophy (wasting) of muscles. It is possible to reverse the effects of marasmus with an adequate diet.

Kwashiorkor (a native word in Ghana meaning "displaced child") is malnutrition resulting from protein deficiency. It is common among children in all poor areas of the world. The main symptoms include edema (fluid retention and swelling); dry, brittle hair; apathy; stunted growth; and sometimes mental retardation. One of the most typical features of kwashiorkor is a pronounced swelling of the abdomen (Figure 18.1). Kwashiorkor can be treated by gradually restoring a balanced diet.

Figure 18.1 Kwashiorkor. Millions of children suffer from this disease, which is caused by severe protein deficiency. Note the characteristic swollen belly, which results from fluid retention. Photographed in Haiti.

People who eat food in excess of that required are **overnourished**. Generally, a person suffering from overnutrition has a diet high in saturated (animal) fats, sugar, and salt. Overnutrition results in obesity, high blood pressure, and an increased likelihood of such disorders as diabetes and heart disease. In nutrition experiments, overnutrition in rodents resulted in a higher incidence of cancer compared with rodents fed a calorie-restricted diet, but evidence that links overnutrition to cancer in humans is sparse. Many human studies show a correlation between diets high in animal fat and red meat and certain kinds of cancer (colon and prostate). Overnutrition is most common among people in highly developed nations, such as the United States, where WHO estimated that 55% of all adults were overweight in the mid-1990s. Overnutrition is also emerging in some developing countries, particularly in urban areas. As people in developing countries earn more money, their diets shift from primarily cereal grains to more processed foods and livestock products.

Producing Enough Food

Producing enough food to feed the world's people is the largest challenge in agriculture today, and the challenge grows more difficult each year because the human population is continually expanding. As seen in Figure 18.2, annual grain production increased from 1970 to 2001. However, because the world population increased by more than 2 billion during that period, the amount of grain *per person* has not changed appreciably.

Some demographers project an *annual* increase in the human population of as much as 80 million for the next 20 years. Almost all of this growth will occur in developing countries. As of 2001, 1.94 billion metric tons (2.14 billion tons) of grains such as wheat, corn, rice, and barley were required to feed the world's population for 1 year. Each year, an additional 25.3 million metric tons (27.9 million tons) of grain must be produced over the previous year's production to account for the projected increase in population.

During the 1990s and early 2000s, world increases in food production barely kept pace with population growth. (We will ignore for the moment the fact that hunger still persists in many places in the world.) Grain production increased from 247 kg per person in 1950 to its highest level, 342 kg per person, in 1984. Since then, it has declined to a 2001 level of 299 kg per person. Annual world grain production was lower than consumption for 3 consecutive years, beginning in 2000.

Global food production can be increased in the short term, although whether this increase is sustainable is questionable. The long-term solution to the food supply problem is control of human population growth.

Famines

Crop failures caused by drought, war, flood, or some other catastrophic event may result in **famine**, a severe food shortage. Throughout human history, famine has struck one or more regions of the world every few years. The developing nations of Africa, Asia, and Latin America are most at risk. The worst African famine in history, which was caused in part by widespread drought, occurred from 1983 to 1985. Hardest hit were Ethiopia and Sudan, in which 1.5 million people died of starvation. The people living in this region lacked sufficient money to purchase food and did not have stored food reserves to protect them against several years of crop failures. More recently, drought and civil unrest in Somalia resulted in famine for an estimated 2 million Somalis. In 1993 the United Nations sent soldiers from 11 nations on a

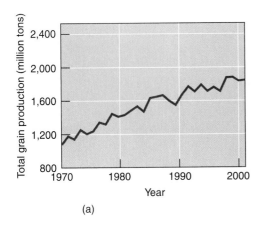

(a)

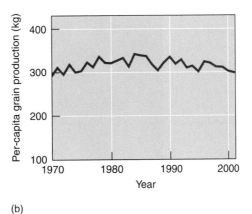

(b)

Figure 18.2 Total world grain production and grain production per person, 1970 to 2001.
(a) Total world grain production increased from 1.1 billion tons in 1970 to 1.8 billion tons in 2001. (b) The amount of grain produced per person has remained more or less stable, at about 300 kg per person.

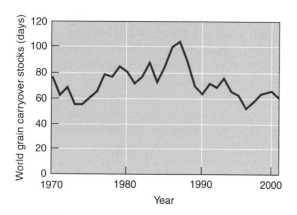

Figure 18.3 World grain carryover stocks, 1970 to 2001.
World grain carryover stocks, which are expressed as the number of days' supply on hand, should be at least 70 days' worth to provide both a cushion against poor harvests and stability in world grain prices.

humanitarian mission to Somalia to stop warring factions from stealing food from relief missions. More than 2 million North Koreans died from starvation or hunger-related illnesses during a famine in the late 1990s. The North Korean famine was caused by several years of floods and drought that destroyed its economy and collective farming system.

Famines get a great deal of media attention because of the huge and obvious amount of human suffering they cause. However, many more people die from undernutrition and malnutrition than from the starvation associated with famine.

Maintaining Grain Stockpiles

World grain carryover stocks are the amounts of rice, wheat, corn, and other grains remaining from previous harvests, as estimated at the start of a new harvest. World grain carryover stocks provide a measure of world **food security**, a goal in which all people have access at all times to adequate amounts and kinds of food needed for healthy, active lives. Stockpiles of grain have decreased each year since their all-time high in 1987 of 422 million metric tons (466 million) tons. This amount would have been enough, assuming it was evenly distributed, to feed the world's people for 104 days. The amount of grain stockpiled in 2001 would feed the world's people for only 60 days (Figure 18.3). According to the United Nations, the 2001 stockpile was below the minimum amount of 70 days' supply needed to safeguard world security. When food is scarce and prices increase, the risk of political instability is a real concern in poor nations. Some experts think that it will take at least two to three bountiful harvests to rebuild stocks. Others are more pessimistic about the low grain stocks and think that they indicate the beginning of a period of agricultural scarcity and higher food prices. Supporting this view is the fact that world grain carryover stocks reached 70 days or higher only two times since 1990 (in 1991 and 1993).

There are two main reasons that world grain stocks have dropped in the past few years. One, weather conditions such as heat and drought have caused poor harvests. The 1998 wheat harvest in Russia was 39% less than the previous year's because of bad weather and internal economic problems. Two, consumption of beef, pork, poultry, and eggs has increased in China and other developing countries, where some people are becoming more affluent and can afford to diversify their diets (Figure 18.4). Animal products account for 40% of the kilocalories people consume in highly developed countries and only 5% of the kilocalories consumed in developing countries. Table 18.1 compares meat consumption in India, China, Italy, and the United States.

Figure 18.4 Meat market in Guangzhou (Canton), China. As their incomes rise, people in China and many other developing countries are eating more meat and meat products. Because livestock animals eat grains, the increased demand for meat has contributed to the decline in world grain carryover stocks. According to Norman Myers, a scientist at Oxford University, if each of China's people were to eat just one extra chicken per year, the grain required to raise those chickens would equal all of the grain exports of Canada. (Canada is the world's second largest exporter of grain.)

Table 18.1	Annual Per-Capita Consumption of Meat in Selected Countries			
Country	Beef (kg)	Pork (kg)	Poultry (kg)	Mutton (kg)
India	1	0.4	1	1
China	4	30	6	2
Italy	26	33	19	2
United States	45	31	46	1

The increased consumption of meat and meat products has prompted a surge in grain used to feed the world's 45 billion livestock animals. The U.S. Office of Technology calculates that every kilogram of chicken meat requires an average input of 2.7 kg of livestock feed grain. Similarly, 1 kg of pork meat requires 6 kg of grain, and 1 kg of beef meat requires 7 kg of grain.

Poverty and Food: Making Food Affordable for the Poor

Despite the fact that enough food is currently produced to feed the world's people with an adequate but not generous diet, the harsh reality is that food is not shared equally by all people. The main cause of undernutrition and malnutrition is poverty. The world's poorest people—those living in developing countries in Asia, Africa, and Latin America—do not own land on which to grow food and do not have sufficient money to purchase food. Over 1.3 billion people in developing countries have incomes equivalent to less than $1 dollar a day and are so poor that they cannot afford to eat enough food or enough of the right kinds of food.

Worldwide, hunger is more common in rural areas than in urban areas. Infants, children, and the elderly are more susceptible to poverty and chronic hunger. The greatest proportion of chronically hungry people is in Africa. The largest number of hungry people is in Asia. Poverty and hunger are not restricted to developing nations, however. Poor hungry people also are found in the United States, European countries, and Australia.

Economic and Cultural Effects on Human Nutrition

Despite the fact that the major food-producing countries currently produce enough grain to feed the rest of the world, there is still an enormous economic problem. It costs money to produce, store, transport, and distribute food. Asian, African, and Latin American countries, which have the greatest need for imported food, are least able to pay for it. On the other hand, the food-producing nations cannot afford simply to give food away and absorb the costs indefinitely. In addition, government inefficiency and bureaucratic red tape can add to food

problems, sometimes making it difficult to distribute the food to the hungriest people and to ensure that those who need it get it instead of those who do not need it. (There have been instances when dishonest government officials sell food intended for hungry people for personal profit.)

Thus, getting food to the people who need it is largely an economic problem. The solution is obvious, although not necessarily easy to implement. Food must be produced in the area where it is to be consumed. If food production matches market demands in developing countries, not only are people fed, but also economic growth occurs as agriculture generates incomes and employment.

Cultural acceptance of food is another important food problem. If, for example, a particular group of people have been eating rice for hundreds of generations, it is unlikely that they will eat corn even if they are very hungry. This may sound strange, but think how reluctant you might be to eat foods such as insect grubs or dog meat, which may be exotic to some but are common fare in parts of Africa and Asia (Figure 18.5). It is

Figure 18.5 Insect grubs as food. Mopani, a high-protein food in South Africa, are caterpillars of the emperor moth.

human nature to be suspicious of foods to which we are unaccustomed.

World Food Problems: A Summary

World food problems are many, as are their solutions. We need to increase the sustainable production of food, assist overall economic development, improve food distribution, and overcome cultural barriers to acceptance of foods. Highly developed nations can help developing countries become agriculturally self-sufficient by providing economic assistance and technical aid. But the ultimate solution to world hunger is tied to achieving a stable population in each nation at a level that it can support.

Having a general grasp of world food problems, we now consider some of the complex challenges facing agriculture today. If you were to travel around the world, you would find many different kinds of agriculture and types of food. Despite this diversity, there has been an overall trend toward greater uniformity in the plants and animals we eat; humans have come to rely on fewer and fewer types of plants and animals for the bulk of food production. There has also been an overall trend toward uniformity in agricultural practices, as farmers in developing nations have adopted techniques used in highly developed countries.

◼ PLANTS AND ANIMALS THAT STAND BETWEEN PEOPLE AND STARVATION

Although we do not know precisely how many different plant species exist, biologists estimate that there are about 270,000. Of these, slightly more than 100 provide about 90% of the food that humans consume, either directly or indirectly. (Humans consume foods indirectly when cereal grains are used to feed livestock that humans eat as meat.) Just 15 species of plants provide the bulk of food for humans (Table 18.2). Of these plants, just three—rice, wheat, and corn—provide about half of the calories that people consume. Our dependence on so few species of plants for the bulk of our food puts us in an extremely vulnerable position. Should disease or some other factor wipe out one of the important food crops, humans might be threatened by severe famine. There are tens of thousands of kinds of plants that have been used as sources of food at one time or another. Many of these doubtless could be developed into important sources of food. We need to identify them, find out how to use them, and study their cultivation requirements.

Animals provide us with foods that are particularly rich in protein. These foods include fishes, shellfish, meat, eggs, milk, and cheese. Cows, sheep, pigs, chickens, turkeys, geese, ducks, goats, and water buffalo are the most important types of about 80 species of livestock. Although nutritious, livestock is an expensive source of

Table 18.2	The 15 Most Important Food Crops in Terms of Production

Plant Crop	Type of Crop	2001 World Production (1,000 metric tons)
Sugarcane	Sugar plant (stem)	1,254,856
Corn (maize)	Cereal grain	609,181
Rice, paddy	Cereal grain	592,831
Wheat	Cereal grain	582,691
White potato	Ground crop (tuber)	308,194
Cassava (manioc)	Ground crop (root)	178,868
Soybean	Legume	176,638
Barley	Cereal grain	141,219
Sweet potato	Ground crop (root)	135,918
Sorghum	Cereal grain	58,149
Peanuts (ground nuts)	Legume	35,096
Oats	Cereal grain	27,278
Rye	Cereal grain	22,717
Beans, dry	Legume	16,772
Peas, dry	Legume	10,511

food because animals are inefficient converters of plant food. Of every 100 calories of plant material a cow consumes, it burns off approximately 86 in its normal metabolic functioning. That means that only 14 calories out of 100 (14%) are stored in the cow to be consumed by humans (see Figures 4.13 and 4.14, which show pyramids of biomass and energy). Meat consumption is high in affluent societies, so large portions of the crops grown in highly developed countries are used to produce livestock animals for human consumption. Almost half of the cereal grains grown in highly developed countries are used to feed livestock (see "You Can Make a Difference: Vegetarian Diets").

◼ THE PRINCIPAL TYPES OF AGRICULTURE

Agriculture can be roughly divided into two types: industrialized agriculture and subsistence agriculture (see "Mini-Glossary: The Principal Types of Agriculture"). Most farmers in highly developed countries and some in developing countries practice **industrialized agriculture,** also called **high-input agriculture.** It relies on large inputs of capital and energy, in the form of fossil fuels, to produce and run machinery, irrigate crops, and produce agrochemicals, such as commercial inorganic fertilizers and pesticides (Figure 18.6). Industrialized agriculture produces high **yields** (the amount of a food crop produces per unit of land), enabling forests and other natural areas to remain wild instead of being converted to agricultural land. The productivity of industrialized agriculture has not been without costs, however.

YOU CAN MAKE A DIFFERENCE

Vegetarian Diets

A vegetarian is a person who does not eat the flesh of any animal, including that of fish and poultry. People embrace vegetarian diets for many reasons. Balanced vegetarian diets provide good nutrition without high levels of saturated fats or cholesterol, both of which cause health problems such as heart disease and obesity. Some studies in the United States indicate that people who are vegetarians live longer, healthier lives than nonvegetarians.

Some people become vegetarians because they are morally or philosophically opposed to killing animals, even for food. Certain religious groups, notably Hindus and Seventh Day Adventists, exclude animal products from their diets. Other people convert to vegetarianism because of a sense of responsibility for land use and its wide repercussions. Fewer plants are required to support vegetarians than to support meat eaters. The amount of usable energy in the food chain is decreased by approximately 90% by adding an additional level—that is, the animals we eat—to the chain (see discussion of energy pyramids in Chapter 4). The actual percentage varies because all animals are not alike in their efficiency at turning food into meat. Simply stated, if everyone were to become a vegetarian, much more food would be available for human consumption.

Vegetarians are divided into four groups—lacto-ovo vegetarians, lacto vegetarians, ovo vegetarians, and vegans—based on whether they eat milk and/or egg products. *Lacto-ovo vegetarians* eat milk, eggs, and foods made from milk and eggs. *Lacto vegetarians* do not eat eggs but they eat milk and milk products such as cheese, yogurt, and butter. *Ovo vegetarians* do not eat dairy, but eggs are allowed. *Vegans* exclude milk and eggs in all forms from their diets.

Some people are reluctant to switch to a vegetarian diet because they fear they will not get enough protein (plant foods generally have a lower percentage of protein than do animal foods). However, meat eaters in highly developed countries usually consume much more protein than they need. The problem with a vegetarian diet is usually not lack of protein but obtaining the proper balance of essential amino acids. It is relatively easy for lacto-ovo vegetarians, lacto vegetarians, and ovo vegetarians to plan a healthy diet because milk and eggs contain all the essential amino acids. Milk is also rich in calcium, which is important for strong bones and teeth. Both milk and eggs contain vitamin B_{12}, which helps form red blood cells. Vegans, however, must plan their diets carefully to obtain a proper balance of amino acids, as well as enough calcium and vitamin B_{12}. Sesame seeds, broccoli, spinach, and other leafy green vegetables are rich in calcium, so these foods must be included in the vegetarian diet every day. Because vitamin B_{12} is found almost exclusively in animal products, most vegans take vitamin B_{12} supplements.

A nutritious vegetarian diet includes a combination of foods that contains all the essential amino acids. A meal of rice and beans or corn and beans provides the proper complement of essential amino acids Cookbooks and other references contain menus and recipes that give the vegetarian diet adequate amounts of high-quality protein. The following list provides an overview of combinations of foods that offer a proper balance of essential amino acids. At least one food from each column should be consumed at every meal.

Column I	Column II
Grains Barley, corn, oats, rice, rye, wheat	**Legumes** Peas, black-eyed peas, chick peas, black beans, fava beans, kidney beans, lima beans, pinto beans, mung beans, navy beans, soybeans (usually eaten as bean curd, or tofu)
Nuts and seeds Almonds, beechnuts, Brazil nuts, cashews, filberts, pecans, pumpkin seeds, sunflower seeds, walnuts	**Dairy products** Cheese, cottage cheese, eggs, milk, yogurt

Several problems, such as soil degradation and increases in pesticide resistance in agricultural pests, are caused by industrialized agriculture; we discuss these and other problems later in the chapter.

Most farmers in developing countries practice **subsistence agriculture**, the production of enough food to feed oneself and one's family, with little left over to sell or reserve for hard times. Subsistence agriculture, too, requires a large input of energy, but from humans and draft animals rather than from fossil fuels.

Some types of subsistence agriculture require large tracts of land. **Shifting cultivation** is a form of subsistence agriculture in which short periods of cultivation are followed by longer periods of fallow (land left uncultivated) in which the land reverts to forest. Shifting cultivation supports relatively small populations. **Slash-and-burn agriculture** is one of several distinct type of shifting cultivation that involves clearing small patches of tropical forest to plant crops (see Chapter 17). Because tropical soils lose their productivity very quickly when they are cultivated (see Chapter 14), farmers using slash-and-burn agriculture must move from one area of forest to another every 3 years or so, so slash-and-burn agriculture is land-intensive. **Nomadic**

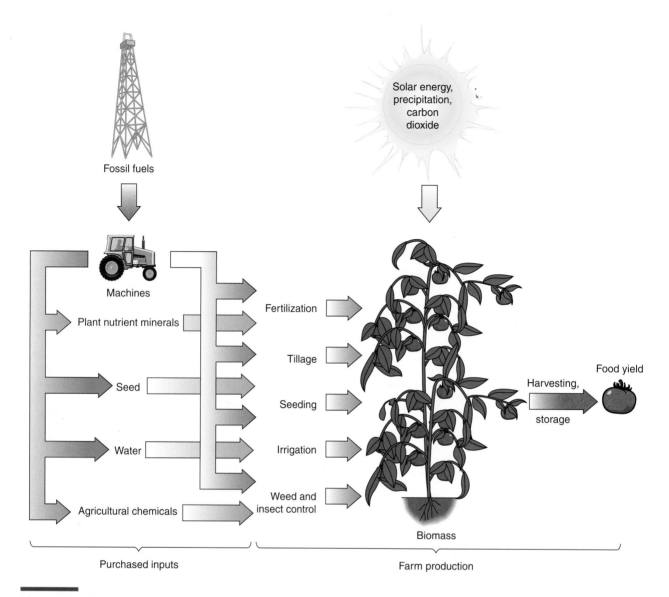

Figure 18.6　Energy inputs in industrialized agriculture. Fossil fuel energy inputs occur at virtually all stages in agricultural production.

herding, in which livestock is supported by land that is too arid for successful crop growth, is another type of land-intensive subsistence agriculture. Nomadic herders must continually move their livestock to find adequate food for them.

Intercropping is a form of intensive subsistence agriculture that involves growing a variety of plants simultaneously on the same field. When certain crops are grown together, they produce higher yields than when they are grown as **monocultures**. One reason for higher yields is that different pests are found on each crop, and intercropping discourages the buildup of any single pest species to economically destructive levels. Native Americans practiced intercropping when they

planted corn, bean, and squash seeds in the same mound of soil. Because the root systems of these plants grow to different depths, they do not compete with one another for water and essential minerals. In addition, the protein-rich bean crop fixes nitrogen that fertilizes the corn and squash plants naturally. **Polyculture** is a complex type of intercropping in which several kinds of plants that mature at different times are planted together. In polyculture that is practiced in the tropics, fast- and slow-maturing crops are often planted together so different crops are harvested throughout the year. Vegetable crops and cereal grains, which mature first, might be planted with papayas and bananas, which mature later.

The Principal Types of Agriculture

industrialized agriculture: Modern agricultural methods, which require a large capital input and less land and labor than traditional agricultural methods. Industrialized agriculture uses large inputs of energy (from fossil fuels), mechanization, water, and agrochemicals (fertilizers and pesticides) to produce large crop and livestock yields; also called *high-input agriculture*.

subsistence agriculture: Traditional agricultural methods, which are dependent on labor and a large amount of land to produce enough food to feed oneself and one's family, with little left over to sell or reserve for hard times. Subsistence agriculture uses humans and draft animals as its main source of energy.

shifting cultivation: A traditional form of subsistence agriculture in which short periods of cultivation are followed by longer periods of fallow (land left uncultivated), during which time the natural ecosystems may become reestablished. Slash-and-burn agriculture is an example of shifting cultivation.

slash-and-burn agriculture: A type of shifting cultivation in tropical forests in which a patch of vegetation is burned, leaving nutrient minerals in the ash. Crops are planted for a few years until the soil is depleted of nutrient minerals, after which the land must lie fallow for many years to recover.

nomadic herding: A traditional grazing method in which nomadic herders wander freely over rangelands in search of good grazing for their livestock.

intercropping: A form of intensive subsistence agriculture that involves growing several crops simultaneously on the same field.

polyculture: A type of intercropping in which several kinds of plants that mature at different times are planted together.

THE EFFECT OF DOMESTICATION ON GENETIC DIVERSITY

Wild plant and animal populations usually have great genetic diversity—that is, a lot of variation in their *genes*, which are units of hereditary information that specify certain traits (see Figure 16.1 for a visual demonstration of genetic diversity in corn). **Genetic diversity** contributes to species' long-term survival by providing the variation that enables each population to adapt to changing environmental conditions. When plants and animals are **domesticated**, much of this genetic diversity is lost, because the farmer selects for propagation only those plants and animals with the most desirable agricultural characteristics. At the same time, other traits that are not of obvious value to humans are selected against. Hence, many of the high-yielding crops produced by modern agriculture are genetically uniform. Most of the vegetable crops grown in the United States are of only a few different varieties. Likewise, dairy cattle and poultry in the United States have low genetic diversity. The familiar black-and-white Holsteins comprise 91% of U.S. dairy cattle, and white leghorns produce almost all the white eggs consumed in the United States.

The loss of genetic diversity that accompanies modern agriculture usually does not prove disastrous to crop plants, because they do not have to survive in the wild under natural conditions. Under cultivation, they are watered, fertilized, and protected as much as possible from pests, including weeds, insects, and disease organisms. Domesticated animals are also protected from the challenges of nature.

However, the lower genetic diversity of domesticated plants and animals increases the likelihood that they will succumb to new strains of disease-causing organisms. These organisms, which include bacteria, fungi, and viruses, evolve quite rapidly because of their high reproductive rates. When a disease breaks out in a domesticated plant or animal population, the entire uniform population is susceptible; thus, the loss is greater than it would be in a natural, varied population, in which at least some individuals would contain genes to resist the disease-causing organism.

The Global Decline in Domesticated Plant and Animal Varieties

Although domestication contributes in general to less genetic diversity than is found in wild relatives, the many farmer-breeders around the world who have selected for specific traits have developed many local varieties of each domesticated plant and animal. French farmers, for example, developed 200 different breeds of cattle during the 18th and 19th centuries. Each traditional variety represents the legacy of the hundreds of farmers who developed it over the centuries. A traditional variety is adapted to the climate where it was bred and contains a unique combination of traits conferred by its unique combination of genes.

A global trend is currently underway to replace the many local varieties of a particular crop or domesticated farm animal with just a few kinds. According to a 1999 Worldwatch Institute report, Mexican farmers are currently growing only about 20% of the corn varieties that they grew in the 1930s. When farmers abandon their traditional varieties in favor of more modern ones, which are bred for uniformity and maximum production, the former varieties frequently face extinction (Figure 18.7). This represents a great loss in genetic diversity, because each variety's characteristic combination of genes gives it distinctive nutritional value, size, color, flavor, resistance to disease, and adaptability to different climates and soil types.

The gene combinations of local varieties are potentially valuable to agricultural breeders because they can be transferred to other varieties, either by traditional breeding methods or by genetic engineering (discussed later in this chapter). U.S. wheat and barley crops were

Figure 18.7 Dutch Belted cow. The USDA recognizes Dutch Belts as a viable dairy breed that produces a high-quality, flavorful milk. The breed is endangered in the United States, with a population of about 200 purebred cows, and rare in Canada. The Netherlands was the original source of this variety, which was known by the 1600s, but the American Dutch Belt is genetically closer to the original breed than the few Dutch Belts still remaining in the Netherlands. Photographed in Wisconsin.

infested in the late 1980s by Russian aphids (aphids are tiny insects that suck out the juices of plant leaves and stems). Over several years, aphid-resistant varieties of wheat and barley were developed using genes of several wheat and barley varieties from the Middle East.

To preserve older, more diverse varieties of plants, many countries are collecting germplasm. **Germplasm** is any plant or animal material that may be used in breeding. It includes seeds, plants, and plant tissues of traditional crop varieties, and the sperm and eggs of traditional livestock breeds. The International Plant Genetics Resources Institute in Rome, Italy, is the scientific organization that oversees plant germplasm collections worldwide. National germplasm collections range in size from the U.S. National Plant Germplasm System in Colorado, which holds almost half a million different varieties, to the national gene bank in the African country Malawi, which holds about 8,000 varieties of native crops and fruits. (See section on seed banks in Chapter 16 for more information about plant germplasm collections.) Several private organizations—such as Seeds of Change in New Mexico and Seed Savers Exchange in Iowa—save, propagate, and distribute traditional crop varieties.

Who actually owns germplasm, particularly crop diversity, has been a contentious international issue for about 20 years. Developing nations, which possess much of the world's crop genetic diversity, have maintained that they own the germplasm found in their countries. However, plant breeders want free access to that germplasm, and agricultural firms want the right to patent any crop improvements they make using germplasm. In 2001 the FAO adopted the **International Treaty on Plant Genetic Resources for Food and Agriculture**. It limits the genetic materials that agricultural companies are allowed to patent and affirms the right of farmers to save, use, exchange, and sell farm-saved seeds. Germplasm is to be made available to plant breeders and agricultural companies in exchange for royalty payments that will be used for conservation and breeding programs. The treaty goes into effect after 40 countries have ratified it. As of 2002, 53 countries had signed the treaty and 7 countries had ratified it. The United States has neither signed nor ratified the treaty.

INCREASING CROP YIELDS

Until the 1940s, agricultural yields among various countries, both highly developed and developing, were generally equal. However, advances made by research scientists have caused a dramatic increase in food production in highly developed countries (Figure 18.8). Greater knowledge of plant nutrition has resulted in fertilizers that promote high yields. The use of pesticides to control insects, weeds, and disease-causing organisms has also improved crop yields. Selective breeding programs have resulted in agricultural plants with more desirable features. Breeders developed wheat plants with larger, heavier grain heads

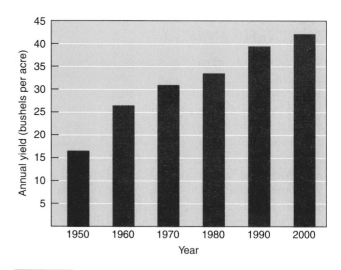

Figure 18.8 Average U.S. wheat yields, 1950 to 2000. Each year shown is actually an average of 3 years to minimize the effects that poor weather conditions might have in a given year. Similar increases in yield occurred in other grain crops.

(for higher yield). Because of the weight of the heads, other traits were gradually incorporated into wheat, such as shorter, thicker stalks, which prevent the plants from falling over during storms.

CASE·IN·POINT The Green Revolution

By the middle of the 20th century, serious food shortages occurred in many developing countries, and it was widely recognized that additional food supplies were needed to feed their growing populations. The development and introduction during the 1960s of high-yielding varieties of wheat and rice to Asian and Latin American countries gave these nations the chance to provide their people with adequate supplies of food. But the high-yielding varieties required intensive cultivation methods, including the use of commercial inorganic fertilizers, pesticides, and mechanized machinery, in order to realize their potential. These agricultural technologies were passed from highly developed nations to developing nations.

The production of more food per acre of cropland by using modern cultivation methods and the new, high-yielding varieties of certain staple crops has been called the **green revolution**. Norman Borlaug, a U.S. scientist who worked on wheat in Mexico during the 1940s and 1950s, is credited with beginning the green revolution. Borlaug's introduction of a short-stemmed, hybrid strain of "miracle wheat" to Mexico and countries in South Asia was followed by the development by other plant breeders of high-yielding varieties of rice and other grain crops (Figure 18.9). For his contributions, Borlaug was awarded the Nobel Prize in 1970.

Some of the success stories of the green revolution have been remarkable. During the 1920s, Mexico produced less than 700 kg (0.77 ton) of wheat per hectare annually. During the green revolution years that began in 1965, Mexico's annual wheat production rose to more than 2,400 kg (2.65 tons) per hectare. Indonesia used to import more rice than any other country in the world. Today Indonesia produces not only enough rice to feed its people, but also some for export.

Problems with the Green Revolution Critics of the green revolution argue that it has made developing countries dependent on imported technologies, such as agrochemicals and tractors, at the expense of traditional agriculture. The two most important problems associated with higher crop production are the high energy costs that are built into this form of agriculture and the serious environment problems caused by the intensive use of commercial inorganic fertilizers and pesticides. The environmental impacts of intensive agriculture, including the green revolution, are discussed later in the chapter.

The production of agricultural chemicals requires fossil fuels, which are nonrenewable resources (see

Tall conventional plant   **Improved high-yielding plant**   **Low-tillering ideotype (new plant type)**

Figure 18.9 Development of high-yielding rice varieties. Traditional rice plants on the *left* are taller and do not yield as much grain (clusters at top of plant) as the more modern varieties shown in the *middle* and on the *right*. The rice plant in the middle was developed during the 1960s by crossing a high-yielding, disease-resistant variety with a dwarf variety to prevent the grain-heavy plants from lodging, or falling over. Improvements since the green revolution have been modest, as the rice variety (*right*), developed during the 1990s, shows. Some researchers think rice and certain other genetically improved crops are near their physical limits of productivity.

Chapter 10). Commercial inorganic fertilizers require a great deal of energy to produce, so their production costs are tied closely to the price of energy. Significant amounts of fossil fuels are required to build and provide power for farm equipment, such as tractors and combines. The installation and operation of irrigation systems, including construction of dams and canals, also require a substantial energy input. In general, it takes three times more energy per hectare to produce corn by irrigation than it does to produce the same amount of corn under rain-fed conditions. (The energy use in this calculation is based on pumping groundwater from a depth of 30 m.)

Rice and wheat are not the only crops to have been improved by the green revolution. High-yielding varieties of crops such as potatoes, barley, and corn have also been developed. Nonetheless, many important food crops remain to be improved by selective breeding and scientific research. People in Africa eat sorghum, millet, cassava, and sweet potatoes, none of which has been greatly improved by green revolution technology. Also, the green revolution has benefited large landowners but not subsistence farmers, who represent a substantial segment of the agricultural community in most developing nations. Subsistence farmers need improved crops that respond to labor-intensive agriculture (human and animal labor) and do not require large outlays of energy and capital.

Increasing Crop Yields in the Post–Green Revolution Era

In 1999, the International Food Policy Research Institute projected that the world demand for rice, wheat, and corn will increase by 40% between 2000 and 2020. This increase in grain production will be needed to feed the increasing human population and to satisfy the appetites of increasing numbers of affluent people who can afford to buy meat.

This challenge cannot be met by increasing the amount of land under cultivation, as the best arable lands are already being cultivated. Projected freshwater shortages, increasing costs of agricultural chemicals, and deteriorating soil quality caused by intensive agricultural techniques may further constrain productivity. Also, as Figure 18.9 demonstrates, recent progress in coaxing more grain out of crops that were genetically improved during the green revolution has resulted in diminishing returns. Grain yields have continued to rise since the 1960s, but in recent years, the rates of increase have not been as great as they were in earlier years.

Despite these problems, most plant geneticists think that we will be able to produce enough food in the 21st century if countries spend more money to support a concerted scientific effort to develop improved crops. Many scientists think that genetic engineering, which will be discussed shortly, will be able to assist in the endeavor to breed more productive varieties. In addition to developing improved crop varieties, modern agricultural methods, such as water-efficient irrigation, will have to be introduced to developing countries that do not currently have them if we are going to continue to increase crop yields.

Efforts are under way to increase food security in low-income, food-deficient countries. During the 1990s the FAO initiated a special program for farmers in 19 different nations, most of which are in Africa. Participating farmers are given genetically improved seeds, commercial inorganic fertilizers, and pesticides and are trained in improved agricultural techniques. These farmers then provide demonstrations to neighboring farmers on how to increase and diversify food production, reduce water use, control pests, and protect the soil and other natural resources. In Ethiopia this program doubled grain yields at demonstration sites and resulted in substantial profits after the costs of the seeds and agricultural chemicals were subtracted.

◼ INCREASING LIVESTOCK YIELDS

Animal production has been increased by the use of hormones and antibiotics, but the use of both is controversial. Although U.S. and Canadian farmers use **hormones**, usually administered by ear implants, the European Union (EU) currently bans all imports of hormone-treated beef because of health concerns for human consumers. They cite a few studies that suggest that these hormones or their breakdown products, both of which are found in trace amounts in meat and meat products, could cause cancer or affect the growth of young children. In 1999, however, an international scientific committee organized by the FAO and WHO examined the hormone issue in detail. They concluded that the traces of hormones found in beef are safe because they are so low compared to normal hormone concentrations in the human body. However, they did not consider if excess hormones that are excreted in cattle feces could end up in other foods or drinking water. Critics of the EU ban contend that it is a move to reduce competition and protect the EU beef industry. These critics note that many foods, such as ice cream, peas, butter, wheat germ, and soybean oil, have natural hormone levels higher than hormone-implanted beef.

Modern agriculture has embraced the routine addition of low doses of **antibiotics** to the feed for pigs, chickens, and cattle. These animals typically gain 4% to 5% more weight than animals that do not receive antibiotics, presumably because they must expend less energy to fight infections. According to the *New England Journal of Medicine*, 40% of the 25,000 tons of antibiotics produced annually in the United States is used in livestock operations, particularly those in which large numbers of animals are confined in small areas. Most of these antibiotics are administered continuously to healthy animals.

Several studies link the indiscriminate use of antibiotics in humans and livestock to bacteria that are increasingly resistant to antibiotics. The development of bacterial resistance to antibiotics is an example of **evolution**. Bacteria are continually evolving, even inside the bodies of human and animal hosts. When an antibiotic is used to treat a bacterial infection, a few bacteria may survive because they are genetically resistant to the antibiotic,[1] and they pass these genes to future generations. As a result, the bacterial population contains a larger percentage of antibiotic-resistant bacteria than before. The Worldwatch Institute reports that antibiotic resistance has been documented in more than 20 different kinds of potentially harmful bacteria (the tuberculosis bacterium is one example), and some bacterial strains are resistant to every antibiotic known, a total of more than 100 drugs.

Because there is increasing evidence that the use of antibiotics in agriculture reduces their medical effective-

---

[1] Bacteria develop genetic resistance through mutations (spontaneous changes in their genetic material) and by acquiring new genes from other bacteria.

ness for humans, WHO has recommended that livestock receive antibiotics that are different from the ones used in human medicine. Many European countries have stopped administering low doses of antibiotics for growth promotion in livestock. However, the United States and many other countries continue the practice.

FOOD SAFETY

Whether consumers buy food at a grocery store and prepare it at home or eat in a restaurant, they must have confidence that the food is safe and will not make them sick. Food safety has become an increasingly complex issue, particularly because much of our food comes from other countries. Outbreaks of illness caused by contaminated food are fairly common in the United States. According to the Centers for Disease Control and Prevention, each year in the United States there are about 76 million cases of food-borne illness. Of these, 325,000 require hospitalization, and 5,000 deaths occur. These numbers may make you think that food safety problems are at an all-time high. However, the risk of food-borne illness is minor compared to the risks involved, for example, in eating the kinds of foods that contribute to obesity, cardiovascular disease, or cancer. (Recall the discussion in Chapter 2 of the discrepancy between real and perceived risk.)

The USDA addresses the safety of meat, poultry, and food products that contain meat or poultry. The U.S. Food and Drug Administration (USFDA) is responsible for the safety of all other foods, such as fresh produce, eggs, milk, seafood, canned foods, and processed foods that do not contain meat. Health experts say that food-borne bacteria, viruses, and parasites pose the greatest risk to the safety of our food supply. In comparison, the food-safety threat posed by food additives, pesticide residues, and unlabelled allergenic substances, while of concern, is relatively unimportant.

Food-borne contaminants can be introduced at several stages in the food cycle, including production, processing, distribution, and preparation. Consider some of the ways that chicken can become contaminated. Bacteria such as *Salmonella* and *Campylobacter* in fecal matter may infect live chickens, which are typically crowded together in small pens. When the chickens are slaughtered at the slaughterhouse, contaminated mechanical pluckers may deposit bacteria on the skin as the feathers are plucked. Chicken is usually packaged at the processing plant, but if juices leak out of the package during shipment, they may contaminate other packages. If the chickens are transported or stored at improper temperatures, the bacteria will proliferate. Harmful bacteria are killed if chicken is cooked until the juices run clear. However, it is easy to cross-contaminate other foods such as vegetables and fruits if they are prepared, for example, on the same cutting board that was used to cut up the chicken. All

kitchen surfaces that come into contact with the raw juices of a chicken should be thoroughly washed. Good hygiene, including washing hands frequently during food preparation, is also important.

According to the National Center for Food Safety and Technology, the U.S. food supply is safer today than in the past. Enhanced surveillance occurs during all stages of food production, which accounts in part for the 23% decline observed in bacterial food-borne illness since 1996. Because it is possible to track food products around the country, it is easier to link two outbreaks in different places to the same contaminant in the same food product. Food companies can quickly identify the lot of food that is contaminated and where it was shipped, so they can issue food recalls and warnings.

Food Processing and Food Additives

Most of the food we eat is not harvested and used directly. Instead, after harvest it is processed, then offered for sale in grocery stores. There are two aspects to food processing. The first encompasses procedures such as drying, freezing, canning, pasteurizing, curing, irradiating (see Chapter 22), and refrigerating food to retard spoilage. The second involves using **food additives**—chemicals that enhance the taste, color, or texture of the food; improve its nutrition; reduce spoilage and prolong shelf life; or maintain the food's consistency.

Sugar and salt are the two most common food additives. Although they are added to food primarily to make it taste better, in large amounts sugar and salt can be used to help preserve food—for example, fruit jelly and salted meat. Salt and sugar have been used as preservatives for centuries.

Essential amino acids and vitamins are sometimes added to foods to make them more nutritious than they would be naturally or to replace nutrients that are lost during processing. Both natural and synthetic **coloring agents** are used to make food visually appealing. Sodium propionate and potassium sorbate are two examples of **preservatives**, which are chemicals added to food to retard the growth of bacteria and fungi that cause food spoilage. Food can also spoil when lipids undergo oxidation. Food additives that prevent oxidation are called **antioxidants** and include butylated hydroxyanisole (BHA) and butylated hydroxytoluene (BHT).

Nitrates (compounds containing NO_3^-) and **nitrites** (compounds containing NO_2^-) are used as food additives and for curing meats. Ingested nitrites are capable of reacting with other chemicals in food and in tobacco to form **N-nitroso compounds** (related nitrogen-containing compounds) in the stomach. Some *N*-nitroso compounds cause cancer. *N*-nitrosodimethylamine is a potent cancer-causing substance found in cheeses, bacon and other cured meats, smoked and salted fishes, beer, and tobacco smoke.

Table 18.3	Food Additives That May Be Harmful	
Food Additive	*Food*	*Possible Effect*
BHA, BHT	Oils, potato chips, chewing gum	May be carcinogenic; allergic reactions in some
Citrus red dye #2	Skin of some oranges	May be carcinogenic
Nitrates, nitrites	Bacon, corned beef, hot dogs, ham, smoked fish, luncheon meats	Formation of carcinogenic *N*-nitroso compounds
Red dyes #3, 8, 9, 19, 37	Cherries (maraschino, fruit cocktail, candy)	May be carcinogenic
Sulfites (sulfur dioxide, sodium sulfide, sodium bisulfite)	Dried fruit, canned and frozen vegetables, some beverages (e.g., wine), bread, salad dressings, and more	Severe allergic reactions

Protection of the Consumer   The USFDA is charged with the responsibility of monitoring food additives. In 1958, regulations that require any new food additive to undergo extensive toxicity testing by the manufacturer were passed. The USFDA then evaluates the results of such tests to determine whether the additive is safe.

Additives that were in use prior to 1958, however, do not have to undergo such testing. In 1959, the FDA designated these chemicals "generally recognized as safe" and made up a list of them—usually called the GRAS (pronounced *grass*) list. All substances on the GRAS list have subsequently been reviewed. Some have been banned, including cyclamates (used for sweetening) and brominated vegetable oil. BHA, BHT, and several other substances on the GRAS list are undergoing further tests.

The Center for Science in the Public Interest has compiled a list of food additives that may be harmful and should be avoided or eaten sparingly (Table 18.3). These include the preservatives BHA and BHT, several coloring agents (especially the red dyes), and nitrates and nitrites.

Are Food Additives Bad?   There are two opposing viewpoints about the safety of our food. Some people are very concerned about the large number and amounts of food additives in processed food. Their main fear is that some additives may cause cancer. Supporters of this view worry that although the risk of developing cancer from exposure to each individual chemical may be quite small, the effects of hundreds of different food additives added together may be significant. Also, there is concern that these chemicals may interact **synergistically**—that is, their total effect may be greater than the sum of their individual effects.

Others think that the health hazards from food additives are greatly exaggerated. They think it is important to put concerns about food additives and even pesticide residues in proper perspective. Everyone acknowledges that the chemicals in our foods do not pose anywhere near the threat of cancer that smoking does (see discussion of risk assessment in Chapter 2). Moreover, much larger quantities of *natural* cancer-causing substances are present in food. Nature is not benevolent; plants have

evolved natural chemical defenses to discourage insects from consuming them. Some of these compounds, which can be present in large amounts, are toxins; some cause cancer. Thus, some experts contend that food additives pose a much smaller threat to humans than do the natural, unavoidable chemicals already present in food.

■ THE ENVIRONMENTAL IMPACTS OF AGRICULTURE

The practices of industrialized agriculture have resulted in several environmental problems that impair the ability of nonagricultural terrestrial and aquatic ecosystems to provide essential **ecosystem services**. This raises questions about the sustainability of intensive agriculture.

The agricultural use of fossil fuels and pesticides produces air pollution. Untreated animal wastes and agricultural chemicals such as fertilizers and pesticides cause water pollution that can reduce biological diversity, harm fisheries, and lead to outbreaks of nuisance species (see envirobrief on harmful algal blooms in Chapter 21). According to the Environmental Protection Agency, agricultural practices are the single largest cause of surface-water pollution in the United States. Water pollution from agriculture is particularly significant in Midwestern states such as Iowa, Wisconsin, and Illinois. Some of these contaminants flow into the Mississippi River and, from there, into the Gulf of Mexico (see Chapter 21). Some agricultural chemicals have been detected in water deep underground, as well as in surface waters. Nitrates from animal wastes and commercial inorganic fertilizers are probably the most widespread groundwater contaminant in agricultural areas.

Industrialized agriculture has favored the replacement of traditional family farms by large agribusiness conglomerates (see Chapter 17). In the United States, most cattle, hogs, and poultry are now grown in feedlots and livestock factories (Figure 18.10). In livestock factories, thousands of animals are confined to small pens in buildings the size of football fields. Such large concentra-

Figure 18.10 Hog factory in Ohio. The hogs remain indoors and are fed and watered by machine at timed intervals throughout the day. Although livestock factories are very efficient and produce meat relatively inexpensively, they also cause environmental problems such as sewage disposal. Some livestock factories produce as much sewage as a small city, yet the wastes are currently not covered by water-quality laws and therefore do not go to a sewage treatment plant.

tions of animals create many environmental problems, including air and water pollution. The quantity of manure produced by several hundred thousand pigs in one livestock factory causes a severe waste disposal problem. At hog factories, the manure is often stored in deep lagoons that have the potential to pollute the soil, surface water, and groundwater. This happened during Hurricane Fran in 1996, when manure from 22 large animal waste lagoons in North Carolina spilled onto the floodplain and into streams, causing major fish kills. People living near livestock factories dislike the odor, which often exceeds federal and state guidelines for emissions and causes their property values to decline.

Many insects, weeds, and disease-causing organisms have developed or are developing resistance to pesticides. Pesticide resistance forces farmers to apply progressively larger quantities of pesticides (Figure 18.11; also see Chapter 22). Residues of pesticides contaminate our food supply and reduce the number and diversity of beneficial microorganisms in the soil. Fishes and other aquatic organisms are sometimes killed by pesticide runoff into lakes, rivers, and estuaries.

Degradation is the natural or human-induced process that decreases the future ability of the land to support crops or livestock animals. Soil erosion, which is exacerbated by large-scale mechanized operations, causes a decline in soil fertility, and the sediments lost because of erosion damage water quality. The USDA estimates that about one fifth of U.S. cropland is vulnerable to soil erosion damage, and soil erosion is an even greater problem in some developing nations. Examples of other types of degradation are compaction of soil by heavy farm

machinery and waterlogging and salinization of soil caused by improper irrigation methods.

Crop production requires enormous amounts of water. According to Lester Brown of the Worldwatch Institute, it takes 1,000 tons of water to produce 1 ton of grain. Worldwide, irrigation consumes about almost 70% of the total fresh water that humans withdraw from aquifers and surface waters. Some agricultural regions remove water from aquifers faster than it is recharged by precipitation, lowering water tables. The huge Ogallala Aquifer under Nebraska, Kansas, Colorado, and other states is one of the best known examples of overdrawing an aquifer for agriculture (see Chapter 13). Most of the water in the Ogallala is ancient, having been left by melting glaciers at the end of the last Ice Age. Thus, the Ogallala Aquifer is largely a nonrenewable resource. Lack of water increasingly affects agricultural yields in parts of Africa, Australia, China, India, and the former Soviet Union. As a result of mismanagement of water resources, vast areas of irrigated land have become too waterlogged or too salty to grow crops (see Chapter 21).

Clearing grasslands and forests and draining wetlands to grow crops have resulted in **habitat fragmentation** and losses that reduce biological diversity (see Chapter 16). Many species have become endangered or threatened as a result of habitat loss caused by agriculture. The most dramatic example of habitat loss in North America is tallgrass prairie, more than 90% of which has been converted to agriculture.

Figure 18.11 Colorado potato beetles. As a result of being exposed to heavy applications of pesticides over the years, Colorado potato beetles are resistant to most insecticides that are registered for use on potatoes. Photographed in Assawoman, Virginia.

Using More Land for Cultivation

You may be surprised to learn that the total amount of the world's land devoted to agriculture is about the same as it was 50 years ago. In the temperate areas of the world, almost all the fertile land with an adequate supply of water is used for agriculture. Although in some regions the loss of prime farmland to urbanization is of concern, in temperate areas very little prime farmland not already under cultivation remains. Many tropical areas, on the other hand, have little prime agricultural land to start with. Those tropical areas with the greatest potential for cultivation and cropland expansion are in Latin America and sub-Saharan Africa.

The United States had agricultural surpluses during the 1980s. One reason is that farmers brought large amounts of previously unused land into production. Unfortunately, much of this land was marginal as agricultural land, because it was prone to soil erosion caused by intermittent floods or frequent droughts (and therefore wind erosion when the ground cover was removed). Harvesting crops from highly erodible land is ecologically unsound and cannot be done indefinitely. Some of the marginal farmlands in the United States have been retired from use (see discussion of the Conservation Reserve Program in Chapter 14).

Other countries have paid a high price for cultivating land highly prone to erosion. The former Soviet Union began to cultivate a large area of marginal land during the 1950s. Although the initial production of cereal crops was high, by the 1980s much of this land had to be abandoned. The annual per-capita food production in Haiti, a very poor Caribbean nation, is half what it was in 1950 as a result of both population growth and lower production on eroded soils.

From the 1960s to the 21st century, the amount of dry land being irrigated for agriculture was greatly increased. About 70% of the world's total irrigated land is found in Asia, and the amount of irrigated land there continues to expand each year. Since 1995, however, the amount of irrigated land in the rest of the world has remained steady. In Europe and Oceania, the amount of irrigated land has actually decreased. This change is due to the increasing cost of irrigation, the depletion of aquifers, the abandonment of salty soil, and the diversion of irrigation water to residential and industrial uses.

■ SOLUTIONS TO AGRICULTURAL PROBLEMS

Food production poses an environmental quandary. The green revolution and industrialized agriculture have unquestionably met the food requirements of most of the human population even as it has more than doubled since 1960. But we have had to pay for food gains with serious environmental problems, and we do not know if industrialized agriculture is sustainable for more than a few decades. To compound the issue, we must continue to increase food production to feed the growing human population, but the resulting damage to the environment may lessen our chances of increasing food production in the future.

Fortunately, the dilemma is not as hopeless as it seems. Farming practices and techniques exist that can ensure a sustainable agricultural output, preferably at yields comparable to industrialized agriculture. Farmers who practice industrialized agriculture can adopt these alternative agricultural methods, which cost less and are less damaging to the environment. Advances are also being made in sustainable subsistence agriculture. In addition, genetic engineering promises greater productivity and more nutritious foods.

A Move Toward Sustainable Agriculture, a Substitute for Industrialized Agriculture

More and more farmers are trying forms of agriculture that cause fewer environmental problems than industrialized agriculture. **Sustainable agriculture**, also called **alternative** or **low-input agriculture**, relies on beneficial biological processes and environmentally friendly chemicals (those that disintegrate quickly and do not persist as residues in the environment). In sustainable agriculture, certain modern agricultural techniques are carefully combined with traditional farming methods from agriculture's past. The sustainable farm consists of field crops, trees that bear fruits and nuts, small herds of livestock, and even tracts of forest. Such diversification protects the farmer against unexpected changes in the marketplace. The breeding of disease-resistant crop plants and the maintenance of animal health rather than the continual use of antibiotics to prevent disease are important parts of sustainable agriculture. Water and energy conservation are also practiced in sustainable agriculture.

Instead of using large quantities of chemical pesticides, sustainable agriculture controls pests by enhancing natural predator–prey relationships. Apple growers in Maryland monitor and encourage the presence of ladybird beetles in their orchards because these insects feed voraciously on European red mites, a major pest of apples. As a general rule, sustainable agriculture tries to maintain biological diversity on farms as a way to minimize pest problems. Providing hedgerows (rows of shrubs) between fields provides a habitat for birds and other insect predators.

Crop selection also helps control pests without heavy pesticide use. In parts of Oregon, apples can be grown without major pest problems, but insects often infest peaches, whereas in western Colorado, apples have major pest problems but peaches do well. Therefore, apples would be the preferred crop for sustainable agriculture in Oregon, as would peaches in Colorado.

An important goal of sustainable agriculture is to preserve the quality of agricultural soil. Crop rotation, conservation tillage, and contour plowing help control erosion and maintain soil fertility (see Figures 14.15b and 14.16a). Sloping hills that are converted to mixed-grass pastures erode less than do hills planted with field crops, thereby conserving the soil and supporting livestock.

Animal manure added to soil decreases the need for high levels of commercial inorganic fertilizers and cuts costs. Also, using biological nitrogen fixation to convert atmospheric nitrogen into a form that can be used by plants lessens the need for nitrogen fertilizers (see Chapter 6). A 15-year study of the effects of different agricultural practices on soil fertility was reported in 1998 in the journal *Nature*. The researchers grew corn under three sets of conditions: (1) application of manure and crop rotations between corn and soybeans (a nitrogen-fixing legume); (2) crop rotations between corn and soybeans; and (3) a control in which corn and soybeans were grown in separate fields using intensive agricultural methods. The results of this experiment indicated that the average corn yields were essentially the same in the three sets of conditions. Soil fertility, as measured by the amounts of soil organic matter and nitrogen, improved somewhat in the legume system (2) and markedly in the manure–legume system (1). Soil fertility in the control fields remained unchanged or declined. Also, 60% more nitrate leached into groundwater in the control as compared to the other two systems. The study concluded that a combination of manure and crop rotations with legumes is environmentally superior to intensive agricultural methods that use commercial inorganic fertilizers to supply nitrogen.

As you can see from this discussion, sustainable agriculture is not a single program but a series of programs that are adapted for specific soils, climates, and farming requirements. Some sustainable farmers—those who practice **organic agriculture**—use no pesticide chemicals, whereas other sustainable farmers use a system of **integrated pest management** (**IPM**). In IPM a limited use of pesticides is incorporated with such practices as crop rotation, continual monitoring for potential pest problems, use of disease-resistant varieties, and biological pest controls (see Chapter 22).

Making Subsistence Agriculture Sustainable and More Productive

Traditional slash-and-burn agriculture is sustainable as long as there are few farmers and large areas of rain forest. Because relatively small patches of forest are cleared for raising crops, the trees quickly return when the land is abandoned and the farmer has moved on to clear another plot of forest. If the abandoned land lies fallow for a period of 20 to 100 years, the forest recovers to the point where subsistence farmers can again clear the forest for planting. Burning the trees releases nutrients into the soil

so crops can again be grown there. Today, however, too many people practice slash-and-burn agriculture, and as a result, more and more tropical forests are being destroyed (see Chapter 17). Also, because so many people are trying to grow crops on rainforest land, the soil does not lie uncultivated between farming cycles long enough for it to recover.

Some researchers have been seeking ways to make former rainforest land retain its productivity for longer periods than are usual in shifting cultivation. Consider Papua New Guinea, a small island nation in which approximately 80% of the people are subsistence farmers. Research scientists in this country have developed methods to deal with some of the most troublesome problems associated with shifting cultivation: soil erosion, declining fertility, and attacks by insects and diseases. Their research, which is part of the Shifting Agriculture Improvement Program in Papua New Guinea, has helped forest plots remain productive for longer periods of time. Heavy mulching with organic material, such as weed and grass clippings, has lessened soil infertility and erosion. The composted mulch is then piled into rows that follow the contours of the land, further reducing erosion. Several crops are planted together, reducing insect damage. One of the crops is always a legume (such as beans), which helps restore nitrogen fertility to the soil. An extension program demonstrates these methods to farmers and distributes a book, *Subsistence Agriculture Improvement Manual*, to help educate rural farmers.

Genetic Engineering

The ability to take a specific gene from a cell of one kind of organism and place it into a cell of an unrelated organism, where it is expressed, is called **genetic engineering**. Genetic engineering has begun to revolutionize medicine and has great potential to improve agriculture as well.

The goals of genetic engineering in agriculture are not new. Using traditional breeding methods, farmers and scientists have developed desirable characteristics in crop plants and agricultural animals for centuries. It takes time to develop such genetically improved organisms, however. Using traditional breeding methods, it might take 15 years or more to incorporate genes for disease resistance into a particular crop plant. Genetic engineering has the potential to accomplish the same goal in a fraction of that time.

Genetic engineering differs from traditional breeding methods in that desirable genes from *any* organism can be used, not just those from the species of the plant or animal that is being improved. If a gene for disease resistance found in petunias would be beneficial in tomatoes, the genetic engineer can splice the petunia gene into the tomato plant (Figure 18.12). This could never be done by traditional breeding methods, because petunias and tomatoes belong to separate groups of plants and do not interbreed.

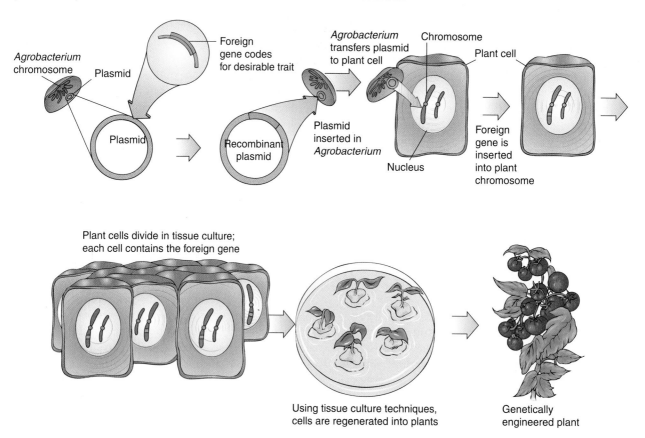

Figure 18.12 Genetic engineering. This example of genetic engineering uses a plasmid, which is a small circular molecule of DNA (genetic material) found in many bacteria. The plasmid of a bacterium called *Agrobacterium* can be used to introduce desirable genes from another organism into a plant. After the foreign DNA is spliced into the bacterial plasmid, the plasmid is inserted into the bacterium, which is then used to infect plant cells in culture. The foreign gene is inserted in the plant's chromosome, and genetically modified plants are then produced from the cultured plant cells.

Genetic engineering may produce food plants that would be more nutritious because they would contain all the essential amino acids. (Currently, no single food crop has this trait.) Genetic engineering may produce crops rich in beta-carotene, which the body uses to make vitamin A. According to WHO, 250 million of the world's children are at risk of vitamin A deficiency, which can cause poor vision, protein deficiency (vitamin A helps the body absorb and use amino acids), and an impaired immune system. In 2000, an international team of scientists reported that they had successfully engineered rice grains to produce beta-carotene. This feat has great potential for improved world health because about half the world's people eat rice as their staple food, and rice is a poor source of many vitamins, including vitamin A.

Crop plants resistant to insect pests, viral diseases, drought, heat, cold, herbicides, or salty or acidic soils are being developed. Scientists have inserted a gene from a virus into yellow squash and zucchini, thereby making them resistant to viral diseases that yellow their leaves and reduce crop yields. As another example, researchers at the USDA have identified a gene in rye that codes for a protein that prevents plant roots from absorbing aluminum, a metal that is more soluble in acidic soils and often reaches toxic levels in plants grown in acidic soils. (Aluminum is a natural component of the inorganic minerals of soils, but it is normally insoluble and therefore cannot be absorbed by plant roots.) Acidic soils are widespread in the tropics; in Latin America, 51% of all soils are acidified. Incorporation of the anti-aluminum gene into crop plants such as wheat would allow them to be grown in areas where they do not currently grow.

Genetic engineering has been used to develop more productive farm animals, including rapidly growing hogs and fishes. Perhaps the greatest potential contribution of genetic engineering in the animal arena, however, is in the production of vaccines against disease organisms that harm agricultural animals. For example, genetically engineered vaccines to protect cattle against a deadly viral disease known as rinderpest, which has reached epidemic proportions in parts of Asia and Africa, have been developed. Several hundred private genetic engineering firms, as well as thousands of scientists in colleges, universities, and government research labs around the world, are involved in agricultural genetic engineering. A great deal of research must be done before most of the envisioned benefits from genetic engineering are realized. Genetic engineering has already transformed agriculture, however. In 2000, **genetically modified (GM)** crops were cultivated on almost 30.3 million hectares (75 million acres) in the United States, the world's top producer of GM crops. Most of these crops are regulated by the USFDA, which applies the same policies and rules on health risk assessment to genetically modified crops as it does to new crops developed by traditional breeding methods.

The Safety of Genetic Engineering While acknowledging that the potential uses of genetic engineering are important and beneficial, the scientists who developed genetic engineering recognized that someone might engineer an organism that would cause ecological or health problems. Scientists therefore proposed stringent guidelines for making the new technology safe. Recent history has failed to bear out most of such concerns. Millions of experiments have demonstrated that genetic engineering experiments can be carried out safely. Today, most scientists recognize the practical importance of genetic engineering and generally agree that the potential threat to humans and the environment was overestimated.

Although many regulations have been relaxed, strict guidelines still exist in areas of genetic engineering research in which there are unanswered questions about possible effects on the environment. Much research is currently being conducted to assess the effects of introducing GM organisms into natural environments—for example, agricultural strains of plants whose seeds or pollen might spread in an uncontrolled manner. Carefully conducted tests have shown that GM organisms are not dangerous to the environment simply because they are genetically modified. However, it is important to assess the biology of each genetically modified organism. In this way scientists will be able to determine if it has characteristics that might cause an environmental hazard under certain conditions.

The **Biosafety Protocol**, an outgrowth of the 1992 U.N. Convention on Biological Diversity, lessens the threat of gene transfer from GM organisms to their wild relatives by providing appropriate procedures in the handling and use of GM organisms. The Biosafety Protocol does not go into effect until at least 50 nations ratify it. As of 2002, 36 nations had ratified the protocol, and an additional 67 nations had signed it and were considering ratification. (The United States has neither signed nor ratified the protocol.)

The Backlash Against Genetically Modified Foods During the late 1990s and early 2000s, opposition to genetically engineered crops increased in many highly developed countries in Europe as well as elsewhere. In 1999 the EU placed a moratorium on virtually all approvals of GM crops. The EU also refused to buy U.S. corn because it might be genetically modified. Some of the opposition to genetically engineered crops is based on fears that do not have any scientific foundation; some opposition may be the result of economic considerations, such as protecting the market for home-grown foods by banning imports; and some opposition is based on legitimate scientific concerns.

One concern is that the inserted genes could spread to weeds or wild relatives of crop plants and possibly harm natural ecosystems in the process. Scientists recognize this concern as legitimate and need to take special precautions to avoid this possibility.

Some consumers are concerned that GM foods might be hazardous to human health. Critics worry that some consumers might develop food allergies. Scientists also recognize this concern and routinely screen new GM crops for allergenicity. In the mid-1990s, for example, an agricultural company was engineering a soybean variety to be more nutritious by inserting a gene from Brazil nuts. During preliminary safety tests before the soybean variety was to be made commercially available, scientists determined that people allergic to Brazil nuts were also allergic to the GM soybean. As a result, the soybean variety was not placed on the market.

Should Foods from Genetically Modified Crops and Livestock Be Labeled? The question of whether or not to label foods containing GM organisms is controversial. Some consumers want such labels because they view genetic engineering as "unnatural." However, the position of the USFDA and most scientists is that such labeling would be counterproductive, in the sense that it would increase public anxiety over a technology that is essentially the same as conventional breeding methods. Labeling would also be expensive because the GM food would have to be kept separate from other foods during planting, harvesting, processing, and distribution; thus, production costs would be higher.

The scientific consensus is that the risks associated with consuming food derived from GM varieties are the same as those associated with consuming food derived from new varieties produced by traditional genetic techniques. Dozens of new plant varieties produced by traditional breeding methods enter the marketplace every year, from corn and rice to pumpkins and tomatoes; all are safe, and none are labeled. In 1996 the U.S. Court of Appeals upheld the USFDA view that labeling should not be required just because some consumers want to know.

Genetic Engineering and Edible Vaccines

Although reports of genetically modified crops often set off alarms, current agricultural and medical developments may give it a friendlier image. The new research field of "pharming" strives to genetically engineer animals to produce pharmaceuticals for human use. At the University of Massachusetts, scientists are developing GM calves in the hope of eventually producing females whose milk would contain certain desired medicines. In another cross of agriculture and medicine, molecular biologists are developing GM plants that deliver edible vaccines, including a tested anti-diarrhea vaccine and a promising cholera vaccine, both of which are contained in raw potato. Other promising pharmaceuticals from GM crops include an extract that prevents tooth decay and a vaccine against ulcers. Bananas, which are more palatable than raw potatoes, are also being genetically manipulated to deliver vaccines. Edible vaccines hold particular promise for developing countries, as they could be grown where most needed, and they would not require constant refrigeration or a supply of clean needles.

FISHERIES OF THE WORLD

The ocean contains a valuable food resource. About 90% of the world's total marine catch is fishes, with clams, oysters, squid, octopus, and other mollusks representing an additional 6% of the total catch. Crustaceans, including lobsters, shrimp, and crabs, make up about 3%, and marine algae constitute the remaining 1% (Figure 18.13).

Fishes and other seafood are highly nutritious because they contain high-quality protein (protein with a good balance of essential amino acids) that is easily digestible. Humans obtain approximately 5% of the total protein in their diet from fishes and other seafood; the rest is obtained from milk, eggs, meat, and plants. However, in certain countries, particularly in developing nations that border the ocean, seafood makes a much larger contribution to the total protein in the human diet.

Fleets of fishing vessels obtain most of the world's marine catch. In addition, numerous fishes are captured in shallow coastal waters and inland waters. According to the FAO, the world annual fish harvest increased substantially from 1950 (19 million tons) to 1996, when a record 93 million tons were caught. In 1998, the latest date available, the world fish catch was 86 million tons.

Problems and Challenges for the Fishing Industry

No nation lays legal claim to the open ocean. Consequently, resources in the ocean are more susceptible to overuse and degradation than are resources on the land, which individual nations own and for which they feel responsible (see "Case in Point: The Tragedy of the Commons" in Chapter 2).

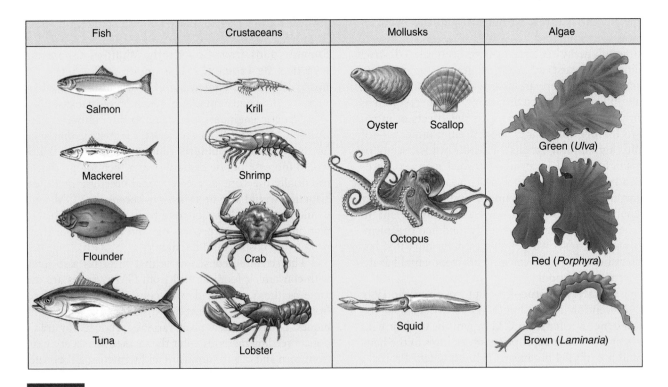

Figure 18.13 Major types of seafood. Fishes, crustaceans, mollusks, and algae comprise the ocean's food resources.

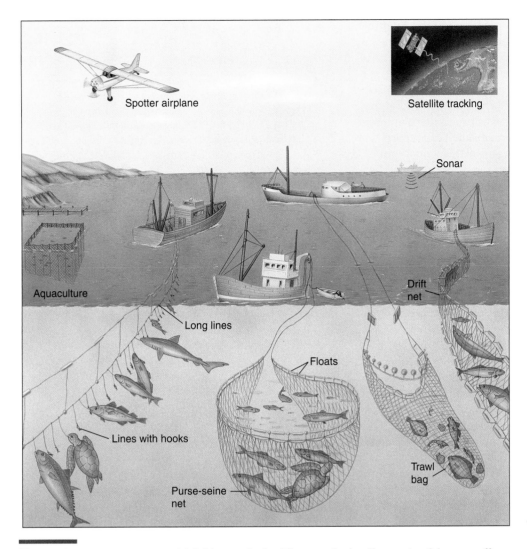

Figure 18.14 Modern commercial fishing methods. These methods of harvesting fish are so effective that many fish species have become commercially extinct. Sea turtles, dolphins, seals, whales, and other aquatic organisms are accidentally caught and killed in addition to the target fish. The depth of longlines can be adjusted to catch open-water fishes such as sharks and tuna or bottom fishes such as cod and halibut. Purse seines are used to catch anchovies, herring, mackerel, tuna, and other fishes that swim near the water's surface. Trawls are used to catch cod, flounder, red snapper, scallops, shrimp, and other fishes and shellfish that live on or near the ocean floor. Drift nets are used to catch salmon, tuna, and other fishes that swim in open waters.

The most serious problem for marine fisheries is that many marine species have been overharvested to the point that their numbers are severely depleted. Each fish species has a maximum sustainable harvest level; if a particular species is overharvested, its numbers drop, and harvest is no longer economically feasible.

According to the FAO, 62% of the world's fish stocks are in urgent need of management action. There are two reasons why fisheries have experienced such pressure: One, the growing human population requires protein in its diets, leading to a greater demand; and two, technological advances in fishing gear have made it possible to fish so efficiently that every single fish is often removed from an area.

Sophisticated fishing equipment includes sonar, radar, computers, airplanes, and even satellites to locate fish schools (Figure 18.14). Some boats set out *longlines*, which are fishing lines with thousand of baited hooks; each longline is up to 128 km (80 mi) long. *Purse-seine nets* are huge nets, as long as 2,000 m (1.25 mi), that are set out by small powerboats to encircle large schools of tuna and other fishes; after the fishes are completely surrounded, the bottom of the net is closed to trap them. A *trawl net* is a weighted, funnel-shaped net that is pulled along the bottom of the ocean to catch bottom-feeding fishes and shrimp; as much as 27 metric tons (30 tons) of fishes, shrimp, and other seafood can be caught in a single net. Trawl nets also destroy the ocean floor habitat.

Drift nets are plastic nets up to 64 km (40 mi) long that entangle thousands of fishes and other marine organisms; although drift nets have been banned by most countries, they continue to be used illegally.

Fishermen tend to concentrate on a few fish species with high commercial value, such as menhaden, salmon, tuna, and flounder, while other fish species are unintentionally caught and then discarded. The FAO reports that about 25% of all marine organisms caught—some 27 million metric tons (30 million tons)—are dumped back into the ocean. Most of these unwanted fishes, dolphins, and sea turtles, collectively known as **bycatch**, are dead or soon die because they are crushed by the fishing gear or are out of the water too long. The United States and other countries are trying to significantly reduce the amount of bycatch or develop uses for it.

In response to overharvesting, many nations have adopted a policy of **ocean enclosure**, which puts the organisms within 320 km (200 mi) of land under the jurisdiction of the country bordering the ocean. Open enclosure is supposed to prevent overharvesting by allowing nations to regulate the amounts of fishes and other seafood harvested from their waters. However, many countries have a policy of **open management**, in which all fishing boats of that country are given unrestricted access to fishes in national waters.

Marine fisheries of the United States are regulated by the **Magnuson Fishery Conservation Act**, which went into effect in 1977. It established eight regional fishery management councils, each of which developed a management plan for its region. Until 1996, the act was not particularly successful because managers were often pressured to set quotas too high, and the National Marine Fisheries Service estimated that more than one third of U.S. fish stocks were being fished at higher levels than could be sustained. In 1996 the act was reauthorized as the **Magnuson–Stevens Fishery Conservation and Management Act**. It required the regional councils and the National Marine Fisheries Service to protect "essential fish habitat" for more than 600 fish species, reduce overfishing, rebuild the populations of overfished species, and minimize bycatch. Overfishing will be reduced by using several management tools, including quotas, restrictions of certain types of fishing gear, limits on the number of fishing boats, and closure of fisheries during spawning periods.

One piece of good news for fisheries was reported in the journal *Science* in 1995. Fisheries biologists studied the population data of 128 depleted fish stocks and concluded that 125 of them can recover if fishing is carefully managed. Prior to this study, fisheries biologists had been concerned that many fish populations had declined so much that they would not be able to recover even if fishing were discontinued. Although this study provides good news about fish populations in the long term, it does not mean that recovery will occur rapidly. It may require the establishment of networks of

---

ENVIROBRIEF

Labels for Dolphin-Safe Tuna

Since the early 1990s, consumers have relied on the "dolphin safe" logo on canned tuna labels to certify that no dolphins were harmed, or were even present, when the tuna were caught. In a change of policy, the U.S. Department of Commerce in 1999 approved a new dolphin-safe label standard for tuna caught in the Eastern Tropical Pacific Ocean. Under the new guidelines, as requested by the International Dolphin Conservation Program, the dolphin-safe label will still be assigned if dolphins are present when tuna are caught, as long as no dolphins are seriously injured or killed in the process. The new standard assumes that netting dolphins does not significantly harm them.

Deaths to dolphins from tuna fishing in the Eastern Tropical Pacific have declined dramatically, by 98.5% since 1992. Several environmental organizations, such as Greenpeace, World Wildlife Fund, and the Center for Marine Conservation, support the new dolphin-safe standard, as they think sufficient progress is being made to curtail dolphin mortality. Critics of the new label—including, among others, Earth Island Institute, ASPCA, Sierra Club, and Defenders of Wildlife—argue that the logo will now be meaningless, even dishonest, to consumers, and that the less restrictive fishing methods will indeed hurt sensitive dolphin stocks. These groups think the decision favors international trade over science, as many nations fish the Eastern Tropical Pacific tuna stocks.

"no-take" reserves and a substantial reduction of fishing fleets. Furthermore, it may take years for some commercially important fish populations to rebound. All of this translates into severe economic hardship for the fishing industry.

Pollution and Deteriorating Habitat One of the great paradoxes of human civilization is that the same ocean that is used to provide food to a hungry world is also used as a dumping ground. Pollution increasingly threatens the world's fisheries. Everything from accidental oil spills to the deliberate dumping of litter pollutes the water. Heavy metals such as lead, mercury, and cadmium are widely used in industry but enter the environment and find their way into aquatic food webs, where they are highly toxic to both fishes and humans who eat fish.

Between 60% and 80% of all commercially important ocean fishes spend at least part of their lives in coastal areas. Tidal marshes, mangrove swamps, estuaries, and the like serve as spawning areas, nurseries, and feeding grounds. Coastal areas are also in high demand for recreational and residential development, and many coastal waters are polluted. The World Resources Institute estimates that about 80% of global ocean pollution comes from human activities on land. Stormwater runoff from cities, agricultural areas, and roads is the single largest source of ocean pollution. As development and pollution degrade coastal ecosystems, the habitats of young marine animals are undermined, contributing to further depletion of fish stocks already suffering from overfishing. (Chapter 7 discusses the impact of human activities on the ocean, and Chapter 17 discusses coastlines.)

Aquaculture: Fish Farming

Aquaculture, the rearing of aquatic organisms, is more closely related to agriculture on land than it is to the fishing industry just described (Figure 18.15). Aquaculture is carried out both in fresh water and marine water near the shore; the cultivation of marine organisms is sometimes called **mariculture**. To optimize the quality and productivity of their "crops," aquaculture farmers control the diets, breeding cycles, and environmental conditions of their ponds or enclosures. Aquaculturists try to reduce pollutants that might harm the organisms they are growing, and they keep them safe from potential predators.

Although aquaculture is an ancient practice that probably originated in China several thousand years ago, its enormous potential to provide food has only recently been appreciated. Aquaculture can contribute variety to the diets of people in highly developed countries. Inhabitants of developing nations can benefit even more from aquaculture: It may provide them with much-needed protein and even serve as a source of foreign exchange when they export such delicacies as aquaculture-grown shrimp.

According to the FAO, world aquaculture production of finfish and shellfish reached 36.1 million tons in 2000 (latest data available). Other important aquaculture crops include seaweeds, oysters, mussels, clams, lobsters, and crabs. Currently, aquaculture is the fastest growing type of food production in the world, and one out of every three fish destined for human consumption comes from fish farms. The nation with the largest aquaculture harvest is China, which in 2000 accounted for about 68% of total world production.

Aquaculture is a $900 million–a-year industry in the United States, where it accounts for 6% of all fish and seafood consumed. All striped bass and rainbow trout available at U.S. retail markets are produced by aquaculture, as are more than half the fresh salmon served in the United States. Catfish, tilapia, salmon, shrimp, and oysters are the most important types of seafood grown by aquaculture in the United States.

Aquaculture differs from fishing in several respects. For one thing, although the highly developed nations harvest more fishes from the ocean, the developing nations produce much more seafood by aquaculture. One reason for this is that developing nations have an abundant supply of cheap labor, which is a requirement of aquaculture because it, like land-based agriculture, is labor-intensive. Another difference between fishing and aquaculture is that the limit on the size of a catch in fishing is the size of the natural population, whereas the limit on aquacultural production is largely the size of the area in which they are grown.

In addition to being done inland, aquaculture is practiced in estuaries and in the ocean near the shore. Therefore, other uses of coastlines compete with aquaculture for available space. Developing countries that grow shrimp by aquaculture cut down the coastal mangroves that provide so many important environmental benefits (Figure 18.16; also see Chapter 7). Many marine fishes breed in the tangled roots of mangroves, and there are concerns that an expansion of shrimp farming could contribute to a decline in marine fish populations.

Because so many fishes are concentrated in a relatively small area, aquaculture produces wastes that can pollute the adjacent water and harm other organisms. Aquaculture also causes a net loss of wild fish because many of the fishes that are farmed are carnivorous. Sea bass and salmon, for example, eat up to 5 kg of wild fish to gain 1 kg of weight.

Although interest in aquaculture is increasing worldwide, several factors are slowing its expansion. Setting up and running an aquaculture facility is expensive. Also,

Figure 18.15 Aquaculture. Shown are tilapia at a large aquaculture facility in Hawaii. Aquaculture has become increasingly important in providing the world's seafood.

Figure 18.16 Aquaculture in a coastal mangrove forest. The dykes enclose shrimp ponds. Aquaculture of shrimp is the single largest factor responsible for mangrove habitat losses worldwide. Photographed in Borneo.

scientific research is necessary in order to make aquaculture of certain organisms profitable. An organism's requirements for breeding must be established, and ways to control excessive breeding must be available so that the population does not overbreed and produce many stunted individuals rather than fewer large ones. The population must be continually monitored for diseases, which have a tendency to spread rapidly in the crowded conditions that are characteristic of aquaculture.

One of the most important limits on aquaculture's potential is the receptivity of animals to the domestication process itself. Land animals such as cows, pigs, and

sheep were domesticated over a period of thousands of years. During this time, there were undoubtedly failed attempts to domesticate other animals, which for one reason or another could not be domesticated. The same is true of aquaculture: It is not simply a matter of observing a need for more tuna, for instance, and therefore opening an aquaculture facility that produces tuna. The organisms that are to be produced profitably by aquaculture must have certain traits that make their domestication possible. Aquatic organisms that are social by nature and do not exhibit territoriality or aggressive behavior are possible candidates for domestication.

SUMMARY WITH SELECTED KEY TERMS

I. Humans require a balanced diet that includes **carbohydrates**, **proteins**, and **lipids** in addition to **vitamins**, **minerals**, and water.

A. People who consume fewer calories than they need are **undernourished**, whereas people who consume enough calories but whose diets are lacking in some specific nutrients are **malnourished**.

B. People who consume food in excess of that required are **overnourished**.

II. About 828 million people lack access to the food needed to be healthy and to lead productive lives.

A. The two regions of the world with the greatest food insecurity are South Asia and sub-Saharan Africa.

B. Crop failures caused by drought, war, flood, or some other catastrophe may result in severe food shortages called **famines**.

C. The greatest challenge in agriculture today is producing enough food to feed the world's population.

   1. **World grain carryover stocks** are the amounts of rice, wheat, corn, and other grains remaining from previous harvests, as estimated at the start of a new harvest.

   2. World grain carryover stocks provide a measure of **food security**, a goal in which all people have access at all times to adequate amounts and kinds of food needed to lead healthy, active lives.

   3. Annual world grain production was lower than consumption for 3 consecutive years, beginning in 2000.

D. Providing enough food for all people is complicated by poverty, problems of distributing food where it is needed, and cultural acceptance of nutritious but unfamiliar foods.

E. The long-term solution to the problem of producing adequate food is the stabilization of the human population.

III. Although thousands of plant species are edible, today only about 100 plants provide about 90% of the food that humans consume.

A. Rice, wheat, and corn provide about half of the calories that people consume.

B. Our dependence on so few species of plants for the bulk of our food makes us vulnerable should disease or some other factor wipe out one of our important food crops.

IV. Agricultural trends today exhibit greater uniformity, in both the foods we eat and agricultural techniques.

A. Most farmers in highly developed countries and some in developing countries rely on **industrialized agriculture**.

    1. Industrialized agriculture produces high **yields**, amounts of food produced per unit of land.

    2. Industrialized agriculture requires a lot of energy (i.e., fossil fuels) to produce commercial inorganic fertilizers and other chemicals, power farm machinery, and operate irrigation systems.

B. Most farmers in developing countries still practice **subsistence agriculture**, the production of enough food to feed oneself and one's family, with little left over to sell or reserve for hard times.

    1. **Shifting cultivation**, including **slash-and-burn agriculture**, and **nomadic herding** are examples of subsistence agriculture that are very land-intensive.

    2. **Polyculture** is a type of **intercropping** in which several crops that mature at different times are grown together.

    3. In subsistence agriculture, human and animal energy is used instead of fossil fuels.

V. When plants and animals are **domesticated**, much of the **genetic diversity** found in wild populations is lost.

A. Agriculture protects domesticated plants and animals as much as possible from pests and diseases.

B. Globally, a few agricultural varieties are replacing the hundreds of varieties developed by farmer-breeders over the centuries.

VI. The **green revolution**, in which new, high-yielding varieties are intensively cultivated with mechanized machinery, commercial inorganic fertilizers, and pesticides, began in the 1960s.

A. The green revolution gave Latin American and Asian countries the chance to produce adequate supplies of food.

B. Africa has not benefited much from the green revolution because its common crops have not been greatly improved by green revolution technology.

VII. The use of **hormones** and **antibiotics** has increased animal production.

A. The European Union bans imports of hormone-fed beef, ostensibly because of health concerns. The U.N. Food and Agricultural Organization and World Health Organization have said that hormone-fed livestock do not pose any health concerns.

B. Several studies link the indiscriminate use of antibiotics in humans or livestock to bacteria that are increasingly resistant to antibiotics.

VIII. Food safety is a complex issue, particularly because much of our food comes from other countries.

A. The U.S. Department of Agriculture and the U.S. Food and Drug Administration address the safety of the foods we eat.

B. Food-borne contaminants such as bacteria can be introduced into the food cycle during production, processing, distribution, and preparation.

C. The U.S. food supply is safer today than in the past.

D. **Food additives** are chemicals that enhance the taste, color, or texture of food; improve its nutrition; reduce spoilage and prolong shelf life; or maintain the food's consistency. **Coloring agents**, **preservatives**, and **antioxidants** are examples of food additives.

IX. In industrialized agriculture, environmental costs are high.

A. Soil erosion causes a decline in soil fertility as well as downstream sediment pollution.

B. Agricultural chemicals such as pesticides and commercial inorganic fertilizers cause air, water, and soil pollution.

    1. Both surface waters and groundwater have been adversely affected.

    2. Soil microorganisms are sensitive to pesticides, as are fishes and other aquatic organisms, which are exposed to pesticide runoff from fields.

    3. Many insects, weeds, and disease-causing organisms have developed resistance to pesticides, forcing farmers to apply larger quantities.

C. Irrigation consumes huge quantities of fresh water.

D. Expanding the amount of agricultural land has resulted in **habitat fragmentation** and losses that reduce biological diversity.

X. The challenges confronting agriculture are being met in a variety of ways.

A. Methods are being developed to make industrialized agriculture sustainable. **Sustainable agriculture** relies on beneficial biological processes and environmentally friendly chemicals.

    1. **Organic agriculture** uses no pesticides.

    2. **Integrated pest management** uses such practices as crop rotation, disease-resistant varieties, and biological pest controls to limit the use of pesticides.

B. Methods are being developed to make subsistence agriculture sustainable and more productive.

C. **Genetic engineering**, the transfer of specific genes from one species to another, is the high-technology answer to some of agriculture's challenges.

    1. Genetic engineering may produce food plants that are more nutritious, resistant to insect pests and viral diseases, or tolerant of drought, heat, cold, herbicides, or salty soil.

    2. Strict guidelines exist in areas of genetic engineering research in which there are unanswered questions about possible effects on the environment or health.

XI. The ocean contains a valuable food resource.

A. Currently, 62% of the world's fish stocks are either fully exploited, overexploited, or depleted because of the growing human population and technological advances in fishing gear.

1. About 25% of all marine organisms caught are unwanted **bycatch** that are discarded.

2. In response to overharvesting, many nations have adopted a policy of **ocean enclosure**, which puts the ocean up to 320 km from land under the jurisdiction of the country bordering the ocean.

3. Many countries have a policy of **open management**, which allows all fishing boats of that country unrestricted access to fishes in national waters.

B. **Aquaculture**, the rearing of aquatic organisms, is supplementing traditional fishing in the supply of high-quality protein from seafood.

1. Aquaculture is like agriculture, whereas fishing is like hunting.

2. Currently, aquaculture is the fastest growing type of food production in the world, and one out of every three fish destined for human consumption comes from fish farms.

3. Although aquaculture has an enormous potential to provide food, it causes environmental problems, such as loss of coastlines and water pollution.

THINKING ABOUT THE ENVIRONMENT

1. How does Walnut Acres, the farm described in the chapter introduction, mimic a natural ecosystem?

2. Distinguish between *undernutrition*, *malnutrition*, and *overnutrition*.

3. What age group in humans is usually most affected by undernutrition, malnutrition, and famine? Why?

4. What are world grain carryover stocks?

5. Why is population control the most fundamental solution to world food problems? Explain your answer.

6. Why does decreased genetic diversity in farm plants and animals increase the likelihood of economic disaster from disease?

7. Distinguish between *industrialized agriculture* and *subsistence agriculture*.

8. Describe *shifting cultivation*, *nomadic herding*, and *intercropping*.

9. Distinguish between *shifting cultivation* and *slash-and-burn agriculture*.

10. Distinguish between *intercropping* and *polyculture*.

11. What is the green revolution? Describe the benefits and problems associated with the green revolution.

12. Describe current food safety issues.

13. What are the major environmental problems associated with industrialized agriculture?

14. Describe the environmental problems associated with farming each of these areas: tropical rain forests; hillsides; arid regions.

15. What is sustainable agriculture? Give at least three examples of ways that industrialized agriculture could be made more sustainable.

16. List the pros and cons of genetically engineering crops.

17. Some scientists are genetically engineering herbicide resistance into crop plants so that when they apply herbi-

cides, only the weeds die and not the crops. How might this specific example of genetic engineering have negative impacts on the environment?

18. Explain why aquaculture is more like agriculture than it is like traditional fishing.

19. What are some of the harmful environmental effects associated with aquaculture?

*20. It takes 7 kg of livestock grain such as corn to produce 1 kg of beef; 6 kg of livestock grain to produce 1 kg of pork; and 2.7 kg of livestock grain to produce 1 kg of poultry. If corn sells for $150 a ton, calculate how much it costs to produce 1 kg of each kind of meat. (One ton equals 907.2 kg.)

*21. Complete Table 18.A on world grain production in 1970 and 2001. What does this information tell you about how the average yield (amount of grain harvested per unit area of land) changed between 1970 and 2001?

Table 18.A World Grain Production, 1970 and 2001

	1970	2001	Percent Increase
Agricultural land (million hectares)	663	684	—
Grain production (million tons)	1,079	1,843	—
Tons of grain per hectare	—	—	—

* Solutions to questions preceded by an asterisk appear in Appendix VII.

TAKE A STAND

Visit our Web site at **http://www.wiley.com/college/raven** (select Chapter 18 from the Table of Contents) for links to more information about issues involving genetically modified (GM) foods. Consider the opposing views of those who support and those who oppose the development and use of GM foods and debate the issue with your classmates. You will find tools to help you organize your research, analyze the data, think critically about the issues, and construct a well-considered argu-ment. Take a Stand activities can be done individually or as part of a team, as oral presentations, written exercises, or Web-based (e-mail) assignments.

Additional on-line materials relating to this chapter, including Student Quizzes, Activity Links, Useful Web Sites, Flash Cards, and more, can also be found on our Web site.

SUGGESTED READING

Ackerman, J. "Food: How Altered?" *National Geographic*, Vol. 201, No. 5 (May 2002). An excellent overview of the promise and debate over genetic engineering.

Ackerman, J. "Food: How Safe?" *National Geographic*, Vol. 201, No. 5 (May 2002). An excellent overview of world food safety issues.

Brown, L.R. "Eradicating Hunger: A Growing Challenge." In *State of the World 2001*, New York: W.W. Norton & Company (2001). An in-depth analysis of why addressing the problem of eradicating hunger is such a complex issue.

Cowley, G. "Certified Organic." *Newsweek* (September 30, 2002). New federal rules will define what "organic foods" actually are.

Gardner, G., and B. Halweil. "Escaping Hunger, Escaping Excess." *WorldWatch*, Vol. 13, No. 4 (July–August 2000). Both poor and affluent countries are faced with the problem of malnutrition.

Menzel, P. "What's For Dinner?" *Smithsonian* (January 2002). A fascinating photo essay of what people around the world eat.

Milius, S. "Carnivorous Fish Nibble at Farming Gain." *Science News*, Vol. 158 (July 1, 2000). Farmed fish that are carnivorous consume large quantities of wild fish.

Mlot, C. "Antidotes for Antibiotic Use on the Farm." *BioScience*, Vol. 50, No. 11 (November 2000). Scientists are trying to find alternatives to heavy antibiotic use in agriculture.

Raloff, J. "Downtown Fisheries?" *Science News*, Vol. 157 (May 13, 2000). Fish farming has potential even in inner cities.

Raloff, J. "Hormones: Here's the Beef." *Science News*, Vol. 161 (January 5, 2002). The practice of giving hormones to livestock causes environmental concerns.

Tuxill, J. "The Biodiversity That People Made." *WorldWatch*, Vol. 13, No. 3 (May–June 2000). Genetic diversity in the world's crops is a valuable resource.

Winslow, R., and P. Landers. "Obesity: A World-wide Woe." *Wall Street Journal* (July 1, 2002). Worldwide, the number of obese people is increasing at an alarming rate.

Chattanooga, Tennessee. Chattanooga's air quality has improved dramatically during the past several decades.

Air Pollution

Learning Objectives

After you have studied this chapter you should be able to

1. List the seven major classes of air pollutants and describe their characteristics and effects.

2. Relate the adverse health effects of specific air pollutants and explain why children are particularly susceptible to air pollution.

3. Describe industrial smog, photochemical smog, temperature inversions, urban heat islands, and dust domes.

4. Summarize the effects of the Clean Air Act on U.S. air pollution.

5. Contrast air pollution in highly developed and developing countries.

6. Describe the global distillation effect and tell where it commonly occurs.

7. Summarize the sick building syndrome.

8. Describe the physiological effects of noise pollution on the human body.

During the 1960s the federal government gave Chattanooga, Tennessee, the dubious distinction of having the worst air pollution in the United States. The air was so dirty in this manufacturing city that sometimes people driving downtown had to turn on their headlights in the middle of the day. The orange air soiled their white shirts so quickly that many businessmen brought extra ones to work. To compound the problem, mountains that surround the city kept the pollutants produced by its inhabitants from dispersing.

Today the air in this scenic midsized city of 200,000 people is clean, and Chattanooga ranks high among U.S. cities in terms of air quality. Efforts by city and business leaders are credited with transforming Chattanooga's air. Soon after the passage of the federal Clean Air Act of 1970, the city established an air pollution control board to enforce regulations controlling air pollution. New local regulations allowed open burning by permit only, placed limits on industrial odors and particulate matter (dust and ash), outlawed visible automotive emissions, and set a 4% cap on sulfur content in fuel, which controlled the production of sulfur oxides. Businesses complied by installing expensive air pollution control devices. The city started an emissions-free electric bus system that was carrying more than 1 million passengers per year by the late 1990s. Chattanooga also decided to recycle its solid waste rather than build an incinerator that would have produced emissions.

By 1972, the measures had proven so effective that Chattanooga drew national attention for its cleanup effort. The National Air Pollution Control Association awarded Chattanooga first place in its annual Cleaner Air Week ceremonies. The *Wall Street Journal*, the *New York Times*, ABC-TV, CBS-TV, and *U.S. News and World Report* did stories covering the success.

In 1984, Chattanooga was officially designated *in attainment* for particulate matter, one of the pollutants regulated by the Environmental Protection Agency (EPA); this designation meant that particulate levels had been below the federal health standard for 1 year. The city reached attainment status for ozone in 1989. Since then, the levels for all seven air pollutants regulated by the EPA have been better in Chattanooga than federal standards require. (Federal standards for ozone and particulate matter are currently being revised, so Chattanooga and other U.S. cities will soon have to comply with stricter attainment levels.)

In the early 2000s, Chattanoogans continued to move their city toward environmental sustainability. The city plans to convert a run-down business district into a community in which people live near their places of work. Businesses located in this district will form an industrial ecosystem in which the wastes of one business will be used as raw materials by another business (see Chapter 15).

Chattanooga's air quality is an environmental success story that could be emulated by other U.S. cities. The air we breathe is often dirty and contaminated with pollutants, particularly in urban areas. Air pollution also extends indoors, and the air we breathe at home, at work, and in our automobiles may be more polluted than the air outdoors. Because air pollution causes a great many health and environmental problems, most highly developed nations and many developing nations have established air quality standards for numerous pollutants.

THE ATMOSPHERE AS A RESOURCE

The atmosphere is a gaseous envelop surrounding Earth (see Figure 6.10). Excluding water vapor and air pollutants, four gases comprise the atmosphere: nitrogen (N_2, 78.08%), oxygen (O_2, 20.95%), argon (Ar, 0.93%), and carbon dioxide (CO_2, 0.04%). The two atmospheric gases most important to humans and other organisms are carbon dioxide and oxygen. During photosynthesis, plants, algae, and certain bacteria use carbon dioxide to manufacture sugars and other organic molecules. During cellular respiration, most organisms use oxygen to break down food molecules and supply themselves with chemical energy. Nitrogen gas is an important component of the nitrogen cycle (see Chapter 6). The atmosphere performs additional **ecosystem services**, namely, blocking the surface of Earth from much of the ultraviolet radiation (UV) coming from the sun, moderating the climate,

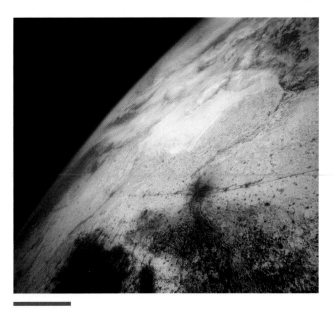

Figure 19.1 The atmosphere. The "ocean of air" is an extremely thin layer compared to the size of Earth. In this photo, shot in a low-Earth orbit from the space shuttle *Endeavor*, the atmosphere is a thin blue layer that separates the planet from the blackness of space. (The Trans-Siberian Railway, the world's longest, appears as a dark line against a background of snow.)

and redistributing water in the hydrologic cycle (see Chapters 6 and 20).

Living at the interface of the Earth's surface and atmosphere, humans think of the atmosphere as an unlimited resource, but we should reconsider. Ulf Merbold, a German space shuttle astronaut, felt very differently about the atmosphere after viewing it in space (Figure 19.1). "For the first time in my life, I saw the horizon as a curved line. It was accentuated by the thin seam of dark blue light—our atmosphere. Obviously, this was not the 'ocean' of air I had been told it was so many times in my life. I was terrified by its fragile appearance."

TYPES AND SOURCES OF AIR POLLUTION

Air pollution consists of gases, liquids, or solids present in the atmosphere in high enough levels to harm humans, other organisms, or materials. Although air pollutants can come from natural sources—as when lightning causes a forest fire or a volcano erupts—human activities release many kinds of substances into the atmosphere and make a major contribution to global air pollution. Some of these substances are harmful when they precipitate (form a solid) and settle on land and surface waters, whereas other substances are harmful because they alter the chemistry of the atmosphere. From the standpoint of human health, probably more significant than the overall human contribution of air pollution is the fact that much

Table 19.1   Major Air Pollutants

Pollutant	Composition	Primary or Secondary	Characteristics
Particulate matter			
Dust	Variable	Primary	Solid particles
Lead	Pb	Primary	Solid particles
Sulfuric acid	H_2SO_4	Secondary	Liquid droplets
Nitrogen oxides			
Nitrogen dioxide	NO_2	Primary	Reddish brown gas
Sulfur oxides			
Sulfur dioxide	SO_2	Primary	Colorless gas with strong odor
Carbon oxides			
Carbon monoxide	CO	Primary	Colorless, odorless gas
Carbon dioxide*	CO_2	Primary	Colorless, odorless gas
Hydrocarbons			
Methane	CH_4	Primary	Colorless, odorless gas
Benzene	C_6H_6	Primary	Liquid with sweet smell
Ozone	O_3	Secondary	Pale blue gas with sweet smell
Air toxics			
Chlorine	Cl_2	Primary	Yellow-green gas

* Discussed in Chapter 20.

of the air pollution released by humans is concentrated in densely populated urban areas.

Although many different air pollutants exist, we will focus our attention on the seven most important types from a regulatory perspective: particulate matter, nitrogen oxides, sulfur oxides, carbon oxides, hydrocarbons, ozone, and air toxics (Table 19.1 and "Mini-Glossary: Air Pollutants"). Air pollutants are often divided into two categories, primary and secondary (Figure 19.2). **Primary air pollutants** are harmful chemicals that enter directly into the atmosphere. The major ones are carbon oxides, nitrogen oxides, sulfur dioxide, particulate matter, and hydrocarbons. **Secondary air pollutants** are harmful chemicals that form from other substances that have been released into the atmosphere. Ozone and sulfur trioxide are secondary air pollutants because both are formed by chemical reactions that take place in the atmosphere.

Major Classes of Air Pollutants

Particulate matter consists of thousands of different solid and liquid particles that are suspended in the atmosphere. **Solid particulate matter** is generally referred to as *dust*, whereas liquid suspensions are commonly called *mists*. Particulate matter includes many things that can be pollutants, such as soil particles, soot, lead, asbestos, sea salt, and sulfuric acid droplets. Particulate matter reduces visibility by scattering and absorbing sunlight. Urban areas receive less sunlight than rural areas, partly as a result of greater quantities of particulate matter in the air. Particulate matter corrodes metals, erodes buildings and works of sculpture when the air is humid, and soils clothing and draperies.

All particulate matter eventually settles out of the atmosphere, but microscopic particles, some of which are especially harmful to humans, can remain suspended in the atmosphere for weeks or even years. Traces of hundreds of different chemicals bind to these microscopic particles; inhaling the particles introduces the chemicals,

MINI-GLOSSARY

Air Pollutants

primary air pollutants: Pollutants emitted directly into the atmosphere.

secondary air pollutants: Air pollutants formed from primary air pollutants as a result of chemical reactions taking place in the atmosphere.

particulate matter: Fine solids or liquid droplets suspended in the air; includes grit, dust, fumes, aerosols, and smoke.

nitrogen oxides: A collective term for several gases, of which nitrous oxide, nitric oxide, and nitrogen dioxide are most important.

sulfur oxides: A collective term for two important gases—sulfur dioxide and sulfur trioxide.

carbon oxides: A collective term for the gases carbon monoxide and carbon dioxide.

hydrocarbons: Organic compounds that contain the elements carbon and hydrogen.

ozone: A form of oxygen, each molecule of which consists of three oxygen atoms. Produced naturally in the stratosphere, ozone in the troposphere is a secondary air pollutant produced by photochemical reactions.

hazardous air pollutants: Potentially harmful air pollutants that may pose health risks to people exposed to them.

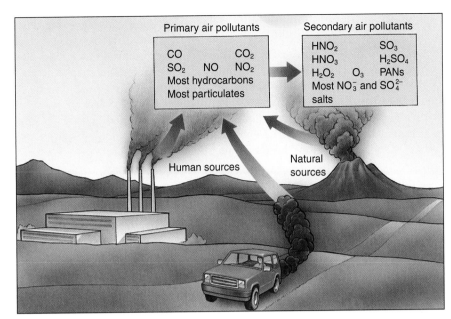

Primary air pollutants

CO   CO₂
SO₂   NO   NO₂
Most hydrocarbons
Most particulates

Secondary air pollutants

HNO₂   SO₃
HNO₃   H₂SO₄
H₂O₂   O₃   PANs
Most NO₃⁻ and SO₄²⁻
salts

Human sources

Natural sources

Figure 19.2 Primary and secondary air pollutants. Primary air pollutants are emitted, unchanged, from a source directly into the atmosphere, whereas secondary air pollutants are produced from chemical reactions involving primary air pollutants.

some of which are toxic, into the body. Microscopic particles are considered more dangerous than larger particles because they are inhaled more deeply into the lungs. The EPA samples microscopic particulate matter at 1,000 locations around the United States so that they can better understand its composition, which varies with location and season.

Nitrogen oxides are gases produced by the chemical interactions between nitrogen and oxygen when a source of energy, such as combustion of fuels, produces high temperatures. Collectively known as NO_x, nitrogen oxides consist mainly of nitric oxide (NO), nitrogen dioxide (NO_2), and nitrous oxide (N_2O). Nitrogen oxides inhibit plant growth and, when breathed, aggravate health problems such as asthma. They are involved in the production of photochemical smog and acid deposition (when nitrogen dioxide reacts with water to form nitric acid and nitrous acid). Nitrous oxide is associated with global warming (nitrous oxide traps heat in the atmosphere and is therefore known as a **greenhouse gas**). Nitrous oxide also depletes ozone in the stratosphere. Nitrogen oxides cause metals to corrode and textiles to fade and deteriorate.

Sulfur oxides are gases produced by the chemical interactions between sulfur and oxygen. Sulfur dioxide (SO_2), a colorless, nonflammable gas with a strong, irritating odor, is a major sulfur oxide emitted as a primary air pollutant. Another major sulfur oxide is sulfur trioxide (SO_3), a secondary air pollutant that forms when sulfur dioxide reacts with oxygen in the air. Sulfur trioxide, in turn, reacts with water to form another secondary air pollutant, sulfuric acid. Sulfur oxides are very important in acid deposition, and they corrode metals and damage stone and other materials. Sulfuric acid and sulfate salts that are produced in the atmosphere from sulfur oxides

damage plants and irritate the respiratory tracts of humans and other animals.

Carbon oxides are the gases carbon monoxide (CO) and carbon dioxide (CO_2). Carbon monoxide, a colorless, odorless, and tasteless gas produced in the largest quantities of any atmospheric pollutant except carbon dioxide, is poisonous and reduces the blood's ability to transport oxygen. Carbon dioxide, also colorless, odorless, and tasteless, is a greenhouse gas; its buildup in the atmosphere is associated with global warming.

Hydrocarbons are a diverse group of organic compounds that contain only hydrogen and carbon; the simplest hydrocarbon is methane (CH_4). Small hydrocarbon molecules are gaseous at room temperature. Methane is a colorless, odorless gas that is the principal component of natural gas. (The odor of natural gas comes from sulfur compounds that are deliberately added so that humans can indirectly detect the presence of the explosive methane gas by smelling the sulfur-containing compounds.) Medium-sized hydrocarbons such as benzene (C_6H_6) are liquids at room temperature, although many are *volatile* and evaporate readily. The largest hydrocarbons, such as they waxy substance paraffin, are solids at room temperature. The many different hydrocarbons have a variety of effects on human and animal health. Some appear to cause no adverse effects, some injure the respiratory tract, and others have been demonstrated to cause cancer. All except methane are important in the production of photochemical smog. Methane is a potent greenhouse gas that is linked to global warming.

Ozone (O_3) is a form of oxygen considered a pollutant in one part of the atmosphere but an essential component in another. In the stratosphere, which extends from 10 to 45 km (6.2 to 28 mi) above the Earth's surface, oxygen reacts with UV coming from the sun to form ozone. Stratospheric ozone prevents much of the solar UV from penetrating to the Earth's surface. Unfortunately, certain human-made pollutants (chlorofluorocarbons, or CFCs) react with stratospheric ozone, breaking it down into molecular oxygen, O_2.

Unlike stratospheric ozone, ozone in the troposphere—the layer of atmosphere closest to the Earth's surface—is a human-made air pollutant. (Ground-level ozone does not replenish the ozone that has been depleted from the stratosphere because it breaks down to form oxygen long before it drifts up to the stratosphere.) Ozone in the troposphere is a secondary air pollutant that forms when sunlight catalyzes reactions between nitrogen oxides and volatile hydrocarbons. The most harmful component of photochemical smog, ozone reduces air visibility and causes health problems. Ozone also stresses plants and reduces their vigor, and chronic (of long duration) ozone exposure lowers crop yields (Figure 19.3). Ozone has been observed to reduce soybean yields by as much as 35% (the average reduction from exposure to ozone is 15%), according to Dr. Loucks at Miami University in Ohio. Chronic exposure to ozone is a possible contributor to forest decline, and ground-level ozone is a greenhouse gas associated with global warming.

Most of the hundreds of other air pollutants—such as chlorine, lead, hydrochloric acid, formaldehyde, radioactive substances, and fluorides—are present in very low concentrations, although it is possible to have high local concentrations of specific pollutants. Some of these air pollutants, known as **hazardous air pollutants**, or **air toxics**, are potentially harmful and may pose long-term health risks to people who live and work around chemical factories, incinerators, or other facilities that produce or use them. To limit the release of more than 180 hazardous air pollutants, the Clean Air Act Amendments of 1990 (discussed later in the chapter) regulate emissions of both large and small businesses, such as bakeries, distilleries, dry cleaners, furniture makers, gasoline service stations, hospitals, auto paint shops, and print shops.

Sources of Outdoor Air Pollution

The two main human sources of primary air pollutants are transportation (mobile sources) and industries (stationary sources) (Figure 19.4). Automobiles and trucks, known as *mobile sources*, release significant quantities of nitrogen oxides, carbon oxides, particulate matter, and hydrocarbons as a result of the combustion of gasoline. Although diesel engines in trucks, buses, trains, and ships consume less fuel than other types of combustion engines, they produce more air pollution. One heavy-duty truck emits as much particulate matter as 150 automobiles, whereas one diesel train engine produces, on average, 10 times the particulate matter of a diesel truck. Pollutants from engines used in outboard motorboats, jet skis, and other mobile sources cause both air and water pollution. According to the California Air Resources Board, a 2-hour ride on a 100-horsepower Jet Ski produces as much pollution as driving 139,000 miles in a 1998 automobile.

Electric power plants and other industrial facilities, known as *stationary sources*, emit most of the particulate matter and sulfur oxides released in the United States; they also emit sizable amounts of nitrogen oxides, hydrocarbons, and carbon oxides. The combustion of fossil fuels, especially coal, is responsible for most of these emissions (see Chapter 10). The top three industrial sources of toxic air pollutants—that is, chemicals released into the air that are fatal to humans at specified concentrations—are the chemical industry, the metals industry, and the paper industry.

Not all air pollution is generated by human activities. On a hot summer day in the Blue Ridge Mountains, which are part of the Appalachians, a blue haze hangs over the forested hills. This haze is caused by hydrocarbon emissions that are emitted from tree leaves. Many plants produce a variety of hydrocarbons in response to heat. The hydrocarbon isoprene, for example, is thought to help protect leaves from high temperatures. However, isoprene and other hydrocarbons are volatile and evaporate into the atmosphere, where they affect atmospheric chemistry. These hydrocarbons, which remain in the atmosphere for about 6 hours before breaking down, are highly reactive and contribute to ozone formation, a key ingredient in photochemical smog (discussed shortly). Carbon monoxide, another important air pollutant, is one of the breakdown products of isoprene. The contri-

Figure 19.3 Ozone damage. Compare the soybean leaf grown in clean air (*right*) with the one damaged by ozone (*left*). Note the darkened, dead tissue along the leaf's margin and the buckling of leaf tissue between the veins. The ozone-damaged leaf is also lighter green, indicating it does not have as much chlorophyll. Plants exposed to ozone exhibit other symptoms, including reduced root growth and a lowered productivity. On average, soybeans exposed to ozone pollution exhibit a 15% reduction in yield.

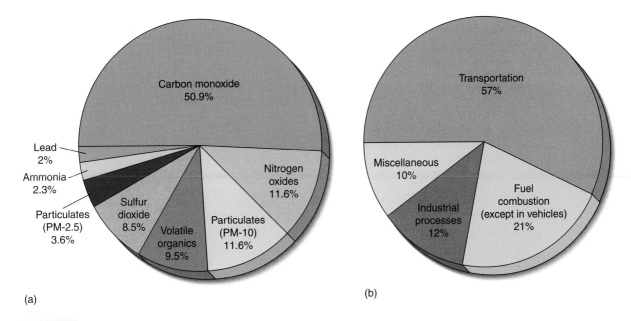

(a)

(b)

Figure 19.4 Primary air pollutants and sources. (a) Air pollutants in the United States, 2000. The concentrations of nitrogen oxides and volatile organics are an indirect measure of ozone, which is a secondary air pollutant formed in their presence. (b) Pollution sources in 1998 (latest data available). Note that transportation and industrial fuel combustion (such as electric power plants) are major contributors of pollutants.

bution of biologically generated hydrocarbon emissions in southeastern cities such as Atlanta, which is heavily wooded, is substantial (Figure 19–5).

EFFECTS OF AIR POLLUTION

Air pollution injures organisms, reduces visibility, and attacks and corrodes materials such as metals, plastics, rubber, and fabrics. The respiratory tracts of animals, including humans, are particularly harmed by air pollutants, which also worsen existing medical conditions such as chronic lung disease, pneumonia, and cardiovascular problems. Most forms of air pollution reduce the overall productivity of crop plants, and when combined with other environmental stressors, such as low winter temperatures or prolonged droughts, air pollution causes plants to decline and die. Air pollution is involved in acid deposition, global temperature changes, and stratospheric ozone depletion (all discussed in Chapter 20).

Air Pollution and Human Health

Generally speaking, exposure to low levels of pollutants such as ozone, sulfur oxides, nitrogen oxides, and particulate matter irritates the eyes and causes inflammation of the respiratory tract (Table 19.2). Evidence exists that many air pollutants also suppress the immune system, increasing susceptibility to infection. In addition, evidence continues to accumulate indicating that exposure

to air pollution during respiratory illnesses may result in the development later in life of chronic respiratory diseases, such as emphysema and chronic bronchitis. In emphysema, the air sacs (alveoli) in the lungs become irreversibly distended, decreasing the efficiency of respiration and causing breathlessness and wheezy breathing. Chronic bronchitis is a disease in which the air passages

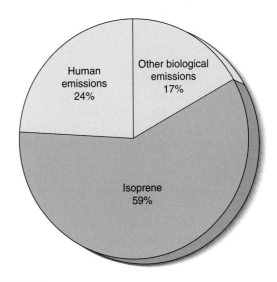

Figure 19.5 Hydrocarbon emissions in Atlanta, Georgia. Isoprene, which is produced by plants, and other biological emissions contribute about three fourths of all hydrocarbon emissions in the area.

Table 19.2 | **Health Effects of Several Major Air Pollutants**

Pollutant	Source	Effects
Particulate matter	Industries, motor vehicles	Aggravates respiratory illnesses; long-term exposure may cause increased incidence of chronic conditions such as bronchitis
Sulfur oxides	Electric power plants and other industries	Irritate respiratory tract; same effects as particulates
Nitrogen oxides	Motor vehicles, industries, heavily fertilized farmland	Irritate respiratory tract; aggravate respiratory conditions such as asthma and chronic bronchitis
Carbon monoxide	Motor vehicles, industries	Reduces blood's ability to transport oxygen; headache and fatigue at lower levels; mental impairment or death at high levels
Ozone	Formed in atmosphere (secondary air pollutant)	Irritates eyes; irritates respiratory tract; produces chest discomfort; aggravates respiratory conditions such as asthma and chronic bronchitis

(bronchi) of the lungs become permanently inflamed, causing breathlessness and chronic coughing.

Health Effects of Specific Air Pollutants Both sulfur dioxide and particulate matter irritate the respiratory tract and, because they cause the airways to constrict, actually impair the lungs' ability to exchange gases. People suffering from emphysema and asthma, a disease in which breathing is wheezy and labored because of airway constriction, are very sensitive to sulfur dioxide and particulate pollution. Nitrogen dioxide also causes airway constriction and, in people suffering from asthma, an increased sensitivity to pollen and dust mites (microscopic animals found in household dust).

One of the largest studies ever conducted on the effects of particulate pollution on human health was published in 1995. This landmark Harvard study tracked the health records of more than 500,000 people in 151 U.S. cities from 1982 to 1989. The study compared mortality data with the amount of microscopic particulate matter at each location. Because so many people were included in the study, scientists were able to discount the effects of tobacco, alcohol, poverty, and other factors that are related to death rates. The study found that people who live and work in the country's most polluted cities are 15 to 17% more likely to die prematurely than those living in U.S. cities with the cleanest air.

Carbon monoxide binds irreversibly with iron in the blood's hemoglobin, eliminating its ability to transport oxygen. At medium concentrations, carbon monoxide causes headaches and fatigue. As the concentration of carbon monoxide increases, reflexes slow down and drowsiness occurs; at a certain high level, carbon monoxide causes death. People at greatest risk from carbon monoxide include pregnant women, infants, and those with heart or respiratory diseases. A four-year study in seven U.S. cities—Chicago, Detroit, Houston, Los Angeles, Milwaukee, New York, and Philadelphia—linked carbon monoxide concentrations in the air to increases in hospital admissions for congestive heart failure.

Ozone and the volatile compounds in smog are irritants that cause a variety of health problems, including burning eyes, coughing, and chest discomfort. Ozone also brings on asthma attacks and suppresses the immune system. A 2002 study at the University of California, Los Angeles concluded that pregnant women exposed to high levels of ozone and carbon monoxide are three times more likely to give birth to infants with serious heart defects.

The health effects of about 150 hazardous air pollutants produced by motor vehicles, businesses, and industries have not been widely studied, although long-term exposure to certain air toxics has been linked to cancer. The Environmental Defense Fund estimates that 360 people out of every million Americans develop cancer as a result of air toxics, although the cancer rate varies widely from one place to another. To put the cancer risk from air toxics into perspective, however, note that almost 150,000 of every million cigarette smokers will die of cancer.

Children and Air Pollution As is true of essentially all environmental stressors, air pollution is a greater health threat to children than it is to adults. The lungs continue to develop throughout childhood, and air pollution can restrict lung development. In addition, a child has a higher metabolic rate than an adult and therefore needs more oxygen. To obtain this oxygen, a child breathes more air—about two times as much air per pound of body weight as an adult. This means that a child also breathes more air pollutants into the lungs. A 1990 study in which autopsies were performed on 100 Los Angeles children who died for unrelated reasons found that more than 80% had subclinical lung damage, which is lung disease in its early stages, before clinical symptoms appear. (Los Angeles has some of the worst air quality in the world.)

A 10-year study of about 5,000 children in 12 communities in Southern California examined the effects on children's developing lungs of chronic exposure to air pollution. Results from this study, which ended in 2001, indicate that children who live in high ozone areas and participate in sports are more likely to develop asthma than children who live there but do not participate in sports. In addition, results indicate that children who breathe the most polluted air (higher concentrations of nitrogen dioxide, particulate matter, and acid vapor) have less lung growth than children who breathe cleaner air. If the children moved to areas with less particulate air pollution, their lung development increased, but if they moved to areas with worse particulate air pollution, their lung development decreased.

Smoking   Smoking, which causes serious diseases such as lung cancer, emphysema, and heart disease, is responsible for the premature deaths of nearly half a million people in the United States each year. Cigarette smoking annually causes about 120,000 of the 140,000 deaths from lung cancer in the United States. Smoking also contributes to heart attacks and strokes and to cancer of the bladder, mouth, throat, pancreas, kidney, stomach, voice box, and esophagus.

Cigarette smoke is a "portable" mixture of air pollutants that includes hydrocarbons, carbon dioxide, carbon monoxide, particulate matter, cyanide, and a small amount of radioactive materials that come from the fertilizer used to grow the tobacco plants. Smokers move about, exhaling tobacco smoke into the air we all must breathe. Passive smoking, which is nonsmokers' chronic breathing of smoke from cigarette smokers, also increases the risk of cancer. Tobacco smoke has been linked to annual deaths from lung cancer of about 3000 nonsmokers in the United States. In addition, passive smokers suffer more respiratory infections, allergies, and other chronic respiratory diseases than other nonsmokers. Passive smoking is particularly harmful to infants and young children, pregnant women, the elderly, and people with chronic lung disease. When parents of infants smoke, the infant has double the chance of pneumonia or bronchitis in its first year of life.

There is good news and bad news about smoking. The good news is that fewer people in highly developed nations are smoking. A poll taken in the mid-1990s found that about 25% of U.S. adults said they were currently smoking compared with a peak of 41% in the mid-1960s. In addition, cigarette production in the United States declined by nearly 6% from 1996 to 1998 (latest data available). Smoking has also declined in Japan and in most European countries.

The bad news is that more and more people are taking up the habit in China, Brazil, Pakistan, and other developing nations (Figure 19.6). In some countries, the smoking habit costs as much as 20% of a worker's annual income. Tobacco companies in the United States pro-

Figure 19.6 A young boy smokes in Kathmandu, Nepal. U.S. tobacco companies export the cigarette habit abroad to compensate for a lower consumption in the United States.

mote smoking abroad, and a substantial portion of our tobacco crop is exported. Cigarette sales in developing countries have increased by 80% since 1990. The World Health Organization (WHO) estimates that worldwide, 3 million people die each year of smoking-related causes, and it wants a global ban on tobacco advertising.

Although fewer U.S. citizens are smoking, certain groups in our society still have high numbers of tobacco addicts, including certain minority groups and those with the least education. A need exists to continue educating these groups, as well as all young people (more than 1 million U.S. children and teenagers take up smoking each year), about the dangers of smoking before they become addicted.

URBAN AIR POLLUTION

Air pollution that is localized in urban areas, where it reduces visibility, is often called **smog**. The word *smog* was coined at the beginning of the 20th century for the smoky fog that was so prevalent in London because of coal combustion. Today there are several different types of smog. Traditional London-type smog—that is, smoke pollution—is sometimes called **industrial smog**. The principal pollutants in industrial smog are sulfur oxides and particulate matter. The worst episodes of industrial smog typically occur during winter months, when combustion of household fuel such as heating oil or coal is high. In December 1952, 4,000 Londoners died in the world's worst industrial smog incident. Because of air quality laws and pollution control devices, industrial smog is generally not a significant problem in highly developed countries today, but it is often serious in many communities and industrial regions of developing countries.

Another important type of smog is **photochemical smog**. This brownish orange haze is called *photochemical*

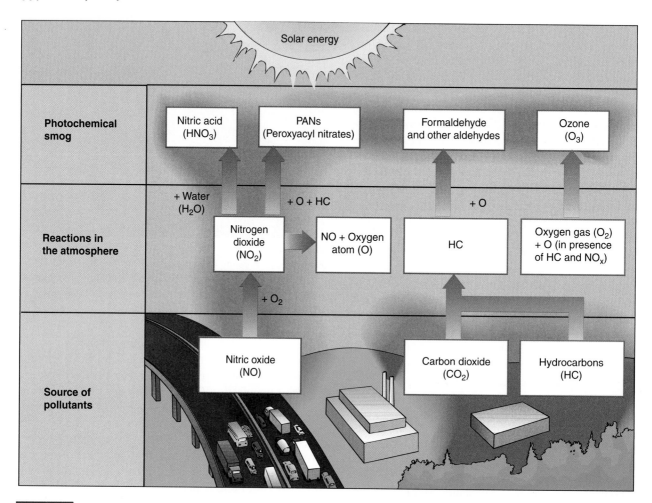

Figure 19.7 Composition of photochemical smog. Photochemical smog is a complex mixture that includes ozone, peroxyacyl nitrates (PANs), nitric acid, and organic compounds such as formaldehyde.

because light—that is, sunlight—initiates several chemical reactions that collectively form the ingredients in photochemical smog. First noted in Los Angeles in the 1940s, photochemical smog is generally worst during the summer months. Both nitrogen oxides and hydrocarbons are involved in its formation. One of the photochemical reactions occurs among nitrogen oxides (largely from automobile exhaust), volatile hydrocarbons, and oxygen in the atmosphere to produce ozone; this reaction requires solar energy (Figure 19.7). The ozone formed in this way then reacts with other air pollutants, including hydrocarbons, to form more than 100 different secondary air pollutants (peroxyacyl nitrates, or PANs, for example) which can injure plant tissues, irritate eyes, and aggravate respiratory illnesses in humans.

The main human source of the ingredients for photochemical smog is the automobile, but bakeries and dry cleaners are also significant contributors. When bread is baked, yeast byproducts that are volatile hydrocarbons are released to the atmosphere where solar energy powers their interactions with other gases to form ozone.

The volatile fumes from dry cleaners also contribute to photochemical smog.

How Weather and Topography Affect Air Pollution

Variation in temperature during the day usually results in air circulation patterns that help to dilute and disperse air pollutants. As the sun increases surface temperatures, the air near the ground is warmed. This heated air expands and rises to higher levels in the atmosphere (warm air is lighter and more buoyant than cool air), causing a low-pressure area near the ground. The surrounding air then moves into the low-pressure area. Thus, under normal conditions, air circulation patterns prevent toxic pollutants from increasing to dangerous levels near the ground.

However, during periods of **temperature inversion**, also called **thermal inversion**, the air near the ground is colder than the air at higher levels, and polluting gases and particulate matter remain trapped in high concentrations

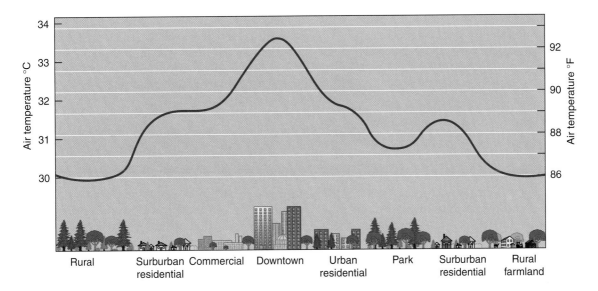

Figure 19.8 **Urban heat island.** This figure shows how temperatures might vary on a summer afternoon. The city stands out as a heat island against the surrounding rural areas.

close to the ground, where people live and breathe. Temperature inversions usually persist for only a few hours before being broken up by solar heating that warms the air near the ground. Sometimes, however, atmospheric stagnation caused by a stalled high-pressure air mass allows a temperature inversion to persist for several days.

Certain types of topography (surface features) increase the likelihood of temperature inversions. Cities located in valleys, near the coast, or on the leeward side of mountains (the side toward which the wind blows) are prime candidates for temperature inversions. The Los Angeles Basin, for example, is a plain that lies between the Pacific Ocean and mountains to the north and east. During the summer, the sunny climate produces a layer of warm, dry air at upper elevations. However, a region of upwelling occurs just off the Pacific coast, bringing cold ocean water to the surface and cooling the ocean air. As this cool air blows inland over the basin, the mountains block its movement further. Thus, a layer of warm, dry air overlies cool air at the surface, producing a temperature inversion.

Urban Heat Islands and Dust Domes

Streets, rooftops, and parking lots in areas of high population density absorb solar radiation during the day and radiate heat into the atmosphere at night. Heat released by human activities such as fuel combustion is also highly concentrated in cities. The air in urban areas is therefore warmer than the air in the surrounding suburban and rural areas. Such localized heat buildup is known as an **urban heat island** (Figure 19.8).

Urban heat islands affect local air currents and weather conditions, particularly by increasing the number of thunderstorms over the city during summer months.

The uplift of warm air over the city produces a low-pressure cell that draws in cooler air from the surroundings. As the heated air rises, it cools, causing water vapor to condense into clouds and producing thunderstorms.

Urban heat islands also contribute to the buildup of pollutants, especially particulate matter, in the form of **dust domes** over cities (Figure 19.9a). Pollutants concentrate in a dust dome because convection (that is, the vertical motion of warmer air) lifts pollutants into the air,

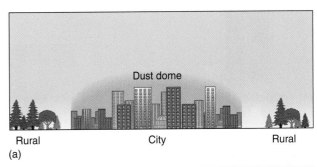

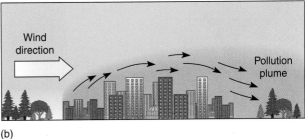

Figure 19.9 **Dust dome.** (a) A dust dome of pollutants forms over a city when the air is somewhat calm and stable. (b) When wind speeds increase, the pollutants move downwind from the city.

where they remain because of somewhat stable air masses produced by the urban heat island. If wind speeds increase, the dust dome moves downwind from the city, and the polluted air spreads over rural areas (Figure 19.9b).

CONTROLLING AIR POLLUTANTS

Many of the measures we have already discussed for energy efficiency and conservation (see Chapter 12) also help to reduce air pollution. Smaller, more fuel-efficient automobiles produce fewer emissions, for example. Appropriate technologies exist to control all the forms of air pollution discussed in this chapter except carbon dioxide.

Smokestacks that have been fitted with electrostatic precipitators, fabric filters, scrubbers, or other technolo-

gies remove particulate matter (Figure 19.10; also see Chapters 10, 15, and 23 for additional discussion of these air pollution control devices). In addition, particulate matter is controlled by careful land-excavating activities, such as sprinkling water on dry soil that is being moved during road construction.

Several methods exist for removing sulfur oxides from flue (chimney) gases, but it is often less expensive simply to switch to a low-sulfur fuel such as natural gas or even to a non–fossil fuel energy source such as solar energy. Sulfur can also be removed from fuels before they are burned, as in coal gasification (see Chapter 10).

Reduction of combustion temperatures in automobiles lessens the formation of nitrogen oxides. Use of mass transit helps reduce automobile use, thereby decreasing nitrogen oxide emissions. Nitrogen oxides produced during high-temperature combustion processes in industry

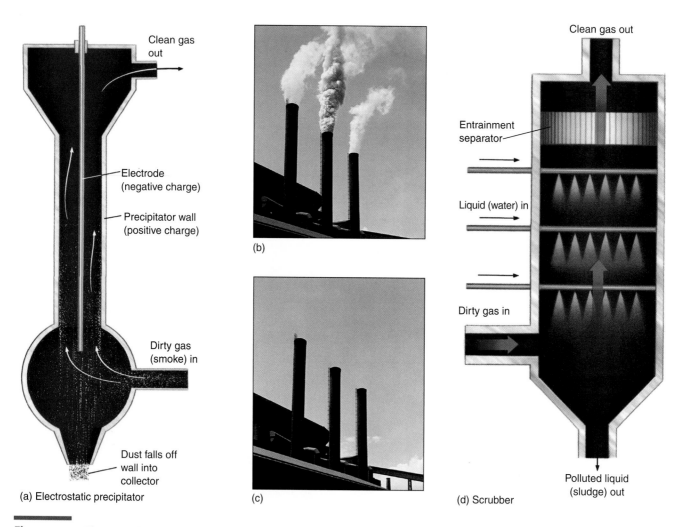

(a) Electrostatic precipitator

(b)

(c)

(d) Scrubber

Figure 19.10 Electrostatic precipitator and scrubber. (a) In an electrostatic precipitator, the electrode imparts a negative charge to particulates in the dirty gas. These particles are attracted to the positively charged precipitator wall, and then fall off into the collector. (b, c) A comparison of emissions from a Delaware Valley steel mill with the electrostatic precipitator turned off (b) and on (c). (d) In a scrubber, mists of water droplets trap particulates in the dirty gas. The toxic dust produced by electrostatic precipitators and the polluted sludge produced by scrubbers must be safely disposed of or they will cause soil and water pollution.

New Mowers for Cleaner Air

Small spark-ignition (SI) engines—those used in lawn-mowers and other lawn equipment—produce about 10% of U.S. hydrocarbon emissions attributed to mobile sources, and also release nitrogen oxides. Both hydrocarbons and nitrogen oxides contribute to the production of ground-level ozone, the primary component of photochemical smog. Nitrogen oxides also contribute to the formation of acid rain.

The EPA set emissions standards for new SI engines, beginning with model year 1997, for both handheld (leaf blowers, chain saws, trimmers, etc.) and non-handheld (mowers, etc.) equipment. These standards were expected to reduce hydrocarbon emissions from these engines by 32%. Now, as a result of new emissions-control technology, the EPA is implementing much more stringent "phase 2" regulations for mowers and other non-handheld SI equipment. The new standards, to be phased in between 2001 and 2007, will reduce hydrocarbon and nitrogen oxide emissions by an additional 59% beyond the 1997 standards. Emissions for handheld SI equipment are also scheduled for tighter regulation. Emission standards are to be phased in gradually to give industry, especially small manufacturers, greater flexibility in adopting new engine designs and technologies. When final standards are met, lawn equipment will be slightly more expensive, but fuel savings with the new engines—a decrease in fuel consumption of about 15%—may actually result in a net savings to the consumer. Until the new equipment is available, lawnmower owners can best reduce toxic emissions by avoiding gasoline spills when refilling gas cans and equipment tanks, and by regularly maintaining their engines.

year. The good news is that overall, air quality has improved since 1970. This improvement in air quality has been largely due to the U.S. Clean Air Act.

The Clean Air Act

The **Clean Air Act** was first passed in 1970 and has been updated and amended twice since then, in 1977 and 1990. This law authorizes the EPA to set limits on the amount of specific air pollutants that are permitted everywhere in the United States. Individual states are responsible for meeting deadlines to reduce air pollution to acceptable levels. States may pass stronger pollution controls than the EPA authorizes, but they cannot mandate weaker limits than those stipulated in the Clean Air Act.

The EPA, which oversees the Clean Air Act, has focused on six air pollutants (lead, particulate matter, sulfur dioxide, carbon monoxide, nitrogen oxides, and ozone) and established maximum acceptable concentrations for each. The most dramatic improvement has been in the amount of lead in the atmosphere, which showed a 98% decrease between 1970 and 2000, primarily because of the switch from leaded to unleaded gasoline. Atmospheric levels of the other pollutants, with the exception of nitrogen oxides, have also been reduced (Figure 19.11). For example, between 1970 and 2001, sulfur dioxide emissions declined by 44%. During this same period, the U.S. gross

can be removed from smokestack exhausts. The release of nitrogen oxides from cultivated fields to which nitrogen fertilizers have been applied is reduced significantly when no-tillage is practiced (see Chapter 14).

Modification of furnaces and engines to provide more complete combustion helps control the production of both carbon monoxide and hydrocarbons. Catalytic afterburners, used immediately following combustion, oxidize most unburned gases. The use of catalytic converters to treat auto exhaust can reduce carbon monoxide and volatile hydrocarbon emissions by about 85% over the life of the car. Careful handling of petroleum and hydrocarbons, such as benzene, reduces air pollution from spills and evaporation.

AIR POLLUTION IN THE UNITED STATES

There is bad and good news about air pollution in the United States. The bad news is that many locations throughout the country still have unacceptably high levels of one or more air pollutants. Moreover, most health experts estimate that air pollution causes the premature deaths of thousands of people in the United States each

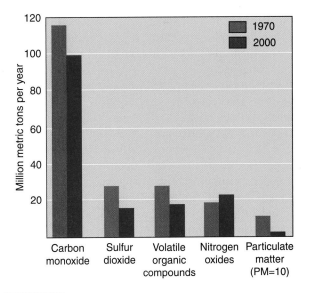

Figure 19.11 Emissions in the United States, 1970 and 2000. Carbon monoxide, sulfur dioxide, volatile organic compounds (many of which are hydrocarbons), and particulate matter showed decreases, and only nitrogen oxides did not decline. PM-10 applies to particles less than or equal to 10 μm (10 micrometers) in size. Since 1990, the EPA has also monitored PM-2.5, which are very small particles less than or equal to 2.5 μm.

Table 19.3	U.S. Urban Areas with The Worst Air Quality (Ozone Nonattainment Areas), 2002

Extreme (2010)*

Los Angeles South Coast Air Basin, California

Very Severe (2007)*

Chicago, Gary and Lake County, Illinois–Indiana
Houston, Galveston, and Brazoria, Texas
Milwaukee and Racine, Wisconsin
New York City, Northern New Jersey, and Long Island, New York–New Jersey–Connecticut
Southeast Desert, California

Severe (2005)*

Baltimore, Maryland
Philadelphia, Wilmington, Trenton, Pennsylvania–New Jersey–Delaware–Maryland
Sacramento, California
San Joaquin Valley, California
Ventura Country (between Santa Barbara and Los Angeles), California

*The year in which the nonattainment areas should be compliant.

domestic product increased 161%, energy consumption increased 42%, and vehicle miles increased 149%.

Although air quality has been gradually improving, more than 154 million metric tons (170 million tons) of pollutants are emitted into U.S. air each year. The atmosphere in many urban areas still contains higher levels of pollutants than are recommended on the basis of health standards, and photochemical smog continues to be a major problem in many metropolitan areas (Table 19.3). In EPA language, these cities are classified as *nonattainment areas for one or more criteria air pollutants.* Nonattainment areas are classified on the basis of how badly polluted they are. This classification ranges from marginal (somewhat polluted and therefore relatively easy to clean up) to extreme (so polluted that it will take many years to clean up the air). The EPA estimates that some 90 million Americans currently live in nonattainment areas.

The Clean Air Act of 1970 and its amendments in 1977 and 1990 required progressively stricter controls of motor vehicle emissions. The provisions of the Clean Air Act Amendments of 1990 include the development of "superclean" cars, which emit lower amounts of nitrogen oxides and hydrocarbons, and the use of cleaner-burning gasoline in the most polluted cities in the United States. These changes were phased in gradually by the year 2000. More recent automobile models do not produce as many pollutants as older models (unless the pollution control devices of the recent models have been deliberately tampered with). Yet despite the increasing percentage of newer automobile models on the road, air quality has not improved in some areas of the United States because of the large increase in the *number* of cars being driven.

The Clean Air Act Amendments of 1990 focus on industrial airborne toxic chemicals in addition to motor vehicle emissions. Between 1970 and 1990, the airborne emissions of only seven toxic chemicals were regulated. In comparison, the Clean Air Act Amendments of 1990 required a 90% reduction in the atmospheric emissions of 189 toxic chemicals by 2003. To comply with this requirement, both small businesses such as dry cleaners and large manufacturers such as chemical companies had to install pollution control equipment if they had not already done so.

Changes to the Clean Air Act in 1997   New scientific evidence that accumulated since the passage of the Clean Air Act Amendments of 1990 suggested that the standards the EPA had originally established for ground-level ozone and particulate matter were not strict enough to protect the health of U.S. citizens. The Clean Air Act Amendments of 1990 limited the emission of particulate matter with particle sizes less than or equal to 10 μm (10 micrometers), designated PM-10. Because of concern over the potential health effects of microscopic particulate matter, the EPA proposed separate standards on the emission of particles less than or equal to 2.5 μm (2.5 micrometers), designated PM-2.5. The EPA also revised ozone standards.

The revised standards provoked an outcry from industries such as the Chemical Manufacturers Association, the American Trucking Association, and other industry and state groups, some of which challenged the new standards in court. In 1999 the U.S. Court of Appeals decided in favor of the industry groups because it said the EPA had not demonstrated how it had decided on the limits it set.

ENVIROBRIEF

Commuting for Clean Air

Although new cars release much lower levels of toxic emissions than in the past, air pollution from increasingly crowded highways remains a threat to the environment and to human health. The percentage of licensed drivers in the United States is growing more rapidly than the population, and the EPA estimates that as many as half of all cancers associated with outdoor air pollution are attributable to emissions from cars and other mobile sources. Commuter Choice programs are voluntary efforts by employers to encourage employees to choose commutes that will reduce air pollution and highway congestion. These programs, which are tied to 1998 changes in federal tax law, potentially benefit everyone. Employees are offered tax-free transit, vanpool, or parking benefits, or they are permitted to set aside their own pre-tax income to pay for similar commuting options. Employers receive tax savings and a new tool to enhance employee recruitment. State and local governments can apply emission reductions gained through Commuter Choice programs toward meeting air quality standards specified in the federal Clean Air Act. The ultimate gains associated with successful Commuter Choice programs are better air quality and reduced consumption of fossil fuels.

The EPA appealed the decision at the Supreme Court. In 2001 the U.S. Supreme Court upheld the way the EPA set its air-quality standards. The court also reaffirmed that the EPA must set these standards on the basis of health considerations, not cost considerations.

Other Ways to Improve Air Quality

Reducing the sulfur content in gasoline from its current average of 330 parts per million (ppm) to 30 ppm or lower would significantly reduce air pollution. (**Parts per million** is the number of molecules of a particular pollutant found in a million molecules of air, water, or some other material.) Sulfur clogs catalytic converters so that they cannot effectively remove emissions from automobile exhaust. The technology to remove sulfur from gasoline exists, but it would be expensive to implement because many oil refineries would have to be modernized. The American Petroleum Institute estimates that the reduction in sulfur in U.S. gasoline would add $0.03 to $0.06 to the cost of each gallon of gasoline, but the EPA's estimate is much lower, $0.01 to $0.02 cents per gallon. The EPA says the overall improvement in the air would be equivalent to removing 50 million cars from U.S. roads each year and would result in a more than $16 billion savings in health benefits. California, which has long been a leader in environmental quality, has imposed a sulfur limit in gasoline of 40 ppm, as have Canada and the European Union.

Minivans, sport utility vehicles (SUVs), and light pickup trucks, which account for almost 50% of new passenger vehicles, currently do not have the same federal emissions standards as ordinary automobiles. Many of these larger vehicles produce more than twice the pollution of a car. In 1998 California issued strict pollution standards for these vehicles, requiring that they emit the same amount of pollution as automobiles; these standards go into effect with the 2004 model year. In 1999 the federal government, which often follows California's actions, announced similar regulations for the entire country. The federal regulations also call for cutting tailpipe emissions of nitrogen oxides to 0.7 gram per mile in regular automobiles by the 2004 model year; sport utility vans and minivans will have to comply with this requirement by the 2007 model year.

Many states, such as California, New York, New Jersey, and Connecticut, now require diesel trucks and buses to undergo emissions tests similar to those that these states have required for automobiles for many years. Diesel exhaust, particularly particulate matter, is viewed as a toxic pollutant. Public health advocates say that breathing diesel exhaust may contribute to asthma, an increased incidence of lung cancer, and other lung diseases.

CASE·IN·POINT Los Angeles

Los Angeles, California, has some of the worst smog in the world. Its location, combined with a sunny climate, is con-

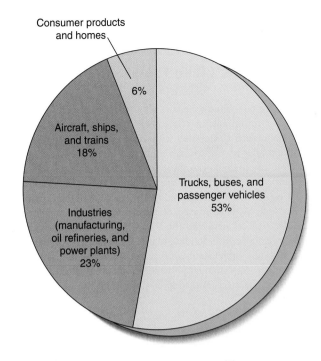

Figure 19.12 Sources of smog in Los Angeles, California. Trucks, buses, and passenger vehicles account for more than half of the emissions that produce smog. Data are from 1997.

ducive to the formation of stable temperature inversions that trap photochemical smog near the ground. The large number of passenger vehicles, heavy-duty trucks, and buses are the source of more than half of the emissions that produce smog in Los Angeles (Figure 19.12).

In 1969, California became the first state to enforce emission standards on motor vehicles, largely because of air pollution problems in Los Angeles. Today Los Angeles has stringent smog controls that regulate everything from low-emission alternative fuels (such as compressed natural gas) for buses to lawn mower emissions to paint vapors. Emissions from large industrial and manufacturing sources, including oil refineries and power plants, have been reduced significantly by installing the cleanest emission reduction equipment available. California has no coal-fired power plants, and most of its power plants burn natural gas. To help reduce the amount of automobile emissions, Los Angeles also has a subway, although it is limited in length. Future pollution reductions will come in part from requiring auto manufacturers to sell ultra-low-emission autos (see "Meeting the Challenge: Clean Cars, Clean Fuels").

After several decades devoted to improving its air quality, Los Angeles now has the cleanest skies since the 1950s. Despite impressive progress, however, the city's air is still among the dirtiest in the United States. Los Angeles still exceeds federal air quality standards on more days than almost any other metropolitan region in the United States. (In 2000, Los Angeles, exceeded air

MEETING THE CHALLENGE

Clean Cars, Clean Fuels

Performance and style have almost always been higher priorities for automobile manufacturers than reducing environmental pollutants. Concerned with an ever increasing number of automobiles on California's congested highways—automobiles that burn too much gasoline and that spew noxious emissions—California decided to legislate a clean, efficient car. A California law passed in 1990 mandated that at least 2% of the cars sold in that state by 1998 must have ultra-low tailpipe emissions; by 2003, 10% of new cars must be ultra-low-emission vehicles. Because more automobiles are bought in the state of California than in any other state, auto manufacturers could not ignore the California mandate. In addition, New York and other northeastern states also adopted California's ultra-low-emission standards.

To develop the new technologies needed for clean cars, the Clinton administration announced in 1993 a 10-year Partnership for a New Generation of Vehicles (PNGV) program among the "Big Three" carmakers—Chrysler, General Motors, and Ford—and government agencies such as the Commerce Department and the EPA. In 2002 the Bush administration announced a new cooperative program, known as Freedom Cooperative Automotive Research (FreedomCAR) to replace PNGV. Both programs have the same overall goal of producing a clean car with the same performance and cost of today's automobiles but with much greater efficiency. Improved fuel efficiency translates into lower emissions and the ability to meet stringent clean-air standards. However, PNGV invested money and research in hybrid cars, whereas FreedomCAR promotes hydrogen as the primary fuel for cars and trucks.

Let us examine some of the new-car designs being considered by carmakers. An engine that runs on electricity is much cleaner and quieter than a gasoline engine. However, electric cars are cleaner than gasoline powered cars only when the source of their electricity is natural gas or solar energy. If electricity comes from a coal-fired power plant, electric cars actually produce *more* emissions than gasoline-powered cars; these emissions are produced at the power plant rather than coming out of a tailpipe. According to consumers, the biggest drawback to electric cars is that they must be recharged for up to 8 hours every 100 miles or less. Short, local trips are not a problem for an electric car, but long trips are out of the question. Also, U.S. consumers have little incentive to buy electric cars, which are more expensive because of their new technology, as long as the price of gasoline is affordable. As a result, U.S. auto makers have discontinued electric-vehicle programs.

Hybrid cars that use a combination of gasoline and electricity are a very promising new technology. The hybrid car starts with the gasoline engine but uses the electric motor alone when driving at low speeds or when idling in traffic. At normal driving speeds, both engines contribute power, and when the car is slowing down, the wheels power the electric generator to store electricity in the battery for future use. Both Honda and Toyota currently sell hybrid cars in the United States. The Honda Insight is a two-seater that has fuel economy ratings of 51 miles per gallon (city) and 68 miles per gallon (highway). The Toyota Prius, a sedan, is rated at 45 miles per gallon (city) and 52 miles (highway). General Motors, Ford, and Daimler-Chrysler are currently working on hybrid SUVs.

Liquid hydrogen is an extremely clean fuel (the only waste product is water). Some car designs have fuel cells that combine stored hydrogen with oxygen from the air to produce electricity. (See Figure 12–7b for a diagram of a hydrogen fuel cell.) Fuel cells do not produce the harmful emissions that conventional gasoline and diesel engines do. Fuel cells are also superior performers, in part because they do not waste fuel by idling when a vehicle is stopped in traffic. Mercedes-Benz was the first carmaker to unveil a prototype fuel-cell-powered minivan, in 1995, but most major auto manufacturers are doing research and development on fuel cells for automobiles. Some use pressurized hydrogen for fuel, whereas others are experimenting with hydrogen extracted from methanol or gasoline.

The widespread adoption of pressurized hydrogen fuel would require an infrastructure of service stations that would allow motorists to fill up with liquid hydrogen. To get around the need for hydrogen filling stations, some auto manufacturers are experimenting with on-board "reformers" that use gasoline or methanol to produce hydrogen that is then fed into the fuel cell. Although such "reformer" autos would produce emissions, the quantity of emissions would be a fraction of those emitted by today's gasoline- and diesel-powered motor vehicles.

Currently, hydrogen for non- "reformer" fuel cells is produced from natural gas, and although natural gas is the cleanest of fossil fuels, it does produce some emissions when burned. Therefore, when the production of hydrogen is taken into account, hydrogen fuel does not produce zero emissions. Futurists look ahead to the time in the not-so-distant future when solar hydrogen will power vehicles (see Chapter 12). Solar hydrogen fuel is produced when solar energy splits water molecules to produce hydrogen. Cars powered by solar hydrogen will require such extensive modification of existing designs, however, that they could not be a viable alternative until well into the 21st century.

quality standards on 45 days; only Riverside-San Bernardino, Bakersfield, and Fresno, all in California, exceeded air quality standards on more days than Los Angeles did.) At its current rate of improvement, however, Los Angeles should attain federal clean-air standards by 2010.

AIR POLLUTION IN DEVELOPING COUNTRIES

As developing nations become more industrialized, they also produce more air pollution. The leaders of most developing countries believe they must become industrialized rapidly to compete economically with highly developed countries. Environmental quality is usually a low priority in the race to develop. Outdated technologies are often adopted, and air pollution laws, where they exist, are not enforced. Thus, air quality is deteriorating rapidly in many developing nations.

Shenyang and neighboring cities in China have so many smokestacks belching coal smoke (coal is burned to heat many homes) that residents can see the sun only a few weeks of the year (Figure 19.13). The rest of the time residents are choked in a haze of orange-colored coal dust. In other developing countries, such as India and Nepal, wood or animal dung is burned indoors, often in poorly designed stoves with little or no outside ventilation, thereby exposing residents to serious indoor air pollution.

The growing number of automobiles in developing countries is also contributing to air pollution, particularly in urban areas. Many vehicles in these countries are 10 or

more years old and have no pollution control devices. Motor vehicles produce about 60% to 70% of the air pollutants in urban areas of Central America, and 50% to 60% in urban areas of India. During the 1990s the most rapid proliferation of motor vehicles worldwide occurred in Latin America, Asia, and Eastern Europe.

Lead pollution from heavily leaded gasoline is an especially serious problem in developing nations. The gasoline refineries in these countries are generally not equipped to remove lead from gasoline. (The same situation occurred in the United States until federal law mandated that the refineries upgrade their equipment.) In Cairo, for example, children's blood lead levels are more than two times higher than the level considered at-risk in the United States. Lead can retard children's growth and cause brain damage.

According to a 1999 study funded by the WHO, the five worst cities in the world in terms of exposing children to air pollution are Mexico City, Mexico; Beijing, China; Shanghai, China; Tehran, Iran; and Calcutta, India. The study determined that respiratory disease is now the leading cause of death for children worldwide. More than 80% of these deaths occur in young children (under the age of 5) who live in cities in developing countries.

CASE-IN-POINT Mexico City

Mexico City, the world's second largest city, has the second most polluted air of any major metropolitan area in the world (Figure 19.14). (According to 2001 data from the World Bank, the air quality in Beijing is the world's worst.) Average visibility has dropped from 11 km (7 mi) in the 1940s, when surrounding snow-capped volcanoes were commonly seen, to 1.6 km (1 mi). Mexico City's air pollution is due in part to its large population growth in the past several decades (according to the U.N. Population Division, Mexico City has grown from 5.4 million in 1960 to 18.1 million in 2000) and in part to its location. Mexico City is in a bowl-shaped valley that is ringed on three sides by mountains; winds coming in from the open northern end are trapped in the valley. Air quality is at its worst from October to January, largely as a result of temperature inversions caused by seasonal variations in atmospheric conditions.

The city has more than 3 million passenger vehicles, 360 gasoline stations, and about 36,000 businesses, which the Mexican government says spew 3.94 million metric tons (4.35 million tons) of pollutants into the air each year. Mexican gasoline contains a lot of contaminants, and the average automobile is 10 years old and therefore produces more pollutants. The air also contains particles of dried fecal matter from the millions of gallons of sewage dumped onto land near the city. In addition, liquefied petroleum gas, which is the major source of energy for cooking and heating in Mexico City, escapes unburned into the atmosphere from thousands of leaks, increasing the level of hydrocarbons in the city's air. Sim-

Figure 19.13 Air pollution in China. Coal smoke pollutes the air above workers' houses in Liaoning Province, China. All forms of pollution are increasing threats as China becomes industrialized.

Figure 19.14 Smog in Mexico City, Mexico. The Mexican flag blows in an extremely polluted breeze. Mexico City has the dubious distinction of having some of the worst air quality in the world. Photographed on March 31, 1998.

ply breathing the air in Mexico City is equivalent to smoking two packs of cigarettes a day.

During the 1990s Mexico embarked on an ambitious plan to improve Mexico City's air quality. It spent more than $5 billion to replace old buses, taxis, delivery trucks, and cars with cleaner vehicles, such as those with catalytic converters; Mexico also switched to unleaded gasoline and reforested some of the nearby hillsides to reduce particulate matter produced by wind erosion. Driving restrictions apply when the air quality is particularly poor, and exhaust emissions are periodically checked on autos.

In addition, Pemex, Mexico's national oil company, has upgraded its refineries and increased gas imports from the United States, which produces a cleaner fuel. An old, polluting oil refinery within the city limits was closed, and several large industries installed pollution control devices. All of these changes translate into cleaner air. In the early 2000s, Mexico City's air pollution, although still unacceptably high, appears to be gradually improving.

LONG-DISTANCE TRANSPORT OF AIR POLLUTION

Certain hazardous air pollutants are distributed globally by atmospheric transport. Persistent compounds, such as polychlorinated biphenyls (PCBs, industrial compounds) and dichloro-diphenyl-trichloroethane (DDT, a pesticide), may be restricted from use or even banned by many countries. *Persistent compounds* are those that do not readily break down, and so accumulate in the environment. Yet because they are volatile, they move through the atmosphere from warmer developing countries where

they are still used to colder highly developed nations, where they condense and are deposited on land and surface water. The process in which volatile chemicals evaporate from land as far away as the tropics and are transported by winds to higher latitudes, where they condense and fall to the ground, is known as the **global distillation effect** (Figure 19.15).

A 1995 study reported that many industrialized countries continue to be highly contaminated by persistent compounds despite their restricted use. The effect is more pronounced where it is colder—that is, at higher latitudes and higher elevations. Dangerous levels of certain persistent toxic compounds have been measured in the Yukon (northwestern Canada) and in other pristine arctic regions. These chemicals enter food webs and become concentrated in the body fat of animals at the top of the food chain (see discussion of biological magnification in Chapter 22). Fishes, seals, polar bears, and arctic people such as the Inuit are particularly vulnerable. When an Inuit consumes a single bite of raw whale skin, she ingests more PCBs than scientists think should be consumed in a week. The level of PCBs in the breast milk of that Inuit woman is five times higher than in the milk of women who live in southern Canada.

In 1996, concern over protecting the Arctic led to the formation of the Arctic Council, consisting of Canada, the United States, Russia, Finland, Norway, Sweden, Denmark, and Iceland. Since then scientists from the Arctic Monitoring and Assessment Program, based in Norway, conducted additional environmental studies that reinforce earlier studies. In 2001, the **Stockholm Convention on Persistent Organic Pollutants** was adopted. Its goal is to phase out the use of at least 12 persistent toxic chemicals, including PCBs, dioxins and furans (chemical contaminants), and DDT and eight other pesticides. The convention goes into

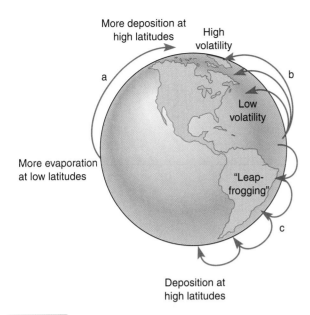

Figure 19.15 Global distillation effect. Volatile chemicals from lower latitudes evaporate into the atmosphere and are transported toward the poles, where they condense and fall to the ground. (a) Long-distance atmospheric transport occurs in part because evaporation exceeds deposition onto the land and ocean at low latitudes, whereas the opposite (deposition exceeds evaporation) is true at high latitudes. (b) The more volatile the chemical, the farther it travels before being deposited onto the land and ocean. (c) Some chemicals move to higher latitudes by repeatedly evaporating and settling ("leapfrogging"), sometimes taking several decades before being permanently deposited in colder regions.

effect when 50 countries have ratified it; ratification had just begun when this book went to press. (For additional discussion of the proposed ban of these chemicals, see the section on the global ban of persistent organic pollutants in Chapter 22.)

Movement of Air Pollution over the Ocean

The Indian Ocean has generally been considered one of the world's cleanest areas because the countries that surround it are not heavily industrialized. During a 6-week study in the winter of 1999, however, scientists from the National Science Foundation and the Scripps Institution of Oceanography reported that a large portion of the Indian Ocean was covered by hazy, polluted air. The pollution extended over 9.5 million km^2 (3.8 million mi^2), an area the size of the United States. It is thought that prevailing winds during the winter monsoon blow the pollution, which includes particulate matter and sulfur droplets, from the Indian subcontinent, China, and Southeast Asia. The hazy area is expected to increase as industry develops in these areas.

Examples of pollution traveling from one continent to another have not been well documented until recently.

Certain atmospheric conditions (that is, a low-pressure system over the Aleutians and a high-pressure system near Hawaii) cause a strong wind toward North America that allows air pollution from Asia to cross the Pacific Ocean. In 1997 scientists from the University of Washington detected carbon monoxide, particulate matter, and PANs in the atmosphere over the western United States. Computer models suggested that these pollutants had been produced in Asia 6 days earlier. In 1998 more definitive evidence of pollution from Asia affecting air quality in North America occurred when a major dust storm in China produced a visible cloud of particulate matter that was tracked by satellite across the Pacific Ocean. The polluted air was analyzed when it reached the United States a few days later and found to contain arsenic, copper, lead, and zinc from ore smelters in Manchuria.

It is also thought that air pollution generated in North America sometimes travels across the Atlantic Ocean to Europe. However, definitive evidence of a North American–European connection has not been demonstrated.

INDOOR AIR POLLUTION

The air in enclosed places such as automobiles, homes, schools, and offices may have significantly higher levels of air pollutants than the air outdoors. In congested traffic, levels of harmful pollutants such as carbon monoxide, benzene, and airborne lead may be several times higher inside an automobile than in the air immediately outside. The concentrations of certain indoor air pollutants may be five times greater than outdoors. Indoor pollution is of particular concern to urban residents because they may spend as much as 90% to 95% of their time indoors.

Because illnesses caused by indoor air pollution usually resemble common ailments such as colds, influenza, or upset stomachs, they are often not recognized. The most common contaminants of indoor air are radon (discussed shortly), cigarette smoke, carbon monoxide, nitrogen dioxide (from gas stoves), formaldehyde (from carpeting, fabrics, and furniture), household pesticides, cleaning solvents, ozone (from photocopiers), and asbestos (Figure 19.16). In addition, viruses, bacteria, fungi (yeasts, molds, and mildews), dust mites, pollen, and other organisms or their toxic parts are important forms of indoor air pollution that are often found in heating, air conditioning, and ventilation ducts.

Health officials are paying increasing attention to the **sick building syndrome**, the presence of air pollution inside office buildings that can cause eye irritations, nausea, headaches, respiratory infections, depression, and fatigue. The Labor Department estimates than more than 20 million employees are exposed to health risks from indoor air pollution. The EPA estimates that the annual medical costs for treating the health effects of indoor air pollution in the United States exceed $1 billion. When

Figure 19.16 Indoor air pollution. Homes may contain higher levels of toxic pollutants than outside air, even near polluted industrial sites.

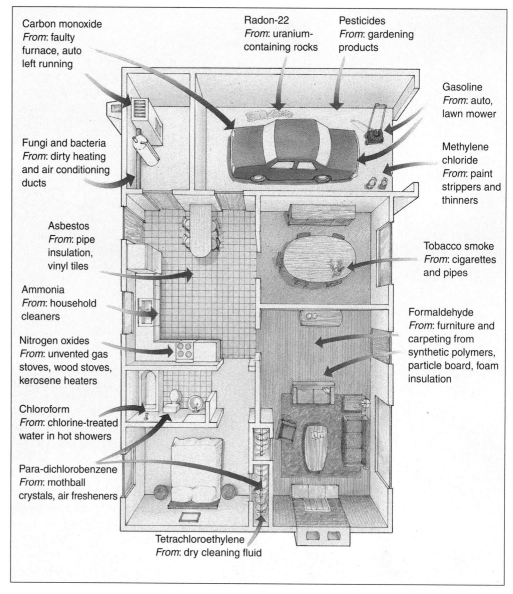

Carbon monoxide
From: faulty furnace, auto left running

Fungi and bacteria
From: dirty heating and air conditioning ducts

Asbestos
From: pipe insulation, vinyl tiles

Ammonia
From: household cleaners

Nitrogen oxides
From: unvented gas stoves, wood stoves, kerosene heaters

Chloroform
From: chlorine-treated water in hot showers

Para-dichlorobenzene
From: mothball crystals, air fresheners

Radon-22
From: uranium-containing rocks

Pesticides
From: gardening products

Gasoline
From: auto, lawn mower

Methylene chloride
From: paint strippers and thinners

Tobacco smoke
From: cigarettes and pipes

Formaldehyde
From: furniture and carpeting from synthetic polymers, particle board, foam insulation

Tetrachloroethylene
From: dry cleaning fluid

lost work time and diminished productivity are added to health care costs, the total annual cost to the economy may be as much as $50 billion. Fortunately, most building problems are relatively inexpensive to alleviate.

Indoor Air Pollution and the Asthma Epidemic

Asthma was considered a rare disease until the middle of the 20th century, and it is far more common in industrialized nations than in developing countries. Since 1970 the number of people in the United States who suffer from asthma has doubled, to more than 15 million; 9 million asthma sufferers are children. Health officials are worried by this trend, which is due in part to indoor air pollution. Indoor exposure to different air pollutants has been demonstrated to contribute to the development and exacerbation of asthma. Exactly what pollutant(s) is/are causing the increase in asthma is unknown, although some evidence suggests that exposure to allergens (sub-

stances that stimulate an allergic reaction) such as dust mites and cockroaches is a major cause.

Radon

The most serious indoor air pollutant is probably **radon**, a colorless, tasteless radioactive gas produced naturally during the radioactive decay of uranium in the Earth's crust. Radon seeps through the ground and enters buildings, where it sometimes accumulates to dangerous levels (Figure 19.17). Although radon is also emitted into the atmosphere, it gets diluted and dispersed and is of little consequence outdoors.

Radon and its decay products emit alpha particles, a form of ionizing radiation that is very damaging to tissue but cannot penetrate very far into the body. Consequently, radon can harm the body only when it is ingested or inhaled. The radioactive particles lodge in the tiny passages of the lungs and damage surrounding

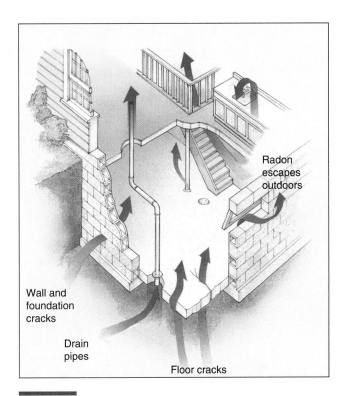

Radon
escapes
outdoors

Wall and
foundation
cracks

Drain
pipes

Floor cracks

Figure 19.17 How radon infiltrates a house. Cracks in basement walls or floors, openings around pipes, and pores in concrete blocks provide some of the entries for radon.

tissue. There is compelling evidence, based primarily on several studies of uranium miners, that inhaling large amounts of radon increases the risk of lung cancer. Several other studies suggest that people who are exposed to relatively low levels of radon over an extended time are also at risk for lung cancer. In 1998 the National Research Council of the National Academy of Sciences released an extensive evaluation of radon's effects on human health. It estimated that residential exposure to radon causes 12% of all lung cancers—between 15,000 and 22,000 lung cancers annually. Cigarette smoking exacerbates the risk from radon exposure; about 90% of radon-related cancers occur among current or former smokers.

According to the EPA, about 6% of U.S. homes have high enough levels of radon to warrant corrective action—that is, a radon level above 4 picocuries per liter of air. (As a standard of reference, outdoor radon concentrations range from 0.1 to 0.15 picocuries per liter of air worldwide.) The highest radon levels in the United States have been found in homes on a geological formation called the **Reading Prong**, which runs across southeastern Pennsylvania into northern New Jersey and New York. Iowa appears to have the most pervasive radon problem, where 71% of the homes tested in 1989 had radon levels high enough to warrant corrective action.

Ironically, efforts to make our homes more energy-efficient have increased the hazard of indoor air pollu-

tants, including radon. Drafty homes waste energy but allow radon to escape outdoors so it does not build up inside. Every home should be tested for radon because levels vary widely from home to home, even in the same neighborhood. Testing is inexpensive, and corrective actions are reasonably priced. Radon concentrations in homes can be minimized by sealing basement concrete floors and by ventilating crawl spaces and basements.

Asbestos

Asbestos is a natural mineral that does not burn or conduct heat or electricity. These properties make asbestos valuable in construction and industry for fire-retardant materials, electrical insulation, and roofing and pipe insulation. It is also used in automobile brake linings.

Some kinds of asbestos have the ability to separate into long, thin fibers, which are so small and lightweight that they can remain suspended in air almost indefinitely. Asbestos fibers are easily inhaled into the lungs. The mucous membranes of the respiratory system remove some of the asbestos, but if more is inhaled than the body can handle, some of it reaches and lodges in the lungs. Definite links exist between exposure to high levels of asbestos and several serious diseases, including lung cancer and **mesothelioma**, a rare and almost always fatal cancer of the body's internal linings. The lag time between exposure to high levels of asbestos and development of cancer is anywhere from 20 to 40 years.

The risk of developing cancer from exposure to asbestos fibers is extremely small for most people. The EPA estimates that between 3,000 and 12,000 U.S. citizens die each year from asbestos-related cancer. Construction workers, firefighters, and building custodians are more likely than others to be exposed to asbestos fibers. Smoking also greatly increases the risk of developing disease from asbestos exposure.

Although health concerns about asbestos had been known since the 1920s, it was not until 1979 that the EPA began regulating its use. During the 1980s concern increased about the widespread presence of asbestos in schools, public and commercial buildings, and homes. Federal legislation required the removal of asbestos from all school buildings. By 1990 about $6 billion had been spent removing asbestos from schools, but by the early 1990s scientists and physicians generally agreed that the threat of asbestos had been greatly exaggerated. The presence of asbestos in a building is not a major health risk unless the asbestos is exposed and crumbling. A much greater danger probably comes from the fibers that are released to the air when asbestos is removed improperly. By the mid-1990s, most experts recommended that asbestos remaining in schools and other public buildings should be left alone. It is both safer and less expensive to seal the asbestos by painting it or covering it up instead of trying to remove it.

NOISE POLLUTION

Sound is caused by vibrations in the air (or some other medium) that reach the ears and stimulate a sensation of hearing. Sound is called **noise pollution** when it becomes loud or disagreeable, particularly when it results in physiological or psychological harm. Of the 28 million U.S. citizens with hearing impairments, as many as 10 million can attribute their impairments at least in part to noise pollution.

Most of the noise that travels in the atmosphere is of human origin. Vehicles, from trains to powerboats to snowmobiles, produce a great deal of noise. Power lawn mowers, jets flying overhead, chain saws, jackhammers, cars with stereos booming, and heavy traffic are just a few examples of the outside noise that assails our ears. Indoors, dishwashers, trash compactors, washing machines, televisions, and stereos add to the din.

The **intensity** (loudness) of sound is measured relative to a reference sound that is so low it is almost inaudible to the human ear. Relative loudness is expressed numerically using the **decibel** (**db**) scale or a modified decibel scale called the **decibel-A** (**dbA**) scale, which takes into account high-pitched sounds to which the human ear is more sensitive (Table 19.4).

Table 19.4 The Decibel-A Scale (dbA)

dbA	Example	Perception/ General Effects
0		Hearing threshold
10	Rustling leaves, breathing	Very quiet
20	Whisper	Very quiet
40	Library	Quiet
50	Quiet neighborhood (daytime)	Quiet to moderately loud
70	Vacuum cleaner, television	Moderately loud
80	Washing machine, typical factory	Very loud, intrusive
90	Motorcycle at 8 m	Very loud; impaired hearing with prolonged exposure
100	Jet flyover at 300 m	Very loud to uncomfortably loud
110	Rock band, boom box held close to ear	Uncomfortably loud
120	Chain saw	Uncomfortably loud to painfully loud
150	Jet at takeoff	Painfully loud to ruptured eardrum

Effects of Noise

Prolonged exposure to noise damages hearing. The part of the inner ear that perceives sound is the **cochlea**, a spiral tube that resembles a snail's shell. Inside the cochlea are approximately 24,000 **hair cells** that detect differences in pressure caused by sound waves. When the hairlike projections of the hair cells move back and forth in response to sound, they initiate nerve impulses that the auditory nerve transmits to the brain. Loud, high-pitched noise injures the hair cells in the cochlea. Because the body does not replace injured hair cells, prolonged exposure to loud noise results in permanent hearing impairment.

In addition to hearing loss, noise produces several other physiological effects in the body. It increases the heart rate, dilates the pupils, and causes muscle contraction. Evidence exists that prolonged exposure to high levels of noise causes a permanent constriction of blood vessels, which can increase the blood pressure, thereby contributing to heart disease. Other physiological effects associated with noise pollution include migraine headaches, nausea, dizziness, and gastric ulcers. Noise pollution also causes psychological stress.

Controlling Noise

Obviously, producing less noise can reduce noise pollution. This can be accomplished in a variety of ways, from restricting the use of sirens and horns on busy city streets to engineering motorcycles, vacuum cleaners, jackhammers, and other noisy devices so that they produce less noise. The engineering approach is technologically feasible but is often avoided because consumers associate loud noise with greater power.

Putting up shields between the noise producer and the hearer can also help control noise pollution. Examples of sound shields include the noise barriers erected along heavily traveled highways and the noise-absorbing material installed around dishwashers. Earplugs are an effective way to protect oneself from unwanted noise; they differ from the preceding examples in that they shield the receiver rather than the noise producer.

ELECTROMAGNETISM: A CONTINUING CONTROVERSY

Scientists have been investigating a possible link between health problems and invisible electric and magnetic fields (EMFs) in the atmosphere. EMFs are associated with such common items as power lines, electric blankets, video displays, and microwave ovens. We are subjected to EMFs outdoors (high-voltage transmission lines) as well as indoors (when electricity is used in household appliances, electrical wiring, and light fixtures).

Several studies have suggested that children who live near power lines may have an increased risk of leukemia. In addition, a study of men who died from brain tumors found that a disproportionate number of them had careers in which they would have received higher-than-normal exposures to EMFs (such as electricians, telephone workers, and electronics engineers). Some laboratory research has suggested that exposure to weak EMFs alters cells.

Because of public concern about possible health hazards from EMFs, Congress requested the National Research Council to decide if protective regulations were needed. Their report, released in 1996 after more than 500 studies were reviewed, concluded that EMFs do not pose significant health risks. In 1997 a study by the National Cancer Institute agreed that there was no link between exposure to EMFs and childhood leukemia. However, the controversy was revived in 1998 when the National Institute of Environmental Health Sciences (part of the U.S. National Institutes of Health) convened an expert panel to review recent studies. The panel concluded that EMFs should be designated a "possible human carcinogen" because EMFs generated by power-line frequencies might cause leukemia in children and because occupational exposure to EMFs might cause leukemia in adults. The panel said the data were inadequate to evaluate any links between EMFs and other cancers, such as brain cancer and breast cancer. Health experts stress, however, that the risk of cancer from exposure to EMFs is very low when compared to other cancer-causing exposures, such as smoking.

SUMMARY WITH SELECTED KEY TERMS

I. Air pollution consists of gases, liquids, or solids present in the atmosphere in high enough levels to harm humans, other organisms, or materials.

A. Air pollution comes from both natural sources and human activities.

  1. The two main sources of human-generated air pollution are motor vehicles (mobile sources) and industry (stationary sources).

  2. Trees and other plants produce hydrocarbons in response to heat. The contribution of biologically generated hydrocarbon emissions in some areas, such as the Blue Ridge Mountains, is substantial.

B. Primary air pollutants are harmful chemicals that enter directly into the atmosphere. **Secondary air pollutants** are harmful chemicals that form from other substances that have been released into the atmosphere.

C. All major forms of human-produced air pollution except carbon dioxide can be prevented or controlled with current technologies, although control may involve considerable expense.

II. The main classes of air pollutants produced by human activities are particulate matter, nitrogen oxides, sulfur oxides, carbon oxides, ozone, hydrocarbons, and hazardous air pollutants.

A. Particulate matter is solid particles and liquid droplets suspended in the atmosphere. **Solid particulate matter** is commonly called dust. Microscopic particulate matter is more dangerous to human health because it can be inhaled more deeply into the lungs than larger particulates. Particulate matter corrodes metals, erodes buildings, and soils fabrics.

B. Nitrogen oxides, such as nitric oxide (NO), nitrogen dioxide (NO_2), and nitrous oxide (N_2O), are gases produced by chemical interactions between nitrogen and oxygen. Nitrogen oxides are associated with photochemical smog and acid deposition. Nitrous oxide is a potent **greenhouse gas** that is associated with global warming as well as stratospheric ozone depletion. Nitrogen oxides also corrode metals and fade textiles.

C. Sulfur oxides, such as sulfur dioxide (SO_2) and sulfur trioxide (SO_3), are gases produced by chemical interactions between sulfur and oxygen. Sulfur oxides are associated with acid deposition, and they corrode metals and damage stone and other materials.

D. Carbon oxides, such as carbon monoxide (CO) and carbon dioxide (CO_2), are gases produced by chemical interactions between carbon and oxygen. Carbon monoxide is poisonous, and carbon dioxide is a greenhouse gas.

E. Hydrocarbons may be solids, liquids, or gases, depending on their molecular size. Hydrocarbons are associated with photochemical smog; methane is a greenhouse gas. Some hydrocarbons are dangerous to human health.

F. Ozone (O_3) is a secondary air pollutant in the lower atmosphere (troposphere) but an essential part of the stratosphere. Tropospheric ozone forms when sunlight catalyzes a reaction between nitrogen oxides and volatile hydrocarbons. Ozone reduces air visibility, causes health problems, and stresses plants. Tropospheric ozone is also a greenhouse gas.

G. Some of the hundreds of other kinds of air pollutants are collectively called **hazardous air pollutants**, or **air toxics**, because they are potentially harmful and may pose long-term health risks to people who are exposed to them. Chlorine, lead, hydrochloric acid, formaldehyde, radioactive substances, and fluorides are examples.

III. The effect of air pollution on human health is a concern.

A. In general, air pollutants irritate the eyes, inflame the respiratory tract, and suppress the immune system.

  1. Sulfur dioxide, particulate matter, and nitrogen dioxide constrict airways, impairing the lungs' ability to exchange gases.

  2. Carbon monoxide combines with hemoglobin and reduces its ability to transport oxygen. Carbon monoxide poisoning can cause death.

3. Exposure to air pollution may result in the development of chronic respiratory diseases such as emphysema and chronic bronchitis.

B. Adults at greatest risk from air pollution include those with heart and respiratory diseases.

C. Air pollution is a greater health threat to children than it is to adults, in part because air pollution impedes lung development. Children with weaker lungs are more likely to develop respiratory problems, including chronic respiratory diseases.

D. Smoking is responsible for the premature deaths of nearly half a million people in the United States each year.

  1. Smoking is linked to lung cancer, emphysema, heart disease, strokes, and several other kinds of cancer.

  2. Passive smoking is particularly harmful to infants, young children, pregnant women, the elderly, and people with chronic lung diseases.

IV. Air pollution is often localized in urban areas.

A. **Industrial smog** is composed primarily of sulfur oxides and particulate matter.

B. The formation of a brownish orange haze known as **photochemical smog** involves a complex series of chemical reactions among nitrogen oxides, volatile hydrocarbons, ozone, and sunlight.

C. Certain climates and topographies cause **temperature inversions**, in which the lower layers of air are cooler than higher layers. Temperature inversions that persist in congested urban areas cause air pollutants to accumulate to dangerous levels.

D. An area of local heat production associated with high population density is known as an **urban heat island**. Heat islands affect air currents and can cause pollutants to accumulate in **dust domes** over cities.

V. Air quality in the United States has slowly improved since passage of the **Clean Air Act** in 1970. This law authorizes the EPA to set limits on how much of specific air pollutants are permitted in the United States.

A. The most dramatic improvement has been the decline in the amount of lead in the atmosphere, although levels of sulfur oxides, ozone, carbon monoxide, volatile organic compounds (many of which are hydrocarbons), and particulate matter have also been reduced.

B. Nitrogen oxides have increased slightly since the passage of the Clean Air Act.

C. Despite decades of progress in improving its air quality, Los Angeles still has some of the worst air pollution in the United States. At its current rate of improvement, Los Angeles should meet federal clean-air standards by 2010.

VI. Air quality is deteriorating in developing nations.

A. Rapid industrialization, a growing number of automobiles in developing countries, and lack of emissions standards are contributing to air pollution, particularly in urban areas.

B. Mexico City has some of the worst air pollution in the world.

VII. The **global distillation effect** is the process in which volatile chemicals evaporate from land as far away as the tropics and are transported by winds to higher latitudes, where they condense and fall to the ground. Volatile chemicals contaminate some remote arctic regions as a result of the global distillation effect.

VIII. Automobiles, homes, schools, and offices may have significantly higher levels of air pollutants than the air outdoors.

A. Eye irritations, nausea, headaches, respiratory infections, depression, and fatigue caused by indoor air pollution are collectively known as the **sick building syndrome**.

B. The most serious indoor air pollutant is **radon**, a radioactive gas that is produced naturally during the radioactive decay of uranium in the Earth's crust.

C. Although loose asbestos fibers in indoor air are an extremely dangerous threat to human health, asbestos is usually not in this form and thus is normally not a major health hazard.

IX. Noise pollution has the potential not only to cause hearing impairment but also to alter many physiological processes in the body and cause psychological stress.

X. Electric and magnetic fields produced by power lines, electrical wiring, and electrical appliances may be hazardous to human health and should be evaluated further.

THINKING ABOUT THE ENVIRONMENT

1. The atmosphere of Earth has been compared to the peel covering an apple. Explain the comparison.

2. List the seven main kinds of air pollutants and briefly describe their sources and effects, including health effects.

3. Distinguish between primary and secondary air pollutants.

4. Distinguish between mobile and stationary sources of air pollution.

5. Briefly describe the health effects of exposure to various air pollutants.

6. Why is air pollution a greater threat to children than it is to adults?

7. The World Health Organization would like to seek a worldwide ban on smoking in public places. Do you agree or disagree? Why?

8. Why might it be more effective to control photochemical smog in heavily wooded areas like the Atlanta metropolitan area by reducing nitrogen oxides than by reducing volatile hydrocarbons?

9. Which is a more stable atmospheric condition, cool air layered over warm air or warm air layered over cool air? Explain. Which condition is a temperature inversion?

10. What are urban heat islands? What are dust domes?

11. What is the U.S. Clean Air Act and how has it reduced outdoor air pollution?

12. What does the Environmental Protection Agency mean by nonattainment areas for one or more criteria air pollutants?

13. What two air pollutants do the 1997 provisions of the Clean Air Act target?

14. Is air pollution worse in highly developed nations or in developing countries? Why?

15. What is the global distillation effect? What kinds of air pollutants are involved in the global distillation effect? Where do these pollutants get permanently deposited? Why?

16. One of the most effective ways to reduce the threat of radon-induced lung cancer is to quit smoking. Explain.

17. How have energy conservation efforts contributed to indoor air pollution and the sick building syndrome?

18. Why are some places considering restricting the loudness of music in public places, including nightclubs and concerts, to 85 dbA?

19. Why are noise pollution and electric and magnetic fields considered forms of air pollution?

*20. Graph the data in Table 19.A on world sulfur emissions from fossil fuel combustion, 1950 to 1994 (latest data available). What long-term trend does the graph indicate?

Table 19.A	World Sulfur Emissions, 1950 to 1994

Year	*World Sulfur Emissions* (million tons)
1950	30.2
1960	47.3
1970	56.7
1980	59.1
1990	70.1
1994	70.7

* The solution to this question appears in Appendix VII.

TAKE A STAND

Visit our Web site at **http://www.wiley.com/college/raven** (select Chapter 19 from the Table of Contents) to learn more about the growing number of clean car choices. You will find tools to help you organize your research, analyze the data, think critically about the issues, and construct a well-considered argument. Take a Stand activities can be done individually or in a team, as oral presentations, written exercises, or Web-based (e-mail) assignments.

Additional on-line materials relating to this chapter, including Student Quizzes, Activity Links, Useful Web Sites, Flash Cards, and more, can also be found on our Web site.

SUGGESTED READING

Conlin, M. "Is Your Office Killing You?" *Business Week* (June 5, 2000). Indoor air pollution can cause serious illnesses.

Crutzen, P.J., and V. Ramanathan. "The Ascent of Atmospheric Sciences." *Science*, Vol. 290 (October 13, 2000). This article provides an excellent overview of the history of atmospheric sciences as well as some of the problems scientists are working on today.

Dalton, R. "Which Way to Energy Utopia?" *Nature*, Vol. 414 (December 13, 2001). Examines the challenges associated with transferring from gasoline-powered to hydrogen-powered motor vehicles.

Gorman, J. "The Air That's Up There." *Science News*, Vol. 161 (June 1, 2002). Studying air pollutants is a challenging endeavor.

Maio, P. "Driving Force." *Wall Street Journal* (September 13, 1999). Examines the potential of fuel-cell–powered cars.

Pyne, S. "Small Particles Add Up to Big Disease Risk." *Science*, Vol. 295 (March 15, 2002). A discussion of the dangers of PM-2.5.

Raloff, J. "Hey, Polluters! This Billboard's For You." *Science News*, Vol. 157 (March 11, 2000). As you drive by a new electronic billboard, it tells you if your car is emitting excessive air pollutants.

Raloff, J. "Ill Winds." *Science News*, Vol. 160 (October 6, 2001). Air pollutants can be carried from one country to another and from one continent to another.

Raloff, J. "U.S. Smog Limit Permits Subtle Lung Damage." *Science News*, Vol. 157 (May 13, 2000). A European study indicates that ozone at current levels currently considered safe may harm human health.

Weiner, T. "Terrific News in Mexico City." *New York Times* (January 5, 2001). Although it's still a major concern, Mexico City's air is cleaner now than it has been in years.

Wilkening, K.E., L.A. Barrie, and M. Engle. "Trans-Pacific Air Pollution." *Science*, Vol. 290 (October 6, 2000). Pollution produced in Asia can cross the Pacific Ocean and contaminate North America.

Wolkomir, R., and Wolkomir, J. "Noise Busters." *Smithsonian* (March 2001). How scientists study noise pollution.

Reforestation project in Guatemala.

Regional and Global Atmospheric Changes

Learning Objectives

After you have studied this chapter you should be able to

1. Describe the greenhouse effect and list the five main greenhouse gases.

2. Discuss the ramifications of some of the potential effects of global warming, including rising sea level, changes in precipitation patterns, effects on organisms, effects on human health, and effects on agriculture.

3. Give examples of several ways to mitigate and adapt to global warming.

4. Describe the importance of the stratospheric ozone layer and distinguish between tropospheric and stratospheric ozone.

5. Explain how ozone depletion takes place and relate some of the harmful effects of ozone depletion.

6. Relate how the international community is working to protect the ozone layer.

7. Explain how acid deposition develops and relate some of the effects of acid deposition.

8. Describe how North American lakes are affected by interactions among global warming, ozone depletion, and acid deposition.

uring the 1990s, farmers in Guatemala planted 52 million trees, paid for by Applied Energy Services (AES), a company that recently built a coal-burning power plant in Connecticut. This project was undertaken to mitigate the adverse effects of carbon dioxide, an air pollutant produced in large quantities when coal and other fossil fuels are burned. (Carbon dioxide and certain other atmospheric pollutants trap solar heat in the atmosphere and may cause the Earth's climate to warm.) Expectations are that as the trees photosynthesize, they will remove the same amount of carbon dioxide from the air that the new power plant will release into the air during its 40-year life span.

The tree-planting venture is actually a continuation of an established reforestation project that was initiated in 1975 by the National Forestry Institute and the Peace Corps. In 1986 the Cooperative for American Relief Everywhere (CARE), an international development organization, became involved. The goal of this project is to help poor farmers improve their degraded lands through reforestation and soil conservation practices. However, CARE's funding was scheduled to expire in 1989. AES provided $2 million to continue the project, which is staffed by Peace Corps volunteers and involves about 40,000 Guatemalan landholders.

This reforestation project makes use of **agroforestry,** in which both forestry and agricultural techniques are used to improve degraded areas. In agroforestry, crops are often planted between the rows of tree seedlings. The trees grow for many years and provide several environmental benefits, such as reducing soil erosion, regulating the release of rainwater into groundwater and surface waters, and providing habitat for the natural enemies of crop pests. Over time, the degraded land slowly improves. When the trees are so tall that they shade out the crops, the forest provides the farmers with food (such as fruits and nuts), fuelwood, lumber, and other forestry products.

AES is not the only utility company to try to offset U.S. carbon dioxide emissions by planting trees in another country. American Electric Power Company, one of the largest coal-burning plants in the United States, is currently spending $5.5 million on a reforestation project in Bolivia.

Although these utilities are to be commended for an innovative approach to the environmental problem of global warming, planting trees is only a partial solution. It has been calculated that an area larger than the United States would have to be reforested *each year* to offset both the world's annual emissions of carbon dioxide and the deforestation that is currently occurring (see Chapter 17). In order to reduce the possibility of disastrous consequences from accelerated global warming during the 21st century, other ways must be found to reduce our carbon dioxide emissions.

In this chapter we examine how air pollutants cause three serious regional and global changes: global warming, depletion of the stratospheric ozone shield, and acid deposition (commonly called acid rain).

GLOBAL WARMING

The Earth's average temperature is based on daily measurements taken at several thousand land-based meteorological stations around the world, as well as data from weather balloons, orbiting satellites, transoceanic ships, and hundreds of sea-surface buoys with temperature sensors. Data indicate that Earth's average surface temperature in 2001 (latest data available) was the second highest since the mid-1800s. The three warmest years on record were 1998, 2001, and 1997. According to the National Oceanic and Atmospheric Administration (NOAA), global temperatures in 1998 may have been the highest in the last 1,200 years. (Although reliable records have been kept only since the mid-19th century, scientists can reconstruct earlier temperatures using indirect climate evidence contained in tree rings, lake and ocean sediments, ancient ice, and coral reefs.) The last two decades of the 20th century were its warmest (Figure 20.1).

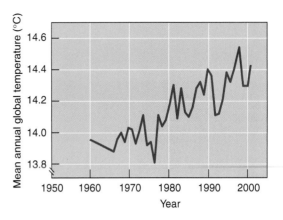

Figure 20.1 Mean annual global temperature, 1960 to 2001. Data are presented as surface temperatures (°C) for 1960, 1965, and every year thereafter. The measurements, which naturally fluctuate, clearly show the warming trend of the last several decades. The warmest year on record was 1998, and the second warmest year was 2001. The eight warmest years occurred from 1990 to 2001. (The dip in global temperatures in the early 1990s, which was caused by the eruption of Mount Pinatubo in 1991, is discussed later in the chapter.)

Other evidence also suggests an increase in global temperature. Several studies have documented that spring in the Northern Hemisphere now comes about 6 days earlier than it did in 1959, and autumn is delayed by 5 days. (Spring is determined by when buds of specific plants open, and autumn by when leaves of specific plants turn color and fall.) Since 1949, the United States has experienced an increased frequency of extreme heat stress events, which are extremely hot, humid days during summer months; medical records indicate that heat-related deaths among elderly and other vulnerable people increase during these events. In the past few decades the sea level has risen slightly, glaciers worldwide have retreated, and extreme weather events such as severe rainstorms have occurred with increasing frequency in certain regions.

Scientists around the world have been researching global warming for the past 50 years. As the evidence has accumulated, those most qualified to address the issue have reached a strong consensus that the 21st century will experience significant climate change and that human activities will have been at least partly responsible for this change.

In response to this growing scientific consensus, governments around the world organized the U.N. Intergovernmental Panel on Climate Change (IPCC). With input from hundreds of climate experts, the IPCC provides the most definitive scientific statement about global warming. The IPCC reviews all the published literature over the previous 5 or so years and determines the current state of both knowledge and uncertainty as it relates to global climate change. The IPCC Third Assessment

Report, issued in January 2001, concluded that human-produced air pollutants have caused most of the climate warming observed over the last 50 years. The IPCC report projected a 1.4° to 5.8°C (2.5° to 10.4°F) increase in global temperature by the year 2100, although the warming will probably not be uniform from region to region. Thus, Earth may become warmer during the 21st century than it has been for at least the last 10,000 years, based on paleoclimate data.

Almost all climate experts agree with the IPCC's assessment that the warming trend has begun and will continue throughout the 21st century. Since the IPCC's second assessment in 1995, a huge body of evidence has accumulated that implicates a growing human influence on the global climate system.

The third IPCC report now has enough data to allow certain projections for the 21st century. During the 21st century, the report estimates a very likely (90% to 99%) chance that we will observe higher maximum temperatures and more hot days over nearly all land areas. The same level of confidence (90% to 99%) exists for higher minimum temperatures, fewer frost days, fewer cold days, an increase in the heat index, and more intense precipitation events over many areas. There is a 66% to 90% chance that the continental interiors in the midlatitudes will experience an increased risk of drought and that some coastal areas will experience stronger hurricanes.

The Causes of Global Warming

Carbon dioxide (CO_2) and certain other trace gases, including methane (CH_4), nitrous oxide (N_2O), chlorofluorocarbons (CFCs), and tropospheric ozone (O_3), are accumulating in the atmosphere as a result of human activities (Table 20.1).[1] The concentration of atmospheric carbon dioxide has increased from about 288 parts per million (ppm) approximately 200 years ago (before the Industrial Revolution began) to 371 ppm in 2001 (Figure 20.2). Burning carbon-containing fossil fuels—coal, oil, and natural gas—accounts for most human-made carbon dioxide. Carbon dioxide is also released by land conversion, such as when tracts of tropical forests are logged or burned. (The burning itself releases CO_2 into the atmosphere, and because trees normally remove CO_2 from the atmosphere during photosynthesis, tree removal prevents this process.) Scientists estimate that by 2050 the concentration of atmospheric CO_2 will be double what it was in the 1700s.

The levels of the other trace gases associated with global warming are also rising. Every time you drive your car, the combustion of gasoline in the car's engine

Table 20.1	Increase in Selected Atmospheric Greenhouse Gases Preindustrial to Present	
Gas	**Estimated Preindustrial Concentration**	**Present Concentration**
Carbon dioxide	288 ppm*	370.9 ppm†
Methane	848 ppb**	1,783 ppb‡
Nitrous oxide	285 ppb	315 ppb‡
Chlorofluorocarbon-12	0 ppt***	541 ppt‡
Chlorofluorocarbon-11	0 ppt	262 ppt‡

* ppm = parts per million.
** ppb = parts per billion.
*** ppt = parts per trillion.
† 2001 annual average.
‡ 1999 values.

releases not only CO_2 but also nitrous oxide and triggers the production of tropospheric ozone. Nitrous oxide is also produced by various industrial processes land-use conversion, and the use of fertilizers. CFCs, which are discussed later in the chapter as they relate to depletion

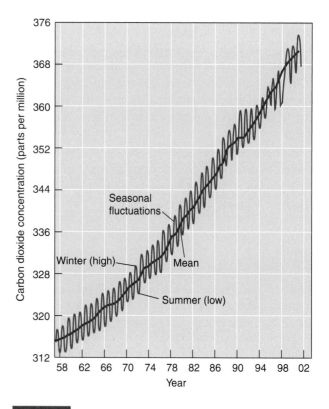

Figure 20.2 Carbon dioxide (CO_2) in the atmosphere, 1958 to 2001. Note the steady increase in the concentration of atmospheric CO_2 since 1958, when measurements began at the Mauna Loa Observatory in Hawaii. This location was selected because it is far from urban areas where factories, power plants, and motor vehicles emit CO_2. The seasonal fluctuations correspond to winter (a high level of CO_2), when plants are not actively growing and absorbing CO_2, and summer (a low level of CO_2), when plants are growing and absorbing CO_2.

[1] Additional greenhouse gases include carbon tetrachloride, methyl chloroform, chlorodifluoromethane (HCFC-22), sulfur hexafluoride, trifluoromethyl sulfur pentafluoride, fluoroform (HFC-23), and perfluoroethane.

of the stratospheric ozone layer, are refrigerants that are released into the atmosphere from old, leaking refrigerators and air conditioners. Decomposition of carbon-containing organic material by anaerobic bacteria in moist places as varied as rice paddies, sanitary landfills, and the intestinal tracts of cattle and other large animals (humans included) is a major source of methane. Water vapor is also a greenhouse gas, and it has a **positive feedback** on the climate that may amplify warming. (Warmer air temperatures cause greater evaporation from the ocean, which in turn causes warmer air temperatures.)

Global warming occurs because these gases absorb infrared radiation—that is, heat—in the atmosphere. This absorption slows the natural heat flow into space, warming the lower atmosphere. Some of the heat from the lower atmosphere is transferred to the ocean and raises its temperature as well. This retention of heat in the atmosphere is a natural phenomenon that has made Earth habitable for its millions of species. However, as human activities increase the atmospheric concentration of these gases, the atmosphere and ocean may continue to warm, and the overall global temperature may rise.

Because CO_2 and other gases trap the sun's radiation somewhat like glass does in a greenhouse, the natural trapping of heat in the atmosphere is referred to as the **greenhouse effect**, and the gases that absorb infrared radiation are known as **greenhouse gases**. The additional warming that may be produced by increased levels of gases that absorb infrared radiation is known as the **enhanced greenhouse effect** (Figure 20.3).

Although current rates of fossil fuel combustion and deforestation are high, causing the CO_2 level in the atmosphere to increase markedly, scientists think the warming trend will be slower than the increasing level of CO_2 might indicate. The reason is that water requires more heat to raise its temperature than gases in the atmosphere do (recall the discussion of the high heat capacity of water in Chapter 13). As a result, the ocean takes longer than the atmosphere to absorb heat. Climate scientists think that warming will probably be more pronounced in the second half of the 21st century than in the first half.

Other Pollutants Cool the Atmosphere One of the complications that makes the rate and extent of global warming difficult to predict is that other air pollutants, known as atmospheric aerosols, tend to cool the atmosphere. **Aerosols**, which come from both natural and human sources, are tiny particles that are so small that they remain suspended in the atmosphere for days, weeks, or even months. Sulfur haze is an aerosol that cools the planet by reflecting sunlight back into space, away from the Earth.[2] This effect reduces the amount of

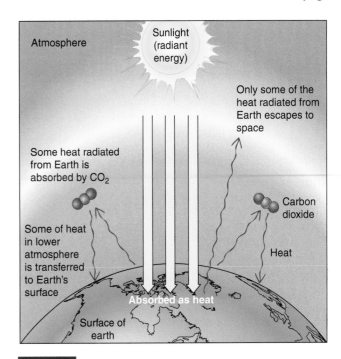

Figure 20.3 Enhanced greenhouse effect. The buildup of carbon dioxide (CO_2) and other greenhouse gases warms the atmosphere by absorbing some of the outgoing infrared (heat) radiation. Some of the heat in the warmed atmosphere is transferred back to Earth's surface, warming the land and ocean.

solar energy that reaches the Earth's surface, thereby cooling the atmosphere. Climate scientists have evidence that sulfur haze significantly moderates warming in industrialized parts of the world. Atmospheric cooling that occurs where and when aerosol pollution is greatest is known as the **aerosol effect**.

Sulfur emissions that produce sulfur haze come from the same smokestacks that pour forth carbon dioxide. (For world sulfur emissions, see Table 19.A at the end of Chapter 19.) In addition, volcanic eruptions can inject sulfur-containing particles into the atmosphere. The explosion of Mount Pinatubo in the Philippines in June 1991 was the largest volcanic eruption in the 20th century (Figure 20.4). The force of this eruption injected massive amounts of sulfur into the stratosphere (the layer of the atmosphere above the troposphere), where sulfur tends to remain longer than it does when emitted into the troposphere. Because sulfur in the stratosphere reduces the amount of sunlight that reaches the planet, this eruption caused the Earth to enter a temporary period of global cooling. Compared to the rest of the 1990s, 1992 and 1993 global temperatures were relatively cool (see Figure 20.1).

Human-produced sulfur emissions should not be viewed as a panacea for human-enhanced global warming, despite their cooling effect. For one thing, sulfur emissions are produced in heavily populated industrial areas, primarily in the Northern Hemisphere. Because they do not remain in the atmosphere for long, they do

---

[2] Complicating the cooling effect of most aerosols is the fact that soot particles from the incomplete combustion of fuels exert a *warming* effect on the atmosphere.

Figure 20.4 Eruption of Mount Pinatubo in June 1991. In the largest volcanic eruption in the 20th century, Mount Pinatubo spewed massive amounts of sulfur into the atmosphere.

not disperse globally. Thus, sulfur pollution may cause regional, but not global, cooling.

Overall, the greenhouse gases will win out over sulfur haze. Greenhouse gases remain in the atmosphere for hundreds of years, whereas human-produced sulfur emissions remain for only days, weeks, or months. And carbon dioxide and other greenhouse gases help warm the planet 24 hours a day, whereas sulfur haze cools the planet only during the daytime. In addition, because sulfur emissions are a respiratory irritant and cause acid deposition (discussed later in this chapter), most nations are trying to *reduce* their sulfur emissions, not maintain or increase them.

Developing Climate Models

Climate is affected by many interacting factors, such as winds, clouds, ocean currents, and albedo (see Chapter 6). Some of these factors enhance global warming and others reduce it. Because the interactions among the atmosphere, the ocean, and the land are too complex and too large to study in a laboratory, climate scientists develop simulation **models** by using powerful computers. Such models calculate complex equations that represent the overall effect of competing factors to describe the current climate situation in numerical terms (Figure 20.5). Models are tested against observations of current and past climate conditions. Models can also be used to suggest the future consequences ("what-if" scenarios) of climate warming on the biosphere and its life support systems.

A climate model is only as good as the data and assumptions on which it is based. Although global

warming models have been increasingly refined in recent years, some uncertainties remain. Consider the uncertain impact of clouds on climate. Global warming will probably cause greater evaporation from the Earth's surface, resulting in more clouds. If global warming causes more low-lying clouds to form, they may block some sunlight and decrease the warming trend (that is, **negative feedback**). On the other hand, if global warming causes more high, thin cirrus clouds to form, they might intensify the warming (that is, positive feedback). As new data about these and other uncertainties become available, they are used to make the models' predictions more precise.

Most climate models examine what the climate will be like in a few decades or a century from now. One model, which was developed by Jerry Mahlman of Princeton University, has examined the implications of climate warming 5 centuries from now. This model assumes that human conservation measures will cause greenhouse gas emissions to stabilize at 2050 levels (that is, at double the carbon dioxide level of the preindustrial world). It presents a dramatic scenario of a warmer climate that future generations will have to live with. Among other changes, the sea level would rise enough to inundate the southern end of Florida from Key Largo to Ft. Lauderdale. Average summer temperatures in the southeastern and mideastern states (north to Philadelphia) would increase from the current 27°C (80°F) to 31°C (87°F), but because this warm air would be able to hold more moisture, the average temperature would feel more like 36°C (97°F).

ENVIROBRIEF

Developing Countries and Carbon Emissions

There has been a general expectation that as the economies of developing countries progress, these nations will follow the path of industrialized countries and consume more fossil fuels—and thus emit more greenhouse gases. Most of the increase in emissions would be a direct result of providing basic human needs for increasing populations. Because of their rapid population growth, developing nations are predicted to release higher levels of greenhouse gases than industrialized nations by 2020. This scenario is not unfolding exactly as predicted, however, because some developing nations—such as China, India, Mexico, Saudi Arabia, South Africa, and Brazil—are making surprising progress toward controlling greenhouse gas emissions. Efforts to reduce carbon emissions include increasing energy efficiency, applying governmental limitations on fossil fuel use, and developing alternative energy programs. Increased control of greenhouse gas emissions in developing countries is not the result of specific climate change policies but instead is a secondary benefit of these nations' efforts to meet their own social, economic, and health needs. Air pollution arising from the use of fossil fuels creates serious public health problems in developing countries, so curbing fossil fuel consumption creates healthier living conditions locally while improving global air quality.

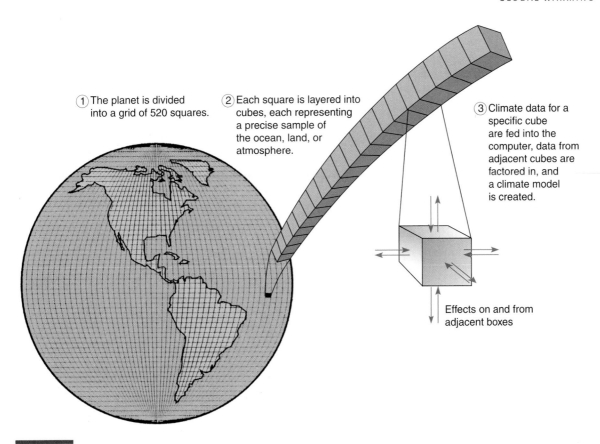

① The planet is divided into a grid of 520 squares.

② Each square is layered into cubes, each representing a precise sample of the ocean, land, or atmosphere.

③ Climate data for a specific cube are fed into the computer, data from adjacent cubes are factored in, and a climate model is created.

Effects on and from adjacent boxes

Figure 20.5 Building a climate model. Modelers place a grid over the planet's surface, dividing it into about 520 squares. Each square of the grid is layered into three-dimensional cubes that extend down into the ocean (about 20 cubes deep) and up into the atmosphere (about 18 cubes high). Each layer's program contains conditions such as seasonal changes in sunlight, temperature, air pressure, currents or winds, and water vapor that affect climate in that particular cube. These conditions are allowed to interact within the layered cubes, using formulas that take into account basic physical laws such as the first law of thermodynamics. Conditions within neighboring cubes are also factored into the calculations. The computer continually performs these calculations, either alone or with an added variable such as increasing greenhouse gas emissions over a period of simulated days, months, or years.

Climate models present humans with a potential ethical dilemma. Do we wait for greater scientific certainty and continue on our present course of using fossil fuel energy to power economic development (which many view as beneficial, at least in the short term)? Or do we make serious cuts in present energy use, which may curtail economic development, to avoid some of the problems we could inflict on humans who will live generations from now? How much are we willing to sacrifice for benefits that may not be realized in our lifetimes?

These questions may not be as difficult to address as they appear at first glance. According to the Worldwatch Institute, the global economy grew by 6% to 8% during 1998 and 1999, yet CO_2 emissions held steady during that 2-year period. Consider China. In 1998, China's economy grew by 7.2%, but CO_2 emissions decreased by 3.7% as many of its industries burned less coal. (The combustion of coal releases more CO_2 per unit of heat produced than other fossil fuels.) Analysts say such reports are evidence that economic growth is in the process of uncoupling from CO_2 production, which means it may be easier for nations to agree to emissions cuts. Moreover, a serious effort to replace fossil fuels with climate-friendly energy sources has the potential to benefit the global economy, as millions of new jobs would probably be created.

The Effects of Global Warming

We now consider some of the observed and potential effects of global warming, including changes in sea level, precipitation patterns, organisms, human health, and agriculture. In addition to these effects, unanticipated surprises will probably occur as a result of climate change. We consider one such "surprise," the possibility that climate warming could disrupt large ocean currents.

Melting Ice and Rising Sea Levels As the overall temperature of the Earth increases, there could be a major thawing of glaciers and the polar ice caps. During the Antarctic summer in 2002, most of the Larsen B ice shelf, an area roughly the size of Rhode Island, broke off the Antarctic Peninsula. (Antarctic ice shelves are thick sheets of floating ice that are fed mainly by glaciers that flow off the land.) This loss of ice coincided with a decades-long trend of atmospheric warming in the Antarctic.

A study from the 1970s to the 1990s showed that the area of ice-covered ocean in the Arctic has also retreated. The southern edge of arctic ice that was found between 71°N and 72°N latitude during the 1970s has retreated northward to 75°N. Sonar measurements from naval submarines operating under the ice indicate that the remaining arctic ice pack has thinned rapidly, losing 40% of its volume in less than 3 decades. A 1993 to 1998 analysis of the Greenland ice sheet, the world's second largest land-bound ice (Antarctica is the largest), determined that Greenland lost about 8.3 km³ (2 mi³) of ice each year during that period.

Mountain glaciers around the world are melting at accelerating rates. Qori Kalis Glacier, the largest glacier in the Peruvian Andes, is currently retreating by about 30.5 m (100 ft) a year. The Gangotri Glacier in India is retreating at a similar rate. According to the Worldwatch Institute, 100 of Glacier National Park's 150 glaciers have melted since 1850; the remaining 50 are retreating so rapidly that they will probably be completely gone by 2030. Melting water from many retreating glaciers have formed unstable lakes behind moraine (a mass of rocks, gravel, and sand formed at the end of glacier). Areas at lower elevations are at risk from these lakes, which have the potential to burst, sending floodwater downward.

In addition to sea-level rise caused by the retreat of glaciers and thawing of polar ice, the sea level will probably rise because of thermal expansion of the warming ocean. Water, like other substances, expands as it warms. The IPCC reports that during the 20th century, the sea level rose by 0.1 to 0.2 m (4 to 8 in.), most of it due to thermal expansion.

The IPCC estimates that the sea level will rise by an additional 48 cm (19 in.) by 2100. Such a rise in sea level would flood low-lying coastal areas such as southern Louisiana and South Florida. Coastal areas that are not inundated will be more likely to suffer erosion and other damage from more frequent and more intense weather events such as hurricanes. These likely effects are certainly a cause for concern, particularly because about two thirds of the world's population lives within 150 km (93 mi) of a coastline.

In 1999 two uninhabited islands (Tebua Tarawa and Abanuea) in the South Pacific were submerged under rising water. In 2001, the 11,000 residents of nearby Tuvalu announced that they must evacuate because the rising sea level has caused lowland flooding, harming their water supply and food production. (As this book went to press, no country had agreed to accept the Tuvaluans.) Small island nations such as the Maldives, a chain of 1,200 islands in the Indian Ocean, are considered highly vulnerable to a rise in sea level. About 80% of the Maldives is lower than 1 m (39 in.) above sea level, and the country's highest point is only 2 m above sea level. As sea levels rise, storm surges could easily sweep over entire islands. Other countries that are vulnerable to a rise in sea level—such as Bangladesh, Egypt, Vietnam, and Mozambique—have dense populations living in low-lying river deltas. A rising sea level could cause Bangladesh to lose as much as 18% of its land, displacing millions of people in this densely populated nation.

Melting ice that drains into the ocean can raise the sea level, but what about land-bound melting ice? Evidence indicates that **permafrost**, the permanently frozen subsoil characteristic of the tundra and boreal forests of Alaska, Canada, Russia, China, and Mongolia, is melting. Permafrost provides the foundation on which tundra plants and forest trees are anchored and on which houses and roads are built. As the permafrost melts, this foundation collapses. Near Fairbanks, Alaska, hundreds of homes and telephone poles are sinking at odd angles into the ground (Figure 20.6.)

Changes in Precipitation Patterns Precipitation patterns that occurred thousands of years ago when the Earth was warmer have been used to develop computer models of weather changes as global warming occurs.

Figure 20.6 Melting permafrost. The telephone poles tilt as the permafrost in which they are anchored melts. Melting permafrost causes ground subsidence, erosion, and landslides.

Arctic Warming and the Native Cultures of the North

The indigenous people of Alaska's and Canada's far north, the Eskimo Inuit, pursue a way of life dictated by the frigid climate. Effects of global warming are changing the Inuit traditional existence. Many populations of wildlife, which the Inuit harvest for food, have been reduced or displaced. Other changes that threaten subsistence livelihoods include reduced snow cover, shorter river ice seasons, and thawing of permafrost. Warmer temperatures also increase the risk of contaminating water supplies, as bacteria move more freely through thawed soil. Larger thawed areas could also lead to the collapse of bridges, buildings, roads, and oil pipelines.

Glimpses of these observed and potential changes are gained from the region's natives, who report changes thought to be linked to warmer temperatures: drying tundra; thinner and retreating sea ice; warmer winters; and changes in the numbers, distribution, and migration of some wildlife species. Scientifically obtained climate change data support these observations. Scientists analyzing lake-sediment cores to explore climate changes for the past 400 years note that the greatest warming trend occurred from 1840 to the late 20th century. Although some of the trend is attributed to long-term natural changes, an abrupt rise in atmospheric temperatures coincides with rapid 20th-century increases in emissions of greenhouse gases such as carbon dioxide. With temperatures projected to rise even faster in the 21st century, scientists caution that the relatively undisturbed Arctic may be particularly susceptible to the effects of climate change.

These simulations indicate that precipitation patterns will change, causing some areas to have more frequent droughts (Figure 20.7). At the same time, heavier snow and rainstorms may cause more frequent flooding in other areas.

Changes in precipitation patterns could affect the availability and quality of fresh water in many locations. It is projected that areas that are currently arid or semi-arid, such as the Sahel region just south of the Sahara Desert (see Figure 6.21), will have the most troublesome water shortages as the climate changes. Closer to home, water experts predict water shortages in the American West because warmer winter temperatures will cause more precipitation to fall as rain rather than snow; melting snow currently provides 70% of stream flows in the West during summer months.

The frequency and intensity of storms over warm surface waters may also increase. In 1998, NOAA developed a computer model examining how global warming might affect hurricanes. When the model was run with a sea surface temperature 2.2°C warmer than today, more intense hurricanes resulted. (The question of whether hurricanes will be more *frequent* in a warmer climate remains uncertain.) Changes in storm frequency and intensity are expected because as the atmosphere warms, more water evaporates, which in turn releases more energy into the atmosphere (recall the discussion of water's heat of vaporization in Chapter 13). This energy generates more powerful storms.

As we discussed in Chapter 6, the El Niño–Southern Oscillation (ENSO), the periodic warming (El Niño) and cooling (La Niña) of the tropical Pacific Ocean, affects precipitation and other aspects of the entire global climate system. Until recently climate scientists were unable to predict if human-induced global warming would affect ENSO. The IPCC's analysis of complex computer models indicates greater extremes of drying and heavy rainfall during El Niño events. Scientists are still uncertain if El Niño events will occur more frequently with global warming.

Effects on Organisms   An increasing number of studies report measurable changes in the biology of plant and animal species as a result of climate warming. Such effects range from earlier flowering times for plant species to migrations of aquatic species. Changes are also

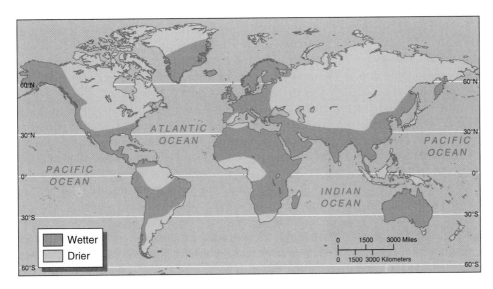

Figure 20.7 Effect of climate warming on precipitation. This map, based on precipitation patterns that occurred thousands of years ago when the Earth was warmer, shows one possible scenario of how precipitation might be altered by warmer global temperatures.

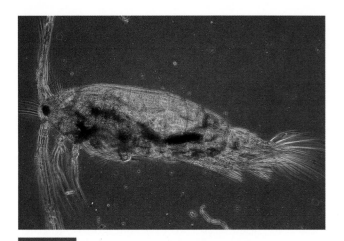

Figure 20.8 Climate warming and zooplankton. A 40-year study of the California Current discovered that zooplankton, such as this copepod (shown magnified 100×), respond dramatically to very small temperature changes. Zooplankton populations have declined by 80% in that time, the result of a slight warming of surface waters off the coast of California. Similarly, a recent long-term study (1946 to 2002) of changes in the species composition of copepods in the North Atlantic reported a decrease in the number of colder-water species.

evident in many populations, communities, and ecosystems. Other human-induced factors, such as pollution and changes in land use, will exacerbate threats posed by climate change. Here we report on the results of several of the hundreds of studies that have been conducted.

Researchers determined that populations of zooplankton in the California Current, which flows from Oregon southward along the California coast, had declined by 80% since 1951, apparently because the current has warmed slightly (Figure 20.8). The decline in zooplankton in the California Current has affected the entire food web there, and populations of plankton-eating fishes and seabirds have also declined.

As temperatures have risen in the waters around Antarctica during the past 2 decades, a similar decline in shrimplike krill has contributed to a reduction in Adélie penguin populations (see "Case in Point: How Humans Have Affected the Antarctic Food Web" in Chapter 4). Because there are fewer krill, the birds are not able to get enough food. Warmer temperatures in Antarctica—during the past 50 years the average annual temperature on the Antarctic Peninsula has increased 2.6°C (5°F)—have also contributed to reproductive failure in Adélie penguins. The birds normally lay their eggs in snow-free rocky outcrops, but the warmer temperatures have caused more snowfall (recall that warmer air can hold more moisture), which melts when the birds incubate the eggs to become cold pools of slush that kill the developing chick embryos.

The first scientific evidence that a species shifted its geographical range, probably in response to climate warming, was published in 1996. A biologist at the University of Texas compared the current geographical range of a western butterfly (the Edith's checkerspot butterfly) to previously recorded observations. She found that the butterfly has disappeared in the southern parts of its range and shifted about 160 km (about 100 mi) northward by establishing new colonies there. In 1999 similar studies conducted in Europe found that 22 butterfly species out of 35 species examined had also shifted their ranges northward, anywhere from 32 to 240 km (20 to 150 mi). Scientists studying birds in Great Britain also reported in 1999 that during the past 20 years the ranges of several dozen species had moved northward by an average of 19 km (12 mi).

Each species reacts to changes in temperature differently. As warming accelerates in the 21st century, some species will undoubtedly become extinct, particularly those with narrow temperature requirements, those confined to small, specialized habitats, and those living in fragile ecosystems. Other species may survive in greatly reduced numbers and ranges.

Ecosystems considered to be at greatest risk of loss of species in the short term are polar seas, coral reefs, mountain ecosystems, coastal wetlands, and tundra. Water temperature increases of 1° to 2°C can cause coral bleaching that contributes to the destruction of coral reefs. In 1998 scientists documented the most geographically extensive and most severe epidemic of coral bleaching ever observed (see section on coral reefs in Chapter 7). According to an ocean scientist at the University of Georgia, about 10% of the world's corals died that year, in many cases from infections caused by viruses, bacteria, or fungi. In 1998 tropical waters were the warmest ever recorded, and scientists suspect that the warmer temperatures stress the corals, making them more susceptible to disease-causing organisms that healthy corals are normally resistant to.

Biologists generally agree that global warming will have an especially severe impact on plants because they cannot move about when environmental conditions change. Although seeds are dispersed by wind and animals, sometimes over long distances, seed dispersal has definite limitations in terms of the speed of migration. During past climate warmings, such as during the glacial retreat that took place some 12,000 years ago, tree species are thought to have migrated from 4 to 200 km (2.5 to 124 mi) per century. If the Earth warms by the projected 1.4° to 5.8°C during the 21st century, the ideal ranges for some temperate tree species (that is, the environment where they grow best) may shift northward as much as 483 km (300 mi) (Figure 20.9). Moreover, soil characteristics, water availability, competition with other plant species, and **habitat fragmentation** all affect the rate at which plants can move into a new area.

Some species will come out of global warming as winners, with greatly expanded numbers and range. Those organisms considered most likely to prosper include certain weeds, insect pests, and disease-carrying

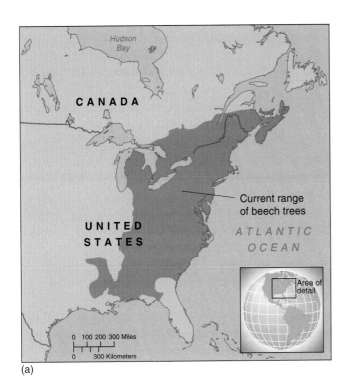

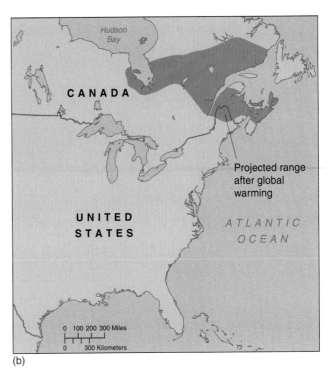

Figure 20.9 Climate change and beech trees in North America. (a) Present geographical range of American beech trees. These large shade trees produce edible beechnuts that are an important food source for squirrels, raccoons, bears, game birds, and many other kinds of wildlife. (b) One projected range of beech trees after global warming occurs. This figure is based on a model of climate change developed by the U.S. Geophysical Fluid Dynamics Laboratory (GFDL). According to the GFDL scenario of changed temperature and precipitation patterns, beeches would continue to grow in parts of Maine and Nova Scotia, and they would expand into Quebec and northern Ontario.

organisms that are already common in a wide range of environments. Biologists note that a 3°C (5.4°F) increase in average surface temperature could allow the Mediterranean fruit fly, an economically important insect pest (see Chapter 22), to expand its range into northern Europe.

Effects on Human Health Currently, most evidence that links climate warming to disease outbreaks is circumstantial, so scientists are reluctant to ascribe cause-and-effect relationships. Nonetheless, data linking climate warming and human health problems are accumulating.

Scientists hypothesize that the increase in CO_2 in the atmosphere and the resultant more frequent and more severe heat waves during summer months will cause an increase in the number of heat-related illnesses and deaths, particularly of elderly people. During the summer 1998 heat wave in Dallas, Texas, more than 100 people died from heatstroke and other heat-related illnesses. India recorded almost 1,300 deaths caused by high temperatures during the 1998 summer heat wave.

Human health may also be affected indirectly by climate warming. Mosquitoes and other disease carriers could expand their range into the newly warm areas and

spread malaria, dengue fever, schistosomiasis, and yellow fever (Table 20.2). As many as 50 million to 80 million additional cases of malaria could occur annually in tropical, subtropical, and temperate areas. According to the World Health Organization, during 1998, the warmest year on record, the incidence of malaria, Rift Valley fever, and cholera surged in developing countries.

Highly developed countries are less vulnerable to such disease outbreaks because of better housing (which keeps mosquitoes outside), medical care, pest control, and public health measures such as water treatment plants. Texas reported a few cases of dengue fever in the late 1990s, whereas nearby Mexico had thousands of cases during that period.

Effects on Agriculture Global warming will increase problems for agriculture, which is already beset with the challenge of providing enough food for a hungry world without doing irreparable damage to the environment. The rise in sea level may cause water to inundate river deltas, which are some of the world's best agricultural lands. Certain agricultural pests and disease-causing organisms will probably proliferate and reduce yields. As mentioned earlier in the chapter, scientists think that

Table 20.2   Likelihood of Altered Distribution of Infectious Diseases as a Result of Climate Change

Disease	Vector	Present Distribution	Likelihood of Distribution Changes with Climate Warming
Malaria	Mosquito	Tropics, subtropics	Highly likely
Dengue fever	Mosquito	Tropics, subtropics	Very likely
Schistosomiasis	Water snails	Tropics, subtropics	Very likely
Yellow fever	Mosquito	Tropical South America and Africa	Very likely
Onchocerciasis	Blackfly	Africa, Latin America	Very likely
Lymphatic filariasis	Mosquito	Tropics, subtropics	Likely
Leishmaniasis	Phlebotomine sandfly	Asia, southern Europe, Africa, Americas	Likely
American trypanosomiasis	Triatomine bug	Central and South America	Likely
African trypanosomiasis	Tsetse fly	Tropical Africa	Likely
Dracunculiasis	Copepod	Central and West Africa	Unknown

global warming will also increase the frequency and duration of droughts, a problem that will be particularly serious for countries with limited water resources. It is also likely that warmer temperatures will result in decreases in soil moisture in many agricultural soils (warmer temperatures cause increased evaporation).

Another effect of climate warming on agriculture involves nighttime temperatures, which have generally increased more than daytime temperatures since 1950, when measurements began. In 1999 a study at the National Science Foundation's Long-Term Ecological Research Site in northeastern Colorado found that warmer nighttime temperatures were linked to changes in the types and distribution of grasses on the prairie. Most notably, buffalo grass, an important food for cattle and other livestock, has declined in abundance and been largely replaced by weeds and non-native grasses. Range scientists point out that buffalo grass can withstand continuous grazing, but the invading plants that are replacing buffalo grass may be more sensitive to grazing pressures. Ecologists suggest that such changes in the structure and dynamics of rangeland ecosystems worldwide could have a profound effect on livestock production.

On a regional scale, current models of modest warming forecast that agricultural productivity will increase in some areas and decline in others. Models suggest that Canada and Russia may be able to increase their agricultural productivity in a warmer climate, whereas tropical and subtropical regions where many of the world's poorest people live will be hardest hit by declining agricultural productivity. Central America and Southeast Asia may experience some of the greatest declines in agricultural productivity.

In addition to all of these potential impacts of climate warming on agriculture, there is a likelihood that modern agricultural methods may have to be altered so they are less reliant on CO_2-producing fossil fuels (see Chapter 18). The manufacture of fertilizers, pesticides, and other agricultural chemicals requires a huge input of

energy from fossil fuels, as does the production and use of modern farm equipment. It is not clear how the current reliance of modern agriculture on fossil fuels will affect future agriculture.

Surprises That May Be Caused by Global Warming   Our current knowledge of the many complexities of global climate is so incomplete that unanticipated effects from a globally warmed world will undoubtedly occur. One recently proposed surprise may be a disruption of the *ocean conveyor belt*, which transports heat around the globe (see Figure 6.15). You may recall from Chapter 6 that the ocean conveyor belt delivers heat from the tropics into the northern part of the North Atlantic Ocean. Some of this heat is transferred to the atmosphere, thereby warming Europe and adjacent lands by as much as 10°C (18°F). As the warm North Atlantic water transfers heat to the atmosphere, it cools, sinks, and flows southward. The cooler, sinking water carries some of the CO_2 from the atmosphere deep into the ocean where, by mechanisms that are incompletely understood, much of the carbon is sequestered (stored).

Models based on the behavior of the ocean conveyor belt during past episodes of climate warming—immediately following the ice ages, for example—suggest that warming could weaken or even shut down the ocean conveyor belt. Data reported at a 2002 ocean science conference suggest that such a change may be underway. No one knows for sure, but changes in the ocean conveyor belt could cause major cooling in Europe and greater warming in the tropics. In addition, a weakened ocean conveyor belt would not be able to sequester as much carbon in the ocean, leading to a positive feedback loop. Less CO_2 stored in the ocean would mean more CO_2 in the atmosphere, which would cause additional atmospheric warming, which in turn would cause the ocean conveyor belt to weaken even further. Thus, once the feedback has begun, it will be difficult for human intervention to reverse the feedback.

Given our limited knowledge at this time, climate scientists are reluctant to do more than make educated guesses about the interactions between climate warming and the ocean conveyor belt. But scientists warn that climate warming may make us vulnerable to this or other as yet unknown surprises, possibly with disastrous consequences. Many researchers think the uncertainties associated with global warming, particularly unknown feedback loops that may accelerate global warming, make a strong case for immediate action to deal with the issue rather than waiting until we have more knowledge about the costs and consequences of global warming.

International Implications of Global Warming

Dealing with global warming is complicated by social, economic, and political factors that vary from one country to another. How will the global community deal with the environmental refugees produced by the impacts of global warming, such as extreme weather events that lead to agricultural failures? Where will they go? Who will help them to resettle? It will be difficult for all countries to develop a consensus on dealing with global warming, partly because global warming will clearly have greater impacts on some nations than on others. However, all nations must cooperate if we are to effectively address global warming and its impacts.

Highly Developed Nations Versus Developing Nations

Although greenhouse gases are produced primarily by the highly developed countries, their *rate* of production by certain developing countries is rapidly increasing. Furthermore, many developing nations may experience the greatest impacts of global warming. Because developing countries have less technical expertise and fewer economic resources, they are the very ones that are least able to respond to the challenges of global warming. Consider Bangladesh. In 2000, Mrs. Sajeeda Choudhury, the Bangladeshi Environment Minister, estimated that rising sea levels may displace perhaps 20 million people from coastal areas of Bangladesh. Because there is not enough land at higher elevations in Bangladesh to accommodate these refugees, she called on highly developed countries to rethink their immigration policies.

Tensions have mounted among nations, especially between the highly developed and developing countries, over their differing self-interests. Most developing countries see increased use of fossil fuels as their route to industrial development and resist pressure from highly developed nations to decrease fossil fuel consumption. Developing countries such as India also ask why they should have to take actions to curb CO_2 emissions when the rich industrialized nations historically have been the main cause of the problem. Since 1950, the 20% of the world's population living in highly developed nations have produced 74% of the CO_2 emissions. Currently,

highly developed countries produce about five or six times more CO_2 emissions per person than developing countries (Figure 20.10).

In contrast, highly developed countries argue that the booming economic growth and much greater number of people living in developing countries threatens to overwhelm the world with carbon dioxide emissions as the developing countries become industrialized. According to the U.S. Department of Energy, CO_2 emissions from developing countries will surpass those from highly developed countries by 2020, assuming current trends in fossil fuel consumption continue. Developing countries respond that even when they are producing half of the world's CO_2 emissions, it will still be unequal because 80% of the world's population living in developing countries will be producing only half the emissions.

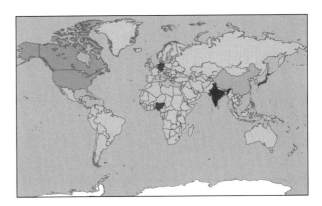

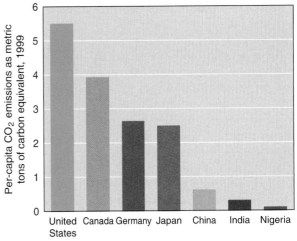

Figure 20.10 Per-capita carbon dioxide (CO_2) emission estimates for selected countries, 1999. Note that industrialized nations currently produce a disproportionate share of CO_2 emissions. As developing nations such as China, India, and Nigeria industrialize, however, their per-capita CO_2 emissions increase. (In conformance with scientific practice, CO_2 emissions are quantified by their weight in carbon. One metric ton of carbon equivalent equals 3.67 metric tons of CO_2.)

International Efforts to Reduce Greenhouse Gas Emissions Despite all the posturing, the international community recognizes that it must stabilize CO_2 emissions. At least 174 nations, including the United States, signed the U.N. Framework Convention on Climate Change developed at the 1992 Earth Summit (see Chapter 1). Its ultimate goal was to stabilize greenhouse gas concentrations in the atmosphere at levels low enough to prevent dangerous human influences on the climate. However, the details of how that goal was to be accomplished were left to future conferences.

At the 1996 U.N. Climate Change Convention held in Geneva, Switzerland, highly developed countries agreed to establish legally binding timetables to cut emissions of greenhouse gases. These timetables were decided at a meeting of representatives from 160 countries held in Kyoto, Japan, in 1997. By October 2002, 95 countries had ratified the so-called **Kyoto Protocol**. This extremely complex international treaty, which is legally binding, provides operational rules on reducing greenhouse gas emissions. It is noteworthy that Australia, Canada, Russia, and the United States have not yet ratified the Kyoto Protocol, which does not enter into force until enough industrialized countries to represent 55% of 1990 global emissions of greenhouse gases have ratified it. (As of October 2002, enough highly developed countries, including Europe and Japan, had ratified the protocol to represent 37% of 1990 global emissions.)

The United States signed the Kyoto Protocol in 1998, but the administration of President George W. Bush withdrew the United States from that commitment in 2001 on the grounds that it would be an unacceptable economic burden. (To curb their production, greenhouse gas emissions will have to be taxed or regulated in some other manner, such as tradable emissions permits.) However, when the costs of climate warming—such as damage to agriculture and human health—are considered, many climate experts point out that it is economically *beneficial* to enact the Kyoto Protocol. Most analysts think the Kyoto Protocol will accomplish little without the full participation of the United States. In place of the Kyoto Protocol, President Bush's strategy was to have U.S. companies voluntarily reduce their emissions. To reduce greenhouse gas emissions, Bush wanted U.S. citizens and companies to use fossil fuels more efficiently and, where possible, to adopt renewable energy in place of fossil fuels (see Chapter 12).

Even if all countries, including the United States, were to implement the Kyoto Protocol fully, it would not prevent the continuing buildup of atmospheric greenhouse gases, only slow their rate of buildup. The international community recognizes that the Kyoto Protocol is only the first step in addressing climate warming. However, the protocol provides the legal framework whereby countries can agree to make additional cuts in greenhouse gases in the future. Perhaps more important, the reduced rate of greenhouse gas buildup provided by full participation in the Kyoto Protocol would give us more time to deal with global climate change.

How We Can Deal with Global Warming: Specifics

Despite certain gaps in our knowledge of global warming, our present understanding of the changing global climate and its potential impact on human society and other species gives us many legitimate reasons to develop strategies that will deal with this problem. Even if we were to immediately stop polluting the atmosphere with greenhouse gases (which we cannot), global temperature would continue to increase for many decades because of the greenhouse gases that have accumulated during the past 100 years. The global climate system is very slow to respond to changes in greenhouse gas concentrations, and it cannot be turned off quickly.

All greenhouse gases will have to be dealt with, but we focus on CO_2 because it is produced in the greatest quantities and has the largest total effect (about 60%) of all the greenhouse gases. Carbon dioxide has an atmospheric lifetime of more than 1 century, so the emissions we are producing today will still be around in the 22nd century. The amount and severity of global warming depend on how much additional greenhouse gas emissions we add to the atmosphere. Many studies make the assumption that we will be able to stabilize atmospheric CO_2 at 550 ppm, which is roughly twice the concentration of atmospheric CO_2 that scientists estimate existed in the preindustrial world and about 50% higher than the CO_2 currently in the atmosphere.

There are two basic ways to attempt to manage global warming: mitigation and adaptation. *Mitigation* is the moderation or postponement of global warming, which buys us time to pursue other, more permanent solutions that stop or reverse global warming. Mitigation also gives us time to understand more fully how global warming operates so we can avoid some of its worst consequences. *Adaptation* is responding to changes brought about by global warming. Creating strategies to adapt to climate warming implies an assumption that global warming is unavoidable.

Mitigation of Global Warming Because climate warming is essentially an energy issue, the development of alternatives to fossil fuels offers a solution to warming caused by CO_2 emissions. Alternatives to fossil fuels are also necessary, given that fossil fuels are present in limited amounts (see Chapter 10). Some alternatives to fossil fuels—solar energy and nuclear energy—were discussed in Chapters 11 and 12.

Combustion of fossil fuels with its accompanying CO_2 emissions is largely a function of human population size and level of consumption: The larger the population and the greater the degree of affluence, with its accompanying higher levels of consumption, the larger the global emissions of greenhouse gases. Obviously, it will be easier to stabilize CO_2 emissions if population growth is slowed. Yet many scientists and policy makers do not include population control as one of the strategies to mitigate climate warming. The human population, including socioeconomic factors that influence population growth, is discussed in Chapters 8 and 9.

Increasing the energy efficiency of automobiles and appliances—and thus reducing the output of CO_2—would help mitigate global warming. This could be accomplished by establishing regulatory programs that increase minimum energy efficiency standards and by negotiating agreements with industry. Energy-pricing strategies, such as carbon taxes and the elimination of energy subsidies (see Chapter 10), are examples of other policies that could help mitigate global warming. Most experts think that by using existing technologies and developing such policies, greenhouse gas emissions can be significantly reduced with little cost to society.

Other mitigation strategies being considered are planting trees, carbon management, and fertilizing the ocean with iron. Planting trees is a low-tech, relatively inexpensive strategy, whereas carbon management and fertilizing the ocean with iron require the development of new technologies and extensive testing.

Sequestering Carbon in Trees

As we discussed in the chapter introduction, one way to mitigate global warming involves removing atmospheric carbon dioxide from the air by planting and maintaining forests. Like other green plants, trees incorporate the carbon into organic matter in leaves, stems, and roots through the process of photosynthesis. Because trees typically live for 100 or more years, the carbon in their roots and stems remains sequestered away from the atmosphere for a relatively long time. Although estimates vary widely, a reasonable estimate is that trees could remove 10% to 15% of the excess CO_2 in the atmosphere, but it would require enormous plantings (recall the discussion of this topic in the chapter introduction). Scientists who have studied sequestering carbon in trees think this proposal might provide short-term benefits for the climate. However, it is no substitute for cutting emissions of greenhouse gases.

Carbon Management

In addition to taking steps to curb greenhouse gas emissions, many countries are investigating **carbon management**, ways to separate and capture the CO_2 produced during the combustion of fossil fuels and then sequester it away from the atmosphere.

Several power plants currently capture CO_2 emitted in their flue gases, but the technology is very new. Technological innovations that can more efficiently trap the CO_2 being emitted from smokestacks would help prevent global warming and yet allow us to continue using fossil fuels (while they last) for energy. Government incentives, such as providing research grants for the development of such technologies, most likely will be necessary to inspire innovations. Several nations have imposed taxes on greenhouse gases; the taxes motivate industrial emitters to improve efficiency and to develop CO_2 capturing technologies.

Carbon could be sequestered in geological formations or depleted oil or natural gas wells on the land or injected as a liquid into the ocean depths. Because of its vastness, the ocean is considered to have great potential as a repository for CO_2. Carbon management is a new, unproven technology, however, and its potential ecological effects will have to be studied fully before it is adopted on a large scale. Injecting large quantities of CO_2 could change the ocean's chemistry and thereby harm the organisms living there. Much of this CO_2 would eventually find its way back to the atmosphere, but ocean chemists estimate the underwater reservoir would remain for several hundred years before appreciable amounts returned to the atmosphere. Norway, Japan, the United States, Canada, and Australia are among the countries most interested in developing carbon management.

Fertilizing the Ocean with Iron

In 1987 an oceanographer proposed a novel way to remove some CO_2 from the atmosphere, by fertilizing parts of the ocean with iron, which is an essential element required by all forms of life. About 20% of the ocean's surface is rich in nutrients such as nitrogen but very low in iron (that is, iron is a **limiting factor**, as discussed in Chapter 5). These areas of the ocean also have very low numbers of phytoplankton, which are microscopic algae that form the base of the ocean's food web. The scientist reasoned that the addition of iron would, in effect, fertilize the ocean, stimulating large numbers of phytoplankton to grow. As they photosynthesized, the phytoplankton would remove CO_2 from the water, which would be replenished by CO_2 from the atmosphere, thereby lowering the atmospheric CO_2 level. When the phytoplankton and zooplankton that eat the phytoplankton die, many sink to the ocean floor. Also, zooplankton droppings, which contain carbon, sink. Thus, some carbon would be stored on the ocean floor, far from the atmosphere (Figure 20.11).

The iron hypothesis was first tested in 1995, when small amounts of dissolved iron were added to a 64-km^2 (26-m^2) area of the equatorial Pacific Ocean. It stimulated a marked increase in the number of phytoplankton, so many that the ocean temporarily turned from blue to green. Although the first test of the iron hypothesis was successful, a similar study in 2000 in the polar Southern Ocean found that the downward export of carbon to deep ocean waters did not increase when the ocean was fertilized with iron. Most scientists reject this method of removing CO_2 from the atmosphere because we do not understand the ecological consequences of such an action.

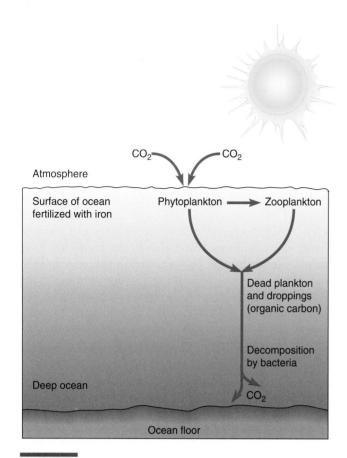

semitropical crops to determine the best ones to substitute for traditional crops as the climate warms. Large lumber companies are currently developing drought-resistant strains of trees, because the trees planted today will be harvested in the last half of the 21st century, when global warming may already be well advanced.

Adaptation to global warming is currently under study at 15 locations around the United States. Each study group consists of scientists and various representatives of municipal governments, federal agencies, local businesses, and community organizations. One of the potential problems identified in the New York City study group involves its sewer system. The waterways for storm runoff normally close during high tides to prevent salt water from the Atlantic Ocean from backing into the system. As the sea level rises in response to global warming, the waterways will also have to be shut during many low tides, which will greatly increase the risk of flooding during storms. (When the waterways are closed, excess water does not drain away.) City planners will therefore have to rebuild the storm runoff system, which is a very expensive proposition, or find some other way to prevent flooding. Evaluating such problems and finding and implementing solutions now will ease future stresses of climate warming.

Figure 20.11 Iron fertilization hypothesis. Phytoplankton, which are the base of the marine food web, photosynthesize, converting carbon dioxide (CO_2) to the carbon in organic molecules. Most of the carbon is converted back to CO_2 during cellular respiration by phytoplankton, zooplankton, and other members of the food web. However, some of the organic carbon (*red arrows*) sinks to the deep ocean, where it is decomposed by bacteria.

Adaptation to Global Warming Because the overwhelming majority of climate experts think that human-induced global warming is inevitable, government planners and social scientists are developing strategies to help various regions and sectors of society adapt to climate warming. One of the most pressing issues is rising sea level. People living in coastal areas could be moved inland, away from the dangers of storm surges. This solution would have high societal and economic costs, however. An alternative, also extremely expensive, is the construction of dikes and levees to protect coastal land. Rivers and canals that spill into the ocean would have to be channeled to prevent saltwater intrusion into fresh water and agricultural land (see Chapter 13). The Dutch, who have been building dikes and canals for several hundred years, have offered their technical expertise to several developing nations that are particularly threatened by a rise in sea level.

We also need to adapt to shifting agricultural zones. Many countries with temperate climates are evaluating

OZONE DEPLETION IN THE STRATOSPHERE

You may recall from Chapter 19 that ozone (O_3) is a form of oxygen that is a human-made pollutant in the troposphere but a naturally produced, essential component in the stratosphere, which encircles our planet some 10 to 45 km (6 to 28 mi) above the surface. The stratosphere contains a layer of ozone that shields the surface from much of the ultraviolet radiation coming from the sun (Figure 20.12).

Ultraviolet (UV) radiation is the name given to that part of the electromagnetic spectrum with wavelengths just shorter than visible light; it is a high-energy form of radiation that can be lethal to organisms in larger quantities. Should ozone disappear from the stratosphere, Earth would become uninhabitable for most forms of life. Scientists divide UV radiation into three bands: UV-A (with wavelengths of 320 to 400 nm), UV-B (280 to 320 nm), and UV-C (200 to 280 nm). (A nanometer, abbreviated *nm*, is one billionth of a meter.) The shorter the wavelength, the more energetic and thus the more dangerous UV radiation is. Fortunately, oxygen and ozone in the atmosphere absorb all incoming UV-C (the most lethal wavelengths), and none reaches the surface of Earth. The ozone layer also absorbs most, but not all, incoming UV-B radiation, so some UV-B reaches the surface of Earth. UV-A is not affected by ozone, and most reaches the surface.

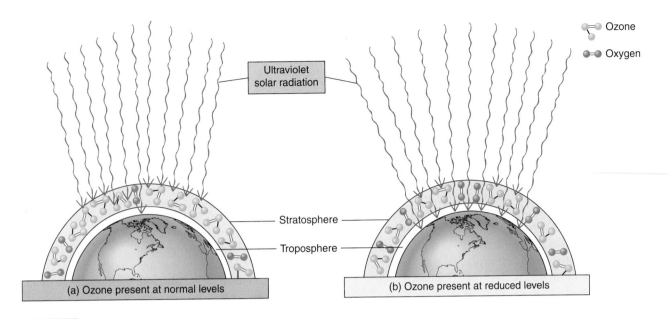

Ozone

Oxygen

Ultraviolet
solar radiation

Stratosphere

Troposphere

(a) Ozone present at normal levels

(b) Ozone present at reduced levels

Figure 20.12 Stratospheric ozone layer. (a) Stratospheric ozone absorbs about 99% of incoming solar ultraviolet (UV) radiation, effectively shielding the surface. (b) When stratospheric ozone is present at reduced levels, more high-energy UV radiation penetrates the atmosphere to the surface, where its presence harms organisms.

A slight thinning in the ozone layer over Antarctica forms naturally for a few months each year. In 1985, however, the thinning was first observed to be greater than it should have been if natural causes were the only factor inducing it. This increased thinning, which occurs each September, is commonly referred to as the "ozone hole" (Figure 20.13). There, ozone levels decrease by as much as 70% each year. During the 1990s the ozone-thinned area continued to grow, and by 1998 it had reached the record size of 27.5 million km² (11 million mi²), which is larger than the North American continent. In 2000, the ozone-thinned area was even larger—28.3 million km² (11.3 million mi²). A smaller thinning has also been detected in the stratospheric ozone layer over the Arctic.

Probably the most disquieting news is that world levels of stratospheric ozone have been decreasing for several decades. According to the National Center for Atmospheric Research, ozone levels over Europe and North America have dropped by almost 10% since the 1970s.

The Causes of Ozone Depletion

Both chlorine- and bromine-containing substances catalyze ozone destruction. The primary chemicals responsible for ozone loss in the stratosphere are a group of versatile chlorine-containing compounds called **chlorofluorocarbons (CFCs)**. Chlorofluorocarbons have been used as propellants for aerosol cans, as coolants in air conditioners and refrigerators (for example, Freon), as foam-blowing agents for insulation and packaging (for example, Styrofoam), and as solvents.

Additional compounds that also attack ozone include halons, methyl bromide, methyl chloroform, carbon tetrachloride, and nitrous oxide. Halons, which contain both bromine and chlorine, are used as fire retardants.

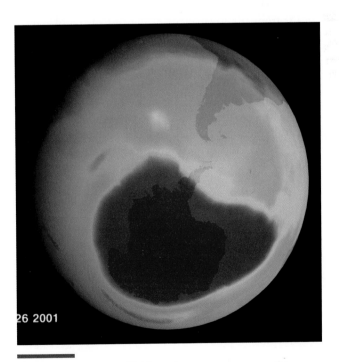

26 2001

Figure 20.13 Ozone thinning. A computer-generated image of part of the Southern Hemisphere, taken on September 26, 2001, reveals the ozone thinning (*purple area over Antarctica*). The ozone-thin area is not stationary but moves about as a result of air currents.

Methyl bromide is a pesticide widely used as a soil and crop fumigant; it kills a variety of pests, such as nematodes (parasitic worms), fungi, and weed seeds (see Table 22.1). Methyl chloroform is an industrial solvent. Carbon tetrachloride, another solvent, is used in many industrial processes, including the manufacture of pesticides and dyes. Nitrous oxide is released from the burning of fossil fuels, particularly coal, and from the breakdown of nitrogen fertilizers in the soil.

The scientific evidence linking CFCs and other human-made compounds to stratospheric ozone destruction includes thousands of laboratory measurements, atmospheric observations, and calculations by computer models. In 1995 the Nobel Prize in chemistry was awarded to Sherwood Rowland, Mario Molina, and Paul Crutzen, the scientists who first explained the connection between the thinning ozone layer and chemicals such as CFCs. This Nobel Prize was the first one ever given for work in environmental science.

After release into the lower troposphere, CFCs and other chlorine-containing compounds slowly drift up to the stratosphere, where UV radiation breaks them down, releasing chlorine. Similarly, bromine is released by the breakdown of halons and methyl bromide. The hole in the ozone layer that was discovered over Antarctica occurs annually between September and November (spring in the Southern Hemisphere). At this time, two important conditions are present: Sunlight returns to the polar region, and the *circumpolar vortex*, a mass of cold air that circulates around the southern polar region and isolates it from the warmer air in the rest of the planet, is well developed. The cold air causes polar stratospheric clouds to form; these clouds contain ice crystals to which chlorine and bromine adhere, making them available to destroy ozone. The sunlight catalyzes the chemical reaction by which chlorine or bromine breaks ozone molecules apart, converting them into oxygen molecules. The chemical reaction by which ozone is destroyed does not alter the chlorine or bromine, and therefore a single chlorine or bromine atom can break down many thousands of ozone molecules. The chlorine and bromine remain in the stratosphere for many years. When the circumpolar vortex breaks up, the ozone-depleted air spreads northward, diluting ozone levels in the stratosphere over South America, New Zealand, and Australia.

Human-produced pollution is not the only cause of ozone depletion. Volcanic eruptions such as that of Mount Pinatubo in 1991 also accelerate ozone loss because the sulfur aerosols they spew into the atmosphere speed the breakdown of CFCs and halons.

The Effects of Ozone Depletion

With depletion of the ozone layer, higher levels of UV radiation reach the surface of the Earth. A study conducted in Toronto from 1989 to 1993 showed that wintertime levels of UV-B increased by more than 5% each

year as a result of lower ozone levels. A study conducted in New Zealand showed that summertime levels of peak UV-B radiation were 12% higher during the 1998–99 summer than they had been during similar periods a decade earlier.

Excessive exposure to UV radiation is linked to several health problems in humans, including eye cataracts, skin cancer, and weakened immunity. The lens of the eye contains transparent proteins that are replaced at a very slow rate. Exposure to excessive UV radiation damages these proteins, and over time, the damage accumulates so that the lens becomes cloudy, forming a cataract. Cataracts can be cured by surgery, but millions of people in developing nations cannot afford the operation and so remain partially or totally blind.

Most cases of skin cancer are caused by excessive, chronic exposure to UV radiation. Ultraviolet B radiation can cause mutations, or changes, in the deoxyribonucleic acid (DNA) residing in skin cells. Over time such changes gradually accumulate and may lead to skin cancer. Globally, about 2.2 million cases of skin cancer occur each year. Malignant melanoma, the most dangerous type of skin cancer, is increasing faster than any other type of cancer. Some forms of malignant melanoma spread rapidly through the body and may cause death a few months after diagnosis.

Scientists are concerned that increased levels of UV radiation may disrupt ecosystems. Recall from Chapter 4 that the productivity of Antarctic phytoplankton, the microscopic drifting algae that are the base of the Antarctic food web, has declined from increased exposure to UV radiation. Research shows that surface UV-B inhibits photosynthesis in these phytoplankton. Direct damage to natural populations of Antarctic fish has also been documented. Increased DNA mutations in ice-fish eggs and larvae (young fish) were matched to increased levels of UV radiation. Researchers are currently studying if these mutations lessen the animals' ability to survive. The possible link between the widespread decline of amphibian populations and increased UV radiation is discussed in Chapter 16. Because organisms are interdependent, the negative effect on one species has ramifications throughout the ecosystem.

There is also concern that high levels of UV radiation may damage crops and forests, but the effects of UV-B radiation on plants are very complex and have not been adequately studied. Plants interact with a large number of other species in both natural ecosystems and agricultural ecosystems, and the impacts of UV radiation on each of these organisms affect plants indirectly. Exposure to higher levels of UV-B radiation appears to increase wheat yields by inhibiting fungi that cause disease in wheat. On the other hand, exposure to higher levels of UV radiation appears to decrease cucumber yields by increasing their incidence of disease. A 2000 study of tobacco and *Arabidopsis* plants showed that exposure to UV-B radiation caused changes in both plants' DNA.

CASE·IN·POINT ## Facilitating the Recovery of the Ozone Layer

In 1978 the United States, the world's largest user of CFCs, banned the use of CFC propellants in products such as antiperspirants and hair sprays. Although this ban was a step in the right direction, it did not solve the problem. Most nations did not follow suit, and besides, propellants represented only the tip of the iceberg in terms of CFC use.

In 1987, representatives from many countries met in Montreal to sign the **Montreal Protocol**, an agreement that originally stipulated a 50% reduction of CFC production by 1998. Since then, 183 countries have ratified this agreement. Despite this effort, the environmental news about CFCs continually worsened in the early 1990s. After scientists reported that decreases in stratospheric ozone occurred over the heavily populated midlatitudes of the Northern Hemisphere in all seasons, the Montreal Protocol was modified to include stricter measures to limit CFC production.

Industrial companies that manufacture CFCs quickly developed substitutes, such as hydrofluorocarbons (HFCs) and hydrochlorofluorocarbons (HCFCs) (see "Meeting the Challenge: Business Leadership in the Phaseout of CFCs"). HFCs do not attack ozone, although they are potent greenhouse gases. HCFCs attack ozone but are not as destructive as the chemicals they are replacing. Although production of HFCs and HCFCs has increased rapidly, these chemicals are transitional substances that will be used only until industry develops nonfluorocarbon substitutes.

CFC, carbon tetrachloride, and methyl chloroform production was completely phased out in the United States and other highly developed countries in 1996, except for a relatively small amount exported to developing countries (Table 20.3). Existing stockpiles could be used after the deadline, however. Developing countries are on a different timetable and will phase out CFC use by 2005. Methyl bromide will be phased out in highly developed countries, which are responsible for 80% of the global use of that chemical, by 2005. HCFCs will be phased out in 2030.

Measurements taken in the lower atmosphere between 1994 and 1998 showed that the amount of ozone-depleting chemicals had slowly decreased. Satellite measurements taken in 1997 provided the first evidence that the levels of ozone-depleting chemicals were starting to decline in the stratosphere. However, two chemicals—CFC-12 and halon-1211—may have increased and therefore still represent a threat to ozone recovery. Although highly developed countries no longer manufacture CFC-12, it continues to leak into the atmosphere from old refrigerators and vehicle air conditioners discarded in those countries. Moreover, developing countries such as China, India, and Mexico have increased their production of CFC-12. Although China and Korea currently produce 95% of the world's halon-1211, the U.N. Environment Program is working with these countries on an accelerated phaseout of halon use by 2006 (the Montreal Protocol says developing countries must phase out halons by 2010). An international fund known as the Montreal Multilateral Fund is available to help developing countries during their transition from ozone-depleting chemicals to safer alternatives.

Unfortunately, CFCs are extremely stable, and those being used today probably will continue to deplete stratospheric ozone for at least 50 years. Scientists expect human-exacerbated ozone thinning to reappear over Antarctica each year, although the area and degree of thinning will gradually decline over time, until full recovery takes place sometime after 2050.

Smuggling CFCs The U.S. government imposed a tax on CFCs, and the tax, along with a diminishing supply of CFCs, caused prices to escalate during the 1990s. One of the purposes of the tax was to encourage consumers to switch to CFC substitutes. An automobile's air conditioning system can be modified to use HFCs for between $50 and $400, but consumers might not consider making these modifications if their existing air conditioners remained inexpensive to maintain.

The increased cost of CFCs led to a thriving black market. Beginning in the mid-1990s, thousands of tons of CFCs were smuggled into the United States each year. Black marketers buy CFCs in countries such as Russia, India, and China and sell them in the United States for less than the retail price. According to the U.S. Customs Service, CFCs are the second-largest illegal import in Miami (the largest is cocaine). The availability of smuggled CFCs provides a cheaper alternative than modifying air conditioners to use HFCs. Therefore, fewer consumers switched to CFC alternatives and, consequently, damage to the ozone layer continues. ∎

Table 20.3 Estimated Annual World CFC* Production for Selected Years, 1950 to 2000

Year	Total CFCs (metric tons)
1950	41,187
1955	83,869
1960	149,142
1965	312,889
1970	559,235
1975	695,041
1980	767,841
1985	917,291
1990	658,325
1995	145,612
2000	36,124

* CFC = chlorofluorocarbon.

MEETING THE CHALLENGE

Business Leadership in the Phaseout of CFCs

As scientific evidence began to accumulate linking chlorofluorocarbons (CFCs) to the thinning of the ozone layer, some business leaders predicted that the economy would suffer and our standard of living would decline when CFCs were banned. Other companies viewed the impending phaseout of CFCs as an opportunity—a chance to achieve a competitive advantage by being among the first to adopt environmentally friendly products. These companies moved quickly and voluntarily to restrict their use of CFCs. Here we examine how three companies met the challenge: McDonald's, Nortel (formerly Northern Telecom), and York International. We then discuss how the chemical industry worked to develop safe, effective alternatives to CFCs.

Amid growing public concern over the effect of CFCs on the ozone layer, a group of schoolchildren petitioned McDonald's to stop using plastic-foam food containers made with CFCs as the foam-blowing agent. In August 1987, McDonald's told its suppliers of disposable foam packaging that they had 18 months to come up with alternatives. McDonald's announcement not only generated a great deal of positive publicity for the company but also galvanized the foam packaging industry, which voluntarily ended all use of CFCs by the end of 1988. Today all foam packaging made in the United States is completely ozone-friendly.

Nortel is a major electronics firm that used CFCs to clean circuit boards and other electronic components. In 1988, the company's engineers met with government officials to develop a planned phaseout of CFCs. They decided to eliminate all use of CFCs in 3 years ("free in three"). Nortel tested possible CFC substitutes. It also tried to reduce CFC consumption at its factories, which competed with one another to be the first to "get to zero." Meanwhile, Nortel worked within the electronics industry to modify the production process so that the need for CFCs and other cleaning solvents was eliminated altogether.

Nortel's leadership not only made it one of the first electronics companies to stop using CFCs, but gave it unanticipated benefits because the new manufacturing processes turned out to be cheaper and more effective.

Manufacturers of refrigeration and air conditioning systems, including "chillers" for larger buildings, were initially skeptical that they could eliminate CFCs, which were used as refrigerants, without reducing energy efficiency. Nonetheless, the industry began an aggressive search for alternatives to CFCs. York International was the first company in the chiller industry to develop an alternative, HCFC-123. Other companies followed suit and stopped selling CFC-based chillers in the United States by 1993. Many existing CFC chillers in buildings have been retrofitted or replaced by the new chillers, which are more energy efficient and therefore save owners money in electricity costs. The chiller story is not an unqualified success, however, because most new chillers still rely on hydrofluorocarbons or hydrochlorofluorocarbons. Carrier is the first manufacturer to offer a chlorine-free refrigerant (Puron) in its air conditioners, beginning in 1996.

The international chemical industry worked hard to ensure that deadlines in the global phaseout of CFCs could be met. Seventeen major chemical companies collectively established two programs, the Alternative Fluorocarbons Environmental Acceptability Study (AFEAS) and the Program for Alternative Fluorocarbon Toxicity Testing (PAFT). AFEAS was created to provide data on the potential environmental effects of CFC alternatives, and PAFT, to provide information about potential effects on human health. Both programs relied on international cooperation among scientists from academic institutions, government research programs, and chemical companies. This collaborative effort enabled CFCs to be phased out more rapidly than would usually be the case when new chemicals have to be subjected to environmental and toxicity testing.

■ ACID DEPOSITION

What do fishless lakes in the Adirondack Mountains, recently damaged Mayan ruins in southern Mexico, and dead trees in the Czech Republic have in common? The answer is that these damages are the result of acid precipitation or, more properly, **acid deposition**. Acid deposition is a type of air pollution. It includes sulfuric and nitric acids in precipitation (sometimes called **wet deposition**) as well as dry, sulfuric acid– and nitric acid–containing particles that settle out of the air (sometimes called **dry deposition**).

Acid deposition is not a new phenomenon. It has been around since the Industrial Revolution began. Robert Angus Smith, a British chemist, coined the term *acid rain* in 1872 after he noticed that buildings in areas with heavy industrial activity were being worn away by rain. Acid precipitation, including acid rain, sleet, snow, and fog, poses a serious threat to the environment. Industrialized countries in the Northern Hemisphere have been hurt the most, especially the Scandinavian countries, central Europe, Russia, and North America. In the United States alone, the damage from acid deposition has been estimated at $10 billion each year. More recently, with the realization that acid deposition now occurs in many temperate and tropical countries that are rapidly becoming industrialized, acid deposition has been recognized as a global problem. For example, Chinese scientists reported in 1998 that 40% of their country is affected by acid rain.

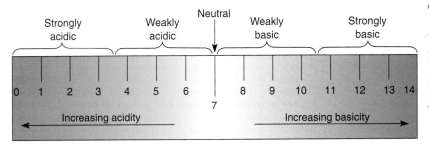

Figure 20.14 The pH scale. This scale, which includes values between 0 and 14, expresses the concentrations of acidic and basic solutions. Pure water, which is neutral, has a pH of 7. Each decrease of 1 pH unit represents a tenfold increase in acidity.

Measuring Acidity

The **pH scale**, which runs from 0 to 14, expresses the relative degree of acidity or basicity of a substance (Figure 20.14; also see Appendix I). A pH of 7 is neither acidic nor basic, whereas a pH less than 7 indicates an acidic solution. The pH scale is logarithmic, so a solution with a pH of 6 is 10 times more acidic than a solution with a pH of 7. Likewise, a solution with a pH of 5 is 10 times more acidic than a solution with a pH of 6 and 100 times more acidic than a solution with a pH of 7. A solution with a pH greater than 7 is basic, or alkaline.

For purposes of comparison, distilled water has a pH of 7, tomato juice has a pH of 4, vinegar has a pH of 3, and lemon juice has a pH of 2. Normally, rainfall is slightly acidic (with a pH from 5 to 6) because CO_2 and other naturally occurring compounds in the air dissolve in rainwater, forming dilute acids. However, the pH of precipitation in the northeastern United States averages 4 and is often 3 or even lower.

How Acid Deposition Develops

Acid deposition occurs when sulfur dioxide and nitrogen oxides are released into the atmosphere (Figure 20.15; also see Chapter 19). Motor vehicles are a major source of nitrogen oxides. Coal-burning power plants, large smelters, and industrial boilers are the main sources of sulfur dioxide emissions and produce substantial amounts of nitrogen oxides as well. Sulfur dioxide and nitrogen oxides, released into the air from tall smokestacks, can be carried long distances by winds. Tall smokestacks allow England to "export" its acid deposition problem to the Scandinavian countries, and the midwestern United States to "export" its acid emissions to New England and Canada.

During their stay in the atmosphere, sulfur dioxide and nitrogen oxides react with water to produce dilute solutions of sulfuric acid (H_2SO_4), nitric acid (HNO_3), and nitrous acid (HNO_2). Acid deposition returns these acids to the ground, causing the pH of surface waters and soil to decrease.

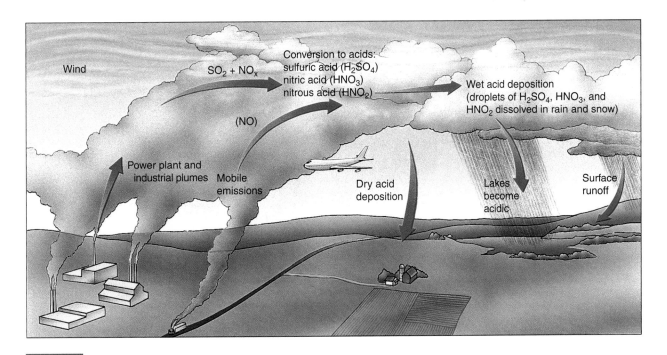

Figure 20.15 Acid deposition. Sulfur dioxide and nitrogen oxide emissions react with water vapor in the atmosphere to form acids that return to the surface as either dry or wet deposition.

Figure 20.16 Acid rain damage. The pockmarks in the White Lee marble on this building in Washington, D.C. occurred when acid rain reacted with some of the minerals in the marble. Over time, these marks will get larger and deeper.

The Effects of Acid Deposition

Acid deposition corrodes metals and building materials (Figure 20.16). It eats away at important monuments, such as the Washington Monument in Washington, D.C., and the Statue of Liberty in New York harbor. Historic sites in Venice and Rome and all across Europe are being worn away by acid deposition. Emissions from uncapped Mexican oil wells in the Gulf of Mexico cause acid deposition that has been tied to the destruction of ancient Mayan ruins in southern Mexico.

The link between acid deposition and declining aquatic animal populations is well established (Figure 20.17). Field investigations were conducted from 1984 to 1987 by the Adirondack Lakes Survey Corporation (ALSC), a nonprofit group that works with various universities, the Environmental Protection Agency (EPA), and various state and local organizations. Of the 1,469 Adirondack lakes and ponds examined, 352 were found to have waters with pH values of 5.0 or less, and 346 of those acidified lakes and ponds had no fish populations. Toxic metals such as aluminum become soluble in acidic lakes and streams and enter food webs. This increased concentration of toxic metals may explain how acidic water adversely affects fishes.

During the 1990s the ALSC began a long-term monitoring program to evaluate the effectiveness of the Clean Air Act Amendments of 1990. Preliminary results indicate that sulfate levels have declined by 15% to 23% in 92% of the lakes being monitored.

Although fishes have received the lion's share of attention in regard to the acid deposition issue, other animals are also adversely affected. Several studies of the effects of acid deposition on birds found that birds living in areas with pronounced acid deposition were much more likely to lay eggs with thin, fragile shells that break

Figure 20.17 Sensitivity to low pH caused by acid deposition. Some aquatic organisms, such as the freshwater mussel, can tolerate very little acid, whereas others, such as the water boatman, have a wide pH tolerance. As the figure shows, freshwater mussels are not found in water with a pH lower than 6.0. The darker color of the mussel at a pH of 6.5 indicates that mussels are more common at this pH than at a pH of 6.0.

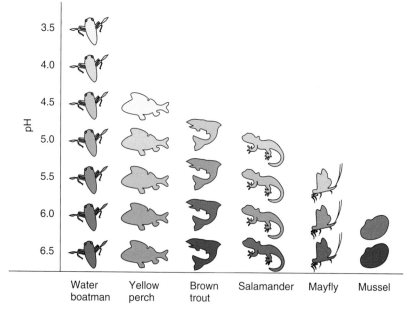

or dry out before the chicks hatch. The inability to produce strong eggshells was attributed to reduced calcium in the birds' diets. Calcium is unavailable to them because in acidic soils it becomes soluble and is washed away, with little left for absorption by plant roots. A smaller amount of calcium in plant tissues means a smaller amount of calcium in the insects and snails that eat the plants. Therefore, less calcium is available to the birds that eat these insects and snails.

Acid Deposition and Forest Decline   Forest surveys in the Black Forest of southwestern Germany indicate that up to 50% of trees in the areas surveyed are dead or severely damaged. (Damage is determined by monitoring the loss of leaves and needles from trees.) The same is true for trees in many other European forests. More than half of the red spruce trees in the mountains of the northeastern United States have died since the mid-1970s, and

sugar maples in eastern Canada and the United States are also dying. Beginning in the late 1990s, the Northern Hardwood Damage Survey has observed and mapped widespread tree death at higher elevations of the Appalachian Mountains from Georgia to Maine.

Many trees that are not dead are exhibiting symptoms of **forest decline**, characterized by gradual deterioration and eventual death of trees. The general symptoms of forest decline are reduced vigor and growth, but some plants exhibit specific symptoms, such as yellowing of needles in conifers. Forest decline is more pronounced at higher elevations, possibly because most trees growing at high elevations are at the limits of their normal range and are therefore less vigorous and more susceptible to stressors of any type.

Many factors can interact to decrease the health of trees (Figure 20.18), and no single factor alone accounts for the recent instances of forest decline. Although acid

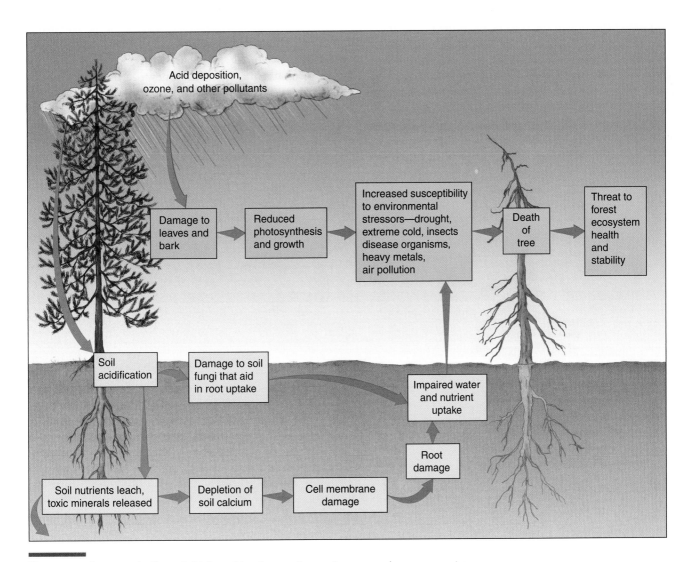

Figure 20.18 Forest decline. Acid deposition is one of several stressors that appear to interact, contributing to the decline and death of trees. Red spruce, which are represented here, are dying throughout the Appalachian Mountains.

deposition correlates well with areas that are experiencing tree damage, it is thought to be only partly responsible. Several other human-induced air pollutants have also been implicated, including tropospheric (surface-level) ozone and toxic heavy metals such as lead, cadmium, and copper. Power plants, ore smelters, refineries, and motor vehicles produce these pollutants in addition to the sulfur and nitrogen oxides that interact to form acid deposition. Insects and weather factors such as drought and severe winters—cold and wind can injure susceptible plants— may also be important. To complicate matters further, the actual causes of forest decline may vary from one tree species to another and from one location to another. Thus, forest decline appears to result from the combination of multiple stressors—acid deposition, tropospheric ozone, UV radiation (which is more intense at higher altitudes), insect attack, drought, and so on. When one or more stressors weakens a tree, then an additional stressor, such as air pollution, may be decisive in causing its death.

One way in which acid deposition harms plants is well established: Acid deposition alters the chemistry of soils (Figure 20.19), which affects the development of plant roots as well as their uptake of dissolved minerals and water from soil. Essential plant minerals such as calcium and potassium wash readily out of acidic soil, whereas others, such as nitrogen, become available in larger amounts. Also, heavy metals such as manganese and aluminum dissolve in acidic soil, becoming available for absorption in toxic amounts. A study completed in 1989 in Central Europe, which has experienced greater forest damage than North America, found a strong correlation between forest damage and soil chemistry altered by acid deposition.

The Politics of Acid Deposition

One of the factors that makes acid deposition so hard to combat is that it does not occur only in the locations where the gases that cause it are emitted. Acid deposition does not recognize borders between states or countries; it is entirely possible for sulfur and nitrogen oxides that were released in one spot to return to Earth's surface hundreds of kilometers from their source.

The United States has wrestled with this issue. Several states in the Midwest and East—Illinois, Indiana Missouri, Ohio, Pennsylvania, Tennessee, and West Virginia—produce between 50% and 75% of the acid deposition that contaminates New England and southeastern Canada. When legislation was formulated to deal with the problem, however, arguments ensued about who should pay for the installation of expensive air pollution devices to reduce emissions of sulfur and nitrogen oxides.

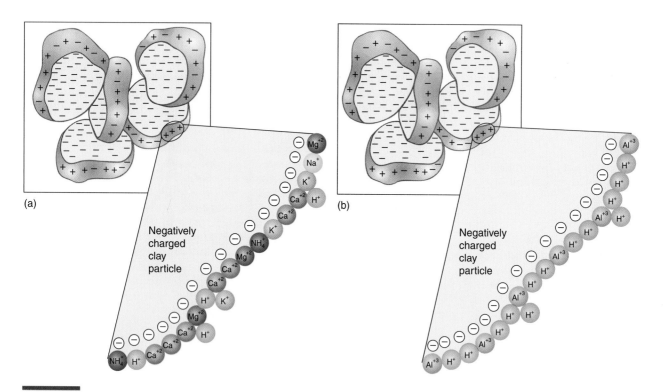

Figure 20.19 Effect of acid on soil chemistry. The surfaces of soil particles are negatively charged. (a) In normal soil, positively charged mineral ions such as calcium and magnesium are attracted to the negatively charged soil particles. (b) In acidified soil, hydrogen ions from acid deposition displace the positively charged mineral ions. In addition, aluminum ions released when the soil becomes acidified also adhere to the soil particles.

Should the states emitting the gases be required to pay all the expenses to clean up the air, or should some of the cost be absorbed by the areas that stand to benefit most from a reduction in pollution?

Pollution abatement issues are quite complex within one country, but are magnified even more in international disputes. For example, England, which has large reserves of coal, uses coal to generate electricity. Gases from coal-burning power plants in England move eastward with prevailing winds and return to the surface as acid deposition in Sweden and Norway. Similarly, emissions from mainland China produce acid deposition in Japan, Taiwan, and North and South Korea.

Facilitating Recovery from Acid Deposition

Although the science and the politics surrounding acid deposition are complex, the basic concept of control is straightforward: Reducing emissions of sulfur dioxide and nitrogen oxides curbs acid deposition. Simply stated, if sulfur dioxide and nitrogen oxides are not released into the atmosphere, they cannot come down as acid deposition. Installation of scrubbers in the smokestacks of coal-fired power plants (see Figure 19.10d) and use of clean-coal technologies to burn coal without excessive emissions effectively diminish acid deposition (see Chapter 10). In turn, a decrease in acid deposition prevents surface waters and soil from becoming more acidic than they already are.

According to a 1999 report from the National Acid Precipitation Assessment Program, a consortium of federal agencies, rainfall in parts of the Midwest, Northeast, and Mid-Atlantic regions is less acidic than it was a few years ago, the result of cleaner-burning power plants and the use of reformulated gasoline. Many power plants in the Ohio Valley switched from high-sulfur coal to western low-sulfur coal. However, solving one environmental problem often creates others. While the move to western low-sulfur coal reduced sulfur emissions, it contributed to the problem of global warming. (Because western coal has a lower heat value than eastern coal, more of it must be burned to generate a given amount of electricity. Therefore, more CO_2 is emitted.) Western coal also contains higher levels of mercury and other trace metals, so burning more of it adds more of these hazardous pollutants to the air.

Despite the fact that the United States, Canada, and many European countries have reduced emissions of sulfur, acid precipitation remains a serious problem. Acidified forests and bodies of water have not recovered as quickly as hoped. Trees in the U.S. Forest Service's Hubbard Brook Experimental Forest in New Hampshire, an area damaged by acid deposition, have grown very little since 1987, when emissions started to decline. Also, many northeastern streams and lakes, such as those in New York's Adirondack Mountains, remain acidic. A primary reason for the slow recovery is probably that the past 30 or more years of acid rain have profoundly altered soil chemistry in many areas (as discussed earlier). Essential plant minerals such as calcium and magnesium have been leached from forest and lake soils. Because soils take hundreds or even thousands of years to develop, it may be decades or centuries before they recover from the effects of acid rain.

Another factor that may be slowing recovery is that although sulfur emissions have declined, nitrogen oxide emissions have not. In fact, nitrogen oxide emissions increased slightly during the 1990s, in part because there are more motor vehicles being driven longer distances than a decade ago (see Figure 19.11). Many scientists are convinced that ecosystems will not recover from acid rain damage until substantial reductions in nitrogen oxide emissions occur. Nitrogen oxide emissions are harder to control than sulfur dioxide emissions because motor vehicles produce a substantial portion of nitrogen oxides. Engine improvements may help reduce nitrogen oxide emissions, but as the population continues to grow, the engineering gains will probably be offset by an increase in the number of motor vehicles. Thus, dramatic cuts in nitrogen oxide emissions will probably require expensive solutions such as increased mass transit, which would reduce the number of miles people drive, or switching from gasoline-powered cars to clean cars (see "Meeting the Challenge: Clean Cars, Clean Fuels" in Chapter 19).

To add to the complexity of the acid deposition problem, new research has implicated other pollutants, such as volatile organic compounds and ammonia, in the chemical reactions that cause acidification. Most scientists now say that air pollution controls that are stricter than those mandated by the Clean Air Act Amendments of 1990 will have to be initiated if forests and lakes are to recover in the near future.

LINKS AMONG GLOBAL WARMING, OZONE DEPLETION, AND ACID DEPOSITION

Most environmental studies examine a single issue, such as acid deposition, global warming, or ozone depletion. Canadian researchers at the Experimental Lakes Area in Ontario, Canada, decided to take a different approach and explore the interactions of all three environmental problems simultaneously. In 1996, the scientists reported that North American lakes may be more susceptible to damage from UV radiation than the thinning of the ozone layer alone would indicate. It appears that the combined effects of acid deposition and climate warming increase how far UV radiation penetrates lake water. Some of the possible effects of increased UV penetration are to disrupt photosynthesis in algae and aquatic plants and to cause sunburn damage (skin lesions) in fishes.

How far UV radiation penetrates lake water is related to the presence of dissolved organic compounds.

This organic material, which is present even in the clearest, least polluted lakes, comes from the decomposition of dead organisms. Dissolved organic material acts like a sunscreen by absorbing UV radiation so that it can penetrate the water only few inches.

A warmer climate increases evaporation, which reduces the amount of water flowing into a lake from the surrounding watershed. Because most of a lake's dissolved organic compounds have been washed into the lake, even a slightly drier climate reduces how much organic material is present in the lake. Therefore, UV radiation penetrates deeper into lake water as a result of climate warming.

Acid deposition also affects the amount of dissolved organic compounds in a lake. The presence of acid causes organic matter to clump together and settle on the lake floor. The removal of organic material from the water allows UV radiation to penetrate farther.

Climate Warming May Hurt Ozone Recovery

Several studies have reported a link between human-induced climate warming and polar ozone depletion.

Greenhouse gases that warm the troposphere induce stratospheric cooling, presumably because heat trapped in the troposphere is not available to warm the stratosphere. Temperatures in the stratosphere have been cooling for the past several years, and, as discussed earlier in this chapter, these lower temperatures provide better conditions for ozone-depleting chemicals to attack stratospheric ozone. The record ozone holes over Antarctica in 1998 and 2000 were attributed to cooler stratospheric temperatures in the circumpolar vortex. Some scientists now speculate that if the cooling trend in the stratosphere continues, recovery of the ozone layer may be delayed. This means that climate warming could prolong ozone depletion in the stratosphere despite the success of the Montreal Protocol.

Scientists now know that environmental problems cannot be studied as separate issues because they often interact in surprisingly subtle ways. As acid deposition, global warming, and ozone depletion are studied further, it is likely that other synergistic interactions among them will be discovered.

SUMMARY WITH SELECTED KEY TERMS

I. On the basis of numerous scientific studies, in 2001, the IPCC, a United Nations panel of experts, concluded that human-produced air pollutants have caused most of the recent climate warming.

A. Most climate scientists think that the world will continue to warm during the 21st century. Scientists are less certain about what regional patterns may emerge.

B. Carbon dioxide (CO_2) emissions from the combustion of fossil fuels cause the atmosphere to retain heat (infrared radiation), which warms the Earth.

1. The natural trapping of heat in the atmosphere is referred to as the **greenhouse effect**, and the gases that absorb infrared radiation are known as **greenhouse gases**.

2. The additional warming that may be produced by increased levels of gases that absorb infrared radiation is known as the **enhanced greenhouse effect**.

3. Other greenhouse gases are methane, nitrous oxide, chlorofluorocarbons (CFCs), and tropospheric (surface-level) ozone.

4. Air pollutants known as atmospheric **aerosols** tend to cool the atmosphere, but it is generally agreed that the effects of greenhouse gases will override the **aerosol effect**.

C. During the 21st century, global warming will probably cause a rise in sea level, changes in precipitation patterns extinction of many species, and problems for agriculture. It could result in the displacement of millions of people, thereby increasing international tensions.

D. The challenge of global warming can be met by mitigation (slowing down the rate of global warming) and adaptation (making adjustments to live with global warming).

1. Representatives from 160 countries developed the **Kyoto Protocol**, which established timetables for countries to reduce emissions of CO_2, methane, and nitrous oxide. This international treaty does not become binding until more industrialized countries ratify it.

2. Mitigation includes the development of alternatives to fossil fuels, controlling the human population, increasing energy efficiency of automobiles and appliances, planting and maintaining forests, and **carbon management**, ways to separate and capture CO_2 and sequester it away from the atmosphere.

II. The ozone layer in the stratosphere helps to shield the Earth from harmful **ultraviolet (UV) radiation**.

A. The total amount of ozone in the stratosphere is slowly declining globally, and large ozone holes develop over Antarctica and the Arctic each year.

1. **Chlorofluorocarbons** (**CFCs**), chlorine-containing compounds that have been used as refrigerants, foam-blowing agents, solvents, and propellants for aerosol cans, are responsible for attacking the ozone layer.

2. Additional compounds that also attack ozone include halons, methyl bromide, methyl chloroform, carbon tetrachloride, and nitrous oxide.

3. Thinning of the ozone layer over Antarctica occurs when the circumpolar vortex, a mass of cold air that circulates around the southern polar region, isolates the cold air from the rest of the planet's warmer air.

  a. The cold causes ice crystals to form in stratospheric clouds.

  b. Chlorine and other chemicals adhere to the ice particles and, from there, attack ozone.

  c. Sunlight catalyzes the chemical reaction by which chlorine or bromine breaks ozone molecules apart, converting them into oxygen molecules.

B. With depletion of the ozone layer, higher levels of UV radiation reach Earth's surface.

  1. In humans, excessive exposure to UV radiation causes cataracts, weakened immunity, and skin cancer.

  2. Increased levels of UV radiation may disrupt ecosystems, such as the Antarctic food web, because the negative effect on one species has ramifications throughout the ecosystem.

  3. The effects of increased levels of UV radiation on plants are very complex and require additional study.

C. International negotiations known as the **Montreal Protocol** resulted in an agreement to phase out CFC production.

III. Acid deposition, commonly called acid rain, is a serious regional problem caused by sulfur and nitrogen oxides.

A. Acid deposition occurs when sulfur dioxide and nitrogen oxides are released into the atmosphere.

  1. These pollutants react with water to produce sulfuric acid, nitric acid, and nitrous acid.

  2. Acid deposition returns these acids to surface waters and soil.

B. Acid deposition kills aquatic organisms and may contribute to forest decline by changing soil chemistry.

  1. Although some countries have reduced emissions of sulfur, it may be decades or centuries before their acidified forests and bodies of water recover.

  2. Acid deposition also attacks materials such as metals and stone.

C. Acid deposition has international implications because sulfur dioxide and nitrogen oxides released in one area can return to the surface hundreds of kilometers from their source.

IV. In some instances, global warming, ozone depletion, and acid deposition interact. The combined effects of acid deposition and climate warming make North American lakes more susceptible to damage from UV radiation caused by the thinning of the ozone layer.

THINKING ABOUT THE ENVIRONMENT

  1. What is the enhanced greenhouse effect, and how does it affect global climate?

  2. What are greenhouse gases? List the main greenhouse gases.

  3. On the basis of what you know about the nature of science, at this time can we say with absolute certainty that the global warming hypothesis (the increased production of greenhouse gases is causing global warming) is correct? Why or why not?

  4. Explain how global warming might affect each of the following: sea level, precipitation patterns, living organisms, human health, and agriculture.

  5. Austrian biologists who study plants growing high in the Alps found that plants adapted to cold-mountain conditions migrated up the peaks as fast as 3.7 m a decade during the 20th century, apparently in response to climate warming. Assuming that warming continues during the 21st century, what will happen to the plants if they reach the tops of the mountains?

  6. Discuss and give examples of the two overall approaches to global warming: mitigation and adaptation.

  7. Why will adaptation to global warming be easier for highly developed nations than for developing nations? Why will it be more difficult?

  8. Insurance companies that provide policies for hurricanes and other natural disasters are thinking of shifting hundreds of millions of dollars of their investments from fossil fuels to solar energy. On the basis of what you have learned in this chapter, explain why insurance companies consider such an investment shift to be in their best interest.

  9. Some environmentalists contend that the wisest way to "use" fossil fuels is to leave them in the ground. How would this affect air pollution? Global warming? Energy supplies?

  10. What is the stratospheric ozone layer? How does it protect life on Earth?

  11. Distinguish between the benefits of the ozone layer in the stratosphere and the harmful effects of ozone at ground level.

  12. What is stratospheric ozone depletion? How does it occur?

  13. Discuss at least two harmful effects of stratospheric ozone depletion.

  14. What is the Montreal Protocol?

  15. What is acid deposition? What are the causes of acid deposition?

  16. Discuss the harmful effects of acid deposition on materials, aquatic organisms, and soils.

  17. Discuss some of the possible causes of forest decline. How might these factors interact to speed the rate of decline?

  18. Explain how acid deposition, climate warming, and ozone depletion interact synergistically to increase the penetration of ultraviolet radiation in North American lakes.

*19. The pH of clouds in Vermont mountains has been measured at 3. The clouds are how many times more acidic than pure water?

*20. Using the data in Table 20.A, calculate the mean global temperature in degrees Celsius for the 1970s, 1980s, 1990s, and early 2000s:

Mean global temperature, 1970s _____

Mean global temperature, 1980s _____

Mean global temperature, 1990s _____

Mean global temperature, early 2000s _____

Write a statement to summarize what the mean temperatures you have calculated demonstrate.

* Solutions to questions preceded by asterisks appear in Appendix VII.

Table 20.A	Mean Global Temperatures, 1970 to 2001				
Year	**Temp. (°C)**	**Year**	**Temp. (°C)**	**Year**	**Temp. (°C)**
1970	14.02	1981	14.21	1992	14.14
1971	13.89	1982	14.06	1993	14.15
1972	14.00	1983	14.25	1994	14.25
1973	14.13	1984	14.07	1995	14.37
1974	13.89	1985	14.03	1996	14.23
1975	13.94	1986	14.12	1997	14.39
1976	13.86	1987	14.27	1998	14.54
1977	14.11	1988	14.29	1999	14.30
1978	14.02	1989	14.18	2000	14.30
1979	14.10	1990	14.36	2001	14.43
1980	14.16	1991	14.31		

TAKE A STAND

Visit our Web site at **http://www.wiley.com/college/raven** (select Chapter 20 from the Table of Contents) for links to more information about the Kyoto Protocol. Consider the opposing views of proponents and opponents of the protocol, and debate the issue with your classmates. You will find tools to help you organize your research, analyze the data, think critically about the issues, and construct a well-considered argument. Take a Stand activities can be done individually or in a team, as oral presentations, written exercises, or Web-based (e-mail) assignments.

Additional on-line materials relating to this chapter, including Student Quizzes, Activity Links, Useful Web Sites, Flash Cards, and more, can also be found on our Web site.

SUGGESTED READING

Armstrong, S. "Ask the Experts." *New Scientist* (November 3, 2001). The author interviewed some Canadian Inuits, who are dealing with some of the consequences of climate warming.

Broecker, W.S. "Glaciers That Speak in Tongues and Other Tales of Global Warming." *Natural History* (October 2001). Glacier retreats and other effects of global warming are examined.

Brown, K. "Homeless." *New Scientist* (May 13, 2000). Global warming threatens many of the world's organisms and the ecosystems of which they are a part.

Cifuentes, L., et al. "Hidden Health Benefits of Greenhouse Gas Mitigation." *Science*, Vol. 293 (August 17, 2001). The same actions that reduce the buildup of greenhouse gases will also provide immediate benefits to human health.

Falkowski, P.G. "The Ocean's Invisible Forest." *Scientific American*, Vol. 287, No. 2 (August 2002). Examines the potential of the iron fertilization hypothesis to reduce global carbon dioxide (CO_2) levels by using phytoplankton.

Huq, S. "Climate Change and Bangladesh." *Science*, Vol. 294 (November 23, 2001). This editorial examines the potentially catastrophic effects of climate change on Bangladesh.

Krajick, K. "Long-Term Data Show Lingering Effects From Acid Rain." *Science*, Vol. 292 (April 13, 2001). The United States may need more stringent air pollution controls to help ecosystems recover from the effects of acid deposition.

McGeehin, M.A., and M. Mirabelli. "The Potential Impacts on Climate Variability and Change on Temperature-Related Morbidity and Mortality in the United States." *Environmental Health Perspectives*, Vol. 109, Supplement 2 (May 2001). One of several articles in this supplement that deal with the potential health consequences of climate variability and change.

Petit, C.W. "Perilous Waters." *U.S. News and World Report* (April 1, 2002). A climate surprise involving the ocean conveyor belt may be starting in the North Atlantic.

Schrope, M. "Consensus Science, or Consensus Politics?" *Nature*, Vol. 412 (July 12, 2001). Examines the workings of the Intergovernmental Panel on Climate Change.

Tangley, L. "High CO_2 Levels May Give Fast-Growing Trees an Edge." *Science*, Vol. 292 (April 6, 2001). When the atmosphere contains higher levels of CO_2, loblolly pines grow faster and reproduce earlier, potentially enabling them to outcompete other tree species.

Watson, T., and J. Weisman. "6 Ways to Combat Global Warming." *USA Today* (July 16, 2001). Examines what, if anything, could be done to combat global warming.

Checking the level of bacteria in the wastewater treatment system in Arcata, California. This constructed wetland is a successful way to treat sewage in a small community like Arcata.

21

Water and Soil Pollution

Learning Objectives

After you have studied this chapter you should be able to

1. List and briefly define eight categories of water pollutants.

2. Discuss how sewage is related to eutrophication, biochemical oxygen demand (BOD), and dissolved oxygen.

3. Distinguish between *oligotrophic* and *eutrophic lakes* and explain how humans induce artificial eutrophication.

4. Contrast *point source pollution* and *nonpoint source pollution*.

5. Describe how most drinking water is purified in the United States and discuss the chlorine dilemma.

6. Distinguish among *primary*, *secondary*, and *tertiary treatments* for wastewater.

7. Compare the goals of the Safe Drinking Water Act and the Clean Water Act.

8. Define *soil pollution* and briefly discuss the specific problem of salinization.

Sixteen thousand people inhabit the town of Arcata on the coast of northern California. Home to Humboldt State University, this small college town has an international reputation for ecological innovation. Faced with helping to finance a $25 million regional wastewater treatment plant in 1975, Arcata decided in 1978 to pioneer a low-tech, natural approach. The city restored and constructed a series of freshwater wetlands in a former industrial area and then routed the wastewater through these wetlands. The wetland wastewater treatment plant was completed in 1986 at a cost of $7 million. In 1987 Arcata received the Innovations in Government Award from the Ford Foundation/Harvard University John F. Kennedy School of Government for its wastewater treatment approach.

Arcata's treatment of wastewater initially follows the steps used in most municipalities. The solid contaminants are allowed to settle out, and the dissolved organic wastes are biologically degraded and then treated with chlorine to remove disease-causing agents. However, conventional treatment does not remove other pollutants, such as nitrogen and phosphorus, because it is too expensive; such pollutants are usually left in the treated wastewater. Unfortunately, when treated wastewater is discharged into rivers, streams, or the ocean, these contaminants sometimes cause problems.

Arcata developed a way to remove such contaminants from treated water for a fraction of the cost of a normal advanced treatment plant and, at the same time, to increase the amount of ecologically important wetlands in the town's vicinity. The town hired biologists, who worked with city engineers to develop a series of six marshes that occupy about 62 hectares (154 acres). Essentially, Arcata uses cattails, bulrushes, and other marsh plants to remove the contaminants by absorbing and assimilating them, thus cleaning the wastewater. Algae, fungi, and bacteria living in the marsh also feed on these contaminants. After water spends some time being purified in the series of marshes, it is pumped to a treatment center, where it is chlorinated to kill bacteria, treated with sulfur dioxide to remove any remaining chlorine, and finally released into nearby Humboldt Bay.

The highly productive marsh ecosystem, known as the Arcata Marsh and Wildlife Sanctuary, provides wildlife habitat for many organisms, such as fishes, muskrats, raccoons, and river otters. Thousands of birds reside in the wetlands permanently or temporarily, and more than 200 species of birds such as ducks, coots, herons, egrets, grebes, and osprey can be observed in the sanctuary. For human recreation the sanctuary has 7.2 km (4.5 mi) of trails that meander through the wetlands.

To date, at least 800 towns and cities around the world have followed Arcata's example and built wetlands to treat wastewater. Although most of the constructed wetlands for wastewater treatment are found in small coastal communities, even large urban cities such as Phoenix and Orlando have made use of this approach. Orlando, Florida, restored a 486-hectare (1,200-acre) wetland that had been drained and used as a cow pasture since the late 1800s. This wetland now removes phosphorus and nitrogen contaminants from 49 million L (13 million gal) of treated city wastewater each day.

As we saw in Chapter 13, water is required by all organisms for their survival. However, having water of good quality is just as important as having enough water. When water is used in households and industries, as well as when it runs off agricultural fields and pastures, it contains a pollution load of feces, urine, industrial liquid wastes, fertilizers, and pesticides. Water is used over and over—as for example, when a downstream town uses river water that an upstream city first used. Therefore, wastewater treatment is an important part of **sustainable water use.**

Because soil and water are intimately connected, substances that pollute water frequently pollute the soil as well. This chapter discusses some of the pollutants found in water and soil and how we can improve water and soil quality. (Ocean pollution is discussed in Chapters 7 and 18.) Although the United States has made visible progress in cleaning up its water since the early 1980s, much remains to be done. In some areas, water quality has actually deteriorated, whereas in other areas, strong cleanup efforts have allowed us only to hold our ground, not make any gains.

TYPES OF WATER POLLUTION

Water pollution consists of any physical or chemical change in water that adversely affects the health of humans and other organisms. Water pollution is a global problem that varies in magnitude and type of pollutant from one region to another. In many locations, particularly in developing countries, the main water pollution issue is lack of disease-free drinking water.

Water pollutants can be divided into eight categories: sewage, disease-causing agents, sediment pollution, inorganic plant and algal nutrients, organic compounds, inorganic chemicals, radioactive substances, and thermal pollution (see "Mini-Glossary: Water Pollutants"). We now examine each of these types of water pollution.

Sewage

Sewage is the release of wastewater from drains or sewers (from toilets, washing machines, and showers) and includes human wastes, soaps, and detergents. The release of sewage into water causes several pollution problems. First, because it carries disease-causing agents, water polluted with sewage poses a threat to public health (see the next section, on disease-causing agents). Sewage also generates two serious environmental problems in water, enrichment and oxygen demand. **Enrichment**, the fertilization of a body of water, is caused by the presence

---

MINI-GLOSSARY

Water Pollutants

sewage: Wastewater carried off by drains or sewers; contains human wastes, soaps, and detergents.

disease-causing agents: Infectious agents such as viruses and bacteria that come from the wastes of infected individuals.

sediment pollution: Excessive amounts of soil particles that enter the water as a result of erosion.

inorganic plant and algal nutrients: Nitrogen, phosphorus, and other substances that stimulate plant and algal growth; from animal wastes, plant residues, and fertilizer runoff.

organic compounds: Carbon-containing chemicals that are usually synthetic and often toxic to aquatic organisms.

inorganic chemicals: Contaminants, such as acids, salts, and heavy metals, that contain elements other than carbon.

radioactive substances: Wastes from mining, refining, and use of radioactive metals.

thermal pollution: Heated water produced during certain industrial processes.

of high levels of plant and algal nutrients such as nitrogen and phosphorus. Because these nutrients get into waterways not only from sewage but also from other sources, we will consider enrichment later in the chapter and confine this discussion to oxygen demand in water.

Sewage and other organic materials are decomposed into carbon dioxide (CO_2), water, and similar inoffensive materials by the action of microorganisms. This degradation process, known as **cellular respiration**, requires the presence of oxygen. Dissolved oxygen is also used by most organisms, including fishes, that live in healthy aquatic ecosystems. But oxygen has a limited ability to dissolve in water, and when an aquatic ecosystem contains high levels of sewage or other organic material, the decomposing microorganisms use up most of the dissolved oxygen, leaving little for fishes or other aquatic animals. At extremely low oxygen levels, fishes and other animals leave or die.

Sewage and other organic wastes are measured in terms of their **biochemical oxygen demand (BOD)**, also known as **biological oxygen demand**, the amount of oxygen needed by microorganisms to decompose the wastes into carbon dioxide, water, and minerals. BOD is usually expressed as milligrams of dissolved oxygen per liter of water for a specific number of days at a given temperature. A large amount of sewage in water generates a high BOD, which robs the water of dissolved oxygen (Figure 21.1). When dissolved oxygen levels are low, anaerobic (without oxygen) microorganisms also produce compounds that have very unpleasant odors, further deteriorating water quality.

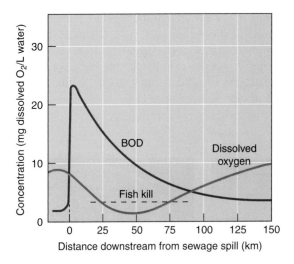

Figure 21.1 Effect of sewage on dissolved oxygen and biochemical oxygen demand (BOD). Note the initial oxygen depletion and increasing BOD close to the sewage spill (at distance 0). The stream gradually recovers as the sewage is diluted and degraded. As indicated on the graph, fishes cannot live in water that contains less than 4 mg of dissolved oxygen per liter of water.

Disease-Causing Agents

Disease-causing agents are infectious organisms that cause diseases; they come from the wastes of infected individuals. Municipal wastewater usually contains many bacteria, viruses, protozoa, parasitic worms, and other infectious agents that cause human or animal diseases (Table 21.1). Typhoid, cholera, bacterial dysentery, polio, and infectious hepatitis are some of the more common diseases caused by bacteria or viruses that are transmissible through contaminated food and water. However, many human diseases such as acquired immunodeficiency syndrome (AIDS) are not transmissible through water.

The vulnerability of our public water supplies to waterborne disease-causing agents was dramatically demonstrated in 1993 when a microorganism known as *Cryptosporidium* contaminated the water supply in the greater Milwaukee area. About 370,000 people developed diarrhea, making it the largest outbreak of a waterborne disease ever recorded in the United States, and several people with weakened immune systems died. Smaller outbreaks of contamination by salmonella and other bacteria have occurred in several cities and towns since the Milwaukee episode. In 2000, the first waterborne outbreak in North America of the deadly strain of *Escherichia coli* (0157:H7) occurred in an Ontario town (Walkerton) in Canada. Several people were killed, and several thousand became sick. Prior to this outbreak, this *E. coli* strain was thought to be transmitted almost exclusively through contaminated food. These and similar outbreaks have triggered additional concerns about the safety of our drinking water.

Monitoring for Sewage Because sewage-contaminated water is a threat to public health, periodic tests are made for the presence of sewage in our water supplies. Although many different microorganisms thrive in sewage, the common intestinal bacterium, *E. coli*, is typically used as an indication of the amount of sewage present in water and as an indirect measure of the presence of disease-causing agents. *E. coli* is perfect for monitoring sewage because it is not present in the environment except from human and animal feces, where it is found in large numbers.

To test for the presence of *E. coli* in water, the **fecal coliform test** is performed (Figure 21.2). A small sample of water is passed through a filter to trap all bacteria. The filter is then transferred to a petri dish that contains nutrients. After an incubation period, the number of greenish colonies present indicates the number of *E. coli*. Safe drinking water should contain no more than 1 coliform bacterium per 100 mL of water (about $1/2$ cup), safe swimming water should have no more than 200 per 100 mL of water, and general recreational water (for boating) should have no more than 2,000 per 100 mL. In contrast, raw sewage may contain several million coliform bacteria

Table 21.1	Some Human Diseases Transmitted by Polluted Water		
Disease	**Infectious Agent**	**Type of Organism**	**Symptoms**
Cholera	*Vibrio cholerae*	Bacterium	Severe diarrhea, vomiting; fluid loss of as much as 20 quarts per day causes cramps and collapse
Dysentery	*Shigella dysenteriae*	Bacterium	Infection of the colon causes painful diarrhea with mucus and blood in the stools; abdominal pain
Enteritis	*Clostridium perfringens,* other bacteria	Bacterium	Inflammation of the small intestine causes general discomfort, loss of appetite, abdominal cramps, and diarrhea
Typhoid	*Salmonella typhi*	Bacterium	Early symptoms include headache, loss of energy, fever; later, a pink rash appears along with (sometimes) hemorrhaging in the intestines
Infectious hepatitis	Hepatitis virus A	Virus	Inflammation of liver causes jaundice, fever, headache, nausea, vomiting, severe loss of appetite, muscle aches, and general discomfort
Poliomyelitis	Poliovirus	Virus	Early symptoms include sore throat, fever, diarrhea, and aching in limbs and back; when infection spreads to spinal cord, paralysis and atrophy of muscles
Cryptosporidiosis	*Cryptosporidium* sp.	Protozoon	Diarrhea and cramps that last up to 22 days
Amoebic dysentery	*Entamoeba histolytica*	Protozoon	Infection of the colon causes painful diarrhea with mucus and blood in the stools; abdominal pain
Schistosomiasis	*Schistosoma* sp.	Fluke	Tropical disorder of the liver and bladder causes blood in urine, diarrhea, weakness, lack of energy, repeated attacks of abdominal pain
Ancylostomiasis	*Ancylostoma* sp.	Hookworm	Severe anemia, sometimes symptoms of bronchitis

per 100 mL of water. Although most strains of coliform bacteria do not cause disease, the fecal coliform test is a reliable way to indicate the likely presence of pathogens, or disease-causing agents, in water.

When dangerous levels of fecal coliform bacteria are discovered in a stream or other body of water, it is important to determine the source of contamination so it can be cleaned up. Finding the source is not always easy because coliform bacteria reside in the intestinal tracts of many animals. The contamination could be coming from humans wastes, such as from septic systems that are not operating effectively (discussed later in the chapter); from animal feedlots; or even from the droppings of raccoons, birds, and other wildlife. A new field of science, known as **bacterial source tracking** (**BST**), attempts to make the proper identification. Bacterial source tracking uses some of the latest techniques in molecular biology to determine subtle differences in strains of *E. coli* on the basis of their animal host. Although this science is still in its infancy, it has successfully identified the source of coliform bacteria in several cases, such as Virginia's Four Mile Run near Washington, D.C.

Figure 21.2 Fecal coliform test. This test is used to indicate the likely presence of disease-causing agents in water. A water sample is first passed through a filtering apparatus. (a) The filter disk is then placed on a medium that supports coliform bacteria for a period of 24 hours. (b) After incubation, the number of bacterial colonies is counted. Each colony of *Escherichia coli* arose from a single coliform bacterium in the original water sample.

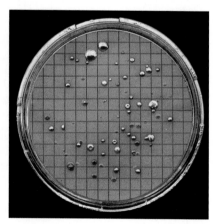

(a)   (b)

Sediment Pollution

Clay, silt, sand, and gravel are sediments that can be suspended and carried in water. When a river flows into a lake or ocean, its flow velocity decreases, and the sediments often settle out. Over time, as sediments accumulate, new land is formed. A river delta is a flat, low-lying plain created from these sediments. River deltas, with their abundant wildlife and waterways for trade routes, have always been important settlement sites for people. Today, river deltas are among the most densely populated areas in the world. Sediments are also deposited on land when a river overflows its banks during a flood.

Sediment pollution consists of excessive amounts of suspended soil particles that eventually settle out and accumulate on the bottom of a body of water. Sediment pollution comes from erosion of agricultural lands, forest soils exposed by logging, degraded stream banks, overgrazed rangelands, strip mines, and construction.

Control of soil erosion, which is discussed in detail in Chapter 14, reduces sediment pollution in waterways.

Sediment pollution causes problems by reducing light penetration, covering aquatic organisms, bringing insoluble toxic pollutants into the water, and filling in waterways. When sediment particles are suspended in the water, they make the water turbid (cloudy), which in turn decreases the distance that light can penetrate. Because the base of the food web in an aquatic ecosystem consists of photosynthetic algae and plants that require light for photosynthesis, turbid water lessens the ability of producers to photosynthesize. Extreme turbidity also reduces the number of photosynthesizing organisms, which in turn causes a decrease in the number of aquatic organisms that feed on the primary producers (Figure 21.3). Sediment that settles out of the water and forms a layer over coral reefs or shellfish beds can clog the gills and feeding structures of many aquatic animals.

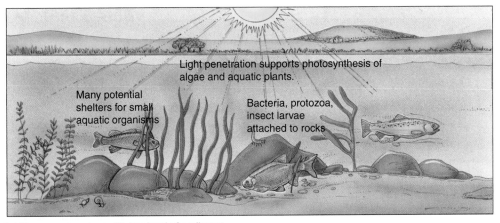

(a) Stream ecosystem with low level of sediment

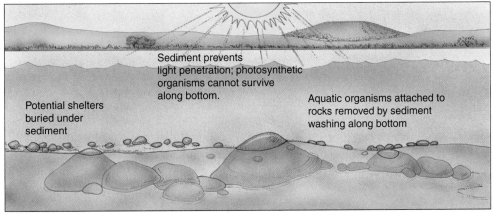

(b) Same stream with high level of sediment

Figure 21.3 Sediment pollution. (a) A stream ecosystem without sediment pollution can support diverse aquatic life. (b) How the same ecosystem would appear after prolonged exposure to heavy sediment pollution.

Sediments adversely affect water quality by carrying toxic chemicals, both inorganic and organic, into the water. The sediment particles provide surface area to which some insoluble, toxic compounds adhere, so when sediments get into water, so do the toxic chemicals. Disease-causing agents can also be transported into water via sediments.

On the basis of the growing recognition that most of the toxic pollutants in water are stored in and released from sediments, the U.S. Environmental Protection Agency (EPA) evaluated data from more than 21,000 sampling stations in 1,363 U.S. watersheds. The resulting survey, which was released in 1998, determined that sediments in 7% of watersheds are so seriously contaminated with toxic pollutants that eating fishes from those waterways would potentially threaten human health.

When sediments settle out of solution, they fill in waterways. This problem is particularly serious in reservoirs and in channels through which ships must pass. Thus, sediment pollution may adversely affect the shipping industry.

Inorganic Plant and Algal Nutrients

Inorganic plant and algal nutrients are chemicals such as nitrogen and phosphorus that stimulate the growth of plants and algae. Inorganic plant and algal nutrients are essential for the normal functioning of healthy ecosystems but are harmful in larger concentrations. Nitrates and phosphates come from such sources as human and animal wastes, plant residues, atmospheric deposition, and fertilizer runoff from agricultural and residential land. Inorganic plant and algal nutrients encourage excessive growth of algae and aquatic plants. Although algae and aquatic plants are the base of the food web in aquatic ecosystems, their excessive growth disrupts the natural balance between producers and consumers and causes other problems, including enrichment, bad odors, and a high BOD. The high BOD occurs when the excessive numbers of algae die and are decomposed by bacteria.

The "Dead Zone" in the Gulf of Mexico　Every spring and summer, fertilizer runoff from Midwestern fields and manure runoff from livestock operations in such states as Iowa, Wisconsin, and Illinois eventually find their way into the Mississippi River and, from there, into the Gulf of Mexico. The amount of such runoff is considerable. According to the 1998 Senate Agriculture Hearings, livestock produce approximately 20 times the feces and urine that humans do in the United States, yet the wastes are currently not covered by water quality laws and therefore do not go to sewage treatment plants (see Chapter 18).

These nutrients are largely responsible for a huge "dead zone" in the Gulf of Mexico that is nearly as large as the state of New Jersey—more than 17,500 km² (7,000 mi²) in area. The dead zone extends from the seafloor up into the water column, sometimes to within a few meters of the surface. Floods and droughts change its size and shape. It generally persists from March or April, as snowmelt and spring rains flow from the Mississippi River into the gulf, to September. It is most severe in June, July, and August. Overall, the dead zone appears to be growing. In 2001, for example, it was more than 20,000 km² (8000 mi²), in size, which is the largest dead zone in the Gulf of Mexico since measurements began in the 1980s.

Other than bacteria that thrive in oxygen-free environments, no life exists in the dead zone. The water does not contain enough dissolved oxygen to support fishes, shrimp, or other aquatic organisms. Fishes, shrimp, and other active swimmers avoid the area, but bottom dwellers such as sea stars, brittle stars, worms, and clams suffocate and die.

This oxygen-free condition, known as **hypoxia**, occurs when algae grow rapidly because of the presence of nutrients such as nitrates in the water. When these algae die, they sink to the bottom and are decomposed by bacteria, which deplete the water of dissolved oxygen, leaving too little for other sea life. Although hypoxia has

---

ENVIROBRIEF

Harmful Algal Blooms

When some pigmented marine algae experience population explosions or *blooms*, their great abundance frequently colors the water orange, red, or brown. Known as **red tides,** these blooms may cause serious environmental harm and threaten the health of humans and animals. Some of the algal species that form red tides produce toxins that attack the nervous systems of fishes, leading to massive fish kills. Water birds such as cormorants suffer and sometimes die when they eat the contaminated fishes. The toxins also work their way up the food web to marine mammals and people. In 1997, more than 100 monk seals, one third of the endangered species' total numbers, died from algal toxin poisoning off the West African coast. Humans may also suffer if they consume algal toxins in shellfish or fishes. Even nontoxic algal species may wreak havoc when they bloom, as they shade aquatic vegetation and upset food web dynamics.

No one knows what triggers red tides, which are becoming more common and more severe, but many experts think the blame lies with coastal pollution. Wastewater and agricultural runoff to coastal areas contains increasingly larger quantities of nitrogen and phosphorus, two nutrients that stimulate algal growth. Changes in ocean temperatures, such as those attributed to global warming, may also trigger algal blooms. In addition, a possible connection exists between red tide outbreaks in Florida's coastal waters and the arrival of dust clouds from Africa. These dust clouds, which sometimes blow across the Atlantic Ocean, enrich the water with iron and appear to trigger algal blooms.

Few control measures are in place to prevent the blooms or to end them when they occur. However, newer technologies like satellite monitoring and weather-tracking systems allow better prediction of conditions likely to stimulate blooms.

been reported in more than 60 coastal areas around the world, the dead zone in the Gulf of Mexico is one of the largest. (Dead zones in the Black Sea and the Baltic Sea are reportedly larger.)

In 1999 the EPA released recommendations to deal with the problem because of its threat to commercial fisheries and to the marine environment in general. At the same time, the EPA tried to balance the needs of farmers because using less fertilizer without modifying other farming practices could reduce crop yields and raise food prices. (The Mississippi River drains all or part of 31 states and two Canadian provinces, and its watershed contains more than half of all U.S. farms.) The challenge is to modify farming methods so that 20% less fertilizer is needed, because then most of the fertilizer that is applied would remain on the farmers' fields.

Other potential inorganic nutrient sources, such as sewage treatment plants and airborne nitrogen oxides from automobile emissions, will also have to be addressed. Restoring former wetlands in the Mississippi River watershed would help reduce the nitrate load from fertilizers entering the Gulf of Mexico. (Recall from the chapter introduction that wetlands retain nutrients such as nitrogen.) The EPA recognizes that the dead zone problem is immense in scope and will take billions of dollars and decades of effort to fix.

Organic Compounds

Organic compounds are chemicals that contain carbon atoms. Most of the thousands of organic compounds found in water are synthetic chemicals that are produced by human activities; these include pesticides, solvents, industrial chemicals, and plastics. (Several examples of toxic organic compounds sometimes found in polluted water are given in Table 21.2.)

Some organic compounds find their way into surface water and groundwater by seeping from landfills. Others, such as pesticides, leach downward through the soil into groundwater or get into surface water by runoff from farms and residences. Many organic compounds are dumped directly into waterways by industries.

The most comprehensive study of synthetic organic pollutants in U.S. waterways was completed in 2002. Researchers from the U.S. Geological Survey collected samples of water from 139 streams in 30 states. Most of the streams were downstream from cities and agricultural areas such as dairies and pig farms. The scientists tested the water samples for 95 different organic compounds, such as antibiotics, ibuprofen, acetaminophen, insect repellents, antimicrobial substances, fragrances, caffeine, and steroids such as hormones from birth control pills and hormone therapy. Low concentrations—in the range of parts per billion—of 82 of these organic compounds were detected. At least one organic chemical was present in 80% of the streams, whereas more than one third of the water samples contained traces of 10 or more organic compounds. The effects on human health of ingesting drinking water containing traces of these chemicals are generally unknown. However, the hormone contaminants have been demonstrated to cause problems in many aquatic organisms (see Chapter 1). Of the organic chemicals that were tested, 33 are suspected to have hormonal effects.

There are several ways to control the presence of organic compounds in our water. Everyone, from individual homeowners to large factories, should take care to prevent organic compounds from ever finding their way into water. Alternative organic compounds, which are less toxic and degrade more readily so that they are not as persistent in the environment, can be developed and used. Also, tertiary water treatment, considered later in this chapter, effectively eliminates many synthetic organic compounds in water.

Inorganic Chemicals

Inorganic chemicals are contaminants that contain elements other than carbon; examples include acids, salts, and heavy metals. Inorganic chemicals do not easily degrade,

Table 21.2	Some Synthetic Organic Compounds Found in Polluted Water
Compound	*Some Reported Health Effects*
Aldicarb (pesticide)	Attacks nervous system
Benzene (solvent)	Associated with blood disorders (bone marrow suppression); leukemia
Carbon tetrachloride (solvent)	Possibly causes cancer; liver damage; may also attack kidneys and vision
Chloroform (solvent)	Possibly causes cancer
Dioxins (TCDD) (chemical contaminants)	Some cause cancer; may harm reproductive, immune, and nervous systems
Ethylene dibromide (EDB) (fumigant)	Probably causes cancer; attacks liver and kidneys
Polychlorinated biphenyls (PCBs) (industrial chemicals)	Attack liver and kidneys; possibly cause cancer
Trichloroethylene (TCE) (solvent)	Probably causes cancer; induces liver cancer in mice
Vinyl chloride (plastics industry)	Causes cancer

or break down. Therefore, when they are introduced into a body of water, they remain there for a long time. Many inorganic chemicals find their way into both surface water and groundwater from sources such as industries, mines, irrigation runoff, oil drilling, and urban runoff from storm sewers. Some of these inorganic pollutants are toxic to aquatic organisms. Their presence may make water unsuitable for drinking or other purposes.

Here we consider the heavy metals lead and mercury, two inorganic chemicals that sometimes contaminate water and accumulate in the tissues of humans and other organisms (see the discussion of bioaccumulation and biological magnification in Chapter 22). Arsenic, another heavy metal, is discussed later in this chapter.

Lead   People used to think of lead poisoning as affecting only inner-city children who ate paint chips that contained lead. Lead-based paint was banned in the United States in 1978, but the EPA estimates that more than three fourths of U.S. homes still contain some lead-based paint. Although lead-based paint remains an important source of lead poisoning in children, lead lurks in many other places in the environment as well.

Lead-containing anti-knock agents in gasoline were outlawed in the United States in 1986; prior to 1986 lead dust was released into the atmosphere when the fuel was burned, and that lead still contaminates the soil, particularly in inner cities near major highways. Thus, children living in the inner city may be at risk when they play outdoors in their schoolyards and backyards. Lead also contaminates the soil, surface water, and groundwater when incinerator ash is dumped into ordinary sanitary landfills. It may be spewed into the atmosphere from old factories that lack air pollution control devices. We ingest additional amounts of lead from pesticide and fertilizer residues on produce, from food cans that are soldered with lead, and even from certain types of dinnerware on which our food is served. Low amounts of lead also originate from natural sources such as volcanoes and wind-blown dust.

According to the EPA, in the mid-1990s more than 10% of all large and medium-sized municipal water supplies contained lead levels that exceeded the maximum permitted by the Safe Drinking Water Act (discussed later in this chapter). In addition, tap water often contains higher levels of lead than are in municipal water supplies; the extra lead comes from the corrosion of old lead water pipes or of lead solder in newer pipes.

Millions of U.S. residents, many of them children, have damaging levels of lead in their bodies. According to the Agency for Toxic Substances and Disease Registry in Atlanta, 17% of U.S. children—about 3 million to 4 million children—have blood lead levels that exceed 15 micrograms (μg) per deciliter of blood. A blood lead level above 10 μg per deciliter is considered dangerous.

The three groups of people at greatest risk from lead poisoning are middle-aged men, pregnant women, and young children. Middle-aged men with high levels of lead are more likely to develop **hypertension**, or high blood pressure. High lead levels in pregnant women increase the risk of miscarriages, premature deliveries, and stillbirths. Children with even low levels of lead in their blood may suffer from a variety of mental and physical impairments, including partial hearing loss, hyperactivity, attention deficit, lowered IQ, and learning disabilities. A 1996 study by the University of Pittsburgh School of Medicine reported a link between male juvenile delinquency and high bone lead concentrations. (Because lead accumulates in bones, bone lead levels more accurately reflect long-term exposure than do blood lead levels.) A 2001 study by the American Medical Association reported a link between murder rates and lead levels in the air.

Mercury   Mercury is a metal that can vaporize at room temperatures. This characteristic poses special environmental challenges when dealing with mercury. Small amounts of mercury occur naturally in the environment, but most mercury pollution is caused by human activities. According to a 1997 EPA report to Congress, coal-fired power plants release the largest amount (33%) of mercury into the environment. Coal contains traces of mercury that vaporize and are released into the atmosphere with the flue gases when the coal is burned. This mercury then moves from the atmosphere to the water via precipitation. The technology exists to control mercury emissions from coal-burning power plants, but it is very expensive and is currently not mandated by the Clean Air Act. Also, the mercury trapped by such devices would have to be properly disposed of in a hazardous waste landfill (see Chapter 23) or it could recontaminate the environment. Environmentalists recommend that in the short term, utilities should burn natural gas instead of coal; the switch to natural gas would reduce not only mercury pollution but also carbon dioxide emissions, which are associated with global climate warming (see Chapter 20).

Municipal waste incinerators account for 18% of mercury released into the environment, and medical waste incinerators for an additional 10%. This mercury is released when incinerators burn materials containing mercury. Fluorescent lights and thermostats are examples of municipal waste that contain mercury, whereas thermometers and blood-pressure cuffs are examples of medical waste.

Significant amounts of mercury are also released into the environment during the smelting of metals such as lead, copper, and zinc. Mercury is used in a variety of industrial processes, such as chemical plants that manufacture chlorine and caustic soda. Some of this mercury vaporizes, thereby entering the atmosphere. In addition, when industries release their wastewater, some metallic mercury may enter natural bodies of water along with the wastewater. Mercury sometimes enters water by precipi-

Figure 21.4 Mercury pollution. Because mercury readily enters the food web, fish advisories are posted where mercury pollution is a problem. Photographed in the Florida Everglades.

tation after household trash containing batteries, paints, and plastics has been burned in incinerators.

Once in a body of water, mercury settles into the sediments and is converted by bacteria to methyl mercury compounds, a more toxic form that readily enters the food web (Figure 21.4). Mercury accumulates in the muscles of tuna, swordfish, sharks, and marine mammals—the top predators of the open ocean. Human exposure to mercury is primarily by eating fishes and marine mammals containing high levels of mercury. At least 40 states in the United States have released health advisories on the consumption by humans of mercury-tainted seafood.

Methyl mercury compounds remain in the environment for a long time and are highly toxic to organisms, including humans. Mercury exposure to developing fetuses in pregnant women has been linked to a variety of conditions, such as mental retardation, cerebral palsy, and developmental delays. Prolonged exposure to methyl mercury compounds causes kidney disorders and severely damages the nervous and cardiovascular systems. Methyl mercury compounds are unusual in that they are able to cross the body's blood–brain barrier (many materials do not pass from the blood to the cerebrospinal fluid and brain). Low levels of mercury in the brain cause neurological problems such as headache, depression, and quarrelsome behavior.

Radioactive Substances

Radioactive substances contain atoms of unstable isotopes that spontaneously emit radiation. Radioactive substances can get into water from several sources, including the mining and processing of radioactive minerals such as uranium and thorium. Many industries use radioactive

substances; although nuclear power plants and the nuclear weapons industry use the largest amounts, medical and scientific research facilities also employ them. It is possible for radiation to inadvertently escape from any of these facilities, polluting the air, water, and soil. Accidents at nuclear power plants can release into the atmosphere large quantities of radiation, which eventually contaminate soil and water. Radiation from natural sources can also pollute groundwater.

Since the mid-1980s, low levels of radioactive substances have been discovered in the wastewater of several sewage treatment plants in the United States. The EPA reports that radioactive materials may concentrate in sludge (a slimy solid mixture formed during the treatment of sewage). Guidelines from the EPA help municipal sewage treatment plants to identify radioactive materials in sewage sludge and, when present, to reduce or eliminate the contamination.

Radon   Radon is a naturally occurring radioactive gas. It is produced in Earth's crust and increases the risk of lung cancer when it is inhaled over long periods of time (see Chapter 19). It is also possible for radon to temporarily dissolve in groundwater. When you shower, wash dishes, or wash clothes with radon-contaminated well water, the radon gets into the air in your home. In addition, the EPA is concerned about possible health effects caused by long-term exposure to radon in drinking water. However, the exposure to radon in water is generally quite small (about 1% to 2% of total radon exposure), compared with the exposure in air that seeps into houses from the ground (see Figure 19-17).

Thermal Pollution

Thermal pollution occurs when heated water produced during certain industrial processes is released into waterways. Many industries, such as steam-generated electric power plants, use water to remove excess heat from their operations. Afterward, the heated water is allowed to cool a little before it is returned to waterways, but its temperature is still warmer than it was originally. The result is that the waterway is warmed slightly.

A rise in temperature of a body of water has several chemical, physical, and biological effects. Chemical reactions, including decomposition of wastes, occur faster, depleting the water of oxygen. Moreover, less oxygen dissolves in warm water than in cool water (Table 21.3), and the amount of oxygen dissolved in water has important effects on aquatic life. When the level of dissolved oxygen is lowered due to thermal pollution, a fish responds by ventilating its gills more frequently to obtain enough oxygen. Gill ventilation, however, is an activity that also requires an increased consumption of oxygen. This situation puts a great deal of stress on the fish as it tries to obtain a greater supply of oxygen from a smaller supply dissolved in the water.

Table 21.3	Dissolved Oxygen in Water at Various Temperatures

Temperature ($°C$)	Dissolved Oxygen Capacity* ($g\ O_2$ per L H_2O)
0	0.0141
10	0.0109
20	0.0092
25	0.0083
30	0.0077
35	0.0070
40	0.0065

* Data for water in contact with air at 760 mm mercury pressure.

There may be other subtle changes in the activities and behavior of aquatic organisms in thermally polluted water, because temperature affects reproductive cycles, digestion rates, and respiration rates. At warmer temperatures, fishes require more food to maintain body weight. They also typically have shorter life spans and smaller populations. In cases of extreme thermal pollution, fishes and other aquatic organisms die.

EUTROPHICATION: AN ENRICHMENT PROBLEM IN AQUATIC ECOSYSTEMS

Lakes, estuaries, and slow-flowing streams that have minimal levels of nutrients are said to be unenriched, or **oligotrophic**. An oligotrophic lake has clear water and supports small populations of aquatic organisms (Figure 21.5a). **Eutrophication** is the enrichment of a lake, estuary, or slow-flowing stream by inorganic plant and algal nutrients such as phosphorus; a body of water that is enriched is said to be **eutrophic**. The enrichment of water results in an increased photosynthetic productivity. Thus, the water in a eutrophic lake is cloudy and usually resembles pea soup because of the presence of vast numbers of algae and cyanobacteria that are supported by the nutrients (Figure 21.5b).

Although eutrophic lakes contain large populations of aquatic animals, they are different kinds of organisms from those predominant in oligotrophic lakes. For example, an unenriched lake in the northeastern United States may contain pike, sturgeon, and whitefish. All three are found in the deeper, colder part of the lake, where there is a higher concentration of dissolved oxygen. In eutrophic lakes, on the other hand, the deeper, colder levels of water are depleted of dissolved oxygen because when the excessive numbers of algae die, they settle to the lake's bottom and stimulate an increased amount of decay. Microorganisms that decompose the dead algae use up much of the lake's dissolved oxygen in the process. Thus, there is a high BOD caused by decomposition on the lake floor, and fishes such as pike, sturgeon, and whitefish die out and are replaced by warm-water fishes,

such as catfish and carp, that can tolerate lesser amounts of dissolved oxygen (Figure 21.6).

Over vast periods, oligotrophic lakes, estuaries, and slow-moving streams become eutrophic naturally. As natural eutrophication occurs, these bodies of water are slowly enriched and grow shallower from the immense number of dead organisms that have settled in the sediments over a long period. Gradually, plants such as water

(a)

(b)

Figure 21.5 Oligotrophic and eutrophic lakes. (a) Crater Lake in Oregon, like other oligotrophic lakes, is low in nutrients. (b) Eutrophic lakes and ponds, such as this one in western New York, are often covered with slimy, smelly mats of algae and cyanobacteria.

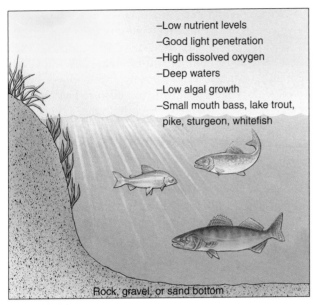

-Low nutrient levels
-Good light penetration
-High dissolved oxygen
-Deep waters
-Low algal growth
-Small mouth bass, lake trout, pike, sturgeon, whitefish

Rock, gravel, or sand bottom

(a) Oligotrophic lake

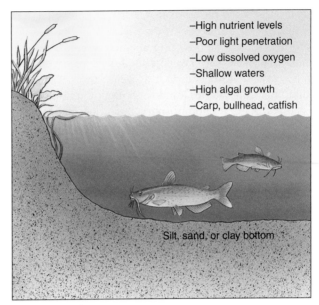

-High nutrient levels
-Poor light penetration
-Low dissolved oxygen
-Shallow waters
-High algal growth
-Carp, bullhead, catfish

Silt, sand, or clay bottom

(b) Eutrophic lake

Figure 21.6 Characteristics of oligotrophic and eutrophic lakes. (a) An oligotrophic lake has a low level of inorganic plant and algal nutrients. (b) A eutrophic lake has a high level of these nutrients.

lilies and cattails take root in the nutrient-rich sediments and begin to fill the shallow waters, forming a marsh.

However, eutrophication can be markedly accelerated by human activities. This fast, human-induced process is usually called **artificial eutrophication**, or **cultural eutrophication**, to distinguish it from natural eutrophication. Artificial eutrophication results from the enrichment of aquatic ecosystems by inorganic plant and algal nutrients that are found predominantly in sewage and fertilizer runoff.

Controlling Artificial Eutrophication

Water, sunlight, carbonate (dissolved carbon dioxide), nitrogen, phosphorus, and certain other inorganic nutrients are the main requirements for algal growth, which is limited by the essential material that is in shortest supply (recall the discussion of limiting factors in Chapter 5). Because phosphorus is often the limiting factor in freshwater ecosystems, the most effective way to slow artificial eutrophication is usually to limit the amount of phosphorus entering the aquatic system. When treated sewage was no longer dumped in Lake Washington, the phosphorus level in the lake declined by about 75% and there was a corresponding drop in algal growth (see Figure 2.10b and discussion in Chapter 2). Nitrogen (as nitrates) has also been implicated in eutrophication. In Sweden, eutrophication of the Kattegat, the narrow waterway between Denmark and Sweden, has been linked to excessive nitrogen runoff from agriculture.

SOURCES OF WATER POLLUTION

Water pollutants come from both natural sources and human activities. For example, some of the mercury that contaminates the biosphere comes from natural sources in Earth's crust; the remainder is emitted by human activities. Nitrate pollution has both natural and human sources—the nitrate that occurs in soil and the inorganic fertilizers that are added to it, respectively. Although natural sources of pollution are sometimes of local concern (see discussion of arsenic later in the chapter), pollution caused by human activities is generally more widespread.

The sources of water pollution are classified into two types, point source pollution and nonpoint source pollution. **Point source pollution** is discharged into the environment through pipes, sewers, or ditches from specific sites such as factories or sewage treatment plants. Point source pollution is relatively easy to control legislatively, but accidents can still occur. A cyanide spill contaminated the Tisza and Danube Rivers in Europe in 2000, killing millions of fishes and shutting off downstream water supplies. The cyanide was in a holding basin at a gold mine in Romania, but heavy snow and rain caused it to overflow into the river. Although cyanide disperses to nonlethal levels rapidly, the heavy metals that were also present in the holding basin will contaminate the rivers and Black Sea, which they empty into, for many years.

Nonpoint source pollution, also called **polluted runoff**, is caused by land pollutants that enter bodies of water over large areas rather than at a single point. Non-point source pollution occurs when precipitation moves over and through the soil, picking up and carrying away pollutants that eventually are deposited in lakes, rivers, wetlands, groundwater, estuaries, and the ocean. Although nonpoint sources are diffuse, their cumulative effect is often huge. Nonpoint source pollution includes agricultural runoff (such as fertilizers, pesticides, live-stock wastes, and salt from irrigation), mining wastes (such as acid mine drainage), municipal wastes (such as inorganic plant and algal nutrients), and construction sediments. Soil erosion from fields, logging operations, eroding stream banks, and construction sites is a major cause of nonpoint source pollution.

Three major sources of human-induced water pollution that we now examine in greater detail are agriculture, municipalities (that is, domestic activities), and industries.

Water Pollution from Agriculture

According to the EPA, agriculture is the leading source of water quality impairment of surface waters nationwide. For example, 72% of the water pollution in rivers is attributed to agriculture. Agricultural practices produce several types of pollutants that contribute to nonpoint source pollution (Figure 21.7). Fertilizer runoff causes water enrichment. Animal wastes and plant residues in waterways produce high BODs and high levels of suspended solids as well as water enrichment. Amid growing concern about polluted runoff from animal wastes, the EPA unveiled a largely voluntary program in 1998 that asks the 450,000 livestock operations in the United States to develop Comprehensive Nutrient Management Plans by 2008. These

Figure 21.7 Water pollution from agriculture. The runoff from feedlots and fields where cattle graze contributes animal wastes to water. Photographed in Wisconsin

plans are to consider safe ways to handle and store manure so that it does not become polluted runoff.

Chemical pesticides used in agriculture may leach into the soil and from there into water. These chemicals are highly toxic and can adversely affect human health as well as the health of aquatic organisms. The National Water Quality Assessment Program, an ongoing study of pesticides and their degradation products, indicates that pesticides are widespread in U.S. rivers, streams, and groundwater. More than 95% of the river and stream samples and almost 50% of groundwater samples contained at least one pesticide. Many water samples contained a mixture of pesticides. More than 50% of all stream samples contained five or more pesticides, and about 10% of all streams contained 10 or more pesticides.

Soil erosion from fields and rangelands causes sediment pollution in waterways. In addition, some agricultural chemicals that are not very soluble in water such as certain pesticides, find their way into waterways by adhering to sediment particles. Thus, soil conservation methods not only conserve the soil but also reduce water pollution.

Municipal Water Pollution

Although sewage is the main pollutant produced by cities and towns, municipal water pollution also has a nonpoint source: urban runoff from storm sewers (Figure 21.8; also see "You Can Make a Difference: Preventing Water Pollution"). The water quality of urban runoff from city streets is often worse than that of sewage. Urban runoff carries salt from roadways, untreated garbage, construction sediments, and traffic emissions (via rain that washes pollutants out of the air). It often may contain such contaminants as asbestos, chlorides, copper, cyanides, grease, hydrocarbons, lead, motor oil, organic wastes, phosphates, sulfuric acid, and zinc.

Some 1,100 cities across the United States, such as New York, San Francisco, Pittsburgh, and Boston, have a **combined sewer system** in which human and industrial wastes are mixed with urban runoff from storm sewers before flowing into the sewage treatment plant. A problem arises when there is heavy rainfall or a large snowmelt because even the largest sewage treatment plant can only process a given amount of wastewater each day. When too much water enters the system, the excess, known as **combined sewer overflow**, flows into nearby waterways without being treated. Combined sewer overflow, which contains raw sewage, has been illegal since passage of the Clean Water Act of 1972 (to be discussed shortly), but cities have only recently begun to address the problem. According to the EPA, 1.2 trillion gallons of combined sewer overflow are discharged into U.S. waterways every year.

Some cities, such as St. Paul, Minnesota, have installed two separate sewers, one for sewage and indus-

Figure 21.8 Urban runoff. The largest single pollutant in urban runoff is organic waste, which removes dissolved oxygen from water as it decays. Fertilizers cause excessive algal growth, which further depletes the water of oxygen, harming aquatic organisms. Other everyday pollutants include used motor oil, which is often poured into storm drains, and heavy metals. These pollutants may be carried from storm drains on streets to streams and rivers.

trial wastes and one for urban runoff. However, such an installation is expensive and requires that every street be dug up. Other cities, such as Birmingham, Michigan, have kept their combined sewer systems but installed huge retention basins to hold the overflow until it can be treated. The one in Birmingham, which was installed in 1998, holds 20.8 million L (5.5 million gal). Such tanks are less expensive to install than separate sewer systems, but there are concerns that after several days of heavy rain or snow, the tank itself could overflow. Thus far in its operation, the Birmingham retention basin has overflowed less than 6 times a year, as compared to an average of 60 times a year that Birmingham experienced combined sewer overflow before the retention basin was installed.

Industrial Wastes in Water

Different industries generate different types of pollutants. Food processing industries produce organic wastes that are readily decomposed but have a high BOD. Pulp and paper mills produce toxic compounds and sludge, although the industry has begun to adopt new manufacturing methods, such as the production of paper without the use of chlorine as a bleaching agent, that produce significantly less toxic effluents.

Many industries in the United States treat their wastewater with advanced treatment methods. The electronics industry produces wastewater containing high levels of heavy metals such as copper, lead, and manganese but uses special techniques such as ion exchange and electrolytic recovery to reclaim those heavy metals. Plates with commercial value are produced from the recovered metals that would otherwise have become a component of hazardous sludge. Although U.S. industries do not usually dump highly toxic wastes into water, disposal is still sometimes a problem.

GROUNDWATER POLLUTION

Roughly half the people in the United States obtain their drinking water from groundwater, which is also

YOU CAN MAKE A DIFFERENCE

Preventing Water Pollution

Although individuals produce very little water pollution, the collective effect of municipal water pollution, even in a small neighborhood, can be quite large. There are many things you can do to protect surface waters and groundwater from water pollution. Here are some specific dos and don'ts that you should adopt.

1. Many household chemicals, such as oven cleaners, mothballs, drain cleaners, and paint thinners, are quite toxic. Use such products sparingly, and try to substitute less hazardous chemicals wherever possible. When disposing of unwanted hazardous household chemicals, contact the solid waste management office in your county for information about hazardous waste collection centers in your area. Never put these chemicals down a drain or toilet, because they may disrupt your septic system or contaminate sewage sludge produced at municipal sewage treatment facilities. Also, never pour these chemicals on the ground because they may contaminate runoff when it rains. Here are some safer alternatives (note, however, that these chemicals are not nontoxic; they are simply less toxic than many commercial products):
   a. Ammonia, to clean appliances and windows.
   b. Bleach, to disinfect. Never mix ammonia and bleach, however, because the mixture releases toxic chlorine gas.
   c. Borax, to remove stains and mildew.
   d. Baking soda, to remove stains, deodorize, and clean household utensils.
   e. Mineral oil, to polish furniture and wax floors.
   f. Vinegar, to clean surfaces, polish metals, and remove stains and mildew.
2. Never throw unwanted medicines down the toilet. Several studies have shown than traces of a variety of drugs are showing up in tap water.
3. Never pour used motor oil or antifreeze down storm drains or on the ground. Recycle these chemicals by dropping them off at a service station or local hazardous waste collection center.
4. Pick up pet waste and dispose of it in the garbage or toilet. If left on the ground, it eventually washes into water-

ways where it can contaminate shellfish and enrich the water.
5. Drive less. The air pollution emitted by automobiles eventually finds its way into groundwater and surface water. Also, toxic metals and oil byproducts that are deposited on the road by automobiles—the average automobile annually leaks more than 1 quart of petroleum products onto roads and parking lots—are washed into surface waters by precipitation.
6. If you are a homeowner, replace some of your grass lawn with trees, shrubs, and ground covers, which absorb up to 14 times more precipitation and require little or no fertilizer. To reduce erosion, use mulch to cover bare ground (see Chapter 14).
7. Use fertilizer sparingly because excess fertilizer leaches into groundwater or waterways. If you hire a professional lawn care service to apply fertilizer and pesticides, use one that monitors and applies chemicals when they are needed. Many of these companies do calendar spraying, applying chemicals every few months regardless of need.
8. Never apply fertilizer near a body of water. Always allow a buffer zone of at least 20 to 40 feet.
9. Make sure that gutters and downspouts drain onto water-absorbing grass or graveled areas instead of paved surfaces.
10. Clean up spilled oil, brake fluid, and antifreeze and sweep sidewalks and driveways instead of hosing them off. Dispose of the dirt properly; do not sweep the dirt into gutters or storm drains.
11. Likewise, do not let grass clippings or leaves wash into gutters or storm drains.
12. Use pesticides sparingly, both indoors and outdoors. Dispose of unwanted pesticides at hazardous waste collection centers.
13. Replace paved driveways and sidewalks with porous surfaces, such as interlocking bricks or stones, and build wood decks instead of concrete patios. These features will allow precipitation to seep into the ground, thereby decreasing runoff.

---

withdrawn for irrigation and industry. In recent years attention has been drawn to the quality of the nation's groundwater, which can become contaminated in several ways. The most common pollutants, such as pesticides, fertilizers, and organic compounds, can seep into groundwater from municipal sanitary landfills, underground storage tanks, backyards, golf courses, and intensively cultivated agricultural lands (Figure 21.9). More than 250,000 underground petroleum storage

tanks are thought to be leaking at service stations in the United States.

Nitrates sometimes contaminate shallow groundwater—30.5 m (100 ft) or less from the surface—with fertilizer being the most common source. High nitrate levels are a concern in some rural areas, where 80% to 90% of the residents use shallow groundwater for drinking water. When nitrates get into the human body, they are converted to nitrites, which reduce the blood's

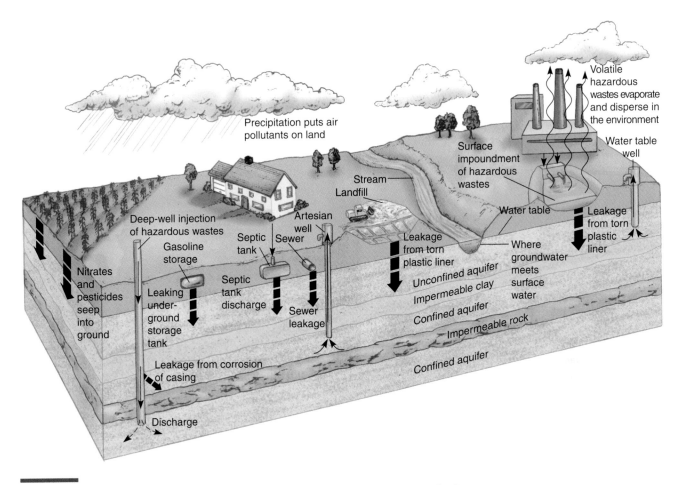

Figure 21.9 Sources of groundwater contamination. Agricultural practices, sewage (both treated and untreated), landfills, industrial activities, and septic systems are some of the sources of groundwater pollution. Once groundwater is contaminated, it does not readily cleanse itself by natural processes. We do not know the degree of groundwater contamination because access is difficult and because contaminants do not readily disperse (groundwater moves very slowly). (Figure is not drawn to scale.)

ability to transport oxygen. This condition is one of the causes of cyanosis (the "blue baby" syndrome), a serious disorder in very young children. The level of nitrate in drinking water is monitored from municipal systems, so it is generally not of concern. If you drink well water, however, you should probably have its nitrate level checked periodically.

Contamination of groundwater is a relatively recent environmental concern. People used to think that the underlying soil and rock through which surface water must seep in order to become groundwater filtered out any contaminants, thereby ensuring the purity of groundwater. This assumption proved false when the quality of groundwater began to be monitored and contaminants were discovered at certain sites. It appears that the natural capacity of soil and rock to remove pollutants from groundwater varies widely from one area to another.

Currently, most of the groundwater supplies in the United States are of good quality and do not violate standards established to protect human health. There

are some local problems that have led to well closures and raised public health concerns. For example, in 1996 Santa Monica, California, closed 7 of its 11 municipal wells when methyl tertiary butyl ether. (MTBE), a gasoline additive that reduces tailpipe emissions, was detected in the groundwater. MTBE, which has been added to gasoline in the United States since about 1979, is suspected of causing cancer. The groundwater became polluted primarily from leaking underground storage tanks that contained petroleum. A partial settlement was reached in 2002 in which some of the oil companies involved agreed to pay to construct and operate a water treatment facility to clean up the contamination.

Cleanup of polluted groundwater is very costly, takes years, and in some cases is not technically feasible. Compounding the cleanup problem is the challenge of safely disposing of the toxic materials removed from groundwater, which, if not handled properly, could contaminate groundwater once again.

■ IMPROVING WATER QUALITY

Water quality can be improved by removing contaminants from the water supply before and after it is used (Figure 21.10). Technology can assist in both processes.

Purification of Drinking Water

The United States has nearly 60,000 municipal water facilities that serve 232 million people. Surface-water sources of municipal water supplies include streams, rivers, and lakes. Often artificial lakes, called **reservoirs**, are produced by building a dam across a river or stream. Reservoirs allow water to be accumulated and stored when there is an adequate supply for use during periods of drought.

In the United States, most municipal water supplies are treated before being used so that the water is safe to drink. Water that is turbid is treated with a chemical (aluminum sulfate) that causes the suspended particles to clump together and settle out. The water is then filtered through sand to remove remaining suspended materials as well as many microorganisms. A few cities, such as Cincinnati, also pump the water through activated carbon granules to remove much of the organic compounds dissolved in the water.

In the final purification step before distribution in the water system, the water is disinfected to kill any remaining disease-causing agents. The most common way to disinfect water is by adding chlorine. A small amount of chlorine is left in the water to provide protection during its distribution through many kilometers of pipes. Other disinfection systems use ozone or ultraviolet (UV) radiation in place of chlorine.

The Chlorine Dilemma   During the 19th century, drinking water supplies in the United States were often contaminated by waterborne, disease-causing organisms. The discovery that chlorine kills these organisms allowed 20th-century Americans to drink water with little fear of contracting typhoid, cholera, or dysentery. The addition of chlorine to our drinking water supply has undoubtedly saved millions of lives.

However, chlorine byproducts, formed when chlorine reacts with organic matter in treated wastewater, have been tentatively linked to several kinds of cancer (rectal, pancreatic, and bladder), an increased risk of miscarriages, and possibly rare birth defects. As a result, use of chlorine to disinfect drinking water has triggered a national debate over the costs and benefits of chlorinating water. The concern is whether there is a long-term hazard from low levels of chlorine in drinking water.

Because there are few viable alternatives to chlorination, the EPA was initially reluctant to reduce the level of chlorine permissible in drinking water, despite the evidence of potential risks. The EPA did not want what happened in Peru to occur in the United States. In 1991 a

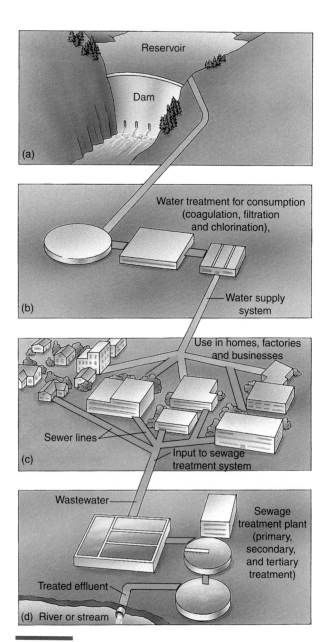

Figure 21.10 Water treatment for municipal use. (a) The water supply for a town may be stored in a reservoir, as shown, or obtained from groundwater. (b) The water is treated before use so that it is safe to drink. During coagulation chemicals are added that clump with suspended particles. Filtration is the use of screens or other devices to remove the clumps of suspended materials produced by coagulation. Chlorination destroys microorganisms and improves the odor and taste of water. (c) After use, the wastewater is collected by municipal sewer lines. (d) The quality of the wastewater is fully or partially restored by sewage treatment (primary, secondary, and sometimes tertiary treatment) before the treated effluent is dispersed into a nearby body of water.

terrible cholera epidemic swept much of Peru, infecting more than 300,000 people and killing at least 3,500. This outbreak occurred when Peruvian officials decided to stop chlorinating much of the country's drinking water after they learned of the slightly increased cancer risk due to chlorination. Peru has since resumed chlorinating its drinking water.

After a detailed review of current scientific evidence linking chlorine to cancer, the EPA proposed in 1994 that water treatment facilities reduce the maximum permissible level of chlorine in drinking water. One alternative to chlorination is to use chloramine, a disinfectant produced by combining chlorine with ammonia; chloramine does not form potentially harmful byproducts. Another alternative is to filter water through activated carbon granules, as is done in Cincinnati; one-third less chlorine then needs to be used in the final step. Another alternative, the use of UV disinfection, has been widely adopted in Europe as an alternative to chlorination.

Fluoridation   The addition of small amounts of fluoride to most municipal drinking water has been practiced since the mid-1940s to help prevent tooth decay. Fluoride has also been added to many toothpastes since that time, for the same reason.

This practice has been controversial, with opponents questioning the safety and effectiveness of fluoride and supporters saying it is completely safe and very effective in preventing decay. More than 40 years of research have failed to link fluoridation to cancer, kidney disease, birth defects, or any other serious medical condition. Most dental health officials think fluoride is the main reason

for the 50% to 60% decrease in tooth decay observed in children during the past several decades. This observation is based on comparisons of cavity rates in schoolchildren between cities with fluoridation and without.

As of 2002, 66% of U.S. public water supplies were fluoridated. Currently, fluoridation is more common in the eastern half of the country than in the western half, although California mandated fluoridation in 1995.

Municipal Sewage Treatment

Wastewater, including sewage, usually undergoes several treatments at a sewage treatment plant to prevent environmental and public health problems. The treated wastewater is then discharged into rivers, lakes, or the ocean.

Primary treatment removes suspended and floating particles, such as sand and silt, by mechanical processes such as screening and gravitational settling (Figure 21.11, left side). The solid material that settles out at this stage is known as **primary sludge**. Primary treatment, however, does little to eliminate the inorganic and organic compounds that remain suspended in the wastewater. The wastewater treatment facilities for about 10.8% of the U.S. population have primary treatment only.

Secondary treatment uses microorganisms to decompose the suspended organic material in wastewater (Figure 21.11, right side). One of the several types of sec-

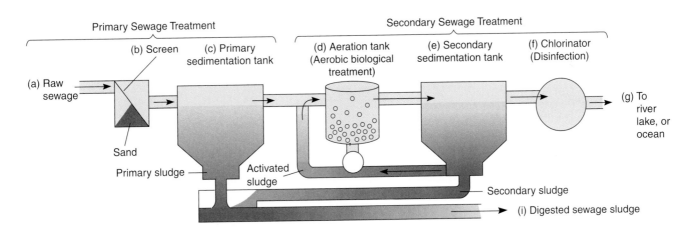

Figure 21.11 Primary and secondary sewage treatment. (a) Raw sewage enters the water treatment plant via the municipal sewage system. (b) Large debris, such as wood, metals, and plastics are removed, and sand settles to the bottom. (c) In the primary sedimentation tank, suspended solids and floating particles sink to the bottom. (d) Aeration tanks mix the partially treated wastewater with air (oxygen) to support bacteria that consume the suspended organic wastes. (e) The cleanest waste is taken from the surface and placed in a secondary sedimentation tank, where any remaining suspended particles settle to the bottom. (f) The cleanest water is taken from the surface and disinfected by chlorination or ultraviolet light to kill any disease-causing bacteria that may be present. (g) The treated water is discharged to a river or other natural water source. (h) Sludge is drawn from the bottom of the primary and secondary sedimentation tanks and pumped to a digester, where bacteria consume the organic wastes. (The activated sludge process is represented.) (i) The digested sewage sludge is disposed of in a sanitary landfill, incinerated, or converted into fertilizer.

ondary treatment is *trickling filters*, in which wastewater trickles through aerated rock beds that contain bacteria and other microorganisms, which degrade the organic material in the water. In another type of secondary treatment known as the *activated sludge process*, wastewater is aerated and circulated through bacteria-rich particles; the bacteria degrade suspended organic material. After several hours, the particles and microorganisms are allowed to settle out, forming **secondary sludge**, a slimy mixture of bacteria-laden solids. Water that has undergone primary and secondary treatment is clear and free of organic wastes such as sewage. The wastewater treatment facilities for about 62% of the U.S. population have both primary and secondary treatments.

Even after primary and secondary treatments, wastewater still contains pollutants, such as dissolved minerals, heavy metals, viruses, and organic compounds (Figure 21.12). Advanced wastewater treatment methods, also known as **tertiary treatment**, include a variety of biological, chemical, and physical processes. Tertiary treatment must be employed to remove phosphorus and nitrogen, the nutrients most commonly associated with enrichment. Tertiary treatment can also be used to purify wastewater so that it can be reused in communities where water is scarce. The wastewater treatment facilities for about 26.7% of the U.S. population have primary, secondary, and tertiary treatments.

Disposal of Sludge   A major problem associated with wastewater treatment is disposal of the primary and secondary sludge that is formed during primary and secondary treatments. Five possible ways to handle sludge are anaerobic digestion, application to soil as a fertilizer, incin-

eration, ocean dumping, and disposal in a sanitary landfill. In anaerobic digestion, the sludge is placed in large circular digesters and kept warm (about 35°C, or 95°F), which allows anaerobic bacteria to break down the organic material into gases such as methane and CO_2. The methane can be trapped and burned to heat the digesters.

After a few weeks of digestion, the sludge resembles humus and can be used as a fertilizer. It has the advantage of being rich in plant nutrients, although sometimes it contains too many heavy metals from industrial effluents to be used commercially. This happens when sewer systems mix industrial waste, which may contain toxic substances, with household waste. Farmers have long used sludge to fertilize hay and feed-grain crops. However, many farmers have been reluctant to use it on crops for direct human consumption because consumers might not purchase the food grown in sludge out of concern that it may pose a threat to human health. In 1996, however, the National Research Council, a society of distinguished scholars organized by the National Academy of Sciences, announced that properly treated sludge could be used safely to fertilize food crops.

Although sludge can be used to condition soil, it is generally treated as a solid waste. Dried sludge is often incinerated, which may contribute to air pollution, although sometimes the heat produced by this process is used constructively. Coastal cities such as New York traditionally dumped their sludge into the ocean. However, in 1988 the U.S. Congress passed the **Ocean Dumping Ban Act**, which barred ocean dumping of sludge and industrial waste, beginning in 1991. Alternatively, sludge can be disposed of in sanitary landfills (see Chapter 23). As landfill space becomes more costly, many cities are looking for

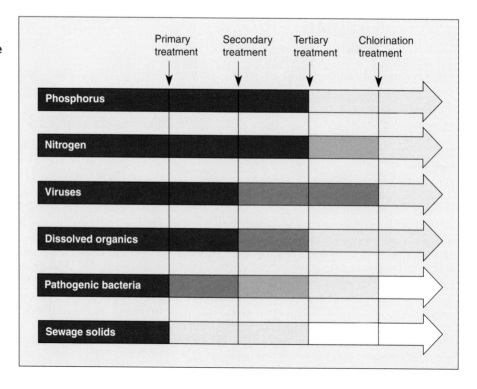

Figure 21.12 Effectiveness of primary, secondary, and tertiary sewage treatments. This figure shows the relative concentrations of various water pollutants after each treatment. The most intense color represents the greatest concentration of a given pollutant, whereas white means that all of that specific pollutant has been removed. Note how ineffective secondary treatment is in removing certain contaminants, such as phosphorus and nitrogen. Also note that even after tertiary treatment, some pollutants are still detectable, although at greatly reduced levels.

other ways to handle sludge. Houston plans to use a special process to reduce its sludge to sterile ash with approximately 5% the original volume, thus cutting the cost of disposal. Meanwhile, Texas is testing sludge ash as a paving material.

Individual Septic Systems

Many private residences, particularly in rural areas, use individual septic systems for sewage disposal instead of using municipal sewage treatment. Household sewage is piped to the septic tank, where particles settle to the bottom (Figure 21.13a). Grease and oils form a scummy layer at the top, where bacteria decompose much of it. Wastewater containing suspended organic and inorganic material then flows into the drain field through a network of small, perforated pipes set in trenches of gravel or crushed stone (Figure 21.13b). The drain field is located just below the soil's surface, and bacteria decompose the remaining organic material in the well-aerated soil. The purified wastewater then percolates into the groundwater or evaporates from the soil.

Septic tank systems require care in order to operate properly. Household chemicals such as bleach and drain cleaners should be used sparingly because they could kill the bacteria that break down the organic wastes. A kitchen garbage disposal should not be used because it could overload the system. Every 2 to 5 years, depending on use, the sludge that collects at the bottom of the septic

Something Fishy Near Sewage Treatment Plants

Endocrine disrupters are chemicals that mimic natural hormones and interfere with normal development processes. Several studies of aquatic animals have revealed sexual abnormalities and reproductive disorders that may be caused by environmental chemicals' mimicking natural hormones in the animals' bodies. Recent work in Britain suggested a link between hormonal defects in fishes and normal concentrations of treated sewage effluent in rivers near sewage treatment plants. Male fishes were feminized, born half-male and half-female. The results indicate that hormonal damage can occur in apparently healthy habitats that meet criteria established by environmental laws. Long-term endocrine disruption could dramatically alter ecosystems, as sterilization of males might eventually lead to the elimination of entire fish populations. The culprit could be any of hundreds of chemicals, though a likely suspect is *nonylphenol*, commonly found in cleaning products, detergents, and industrial solvents. Other studies in the United States as well as Britain linked the presence of natural human hormones in treated sewage effluent—probably from women's urine—to apparent endocrine disruption in fishes. Higher incidences of feminization in Britain may reflect its greater human population densities.

tank is removed and taken to a municipal sewage treatment plant for proper disposal. If a septic system is not maintained properly, it can malfunction or overflow, releasing bacteria and nutrients into groundwater or waterways.

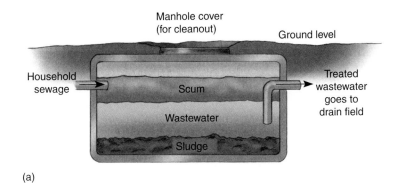

(a)

Figure 21.13 Septic tank system. Many private homes in rural areas use a septic tank and drainage field. Both are located underground. (a) The septic tank works much like primary treatment in municipal sewage treatment. Sewage from the house is piped to the septic tank, where particles settle to the bottom. (b) Wastewater containing suspended organic and inorganic material flows into the drain field and gradually seeps into the soil.

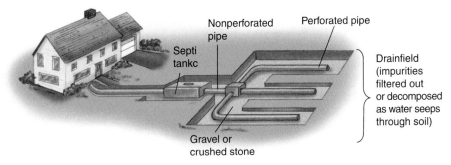

(b)

MEETING THE CHALLENGE

Using Citizen Watchdogs to Monitor Water Pollution

Insufficient staff and lack of funds prevent well-intentioned government agencies from effectively monitoring and enforcing environmental laws such as the Safe Drinking Water Act. For example, the San Francisco Bay Conservation and Development Commission is California's federally designated management agency charged with patrolling 1,600 km (1,000 mi) of shoreline and 1,500 km^2 (600 mi^2) of water to ascertain that no one is illegally polluting San Francisco Bay. In addition, it handles hundreds of cases arising from its monitoring activities. The agency is severely understaffed and unable to adequately protect the bay.

San Francisco Bay, like most other aquatic ecosystems, is endangered by the combined impact of many different pollution sources rather than from a single disaster, such as occurred when the *Exxon Valdez* spilled oil in Alaska (see Chapter 10). Thus, continual monitoring is necessary to ensure that many small polluters do not collectively do irreparable harm to the bay.

A growing number of private citizens have become actively involved in monitoring and enforcing environmental laws to protect waterways in their communities. Provisions in the Clean Water Act, the Safe Drinking Water Act, and other key environmental laws allow citizens to file suit when the government does not enforce the laws. Citizen action groups also pressure firms to clean up.

In San Francisco Bay, some 300 citizen watchdogs called Bay Keepers monitor the bay from boats, airplanes, and helicopters. Law students at local universities advise the keepers on issues of litigation. The San Francisco Bay Keeper program is modeled after the Hudson River Keeper Program, which first organized in 1966 as a coalition of commercial and recreational fisherman who wanted to reclaim the Hudson River from its polluters. In 1983, using money obtained from successful lawsuits against polluters, the Hudson River Keeper program hired its first full-time River Keeper. A network of community fishermen and environmentalists inform the River Keeper of any suspicious activity on the river. The River Keeper monitors water quality in a boat, attends board meetings, educates the public, and employs litigation as a last resort. The River Keeper position is described as part investigator, scientist, lawyer, lobbyist, and public relations agent.

At least 38 River, Sound, Bay, Inlet, Channel, and Coast Keeper groups have organized across the United States, from the Cook Inlet Keeper in Alaska to the Pensacola Gulf Coast Keeper in Florida. The umbrella organization for keeper groups is the Water Keeper Alliance, which organized in 1992. Its philosophy is based on the idea that daily vigilance by citizens is required to protect a community's natural resources. The Water Keeper Alliance helps new keeper programs organize, both in the United States and other countries. Keeper programs are being established in Belize, the Czech Republic, the Dominican Republic, Mexico, the Philippines, and Poland.

LAWS CONTROLLING WATER POLLUTION

Many governments have passed legislation aimed at controlling water pollution. Point source pollutants lend themselves to effective control more readily than nonpoint source pollutants. Governments generally control point source pollution in one of two ways, either by imposing penalties on polluters (a common approach in the United States) or by taxing polluters to pay for the cleanup (common in Japan).

Although most countries have passed laws to control water pollution, monitoring and enforcement are difficult, even in highly developed countries. Typically, too few resources are allotted for enforcement. For example, in 1996 the 70 enforcement agents in the Office of Drinking Water of the EPA were expected to handle more than 80,000 complaints about drinking water safety. (See "Meeting the Challenge: Using Citizen Watchdogs to Monitor Water Pollution.")

The United States has attempted to control water pollution through legislation since the passage of the **Refuse Act** of 1899, which was intended to reduce the release of pollutants into navigable rivers. The two federal laws that have the most impact on water quality today are the Safe Drinking Water Act and the Clean Water Act.

Safe Drinking Water Act

Prior to 1974, individual states set their own standards for drinking water, which of course varied a great deal from state to state. In 1974 the **Safe Drinking Water Act** was passed, which set uniform federal standards for drinking water in order to guarantee safe public water supplies throughout the United States. This law required the EPA to determine the **maximum contaminant level**, which is the maximum permissible amount of any water pollutant that might adversely affect human health. The EPA oversees the states to ensure that they adhere to the maximum contaminant levels for specific water pollutants. The EPA noted in 1998 that 40,000 water systems reported violations of public safe drinking water laws for that year; this number, while large, represents less than 25% of U.S. water systems. Of these, 9,600 water systems (about 6%) had significant violations.

Most water suppliers take few or no steps to prevent the contamination of the watershed or groundwater that they draw from. The vast majority of water utilities do not use modern water-treatment technologies such as activated carbon granules or UV disinfection to reduce chemical contamination by pesticides, arsenic, and chlorine disinfection byproducts. Also, the average water pipe in the United States is 100 or more years old before it is replaced. Many aging pipes are cracked, which permits contaminated water to seep into them and increases the risk of waterborne diseases.

The Safe Drinking Water Act was amended in 1986 and in 1996. The 1996 version requires municipal water suppliers to tell consumers what contaminants are present in their city's water and if these contaminants pose a health risk. The law also requires the EPA to review risks posed by radon and arsenic in drinking water and to revise its drinking water standards for each contaminant accordingly.

Clean Water Act

The quality of rivers, lakes, aquifers, estuaries, and coastal waters in the United States is most affected by the **Clean Water Act**. Originally passed as the Water Pollution Control Act of 1972, it was amended and renamed the Clean Water Act of 1977; additional amendments were made in 1981 and 1987. Congress should reauthorize the Clean Water Act sometime in the early 2000s. The Clean Water Act has two basic goals: to eliminate the discharge of pollutants in U.S. waterways and to attain water quality levels that make these waterways safe to fish and swim in. Under the provisions of this act, the EPA is required to set up and monitor **national emission limitations**, which are the maximum permissible amounts of water pollutants that can be discharged from sewage treatment plants, factories, and other point sources.

Overall, the Clean Water Act has been effective at improving the quality of water from point sources, despite the relatively low fines it imposes on polluters. It is not hard to identify point sources, which must obtain permits from the **National Pollutant Discharge Elimination System** to discharge untreated wastewater.

According to the EPA, nonpoint source pollution is a major cause of water pollution. However, nonpoint source pollution is much more difficult and expensive to control than point source pollution. The 1987 amendments to the Clean Water Act expanded the National Pollutant Discharge System to include nonpoint sources, such as sediment erosion from construction sites.

To date, U.S. environmental policies have failed to effectively address nonpoint source pollution, which could be reduced by regulating land use, agricultural practices, and many other activities. The problem is that such regulation would require the interaction and cooperation of many government agencies, environmental organizations, and private citizens. Such coordination is enormously challenging but necessary if nonpoint source pollution is to be reduced.

The United States has improved its water quality in the past several decades, thereby demonstrating that the environment can recover rapidly once pollutants are eliminated. But much remains to be done. The EPA's 2002 National Water Quality Inventory indicated that water pollution has increased in U.S. rivers, lakes, estuaries, and coastal areas in recent years. According to the report, which was based on 2000 data submitted by the states, 39% of the nation's rivers, 45% of its lakes, and 51% of its estuaries were too polluted for swimming, fishing, or drinking.

Laws That Protect Groundwater

Several federal laws attempt to control groundwater pollution. The Safe Drinking Water Act contains provisions to protect underground aquifers that are important sources of drinking water. In addition, underground injection of wastes is regulated by the Safe Drinking Water Act in an effort to prevent groundwater contamination. The **Resource, Conservation, and Recovery Act** deals with the storage and disposal of hazardous wastes and helps prevent groundwater contamination (see Chapter 23). Several miscellaneous laws related to pesticides, strip mining, and cleanup of abandoned hazardous waste sites also indirectly protect groundwater.

The many laws that directly or indirectly affect groundwater quality were passed at different times and for different reasons. These laws provide a disjointed and, at times, inconsistent protection of groundwater. The EPA makes an effort to coordinate all these laws, but groundwater contamination still occurs.

CASE-IN-POINT ## Water Pollution in the Great Lakes

The five Great Lakes of North America—Lakes Superior, Michigan, Huron, Erie, and Ontario—formed about 10,500 years ago when the melting waters of retreating glaciers drained into the lake basins that were carved from river valleys by the glaciers. The Great Lakes, which are connected to one another, collectively hold about one fifth of the world's fresh surface water (Figure 21.14). The combined area of the Great Lakes is 244,000 km^2 (94,200 mi^2).

Industrial wastes, sewage, fertilizers, and other pollutants have contaminated the Great Lakes since the mid-1800s. More than 33 million people live in the Great Lakes watershed, an area that is home to agriculture, trade, industry, and tourism. Important industrial cities, such as Duluth, Milwaukee, Chicago, Cleveland, Erie, Buffalo, and Toronto, are located along the lake shorelines. At least 38 million people obtain their drinking water from the Great Lakes.

During the 1960s, pollution in the Great Lakes became a highly visible problem, particularly in Lake

Figure 21.14 Great Lakes and their drainage basin. Although many small streams empty onto the Great Lakes, they drain a relatively small area. Lake Superior is the largest body of fresh water in the world; it is also the least polluted of the Great Lakes. Lake Michigan is the only Great Lake located entirely within the United States; all others are located in both the United States and Canada. Lake Huron is the second least polluted of the Great Lakes. Water in the Great Lakes flows eastward and eventually drains into the Atlantic Ocean via the St. Lawrence River. The two easternmost lakes, Lake Erie and Lake Ontario, therefore contain their own pollution plus pollution from the other three lakes.

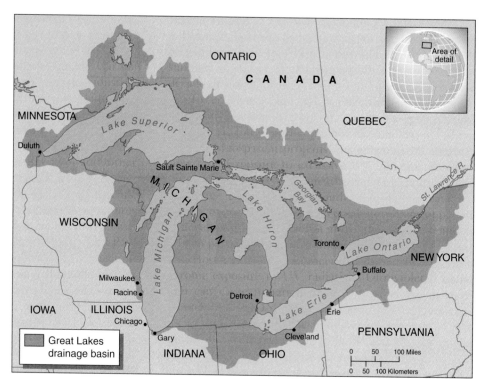

Erie, Lake Ontario, and shallow parts of Lake Huron and Lake Michigan. (Lake Erie is the most polluted of the Great Lakes because it has the largest human population on its shores. It is also the shallowest of the Great Lakes and therefore contains the smallest volume of water.) Thousands of toxic chemicals polluted the lakes. Eutrophication was pronounced, bacterial counts were high enough to be a health hazard, and fish kills were common. Birth defects, such as missing brains, internal organs located outside the body, and deformed feet and wings, were observed in almost 50% of the animal species studied.

Canada and the United States have cooperated to improve the condition of the Great Lakes. In 1972, a joint pollution control program was enacted. Since then, $20 billion has been spent to clean up the lakes. The **Great Lakes Toxic Substance Control Agreement** was signed by the eight Great Lakes states in 1986 and by the two Great Lakes Canadian provinces in 1988. This legislation is designed to reduce pollution in the lakes by developing coordinated programs among the eight states and two provinces.

As of the early 2000s, the Great Lakes are in better shape than they have been in recent memory. The U.S. EPA and Environment Canada report that water quality and human health are slowly improving in the Great Lakes region. Levels of dichlorodiphenyltrichloroethane (DDT) in women's breast milk have declined since 1967, as have levels of PCBs in trout, Coho salmon, and herring gulls. Some animal populations,

such as double-crested cormorants and bald eagles, have rebounded.

Many problems remain, however. Forty-three shoreline areas are so polluted that they are designated "areas of concern" by the EPA and Environment Canada. Zebra mussels, sea lampreys, and at least 160 other invasive aquatic species have proliferated and threaten native species (recall the discussion of invasive species in Chapter 1). Shoreline development continues to encroach on natural areas and contributes to flooding and shoreline erosion during storms. Pesticides and fertilizers from suburban lawns pollute the water.

Persistent toxic compounds remain in the lakes, many of them coming from air pollution, such as the spraying of insecticides and the incineration of wastes, or from contaminated lake sediments. Mercury contamination is on the increase. Levels of PCBs in many parts of Lakes Ontario, Erie, and Michigan are still 100 times higher than the level considered safe for human health. The water itself is safe to drink, but eating fish with concentrated levels of PCBs damages human health. The presence of certain toxic chemicals is causing hormonal changes that are linked to reproductive failures, abnormal development, and abnormal behaviors in some fishes, birds, and mammals (recall the discussion of endocrine disrupters in Chapter 1). Because persistent toxic chemicals have accumulated in food webs (Table 21.4), fish consumption advisories are issued warning people who eat Great Lakes fishes of potential health problems.

Table 21.4	Major Contaminants That Persist in the Great Lakes

Chemicals Usually Found*

Mercury (heavy metal)
Dieldrin (pesticide)
PCBs (polychlorinated biphenyls—industrial chemicals)
Chlordane (pesticide)
DDT (dichlorodiphenyltrichloroethane—pesticide)
Dioxins (chemical contaminants)
Furans (chemical contaminants)

* Based on contaminant monitoring of fish tissues at multiple sites in the Great Lakes and tributary river mouths.

WATER POLLUTION IN OTHER COUNTRIES

In the late 1970s, only about one third of the world's people had access to safe drinking water, and most of these lived in highly developed countries. Since then, tremendous progress has been made in providing clean water. Yet, according to the World Health Organization, an estimated 1.4 billion people still do not have access to safe drinking water, and about 2.9 billion people do not have access to adequate sanitation systems; most of these people live in rural areas of developing countries. Worldwide, at least 250 million cases of water-related illnesses occur each year, with 5 million or more of these resulting in death.

Municipal water pollution from sewage is a greater problem in developing countries, many of which lack water treatment facilities, than in highly developed nations. Sewage from many densely populated cities in Asia, Latin America, and Africa is dumped directly into rivers or coastal harbors.

Almost every nation in the world faces problems of water pollution. For an international perspective on water pollution, we now examine some specific issues in South America, Europe, Asia, and Africa.

Lake Maracaibo, Venezuela

Lake Maracaibo in Venezuela is the largest lake in South America (Figure 21.15). It receives fresh water from several rivers, and water flows from it into the Caribbean Sea. Larger than the state of Connecticut, Lake Maracaibo exemplifies most lakes and inland seas around the world. Lake Maracaibo is suffering from the effects of oil pollution and human wastes as well as contamination by farms and factories. About 10,000 oil wells have been drilled to tap the oil and natural gas reserves under the shallow lake. An underwater network of old oil pipes about 15,400 km (9,600 mi) long leaks oil into the lake. Fertilizers and other agricultural chemicals drain into the lake from nearby farms, providing the nutrients for an overgrowth of algae. Until recently, raw sewage from the city of Maracaibo's 1.4 million people and many smaller communities was discharged directly into the water, contributing to the nutrient overload. Modern sewage-treatment facilities were installed during the 1990s to take care of human wastes, but Lake Maracaibo still needs to have its other pollution problems addressed.

Po River, Italy

The Po River, which flows across northern Italy, empties into the Adriatic Sea. The Po is Italy's equivalent of

Figure 21.15 Lake Maracaibo, Venezuela. Drilling platforms are located throughout the lake, and pipelines join the oil wells to refineries on land. Shown are storage and export facilities in the foreground and the town of Cabimas in the background. An oil tanker is docked at the leftmost pier.

the Mississippi River, and it is heavily polluted. Many cities, including Milan with 1.3 million residents, dump their treated and untreated sewage into the Po. Industry is responsible for half the pollutants that enter the Po. Italian agriculture, including large poplar plantations, relies heavily on chemicals and is responsible for massive amounts of nonpoint source pollution. Soil erosion has resulted in so much sediment deposition at the mouth of the river that the Po River delta is advancing into the Adriatic Sea by about 81 hectares (200 acres) each year.

More than 16 million people—almost one third of all Italians—live in the Po River basin. The health of many Italians is threatened because the Po is the source of their drinking water. In addition, pollution from the Po has jeopardized tourism and fishing in the Adriatic Sea. Pollution has closed swimming at some beaches, for example. Although Italy recognizes the problems of the Po and would like to do something about them, the improvement of water quality will be very difficult to implement because the river is under the jurisdiction of dozens of local and regional governments. The cleanup of the Po will require the implementation of a national plan over a period of several decades.

Ganges River, India

The Ganges River is a holy river that symbolizes the spirituality and culture of the Indian people. The river is also widely used for bathing and washing clothes (Figure 21.16). It is highly polluted. Little of the sewage and industrial waste produced by the 350 million people who live in the Ganges River basin is treated. Another major source of contamination is the 35,000 human bodies per year that are cremated in the open air in Varanasi, the holy city of the Hindus. (The Hindus cremate the body to free the soul; dumping the ashes into the Ganges increases the chances of the soul getting into heaven.) Bodies that are incompletely burned are also dumped into the Ganges River, where their decomposition adds to the BOD of the river. In addition, people who cannot afford cremation costs for their dead dump human remains into the river.

The Indian government has initiated the Ganga Action Plan, an ambitious cleanup project that includes construction of water treatment plants in 29 large cities in the river basin. In addition, 32 electric crematoriums are being set up along the banks. However, although about $100 million has been spent constructing sewage treatment plants in major cities along the Ganges, most are not completed or are not working effectively. Raw sewage is still being discharged into the river by most of the 29 cities in the river basin. Costs have escalated, and many delays in the Ganga Action Plan have occurred. According to critics, one of the reasons the government plan has not worked is that it has not tried to get people involved at the community level. People along the

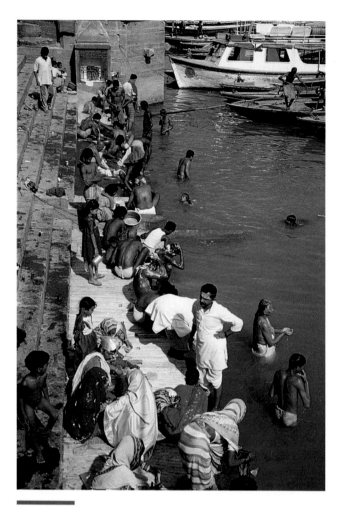

Figure 21.16 Ganges River. Bathing and washing clothes in the Ganges River are common practices in India. The river is contaminated by raw sewage discharged directly into the river at many different locations.

Ganges have not been informed about what the government is trying to do, why it is important, and how they can help.

Kwale, Kenya

For many Africans, serious health problems are caused by surface water supplies that are contaminated with disease-causing organisms. A hand pump project sponsored by the World Bank and the U.N. Development Program was meant to address the water safety problem for many Kenyans. For example, Kwale, Kenya, had been the site of cholera and diarrhea outbreaks in the past. With the installation of village wells with hand pumps, it was hoped that groundwater that is clean and healthful would be available to the inhabitants of Kwale. Unfortunately, an acute drought lowered water levels in the early 2000s, drying up many of the wells. Also, pit latrines contaminated the water in many of the wells, leading to an outbreak of cholera in 2002.

Arsenic Poisoning in Bangladesh

The same kind of water development program that was implemented in Kenya was put into practice in Bangladesh. During the 1980s, world health organizations, concerned about illnesses caused by drinking contaminated surface water in Bangladesh, funded the installation of more than 2.5 million wells with hand pumps. Tragically, the groundwater in many of these wells is contaminated with high levels of naturally occurring arsenic. Tens of thousands of people now suffer from chronic arsenic poisoning, which initially causes skin lesions, particularly on the hands and feet. Eventually, when the cumulative dose is large enough, arsenic poisoning leads to death from cancer. World health authorities do not know how many people are at risk. Estimates range from hundreds of thousands to as many as 70 million.

International agencies are currently providing funding to test all the wells. Meanwhile, the Bangladeshi people continue to drink the groundwater because there is no other option for them. Scientists at Stevens Institute of Technology have developed an inexpensive bucket filtration system to remove arsenic from well water. A pilot program is underway in Bangladesh to test the system. Safe disposal of the arsenic-rich sediments produced by the filtration system would have to be arranged should this procedure be widely adopted.

SOIL POLLUTION

Soil pollution is any physical or chemical change in soil that adversely affects the health of plants and other organisms living in and on it. Soil pollution is important not only in its own right but because so many soil pollutants tend to move into surface water, groundwater, and air. For example, selenium, an extremely toxic natural element that is found in many western soils, leaches off irrigated farmlands and poisons nearby lakes, ponds, and rivers, causing death and deformity in thousands of migratory birds and other organisms annually.

Except for salts, petroleum products, and heavy metals, most soil pollutants originate as agricultural chemicals, including fertilizers and pesticides. Many of these agricultural chemicals persist in the soil by seeping into tiny cracks, called micropores, and adhering to soil particles. It appears that the soil may act like a reservoir that stores contaminants and continuously releases them to surface water and groundwater, as well as topsoil, over a long period.

Some heavy metals have been slowly accumulating on farmland soils. One of the most toxic is cadmium, which is present in trace amounts in certain fertilizers. A buildup of cadmium in soil would probably not be occurring if we were not applying so much fertilizer to increase food production. Fortunately, there is a growing interest in sustainable farming practices that reduce the need for large chemical applications to the soil (see Chapter 18). **Sustainable agriculture** may help solve the dual problems of producing enough food and preventing environmental contamination.

Irrigation and Salinization of the Soil

Soils found in arid and semiarid regions often contain high natural concentrations of inorganic compounds as mineral salts. In these areas, the amount of water that drains into lower soil layers is minimal because the little precipitation that falls quickly evaporates, leaving behind the salt. In contrast, humid climates have enough precipitation to leach salts out of soils and into waterways and groundwater.

Irrigation of agricultural fields often results in their becoming increasingly saline (salty), an occurrence known as **salinization** (see Figure 13.10). Irrigation water contains small amounts of dissolved salts (recall from Chapter 13 that fresh water contains some salt). The continued application of such water, season after season, year after year, leads to the gradual accumulation of salt in the soil. When the water evaporates, the salts are left behind, particularly in the upper layers of the soil, which are the layers that are most important for agriculture. Given enough time, the salt concentration can rise to such a high level that plants are poisoned or their roots dehydrated. Also, when irrigated soil becomes waterlogged, capillary movement may carry salts from groundwater to the soil surface, where they are deposited as a crust of salt.

How Soil Salinization Affects Plants Most plants cannot obtain all the water they need from saline soil. The cause is a water balance problem that exists because water always moves from an area of higher concentration (of water) to an area of lower concentration. (The concentration of water is determined by the amount of dissolved materials in it. For example, a solution containing 10% dissolved salt has 90% water.) Under normal conditions, the dissolved materials in plant cells give them a lower concentration of water than that in soil. As a result, water moves from the soil into plant roots. When soil water contains a large quantity of dissolved salts, however, its concentration of water may be lower than that in the plant cells; if so, water consequently moves *out* of the plant roots and into the saline soil (Figure 21.17).

Obviously, most plants cannot survive under these conditions. Plant species that thrive in saline soils have special adaptations that enable them to tolerate the high amount of salt. Most crops, unless they have been genetically selected to tolerate high salt, are not productive in saline soil.

Soil Remediation

Until relatively recently, the only sure way to remove contaminants from soil was to excavate (dig up) the soil

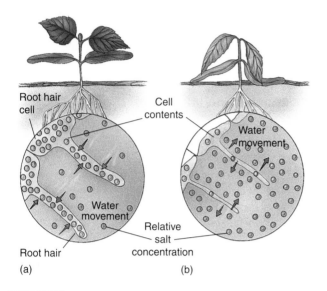

Root hair
cell

Cell
contents

Water
movement

Water
movement

Relative
salt
concentration

Root hair

(a)   (b)

Figure 21.17 Effect of salinized soil on plants. (a) Normally, the water concentration inside plant cells is lower than that in the soil, resulting in a net movement of water (*blue arrows*) into root cells. (b) When soil contains a high amount of salt, its relative water concentration can be lower than the water concentration inside cells. This causes water to move out of roots into the soil, even when the soil is wet.

from an entire field and incinerate it. This solution, besides being impractical, killed all beneficial soil organisms. As a result of research in **soil remediation**, other techniques are available to clean up contaminated soil. Four of these techniques are dilution, vapor extraction, bioremediation, and phytoremediation.

Dilution involves running large quantities of water through contaminated soil in order to leach out pollutants such as excess salt. Although this sounds like a straightforward process, it is extremely difficult, and in many cases impossible, to accomplish. Many soils do not have good drainage properties, so adding lots of water simply causes them to become waterlogged. Dilution also results in polluted water, which poses a disposal problem and is expensive when large volumes of soil have been contaminated. Moreover, it is not feasible in arid and semiarid regions because water is in short supply.

In **vapor extraction**, air is injected into or pumped through soil to remove organic compounds that are volatile (evaporate quickly). Vapor extraction, which is usually done for 6 months to a year, does not require soil excavation and has been used successfully thousands of times. The oxygen in the injected or pumped air also helps to clean up the soil by stimulating certain microorganisms to degrade nonvolatile organic compounds.

Soil remediation specialists are increasingly using microorganisms to degrade organic contaminants such as oil and sludge. The use of bacteria and other microorganisms to clean up soil and water pollution, known as **bioremediation**, is discussed further in Chapter 23.

Plants can clean up polluted soils, particularly those contaminated by salts or heavy metals such as cobalt and zinc. The roots of certain plants absorb these contaminants, which subsequently accumulate in the stems and leaves (see Table 23–2). The use of plants to clean up polluted soil, known as **phytoremediation**, is a relatively inexpensive soil remediation technique, although it requires several to many years before the soil is clean.

SUMMARY WITH SELECTED KEY TERMS

I. Water pollution consists of any physical or chemical change in water that adversely affects the health of humans and other organisms. Wastewater treatment is an important part of **sustainable water use**.

A. Sewage, wastewater carried off by drains or sewers, contributes to **enrichment** (fertilization of water) and produces an oxygen demand as it is decomposed.

  1. Microorganisms use the process of **cellular respiration** to break down sewage into carbon dioxide, water, and similar materials.

  2. Biochemical oxygen demand (BOD) is the amount of oxygen needed by microorganisms to decompose sewage and other organic wastes.

B. Disease-causing agents, such as bacteria, viruses, protozoa, and parasitic worms, are transmitted in sewage.

  1. The common intestinal bacterium *Escherichia coli* is used as an indication of the amount of sewage present in water and as an indirect measure of disease-causing organisms.

  2. The **fecal coliform test** determines the presence of *E. coli* in water.

C. Sediment pollution, which is caused primarily by soil erosion, increases water turbidity, thereby reducing photosynthetic productivity in the water.

D. Inorganic plant and algal nutrients, such as nitrogen and phosphorus, contribute to enrichment.

  1. Fertilizer runoff from agricultural and residential land is a major contributor of inorganic plant and algal nutrients.

  2. Fertilizer runoff from Midwestern fields is carried by the Mississippi River to the Gulf of Mexico, where it causes an oxygen-free condition known as **hypoxia** over a large area.

E. Many **organic compounds**, which contain carbon atoms, are synthetic and do not decompose readily. Some of these, such as pesticides, solvents, and industrial chemicals, are quite toxic to organisms.

F. Inorganic chemicals, contaminants that contain elements other than carbon, include toxins such as lead and mercury.

   1. Small amounts of lead occur naturally in the environment, but most lead contamination can be traced to human activities. Children with low levels of lead in their blood may suffer from partial hearing loss, hyperactivity, attention deficit, lowered IQ, and learning disabilities.

   2. Mercury is used in many industrial processes and is also released during the combustion of coal. Once in a body of water, mercury is converted to methyl mercury, which readily enters the food web and accumulates in fishes. Methyl mercury compounds are highly toxic.

G. Radioactive substances include the wastes from mining, refining, and using radioactive metals. Radioactive substances may concentrate in sewage sludge.

H. Thermal pollution occurs when heated water, produced during many industrial processes, is released into waterways.

II. Eutrophication, the nutrient enrichment of **oligotrophic** lakes, estuaries, or slow-moving streams, results in high photosynthetic productivity, which supports an overpopulation of algae.

A. Eutrophic bodies of water tend to fill in rapidly as dead organisms settle to the bottom. Eutrophication also kills fishes and causes a decline in water quality as large numbers of algae die and decompose rapidly.

B. Artificial eutrophication, also called **cultural eutrophication**, is enrichment of an aquatic ecosystem that occurs at an accelerated rate because of human activities such as fertilizer runoff and sewage.

III. Water pollutants come from both natural sources and human activities.

A. Pollution that enters the water at specific sites, such as pipes from industrial or sewage treatment plants, is called **point source pollution**.

B. Nonpoint source pollution, also called **polluted runoff**, comes from the land rather than from a single point of entry.

C. Three major sources of human-induced water pollution are agriculture, municipalities (sewage and urban runoff), and industries.

IV. Pollutants that seep from sanitary landfills, underground storage tanks, and agricultural operations can contaminate groundwater.

A. Currently, most of the groundwater supplies in the United States are of good quality, although there are some local problems.

B. Cleanup of polluted groundwater is very costly, takes years, and in some cases is not technically feasible.

V. In the United States, most municipal water supplies are treated before being used so that the water is safe to drink.

A. Water is usually treated with aluminum sulfate to cause suspended particles to clump and settle out, filtered through sand, and disinfected by adding chlorine.

B. Because there is concern over whether low levels of chlorine in drinking water pose a health hazard, the Environmental Protection Agency (EPA) has proposed that water treatment facilities reduce the maximum permissible level of chlorine in drinking water.

C. Many municipalities' drinking water contains small amounts of added fluoride to help prevent tooth decay.

VI. Wastewater treatment may include **primary treatment** (the physical settling of solid matter), **secondary treatment** (the biological degradation of organic wastes), and **tertiary treatment** (the removal of special contaminants such as organic chemicals, nitrogen, and phosphorus).

A. The slimy mixture of bacteria-laden solids that settles out during sewage treatment is called **primary sludge** (formed during primary treatment) and **secondary sludge** (formed during secondary treatment).

B. One of the most pressing problems of wastewater treatment is disposal of the sludge that results from primary and secondary treatments.

VII. Industrial wastes, sewage, fertilizers, and other pollutants have contaminated the Great Lakes since the mid-1800s. During the 1960s, pollution in the Great Lakes became a highly visible problem.

A. Since 1972 Canada and the United States have worked together to reduce pollution in the Great Lakes. The **Great Lakes Toxic Substance Control Agreement** is designed to reduce pollution in the Great Lakes by developing coordinated programs among the states and provinces that border the lakes.

B. Many problems remain, such as invasive species and the presence of persistent toxic compounds.

VIII. Laws attempt to control water pollution. Monitoring and enforcement are difficult, however.

A. The **Safe Drinking Water Act** requires the EPA to establish **maximum contaminant levels** for water pollutants that might affect human health.

B. The quality of rivers, lakes, aquifers, estuaries, and coastal waters in the United States is most affected by the **Clean Water Act**, which requires the EPA to establish national emission limitations for wastewater that is discharged into U.S. surface waters.

C. Legislation has been more effective in controlling point source pollution than in controlling nonpoint source pollution.

D. The many laws that address groundwater pollution operate in isolation from one another and often at cross-purposes.

IX. Worldwide, about 1.4 billion people do not have access to safe drinking water. About 2.9 billion people do not have access to adequate sanitation systems.

X. Soil pollution is any physical or chemical change in soil that adversely affects the health of plants and other organisms living in and on it.

A. Soil pollution is important not only in its own right but because so many soil pollutants tend to move into surface water, groundwater, and air.

B. The chief soil pollutants are salts, petroleum products, heavy metals, and agricultural chemicals.

C. **Salinization**, a common problem in irrigated arid and semiarid regions, makes soil unfit for growing most crops. It

is extremely difficult to remove excess salts from salinized soils.

D. A variety of techniques are used for **soil remediation**, which is cleaning up contaminated soil.

THINKING ABOUT THE ENVIRONMENT

1. What is water pollution? Why is wastewater treatment an important part of sustainable water use?

2. What are the eight main groups of water pollutants? Give an example of each type.

3. Explain why untreated sewage may kill fishes when it is added directly to a body of water.

4. How do Midwestern farmers threaten the livelihood of fishermen in the Gulf of Mexico?

5. Distinguish between oligotrophic and eutrophic lakes. What causes artificial eutrophication?

6. Tell whether each of the following represents point source pollution or nonpoint source pollution: fertilizer runoff from farms, thermal pollution from a power plant, urban runoff, sewage from a ship, erosion sediments from deforestation.

7. Compare the potential pollution problems of groundwater and surface water that are used as sources of drinking water.

8. How is most drinking water purified in the United States?

9. Why is chlorine added to drinking water? Why does the Environmental Protection Agency recommend that public water treatment facilities find alternatives to chlorine?

10. Distinguish among *primary sewage treatment*, *secondary sewage treatment*, *tertiary sewage treatment*, and *septic tanks*.

11. What are the main goals of the Safe Drinking Water Act? The Clean Water Act?

12. Is the Clean Water Act related in any way to the quality of public drinking water in the United States? Explain your answer.

13. What is soil pollution?

14. The United States has a Clean Air Act and a Clean Water Act. Should we also have a Clean Soil Act? Present at least one argument for and one argument against such legislation.

15. Explain why saline soils are physiologically dry for plants even though they may be physically wet.

*16. The water pollution from a slaughterhouse in which one ton of cattle was slaughtered each day produced a daily biochemical oxygen demand (BOD) of 10.5 kg. Calculate the human population equivalent of this amount of BOD—that is, the number of people this amount of water pollution is equivalent to. (Each person produces an average of 0.07 kg of BOD daily.)

*17. An asphalt parking lot measures 100 m by 500 m. Calculate its area. If 2 cm of rain fall on the parking lot during a heavy rainstorm, how much water (in cubic meters) will run off? Where does the runoff go?

* Solutions to questions preceded by an asterisk appear in Appendix VII.

TAKE A STAND

Visit our Web site at **http://www.wiley.com/college/raven** (select Chapter 21 from the Table of Contents) for links to more information about the complexities of the water chlorination issue. Consider the opposing views of those who support and those who oppose the chlorination of water, and debate the issue with your classmates. You will find tools to help you organize your research, analyze the data, think critically about the issues, and construct a well-considered argument. Take a

Stand activities can be done individually or as part of a team, as oral presentations, written exercises, or Web-based (e-mail) assignments.

Additional on-line materials relating to this chapter, including Student Quizzes, Activity Links, Useful Web Sites, Flash Cards, and more, can also be found on our Web site.

SUGGESTED READING

Becker, E. "Farms Face Scrutiny For Water Pollution." *New York Times* (February 10, 2002). Animal wastes and fertilizers from farms across the United States can pollute the coastal waters.

Clarke, T. "Bangladeshis to Sue Over Arsenic Poisoning." *Nature*, Vol. 413 (October 11, 2001). The arsenic crisis in Bangladesh may result in lawsuits that discourage aid agencies from working in developing countries.

Harder, B. "A Confluence of Contaminants." *Science News*, Vol. 161 (March 23, 2002). Summarizes the U.S. Geological Survey study of organic pollutants in 139 streams in 30 states.

Malakoff, D. "Microbiologists on the Trail of Polluting Bacteria." *Science*, Vol. 295 (March 29, 2002). Examines the potential of bacterial source tracking in identifying the source of coliform contamination of waterways.

Mielke, H.W. "Lead in the Inner Cities." *American Scientist*, Vol. 87 (January–February 1999). Research shows that lead poisoning in inner-city children is caused more from lead contamination of the soil, the product of leaded gasoline used decades ago, than from lead-based paints.

Motavalli, J. "Heavy Metal Harm." *E Magazine* (May–June 2002). All about mercury pollution—where it comes from, what it can do, and how to deal with it.

Raloff, J. "Even Low Lead in Kids Has a High IQ Cost." *Science News*, Vol. 159 (May 5, 2001). Lead toxicity can permanently impair brain development in young children.

Sampat, P. "Groundwater Shock: The Polluting of the World's Major Freshwater Stores." *World Watch*, Vol. 13, No. 1 (January–February 2000). Contamination of major aquifers is increasing.

Sampat, P. "Uncovering Groundwater Pollution." *State of the World 2001* by L.R. Brown et al. New York: W. W. Norton & Company (2001). Groundwater pollution is a serious problem around the world.

Simpson, S. "Shrinking the Dead Zone." *Scientific American*, Vol. 285, No. 1 (July 2001). The annual hypoxia problem in the Gulf of Mexico needs to be addressed.

Zwingle, E. "Po: River of Pain and Plenty." *National Geographic*, Vol. 201, No. 5, May 2002. This article highlights the importance and challenges facing Italy's longest river.

Salad vac. This machine controls certain insect pests without the use of chemical pesticides. Photographed in Salinas, California.

The Pesticide Dilemma

Learning Objectives

After you have studied this chapter you should be able to

1. Define *pesticide*, distinguish among various types of pesticides such as insecticides and herbicides and describe the major groups of insecticides and herbicides.

2. Relate the benefits of pesticides.

3. Summarize the problems associated with pesticide use, including development of genetic resistance; creation of imbalances in the ecosystem; persistence, bioaccumulation, and biological magnification; and mobility in the environment.

4. Discuss pesticide risks to human health, including short-term effects, long-term effects, pesticides as endocrine disrupters, and risks to children.

5. Describe alternative ways to control pests, including cultivation methods, biological controls, reproductive controls, pheromones and hormones, genetic controls, quarantine, integrated pest management, and irradiating foods.

6. Briefly summarize the three U.S. laws that regulate pesticides: the Food, Drug, and Cosmetics Act; the Federal Insecticides, Fungicide, and Rodenticide Act; and the Food Quality Protection Act.

7. Describe the purpose of the Stockholm Convention on Persistent Organic Pollutants.

Picture a giant vacuum cleaner slowly moving over rows of strawberry or vegetable crops and sucking insects off the plants. Such a machine, given names like "salad vac" and "bug vac," was invented by an entomologist (a biologist who studies insects) as a substitute for the chemical poisons we call pesticides. Each vacuuming eliminates the need for one application of pesticide.

More than forty California growers use the farm-sized vacuum cleaner to remove and kill lygus bugs, leafhoppers, Colorado potato beetles, and other insect pests from their strawberries and other crops. Some beneficial insects are also removed and killed by the bug vac, but fewer than would be killed if chemical insecticides were sprayed. Increasing public concern about pesticide residues on food has caused farmers to look seriously at other ways to control pests—even by vacuuming insects.

In this chapter we examine the types and uses of pesticides, their benefits, and their disadvantages. Pesticides have saved millions of lives by killing insects that carry disease and by increasing the amount of food we grow. Modern agriculture depends on pesticides to produce blemish-free fruits and vegetables at a reasonable cost to farmers (and therefore to consumers).

However, pesticides also cause environmental and health problems, and it appears that in many cases their harmful effects outweigh their benefits. Pesticides rarely affect the pest species alone, and the balance of nature, such as predator–prey relationships, is upset. Certain pesticides concentrate at higher levels of the food chain.

Humans who apply and work with pesticides may be at risk for pesticide poisoning (short term) and cancer (long term), and people who eat traces of pesticide on food are concerned about the long-term effects. In this chapter we also consider some alternatives to pesticides and discuss the pesticide laws that are supposed to protect our health and the environment.

WHAT IS A PESTICIDE?

Any organism that interferes in some way with human welfare or activities is called a **pest**. Some weeds, insects, rodents, bacteria, fungi, nematodes (microscopic worms), and other pest organisms compete with humans for food; other pests cause or spread disease. The definition of *pest* is subjective; a mosquito may be a pest to you, but it is not a pest to the bat or bird that eats it. People try to control pests, usually by reducing the size of the pest population. Toxic chemicals called **pesticides** are the most common way of doing this, particularly in agriculture. Pesticides can be grouped by their target organisms—that is, by the pests they are supposed to eliminate. Thus, **insecticides** kill insects, **herbicides** kill plants, **fungicides** kill fungi, and **rodenticides** kill rodents such as rats and mice.

Agriculture is the sector that uses the most pesticides worldwide—approximately 85% of the estimated 2.6 million metric tons (2.9 million tons) used each year. Highly developed countries use about three fourths of all pesticides, but pesticide use is increasing most rapidly in developing countries.

The "Perfect" Pesticide

The ideal pesticide would be a **narrow-spectrum pesticide** that would kill only the organism for which it was intended and not harm any other species. The perfect pesticide would also be readily broken down, either by natural chemical decomposition or by biological organisms, into safe materials such as water, carbon dioxide, and oxygen. The ideal pesticide would stay exactly where it was put and would not move around in the environment.

Unfortunately, there is no such thing as an ideal pesticide. Most pesticides are **broad-spectrum pesticides**, which kill a variety of organisms, including some that are beneficial, in addition to the target pest. Some pesticides do not degrade readily or else break down into compounds that are as dangerous as, if not more dangerous than, the original pesticide. And most pesticides move around a great deal throughout the environment.

First-Generation and Second-Generation Pesticides

Before the 1940s, pesticides were of two main types, inorganic compounds (also called minerals) and organic

Figure 22.1 Pesticide derived from plants. Chrysanthemum flowers, shown here as they are harvested in Rwanda, are the source of the insecticide pyrethrin. Botanicals are chemicals from plants that can be used as pesticides.

compounds. Inorganic compounds that contain lead, mercury, and arsenic are extremely toxic to pests but are not used much today, in part because of their chemical stability in the environment. Natural processes do not degrade inorganic compounds, which therefore persist and accumulate in the soil and water. This accumulation poses a threat to humans and other organisms, which, like the target pests, are susceptible to poisoning by inorganic compounds.

Plants, which have been fighting pests longer than humans, have developed several natural organic compounds that are poisonous, particularly to insects. Such plant-derived pesticides are called **botanicals**. Examples of botanicals include nicotine from tobacco, pyrethrin from chrysanthemum flowers (Figure 22.1), and rotenone from roots of the derris plant, all of which are used to kill insects. Botanicals are easily degraded by microorganisms and therefore do not persist for long in the environment. However, they are highly toxic to aquatic organisms and to bees (beneficial insects that pollinate crops).

Synthetic botanicals are human-made insecticides produced by chemically modifying the structure of natural botanicals. An important group of synthetic botanicals are the pyrethroids, which are chemically similar to pyrethrin. Pyrethroids do not persist in the environment; they are slightly toxic to mammals and bees but very toxic to fishes. Allethrin is an example of a pyrethroid.

In the 1940s a large number of synthetic organic pesticides began to be produced. Earlier pesticides, both inorganic compounds and botanicals, are called **first-generation pesticides** to distinguish them from

Figure 22.2 Early use of DDT.
Pesticides such as DDT (dichlordiphenyltri-chloroethane), shown as it was sprayed to control mosquitoes at New York's Jones Beach State Park in 1945, were used in ways that would be unacceptable now. The sign on the truck reads, in part, "D.D.T. Powerful insecticide. Harmless to humans." The harmful environmental effects of DDT were not known until many years later.

the vast array of synthetic poisons in use today, called **second-generation pesticides**. The insect-killing ability of **dichlorodiphenyltrichloroethane** (**DDT**), the first of the second-generation pesticides, was recognized in 1939 (Figure 22.2). There are currently about 20,000 registered commercial pesticide products, consisting of combinations of about 675 active chemical ingredients. Table 22.1, which summarizes the characteristics of many of the pesticides discussed in the chapter, provides a useful overview of the varied effects different pesticides have on mammals, including humans, and on the environment.

The Major Groups of Insecticides

Insecticides, the largest category of pesticides, are usually classified into groups based on chemical structure. Three of the most important groups of second-generation insecticides are the chlorinated hydrocarbons, organophosphates, and carbamates.

DDT is an example of a **chlorinated hydrocarbon**, an organic compound containing chlorine. After DDT's insecticidal properties were recognized, many more chlorinated hydrocarbons were synthesized as pesticides. Generally speaking, chlorinated hydrocarbons are broad-spectrum insecticides. Most are slow to degrade and therefore persist in the environment (even inside organisms) for many months or even years. They were widely used from the 1940s until the 1960s, but since then many have been banned or their use has largely been restricted, mainly because of problems associated with their persistence in the environment and impacts on humans and wildlife. Three chlorinated hydrocarbons still in use in

the United States are endosulfan, lindane, and methoxychlor. Many people first became aware of the problems with pesticides in 1963, when Rachel Carson published her book *Silent Spring* (see Chapter 3).

Organophosphates, organic compounds that contain phosphorus, were developed during World War II as an outgrowth of German research on nerve gas. Organophosphates are more poisonous than other types of insecticides, and many are highly toxic to birds, bees, and aquatic organisms. The toxicity of many organophosphates in mammals, including humans, is comparable to that of some of our most dangerous poisons—arsenic, strychnine, and cyanide. Organophosphates do not persist in the environment as long as chlorinated hydrocarbons do. As a result, organophosphates have generally replaced the chlorinated hydrocarbons in large-scale uses such as agriculture, although many are not widely available to consumers because of their high level of toxicity. Methamidophos, dimethoate, and malathion are three examples of organophosphates.

Carbamates, the third group of insecticides, are broad-spectrum insecticides derived from carbamic acid. Carbamates are generally not as toxic to mammals as the organophosphates, although they still show broad, nontarget toxicity. Two common carbamates are carbaryl and aldicarb.

The Major Kinds of Herbicides

Chemicals that kill or inhibit the growth of unwanted vegetation such as weeds in crops or lawns are called herbicides. Like insecticides, herbicides can be classified into groups on the basis of chemical structure, but this method is cumbersome because there are at least 12 different chemical groups that are used as herbicides. It is easier to group herbicides according to how they act and what they kill. **Selective herbicides** kill only certain types of plants, whereas **nonselective herbicides** kill all vegetation. Selective herbicides can be further classified according to the types of plants they affect. **Broad-leaf herbicides** kill plants with broad leaves but do not kill grasses; **grass herbicides** kill grasses but are safe for most other plants.

Two common herbicides with similar structures are 2,4-dichlorophenoxyacetic acid (2,4-D) and 2,4,5-trichlorophenoxyacetic acid (2,4,5-T). Both were developed in the United States in the 1940s. These broad-leaf herbicides are similar in structure to a natural growth hormone in plants and therefore disrupt the plants' natural growth processes; they kill plants such as dandelions but do not harm grasses. You may recall from Chapter 18 that many of the world's important crops, such as wheat, corn, and rice, are cereal grains, which are grasses. Both 2,4-D and 2,4,5-T can be used to kill weeds that compete with these crops, although 2,4,5-T is no longer used in the United States. The Environmental Protection Agency (EPA) banned most uses of 2,4,5-T in 1979 because of possible harmful side effects to humans that became apparent after its use in the Vietnam War.

CASE·IN·POINT The Use of Herbicides in the Vietnam War

One of the controversial aspects of the Vietnam War was the defoliation program carried on by the United States in South Vietnam. From 1962 to 1971, the United States sprayed more than 12 million gallons of herbicides over about 4.5 million acres of South Vietnam to kill vegetation around military bases and to expose hiding places and destroy crops planted by the Vietcong and North Vietnamese troops (Figure 22.3). The three mixtures of herbicides used were designated Agent White, Agent Blue, and Agent Orange.

The negative impacts of these herbicides on the environment are still being felt today. It is estimated that 14% of the ecologically important mangrove forests of South Vietnam were destroyed. Shrubs and wild grasses have replaced the vast tracts of denuded forestlands, and the forests may take many decades to return. Approximately 30% of the nation's commercially valuable hardwood forests were killed, and bamboo and weedy grasses have replaced them.

In addition to the ecological damage, the herbicide sprays appear to have caused health problems in the Vietnamese people and in some members of the U.S. military who were exposed to them in the Vietnamese jungles. Agent Orange, a mixture of two herbicides (2,4-D and 2,4,5-T), also contained minute amounts of **dioxins**, a group of mildly to very toxic chemical compounds formed during the manufacture of these herbicides. The

dioxins in Agent Orange were reportedly 1,000 times more toxic than were those in domestic herbicides. (The contamination of Agent Orange by dioxins was discovered after the war.)

High doses of dioxins have been shown to cause birth defects in animals. According to the Vietnamese government, about 500,000 Vietnamese people have died or contracted serious illnesses as a result of exposure of themselves, their parents, or their grandparents to Agent Orange. Several researchers have noted that the number of birth defects, stillbirths, female reproductive disorders, and certain adult soft-tissue cancers increased dramatically in Vietnam following the war, particularly in areas where the herbicide was heavily sprayed or where it was stored and subsequently leaked into the soil and bodies of water. Although medical researchers caution that it is very difficult to prove a direct cause-and-effect relationship between Agent Orange and these medical problems, statistical evidence suggests that Agent Orange is at least partly responsible.

The level of dioxin in the breast milk of Vietnamese mothers has been measured at 1,800 parts per trillion. In comparison, the breast milk of U.S. mothers contains an average of 4 parts per trillion of dioxin. The high level of dioxin in human milk indicates that dioxin has contaminated Vietnam's food chain, including fish. Vietnamese people who eat the most fish have the highest levels of dioxins in their blood.

As a group, U.S. veterans who were exposed to high levels of Agent Orange have more health problems than do other veteran groups. The U.S. Department of Veterans Affairs currently recognizes 10 medical conditions linked to dioxin exposure. These include a variety of soft-

Figure 22.3 Herbicide spraying during the Vietnam War. This historic photo, taken in 1966, shows a forested area near a South Vietnamese highway.

| Table 22.1 | Characteristics of Selected Pesticides |

Type	Example	Toxicity to Mammals/ Regulatory Status*	Ecological Effects	Persistence/ Bioaccumulation
Insecticides				
Chlorinated hydrocarbons	Endosulfan	Highly toxic; restricted-use pesticide**	Highly toxic to birds and fishes	Moderately persistent; bioaccumulates
	Lindane	Moderately toxic; some formulations are restricted use	Highly toxic to fishes and aquatic invertebrates; thins egg shells of birds	Highly persistent; bioaccumulates
	Methoxychlor	Slightly toxic; general-use pesticide	Slightly toxic to birds; highly toxic to fishes and aquatic invertebrates	Relatively low persistence; does not bioaccumulate to any significant degree
	Dichlorodiphenyl-trichloroethane (DDT)	Moderately to slightly toxic; banned from use in the United States in 1972	Chronic exposure in birds may cause eggshell thinning and embryo death; highly toxic to many aquatic invertebrates and fish species	Highly persistent; bioaccumulates significantly
Organophosphates	Methamidophos	Highly toxic; restricted-use pesticide	Highly toxic to birds, bees, and aquatic organisms	Relatively low persistence; does not bioaccumulate
	Dimethoate	Moderately toxic; general-use pesticide	Moderately to highly toxic to birds, fishes, aquatic invertebrates, and bees	Relatively low persistence; does not bioaccumulate
	Malathion	Slightly toxic; general-use pesticide	Highly toxic to bees and aquatic invertebrates; moderately toxic to birds; range of toxicities for fishes	Relatively low persistence; bioaccumulation demonstrated in brown shrimp
Insecticides				
Carbamates	Aldicarb	Highly toxic; restricted-use pesticide	Highly toxic to birds; moderately toxic to fishes; not toxic to bees	Relatively low persistence; does not bioaccumulate
	Carbaryl	Slightly, moderately, or highly toxic, depending on formulation; general-use pesticide	Moderately toxic to aquatic organisms; highly toxic to bees and other beneficial insects	Relatively low persistence; bioaccumulates in acidic waters
Botanicals	Pyrethrin-I, Pyrethrin-II	Slightly, moderately, or highly toxic, depending on formulation; most are general-use pesticides	Highly toxic to aquatic organisms and bees; slightly toxic to birds; more toxic in acidic waters	Relatively low persistence; does not bioaccumulate
Synthetic botanicals (pyrethroids)	Allethrin	Slightly toxic; general-use pesticide	Toxic to fishes; slightly toxic to bees; not toxic to birds	Data not available

(continues)

tissue cancers, skin diseases, urological disorders, and birth defects. However, the U.S. government has not agreed to compensate Vietnamese people with similar health problems. ■

BENEFITS OF PESTICIDES

Each day a war is waged as farmers, struggling to produce bountiful crops, battle insects and weeds. Similarly, health officials fight their own war against the ravages of human diseases transmitted by insects. One of the most effective weapons in the arsenals of farmers and health officials is the pesticide.

Disease Control

Insects transmit several devastating human diseases. Fleas and lice carry the microorganism that causes typhus in humans. Malaria, which is also caused by a microor-

Table 22.1 (Continued)

Type	Example	Toxicity to Mammals/ Regulatory Status*	Ecological Effects	Persistence/ Bioaccumulation
Herbicides				
Triazine	Cyanazine	Moderately toxic; causes a variety of birth defects in rats; restricted-use pesticide	Slightly to moderately toxic to birds and aquatic organisms; not toxic to bees	Low to moderate persistence; has been found in groundwater
	Atrazine	Slightly to moderately toxic; restricted-use pesticide (because it has the potential to contaminate groundwater)	Not toxic to birds or bees; slightly toxic to aquatic organisms	Highly persistent in soil; soluble in water and a common river contaminant in agricultural areas; does not bioaccumulate
Phenoxy compound	2,4-D	Slightly to moderately toxic; causes birth defects in rats at high doses; general-use pesticide	Slightly to moderately toxic to birds; some formulations are highly toxic to fishes	Low persistence in soil and water; does not bioaccumulate
Thiocarbamate	Butylate	Slightly toxic; general-use pesticide	Moderately toxic to fishes; not toxic to birds, bees, or other wildlife	Low to moderate persistence in soil; low to moderate bioaccumulation in fishes
Analine	Alachlor	Slightly toxic; potential to cause cancer in rats; restricted-use pesticide	Moderately toxic to fishes; slightly toxic to some birds and aquatic invertebrates; not toxic to bees	Low persistence; does not bioaccumulate
Dinitroanaline compound	Trifluralin	Slightly toxic; general-use pesticide	Not toxic to birds or bees; highly toxic to fishes and aquatic invertebrates	Moderate to high persistence in soil; moderate bioaccumulation
Fungicides				
Phthalimide	Captan	Very slightly toxic; general-use pesticide	Not toxic to birds or bees; highly toxic to fishes	Low persistence; low to moderate bioaccumulation
Fumigants				
—	Methyl bromide	Highly toxic gas; restricted-use pesticide	Potential to destroy ozone; moderately toxic to aquatic organisms; not toxic to bees	Moderately persistent in soil; evaporates into atmosphere; does not bioaccumulate

* The information in this table does not replace or supersede the information on the pesticide product or other regulatory requirements.
** Restricted use pesticides may be purchased and used only by certified applicators.

ganism, is transmitted to millions of humans each year by female *Anopheles* mosquitoes (Figure 22.4). According to the World Health Organization (WHO), approximately 300 million to 500 million people currently suffer from malaria, and as many as 2.7 million people, mostly children in developing countries, die from the disease each year. Because there are only a few antimalarial drugs available to treat malaria, the focus of controlling this disease is on killing the mosquitos that carry it.

Pesticides, particularly DDT, have helped control the population of mosquitoes, thereby reducing the incidence of malaria. Consider Sri Lanka. In the early 1950s, more than 2 million cases of malaria were reported in Sri Lanka each year. When spraying of DDT was initiated to control mosquitoes, malaria cases dropped to almost zero. When DDT spraying was discontinued in 1964, malaria reappeared almost immediately. By 1968, its annual incidence had increased to greater than 1 million

Figure 22.4 Location of malaria.
Insecticides sprayed to control mosquitoes in these locations have saved millions of lives.

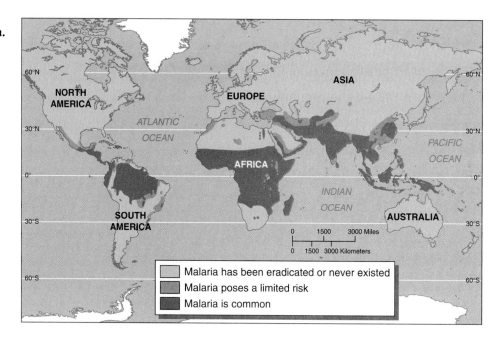

cases per year. Despite the negative effects of DDT on the environment and organisms, the Sri Lankan government decided to begin spraying DDT once again in 1968. Today, DDT is still used in at least 20 tropical countries to control mosquitoes.

Crop Protection

Although exact assessments are difficult to make, it is widely estimated that more than one third of the world's crops are eaten or destroyed by pests. Given our expanding population and world hunger, it is easy to see why control of agricultural pests is desirable.

Pesticides reduce the amount of a crop that is lost through competition with weeds, consumption by insects, and diseases caused by plant **pathogens** (microorganisms, such as fungi and bacteria, that cause disease). Although many insect species are beneficial from a human viewpoint (two examples are honeybees, which pollinate crops, and ladybugs, which prey on crop-eating insects), a large number are considered pests. Of these, about 200 species have the potential to cause large economic losses in agriculture. For example, the Colorado potato beetle is one of many insects that voraciously consume the leaves of the potato plant, reducing the plant's ability to produce large tubers for harvest (see Figure 18.11).

Serious agricultural losses are minimized in the United States and other highly developed nations primarily by the heavy application of pesticides. Pesticide use is usually justified economically, in that farmers save an estimated $3 to $5 in crops for every $1 that they invest in pesticides. In developing countries where pesticides are not used in appreciable amounts, the losses due to agricultural pests can be considerable.

Why are agricultural pests found in such great numbers in our fields? Part of the reason is that agriculture is usually a **monoculture**; that is, only one variety of one crop species is grown on large tracts of land. The cultivated field thus represents a very simple ecosystem. In contrast, forests, wetlands, and other natural ecosystems are extremely complex and contain many different species, including predators and parasites that control pest populations and plant species that are not used by pests for food. A monoculture reduces the dangers and accidents that might befall a pest as it searches for food. A Colorado potato beetle in a forest would have a hard time finding anything to eat, but a 500-acre potato field is like a big banquet table set just for the pest. It eats, prospers, and reproduces. In the absence of many natural predators and in the presence of plenty of food, the population thrives and grows, and more of the crop becomes damaged.

PROBLEMS ASSOCIATED WITH PESTICIDE USE

Although pesticides have their benefits, they are accompanied by several problems. For one thing, many pest species evolve a resistance to pesticides after repeated exposure to them. Also, pesticides affect numerous species in addition to the target pests, generating imbalances in the ecosystem (including agricultural fields) and posing a threat to human health. And, as mentioned earlier, the ability of some pesticides to resist degradation and to readily move around in the environment causes even more problems for humans and other organisms.

Evolution of Genetic Resistance

The prolonged use of a particular pesticide can cause a pest population to develop genetic resistance to the pesticide. **Genetic resistance** is any inherited characteristic that decreases the effect of a pesticide on a pest.

In the 50 years during which pesticides have been widely used, at least 520 species of insects and mites have evolved genetic resistance to certain pesticides (Figure 22.5). Many pests now have multiple resistance to several pesticides, and at least 17 species, such as diamondback moths and palm thrips, are resistant to *all* major classes of insecticides that farmers are legally allowed to use on them. Insects are not the only pests to evolve genetic resistance; at least 84 weed species are currently resistant to certain herbicides. Some weeds, such as annual ryegrass and canary grass, are resistant to all available herbicides.

How does genetic resistance to pesticides occur? Every time a pesticide is used to control a pest, some survive. The survivors, because of certain genes they already possess, are genetically resistant to the pesticide, and they pass on this trait to future generations. Thus, **evolution**—any cumulative genetic change in a population of organisms—occurs, and subsequent pest populations contain larger percentages of pesticide-resistant pests than before.

Insects and other pests are constantly evolving. The short generation times (the period between the birth of one generation and that of another) and large populations that are characteristic of most pests favor rapid evolution, which allows the pest population to quickly adapt to the pesticides used against it. As a result, an insecticide that kills most of an insect population becomes less effective after prolonged use because the survivors and their offspring are genetically resistant.

Manufacturers of chemical pesticides have often responded to genetic resistance by recommending that the pesticide be applied more frequently or in larger doses. Alternatively, they recommend switching to a new, often more expensive, pesticide. These responses result in a predicament that has come to be known as the **pesticide treadmill**, in which the cost of applying pesticides increases while their effectiveness decreases. Over time, the pesticide treadmill results in increased pesticide use, higher production costs, and declines in crop yields.

Resistance Management **Resistance management** is a relatively new approach to dealing with genetic resistance. It involves efforts to delay the evolution of genetic resistance in insect pests or weeds so that the period of time in which a pesticide is useful is maximized. Strategies of resistance management vary depending on the pest species involved.

One strategy of resistance management for insect pests is to maintain a nearby "refuge" of untreated plants where the insect pest can avoid being exposed to the insecticide. Those insects that live and grow in the refuge remain susceptible to the insecticide. When susceptible insects migrate into the area being treated with insecticide, they mate with, and thus mix genes with, the genetically resistant population. This interbreeding delays the development of genetic resistance in the population as a whole.

Avoiding repeated use of the same herbicide on the same field is one strategy of resistance management that slows the development of weed resistance to herbicides. After herbicides are applied, the field should be scouted to see if any weed plants survived the herbicide application. These weeds are resistant to the herbicide, and they should be removed from the field before they flower. Cultural methods that prolong the usefulness of herbicides include planting seed that is certified to be free of weed seeds and mechanically pulling weeds.

Imbalances in the Ecosystem

One of the worst problems associated with pesticide use is that pesticides affect species other than the pests for which they are intended. Beneficial insects are killed as effectively as pest insects. In a study of the effects of spraying the insecticide dieldrin to kill Japanese beetles, scientists found a large number of dead animals in the treated area, such as various birds, rabbits, ground squirrels, cats, and beneficial insects. (Use of dieldrin in the United States has since been banned.) Pesticides do not

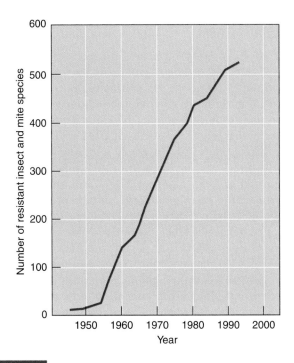

Figure 22.5 Genetic resistance. There has been a dramatic increase in the number of insect species exhibiting genetic resistance to insecticides. More than 520 insect and mite species have evolved resistance to insecticides.

Figure 22.6 Insect pests, insect predators, and pesticides. Natural population fluctuations are controlled by a variety of factors, including the presence of predators and parasites. When pesticides are applied, the predators and parasites of the pest species are also affected. The effect on predator/parasite populations disrupts the normal interactions between species and can cause a huge increase in the population of the pest species a short time after the pesticide is applied.

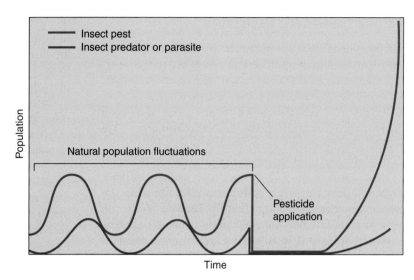

have to kill organisms to harm them. Quite often the stress of carrying pesticides in its body makes an organism more vulnerable to predators, diseases, or other stressors in its environment.

Because the natural enemies of pests often starve or migrate in search of food after pesticide has been sprayed in an area, pesticides are indirectly responsible for a large reduction in the populations of these natural enemies. Pesticides also kill natural enemies directly, because predators consume a lot of the pesticide by consuming the pests. After a brief period, the pest population rebounds and gets larger than ever, partly because no natural predators are left to keep its numbers in check (Figure 22.6).

Despite a 33-fold increase in pesticide use in the United States since the 1940s, crop losses due to insects, diseases, and weeds have not changed appreciably (Table 22.2). Increasing genetic resistance to pesticides and the destruction of the natural enemies of pests provide a partial explanation. Changes in agricultural practices are also to blame; for example, crop rotation, a proven way of controlling certain pests, is not practiced as much today as it was several decades ago (see Chapter 14).

Creation of New Pests   In some instances, the use of a pesticide has resulted in a pest problem that did not exist before. Creation of new pests—that is, turning minor pest organisms into major pests—is possible because the

natural predators, parasites, and competitors of a certain pest may be largely killed by a pesticide, allowing the pest's population to rebound. The use of DDT to control certain insect pests on lemon trees was documented as causing an outbreak of a scale insect (a sucking insect that attacks plants) that had not been a problem before spraying (Figure 22.7). In a similar manner, the European red mite became an important pest on apple trees in the northeastern United States, and beet armyworms became an important pest on cotton, both after the introduction of pesticides.

Persistence, Bioaccumulation, and Biological Magnification

The effects of DDT on many bird species first demonstrated certain problems of chlorinated hydrocarbon pesticide use. Falcons, pelicans, bald eagles, ospreys, and many other birds are very sensitive to traces of DDT in their tissues. A substantial body of scientific evidence indicates that one of the effects of DDT on these birds is that they lay eggs with extremely thin, fragile shells that usually break during incubation, causing the chicks' deaths. After 1972, the year DDT was banned in the United States, the reproductive success of many birds improved (Figure 22.8).

The impact of DDT on birds is the result of three characteristics of DDT: its persistence, bioaccumulation, and biological magnification. Some pesticides, particularly chlorinated hydrocarbons, are extremely stable in the environment and may take many years to be broken down into less toxic forms. The **persistence** of synthetic pesticides is a result of their novel (not found in nature) chemical structures. Natural decomposers such as bacteria have not yet evolved ways to degrade synthetic pesticides, so they accumulate in the environment and in the food web.

When a pesticide is not metabolized (broken down) or excreted by an organism, it is simply stored, usually in

Table 22.2	Percentage of Crops Lost Annually to Pests in the United States		
Period	**Insects**	**Diseases**	**Weeds**
1989–1999	13.0	12.0	12.0
1974	13.0	12.0	8.0
1951–1960	12.9	12.2	8.5
1942–1951	7.1	10.5	13.8

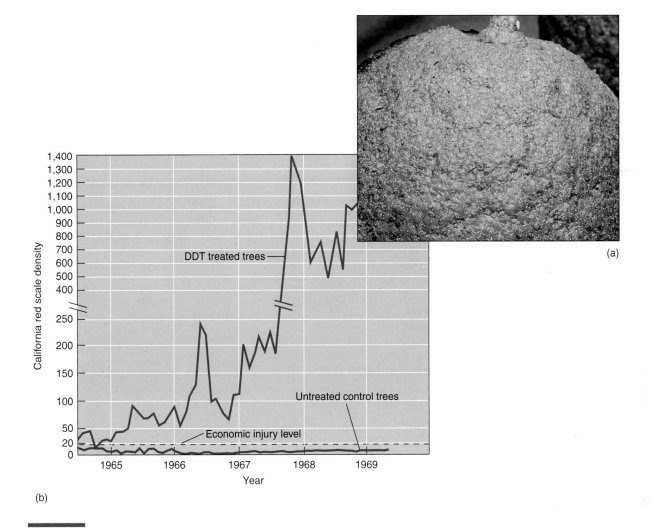

(a)

(b)

Figure 22.7 Pesticide use and new pest species. An infestation of red scale insects on lemons occurred after DDT (dichlorodiphenyltrichloroethane) was sprayed to control a different pest. Prior to DDT treatment, red scale did not cause significant economic injury to citrus crops. (a) Red scale on green (unripened) citrus fruit. (b) A comparison of red scale populations on DDT-treated trees (*blue line*) and untreated trees under biological control (*red line*).

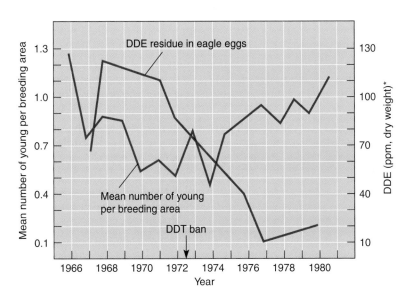

Figure 22.8 Effect of DDT on birds. A comparison of the number of successful bald eagle offspring with the level of DDT (dichlorodiphenyltrichloroethane) residues in their eggs. Note that reproductive success improved after DDT levels decreased. (DDE = dichlorodiphenyldichloroethylene.)

fatty tissues. Over time, the organism may accumulate high concentrations of the pesticide. The buildup of a persistent pesticide in an organism's body is known as **bioaccumulation** or **bioconcentration**.

Organisms at higher levels on food webs tend to have greater concentrations of bioaccumulated pesticide stored in their bodies than those lower on food webs. The increase in pesticide concentrations as the pesticide passes through successive levels of the food web is known as **biological magnification** or **biological amplification**.[1]

As an example of the concentrating characteristic of persistent pesticides, consider a food chain studied in a Long Island salt marsh that was sprayed with DDT over a period of years for mosquito control: algae and plankton → shrimp → American eel → Atlantic needlefish → ring-billed gull (Figure 22.9). The concentration of DDT in water was extremely dilute, on the order of 0.00005 parts per million (ppm). The algae and other plankton contained a greater concentration of DDT, 0.04 ppm. Each shrimp grazing on the plankton concentrated the pesticides in its tissues to 0.16 ppm. Eels that ate shrimp laced with pesticide had a pesticide level of 0.28 ppm, and needlefish that ate eels contained 2.07 ppm of DDT. The top carnivores, ring-billed gulls, had a DDT level of 75.5 ppm from eating contaminated fishes. Although this example involves a bird at the top of the food chain, it is important to recognize that *all* top carnivores, from fishes to humans, are at risk from biological magnification. Because of this risk, currently approved pesticides have been tested to ensure they do not persist and accumulate in the environment.

Mobility in the Environment

Another problem associated with pesticides is that they do not stay where they are applied but tend to move through the soil, water, and air, sometimes long distances (Figure 22.10). Pesticides that are applied to agricultural lands and then wash into rivers and streams when it rains can harm fishes. If the pesticide level in their aquatic ecosystem is high enough, the fishes may be killed. If the level is sub-lethal (that is, not enough to kill the fishes), the fishes may still suffer from undesirable effects such as bone degeneration. These effects may decrease their competitiveness and increase their chances of being preyed upon.

Pesticide mobility is also a problem for humans. In 1994 the Environmental Working Group (EWG), a private environmental organization, analyzed herbicides in drinking water by evaluating 20,000 water tests performed by state and federal government inspectors.

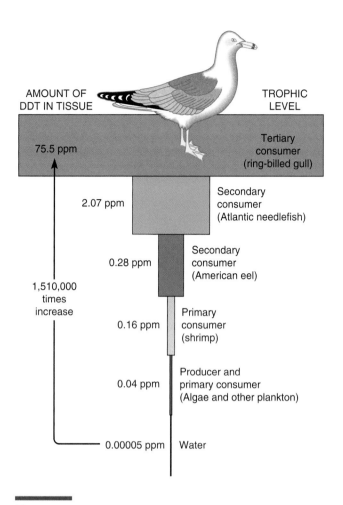

Figure 22.9 Biological magnification of DDT in a Long Island salt marsh. Note how the level of DDT (dichlorodiphenyl-trichloroethane), expressed as parts per million, increased in the tissues of various organisms as DDT moved through the food chain from producers to consumers. The ring-billed gull at the top of the food chain had approximately 1.5 million times more DDT in its tissues than the concentration of DDT in the water.

Their study revealed that 14.1 million U.S. residents drink water that contains traces of five widely used herbicides. Because these herbicides are often used on corn and soybeans, the study focused on the Midwestern states where these crops are commonly grown. The study concluded that 3.5 million people living in the Midwest face a slightly elevated cancer risk because of their exposure to the herbicides. The EPA recently reduced the use of five of the herbicides (alachlor, metachlor, atrazine, cyanazine, and simazine).

From 1996 to 1998 the EWG conducted a different study in California that revealed pesticide mobility in the atmosphere. They collected nearly 100 air samples in several California counties and submitted them to a certified laboratory for analysis. Almost two thirds of the samples contained small amounts of pesticides that had drifted from farm fields.

---

[1] Other toxic substances besides pesticides may exhibit bioaccumulation and biological magnification, including radioactive isotopes, heavy metals such as mercury, flame retardants known as polybrominated diphenyl ethers (PBDEs), and industrial chemicals such as PCBs (polychlorinated biphenyls); see Chapters 11, 21, and 23.

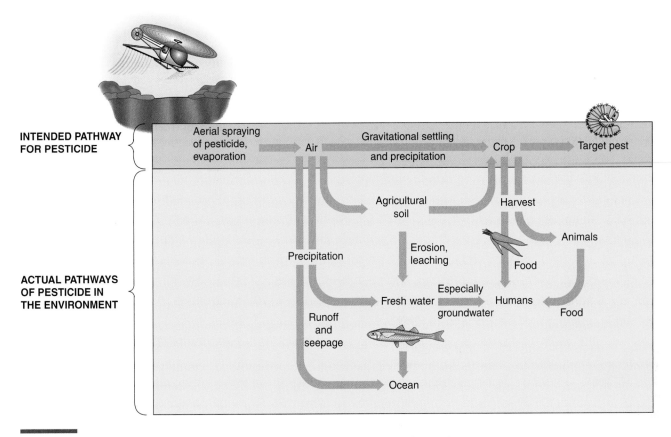

INTENDED PATHWAY FOR PESTICIDE

ACTUAL PATHWAYS OF PESTICIDE IN THE ENVIRONMENT

Figure 22.10 Intended versus actual pathways of pesticides. The intended pathway of pesticides in the environment is shown at the top of the figure, while the actual pathways are shown at the bottom.

RISKS OF PESTICIDES TO HUMAN HEALTH

Exposure to pesticides, which is greater than most people realize and often occurs without their knowledge, can also damage human health. Pesticide poisoning caused by short-term exposure to high levels of pesticides can result in harm to organs and even death, whereas long-term exposure to lower levels of pesticides can cause cancer. There is also concern that exposure to trace amounts of certain pesticides can disrupt the human endocrine (hormone) system. Children are considered to be at greater risk from exposure to household pesticides than adults.

Short-Term Effects of Pesticides

Pesticides poison approximately 67,000 people in the United States each year. Most of these are farm workers or others whose occupations involve daily contact with large quantities of pesticides. A person with a mild case of pesticide poisoning may exhibit symptoms such as nausea, vomiting, and headaches. More serious cases, particularly organophosphate poisonings, may result in

permanent damage to the nervous system and other body organs. Although the number is low in the United States, people do die from overexposure to pesticides. Almost any pesticide can kill a human if the dose is large enough.

The WHO estimates that globally more than 3 million people are poisoned by pesticides each year; of these, about 220,000 die. The incidence of pesticide poisoning is highest in developing countries, in part because they often use dangerous pesticides that have been banned or greatly restricted by highly developed nations. Also, pesticide users in developing nations often are not trained in the safe handling and storage of pesticides, and safety regulations are generally more lax there.

CASE·IN·POINT The Bhopal Disaster

In December 1984, the world's worst industrial accident occurred at a Union Carbide pesticide plant in the Indian city of Bhopal. As much as 36 metric tons (40 tons) of methyl isocyanate (MIC) gas, which is used to produce carbamate pesticides, erupted from an underground storage tank after water leaked in and caused an explosive chemical reaction. Some of the MIC, which is itself highly toxic, converted in the atmosphere to hydrogen cyanide, which is even more deadly. The toxic cloud settled over 78 km² (30 mi²), exposing up to 600,000 people.

According to official counts, about 2,500 people were killed outright from exposure to the deadly gas, and at least 2,500 have died since then. An international group of medical specialists (the International Medical Commission on Bhopal) estimated in 1996 that between 50,000 and 60,000 survivors had serious respiratory, ophthalmic, intestinal, reproductive, and neurological problems. Young women who were exposed to the gas were unable to marry because it was widely assumed they were sterile. Many also suffer from psychological disorders, such as post-traumatic stress and pathological guilt over being unable to save loved ones.

Union Carbide agreed in 1989 to pay $470 million in compensation and spend more than $100 million to build a hospital for victims. The Indian government is still disbursing the $470 million paid by Union Carbide. Most people who have been compensated have received $500 each.

Meanwhile, cleanup of contaminated land and groundwater in the vicinity of the former pesticide plant remains to be addressed. Tests conducted in 1999 by Greenpeace International and several Bhopal-based organizations revealed contamination of land and drinking water supplies with heavy metals and persistent organochlorine compounds. People living in communities surrounding the former plant site may be at risk from these hazardous chemicals, particularly in their drinking water. ◼

Long-Term Effects of Pesticides

Many studies of farm workers and workers in pesticide factories, both of whom are exposed to low levels of pesticides over many years, show an association between cancer and long-term exposure to pesticides. A type of lymphoma (a cancer of the lymph system) has been associated with the herbicide 2,4-D. Other pesticides have been linked to a variety of cancers, such as leukemia and cancers of the brain, lungs, and testicles. Although the issue of whether certain pesticides cause breast cancer remains unresolved, researchers have noted a correlation between a high level of one or more pesticides in the breast's fatty tissue and cancer.

Long-term exposure to at least one pesticide may have resulted in sterility in thousands of farm workers on banana and pineapple plantations. More than 26,000 workers in 12 countries sued their employers or pesticide manufacturers, and most of the defendants have paid settlements but not admitted liability.

A 2001 study that compared records of pesticide applications in 1984 in California's Central Valley with state health records for the same area showed that pregnant women who live near pesticide applications have higher rates of miscarriage. Other studies have indicated that the children of agricultural workers are at greater risk for birth defects, particularly stunted limbs. There is also evidence that exposure to pesticides may compromise the body's ability to fight infections.

Long-term exposure to pesticides may increase the risk of Parkinson's disease, which afflicts about 1 million people in the United States. When researchers injected rats with repeated doses of rotenone, a plant-derived pesticide, the rats developed the symptoms of Parkinson's disease, including difficulty in walking, tremors, and abnormal protein deposits in their brains. Many pesticides have chemical structures similar to rotenone, which suggests that other pesticides may increase the risk of Parkinson's disease as well. It is important to remember that this connection between pesticide exposure and Parkinson's disease is very tentative and will require additional research to prove or disprove. To date, no study has connected Parkinson's disease in humans with exposure to a specific pesticide.

Pesticides as Endocrine Disrupters

In the 1990s, many scientific articles were published that linked certain pesticides and other persistent toxic chemicals with reproductive problems in animals (Table 22.3; see Figure 1.5 and Chapter 1 discussion on endocrine disrupters). River otters exposed to synthetic chemical pollutants were found to have abnormally small penises. Female sea gulls in Southern California exhibited behavioral aberrations by pairing with one another rather than with males during the mating season. In many cases scientists have been able to reproduce the same abnormal symptoms in the laboratory, providing support that the defects are caused by certain persistent chemicals.

However, it was the suggestion in the 1996 book, *Our Stolen Future*, by Theo Colburn, Dianne Dumanoski, and John Myers, that persistent toxic chemicals in the environment are disrupting *human* hormone systems that ignited a barrage of media attention. Theo Colburn, a senior scientist at the World Wildlife Fund, hypothesized that ubiquitous chemicals in the environment are linked to disturbing

Table 22.3 Some Pesticides That Are Known Endocrine Disrupters*

Pesticide	General Information
Atrazine	Herbicide; still used
Chlordane	Insecticide; banned in United States in 1988
DDT (dichlorodiphenyl-trichloroethane)	Insecticide; banned in United States in 1972
Endosulfan	Insecticide; still used
Kepone	Insecticide; banned in United States in 1977
Methoxychlor	Insecticide; still used

* Based on experimental research with laboratory animals.

trends in human health. These include increases in breast and testicular cancer, increases in male birth defects, and decreases in sperm counts (see Chapter 1).

Although a few studies suggest that certain chemicals in the environment may affect human health, a direct cause-and-effect relationship between these chemicals and adverse effects on the human population remains to be established. Some scientists think the potential danger is so great, however, that persistent chemicals such as DDT should be internationally banned immediately. Other scientists are more cautious in their assessment of the danger but still think the problem should not be disregarded.

Our Stolen Future ignited public concern and triggered scientific investigations by universities, governments, and industries on how synthetic chemicals interfere with the actions of human hormones. It will take years before these studies can tell us if these chemicals are acting as endocrine disrupters in humans. Meanwhile, an international effort is currently underway to ban all production and use of nine pesticides suspected of being endocrine disrupters (see section on the global ban of persistent organic pollutants later in this chapter).

Pesticides and Children

In recent years, there has been increased attention to the health effects of household pesticides on children because it appears that household pesticides are a greater threat to children than to adults. For one thing, children tend to play on floors and lawns, where they are exposed to greater concentrations of pesticide residues. Also, children may be more sensitive to pesticides because their bodies are still developing. Several preliminary studies suggest that exposure to household pesticides may cause brain cancer and leukemia in children, but more research must be done before any firm conclusions can be made.

The EPA estimates that 84% of U.S. homes use pesticide products, such as pest strips, bait boxes, bug bombs, flea collars, pesticide pet shampoos, aerosols, liquids, and dusts. Several thousand different household pesticides are manufactured, and these contain over 300 active ingredients and more than 2,500 inert ingredients (discussed later in the chapter). Poison control centers in the United States annually receive more than 130,000 reports of exposure and possible poisoning from household pesticides. More than half of these incidents involve children.

There is also concern about the ingestion by children of pesticide residues on food. The National Research Council published a 3-year study in 1993 called *Pesticides in the Diets of Infants and Children.* It called for additional research on how pesticide residues on food affect the young because it is not known if infants and children are more, or less, susceptible than adults. Also, because current pesticide regulations are intended to protect the health of the general population, the report

stated that infants and children may not be adequately protected.

Recent research supports an emerging hypothesis that exposure to pesticides may affect the development of intelligence and motor skills of infants, toddlers, and preschoolers. One study, published in *Environmental Health Perspectives* in 1998, compared two groups of rural Yaqui Indian preschoolers. These two groups, both of which live in northwestern Mexico, shared similar genetic backgrounds, diets, water mineral contents, cultural patterns, and social behaviors. The main difference between the two groups was their exposure to pesticides: One group lived in a farming community where pesticides were used frequently (45 times per crop cycle) and the other in an adjacent nonagricultural area where pesticides were rarely used. When asked to draw a person, most of the 17 children from the low-pesticide area drew recognizable stick figures, whereas most of the 34 children from the high-pesticide area drew meaningless lines and circles (Figure 22.11). Additional tests of simple mental and physical skills revealed similar striking differences between the two groups of children.

Drawings of a person
(by 4-year-olds)

Foothills   Valley

54 mo.   54 mo.
(a) female   (b) female

Figure 22.11 Effect of pesticide exposure on preschoolers. A study in Sonora, Mexico, found that Yaqui Indian preschoolers varied in their motor skills on the basis of the degree of pesticide exposure. The children were asked to draw a person. Two representative pieces of art are shown, both drawn by $4^{1}/_{2}$-year-old girls. (a) Most preschoolers who received little pesticide exposure were able to draw a recognizable stick figure. (b) Most preschoolers who lived in an area where agricultural pesticides were widely used could only draw meaningless lines and circles.

ALTERNATIVES TO PESTICIDES

Given the many problems associated with pesticides, it is clear that they are not the final solution to pest control. Fortunately, pesticides are not the only weapons in our arsenal. Alternative ways to control pests include cultivation methods, biological controls, reproductive controls, pheromones and hormones, genetic controls, quarantine, and irradiation. A combination of these methods in agriculture, often including a limited use of pesticides as a last resort, is known as integrated pest management (IPM). IPM is the most effective way to control pests. (Also see the Chapter 18 introduction for a discussion of the growing popularity of organic foods, which are grown in the absence of pesticides and commercial fertilizers.)

Using Cultivation Methods to Control Pests

Sometimes agricultural practices can be altered in such a way that a pest is adversely affected or discouraged from causing damage. Although some practices, such as the insect vacuum mentioned at the beginning of the chapter, are relatively new, other cultivation methods that discourage pests have been practiced for centuries.

One way to reduce damage by crop pests is by *interplanting* mixtures of plants, such as by alternating rows of different plants. When corn was interplanted with molasses grass in an experiment in Kenya, only about 5% of the corn crop was damaged, as compared with about 39% damage in the control field of corn. Molasses grass attracts wasps that lay their eggs inside stem borers, insects that destroy the stems of corn plants.

A technique that has been used with success in alfalfa crops is *strip cutting*, in which only one segment of the crop is harvested at a time. The unharvested portion of the crop provides an undisturbed habitat for natural predators and parasites of the pest species. The same type of benefit is derived from keeping strips of unplowed plants (that is, wild plants, including weeds) along the margins of fields. A German study found that pest mortality was about 50% higher in oilseed rape fields with margin strips of other plants, as compared to fields without margins. The margins provided a refuge for three parasites of the pollen beetle, which is a significant pest on oilseed rape plants.

The proper timing of planting, fertilizing, and irrigating promotes healthy, vigorous plants that are more able to resist pests because they are not being stressed by other environmental factors. The rotation of crops also helps control pests. When corn is not planted in the same field for 2 years in a row, the corn rootworm is effectively controlled.

Biological Controls

Biological controls involve the use of naturally occurring disease organisms, parasites, or predators to control pests. As an example, suppose that an insect species is accidentally introduced into a country where it was not found previously, and becomes a pest. It might be possible to control this pest by going to its native country and identifying an organism that is an exclusive predator or parasite of the pest species. That predator or parasite, if successfully introduced, may be able to lower the population of the pest species so that it is no longer a problem.

Cottony-cushion scale provides an example of successfully using one organism to control another. The cottony-cushion scale is a small insect that sucks the sap from the branches and bark of many fruit trees, including citrus trees (Figure 22.12). It is native to Australia but was accidentally introduced to the United States in the 1880s. A U.S. entomologist went to Australia and returned with several possible biological control agents. One, the vedalia beetle, was found to be very effective in controlling scale, which it eats voraciously and exclusively. Within 2 years of its introduction, the vedalia beetle had significantly reduced the cottony-cushion scale in citrus orchards. Today both the cottony-cushion scale and the vedalia beetle are present in very low numbers, and the scale is not considered an economically important pest.

More than 300 species have been introduced as biological control agents to North America. The Agricultural Research Service of the U.S. Department of Agriculture (USDA) is currently investigating possible biological controls for about a dozen other insect and weed pests. Although some examples of biological control are quite spectacular, finding an effective parasite or predator is very difficult. And just because a parasite or predator has been identified does not mean it can become successfully established in a new environment. Slight variations in environmental conditions such as temperature and moisture can alter the effectiveness of the biological control organism in its new habitat.

Care must also be taken to ensure that the introduced control agent does not attack unintended hosts and become a pest itself. To guard against this when the control agent is an insect, scientists put the insects in cages with samples of important crops, ornamentals, and native plants to determine if the insects will eat the plants when they are starving.

Despite such tests, organisms introduced as biological controls sometimes cause unintended problems in their new environment, and once they are introduced, they cannot be recalled. A weevil was introduced in 1968 to control the Eurasian musk thistle, a noxious weed that arrived in North America in the mid-19th century. Since then, the weevil has expanded its host range to include a North American thistle that is listed as a threatened species. The weevils significantly reduce seed production in the native thistles.

Insects are not the only biological control agents. Bacteria and viruses that harm insect pests have also been used successfully as biological controls. *Bacillus popilliae*,

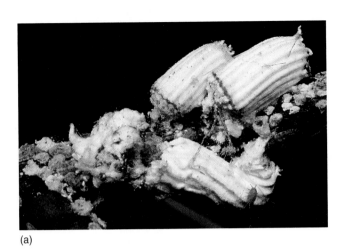

(a)

(b)

Figure 22.12 Biological control of cottony-cushion scale. (a) The cottony-cushion scale is an insect that attacks the stems and bark of several important crops. The white, ridged mass on top of each female contains up to 800 eggs. The males are reddish brown. (b) A vedalia beetle larva feeds on cottony-cushion scale. Both adults and larvae of the vedalia beetle control the cottony-cushion scale.

which causes milky spore disease in insects, can be applied as a dust on the ground to control the larval (immature) stage of Japanese beetles. The common soil bacterium *Bacillus thuringiensis*, or *Bt*, produces a natural pesticide that is toxic to some insects, such as the cabbage looper, a green caterpillar that damages many vegetable crops, and the corn earworm. When eaten by insect larvae, *Bt* toxin damages the intestinal tract, killing the young insect. The toxin does not persist in the environment and is not known to harm mammals, birds, or other noninsect species.

Pheromones and Hormones

Pheromones are natural substances produced by animals to stimulate a response in other members of the same species. Pheromones are commonly called sexual attractants because they are often produced to attract members of the opposite sex for mating. Each insect species produces its own specific pheromone, so once the chemical structure is known, it is possible to make use of pheromones to control individual pest species. Pheromones have been successfully used to lure insects such as Japanese beetles to traps, where they are killed (Figure 22.13). Alternatively, pheromones can be released into the atmosphere to confuse insects so that they cannot locate mates.

Insect **hormones** are natural chemicals produced by insects to regulate their own growth and metamorphosis, which is the process by which an insect's body changes through several stages to an adult. Specific hormones must be present at certain times in the life cycle of the insect; if they are present at the wrong time, the insect develops abnormally and dies. Many such insect hormones have been identified, and synthetic hormones with similar structures have been made. Entomologists are actively pursuing the possibility of using these substances to control insect pests.

A synthetic version of the insect hormone ecdysone, which causes molting, was the first hormone to be approved for use. Known as MIMIC, the hormone triggers abnormal molting in insect larvae (caterpillars) of moths and butterflies. Since MIMIC also affects some beneficial insects, its use has risks.

Figure 22.13 Pheromone traps. A Japanese beetle trap uses a sexual attractant to lure Japanese beetles, which fall into the bag and die. Such insect traps draw large numbers of Japanese beetles.

Reproductive Controls

Like biological controls, reproductive controls of pests involve the use of organisms. Instead of using another species to reduce the pest population, however, reproductive control strategies suppress pests by sterilizing some of its members. In the **sterile male technique**, large numbers of males are sterilized in a laboratory, usually with radiation or chemicals. Males are sterilized rather than females because male insects of species selected for this type of control mate several to many times, but females of that species mate only once. Thus, releasing a single sterile male may prevent successful reproduction by several females, whereas releasing a single sterile female would prevent successful reproduction by only that female.

The sterilized males are released into the wild, where they reduce the reproductive potential of the pest population by mating with normal females, who then lay eggs that never hatch. As a result, of course, the population of the next generation is much smaller.

One disadvantage of the sterile male technique is that it must be carried out continually to be effective. If

sterilization is discontinued, the pest population rebounds to a high level in a few generations (which, you will recall, are very short). The procedure is also expensive, as it requires the rearing and sterilization of large numbers of insects in a laboratory or production facility. During the 1990 Mediterranean fruit fly (medfly) outbreak in California, as many as 400 million sterile male medflies were released each week. Such extraordinary measures are taken because the medfly is so destructive. The adult medfly lays its eggs on 250 different fruits and vegetables. When the eggs hatch, the maggots feed on the fruits and vegetables and turn them into a disgusting mush.

Genetic Controls

Traditional selective breeding has been used to develop many varieties of crops that are genetically resistant to disease organisms or insects. Traditional breeding of crop plants typically involves identifying individual plants that are in an area where the pest is common but that do not appear to be damaged by the pest. These individuals are then crossed with standard crop varieties in an effort to produce a pest-resistant version. It may take as long as 10 to 20 years to develop a resistant crop variety, but the benefits are usually worth the time and expense.

Although traditional selective breeding has resulted in many disease-resistant crops and decreased the use of pesticides, there are potential problems. Fungi, bacteria, and other plant pathogens evolve rapidly. As a result, they can quickly adapt to the disease-resistant host plant, meaning that the new pathogen strains can cause disease in the formerly disease-resistant plant variety. Plant breeders, then, are in a continual race to keep one step ahead of plant pathogens.

Genetic engineering offers great promise in breeding pest-resistant plants more quickly (see Chapter 18). For example, a gene from the soil bacterium *Bt* (already discussed in the section on biological controls) has been introduced into several plants, such as cotton. Caterpillars that eat cotton leaves from these **genetically modified (GM)** plants die or exhibit stunted growth.

CASE-IN-POINT *Bt,* Its Potential and Problems

Bt has been marketed since the 1950s, but it was not sold on a large scale until recently, mainly because there are many different varieties of the *Bt* bacterium, and each variety produces a slightly different protein toxin. Each is toxic to only a small group of insects. For example, the *Bt* variety that works against corn borers would not be effective against Colorado potato beetles. As a result, *Bt* was not economically competitive against chemical pesticides, each of which could kill many different kinds of pests on many different crops.

Genetic engineers have greatly increased the potential of *Bt*'s toxin as a natural pesticide by modifying the gene coding for the toxin so that it affects a wider range of insect pests. They then inserted the *Bt* gene that codes for the toxin into at least 18 crop species, including corn, potato, and cotton. *Bt* corn, one of the first genetically modified (GM) crops, has been engineered to produce a continuous supply of toxin, which provides a natural defense against insects such as the European corn-borer. Likewise, *Bt* tomato and *Bt* cotton are more resistant to pests such as the tomato pinworm and the cotton boll-worm. Early ecological risk assessment studies, performed before EPA approved their use, indicated that genetically modified crops are essentially like their more conventional counterparts that are produced by selective breeding. Genetically modified crops do not become invasive pests or persist in the environment longer than crops that have not been genetically modified. Significantly fewer pesticide applications are needed for GM crops with the *Bt* gene.

The future of the *Bt* toxin as an effective substitute for chemical pesticides is not completely secure, however. Beginning in the late 1980s, several farmers began to notice that chemically applied *Bt* was not working as well against the diamondback moth as it had in the past. All of the farmers who reported this reduction in effectiveness had used *Bt* frequently and in large amounts on their fields. Also, in 1996, many farmers growing *Bt* cotton reported that their crop succumbed to the cotton bollworm, which *Bt* is supposed to kill. It appears that certain insects, such as the diamondback moth and possibly the cotton bollworm, may have developed resistance to this natural toxin in much the same way that they develop resistance to chemical pesticides. If *Bt* continues to be used in greater and greater amounts, it is likely that more insect pests will develop genetic resistance to it, greatly reducing *Bt*'s potential as a natural pesticide. Scientists are studying resistance management strategies to curb pest resistance to genetically engineered crops that make *Bt*.

Unintended Ecological Risks of Genetically Modified Crops

One of the concerns about growing GM crops is that they might do some unintentional harm to the environment (see Chapter 18 discussion of genetic engineering). In 1999 biologists from Cornell University published results that suggested that GM crops containing the *Bt* gene might harm the monarch butterfly. The primary food of monarch larvae (caterpillars) is milkweed, a weed that often grows along the margins of cornfields. Corn is a wind-pollinated plant that disperses its pollen up to 60 m (197 ft). Corn pollen lands on other plants near cornfields and can be eaten by the organisms that consume these plants. The biologists conducted a laboratory experiment in which they fed monarch larvae milkweed leaves dusted with *Bt* corn pollen. In a few days these larvae ate less, grew more slowly, and had higher mortality rates than control larvae fed milkweed leaves without *Bt* corn pollen.

Supporters of GM crops point out that the chemical pesticides that were sprayed before *Bt* was introduced into corn would certainly have killed many more nontarget organisms, including large numbers of monarch larvae. Moreover, because the researchers did not perform field studies, we cannot assume that the monarch population is at risk from the increasing acreage of *Bt* corn being planted in the United States. (In 2000, 6.2 million hectares, or 15.3 million acres, of *Bt* corn were planted in the United States; this represents about 20% of the U.S. corn crop.) A 1999 study suggests that monarchs in the wild are not at significant risk from *Bt* corn. ■

Quarantine

Governments attempt to prevent the importation of foreign pests and diseases by practicing **quarantine**, or restriction of the importation of exotic plant and animal material that might harbor pests. If a foreign pest is accidentally introduced, quarantine of the area where it is detected helps prevent its spread. If a foreign pest is detected on a farm, the farmer may be required to destroy the entire crop.

Quarantine is an effective, although not foolproof, means of control. The USDA has blocked the accidental importation of medflies on more than 100 separate occasions. On the few occasions when quarantine failed and these insects successfully passed into the United States, millions of dollars' worth of crop damage has been incurred in addition to the millions spent to eradicate the pests. Eradication efforts include the use of helicopters to spray the insecticide malathion over hundreds of square kilometers and the rearing and releasing of millions of sterile males to breed the medfly out of existence.

Many experts think that the repeated finds of medflies in California indicate that, rather than being accidentally introduced each year, the medfly has become established in the state. If so, there are potentially disastrous consequences for California's $18 billion agricultural economy. Other countries could stop importing California produce or require expensive inspections and treatments of every shipment in order to prevent the importation of the medfly into their countries.

Integrated Pest Management

Many pests cannot be controlled effectively with a single technique; a combination of control methods is often more effective. **Integrated pest management (IPM)** combines the use of a variety of biological, cultivation, and pesticide controls that are tailored to the conditions and crops of an individual farm (Figure 22.14). Biological and genetic controls, including GM crops designed to resist pests, are used as much as possible, and conventional pesticides are used sparingly and only when other methods fail (see "Meeting the Challenge: Reducing Agricultural Pesticide Use by 50% in the United

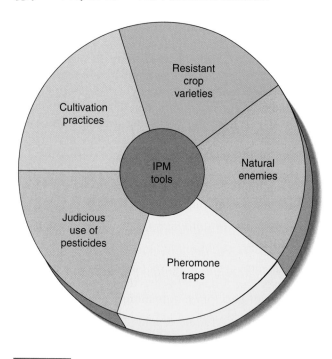

Figure 22.14 Tools of integrated pest management (IPM).
Cultivation practices are agricultural practices, such as crop rotation, that reduce the pest populations for a particular crop from one season to the next. Where available, resistant crop varieties are grown to protect against important pests. IPM encourages conditions for natural enemies of pests. Pheromone traps keep pest populations at acceptable levels because they lure and destroy male insects, thereby reducing the insects' rate of reproduction. Pesticides are applied only after available alternatives have been tried and the pest population is large enough to cause economic damage.

States"). When pesticide methods are required, the least toxic pesticides that are effective are applied in the lowest possible quantities. Thus, IPM allows the farmer to control pests with a minimum of environmental disturbance and often at a minimal cost.

In order to be effective, IPM requires a thorough knowledge of the life cycles, feeding habits, travel, and nesting habits of the pests as well as all of their interactions with their hosts and other organisms. The timing of planting, cultivation, and treatments with biological controls is critical and is determined by carefully monitoring of the concentration of pests. Integrated pest management also optimizes natural controls by using agricultural techniques that discourage pests. Integrated pest management is an important part of **sustainable agriculture** (see Chapter 18).

There are two fundamental premises associated with IPM. First, IPM is the *management* rather than the eradication of pests. Farmers who have adopted the principles of IPM allow a low level of pests in their fields and accept a certain amount of economic damage caused by the pests. These farmers do not spray pesticides at the first

sign of a pest. Instead, they periodically sample the pest population in the field to determine when the pest population reaches an *economic injury threshold* at which the benefit of taking action (such as the judicious use of pesticides) exceeds the cost of that action.

Second, IPM requires that farmers be educated so they can adopt the strategies that will work best in their particular situations. Managing pests is much more complex than trying to eradicate them. The farmer must know what pests to expect on each crop, and what steps can be taken to minimize their effects. The farmer must also know what beneficial species will assist in controlling the pests, and how these beneficial species can be encouraged.

Cotton, which is attacked by many insect pests, has responded well to IPM. Cotton has the heaviest insecticide application of any crop: Although only about 1% of agricultural land in the United States is used for this crop, cotton accounts for almost 50% of all the insecticides used in agriculture! By applying simple techniques such as planting a strip of alfalfa adjacent to the cotton field, the need for chemical pesticides is lessened. Lygus bugs, a significant pest of cotton, move from the cotton field to the strip of alfalfa, which they prefer as a food. Thus, less damage is done to the cotton plants.

Adoption of IPM by U.S. farmers has steadily increased since the 1960s, but the overall proportion of

---

ENVIROBRIEF

Organic Cotton

Although organic foods are doing well in the United States, consumers have traditionally been unwilling to pay the premium price placed on clothes made from pesticide- and fertilizer-free cotton. Now, however, the organic cotton industry, backed by several of the country's largest garment manufacturers, is slowly making gains. Patagonia, Inc., which is the most enthusiastic backer of organic cotton, shifted its entire cotton line to organic fibers in 1996. More recently, organic cotton farmers made an unusual deal with Levi-Strauss, Nike, and the Gap. These companies agreed to buy organic cotton and mix it with ordinary cotton. Shoppers cannot buy all-organic clothing, but the clothing companies can tout their environmental commitment, and organic cotton farmers have gained a tremendous foothold in the industry. Growing interest in personal care products made from organic cotton (cotton balls, ear swabs, tampons, etc.) has also boosted the U.S. industry, which planted an estimated 11,500 acres of organic cotton across seven states in 2001.

These changes couldn't come soon enough, as runoff from cotton farming is the most significant source of water pollution in some states, and globally, more insecticides are used to grow cotton than any other crop. The European Union is financing a $12.7 million effort to grow environment-friendly cotton in Asia, including China, India, and Pakistan, which together produce 46% of the world's cotton. Farmers there will be trained in integrated pest management techniques. The project's goals are to cut pesticide use by half, while increasing cotton production.

MEETING THE CHALLENGE

Reducing Agricultural Pesticide Use by 50% in the United States

Pesticides have benefited farmers (by increasing agricultural productivity) and consumers (by lowering food prices). But pesticide use has had its price—not necessarily in economic terms, for it is difficult to assign monetary values to many of its effects—but in terms of health problems and damage to agricultural and natural ecosystems. Society is increasingly concerned about pesticide use.

Is it feasible to ban all pesticides known to cause cancer in laboratory animals? Probably not, at least not now. In many cases, substitute pesticides either do not exist, are less effective, or are considerably more expensive. A pesticide ban would increase food prices, although estimates on the magnitude of that increase vary considerably from one study to another. A pesticide ban would also cause considerable economic hardship for certain growers, although other farmers might benefit. For example, some insects are more troublesome in certain areas than in others. A farmer growing a crop in an area where an insect was very harmful might not be able to afford the crop losses that would occur without the use of a banned pesticide. Growers in areas where the insect was less of a problem could then increase their production of that crop, benefiting financially from the first farmer's loss.

Since it is impractical to ban large numbers of pesticides right now, is there another way to provide greater protection to the environment and human health without reducing crop yields? Governments in Sweden, Denmark, the Netherlands, and the Canadian Province of Ontario think so. Sweden achieved a 50% reduction in pesticide use in 1992 and is now on a second program to reduce pesticide use by another 50%. Denmark, the Netherlands, and the Province of Ontario also implemented similar programs to reduce

pesticide use by 50% during the 1990s. As of 1999, Denmark had achieved this goal, and the Netherlands and Ontario were still working toward it. Strategies to reduce pesticide use include removing subsidies that encourage pesticide use, applying pesticide only when needed, using improved application equipment, and adopting integrated pest management (IPM) practices.

Too often pesticides are applied unnecessarily to prevent a possible buildup of pests. **Calendar spraying** is the regular use of pesticides regardless of whether pests are a problem or not. Pesticide use can be decreased by continually monitoring pests so that pesticides are applied only when pests become a problem; this technique is known as **scout-and-spray**. A 1991 study at Cornell University determined that a monitoring program might reduce the use of insecticides on cotton by 20% for example.

The use of aircraft to apply pesticide is extremely wasteful because 50% to 75% of the pesticide does not reach the target area, but instead drifts in air currents until it settles on soil or water (recall Figure 22.10). Pesticides applied on land with traditional methods also drift in air currents. Advances in the design of equipment for applying pesticides could reduce pesticide use considerably. A recently developed rope-wick applicator reduces herbicide use on soybean fields by approximately 90%.

Pesticide use can also be reduced considerably through alternative pest control strategies. The widespread adoption of IPM would make it feasible, based on current available technologies, for the United States to reduce pesticide use by 50% within a 5- to 10-year period, assuming that congress passed the necessary legislation (similar to what Sweden and Denmark passed). The economic results of such a reduction would benefit farmers and consumers.

---

farmers using IPM is still small. One reason that IPM is not more widespread is that the knowledge required to use pesticides is relatively simple compared to the complex, sophisticated knowledge needed to use IPM.

Integrated pest management has been most successful in controlling insect pests. Scientists are now trying to develop IPM techniques to effectively control weeds with a minimal use of herbicides. There is also a widely recognized need to develop IPM techniques to reduce pesticide use in urban and suburban environments.

CASE-IN-POINT **Integrated Pest Management in Asia**

Beginning in the 1950s, rice farmers in many Asian countries applied pesticides on their paddies several times during each growing season. They were encouraged to do so by both scientists and chemical pesticide companies. In the 1980s and 1990s, however, a quiet revolution against the indiscriminate use of pesticides occurred. The International Rice Research Institute (IRRI), the world's largest scientific establishment devoted to rice cultivation, promoted this revolution.

The IRRI has gone into different rice-growing areas in Asia and asked local farmers to conduct a group experiment. Some farmers were asked to not spray *any* pesticide during the first 40 days of the growing season. These farmers, who received training in IPM, were also told to use the more traditional cultivation practices that were abandoned when pesticides were introduced. Other farmers were asked to spray pesticides on their rice crops as usual, including the two, three, or more "preventative" treatments they usually apply during the early part of the growing season; this group served as the experiment's control. When the crops were harvested, the rice yields on treated and untreated fields were compared. Many farmers were astonished to learn that untreated fields had yields as large as or larger than those of treated fields (Figure 22.15).

The prevailing agricultural philosophy—to wipe out pests while their populations are low by repeatedly applying pesticides throughout the growing season—is expensive, damaging to the environment, and, as the rice farmers discovered, ineffective. Farmers reported seeing many more natural enemies of rice pests, such as spiders, frogs, and carnivorous beetles, in the untreated fields than in fields that were regularly sprayed with pesticides. Predator populations, which were normally held in check by pesticides, had a chance to grow when paddies were untreated.

The Indonesian government helped encourage the widespread adoption of IPM, in part by phasing out pesticide subsidies. (Many governments subsidize pesticide use on the assumption that pesticide use boosts crop production. These subsidies, which make pesticides cheaper and more available, encourage the inefficient use or overuse of pesticides and discourage the adoption of alternative pest control measures.) Between 1985 and 1990, Indonesian pesticide subsidies declined from $141 million U.S. dollars to zero, according to the World Bank. Pesticide production and importation declined by 58%, whereas rice production rose by 14% during that time period.

Irradiating Foods

It is possible to prevent insects and other pests from damaging harvested food without using pesticides. The food can be harvested and then exposed to ionizing radiation (usually gamma rays from cobalt 60), which kills many microorganisms, such as *Salmonella*, a bacterium that causes food poisoning. Numerous countries, such as Canada, much of Western Europe, Japan, Russia, and Israel, extend the shelf lives of foods with irradiation. The U.S. Food and Drug Administration (FDA) approved this process for fruits and vegetables as well as fresh poultry in 1986, and the first irradiated food was sold in the United States in 1992. In 2000, the USDA approved the irradiation of additional raw meats, such as ground beef, steaks, and pork chops, to eliminate bacterial contamination.

The irradiation of foods is somewhat controversial. Some consumers are concerned because they fear that irradiated food is radioactive. This is not true. In the same way that you do not become radioactive when your teeth are x-rayed, food does not become radioactive when it is irradiated. Therefore, you are not exposed to radiation when you eat irradiated foods. Nevertheless, many U.S. consumers refuse to purchase food that is labeled as having been irradiated.

Critics of irradiation are concerned because irradiation forms traces of certain chemicals called free radicals, some of which have been demonstrated to be carcinogenic in laboratory animals. They also point out that we do not know the long-term effects of eating irradiated foods. Proponents of irradiation argue that free radicals normally occur in food and are also produced by cooking processes such as frying and broiling. They assert that more than 1,000 investigations of irradiated foods, conducted worldwide for more than 3 decades, have demonstrated that it is safe. Furthermore, irradiation lessens the need for pesticides and food additives.

Figure 22.15 Rice production and pesticide use in Indonesia, 1972 to 1990. The decline in pesticide use in the late 1980s and early 1990s did not cause a decrease in rice yields. Instead, rice production increased during the 4 years following the new policy. Indonesia was the first Asian country to widely embrace integrated pest management (IPM). It phased out pesticide subsidies, banned the use of dozens of pesticides on rice, and trained more than 200,000 farmers in IPM techniques.

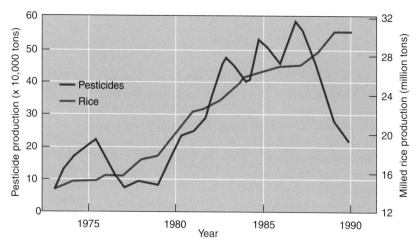

LAWS CONTROLLING PESTICIDE USE

The federal government has passed several laws to regulate pesticides in the interest of protecting human health and the environment. These include the Food, Drug, and Cosmetics Act; the Federal Insecticide, Fungicide, and Rodenticide Act; and the Food Quality Protection Act. The EPA and, to a lesser extent, the FDA and the USDA oversee the implementation of these laws.

Food, Drug, and Cosmetics Act

The **Food, Drug, and Cosmetics Act (FDCA)**, passed in 1938, recognized the need to regulate pesticides found in food but did not provide a means of regulation. The FDCA was made more effective in 1954 with the passage of the **Pesticide Chemicals Amendment**. This amendment, also called the **Miller Amendment**, required the establishment of acceptable and unacceptable levels of pesticides in food.

An amended FDCA, passed in 1958, contained an important section known as the **Delaney Clause**, which stated that no substance capable of causing cancer in test animals or in humans would be permitted in processed food. Processed foods are prepared in some way, such as frozen, canned, dehydrated, or preserved, before being sold. The Delaney Clause recognized that pesticides tend to concentrate in condensed processed foods, such as tomato paste and applesauce.

The Delaney Clause, although commendable for its intent, had two inconsistencies. First, it did not cover pesticides on raw foods such as fresh fruits and vegetables, milk, meats, fish, and poultry. As an example of this double standard, residues of a particular pesticide might be permitted on fresh tomatoes but not in tomato ketchup. Second, because the EPA lacked sufficient data on the cancer-causing risks of pesticides that have been used for a long time, the Delaney Clause applied only to pesticides that were registered after strict tests were put into effect in 1978. This situation gave rise to one of the paradoxes of the Delaney Clause. There were cases in which a newer pesticide that posed minimal risk was banned because of the Delaney Clause, but an older pesticide, which the newer one was to have replaced, was still used despite the fact that it was many times more dangerous.

When the Delaney Clause was passed, the technologies for detecting pesticide residues could only reveal high levels of contamination. Modern scientific techniques are so sensitive, however, that it is almost impossible for any processed food to meet the Delaney standard. As a result, the EPA found it difficult to enforce the strict standard required by the Delaney Clause. In 1988 the EPA began granting exceptions that permitted a "negligible risk" of one case of cancer in 70 years for every 1 million people. However, the EPA was taken to court because of its failure to follow the Delaney Clause as cur-rently written, and the U.S. courts decided in 1994 that no exceptions could be granted unless Congress modified the Delaney Clause. (One of the key provisions of the 1996 Food Quality Protection Act, discussed shortly, did just that.)

Federal Insecticide, Fungicide and Rodenticide Act

The **Federal Insecticide, Fungicide, and Rodenticide Act (FIFRA)** was originally passed in 1947 to regulate the effectiveness of pesticides—that is, to prevent people from buying pesticides that did not work. FIFRA has been amended over the years to require testing and registration of the active ingredients of pesticides. Also, any pesticide that does not meet the tolerance standards established by the FDCA must be denied registration by FIFRA.

In 1972 the EPA was given the authority to regulate pesticide use under the terms of the FDCA and FIFRA. Since that time, the EPA has banned or restricted the use of many chlorinated hydrocarbons. In 1972 the EPA banned DDT for almost all uses. Aldrin and dieldrin were outlawed in 1974 after more than 80% of all dairy products, fish, meat, poultry, and fruits were found to contain residues of these insecticides. The banning of kepone occurred in 1977 and of chlordane and heptachlor in 1988.

A 2-year study by the National Research Council concluded in 1987 that U.S. laws regarding pesticide residues in food were not adequate to protect the public from cancer-causing pesticides (Table 22.4). It included several recommendations that were made into law—an amended FIFRA—in 1988. The 1988 law required

Table 22.4	Worst-Case Estimates of Risk of Cancer from Pesticide Residues on Food
Food	**Cancer Risk***
Tomatoes	$8.75 \times 10^{-4\dagger}$
Beef	6.49×10^{-4}
Potatoes	5.21×10^{-4}
Oranges	3.76×10^{-4}
Lettuce	3.44×10^{-4}
Apples	3.23×10^{-4}
Peaches	3.23×10^{-4}
Pork	2.67×10^{-4}
Wheat	1.92×10^{-4}
Soybeans	1.28×10^{-4}

* Please note that these figures are worst-case estimates. Four assumptions were made in arriving at these values: (1) the entire U.S. crop (of tomatoes, for example) is treated (2) with *all* pesticides that are registered for use on that crop; (3) the pesticides are applied the maximum number of times (4) at the maximum rate, or amount, each time.
† As an example of how to interpret these figures, tomatoes are estimated to cause an average of 8.75 deaths from cancer for every 10,000 people.

reregistration of older pesticides, which subjected them to the same toxicity tests that new pesticides face.

The 1988 law had its critics. Although it was stricter than previous legislation, it represented a compromise between agricultural interests, including pesticide manufacturers, and those opposed to all uses of pesticides. The new law did not address a very important issue, the contamination of groundwater by pesticides. It also failed to address the establishment of standards for pesticide residues on foods and the safety of farm workers who are exposed to high levels of pesticides.

FIFRA also did not require pesticide companies to disclose the inert ingredients in their formulations. Many pesticide products contain as much as 99% inert ingredients, which are not supposed to have active properties against the target organism. The National Coalition for Alternatives to Pesticides has examined the EPA's chemical ingredient database and determined that 394 of the more than 2,500 chemicals listed as "inert" were once listed as active ingredients. Many inert ingredients are generally recognized as safe (examples include pine oil, ethanol, silicone, and water). Others are known to be toxic (examples include asbestos, benzene, formaldehyde, lead, and cadmium). In fact, more than 200 inert ingredients are classified as hazardous air and water pollutants, and 21 are known or suspected to cause cancer.

Food Quality Protection Act

The **Food Quality Protection Act** of 1996 amended both the FDCA and FIFRA. It revised the Delaney Clause by establishing identical pesticide residue limits—those that pose a negligible risk—for both raw produce and processed foods. The law requires that the increased susceptibility of infants and children to pesticides be considered when establishing pesticide residue limits for some 9,700 pesticide uses on specific crops. The pesticide limits are established for all health risks, not just cancer. For example, the EPA must develop a program to test pesticides for endocrine-disrupting properties. The EPA has until 2006 to reevaluate pesticide limits, which must include the cumulative effect of all exposures, from household bug sprays to traces in drinking water and food. Another key provision of the Food Quality Protection Act is that it reduces the time it takes to ban a pesticide considered dangerous, from 10 years to 14 months.

■ THE MANUFACTURE AND USE OF BANNED PESTICIDES

Some U.S. companies manufacture pesticides that have been banned or heavily restricted in the United States and export them to developing countries, particularly in Asia, Africa, and Latin America. International trade of banned or restricted pesticides is notoriously difficult to

monitor. The Foundation for Advancements in Science and Education began documenting the extent of the trade in hazardous pesticides in 1991 by examining customs records from U.S. ports. The latest available data indicate that during 1995 and 1996, U.S. companies exported 9,524 metric tons (10,504 tons) of pesticides forbidden from use in the United States. This amount includes pesticides that are considered too dangerous to use in the United States as well as pesticides that have never been evaluated by the EPA. Other highly developed nations also export banned pesticides.

The U.N. Food and Agriculture Organization (FAO) is attempting to help developing nations become more aware of dangerous pesticides. It established a "red alert" list of more than 50 pesticides that have been banned in five or more countries. The FAO further requires that the manufacturers of these pesticides inform importing countries about why such pesticides have been banned. The United States supports these international guidelines and exports banned pesticides only with the informed consent of the importing country. However, many foreign farmers never receive these guidelines or any training on the safe storage and application of pesticides.

Another concern relating to banned pesticides is that unwanted stockpiles of leftover, deteriorated pesticides are accumulating, particularly in developing countries. The United Nations estimates there are more than 100,000 tons of these obsolete pesticides in developing countries. They are often stored in drums at waste sites in the countryside because developing countries have few or no hazardous waste disposal facilities. Over time, chemicals leach from such waste sites into the soil, and from there they get into waterways and groundwater (see Chapter 23).

The Importation of Food Tainted with Banned Pesticides

The fact that many dangerous pesticides are no longer being used in the United States is no guarantee that traces of those pesticides are not in our food. Although many pesticides have been restricted or banned in the United States, they are widely used in other parts of the world. Much of our food—some 1.2 million shipments annually—is imported from other countries, particularly those in Latin America. Some produce contains traces of banned pesticides such as DDT, dieldrin, chlordane, and heptachlor.

It is not known how much of the food coming into the United States is tainted with pesticides. The FDA monitors toxic residues on incoming fruits and vegetables, but it is able to inspect only about 1% of the food shipments that enter the United States each year. In addition, the General Accounting Office reports that some food importers illegally sell food after the FDA has found it to be tainted with pesticides. When caught, these companies face fines that are not severe enough to discourage such practice.

The Global Ban of Persistent Organic Pollutants

The **Stockholm Convention on Persistent Organic Pollutants**, which was adopted in 2001, is an important international treaty that seeks to protect human health and the environment from the 12 most toxic chemicals on Earth (Table 22.5). Classified as **persistent organic pollutants**, or **POPs**, these chemicals, 9 of which are pesticides, bioaccumulate in organisms and can travel thousands of kilometers through air and water to contaminate sites far removed from their source (see discussion of long-distance transport of air pollution in Chapter 19). Some POPs also disrupt the endocrine system, cause cancer, or adversely affect the developmental processes of organisms.

The Stockholm Convention requires countries to develop plans to eliminate the production and use of intentionally produced POPs. A notable exception to this requirement is that DDT can still be produced and used to control mosquitoes that carry the malaria pathogen in countries where no affordable alternatives exist. (DDT is inexpensive, and many of these countries cannot afford safer alternatives.) The treaty also specifies that efforts should be undertaken to eliminate, where feasible, the unintentional production of industrial by-products such as dioxins and furans.

CHANGING ATTITUDES

Heavy pesticide use can be attributed partly to the consumer, who has come to expect perfect, unblemished produce. There is no question that pesticides help farm-

Figure 22.16 Organically grown produce in a supermarket. If consumers are willing to pay extra for food that is grown in the absence of pesticides, then farmers will grow more pesticide-free crops.

ers grow crops that are more visually appealing. However, consumers must ask themselves, would they rather buy apples that are smaller and have an occasional blemish or worm but are pesticide-free, or apples that are free from all imperfections but contain traces of pesticides? Until consumers change their attitudes and demand produce that is grown without pesticides, farmers will continue to use pesticides (Figure 22.16).

Many farmers are exploring alternatives to pesticides on their own because they come into contact with pesticides on a regular basis. These farmers are aware of the dangers and problems associated with pesticide use and are concerned for their own safety and that of their families. They do not want to inhale pesticide or let it settle on their skin when they are applying it. They do not want to drink water or eat food with traces of pesticide any more than the typical consumer does.

Pesticide Risk Assessment

Although the effects of long-term exposure to low levels of carcinogenic pesticides should be of concern to all informed consumers, panic, which is often fueled by sensational news reports, is not justified. It is important to keep a balanced perspective when considering pesticides. The threat of cancer from consuming pesticide residues on our food is quite small compared to the threat of cancer from smoking cigarettes or from overexposure to ultraviolet radiation from the sun (see Chapter 2 discussion of risk assessment).

However, it is difficult for consumers to make informed decisions about the risks of pesticide residues on food, because we have no way of knowing what kinds of pesticides have been used on the foods we eat. A requirement that all foods list such chemicals would go a long way toward helping us assess risks, as well as discouraging heavy pesticide use.

Table 22.5	Persistent Organic Pollutants: The "Dirty Dozen"
Persistent Organic Pollutant	*Use*
Aldrin	Insecticide
Chlordane	Insecticide
DDT (dichlorodiphenyl-trichloroethane)	Insecticide
Dieldrin	Insecticide
Endrin	Rodenticide and insecticide
Heptachlor	Fungicide
Hexachlorobenzene	Insecticide; fire retardant
Mirex™	Insecticide
Toxaphene™	Insecticide
PCBs (polychlorinated biphenyls)	Industrial chemical
Dioxins	By-product of certain manufacturing processes
Furans (dibenzofurans)	By-product of certain manufacturing processes

SUMMARY WITH SELECTED KEY TERMS

I. Pesticides are toxic chemicals that are used to kill pests such as insects (**insecticides**), weeds (**herbicides**), fungi (**fungicides**), and rodents (**rodenticides**).

A. The ideal pesticide would be a **narrow-spectrum pesticide** that kills only the target organism.

B. Most pesticides are **broad-spectrum pesticides**, which kill a variety of organisms besides the target organism.

II. Insecticides are classified into groups based on their chemical structure.

A. Chlorinated hydrocarbons are organic compounds containing chlorine. **Dichlorodiphenyltrichloroethane, or (DDT)** is a broad-spectrum chlorinated hydrocarbon that is highly persistent in the environment; its use has been banned in the United States and many other countries.

B. Organophosphates are organic compounds that contain phosphorus. As a group, they are highly toxic to many organisms, although they do not persist in the environment as long as chlorinated hydrocarbons do.

C. Carbamates are derived from carbamic acid. As a group, they show broad, nontarget toxicity, although they have a low persistence in the environment.

III. Herbicides may be classified as either selective or nonselective.

A. Selective herbicides kill only certain types of plants. Selective herbicides include **broad-leaf herbicides**, which kill plants with broad leaves but do not kill grasses, and **grass herbicides**, which kill grasses but are safe for most other plants.

B. Nonselective herbicides kill all vegetation.

IV. Pesticides provide important benefits to humans.

A. Pesticides help prevent diseases such as malaria that are transmitted by insects.

B. Pesticides reduce crop losses from pests, thereby increasing agricultural productivity.

1. Pesticides reduce competition with weeds, consumption by insects, and diseases caused by plant **pathogens** such as certain fungi and bacteria.

2. Agricultural fields are **monocultures** that provide abundant food for pest organisms. Many natural predators are absent from monocultures.

V. Pesticide use has caused environmental problems.

A. Pests have evolved **genetic resistance**, inherited characteristics that decrease the effect of the pesticide on the pest. **Resistance management** is a series of techniques used to delay the **evolution** of genetic resistance in the population so that the period in which an insecticide is useful can be maximized.

B. Pesticides affect species other than those for which they are intended, causing imbalances in ecosystems. In some instances the use of a pesticide has resulted in a pest problem that did not exist before.

C. Other problems associated with some types of pesticides are persistence, bioaccumulation, and biological magnification.

1. A pesticide that demonstrates **persistence** takes a long time to be broken down into less toxic forms.

2. Bioaccumulation is the buildup of a persistent pesticide in an organism's body.

3. Biological magnification is an increase in pesticide concentration as the pesticide passes through successive levels of the food web.

D. Many pesticides move through the soil, water, and air, sometimes for long distances.

VI. Some serious health problems are associated with pesticide use.

A. Humans may be poisoned by exposure to a large amount of pesticide.

B. Lower levels of many pesticides may pose a long-term threat of cancer.

C. Certain persistent pesticides may interfere with the actions of natural hormones.

D. Pesticides are a greater threat to children than to adults. For example, pesticides may affect the development of intelligence and motor skills in infants and young children.

VII. Alternative methods exist to reduce the need for pesticides in agriculture.

A. Cultivation techniques such as strip cutting, interplanting, and crop rotation are effective in controlling pests.

B. Biological controls involve the use of naturally occurring disease organisms, parasites, or predators to control pests.

C. Reproductive controls include reducing the pest population by sterilizing some of its members (the **sterile male technique**).

D. Pheromones, natural substances produced by animals to stimulate a response in other members of the same species, can be used to lure insects to traps or to confuse insects so they cannot locate mates.

E. Insect hormones are natural substances produced by insects to regulate their own growth and metamorphosis; a hormone that is present at the wrong time in an insect's life cycle disrupts its normal development.

F. Genetic control of pests involves producing varieties of crops and livestock animals that are genetically resistant to pests. Some **genetically modified** (**GM**) crops contain the *Bacillus thuringiensis* (*Bt*) gene that codes for a toxin used against insect pests.

G. Quarantine involves restricting the importation of exotic plant and animal material that might harbor pests. If a foreign pest is accidentally introduced, quarantine of the area where it is detected helps prevent its spread.

H. Integrated pest management (**IPM**) is a combination of agricultural methods that emphasizes biological controls and cultivation methods along with the judicious use of pesticides.

1. IPM is the management rather than the eradication of pests.

2. IPM requires that farmers be educated so they can adopt the strategies that will work best in a given pest situation.

I. Irradiation of food can be used to control pests after food has been harvested.

VIII. Several U.S. laws regulate pesticides in the interest of protecting human health and the well-being of the environment.

A. The **Food, Drug, and Cosmetics Act (FDCA)**, as originally passed in 1938, recognized the need to regulate pesticides in food but did not provide a means of regulation.

  1. The **Miller Amendment**, passed in 1954, required the establishment of acceptable and unacceptable levels of pesticides in food.

  2. The **Delaney Clause**, passed in 1958, stated that no substance capable of causing cancer in laboratory animals or in humans would be permitted in processed food.

    a. The Delaney Clause overlooked pesticides on raw foods.

    b. It applied only to pesticides that were registered after strict tests were put into effect in 1978.

B. The **Federal Insecticide, Fungicide, and Rodenticide Act (FIFRA)**, originally passed in 1947, has been amended over the years to require testing and registration of the active ingredients of pesticides. The 1988 version required the reregistration of older pesticides, which subjected them to the same toxicity tests that new pesticides face.

C. The **Food Quality Protection Act** of 1996 amended both the FDCA and FIFRA. It revised the Delaney Clause by establishing identical pesticide residue limits—those that pose a negligible risk—for both raw and processed foods.

THINKING ABOUT THE ENVIRONMENT

1. What is a pest? What is a pesticide?

2. Distinguish among *insecticides, herbicides, fungicides,* and *rodenticides.*

3. Explain the difference between *narrow-spectrum* and *broad-spectrum pesticides.*

4. Describe the general characteristics of each of the following groups of insecticides: chlorinated hydrocarbons, organophosphates, and carbamates.

5. Explain two benefits of pesticide use.

6. What is the dilemma referred to in the title of this chapter?

7. Overall, do you think the benefits of pesticide use outweigh its disadvantages? Give at least two reasons for your answer.

8. Sometimes pesticide use can increase the damage done by pests. Explain.

9. How is the buildup of insect resistance to pesticides similar to the increase in bacterial resistance to antibiotics?

10. Distinguish among *persistence, bioaccumulation,* and *biological magnification.*

11. How do pesticides move around in the environment?

12. What is the pesticide treadmill?

13. Describe some of the effects of pesticides on human health.

14. Describe the pesticide disaster that occurred at Bhopal, India.

15. Biological control is often much more successful on a small island than on a continent. Offer at least one reason why this might be the case.

16. It is more effective to use the sterile male technique when an insect population is small than when it is large. Explain.

17. Define *integrated pest management (IPM)*. List five tools of IPM, and give an example of each.

18. Is a major goal of IPM to eradicate the pest species? Explain your answer.

19. Which of the following uses of pesticides do you think are most important? Which are least important? Explain your views.

  a. Keeping roadsides free of weeds

  b. Controlling malaria

  c. Controlling crop damage

  d. Producing blemish-free fruits and vegetables

20. List three laws by which pesticides are regulated in the United States. Summarize the goals of each law.

21. What is the Stockholm Convention on Persistent Organic Pollutants?

22. Why is pesticide misuse increasingly viewed as a *global* environmental problem?

*23. A water sample was measured and found to have 0.00005 ppm of DDT. The plankton living in the water were then measured and found to have 800 times that amount of DDT. What was the concentration of DDT, in ppm, in the plankton?

*24. Worldwide, the pesticide industry sold $27.8 billion in pesticides in 1994. Preliminary data indicate that sales in 1998 were $34 billion. Calculate the average percent *annual* increase in pesticide sales from 1994 to 1998.

*Solutions to questions preceded by an asterisk appear in Appendix VII.

TAKE A STAND

Visit our Web site at **http://www.wiley.com/college/raven** (select Chapter 22 from the Table of Contents) for links to more information about issues surrounding the genetic engineering of the *Bt* gene into crops. Consider the opposing views of those who support and those who oppose the use of genetic engineering for this purpose and debate the issue with your classmates. You will find tools to help you organize your research, analyze the data, think critically about the issues, and construct a well-considered argument. Take a Stand activities can be done individually or as part of a team, as oral presentations, written exercises, or Web-based (e-mail) assignments.

Additional on-line materials relating to this chapter, including Student Quizzes, Activity Links, Useful Web Sites, Flash Cards, and more, can also be found on our Web site.

SUGGESTED READING

Carson, R.L. *Silent Spring.* Boston: Houghton Mifflin, 1962. The classic book that first alerted the public to the environmental dangers of pesticide use.

Colburn, T., D. Dumanoski, and J.P. Myers. *Our Stolen Future.* New York: Dutton, 1996. Drawing on work from many fields, this controversial book records the rapidly unfolding story of commonly used chemicals that disrupt the activities of hormones.

Dickie, P. "Use as Directed: How Much Do We Know About Home Pest Control?" *Living Planet* (summer 2001). A quick consumer course on home pesticide use.

Dreyfuss, R. "Apocalypse Still." *Mother Jones* (February 1, 2000). A well-researched article on the effects on the Vietnamese people and land of Agent Orange use during the Vietnam War.

Helmuth, L. "Pesticide Causes Parkinson's in Rats." *Science,* Vol. 290 (November 10, 2000). This news article reports on a study that tentatively implicates pesticide exposure to the development of Parkinson's disease.

Holcomb, B. "How Safe Is Your Dinner?" *Good Housekeeping* (March 2000). This short article answers many of the questions consumers have about irradiated food.

Knight, J. "Alien Versus Predator." *Nature,* Vol. 412 (July 12, 2001). The use of biological controls is gaining in popularity.

McGinn, A.P. "Malaria, Mosquitoes, and DDT." *WorldWatch,* Vol. 15, No. 3 (May–June 2002). The war against malaria and the mosquitoes that transmit it is relentless.

Obrycki, J.J., J.E. Losey, O.R. Taylor, and L.C.H. Jesse. "Transgenic Insecticidal Corn: Beyond Insecticidal Toxicity and Ecological Complexity." *BioScience,* Vol. 51, No. 5 (May 2001). Don't let the title scare you away from this informative article on *Bt* crops.

Raloff, J. "The Case for DDT." *Science News,* Vol. 158 (July 1, 2000). The use of the environmental pollutant DDT saves lives.

Strong, D.R., and R.W. Pemberton. "Biological Control of Invading Species—Risk and Reform." *Science,* Vol. 288 (June 16, 2000). Biological controls can be used successfully against invasive species, but they are not a panacea.

Yoon, C.K. "When Biological Control Gets Out of Control." *New York Times* (March 6, 2001). A fascinating account of a biological control agent that is now eating many nontarget species.

Obsolete computer equipment.

Solid and Hazardous Wastes

Learning Objectives

After you have studied this chapter you should be able to

1. Distinguish between *municipal solid waste* and *nonmunicipal solid waste*.

2. Describe the features of a modern sanitary landfill and relate some of the problems associated with sanitary landfills.

3. Describe the features of a mass burn incinerator and relate some of the problems associated with incinerators.

4. Summarize how source reduction, reuse, and recycling help reduce the volume of solid waste.

5. Define *hazardous waste* and briefly characterize representative hazardous wastes (dioxins, PCBs, and radioactive wastes).

6. Contrast the Resource Conservation and Recovery Act and the Comprehensive Environmental Response, Compensation, and Liability Act (the Superfund Act).

7. Discuss the scope, accomplishments, and shortcomings of the Superfund Program.

8. Define *environmental justice* and discuss the issue of international waste management as it relates to environmental justice.

I n the United States and other highly developed countries, the average computer is replaced every 2 years, not because it is broken but because rapid technological developments and new generations of software make it obsolete. Old computers may still be in working order, but they have no resale value and are even difficult to give away. As a result, they frequently are thrown away with the trash and end up in landfills or incinerators. According to one estimate, 5 million to 7 million tons of computer and electronic waste were discarded in the United States in 1998 (latest data available).

This disposal represents a huge waste of the high-quality plastics and metals (aluminum, copper, tin, nickel, palladium, silver, and gold) that make up computers. Computers also contain the toxic heavy metals lead, cadmium, mercury, and chromium, which could potentially leach from landfills into soil and groundwater. Most computers contain from 3 to 8 pounds of lead, for example.

In addition, there is a market for old computers because many groups, such as nonprofit organizations, schools, and people with disabilities, cannot afford the current generation of computers but could benefit from inexpensive, entry-level computers. This situation has provided a business opportunity for certain companies.

For example, Technology Recycling in Denver, Colorado, reclaims and resells old computers, monitors, fax machines, printers, and other electronic equipment they obtain from businesses and individuals. They also receive obsolete equipment from computer and telecommunications equipment manufacturers. The company charges a fee to pick up, haul, dismantle, and recycle old computer systems, which are sorted, tested, and, where possible, repaired. Computer equipment beyond economic repair is broken down for individual components or stripped for recycling of metals, plastics, and glass. For example, the plastics in broken components are recycled to make such items as park benches and shelving.

Technology Recycling is not the only U.S. company in the electronics recycling business. Because dozens of entrepreneurs are entering the field, the International Association of Electronics Recyclers, Inc., the trade association for the electronics recycling industry, is developing environmental standards and standard business practices for recycling electronics.

Other inventive ways to deal with obsolete computers are being tried. Conigliaro Industries, which received grants from the American Plastics Council and the state of Massachusetts Department of Environmental Protection, has demonstrated that the plastic parts from computers and other electronic equipment can be converted into an asphalt-like material for filling potholes. The state of Massachusetts is also removing and recycling valuable metals from computers' circuit boards.

Currently, about 11% of discarded computer and electronic components are recycled in the United States. Although companies like Technology Recycling and Conigliaro Industries handle obsolete computers in the United States, most U.S. computers to be recycled are shipped overseas to developing countries such as India, Pakistan, and China. There the computers are disassembled, often by methods that are potentially dangerous to the workers taking them apart. Other highly developed countries have been more progressive than the United States in dealing with their computer waste. The European Parliament approved a plan in 2001 that requires electronics manufacturing companies to take back discarded electronic waste and remove some of the most toxic chemicals. Similarly, Japan has passed an electric appliance recycling law that will soon include computers.

Like it or not, humans produce solid waste, and it is our responsibility to deal with our waste in safe, cost-effective, and environmentally sensitive ways. There is no single solution to the challenge of solid waste disposal today. A combination of source reduction, reuse, recycling, composting, burning, and burying in sanitary landfills is currently the optimal way to manage solid waste. In this chapter we examine the problems and the opportunities associated with the management of solid waste. We then examine hazardous waste, which contains toxic materials and requires special disposal.

THE SOLID WASTE PROBLEM

The United States generates more solid waste, per capita, than any other country. (Canada is a close second.) Every man, woman, and child in the United States produces an average of 2.0 kg (4.5 lb) of solid waste per day. This amount corresponded to a total of 210 million metric tons (232 million tons) in 2000 (Figure 23.1). And the problem worsens each year as the U.S. population increases.

The solid waste problem has been made abundantly clear by several highly publicized instances of garbage barges wandering from port to port and from country to country, trying to find someone willing to accept their cargo. In 1987, a garbage barge from Islip, New York, was towed by the tugboat *Break of Dawn* to North Carolina. When North Carolina refused to accept the solid waste, the *Break of Dawn* set off on a journey of many months. In total, six states and three countries rejected the waste, which was eventually returned to New York to be incinerated.

Another example of our solid waste problem is the story of the *Khian Sea*, a Bahamian ship that was hired by the city of Philadelphia in 1986. It was to transport 14,000 tons of incinerated ash to the Bahamas, but before it got to port, the Bahamas said no because they worried that the ash might contain toxic chemicals. For the next few years, the barge tried unsuccessfully to dump the ash in Puerto Rico, the Dominican Republic, the Netherlands Antilles, Honduras, and Guinea-Bissau on the western coast of Africa. In 1988, about 4,000 tons were dumped on a beach in Haiti, which demanded that it be reloaded onto the barge. The barge took off without reloading the ash and then disappeared for several months, only to reappear in Singapore under a new name and minus its cargo, which was dumped somewhere in the Indian Ocean. In 2002 the Pennsylvania Department of Environmental Protection transported the ash remaining in Haiti back to Pennsylvania for permanent disposal.

Waste generation is an unavoidable consequence of prosperous, high-technology, industrial economies. It is a problem not only in the United States but also in other highly developed nations. Many products that have the potential to be repaired, reused, or recycled are simply thrown away. Others, including paper napkins and disposable diapers, are designed to be used once and then discarded. Packaging, which not only makes a product more attractive and therefore more likely to sell, but protects it and keeps it sanitary, also contributes to waste. Nobody likes to think about solid waste, but the fact is that it is a concern of modern soci-

ety—we keep producing it, and places to dispose of it safely are dwindling in number.

Types of Solid Waste

Municipal solid waste consists of solid materials discarded by homes, office buildings, retail stores, restaurants, schools, hospitals, prisons, libraries, and other commercial and institutional facilities. Municipal solid waste is a heterogeneous mixture composed primarily of paper and paperboard, yard waste, plastics, metals, wood, food waste, glass, and other materials such as rubber, leather, and textiles (Figure 23.2). The proportions of the major types of solid waste in this mixture change over time. Today's solid waste contains more paper and plastics than in the past, whereas the amounts of glass and steel have declined.

Municipal solid waste is a relatively small portion of all the solid waste produced. **Nonmunicipal solid waste**, which includes wastes from mining (mostly waste rock, about 75% of the total solid waste production), agriculture (about 13%), and industry (about 9.5%), is produced in substantially larger amounts than municipal solid waste (about 1.5%). Thus, most solid waste generated in the United States is from nonmunicipal sources.

◼ DISPOSAL OF SOLID WASTE

Solid waste has traditionally been regarded as material that is no longer useful and that should be disposed of. There are four ways to get rid of solid waste: dump it, bury it, burn it, or compost it.

Open Dumps

The old method of solid waste disposal was dumping. Open dumps were unsanitary, malodorous places in which disease-carrying vermin such as rats and flies proliferated. Methane gas was released into the surrounding air as microorganisms decomposed the solid waste, and

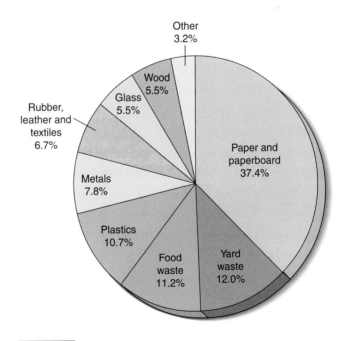

Figure 23.2 **Composition of municipal solid waste.** Municipal solid waste is composed of paper, yard waste, food scraps, plastics, metals, glass, wood, and other materials such as rubber and leather.

fires polluted the air with acrid smoke. Liquid that oozed and seeped through the solid waste heap ultimately found its way into the soil, surface water, and groundwater. Hazardous materials that were dissolved in this liquid often contaminated soil and water.

Sanitary Landfills

Open dumps have been replaced by **sanitary landfills**, which receive 55.4% of the solid waste generated in the United States today (Figure 23.3). Sanitary landfills differ from open dumps in that the solid waste is placed in a hole, compacted, and covered with a thin layer of soil every day (Figure 23.4). This process reduces the number of rats and other vermin usually associated with solid waste, lessens the danger of fires, and decreases the amount of odor. If a sanitary landfill is operated in accordance with solid waste management-approved guidelines, it does not pollute local surface and groundwater. Safety is ensured by the presence of layers of compacted clay and plastic sheets at the bottom of the landfill, which prevent liquid waste from seeping into groundwater. Newer landfills possess a double liner system (plastic, clay, plastic, clay) and use sophisticated systems to collect **leachate** (liquid that seeps through the solid waste) and gases that form during decomposition.

Sanitary landfills charge "tipping fees" to accept solid waste. This money helps offset the landfill's operating costs. It also enables the jurisdiction to charge lower property taxes for homes and businesses located in close proximity to the sanitary landfill. Tipping fees vary widely from one state to another: Sanitary landfills in

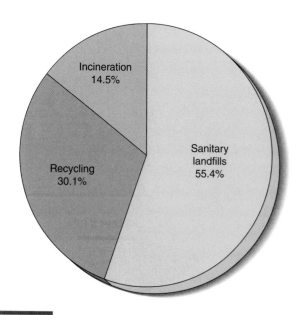

Figure 23.3 Disposal of municipal solid waste in 2000. The recovery of waste from recycling increased in both percentage and tons during the 1990s. In 1990, the percentage of recycling was 16.2%, whereas in 2000, it was 30.1%.

Wyoming currently charge tipping fees of $10 per ton of solid waste, whereas Vermont landfills charge $65 per ton. Some jurisdictions cannot handle all their waste, and so they export their solid waste to nearby states with lower tipping fees. For example, sanitary landfills in Ohio and Indiana accept solid waste from New York and New Jersey.

The location of an "ideal" sanitary landfill is based on a variety of factors, including the geology of the area, soil drainage properties, and the proximity of nearby bodies of water and wetlands. The landfill should be far enough away from centers of dense population to be inoffensive but close enough so as not to require high transportation costs. Landfill designs should take into account an area's climate, such as rainfall, snowmelt, and the likelihood of flooding.

Problems Associated with Sanitary Landfills   Although the operation of sanitary landfills has improved over the years with the passage of stricter and stricter guidelines, very few landfills are ideal. Most sanitary landfills in operation today do not meet current legal standards for new landfills.

One of the problems associated with sanitary landfills is the production of methane gas by microorganisms that decompose organic material anaerobically (in the absence of oxygen). This methane may seep through the solid waste and accumulate in underground pockets, creating the possibility of an explosion. It is even possible for methane to seep into basements of nearby homes, which is an extremely dangerous situation. Conventionally, landfills collected the methane and burned it off in flare systems. A growing number of landfills, however, have begun to use the methane for gas-to-energy projects. About 115 landfills in the United States currently use methane gas to generate electricity, and these facilities collectively produce enough electricity for about 500,000 homes.

Another problem associated with sanitary landfills is the potential contamination of surface water and groundwater by leachate that seeps from unlined landfills or through cracks in the lining of lined landfills. New York's Fresh Kills Landfill, which was the largest in the United States before it closed in 2002, continues to produce an estimated 1 million gallons of leachate each day. Because household trash contains toxic chemicals such as heavy metals, pesticides, and organic compounds that can be carried into groundwater and surface water, the leachate must be collected and treated, even though the sanitary landfill has closed.

Landfills are not a long-term remedy for waste disposal, because they are filling up. During the period from 1978 to 2000, the number of U.S. landfills in operation decreased from 20,000 to 1,967. Many of the landfills closed because they had reached their capacity. Other landfills closed because they did not meet state or federal environmental standards.

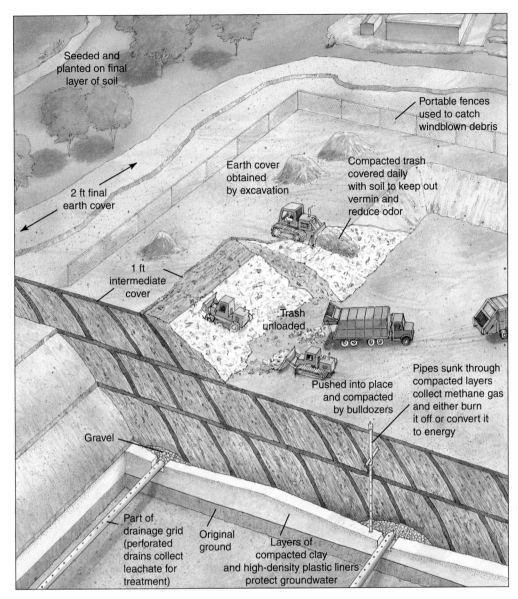

Figure 23.4 Sanitary landfill. Solid waste is spread in a thin layer, compacted into small sections, called cells, and covered with soil. Sanitary landfills constructed today have protective liners of compacted clay and high-density plastic and sophisticated leachate collection systems that minimize environmental problems such as contamination of groundwater.

Labels in figure:

Seeded and planted on final layer of soil

Portable fences used to catch windblown debris

2 ft final earth cover

Earth cover obtained by excavation

Compacted trash covered daily with soil to keep out vermin and reduce odor

1 ft intermediate cover

Trash unloaded

Pushed into place and compacted by bulldozers

Pipes sunk through compacted layers collect methane gas and either burn it off or convert it to energy

Gravel

Part of drainage grid (perforated drains collect leachate for treatment)

Original ground

Layers of compacted clay and high-density plastic liners protect groundwater

---

Fewer new sanitary landfills are being opened to replace the closed ones. The reasons for this are many and complex. Many desirable sites have already been taken. Also, people living near a proposed site are usually adamantly opposed to a landfill near their homes. Recall from Chapter 11 that the opposition of people to the location of hazardous facilities near their homes is known as the **not-in-my-backyard**, or **NIMBY**, response. This attitude is partly the result of past problems with landfills, everything from offensive odors to dangerous contamination of drinking water. It is also caused by the fear that property values will be lowered by a nearby facility.

Closing a Sanitary Landfill   Considerable expense is involved in closing a sanitary landfill once it is full. The possibility of groundwater pollution and gas explosions remains for a long time, so the Environmental Protection Agency (EPA) currently requires owners to continuously

monitor a landfill for 30 years after the landfill is closed. In addition, no homes or other buildings can be built on a closed sanitary landfill for many years.

The Special Problem of Plastic   The amount of plastic in our solid waste is growing faster than any other component of municipal solid waste. More than half of this plastic is from packaging. Plastics are chemical **polymers** that are composed of chains of repeating carbon compounds. The properties of the many types of plastics—polypropylene, polyethylene, and polystyrene, to name a few—differ on the basis of their chemical compositions.

One of the characteristics of most plastics is that they are chemically stable and do not readily break down, or decompose. This characteristic, which is essential in the packaging of certain products, such as food, causes long-term problems: Most plastic debris disposed of in sanitary landfills will probably last for centuries.

In response to concerns about the volume of plastic waste, some places have actually banned the use of certain types of plastic, such as the polyvinyl chloride employed in packaging. Special plastics that have the ability to degrade or disintegrate have been developed. Some of these, however, are **photodegradable**; they break down only after being exposed to sunlight, which means they will not break down in a sanitary landfill. Other plastics are **biodegradable**—that is, decomposed by microorganisms such as bacteria. Whether biodegradable plastics actually break down under the conditions found in a sanitary landfill is not yet clear, although preliminary studies indicate that they probably do not. Many factors, such as temperature, amount of oxygen present, and composition of the microbial community, affect biodegradation of plastics and other organic materials. (Other waste management options for plastic will be discussed later in this chapter.)

The Special Problem of Tires

One of the most difficult materials to manage is rubber. Discarded tires—about 273 million each year in the United States—are made of vulcanized rubber, which cannot be melted and reused for tires. More than 800 million old tires have accumulated in tire dumps, as well as along roadsides and in vacant lots. According to the EPA, 24% of the scrap tires generated in 2001 were disposed of in landfills, stockpiled, or illegally dumped. Disposal of tires in sanitary landfills is a real problem because tires, being relatively large and light, have a tendency to move upward through the accumulated solid waste. After a period, they work their way to the surface of the landfill. These tires are a fire hazard, creating fires that are difficult to extinguish, and collect rainwater, providing a good breeding place for mosquitoes. Accordingly, most states either ban tires from sanitary landfills or require that they be shredded to save space and prevent water from pooling in them. (Other waste management options for tires will be discussed later in the chapter.)

Incineration

When solid waste is incinerated, two positive things are accomplished. First, the volume of solid waste is reduced by up to 90%. The ash that remains is, of course, much more compact than solid waste that has not been burned. Second, incineration produces heat that, if properly channeled, can produce steam to warm buildings or generate electricity. In 2000 the United States had 102 waste-to-energy incinerators, which burned 14.5% of the nation's solid waste. (In comparison, only 1% of U.S. solid waste was incinerated in 1970.) Waste-to-energy incinerators produce substantially less carbon dioxide emissions than equivalent power plants that burn fossil fuels (Figure 23.5). (Recall from Chapter 20 that carbon dioxide is a potent greenhouse gas.)

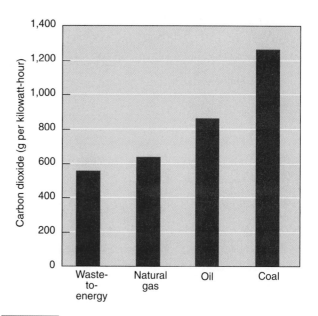

Figure 23.5 Carbon dioxide emissions per kilowatt-hour of electricity generated. Waste-to-energy incinerators release less carbon dioxide into the atmosphere than do equivalent power plants that burn fossil fuels.

Some materials are best removed from solid waste before incineration occurs. Glass does not burn, and when it melts, it is difficult to remove from the incinerator. Although food waste burns, its high moisture content often decreases the efficiency of incineration, so it is better to remove it before incineration. Removal of batteries, thermostats, and fluorescent lights is desirable because it eliminates most mercury emissions produced during combustion. The best materials for incineration are paper, plastics, and rubber.

Paper is a good candidate for incineration because it burns readily and produces a great amount of heat. Several studies have examined the economic and environmental costs and benefits of various waste paper management options. Many of these studies have concluded that waste-to-energy incineration is better than recycling, which in turn is better than disposal in a sanitary landfill.[1] One potential environmental complication with burning paper is the presence in the ink and paper of toxic compounds, which might be emitted during incineration. Some types of paper release dioxins into the atmosphere when burned; dioxins are toxic compounds that are discussed later in the chapter.

Plastic produces a lot of heat when it is incinerated: One kilogram of plastic waste yields almost as much heat

---

[1] The 50 or more studies do not reach unanimous conclusions because economists do not agree about the cash value that should be applied for environmental benefits and costs. For example, estimates for the environmental cost of emitting 1 kg of carbon dioxide range from $1 to more than $50.

Figure 23.6 Tires that will be burned to generate electricity. This mountain in Westley, California, contains 4 million to 6 million old tires. The power plant that burns them supplies electricity to 3,500 homes. (The person wearing red gives a sense of scale.)

as a kilogram of fuel oil. As with paper, there is concern about some of the pollutants that might be emitted during the incineration of plastic. Polyvinyl chloride, a common component of many plastics, may release dioxins and other toxic compounds when incinerated.

One of the best uses for old tires is incineration, because burning rubber produces much heat. Some electric utilities in the United States and Canada are burning tires instead of or in addition to coal (Figure 23.6). Tires produce as much heat as coal and often generate less pollution. In 2001, 42% of all discarded tires were incinerated.

Types of Incinerators   The three types of incinerators are mass burn, modular, and refuse-derived fuel. **Mass burn incinerators** are large furnaces that burn all solid waste except for unburnable items such as refrigerators. Most mass burn incinerators are large and are designed to recover the energy produced from combustion (Figure 23.7). **Modular incinerators** are smaller incinerators that burn all solid waste. They are assembled at factories and so are less expensive to build. In **refuse-derived fuel incinerators**, only the combustible portion of solid waste is burned. First, noncombustible materials such as glass and metals are removed by machine or by hand. The remaining solid waste, including plastic and paper, is shredded or shaped into pellets and burned.

Problems Associated with Incineration   The combustion of any fuel, whether it is coal or municipal solid waste, yields some air pollution. The possible production

of toxic air pollutants is the main reason people oppose incineration. Incinerators can pollute the air with carbon monoxide, particulates, heavy metals such as mercury, and other toxic materials unless expensive air pollution control devices are used. Such devices include **lime scrubbers**, towers in which a chemical spray neutralizes acidic gases, and **electrostatic precipitators**, which give ash a positive electrical charge so that it adheres to negatively charged plates rather than going out the chimney (see Figure 19.10 for diagrams of a scrubber and an electrostatic precipitator).

Incinerators produce large quantities of ash that must be disposed of properly. Two kinds of ash are produced, bottom ash and fly ash. **Bottom ash**, also known as **slag**, is the residual ash left at the bottom of the incinerator when combustion is completed. **Fly ash** is the ash from the flue (chimney) that is trapped by air pollution control devices. Fly ash usually contains more toxic materials, including heavy metals and possibly dioxins, than bottom ash.

Currently, both types of incinerator ash are best disposed of in specially licensed hazardous waste landfills (discussed later in this chapter). What happens to the toxic materials in incinerator ash when it is placed in an average sanitary landfill is unknown, but there are concerns that the toxic materials could contaminate groundwater. In 1994 the Supreme Court considered a case involving a Chicago waste-to-energy incinerator that was disposing of its ash, which contained very high levels of lead and cadmium, in a regular sanitary landfill. The court ruled that incinerator ash must be tested regularly

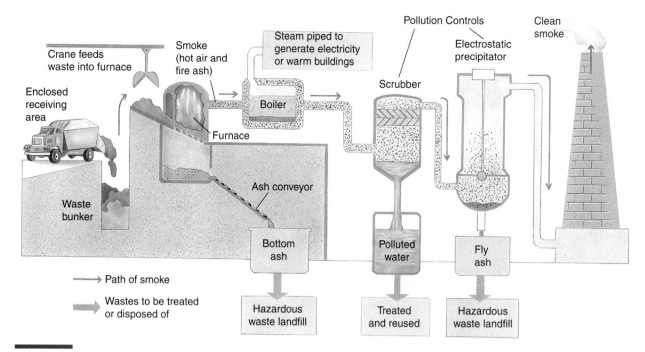

Figure 23.7 Mass burn, waste-to-energy incinerator. Modern incinerators have pollution control devices such as lime scrubbers and electrostatic precipitators to trap dangerous and dirty emissions.

for toxic materials. When the hazardous materials in incinerator ash exceed acceptable levels, the ash must be disposed of in licensed hazardous waste sites.

Site selection for incinerators, like that for sanitary landfills, is controversial. People may recognize the need for an incinerator, but they do not want it near their homes. Another drawback of incinerators is their high cost. Prices have been escalating because costly pollution control devices are now required.

Composting

Yard waste, such as grass clippings, branches, and leaves, is a substantial component of municipal solid waste (see Figure 23.2). As space in sanitary landfills becomes more limited, other ways to dispose of yard waste are being developed and implemented. One of the best ways is to convert organic waste into soil conditioners such as compost or mulch (see "You Can Make a Difference: Practicing Environmental Principles," in Chapter 14). Food scraps, sewage sludge, and agricultural manure are other forms of solid waste that can be used to make compost. Compost and mulch can be used for landscaping in public parks and playgrounds or as part of the daily soil cover at sanitary landfills. Compost and mulch can also be sold to gardeners.

Composting as a way to manage solid waste first became popular in Europe. Several municipalities in the United States have composting facilities as part of their comprehensive solid waste management plans, and more

than 20 states have banned yard waste from sanitary landfills. This trend is likely to continue, making composting even more desirable (see "Meeting the Challenge: Municipal Solid Waste Composting," in Chapter 14). The United States currently has 3,260 municipal composting programs that recycle about 56.9% of yard wastes (Figure 23.8).

Composting Food Waste at Rikers Island

New York City's Rikers Island, site of 10 jails and home to 10,000 inmates, serves as the city's surprising frontrunner in innovative recycling, composting, and gardening. Prisoners compost 6 tons of food scraps each day, in a facility with a roof made of photovoltaic cells. The composting produces high-grade fertilizer used in the prison's garden and farm programs. At the city's largest community garden, in 1998 inmates raised 30,000 pounds of fresh vegetables for prison meals, which would have cost more than $8,000. Inmates are also offered classroom instruction on basic plant biology and horticulture.

Staten Island's Fresh Kills landfill closed in 2002, and New York City faces strong opposition to its efforts to export more of its trash to other states. Composting programs like that at Rikers Island could provide at least a partial solution to some of the city's garbage woes. If the program is proven economical at Rikers Island, additional composting centers could be established elsewhere in the city, making New York potentially capable of one day recycling all of its food wastes, which amounts to about one fourth of the city's total garbage.

Figure 23.8 Municipal composting. Inside this steel tube, bacteria break down the organic wastes in the presence of water and air. After about a week of processing, the organic material is converted into compost, a fertile soil conditioner. Photographed at the recycling center in Columbia County, Wisconsin.

■ WASTE PREVENTION

Given the problems associated with sanitary landfills and incinerators, it makes sense to do whatever we can to lessen the need for these waste disposal methods. The three goals of waste prevention, in order of priority, are: (1) *reduce* the amount of waste as much as possible; (2) *reuse* products as much as possible; and (3) *recycle* materials as much as possible.

Reducing the amount of waste includes purchasing products that have less packaging and that last longer or are repairable. Consumers can also reduce waste by decreasing their consumption of products. Before deciding to purchase a product, a consumer should ask, "Do I really *need* this product, or do I merely *want* it?" Many U.S. consumers have participated actively for more than a decade in efforts to convert their throwaway economy into a waste prevention economy. Individual efforts have focused mainly on recycling, however, and more needs to be done in the areas of waste reduction and reuse. Reuse is considered a lower priority than reducing waste because reuse of products requires more energy and produces more pollution than not having the product in the first place. Reuse saves energy and reduces pollution significantly as compared to recycling, however.

We already discussed dematerialization, reuse, and recycling in Chapter 15, in the context of resource conservation. Now we examine the impact of these practices on solid waste (see "Meeting the Challenge: Reusing and Recycling Old Automobiles"). We then consider the fee-per-bag approach that many communities have adopted to encourage waste reduction.

Reducing the Amount of Waste: Source Reduction

Industries can often design and manufacture their products in ways that decrease the volume of solid waste and the amount of hazardous materials in the solid waste stream. Such a strategy, known as **source reduction**, is the most underutilized aspect of waste management. Source reduction is accomplished in a variety of ways, such as substituting raw materials that introduce less waste during the manufacturing process and reusing and recycling wastes at the plants where they are generated.

ENVIROBRIEF

An Industrial Ecosystem

The J. R. Simplot Company supplies about half of the french fries sold by McDonald's each year. Jack Richard Simplot, the founder of the Idaho-based firm, has found a profitable use for the company's waste with an extraordinary waste reduction system. Because french fries are a quality-controlled product, approximately 50% of the potatoes processed by Simplot's company ended up as waste, until he learned that the peelings could be mixed with grain and hay to feed cattle. Company potato residues feed about 400,000 animals each year. But that is only the beginning. Potato particles contained in processing water are anaerobically digested to produce methane for power plants, and the nutrient-bearing process water is used to irrigate and fertilize agricultural crops. Energy in the form of fuel-grade ethanol is also derived from processing waste and cull potatoes. The sludge from the ethanol process is converted to low cost, high-protein livestock feed. Manure from the cattle feedlots is applied to farmland to help nourish the hay and grain crops used to feed the cattle.

MEETING THE CHALLENGE

Reusing and Recycling Old Automobiles

About 11 million cars and trucks are discarded each year in the United States. Although by weight about 75% of a retired car is easily reused (as secondhand parts) or recycled (as scrap metal), the remaining 25% is not easy to recycle and usually ends up in sanitary landfills. Approximately 2,000 companies are involved in automotive recycling in the United States.

Because automobiles typically contain about 600 different materials—glass, metals, plastics, fabrics, rubber, foam, leather, and so on—identifying ways to reuse or recycle old parts is a complex problem. Economics is an important aspect of the problem, as reuse and recycling companies must make money in their recycling endeavors.

How does a car disassembly factory work? Workers typically begin disassembling a used car by draining all fluids—such as antifreeze, gasoline, transmission fluid, Freon, oil, and brake fluid—and recycling the fluids or processing them for disposal. Reusable parts and components, such as the engine, tires, and battery, are then removed, cleaned, tested, and inventoried before being sold as used parts. Body shops, new and used car dealers, repair shops, and auto and truck fleets are the main buyers of used parts. In addition to being reused, some parts are disassembled for their materials. Catalytic converters are disassembled because they contain valuable amounts of platinum and rhodium.

An automotive recycling facility then sends the remaining vehicle "shell" to a scrap processor for "scrapping." At the scrap processing facility, a giant machine shreds the entire automobile into pieces, each the size of a jar lid. Magnets and other machines sort the pieces into piles of steel, iron, copper, aluminum, and "fluff," which consists of the remaining materials, such as plastic, rubber, upholstery, and glass.

About 37% of the iron and steel scrap reprocessed in the United States comes from old automobiles. Recycling iron and steel saves energy and reduces pollution. According to the Environmental Protection Agency, recycling scrap iron and steel produces 86% less air pollution and 76% less water pollution than mining and refining an equivalent amount of iron ore.

Recycling plastic is one of the biggest challenges in auto recycling. Plastic is lightweight, and, as a result, auto makers are using more and more plastic to improve fuel efficiency. Also, no industry standards for plastic parts currently exist, so the kinds and amounts of plastic from which cars are made vary a great deal. As many as 15 different plastics comprise some dashboards, and because many of these plastics are chemically incompatible, they cannot be melted together for recycling.

Auto manufacturers around the world have begun to address the challenge of reusing and recycling old cars. Toyota has developed a way to recover urethane foam and other shredded materials. This "fluff" is used to make silencer padding. The Chrysler Corporation is developing a Composite Concept Vehicle that will have body sections that are completely recyclable. Honda, Mercedes-Benz, Peugeot, Toyota, Volkswagen, and Volvo are a few of the auto manufacturers that have started to design cars so that each part of an old automobile can be reconditioned or recycled.

Legislators are watching automakers' efforts, perhaps with the idea of someday devising legislation that would require U.S. manufacturers to accept scrapped autos from consumers after they have outlived their usefulness. Such a requirement exists in the European Union, which has mandated that by 2015, 95% of each discarded car must be recoverable. When they design new car models, European auto manufacturers now take into account the car's entire life cycle, including when it is taken out of service and discarded. This concept of **product stewardship**, in which manufacturers assume responsibility for their products from cradle to grave, has also been successfully applied to other businesses. Product stewardship encourages optimal recycling when products are returned to manufacturers.

Innovations and product modifications can reduce the waste produced after a product has been used by the consumer. Dry-cell batteries, for example, contain much less mercury today than they did in the early 1980s. The 35% reduction in the weight of aluminum cans since the 1970s provides another example of source reduction.

The **Pollution Prevention Act** of 1990 was the first U.S. environmental law to focus on the reduced generation of pollutants at their point of origin rather than the reduction of pollutants or repair of damage caused by such substances. The act was written to increase the adoption of cost-effective source reduction measures. It required the EPA to develop source reduction models, and it required manufacturing facilities to report to the EPA annually on their source reduction and recycling activities.

Dematerialization, the progressive decrease in the size and weight of a product as a result of technological improvements, is discussed in Chapter 15. *Dematerialization results in source reduction only if the new product is as durable as the one it replaced.* If smaller, lighter products have shorter life spans so that they have to be thrown out and replaced more often, source reduction is not accomplished.

Reusing Products

One example of reuse is refillable glass beverage bottles. Years ago, refillable beverage bottles were used a great deal in the United States (see Chapter 15). Today their use is rare, although about 11 states still have them. In order for a glass bottle to be reused, it must be considerably thicker (and therefore heavier) than one-use bottles. Because of the increased weight, transportation costs are higher. In the past, reuse of glass bottles made economic sense because there were many small bottlers scattered across the United States, helping to minimize transportation costs. Today there are approximately one tenth as many bottlers. Because of the centralization of bottling beverages, it is economically difficult to go back to the days of refillable bottles. The price of beverages might have to increase to absorb increased transportation costs and the energy used to sterilize dirty bottles.

Although the quantity of reusable glass bottles in the United States has declined, certain countries still reuse glass to a large extent. In Japan, almost all beer and sake bottles are reused as many as 20 times. Bottles in Ecuador may remain in use for 10 years or longer. European countries such as Denmark, Finland, Germany, the Netherlands, Norway, Sweden, and Switzerland have passed legislation that promotes the refilling of beverage containers. Parts of Canada and about 11 states in the United States also have deposit laws.

Can Refillable Bottles Make a Comeback in the United States? Some beverage producers have recently reversed the trend toward disposable bottles. Some dairies have started delivering milk directly to consumers in bottles made of acrylic plastic that can be reused 50 to 100 times before being disposed of. A few breweries have also switched from disposable back to refillable bottles.

Despite these examples of a shift toward refillables, however, it is unlikely that the United States will switch to refillable bottles on a large scale unless legislators and consumers take action. If a federal deposit law—a national "bottle bill"—were passed, consumers would return bottles to get the small deposit, such as a dime or quarter per bottle. Other legislation, such as a requirement that bottlers sell a certain percentage of their product in refillable bottles, would also encourage reuse.

Recycling Materials

It is possible to collect and reprocess many materials found in solid waste into new products of the same or different type. Recycling is preferred over landfill disposal because it conserves our natural resources and is more environmentally benign. Every ton of recycled paper saves 17 trees, 7,000 gallons of water, 4,100 kilowatt-hours of energy, and 3 cubic yards of landfill space. Recycling also has a positive effect on the economy by generating jobs and revenues (from selling the recycled materials). However, recycling does have environmental costs. It uses energy (as does any human activity) and generates pollution (as does any human activity). For example, the de-inking process in paper recycling requires energy and produces a toxic sludge that contains heavy metals.

The many different materials in municipal solid waste must be separated from one another before recycling can be accomplished. It is easy to separate materials such as glass bottles and newspapers, but the separation of materials in items with complex compositions is difficult. Some food containers are composed of thin layers of metal foil, plastic, and paper, and trying to separate these layers is a daunting prospect, to say the least.

The number of communities with recycling programs increased remarkably during the 1990s but leveled off in the early 2000s. By 1997, there were more than 8,800 curbside recycling programs in the United States. The average U.S. family of four now recycles over 454 kg (1,000 lb) of aluminum and steel cans, plastic bottles, glass containers, newspapers, and cardboard each year (see Figure 23.1). Some 65 million people now live in areas with curbside-collection recycling programs. Other communities collect solid waste as a mixture and take it to special resource recovery facilities, where it is either hand-sorted or separated using a variety of technologies, including magnets, screens, and conveyor belts. The United States currently recycles 30.1% of its municipal solid waste, which is a higher percentage than any other highly developed nation (Figure 23.9).

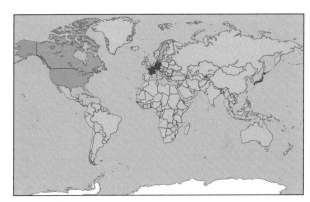

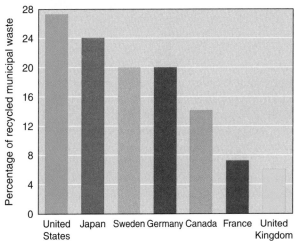

Figure 23.9 Recycling of municipal solid waste in selected highly developed countries. Data from the early 1990s are the latest available for all countries shown.

Most people think that recycling involves merely separating certain materials from the solid waste stream, but that is only the first step in recycling. In order for recycling to work, there must be a market for the recycled goods, and the recycled products must be used in preference to virgin products. Prices paid by processors for old newspapers, used aluminum cans, used glass bottles, and the like vary significantly from one year to the next, largely depending on the demand for recycled products.

Paper The United States currently recycles about 45% of its paper and paperboard. Many highly developed countries have greater recycling rates. Denmark recycles 97% of its paper. Part of the reason paper is not recycled more in the United States is that many older paper mills are not equipped to process waste paper. The number of mills that can handle waste paper increased during the 1990s, however, in part because of consumer demand. Most of the new mills in the United States are located near cities so they can take advantage of a local supply of scrap paper. For example, the small but densely populated state of New Jersey now has 13 paper mills that use scrap paper exclusively.

In addition to a slow increase in paper recycling in the United States, there is a growing demand for U.S. waste paper in other countries. Mexico imports a large quantity of waste paper and cardboard from the United States. Used paper is also in great demand in China, Taiwan, and Korea. The United States currently exports about 19% of its recovered paper.

As mentioned earlier, paper recycling does not work unless there is a market for recycled paper products. Sometimes the demand can be created by legislation. In Toronto, Canada, the city council passed a law that daily newspapers must contain at least 50% recycled fiber, or the publishers cannot have vending boxes on city streets. President Clinton issued an executive order, effective in 1998, that all paper purchased by the federal government must contain 30% postconsumer (that is, recycled) content.

Glass Glass is another component of solid waste that is appropriate for recycling. The United States currently recycles about 26.3% of its glass containers. Recycled glass costs less than glass made from virgin materials. Glass food and beverage containers are crushed to form **cullet**, which can be melted and used by glass manufacturers to make new products without any special adaptations in their factories. Although cullet is much more valuable when glass containers of different colors are separated before being crushed, cullet made from a mixture of colors can be used to make glasphalt, a composite of glass and asphalt that makes an attractive roadway (Figure 23.10).

Aluminum Although the recycling of aluminum has declined since peaking in the 1990s (see Chapter 15), it is still one of the best success stories in U.S. recycling, largely because of economic factors (Figure 23.11). Making a new aluminum can from a recycled one requires a fraction of the energy it would take to make a new can from raw metal. Because energy costs for new cans are high, there is strong economic incentive to recycle aluminum. According to the EPA, in 2000 almost 55% of aluminum beverage cans—more than 49 billion—were recycled, saving about 15.5 million barrels of oil.

Metals Other Than Aluminum Other recyclable metals include lead, gold, iron and steel, silver, and zinc. One of the obstacles to recycling metal products discarded in municipal solid waste is that their metallic compositions are often unknown. It is also difficult to extract metal from products, such as stoves, that contain other mate-

Figure 23.10 Glasphalt. Shown is a closeup of a Baltimore street paved with glasphalt, which contains a mixture of broken glass from different-colored containers. At night the road sparkles, as light from automobile headlights reflects off the pieces of glass.

Figure 23.11 Recycled aluminum.
These stacks of aluminum ingots are made from recycled aluminum beverage cans. Photographed at the Reynolds Metals Company in Sheffield, Alabama.

rials besides metal (plastic, rubber, or glass, for instance). In contrast, any waste metal produced at factories is recycled easily because its composition is known.

The economy has a large influence on whether metal is recycled or discarded. Greater recycling occurs when the economy is strong than when there is a recession. Thus, although the supply of metal waste is fairly constant, the amount of recycling varies from year to year.

One exception to this generalization is steel. Before the 1970s, almost all steel was produced in large mills that processed raw ores. Starting in the 1970s and continuing to the present, however, "mini-mills" that produce steel products from up to 100% scrap became increasingly important. These mills are located near many cities in the United States so they can process local scrap more profitably (because they do not have to pay to transport the scrap long distances). Mini-mills usually have electric arc furnaces that are energy-efficient and less polluting than the furnaces in old steel mills. According to the Institute of Scrap Recycling Industries, new steel products contain an average of 56% recycled scrap steel.

Plastic   Less than 20% of plastic is recycled. One reason plastic is not recycled much is that depending on the economic situation, it is sometimes less expensive to make it from raw materials (petroleum and natural gas) than to recycle it. In other words, plastic recycling—indeed, all recycling—is driven by economics. Some local and state governments support or require the recycling of plastic.

Polyethylene terphthalate (PET), the plastic used in soda bottles, is recycled more than any other plastic. According to the EPA, 34.9% of the 9.5 billion PET bottles sold annually are recycled to make such diverse products as carpeting, automobile parts, tennis ball felt, and polyester cloth (Figure 23.12); it takes about 25 plastic bottles to make one polyester sweater.

Polystyrene (one form of which is Styrofoam) is an example of a plastic that has great recycling potential but is currently not recycled appreciably. Cups, tableware, and packaging materials that are made of polystyrene can be recycled into a variety of products, such as coat hangers, flower pots, foam insulation, and toys. Because approximately 2.3 billion kg (5 billion lb) of polystyrene are produced each year, large-scale recycling would make a major dent in the amount of polystyrene that ends up in landfills.

One of the challenges associated with recycling plastic is that there are many different kinds. Forty-six different plastics are common in consumer products, and many products contain multiple kinds of plastic. A plastic ketchup bottle, for example, may have up to six layers of different plastics bonded together. To allow for effective recycling of high-quality plastic, the different types must be meticulously sorted or separated. If two or more resins are recycled together, the resultant plastic is of lower quality.

Low-quality plastic mixtures are used to make a construction material similar to wood. This "plastic lumber" is particularly useful for outside products, such as fence posts, planters, highway retaining walls, picnic tables, and park benches, because of its durability.

Figure 23.12 Recycled plastic. A wide variety of products can be made from recycled polyethylene terphthalate (PET; shown) and other plastics.

Tires As mentioned earlier in the chapter, about 273 million tires are discarded in the United States each year. Given the huge quantity of tires available for recycling, the number of products that can be made from old tires is limited. Uses for old tires include retread tires; playground equipment; trashcans, garden hoses, and other consumer products; reef building on coastlines; and rubberized asphalt for pavement. Some research in product development is being carried out, and almost all states have established tire-recycling programs. According to the EPA, 9% of tires are currently recycled to make other products.

The Fee-Per-Bag Approach

Consumers have traditionally paid the same amount in taxes or collecting fees whether they put out 1 bag or 20 bags of solid waste. Many communities have reduced the volume of solid waste and encouraged reuse and recycling by charging households for each container of solid waste—the **fee-per-bag approach**.

Two environmental economists examined the weight and volume of weekly solid waste and recycling in Charlottesville, Virginia, before and after a charge of $0.80 per 32-gallon bag of solid waste was implemented in 1992.

They determined that charging consumers for each bag of solid waste that they discard is an effective way to reduce the volume of solid waste. The average Charlottesville household reduced the weight of its solid waste by 14% and the volume of its solid waste by 37%. Increased recycling accounted for much of the decrease in solid waste. Similar results were obtained in a 1996 study of nine communities in California, Illinois, and Michigan. However, the fee-per-bag approach has raised concerns about people illegally disposing of solid waste by burning or dumping along back roads or in vacant lots.

HAZARDOUS WASTE

Toxic waste has held national attention since 1977, when it was discovered that toxic waste from an abandoned chemical dump had contaminated homes and possibly people in Love Canal, a small neighborhood on the edge of Niagara Falls, New York. Lois Gibbs, a housewife in Love Canal, led a successful crusade to evacuate the area after she discovered what seemed to be a high number of serious illnesses, particularly among children, in the neighborhood. (Today, Ms. Gibbs is executive director for the Center for Health, Environment, and Justice.) As a result of the publicity, Love Canal became synonymous with chemical pollution caused by negligent hazardous waste management. In 1978, it became the first location ever declared a national emergency disaster area because of toxic waste; some 700 families were evacuated. The Love Canal episode resulted in the passage of the federal Superfund law, which held polluters accountable for the cost of cleanups (discussed shortly).

From 1942 to 1953, a local industry, Hooker Chemical Company, disposed of 19,945 metric tons (22,000 tons) of toxic chemical waste in the 914-m-long (3,000-ft-long) Love Canal. When the site was filled, Hooker added topsoil and donated the land to the local board of education. A school and houses were built on the site, which began oozing toxic waste several years later. Over 300 chemicals, many of them carcinogenic, have been identified in Love Canal's toxic waste.

Tons of contaminated soil were removed during the cleanup that followed, but because the canal was so huge, the federal government decided to contain the waste and construct drainage trenches to prevent toxic wastes from leaking from the site. In 1990, after almost 10 years of cleanup, the EPA and the New York Department of Health declared the area safe to be resettled. The nearby housing area was renamed Black Creek Village. Today, the canal is a 40-acre mound covered by clay and surrounded by a chain-link fence and warning signs.

Whether the site caused adverse health effects in residents remains unanswered because of a lack of reliable scientific data. On average, residents of Love Canal seem to have had more health problems, from miscarriages and birth defects to psychological disorders.

The Love Canal disaster generated an immediate concern about the hazardous waste in thousands of old landfills, dumps, and junkyards across the United States, and that concern has been with us in one form or another ever since. A 1997 California study reported that women living within a quarter mile of untreated hazardous waste sites (Superfund sites, discussed shortly) were at greater risk of having babies with serious birth defects, such as malformed neural tubes and defective hearts.

Other countries have the same problems with toxic waste management. How shall we deal with the bewildering array of hazardous waste that is continually generated and released in ever increasing amounts into the environment by mining, industrial processes, incinerators, military activities, and thousands of small businesses? How do we clean up the hazardous materials that have already contaminated our world?

Types of Hazardous Waste

Any discarded chemical that threatens human health or the environment is known as **hazardous** or **toxic waste**. Hazardous waste, which accounts for about 1% of the solid waste stream in the United States, includes chemicals that are dangerously reactive, corrosive, explosive, or toxic (Table 23.1). The chemicals may be solids, liquids, or gases.

More than 700,000 different chemicals are known to exist. How many are hazardous is unknown, because most have never been tested for toxicity, but without a doubt, hazardous substances number in the thousands.

According to the EPA report, *Chemical Hazard Data Availability Study*, only 7% of the 3,000 chemicals used in large quantities (more than 500 tons annually) in U.S. commerce have undergone comprehensive studies of potential health and environmental effects. Hazardous chemicals include a variety of acids, dioxins, abandoned explosives, heavy metals, infectious waste, nerve gas, organic solvents, polychlorinated biphenyls (PCBs) pesticides, and radioactive substances. Many of these chemicals have already been discussed, particularly in Chapters 1, 11, 19, 21, and 22, which examine endocrine disrupters, radioactive waste, air pollution, water and soil pollution, and pesticides. Here we discuss dioxins, PCBs, and radioactive and toxic wastes produced during the Cold War.

Dioxins   **Dioxins** are a group of 75 similar chemical compounds that are formed as unwanted by-products during the combustion of chlorine compounds. Some of the known sources of dioxins are medical waste and municipal waste incinerators, iron ore mills, copper smelters, cement kilns, metal recycling, coal combustion, pulp and paper plants that use chlorine for bleaching, and chemical accidents. Incineration of medical and municipal wastes accounts for 70% to 95% of known human emissions of dioxins. In Japan, where nearly 75% of its solid waste is burned in incinerators, the air contains nearly 10 times the amount of dioxins found in other highly developed countries. Hospital waste incinerators are probably the largest dioxin polluters because they are so numerous (there are more than 6,000 of them in the

---

Table 23.1   **Examples of Hazardous Waste**

Hazardous Material	*Some Possible Sources*
Acids	Ash from power plants and incinerators; petroleum products
CFCs (chlorofluorocarbons)	Coolant in air conditioners and refrigerators
Cyanides	Metal refining; fumigants in ships, railway cars, and warehouses
Dioxins	Emissions from incinerators and pulp and paper plants
Explosives	Old military installations
Heavy metals	Paints, pigments, batteries, ash from incinerators, sewage sludge with industrial waste, improper disposal in landfills
Arsenic	Industrial processes, pesticides, additives to glass, paints
Cadmium	Rechargeable batteries, incineration, paints, plastics
Lead	Lead-acid storage batteries, stains and paints; TV picture tubes and electronics discarded in landfills
Mercury	Coal-burning power plants; paints, household cleaners (disinfectants), industrial processes, medicines, seed fungicides
Infectious waste	Hospitals, research labs
Nerve gas	Old military installations
Organic solvents	Industrial processes; household cleaners, leather, plastics, pet maintenance (soaps), adhesives, cosmetics
PCBs (polychlorinated biphenyls)	Older appliances (built before 1980); electrical transformers and capacitors
Pesticides	Household products
Radioactive waste	Nuclear power plants, nuclear medicine facilities, weapons factories

United States), and they generally have unsophisticated pollution controls. Motor vehicles, barbecues, and cigarette smoke also emit dioxins, although their contribution is minor. Forest fires and volcanic eruptions are natural sources of dioxins. (Recall from "Case in Point: The Use of Herbicides in the Vietnam War" in Chapter 22 that dioxins are also formed during the production of certain pesticides.)

Dioxins are emitted in smoke and then settle on plants, the soil, and bodies of water; from there they are incorporated into the food web. When humans and other animals ingest dioxins, they are stored and accumulate in their fatty tissues (see discussion of bioaccumulation and biological magnification in Chapter 22). Humans are primarily exposed when they eat contaminated meat, dairy products, and fish. Because dioxins are so widely distributed in the environment, virtually everyone has dioxins in their body fat.

Just how dangerous dioxins are to humans is somewhat controversial. Dioxins are known to cause several kinds of cancer in laboratory animals, but the data are conflicting on their cancer-causing ability in humans. A 1997 follow-up study of residents of Sevesco, Italy, who were exposed to high levels of dioxin after a chemical accident in 1976 revealed a statistically significant increase in cancer deaths. According to a 2001 EPA report on the health effects of dioxins, dioxins probably cause several kinds of cancer in humans.

The 2001 EPA report also raised other concerns about the effects of dioxins on the human reproductive, immune, and nervous systems. Dioxins may delay fetal development and cause cognitive damage, cause endometriosis in women (the growth of uterine tissue in abnormal locations in the body), and decrease sperm production in men. Dioxins have also been linked to an increased risk of heart disease. Because human milk contains dioxins, nursing infants, who feed almost exclusively on milk, are considered particularly at risk. In 1998 a panel of 40 scientists urged the World Health Organization's International Agency for Research on Cancer to reduce its recommended *maximum daily intake* of dioxins to one tenth of what had been formerly recommended. WHO also classified the dioxin TCDD (tetrachlorodibenzodioxin) as a known human carcinogen.

PCBs Polychlorinated biphenyls (**PCBs**) are a group of 209 industrial chemicals composed of carbon, hydrogen, and chlorine. They are clear or light yellow, oily liquids or waxy solids. PCBs were manufactured in the United States between 1929 and 1979 for a wide variety of uses. They have been employed as cooling fluids in electrical transformers, electrical capacitors, vacuum pumps, and gas-transmission turbines. PCBs were also used in hydraulic fluids, fire retardants, adhesives, lubricants, pesticide extenders, inks, and other materials.

The first evidence that PCBs were dangerous appeared in 1968, when hundreds of Japanese who ate rice bran oil that had accidentally been contaminated with PCBs experienced serious health problems, including liver and kidney damage. A similar mass poisoning, also attributed to PCB-contaminated rice oil, occurred in Taiwan in 1979. Since then, toxicity tests conducted on animals indicate that PCBs harm the skin, eyes, reproductive organs, and gastrointestinal system. PCBs are also endocrine disrupters because they interfere with hormones released by the thyroid gland. Several studies have demonstrated that children exposed to PCBs before birth have certain intellectual impairments, such as poor reading comprehension, memory problems, and difficulty paying attention. PCBs may also be carcinogenic; they have been shown to cause liver cancer in rats. A 1997 study in Sweden reported that cancer patients tend to have higher PCB concentrations in their body fat than healthy people. A 1997 study by the U.S. National Cancer Institute found a correlation between high levels of PCBs in blood fat and the incidence of non-Hodgkin's lymphomas, cancers that attack the body's immune system.

Some of the properties of PCBs that make them so useful in industry also make them dangerous to organisms and difficult to eliminate from the environment. PCBs are chemically stable and resist chemical and biological degradation. Like DDT (dichlorodiphenyltrichloroethane) and dioxins, PCBs accumulate in fatty tissues and are subject to biological magnification in food webs. Research indicates that the general human population is mainly exposed to PCBs by eating food that became contaminated through biological magnification. One way that PCBs enter aquatic food webs is by benthic invertebrates that live in contaminated sediments. (PCBs tend to bind to organic particles in aquatic sediments.) Small fish eat these invertebrates, and as larger fish eat the smaller fish, the PCBs bioaccumulate. Human populations whose diets consist primarily of fish and marine mammals, such as the Inuit of northern Canada, are exposed to very large amounts of PCBs.

Prior to their banning by the EPA in the 1970s, PCBs were dumped in large quantities into landfills, sewers, and fields. Such improper disposal is one of the reasons PCBs are still a threat today. Also, when sealed electrical transformers and capacitors leak or catch fire, PCB contamination of the environment occurs.

One of the most effective ways to destroy PCBs is by high-temperature incineration. However, incineration is not practical for the removal of PCBs that have leached into the soil and water, because, among other difficulties, the cost of incinerating large quantities of soil is prohibitively high. Another way to remove PCBs from soil and water is to extract them by using solvents. This method is undesirable for two reasons: First, the solvents themselves are hazardous chemicals, and second, extraction is costly because wells must be dug to collect the chemically contaminated leachate in order to decontaminate it.

More recently, researchers have discovered several bacteria that can degrade PCBs at a fraction of the cost of incineration. One of the limitations of this method is that if PCB-eating bacteria are sprayed on the surface of the soil, they cannot decompose the PCBs that have already leached deep into the soil or groundwater systems. Although these microorganisms show promise in removing PCBs from the environment, additional research will have to be conducted to make the biological degradation of PCBs practical. (Additional discussion about using bacteria to break down hazardous waste is found later in this chapter.)

CASE·IN·POINT Hanford Nuclear Reservation

U.S. nuclear weapons facilities are not actively manufacturing nuclear weapons anymore, but they present us with a greater challenge, cleaning up and disposing of radioactive and toxic wastes that have accumulated at numerous sites around the United States since the 1940s. Every step in the production of nuclear warheads generated radioactive and chemical wastes. We focus our discussion on the Hanford Nuclear Reservation, a 1,400-km^2 (560-mi^2) area on the Columbia River in south central Washington state (Figure 23.13a). Hanford is the largest, most seriously contaminated site in the U.S. nuclear weapons infrastructure.

The sheer immensity of the cleanup task at Hanford is daunting. Tons of highly radioactive solid and liquid wastes were stored or dumped into trenches, pits, tanks, ponds, and underground cribs—a total of 1,377 waste sites. (These methods of disposal were considered acceptable at the time.) For example, two concrete pools of water store more than 100,000 spent fuel rods. As they corrode, the rods release highly radioactive uranium, plutonium, cesium, and strontium into the water. Because these pools are leaking, soil and groundwater have been contaminated, and the Columbia River is in danger.

The Columbia River is also threatened by 54 million gallons of toxic chemical and radioactive liquid wastes stored in 177 tanks. Liquids in some of these tanks are so reactive that they boiled for years from the heat of their own radioactivity or chemical activity, but most of them are now covered by semisolid crusts that formed from chemical reactions within the mixtures. Some of the tanks are potentially explosive: Chemical reactions in some of the tanks produce hydrogen and other toxic gases. Vents have been installed to allow the gas to escape and reduce the danger of ignition. Many of these tanks are also leaking their poisons into the ground.

Cleanup of this toxic waste nightmare is complicated by the fact that the *extent* of radioactive pollution and *kinds* of toxic mixtures present are not well known. As a result, scientists and engineers are assessing the damage, prioritizing the cleanup process, and determining how

(a)

(b)

Figure 23.13 Hanford Nuclear Reservation. (a) Location of Hanford along the Columbia River in Washington state. (b) A worker at the Waste Receiving and Processing Facility crushes drums of low-level nuclear waste. The crushed drums will then be packed into larger drums for permanent burial in the hazardous waste landfill at Hanford.

best to proceed for each type of contamination (Figure 23.13b). Most experts believe the cleanup, which is under the direction of the U.S. Department of Energy (DOE) will take decades to complete and cost hundreds of billions of tax dollars.

After the cleanup is finished, Hanford will remain hazardous for hundreds or even thousands of years, in part because we do not have the technologies to address the widespread soil contamination. The DOE will have to establish and maintain a long-term program at the site to limit human exposure to remaining hazards. ■

■ MANAGEMENT OF HAZARDOUS WASTE

Humans have the technology to manage toxic waste in an environmentally responsible way, but it is extremely expensive. Although great strides have been made in educating the public about the problems posed by hazardous waste, we have only begun to address many of the issues of hazardous waste disposal. No country currently has an effective hazardous waste management program, but several European countries have led the way by producing smaller amounts of hazardous waste and by using fewer hazardous substances.

Chemical Accidents

When a chemical accident occurs in the United States, whether at a factory or during the transport of hazardous chemicals, the EPA's Emergency Response Notification System (ERNS) is notified. More than 60% of all chemical accidents reported to ERNS involve oil, gasoline, or other petroleum spills. The remaining accidents involve some 1,032 other hazardous chemicals, such as PCBs, ammonia, sulfuric acid, and chlorine. In 1998, the ERNS database indicated that 16,958 metric tons (18,704 tons) of toxic chemicals were released into the environment during reported accidents. The three states with the greatest number of toxic chemical accidents in 1998 were Texas (1,392 accidents), Louisiana (644 accidents), and California (215 accidents).

Chemical safety programs have traditionally stressed accident mitigation and adding safety systems to existing procedures. More recently, however, industry and government agencies have stressed accident prevention through the **principle of inherent safety**, in which industrial processes are redesigned to involve less toxic materials so that dangerous accidents are prevented. The principle of inherent safety is an important aspect of source reduction discussed earlier in the chapter.

Current Management Policies

Currently, two federal laws dictate how hazardous waste should be managed: (1) the Resource Conservation and Recovery Act, which is concerned with managing hazardous waste that is being produced now, and (2) the Superfund Act, which provides for the cleanup of abandoned and inactive hazardous waste sites.

The **Resource Conservation and Recovery Act (RCRA)** was passed in 1976 and amended in 1984. Among other things, RCRA instructs the EPA to identify which waste is hazardous and to provide guidelines and standards to states for hazardous waste management programs. RCRA bans hazardous waste from land disposal unless it has been treated to meet EPA's standards of reduced toxicity. In 1992 the EPA initiated a major reform of RCRA to expedite cleanups and streamline the permit system to encourage hazardous waste recycling.

In 1980 the **Comprehensive Environmental Response, Compensation, and Liability Act (CERCLA)**, commonly known as the **Superfund Act**, established a program to tackle the huge challenge of cleaning up abandoned and illegal toxic waste sites across the United States. At many of these sites, hazardous chemicals have migrated deep into the soil and have polluted groundwater. The greatest threat to human health from toxic waste sites comes from drinking water laced with such contaminants.

Cleaning Up Existing Toxic Waste: The Superfund Program

The federal government estimates that the United States has more than 400,000 hazardous waste sites with leaking chemical storage tanks and drums (both above and below ground), pesticide dumps, and piles of mining waste (Figure 23.14a). This estimate does not include the hundreds or thousands of toxic waste sites at military bases and nuclear weapons facilities.

By 2002, 11,330 sites were in the CERCLA inventory, which means that they have been identified by the EPA as qualifying for cleanup (Figure 23.14b). (Since 1980, almost 33,000 sites have been cleaned up and removed from the CERCLA inventory.) These sites are not identified according to any particular criteria. Some are dumps that local or state officials have known about for years, whereas concerned citizens identify others. The sites in the inventory are evaluated and ranked to identify those with extremely serious hazards. The ranking system uses data from preliminary assessments, site inspections, and expanded site inspections, which include contamination tests of soil and groundwater and sampling of toxic waste.

The sites that pose the greatest threat to public health and the environment are placed on the **Superfund National Priorities List**, which means that the federal government will assist in their cleanup. As of 2002, 1,234 sites were on the National Priorities List. One fifth of these sites are open dumps or sanitary landfills that operated before state and federal regulations were imposed and that received solid waste that would be considered hazardous today. The five states with the greatest number of sites on the national priorities list are New Jersey (113 sites as of October 2002), California (96 sites), Pennsylvania (94 sites), New York (90 sites), and Michigan (67 sites).

As of 2002, 259 hazardous waste sites had been cleaned up enough to be deleted from the National Priorities List, and 656 other sites had been partially corrected. These sites, designated "construction completes," have had extensive cleanups but remain on the Superfund list because of continued problems with groundwater contamination. The average cost of cleaning up a site is $20 million.

(a)

(b)

Figure 23.14 Cleaning up hazardous waste. (a) Toxic waste in deteriorating drums at a site near Washington, D.C. The metal drums in which much of the waste is stored have corroded and started to leak. Old toxic waste dumps are commonplace around the United States. (b) Cleanup of a hazardous waste site near Minneapolis, Minnesota. Removal and destruction of the wastes are complicated by the fact that usually nobody knows what chemicals are present.

One reason for the urgency about cleaning up the sites on the National Priorities List is their locations. Most were originally in rural areas on the outskirts of cities. With the growth of cities and their suburbs, however, many of the dumps are now surrounded by residential developments. About 27 million U.S. citizens, including over 4 million children, live within 6.4 km (4 mi) of one or more Superfund sites.

The Superfund program got off to a slow start, in part because the EPA, which was authorized to administer it, experienced severe budget cuts. An amendment passed in 1986 provided the EPA with more money and instructions to work faster. From 1981 to 2001, Congress appropriated $12.6 billion to implement remedial and removal programs of the Superfund Act, and $1.3 billion were authorized for 2000. The Superfund Act, which expired in 1994, has been under review by Congress and should be amended and reauthorized sometime in the 2000s. The tax on oil and chemical companies that helps to fund the Superfund program expired in 1995. Congress has not reinstated the tax, and taxpayer revenues now pay for cleaning up "orphan sites," where the polluter cannot be identified or cannot pay.

Because the federal government cannot assume major responsibility for cleaning up every old dumping ground in the United States, the current landowner, prior owners, anyone who has dumped waste on the site, and anyone who has transported waste to a particular site share cleanup costs. For some sites, many different parties are considered liable for cleanup costs. The cleanup process has been mired in litigation, mostly by companies, charged with polluting, suing each other. Despite the urgency of cleaning up sites on the National Priorities List, it will take many years to complete the job.

Although critics decry the slow pace and high cost of cleaning up Superfund sites, the very existence of CERCLA has been a deterrent to further polluting. Compa-

ENVIROBRIEF

Transforming the Rocky Mountain Arsenal

One of the most heavily contaminated Superfund sites in the United States is being treated to an environmental makeover. For more than 40 years the Rocky Mountain Arsenal, just northeast of Denver, was the site of the U.S. Army's manufacture and storage of nerve gases, mustard gas, and other toxic chemical weapons as well as napalm and rocket fuel for the U.S. Air Force. Shell Oil and other companies also used the site to manufacture pesticides for agriculture and gardening.

Production of chemicals and weapons ceased long ago, and a 14-year, $2 billion remediation effort was launched in 1996. The project is cleaning contaminated land as much as possible so that most of the arsenal's 17,000 acres can be converted into a national wildlife refuge. Although residues from more than 600 chemicals have been found on the arsenal, it has always served as a haven to wildlife, probably because much of the site is relatively undisturbed. Contaminants such as dieldrin and other organochlorine pesticides create the most concern, as they persist in soil and groundwater for decades. Studies of many of the wildlife species on the site indicate that while some animals are relatively unaffected by contaminants, rodents, smaller birds such as kestrels, and aquatic insects suffer a great deal more.

Much of the site's cleanup is completed or under way: buildings cleaned and demolished, chemical waste basins capped with crushed concrete, landfills with multiple linings constructed to contain contaminated soil or nerve gas and pesticide wastes, and groundwater captured and treated extensively. Though many observers question the likelihood that the Rocky Mountain Arsenal will ever be truly clean, others accept the cleanup as a compromise resulting in a beautiful—if not pristine—wildlife habitat.

nies that produce hazardous waste are now fully aware of the costs of liability and cleanup, and therefore are more likely to take steps to ensure that their hazardous wastes are disposed of properly.

The Biological Treatment of Hazardous Contaminants

A variety of methods are employed to clean up soil contaminated with hazardous waste (recall the section on soil remediation in Chapter 21). Because most of these processes are prohibitively expensive, innovative approaches such as bioremediation and phytoremediation are being developed to deal with hazardous waste. **Bioremediation** is the use of bacteria and other microorganisms to break down hazardous waste, whereas **phytoremediation** is the use of plants to absorb and accumulate toxic materials from the soil (see Chapter 21). (The prefix *phyto-* comes from the Greek word meaning "plant.")

To date, more than 1,000 different species of bacteria and fungi have been used to clean up various forms of pollution. Bioremediation takes a little longer to work than traditional hazardous waste disposal methods, but it accomplishes the cleanup at a fraction of the cost. In bioremediation, the contaminated site is exposed to an army of microorganisms, which gobble up the poisons such as petroleum and other hydrocarbons and leave behind harmless substances such as carbon dioxide, water, and chlorides. Bioremediation encourages the natural processes by which bacteria consume organic molecules such as hydrocarbons. During bioremediation, conditions at the hazardous waste site are modified so that the desired bacteria will thrive in large enough numbers to be effective. Environmental engineers might pump air through the soil (to increase its oxygen level) and add a few soil nutrients such as phosphorus or nitrogen. They might also install a drainage system to pipe any contaminated water that leached through the soil back to the surface for another exposure to the bacteria.

In phytoremediation, plant species that are known to remove specific toxic materials from the soil are grown at a contaminated site. As the roots penetrate the soil, they selectively absorb the toxins, which accumulate in both root and shoot tissues. Later, the plants are harvested and disposed of in a hazardous waste landfill. Alternatively, the plants break down a toxic chemical into more benign chemicals.

The field of phytoremediation is in its infancy, but already, specific plants have been identified that remove such hazardous materials as trinitrotoluene (TNT), radioactive strontium and uranium, selenium, lead, and other heavy metals (Table 23.2). Researchers in England have also identified three species of marsh plants that degrade herbicides and other pesticides; in this case, the actual organisms that attack the pesticide compounds are bacteria that live in symbiotic association with the plants' roots. Some researchers are working to genetically engineer certain plants to do an even better job of accumulating toxins.

Phytoremediation is much cheaper than conventional methods to clean up hazardous waste sites, but it does have limitations. Plants cannot remove contaminants that are present in the soil deeper than their roots normally grow. There is also concern that insects and other animals might eat the plants, thereby introducing the toxins into the food web.

Managing the Toxic Waste We Are Producing Now

Many people think, incorrectly, that the establishment of the Superfund has eliminated the problem of toxic waste. The Superfund deals only with hazardous waste produced in the past. It does nothing to eliminate the large amount of toxic waste that continues to be produced today. There are three ways to manage hazardous waste: (1) source reduction, (2) conversion to less hazardous materials, and (3) long-term storage.

Table 23.2 Selected Examples of Phytoremediation

Plant	Materials That Plant Absorbs and Accumulates	Location Under Study
Indian mustard	Strontium-90	Near Chernobyl nuclear power plant in Ukraine
	Cadmium	Abandoned lead mine in England
	Chromium	Contaminated soil from an old chromium smelter in New Jersey
Sunflower	Strontium-90, cesium-137	Former Department of Energy facility in Ashtabula, Ohio; near Chernobyl nuclear power plant
Cattail	Selenium	Oil refinery in Point Richmond, California
Parrot feather, stonewort	TNT (trinitrotoluene)	Environmental Protection Agency project in Athens, Georgia
Twist flower	Nickel	Bureau of Mines test site in California
Brake fern	Arsenic	Contaminated soil in central Florida

As with municipal solid waste, the most effective of the three is source reduction—that is, reducing the amount of hazardous materials used in industrial processes and substituting less hazardous or nonhazardous materials for hazardous ones. **Environmental chemistry**, also known as **green chemistry**, is an increasingly important subdiscipline of chemistry in which commercially important chemical processes are redesigned to significantly reduce environmental harm. For example, chlorinated solvents are widely used in electronics, dry cleaning, foam insulation, and industrial cleaning. It is sometimes possible to accomplish source reduction by substituting a less hazardous water-based solvent for a chlorinated solvent. Substantial source reduction of chlorinated solvents can also be realized by reducing solvent emissions. Most chlorinated solvent pollution gets into the environment by evaporation during industrial processes. Installing solvent-saving devices not only benefits the environment but also provides economic gains, because smaller amounts of chlorinated solvents must be purchased. No matter how efficient source reduction becomes, however, it will never entirely eliminate hazardous waste.

The second best way to deal with hazardous waste is to reduce its toxicity. This can be accomplished by chemical, physical, or biological means, depending on the nature of the hazardous waste. One way to detoxify organic compounds is by high-temperature incineration. The high heat of combustion reduces these dangerous compounds, such as pesticides, PCBs, and organic solvents, into safe products such as water and carbon dioxide. The incineration ash is hazardous, however, and must be disposed of in a landfill designed specifically for hazardous materials. One method to reduce the toxicity of hazardous waste is incineration using a plasma torch, which produces such high temperatures (up to 10,000°C) that hazardous waste is almost completely converted to harmless gases. In comparison, conventional incinerators produce temperatures no higher than 2,000°C.

Hazardous waste that is produced in spite of source reduction and that is not completely detoxified must be placed in long-term storage. Landfills that are equipped to store hazardous substances have many special features, including several layers of compacted clay and high-density plastic liners at the bottom of the landfill to prevent leaching of hazardous substances into surface water and groundwater. Liquid that percolates through a hazardous waste landfill is collected and treated to remove contaminants. The entire facility and nearby groundwater deposits are carefully monitored to make sure there is no leaking. Only solid chemicals (not liquids) that have been treated to detoxify them as much as possible are accepted at hazardous waste landfills. These chemicals are placed in sealed barrels before being stored in the hazardous waste landfill.

Very few facilities are certified to handle toxic waste (Figure 23.15). In 2001 there were only 23 commercial hazardous waste landfills in the United States, although many larger companies were licensed to treat their hazardous waste on-site. As a result, much of our hazardous waste is still placed in sanitary landfills, burned in incinerators that lack the required pollution control devices, or discharged into sewers.

ENVIRONMENTAL JUSTICE

Opposition to the environmental inequities faced by low-income minority communities in both rural and urban

Figure 23.15 Hazardous waste landfill. These landfills are subject to strict environmental criteria and design features. For example, they are located as far as possible from streams, wetlands, and residences. The soil must contain a lot of clay, and no accessible aquifers can be located at the site. The bottom of the hazardous waste landfill has several feet of compacted clay covered by three high-density plastic liners (one is shown). A leak detection system is installed underneath the clay, and a drain system located above the plastic liners allows liquid leachate to collect in a basin where it can be treated. Barrels of hazardous waste are then placed on the liners and covered with soil. Photographed near Detroit, Michigan.

areas has increased in recent years. Many studies indicate that poor minority neighborhoods are more likely to have hazardous waste facilities, sanitary landfills, sewage treatment plants, and incinerators in their neighborhoods. A 1990 study at Clark Atlanta University found that six of Houston's eight incinerators were located on less expensive land in predominantly black neighborhoods. Such communities often have a limited involvement in the political process and may not even be aware of their exposure to higher levels of pollutants.

Because people in low-income communities frequently lack access to sufficient health care, they may not be treated adequately for exposure to environmental hazards. The high incidence of asthma in many minority communities is an example of a health condition that may be caused or exacerbated by exposure to environmental pollutants. Few studies have been conducted to examine how environmental pollutants interact with other socioeconomic factors to cause health problems. Those studies that have been done often fail to conclusively tie exposure to environmental pollutants to the health problems of poor and minority communities. A 1997 health study of residents in San Francisco's Bayview–Hunters Point area found that hospitalizations for chronic illnesses were almost four times higher than the state average. Bayview–Hunters Point is heavily polluted. It has 700 hazardous waste facilities, 325 underground oil storage tanks, and two Superfund sites. Yet the study did not demonstrate that an increased exposure to toxic pollutants was the primary cause of illnesses in the residents. Although anecdotal evidence abounds, there is currently little hard scientific evidence that shows to what extent a polluted environment is responsible for the disproportionate health problems of poor and minority communities. Lead contamination is a notable exception (see Chapter 21 discussion of lead).

In addition to concerns about pollution in their neighborhoods, low-income communities may not receive equal benefits from federal cleanup programs. A 1992 paper published in the *National Law Journal* reported that toxic waste sites in white communities were cleaned up faster and better than were those in communities of blacks, Hispanics, or other minorities.

Environmental Justice and Ethical Issues

There is an increasing awareness that environmental decisions such as where to locate a hazardous waste landfill have important ethical dimensions. The most basic ethical dilemma centers on the rights of the poor and disenfranchised versus the rights of the rich and powerful. Whose rights should have priority in these decisions? The challenge is to find and adopt just solutions that respect all groups of people, including those yet to be born. **Environmental justice** means that every citizen, regardless of age, race, gender, social class, or other factor, is entitled to adequate protection from environmental hazards. Viewed ethically, environmental justice is a fundamental human right. Although we may never be able to completely eliminate environmental injustices of the past, we have a moral imperative to prevent them today so that the negative effects of pollution do not disproportionately affect any segment of society.

In response to these concerns, a growing environmental justice movement has emerged at the grassroots level as a strong motivator for change. Advocates of the environmental justice movement are calling for special efforts to clean up hazardous sites in low-income neighborhoods, from inner-city streets to Indian reservations. Many environmental justice groups base their demands on the inherent "rightness" of their position. Other groups want science to give their demands legitimacy. Many advocates cite the need for more research on human diseases that may be influenced by environmental pollutants.

Mandating Environmental Justice at the Federal Level

In 1994, President Clinton signed an executive order requiring all federal agencies to develop strategies and policies to ensure that their programs do not discriminate against poor and minority communities when decisions are made about where hazardous facilities are to be located. The National Environmental Justice Advisory Council, also established in 1994, provides grants to help low-income communities around the United States identify and address local environmental problems.

The first response to President Clinton's initiative occurred in 1997 when the Nuclear Regulatory Commission (NRC) rejected a request to build a uranium processing plant near two minority neighborhoods in northern Louisiana. The commission decided that racial considerations had been a factor in site selection because the applicant had ruled out all potential sites near predominantly white neighborhoods. This decision by the NRC is significant because it sends a message that the U.S. government will protect the rights of vulnerable members of society.

Environmental Justice and International Waste Management

Environmental justice applies to countries as well as to individuals. Although there are ways to reduce and dispose of waste in an environmentally sound manner, industrialized countries have sometimes chosen to send their waste to other countries. (As industrialized nations develop more stringent environmental standards, disposing of hazardous waste at home becomes much more expensive than sending it to a developing nation for disposal.) Some waste is exported for legitimate recycling, but some has been exported strictly for disposal.

The export of both solid and hazardous wastes by the United States, Japan, and the European Economic Community is one of the most controversial aspects of waste management today. Usually the recipients of this waste are developing nations in Africa, Central and South America, and the Pacific Rim of Asia, although Eastern and Central Europe and countries of the former Soviet Union are also important sites. The governments, industries, and citizens of these countries are often inexperienced and ill equipped to contain and monitor such materials. As a result, the waste often causes the same types of environmentally hazardous sites in developing nations that industrialized nations are trying to clean up at home.

In 1989 the U.N. Environment Program developed a treaty known as the **Basel Convention** to restrict the international transport of hazardous waste. It allows countries to export hazardous waste only with prior consent of the importing country as well as of any countries that the waste passes through in transit. The Basel Convention became official in 1992, and 152 countries had ratified it as of 2002. (The United States signed the treaty in 1989 but has not yet ratified it because Congress has not passed the necessary legislation.) In 1995 the Basel Convention was amended to ban the export of *any* hazardous waste between industrial and developing countries; as of 2002, 31 countries had ratified the 1995 amendment.

INTEGRATED WASTE MANAGEMENT

The most effective way to deal with solid and hazardous waste is with a combination of techniques. In **integrated waste management**, a variety of options that minimize waste, including the three R's of waste prevention (*reduce*, *reuse*, and *recycle*), are incorporated into an overall waste management plan (Figure 23.16). Even on a large scale, recycling and source reduction will not entirely eliminate the need for disposal facilities such as incinerators and landfills. They will, however, substantially reduce the amount of solid waste requiring disposal in incinerators and landfills.

CHANGING ATTITUDES

In the United States we have become accustomed to the convenience of a throwaway society. The products we purchase rarely, if ever, have disposal expenses built into their prices, so we are not aware of their true cost to our communities. And many municipalities charge all citizens

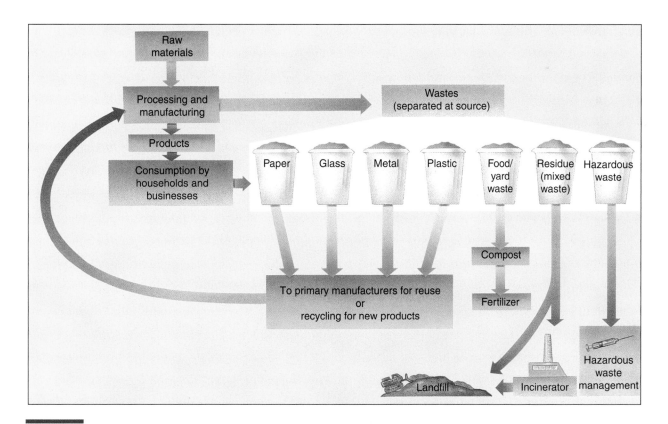

Figure 23.16 Integrated waste management. Source reduction, reuse, recycling, and composting are part of integrated waste management, in addition to incineration and disposal in landfills.

equally for trash collection, regardless of whether they generate a lot or a little solid waste.

If all other factors are equal, most people in the United States do not select a particular product because it produces less waste than another. Most of us like the neatness, squeezability, and unbreakability of plastic ketchup bottles, and we prefer the convenience of disposable diapers over cloth diapers. We value attractive packaging. In contrast, Europeans are generally more willing than Americans to accept new products or changes in packaging that produce less solid waste. (European countries such as Germany, Spain, and France generate about 2 lb of municipal solid waste per person per day, in comparison to the United States' 4.5 lb per person per day.) Europeans' recycling technologies are also more advanced.

Voluntary Simplicity

An increasing number of people in the United States and other highly developed nations have embraced the concept of *voluntary simplicity*, which recognizes that individual happiness and quality of life are not necessarily linked to the accumulation of material goods. People who embrace voluntary simplicity recognize that a person's values and character define that individual more than how many things he or she owns. Voluntary simplicity requires a change in behavior as people purchase and use fewer items than they might have formerly.

Many individuals, businesses, and communities are developing innovative ways to make do with less. Some communities have formed "tool libraries" from which people can check out power and hand tools for a minimal fee. Such organized sharing reduces the requirement for every household to own a complete set of garden and maintenance tools. Other items, such as luggage, household appliances, cookbooks, and recreational equipment, could also be shared or rented at low cost within a community.

Thousands of people in Germany, Switzerland, the Netherlands, Austria, Denmark, Sweden, Italy, and Ireland have embraced the principle of voluntary simplicity by signing up with car sharing programs. Such programs, which are designed for people who use a car once every day or so, offer an economical alternative to individual ownership. Car sharing also may reduce the number of cars manufactured. Studies show that most car sharers drive significantly fewer miles than they did before they joined the program.

■ SUMMARY WITH SELECTED KEY TERMS

I. Solid waste increases in quantity each year.

A. Municipal solid waste consists of solid materials discarded by homes, office buildings, retail stores, restaurants, schools, hospitals, prisons, libraries, and other commercial and institutional facilities.

B. Nonmunicipal solid waste includes wastes from mining, agriculture, and industry.

II. Traditionally, municipal solid waste has been disposed of in open dumps and, more recently, in **sanitary landfills**.

A. The location of a sanitary landfill must take into account the geology of the area, soil drainage properties, the proximity of nearby surface waters and wetlands, and distance from population centers.

B. Despite design features such as high-density plastic liners and leachate collection systems, most sanitary landfills have the potential to contaminate soil, surface water, and groundwater.

III. Incineration reduces municipal solid waste to a fraction of its original volume, and the heat produced during incineration can be used to warm buildings or generate electricity.

A. Mass burn incinerators are large furnaces that burn all solid waste except for unburnable items such as refrigerators. Most mass burn incinerators recover the energy produced from combustion.

B. Modular incinerators are smaller incinerators that burn all solid waste.

C. Refuse-derived fuel incinerators burn the combustible portion of solid waste. Noncombustible materials such as glass and metals are removed, and the remaining solid waste is shredded or shaped into pellets before being burned.

D. One of the drawbacks of incineration is the great expense of installing pollution control devices on the incinerators.

   1. Lime scrubbers produce a chemical spray that neutralizes acidic gases. **Electrostatic precipitators** give ash a positive electrical charge so it adheres to negatively charged plates.

   2. These controls make the gaseous emissions from incinerators safe (as far as we know) but make the ash that remains behind more toxic.

E. Incinerators produce two kinds of ash, **bottom ash** (the residual ash left at the bottom of the incinerator) and **fly ash** (ash from the flue, or chimney). Fly ash is more toxic than bottom ash.

IV. Composting reduces the amount of organic waste, particularly yard waste (grass clippings, branches, and leaves), in the solid waste stream.

V. The three goals of waste prevention are to reduce the amount of waste as much as possible, reuse products as much as possible, and recycle materials as much as possible.

A. Reducing the amount of waste includes purchasing fewer products and purchasing products that have less packaging and that last longer or are repairable.

B. Source reduction has great potential to reduce the volume of solid waste.

1. Source reduction can be accomplished by substituting raw materials that introduce less solid or hazardous waste during the manufacturing process and by reusing and recycling solid and hazardous wastes at the plants where they are generated.

2. Consumers can practice source reduction by decreasing their consumption of products and by purchasing durable products that are designed to generate less solid waste.

C. One example of reuse is refillable glass beverage bottles. Reuse of refillable bottles is not as widespread in the United States as it was in the past. Other countries reuse containers to a greater extent than the United States in order to conserve resources as well as reduce solid waste.

D. Recycling involves collecting and reprocessing materials into new products.

1. Many communities are recycling paper, glass, metals, and plastic.

2. In **product stewardship**, manufacturers assume responsibility for their products from cradle to grave.

E. Many communities have reduced the weight and volume of solid waste by the **fee-per-bag approach** in which households are charged for each container of solid waste that they put out.

VI. **Hazardous wastes** are solids, liquids, and gases that pose a real or potential threat to the environment or to human health.

A. Dioxins are hazardous chemicals formed as unwanted by-products during the combustion of many chlorine compounds.

B. Polychlorinated biphenyls (**PCBs**) are hazardous, oily, industrial chemicals composed of carbon, hydrogen, and chlorine.

C. Radioactive and chemical wastes have accumulated at numerous U.S. nuclear weapons facilities around the United States. The largest, most seriously contaminated site is Hanford Nuclear Reservation in Washington state.

VII. Management of hazardous waste is an ongoing challenge.

A. Industry and government agencies are increasingly stressing the **principle of inherent safety**, in which industrial processes are redesigned to involve less toxic materials so that dangerous accidents are prevented.

B. In the past, hazardous waste was dumped, along with other solid waste, in open dumps and sanitary landfills.

1. As a result, we are faced with the daunting prospect of cleaning up old toxic waste dumps and the soil and water they have polluted.

2. The **Comprehensive Environmental Response, Compensation, and Liability Act**, also known as the **Superfund Act**, addresses the challenge of cleaning up abandoned and illegal toxic waste sites in the United States.

3. The sites that pose the greatest threat to public health and the environment are placed on the **Superfund National Priorities List**.

C. Hazardous waste continues to be produced and must be disposed of safely.

1. Source reduction is the best way to reduce hazardous waste. **Environmental chemistry** is a subdiscipline in chemistry in which commercially important chemical processes are redesigned with source reduction in mind.

2. High-temperature incineration is one way to detoxify hazardous organic compounds such as pesticides, PCBs, and organic solvents.

3. Hazardous waste must be stored in special landfills designed to minimize the risk of environmental contamination. In the United States, only 23 commercial facilities are currently certified to handle hazardous waste.

VIII. Opposition to environmental inequities faced by low-income communities has resulted in the **environmental justice** movement, which believes that every citizen, regardless of age, race, gender, or social class, is entitled to adequate protection from environmental hazards.

IX. There is no single solution to the problem of solid and hazardous wastes.

A. The best approach is to use **integrated waste management**, which is a combination of techniques including source reduction, reuse, recycling, and composting, to reduce the amount of waste to be discarded.

B. Combustible materials that remain can then be incinerated to further reduce the volume of solid waste. Incinerators must be equipped with pollution control devices to ensure that the pollutants are not transferred from one medium (solid waste) to another (air or water pollution).

C. The remaining solid waste, a small fraction of the original solid waste, should be disposed of in an environmentally safe sanitary landfill.

THINKING ABOUT THE ENVIRONMENT

1. What is solid waste? Distinguish between *municipal solid waste* and *nonmunicipal solid waste*.

2. Describe the main features of a sanitary landfill.

3. Describe the main features of a mass burn incinerator.

4. Compare the advantages and disadvantages of disposing of waste in sanitary landfills and by incineration.

5. What is source reduction? Explain how source reduction, reuse, and recycling reduce the volume of solid waste.

6. List what you think are the best ways to treat each of the following types of solid waste, and explain the benefits of the processes you recommend: paper, plastic, glass, metals, food waste, yard waste.

7. How do industries such as Goodwill, which accepts donations of clothing, appliances, and furniture for resale, affect the volume of solid waste?

8. It could be argued that a business that collects and sells its waste paper is not really recycling unless it also buys products made from recycled paper. Explain.

9. Why is creating a demand for recycled materials sometimes referred to as "closing the loop"?

10. What is hazardous waste?

11. What are dioxins, and how are they produced? What harm do they cause?

12. What are PCBs, and what harm do they cause?

13. Suppose hazardous chemicals were suspected to be leaking from an old dump near your home. Outline the steps you would take to (1) have the site evaluated to determine if there is a danger and (2) mobilize the local community to get the site cleaned up.

14. Briefly contrast the Resource Conservation and Recovery Act with the Comprehensive Environmental Response, Compensation, and Liability Act.

15. What are the goals, strengths, and weaknesses of the Superfund program?

16. Define *environmental justice*.

17. The Organization for African Unity has vigorously opposed the export of hazardous waste from industrialized countries to developing nations. They call this practice "toxic terrorism." Explain.

18. What is integrated waste management? Why must a sanitary landfill always be included in any integrated waste management plan?

19. Compare integrated pest management, discussed in Chapter 22, to integrated waste management. How does each reduce potential damage to the environment?

*20. In 1999, new steel products contained, on average, 56% recycled steel. If a total of 98.1 million tons of steel was produced in 1999, how many million tons were scrap (recycled)?

* The solution to this question appears in Appendix VII.

TAKE A STAND

Visit our Web site at **http://www.wiley.com/college/raven** (select Chapter 23 from the Table of Contents) for links to more information about whether recycling programs should be required components of municipal-waste-management plans. You will find tools to help you organize your research, analyze the data, think critically about the issues, and construct a well-considered argument. Take a Stand activities can be done individually or as part of a team, as oral presentations, written exercises, or Web-based (e-mail) assignments.

Additional on-line materials relating to this chapter, including Student Quizzes, Activity Links, Useful Web Sites, Flash Cards, and more, can also be found on our Web site.

SUGGESTED READING

Atkin, R. "The Mixed Bag of Community Recycling." *Christian Science Monitor* (April 17, 2002). Examines the current status of recycling in the United States.

Bramwell, P. "Sum of the Parts: How Long Should Cars Last?" *Living Planet* (Fall 2001). Examines cars' life spans and what happens to them when they are at the end of their usefulness.

Hileman, B. "Reassessing Dioxins." *Chemical and Engineering News* (May 28, 2001). The EPA's long-awaited report on the health risks of dioxins is very controversial.

Hilts, P.J. "Dioxin in Arctic Circle Is Traced to Sources Far to the South." *New York Times* (October 17, 2001). The Inuit of Canada carry a heavy dioxin load in their bodies, despite the fact that they live far from any sources of dioxins.

Parks, K. "From Landfills to Links." *St. Petersburg Times* (April 16, 2001). Highlights a company that converts landfills into golf courses and other developed areas.

Poliakoff, M., and P. Anastas. "A Principled Stance." *Nature*, Vol. 413 (September 20, 2001). The authors think green chemistry has great potential.

Probst, K., and A. Lowe. "The $200 Billion Question: Does Anyone Care about Cleaning Up the Nation's Nuclear Weapon's Sites?" *Resources*, Issue 138 (Winter 2000). Former nuclear weapons manufacturing sites are highly contaminated.

Raloff, J. "Landfills Make Mercury More Toxic." *Science News*, Vol. 160 (July 7, 2001). Landfill disposal of products containing mercury alters the mercury to a more toxic form that is more readily released into the atmosphere.

Raloff, J. "Memory Problems Linked to PCBs in Fish." *Science News*, Vol. 159 (June 16, 2001). A new study links subtle memory problems to eating Great Lakes fish contaminated with PCBs.

Revkin, A. "New Pollution Tool: Toxic Avengers with Leaves." *New York Times* (March 6, 2001). Highlights the relatively new field of phytoremediation.

Schultz, J. "Waste Not, Want Not." *Far Eastern Economic Review* (September 20, 2001). A Tokyo hotel composts its garbage and turns a tidy profit in the process.

Zwillich, T. "A Tentative Comeback for Bioremediation." *Science*, Vol. 289 (September 29, 2000). The Department of Energy is using bacteria to consume hazardous wastes.

People enjoy an outdoor concert in Madison, Wisconsin. Tomorrow's world will require that we revalue ourselves according to a different set of ideas. Quality of life will have to become a better measure of success than wealth and material possessions.

Tomorrow's World

Learning Objectives

After you have studied this chapter you should be able to

1. Define *sustainability* and explain the very different reasons why people in highly developed countries and developing countries are not living sustainably.

2. Discuss how the natural environment is linked to sustainable development.

3. Explain why we should respect and care for the community of life.

4. Describe some of the challenges confronting our efforts to improve the quality of human life worldwide.

5. List two major steps we must take to stay within Earth's carrying capacity.

6. Describe the role of education in changing personal attitudes and practices that affect the environment.

7. Discuss at least two important environmental goals that can be accomplished most effectively at national and international levels.

8. Write a one-page essay describing what kind of world you want to leave for your children.

You have been presented with a broad overview of the environment, including both the principles necessary to understand how ecosystems operate and a detailed review of the individual issues that are having an important influence on today's world. Now we wish to speak directly to you, not only as students but also as citizens. Our world is in very serious trouble today, presenting us with an enormous challenge. What follows represents where we stand, our opinions on what must be done to meet today's critical environmental issues. There are reasonable people who hold other views, but we believe that we would not be doing our job as responsible teachers if we did not present our views for you. Read, think, discuss, and come to your own conclusions. And then, if our world is to have a peaceful, prosperous, and sustainable future, act.

A STRATEGY FOR SUSTAINABLE LIVING

We have organized our comments around the nine principles for sustainable living presented in *Caring for the Earth*, a report published in 1991 by the World Conservation Union (IUCN), the United Nations Environment Program (UNEP), and the World Wide Fund for Nature (WWF). In preparing this chapter, we have paraphrased much of what is included in that excellent publication. Much discussion in the preceding chapters of this book has laid the foundation for understanding the principles and problems involved, and the framework in *Caring for the Earth* is a convenient way to organize an agenda for action.

Major Environmental Challenges

The problems we face are enormous and based in part on the large disparities in the standards of living experienced in different parts of the world. We who live in the United States at the beginning of the 21st century enjoy an abundance of what the world has to offer. In fact, we are collectively the wealthiest people who have ever existed, with the highest standard of living (shared with a few other rich countries). Because the United States, with 4.6% of the world's people, controls approximately 25% of the world's economy, it is obvious that we depend on many other nations for our prosperity. In our actions, however, we often seem to miss this relationship and to underestimate our effects on the environment that supports us.

Throughout the world, nearly one in four people lives in extreme poverty, and one in eight cannot get enough food to make it possible for their bodies to grow and function properly (Figure 24.1). Each year, many millions of children die of malnutrition and disease. Despite the magnitude of these problems, which impact everyone on Earth directly or indirectly, we find it difficult to take seriously the need to alter the basic activities by which we use the world's natural resources. Only such a strategy, however, will bring us safely into the kind of future that we all want for our children and grandchildren.

Throughout the text, we have demonstrated that humans are not managing the world's resources sustainably. Instead, we are depleting them rapidly. During the second half of the 20th century, as the world population grew from 2.5 billion to 6 billion people, we lost about one fourth of the world's topsoil. Because of salinization, desertification, urban sprawl, erosion, and other factors, we are feeding well over twice as many people as in 1950 on only 80% of the agricultural lands that were being cultivated then. Global warming is underway, with greater than one-sixth more carbon dioxide (CO_2), the most important greenhouse gas, having been added to

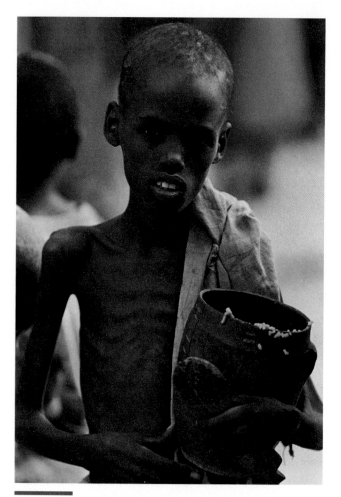

Figure 24.1 Undernourished child in Somalia. This undernourished boy is receiving food at a Red Cross food distribution center.

the total amount in the Earth's atmosphere since 1950. The depletion of the stratospheric ozone layer that protects us from carcinogenic ultraviolet B (UV-B) rays has been substantial. Collectively, we are using more than 30% of the energy ultimately derived from the sun and transformed through the process of photosynthesis into a form that supports all living organisms on Earth, including ourselves (see section on human impact on net primary productivity in Chapter 4). Since 1950, we have cut about a third of the forests that were in existence then, without replacing them. Most seriously of all, we are driving the world's species of plants, animals, fungi, and microorganisms to extinction at a rate thousands of times the rate that has been in operation for the past 65 million years. As many as two thirds of all terrestrial species may vanish by the end of the 21st century.

If the world's resources were considered a bank account, we could be said to be living off the principal, and not off the interest. Living off the interest alone, without touching the principal, would be sustainable. Our way of living, however, is clearly unsustainable. We exist as we do by rapidly exhausting the quantity and quality of the natural resources that will still be available to people in the future.

Living Sustainably

Sustainability is a concept that people have been discussing for many years. "Our Common Future," the Report of the World Commission on Environment and Development (1987), presented the closely related concept of **sustainable development**. It was defined as development that meets the needs of the present without compromising the ability of future generations to meet their own needs. The authors of the report pointed out that the concept of *needs* includes meeting the needs of the world's poor, because unless their needs are met, there can be no overall sustainability. "Our Common Future" also pointed out that the environment's ability to meet present and future needs is directly related to the state of technology and social organization existing at a given time and in a given place. The number of people existing, their degree of affluence (that is, their level of consumption), and their choices of technology all interact to produce the total effect of a given society, or of society at large, on the sustainability of the environment.

When the [20th] century began, neither human numbers nor technology had the power to radically alter planetary systems. As the century closes, not only do vastly increased human numbers and their activities have that power, but major unintended changes are occurring in the atmosphere, in soils, in waters, among plants and animals, and in the relationships among all of these. The rate of change is outstripping the ability of scientific disciplines and our current capabilities to assess and advise. It is frustrating the attempts of political and economic institutions, which evolved in a different, more fragmented world, to adapt and cope.

Even with the use of the best technologies we could imagine, the productivity of Earth still has its limits, and the extent of our use of it cannot be expanded indefinitely. To live within these limits, population growth must be held at a level that can be sustained, and the wealthy must first stabilize their use of natural resources, and then reduce this use to a level that can be maintained. The world does not contain nearly enough resources to sustain everyone at the level of consumption that is enjoyed in the United States, Europe, and Japan (Figure 24.2). Suitable strategies, however, do exist to reduce these levels of consumption without concurrently reducing the real quality of life.

To accomplish this reduction of consumption at individual, community, national, and finally global levels, we must act on the basis of an understanding that development can take place only within the limits of the ecosystems themselves. Once we accept that principle, we realize that conservation and development are not only compatible, but that they must be so. Although we cannot foresee the consequences of particular kinds of development with full accuracy, all development inevitably must take place within the carrying capacity of the ecosystems that support it.

Kai N. Lee is a very thoughtful ecologist who is a professor at Williams College. In his book *Compass and Gyroscope: Integrating Science and Politics for the Environment*, Lee puts it this way:

Against this background it is possible to see that sustainable development is not a goal, not a condition likely to be attained on Earth as we know it. Rather, it is more like freedom or justice, a direction in which we strive, along which we search for a life good enough to warrant our comforts. Freedom and justice are easily taken for granted, although many have died in their pursuit and defense. A more materialist goal such as sustainability is harder to imbue with romance and ideology. But the enormous

Figure 24.2 Consumption in the United States. This "typical" family of four lives in Pearland, Texas. All their material possessions were removed from the house and garage and placed in the driveway and street to emphasize the large amount of natural resources consumed by people living in highly developed countries.

changes our species has wrought leave us a difficult choice: either to accept our humanity in the company of the whole human race and the natural world we jointly share, or to concede that being human is too difficult for the richest, most advanced beings in history.

In the following nine sections, we now present the principles involved in building a sustainable society, along with some of the actions we think are critical for making these principles a reality.

PRINCIPLE 1: BUILDING A SUSTAINABLE SOCIETY

Sustainable development must provide real improvements in the quality of human life. At the same time it must maintain the life-support systems on which our lives, and the lives of all other species, are based. It is not possible to ignore the functions of these biological and physical systems in achieving economic development; if we do ignore them, we ultimately degrade the quality of human life as well. The development strategies most suitable for a given locality depend on the complex mix of biological, physical, and human factors that are in operation there, and we must strive to understand these if our decisions are to be rational. Preserving Earth's biological diversity makes the whole system stronger, and the more variety that is preserved, the more interesting and beautiful the world is. Biological diversity and human cultural diversity are intertwined: They are in fact two sides of the same coin (Figure 24.3).

To build and maintain a sustainable society, it is necessary to preserve the productive natural systems that support us. Renewable resources such as forests and fisheries must be treated in ways that ensure their long-term productivity. Their capacity for renewal must be understood and respected. The need to conserve nonrenewable resources such as oil and natural gas, or minerals, is obvious. Somewhat unexpectedly, however, it is the potentially renewable resources—productive living systems such as forests, shrublands, grazing lands, and swamps—that have been so badly damaged over the past 200 years (Figure 24.4). In addition, discoveries of new supplies of nonrenewable resources have often given us the illusion that they are inexhaustible.

The basic message is that whatever actions we take, they must be taken within the **carrying capacity** of individual ecosystems, and ultimately that of Earth itself. However, the transition to a sustainable world depends on changing individual attitudes and practices in intelligent ways based on scientific information. Throughout the world, communities are finding that sustainable activities preserve their local environments and improve their standards of living. Getting the best available information to rural communities is essential to help the people who live there make the most appropriate choices for

Figure 24.3 Biological diversity and human cultural diversity in Brazil. Humans in increasing numbers are encroaching on the traditional environments of organisms such as this scarlet macaw and of native peoples such as this Amazonian boy.

themselves. Once attitudes have shifted and been nurtured at the individual and community levels, then national strategies can be devised that safeguard the ecological capital of each country for the future. And finally, because many environmental problems extend across national boundaries, alliances among countries are necessary to preserve the lands and waters on which our livelihoods depend.

PRINCIPLE 2: RESPECTING AND CARING FOR THE COMMUNITY OF LIFE

At the most basic level, we are part of the web of life on Earth, having evolved within it and as part of it. We are entirely dependent on that web, with all of its complex interactions, for our survival and for anything that we wish to attain individually or collectively in the future. In view of our utter dependency on the web of life, many people have raised the question as to whether it is morally acceptable for us to continue destroying it so rapidly (Figure 24.5). Can we justify, in ethical terms, the fact that we are driving species to extinction that are, as far as we know, our only living companions in the universe? Clearly, our individual rights as humans are ultimately dependent on our respect for one another and for the natural environment that surrounds and supports us. There are no exemptions from this principle.

Pragmatically, we have a clear interest in protecting Earth's biological diversity and managing it sustainably. We obtain from living organisms all of our food, most of our medicines, much building and clothing materials, biomass for energy, and many other products that are of fundamental importance to us. In addition, communities of organisms and ecosystems provide an enormous array of **ecosystem services** without which we would be unable to survive. These include the protection of watersheds and topsoils; the development of fertile agricultural lands; the determination of local climate and, through

Figure 24.5 Endangered green sea turtle in the Atlantic Ocean. After mating in shallow water, the females lay their eggs on beaches from Cape Canaveral to Palm Beach. Coastal development is one of the ways humans affect sea turtles adversely.

processes such as carbon sequestration, of global climate; and the maintenance of habitats for beneficial animals and plants. In view of these ecosystem services, the ultimate reason for caring for the community of life on Earth is a selfish one. The interdependence of humans with nature depends on living sustainably on Earth.

While these words may sound obvious, almost no human society lives according to the underlying principle, although some nations and some localities obviously do better than others.

PRINCIPLE 3: IMPROVING THE QUALITY OF HUMAN LIFE

The ultimate goal of development is to improve the quality of human life, making it possible for humans throughout the world to enjoy long, healthy, and fulfilling lives. A serious complication lies in the fact that the distribution of the world's resources is very unequal. Those who live in highly developed countries, a rapidly shrinking 19% of the global population, control about 79% of the world's finances, as measured by summing gross domestic products. The per-capita GNI PPP (gross national income purchasing power parity) in highly developed countries in 2002 was about $22,060; the per-capita GNP PPP of the other 80% of the people in the world was about $3,580. At present, nearly 1.4 billion people, almost one

Figure 24.4 Damage to potentially renewable resources in Kenya. Extensive withdrawal of water has contributed to erosion in this once-fertile area. Careful stewardship of the land can prevent such damage.

Figure 24.6 Children gathering firewood in Ethiopia.
Poverty forces children to help contribute to the family's livelihood.

fourth of the world's population, live on less than $1 per day.

Among the world's poor people, roughly one in seven receives less than 80% of the minimum caloric intake recommended by the United Nations, so that their brains and bodies are unable to develop properly. For many of the world's women and children, life is an endless struggle for survival, centering on the daily requirements for firewood, clean water, and food (Figure 24.6). Such poverty, along with the enormous pressures that human population growth, consumption rates, and application of technologies are putting on the world's productive capacity, are global problems that cannot be solved without modifying the standard of living enjoyed in highly developed countries.

Not only does failing to confront the problem of poverty around the world continue to make it impossible to attain global sustainability but it also is morally indefensible. Faced with the facts, most people would find unacceptable the deaths of 35,000 babies each day, most of which could have been avoided by access to adequate supplies of food or basic medical techniques and supplies. For us to allow so many to starve, to go hungry, and to live in absolute poverty is to threaten the future of the

global ecosystem that sustains us all. Everyone must have a reasonable share of Earth's productivity, or our civilization will eventually come unraveled. As U.S. President Franklin Delano Roosevelt said so well in his second inaugural address in 1937, "The test of our progress is not whether we add more to the abundance of those who have much; it is whether we provide enough for those who have too little."

Improving the quality of life in lower-income countries will require increasing economic growth so that issues of health, nutrition, and education can be addressed adequately. The role of women requires special attention, since women are often disproportionately disadvantaged in poor countries, and the improvement of their status can make a significant contribution to the stability and prosperity of those communities. As Nafis Sadik, head of the U.N. Population Fund has pointed out, women hold a paradoxical place in many societies. As part of their traditional duties as mothers and wives, they are expected to bear the whole responsibility for childcare; at the same time, they are often expected to contribute significantly to the family income by direct labor (Figure 24.7). In many developing countries

Figure 24.7 A woman harvesting tomatoes in Mali. In addition to caring for their children, women are the farmers in much of Africa.

women have few rights and little legal ability to protect their property, their rights to their children, their income, or anything else. Improving the status of women is a crucial aspect of development.

In addition, raising the standard of living for poor countries will require that special attention be given to the poorest segments of all populations—that is, to those without much hope. In this context, the universal education of children and the reduction of illiteracy are of critical importance in raising and maintaining appropriate standards of living in every country.

Finally, one of the greatest barriers to equalizing the gap between highly developed and developing nations is the lack of trained professionals in the latter group. Only about 10% of the world's scientists and engineers live in less developed countries. Considering that a majority of those scientists and engineers live in only four nations (China, Brazil, Mexico, and India), it is apparent that for most developing countries, trained technical personnel are virtually lacking. Therefore, how are these countries to decide, on the basis of their limited knowledge, whether to join international agreements concerning the environment or how best to manage their own natural resources? The training and employment of professional scientists and engineers in developing countries is a matter of high priority.

PRINCIPLE 4: CONSERVING EARTH'S VITALITY AND BIOLOGICAL DIVERSITY

Development will succeed only if it is carried out in such a way as to maintain the sustainable productivity of the biosphere. A human population of more than 6 billion people, using an estimated 40% of land-based net terrestrial photosynthetic productivity (see Chapter 4) and approximately 55% of accessible, renewable supplies of fresh water, puts enormous and unprecedented pressures on the life-support systems that nature provides and on the sustainability of renewable resources. Many of the life-support systems adversely affected by human activities were reviewed in Parts 5 and 6 of this book. The negative impacts discussed there affect the ecological processes that maintain topsoil, the atmosphere, global climate, the hydrologic cycle, and the other natural processes that maintain Earth's vitality.

With respect to biological diversity, some 80% of the species of plants, animals, fungi, and microorganisms on which we all depend occur in developing countries. How will these relatively poor countries be able to sustainably manage and conserve these precious resources? Biological diversity is an intrinsically local problem: Each nation must address it for the sake of its own people's future, as well as for the world at large. The problem is made more complicated by the fact that an estimated five sixths of all kinds of organisms have not yet been recognized and described scientifically. Biological diversity, like most problems of sustainable development, can be addressed adequately only if we help to increase the number of scientists and engineers in developing countries.

The science of restoration ecology must be pursued aggressively, and natural communities must be restored where possible throughout the world (Figure 24.8). Natural areas, as well as significantly modified ecosystems, must be protected to the greatest extent possible, with a special emphasis on preserving the biological diversity that they contain. Rich countries must aid the poorer ones if this is to be carried out effectively.

Figure 24.8 Prairie restoration in Wisconsin. In 1935, when restoration began on this tall-grass prairie, it was severely degraded agricultural land. Restoration ecology creates biological habitats and regenerates the soil that was damaged by human activities such as agriculture or mining.

Farmland, grazing land, forest plantations, and similar modified ecosystems must be managed efficiently so that they can produce as much as possible and thus reduce the pressure on natural lands everywhere. This is especially true because worldwide we are growing our food and other agricultural products (such as cotton and biomass) on an area about the size of South America. Very little additional land—land that is not now cultivated—is suitable for agriculture. Obviously, improving agriculture is one of the highest priorities involved in achieving future global sustainability, although in a rich country like the United States, where food is very inexpensive, it is difficult to appreciate this goal properly.

Many improvements in agriculture will be needed to feed the world's people adequately in the future. In general, per-capita grain production has kept pace with human population growth over the past 40 years. However, per-capita grain production and other examples of expanded agricultural productivity have taken place at high environmental costs that are often not sustainable. We need to develop sustainable agricultural systems that are suitable for the different regions of the globe. First, the negative effects of agriculture, including soil and water pollution, air pollution, soil erosion, loss of soil fertility, and aquifer depletion, must be brought under control by an array of different strategies. For example, conservation tillage, in which the topsoil is protected, is one important element of sustainable agriculture. This practice is growing rapidly throughout the world. Precision farming, in which satellite data are used to determine the most appropriate levels of fertilizer application for individual small areas, provides an important key to agriculture of the future. Genetic engineering should be used to develop nutritious crops with the precise combination of characteristics that will be suitable for different local conditions. Integrated pest management, involving the use of beneficial insects and improved cultural practices, with the application of pesticides only as a last resort, needs to be applied more widely.

Each country should develop a comprehensive plan for the protection of its natural areas, with international assistance provided where needed. Biological diversity will not survive without our direct intervention, as 6.2 billion humans have placed so much of Earth's organisms at risk. For our own future good, we must reverse the pressure we are putting on Earth's organisms.

One of the major factors destroying biological diversity throughout the world is chemical pollution, as pointed out so effectively by Rachael Carson in the 1970s. Pollution by damaging chemicals, often used in industry or agriculture, is having a huge negative effect on biological diversity throughout the world. For example, it is estimated that some 65 million birds die each year in the United States as a result of spraying with pesticides. Means must be found to reduce the enormous pesticide burden on Earth, and to achieve pest control by following alternative pathways.

PRINCIPLE 5: KEEPING WITHIN EARTH'S CARRYING CAPACITY

The maximum human population that can be sustained by a given ecosystem, or by the world as a whole, is defined as its *carrying capacity*. The carrying capacity of a given ecosystem is determined ultimately by its ability to absorb wastes and renew itself. In understanding Earth's carrying capacity and designing appropriate development strategies, we should bear in mind the great disparities between living standards and expectations in different areas. The poorer people living in developing countries have far less than their share of the world's resources, and most cannot live sustainably, given their current circumstances. In addition, population growth rates tend to be highest where poverty is most intense. This overall situation is clearly unstable and must be corrected by the determination and adoption of acceptable levels of population and consumption for all regions.

The world population grew from 2.5 billion in 1950 to 6.2 billion in 2002. The United Nations has recently (in 2000) calculated that world population could stabilize at a level of approximately 9.3 billion people by the end of the 21st century. To be able to restrict the global population to this enormous total, however, there must be sustained worldwide attention to family planning. Our population will not automatically reach a level of 9.3 billion people and then stop growing. If we continue to pay consistent attention to overpopulation and devote the resources that are necessary to make family planning available for everyone, we can make the human population stabilize (Figure 24.9). If we do not continue to emphasize family planning measures, we simply will not achieve population stability.

To stay within Earth's carrying capacity, it will be necessary not only to reach a stable population but also to greatly reduce excessive consumption and waste. These factors must be managed in an integrated fashion and be coupled with educational programs everywhere, so that people will have the opportunity to understand that Earth's carrying capacity is not unlimited. New technologies must be developed that will be more environmentally friendly than the ones they will replace.

One of the most valuable tools that can be used to moderate the excessive consumption of resources is an economic one: **full-cost accounting**. Tax subsidies that have been enacted to benefit certain classes of manufacturers or investors often lead to the misstatement of true costs and to the consequent waste of the commodities involved. The varying cost of gasoline around the world provides an excellent example of the ways in which prices can be manipulated. In the United States, the cost of

gasoline is held artificially low (see Table 10.2), and price increases are always regarded with alarm.

PRINCIPLE 6: CHANGING PERSONAL ATTITUDES AND PRACTICES

Any long-term improvement in the condition of the world must start with individuals—our values, attitudes, and practices. Each of us makes a difference, and it is ultimately our collective activities that make the world what it is. In the richer parts of the world, apathy, ignorance, or incentives to wasteful consumption have negative effects. In the less developed countries, survival may be such an overwhelming concern that conservation in any sense of the word may seem irrelevant.

As people adopt new lifestyles, they must be educated so that they understand the reasons for changing practices that may be highly ingrained or traditional. People are generally concerned about the environment, but their concerns do not naturally translate into action. Furthermore, most people believe that individual actions do not really make a difference, whereas in fact, they are the only factors that do so, both individually and collectively. If people have the opportunity to understand the way the natural world functions, they will be able to appreciate their own place in it, and to value actions that are sustainable. Formal education and informal education are both important in bringing about change and in contributing to the sustainable management of resources. Accurate information must be made available widely; the media have an important part to play in such efforts. No national plan can be successful without a clear statement

of its strategic goals, accompanied by a pervading educational program laying out the fundamental reasons why these goals are so important and what individuals can do to help realize them.

In the United States, every citizen will be called on throughout their lives to take actions of many kinds that are based on an understanding of the environment. Many of these are personal choices, such as the amount of water or the kind of transportation to use. In a democracy, there are also many choices made at a local, national, and global level that require environmental knowledge. Should we ratify the Convention on Biological Diversity or remain one of a handful of countries that have failed to do so? Should we support the Kyoto Protocol on global warming? Should we promote world trade—and what kind of trade? How important is it to reduce sulfur emissions from coal-fired plants? To be able to answer these and similar questions, we can read and study. To help our fellow citizens understand these issues, we can

1. Set up environmental curricula in primary and secondary schools and in colleges.

2. Encourage the activities of environmental organizations.

3. Support institutions such as natural history museums, zoos, aquaria, and botanical gardens, all of which promote conservation and sustainability.

4. Encourage the inclusion of relevant material in the programs of churches, social groups, and other institutions that should logically emphasize them.

Most people are interested in the environment—their own local environment—in their own way. We must work together to create conditions in which everyone will have the opportunity to learn and to contribute. That is the very essence of a democratic society.

PRINCIPLE 7: ENABLING COMMUNITIES TO CARE FOR THEIR OWN ENVIRONMENTS

Most communities care deeply about the sustainable management of their own environments, whether they would express their views precisely in that way or not. By acting together, people can be a strong and effective force, regardless of whether their community is wealthy or poor. In the United States and similar highly developed countries, it is often through the ballot box that action takes place, but there is no substitute for direct action in any part of the world. Developing effective local governments that are responsive to the need for sustainable community development is an essential element in achieving overall success.

We often speak of throwing garbage away, but if we stop to think about it, all that enters Earth, with minor exceptions, is sunlight; and all that leaves is some of the radiated heat, ultimately originating from that sunlight. Thus there is truly no "away," no place where our garbage, our carbon dioxide, our pollution—anything that we produce or concentrate—can go (Figure 24.10). Despite this great truth, we often treat Earth as if it had no limits—as if there were plenty of room in which to dispose of things. We forget that by doing so we are cre-

ating a situation like the one that would arise if we simply took into our homes all that we need or wish to consume, and then just left any part of them that we did not use in the home. There would soon be no room for us! And Earth is precisely the same. We must stop clinging to the illusion that we can really throw things "away." That is basically why each community must learn to take care of itself: to make and use at least the great majority of what it consumes locally, and to dispose of it locally.

In addition to approaching local self-sufficiency in production of goods and disposal of wastes, a community must maintain the vitality of its local ecosystems. As the world becomes less diverse and less resilient because of the operation of the many factors we have discussed in this book, it will not be degraded uniformly. Some parts will be richer in biological diversity, healthier, and more enjoyable places to live. The reason that these particular places will stand out will be because individuals and communities will have taken care of them well.

One important ingredient in community action is the way people transmit information. Knowledge and information are increasingly concentrated in cities. The rural people who manage much of the world's biological and physical resources may have an increasingly difficult time in getting access to the information they need, despite the fact that we live in an age when it is easier to transmit information than at any earlier time.

An outstanding example of one way information can be transmitted to rural people is provided by the establishment of Information Villages around Pondicherry, in southeastern India, by the M.S. Swaminathan Research

Figure 24.10 The solid waste produced by an average U.S. family of four in 1 year. The piles of cans, bottles, and newspapers on the left are what the average family now recycles. The trashcans and bags on the right are filled with the solid waste that remains after recycling. Photographed in San Diego, California.

Foundation. In small farming communities, dominated by women (the men are mostly away earning cash at jobs in cities or industrial sites), the foundation has established computer stations that receive their signals through solar-powered antennae. At these stations, which are manned by local volunteers, information about health, agriculture, and other matters that concern the villagers is made available on a continuing basis. The particular kinds of information presented are selected by the people themselves and provided in their own language and in multimedia (for illiterate users).

In general, actions that give communities more control over their own lives are important in enabling them to move their communities toward sustainability. These actions include access to resources and the ability to play a role in managing them, access to education and training, and the right to participate in decision making. If individual nations empower communities in these ways, their local communities will function well, and so, therefore, will the world as a whole.

PRINCIPLE 8: BUILDING A NATIONAL FRAMEWORK FOR INTEGRATING DEVELOPMENT AND CONSERVATION

Attention to the seven principles outlined thus far makes it possible to build an effective national approach, which is essential to attain sustainable development globally. Although the Earth Summit in Rio de Janeiro in 1992 brought together more heads of state than any other meeting in human history, they did not speak much of international cooperation, mainly stressing how effective their own national efforts were. These efforts differ widely in scope and intensity, but all help to form the basis for achieving collective success in creating a sustainable world.

Most of the world's nations have departments or ministries devoted wholly or in part to environmental protection. They address areas of general public concern, such as air and water pollution, the destruction of natural resources, flooding, and similar threats to human life. Few departments of environmental protection are broadly integrated and devoted to the principle of sustainable development overall. Most of these departments mainly react to problems as they occur instead of identifying system-wide opportunities that could prevent these problems from ever occurring.

One of the most important aspects of effective national action regarding sustainable development is cooperation among the different parts of a given government. Attention to environmental matters at a national level is a relatively new phenomenon, and examining programs at this level often reveals that many aspects of sustainable development are dealt with by different parts of a government. Consequently, integration may be poor,

and the effectiveness of the programs hampered. In the United States, for example, the restoration of the Everglades ecosystem in Florida requires the cooperation of many departments, and we are just learning how to do this. The Department of Interior, U.S. Army Corps of Engineers, National Atmospheric and Oceanic Administration in the Department of Commerce, and the state of Florida are all directly concerned in restoration of the Everglades. In addition, the Environmental Protection Agency, National Science Foundation, Department of Housing and Urban Development, Department of Health and Human Services, and others are peripherally involved in the Everglades restoration. The better integrated the common efforts of these departments can be, the better the results. The problems of sustainability are intrinsically complex and require widespread cooperation to be managed properly.

Similar considerations apply when we are weighing the potential environmental consequences of major public works. Many kinds of development, such as constructing new transportation systems or new suburban communities (Figure 24.11), must be judged in light of their overall impact on the quality of life in the region, and for the people who use these facilities or live in these communities. We can undertake such actions sustainably only if we judge, insofar as is possible, the full range of their future consequences.

Because building a national framework that integrates development and conservation is largely a matter of management, the increasing attention to the environment in the training of lawyers, business managers, architects, engineers, and other professionals is very encouraging. Perhaps some of you will take up these or similar careers in which you could make a strong contribution in the environmental area. Reading this book, you will have learned that environmental considerations pervade all aspects of life, and those who are well trained in these matters, in the context of their profession, will have important contributions to make in the future. The legal framework within which environmental considerations are viewed is exceedingly important. The tax structure is another element in achieving effective environmental action. The advancement of scientific knowledge that underlies the development of appropriate steps to maintain a sound environment is also important. Opportunities abound and will be taken up by those with vision and a determination to create a better world for the future.

Just as individuals must commit themselves to new ethical standards if they are to approach the goal of sustainability in their own lives, so nations must establish fully their commitment to the principles of a sustainable society. The extent to which nations accomplish this will ultimately determine their long-term success. Constitutional documents and other statements of national policy should ideally lay out such commitments clearly, and make possible the generation of appropriate laws and suitable economic policies. As Tim Wirth, a former U.S.

Figure 24.11 Suburban sprawl near Las Vegas, Nevada. Before a community permits such development, various departments should carefully consider the potential environmental repercussions.

senator from Colorado, put it, "The economy is a wholly-owned subsidiary of the environment." The ability of nations to recognize this relationship and to act on their recognition of it will have much to do with the quality of life people enjoy throughout the world.

■ PRINCIPLE 9: CREATING A GLOBAL ALLIANCE

Although the necessity of creating a global alliance for sustainable development is obvious—witness the global problems involving the atmosphere, climate change, the ocean and fisheries, forests, and biological diversity—it has proven very difficult in practice (Figure 24.12). Successfully dealing with these global problems depends on national actions that contribute to an integrated global action. Such action supports nations and people throughout the world in their efforts to harmonize human activities with the maintenance and long-term productivity of the Earth's life-support systems. Immigration from areas of want to those of affluence, legal and illegal, demonstrates vividly the need for international coordination, and so do the increasing demands of international trade. These problems cannot simply be ignored, because their effects are pervasive and will be felt regardless of what actions we take in relation to them.

Treaties govern international relationships, and in principle, treaties could become the basis for comprehensive global alliances. Many global treaties exist on environmental matters, ratified by varying numbers of nations; these concern the atmosphere, ocean, fresh water, wetlands, hazardous waste, climate change, and

Figure 24.12 Cod fishing off the coast of Lofoten, Norway. Declining fisheries have been observed worldwide.

| Table 24.1 | Selected International Environmental Treaties |

Treaty	What it Does	U.S. Ratification
Kyoto Protocol	Addresses the problem of global climate change	No
U.N. Fish Stocks Agreement	Regulates marine fishing	Yes
Madrid Protocol	Protection of Antarctica's unique environment	Yes
U.N. Convention of the Law of the Sea	Regulates deep-sea mining	No
Convention on International Trade in Endangered Species of Wild Flora and Fauna	Protects endangered plant and animal species	Yes
U.N. Convention on Biological Diversity	Protects biological diversity	No
Montreal Protocol	Protects the ozone layer	Yes
Stockholm Convention on Persistent Organic Pollutants	Bans or restricts the production and use of 12 extremely toxic chemicals	No
Basel Convention	Restricts the international transport of hazardous waste	No

the protection of biological diversity, among other topics (Table 24.1). Additional treaties should be devised and implemented in areas deserving such attention, and the international system of governance should be enhanced and widely recognized. The document that we have used as a basis for this chapter, *Caring for the Earth*, calls for a universal declaration and covenant on sustainability, which would certainly contribute to progress in creating a global alliance.

Goods, services, and money flow throughout the world and greatly affect the nations among which they move. We have entered a new era of global trade, within which we need to establish new guidelines for national, corporate, and individual behavior, and new understandings of the dynamics of systems that we understand only imperfectly. To give just a single example at this point, the flow of money from developing countries to highly developed countries has exceeded the flow in the other direction for more than 15 years. Former West German Chancellor Willy Brant termed this phenomenon "a blood transfusion from the sick to the healthy." A world that valued sustainability and social justice for everyone would find ways to reverse this flow for the common good. In such a world, debts from the poorest countries would be forgiven more readily than is the case now, and international development assistance would be enhanced.

If each nation accepts the duty to live sustainably, to help other nations to do so, and to support international laws and treaties that promote sustainability, the world as a whole will function increasingly well. At the same time, ways and means must be found to increase the income of poor countries, because an improving economy is generally agreed to be one of the most important factors that leads away from poverty. The strengthening of the United Nations as an effective force for global sustainability, by putting the whole complex of problems that will determine the future of the world at the center of its programs, would contribute greatly to the creation of a sustainable, healthy, peaceful, and prosperous world.

WHAT KIND OF WORLD DO WE WANT?

Perhaps the most important single lesson to have learned from this book is that those of us who live in highly developed countries are at the core of the problems facing the global environment today. Highly developed countries consume a disproportionate share of resources and must act forcefully to reduce their levels of consumption if we are all going to be able to achieve sustainability. We in industrialized countries continue to strive aggressively to increase our high standard of living, from a level that would be considered utopian by most people on Earth. We drain resources from the entire globe and thus contribute to a future in which neither our children nor our grandchildren will be able to live in anything like the affluence that we experience now.

The citizens of highly developed nations often seem to assume implicitly that overall prosperity can be created as a result of science and technology, and that the environment will take care of itself. If we are to improve the world situation, however, we must radically change our view of the world and adopt new ways of thinking, or we will perish together. We must learn to understand, respect, and work with one another, regardless of the differences that exist among us. The most heartening aspect of the situation that we confront is that people, given the motivation, do have the ability to make substantial changes.

At the deepest level, the most critical environmental problems, from which all others arise, are our own attitudes and values. We are totally out of touch and out of balance with the natural world, and until we reconnect and readjust in some significant way, all solutions will be stopgap ones. As a society we don't feel a part of the global ecosystem; we feel separate, above it, and therefore in a position to consume and abuse without thought of consequences. This book has been about consequences. In the last 30 years or so, we have come to recognize the nature of the impact of human activities on the biosphere. We now understand this impact enough to know that we cannot continue to act as we have been acting and expect any sort of viable future for our species. If all we do, as a result of this new knowledge, is to make some shifts in consumer choices and write a few letters, it won't be nearly enough.

Your generation must become the next pioneers. A pioneer is one who ventures into unexplored territory, a process that is simultaneously terrifying and profoundly exciting. The unexplored territory in this case is the development of a truly different way for humans to exist in the world. No models exist for this kind of change. You must forge a new revolution—akin in scope and effect to the Agricultural Revolution or the Industrial Revolution, yet totally different because it must be deliberate. You must help create the political will for it with your numbers and your commitment. You must create the economic power with your thoughtful decisions as both consumers and leaders. You must create social change with your acceptance and respect of the differences among peoples.

This change will require reconnecting with the natural environment. That means, at a personal level, taking opportunities to be outside (even if it is a city park or a backyard garden), to listen to the wind, and to look at the exquisite variety of plants, insects, and other life forms with which we coexist (Figure 24.13). Humans evolved in nature. Our immensely complex and multidimensional brains evolved precisely because we were able to interact with growing things, weather patterns, and other animals. The world we have created now screens us from all that. The sophisticated devices we imagined and manufactured—things such as televisions, computers, and automobiles—have come to define our world. One of

Figure 24.13 Reconnecting with the natural environment in Idaho. These hikers are enjoying part of the U.S. National Wilderness Preservation System.

Figure 24.14 Tomorrow's generation. The choices we make today will determine if future generations will inherit a sustainable world.

your challenges will be to use technology as a tool but not to let it define your interaction with the world.

The new environmental revolution will require that we revalue ourselves and our lives according to a different set of ideals. Wealth and material possessions have come to mean success, at a tremendous cost to the planet. Such ideas are deeply imbedded and extremely compelling and will be difficult to change. You may need to throw off some of the myths of Western culture, such as the belief that "faster" and "more" inevitably mean "better." You will at least be called on to examine those myths very deeply and very thoughtfully and to decide what they mean to you.

It will require reinventing economic constructs, such as building the cost of environmental impact and damage into our accounting systems, which will cause market forces to work in favor of environmental protection. Business activities must involve developing cooperative partnerships with people all over the world and making decisions based on long-term benefits to the environment.

This is an overwhelming responsibility. The choices we make now will have a greater impact on the future than those that any generation has had before (Figure 24.14). Even choosing to do nothing will have profound consequences for the future. At the same time, it is an incredible opportunity. Margaret Mead (1901 to 1978), the noted American anthropologist, once said, "Never doubt that a small group of thoughtful, committed people can change the world; indeed, it's the only thing that ever has!" This is a time in history when the best of human qualities—vision, courage, imagination, and concern—will play a critical role in establishing the nature of tomorrow's world.

SUMMARY WITH SELECTED KEY TERMS

I. We face enormous environmental problems.

A. Nearly one in four people lives in poverty, and nearly one in eight people does not have enough food to grow and function properly.

B. We are not managing the world's resources sustainably.

II. Sustainable development is development that meets the needs of the present without compromising the ability of future generations to meet their needs.

A. Earth's productivity is limited, and our use of it at the current level cannot be extended indefinitely.

B. The world does not contain enough resources to sustain everyone at the level of consumption that is enjoyed in the United States.

C. Sustainable development must allow for the maintenance of the life-support systems on which our lives and the lives of all other species are based.

III. To build a sustainable society, we must provide improvement in the quality of human life and, at the same time, maintain Earth's life-support systems.

A. We must preserve biological diversity.

B. Whatever actions we take must be taken within Earth's **carrying capacity.**

IV. We must respect and care for the community of life.

A. We are part of and dependent on the web of life.

B. Communities of organisms and ecosystems provide important **ecosystem services** that enable us to survive.

V. We must improve the quality of human life for the world's poor.

A. The distribution of the world's resources is very unequal.

B. Highly developed countries represent about 19% of the global population but control about 79% of the world's finances.

C. Failing to confront the problem of poverty makes it impossible to attain global sustainability.

D. The role of women, who are often disproportionately disadvantaged in poor countries, requires special attention.

VI. We must conserve Earth's biological diversity.

A. Most of the biological diversity on which we depend occurs in developing countries.

B. Each country should develop a comprehensive plan for the protection of its natural areas, with international assistance provided where needed.

VII. We must keep within Earth's carrying capacity.

A. The maximum human population that can be sustained by the world is Earth's carrying capacity.

B. The world population was 6.2 billion in 2002.

C. The United Nations projects that the human population will stabilize at about 9.3 billion by the end of the 21st century.

D. We must not only reach a stable population but also greatly reduce excessive consumption and waste.

VIII. We must change our personal attitudes and practices.

A. Each of us makes a difference, and our collective activities make the world what it is.

B. Education helps people understand the reasons for changing highly ingrained or traditional practices.

IX. We must enable communities to care for their own environments.

A. Local governments must become responsive to the need for sustainable development.

B. Actions that give people more control over their own lives enable them to move their communities toward sustainability.

X. We must build a national framework for integrating development and conservation.

A. Most of the world's nations have one or more departments or ministries devoted to environmental protection, but few are devoted to sustainable development overall.

B. At the national level, integration among the different departments involved in environmental matters improves effectiveness.

XI. We must create a global alliance.

A. Many environmental problems are global in scope.

B. Many global treaties exist on environmental matters, although they are not supported by all nations.

THINKING ABOUT THE ENVIRONMENT

1. What is sustainability? How are people in highly developed countries not living sustainably? How are people living in developing countries not living sustainably?

2. How is the natural environment—Earth's living organisms and ecosystems—an essential part of sustainable development?

3. Why it is important to respect and care for the community of life?

4. Describe some of the challenges that make it difficult to improve the quality of human life around the world.

5. Describe two major goals that we must reach to stay within Earth's carrying capacity.

6. How can education change personal attitudes and practices as they relate to the environment?

7. Give an example of an environmental goal that is accomplished most effectively (1) at the local level, (2) at the national level, and (3) at the international level.

8. Discuss four specific environmental goals that you hope are achieved in your lifetime.

TAKE A STAND

Visit our Web site at **http://www.wiley.com/college/raven** (select Chapter 24 from the Table of Contents) for links to more information about actions that contribute to community sustainability. Examine three examples of such actions, and decide with your classmates which one you could implement in your college community. You will find tools to help you organize your research, analyze the data, think critically about the issues, and construct a well-considered plan. Take a Stand activities can be done individually or as part of a team, as oral presentations, written exercises, or Web-based (e-mail) assignments.

Additional on-line materials relating to this chapter, including Student Quizzes, Activity Links, Useful Web Sites, Flash Cards, and more, can also be found on our Web site.

SUGGESTED READING

Earle, S.A. *Sea Change: A Message of the Oceans.* New York: G.P. Putnam's Sons (1995). Why we should safeguard Earth's largest and most vital natural resource.

Ehrlich, P.R., A.H. Ehrlich, and G.C. Daily. *The Stork and the Plow.* New York: G. P. Putnam's Sons (1995). Insights into the population-consumption problem.

Harte. J. *The Green Fuse: An Ecological Odyssey.* Berkeley: University of California Press (1993). A thoughtful narrative on ecology.

McKibben, B. *Hope, Human and Wild.* New York: Little, Brown (1995). Explores how we confront our problems and find solutions to them.

Quammen, D. *The Song of the Dodo.* New York: Scribner (1996). A great adventure and, at the same time, a wake-up call about the extinction of species.

Safina, C. *Song for the Blue Ocean.* New York: Henry Holt (1998). Examines the connections between the ocean and human survival.

Solbrig, O.T., and D.J. Solbrig. *So Shall You Reap: Farming and Crops in Human Affairs.* Washington, D.C.: Island Press (1994). Reviews agriculture's importance to history and how farming has taken its toll on the physical world.

Trefil, J. *A Scientist in the City.* New York: Doubleday (1994). Examines the promises and consequences of technology in urban living.

Review of Basic Chemistry

Elements

All matter, living and nonliving, is composed of chemical **elements**, substances that cannot be broken down into simpler substances by chemical reactions. There are 92 naturally occurring elements, ranging from hydrogen (the lightest) to uranium (the heaviest). In addition to the naturally occurring elements, about 20 elements heavier than uranium have been made in laboratories by bombarding elements with subatomic particles.

Instead of writing out the name of each element, chemists use a system of abbreviations called **chemical symbols**—usually the first one or two letters of the English or Latin name of the element. The symbol stands for one atom of the element. For example, O is the symbol for one atom of oxygen, C for one atom of carbon, Cl for one atom of chlorine, and Na for one atom of sodium (its Latin name is *natrium*). Note that the first letter of a chemical symbol is capitalized, and the second letter (if there is one) is not.

Table A.1	Symbols for Some Elements That Are Important in Environmental Science		
Element	**Symbol**	**Element**	**Symbol**
Aluminum	Al	Lithium	Li
Bromine	Br	Magnesium	Mg
Calcium	Ca	Mercury	Hg
Carbon	C	Nitrogen	N
Chlorine	Cl	Oxygen	O
Copper	Cu	Phosphorus	P
Fluorine	F	Plutonium	Pu
Helium	He	Potassium	K
Hydrogen	H	Sodium	Na
Iodine	I	Sulfur	S
Iron	Fe	Uranium	U
Lead	Pb	Zinc	Zn

Atoms The **atom** is the smallest subdivision of an element that retains the characteristic chemical properties of that element. Atoms are unimaginably small, much smaller than the tiniest particle visible under a light microscope.

An atom is composed of smaller components called **subatomic particles**—protons, neutrons, and electrons. **Protons** have a positive electrical charge; **neutrons** are uncharged particles with about the same mass as protons. Protons and neutrons make up almost all the mass of an atom and are concentrated in the atomic nucleus. **Electrons** have a negative electrical charge and an extremely small mass (only about 1/1800 of the mass of a proton). The electrons, which behave like waves as well as particles, spin in the space surrounding the atomic nucleus.

Table A.2	Some Properties of Subatomic Particles	
Subatomic Particle	**Electric Charge**	**Location**
Proton	Positive, +1	Nucleus
Neutron	Neutral	Nucleus
Electron	Negative, –1	Outside nucleus

Each kind of element has a fixed number of protons in the atomic nucleus. This number, called the **atomic number**, determines the chemical identity of the atom. The total number of protons plus neutrons in the atomic nucleus is termed the **atomic mass**. For example, the element oxygen has eight protons and eight neutrons in the nucleus; it therefore has an atomic number of 8 and an atomic mass of 16: $^{16}_{8}O$.

When an atom is uncombined, it generally contains the same number of electrons as protons. Some kinds of chemical combinations and certain other circumstances change the number of electrons, but chemical reactions do not affect anything in the atomic nucleus. Because electrons and protons have equal but opposite charges, an uncombined atom is electrically neutral.

The Periodic Table Elements are organized in a **periodic table** (Figure A.1). The periodic table provides information about the each element's chemical behavior as well as its atomic mass and the structure of its atoms. The box for each element shows the name, atomic number, symbol, and atomic mass.

Examine the element magnesium, with an atomic number of 12. The atomic number indicates that magnesium has 12 protons in each atomic nucleus and 12 electrons outside the nucleus. The atomic mass of magnesium, 24.305 atomic mass units (amu), is the average of the masses of its naturally occurring isotopes (discussed shortly). The number of neutrons in the typical magnesium atom is obtained by subtracting the atomic number from the atomic mass: 24 – 12 = 12 neutrons.

Notice that the elements are arranged on the periodic table in order of increasing atomic number. The chemical properties of the elements are a regularly repeated function of their atomic numbers. For example, the elements in the far right column are called **noble gases**. These elements—helium, neon, argon, krypton, xenon, and radon—have similar chemical properties. All are colorless, odorless gases that form very few compounds with other elements.

The periodic table of the elements

1	2	3	4	5	6	7	8	9	10	11	12	13	14	15	16	17	18
Hydrogen 1 **H** 1.0079																	Helium 2 **He** 4.0026
Lithium 3 **Li** 6.941	Beryllium 4 **Be** 9.0122											Boron 5 **B** 10.811	Carbon 6 **C** 12.011	Nitrogen 7 **N** 14.007	Oxygen 8 **O** 15.999	Fluorine 9 **F** 18.998	Neon 10 **Ne** 20.180
Sodium 11 **Na** 22.990	Magnesium 12 **Mg** 24.305											Aluminium 13 **Al** 26.982	Silicon 14 **Si** 28.086	Phosphorus 15 **P** 30.974	Sulfur 16 **S** 32.065	Chlorine 17 **Cl** 35.453	Argon 18 **Ar** 39.948
Potassium 19 **K** 39.098	Calcium 20 **Ca** 40.078	Scandium 21 **Sc** 44.956	Titanium 22 **Ti** 47.867	Vanadium 23 **V** 50.942	Chromium 24 **Cr** 51.996	Manganese 25 **Mn** 54.938	Iron 26 **Fe** 55.845	Cobalt 27 **Co** 58.933	Nickel 28 **Ni** 58.693	Copper 29 **Cu** 63.546	Zinc 30 **Zn** 65.39	Gallium 31 **Ga** 69.723	Germanium 32 **Ge** 72.61	Arsenic 33 **As** 74.922	Selenium 34 **Se** 78.96	Bromine 35 **Br** 79.904	Krypton 36 **Kr** 83.80
Rubidium 37 **Rb** 85.468	Strontium 38 **Sr** 87.62	Yttrium 39 **Y** 88.906	Zirconium 40 **Zr** 91.224	Niobium 41 **Nb** 92.906	Molybdenum 42 **Mo** 95.94	Technetium 43 **Tc** [98]	Ruthenium 44 **Ru** 101.07	Rhodium 45 **Rh** 102.91	Palladium 46 **Pd** 106.42	Silver 47 **Ag** 107.87	Cadmium 48 **Cd** 112.41	Indium 49 **In** 114.82	Tin 50 **Sn** 118.71	Antimony 51 **Sb** 121.76	Tellurium 52 **Te** 127.60	Iodine 53 **I** 126.90	Xenon 54 **Xe** 131.29
Caesium 55 **Cs** 132.91	Barium 56 **Ba** 137.33	57-70 *	Hafnium 72 **Hf** 178.49	Tantalum 73 **Ta** 180.95	Tungsten 74 **W** 183.84	Rhenium 75 **Re** 186.21	Osmium 76 **Os** 190.23	Iridium 77 **Ir** 192.22	Platinum 78 **Pt** 195.08	Gold 79 **Au** 196.97	Mercury 80 **Hg** 200.59	Thallium 81 **Tl** 204.38	Lead 82 **Pb** 207.2	Bismuth 83 **Bi** 208.98	Polonium 84 **Po** [209]	Astatine 85 **At** [210]	Radon 86 **Rn** [222]
Francium 87 **Fr** [223]	Radium 88 **Ra** [226]	89-102 **	Rutherfordium 104 **Rf** [261]	Dubnium 105 **Db** [262]	Seaborgium 106 **Sg** [266]	Bohrium 107 **Bh** [264]	Hassium 108 **Hs** [269]	Meitnerium 109 **Mt** [268]	Ununnilium 110 **Uun** [271]	Unununium 111 **Uuu** [272]	Ununbium 112 **Uub** [277]		Ununquadium 114 **Uuq** [289]				

(Group 3, row 6: Lutetium 71 **Lu** 174.97; Group 3, row 7: Lawrencium 103 **Lr** [262])

*Lathanoids

Lanthanum 57 **La** 138.91	Cerium 58 **Ce** 140.12	Praseodymium 59 **Pr** 140.91	Neodymium 60 **Nd** 144.24	Promethium 61 **Pm** [145]	Samarium 62 **Sm** 150.36	Europium 63 **Eu** 151.96	Gadolinium 64 **Gd** 157.25	Terbium 65 **Tb** 158.92534	Dysprosium 66 **Dy** 162.50	Holmium 67 **Ho** 164.93	Erbium 68 **Er** 167.26	Thulium 69 **Tm** 168.93	Ytterbium 70 **Yb** 173.04

**Actinoids

Actinium 89 **Ac** [227]	Thorium 90 **Th** 232.04	Protactinium 91 **Pa** 231.04	Uranium 92 **U** 238.03	Neptunium 93 **Np** [237]	Plutonium 94 **Pu** [244]	Americium 95 **Am** [243]	Curium 96 **Cm** [247]	Berkelium 97 **Bk** [247]	Californium 98 **Cf** [251]	Einsteinium 99 **Es** [252]	Fermium 100 **Fm** [257]	Mendelevium 101 **Md** [258]	Nobelium 102 **No** 259.1011

Figure A.1 The periodic table of the elements.

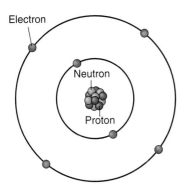

Figure A.2 A carbon atom. In this drawing, known as a Bohr model, electrons move in orbitals that correspond to energy levels. Although Bohr models do not depict electronic configurations accurately (electrons do not actually circle the nucleus in fixed concentric pathways), we use Bohr models because of their simplicity.

The Electronic Configuration of an Atom The way electrons are arranged around an atomic nucleus is referred to as the atom's **electronic configuration**. Knowing the approximate locations of electrons enables chemists to predict how atoms can combine to form different types of chemical compounds.

An atom may have several **energy levels**, or **electron shells**, where electrons are located. The lowest energy level is the one closest to the nucleus (Figure A.2). Only two electrons can occupy this energy level. The second energy level can accommodate a maximum of eight electrons. Although the third and outer shells can each contain more than eight electrons, they are most stable when only eight are present. We may consider the first shell complete when it contains two electrons and every other shell complete when it contains eight electrons.

The energy levels correspond roughly to physical locations of electrons, called **orbitals**. There may be several orbitals within a given energy level. Electrons are thought to whirl around the nucleus in an unpredictable manner, now close to it, now farther away. Orbitals represent the places where electrons are most probably found.

Isotopes **Isotopes** are atoms of the same element that contain the same number of protons but different numbers of neutrons. Isotopes, therefore, have the same atomic number but different atomic mass numbers. The three isotopes of hydrogen contain zero, one, and two neutrons, respectively. Elements usually occur in nature as mixtures of isotopes.

Table A.3	Mixture of Isotopes in Naturally Occurring Oxygen

Isotope	Percentage of Atoms
$^{16}_{8}O$	99.759
$^{17}_{8}O$	0.037
$^{18}_{8}O$	0.204

All isotopes of a given element have essentially the same chemical characteristics. Some isotopes with excess neutrons

Figure A.3 The water molecule. Each molecule of water consists of two hydrogen atoms bonded to one oxygen atom.

are unstable and tend to break down, or decay, to a more stable isotope (usually of a different element). Such isotopes are termed **radioisotopes** because they emit high-energy radiation when they decay (see Chapter 11).

Molecules Two or more atoms may combine chemically to form a **molecule**. When two atoms of oxygen combine, for example, a molecule of oxygen is formed. Different kinds of atoms can combine to form **chemical compounds**. A chemical compound is a substance that consists of two or more different elements combined in a fixed ratio. Water is a chemical compound in which each molecule consists of two atoms of hydrogen combined with one atom of oxygen (Figure A.3).

Chemical Bonds The number and arrangement of electrons in the *outermost* energy level (electron shell) primarily determine the chemical properties of an element. In a few elements (the noble gases), the outermost shell is filled. These elements are chemically inert, meaning that they will not readily combine with other elements. The electrons in the outermost energy level of an atom are referred to as **valence electrons**. The valence electrons are chiefly responsible for the chemical activity of an atom. When the valence (outer) shell of an atom contains fewer than eight electrons (the stable number for the valence shell of most atoms), the atom tends to lose, gain, or share electrons to achieve an outer shell of eight. (The valence shells of the lightest elements, hydrogen and helium, are full when they contain two electrons.)

The elements in a compound are always present in a certain proportion. This reflects the fact that atoms are attached to each other by chemical bonds in a precise way to form a compound. A **chemical bond** is the attractive force that holds two atoms together. Each bond represents a certain amount of potential chemical energy. The atoms of each element form a specific number of bonds with the atoms of other elements—a number dictated by the number of valence electrons (Figure A.4).

Ions Some atoms have the ability to gain or lose electrons. Because the number of protons in the nucleus remains unchanged, the loss or gain of electrons produces an atom with a net positive or negative charge. Such electrically charged atoms are termed **ions** (Figure A.5).

Chemical Formulas

A **chemical formula** is a shorthand method for describing the chemical composition of a molecule. Chemical symbols are used to indicate the types of atoms in the molecule, and subscript numbers are used to indicate the number of each atom present. The chemical formula for molecular oxygen, O_2, tells

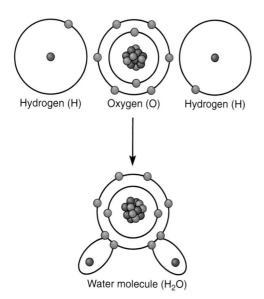

Hydrogen (H)   Oxygen (O)   Hydrogen (H)

Water molecule (H₂O)

Figure A.4 Electronic configuration of a water molecule. A molecule of water is formed when two hydrogen atoms share their valence electrons with the valence electrons of oxygen.

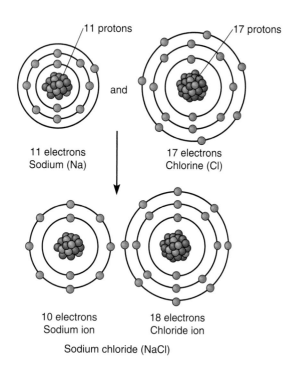

11 protons   17 protons

and

11 electrons   17 electrons
Sodium (Na)   Chlorine (Cl)

10 electrons   18 electrons
Sodium ion   Chloride ion

Sodium chloride (NaCl)

Figure A.5 Formation of sodium and chloride ions. Ions gain or lose one or more valence electrons. A sodium atom becomes a positive ion when it donates its single valence electron to chlorine, which has seven valence electrons. With this additional electron, the chlorine atom becomes a negative chloride ion. The attraction between a positively charged sodium ion and a negatively charged chloride ion produces the chemical compound sodium chloride.

us that each molecule consists of two atoms of oxygen. This formula distinguishes it from another form of oxygen, ozone, which has three oxygen atoms and is written O_3. The chemical formula for water, H_2O, indicates that each molecule consists of two atoms of hydrogen and one atom of oxygen. Note that when a single atom of one type is present it is not necessary to write 1; it is not necessary to write H_2O_1.

Chemical Equations

The chemical reactions that occur between atoms and molecules—for example, between methane and oxygen—can be described on paper by means of **chemical equations**.

Reactant A		Reactant B		Product C		Product D
CH_4	+	$2O_2$	$\rightarrow$	CO_2	+	$2H_2O$
Methane		Oxygen		Carbon Dioxide		Water

Methane is broken down in this reaction.

In a chemical reaction, the **reactants** (the substances that participate in the reaction) are written on the left side of the equation and the **products** (the substances formed by the reaction) are written on the right side. The arrow means *yields* and indicates the direction in which the reaction tends to proceed. The number preceding a chemical symbol or formula indicates the number of atoms or molecules reacting. Thus, $2O_2$ means two molecules of oxygen. The absence of a number indicates that only one atom or molecule is present. Thus, the above equation can be translated into ordinary language as "One molecule of methane reacts with two molecules of oxygen to yield one molecule of carbon dioxide and two molecules of water."

Acids and Bases

An **acid** is a compound that ionizes in solution to yield hydrogen ions (H^+)—that is, protons—and negatively charged ions. Acids turn blue litmus paper red and have a sour taste. Hydrochloric acid (HCl) and sulfuric acid (H_2SO_4) are examples of acids.

		in water				
HCl		$\rightarrow$		H^+	+	Cl^-
Hydrochloric acid				Hydrogen ion		Chloride ion

The strength of an acid depends on the degree to which it ionizes in water, releasing hydrogen ions. Thus, HCl is a very strong acid because most of its molecules dissociate, producing hydrogen and chloride ions.

Most bases are substances that yield hydroxide ions (OH^-) and positively charged ions when dissolved in water. Bases turn red litmus paper blue. Sodium hydroxide (NaOH) and aqueous ammonia (NH_4OH) are examples of bases.

		in water				
NaOH		$\rightarrow$		Na^+	+	OH^-
Sodium hydroxide				Sodium ion		Hydroxide ion

Bases react with hydrogen ions and remove them from solution.

pH

Because the concentration of hydrogen or hydroxide ions is usually small, it is convenient to express the degree of acidity or basicity in a solution in terms of **pH**, formally defined as the negative logarithm of the hydrogen ion concentration. The pH scale is logarithmic, extending from 0, which is the pH of a very strong acid, to 14, which is the pH of a very strong base. The pH of pure water is 7, neither acidic nor basic, but neutral. Even though water does ionize slightly, the concentrations of H^+ ions and OH^- ions are exactly equal; each of them has a concentration of 10^{-7}, which is why we say that water has a pH of 7. Solutions with a pH of *less* than 7 are acidic and contain more H^+ ions than OH^- ions. Solutions with a pH *greater* than 7 are basic and contain more OH^- ions than H^+ ions.

Because the scale is logarithmic to base 10, a solution with a pH of 6 has a hydrogen ion concentration that is 10 times greater than a solution with a pH of 7, and is much more acidic. A pH of 5 represents another tenfold increase. Therefore, a solution with a pH of 4 is 10×10, or 100 times more acidic than a solution with a pH of 6.

The contents of most animal and plant cells are neither strongly acidic nor basic but are an essentially neutral mixture of acidic and basic substances. Most life cannot exist if the pH of the cell changes very much.

Graphing

Much of the study of environmental science involves learning about relationships between variables. For instance, there is a definite relationship between the amount of pollution discharged and the cost of the environmental damage caused by the pollution. Often such relationships can be expressed and understood through graphs.

A **graph** is a diagram that expresses a relationship between two or more quantities. In some cases there is a definite cause-and-effect relationship, whereas in others the association is not as direct. Graphic presentation of data may not explain the *reason* for the relationship (as, for example, the worldwide increase in fertilizer use since 1960), but the shape of it can provide clues. A graph puts into visual form abstract ideas or experimental data, so that their relationships become more apparent.

Variables

The related quantities displayed on a graph are called **variables**. The simplest sort of graph uses a system of coordinates or axes to represent the values of the variables. Usually the relative size of a variable is represented by its position along the axis. Numbers along the axis allow the reader to estimate the values.

If the relationship being plotted is one of cause and effect, the variable that expresses the cause is called the **independent variable**. Usually this is represented by the horizontal axis, which is called the *x*-axis. The variable that changes as a result of changes in the independent variables is the **dependent variable**. It is usually represented on the vertical axis, which is called the *y*-axis. The two axes are arranged at right angles to each other and cross at a point called the origin (Figure A.6).

To show the relationship between two variables that are directly related at some specific value, such as point *A* in Figure A.6, the value on the *x*-axis (x_1) is extended vertically, and the corresponding value on the *y*-axis (y_1) is extended horizontally. The relationship between the two variables determines the point *A* at which these lines cross.

If another pair of points (x_2 and y_2) is chosen, their point of intersection on the graph can also be plotted; this is point *B*. A line drawn between points *A* and *B* can then give information about how all other *x*- and *y*-values on this graph should relate to each other.

Types of Relationships

As you may have guessed, this explanation represents a very simple case in which some important assumptions were made. We first assumed that for every *x*-value there was only one *y*-value. We further assumed that all of the *y*-variables were directly related to all of the *x*-variables. This is the simplest kind of relationship that a graph can represent. It is called a **direct relation-**

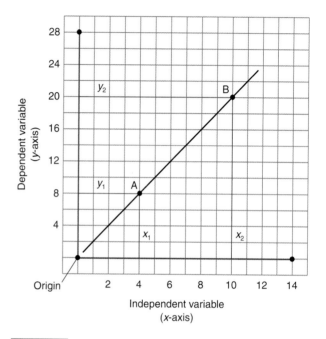

Figure A.6 Typical line graph. Two numbers specify the position of A; these are its *x* and *y* coordinates. The *x* coordinate, 4, is the foot of the perpendicular from A to the *x*-axis (see x_1). The *y* coordinate, 8, is the foot of the perpendicular from A to the *y*-axis (see y_1). Point B is obtained by extending the corresponding values on the *x*- and *y*-axes until they intersect, at $x = 10$, $y = 20$.

ship: The *y*-values get larger as the *x*-values get larger. An example of this type of graph is shown in Figure A.7.

Inverse relationships are also common. In inverse relationships, the *y*-values get *smaller* as the *x*-values get larger (Figure A.8). Most relationships found in environmental science are not simple. Over some ranges, a relationship may be direct or inverse, and then it may change as a wider range of variables is considered.

Take a few moments to flip through the pages of this book. You will see many graphs. Some express simple relationships over their entire range of data, whereas others are more complicated, expressing several relationships at once. In some cases there are several lines on the graph, each describing some aspect of the idea being presented. Some data are presented as bar graphs or pie charts instead of lines to illustrate relationships (Figures A.9 and A.10). Whatever their form, all these graphs are designed to present important relationships in the clearest possible way. When you learn to interpret information presented graphically, you are well on your way to understanding environmental science.

A-6

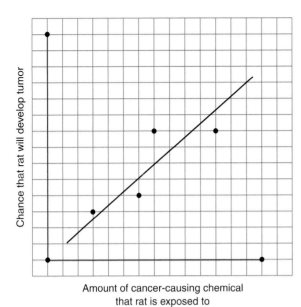

Figure A.7 **Direct relationship.** In a direct relationship, *y* values increase as *x* values increase, producing an upward-sloping straight line. In science, data points often do not fall exactly on a straight line but are scattered about an ideal line, which is determined mathematically and is drawn to show the general relationship.

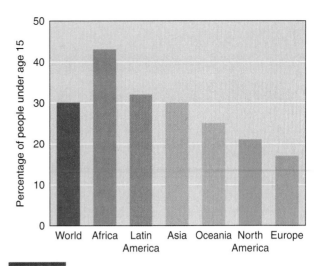

Figure A.9 **Bar graph.** A bar graph has parallel bars to represent the values in a given set of data. Bar graphs are appropriate for comparing discrete values, as for example, when comparing the percentage of people under age 15 in different parts of the world.

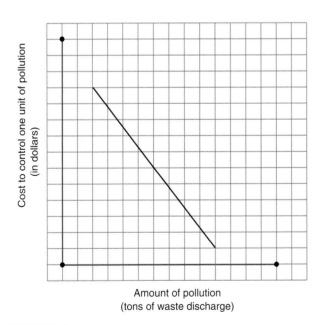

Figure A.8 **Inverse relationship.** In an inverse relationship, *y* values decrease as *x* values increase, producing a downward-sloping line or curve.

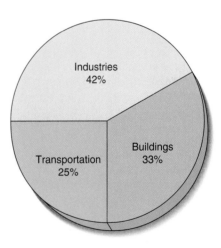

Figure A.10 **Pie chart.** A pie chart, which is always shown as a circle, is a graphical way to show percentages as a whole, such as major energy consumers in the United States. The size of each section of the pie chart is proportional to its share of the whole.

Units of Measure: Some Useful Conversions

SOME COMMON PREFIXES

Prefix and Symbol	Meaning	Example
Giga- (G)	Billion	1 gigaton = 1,000,000,000 tons
Mega- (M)	Million	1 megawatt = 1,000,000 watts
Kilo- (k)	Thousand	1 kilojoule = 1000 joules
Centi- (c)	Hundredth	1 centimeter = 0.01 meter
Milli- (m)	Thousandth	1 milliliter = 0.001 liter
Micro- (μ)	Millionth	1 micrometer = 0.000001 meter
Nano- (n)	Billionth	1 nanometer = 0.000000001 meter
Pico- (p)	Trillionth	1 picocurie = 0.000000000001 curie

LENGTH   Standard Unit = Meter

1 meter (m) = 39.37 in. = 3.28 ft
1 in. = 2.54 cm
1 km = 0.621 mi
1 mi = 1.609 km = 1,609 m
1 nautical mile = 1.15 mi = 1.85 km

VOLUME   Standard Unit = Liter

1 liter (L) = 1,000 cm^3 = 1.057 qt (U.S.)
1 gallon (U.S.) = 3.785 L
1 mi^3 = 4.166 km^3

ENERGY   Standard Unit = Joule

1 joule (J) = 0.24 cal
1 calorie = 4.184 J
1 Calorie = 1,000 calories = 1 kcal
1 kilocalorie = 4.184 kJ
1 British thermal unit = 252 cal
1 kilowatt-hour = 3,600,000 J

PRESSURE   Standard Unit = Pascal

1 bar = 10^5 Pa
1 atm = 1.01 bar = 1.01×10^5 Pa
1 millibar = 0.0145 lb/in^2

AREA   Standard Unit = Square Meter (m^2)

1 hectare = 10,000 m^2 = 0.01 km^2 = 2.471 acres
1 acre = 0.405 hectare
1 km^2 = 100 hectares = 0.386 mi^2
1 mi^2 = 640 acres = 259 hectares = 2.59 km^2

MASS   Standard Unit = Kilogram

1 kilogram (kg) = 2.205 lb = 35.3 oz
1 metric ton = 1,000 kg = 1.103 ton = 2,204.6 lb
1 short ton = 907 kg
1 lb = 453.6 g

ELECTRICAL POWER   Standard Unit = Watt

1 watt (W) = 1 J/second

TEMPERATURE   Standard Unit = Celsius

$°C = (°F - 32) \times {}^5/_9$
$°F = °C \times {}^9/_5 + 32$
$1°C = 1.8°F$

Abbreviations, Formulas, and Acronyms Used in This Text

AES	Applied Energy System
AID	U.S. Agency for International Development
AIDS	acquired immune deficiency syndrome
ALSC	Adirondack Lakes Survey Corporation
ANWR	Arctic National Wildlife Refuge
ARS	USDA Agricultural Research Service
ATP	adenosine triphosphate
BHA	butylated hyroxyanisole
BHT	butylated hydroxytoluene
BLM	Bureau of Land Management
BOD	biological/biochemical oxygen demand
BRD	Biological Resources Discipline
BST	bacterial source tracking
Bt	*Bacillus thuringiensis*
BTU	British thermal unit
CARE	Cooperative for American Relief Everywhere
CERCLA	Comprehensive Environmental Response, Compensation, and Liability Act
CFCs	chlorofluorocarbons
CH_4	methane
CHAART	Center for Health Applications of Aerospace-Related Technologies
CHP	combined heat and power (cogeneration)
CITES	Convention on International Trade in Endangered Species of Wild Flora and Fauna
CO	carbon monoxide
CO_2	carbon dioxide
CO_3^{2-}	carbonate
CRP	Conservation Reserve Program
db	decibel
dbA	decibel-A
DDT	dichlorodiphenytrichloroethane
DMS	dimethyl sulfide
DNA	deoxyribonucleic acid
DOD	U.S. Department of Defense
DOE,	U.S. Department of Energy
ED_{50}	effective dose-50%
EDB	ethylene dibromide
EFI	European Forest Institute
EIS	environmental impact statement
EMFs	electric and magnetic fields
ENSO	El Niño–Southern Oscillation
EPA	U.S. Environmental Protection Agency
ERC	emission reduction credit
ERNS	EPA Emergency Response Notification System

ESA	Endangered Species Act
EU	European Union
EWG	Environmental Working Group
FAO	U.N. Food and Agriculture Organization
FDA	U.S. Food and Drug Administration
FDCA	Food, Drug, and Cosmetics Act
FIFRA	Federal Insecticide, Fungicide, and Rodenticide Act
FWS	U.S. Fish and Wildlife Service
GDP	gross domestic product
GFDL	U.S. Geophysical Fluid Dynamics Laboratory
GHP	geothermal heat pump
GM	genetically modified
GNP	gross national product
GPP	gross primary productivity
GRAS	generally recognized as safe
HCFCs	hydrochlorofluorocarbons
HCPs	habitat conservation plans
HCO_3^-	bicarbonate
HFCs	hydrofluorocarbons
HIV	human immunodeficiency virus
HNO_2	nitrous acid
HNO_3	nitric acid
H_2S	hydrogen sulfide
H_2SO_4	sulfuric acid
IAEA	International Atomic Energy Agency
ICPR	International Commission for Protection of the Rhine
IIASA	International Institute for Applied Systems Analysis
IMF	International Monetary Fund
IPCC	U.N. Intergovernmental Panel of Climate Change
IPM	integrated pest management
IRCA	Immigration Reform and Control Act
IRRI	International Rice Research Institute
IUCN	World Conservation Union (formerly International Union for the Conservation of Nature and Natural Resources)
K	carrying capacity
kcal	kilocalorie
kJ	kilojoule
kWh	kilowatt-hour
LD_{50}	lethal dose-50%
LDC	less developed country

LTER	long-term ecological research		PCBs	polychlorinated biphenyls
medfly	Mediterranean fruit fly		PET	polyethylene terphthalate
MIC	methyl isocyanate		PO_4^{3-}	phosphate
MOX	mixed oxide (form of plutonium)		POET	four variables in urban centers (population, organization, environment, and technology)
mpg	miles per gallon			
mph	miles per hour		POPs	persistent organic pollutants
NAECA	National Appliance Energy Conservation Act		ppb	parts per billion
NASA	National Aeronautics and Space Administration		ppm	parts per million
NDP	net domestic product		PV	photovoltaic
NEPA	National Environmental Policy Act		r	growth rate
NH_3	ammonia		RCRA	Resource Conservation and Recovery Act
NIMBY	not in my backyard		SMCRA	Surface Mining Control and Reclamation Act
NIMTOO	not in my term of office		SO_2	sulfur dioxide
NMFS	National Marine Fisheries Service		SO_3	sulfur trioxide
NNP	net national product		SO_4^{2-}	sulfate
NO	nitric oxide		SO_x	collective name for sulfur oxides
NO_x	collective name for nitrogen oxides		STDs	sexually transmitted diseases
N_2O	nitrous oxide		TCDD	tetrachlorodibenzodioxin (a dioxin)
NO_2	nitrogen dioxide		TCE	trichloroethylene
NO_2^-	nitrite		TNT	trinitrotoluene
NO_3^-	nitrate		2,4-D	2,4-dichlorophenoxyacetic acid
NOAA	National Oceanic and Atmospheric Administration		2,4,5-T	2,4,5-trichlorophenoxyacetic acid
NPP	net primary productivity		U.N.	United Nations
NPS	National Park Service		UNAIDS	U.N. Program on AIDS
NRC	National Research Council		UNCLOS	U.N. Convention on the Law of the Sea
NRCS	National Resources Conservation Service (formerly Soil Conservation Service)		USDA	U.S. Department of Agriculture
			USFS	U.S. Forest Service
NWS	National Weather Service		UV	ultraviolet
O_3	ozone		VOC	volatile organic compound (or chemical)
OH^-	hydroxide ion		WHO	World Health Organization
OPEC	Organization of Petroleum Exporting Countries		WRP	Wetlands Reserve Program
OTEC	ocean thermal energy conversion		WWF	World Wide Fund for Nature (also known as World Wildlife Fund in the United States, Canada, and Australia)
PANs	peroxyacyl nitrates			

How to Make a Difference

Individuals and groups all around the world are beginning to address complex environmental problems. Youths from different countries and cultures are working together to clean up the Mediterranean Sea. Women in India are protecting trees from lumberjacks and planting more trees. Political parties that address environmental issues are gaining increasing political clout throughout Europe. Grassroots environmental groups are organizing in Russia to stop development projects that are environmentally unsound.

In this book we have explored some of the complexities of our environment, and we have examined many problems that are the direct result of human action. Some environmental problems seem almost impossible to address, yet the quality of life and the kind of world we want this place to be, not only for ourselves but for our children and grandchildren, depend largely on the actions we take today.

What You Can Do as an Individual

Throughout this text we have highlighted things you can do as an individual to conserve energy and other natural resources and reduce waste. Although individual savings are small, they represent substantial conservation when combined with other individual contributions. Knowing what you now know about the environment, you should try to break personal habits that harm the environment.

Because you are taking a course in environmental science, you should be well informed on most environmental issues. Make an effort to stay informed and share your knowledge with others. You can help shape environmental policies of the future, but you cannot be effective if you do not know what you are talking about

Most important, should you be particularly concerned about one or more environmental issues, don't just fret over them—get committed. Join appropriate environmental groups whose collective power carries more weight than single voices. Become politically involved on local, state, and national levels. Help to elect leaders who support a sustainable Earth society and to influence those leaders in office to support the environmental causes you advocate. Be persistent in your efforts.

In summary, there are three things you can do to help preserve and improve the environment: (1) Lower your consumption and reduce waste so that your lifestyle reflects the conservation of natural resources; (2) stay informed on environmental topics and share your knowledge; (3) get involved in environmental issues.

Environmental Organizations You Can Join

The following is an incomplete listing of environmental organizations. Some are grassroots groups that depend on their constituents—people like you—for funding and momentum.

Others are private, nonprofit organizations that are funded by foundations or other groups. A few are trade associations or professional societies.

Something most of these groups have in common is that they offer internships for college students. Some groups have organized internship programs, while others have internships available only occasionally. Many offer paid internships, but a few are too small to be able to afford any stipend. You can generally arrange to receive college credit for paid or unpaid internships. When this list was being compiled, a number of the organizations were actively seeking interns, but most environmental organizations do not advertise such positions. Therefore, you must contact the organizations you are most interested in to see if they are currently offering anything of interest to you.

Advocates for Youth, 1025 Vermont Ave. NW, suite 200, Washington, D.C. 20005. (202) 347-5700. www.advocatesforyouth. A nonprofit organization that directs its efforts at the prevention of early childbirth both in the United States and other countries. Internships available.

American Association for World Health, 1825 K Street NW, suite 1208, Washington, D.C. 20006. (202) 466-5883. www.aawhworldhealth.org.

The only private national organization in the United States dedicated to funneling a broad spectrum of critical national and international health information to Americans at the grassroots level. It develops and distributes practical, easy-to-use health education and promotional materials to those community leaders who can most effectively reach U.S. citizens at the local level. Internships available.

American Council for an Energy-Efficient Economy, 1001 Connecticut Avenue NW, suite 801, Washington, D.C. 20036. (202) 429-8873. www.aceee.org.

Nonprofit organization dedicated to advancing energy efficiency as a means of promoting both economic prosperity and environmental protection.

American Farmland Trust, 1200 18th Street NW, suite 800, Washington, D.C. 20036. (202) 331-7300. www.farmland.org.

National, private, nonprofit organization working to stop the loss of productive farmland and promote farming practices that lead to a healthy environment. Offers paid and unpaid internships.

American Fisheries Society, 5410 Grosvenor Lane, Bethesda, MD 20814-2199. (301) 897-8616. www.fisheries.org.

Organization working to improve the conservation and sustainability of fishery resources and aquatic ecosystems. Nonprofit. Offers paid summer internships and graduate fellowships.

American Forests, 910 17th Street NW, suite 600, Washington, D.C. 20006. (202) 955-4500. www.amfor.org.

Nonprofit citizen conservation organization helping people improve the environment through national and international tree planting, forest policy, urban forestry and population programs. Through their Global ReLeaf Program, the right tree species is planted in the right places to reforest damaged public lands here and abroad. Internships available.

American Land Conservancy, 1388 Sutter Street, suite 810, San Francisco, CA 94109. (415) 1749-3010. www.alcnet.org.

Dedicated to the preservation of land and water as enduring public resources, and to the protection and enhancement of our nation's natural, ecological, historical, recreational, and scenic heritage.

American Lung Association, 61 Broadway, 6th floor, New York, NY 10006. (212) 315-8700. www.lungusa.org.

Educates schools and the general community on the health effects of indoor and outdoor air pollution. Local internships available.

American Rivers, Inc., 1025 Vermont Ave NW, suite 720, Washington, D.C. 20005. (202) 347-7550. www.amrivers.org

Dedicated to protecting and restoring the nation's river systems and to fostering a river stewardship ethic. Summer internships are available.

Beyond Pesticides/National Coalition Against Misuse of Pesticides, 701 E Street SE, #200, Washington, D.C. 20003. (202) 543-5450. www.beyondpesticides.org.

Nonprofit organization that serves as a national network committed to pesticide safety and the adoption of alternative pest management strategies that reduce or eliminate a dependency on toxic chemicals.

Center for Environmental Citizenship, 200 G Street NE, suite 300, Washington, D.C. 20002. (202)-547-8435. www.envirocitizen.org.

National nonpartisan organization to encourage college students to be environmental citizens. Dedicated to educating, training and organizing a diverse, national network of young leaders to protect the environment. Six-day summer leadership training program offered.

Center for Environmental Health, 528 61st Street, suite A, Oakland, CA 94609. (510) 594-9864. www.cehca.org.

To protect the public from environmental health hazards and toxic exposure by directly influencing corporate behavior in the public interest, which is done by facilitating communication and cooperation between industry and the public. Internships available for academic credit.

Center for Environmental Information, 55 St. Paul Street, Rochester, NY 14604. (585) 262-2870. www.rochesterenvironment.org.

Provides information, communications services, publications, and educational programs. Offers internships.

Center for Health, Environment and Justice, P.O. Box 6806, Falls Church, VA 22040. (703) 237-2249. www.chej.org. chej@chej.org.

Assists citizens' groups in the fight against toxic polluters and for environmental justice across the country. Internships available.

Center for Plant Conservation, c/o Missouri Botanical Gardens, P.O. Box 299, St. Louis, MO 63166. (314) 577-9450. www.mobot.org/cpc.

Network of 29 U.S. botanical gardens and arboretums that work to conserve plant species native to the United States.

Center for Science in the Public Interest, 1875 Connecticut Ave. NW, suite 300, Washington, D.C. 20009-5728. (202) 332-9110. www.cspinet.org.

Concerned with sustainable agriculture, alcohol, food, and nutrition.

Children's Environmental Health Network, 110 Maryland Avenve NE, suite 511, Washington, D.C. 20002. (202) 543-4033. www.cehn.org.

National project dedicated to promoting a healthy environment and to protecting the fetus and the child from environmental hazards. Internships available at D.C. and California offices.

Clean Water Action Project, 4455 Connecticut Ave. NW, suite A300, Washington, D.C. 20008-2328. (202) 895-0420. www.cleanwateroction.org.

With offices in 16 states, this organization lobbies for clean, safe, and affordable water, prevention of health-threatening pollution, creation of environmentally safe jobs and businesses, and empowerment of people to make democracy work.

Climate Institute, $333^1/_2$ Pennsylvania Ave. SE, Washington, D.C. 20003. (202) 547-0104. www.climate.org

Devoted to helping maintain the balance between climate and life on Earth. Nonprofit. Offers internships.

Concern, Inc., 1794 Columbia Road NW, Washington, D.C. 20009. (202) 328-8160. www.igc.org.

National, nonprofit organization that builds understanding of and support for programs, policies, and practices that are environmentally, economically, and socially sound.

Conservation Fund, 1800 North Kent Street, suite 1120, Arlington, VA 22209. (703) 525-6300. www.conservationfund.org.

Helps local, state, and federal agencies and nonprofit organizations acquire property from willing sellers to protect open space, wildlife habitat, public recreation areas, river corridors, and historic places. Paid and unpaid internships available.

Conservation International, 1919 M Street, suite 600, Washington, D.C. 20036. (202) 912-1000. www.conservation.org.

This group works to conserve Earth's living natural heritage, our global biodiversity, and to demonstrate that human societies are able to live harmoniously with nature. Internships available.

Conservation Law Foundation, 62 Summer Street, Boston, MA 02110. (617) 350-0990. www.clf.org.

Nonprofit, public interest, member-supported, environmental advocacy organization founded in 1966 that uses law, economics, and science to design and implement strategies that conserve natural resources, protect public health, and promote vital communities. Offers summer internships for graduates and undergraduates.

Consumer Federation of America, 1424 16th Street NW, suite 604, Washington, D.C. 20036. (202) 387-6121. www.consumerfed.org.

An advocacy group composed of several hundred consumer organizations that lobby on a broad range of energy and indoor air pollution issues. Internships available.

Cousteau Society, 870 Greenbriar Center, suite 402, Chesapeake, VA 23320. (800) 441-4395. www.cousteausociety.org.

Nonprofit organization that works to protect and improve the quality of life for present and future generations.

Defenders of Wildlife, 1101 14th Street NW, #1400, Washington, D.C. 20005. (202) 682-9400. www.defenders.org.

Works to preserve, enhance, and protect the diversity of wildlife and the habitats critical for their survival. Internships available.

Earth Island Institute, 300 Broadway, suite 28, San Francisco, CA 94133. (415) 788-3666. www.earthisland.org.

Develops and supports projects that counteract threats to the biological and cultural diversity that sustains the environment. Through education and activism these projects promote the conservation, preservation, and restoration of the Earth. Internships available.

Earth Justice, 426 17th Street, 6th floor, Oakland, CA 94612. (510) 550-6700. www.earthjustice.org.

A nonprofit public interest law firm dedicated to protecting natural resources and wildlife and to defending the right of all people to a healthy environment by enforcing and strengthening environmental laws on behalf of hundreds of organizations and communities.

Earth Share, 3400 International Drive NW, suite 2K, Washington, D.C. 20008. (800) 875-3863. (202) 537-7100. www.earthshare.org.

Manages workplace-giving campaigns for its national environmental charities, just as the United Way raises funds for its health and human service charities. In these payroll deduction drives, Earth Share encourages employees to pledge a small amount of each paycheck to help solve environmental problems. Employees may elect to contribute to all member agencies through a gift to Earth Share, or specifically to one or more of them.

Environmental Defense, 257 Park Ave. S, 16th floor, New York, NY 10010. (212) 505-2100. www.edf.org.

Other offices are in Washington, D.C.; Oakland, CA; Boulder, CO; Raleigh, NC; and Austin, TX. A national, nonprofit advocacy group that uses science, economics, and law to develop economically viable solutions to environmental problems. Student internships are offered at all offices.

Environmental Industry Associations, 4301 Connecticut Ave. NW, suite 300, Washington, D.C. 20008. (202) 244-4700. (800) 424-2869. www.envasns.org.

A trade association that works with federal and state governments and provides educational services on solid-waste management and resource recovery.

The Environmental Law Institute, 1616 P Street NW, suite 200, Washington, D.C. 20036. (202) 939-3800. www.eli.org

A nonprofit, grassroots organization involved in finding legal solutions to environmental problems. Internships available.

FINCA International, Inc. (The Foundation for International Community Assistance), 1101 14th Street NW, Washington, D.C. 20005. (202) 682-1510. www.villagebanking.org.

Provides financial services to the world's poorest families so they can create their own jobs, raise household incomes, and improve their standard of living. Internships are available.

Friends of the Earth, 1025 Vermont Ave NW, suite 300, Washington, D.C. 20005. (202) 783-7400 or (877) 843-8687. www.foe.org.

An environmental activist group distinguished by its international approach to global problems such as ozone depletion, drinking water contamination, and hazardous waste disposal, as well as its advocacy work on Capitol Hill. Internships available.

Greenpeace USA, 702 H Street, suite 300, Washington, D.C. 20001. (202) 462-1177. www.greenpeace.org.

An international organization that currently focuses on ocean ecology, hazardous wastes, disarmament, and atmospheric pollution issues. Internships available.

International Planned Parenthood Federation, 120 Wall Street, 9th floor, New York, NY 10005. (212) 248-6400. www.ippf.org.

Headquartered in London, this organization provides technical assistance and financial support to reproductive health organizations, helps facilitate information sharing among its affiliates, and advocates sexual and reproductive rights on a regional and international level in over 120 different countries. Internships available.

Izaak Walton League of America, 707 Conservation Lane, Gaithersburg, MD 20878. (301) 548-0150, (800) IKE-LINE. www.iwla.org.

A grassroots organization concerned with the conservation of wildlife, renewable natural resources, and water quality.

Kids for Saving Earth Worldwide (KSE), P.O. Box 421118, Plymouth, MN 55442. (763) 559-1234. www.kidsforsavingearth.org.

To educate and empower children to help the Earth's environment through action-oriented programs. Preschool through high school environmental educational materials available. Nonprofit. Offers internships.

The Land Institute, 2440 E. Water Well Road, Salina, KS 67401. (785) 823-5376. www.landinstitute.org.

Nonprofit organization that seeks to develop an agriculture that will save soil from being lost or poisoned, while promoting a community life that is both prosperous and enduring.

League of Conservation Voters, 1920 L Street NW, suite 800, Washington, D.C. 20036. (202) 985-8683. www.lcv.org.

A nonpartisan political action group that works to elect pro-environmental candidates to Congress. They also publish a scorecard that rates members of Congress on their environmental votes. Internships available.

National Arbor Day Foundation, 100 Arbor Ave. Nebraska City, NE 68410. (402) 474-5655. www.arborday.org.

Through educational programs, this foundation encourages tree planting and environmental stewardship.

National Audubon Society, 700 Broadway, New York, NY 10003. (212) 979-3000. www.audubon.org.

Also 9 regional offices and 500 local and international chapters. Works to conserve and restore natural ecosystems, focusing on birds and other wildlife, for the benefit of humanity and the Earth's biological diversity. Currently runs over 100 wildlife sanctuaries and nature centers nationwide. Internships available.

National Fish and Wildlife Foundation, 1120 Connecticut Ave NW, suite 900, Washington, D.C. 20036. (202) 857-0166. www.nfwf.org.

Charitable organization with nine regional offices that is dedicated to the conservation of natural resources: fish, wildlife, and plants. The foundation invests in the best possible solutions to natural resource management by awarding challenge grants using its federally appropriated funds to match sector funds. Offers unpaid internships.

National Park Foundation, 11 Dupont Circle, suite 600, Washington, D.C. 20036. (202) 238-4200. www.nationalparks.org.

The official nonprofit partner of the National Park Service, this foundation helps conserve, preserve, and enhance our National Parks for the benefit of the American people. Paid internships are available.

National Parks and Conservation Association, 1300 19th Street NW, Suite 300, Washington, D.C. 20036. (800) 628-7275. www.npca.org.

Acquisition and protection of national parks. Active in environmental issues as they relate to the national parks.

National Resources Defense Council, 40 West 20th Street, New York, NY 10011. (212) 727-2700. www.nrdc.org.

Offices in Washington, D.C.; Los Angeles; and San Francisco. Organization and litigation on a number of environmental issues, including public health, air, and energy issues; nuclear weapons; and land, water, and coastal issues. Internships available.

National Wildflower Research Center, 4801 La Crosse Ave, Austin, TX 78739. (512) 292-4200. www.wildflower.org.

Only national, nonprofit organization that works toward the preservation and reestablishment of native plants and planned landscapes. Emphasis on environmental education. Offers internships.

National Wildlife Federation, 11100 Wildlife Center Drive, Reston, VA 20190. (703) 438-6000. www.nwf.org.

The largest private conservation organization in the world. Conservation education is its primary mission. Its conservation internship program is primarily intended for recent college graduates, but college seniors sometimes qualify. These internships are located throughout the country.

Natural Resources Council of America, 1025 Thomas Jefferson Street NW, suite 109, Washington, D.C. 20007. (202) 333-0411. www.naturalresource.council.org.

A consortium that serves over 75 different environmental groups. Paid and unpaid internships available.

The Nature Conservancy, 4245 North Fairfax Drive, suite 100, Arlington, VA 22203-1606. (800) 628-6860. www.nature.org.

Global preservation of natural diversity. This private, international group preserves habitats and species by saving lands and waters that such species need to survive. Internships available.

North American Association for Environmental Education, 410 Tarvin Road, Rock Spring, GA 30739. (706) 764-2926. www.naaee.org.

Integrated network of professionals in the field of environmental education dedicated to supporting the work of environmental educators around the world.

The Ocean Conservancy, 1725 DeSales NW, suite 600, Washington, D.C. 20009. (202) 429-5609. www.oceanconservancy.org.

Educates the public on topics such as marine pollution issues and marine wildlife sanctuaries. Internships available.

Physicians for Social Responsibility, 1875 Connecticut Ave., suite 1012, Washington, D.C. 20009. (202) 667-4260. www.psr.org.

Focuses on the effects of nuclear war and nuclear weapons on human health.

Planned Parenthood Federation of America, 810 Seventh Avenue, New York, NY 10019. (212) 541-7800. www.plannedparenthood.org.

Addresses the family planning needs of nearly 8 million men and women around the world each year. Internships available.

Population Action International, 1300 19th Street NW, suite 200, Washington, D.C. 20036. (202) 557-3400. www.populationaction.org.

A private, nongovernmental organization that seeks to increase political and financial support for effective population policies and programs grounded in individual rights.

Population Connection, 1400 16th Street NW, suite 320, Washington, D.C. 20036. (202) 332-2200 or 1-800-POP-1956. www.populationconnection.org.

The largest national, nonprofit organization working to slow population growth and achieve sustainable balance between Earth's people and its resources. They seek to protect the environment and ensure a high quality of life for present and future generations. Paid internships available for undergraduate and graduate students.

Population-Environment Balance, Inc., 200 P Street NW, suite 600, Washington, D.C. 20036. (202) 955-5700. www.balance.org

Aims to stabilize the population of the United States and to safeguard the land's carrying capacity. Also supports a responsible immigration policy, increased funding for contraceptive research and availability, and a woman's right to reproductive choice. Offers paid summer internships.

The Population Institute, 107 2nd Street NE, Washington, D.C. 20002. (202) 544-3300. www.populationinstitute.org.

Education, lobbying, and public policy on population growth, particularly in developing nations. Internships available.

Population Reference Bureau, 1875 Connecticut Ave. NW, suite 520, Washington, D.C. 20009-5728. (202) 483-1100. (800) 877-9881. www.prb.org

A private, nonprofit educational organization that disseminates demographic and population information. PRB is a clearinghouse for U.S. and international population matters. Paid summer internships are available.

Population Resource Center, 1725 K Street NW, suite 1102, Washington, D.C. 20006. (202) 467-5030. www.prcdc.org.

Enables policymakers to incorporate the latest research findings in population change into the development of public policy by organizing a variety of educational and informational programs to help policymakers connect demographic, social, and economic change to public policy issues both domestically and internationally. Internships available.

The Public Citizen, 1600 20th Street, NW, Washington, D.C. 20009. (202) 588-1000. www.publiccitizen.org.

Political action group founded by Ralph Nader. Composed of several sister organizations with different missions, including Congress Watch, Critical Mass Energy Project, and the Health Research Group. Internships available.

Rainforest Action Network, 221 Pine Street, suite 500, San Francisco, CA 94104. (415) 398-4404. www.ran.org.

Works to protect the Earth's rain forests and supports the rights of rainforest inhabitants through education, grassroots organizing, and nonviolent direct action. Offers internships.

Rainforest Alliance, 665 Broadway, suite 500, New York, NY 10012. (212) 677-1900, (888) MY-EARTH. www.rainforest-alliance.org.

International nonprofit organization dedicated to the conservation of tropical forests for the benefit of the global community. Summer internships available.

Renew America, 1200 18th Street NW, suite 1100, Washington, D.C. 20036. (202) 721-1545. www.sol.crest.org/renew_america.

A nonprofit organization that coordinates a network of community and environmental groups, businesses, government leaders, and civic activists to exchange ideas and expertise for improving the environment. Unpaid internships available.

Resources for the Future, 1616 P Street NW, Washington, D.C. 20036. (202) 328-5000. www.rff.org.

A nonprofit, nonpartisan think tank that conducts independent research on environmental and natural resource issues. Paid summer internships available.

Sea Shepherd Conservation Society, 22774 Pacific Coast Hwy, Malibu, CA 90265. (310) 456-1141. www.seashepherd.org.

Nonprofit, nongovernmental organization involved with the investigation and documentation of violations of international laws, regulations, and treaties protecting marine wildlife species. Offers internships.

Sierra Club, 85 2nd Street, 2nd floor, San Francisco, CA 94105-3441. (415) 977-5500. www.sierraclub.org.

Interested in conserving the natural environment by influencing public policy decisions. Sierra Club members explore, enjoy, and protect the wild places of the Earth. Internships are offered in the San Francisco office only.

Smithsonian Institution, P.O. Box 37012, ST Building, room 153, MRC 010, Washington, D.C. 20013. www.si.edu.

Sponsors a wide variety of environmental education and research programs.

Student Conservation Association, Inc., P.O. Box 550, Charlestown, NH 03603. (603) 543-1700. www.thesca.org.

Resource assistant program places college students in internship positions at such places as national parks. Opportunities are available throughout the year.

Union of Concerned Scientists, 2 Brattle Square, Cambridge, MA 02238. (617) 547-5552. www.ucsusa.org.

Works to ensure that all people have clean air and energy, as well as safe and sufficient food. Focuses on a future free from threats of global warming and nuclear war. Paid internships available.

Water Environment Federation, 601 Wythe Street, Alexandria, VA 22314-1994. (800) 666-0206. www.wef.org.

Focuses primarily on wastewater treatment and water quality. A few internships are offered, and scholarships are sometimes offered to both undergraduate and graduate students.

The Wilderness Society, 1615 M Street NW, Washington, D.C. 20034. (800) 843-9453. www.tws.org.

Wilderness, parks, and public lands are the focus of this group. Internships available.

Wildlife Habitat Council, 1010 Wayne Ave., suite 920, Silver Spring, MD 20910. (301) 588-8994. www.wildlifehc.org.

A nonprofit, nonlobbying group of corporations, conservation organizations, and individuals dedicated to protecting and enhancing wildlife habitat. Also has four regional offices. Research assistant programs are sometimes available.

Wildlife Management Institute, 1101 14th Street NW, suite 801, Washington, D.C. 20005. (202) 371-1808. www.wildlifemgt.org/wmi.

A private, nonprofit scientific and educational organization committed to the conservation, enhancement, and professional management of North America's wildlife and other natural resources.

Wildlife Society, 5410 Grosvenor Lane, suite 200, Bethesda, MD 20814-2197. (301) 897-9770. www.wildlife.org.

A professional society for wildlife biologists.

Wilderness Watch, P.O. Box 9175, Missoula, MT 59807. (406) 542-2048. www.wildernesswatch.org.

A national nonprofit conservation organization dedicated solely to the preservation and enhancement of American wildernesses and wild and scenic rivers.

World Environment Center, 419 Park Avenue S., suite 500, New York, NY 10016. (212) 683-4700. www.wec.org.

An independent, nonadvocacy, nonprofit organization that advances sustainable development, encourages environmental leadership, and helps improve environmental, health, and safety practices worldwide by providing a platform for collaboration, partnership, and exchange of information between industry, government, nongovernmental organizations, and other sectors.

World Resources Institute, 10 G Street NE, suite 800, Washington, D.C. 20002. (202) 729-7600. www.wri.org.

Provides information, ideas, and solutions to global environmental problems. Paid summer internships available.

World Wildlife Fund (WWF), 1250 24th Street NW, Washington, D.C. 20037. (202) 293-4800. www.wwf.org.

Finds ways to save endangered species, including the acquisition of wildlife habitat. Internships are offered from time to time, primarily for college graduates. (Outside the United States, Canada, and Australia, the World Wildlife Fund is known as the World Wide Fund for Nature and can be found at www.panda.org.)

Worldwatch Institute, 1776 Massachusetts Ave. NW, Washington, D.C. 20036-1904. (202) 452-1999. www.worldwatch.org.

Nonprofit, public-interest research institute concentrating on global and environmental issues. Paid summer internships available.

Get Politically Involved

It is important to let your elected officials know about your concerns. Writing to your senator or congressional representative is an effective way to help influence the formation of policies and laws affecting environmental issues. It also helps to stay in touch when laws that have been passed are being implemented.

How to Write to Elected Officials   When addressing a letter, you should use the appropriate heading. If you are writing to the president of the United States, address your letter to: The President, The White House, 1600 Pennsylvania Avenue NW, Washington, D.C. 20500, Dear Mr. President. A letter to a senator should be addressed to The Honorable _____, Senate Office Building, Washington, D.C. 20510, Dear Senator _____. When writing to your representative, address the letter to The Honorable _____, House Office Building, Washington, D.C. 20515, Dear Representative _____.

You can also contact them via the World Wide Web:
www.whitehouse.gov
www.senate.gov
www.house.gov

The Do's   Letters are more effective if you keep them short (one page maximum) and to the point. Talk about the item of concern as specifically as possible: how it impacts you and others, what your position is on the matter, and why. If possible, make a specific request of the official; that is, let them know what you would like them to do. If you are writing about a particular bill, it would be helpful to know the number or name of the bill. Also, make sure you include your name and return address in case they wish to respond.

The Don'ts   Letters are more effective if you avoid being contentious or rude. Also, don't come across as high and mighty ("as a taxpayer who put you in office…"). Don't introduce more than one issue or allow the letter to ramble on and on.

Federal and International Agencies You Can Contact
The following agencies can be contacted for information about specific environmental bills, laws, and issues.

Agency for International Development, Ronald Reagan Building, Washington, D.C. 20523. (202) 712-4810. www.usaid.gov.

Biological Resources Discipline—USGS, 12201 Sunrise Valley Drive, Reston, VA 20192. (703) 648-4050. www.biology.usgs.gov.

Bureau of the Census, 4700 Silverhill Road, Suitland, MD 20746. (301) 457-1722. www.census.gov.

Bureau of Land Management, U.S. Department of Interior, 1849 C Street NW, Room 406-LS, Washington, D.C. 20240. (202) 452-5125. www.blm.gov.

Bureau of Reclamation, U.S. Department of Interior, 1849 C Street NW, Washington, D.C. 20240-0001. (202) 513-0501. www.usbr.gov.

Congressional Research Service, Library of Congress, 101 Independence Avenue SE, Washington, D.C. 20540. (202) 707-5700. www.crs.loc.gov.

Council on Environmental Quality, 722 Jackson Place NW, Washington, D.C. 20503. (202) 395-5750. www.whitehouse.gov/CEQ.

Department of Agriculture, Washington, D.C. 20250. (202) 720-4197. www.usda.gov.

Department of Energy, 1000 Independence Ave. SW, Washington, D.C. 20585. (800) Dial-DOE. www.energy.gov.

Department of the Interior, 1849 C Street NW, Washington, D.C. 20240. (202) 208-3100. www.doi.gov.

Environmental Protection Agency, Ariel Rios Building, 1200 Pennsylvania Ave. NW, Washington, D.C. 20460. (202) 260-2090. www.epa.gov.

Federal Energy Regulatory Commission, 888 1st Street NE, Washington, D.C. 20426. (866) 208-3372. www.ferc.fed.us.

U.S. Fish and Wildlife Service, 1849 C Street NW, Washington, D.C. 20240. (703) 358-1711. www.fws.gov.

Food and Drug Administration, 5600 Fishers Lane, Rockville, MD 20857. (888) 463-6332. www.fda.gov.

Forest Service, U.S. Department of Agriculture, P.O. Box 96090, Washington, D.C. 20090-6090. (202) 205-8333. www.fs.fed.us.

Geological Survey, U.S. Department of Interior, 12201 Sunrise Valley Drive, Reston, VA 20192. (703) 648-4000. www.usgs.gov.

Government Printing Office, 732 North Capitol Street NW, Washington, D.C. 20401. (202) 512-0000. www.access.gpo.gov.

International Union for the Conservation of Nature and Natural Resources, Rue Mauverney 28, Gland, 1196, Switzerland. 41 (22) 999-0000. www.IUCN.org.

National Academy of Sciences, 500 Fifth Street NW, Washington, D.C. 20001. (202) 334-2000. www.nationalacademies.org/nas/nashome.

National Oceanic and Atmospheric Administration, 14th and Constitution Avenue NW, Washington, D.C. 20230. (202) 482-6090. www.noaa.gov.

National Park Service, U.S. Department of Interior, 1849 C Street NW, Washington, D.C. 20240. (202) 208-6843. www.nps.gov.

National Science Foundation, 4201 Wilson Blvd., Arlington, VA 22230. (703) 292-5111. www.nsf.gov.

Natural Resources Conservation Service, U.S. Department of Agriculture, P.O. Box 2890, Washington, D.C. 20013. (202) 205-0026. www.nrcs.usda.gov.

Nuclear Regulatory Commission, 11545 Rockville Pike, Rockville, MD 20852. (301) 415-7000. www.nrc.gov.

Occupational Safety and Health Administration, U.S. Department of Labor, 200 Constitution Avenue NW, Washington, D.C. 20210. (202) 693-1999. www.osha.gov.

United Nations Environmental Programme, P.O. Box 30552, Nairobi, Kenya, or 2 United Nations Plaza, room DC2-803, New York, NY 10017. (212) 963-8139. www.unep.org.

Green Collar Professions

Careers relating to the environment are varied, and their numbers are ever increasing. The following list of "green-collar" jobs, organized into three general categories, is intended only as a broad sample. It should be noted that many jobs relating to the environment are by nature cross-disciplinary. With a creative approach, you can contribute to the welfare of our environment through almost any career path.

ENVIRONMENTAL PROTECTION (Solid and Hazardous Waste Management, Pollution Control)

air quality engineer (i.e., analyzing and controlling air pollution)

atmospheric scientist (i.e., measuring the chemical composition of the atmosphere)

biostatistician (i.e., measuring the statistical relation between injury or disease and exposure to pollution or toxins)

chemical engineer (i.e., designing systems for chemical waste disposal)

chemist (i.e., testing toxins and their interactions for possible environmental effects)

electrical engineer (i.e., designing energy-efficient power sources)

emergency response specialist (i.e., managing the emergency cleanup of chemical fires or spills)

environmental engineer (i.e., maintaining public water resources)

environmental health scientist (i.e., monitoring the health of ecosystems)

environmental protection specialist (i.e., reviewing government contracts for compliance with Environmental Protection Agency regulations)

geological engineer (i.e., designing environmentally safe sanitary landfills, selecting hazardous waste disposal sites)

hazardous waste manager (i.e., transporting and disposing of hazardous materials)

health physicist (i.e., developing protective measures for radiation exposure)

industrial hygienist (i.e., establishing procedures to minimize worker health risks associated with hazardous materials)

mechanical engineer (i.e., designing machinery to comply with environmental regulations)

meteorologist (i.e., monitoring the effects of atmospheric pollutants on weather patterns)

microbiologist (i.e., studying the effects of toxins on microorganisms as part of natural food chains; discovering microorganisms to be used in bioremediation)

mining engineer (i.e., designing mining tunnels to minimize dangerous methane gas leaks and groundwater contamination)

noise-control specialist (i.e., developing methods of neutralizing noise at heavy-industry manufacturing sites)

nuclear engineer (i.e., improving design safety of nuclear power plants; disposing of nuclear waste)

oceanographer (i.e., testing ocean water for the presence of sludge and industrial wastes)

risk manager (i.e., assessing potential environmental risks of proposed industrial procedures, chiefly for insurance purposes)

safety engineer (i.e., designing environmentally safe manufactured products)

soil scientist (i.e., testing soil for toxins leaked from landfill sites)

solid waste manager (i.e., managing the distribution, transportation, and disposal of solid wastes)

toxicologist (i.e., determining the link between human illness or disease and exposure to toxic substances)

water quality technologist (i.e., testing for contaminants at water treatment plants)

NATURAL RESOURCE MANAGEMENT (Land and Water Conservation, Fishery and Wildlife Management, Forestry, Parks, and Outdoor Recreation)

agricultural engineer (i.e., designing agricultural systems for soil and water conservation)

agronomist (i.e., developing alternatives to chemical fertilizers and pesticides for use in food crop cultivation)

botanist (i.e., identifying new plant species, evaluating wetlands based on the plants in an area, developing conservation plans for endangered or threatened plants)

conservation biologist (i.e., researching wildlife populations and ways to protect them)

ecologist (i.e., studying how natural and human-altered ecosystems work)

fishery biologist (i.e., studying the effects of changing environmental conditions on the survival and growth of fish)

fishery manager (i.e., managing fish hatcheries)

forester/park ranger (i.e., planting and maintaining populations of tree species)

hydrologist/geologist (i.e., managing water resources for agricultural use)

landscape architect (i.e., designing parks, recreational, and other public facilities that preserve or restore natural ecosystems)

petroleum engineer (i.e., ensuring that oil well sites are returned to their original condition after drilling is completed)

wildlife forensic specialist (i.e., providing forensic identification of wildlife parts and products to wildlife law enforcement officers)

wildlife inspector with U.S. Fish and Wildlife Service (i.e., ensuring that wildlife imports and exports comply with U.S. and international wildlife protection laws)

wildlife manager (i.e., restoring or preserving wildlife populations and habitats)

zoologist (i.e., studying the habitat requirements of animal species to prevent their extinction)

COMMUNICATIONS AND PUBLIC AFFAIRS

architect (i.e., designing energy-efficient buildings)

civil engineer (i.e., planning public works projects, including highway and sewage restoration)

computer specialist (i.e., developing and operating computer systems that monitor or simulate environmental problems)

demographer (i.e., researching statistics on human population growth rates to assist efforts to stabilize population growth)

educator (i.e., teaching environmental studies at the primary, secondary, undergraduate, or graduate level, or conducting employee training programs)

environmental lawyer (i.e., specializing in the interpretation of environmental legislation, regulations, and enforcement procedures)

interpretive naturalist (i.e., conducting educational tours in natural settings)

journalist (i.e., reporting on environmental issues)

occupational/environmental physician (i.e., treating patients exposed to radioactive or toxic materials)

technical writer (i.e., preparing environmental impact statements)

urban/community planner (i.e., designing residential areas to preserve natural ecosystems)

Chapter 1, Question 9

$I = P$ (6,000,000,000 people) $\times A$ (0.1 cars/person) $\times T$ (5.4 tons/car/year)

$I =$ **3.24 billion tons of CO_2**

For 2050:

$I =$ 10,000,000,000 people) $\times$ 0.4 cars/person $\times$ 5.4 tons/car/year

$I =$ **21.6 billion tons of CO_2**

Chapter 1, Question 10

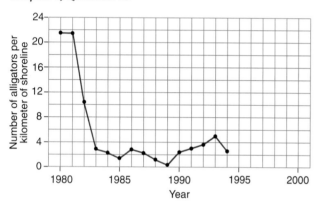

Chapter 2, Question 12

$$\frac{0.05}{1,000,000} = \frac{1}{x}$$
$0.05x = 1,000,000$
$x =$ **20,000,000**

Chapter 2, Question 13

1 kg = 2.205 lb
$$\frac{1}{2.205} = \frac{x}{100}$$
$2.205x = 100$
$x = 45.35$ kg
45.35 kg $\times$ 10.0 mg/kg = 453.5 mg = **0.45 g**

Chapter 2, Question 14

20 mg/kg $\times$ 0.75 kg = **15 mg**

Chapter 3, Question 15

Figure 3.13a: The amount of pollution indicated where the dashed line meets the x-axis is **more** than the economically optimum amount of pollution. The marginal cost of pollution is **higher** than if the pollution level was at its optimum. The cost of pollution abatement is **lower** than if the pollution level was at its optimum.

Figure 3.13b: The amount of pollution is **less** than the economically optimum amount of pollution. The marginal cost of pollution is **lower** than if the pollution level was at its optimum. The cost of pollution abatement is **higher** than if the pollution level was at its optimum.

Chapter 3, Question 16

Curve a corresponds to 1995, and curve b corresponds to 2000. It costs less to clean up a given amount of pollution on curve b than on curve a.

Chapter 4, Question 14

8,833 kcal/m²/year + 11,997 kcal/m²/year = **20,830 kcal/m²/year**

Chapter 4, Question 15

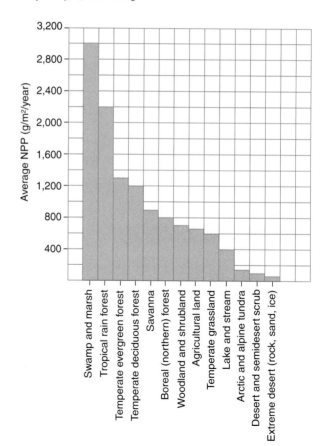

Chapter 5, Question 13

a. **Day 2**
b. **Days 3 to 5**
c. **Day 18**

Chapter 6, Question 15

$°F = °C × 9/5 + 32$
$°F = 245 × 9/5 + 32 = 281 + 32 = -49°F$
$°F = 275 × 9/5 + 32 = 2135 + 32 = -103°F$

Chapter 7, Question 15

$0.44 × 360 = 158.4°$
$0.33 × 360 = 118.8°$
$0.12 × 360 = 43.2°$
$0.10 × 360 = 36.0°$
$00.1 × 360 = 3.6°$

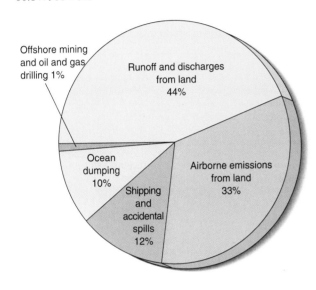

Chapter 7, Question 16

Figure a Temperate deciduous forest. It has the largest total precipitation of all three locations.

Figure b Temperate grassland. It has a moderate amount of rainfall, and the precipitation is somewhat seasonal.

Figure c Temperate desert. It has the least amount of precipitation of all three locations.

Chapter 8, Question 18

$r = b - d$
$r = 2,700/100,000 - 900/100,000$
$r = 0.027 - 0.009 = 0.018 = $ **1.8% per year**

Chapter 8, Question 19

The easiest way to determine the population of India in the year 2005 is to do similar calculations for each year:

$1,049,500,000 × 0.017 = 17,841,500$

$1,049.5$ million $+ 17.8$ million $= $ **1,067.3 million people in 2003**

$1,067.3$ million $× 0.017 = 18,144,100$

$1,067.3$ million $+ 1.8$ million $= $ **1069.1 million people in 2004**

$1,069.1$ million $× 0.017 = 18,174,700$ million

$1,069.1$ million $+ 18.2$ million $= $ **1,087.3 million people in 2005**

$t_d = 70/r = 70/1.7 = 41$ years

$2002 ÷ 41 = $ **the year 2043**

Chapter 9, Question 12

Let x = amount of CO_2 per capita in India
Then $30x$ = amount of CO_2 in the United States
For India, $x × 1,049,500,000 = 1,049,500,000x$
For the United States, $30x × 287,400,000 = 8,622,000,000x$
The United States produces more CO_2 than India.
$8,622,000x/1,049,500,000x = 8.2$
The United States produces 8.2 times more CO_2 pollution than India.

Chapter 10, Question 15

Table 10-A	Per-Capita Gasoline Consumption in the United States, 1950 to 2002			
Year	Gasoline (Million Barrels/ Day)	Gasoline (Million Gallons/ Day)	Population (Millions)	Gasoline Consumption (Gallons/ Day/Person)
1950	2.72	114.24	151.3	0.8
1955	3.66	153.72	165.1	0.9
1960	4.13	173.46	179.3	1.0
1965	4.59	192.78	193.5	1.0
1970	5.78	242.76	203.2	1.2
1975	6.67	280.14	215.5	1.3
1980	6.58	276.36	226.5	1.2
1985	6.83	286.86	238.7	1.2
1990	7.23	303.66	248.7	1.2
1995	7.79	327.18	263.2	1.2
2000	8.36	351.12	275.6	1.3
2002*	9.27	389.34	287.4	1.4

* Latest data available.
Source: Energy Information Administration, U.S. Department of Energy; Population Reference Bureau.

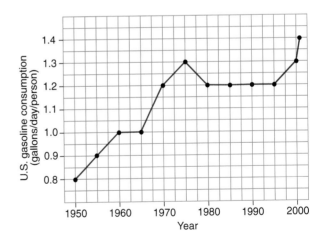

Chapter 11, Question 14

atomic number = **92 = number of protons**
$235 - 92 = $ **143 neutrons**

Chapter 11, Question 15

a. $1/2 × 1,000$ grams = **500 grams**

b. 1 half-life = 500 grams decayed

2 half-lives = 750 grams decayed

$2 \times 250{,}000 =$ **500,000 years**

c. $1{,}000{,}000/250{,}000 = 4$ half-lives

1st half-life: $1{,}000 \text{ g} \rightarrow 500 \text{ g}$

2nd half-life: $500 \text{ g} \rightarrow 250 \text{ g}$

3rd half-life: $250 \text{ g} \rightarrow 125 \text{ g}$

4th half-life: $125 \text{ g} \rightarrow$ **62.5 g remaining**

Chapter 12, Question 12

$2{,}170 - 1{,}930 = 240$
$240/1{,}930 = 0.124 =$ **12.4% in 1991**
$2{,}510 - 2{,}170 = 340$
$340/2{,}170 = 0.157 =$ **15.7% in 1992**
Contine the calculations in this manner to fill the table.

Table 12-A	Global Wind Power Generating Capacity, 1990–2001	
Year	**Capacity (megawatts)**	**Percent Increase Over Preceding Year**
1990	1930	11.6
1991	2170	12.4
1992	2510	15.7
1993	2990	19.1
1994	3,490	16.7
1995	4,780	37.0
1996	6,070	27.0
1997	7,640	25.9
1998	10,150	32.9
1999	13,930	37.3
2000	18,100	29.9
2001	24,800	37.0

Chapter 13, Question 19

1 drop/second $\times$ 60 seconds = 60 drops/minute
60 drops/minute $\times$ 60 minutes = 3,600 drops/hour
3,600 drops/hour $\times$ 24 hours = 86,400 drops/day
86,400 drops/day $\times$ 365 = 31,536,000 drops/year
31,536,000/13,140 = **2400 gallons/year**

Chapter 14, Question 19

1 mm/13 metric tons = 250 mm/x
$x =$ **3,250 metric tons**
$3{,}250 \times 1.103 =$ **3,584.8 tons (English units)**

Chapter 15, Question 16

$$\frac{7{,}291{,}000}{29} = \frac{x}{100}$$
$29x = 729{,}100{,}000$
$x =$ **25,141,379 metric tons**

Chapter 16, Question 19

$1/3 \times \$10$ billion = $3.333 billion = $3,333 million
$3,333 million/365 = **$9.13 million per day**

Chapter 17, Question 21

$0.8\% = 0.008$

14.7 billion hectares $\times$ 0.008 = 0.1176 billion hectares = **117.6 million hectares**

117.6 million hectares $\times$ 2.471 = **290.6 million acres**

Chapter 18, Question 20

$\$150/907.2 \text{ kg} = x/7 \text{ kg}$
$907.2x = 1{,}050$
$x =$ **$1.16 (beef)**
$\$150/907.2 \text{ kg} = x/6 \text{ kg}$
$907.2x = 900$
$x =$ **$0.99 (pork)**
$\$150/907.2 \text{ kg} = x/2.7 \text{ kg}$
$907.2x = 405$
$x =$ **$0.45 (poultry)**

Chapter 18, Question 21

Table 18-A	A World Grain Production, 1970 and 2001		
	1970	**2001**	**Percent Increase**
Agricultural land (million hectares)	663	684	3.2
Grain production (million tons)	1079	1843	70.8
Tons of grain per hectare	1.63	2.70	65.6

The average grain yield increased by 65.6% between 1970 and 2001.

Chapter 19, Question 20

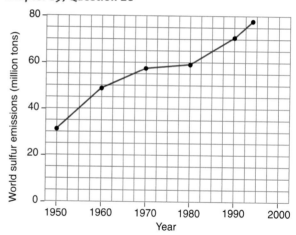

Long term, the graph indicates that world sulfur emissions are increasing.

Chapter 20, Question 19

pH of water = 7; pH of rain = 3
$7 - 3 = 4$
$10^4 =$ **10,000 times more acidic than pure water**

Chapter 20, Question 20

Sum the 10 yearly temperatures for each decade and divide by 10. (For the 2000 temperature sum the years 2000 and 2001 and divide by 2.)

Mean global temperature, 1970s = 13.99°C

Mean global temperature, 1980s = 14.16°C

Mean global temperature, 1990s = 14.30°C

Mean global temperature, 2000–2001 = 14.36°C

Since 1970, the mean global temperature for each decade has been higher than the previous decade.

Chapter 21, Question 16

10.5 kg/ 0.07 kg = **150 people to whom this pollution is equivalent**

Chapter 21, Question 17

100 m × 500 m = **50,000 m^2**

2 cm = 0.02 m

50,000 m^2 × 0.02 m = **1,000 cubic meters of water**

The runoff goes into storm sewers (primarily) and surrounding land (a little).

Chapter 22, Question 23

0.00005 ppm × 800 = **0.04 ppm in plankton**

Chapter 22, Question 24

$34.0 billion – $27.8 billion = $6.2 billion increase, 1994 to 1998

6.2/27.8 = 22.3%, 1994 to 1998

22.3% ÷ 4 yrs = **5.58% average annual increase**

Chapter 23, Question 20

0.56 × 98.1 × 10^6 tons = **54.9 × 10^6 tons, or 54.9 million tons**

Glossary

abiotic Nonliving. Compare *biotic.*

abyssal benthic zone The ocean's benthic environment that extends from a depth of 4000 to 6000 meters. Compare *hadal benthic zone.*

acid A substance that releases hydrogen ions (protons) in water. Acids have a sour taste and turn blue litmus paper red. Compare *base.*

acid deposition A type of air pollution that includes acid that falls from the atmosphere to Earth as precipitation or as dry acidic particles. See *acid precipitation.*

acid mine drainage Pollution caused when sulfuric acid and dangerous dissolved materials such as lead, arsenic, and cadmium wash from coal and metal mines into nearby lakes and streams.

acid precipitation Precipitation that is acidic as a result of both sulfur and nitrogen oxides forming acids when they react with water in the atmosphere; partially due to the combustion of coal; includes acid rain, acid snow, and acid fog.

active solar heating A system in which a series of collection devices mounted on a roof or in a field are used to absorb solar energy. Pumps or fans distribute the collected heat. Compare *passive solar heating.*

acute toxicity Adverse effects that occur within a short period after exposure to a toxicant. Compare *chronic toxicity.*

adaptation An evolutionary modification that improves the chances of survival and reproductive success of a population in a given environment.

adaptive radiation The evolution of a large number of related species from an unspecialized ancestral organism.

additivity A phenomenon in which two or more pollutants interact in such a way that their combined effects are what one would expect given the individual effects of each component of the mixture. Compare *antagonism and synergism.*

aerobic respiration The process by which cells use oxygen to break down organic molecules into waste products (such as carbon dioxide and water) with the release of energy that can be used for biological work.

aerosol Tiny particles of natural and human-produced air pollution that are so small they remain suspended in the atmosphere for days or even weeks.

aerosol effect Atmospheric cooling that occurs as a result of aerosol pollution.

age structure The number and proportion of people at each age in a population.

agroforestry Forestry and agricultural techniques that are used to improve degraded areas; in agroforestry, trees and crops are often planted together.

A-horizon The topsoil; located just beneath the O-horizon of the soil. The A-horizon is rich in various kinds of decomposing organic matter.

air pollution Various chemicals (gases, liquids, or solids) present in high enough levels in the atmosphere to harm humans, other animals, plants, or materials. Excess noise and heat are also considered air pollution.

air toxics See *hazardous air pollutants.*

albedo The proportional reflectance of Earth's surface; glaciers and ice sheets have high albedos and reflect most of the sunlight hitting their surfaces, whereas the ocean and forests have low albedos.

alcohol fuel A liquid fuel such as methanol or ethanol that may be used in internal combustion engines; mixing gasoline with 10% ethanol produces a cleaner-burning mixture known as gasohol.

algae Unicellular or simple multicellular photosynthetic organisms; important producers in aquatic ecosystems.

alpine tundra A distinctive ecosystem located in the higher elevations of mountains, above the tree line; characteristic vegetation includes grasses, sedges, and small tufted plants. Compare *tundra.*

alternative agriculture See *sustainable agriculture.*

altitude The height of a thing above sea level.

amino acids Organic compounds that are linked together to form proteins. See *essential amino acid.*

ammonification The conversion of nitrogen-containing organic compounds to ammonia (NH_3) and ammonium ions (NH_4^+) by certain bacteria (ammonifying bacteria) in the soil; part of the *nitrogen cycle.*

annual A plant that grows, reproduces, and dies in one growing season. Compare *biennial* and *perennial.*

antagonism A phenomenon in which two or more pollutants interact in such a way that their combined effects are less severe than the sum of their individual effects. Compare *addditivity* and *synergism.*

anthracite The highest grade of coal; has the highest heat content and burns the cleanest of any grade of coal. Also known as *hard coal.* Compare *bituminous coal, subbituminous coal,* and *lignite.*

anticline An upward folding of rock layers, or strata.

antioxidant Food additive that prevents oxidation (breakdown) of food molecules.

aquaculture The rearing of aquatic organisms (fishes, seaweeds, and shellfish), either freshwater or marine, for human consumption. See *mariculture.*

aquatic Pertaining to the water. Compare *terrestrial.*

aqueduct A large pipe or conduit constructed to carry water from a distant source.

aquifer The underground caverns and porous layers of underground rock in which groundwater is stored. See *confined aquifer* and *unconfined aquifer.*

aquifer depletion The removal by humans of more groundwater than can be recharged by precipitation or melting snow.

arctic tundra See *tundra.*

arid land A fragile ecosystem in which lack of precipitation limits plant growth. Arid lands are found in both temperate and tropical regions. Compare *semi-arid land.* Also called *desert.*

artesian aquifer See *confined aquifer.*

artificial eutrophication Overnourishment of an aquatic ecosystem by nutrients such as nitrates and phosphates. In artificial eutrophication, the pace of eutrophication is rapidly accelerated due to human activities such as agriculture and discharge from sewage treatment plants. Also called *cultural eutrophication.* See *eutrophication.*

artificial insemination A technique in which sperm collected from a suitable male of a rare species is used to artificially impregnate a female (perhaps located in another zoo in a different city or even in another country).

assimilation The incorporation of a substance into the cells of an organism.

atmosphere The gaseous envelope surrounding Earth.

atom The smallest quantity of an element that retains the chemical properties of that element; composed of protons, neutrons, and electrons.

atomic mass A number that represents the sum of the number of protons and neutrons in the nucleus of an atom. The atomic mass represents the relative mass of an atom. Compare *atomic number*.

atomic number A number that represents the number of protons in the nucleus of an atom. Each element has its own characteristic atomic number. Compare *atomic mass*.

autotroph See *producer*.

Baby Boom The large wave of births that followed World War II, from 1945 to 1962.

background extinction The continuous, low-level extinction of species that has occurred throughout much of the history of life. Compare *mass extinction*.

bacteria Unicellular, prokaryotic microorganisms. Most bacteria are decomposers, but some are autotrophs, and some are parasites.

bacterial source tracking Using molecular biological techniques to identify the source of dangerous bacteria in a stream or other body of water.

base A compound that releases hydroxide ions (OH^-) when dissolved in water. A base turns red litmus paper blue. Compare *acid*.

bellwether species An organism that provides an early warning of environmental damage. Examples include lichens, which are very sensitive to air pollution, and amphibians, which are sensitive to pesticides and other environmental contaminants. Also called *sentinel species*.

benthic environment The ocean floor.

benthos Bottom-dwelling marine organisms that fix themselves to one spot, burrow into the sand, or simply walk about on the ocean floor. Compare *plankton* and *nekton*.

B-horizon The light-colored, partially weathered soil layer underneath the A-horizon; subsoil. The B-horizon contains much less organic material than the A-horizon.

biennial A plant that requires two years to complete its life cycle. Compare *annual* and *perennial*.

bioaccumulation The buildup of a persistent toxic substance, such as certain pesticides, in an organisms's body. Also called *bioconcentration*.

biochemical oxygen demand (BOD) The amount of oxygen needed by microorganisms to decompose (by aerobic respiration) the organic material in a given volume of water. Also called *biological oxygen demand*.

bioconcentration See *bioaccumulation*.

biodegradable A chemical pollutant decomposed (broken down) by organisms or by other natural processes. Compare *nondegradable*.

biodiversity See *biological diversity*.

biodiversity hotspots Relatively small areas of land that contain an exceptional number of endemic species and are at high risk from human activities.

biogas A clean fuel, usually composed of a mixture of gases, whose combustion produces fewer pollutants than either coal or biomass. Biogas is produced from the anaerobic digestion of organic materials.

biogas digester A vat that uses bacteria to break down household wastes, including sewage. The gas produced by the bacteria is burned as a fuel.

biogeochemical cycle Process by which matter cycles from the living world to the nonliving physical environment and back again. Examples of biogeochemical cycles include the *carbon cycle*, *nitrogen cycle*, *phosphorus cycle*, and *sulfur cycle*.

biological amplification See *biological magnification*.

biological control A method of pest control that involves the use of naturally occurring disease organisms, parasites, or predators to control pests. Also called *biological pest control*.

biological diversity The number and variety of organisms; includes *genetic diversity*, *species richness*, and *ecosystem diversity*. Also called *biodiversity*.

biological magnification The increased concentration of toxic chemicals such as PCBs, heavy metals, and certain pesticides in the tissues of organisms at higher trophic levels in food webs. Also called *biological amplification*.

biological oxygen demand (BOD) See *biochemical oxygen demand*.

biological pest control See *biological control*.

biomass (1) A quantitative estimate of the total mass, or amount, of living material. Often expressed as the dry weight of all the organic material that comprises organisms in a particular ecosystem. (2) Plant and animal materials used as fuel.

biome A large, relatively distinct terrestrial region characterized by a similar climate, soil, plants, and animals, regardless of where it occurs on Earth; because it is so large in area, a biome encompasses many interacting ecosystems.

bio-prospecting Investigation of organisms in a forest, coral reef, or other natural area as potential sources of chemical products, including drugs, flavorings, fragrances, and natural pesticides.

bioremediation A method employed to clean up a hazardous waste site that uses microorganisms to break down the toxic pollutants. Compare *phytoremediation*.

biosphere All of Earth's organisms and their interactions with each other, the land, the water, and the atmosphere.

biotic Living. Compare *abiotic*.

biotic pollution The accidental or intentional introduction of a foreign, or exotic, species into an area where it is not native. See *invasive species*.

biotic potential The maximum rate at which a population could increase under ideal conditions. Also called *intrinsic rate of increase*.

birth rate The number of births per 1000 people per year. Also called *crude birth rate* and *natality*.

bituminous coal The most common form of coal; produces a high amount of heat and is used extensively by electric power plants. Also called *soft coal*. Compare *anthracite*, *subbituminous coal*, and *lignite*.

bloom Large algal population caused by the sudden presence of large amounts of essential nutrients (such as nitrates and phosphates) in surface waters.

boreal forest See *taiga*.

botanical A plant-derived chemical used as a pesticide. See *synthetic botanical*.

bottom ash The residual ash left at the bottom of an incinerator when combustion is completed. Also called *slag*. Compare *fly ash*.

breeder nuclear fission A type of nuclear fission in which nonfissionable U-238 is converted to fissionable P-239.

broad leaf herbicide A herbicide that kills plants with broad leaves but does not kill grasses (such as corn, wheat, and rice).

bycatch Unwanted fishes, dolphins, and sea turtles that are caught along with commercially valuable fishes and then dumped, dead or dying, back into the ocean.

calendar spraying The regular use of pesticides regardless of whether pests are a problem or not. Compare *scout-and-spray*.

calorie A unit of heat energy; the amount of heat required to raise 1 g of water 1°C.

cancer A malignant tumor anywhere in the body. Cancers tend to spread throughout the body.

carbamates A class of broad-spectrum pesticides that are derived from carbamic acid.

carbohydrate An organic compound containing carbon, hydrogen, and oxygen in the ratio of 1C:2H:1O. Carbohydrates include sugars and starches, molecules metabolized readily by the body as a source of energy.

carbon cycle The worldwide circulation of carbon from the abiotic environment into organisms and back into the abiotic environment.

carbon management Ways to separate and capture the carbon dioxide produced during the combustion of fossil fuels and then sequester it away from the atmosphere.

carcinogen Any substance that causes cancer or accelerates its development.

carnivore   An animal that feeds on other animals; flesh-eater. Compare *herbivore* and *omnivore*. See *secondary consumer*.

carrying capacity *(K)*   The maximum number of individuals of a given species that a particular environment can support sustainably (long term), assuming there are no changes in that environment. For example, the carrying capacity of a rangeland is the maximum number of animals the rangeland plants can sustain over an indefinite period of time without deterioration of the rangeland.

cell   The basic structural and functional unit of life. Simple organisms are composed of single cells, whereas complex organisms are composed of many cells.

cellular respiration   A process in which the energy of organic molecules is released within cells. *Aerobic respiration* is a type of cellular respiration.

CFCs   See *chlorofluorocarbons*.

chain reaction   A reaction maintained because it forms as products the very materials that are used as reactants in the reaction. For example, a chain reaction occurs during nuclear fission when neutrons collide with other U-235 atoms and those atoms are split, releasing more neutrons, which then collide with additional U-235 atoms.

chaparral   A biome with a Mediterranean climate (mild, moist winters and hot, dry summers). Chaparral vegetation is characterized by small-leaved evergreen shrubs and small trees.

chemosynthesis   The biological process by which certain bacteria take inorganic compounds from their environment and use them to obtain energy and make carbohydrate molecules; light is not required for this process. Compare *photosynthesis*.

chlorinated hydrocarbon   A synthetic organic compound that contains chlorine and is used as a pesticide (for example, DDT) or an industrial compound (for example, PCBs).

chlorofluorocarbons   Human-made organic compounds composed of carbon, chlorine, and fluorine that have a number of industrial and commercial applications but have been banned because they attack the stratospheric ozone layer. Also called *CFCs*.

chlorophyll   A green pigment that absorbs radiant energy for photosynthesis.

C-horizon   The partly weathered layer in the soil located beneath the B-horizon. The C-horizon borders solid parent material.

chronic toxicity   Adverse effects that occur after a long period of exposure to a toxicant. Compare *acute toxicity*.

circumpolar vortex   A mass of cold air that circulates around the southern polar region, in effect isolating it from the warmer air in the rest of the world.

clay   The smallest inorganic soil particles. Compare *silt* and *sand*.

clean-coal technologies   New methods for burning coal that do not contaminate the atmosphere with sulfur oxides and that produce significantly fewer nitrogen oxides.

clearcutting   A forest management technique that involves the removal of all trees from an area at a single time. Compare *seed-tree cutting*, *selective cutting*, and *shelterwood cutting*.

climate   The average weather conditions that occur in a place over a period of years. Includes temperature and precipitation. Compare *weather*.

closed system   A system that does not exchange energy with its surroundings. Compare *open system*.

coal   A black combustible solid found in Earth's crust. Formed from the remains of ancient plants that lived millions of years ago. Used as a fuel. See *fossil fuel*.

coal gasification   The technique of producing a synthetic gaseous fuel (such as methane) from solid coal.

coal liquefaction   The process by which solid coal is used to produce a synthetic liquid fuel similar to oil.

coastal wetlands   Marshes, bays, tidal flats, and swamps that are found along a coastline. See *mangrove forest*, *salt marsh*, and *wetlands*.

cochlea   The part of the ear that perceives sound.

coevolution   The interdependent evolution of two or more species that occurs as a result of their interactions over a long period of time. Flowering plants and their animal pollinators are an example of coevolution because each has profoundly affected the other's characteristics.

cogeneration   An energy technology that involves recycling "waste" heat so that two useful forms of energy (electricity and either steam or hot water) are produced from the same fuel. Also called *combined heat and power (CHP)*.

coloring agent   Natural or synthetic food additives used to make food visually appealing.

combined heat and power   See *cogeneration*.

combined sewer overflow   A problem that arises in a combined sewer system when too much water (from a heavy rainfall or snowmelt) enters the system. The excess flows into nearby waterways without being treated.

combined sewer system   A municipal sewage system in which human and industrial wastes are mixed with urban runoff from storm sewers before flowing into the sewage treatment plant.

combustion   The process of burning by which organic molecules are rapidly oxidized, converting them into carbon dioxide and water with an accompanying release of heat and light.

command and control regulation   Pollution control laws that work by setting pollution ceilings. Examples include the Clean Water and Clean Air Acts. Compare *incentive-based regulation*.

commensalism   A type of symbiosis in which one organism benefits, and the other one is neither harmed nor helped. See *symbiosis*. Compare *mutualism* and *parasitism*.

commercial extinction   Depletion of the population of a commercially important species to the point that it is unprofitable to harvest.

commercial harvest   The collection of commercially important organisms from the wild. Examples include the commercial harvest of parrots (for the pet trade) and cacti (for houseplants).

commercial inorganic fertilizer   See *fertilizer*.

community   An association of different species living together at the same time in a defined habitat with some degree of mutual interdependence. Compare *ecosystem*.

community stability   The ability of a community to withstand environmental disturbances.

compact development   The design of cities so that tall, multiple-unit residential buildings are close to shopping and jobs, all of which are connected by public transportation.

competition   The interaction among organisms that vie for the same resources in an ecosystem (such as food, living, space or other resources). See *interspecific competition* and *intraspecific competition*.

competitive exclusion   The concept that no two species with identical living requirements can occupy the same ecological niche indefinitely. Eventually, one species will be excluded by the other as a result of interspecific (between-species) competition for a resource in limited supply.

compost   A natural soil and humus mixture that improves soil fertility and soil structure.

confined aquifer   A groundwater storage area trapped between two impermeable layers of rock. Also called *artesian aquifer*. Compare *unconfined aquifer*.

conifer   Any of a group of woody trees or shrubs (gymnosperms) that bear needle-like leaves and seeds in cones.

conservation   The sensible and careful management of natural resources. Compare *preservation*.

conservation biology   A multidisciplinary science that focuses on the study of how humans impact organisms and on the development of ways to protect biological diversity; includes *in situ* and *ex situ* conservation.

conservation tillage   A method of cultivation in which residues from previous crops are left in the soil, partially covering it and helping to hold it in place until the newly planted seeds are established. See *no-tillage*. Compare *conventional tillage*.

conservationist   A person who supports the conservation of natural resources.

consumer   An organism that cannot synthesize its own food from inorganic materials and therefore must use the bodies of other organisms as sources of energy and body-building materials. Also called *heterotroph*. Compare *producer*.

consumption   The human use of materials and energy; generally speaking, people in highly developed countries are extravagant consumers, and their use of resources is greatly out of proportion to their numbers.

consumption overpopulation   A situation in which each individual in a population consumes too large a share of resources, that is, much more than is needed to survive. Consumption overpopulation results in pollution, environmental degradation, and resource depletion. See *overpopulation*. Compare *people overpopulation*.

containment building   A safety feature of nuclear power plants that provides an additional line of defense against any accidental leak of radiation.

continental shelf   The submerged, relatively flat ocean bottom that surrounds continents. The continental shelf extends out into the ocean to the point where the ocean floor begins a steep descent.

contour plowing   Plowing that matches the natural contour of the land, that is, the furrows run around rather than up and down a hill; lessens erosion, as on a hillside. See *strip cropping*.

contraceptive   A device or drug used to intentionally prevent pregnancy.

control   An essential part of every scientific experiment in which the experimental variable remains constant. The control provides a standard of comparison in order to verify the results of an experiment. See *variable*.

conventional tillage   The traditional method of cultivation in which the soil is broken up by plowing before seeds are planted. Compare *conservation tillage*.

cooling tower   Part of an electric power generating plant within which heated water is cooled.

coral reef   Structure built from accumulated layers of calcium carbonate ($CaCO_3$); found in warm, shallow seawater. The living portion is composed principally of red coralline algae or of colonies of millions of tiny coral animals.

Coriolis effect   The tendency of moving air or water to be deflected from its path to the right in the Northern Hemisphere and to the left in the Southern Hemisphere. Caused by the direction of Earth's rotation.

corridor   See *wildlife corridor*.

cost–benefit analysis   A mechanism that helps policy-makers make decisions about environmental issues. Compares estimated costs of a particular action with potential benefits that would occur if that action were implemented.

crop rotation   The planting of different crops in the same field over a period of years. Crop rotation reduces mineral depletion of the soil because the mineral requirements of each crop vary.

crude birth rate   See *birth rate*.

crude death rate   See *death rate*.

crude oil   See *petroleum*.

cullet   Crushed glass food and beverage containers that are melted to make new products.

cultural eutrophication   See *artificial eutrophication*.

datum (pl. *data*)   The information, or facts, with which science works and from which conclusions are inferred.

DDT   A chlorine-containing organic compound that has insecticidal properties; because it is slow to degrade and therefore persists in the environment and inside organisms, DDT has been banned in the United States and many other countries.

death rate   The number of deaths per 1000 people per year. Also called *crude death rate* and *mortality*.

debt-for-nature swap   The cancellation of part of a country's foreign debt in exchange for their agreement to protect certain lands (or other resources) from detrimental development.

decibel (dB)   A numerical scale that expresses the relative loudness of sound.

decibel-A (dBA)   A modified decibel scale that takes into account high-pitched sounds to which the human ear is more sensitive.

decommission   The dismantling of an old nuclear power plant after it closes. Compare *entombment*.

decomposer   A heterotroph that breaks down organic material and uses the decomposition products to supply it with energy. Decomposers are microorganisms of decay. Also called *saprotroph*. Compare *detritivore*.

deductive reasoning   Reasoning that operates from generalities to specifics and can make relationship among data more apparent. Compare *inductive reasoning*.

deep ecology   A philosophical worldview that stresses harmony with nature, respect for life, and the belief that humans and all other species have an equal worth. Deep ecologists describe themselves as biocentric (all-life-centered).

deforestation   The temporary or permanent clearance of large expanses of forests for agriculture or other uses.

degradation   The natural or human-induced process that decreases the future ability of the land (or soil) to support crops or livestock animals.

delta   A deposit of sand or soil at the mouth of a river.

demand-side management   A way that electric utilities can meet future power needs, by helping consumers conserve energy and increase energy efficiency. In demand-side management, consumers save money because they use less energy, and utilities save money because they do not have to build new power plants or purchase additional power.

dematerialization   The decrease in the size and weight of a product as a result of technological improvements that occur over time.

demographics   The applied branch of sociology that deals with population statistics; provides information on the populations of various countries or groups of people.

demographic transition   The process whereby a country moves from relatively high birth and death rates to relatively low birth and death rates.

demography   The science of population.

denitrification   The conversion of nitrate (NO_3^-) to nitrogen gas (N_2) by certain bacteria (denitrifying bacteria) in the soil; part of the *nitrogen cycle*.

density-dependent factor   An environmental factor whose effects on a population change as population density changes; density-dependent factors tend to retard population growth as population density increases and enhance population growth as population density decreases.

density-independent factor   An environmental factor that affects the size of a population but is not influenced by changes in population density.

derelict land   Land area that was degraded by mining.

desalination   See *desalinization*.

desalinization   The removal of salt from ocean or brackish (somewhat salty) water. Also called *desalination*.

desert   See *arid land*.

desertification   Degradation of once-fertile rangeland (or tropical dry forest) into nonproductive desert. Caused partly by soil erosion, deforestation, and overgrazing.

detritivore   An organism (such as an earthworm or crab) that consumes fragments of dead organisms. Also called *detritus feeder*. Compare *decomposer*.

detritus   Organic matter that includes dead organisms (such as animal carcasses and leaf litter) and wastes (such as feces).

detritus feeder   See *detritivore*.

deuterium   An isotope of hydrogen that contains one proton and one neutron per atom. Compare *tritium*.

developed country   See *highly developed country*.

developing country   A country not highly industrialized and characterized by a high fertility rate, high infant mortality rate, and low per-capita income. Most developing countries are located in Africa, Asia, and Latin America, and fall into two subcategories: *moderately developed* and *less developed*. Compare *highly developed country*.

dilution A technique of soil remediation that involves running large quantities of water through contaminated soil in order to leach out pollutants such as excess salt.

dioxin Any of a family of mildly to extremely toxic chlorinated hydrocarbon compounds that are formed as by-products in certain industrial processes.

disease A departure from the body's normal healthy state as a result of infectious organisms, environmental stressors, or some inherent weakness.

dispersal The movement of individuals among populations, from one region or country to another. See *immigration* and *emigration*.

distillation A heat-dependent process used to purify or separate complex mixtures. Saltwater or brackish water may be distilled to remove the salt from the water. Compare *reverse osmosis*.

DNA Deoxyribonucleic acid. Present in a cell's chromosomes, DNA contains all of an organism's genetic information.

domesticated Adapted to humans. Describes plants and animals that, during their association with humans, have become so altered from their original ancestors that it is doubtful they could survive and compete successfully in the wild.

dose In toxicology, the amount of a toxicant that enters the body of an exposed organism.

dose-response curve In toxicology, a graph that shows the effect of different doses on a population of test organisms.

doubling time The amount of time it takes for a population to double in size, assuming that its current rate of increase doesn't change.

dragline In coal mining, huge shovel that takes enormous chunks out of a mountain to reach underground coal seams.

drainage basin See *watershed*.

drip irrigation See *microirrigation*.

dry deposition A form of acid deposition in which dry, sulfate-containing particles settle out of the air. Compare *wet deposition*.

Dust Bowl, American A semiarid region of the Great Plains that became desert-like as a result of human mismanagement, extended drought, and severe dust storms during the 1930s.

dust dome A dome of heated air that surrounds an urban area and contains a lot of air pollution. Compare *urban heat island*.

ecological footprint The average amount of land and ocean needed to supply an individual with food, energy, water, housing, transportation, and waste disposal.

ecological niche See *niche*.

ecological pyramid A graphic representation of the relative energy value at each trophic level. See *pyramid of biomass*, *pyramid of energy*, and *pyramid of numbers*.

ecological restoration See *restoration ecology*.

ecological risk assessment The process by which the ecological consequences of human activities are estimated. See *risk assessment*.

ecological succession See *succession*.

ecologically sustainable forest management A new method of forest management that seeks not only to conserve forests for the commercial harvest of timber and nontimber forest products, but to sustain biological diversity, prevent soil erosion, protect the soil, and preserve watersheds that produce clean water. Also called *sustainable forest management* or, simply, *sustainable forestry*.

ecology A discipline of biology that studies the interrelationships between organisms and among organisms and their environment.

economics The study of how people (individuals, businesses, or countries) use their limited economic resources to fulfill their needs and wants. Economics encompasses the production, consumption, and distribution of goods.

ecosystem The interacting system that encompasses a community and its nonliving, physical environment. In an ecosystem, all of the biological, physical, and chemical components of an area form a complex interacting network of energy flow and materials cycling. Compare *community*.

ecosystem diversity Biological diversity that encompasses the variety among ecosystems, such as forests, grasslands, deserts, lakes, estuaries, and oceans. Compare *genetic diversity* and *species richness*.

ecosystem management A conservation focus that emphasizes restoring and sustaining ecosystem quality rather than the conservation of individual species.

ecosystem services Important environmental benefits, such as clean air to breathe, clean water to drink, and fertile soil in which to grow crops, that ecosystems provide.

ecotone The transitional zone where two ecosystems or biomes intergrade.

ecotourism A type of tourism in which tourists pay to observe wildlife in natural settings.

edge effect The ecological phenomenon in which ecotones between adjacent communities often have more kinds of species or greater population densities of certain species than either adjoining community.

effective dose-50% (ED$_{50}$) In toxicology, the dose that causes 50% of a population to exhibit whatever biological response is under study.

E-horizon A heavily leached soil area that sometimes develops between the A- and B-horizons.

electromagnetic spectrum The continuous range of wavelengths of electromagnetic energy; includes radio waves, microwaves, infrared waves, visible light, ultraviolet radiation, x rays, and gamma rays.

electrostatic precipitator An air pollution control device that gives ash a positive electrical charge so that it adheres to negatively charged plates.

El Niño–Southern Oscillation (ENSO) A periodic warming of surface waters of the tropical East Pacific that alters both ocean and atmospheric circulation patterns and results in unusual weather in areas far from the tropical Pacific. Compare *La Niña*.

embryo transfer A technique in which a female of a rare species is treated with fertility drugs, which cause her to produce multiple eggs; some of these eggs are collected, fertilized with sperm, and surgically implanted into a female of a related but less rare species, which later gives birth to offspring of the rare species.

emigration A type of dispersal in which individuals leave a population and thus decrease its size. Compare *immigration*.

emission charge A government policy that controls pollution by charging the polluter for each given unit of emissions; that is, by establishing a tax on pollution.

emission reduction credit (ERC) A waste-discharge permit that can be bought and sold by companies that produce emissions. See *marketable waste-discharge permit*.

endangered species A species whose numbers are so severely reduced that it is in imminent danger of becoming extinct in all or a significant part of its range. Compare *threatened species*.

endemic species Localized, native species that are not found anywhere else in the world.

endocrine disrupter A chemical that interferes with the actions of the endocrine system (the body's *hormones*). Includes certain plastics such as polycarbonate; chlorine compounds such as PCBs and dioxin; the heavy metals lead and mercury; and some pesticides such as DDT, kepone, chlordane, and endosulfan.

energy The capacity or ability to do work.

energy conservation Saving energy by reducing energy use and waste. Carpooling to work or school is an example of energy conservation. Compare *energy efficiency*.

energy efficiency Using less energy to accomplish the same task. Purchasing an appliance such as a refrigerator that uses less energy to keep food cold is an example of energy efficiency. Compare *energy conservation*.

energy flow The passage of energy in a one-way direction through an ecosystem.

energy intensity A statistical estimate of energy efficiency, as for example, a country's or region's total energy consumption divided by its gross natural product.

enhanced greenhouse effect   See *greenhouse effect.*

enrichment   (1) The process by which uranium ore is refined after mining to increase the concentration of fissionable U-235. (2) The fertilization of a body of water, caused by the presence of high levels of plant and algal nutrients such as nitrogen and phosphorus; see *eutrophication*, which is a type of enrichment.

entombment   An option after the closing of an old nuclear power plant in which the entire power plant is permanently encased in concrete. Compare *decommission.*

entropy   A measure of the randomness or disorder of a system.

environment   All the external conditions, both abiotic and biotic, that affect an organism or group of organisms.

environmental chemistry   The subdiscipline of chemistry that deals with redesigning commercially important chemical processes to significantly reduce environmental harm. Also called *green chemistry.*

environmental ethics   A field of applied ethics that considers the moral basis of environmental responsibility and how far this responsibility extends; environmental ethicists try to determine how we humans should relate to nature.

environmental impact statement (EIS)   A statement that accompanies federal recommendations or proposals and is supposed to help federal officials and the public make informed decisions. Required by the National Environmental Policy Act of 1970.

environmentalist   A person who works to solve environmental problems such as overpopulation; pollution of Earth's air, water, and soil; and depletion of natural resources. Environmentalists are collectively known as the environmental movement.

environmental justice   The right of every citizen, regardless of age, race, gender, social class, or other factors, to adequate protection from environmental hazards.

environmental movement   See *environmentalist.*

environmental resistance   Limits set by the environment that prevent organisms from reproducing indefinitely at their biotic potential; includes the limited availability of food, water, shelter, and other essential resources, as well as limits imposed by disease and predation.

environmental science   The interdisciplinary study of how humanity interacts with other organisms and the nonliving physical environment.

environmental stressor   Environmental factors, whether natural or human-induced, that tax an organism's ability to thrive.

environmental sustainability   See *sustainability.*

environmental worldview   A worldview that helps us make sense of how the environment works, our place in the environment, and right and wrong environmental behaviors.

epicenter   The site at Earth's surface that is located directly above an earthquake's focus.

epiphyte   A small organism that grows on another organism but is not parasitic on it. Small plants that live attached to the bark of a tree's branches are epiphytes.

essential amino acid   Any of the ten amino acids that must be obtained in the diet because humans cannot synthesize them from simpler materials.

estuary   A coastal body of water that connects to the ocean, in which fresh water from a river mixes with saltwater from the ocean.

ethanol   A colorless, flammable liquid, C_2H_5OH. Also called *ethyl alcohol.*

ethics   The branch of philosophy that deals with human values. See *environmental ethics.*

ethyl alcohol   See *ethanol.*

euphotic zone   The upper reaches of the pelagic environment, from the surface to a maximum depth of 150 m in the clearest open ocean water; sufficient light penetrates the euphotic zone to support photosynthesis.

eutrophic lake   A lake enriched with nutrients such as nitrates and phosphates and consequently overgrown with plants or algae; water in a eutrophic lake contains very little dissolved oxygen. Compare *oligotrophic lake.*

eutrophication   The enrichment of a lake, estuary, or slow flowing stream by nutrients that cause increased photosynthetic productivity. Eutrophication that occurs naturally is a very slow process in which the body of water gradually fills in and converts to a marsh, eventually disappearing. See *enrichment* and *artificial eutrophication.*

evaporation   The conversion of water from a liquid to a vapor. Also called *vaporization.*

evolution   The cumulative genetic changes in populations that occur during successive generations. Evolution explains the origin of all the organisms that exist today or have ever existed.

ex situ conservation   Conservation efforts that involve conserving biological diversity in human-controlled settings. See *conservation biology.* Compare *in situ conservation.*

exosphere   The outermost layer of the atmosphere, bordered by the thermosphere and interplanetary space.

exotic species   See *biotic pollution.*

exponential population growth   The accelerating population growth that occurs when optimal conditions allow a constant rate of increase over a period of time. When the increase in population number versus time is plotted on a graph, exponential population growth produces a characteristic J-shaped curve.

external cost   A harmful side effect of production or consumption of a product that is borne by people not directly involved in the market exchange for that product; an external cost is usually not reflected in a product's price. Also called *negative externality.*

extinction   The elimination of a species from Earth; occurs when the last individual member of a species dies.

facultative parasite   An organism that is normally saprotrophic but, given the opportunity, becomes parasitic. Compare *obligate parasite.*

fall turnover   A mixing of the lake waters in temperate lakes, caused by falling temperatures in autumn. Compare *spring turnover.*

family planning   Providing the services, including information about birth control methods, to help people have the number of children they want.

famine   Widespread starvation caused by a drastic shortage of food. Famine is caused by crop failures that are brought on by drought, war, flood, or some other catastrophic event.

fault   Fractures in the crust along which rock moves forward and backward, up and down, or from side to side. Fault zones are often found at *plate boundaries.*

fecal coliform test   A water quality test for the presence of *E. coli* (fecal bacteria common in the intestinal tracts of people and animals). The presence of fecal bacteria in a water supply indicates a chance that pathogenic organisms may be present as well.

fee-per-bag approach   A way to reduce the weight and volume of municipal solid waste by charging households for each container of garbage.

fertilizer   A material containing plant nutrients, such as nitrates, phosphates, and potassium, that is put on the soil to enhance plant growth. Organic fertilizers include such natural materials as animal manure, crop residues, bone meal, and compost, whereas commercial inorganic fertilizers are manufactured from inorganic chemical compounds.

first law of thermodynamics   Energy cannot be created or destroyed, although it can be transformed from one form to another. Compare *second law of thermodynamics.*

fission   A nuclear reaction in which large atoms of certain elements are each split into two smaller atoms with the release of a large amount of energy. Compare *fusion.*

flood plain   The area bordering a river that is subject to flooding.

flowing-water ecosystem   A freshwater ecosystem such as a river or stream in which the water flows. Compare *standing-water ecosystem.*

fluidized-bed combustion   A clean-coal technology in which crushed coal is mixed with particles of limestone in a strong air current during combustion; the limestone neutralizes the acidic sulfur compounds produced during combustion.

fly ash The ash from the flue (chimney) that is trapped by electrostatic precipitators. Compare *bottom ash*.

flyway An established route that ducks, geese, and shorebirds follow during their annual migrations.

focus The site, often far below Earth's surface, where an earthquake begins.

food additive A chemical added to food because it enhances the taste, color, or texture of the food, improves its nutrient value, reduces spoilage, prolongs shelf life, or helps maintain its consistency. See *antioxidant* and *preservative*.

food chain The successive series of organisms through which energy flows in an ecosystem. Each organism in the series eats or decomposes the preceding organism in the chain. Compare *food web*.

food security A goal in which all of the world's people have access at all times to adequate amounts and kinds of food needed for healthy, active lives.

food web A complex interconnection of all the food chains in an ecosystem. Compare *food chain*.

forest decline A gradual deterioration (and often death) of many trees in a forest; may be the result of a combination of environmental stressors, such as acid precipitation, toxic heavy metals, and surface-level ozone.

forest edge The often sharp boundary between the forest and surrounding farmlands or residential neighborhoods. Songbird nests near a forest edge are more vulnerable to predation by small mammalian predators and egg-eating birds as well as nest parasitism by brown-headed cowbirds.

forest management See *ecologically sustainable forest management*.

fossil fuel Combustible deposits in Earth's crust. Fossil fuels are composed of the remnants of prehistoric organisms that existed millions of years ago. Examples include *oil (petroleum), natural gas,* and *coal*.

fragmentation See *habitat fragmentation*.

freshwater wetlands Lands that are usually covered by shallow fresh water for at least part of the year and that have a characteristic soil and water-tolerant vegetation; include *marshes* and *swamps*.

frontier attitude The attitude of most Americans during the 1700s and early 1800s that, because the natural resources of North America were seemingly inexhaustible, there was no reason not to conquer and exploit nature as much and as quickly as possible.

fuel cell An electrochemical cell similar to a battery in which fuel cell reactants (hydrogen and oxygen) are supplied from external reservoirs; when hydrogen and oxygen react in a fuel cell, water forms and energy is produced as an electrical current.

fundamental niche The potential ecological niche that an organism could have if there were no competition from other species. See *niche*. Compare *realized niche*.

fungicide A toxic chemical that kills fungi.

fusion A nuclear reaction in which two smaller atoms are combined to make one larger atom with the release of a large amount of energy. Compare *fission*.

Gaia hypothesis The collective name for a series of hypotheses that Earth's organisms adjust the environment to keep it habitable for life.

game farming See *wildlife ranching*.

gas hydrates See *methane hydrates*.

gasohol See *alcohol fuel*.

gene A segment of DNA that serves as a unit of hereditary information.

genetically modified (GM) organism An organism that has had its genes intentionally manipulated.

genetic diversity Biological diversity that encompasses the genetic variety among individuals within a single species. Compare *ecosystem diversity* and *species richness*.

genetic engineering The ability to take a specific gene from one cell and place it into another cell where it is expressed.

genetic resistance An inherited characteristic that decreases the effect of a pesticide on a pest. Over time, the repeated exposure of a pest population to a pesticide causes an increase in the number of individuals that can tolerate the pesticide.

geothermal energy The natural heat within Earth that arises from ancient heat within Earth's core, from friction where continental plates slide over one another, and from decay of radioactive elements; geothermal energy can be used for space heating and to generate electricity.

geothermal heat pump (GHP) A device that heats and cools buildings by taking advantage of the difference in temperature between Earth's surface and subsurface at depths from about 1 to 100 m.

germplasm Any plant or animal material that may be used in breeding; includes seeds, plants, and plant tissues of traditional crop varieties and the sperm and eggs of traditional livestock breeds.

global commons Those resources of our environment that are available to everyone but for which no single individual has responsibility.

global distillation effect The process whereby volatile chemicals evaporate from land as far away as the tropics and are carried by air currents to higher latitudes, where they condense and fall to the ground.

grain stockpiles See *world grain carryover stocks*.

green architecture The practice of designing and building homes with environmental considerations such as energy efficiency, recycling, and conservation of natural resources in mind.

green chemistry See *environmental chemistry*.

green power Electricity produced from renewable sources such as solar power, wind, biomass, geothermal, and small hydroelectric plants.

green revolution The period of time during the 20th century when plant scientists developed genetically uniform, high-yielding varieties of important food crops such as rice and wheat.

greenhouse effect The natural global warming of our atmosphere caused by the presence of carbon dioxide and other greenhouse gases, which trap the sun's radiation somewhat like glass does in a greenhouse. The additional warming that may be produced by increased levels of greenhouse gases that absorb infrared radiation is known as the enhanced greenhouse effect.

greenhouse gases The gases that absorb infrared radiation, which include carbon dioxide, methane, nitrous oxide, chlorofluorocarbons, and tropospheric ozone, all of which are accumulating in the atmosphere as a result of human activities.

gross primary productivity The rate at which energy accumulates in an ecosystem (as biomass) during photosynthesis. Compare *net primary productivity*.

groundwater The supply of fresh water under Earth's surface. Groundwater is stored in underground caverns and porous layers of underground rock called aquifers. Compare *surface water*.

growth rate (r) The rate of change of a population's size, expressed in percent per year. In populations with little or no dispersal, it is calculated by subtracting the death rate from the birth rate. Also called *natural increase* in human populations.

gyre A circular, prevailing wind that generates circular ocean currents.

habitat The local environment in which an organism, population, or species lives.

habitat fragmentation The division of habitats that formerly occupied large, unbroken areas into smaller pieces by roads, fields, cities, and other land-transforming activities.

hadal benthic zone The ocean's benthic environment that extends from 6000 meters to the bottom of the deepest ocean trenches. Compare *abyssal benthic zone*.

hair cell One of a group of cells inside the cochlea of the ear that detects differences in pressure caused by sound waves.

half-life See *radioactive half-life*.

hard coal See *anthracite*.

hazardous air pollutants Air pollutants that are potentially harmful and may pose long-term health risks to people who live and work around chemical factories, incinerators, or other facilities that produce or use them. Also called *air toxics*.

hazardous waste   Any discarded chemical (solid, liquid, or gas) that threatens human health or the environment. Also called *toxic waste*.

herbicide   A toxic chemical that kills plants.

herbivore   An animal that feeds on plants or algae. Compare *carnivore* and *omnivore*. See *primary consumer*.

heterocysts   Oxygen-excluding cells of nitrogen-fixing cyanobacteria.

heterotroph   See *consumer*.

high-grade ore   An ore that contains relatively large amounts of a particular mineral. Compare *low-grade ore*.

high-input agriculture   See *industrialized agriculture*.

high-level radioactive waste   Any radioactive solid, liquid, or gas that initially gives off large amounts of ionizing radiation. Compare *low-level radioactive waste*.

highly developed country   An industrialized country that is characterized by a low fertility rate, low infant mortality rate, and high per-capita income. Highly developed countries include the United States, Canada, Japan, and European countries. Also called *developed country*. Compare *developing country*.

horizons   See *soil horizons*.

hormone   A chemical messenger produced by a living organism in minute quantities to regulate its growth, reproduction, and other important biological functions. Also see *endocrine disrupter*.

hot spot   (1) A rising plume of magma that flows out through an opening in the crust. (2) An area of great species diversity that is at risk of destruction by human activities.

humus   Black or dark brown decomposed organic material.

hydrocarbons   A diverse group of organic compounds that contain only hydrogen and carbon.

hydrogen bond   A bond between water molecules, formed when the negative (oxygen) end of one water molecule is attracted to the positive (hydrogen) end of another water molecule. Hydrogen bonding is the basis for many of water's physical properties.

hydrologic cycle   The water cycle, which includes evaporation, precipitation, and flow to the seas. The hydrologic cycle supplies terrestrial organisms with a continual supply of fresh water.

hydrology   The science that deals with Earth's waters, including the availability and distribution of fresh water.

hydropower   The energy of flowing or falling water; used to generate electricity.

hydrosphere   Earth's supply of water (both liquid and frozen, fresh and salty).

hydrothermal reservoir   Large underground reservoirs of hot water and possibly also steam, depending on the temperature and pressure of the fluid; some of the hot water or steam may escape to the surface, creating hot springs or geysers.

hydrothermal vent   A hot spring on the seafloor where a solution of hot, mineral-rich water rises to the surface. Many hydrothermal vents support thriving communities.

hypertension   High blood pressure.

hypothesis   An educated guess that might be true and is testable by observation and experimentation. Compare *theory*.

hypoxia   Low dissolved oxygen concentrations that occur in many bodies of water when nutrients stimulate the growth of algae that subsequently die and are decomposed by oxygen-using bacteria. Hypoxia often causes too little oxygen for other aquatic life.

illuviation   The deposition of material in the lower layers of soil from the upper layers; caused by leaching.

immigration   A type of dispersal in which individuals enter a population and thus increase its size. Compare *emigration*.

incentive-based regulation   Pollution control laws that work by establishing emission targets and providing industries with incentives to reduce emissions. Compare *command and control regulation*.

indicator species   A species that indicates particular conditions. For example, a large population of *Euglena* indicates polluted water.

inductive reasoning   Reasoning that uses specific examples to draw a general conclusion or discover a general principle. Compare *deductive reasoning*.

industrial ecosystem   A complex web of interactions among various industries, in which "wastes" produced by one industrial process are sold to other companies as raw materials. Finding profitable uses for a company's wastes is an extension of sustainable manufacturing. See *sustainable manufacturing*.

industrial smog   The traditional, London-type smoke pollution, which consists principally of sulfur oxides and particulate matter. Compare *photochemical smog*.

industrialized agriculture   Modern agricultural methods that require a large capital input and less land and labor than traditional methods. Industrialized agriculture uses large inputs of energy (from fossil fuels), mechanization, water, and agrochemicals (fertilizers and pesticides) to produce large crop and livestock yields. Also called *high-input agriculture*. Compare *sustainable agriculture* and *subsistence agriculture*.

infant mortality rate   The number of infant deaths per 1000 live births. (An infant is a child that is in its first year of life.)

infectious disease   A disease caused by a microorganism (such as a bacterium or fungus) or infectious agent (such as a virus). Infectious diseases can be transmitted from one individual to another.

infrared radiation   Electromagnetic radiation with wavelengths longer than visible light but shorter than microwaves. Humans perceive infrared radiation as invisible waves of heat energy.

inherent safety   See *principle of inherent safety*.

inorganic chemical   A chemical that does not contain carbon and is not associated with life. Inorganic chemicals that are pollutants include mercury compounds, road salt, and acid drainage from mines.

inorganic fertilizer   See *fertilizer*.

inorganic plant nutrient   A nutrient such as phosphate or nitrate that stimulates plant or algal growth. Excessive amounts of inorganic plant nutrients, which may come from animal wastes and plant residues as well as fertilizer runoff, can cause both soil and water pollution.

insecticide   A toxic chemical that kills insects.

in situ conservation   Conservation efforts that concentrate on preserving biological diversity in the wild. See *conservation biology*. Compare *ex situ conservation*.

integrated pest management (IPM)   A combination of pest control methods (biological, chemical, and cultivation) that, if used in the proper order and at the proper times, keep the size of a pest population low enough that it does not cause substantial economic loss.

integrated waste management   A combination of the best waste management techniques into a consolidated program to deal effectively with solid waste.

intercropping   A form of intensive subsistence agriculture that involves growing several crops simultaneously on the same field. See *polyculture*.

interspecific competition   Competition between members of different species. See *competition*. Compare *intraspecific competition*.

intertidal zone   The shoreline area between the low tide mark and the high tide mark.

intraspecific competition   Competition among members of the same species. See *competition*. Compare *interspecific competition*.

intrinsic rate of increase   See *biotic potential*.

invasive species   Foreign species whose introduction causes economic or environmental harm.

ionizing radiation   Radiation that contains enough energy to eject electrons from atoms, forming positively charged ions. Ionizing radiation can damage living tissue.

IPAT equation   A model that shows the mathematical relationship between environmental impacts and the forces that drive them (number of people, affluence per person, and environmental effects of technologies used to obtain and consume resources).

isotope   An alternate form of the same element that has a different atomic mass. That is, an isotope has a different number of neutrons but the same number of protons and electrons.

***K* selection**   A reproductive strategy in which a species typically has a large body size, slow development, long life span, and does not devote a large proportion of its metabolic energy to the production of offspring. Compare *r selection.*

kelps   Large brown algae that are common in relatively shallow, cooler temperate marine water along rocky coastlines.

keystone species   A species that is crucial in determining the nature and structure of the entire ecosystem in which it lives; other species of a community depend on or are greatly affected by the keystone species, whose influence is much greater than would be expected by its relative abundance.

kilocalorie   A unit of heat energy equivalent to the energy required to raise the temperature of 1 kilogram of water by 1°C.

kilojoule   A unit of work energy; one kilocalorie equals 4.184 kilojoules.

kinetic energy   The energy of a body that results from its motion. Compare *potential energy.*

krill   Tiny shrimplike animals that are important in the Antarctic food web.

kwashiorkor   Malnutrition, most common in infants and in young children, that results from protein deficiency. Compare *marasmus.*

Kyoto Protocol   An international treaty that stipulates that highly developed countries must cut their emissions of CO_2 and other gases that cause climate warming by an average of 5.2% by 2012; does not go into effect until at least 55% of the countries involved have ratified it.

landscape   A large area that includes one or more ecosystems.

landscape ecology   A sub-discipline in ecology that focuses on connections among ecosystems in a particular area.

La Niña   A periodic occurrence in the eastern Pacific Ocean in which the surface water temperature becomes unusually cool, and westbound trade winds become unusually strong; La Niña often occurs after an El Niño event. Compare *El Niño–Southern Oscillation.*

latitude   The distance, measured in degrees north or south, from the equator.

lava   Magma (molten rock) that reaches Earth's surface.

leachate   The liquid that seeps through solid waste at a sanitary landfill or other waste disposal site.

leaching   The process by which dissolved materials (nutrients or contaminants) are washed away or filtered down through the various layers of the soil.

less developed country   A developing country with a low level of industrialization, a very high fertility rate, a very high infant mortality rate, and a very low per-capita income (relative to highly developed countries). Compare *moderately developed country* and *highly developed country.*

lethal dose-50% (LD$_{50}$)   In toxicology, the dose (usually reported in mg of chemical toxicant per kg of body weight) that is lethal to 50% of a population of test animals.

life index of world reserves   An estimate of the time remaining before the known reserves of a particular nonrenewable resource are expended (at current rates of extraction and use).

lignite   A grade of coal that is brown or brown-black and has a soft, woody texture (softer than bituminous coal). Compare *anthracite, bituminous coal,* and *subbituminous coal.*

lime scrubber   An air pollution control device in which a chemical spray neutralizes acidic gases. See *scrubbers.*

limiting factor   An environmental factor that restricts the growth, distribution, or abundance of a particular population; limiting factors restrict an organism's ecological niche.

limnetic zone   The open-water area away from the shore of a lake or pond that extends down as far as sunlight penetrates (and therefore photosynthesis occurs). Compare *littoral zone* and *profundal zone.*

lipid   A diverse group of organic molecules that are metabolized by cellular respiration to provide the body with a high level of energy; commonly called fats and oils.

liquefied petroleum gas   A mixture of liquefied propane and butane. Liquefied petroleum gas is stored in pressurized tanks.

lithosphere   The soil and rock of Earth's crust.

littoral zone   The shallow-water area along the shore of a lake or pond. Compare *limnetic zone* and *profundal zone.*

loam   An ideal agricultural soil that has an optimum combination of different soil particle sizes: approximately 40% each of sand and silt, and about 20% of clay.

low-grade ore   An ore that contains relatively small amounts of a particular mineral. Compare *high-grade ore.*

low-input agriculture   See *sustainable agriculture.*

low-level radioactive waste   Any radioactive solid, liquid, or gas that gives off small amounts of ionizing radiation. Compare *high-level radioactive waste.*

magma   Molten rock formed within Earth. Compare *lava.*

malnutrition   A condition caused when a person does not receive enough specific essential nutrients in the diet. See *marasmus* and *kwashiorkor.* Compare *overnutrition* and *undernutrition.*

manganese nodule   A small (potato-sized) rock that contains manganese and other minerals. Manganese nodules are common on parts of the ocean floor.

mangrove forest   Swamps of mangrove trees that grow along many tropical coasts.

marasmus   A condition of progressive emaciation that is especially common in children and is caused by a diet low in both total calories and protein. Compare *kwashiorkor.*

marginal cost of pollution   The cost in environmental damage of a unit of pollution emitted into the environment.

marginal cost of pollution abatement   The cost to dispose of a unit of pollution in a nonpolluting way.

mariculture   The rearing of marine organisms (fishes, seaweeds, and shellfish) for human consumption; a subset of aquaculture.

marine snow   Organic debris that drifts into the darkened regions of the oceanic province from the upper, lighted regions.

marketable waste-discharge permit   A government policy that controls pollution by issuing permits allowing the holder to pollute a given amount. Holders are not allowed to produce more emissions than the permit allows. See *emission reduction credit.*

marsh   A wetland that is treeless and dominated by grasses; freshwater marshes are found inland along lakes and rivers, and salt marshes are found in bays and rivers near the ocean or in protected coastal areas. See *salt marsh.*

mass extinction   The extinction of numerous species during a relatively short period of geological time. Compare *background extinction.*

maximum contaminant level   The maximum permissible amount (by law) of a water pollutant that might adversely affect human health.

meltdown   The melting of a nuclear reactor vessel. A meltdown would cause the release of a substantial amount of radiation into the environment. See *reactor vessel.*

mesosphere   The layer of the atmosphere between the stratosphere and the thermosphere. It is characterized by the lowest atmospheric temperatures.

metal   An element that is malleable, lustrous, and a good conductor of heat and electricity. See *mineral.* Compare *nonmetal.*

methane   The simplest hydrocarbon, CH_4, which is an odorless, colorless, flammable gas.

methane hydrates   Reserves of ice-encrusted natural gas located in porous rock in the arctic tundra (under the permafrost) and in the deep ocean sediments of the continental slope and ocean floor. Also called *gas hydrates.*

methanol   A colorless, flammable liquid, CH_3OH. Also called *methyl alcohol.*

methyl alcohol   See *methanol*.

microclimate   Local variations in climate produced by differences in elevation, in the steepness and direction of slopes, and in exposure to prevailing winds.

microirrigation   A type of irrigation that conserves water. In microirrigation, pipes with tiny holes bored into them convey water directly to individual plants. Also called *drip* or *trickle irrigation*.

mineral   (1) An element, inorganic compound, or mixture that occurs naturally in Earth's crust. See *metal* and *nonmetal*. (2) Inorganic nutrients ingested (by animals) or absorbed (by plants) in the form of salts. Compare *vitamin*.

mineral reserve   A mineral deposit that has been identified and is currently profitable to extract. Compare *mineral resource*. See *total resources*.

mineral resource   Any undiscovered mineral deposits or known deposits of low-grade ore that are currently unprofitable to extract. Mineral resources are potential resources that may be profitable to extract in the future. Compare *mineral reserve*. See *total resources*.

model   (1) A formal statement that describes a situation and can be used to understand the present or predict the future course of events. (2) A simulation, using powerful computers, that represents the overall effect of competing factors to describe an environmental situation in numerical terms.

moderately developed country   A developing country with a medium level of industrialization, a high fertility rate, a high infant mortality rate, and a low per-capita income (all relative to highly developed countries). Compare *less developed country* and *highly developed country*.

monoculture   The cultivation of only one type of plant over a large area. Compare *polyculture*.

Montreal Protocol   International negotiations that resulted in a timetable to phase out CFC production.

mortality   See *death rate*.

mulch   Material placed on the surface of soil around the bases of plants. A mulch helps to maintain soil moisture and reduce soil erosion. Organic mulches have the additional advantage of decomposing over time, thereby enriching the soil.

municipal solid waste   Solid waste generated in homes, office buildings, retail stores, restaurants, schools, hospitals, prisons, libraries, and other commercial and institutional facilities. Compare *nonmunicipal solid waste*.

municipal solid waste composting   The large-scale composting of the organic portion of a community's solid waste.

mutation   A change in the DNA (a gene) of an organism. A mutation in reproductive cells may be passed on to the next generation, where it may result in birth defects or genetic disease.

mutualism   A symbiotic relationship in which both partners benefit from the association. See *symbiosis*. Compare *parasitism* and *commensalism*.

mycorrhizae   A mutualistic association between a fungus and the roots of a plant. Most plants form mycorrhizal associations with fungi, which enables plants to absorb adequate amounts of essential minerals from the soil.

narrow-spectrum pesticide   An "ideal" pesticide that only kills the organism for which it is intended and does not harm any other species. Compare *wide-spectrum pesticide*.

natality   See *birth rate*.

national emission limitations   The maximum permissible amount (by law) of a particular pollutant that can be discharged into the nation's rivers, lakes, and oceans from point sources.

National Environmental Policy Act (NEPA)   A U.S. federal law that is the cornerstone of U.S. environmental policy. It requires that the federal government consider the environmental impact of any construction project funded by the federal government.

natural capital   Earth's resources and processes that sustain living organisms, including humans.

natural gas   A mixture of gaseous hydrocarbons (primarily methane) that occurs, often with oil deposits, in Earth's crust. See *fossil fuel*.

natural increase   See *growth rate*.

natural regulation   A controversial park management policy that involves letting nature take its course most of the time, with corrective actions undertaken as needed to adjust for changes caused by pervasive human activities.

natural resources   See *resources*.

natural selection   The tendency of organisms that possess favorable adaptations to their environment to survive and become the parents of the next generation; evolution occurs when natural selection results in genetic changes in a population. Natural selection is the mechanism of evolution first proposed by Charles Darwin.

negative externality   See *external cost*.

negative feedback loop   A mechanism in which a change in some condition triggers a response that counteracts, or reverses, the changed condition. Compare *positive feedback loop*.

nekton   Relatively strong-swimming aquatic organisms such as fish and turtles. Compare *plankton* and *benthos*.

neo-Malthusians   Economists who hold that developmental efforts are hampered by a rapidly expanding population. Compare *pronatalist*.

neritic province   Open ocean from the shoreline to a depth of 200 meters. Compare *oceanic province*.

nest parasitism   A behavior practiced by brown-headed cowbirds and certain other birds in which females lay their eggs in the nests of other bird species and leave all parenting jobs to the hosts.

net primary productivity   Energy that remains in an ecosystem (as biomass) after cellular respiration has occurred. Compare *gross primary productivity*.

niche   The totality of an organism's adaptations, its use of resources, and the lifestyle to which it is fitted. The niche describes how an organism uses materials in its environment as well as how it interacts with other organisms. Also called *ecological niche*. See *fundamental niche* and *realized niche*.

nitrification   The conversion of ammonia (NH_3) and ammonium ions (NH_4^+) to nitrate (NO_3^-) by certain bacteria (nitrifying bacteria) in the soil; part of the *nitrogen cycle*.

nitrogen cycle   The worldwide circulation of nitrogen from the abiotic environment into organisms and back into the abiotic environment.

nitrogen fixation   The conversion of atmospheric nitrogen (N_2) to ammonia (NH_3) by nitrogen-fixing bacteria and cyanobacteria; part of the *nitrogen cycle*.

noise pollution   A loud or disagreeable sound, particularly when it results in physiological or psychological harm.

nomadic herding   A traditional grazing method in which nomadic herdsmen wander freely over rangelands in search of good grazing for their livestock.

nondegradable   A chemical pollutant (such as the toxic elements mercury and lead) that cannot be decomposed (broken down) by organisms or by other natural processes. Compare *biodegradable*.

nonmetal   A mineral that is nonmalleable, nonlustrous, and a poor conductor of heat and electricity. See *mineral*. Compare *metal*.

nonmunicipal solid waste   Solid waste generated by industry, agriculture, and mining. Compare *municipal solid waste*.

nonpoint source pollution   Pollutants that enter bodies of water over large areas rather than being concentrated at a single point of entry. Examples include agricultural fertilizer runoff and sediments from construction. Also called *polluted runoff*. Compare *point source pollution*.

nonrenewable resources   Natural resources that are present in limited supplies and are depleted by use; include minerals such as copper and tin and fossil fuels such as oil and natural gas. Compare *renewable resources*.

nonurban land   See *rural land*.

nonylphenol   A chemical commonly found in cleaning products, detergents, and industrial solvents that may be an endocrine disrupter.

no-tillage   A method of conservation tillage that leaves both the surface and subsurface soil undisturbed. Special machines punch holes in the soil for seeds. See *conservation tillage*. Compare *conventional tillage*.

nuclear energy   The energy released from the nucleus of an atom in a nuclear reaction (fission or fusion) or during radioactive decay.

nuclear fuel cycle   The processes involved in producing the fuel used in *nuclear reactors* and in disposing of radioactive wastes (also called nuclear wastes).

nuclear reactor   A device that initiates and maintains a controlled nuclear fission chain reaction in order to produce energy (electricity).

nutrient cycling   The pathways of various nutrient minerals or elements from the environment through organisms and back to the environment; *biogeochemical cycles* are examples of nutrient cycling.

obligate parasite   An organism that can only exist as a parasite. Compare *facultative parasite*.

oceanic province   That part of the open ocean that is deeper than 200 meters and comprises most of the ocean. Compare *neritic province*.

ocean temperature gradient   The differences in temperature at various ocean depths.

Ogallala Aquifer   A massive groundwater deposit under eight midwestern states, including Nebraska, Kansas, Oklahoma, and Texas.

O-horizon   The uppermost layer of certain soils, composed of dead leaves and other organic matter.

oil   See *petroleum*.

oil sand   See *tar sand*.

oil shales   Sedimentary "oily rocks" that contain a mixture of hydrocarbons known as kerogen; oil shales must be crushed, heated to high temperatures, and refined after they are mined to yield oil.

oligotrophic lake   A deep, clear lake that has minimal nutrients. Water in an oligotrophic lake contains a high level of dissolved oxygen. Compare *eutrophic lake*.

omnivore   An animal that eats a variety of plant and animal material. Compare *herbivore* and *carnivore*.

oncogene   Any of a number of genes that usually are involved in cell growth or division and that cause the formation of a cancer cell when mutated.

open management   A policy in which all fishing boats of a particular country are given unrestricted access to fish in their national waters.

open-pit mining   A type of surface mining in which a giant hole (quarry) is dug to extract iron, copper, stone, or gravel. See *surface mining*. Compare *strip mining*.

open system   A system that exchanges energy with its surroundings. Compare *closed system*.

optimum amount of pollution   The amount of pollution that is economically most desirable. It is determined by plotting two curves, the marginal cost of pollution and the marginal cost of pollution abatement; the point where the two curves meet is the optimum amount of pollution.

ore   Rock that contains a large enough concentration of a particular mineral that it can be profitably mined and extracted.

organic agriculture   Growing crops and livestock without the use of synthetic pesticides or commercial inorganic fertilizers. Organic agriculture makes use of natural organic fertilizers (such as manure and compost) and chemical-free methods of pest control.

organic chemical   A compound that contains the element carbon and is either naturally occurring (in organisms) or synthetic (manufactured by humans). Many synthetic organic compounds persist in the environment for an extended period of time, and some are toxic to organisms.

organic fertilizer   See *fertilizer*.

organophosphate   A synthetic organic compound that contains phosphorus and is very toxic; used as an insecticide.

overburden   Overlying layers of soil and rock over mineral deposits. The overburden is removed during surface mining.

overgrazing   The destruction of an area's vegetation that occurs when too many animals graze on the vegetation, consuming so much of it that it does not recover.

overnutrition   A condition caused by eating food in excess of that required to maintain a healthy body. Compare *malnutrition* and *undernutrition*.

overpopulation   A situation in which a country or geographical area has more people than its resource base can support without damaging the environment. See *people overpopulation* and *consumption overpopulation*.

oxide   A compound in which oxygen is chemically combined with some other element.

ozone   A blue gas, O_3, with a distinctive odor. Ozone is a human-made pollutant in one part of the atmosphere (the troposphere) but a natural and essential component in another (the stratosphere).

parasitism   A symbiotic relationship in which one member (the parasite) benefits and the other (the host) is adversely affected. See *symbiosis*. Compare *mutualism* and *commensalism*.

particulate matter   Solid particles and liquid droplets suspended in the atmosphere.

parts per billion   The number of molecules of a particular substance found in one billion molecules of air, water, or some other material. Abbreviated ppb.

parts per million   The number of molecules of a particular substance found in one million molecules of air, water, or some other material. Abbreviated ppm.

passive solar heating   A system that uses the sun's energy without requiring mechanical devices (pumps or fans) to distribute the collected heat. Compare *active solar heating*.

pathogen   An agent (usually a microorganism) that causes disease.

PCBs   Chlorine-containing organic compounds that enjoyed a wide variety of industrial uses until their dangerous properties, including the fact that they are slow to degrade and therefore persist in the environment, were recognized.

people overpopulation   A situation in which there are too many people in a given geographical area. Even if those people use few resources per person (the minimum amount they need to survive), people overpopulation results in pollution, environmental degradation, and resource depletion. See *overpopulation*. Compare *consumption overpopulation*.

perennial   A plant that lives for more than two years. Compare *annual* and *biennial*.

permafrost   Permanently frozen subsoil characteristic of frigid areas such as the tundra.

persistence   A characteristic of certain chemicals that are extremely stable and may take many years to be broken down into simpler forms by natural processes. Certain pesticides, for example, exhibit persistence and remain unaltered in the environment for years.

persistent organic pollutants (POPs)   A group of persistent, toxic chemicals that bioaccumulate in organisms and can travel long distances through air and water to contaminate sites far removed from their source; some disrupt the endocrine system, cause cancer, or adversely affect the developmental processes of organisms.

pest   Any organism that interferes in some way with human welfare or activities.

pesticide   Any toxic chemical used to kill pests. See *fungicide, herbicide, insecticide,* and *rodenticide*.

pesticide treadmill   A predicament faced by pesticide users, in which the cost of applying pesticides increases because they have to be applied more frequently or in larger doses, while their effectiveness decreases as a result of genetic resistance in the target pest.

petrochemicals   Chemicals, obtained from crude oil, that are used in the production of such diverse products as fertilizers, plastics, paints, pesticides, medicines, and synthetic fibers.

petroleum   A thick, yellow to black, flammable liquid hydrocarbon mixture found in Earth's crust. When petroleum is refined, the mixture is separated into a number of hydrocarbon compounds, including gasoline, kerosene, fuel oil, lubricating oils, paraffin, and asphalt. Also called *crude oil*. See *fossil fuel*.

pH   A number from 0 to 14 that indicates the degree of acidity or basicity (alkalinity) of a substance.

pheromone   A substance secreted by one organism into the environment that influences the development or behavior of other members of the same species.

phosphorus cycle   The worldwide circulation of phosphorus from the abiotic environment into organisms and back into the abiotic environment.

photochemical smog   A brownish orange haze formed by complex chemical reactions involving sunlight, nitrogen oxides, and hydrocarbons. Some of the pollutants in photochemical smog include peroxyacyl nitrates (PANs), ozone, and aldehydes. Compare *industrial smog*.

photodegradable   Breaking down upon exposure to sunlight.

photosynthesis   The biological process that captures light energy and transforms it into the chemical energy of organic molecules (such as glucose), which are manufactured from carbon dioxide and water. Photosynthesis is performed by plants, algae, and several kinds of bacteria. Compare *chemosynthesis*.

photovoltaic (PV) solar cell   A wafer or thin-film device that generates electricity when solar energy is absorbed.

phytoplankton   Microscopic floating algae that are the base of most aquatic food chains. See *plankton*. Compare *zooplankton*.

phytoremediation   A method employed to clean up a hazardous waste site that uses plants to absorb and accumulate toxic materials. Compare *bioremediation*.

pioneer community   The first organisms (such as lichens or mosses) to colonize an area and begin the first stage of ecological succession. See *succession*.

plankton   Small or microscopic aquatic organisms that are relatively feeble swimmers and thus, for the most part, are carried about by currents and waves. Composed of *phytoplankton* and *zooplankton*. Compare *nekton* and *benthos*.

plasma   An ionized gas formed at very high temperatures when electrons are stripped from the gas atoms; plasma is formed during fusion reactions.

plate boundary   Any area where two tectonic plates meet. Plate boundaries are often sites of intense geological activity. See *fault*.

plate tectonics   The theory that explains how Earth's crustal plates move and interact at their boundaries.

point source pollution   Water pollution that can be traced to a specific spot (such as a factory or sewage treatment plant) because it is discharged into the environment through pipes, sewers, or ditches. Compare *nonpoint source pollution*.

polar easterly   A prevailing wind that blows from the northeast near the North Pole or from the southeast near the South Pole.

polluted runoff   See *nonpoint source pollution*.

pollution   Any undesirable alteration of air, water, or soil that harms the health, survival, or activities of humans and other living organisms.

polychlorinated biphenyls   See *PCBs*.

polyculture   A type of intercropping in which several kinds of plants that mature at different times are planted together. See *intercropping*. Compare *monoculture*.

polymer   A compound composed of repeating subunits.

population   A group of organisms of the same species that live in the same geographical area at the same time.

population density   The number of individuals of a species per unit of area or volume at a given time.

population ecology   That branch of biology that deals with the numbers of a particular species that are found in an area and how and why those numbers change (or remain fixed) over time.

population growth momentum   The continued growth of a population after fertility rates have declined, as a result of a population's young age structure; population growth momentum can be either positive or negative but is usually discussed in a positive context.

population momentum   See *population growth momentum*.

positive feedback loop   A mechanism in which a change in some condition triggers a response that intensifies the changing condition. Compare *negative feedback loop*.

potential energy   Stored energy that is the result of the relative position of matter instead of its motion. Compare *kinetic energy*.

poverty   A condition in which people are unable to meet their basic needs for adequate food, clothing, or shelter.

predation   The consumption of one species (the prey) by another (the predator); includes both animals eating other animals and animals eating plants.

preservation   Setting aside undisturbed areas, maintaining them in a pristine state, and protecting them from human activities that might alter the natural state. Compare *conservation*.

preservative   A chemical added to food to retard the growth of bacteria and fungi that cause food spoilage.

prevailing wind   A major surface wind that blows more or less continually.

primary air pollutant   A harmful chemical that enters directly into the atmosphere from either human activities or natural processes (such as volcanic eruptions). Compare *secondary air pollutant*.

primary consumer   An organism that consumes producers. Also called *herbivore*. Compare *secondary consumer*.

primary sludge   A slimy mixture of bacteria-laden solids that settles out from sewage wastewater during primary treatment.

primary succession   An ecological succession that occurs on land that has not previously been inhabited by plants; no soil is present initially. Compare *secondary succession*.

primary treatment   Treating wastewater by removing suspended and floating particles (such as sand and silt) by mechanical processes (such as screens and physical settling). Compare *secondary treatment* and *tertiary treatment*.

principle of inherent safety   Chemical safety programs that stress accident prevention by redesigning industrial processes to involve less toxic materials (so that dangerous accidents are prevented).

producer   An organism (such as a chlorophyll-containing plant) that manufactures complex organic molecules from simple inorganic substances. In most ecosystems, producers are photosynthetic organisms. Also called *autotroph*. Compare *consumer*.

product stewardship   The concept in which manufacturers assume responsibility for their products from cradle to grave; encourages optimal recycling when products are returned to manufacturers.

profundal zone   The deepest zone of a large lake. Compare *limnetic zone* and *littoral zone*.

pronatalists   Those who are in favor of population growth. Compare *neo-Malthusians*.

protein   A large, complex organic molecule composed of amino acid subunits; proteins are the principal structural components of cells.

pyramid of biomass   An ecological pyramid that illustrates the total biomass (for example, the total dry weight of all organisms) at each successive trophic level in an ecosystem. See *ecological pyramid*. Compare *pyramid of energy* and *pyramid of numbers*.

pyramid of energy   An ecological pyramid that shows the energy flow through each trophic level in an ecosystem. See *ecological pyramid*. Compare *pyramid of biomass* and *pyramid of numbers*.

pyramid of numbers   An ecological pyramid that shows the number of organisms at each successive trophic level in a given ecosystem. See *ecological pyramid*. Compare *pyramid of biomass* and *pyramid of energy*.

quarantine   Practice in which the importation of exotic plant and animal material that might be harboring pests is restricted.

r **selection**   A reproductive strategy in which a species typically has a small body size, rapid development, short life span, and devotes a large proportion of its metabolic energy to the production of offspring. Compare *K selection*.

radiation   The emission of fast-moving particles or rays of energy from the nuclei of radioactive atoms.

radioactive atoms   Atoms of unstable isotopes that spontaneously emit radiation.

radioactive decay   The process in which a radioactive element emits radiation and, as a result, its nucleus changes into the nucleus of a different element.

radioactive half-life   The period of time required for one-half of a radioactive substance to change into a different material.

radioisotope   An unstable isotope that spontaneously emits radiation.

radon   A colorless, tasteless, radioactive gas produced during the radioactive decay of uranium in Earth's crust.

rain shadow   An area on the downwind side of a mountain range with very little precipitation. Deserts often occur in rain shadows.

range   The area of Earth in which a particular species occurs.

reactor vessel   A huge steel potlike structure encasing the uranium fuel in a nuclear reactor. The reactor vessel is a safety feature designed to prevent the accidental release of radiation into the environment.

realized niche   The life style that an organism actually pursues, including the resources that it actually uses. An organism's realized niche is narrower than its fundamental niche because of competition from other species. See *niche*. Compare *fundamental niche*.

reclaimed water   Treated wastewater that is reused in some way, such as for irrigation, manufacturing processes that require water for cooling, wetlands restoration, or groundwater recharge.

recycling   Conservation of the resources in used items by converting them into new products. For example, used aluminum cans are recycled by collecting, remelting, and reprocessing them into new cans. Compare *reuse*.

red tide   A red, orange, or brown coloration of water caused by a bloom, or population explosion, of algae; many red tides cause serious environmental harm and threaten the health of humans and animals.

renewable resources   Resources that are replaced by natural processes and can be used forever, provided they are not overexploited in the short term. Examples include fresh water in lakes and rivers, fertile soil, and trees in forests. Compare *nonrenewable resources*.

replacement-level fertility   The average number of children a couple must produce in order to "replace" themselves; the number is greater than two because some children die before reaching reproductive age.

reprocessing   Reusing highly radioactive spent nuclear fuel by chemically separating the unused uranium and plutonium from other radioactive waste products.

reservoir   An artificial lake produced by building a dam across a river or stream; allows water to be stored for use.

resistance management   A relatively new approach to dealing with genetic resistance so that the period of time in which a pesticide is useful is maximized; involves efforts to delay the development of genetic resistance in pests.

resource partitioning   The reduction in competition for environmental resources, such as food, that occurs among coexisting species as a result of each species' niche differing from the others in one or more ways.

resource recovery   The process of removing any material—sulfur or metals, for example—from polluted emissions or solid waste and selling it as a marketable product.

resources   (1) Any parts of the natural environment that are used to promote the welfare of people or other species; examples include clean air, fresh water, soil, forests, minerals, and organisms. Also called *natural resources*. (2) Anything from the environment that meets a particular species' needs.

response   In toxicology, the type and amount of damage caused by exposure to a particular dose.

restoration ecology   The field of science in which the principles of ecology are used to help return a degraded environment as close as possible to its former, undisturbed state. Also called *ecological restoration*.

reuse   Conservation of the resources in used items by using them over and over again. For example, glass bottles can be collected, washed, and refilled again. Compare *recycling*.

reverse osmosis   A desalinization process that involves forcing saltwater through a membrane permeable to water but not to salt. Compare *distillation*.

riparian area   The thin patch of vegetation along the bank of a stream or river that interfaces between terrestrial and aquatic habitats; protects the habitat of salmon, trout, and other aquatic species from sedimentation caused by soil erosion. Also called *riparian buffer*.

riparian buffer   See *riparian area*.

risk   The probability of harm—such as injury, disease, death, or environmental damage—occurring under certain circumstances.

risk assessment   Using statistical methods to quantify the harmful effects on human health or the environment of exposure to a particular danger. Risk assessments are most useful when they are compared with one another. See *ecological risk assessment*.

risk management   Determining whether there is a need to reduce or eliminate a particular risk and, if so, what should be done. Based on data from risk assessment as well as political, economic, and social considerations.

rock   A naturally formed aggregate, or mixture, of minerals; rocks have varied chemical compositions.

rodenticide   A toxic chemical that kills rodents.

runoff   The movement of fresh water from precipitation and snowmelt to rivers, lakes, wetlands, and ultimately, the ocean.

rural land   Sparsely populated areas, such as forests, grasslands, deserts, and wetlands. Also called *nonurban land*.

salinity   The concentration of dissolved salts (such as sodium chloride) in a body of water.

salinity gradient   The difference in salt concentrations that occurs at different depths in the ocean and at different locations in estuaries.

salinization   The gradual accumulation of salt in a soil, often as a result of improper irrigation methods. Most plants cannot grow in salinized soil.

salt marsh   An estuarine wetland dominated by salt-tolerant grasses.

saltwater intrusion   The movement of seawater into a freshwater aquifer located near the coast; caused by groundwater depletion.

salvage logging   A controversial logging method in which trees that are weakened by insects, disease, or fire are harvested. Forestry scientists disprove of salvage logging, particularly in fire-damaged areas, and say it delays or prevents the forest from recovering.

sand   Inorganic soil particles that are larger than clay or silt. Compare *clay* and *silt*.

sanitary landfill   The most common method of disposal of solid waste by compacting it and burying it under a shallow layer of soil.

saprotroph   See *decomposer*.

savanna   A tropical grassland with widely scattered trees or clumps of trees; found in areas of low rainfall or seasonal rainfall with prolonged dry periods.

science   A human endeavor that seeks to reduce the apparent complexity of the natural world to general principles that can be used to make predictions, solve problems, or provide new insights.

scientific method   The way a scientist approaches a problem (by formulating a hypothesis and then testing it by means of an experiment).

scout-and-spray   The use of pesticides that are applied only when pests become a problem; requires continual monitoring. Compare *calendar spraying*.

scrubbers   Desulfurization systems that are used in smokestacks to decrease the amount of sulfur released in the air by 90% or more. One type is a *lime scrubber*.

sea grasses   Flowering plants that grow in quiet, shallow ocean water in temperate, subtropical, and tropical waters.

secondary air pollutant   A harmful chemical that forms in the atmosphere when a primary air pollutant reacts chemically with other air pollutants or natural components of the atmosphere. Compare *primary air pollutant.*

secondary consumer   An organism that consumes primary consumers. Also called *carnivore.* Compare *primary consumer.*

secondary sludge   A slimy mixture of bacteria-laden solids that settles out from sewage wastewater during secondary treatment.

secondary succession   An ecological succession that takes place after some disturbance destroys the existing vegetation; soil is already present. Compare *primary succession.*

secondary treatment   Treating wastewater biologically, by using microorganisms to decompose the suspended organic material; occurs after primary treatment. Compare *primary treatment* and *tertiary treatment.*

second law of thermodynamics   When energy is converted from one form to another, some of it is degraded into a lower quality, less useful form. Thus, with each successive energy transformation, less energy is available to do work. Compare *first law of thermodynamics.*

sediment pollution   Excessive amounts of soil particles that enter the water as a result of erosion.

sedimentation   (1) Letting solids settle out of wastewater by gravity during primary treatment. (2) The process in which eroded particles are transported by water and deposited as sediment on river deltas and the sea floor. If exposed to sufficient heat and pressure, sediments can solidify into sedimentary rock.

seed-tree cutting   A forest management technique in which almost all trees are harvested from an area in a single cutting, but a few desirable trees are left behind to provide seeds for the regeneration of the forest. Compare *clearcutting, selective cutting,* and *shelterwood cutting.*

seismic waves   Vibrations that travel through rock as a result of an earthquake.

selective cutting   A forest management technique in which mature trees are cut individually or in small clusters while the rest of the forest remains intact so that the forest can regenerate quickly (and naturally). Compare *clearcutting, seed-tree cutting,* and *shelterwood cutting.*

semi-arid land   Land that receives more precipitation than a desert but is subject to frequent and prolonged droughts. Compare *arid land.*

sentinel species   See *bellwether species.*

sewage   Wastewater carried off by drains or sewers.

sewage sludge   See *primary sludge* and *secondary sludge.*

shelterbelt   A row of trees planted as a windbreak to reduce soil erosion of agricultural land.

shelterwood cutting   A forest management technique in which all mature trees in an area are harvested in a series of partial cuttings over a period of time. (Typically two or three harvests are made over a decade.) Compare *clearcutting, seed-tree cutting,* and *selective cutting.*

shifting cultivation   A traditional form of subsistence agriculture in which short periods of cultivation are followed by longer periods of fallow (land left uncultivated), during which times the natural ecosystems may become reestablished. See *slash-and-burn agriculture.*

sick building syndrome   Eye irritations, nausea, headaches, respiratory infections, depression, and fatigue caused by the presence of air pollution inside office buildings.

silt   Medium-sized inorganic soil particles. Compare *clay* and *sand.*

sink   In environmental science, the part of the natural environment that receives an input of materials; economies depend on sinks for waste products. Compare *source.*

sinkhole   A large surface cavity or depression when an underground cave roof has collapsed; occurs most frequently when a drought or excessive pumping causes a lowering of the water table.

slag   See *bottom ash.*

slash-and-burn agriculture   A type of shifting cultivation in tropical forests in which a patch of vegetation is burned, leaving nutrient minerals in the ash. Crops are planted for a few years until the soil is depleted of nutrient minerals, after which the land must lie fallow for many years to recover. See *shifting agriculture.*

smelting   Process in which ore is melted at high temperatures to help separate impurities from the molten metal.

smog   Air pollution caused by a variety of pollutants. See *industrial smog* and *photochemical smog.*

soft coal   See *bituminous coal.*

soil   The uppermost layer of Earth's crust, which supports terrestrial plants, animals, and microorganisms. Soil is a complex mixture of inorganic minerals (from the parent material), organic material, water, air, and organisms.

soil erosion   The wearing away or removal of soil from the land; caused by wind and flowing water. Although soil erosion occurs naturally from precipitation and runoff, human activities (such as clearing land) accelerate it.

soil horizons   The horizontal layers into which many soils are organized. May include the O-horizon (surface litter), A-horizon (topsoil), E-horizon, B-horizon (subsoil), and C-horizon (partly weathered parent material).

soil pollution   Any physical or chemical change in soil that adversely affects the health of plants and other organisms living in and on it.

soil profile   A vertical section through the soil, from the surface to the parent material, that reveals the soil horizons.

soil remediation   Use of one or more techniques to remove contaminants from the soil.

soil salinization   See *salinization.*

solar energy   Energy from the sun. Solar energy includes both direct solar radiation and indirect solar energy (such as wind, hydropower, and biomass).

solar thermal electric generation   A means of producing electricity in which the sun's energy is concentrated by mirrors or lenses onto a fluid-filled pipe; the heated fluid is used to generate electricity.

solid waste   See *municipal solid waste* and *nonmunicipal solid waste.*

sound   Vibrations in the air (or some other medium) that reach the ears and stimulate a sensation of hearing.

source   In environmental science, the part of the environment from which materials move; economies depend on sources for raw materials. Compare *sink.*

source reduction   An aspect of waste management in which products are designed and manufactured in ways that decrease not only the volume of solid waste but also the amount of hazardous materials in the solid waste that remains.

species   A group of similar organisms that are able to interbreed with one another but unable to interbreed with other sorts of organisms.

species richness   Biological diversity that encompasses the number of species in an area (or community). Compare *genetic diversity* and *ecosystem diversity.*

spent fuel   The used fuel elements that were irradiated in a nuclear reactor.

spoil bank   A hill of loose rock created when the overburden from a new trench is put into the old (already excavated) trench during strip mining.

spring turnover   A mixing of the lake waters in temperate lakes that occurs in spring as ice melts and the surface water reaches 4°C, its temperature of greatest density. Compare *fall turnover.*

stable runoff   The share of runoff from precipitation available throughout the year. Most geographical areas have a heavy runoff during a few months (the spring months, for example) when precipitation and snowmelt are highest. Stable runoff is the amount that can be depended on every month.

standing-water ecosystem   A body of freshwater that is surrounded by land and whose water does not flow; a lake or pond. Compare *flowing-water ecosystem.*

sterile male technique A method of insect control that involves rearing, sterilizing, and releasing large numbers of males of the pest species.

stewardship The idea that humans share responsibility for the sustainable care and management of our planet.

strata Layers of rock.

Strategic Petroleum Reserve An emergency supply of up to one billion barrels of oil that is stored in underground salt caverns along the coast of the Gulf of Mexico, as mandated by the U.S. Energy Policy and Conservation Act.

stratosphere The layer of the atmosphere between the troposphere and the mesosphere. It contains a thin ozone layer that protects life by filtering out much of the sun's ultraviolet radiation.

stressor See *environmental stressor.*

strip cropping A type of contour plowing that produces alternating strips of different crops that are planted along the natural contours of the land; lessens erosion, as on a hillside. See *contour plowing.*

strip mining A type of surface mining in which a trench is dug to extract the minerals, then a new trench is dug parallel to the old one; the overburden from the new trench is put into the old trench, creating a hill of loose rock known as a *spoil bank.* See *surface mining.* Compare *open-pit mining.*

structural traps Underground geological structures that tend to trap oil or natural gas if it is present.

subbituminous coal A grade of coal, intermediate between lignite and bituminous, that has a relatively low heat value and sulfur content. Compare *anthracite, bituminous coal,* and *lignite.*

subduction The process in which one tectonic plate descends under an adjacent plate.

sublimation The property of water in which it changes from a solid to a vapor without going through the liquid phase; freeze-drying of various products takes advantage of the fact that water sublimates.

subsidence The sinking or settling of land caused by aquifer depletion (as groundwater supplies are removed).

subsidy A form of government support (such as monetary payments, public financing, tax benefits, or tax exemptions) given to a business or institution to promote the activity performed by that group.

subsistence agriculture Traditional agricultural methods that are dependent on labor and a large amount of land to produce enough food to feed oneself and one's family, with little left over to sell or reserve for hard times. Subsistence agriculture uses humans and draft animals as its main source of energy. Compare *high-input agriculture.*

subsurface mining The extraction of mineral and energy resources from deep underground deposits. Compare *surface mining.*

succession The sequence of changes in a plant community over time. Also called *ecological succession.*

sulfide A compound in which an element is combined chemically with sulfur.

sulfur cycle The worldwide circulation of sulfur from the abiotic environment into organisms and back into the abiotic environment.

surface mining The extraction of mineral and energy resources near Earth's surface by first removing the soil, subsoil, and overlying rock strata. See *strip mining* and *open-pit mining.* Compare *subsurface mining.*

surface water Fresh water found on Earth's surface in streams and rivers, lakes, ponds, reservoirs, and wetlands. Compare *groundwater.*

survivorship The probability that a given individual in a population will survive to a particular age; usually presented as a survivorship curve.

sustainability The ability to meet humanity's current needs without compromising the ability of future generations to meet their needs; sustainability implies that the environment can function indefinitely without going into a decline from the stresses imposed by human society on natural systems such as fertile soil, water, and air. Also called *environmental sustainability.*

sustainable agriculture Agricultural methods that rely on beneficial biological processes and environmentally friendly chemicals rather than conventional agricultural techniques. Also called *alternative* or *low-input agriculture.* Compare *industrialized agriculture.*

sustainable consumption The use of goods and services that satisfy basic human needs and improve the quality of life but that also minimize the use of nonrenewable and renewable resources so they are available for future generations.

sustainable development See *sustainable economic development.*

sustainable economic development Economic growth in which natural resources are not overused and excessive pollution is not generated so that our future economic growth is not threatened. See *sustainability.* Also called *sustainable development.*

sustainable forest management See *ecologically sustainable forest management.*

sustainable forestry See *ecologically sustainable forest management.*

sustainable manufacturing A manufacturing system based on minimizing waste by industry. Sustainable manufacturing involves such practices as reuse, recycling, and source reduction. See *industrial ecosystems.*

sustainable soil use The wise use of soil resources, without a reduction in the amount or fertility of soil, so that it is productive for future generations.

sustainable water use The wise use of water resources, without harming the essential functioning of the hydrologic cycle or the ecosystems on which humans depend, so that water is available for future generations.

swamp A wetland that is dominated by trees; freshwater swamps are found inland, and saltwater swamps occur along protected coastal areas. See *mangrove forest.*

symbionts The partners of a symbiotic relationship.

symbiosis An intimate relationship between two or more organisms of different species. See *commensalism, mutualism,* and *parasitism.*

synergism A phenomenon in which two or more pollutants interact in such a way that their combined effects are more severe than the sum of their individual effects. Compare *additivity* and *antagonism.*

synfuel A liquid or gaseous fuel synthesized from coal or other naturally occurring sources and used in place of oil or natural gas. Also called *synthetic fuel.*

synthetic botanical Any of a group of human-made insecticides that are produced by chemically modifying the structure of natural botanicals. See *botanical.*

synthetic fuel See *synfuel.*

taiga A region of coniferous forests (such as pine, spruce, and fir) in the Northern Hemisphere. The taiga is located just south of the tundra and stretches across both North America and Eurasia. Also called *boreal forest.*

tailings Piles of loose rock produced when a mineral such as uranium is mined and processed (extracted and purified from the ore).

tar sand An underground sand deposit permeated with a thick, asphalt-like oil known as bitumen. The bitumen can be separated from the sand by heating. Also called *oil sand.*

temperate deciduous forest A forest biome that occurs in temperate areas where annual precipitation ranges from about 75 cm to 125 cm.

temperate grassland A grassland characterized by hot summers, cold winters, and less rainfall than is found in a temperate deciduous forest biome.

temperate rain forest A coniferous biome characterized by cool weather, dense fog, and high precipitation.

temperature inversion See *thermal inversion.*

terracing A soil conservation method that involves building dikes on hilly terrain in order to produce level, terraced areas for agriculture.

terrestrial Pertaining to the land. Compare *aquatic.*

tertiary treatment Advanced wastewater treatment methods that occur after primary and secondary treatments and include a variety of biological, chemical, and physical processes. Compare *primary treatment* and *secondary treatment.*

theory   An integrated explanation of numerous hypotheses, each of which is supported by a large body of observations and experiments. Compare *hypothesis.*

thermal inversion   A layer of cold air temporarily trapped near the ground by a warmer, upper layer. If a thermal inversion persists, air pollutants may build up to harmful or even dangerous levels. Also called *temperature inversion.*

thermal pollution   Water pollution that occurs when heated water produced during many industrial processes is released into waterways.

thermal stratification   The marked layering (separation into warm and cold layers) of temperate lakes during the summer. See *thermocline.*

thermocline   A marked and abrupt temperature transition in temperate lakes between warm surface water and cold deeper water. See *thermal stratification.*

thermodynamics   The branch of physics that deals with energy and its various forms and transformations. See *first law of thermodynamics* and *second law of thermodynamics.*

thermosphere   The layer of the atmosphere between the mesosphere and the exosphere. Temperatures are very high due to the absorption of x-rays and shortwave ultraviolet radiation.

threatened species   A species in which the population is low enough for it to be at risk of becoming endangered in the foreseeable future throughout all or a significant portion of its range. Compare *endangered species.*

threshold   In toxicology, the maximum dose that has no measurable effect (or the minimum dose that produces a measurable effect).

topography   A region's surface features (such as the presence or absence of mountains and valleys).

total fertility rate   The average number of children born to a woman during her lifetime.

total resources   The combination of a mineral's reserves and resources. See *mineral reserve* and *mineral resource.* Also called *world reserve base.*

toxicant   A chemical with adverse effects on health.

toxicology   The study of harmful chemicals (toxicants) that have adverse effects on health, as well as ways to counteract their toxicity.

toxic waste   See *hazardous waste.*

trade wind   A prevailing tropical wind that blows from the northeast (in the Northern Hemisphere) or from the southeast (in the Southern Hemisphere).

transpiration   The loss of water vapor from the aerial surfaces of plants.

trickle irrigation   See *microirrigation.*

tritium   An isotope of hydrogen that contains one proton and two neutrons per atom. Compare *deuterium.*

trophic level   Each level in a food chain. All producers belong to the first trophic level, all herbivores belong to the second trophic level, and so on.

tropical cyclone   Giant, rotating tropical storms with winds of at least 119 kilometers per hour (74 mph); the most powerful tropical cyclones have wind velocities greater than 250 kilometers per hour (155 mph). Called hurricanes in the Atlantic, typhoons in the Pacific, and cyclones in the Indian Ocean.

tropical dry forest   A tropical forest where enough precipitation falls to support trees, but not enough to support the lush vegetation of a tropical rain forest. Many tropical dry forests occur in areas with pronounced rainy and dry seasons.

tropical rain forest   A lush, species-rich forest biome that occurs in tropical areas where the climate is very moist throughout the year. Tropical rain forests tend to be characterized by old, infertile soils.

troposphere   The atmosphere from Earth's surface to the stratosphere. It is characterized by the presence of clouds, turbulent winds, and decreasing temperature with increasing altitude.

tundra   The treeless biome in the far north that consists of boggy plains covered by lichens and small plants such as mosses. The tundra is characterized by harsh, very cold winters and extremely short summers. Also called *arctic tundra.* Compare *alpine tundra.*

ultraviolet radiation   That part of the electromagnetic spectrum with wavelengths just shorter than visible light; a high-energy form of radiation that can be lethal to organisms in excessive amounts. Also called *UV radiation.*

unconfined aquifer   A groundwater storage area located above a layer of impermeable rock. Water in an unconfined aquifer is replaced by surface water that drains from directly above it. Compare *confined aquifer.*

undernutrition   A condition caused when a person receives fewer calories in the diet than are needed; an undernourished person eats insufficient food to maintain a healthy body. Compare *over-nutrition* and *malnutrition.*

unfunded mandate   Federal requirements imposed on state and local governments without providing a way to pay for the cost of compliance.

upwelling   A rising ocean current that transports colder, nutrient-laden water to the surface.

urban growth   The rate at which a city's population grows.

urban heat island   Local heat buildup in an area of high population density. Compare *dust dome.*

urbanization   The process in which people increasingly move from rural areas to densely populated cities; also involves the transformation of rural areas into urban areas.

UV radiation   See *ultraviolet radiation.*

values   The principles that an individual or society considers important or worthwhile.

vapor extraction   A technique of soil remediation that involves injecting or pumping air into soil to remove organic compounds that are volatile (evaporate quickly).

vaporization   See *evaporation.*

variable   A factor that influences a process. In scientific experiments, all variables are kept constant except for one. See *control.*

vector   An organism that transmits a parasite.

vitamin   A complex organic molecule required in very small quantities for the normal metabolic functioning of living cells. Compare *mineral.*

vitrification   A method of safely storing high-level radioactive liquid wastes in solid form, as enormous glass logs.

water cycle   See *hydrologic cycle.*

water pollution   Any physical or chemical change in water that adversely affects the health of humans and other organisms.

water table   The uppermost level of an unconfined aquifer, below which the ground is saturated with water.

watershed   The area of land drained by a single river or stream. Also called *drainage basin.*

weather   The general condition of the atmosphere (temperature, moisture, cloudiness) at a particular time and place. Compare *climate.*

weathering process   A biological, chemical, or physical process that helps form soil from rock; during weathering, the rock is gradually broken down into smaller and smaller particles.

westerly   A prevailing wind that blows in the mid-latitudes from the southwest (in the Northern Hemisphere) or from the northwest (in the Southern Hemisphere).

Western diseases   A group of noninfectious diseases that are generally more commonplace in industrialized countries, including obesity and heart disease.

wet deposition   A form of acid deposition in which acid falls to Earth as precipitation. Compare *dry deposition.*

wetlands   Lands that are transitional between aquatic and terrestrial ecosystems and are covered with water for at least part of the year.

wide-spectrum pesticide   A pesticide that kills a variety of organisms in addition to the pest against which it is used. Most pesticides are wide-spectrum pesticides. Compare *narrow-spectrum pesticide.*

wilderness   Any area that has not been greatly disturbed by human activities and that humans may visit but do not permanently inhabit.

wildlife corridor   Protected zones that connect unlogged or undeveloped areas; wildlife corridors are thought to provide escape routes and allow animals to disperse so they can interbreed.

wildlife management   An applied field of conservation biology that focuses on the continued productivity of plants and animals; includes the regulation of hunting and fishing and the management of food, water, and habitat for wildlife populations.

wildlife ranching   An alternative use of land in which private landowners maintain herds of wild animals and earn money from tourists, photographers, and sport hunters as well as from animal hides, leather, and meat. Also called *game farming.*

wind   Surface air currents that are caused by the solar warming of air.

wind energy   Harnessing surface air currents caused by the solar warming of air to generate electricity. See *wind farm.*

wind farm   An array of wind turbines for utilizing wind energy by capturing it and converting it to electricity.

wise-use movement   A coalition of several hundred grassroots organizations who think that the government has too many environmental regulations and that federal lands should be used primarily to enhance economic growth.

world grain carryover stocks   The amounts of rice, wheat, corn, and other grains remaining from previous harvests, as estimated at the start of a new harvest; these provide a measure of world food security.

world reserve base   See *total resources.*

worldview   One of many perspectives that is based on a collection of our basic values; worldviews help us make sense of the world, understand our place in it, and determine right and wrong behaviors. See *environmental worldview.*

yield   In agriculture, the amount of a food crop produced per unit of land.

zero population growth   When the birth rate equals the death rate. A population with zero population growth remains the same size.

zooplankton   The nonphotosynthetic organisms—tiny shrimp, larvae, and other drifting animals—that are part of the plankton. See *plankton.* Compare *phytoplankton.*

zooxanthellae   Algae that live inside coral animals and have a mutualistic relationship with them.

Photo Credits

Chapter 1
Opener: ©Barney Taxel, 2001. Photo montage courtesy William McDonough and Partners. Figure 1-1 and 1-2: Earth Images/Stone/Getty Images. Figure 1-4: Jerry Cooke/Photo Researchers. Figure 1-5: Stuart Bauer. Figure 1-7: Jeff Lepore/Photo Researchers. Figure 1-8: Courtesy Tracy Brooks, U.S. Fish and Wildlife Service. Figure 1-9: Courtesy Illinois Department of Conservation. Figure 1-10: Courtesy NASA/Goddard Space Flight Center Scientific Visualization Studio. Figure 1-12: Dario Lopez Mills/AP/Wide World Photos. Figure 1-13: Gary Yeowell/Stone/Getty Images. Page 13: Courtesy Dr. Monty Graham, Dauphin Island Sea Lab, AL.

Chapter 2
Opener: Courtesy L. Beck, NASA Ames Research Center. Figure 2-2: Stephen Ferry/Liaison Agency, Inc./Getty Images. Figure 2-3: Les Stone/Liaison Agency, Inc./Getty Images. Figure 2-5: Courtesy Dr. Patricia A. Crione, EPA. Figure 2-8: Sinclair Stammers/Science Photo Library/Photo Researchers. Figure 2-9: Charlie Ott/Photo Researchers.

Chapter 3
Opener: CC Studio/Photo Researchers. Figure 3-1: Minnesota Historical Society/Corbis Images. Figure 3-2: Bettmann/Corbis Images. Figure 3-3a: National Archives. Figure 3-3b: Wallace Kleck/Terraphotographics/BPS. Figure 3-4: Photo by Tom Coleman. Courtesy Aldo Leopold Foundation, Baraboo, WI. Figure 3-5: ©AP/Wide World Photos. Figure 3-7: Reuters Newmedia, Inc/Corbis Images. Figure 3-8: Albert Copley/Visuals Unlimited. Figure 3-14: Jack Wilburn/Animals Animals. Figure 3-16: TASS/Sovfoto/Eastfoto.

Chapter 4
Opener: Robert Noonan/Photo Researchers. Figure 4-2: Denise Tackett/Tom Stack & Associates. Figure 4-3: Jerry Alexander/Stone/Getty Images. Figure 4-6: Barbara Gerlach/Visuals Unlimited. Figure 4-7: D. Foster, Science VU-WHOI/Visuals Unlimited. Figure 4-11: Peter Johnson/Corbis Images. Figure 4-15: Provided by the SEAWIFS Project, NASA/Goddard Space Flight Center and ORB Image.

Chapter 5
Opener: Harmut Schwarzbach/Still Pictures. Figure 5-4a and c: Wolfgang Kaehler. Figure 5-4b: Glenn N. Oliver/Visuals Unlimited. Figure 5-7a and b: Courtesy Randy Molina, U.S. Forest Service. Figure 5-8: Carlyn Iverson. Figure 5-9: Courtesy USDA, Agricultural Research Service. Figure 5-10: Marilyn Kazmers/Dembinsky Photo Associates. Figure 5-11: Michael Fogden/Animals Animals. Figure 5-12: Chris Newbert/Minden Pictures, Inc. Figure 5-13a: Ed Kanze/Dembinsky Photo Associates. Figure 5-13b: Robert Clay/Visuals Unlimited. Figure 5-15a: Michael Abbey/Photo Researchers.

Chapter 6
Opener: The Image Bank/Getty Images. Figure 6-20: Courtesy NOAA.

Chapter 7
Opener: Courtesy John McColgan, Alaska Fire Service/BLM. Figure 7-4: Warren Garst/Tom Stack & Associates. Figure 7-5: Beth Davidow/Visuals Unlimited. Figure 7-6: David Muench Photography. Figure 7-7: Barbara Miller/Biological Photo Service. Figure 7-8: Annie Griffiths Belt/DRK Photo. Figure 7-9: John Cunningham/Visuals Unlimited. Figure 7-10: Willard Clay/Dembinsky Photo Associates. Figure 7-11: Carlyn Iverson. Figure 7-12a: Frans Lanting/Minden Pictures, Inc. Figure 7-12b: Mark Moffett/Minden Pictures, Inc. Figure 7-16: Gregory G. Dimijian/Photo Researchers. Figure 7-17b: Patti Murray Animals Animals/Earth Scenes. Figure 7-21: Courtesy Dr. David Campbell. Figure 7-22: Norbert Wu/Peter Arnold, Inc. Figure 7-25: Courtesy of Bruce H. Robison, Monterey Bay Aquarium.

Chapter 8
Opener: Malcolm Linton/Liaison Agency, Inc./Getty Images. Figure 8-1: Roger Wilmshurst/Frank Lane Picture/Corbis Images. Figure 8-5: Yvona Momatiuk and John Eastcott/Photo Researchers. Figure 8-6a: Corbis Images. Figure 8-6b: Stephen Dalton/Photo Researchers. Figure 8-10a: Tom McHugh/Photo Researchers. Figure 8-14: Shepard Sherbell/Corbis Images.

Chapter 9
Opener: ©Donna DeCesare. Figure 9-1: Reuters Newmedia, Inc./Corbis Images. Figure 9-3: Peter Ginter/Material World. Figure 9-5: Gunter Marx Photography/Corbis Images. Figure 9-6: Tri-Met photo by Chris Fussell. Figure 9-7: Les Ston/Corbis Sygma. Figure 9-8: Deshakalyan Chowdhury/AFP/Corbis Images. Figure 9-9: Mike Goldwater-Network/Matrix International, Inc. Figure 9-12: E.F. Anderson/Visuals Unlimited. Figure 9-13: Baldev/Corbis Sygma. Figure 9-14: Margaret B. Duda/Photo Researchers.

Chapter 10
Opener: Tony Korody/Corbis Sygma. Figure 10-3: No. GEO85638c, Field Museum of Natural History. Figure 10-5: Kristin Finnegan Photograhy. Figure 10-6: Ken Sherman/AppaLight. Figure 10-7: John Shaw Photography. Figure 10-14a: Chris Wilkins/AFP/Corbis Images. Figure 10-14b: Charles Mason/Black Star. Page 215: Courtesy Chuck Meyers, Office of Surface Mining, Department of the Interior.

Chapter 11
Opener: Courtesy Public Service Electric & Gas Co. Figure 11-3: Courtesy Westinghouse Electric Corp., Commercial Nuclear Fuel Division. Figure 11-7a: Novosti/Liaison Agency, Inc./Getty Images. Figure 11-7b: Patrick Landmann/Liaison Agency, Inc./Getty Images. Figure 11-9: Caroline Penn/Corbis Images. Figure 11-10: Peter Essick/Aurora Photos. Figure 11-12: Alfred J. Hernandez/AP/Wide World Photos.

Chapter 12
Opener: Glen Allison/Stone/Getty Images. Figure 12-4a: L. Lefkowitz/Taxi/Getty Images. Figure 12-5a: Courtesy Solar Design Associates. Figure 12-6 (photo and inset): Energy Conservation Devices, Discover Magazine; inset, United Solar Systems, Discover Magazine. Figure 12-8a: Robert E. Ford/Terraphotographics/BPS. Figure 12-8b: Mike Andrews/Animals Animals/Earth Scenes. Figure 12-9a: Courtesy Marc Sherman. Figure 12-9b: Courtesy U. S. Dept. of Energy. Figure 12-14b: Courtesy Ontario Hydro.

Chapter 13
Opener: Earth Satellite Corporation/Photo Researchers. Figure 13-5: Mark Wagner/Stone/Getty Images. Figure 13-7a: ©Kim D.

Text and Illustration Credits

Chapter 1

Figure 1-3: Population Reference Bureau; **Figure 1-6**: National Marine Fisheries Service; **Table 1-1**: Allan R. Woodward, Florida Fish and Wildlife Conservation Commission; **Figure 1-11**: Dave Keeling and Tim Whorf, Scripps Institute of Oceanography, La Jolla, California. Reprinted by permission of the Carbon Dioxide Information Analysis Center; **page 18, list of 8 principles of deep ecology**: Naess, A. and D. Rothenberg. *Ecology, Community and Lifestyle*. Cambridge, UK:Cambridge University Press (2001); **page 18, Quote in section on environmental worldviews**: Seed, J., Macy, J., Fleming, P, and Naess, A. *Thinking Like a Mountain: Towards a Council of All Beings*. Philadelphia: New Society Publishers (1988); **Envirobrief on coffee**: *World Watch*, Vol. 15 (May/June 2002), *Environmental Review*, Vol. 8 (August 2001), *Christian Science Monitor* (July 10, 1997), *Science*, Vol. 274 (November 22, 1996); **Envirobrief on jellyfish**: *Science*, Vol. 273 (6 July 2001), *Scientific American* (October 2000), online source: http://dockwatch.disl.org/overview.htm.

Chapter 2

Envirobrief on Salton Sea: *Environment* (June 1999); *Science* (April 2, 1999); Cohen, M.J., J.I. Morrison, and E.P. Glenn. *Haven or Hazard: The Ecology and Future of the Salton Sea*, Report of the Pacific Institute for Studies in Development, Environment, and Security (February 1999); **Envirobrief on environmental literacy**: *The New York Times* (April 22, 1997); *Christian Science Monitor* (November 8, 1994); *Environment* (November 1994 and September 1993); **Figure 2-1**: Adapted from Solomon, E.P., L.R. Berg, and D.W. Martin. *Biology*, 5th edition. Philadelphia: Saunders College Publishing (1999); **Figures 2-10 and 2-11**: From Edmondson, W.T. *The Uses of Ecology: Lake Washington and Beyond*. Seattle: University of Washington Press (1991). Copyright 1991 by the University of Washington Press. Adapted and redrawn by permission; **Table 2-1**: Probabilities calculated by author based on latest available data from U.S. Census Bureau. The U.S. population in 1998 was 270.2 million; **Table 2-2**: *Science and Judgment in Risk Assessment*. Washington, D.C.: National Academy Press (1994); **Table 2-3**: Joesten, M.D. and J.L. Wood. *World of Chemistry*, 2nd edition. Philadelphia: Saunders College Publishing, (1996).

Chapter 3

Figure 3-15: Adapted from Figure 11–9, page 348, in Kahn, J.R. The Economic Approach to Environmental and Natural Resources, 2nd edition. Philadelphia: Dryden Press (1998); **Excerpt by Aldo Leopold**, from *A Sand County Almanac: And Sketches Here and There*, by Aldo Leopold, copyright 1949, 1977 by Oxford University Press, Inc. Used by permission of Oxford University Press; **Excerpt from the "Wilderness Essay,"** a letter written by Wallace Stegner to David Pesonen of the U. of California's Wildland Research Center; **Excerpt from *Silent Spring* by Rachel Carson**, Copyright 1962 by Rachel L. Carson, renewed 1990 by Roger Christie. Reprinted by permission of Houghton Mifflin Company. All rights reserved; **Section on natural resources, the environment, and the national income accounts**, Main source: Levin, J. "The Economy and the Environment: Revising the National Accounts." *IMF Survey* (June 4, 1990). Reprinted by permission of the International Monetary Fund, Washington, D.C.; **Envirobrief on trading turtle safety**: *Scientific American* (August 1998); *BioScience* (July 1998); **Envirobrief on how green is your campus**: *E Magazine* (May/June 1999); *World Watch* (May/June 1998); Lerner, S. *Eco-Pioneers*. Cambridge, Massachusetts: MIT Press (1997); **Table 3-1**: Compiled by author.

Chapter 4

Envirobrief on unintended changes in food webs: *BioScience* (November 2001); **Envirobrief on fishing down the food web**: *Environmental Review* (July 1998); *World Watch* (May/June 1998); *Science* (February 6, 1998); **Envirobrief on more is sometimes less**: Source: *Discover* (September 1994); **Table 4-1**: After Whittaker, R.H. *Communities and Ecosystems*, 2nd edition. New York: Macmillan (1975).

Chapter 5: Envirobrief on otters in trouble: *San Francisco Examiner* (May 2, 1999); *New York Times* (January 5, 1999); *Science* (October 16, 1998); **Envirobrief on acorns to lyme disease**: Source: *Science* (February 13, 1998); **Envirobrief on credible science in the biosphere**: Sources: *Nature* (December 9, 1999; August 18, 1994; March 10, 1994); *Science* (May 22, 1998; January 13, 1995; August 19, 1994; March 11, 1994); *Scientific American* (August 1995); **Figure 5-15**: Data for graphs adapted from Gause, G.F. *The Struggle for Existence*. Baltimore: Williams & Wilkins (1934); **Figure 5-16**: Adapted from MacArthur, R.H. "Population Ecology of Some Warblers of Northeastern Coniferous Forests." *Ecology*, Vol. 39 (1958); **Figure 5-17**: Adapted from MacArthur, R.H., and J.W. MacArthur. "On Bird Species Diversity." *Ecology*, Vol. 42 (1961); **Table 5-1, Ecosystem Services**: Adapted from page 527 of *Climate Change Impacts on the United States*, A report of the National Assessment Synthesis Team, U.S. Global Change Research Program, Cambridge University Press (2001).

Chapter 6

Figures 6-2, 6-3, 6-4, 6-5, and 6-6, Values are from Schlesinger, W.H. *Biogeochemistry: An Analysis of Global Change*, 2nd edition. Academic Press, San Diego (1997) and based on several sources; **Figure 6-7**, Adapted from Figure 4.15 in Strahler, A. and A. Strahler. *Physical Geography*, 2nd edition. Hoboken, NJ: John Wiley and Sons, Inc. (2002) and based on data from J.T. Kiehl and K.E. Trenberth, "Eath's Annual Global Mean Energy Budget." *Bull. Am. Met. Soc.*, Vol. 78 (1997). **Figure 6-15**, After Broecker, W.S. "The Great Ocean Conveyor." *Oceanography*, Vol. 4 (1991); **Figure 6-16**, Data adapted from the Climate Analysis Center, National Weather Service, and NOAA, Camp Springs, MD; **Figure 6-18**, From Figure 1-8 in deBlij, H.J. and P.O. Muller. *Geography: Realms, Regions and Concepts*, 10th edition. Hoboken, NJ: John Wiley and Sons, Inc. (2002). After Köppen and Geiger, and updated and modified by several geographers; **Table 6-1**, Schimel et al., "CO_2 in the Carbon Cycle." *Climate Change 1994*. Cambridge, U.K.: Cambridge University Press (1995); **Envirobrief on albedo**, Thompson, G.R., and J. Turk. *Earth Science and the Environment* 2/e. Philadelphia: Saunders College Publishing (1999); **Envirobrief on Earth's rotation**, *Natural History* (September 1996); *Earth* (June 1996).

Chapter 7

Figure 7-1, Adapted from Odum, E.P. *Fundamentals of Ecology*, 3rd edition. Philadelphia: W.B. Saunders Company (1971); **Figure 7-17a**, Adapted from pages 140 to 141 of *AAAS Atlas of Population and Environment*, University of California Press (2000) and based on data from UNEP-WCMC; **Figure 7-18**, Adapted from page 17 of *Resources*, Issue 133 (Fall 1998); **Figure 7-20**, Adapted from Figure 14-1 in Karleskint, G. *Introduction to Marine Biology*. Philadelphia: Harcourt College Publishers (1998); **Figure 7-23**, Adapted from Marangos, J.E., M.P. Crosby, and J.W. McManus. "Coral Reefs and Biodiversity: A Critical and Threatened Relationship." *Oceanography*, Vol. 9 (1996);

Envirobrief on value of nature: *Nature* (October 1, 1998, August 7, 1997, and May 15, 1997); *U.S. News & World Report* (May 26, 1997); *Newsweek* (May 26, 1997); *Science News* (May 17, 1997); *Science* (May 16, 1997); Daily, G., *Nature's Services*. Washington, D.C.: Island Press (1997). **Envirobrief on using goats to fight fires:** *Smithsonian* (October 2001).

Chapter 8

Figure 8-8: After Paynter. Data collected from Kent Island, Maine, 1934 to 1939. Baby gulls were banded to establish identity; **Figure 8-9:** Adapted and redrawn from data by Dr. Rolf O. Peterson, Michigan Technological University; **Figure 8-11, 8-12, 8-17, and 8-18 and Tables 8-1, 8-2, and 8-3:** Data from Population Reference Bureau; **Figure 8-13:** *World Population Prospects, The 2000 Revision*, United Nations Population Division; **Figure 8-19:** OECD, U.S. Bureau of the Census, and Urban Institute; **Envirobrief on U.S. Population:** U.S. Census Bureau; **Envirobrief on the Flood of Refugees:** U.N. High Commissioner for Refugees; *Vital Signs 2001*.

Chapter 9

Table 9-1: *World Urbanization Prospects: The 2001 Revision*. United Nations Population Division; **Table 9-2:** U.N. Statistics Division; **Table 9-3:** Modified from Solomon, Berg, Martin, *Biology*, 6/e, Brooks/Cole, Thomson Learning (2002), **Figure 9-2:** World Bank, Global Development Finance, electronic database, 2001, as cited by Mastny in D.M. Roodman in *Vital Signs 2002*. New York: W.W. Norton & Company (2002); **Figure 9-10:** U.S. Census Bureau; **Figure 9-11:** Population Reference Bureau; **Envirobrief on ecology hits the streets:** *BioScience* (July 2000 and August 1998); *Science News* (April 4, 1998); **Envirobrief on successful shots:** *The World Health Report*, World Health Organization (1997); *WorldWatch* (May-June, 1994); **Envirobrief on reproduction and women:** *2002 World Population Data Sheet*, Population Reference Bureau; *Nature* (March 11, 1999); *St. Petersburg Times* (September 25, 1994).

Chapter 10

Table 10-1: Compiled by author from EIA, U.S. Department of Energy and USGS; **Table 10-2:** EIA, U.S. Department of Energy; **Table 10-3:** U.S. Department of Energy; **Table 10-4:** *National Energy Policy*, 2001, U.S. Department of Energy; **Table 10-A:** EIA, U.S. Department of Energy and Population Reference Bureau; **Figures 10-1, 10-2, 10-4, 10-9, 10-12, 10-13,** EIA, U.S. Department of Energy; **Figure 10-16,** Adapted from Hinrichs, R.A. *Energy*. Philadelphia: Saunders College Publishing (1992); **Envirobrief on energy for China,** *International Energy Outlook 2001*, Department of Energy; *People's Daily* (December 16, 2000); *Science* (March 6, 1998); **Envirobrief on offshore rig debate,** *Christian Science Monitor* (August 22, 1996).

Chapter 11

Table 11-2, Adapted from two sources: Hinrichs, R.A. and M. Kleinbach. *Energy: Its Use and the Environment*, 3rd edition. Philadelphia: Harcourt College Publishers (2002) and *Science for Democratic Action*, Vol. 8, No. 3 (May 2000); **Table 11-3,** International Atomic Energy Agency and EIA, U.S. Department of Energy; **Envirobrief on nuclear waste nightmare:** *New York Times (November 21, 1994 and April 27, 1993)*.

Chapter 12

Figure 12-1: U.S. Department of Energy; **Figure 12-7b and Table 12-1:** Adapted from Hinrichs, R.A. *Energy: Its Use and the Environment*, 2nd ed. Philadelphia: Saunders College Publishing (1996); **Table 12-2:** Princeton Energy Resources International; **Table 12-3:** Calculations by author based on total energy consumption provided by *World Resources 2000-2001*, New York: World Resources Institute (2000); populations and per-capita GDPs provided by Population Reference Bureau; **Table 12-4:** Data from the Rocky Mountain Institute, 1739 Snowmass Creek, Snowmass, CO 81654, (970)927-3851. Reprinted by permission; **Table 12-A:** The Worldwatch Institute. *Vital Signs 2002*.

New York: W.W. Norton & Company (2002); **Envirobrief on starring role in energy efficiency:** Energy Star Web site (EPA/DOE), www.energystar.gov; **Envirobrief on benefits of home energy production:** Green Power Network, Net Metering Web site, www.eren.doe.gov/greenpower/netmetering/index.shtml; *Scientific American* (May 1997).

Chapter 13

Tables 13-1 and 13-2: World Resources Institute and *World Resources 1998-99*. New York: Oxford University Press (1998); **Figure 13-6:** *Control of Water Pollution from Urban Runoff*. Paris: Organization for Economic Cooperation and Development (1986); **Figure 13-11:** Data from U.S. Geological Survey and Water Resources Council; **Figure 13-12:** Environmental Systems Research Institute, Inc. (1999); **Figure 13-13:** U.S. Geological Survey; **Figure 13-16:** Engelbert, E.A., and A. Foley, eds. *Water Scarcity: Impacts on Western Agriculture*. Berkeley: University of California Press (1984). Reprinted by permission. **Figure 13-17:** U.S. Geological Survey; **Figure 13-23a:** California State Water Project; **Figure 13-UN-1:** Data from Kindler, J., and C.S. Russell. *Modeling Water Demands*. Toronto: Academic Press, Inc. (1984) as reported in Gleick, P.H. "Basic Water Requirements for Human Activities: Meeting Basic Needs." Water International, Vol. 21 (1996); **Envirobrief on U.S. water use trends:** Gleick, P.H. *The World's Water 1998-1999: The Biennial Report on Freshwater Resources*. Washington, D.C.: Island Press (1998); **Environbrief on where water conservation is academic:** Gleick, P.H. and A.K. Wong. *Sustainable Use of Water: California Success Stories*. Pacific Institute for Studies in Development, Environment, and Security (1999); **Envirobrief on xeriscaping:** National Xeriscaping Council, Inc.

Chapter 14

Envirobrief on future of Everglades soil: Davis, S.M. and J.C. Ogden. *Everglades: The Ecosystem and Its Restoration*. Delray Beach, FL: St. Lucie Press (1994); **Envirobrief on soil fungi and plant diversity:** *Science News* (December 5, 1998); *Nature* (November 5, 1998).

Chapter 15

Table 15-2: Gardner, G. and P. Sampat. *Mind Over Matter: Recasting the Role of Materials in Our Lives*, Worldwatch Paper 144 (December 1998); **Table 15-3:** Tilton (2002), U.S. Bureau of Mines, and U.S. Geological Survey as reported in MMSD; **Figure 15-3:** Adapted from Joesten, M.D., and J.L. Wood. *World of Chemistry*, Second Edition. Philadelphia: Saunders College Publishing (1996); **Figure 15-7:** CRU International (2001) as reported in MMSD; **Envirobrief on diamonds under the tundra:** Mbendi Mining News (September 7, 2002); *Discover* (December 1994); **Envirobrief on not-so-precious gold:** U.S. Geological Survey (1999); Worldwatch Institute.

Chapter 16

Data in first section on number of species: Data from Reaka-Kudla, M.L., D.E. Wilson, and E.O. Wilson. *Biodiversity II*. Washington, D.C.: Joseph Henry Press (1997); **Table 16-1:** Adapted from Table 2 in Wilcove, D.S. et al. "Quantifying Threats to Imperiled Species in the United States." *BioScience*, Vol. 48, No. 8 (August 1998); **Tables 16-2 and 16-3:** U.S. Fish and Wildlife Service; **Figure 16-8:** Myers, N. et al. "Biodiversity Hotspots for Conservation Priorities." Conservation International; **Envirobrief on pollinators:** *Conservation Biology* (February 1998); *BioScience* (May 1997); **Envirobrief on solving crimes involving organisms:** *Scientific American* (March 1999); *Toronto Star* (January 1, 1997); *Washington Post* (May 2, 1994); **Envirobrief on when one rare species eats another:** *Scientific American* (May 2002); *Science News* (September 1, 2001).

Chapter 17

Table 17-1: U.S. Dept. of Interior, U.S. Dept. of Agriculture, and U.S. Dept. of Defense; **Table 17-2:** National Park Service; **Table 17-3:** "Top Twelve Farm Areas Threatened by Population Growth." American Farmland Trust (1997). Reprinted by permission; **Table 17-**

4: From Box 1.1 in Noss, R.F., M.A. O'Connell, and D.D. Murphy. *The Science of Conservation Planning: Habitat Conservation Under the Endangered Species Act*, Island Press: World Wildlife Fund (1997). Reprinted with permission. **Figure 17-1:** Adapted from the map, "Federal Lands in the Fifty States," produced by the Cartographic Division of the National Geographic Society (October 1996); **Figure 17-8:** Based on Goode Base Map; **Figure 17-11:** Data obtained from Marsh, W.M. and J.M. Grossa, Jr., *Environmental Geography*, 2nd edition, John Wiley & Sons, Inc. (2002); **Figure 17-19:** World Resources Institute and U.N. Food and Agricultural Organization; **Envirobrief on ecologically certified wood:** Brown, L.R., M. Renner, and B. Halweil. *Vital Signs 1999: The Environmental Trends That Are Shaping Our Future*. New York: W.W. Norton & Company (1999); *Christian Science Monitor* (March 31, 1999); *Discover* (March 1999); **Envirobrief on preserving DOE's natural labs:** *Science* (October 23, 1998); *Issues in Science and Technology* (Winter 1997-98, Spring 1998); **Envirobrief on taking cattle to the bank:** www.conservationfund.org/pdf/mou.pdf ; *New York Times* (December 26, 2000).

Chapter 18

Table 18-1: FAO, USDA, and PRB, as reported in Brown, L.R. *Tough Choices: Facing the Challenge of Food Scarcity*. New York: W.W. Norton & Company (1996); **Table 18-2:** World production data from U.N. Food and Agricultural Organization; **Table 18-3:** Data from World Resources Institute, as reported in *AAAS Atlas of Population and Environment*. Berkeley: University of California Press (2000); **Table 18-4:** Center for Science in the Public Interest; **Table 18A and Figure 18-2:** Adapted from USDA data, as reported in Brown, L.R. et al. *Vital Signs 2002*. New York: W.W. Norton & Company (2002); **Figure 18-3:** Adapted from USDA data; **Figure 18-6:** After G.H. Heichel, "Agricultural Production and Energy Resources." *American Scientist*, Vol. 64 (January/February 1976); **Figure 18-8:** USDA; **Envirobrief on genetic engineering and edible vaccines:** *Scientific American* (September 2000); *The Chronicle of Higher Education* (June 5, 1998); *Science News* (February 14, 1998); **Envirobrief on labels for dolphin-safe tuna:** *Environment News Service* (May 4, 1999).

Chapter 19

Quote by Ulf Merbold: From Ulf Merbold of the German Aerospace Liaison. Reprinted by permission; **Envirobrief on buzzing in on air pollution:** *The Christian Science Monitor* (June 10, 1997); **Envirobrief on new mowers for cleaner air:** *EPA Regulatory Announcement*, "New Phase 2 Standards for Small Spark-Ignition Nonhandheld Engines," EPA420-F-99-008 (March 1999); EPA brochure, "Small Engine Emission Standards," EPA420-F-98-025 (August 1998); **Envirobrief on commuting for clean air:** EPA Environmental Fact Sheet, "The Commuter Choice Program: A Way to Save Money and Help the Environment," EPA 420-F-98-029 (December 1998); www.epa.gov/oms/transp; www.myworld.org/programs/transportation; www.metrovanpool.com/commuterchoice.htm; **1995 study on long-distance transport of persistent compounds:** Simonich, S.L. and R.A. Hites. "Global Distillation of Persistent Organochlorine Compounds." *Science*, Vol. 269 (September 29, 1995); **Table 19-3:** Ozone Policy and Strategies Group, Office of Air and Radiation, Environmental Protection Agency; **Table 19-A:** Jane Dignon, Lawrence Livermore National Laboratory; **Figures 19-4, 19-5:** Environmental Protection Agency; **Figure 19-10:** a and d, adapted from Joesten, M.D., and J.L. Wood. *World of Chemistry*, Second Edition. Philadelphia: Saunders College Publishing (1996); **Figure 19-11:** Air Quality Planning and Standards, Office of Air and Radiation, EPA; **Figure 19-12:** South Coast AQMD; **Figure 19-15:** adapted from Wania, F. and D. Mackay. "Tracking the Distribution of Persistent Organic Pollutants." *Environmental Science and Technology*, Vol 30 (1996).

Chapter 20

Figure 20-1 and Table 20-A: Surface Air Temperature Analysis, Goddard Institute for Space Studies, NASA, as reported in *Vital Signs 2002*, The Worldwatch Institute, New York: W.W. Norton and Company (2002); **Figure 20-2:** Dave Keeling and Tim Whorf, Scripps Institution of Oceanography, La Jolla, California; **Figure 20-7:** Kellogg, W.W., and R. Schware. *Climate Change and Society*. Boulder, Colorado: Westview Press (1981). Redrawn by permission of W.W. Kellogg; **Figure 20-9:** Adapted from Davis, M.B., C. Zabinski, R.L. Peters, and T.E. Lovejoy, eds. "Changes in Geographical Range Resulting from Greenhouse Warming: Effects on Biodiversity in Forests." *Global Warming and Biological Diversity*. edited by R.L. Peters, and T.E. Lovejoy. New Haven, Connecticut: Yale University Press (1992); **Figure 20-10:** Greg Marland, Tom Boden (Oak Ridge National Laboratory), and Bob Andres (University of North Dakota); **Table 20-1:** Carbon Dioxide Information Analysis Center, Environmental Sciences Division, Oak Ridge National Laboratory; **Table 20-2:** From Table 4.21 in *Health and Environment in Sustainable Development: Five Years After the Summit*, World Health Organization (1997) and based on an unpublished assessment prepared by A.J. McMichael et al; **Table 20-3:** AFEAS (Alternative Fluorocarbons Environmental Acceptability Study); **Envirobrief on Arctic warming and the native cultures of the North:** Crary, D. "Global Warming Threat to Canada." Associated Press (December 14, 1998); *New Scientist* (November 14, 1998); *Washington Post* (November 14, 1997; *Los Angeles Times* (November 14, 1997; *Science* (November 14, 1997); **Envirobrief on developing countries and carbon emissions:** *Nature* (August 1999); *New Scientist* (January 9, 1999); *World Resources Institute Climate Notes* (July 1997).

Chapter 21

Figure 21-1: Adapted from Joesten, M.D., and J.L. Wood. *World of Chemistry*, 2nd edition. Philadelphia: Saunders College Publishing (1996); **Table 21-2:** Cancer data based on IARC (WHO International Agency for Research on Cancer) list of carcinogens; **Table 21-4:** Office of the Great Lakes, Michigan Department of Environmental Quality; **Envirobrief on harmful algal blooms:** Worldwatch Paper 145, Washington, D.C.: Worldwatch Institute (March 1999); Marine Ecosystems: Emerging Diseases as Indicators of Change. NOAA/NASA Year of the Ocean Special Report (December 1998); **Envirobrief on something fishy near sewage treatment plants:** *Los Angeles Times* (October 22, 1998); *Minneapolis Star Tribune* (April 18, 1998); *Environmental Science and Technology* (January 1, 1998); *Ithaca Journal* (July 30, 1997); *Environmental Health Perspectives* (October 1996).

Chapter 22

Figure 22-4: From "International Travel and Health. Vaccination Requirements and Health Advice: Situation as of January 1, 1992." Geneva: World Health Organization (1992). Reproduced by permission of the World Health Organization; **Figure 22-5:** Data from Mark Whalon, Michigan State University, as reported in Brown, L.R., H. Kane, and D.M. Roodman. *Vital Signs 1994*. New York: W.W. Norton & Co. (1994); **Figure 22-7b:** Adapted from Debach, P. *Biological Control by Natural Enemies*. New York: Cambridge University Press (1974); **Figure 22-8:** Grier, J.W. "Ban of DDT and Subsequent Recovery of Reproduction in Bald Eagles." *Science*, Vol. 218 (December 17, 1982). Redrawn by permission of the American Association for the Advancement of Science; **Figure 22-9:** Based on data from woodwell, G.M., C.F. Worster, Jr., and P.A. Isaacson. "DDT Residues in an East Coast Estuary: A Case of Biological Concentration of a Persistent Insecticide." *Science*, Vol. 156 (May 12, 1967); **Figure 22-11:** Guillette, E.A., M.M. Meza, M.G. Aquilar, A.D. Soto, and I.E. Garcia. "An Anthropological Approach to the Evaluation of Preschool Children Exposed to Pesticides in Mexico." *Environmental Health Perspectives* (May 1998); **Figure 22-14:** Gardner, G. "IPM and the War on Pests." *World Watch*, Vol. 9, No. 2 (March/April 1996). Redrawn by permission of Worldwatch Institute; **Table 22-1:** Adapted by author from Pesticide Information Profiles of the Extension Toxicology Network, maintained and archived at Oregon State University; **Table 22-2:** Adapted from Pimental, D. "Environmental and Economic Effects of Reducing Pesticide Use." *Bioscience*, Vol. 41, No. 6 (June 1991). © by the American

Institute of Biological Sciences; 1999 personal communication with D. Pimental; **Table 22-3:** Compiled by author from various sources, including "Chemically-Induced Alterations in Sexual and Functional Development: The Wildlife-Human Connection," Vol. XXI *Advances in Modern Environmental Toxicology*, edited by Colburn, T. and C. Clement. Princeton: Princeton Scientific Publishing Company (1992); and Colburn, T., D. Dumanoski, and J.P. Myers. *Our Stolen Future*. New York: Dutton (1996); **Table 22-4:** Excerpted from "Worst Case Estimates of Risk of Cancer from Pesticide Residues on Food." From *Regulating Pesticides in Food: The Delaney Paradox*. Washington, D.C.: National Academy Press (1987). © by the National Academy of Sciences, Courtesy of the National Academy Press, Washington, D.C.; **Envirobrief on lethal bug juice:** *Washington Post* (August 9, 1993); **Envirobrief on organic cotton:** *Organic Trade Association*, Issues 10 and 11 (March.April 1999 and May/June 1999); U.N. Food and Agriculture Organization (May 26, 1999); *New York Times* (November 6, 1997).

Chapter 23

Figures 23-2 and 23-3: EPA; **Figure 23-5:** Ecobalance, Inc., and Integrated Waste Services Association; **Figure 23-8:** *Warmer Bulletin*, February 1995, as reported on the EPA Web site; **Table 23-2:** Compiled by author from various sources: *Nature* (February 1, 2001); *Science News* (July 20, 1996); *Environment* (May 1996); *The Wall Street Journal* (August 7, 1995); *Business Week* (May 8, 1995); **Envirobrief on composting food at Rikers Island:** *New York Times* (January 19, 1999); **Envirobrief on an industrial ecosystem:** Personal communication with Mr. Fred Zerza, J.R. Simplot Company (1999); **Envirobrief on transforming the Rocky Mountain Arsenal:** *BioScience* (April 1999).

Chapter 24

Quote from Bruntland Report: *Our Common Future*. World Commission on Environment and Development. Cary, NC: Oxford University Press (1987), p. 22. Reprinted by permission of Oxford University; **Quote from Kai N. Lee:** *Compass and Gyroscope: Integrating Science and Politics for the Environment*. Washington, D.C.: Island Press (1993), p. 200; **9 principles around which the chapter is organized:** *Caring for the Earth*. Published jointly by the World Conservation Union, U.N. Environment Program, and World Wide Fund for Nature (1991).

Index